Water
and Waste-Water
Technology

MARK J. HAMMER

Professor of Civil Engineering
University of Nebraska

JOHN WILEY & SONS, INC.,
New York   London   Sydney   Toronto

Consulting Editors

TOM McCUTCHEN
Central Nebraska Technical College
Hastings, Nebraska

ARLEY GOODENKAUF
Metropolitan Utilities District
Omaha, Nebraska

To my mother
Bertha Grundahl Hammer
and in memory of my father
Herbert Hammer

Library of Congress Cataloging in Publication Data:

Hammer, Mark J   1931–
Water and waste-water technology.

Includes bibliographical references and index.
1. Water-supply engineering. 2. Sewage disposal.
I. Title.

TD345.H25   628.1   75–2000
ISBN 0–471–34726–4

10 9 8 7 6

Preface

This book provides a comprehensive understanding of the technology of municipal water supply and waste-water disposal. The emphasis on art and practice makes it particularly valuable for colleges and technical institutes, public health programs, environmental sciences, and natural resources management. Although it is intended to fill the void left by books now out of date, it does not displace design-oriented approaches that explore the theoretical concepts appropriate to professional sanitary engineering programs and graduate education.

Experience indicates that students benefit from a review of the disciplines that have specific applications in water supply and pollution control. Therefore, the introductory chapters cover fundamentals of chemistry, biology, hydraulics, and hydrology that are unique to sanitary studies. The book is organized in a traditional manner with water distribution and processing separated from waste-water collection and treatment. The final chapters give an overview of advanced waste treatment, water reuse, and land disposal techniques. I have carefully integrated the subject matter so that students can clearly understand the interrelationships between individual unit operations and integration of systems as a whole. In discussing various topics, I specifically included the latest technology: for example, the use of lasers in laying sewer pipe, the phenomenon of eutrophication, the use of high-purity oxygen in waste treatment, the handling of water-treatment-plant sludges, the procedures for determining BOD of industrial wastes, and land disposal. A thorough discussion of the book's content is given in the introduction of Chapter 1.

There are many illustrations that explain fundamental concepts and show modern equipment and facilities. Also, numerous sample calculations help the reader to understand the unit processes being described. Answers are given for all homework problems, mainly to help students who are interested in individual study. Finally, the essential resource material is included in an appendix.

Mark J. Hammer

Acknowledgments

The review recommendations of two technical consultants were incorporated during the development and writing of this textbook. Arley Goodenkauf, Superintendent of Water Purification, Metropolitan Utilities District, Omaha, Nebraska, provided a critique of the chapters on water systems, and Thomas McCutchen, instructor in waste-water technology, Central Technical Community College, Hastings, Nebraska, read the sections relating to waste-water technology. The Metropolitan Utilities District was helpful in contributing data, particularly the data in the chapter on operation of waterworks.

Mrs. Mary Lou Wegener and my wife, Audrey, typed the manuscript and gave invaluable assistance in assembling and proof reading the final draft.

I am grateful to the University of Nebraska and all of my students for motivating and encouraging me to complete this book.

M. J. H.

Contents

chapter 1

Introduction

The hydrologic cycle describes the movement of water in nature. Evaporation from the ocean is carried over land areas by maritime air masses. Vapor from inland waters and transpiration from plants adds to atmospheric moisture that eventually precipitates as rain or snow. Rainfall may percolate into the ground, join surface watercourses, be taken up by plants, or to reevaporate. Groundwater and surface flows drain toward the ocean for recycling.

Man intervenes in the hydrologic cycle, generating man-made water cycles (Figure 1-1). Some communities withdraw groundwater for public supply, but the majority rely on surface sources. After processing, it is distributed to households and industries. Waste water is collected in a sewer system and transported to a plant for treatment prior to disposal. Conventional methods provide only partial recovery of the original water quality. Dilution into a surface watercourse and purification by nature yield additional quality improvement. However, the next city downstream is likely to withdraw the water for a municipal supply before complete rejuvenation. This city in turn treats and disposes of its waste water by dilution. This process of withdrawal and return by successive municipalities in a river basin results in indirect water reuse. During dry weather, maintaining minimum flow in many small rivers relies on the return of upstream waste-water discharges. Thus a man-generated water cycle within the natural hydrologic scheme involves: (1) surface water withdrawal, processing, and distribution, (2) waste-water collection, treatment, and disposal back to surface water by dilution, (3) natural purification in a river, and (4) repetition of this scheme by cities downstream.

Discharge of conventionally treated waste waters to lakes, reservoirs, and estuaries, which act like lakes, accelerates eutrophication. Resulting deterioration of water quality interferes with indirect reuse for public supply and water-based recreational activities. Consequently, advanced waste treatment by either mechanical plants or land disposal techniques has been introduced into the man-made water cycle involving inland lakes and reservoirs.

Installation of advanced treatment systems that reclaim waste water to nearly its original quality has encouraged several cities to consider direct reuse

1

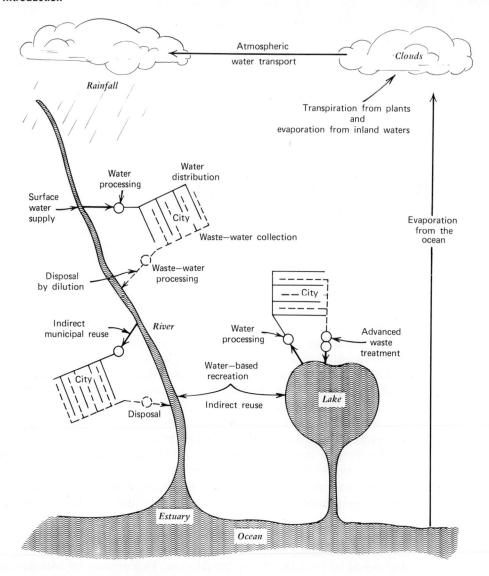

Figure 1-1 Integration of natural and man-generated water cycles.

of water for industrial processing, recreational lakes, irrigation, groundwater recharge, and other applications. However, direct return for a potable water supply is not being encouraged at present because of potential health hazards from viruses and traces of toxic substances that are difficult to detect and may not be removed in water reclamation. Another problem is the buildup of dissolved salts that can be removed only by costly demineralization processes. Nevertheless, it is anticipated, with the increase in demand for fresh water, that direct water reuse by some metropolitan areas will be realistic by the year 2000.

The basic sciences of chemistry, biology, hydraulics, and hydrology are the foundation for understanding water supply and pollution control. Chemical principles find greatest application in water processing, while waste-water treatment relies on biological systems. Knowledge of hydraulics is the key to

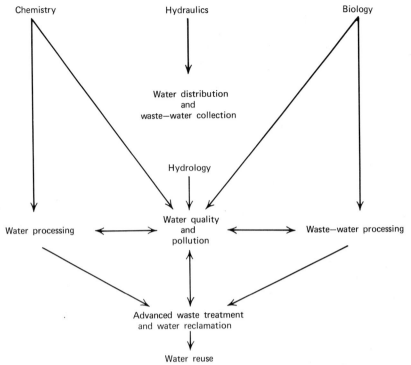

Figure 1-2 Flow diagram relating basic science areas to the disciplines of water and waste-water technology, and the integration of various aspects of water supply and pollution control.

water distribution and waste-water collection. Quality of the water environment is the focus of man's indirect water reuse cycle. As is illustrated in Figure 1-2, chemistry, hydrology, biology, water processing, and waste-water treatment converge to give a perception of water quality and pollution. Finally, insights to future direct water reuse are provided through the technology of advanced waste treatment and water reclamation.

In this book the fundamental principles of science are integrated with different aspects of sanitary technology. Table 1-1 lists prerequisite readings included in this book for various subject areas. The purpose of this table is to help the student correlate information from the text on a particular subject of interest. For example, prior to studying water quality and pollution in Chapter 5, it is recommended that a student review chemistry fundamentals, aquatic organisms from bacteria through fishes, the aquatic food chain, waterborne diseases, and the hydrology of rivers and lakes. Prerequisite readings for water distribution systems are selections from Chapter 4. The instructor who is preparing lecture sequence on this topic may find it advantageous to integrate the principles of hydraulics with the descriptive material from the chapter on water distribution. The advantage is that students can read about real pipe networks while solving simplified pipe flow problems. Conversely, when water processing is taught, it may be most advantageous for the student to review the portions on chemistry listed in Table 1-1 prior to doing the reading assignments in Chapter 7.

Waste-water flows and characteristics are discussed separately in Chapter 9, since this information is needed for both collection systems and waste-water processing. The listing under collection systems also includes applied hydraulics. Conventional municipal waste-water treatment relies on biological

Table 1-1. Listing of Prerequisite Readings for Various Subject Areas in Water and Waste-water Technology

---

Water Quality and Pollution, Chapter 5

2-1 Chemical Elements, Radicals, and Compounds
2-6 Organic Compounds
3-1 Bacteria and Fungi
3-2 Algae
3-3 Protozoa and Multicellular Animals
3-4 Fishes
3-5 Aquatic Food Chain
3-6 Waterborne Diseases
3-7 Indicator Organisms for Water Quality
4-10 Flow in Streams and Rivers
4-11 Hydrology of Lakes and Reservoirs

Water Distribution Systems, Chapter 6

4-1 Water Pressure
4-2 Pressure-Velocity-Head Relationships
4-3 Flow in Pipes Under Pressure
4-4 Flow in Pipe Networks
4-6 Flow Measurement in Pipes
4-8 Centrifugal Pumps

Water Processing, Chapter 7

2-1 Chemical Elements, Radicals, and Compounds
2-2 Chemical Water Analysis
2-3 pH and Chemical Reactions
2-4 Gas Solubility and Alkalinity
2-5 Colloids and Coagulation
2-8 Laboratory Chemical Analyses
4-1 Water Pressure
4-2 Pressure-Velocity-Head Relationships
4-6 Flow Measurement in Pipes
4-8 Centrifugal Pumps

Waste-Water Collection Systems, Chapter 10

4-1 Water Pressure
4-2 Pressure-Velocity-Head Relationships
4-5 Gravity Flow in Circular Pipes
4-7 Flow Measurement in Open Channels
4-8 Centrifugal Pumps
4-9 Amount of Storm Runoff
Chapter 9   Waste-Water Flows and Characteristics

Waste-Water Processing, Chapter 11

2-6 Organic Compounds
2-7 Organic Matter in Wastes
3-1 Bacteria and Fungi
3-2 Algae
3-3 Protozoa and Multicellular Animals
3-7 Indicator Organisms for Water Quality
3-9 Biochemical Oxygen Demand
3-10 Biological Treatment Systems

Table 1-1 (*Cont.*)

processing and, therefore, an understanding of living systems is indispensable. The chapters on operation of systems are not intended to give all-inclusive subject coverage. Since this book deals primarily with art and practice, a great deal of information relative to operation is presented throughout in the book's descriptive material.

Advanced waste treatment incorporates both biological unit operations and chemical processes that are similar to those applied in water treatment. Therefore, to understand the concepts described in Chapter 13, the reader must have adequate knowledge about the handling of both waste water and water. Reuse of reclaimed water employs the most recent treatment technology and requires a comprehensive understanding of both treatment processes and water quality.

chapter 2

Chemistry

This chapter provides basic information about chemistry as it applies to water and waste-water technology. Selected data are compiled and presented as an introduction to the chapters dealing with water quality, pollution, and chemical-treatment processes. For example, the characteristics of common elements, radicals, and inorganic compounds are tabulated. The usual method for presenting chemical water analysis is described, since it is not normally presented in general chemistry textbooks. Sections on chemical reactions, alkalinity, and coagulation emphasize important aspects of applied water chemistry. Since organic chemistry traditionally has not been included in introductory courses, persons practicing in water supply and pollution control are not generally exposed to this area of chemistry. For this reason an introduction on the nomenclature of organic compounds and a brief description of the organic matter in waste water are provided. Finally, the importance, technique, and equipment used in selected laboratory analyses are discussed. Water quality parameters and their characteristics can be understood better when testing procedures are known.

2-1   ELEMENTS, RADICALS, AND COMPOUNDS

The fundamental chemical identities that form all substances are referred to as elements. Each element differs from any other in weight, size, and chemical properties. Names of elements common to water and waste-water technology along with their symbols, atomic weights, common valence, and equivalent weights are given in Table 2-1. Symbols for elements are used in writing chemical formulas and equations.

Atomic weight is the weight of an element relative to that of hydrogen, which has an atomic weight of unity. This weight expressed in grams is called one gram atomic weight of the element. For example, one gram atomic weight of aluminum Al is 27.0 grams. Equivalent or combining weight of an element is equal to atomic weight divided by the valence.

Some elements appear in nature as gases, for example, hydrogen, oxygen,

$Al_2(SO_4)_3$

m.w. → $2(26.98) + 3(32.06 + 4(16)) = 342.14$

Eq weght → $\dfrac{342.14}{6} = 57.02$

Table 2-1. Basic Information on Common Elements

Name	Symbol	Atomic Weight	Common Valence	Equivalent Weight[a]
Aluminum	Al	27.0	3+	9.0
Arsenic	As	74.9	3+	25.0
Barium	Ba	137.3	2+	68.7
Boron	B	10.8	3+	3.6
Bromine	Br	79.9	1−	79.9
Cadmium	Cd	112.4	2+	56.2
Calcium	Ca	40.1	2+	20.0
Carbon	C	12.0	4−	
Chlorine	Cl	35.5	1−	35.5
Chromium	Cr	52.0	3+	17.3
			6+	
Copper	Cu	63.5	2+	31.8
Fluorine	F	19.0	1−	19.0
Hydrogen	H	1.0	1+	1.0
Iodine	I	126.9	1−	126.9
Iron	Fe	55.8	2+	27.9
			3+	
Lead	Pb	207.2	2+	103.6
Magnesium	Mg	24.3	2+	12.2
Manganese	Mn	54.9	2+	27.5
			4+	
			7+	
Mercury	Hg	200.6	2+	100.3
Nickel	Ni	58.7	2+	29.4
Nitrogen	N	14.0	3−	
			5+	
Oxygen	O	16.0	2−	8.0
Phosphorus	P	31.0	5+	6.0
Potassium	K	39.1	1+	39.1
Selenium	Se	79.0	6+	13.1
Silicon	Si	28.1	4+	6.5
Silver	Ag	107.9	1+	107.9
Sodium	Na	23.0	1+	23.0
Sulphur	S	32.1	2−	16.0
Zinc	Zn	65.4	2+	32.7

[a] Equivalent weight (combining weight) equals atomic weight divided by valence.

and nitrogen; mercury appears as a liquid; others appear as pure solids, for instance, carbon, sulfur, phosphorus, calcium, copper, and zinc; and many occur in chemical combination with each other in compounds. Atoms of one element unite with those of another in a definite ratio defined by their valence. Valence is the combining power of an element based on that of the hydrogen atom, which has an assigned value of 1. Thus an element with a valence of 2+ can replace two hydrogen atoms in a compound, or in the case of 2− can react with two hydrogen atoms. Sodium has a valence of 1+ while chlorine has a valence of 1−, therefore, one sodium atom combines with one chlorine atom to form sodium chloride (NaCl), common salt. Nitrogen at a valence of 3− can combine with three hydrogen atoms to form ammonia gas (NH_3). The weight of

Table 2-2.   Common Radicals Encountered in Water

Name	Formula	Molecular Weight	Electrical Charge	Equivalent Weight
Ammonium	NH_4^+	18.0	1+	18.0
Hydroxyl	OH^-	17.0	1−	17.0
Bicarbonate	HCO_3^-	61.0	1−	61.0
Carbonate	CO_3^-	60.0	2−	30.0
Orthophosphate	PO_4^-	95.0	3−	31.7
Orthophosphate, mono-hydrogen	HPO_4^-	96.0	2−	48.0
Orthophosphate, di-hydrogen	$H_2PO_4^-$	97.0	1−	97.0
Bisulfate	HSO_4^-	97.0	1−	97.0
Sulfate	SO_4^-	96.0	2−	48.0
Bisulfite	HSO_3^-	81.0	1−	81.0
Sulfite	SO_3^-	80.0	2−	40.0
Nitrite	NO_2^-	46.0	1−	46.0
Nitrate	NO_3^-	62.0	1−	62.0
Hypochlorite	OCl^-	51.5	1−	51.5

a compound, equal to the sum of the weights of the combined elements, is referred to as molecular weight, or simply mole. The molecular weight of NaCl is 58.4 grams, while one mole of ammonia gas is 17.0 grams.

Certain groupings of atoms act together as a unit in a large number of different molecules. These, referred to as radicals, are given special names, such as, the hydroxyl group (OH^-). The most common radicals in ionized form are listed in Table 2-2. Radicals in themselves are not compounds but join with other elements to form compounds. Data on inorganic compounds common to water and waste-water chemistry are given in Table 2-3. The proper name, formula, and molecular weight are included for all of the chemicals listed. Popular names, for example, alum for aluminum sulfate, are included in brackets. For chemicals used in water treatment, one common use is given; many have other applications not included. Equivalent weights for compounds and hypothetical combinations, for example, $Ca(HCO_3)_2$, involved in treatment are provided.

EXAMPLE 2-1

Calculate the molecular and equivalent weights of ferric sulfate.

Solution

Formula from Table 2-3 is $Fe_2(SO_4)_3$.
Using atomic weight data from Table 2-1

Fe	$2 \times 55.8 = 111.6$
S	$3 \times 32.1 = 96.3$
O	$12 \times 16.0 = \underline{192.0}$

Molecular Weight = 399.9 or 400 grams

Table 2-3. Basic Information on Common Inorganic Chemicals

Name	Formula	Common Usage	Molecular Weight	Equivalent Weight
Activated carbon	C	Taste and odor control	12.0	n.a.[a]
Aluminum sulfate (filter alum)	$Al_2(SO_4)_3 \cdot 14.3H_2O$	Coagulation	600	100
Aluminum hydroxide	$Al(OH)_3$	(Hypothetical combination)	78.0	26.0
Ammonia	NH_3	Chloramine disinfection	17.0	n.a.
Ammonium fluosilicate	$(NH_4)_2SiF_6$	Fluoridation	178	n.a.
Ammonium sulfate	$(NH_4)_2SO_4$	Coagulation	132	66.1
Calcium bicarbonate	$Ca(HCO_3)_2$	(Hypothetical combination)	162	81.0
Calcium carbonate	$CaCO_3$	Corrosion control	100	50.0
Calcium fluoride	CaF_2	Fluoridation	78.1	n.a.
Calcium hydroxide	$Ca(OH)_2$	Softening	74.1	37.0
Calcium hypochlorite	$Ca(ClO)_2 \cdot 2H_2O$	Disinfection	179	n.a.
Calcium oxide (lime)	CaO	Softening	56.1	28.0
Carbon dioxide	CO_2	Recarbonation	44.0	22.0
Chlorine	Cl_2	Disinfection	71.0	n.a.
Chlorine dioxide	ClO_2	Taste and odor control	67.0	n.a.
Copper sulfate	$CuSO_4$	Algae control	160	79.8
Ferric chloride	$FeCl_3$	Coagulation	162	54.1
Ferric hydroxide	$Fe(OH)_3$	(Hypothetical combination)	107	35.6
Ferric sulfate	$Fe_2(SO_4)_3$	Coagulation	400	66.7
Ferrous sulfate (copperas)	$FeSO_4 \cdot 7H_2O$	Coagulation	278	139
Fluosilicic acid	H_2SiF_6	Fluoridation	144	n.a.
Hydrochloric acid	HCl	n.a.	36.5	36.5
Magnesium hydroxide	$Mg(OH)_2$	Defluoridation	58.3	29.2
Oxygen	O_2	Aeration	32.0	16.0
Potassium permanganate	$KMnO_4$	Oxidation	158	n.a.
Sodium aluminate	$NaAlO_2$	Coagulation	82.0	n.a.
Sodium bicarbonate (baking soda)	$NaHCO_3$	pH adjustment	84.0	84.0
Sodium carbonate (soda ash)	Na_2CO_3	Softening	106	53.0
Sodium chloride (common salt)	$NaCl$	Ion-exchanger regeneration	58.4	58.4
Sodium fluoride	NaF	Fluoridation	42.0	n.a.
Sodium hexameta-phosphate	$(NaPO_3)_n$	Corrosion control	n.a.	n.a.
Sodium hydroxide	$NaOH$	pH adjustment	40.0	40.0
Sodium hypochlorite	$NaClO$	Disinfection	74.4	n.a.
Sodium silicate	Na_4SiO_4	Coagulation aid	184	n.a.
Sodium fluosilicate	Na_2SiF_6	Fluoridation	188	n.a.
Sodium thiosulfate	$Na_2S_2O_3$	Dechlorination	158	n.a.
Sulphur dioxide	SO_2	Dechlorination	64.1	n.a.
Sulfuric acid	H_2SO_4	pH adjustment	98.1	49.0
Water	H_2O	n.a.	18.0	n.a.

[a] n.a. = not applicable.

The ferric (oxide iron) atom has a valence of $3+$, thus a compound with 2 ferric atoms has a total electrical charge of $6+$. (Three sulfate radicals have a total of $6-$ charges).

$$\text{Equivalent Weight} = \frac{\text{Molecular Weight}}{\text{Electrical Charge}}$$

$$= \frac{400}{6} = 66.7 \text{ grams per equivalent weight}$$

2-2 CHEMICAL WATER ANALYSIS

When placed in water, inorganic compounds dissociate into electrically charged atoms and radicals referred to as ions. This breakdown of substances into their constituent ions is called ionization. An ion is represented by the chemical symbol of the element, or radical, followed by superscript $+$ or $-$ signs to indicate the number of unit charges on the ion. Consider the following: sodium, Na^+, chloride, Cl^-, aluminum, Al^{+++}, ammonium, NH_4^+, and sulfate, SO_4^-.

Laboratory tests on water, such as those outlined in Section 2-8, determine concentrations of particular ions in solution. Test results are normally expressed as weight of the element or radical in milligrams per liter of water, abbreviated as mg/l. Some books use the term parts per million (ppm), which is for practical purposes identical in meaning to mg/l, since 1 liter of water weighs 1,000,000 milligrams. In other words, 1 mg per liter (mg/l) equals 1 mg in 1,000,000 mg, which is the same as 1 part by weight in one million parts by weight (1 ppm). The concentration of a substance in solution can also be expressed in milliequivalents per liter (meq/l) representing the combining weight of the ion, radical, or compound. Milliequivalents can be calculated from milligrams per liter by

$$meq/l = mg/l \times \frac{valence}{atomic\ weight} = \frac{mg/l}{equivalent\ weight} \tag{2-1}$$

or in the case of a radical or compound the equation reads

$$meq/l = mg/l \times \frac{electrical\ charge}{molecular\ weight} = \frac{mg/l}{equivalent\ weight} \tag{2-2}$$

Equivalent weights for selected elements, radicals, and inorganic compounds are given in Tables 2-1, 2-2, and 2-3, respectively.

A typical chemical water analysis is in Table 2-4. These data can be

Table 2-4. Chemical Analysis of a Surface Water (Values in mg/liter)[a]

Alkalinity (as $CaCO_3$)	108	Lead	0
Alkylbenzene sulfonate		Magnesium	9.9
(detergent)	0.1	Nitrate	2.2
Arsenic	0	pH	7.6
Barium	0	Phenols	0
Bicarbonate	131	Phosphorus	
Cadmium	0	(total inorganic)	0.5
Calcium	35.8	Potassium	3.9
Carbon chloroform extract		Selenium	0
(exotic organic chemicals)	0.04	Silver	0
Chloride	7.1	Sodium	4.6
Chromium	0	Sulfate	26.4
Copper	0.10	Total dissolved solids	220
Cyanide	0	Zinc	0
Fluoride	0.7	Turbidity	5
Iron plus manganese	0.13	Color	5
Iron	0.10	Threshold odor number	1

[a] Tests for Radioactivity and Pesticides not included.

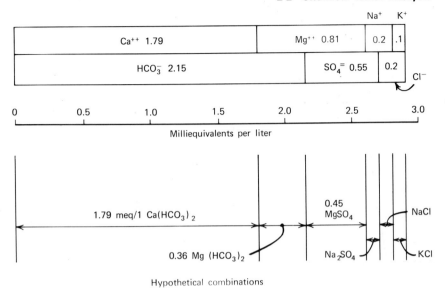

Figure 2-1 Milliequivalents per liter bar graph for water analysis given in Table 2-4.

compared against the chemical characteristics specified by the Drinking Water Standards and Guidelines[1] to determine whether the water is safe for human consumption, or if treatment is required before domestic or industrial use.

Reporting results in milligrams per liter in tabular form is not convenient for visualizing the chemical composition of a water, therefore, results are often expressed in milliequivalents per liter, which permits graphical presentation and a quick check on the accuracy of the analyses for major ions. The sum of the milliequivalents per liter of the cations (positive radicals) must equal the sum of the anions (negative radicals). In a perfect evaluation they would be exactly the same since a water in equilibrium is electrically balanced. Graphical presentation of a water analysis using milliequivalents is performed by plotting the meq/l values to scale, for example, by letting 1 meq/l equal 1 inch. The bar graph shown in Figure 2-1 is based on the water data in Table 2-4. The top row consists of the major cations arranged in the order of calcium, magnesium, sodium, and potassium. The under row is arranged in the sequence of bicarbonate, sulfate, and chloride.

Hypothetical combinations of the positive and negative ions can be developed from a bar graph as in Figure 2-1. These combinations are particularly helpful in considering lime-soda water softening. In Figure 2-1, the carbonate hardness $[Ca(HCO_3)_2 + Mg(HCO_3)_2]$ is 2.15 meq/l and the noncarbonate is 0.45 meq/l ($MgSO_4$).

EXAMPLE 2-2

The results of a water analysis are: calcium 29.0 mg/l, magnesium 16.4 mg/l, sodium 23.0 mg/l, potassium 17.5 mg/l, bicarbonate 171 mg/l (as HCO_3), sulfate 36.0 mg/l and chloride 24.0 mg/l. Convert milligrams per liter concentrations to milliequivalents per liter, list hypothetical combinations, and calculate hardness as mg $CaCO_3$/l for this water.

Solution

Using Eq. 2-1,

Component	mg/l	Equivalent Weight	meq/1
Ca^{++}	29.0	20.0	1.45
Mg^{++}	16.4	12.2	1.34
Na^+	23.0	23.0	1.00
K^+	17.5	39.1	0.45
Total Cations			4.24
HCO_3^-	171	61.0	2.81
$SO_4^=$	36.0	48.0	0.75
Cl^-	24.0	35.5	0.68
Total Anions			4.24

Hypothetical combinations: $Ca(HCO_3)_2$, 1.45 meq/l; $Mg(HCO_3)_2$, 1.34; $NaHCO_3$, 0.02; Na_2SO_4, 0.75; NaCl, 0.23; and KCl, 0.45 meq/l.
Hardness $(Ca^{++} + Mg^{++}) = 2.79$ meq/l
Equivalent Weight $CaCO_3 = 50.0$ (Table 2-3)
By Eq. 2-2,
Hardness $= 2.79 \times 50.0 = 140$ mg/l

2-3   pH AND CHEMICAL REACTIONS

Water (H_2O) dissociates to only a slight degree yielding a concentration of hydrogen ions equal to 10^{-7} mole per liter. Because water yields one hydroxyl (basic) ion for each hydrogen (acid) ion, pure water is considered neutral.

$$H_2O \rightleftarrows H^+ + OH^- \qquad (2\text{-}3)$$

The acidic nature of a water is related to the concentration of hydrogen ions in water solution by use of the symbol pH, where

$$pH = \log \frac{1}{[H^+]} \qquad (2\text{-}4)$$

Since the logarithm of 1 over 10^{-7} is 7, the pH at neutrality is 7.

When an acid is added to water, the hydrogen ion concentration increases resulting in a lower pH number. Conversely, when an alkaline substance is added, the OH^- ions unite with the free H^+, lowering the hydrogen ion concentration causing a higher pH. The pH scale, ranging from 0 to 14, is acid from 0 to 7 and basic from 7 to 14.

Acid range	Basic range
0	7	14

Acid-Base Reactions (Neutralization)

Sulfuric acid added to water dissociates into hydrogen ions (H^+) and sulfate ions ($SO_4^=$) resulting in an acid solution. A basic solution is formed by adding an alkali, such as, NaOH, to water. If both an acid and a base are put in the same water, the H^+ ions combine with the OH^- ions to form water and, if equivalent amounts are added, they neutralize each other forming a salt in solution that is shown in the following equation:

$$H_2SO_4 + 2NaOH = 2H_2O + Na_2SO_4 \qquad (2\text{-}5)$$

A chemical equation presents the formulas for reactants and products, and gives the number of moles of each. It is said to be balanced when the number of atoms for each element on the left side equals those on the right. By using a balanced equation, molecular weight relationships can be used to calculate the quantities of reactants and products. The process of using a balanced chemical equation for making calculations is called stoichiometry. Equivalent weights can be used to determine amounts of reactants that combine in certain types of simple chemical reactions.

Oxidation and Reduction Reactions

Many chemical changes involve the addition of oxygen and/or the change of valence of one of the reacting substances. Oxidation is the addition of oxygen or removal of electrons, and reduction is the removal of oxygen or addition of electrons. A classic oxidation reaction is the rusting of iron by oxygen:

$$4Fe + 3O_2 = 2Fe_2O_3 \qquad (2\text{-}6)$$

The iron is oxidized from Fe to Fe^{+++} while the oxygen is reduced from O to $O^=$.

A practical example of an oxidation-reduction reaction in water treatment is the removal of soluble ferrous iron from solution by oxidation using potassium permanganate. In this reaction (2-7) the iron gains one positive charge while the manganese in the permanganate ion is reduced from a valence of 7+ to a valence of 4+ in manganese dioxide. (Note that an arrow is used instead of an equal sign in Eq. 2-7, since the equation is not balanced. The arrows pointing downward in the right side indicate that these products are solid precipitates.)

$$3Fe^{2+} + Mn^{(7+)}O_4^- \rightarrow 3Fe^{3+} \downarrow + Mn^{(4+)}O_2 \downarrow \qquad (2\text{-}7)$$

Precipitation Reactions

Water is a solvent for most inorganic compounds. Some solids, such as sodium chloride, dissolve readily while others, such as iron, ionize very slowly. Solid calcium carbonate, the common scale found in water pipes, dissociates to calcium and carbonate ions to a degree depending on pH of the water. Water low in calcium ions (soft water) tends to dissolve calcium carbonate scale, whereas a hard water tends to precipitate calcium carbonate on the inside of the pipe. Lime-soda softening involves removal of calcium and magnesium ions from solution by precipitation reactions. In this process, lime slurry $Ca(OH)_2$ is applied to the hard water raising the pH and supplying additional calcium ions. This results in precipitation of the calcium as calcium carbonate (Eq. 2-8), which then can be settled and filtered from the water. Chemical precipitation is also used in the removal of other undesirable ions from solution, such as iron and manganese.

EXAMPLE 2-3

How many pounds of pure sodium hydroxide are required to neutralize an industrial waste with an acidity equivalent to 50 lb of sulfuric acid per million gallons of waste water?

Solution

Using Eq. 2-5
2 moles NaOH neutralize 1 mole H_2SO_4
2×40.0 lb NaOH react with 1×98.1 lb H_2SO_4

$$\text{lb NaOH required} = \frac{80.0 \text{ lb NaOH}}{98.1 \text{ lb } H_2SO_4} \times 50 \text{ lb } H_2SO_4 = 40.8 \text{ lb NaOH/mil gal}$$

Alternate Solution

In neutralization reactions one equivalent of base reacts with one of acid, therefore, by using equivalent weights (Table 2-3) rather than molecular values, the calculation is

$$\text{lb NaOH required} = \frac{40.0 \text{ NaOH}}{49.0 \text{ } H_2SO_4} \times 50 \text{ lb } H_2SO_4 = 40.8 \text{ lb NaOH/mil gal}$$

EXAMPLE 2-4

Theoretically, how many milligrams per liter of reduced iron can be oxidized by one milligrams per liter of potassium permanganate?

Solution

From Eq. 2-7,
$3Fe^{++}$ $(3 \times 55.8 = 167)$ are oxidized for each MnO_4^- reduced ($KMnO_4$ weight is 158)

$$\frac{158 \text{ } KMnO_4}{167 \text{ Fe}} = \frac{1 \text{ mg/l}}{X \text{ mg/l}} \qquad X = 1.06 \text{ mg/l}$$

EXAMPLE 2-5

In precipitation water softening, lime is used to remove calcium hardness by the reaction:

$$CaO + Ca(HCO_3)_2 = 2CaCO_3 \downarrow + H_2O \qquad (2\text{-}8)$$

What dosage of lime with a purity of 78 percent CaO is required to combine with 70 mg/l of calcium?

Solution

1 mole of $Ca(HCO_3)_2$ (162 g) contains 40.1 g of calcium, therefore, 70 mg/l of calcium is equivalent to

$$\frac{162}{40.1} \times 70 = 283 \text{ mg/l } Ca(HCO_3)_2$$

56.1 g of CaO combines with 162 g of $Ca(HCO_3)_2$, therefore 283 mg/l $Ca(HCO_3)_2$ reacts with

$$\frac{56.1}{162} \times 283 = 98.0 \text{ mg/l CaO}$$

For a purity of 78 percent the dosage of commercial lime required is

$$\frac{98.0 \text{ CaO}}{0.78} = 126 \text{ mg/l} = 1050 \text{ lb/mil gal}$$

2-4 GAS SOLUBILITY AND ALKALINITY

Most gases are either soluble in water to some degree or react chemically with water. An exception is methane gas (CH_4), commonly referred to as illuminating gas, which does not interact with water to a measurable extent. (Methane is produced in anaerobic decomposition of waste sludge and may be collected and burned for its heat value.) The two major atmospheric gases, nitrogen and oxygen, while not reacting chemically with water, dissolve to a limited degree. The solubility of each gas is directly proportional to pressure it exerts on the water. At a given pressure, solubility of oxygen varies greatly with water temperature and to a lesser degree with salinity (chloride concentration). Dissolved oxygen saturation values, based on normal atmosphere consisting of 21 percent oxygen, for various temperatures and chloride concentrations are listed in Table 2-5. The table footnote explains how to correct oxygen solubility values for barometric pressure. The change of barometric pressure with altitude depends on the rate of decrease in density of the air. Near the earth's surface, it is sufficiently accurate for most engineering computations to say that pressure decreases at a rate of 25 mm (1.0 in.) of mercury per 1000 ft of increased elevation.

Gases may be formed in solution by the decomposition of organic substances in water. Ammonia biochemically released from nitrogenous matter appears in solution as the ammonium radical if the solution is acidic; however, if the water is basic, it remains as ammonia gas. One technique for removing ammonia nitrogen from waste water uses the principle of raising the pH and then air stripping the ammonia gas. Another gas released from septic waste is hydrogen sulfide (H_2S), identifiable by its rotten egg odor. The SH^- group is biochemically produced in solution, converted to H_2S under reduced conditions, and released from solution as a gas. In sewers this can lead to crown corrosion of the pipe by having the H_2S oxidized to sulfuric acid (H_2SO_4) in the condensation moisture hanging on the interior of the pipe.

Chlorine, the most common oxidizing agent used for disinfection of municipal water supplies in the United States, reacts with water as shown in Eq. 2-9. Liquid chlorine is shipped to treatment plants in pressurized steel cylinders. On pressure release the liquid converts to chlorine gas and is applied to the water. In dilute solution and at pH levels above 4, chlorine combines with water molecules to form hydrochloric acid plus hypochlorous acid which, in turn, yields the hypochlorite ion. The degree of ionization is dependent on pH, with HOCl occurring below 6 and almost complete ionization above pH 9. The pH of a water during disinfection is important, since HOCl is more effective than OCl^- in killing harmful organisms.

$$Cl_2 + H_2O \underset{\text{pH}<4}{\overset{\text{pH}>4}{\rightleftarrows}} HCl + HOCl \underset{\text{pH}<6}{\overset{\text{pH}>9}{\rightleftarrows}} H^+ + OCl^- \qquad (2\text{-}9)$$

Table 2-5. Saturation Values of Dissolved Oxygen in Water Exposed to Water-Saturated Air Containing 20.90 Percent Oxygen Under a Pressure of 760 mm of Mercury[a]

Temperature in °C	Chloride Concentration in Water mg/l			Difference per 100 mg Chloride	Temperature in °C	Vapor Pressure mm
	0	5000	10,000			
	Dissolved Oxygen mg/l					
0	14.6	13.8	13.0	0.017	0	5
1	14.2	13.4	12.6	0.016	1	5
2	13.8	13.1	12.3	0.015	2	5
3	13.5	12.7	12.0	0.015	3	6
4	13.1	12.4	11.7	0.014	4	6
5	12.8	12.1	11.4	0.014	5	7
6	12.5	11.8	11.1	0.014	6	7
7	12.2	11.5	10.9	0.013	7	8
8	11.9	11.2	10.6	0.013	8	8
9	11.6	11.0	10.4	0.012	9	9
10	11.3	10.7	10.1	0.012	10	9
11	11.1	10.5	9.9	0.011	11	10
12	10.8	10.3	9.7	0.011	12	11
13	10.6	10.1	9.5	0.011	13	11
14	10.4	9.9	9.3	0.010	14	12
15	10.2	9.7	9.1	0.010	15	13
16	10.0	9.5	9.0	0.010	16	14
17	9.7	9.3	8.8	0.010	17	15
18	9.5	9.1	8.6	0.009	18	16
19	9.4	8.9	8.5	0.009	19	17
20	9.2	8.7	8.3	0.009	20	18
21	9.0	8.6	8.1	0.009	21	19
22	8.8	8.4	8.0	0.008	22	20
23	8.7	8.3	7.9	0.008	23	21
24	8.5	8.1	7.7	0.008	24	22
25	8.4	8.0	7.6	0.008	25	24
26	8.2	7.8	7.4	0.008	26	25
27	8.1	7.7	7.3	0.008	27	27
28	7.9	7.5	7.1	0.008	28	28
29	7.8	7.4	7.0	0.008	29	30
30	7.6	7.3	6.9	0.008	30	32

[a] Saturation at barometric pressures other than 760 mm (29.92 in.), C'_s is related to the corresponding tabulated values, C_s, by the equation:

$$C'_s = C_s \frac{P-p}{760-p}$$

where C'_s = solubility at barometric pressure P and given temperature, milligrams per liter
C_s = saturation at given temperature from table, milligrams per liter
P = barometric pressure, millimeters
p = pressure of saturated water vapor at temperature of the water selected from table, millimeters

Carbon dioxide, although only about 0.03 percent of atmospheric air, plays a major role in water chemistry, since it reacts readily with water, forming bicarbonate and carbonate radicals. CO_2 may be absorbed from the air or may be produced by bacterial decomposition of organic matter in the water. Once in solution, it reacts to form carbonic acid.

$$CO_2 + H_2O \rightleftharpoons H_2CO_3 \underset{pH\ 4.5}{\rightleftharpoons} H^+ + HCO_3^- \underset{pH\ 8.3}{\rightleftharpoons} H^+ + CO_3^= \quad (2\text{-}10)$$

When the pH of the water is greater than 4.5, carbonic acid ionizes to form bicarbonate which, in turn, is transformed to the carbonate radical if the pH is above approximately 8.3. Carbon dioxide is very aggressive and leads to corrosion of water pipes; therefore, water supplies with low pH are frequently neutralized with a base to reduce pipe corrosion. On the other hand, alkaline waters with $CO_3^=$ ions are hard and form scale by precipitation of $CaCO_3$. Water treatment practice to lower the pH or soften such a water may be advantageous.

The bicarbonate-carbonate character of a water can be analyzed by slowly adding a strong acid solution to a sample of water and reading resultant changes in pH. This process, called titration, is used to measure the alkalinity of a water. Acidity is measured by titrating with a strong basic solution. If two samples of pure water (pH 7) are titrated with sulfuric acid and sodium hydroxide solutions, respectively, the composite titration curve is as shown by Figure 2-2a. Very small initial additions of either titrant results in significant changes in pH, since addition or withdrawal of hydrogen ions is reflected immediately in changing pH readings. Curve b is a titration curve of a water containing a high initial concentration of carbonate ions prepared by adding sodium carbonate to distilled water. When acid is added, the majority of the hydrogen ions from acid combine with the carbonate ions to form bicarbonates (Eq. 2-10). The excess hydrogen ions lower the pH gradually until at pH 8.3 all carbonate radicals have been converted to bicarbonates. Additional hydrogen ions reduce the bicarbonates to carbonic acid below pH 4.5. Stirring of the sample at this time results in release of carbon dioxide gas formed from the original carbonates. Figure 2-2c is the reverse titration of curve b where the addition of a strong base results in conversion of carbonic acid to bicarbonate and, in turn, to carbonate; this follows Eq. 2-10 from left to right. Substances in solution, such as the various ionic forms of carbon dioxide, that offer resistance to change in pH as acids or bases are added, are referred to as buffers. An understanding of buffering action in water and waste-water chemistry is essential, since many chemical reactions in water treatment and biological reactions in waste-water treatment are pH dependent and rely on pH control. Curve d is a typical titration curve of a relatively hard well water having an initial pH of 7.8. Inflections of the curve near pH values 4.5 and 8.3 indicate that the primary buffer is the bicarbonate-carbonate system. Irregularities and deviations from the ideal bicarbonate buffer curve, Figure 2-2b, are frequently observed in actual titrations of water and waste-water samples; these may be caused by any number of buffering compounds frequently found in low concentrations, such as phosphates ($H_2PO_4^-$, $HPO_4^=$, $PO_4^≡$).

Alkalinity of a water is a measure of its capacity to neutralize acids, in other words, to absorb hydrogen ions without significant pH change. Alkalinity is measured by titrating a given sample with 0.02 normal (N) sulfuric acid. (Normality is a method of expressing the strength of a chemical solution. A

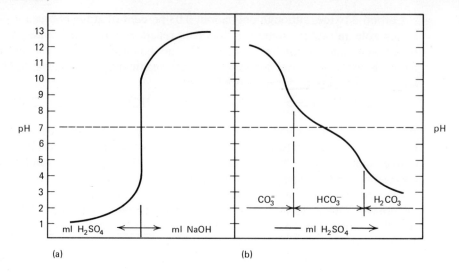

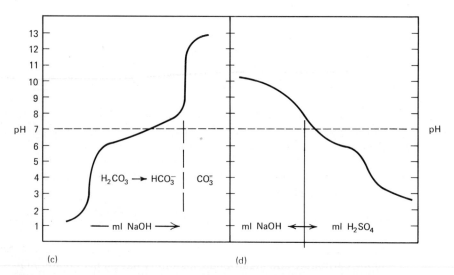

Figure 2-2 Titration curves for pure water (strong acids and bases), carbonate (salt of a weak acid), carbonic acid (weak acid), and typical well water. (a) Pure water. (b) Carbonate in water. (c) Carbonic acid. (d) Well water.

1.00 N solution contains one gram of available hydrogen ions, or its equivalent, per liter of solution. A 0.02 N H_2SO_4 solution contains 0.98 grams of pure sulfuric acid, which is 0.02 times the equivalent weight of sulfuric acid.) For highly alkaline samples, the first step is titrating to a pH of 8.3. The second phase, or first in the case of a water with an initial pH of less than 8.3, is titrating to an indicated pH of 4.5. Colorimetric indicators, chemicals that change color at specific pH values, can be used to determine the end points of these titrations if a pH meter is not used. Phenolphthalein turns from pink to colorless at pH 8.3, and methyl orange changes from orange at pH 4.6 to pink at 4.0. Mixed bromocresol green-methyl red indicator, a suitable substitute for methyl orange, changes from gray to pink at pH 4.8–4.6. That part of the total alkalinity above 8.3 is referred to as phenolphthalein alkalinity. Alkalinity is

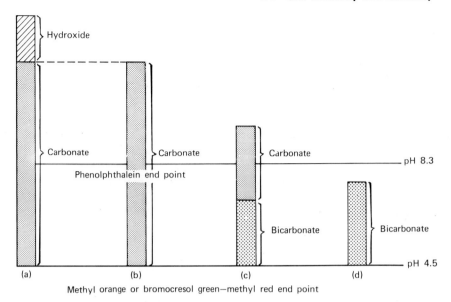

Figure 2-3 Graphical representation of various forms of alkalinity in water samples related to titration end points. Using abbreviations of P for the measured phenolphthalein alkalinity and T for the total alkalinity, the following relationships define the bar graphs in: (a) Hydroxide = $2P - T$, and Carbonate $2(T - P)$; (b) Carbonate = $2P = T$; (c) Carbonate = $2P$, and Bicarbonate = $T - 2P$, and (d) Bicarbonate = T.

conventionally expressed in terms of mg $CaCO_3$ per liter and is calculated by Eq. 2-11. Note, if the water sample is 100 ml and normality of acid 0.02 N, 1 ml of titrant represents 10 mg/l alkalinity.

$$\frac{\text{Alkalinity}}{\text{as mg/l } CaCO_3} = \frac{\text{ml titrant} \times \text{normality of acid} \times 50{,}000}{\text{ml sample}} \qquad (2\text{-}11)$$

Figure 2-3 is a graphical representation of the various forms of alkalinity found in water samples. This diagram can be related to Eq. 2-10. If pH of a sample is less than 8.3, the alkalinity is all in the form of bicarbonate. Samples containing carbonate and bicarbonate alkalinity have a pH greater than 8.3, and titration to the phenolphthalein end point represents one half of the carbonate alkalinity, or conversion to HCO_3^-. If the milliliters of phenolphthalein titrant equals the milliliters of methyl orange titrant, all of the alkalinity is in the form of carbonate. Any alkalinity in excess of this amount is due to hydroxide (OH^-).

EXAMPLE 2-6

What is the dissolved oxygen concentration in water at 18° C, 800 mg/l chloride concentration, and barometric pressure of 660 mm (4000-ft altitude).

Solution

From Table 2-5, saturation at 18° C with 800 mg/l chloride concentration is
$9.5 - 8 \times 0.009 = 9.4$ mg/l
Correcting for a barometric pressure of 660 mm by using the equation in the

footnote to Table 2-5,

$$C'_s = 9.4 \frac{660-16}{760-16} = 8.1 \text{ mg/l}$$

EXAMPLE 2-7

A 100-ml water sample is titrated for alkalinity by using 0.02 N sulfuric acid. To reach the phenolphthalein end point requires 3.0 ml, and an additional 12.0 ml is added for the methyl orange color change. Calculate the phenolphthalein and total alkalinities. Give the ionic forms of the alkalinity present.

Solution

Using Eq. 2-11

$$\frac{\text{Phenolphthalein}}{\text{alkalinity}} = \frac{3.0 \times 0.02 \times 50,000}{100} = 30 \text{ mg/l (as } CaCO_3)$$

$$\frac{\text{Total}}{\text{alkalinity}} = \frac{15.0 \times 0.02 \times 50,000}{100} = 150 \text{ mg/l (as } CaCO_3)$$

Referring to Figure 2-3

Alkalinity as carbonate $(CO_3^=) = 2 \times 30 = 60 \text{ mg/l}$

Alkalinity as bicarbonate $(HCO_3^-) = 150 - 60 = 90 \text{ mg/l}$

2-5   COLLOIDS AND COAGULATION

A large variety of turbidity-producing substances found in polluted waters do not settle out of solution, for example, color compounds, clay particles, microscopic organisms, and organic matter from decaying vegetation or municipal wastes. These particles, referred to as colloids, range in size from 1 to 500 millimicrons and are not visible when using an ordinary microscope. A colloidal dispersion formed by these particles is stable in quiescent water, since the individual particles have such a large surface area relative to their weight that gravity forces do not influence their suspension. Colloids are classified as hydrophobic (water hating) and hydrophilic (water loving).

Hydrophilic colloids are stable because of their attraction for water molecules, rather than the slight charge that they might possess. Typical examples are soap, soluble starch, synthetic detergents, and blood serum. Because of their affinity for water, they are not as easy to remove from suspension as hydrophobic colloids and coagulant dosages of 10 to 20 times the amount used in conventional water treatment are frequently necessary to coagulate hydrophilic materials.

Hydrophobic particles, possessing no affinity for water, are dependent on electrical charge for their stability in suspension. The bulk of inorganic and organic matter in a turbid natural water is of this type. The forces acting on hydrophobic colloids are illustrated in Figure 2-4a. Individual particles are held apart by electrostatic repulsion forces developed by positive ions adsorbed onto their surfaces from solution. An analogy is the repulsive force that exists between the common poles of two bar magnets. The magnitude of the repulsive

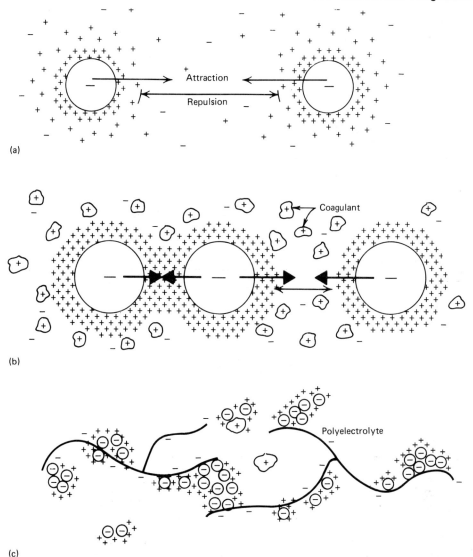

(a)

(b)

(c)

Figure 2-4 Schematic actions of colloids and coagulation. (a) Forces acting on hydrophobic colloids in stable suspension. (b) Compression of the double-layer charge on colloids (destabilization) by addition of chemical coagulants. (c) Agglomeration resulting from coagulation with metal salt and polymer aid.

force developed by the charged double layer of ions attracted to a particle is referred to as the zeta potential.

A natural force of attraction exists between any two masses (Van der Waals force). Random motion of colloids (Brownian movement), caused by bombardment of water molecules, tends to enhance this physical force of attraction in pulling the particles together. However, a colloidal suspension remains dispersed indefinitely when the forces of repulsion exceed those of attraction and the particles are not allowed to contact.

The purpose of chemical coagulation in water treatment is to destabilize suspended contaminants such that the particles contact and agglomerate, forming flocs that drop out of solution by sedimentation. Destabilization of hydrophobic colloids is accomplished by addition of chemical coagulants, for instance, salts of aluminium and iron. Highly charged hydrolyzed metal ions produced by these salts in solution reduce the repulsive forces between colloids by compressing the diffuse double layer surrounding individual particles (Figure 2-4b). With the forces of repulsion suppressed, gentle mixing results in particle contact, and the forces of attraction cause particles to stick to each other, producing progressive agglomeration. Coagulant aids may be used to enhance the process of flocculation. For example, organic polymers provide bridging between particles by attaching themselves to the absorbant surfaces of colloids, building larger flocculated masses (Figure 2-4c).

Removal of turbidity by coagulation depends on the type of colloids in suspension; the temperature, pH, and chemical composition of the water; the type and dosage of coagulants and aids; and, the degree and time of mixing provided for chemical dispersion and floc formation. Although in chemistry the term coagulation means the destabilization of a colloidal dispersion by suppressing the double layer (Figure 2-4a), and flocculation refers to aggregation of the particles, engineers have not traditionally restricted the use of these terms to describe the chemical mechanisms only. More frequently, coagulation and flocculation are associated with the physical processes used in chemical treatment. Mixing, involving violent agitation, is used to dissolve and disperse coagulant chemicals throughout the water being treated. Flocculation is a slow mixing process, following chemical dispersion, during which the destabilized particles form into well-developed flocs of sufficient size to settle. The word coagulation is commonly used to describe the entire mixing and flocculation operation. Actual units used in chemical treatment may be constructed in series (mixing—flocculation—sedimentation) or may be housed in a single compartment tank. An in-line system, as is shown in Figure 7-8, normally provides about one minute for flash mixing, thirty minutes flocculation, and four hours sedimentation, followed by sand filtration to remove nonsettleable matter. A flocculator-clarifier (Figure 7-9) mixes raw water and applied chemicals with previously flocculating solids in the center mixing compartment, thus contacting settleable solids with untreated water and fresh chemicals. Solids settled in the peripheral zone automatically return to the mixing area; excess solids buildup is withdrawn from the bottom for disposal.

2-6   ORGANIC COMPOUNDS

All organic compounds contain carbon atoms connected to each other in chain or ring structures with other elements attached. Major components are

carbon, hydrogen, and oxygen; minor elements include nitrogen, phosphorus, sulfur, and certain metals. Each carbon exhibits four connecting bonds. Organics are derived from nature, for example, plant fibers and animal tissues, produced by synthesis reactions producing rubber, plastics, and the like, and through fermentation processes, for instance, alcohols, acids, antibiotics, and others. In contrast to inorganic compounds, organic substances are usually combustible, high in molecular weight, only sparingly soluble in water reacting as molecules rather than ions, and are a source of food for animal consumers and microbial decomposers.

The molecular formula for an inorganic compound is specific; however, in organic chemistry an empirical formula may represent more than one compound. For example, the formula C_2H_6O may be arranged to represent the following two substances:

$$\begin{array}{ccccc} & H & H & & H & & H \\ & | & | & & | & & | \\ H- & C- & C- & OH & H- & C- & O- & C- & H \\ & | & | & & | & & | \\ & H & H & & H & & H \end{array}$$

The one on the left is ethyl alcohol while the one on the right is dimethyl ether. Compounds having the same molecular formula but that differ in structural formulas are called isomers. Because of isomerism and the frequency of ring structures, organic compounds are normally given as graphic formulas rather than as molecular ones. For ease of printing, formulas not containing rings are frequently written on a single line. For example, ethyl alcohol and dimethyl ether become

$$CH_3CH_2OH \quad \text{and} \quad CH_3-O-CH_3$$

Hydrocarbons

Saturated hydrocarbons (paraffins and alkanes) contain only the elements of carbon and hydrogen with single bonds between carbon atoms. Methane (CH_4), the simplest hydrocarbon, is a gas produced in the anaerobic decomposition of waste sludge. When mixed with approximately 90 percent air, it is highly explosive and can be used as a fuel for gas engines or sludge heaters. Short-chain paraffin hydrocarbons can be obtained from petroleum by fractional distillation. Propane is available in high pressure cylinders for a heating fuel, Eq. 2-12.

$$CH_3CH_2CH_3 + 5O_2 \rightarrow 3CO_2 + 4H_2O + \text{energy} \tag{2-12}$$

Names and formulas of one to four carbon saturated hydrocarbons along with the radicals of these parent compounds are given in Table 2-6,

Unsaturated hydrocarbons (olefins) are distinguished from paraffins by the presence of multiple bonds between some carbon atoms. For example, ethene

Table 2-6. Saturated Hydrocarbons

Name	Formula	Radical	Formula
Methane	CH_4	Methyl	CH_3-
Ethane	CH_3CH_3	Ethyl	CH_3CH_2-
Propane	$CH_3CH_2CH_3$	n-Propyl	$CH_3CH_2CH_2-$
		Isopropyl	$\begin{array}{c} CH_3 \\ \diagdown \\ CH- \\ \diagup \\ CH_3 \end{array}$
n-Butane	$CH_3CH_2CH_2CH_3$	n-Butyl	$CH_3CH_2CH_2CH_2-$

(ethylene) and acetylene are, respectively,

$$\begin{array}{cc} \overset{\displaystyle H}{\underset{\displaystyle |}{}} \ \overset{\displaystyle H}{\underset{\displaystyle |}{}} \\ H-C=C-H \end{array} \qquad H-C\equiv C-H$$

The multiple bonds between carbons displace hydrogen atoms creating units containing fewer hydrogens than could be attached, hence, the term "unsaturated." Natural vegetable oils containing a large number of unsaturated bonds are liquids at room temperature. Popular vegetable shortenings available as solid fats are commercially produced from oils through the process of hydrogenation—the addition of hydrogen gas under controlled conditions. Reducing the number of unsaturated bonds increases the melting point, converting an oil to solid fat.

The parent compound of aromatic hydrocarbons is benzene. It is a six-carbon ring with double bonds between alternate atoms. Frequently the carbon

Benzene

atoms are not shown in writing the graphic formula. Benzene is used in the manufacture of a wide variety of commercial products including explosives, insecticides, certain plastics, solvents, and dyes.

Alcohols

These are formed from hydrocarbons by replacing one or more hydrogen atoms by hydroxyl groups (—OH). Names and formulas of three primary alcohols are listed in Table 2-7. Methanol is manufactured synthetically by a catalytic process from carbon monoxide and hydrogen. It is used extensively in manufacturing organic compounds, such as formaldehyde, solvents, and fuel additives. Ethyl alcohol for beverage purposes is produced by fermentation of a variety of natural materials—corn, wheat, rice, and potatoes. Industrial ethanol is produced from fermentation of waste solutions containing sugars, for example, black strap molasses, a residue resulting from the purification of cane sugar. Propanol has two isomers, the more common being isopropyl alcohol widely used by industry and sold as a medicinal rubbing alcohol. The three primary alcohols listed in Table 2-7 have boiling points less than 100° C, and they are miscible with water.

The derivative of benzene containing one hydroxyl group, known as phenol, has a molecular formula of C_6H_5OH and a graphic representation of

The formula and -ol name ending indicate the characteristics of an alcohol, but phenol, commonly known as carbolic acid, ionizes in water-yielding hydrogen

Table 2-7. Primary Alcohols

Common Name	Proper Name	Formula
Methyl alcohol	Methanol	CH_3OH
Ethyl alcohol	Ethanol	CH_3CH_2OH
n-Propyl alcohol	1-Propanol	$CH_3CH_2CH_2OH$
Isopropyl alcohol	2-Propanol	$\begin{array}{c} CH_3 \\ \diagdown \\ \quad CHOH \\ \diagup \\ CH_3 \end{array}$

ions and exhibits features of an acid. It occurs as a natural component in wastes from coal-gas petroleum industries and in a wide variety of industrial wastes where phenol is used as a raw material. Phenol is a strong toxin that makes these wastes particularly difficult to treat in biological systems. Phenols also impart undesirable taste to water at extremely low concentrations. Drinking water standards set a phenol limit of 0.001 mg/l for this reason.

Aldehydes and Ketones

These are compounds containing the carbonyl group. Formaldehyde, in addition to being a preservative for biological specimens, is used in producing a variety of plastics and resins. Acetone (dimethyl ketone) is a good solvent of fats and is a common cleaning agent for laboratory glassware.

$$\underset{\text{Carbonyl group}}{-\overset{\displaystyle O}{\overset{\|}{C}}-} \qquad \underset{\text{Formaldehyde}}{H-\overset{\displaystyle O}{\overset{\|}{C}}-H} \qquad \underset{\text{Acetone}}{CH_3-\overset{\displaystyle O}{\overset{\|}{C}}-CH_3}$$

Carboxylic Acids

All organic acids contain the carboxyl group written as

$$-\overset{\displaystyle O}{\overset{\|}{C}}-OH \qquad \text{or} \qquad -COOH$$

This is the highest state of oxidation that an organic radical can achieve. Further oxidation results in the formation of carbon dioxide and water.

$$\underset{\text{Hydrocarbon}}{CH_4} \xrightarrow{+O} \underset{\text{Alcohol}}{CH_3OH} \xrightarrow{-2H} \underset{\text{Aldehyde}}{H_2C{=}O} \xrightarrow{+O} \underset{\text{Acid}}{HCOOH} \xrightarrow{+O} CO_2+H_2O$$

Names, formulas, and ionization constants for the five simplest acids are given in Table 2-8. Acids through nine carbon are liquids but those with longer chains are greasy solids and, hence, the common name fatty acid.

Inorganic strong acids and bases approach 100 percent ionization in dilute solutions. In contrast, organic acids are weak and ionize so poorly that it is impractical to express the degree of ionization as a percentage. Furthermore, the amount of dissociation is quite independent of concentration. These observations have led to the use of constants as a mathematical method of expressing degree of ionization. For a typical organic acid, let us consider the dissociation of acetic acid:

$$CH_3-\overset{\displaystyle O}{\overset{\|}{C}}-OH \rightleftarrows CH_3-\overset{\displaystyle O}{\overset{\|}{C}}-O^- + H^+ \qquad (2\text{-}13)$$

Table 2-8. Carboxylic Acids

Common Name	Proper Name	Formula	Ionization Constant, K
Formic	Methanoic	HCOOH	0.000,214
Acetic	Ethanoic	CH_3COOH	0.000,018
Propionic	Propanoic	CH_3CH_2COOH	0.000,014
n-Butyric	Butanoic	$CH_3CH_2CH_2COOH$	0.000,015
Valeric	Pentanoic	C_4H_9COOH	0.000,016
Caproic	Hexanoic	$C_5H_{11}COOH$	Almost insoluble

The ionization (equilibrium) constant is equal to the molar concentration of acetate ion times that of hydrogen ion divided by the molar quantity of molecular acetic acid in solution, Eq. 2-14. The K-value for acetic acid is 0.000,018 at 25° C, which means that in moderately dilute solution only 18 out of every million acetic acid molecules in water dissociate releasing hydrogen ions.

$$K = \frac{[CH_3-\overset{\overset{O}{\|}}{C}-O^-][H^+]}{[CH_3-\overset{\overset{O}{\|}}{C}-OH]} = 0.000,018 \qquad (2\text{-}14)$$

Formic, acetic, and propionic acids have sharp penetrating odors, while butyric and valeric have extremely disagreeable odors associated with rancid fats and oils. Anaerobic decomposition of long-chain fatty acids results in production of two- and three-carbon acids that are then converted to methane and carbon dioxide gas in digestion of waste sludge.

Basic compounds react with acids to produce salts. For example, NaOH reacts with acetic acid to produce sodium acetate. Soaps are salts of long-chain

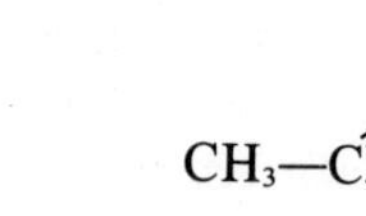

$$CH_3C\overset{\displaystyle \diagup O}{\diagdown O^-Na^+}$$

Sodium Acetate

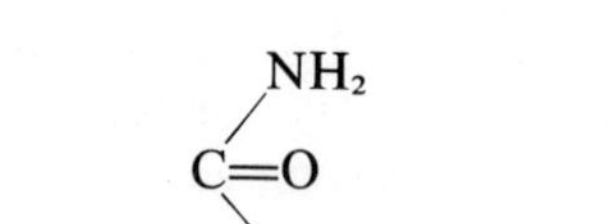

$$CH_3(CH_2)_{13}CH_2C\overset{\displaystyle \diagup O}{\diagdown O^-Na^+}$$

Sodium Palmitate (a soap)

fatty acids. Other derivatives of carboxylic acids include esters, such as ethyl acetate, and amides, for example, urea.

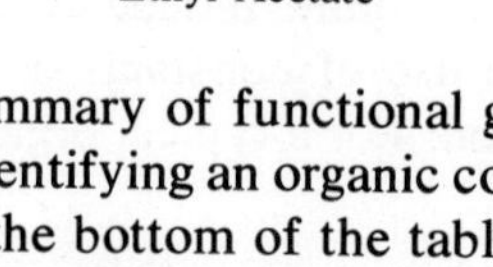

$$CH_3-C\overset{\displaystyle \diagup O}{\diagdown OCH_2CH_3}$$

Ethyl Acetate

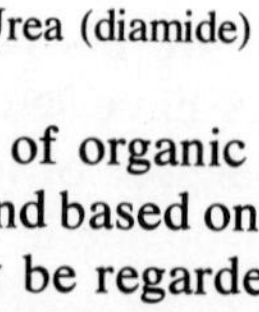

$$C\overset{\displaystyle \diagup NH_2}{\underset{\displaystyle NH_2}{=}O}$$

Urea (diamide)

Table 2-9 is a summary of functional groups of organic compounds. These data are useful in identifying an organic compound based on a formula or name. Amines, shown at the bottom of the table, may be regarded as derivatives of ammonia in which hydrogen atoms have been replaced by carbon chains or rings.

Table 2-9. Functional Groups of Organic Compounds

Functional Group	Type of Compound	Proper Name Ending	Example	Common Name	Proper Name
—C—C—	Saturated hydrocarbon	-ane	CH_3CH_3	Ethane	Ethane
—C=C—	Olefin	-ene	$H_2C=CH_2$	Ethylene	Ethene
—C≡C—	Acetylene	-yne	$H—C≡C—H$	Acetylene	Ethyne
—OH	Alcohol	-ol	$CH_3CH_3—OH$	Ethyl alcohol	Ethanol
—O—	Ether	—	$CH_3CH_2—O—CH_2CH_3$	Ethyl ether	Ethoxyethane
—C—H (O)	Aldehyde	-al	$CH_3—C—H$ (O)	Acetaldehyde	Ethanal
—C— (O)	Ketone	-one	$CH_3—C—CH_3$ (O)	Acetone	Propanone
—C—OH (O)	Carboxylic acid	-oic acid	$CH_3—C—OH$ (O)	Acetic acid	Ethanoic acid
—N (H, H)	Amine	—	$CH_3—NH_2$	Methylamine	Aminomethane

2-7 ORGANIC MATTER IN WASTES

Biodegradable organic matter in municipal waste water is classified into three major categories: carbohydrates, proteins, and fats. Carbohydrates consist of sugar units containing the elements of carbon, hydrogen, and oxygen. A single sugar ring is known as a monosaccharide; few of these occur naturally. Disaccharides are composed of two monosaccharide units. Sucrose, common table sugar, is glucose plus fructose, while the most prevalent sugar in milk is

Sucrose

Lactose

Section of a cellulose molecule

lactose consisting of glucose plus galactose. Polysaccharides, long chains of sugar units, can be divided into two groups: readily degradable starches abundant in potatoes, rice, corn, and other edible plants; and cellulose found in wood, cotton, paper, and similar plant tissues. Cellulose compounds degrade biologically at a much slower rate than starches.

Proteins in the simple form are long strings of amino acids containing carbon, hydrogen, oxygen, nitrogen, and phosphorus. They form an essential part of all living tissue, and constitute a diet necessity for all higher animals. The following is a small section of a protein showing four amino acids connected together.

Fats refer to a variety of biochemical substances that have the common property of being soluble to varying degrees in organic solvents (ether, ethanol, acetone, and hexane) while being only sparingly soluble in water. Because of their limited solubility, degradation by microorganisms is at a very slow rate. A simple fat is a triglyceride composed of a glycerol unit with short- or long-chain fatty acids attached. The formula for glycerol oleobutyropalmitate, found in butter fat, is

The majority of carbohydrates, fats, and proteins in waste water are in the form of large molecules that cannot penetrate the cell membrane of micro-organisms. Bacteria, in order to metabolize high-molecular-weight substances, must be capable of breaking down the large molecules into diffusible fractions for assimilation into the cell. The first step in bacterial decomposition of organic compounds is hydrolysis of carbohydrates into soluble sugars, proteins into amino acids, and fats to short fatty acids. Further aerobic biodegradation results in the formation of carbon dioxide and water. By decomposition in the absence of oxygen, anaerobic digestion, the end products are organic acids, alcohols, and other liquid intermediates as well as gaseous entities of carbon dioxide, methane, and hydrogen sulfide.

Of the organic matter in waste water, 60 to 80 percent is readily available for biodegradation. Several organic compounds, such as cellulose, long-chain saturated hydrocarbons, and other complex compounds, although available as a bacterial food, are considered nonbiodegradable because of the time and environmental limitations of biological waste-water treatment systems. Detergents and pesticides are also a part of the 20 to 40 percent inert fraction of wastes, since they contain organic ring structures that are difficult to decompose. A common group of surfactants are the sulfonated alkyl benzenes (ABS). In the past, ABS units were derived largely from polymers of propylene, and the alkyl group was highly branched. This structure, very resistant to biological attack, resulted in contamination of both surface- and groundwater supplies with an objectionable foaming property. These materials are now made largely from normal (straight-chain) hydrocarbons, and thus the unbranched carbon chains allow ease in bacterial decomposition.

One of the most versatile and effective chlorinated pesticides for the control of objectionable insects is DDT, a chlorinated aromatic compound. An example of a more recent insecticide is endrin. Both of these highly toxic compounds may appear in waters through direct application or through percolation and runoff from treated areas, as well as through industrial wastes from their manufacture. In general, the chlorinated pesticides are the most resistant to biological degradation and may persist for months or years in soils and water, and may tend to concentrate in aquatic plants and animals. They are also resistant to removal by conventional waste-water and water treatment processes.

DDT   Endrin

Although some waste odors are inorganic compounds, for example, hydrogen sulfide gas, many are caused by volatile organic compounds such as mercaptans (organics with SH groups), butyric acid, and others. Industries may produce a variety of medicinal odors in the processing of raw materials. Surface-water supplies plagued with blooms of blue-green algae have fishy or pigpen odors. Actually very little is known about the organic compounds that produce disagreeable odors or their precise origins. Each case of a smelly waste treatment plant or disagreeable taste in a water supply must be investigated separately keeping in mind that the cause may be anaerobic decomposition, industrial chemicals, or growths of obnoxious microorganisms.

2-8 LABORATORY CHEMICAL ANALYSES

Standard Solutions

Solutions of chemical reagents with known concentrations are commonly used in laboratory testing. A 1 molar (1 M) solution contains 1 gram molecular weight of chemical per liter of solution. For example, a molar solution of NaOH would contain 40.0 grams of the pure salt dissolved in a total solution volume of 1 liter. A 1 normal (1 N) solution contains 1 gram equivalent weight of reagent per liter of solution. A normal NaOH solution would contain the same amount of salt as the 1 molar solution, since the molecular weight and equivalent weight are identical (See Table 2-3). However, in the case of sulfuric acid, a 1 M solution would contain 98.1 g/l whereas a 1 N solution would contain 49.0 g/l.

Solution concentrations can also be expressed in terms of milligrams per liter (mg/l), milliequivalents per liter (meq/l), grains per gallon (gpg), pounds per million gallons (lb/mil gal), and pounds per million cubic feet (lb/1,000,000 cu ft). The relationship between milliequivalents per liter and milligrams per liter is given in Eq. 2-1. One mg/l, being 1 part by weight per 1,000,000 parts by weight, is equivalent to 8.34 lb/mil gal, since the weight of 1 gal of water is 8.34 lb. Also since 1 cu ft of water weighs 62.4 lb, 1 mg/l equals 62.4 lb/1,000,000 cu ft. By definition 1 lb contains 7000 grains; hence, 1 mg/l equals 0.0584 gpg, or 1 gpg equals 17.1 mg/l. Equation 2-15 summarizes these concentration factors.

$$1.00 \text{ mg/l} = 8.34 \text{ lb/mil gal} = \frac{62.4}{1,000,000 \text{ cu ft}} = 0.0584 \text{ gpg} \qquad (2\text{-}15)$$

Hydrogen Ion Concentration

pH, defined by Eq. 2-4, is used to express the intensity of an acid or alkaline solution. Measurement of hydrogen ion concentration is most frequently accomplished by using a meter that reads directly in pH units, as is shown in Figure 2-5. A glass electrode, in association with a calomel electrode, dipped into the solution detects hydrogen ions. Prepared standard solutions are used to calibrate the meter.

Colorimetric techniques are available for determining pH. Special indicator solutions when added to water reveal pH by color development. A viewer equipped with a disk of standard colors can be used to estimate pH by the color of a water sample. Alternately special paper can be purchased which varies in color according to pH when dipped into water. Unfortunately, although these techniques require less expensive equipment, their application is limited to clear water samples where high accuracy is not needed. Color and turbidity in the sample can interfere with colorimetric comparisons leading to erroneous readings.

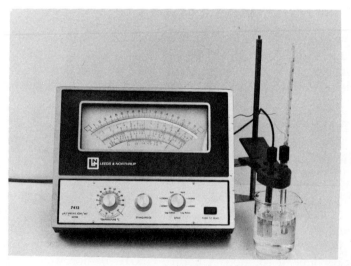

Figure 2-5 A pH meter provided with glass and calomel electrodes for hydrogen ion measurements.

Alkalinity and Acidity

Section 2-4 discusses alkalinity and gives the common forms (Figure 2-3), equation for calculation (Eq. 2-11), and sample computation (Example 2-7). A typical laboratory apparatus for titration analysis is shown in Figure 2-6. The container with a measured sample volume is stirred while a standardized titrant solution in a calibrated burette is dispensed into it. End point of titration is determined either by a colorimetric indicator added to the sample or by use of a glass pH electrode and meter.

The primary need for alkalinity measurements is related to water processing, although this measurement is routinely included in any water analysis. Since lime in water softening and coagulants for turbidity removal react with alkalinity, it is essential that this parameter by monitored in both raw and treated water to insure optimum dosages of treatment chemicals. Buffering action for control of pH in biological systems is the carbon dioxide-bicarbonate system; therefore, alkalinity measurements are performed on aerating waste waters and digesting sludges in evaluating environmental conditions.

Acidity is a measure of carbon dioxide and other acids in solution. Analysis is by titration, similar to that used in the determination of alkalinity. Strong inorganic-acid acidity exists below pH 4.5 while carbon dioxide acidity (carbonic acid) is between pH 4.5 and 8.3 (Figure 2-2c). A measured volume of water sample is titrated with 0.02 N NaOH from the existing pH to a pH of 8.3,

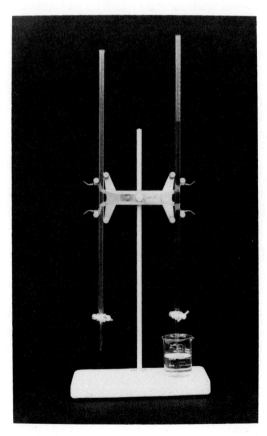

Figure 2-6 Titration apparatus including dispensing burette stand and sample container.

noting the amount of milliliters of titrant used below pH 4.5 and that neutralized between 4.5 and 8.3. Acidity, conventionally expressed in terms of mg $CaCO_3$ per liter, is calculated by Eq. 2-16. Methyl orange acidity is that below pH 4.5, phenolphthalein exists in the pH range of 4.5 to 8.3, and total acidity is the sum of these two values.

$$\frac{\text{Acidity}}{\text{as mg/l } CaCO_3} = \frac{\text{ml titrant} \times \text{normality of base} \times 50,000}{\text{ml sample}} \qquad (2\text{-}16)$$

Hardness

Hardness is caused by multivalent metallic cations; those most abundant in natural waters are calcium and magnesium. Hard waters from both underground and surface supplies are most common in areas having extensive geological formations of limestone. Although satisfactory for human consumption, Ca^{++} and Mg^{++} precipitate soap, reducing its cleansing action, and cause scale [$CaCO_3$ and $Mg(OH)_2$] in water distribution mains and hot-water heaters. Even with the advent of synthetic detergents and methods for dealing with scaling problems, partial softening of extremely hard waters by municipal treatment plants is desirable. Waters with less than 50 mg/l hardness are considered soft, up to 150 mg/l moderately hard, and in excess of 300 mg/l very hard.

The most common testing method for hardness is the EDTA titrimetric method. Disodium ethylenediaminetetraacetate (Na_2EDTA) forms stable complex ions with Ca^{++}, Mg^{++}, and other divalent cations causing hardness, thus removing them from solution. If a small amount of dye, Eriochrome Black T, is added to the water containing hardness ions at pH 10, the solution becomes wine red; in absence of hardness the color is blue. The test procedure uses a titration apparatus like that shown in Figure 2-6. The burette is filled with a standardized EDTA solution for dispensing into a measured water sample containing indicator dye. EDTA added complexes hardness ions until all have been removed from solution and the water color changes from wine red to blue, indicating end of titration, Eq. 2-17.

$$\underset{\text{Wine red color}}{Ca^{++} + Mg^{++} + EDTA} \xrightarrow{pH = 10} \underset{\text{Blue color}}{Ca \cdot EDTA + Mg \cdot EDTA} \qquad (2\text{-}17)$$

Hardness is conventionally expressed in the units of mg/l as $CaCO_3$ and is calculated from the laboratory data by the following:

$$\frac{\text{Hardness as}}{\text{mg/l } CaCO_3} = \frac{\text{ml titrant} \times CaCO_3 \text{ equivalent of EDTA} \times 1000}{\text{ml sample}} \qquad (2\text{-}18)$$

Iron and Manganese

These metals at very low concentrations are highly objectionable in water supplies for domestic or industrial use. Traces of iron and manganese can cause staining of bathroom fixtures, can impart brownish color to laundered clothing, and can affect the taste of water. Groundwaters devoid of dissolved oxygen may contain appreciable amounts of ferrous iron (Fe^{++}) and manganous manganese (Mn^{++}), which are soluble (invisible) forms. When exposed to oxidation, they are transformed to the stable insoluble ions of ferric iron (Fe^{+++}) and manganic manganese (Mn^{++++}), giving water a rust color. Supplies drawn from anaerobic bottom water of reservoirs, or rivers that have contacted natural formations of iron- and manganese-bearing rock, may contain both reduced and oxidized forms, the latter being complexed frequently with organic matter.

The most popular methods of determining iron and manganese in water use colorimetric procedures. Color development as a technique in testing has the major advantage of being highly specific for the ion involved and generally requires minimum pretreatment of the water sample. Concentration of the substance being evaluated is directly related to intensity of the color developed. The simplest method of color comparison is to make up a set of standards by using known concentrations in a series of glass tubes, commonly referred to as Nessler tubes. After the sample has been prepared, it is placed in an identical glass tube and its color is compared visually against the standards. Or, a small comparator (viewing device) can be used to measure color development in a sample against a disk wheel that contains small colored glasses shaded to serve as color standards.

Photoelectric colorimeters of the type shown in Figure 2-7 reduce the human error involved in making color distinctions by eye. In this apparatus, light from an ordinary bulb is transmitted through a lens system and colored glass filter to obtain a monochromatic light beam. Selected colored filters are used in different chemical tests. Percentage of tramsmission of the light beam through

Figure 2-7 Photoelectric colorimeter (filter photometer). (Courtesy of Hach Chemical Company, Ames, Iowa.)

the sample is measured by using a photoelectric cell connected to a galvanometer. Scales developed for each particular chemical analysis can be positioned behind the galvanometer needle such that the result can be read directly in terms of concentration of the ion that is being tested.

The most accurate laboratory apparatus used in colorimetric measurements is the spectrophotometer. It operates in a fashion similar to the photoelectric colorimeter, but the monochromatic light is developed by a precise prism or grating system that allows the selection of any wavelength in the visible spectrum. Figure 2-8 shows a picture and schematic diagram of a spectrophotometer.

The phenanthroline method is preferred for measuring iron in water. The first step is to insure that all iron in solution is in the ferrous state by treating the sample with hydrochloric acid and hydroxolamine as the reducing agent as follows:

$$4Fe^{+++} + 2NH_2OH \xrightarrow{pH = 3.2} 4Fe^{++} + N_2O + H_2O + 4H^+ \qquad (2\text{-}19)$$

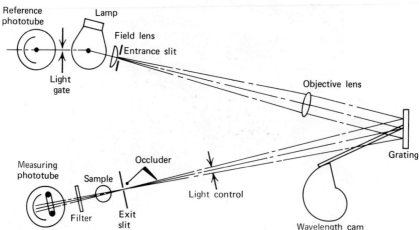

Figure 2-8 Spectrophotometer and optical system schematic. (Courtesy of Bausch and Lomb Inc., Analytical Systems Division.)

Three molecules of 1,10-phenanthroline sequester each atom of ferrous iron to form an orange-red complex.

1,10-phenanthroline   Orange-red complex

A spectrophotometer set for a monochromatic light beam of 510 millimicrons is used to measure the light absorption of the colored solution. Concentration of iron in the sample is determined from a percentage transmission versus iron concentration calibration curve prepared from a series of standard iron solutions previously tested.

Manganese is also determined by a colorimetric procedure—the persulfate method. Chloride interference is overcome by adding mercuric sulfate to form relatively insoluble $HgCl_2$. All manganese in the sample is then converted to permanganate by persulfate in the presence of silver as a catalyst.

$$Mn^{++} + S_2O_8^= \xrightarrow{Ag^+} MnO_4^- + SO_4^= \qquad (2\text{-}21)$$

The concentration of permanganate ion produced and its resultant color represents the amount of manganese in the sample.

Color

Hues in water may result from natural minerals such as iron and manganese, vegetable origins—humus material and tannins—or colored wastes discharged from a variety of industries including mining, refining, pulp and paper, chemicals, and food processing. True color of water is considered to be only that attributable to substances in solution after removal of suspended materials by centrifuging or filtration. In domestic water, color is undesirable aesthetically and may dull clothes or stain fixtures. Stringent color limits are required for water use in many industries—beverage production, dairy and food processing, paper manufacturing, and textiles.

Standard color solutions are composed of potassium chloroplatinate (K_2PtCl_6) tinted with small amounts of cobalt chloride. The color produced by 1 mg/l of platinum in combination with $\frac{1}{2}$ mg/l of metallic cobalt is taken as 1 standard color unit. The yellow-brownish hue produced by these metals in solution is similar to that found in natural waters. Comparison tubes containing standard platinum-cobalt solutions ranging from 0 to 70 color units are used for visual measurements; however, laboratories often employ a colorimeter for readings.

Turbidity

Insoluble particles of soil, organics, microorganisms, and other materials impede the passage of light through water by scattering and absorbing the rays. This interference of light passage through water is referred to as turbidity. In excess of 5 units it is noticeable to the average water consumer and accordingly represents an unsatisfactory condition for drinking water. Turbidity in a typical clear lake water is about 25 units, and muddy water exceeds 100 units.

Figure 2-9 Jackson candle turbidimeter.

One unit of turbidity is 1 mg/l of silica in water suspension. Sometimes standard suspensions are prepared using other materials, for example, diatomaceous earth (fuller's earth), which are then calibrated in silica units. The Jackson candle turbidimeter, consisting of a calibrated glass tube, holder, and candle as shown in Figure 2-9, is the standard instrument for measuring turbidity. With the tube in place over a lighted candle, the water sample is gradually added until the candle flame just disappears from view through the water column in the tube. For accurate determinations below 5 units a Baylis turbidimeter is sometimes used. It consists of a light source and two glass tubes; one holds a standard suspension and the other holds the sample. Light from a bulb passing through the water at right angles is scattered by the turbidity, producing a whitish haze that interfers with the blue color produced at the bottom of the tubes by a piece of cobalt glass. The degree of haziness is proportional to the turbidity, and the concentration is measured by comparison with a series of standard suspensions. Both the Jackson candle and Baylis turbidimeter rely on human vision for judgment in reading. A number of laboratory devices that use a standard light source and photoelectric cell system are available. Portable photoelectric colorimeters are frequently used for field measurements of river and lake waters.

Jar Tests

Effectiveness of chemical coagulation of water or waste water can be experimentally evaluated in the laboratory by using a stirring device as is illustrated in Figure 2-10. The stirrer consists of six paddles capable of variable-speed operation between 0 and 100 rpm. In making tests, 1 liter or more of water is placed in each of the jars or beakers, and is dosed with different amounts of coagulant. After rapid mixing to disperse the chemicals, the samples are stirred slowly for floc formation and then are allowed to settle under quiescent conditions. The jars are mixed at a speed of 60 to 80 rpm for 1 min after adding the coagulant solution and then are stirred at a speed of 30 rpm for 15 min. After stopping the stirrer, the nature and settling characteristics of the floc are observed and are recorded in qualitative terms, as poor, fair, good, or excellent. A hazy sample indicates poor coagulation, while proper coagu-lated water contains floc that are well formed with the liquid clear between particles. The lowest dosage that provides good turbidity removal during a jar test is considered as the first trial dosage in plant operation. Ordinarily a full-scale treatment plant gives better results than a jar test at the same dosage.

For research or special studies, the beakers used in jar testing may be modified to replicate more closely actual mixing units constructed in treatment plants. Figure 2-11 illustrates a container provided with stators mounted on the inside, above and below the paddle, to provide for more thorough mixing. Special provision is made for applying the coagulant dose through a glass tube extending into the jar near the hub of the impeller. The system also includes siphons that can be used to draw off samples from the same depth in each beaker for chemical testing, for example, turbidity measurements during the settling period. Actual treatment plant operations may dictate a change in mixing and settling times to match those being used in water processing. Results from a jar test for coagulation of a turbid river water using alum are given in Table 2-10. The lowest recommended dosage in treating this water is 40 mg/l of coagulant. Since other factors such as temperature, alkalinity, and pH influence coagulation, jar tests can also be run to evaluate these parameters and to determine optimum dosages under differing conditions. For example,

Figure 2-10 Stirring device used in jar tests for chemical coagulation. (Courtesy of Phipps & Bird, Inc.)

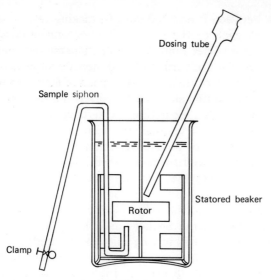

Figure 2-11 Modified container for jar tests including stators, dosing tube, and sampling siphon.

Table 2-10. Results of Jar Test for Coagulation

Jar Number	Aluminum (ml/jar)	Sulfate (mg/l)	Dosage (gpg)	Floc Formation
1	1	10	0.58	None
2	2	20	1.17	Smoky
3	3	30	1.85	Fair
4	4	40	2.34	Good
5	5	50	2.92	Good
6	6	60	3.50	Heavy

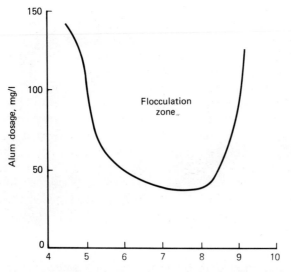

Figure 2-12 Effect of pH on dosage of alum required for good coagulation of a river water.

Figure 2-12 shows the results of a series of jar tests, similar to the one illustrated in Table 2-10, that have been run at different pH values. The plot illustrates the pH of water samples tested versus minimum alum dosage required for good floc formation. In addition to the affect of variation in water quality on chemical dosage, special studies can be conducted to measure optimum application of coagulant aids, such as polymers or activated silica with the primary coagulant.

Fluoride

Excessive fluoride ions in drinking water cause dental fluorosis or mottling of teeth. On the other hand, communities whose drinking water contains no fluoride have a high prevelance of dental caries. Optimum concentrations of fluoride provided in public water supplies, generally in the range of 0.8 to 1.2 mg/l, reduce dental caries to a minimum without causing noticeable dental fluorosis. Several fluoride compounds (Table 2-3) are used in treating municipal water; all of these dissociate readily, yielding the fluoride ion (F^-).

Electrode and colorimetric methods are currently most satisfactory for determining fluoride ion concentration. Both methods are susceptible to interferring substances, for example, chlorine in the colorimetric method and polyvalent cations, such as Al^{+++}, which complex fluoride ions hinder electrode response. Fluoride can be separated from other constituents in water by distillation of hydrofluoric acid (HF) after acidifying the sample with sulfuric acid. Pretreatment separation is performed using a distillation apparatus. Samples not containing significant amounts of interfering ions can be tested directly.

The fluoride ion-activity electrode is a specific ion sensor designed for use with a calomel reference electrode and a pH meter having a millivolt scale. The key element in the fluoride electrode is the laser-type doped single lanthanum fluoride crystal across which a potential is established by the presence of fluoride ions. The crystal contacts the sample solution at one face and an internal reference solution at the other. An appropriate calibration curve must be developed that relates meter reading in millivolts to concentration of fluoride. Fluoride activities dependent on total ionic strength of the sample and the electrode does not respond to fluoride that is bound or complexed. These difficulties are largely overcome by addition of citrate ion to preferentially complex aluminum and by introducing a buffer solution of high total ionic strength to swamp out variations in sample ionic strength.

The colorimetric method is based on the reaction between fluoride and a zirconium-dye lake. (The term "lake" refers to the color produced when zirconium ion is added to SPADNS dye.) Fluoride reacts with the reddish color-dye lake, dissociating a portion of it into a colorless complex anion ($ZrF_6^=$) and the yellow-color dye. As the amount of fluoride is increased, the color becomes progressively lighter and of a different hue. This bleaching action is directly proportional to fluoride ion concentration. Either a spectrophotometer (Figure 2-8) or a filter photometer (Figure 2-7) may be used to measure sample absorbance for comparison against a standard curve.

Chlorine

The most common method for disinfecting water supplies in the United States is by chlorination. As a strong oxidizing agent, it is also used effectively in taste and odor control, and removal of iron and manganese from water supplies.

Chlorine combines with water to form hypochlorous and hydrochloric acids, Eq. 2-9. Hypochlorous acid reacts with ammonia in water to form mono-chloramine (NH_2Cl), dichloramine ($NHCl_2$), and trichloramine (NCl_3), depending on the relative amounts of acid and ammonia. Chlorine, hypochlorous acid, and hypochlorite ion are free chlorine residual, while the chloramines are designated as combined chlorine residual.

The orthotolidine-arsenite (OTA) test differentiates and measures both free and combined chlorine residuals. Orthotolidine is an organic compound oxidized in acid solution by chlorine, chloramines, and other oxidizing agents to produce a yellow-colored compound. The intensity of color development is proportional to the amount of chlorine residual. Free chlorine reacts instantaneously, whereas chloramines react much more slowly. To determine both free and combined residuals, two tests are performed. Total residual is measured by the color developed about 5 min after adding the orthotolidine. In the second test, sodium arsenite solution (a reducing agent) is added within 5 sec after the orthotolidine. This allows time for the free chlorine to react but permits only a small amount of the combined to be oxidized, since arsenite reduces chloramine instantaneously, stopping further action with the orthotolidine. Therefore, the yellow color developed in this case is only due to free chlorine residual. Combined residual is determined by subtracting the free residual of the second test from the total residual of the first. For routine measurements of less than 1 mg/l, a color comparator of the type shown in Figure 2-13 is sufficiently accurate. For more precise determinations, or when

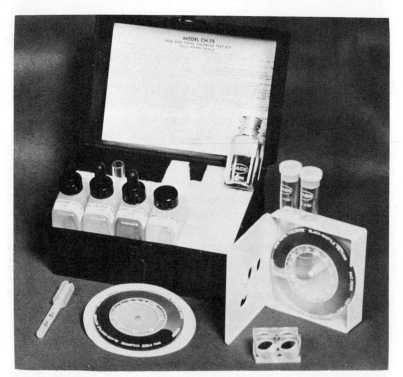

Figure 2-13 Residual chlorine comparator. (Courtesy of Hach Chemical Company, Ames, Iowa.)

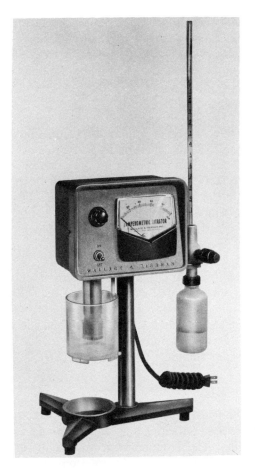

Figure 2-14 Amperometric titrator for determination of chlorine residual in waste water by the iodometric method. (Courtesy of Wallace & Tiernan Division, Pennwalt Corp.)

higher residuals are anticipated, colorimetric determinations can be made by using a filter photometer or spectrophotometer.

Residual chlorine in waste-water samples cannot be reliably determined by the OTA technique because of interference from organic matter. The indirect procedure of the iodometric method is much more precise. A waste-water sample is prepared by adding: a measured volume of standardized phenylarsine oxide, or thiosulfate, solution; excess potassium iodide; and acetate buffer solution to reduce the pH to between 3.5 and 4.2. Chlorine residual oxidizes an equivalent portion of the excess iodide ion to free iodine which, in turn, is immediately converted back to iodide by a portion of the reducing agent. The amount of phenylarsine oxide remaining is then quantitatively measured by titration with a standard iodine solution. In other words, this analysis reverses the end point by determining the unreacted standard phenylarsine oxide in the treated sample using standard iodine, rather than directly titrating the iodine released by the chlorine. Thus contact between the full concentration of liberated iodine and the waste water is avoided. The titration end point may be indicated by using starch which produces a blue color in the presence of free iodine. An alternative, and more accurate, method for observing the end point involves the use of an amperometric titrator (Figure 2-14). The meter needle responds to electrical conductivity of the sample solution being titrated. A deflection of the pointer upscale registers the presence of free iodine and the titration end point. This apparatus can be employed to measure free residual, or to differentiate between monochloramine and dichloramine fractions of combined residual.

Nitrogen

Common forms of nitrogen are: organic, ammonia, nitrite, nitrate, and gaseous nitrogen. Nitrogenous organic matter, such as protein, is essential to living systems. Industrial wastes are often tested for nitrogen and phosphorus content to insure that there are sufficient nutrients for biological treatment. Inorganic nitrogen, principally in the ammonia and nitrate forms, is used by green plants in photosynthesis. Since nitrogen in natural waters is limited, pollution from nitrogen wastes can promote the growth of algae, causing green-colored water. Ammonia is also considered a serious water pollutant because of its toxic effect on fishes.

The test for ammonia nitrogen is a distillation process using the apparatus illustrated in Figure 2-15. The analysis procedure is based on shifting the equilibrium between the ammonium ion and free ammonia, and the release of gaseous ammonia along with the steam which is produced when the water is boiled. After placing the sample in a flask, a buffer solution is added to increase the pH and to shift the equilibrium ammonia to the right in Eq. 2-22.

$$NH_4^+ \underset{\text{acidic}}{\overset{\text{basic}}{\rightleftharpoons}} NH_3 \uparrow + H^+ \tag{2-22}$$

A condenser is then attached to the flask and the mixture is distilled, driving off steam and free ammonia. The steam containing ammonia is condensed and is collected in a boric acid solution. The reaction with boric acid forms ammonium ions, Eq. 2-22 to the left, while producing borate ions from the acid.

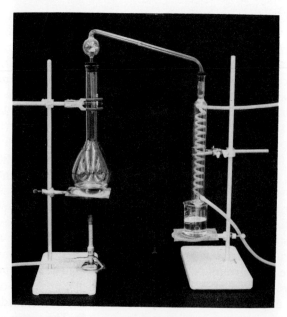

Figure 2-15 Distillation apparatus for ammonia and organic nitrogen tests.

The amount of ammonia in the sample is determined by the quantity of boric acid consumption in the collecting beaker. This may be measured by back titration of the solution with a standard acid to determine the amount of borate ion produced.

Organic nitrogen in a waste is determined by digestion of the organic matter, thus releasing ammonia, and then proceeding with the ammonia nitrogen test. Since the bulk of organic nitrogen is in the form of proteins and amino acids, it can be converted to ammonium ion by boiling in an acidified solution with chemical catalysts, Eq. 2-23. When acid is first added to the waste sample the organics turn black. Complete destruction and conversion after one to two hours of boiling is indicated by a clearing of the liquid. On cooling, a sodium hydroxide reagent is added to raise the pH and to convert the

$$\text{Organic-nitrogen} \xrightarrow[\text{catalysts}]{\text{sulfuric acid}} NH_4^+ \qquad (2\text{-}23)$$

ammonium ion to ammonia. The flask is attached to the connecting bulb of the condenser, and the test is completed by distillation for measurement of the ammonia nitrogen. If both ammonia and organic nitrogen determinations are desired, the sample may be tested first for ammonia and then can be digested and run for organic nitrogen. The sum of these two results is often referred to as total Kjeldahl nitrogen. If separation of organic and ammonia nitrogens is not needed, the total Kjeldahl nitrogen may be determined directly by digesting the entire sample and distilling off both the ammonia nitrogen that originally existed in the sample, as well as that released from digesting the organic nitrogen.

Nitrite and nitrate nitrogen in natural and treated waters are routinely determined by colorimetric techniques. For example, a common test for nitrate is the phenoldisulfonic acid method. The intensity of the yellow color produced by the reaction between this acid and nitrate is directly related to nitrate concentration. The color developed in an unknown specimen can be compared to the hues developed by known concentrations using Nessler tubes, filter photometer, or a spectrophotometer. The nitrite analysis is based on the formation of a reddish-purple color produced between nitrite ion and two organic reagents, sulfanilic acid and 1-naphthylamine hydrochloride. Nitrite and nitrate tests on waste waters are much more difficult to perform because of the high concentrations of interferring substances, such as chlorides and organic matter. *Standard Methods*[2] outlines five tentative methods for nitrate. Each involves special pretreatment of the waste water for removal of turbidity, color, and other inhibiting substances.

The specific nitrogen tests conducted on a natural water, treated water, or waste water depend on the objectives of the study. A drinking water test may require only analysis for nitrate, while a stream pollution survey may be primarily interested in ammonium nitrogen. Total nitrogen in a water is equal to the sum of the organic, ammonia, nitrite, and nitrate nitrogen concentrations. All nitrogen test results are expressed in units of milligrams of N/l.

Phosphorus

The common compounds are orthophosphates ($H_2PO_4^-$, $HPO_4^=$, $PO_4^\equiv$), polyphosphates, such as $Na_3(PO_3)_6$ used in synthetic detergent formulations, and organic phosphorus. All polyphosphates gradually hydrolyze in water to the stable ortho form, while decaying organic matter decomposes biologically to release phosphate. Orthophosphates are, in turn, synthesized back into living animal or plant tissue.

The prime concern in biological waste treatment is insuring sufficient phosphorus to support microbial growth. Although sanitary wastes normally have a surplus of phosphates, some trade wastes may be nutrient deficient because of high carbohydrate or hydrocarbon chemical content. In phosphorus pollution control the primary concern is overfertilization of surface waters, resulting in nuisance growths of algae and aquatic weeds.

A popular procedure for determining orthophosphate is the stannous chloride colorimetric method. Ammonium molybdate and stannous chloride reagents combine with phosphate to produce a blue-colored colloidal suspension. A second technique for orthophosphate is the vanadomolybdophosphoric acid method. Vanadium is used in combination with ammonium molybdate to produce a yellow color with phosphate ion. Color development is quantitatively measured using a spectrophotometer.

Polyphosphates are converted to orthophosphates by acid hydrolysis. Therefore, to analyze for total inorganic phosphate, the water sample is acidified with a strong acid solution, boiled gently for at least 90 minutes, cooled, and neutralized as pretreatment prior to testing for phosphate. The amount of polyphosphate is obtained by subtracting the measured orthophosphate concentration in a sample from the quantity of total inorganic phosphate.

Total phosphorus content of a sample includes ortho, poly, and organic species. Release of phosphate-combined organic matter requires digestion with either perchloric, sulfuric, or nitric acid depending on the type of sample. Following digestion, one of the colorimetric methods can be used to measure the liberated orthophosphate. Organic phosphorus content is computed by subtracting the results of inorganic (ortho + poly) phosphorus tests from the total phosphorus determination.

The classification of phosphorus fractions in a sample includes the physical states of filtrable (dissolved) and particulate, as well as chemical types. Separating dissolved from particulate forms is accomplished by filtration through a 0.45-μ membrane filter. A complete analysis would consist of conducting ortho, acid-hydrolyzable, inorganic phosphate tests on measured portions of unfiltered and filtered sample. Particulate content is calculated by subtracting the filtrable orthophosphate from the total in the whole sample.

The variety of techniques for pretreatment of samples, measurement of phosphorus concentrations, and expression of results frequently leads to confusion. The particular procedure used in analysis should be recorded with test results. The most common shortcoming in presenting phosphorus data in printed literature is failure to document collection technique, storage, and filtration or other pretreatment. The thirteenth edition of *Standard Methods* determines total phosphorus by acid digestion, whereas earlier editions considered that portion available after acid hydrolysis as being the total amount, when in fact this includes only the ortho and poly forms. Another modification involves changing the expression of results. The current *Standard Methods* calculates results of all phosphate tests in terms of milligrams per liter as phosphorus, while previously milligrams per liter as phosphate were used. To convert milligrams per liter as phosphate to milligrams per liter as phosphorus the value given should be divided by 95 (molecular weight of PO_4) and multiplied by 31 (atomic weight of phosphorus). In other words, 1.00 mg/l PO$_4$ equals 0.34 mg/l P, or 1.00 mg/l P equals 3.06 mg/l PO$_4$.

Dissolved Oxygen

Biological decomposition of organic matter uses dissolved oxygen. Levels significantly below saturation values, given in Table 2-5, often occur in polluted surface waters. Since fish and most aquatic life are stifled by a lack of oxygen, dissolved oxygen determination is a principal measurement in pollution surveys. The rate of air supply to aerobic treatment processes is monitored by dissolved oxygen testing to maintain aerobic conditions, and to prevent waste of power by excessive aeration. DO tests are used in the determination of biochemical oxygen demand of a waste water. Small samples of waste are mixed with dilution water and placed in BOD bottles for dissolved oxygen testing at various intervals of time. Oxygen is a significant factor in corrosion of piping systems. Removal of oxygen from boiler feed waters is common practice, and the DO test is the means of control.

The azide modification of the iodometric method is the most common chemical technique for measuring dissolved oxygen. The standard test uses a 300-ml BOD bottle for containing the water sample (Figure 2-16). The chemical reagents used in the test are: manganese sulfate solution, alkali-iodide-azide reagent, concentrated sulfuric acid, starch indicator, and standardized sodium thiosulfate titrant. The first step is to add 2 ml each of the first two reagents to the BOD bottle, restopper with care to exclude air bubbles, and mix by repeatedly inverting the bottle. If no oxygen is present, the manganous ion (Mn^{++}) reacts only with the hydroxide ion to form a pure white precipitate of $Mn(OH)_2$, Eq. 2-24. If oxygen is present some of the Mn^{++} is oxidized to a higher valence (Mn^{++++}) and precipitates as a brown-colored oxide (MnO_2), Eq. 2-25.

$$Mn^{++} + 2OH^- \rightarrow Mn(OH)_2 \downarrow \qquad (2\text{-}24)$$

$$Mn^{++} + 2OH^- + \tfrac{1}{2}O_2 \rightarrow MnO_2 \downarrow + H_2O \qquad (2\text{-}25)$$

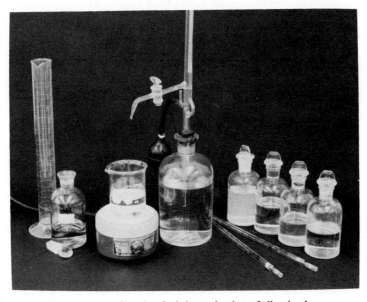

Figure 2-16 Apparatus for chemical determination of dissolved oxygen.

After shaking and allowing sufficient time for all the oxygen to react, the chemical precipitates are allowed to settle leaving clear liquid in the upper portion. Then, 2 ml of concentrated sulfuric acid is added. The bottle is restoppered and mixed by inverting until the suspension is completely dissolved and the yellow color is uniform throughout the bottle. The reaction that takes place with the addition of acid is shown in Eq. 2-26; the manganic oxide is reduced to manganous manganese while an equivalent amount of iodide ion is converted to free iodine. The quantity of I_2^o is equivalent to the dissolved oxygen in the original sample.

$$MnO_2 + 2I^- + 4H^+ \rightarrow Mn^{++} + I_2^o + 2H_2O \qquad (2\text{-}26)$$

A volume of 203 ml, corresponding to 200 ml of the original sample after correction for the loss of sample displaced by reagents additions, is poured from the BOD bottle into a container for titration with 0.0250 N thiosulfate solution. Thiosulfate in the titrant is oxidized to tetrathionate while the free iodine is converted back to iodide ion, Eq. 2-27.

$$2S_2O_3^= + I_2^o \xrightarrow[\text{indicator}]{\text{starch}} S_4O_6^= + 2I^- \qquad (2\text{-}27)$$

Since it is impossible to titrate accurately the yellow-colored iodine solution to a colorless liquid, an end point indicator is needed. Soluble starch in the presence of free iodine produces a blue color. Therefore, after titration to a pale straw color, 1 to 2 ml of starch solution is added and titration is continued to the first disappearance of blue color. If 0.0250 N sodium thiosulfate is used to measure the dissolved oxygen in a volume equal to 200 ml of original sample, 1.0 ml of titrant is equivalent to 1.0 mg/l DO.

Since high concentrations of suspended solids and biological activity of activated-sludge flocs may have high oxygen utilization rates, microbial activity must be stopped at the time of sample collection, and suspended solids must be settled out of solution prior to the iodometric dissolved oxygen test. A common procedure is to use copper sulfate-sulfamic acid inhibitor solution to stop biological activity and to flocculate suspended solids. Recommended sampling procedure is to add 10 ml of inhibitor to a 1 qt, or 1 liter, wide-mouth bottle. For immersion in an aeration tank, the bottle is often placed in a special sampler designed so that the bottle fills from a tube near the bottom and overflows about 25 percent of the bottle capacity. After removing from the sampler, the sample is stoppered and allowed to settle until a clear supernatant can be siphoned into a BOD bottle. DO concentration is then measured by the azide-iodometric method.

Membrane electrodes are available for measurement of dissolved oxygen without chemical treatment of a sample. A dissolved oxygen probe is composed of two solid metal electrodes in contact with a salt solution which is separated from the water sample by a selective membrane (Figure 2-17). The recessed end of the probe containing the metal electrodes is filled with saturated potassium chloride solution and is covered with a polyethylene or teflon membrane held in place by a rubber O-ring. The probe also has a sensor for measuring temperature. The unit inserted in the bottle in Figure 2-17 is designed specifically for measuring dissolved oxygen in nondestructive BOD testing; the same bottle can be measured for oxygen depletion at various time intervals, restoppering between readings. The field probe is submersible and can be lowered into water for DO and temperature measurements in lakes and

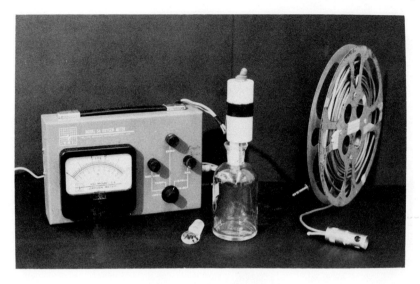

Figure 2-17 Dissolved oxygen meter with the laboratory probe inserted in a BOD bottle. At the right is a field probe with the extension wire wound on a reel.

streams. The same unit attached to a long rod can record dissolved oxygen in aeration tanks. Meters used with the probes have both temperature and dissolved oxygen scales, and can be operated by line power in the laboratory or battery operated in the field. Membrane electrodes may be calibrated by reading against air or, more frequently, a water sample of known DO concentration determined by the iodometric method.

Solids

Suspended and dissolved solids, both organic and inert, are common tests of polluted waters. In drinking water, the maximum recommended total dissolved solids concentration is 500 mg/l. Suspended and volatile solids are common parameters used in defining a municipal or industrial waste water. Operational efficiency of various treatment units is defined by solids removal, for example, suspended solids removal in a settling basin, and volatile solids reduction in sludge digestion.

Total solids, total residue on evaporation, is the term applied to material left in a dish after evaporation of a sample of water or waste water and subsequent drying in an oven. A measured volume of sample is placed in an evaporating dish, usually porcelain, as is shown in Figure 2-18. After evaporation of the water from the dish on a steam bath, the dish is transferred to an oven maintained at 103° C and is dried to a constant weight. The milligrams of total residue is equal to the difference between the cooled weight of the dish and the original weight of the empty dish. Concentration of total residue is calculated using Eq. 2-28.

$$\text{mg/l total residue} = \frac{\text{mg total residue} \times 1000}{\text{ml sample}} \qquad (2\text{-}28)$$

The terms suspended solids and dissolved solids refer to matter that is retained and passed through a filter, respectively. In water analyses they are referred to as nonfiltrable and filtrable residue, whereas in waste water they are

Figure 2-18 Apparatus for solids determinations including a suction flask with attached funnel, filter pads, and drying dish.

called suspended and dissolved solids. In either case, a measured portion of sample is drawn through a glass-fiber filter, retained in a funnel, by applying a vacuum to the suction flask (Figure 2-18). Membrane or paper filters may be used in testing water while only the glass-fiber filter disk is recommended for both water and waste water. Prior to the latest *Standard Methods*,[2] a Gooch crucible (a small porcelain filtering dish with pinholes in the bottom) containing an asbestos fiber mat was used for suspended solids tests. After filtering, the disk is removed from the funnel, dried, and weighed to determine the increase as a result of the residue retained. Calculation of total suspended matter (filtrable residue) uses the same type of formula as Eq. 2-28. Dissolved matter (nonfiltrable residue) is not determined directly but is calculated by subtracting suspended solids concentration from the total residue on evaporation.

Volatile solids are determined by igniting the residue on evaporation, or the filtrable solids, at 550°C in an electric muffle furnace. Water and sludge residues are burned for 1 hr, but waste water samples require only 20 min ignition. Loss of weight on ignition is reported as milligrams per liter of volatile solids, and residue after burning is referred to as fixed solids. The evaporating dish used in volatile solids analysis and the glass-fiber filter disk used for volatile suspended solids must be pretreated by burning in the muffle furnace to determine an accurate initial (empty) weight. Volatile solids in a waste are often interpreted as being a measure of the organic matter. However, this is not precisely true, since combustion of many pure organic compounds leaves an ash and many inorganic salts volatilize during ignition.

Following are typical data and calculations in the analysis of a waste sludge to determine total solids and total volatile solids:

Weight of empty dish on analytical balance		64.532 g
Weight of dish on simple balance	64.5 g	
Weight of dish plus sludge sample	144.7	
Weight of sample	80.2	
Weight of dish plus dry sludge solids		68.950
Weight of sludge solids		4.418

Percentage of total solids $\dfrac{4.42}{80.2} \times 100 = 5.51$ percent

Weight of dish plus ignited solids		65.735
Weight of volatile solids		3.215

Percentage of volatile solids $\dfrac{3.22}{80.2} \times 100 = 4.02$ percent

Percentage of total solids that $\dfrac{3.22}{4.42} \times 100 = 72.8$ percent
 are volatile

Settleable solids are determined by filling a 1-liter conical-shaped container (Imhoff cone) and allowing 1 hr for sedimentation. The tip at the bottom of the cone is scribed with volume markings so that the quantity of settled matter can be read directly in milliliters per liter. If it is desired to express the settleable matter on a weight basis, milligrams per liter, a sample of the waste water is poured into a 1-liter graduated cylinder and is allowed to stand quiescent for 1 hr. Solids analysis run on the initial well-mixed sample, and suspended matter remaining at the midpoint of the graduated cylinder after settling are used in the following calculation:

$$\text{mg/l settleable matter} = \text{mg/l suspended} - \text{mg/l nonsettleable} \qquad (2\text{-}29)$$

Chemical Oxygen Demand

COD is widely used to characterize the organic strength of waste waters and pollution of natural waters. The test measures the amount of oxygen required for chemical oxidation of organic matter in the sample to carbon dioxide and water. The apparatus used in the dichromate reflux method pictured in Figure 2-19, consists of an erlenmeyer flask fitted with a condenser and a hot plate. The test procedure is to add a known quantity of standard potassium dichromate solution, sulfuric acid reagent containing silver sulfate, and a measured volume of sample to the flask. This mixture is refluxed (vaporized and condensed) for 2 hr. Most types of organic matter are destroyed in this boiling mixture of chromic and sulfuric acid, Eq. 2-30. After cooling, washing

$$\text{Organics} + Cr_2O_7^= + H^+ \xrightarrow[Ag^+]{\text{heat}} CO_2 + H_2O + 2Cr^{+++} \qquad (2\text{-}30)$$

down the condenser, and diluting the mixture with distilled water, the dichromate remaining in the specimen is titrated with standard ferrous ammonium sulfate using ferroin indicator. Ferrous ion reacts with dichromate ion as in Eq. 2-31 with an end point color change from blue-green to reddish brown. A blank

$$6Fe^{++} + Cr_2O_7^= + 14H^+ \rightarrow 6Fe^{+++} + 2Cr^{+++} + 7H_2O \qquad (2\text{-}31)$$

sample of distilled water is carried through the same COD testing procedure as

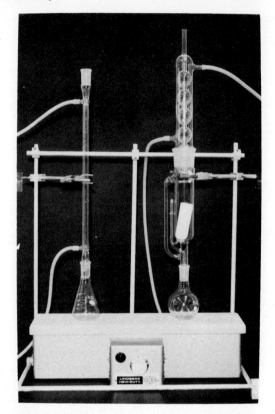

Figure 2-19 Reflux unit for chemical oxygen demand (left) and Soxhlet grease extraction apparatus (right).

the waste-water sample. The purpose of running a blank is to compensate for any error that may result because of the presence of extraneous organic matter in the reagents. COD is calculated using Eq. 2-32; the difference between amounts of titrant used for the blank and the sample is divided by volume of sample and multiplied by normality of the titrant. The 8000 multiplier is to express the results in units of milligrams per liter of oxygen, since 1 liter contains 1000 ml and the equivalent weight of oxygen is 8.

$$\frac{COD}{mg/l} = \frac{(ml\ blank - ml\ sample\ titrant)(normality\ Fe(NH_4)_2(SO_4)_2)8000}{ml\ sample} \quad (2\text{-}32)$$

Grease

A variety of organic substances including hydrocarbons, fats, oils, waxes, and high-molecular-weight fatty acids are collectively referred to as grease. Their importance in municipal and industrial wastes is related to their difficulty in handling and treatment. Because of low solubility, grease separates from water adhering to the interior of pipes and tank walls, reduces biological treatability of a waste, and produces greasy sludge solids difficult to process.

Apparatus for grease determination by the Soxhlet extraction method is shown in Figure 2-19. A sample is prepared for grease extraction by filtering 1 liter of acidified waste water through a filter consisting of muslin cloth overlaid with filter paper supporting diatomaceous-silica filter aid. After drawing the sample through, the filter is removed from the funnel, rolled and placed in an extraction thimble, and is dried in a hot-air oven at 103° C. The boiling flask is carefully weighed and partially filled with either *n*-hexane or trichloro-trifluoroethane. The latter is often selected, since it is nonflammable. After placing the thimble containing the sample into the Soxhlet extension tube, the flask with solvent is fitted on the bottom, and the condenser is placed on top. Heating the hexane drives vapor up through a side arm in the Soxhlet extension tube to the base of the condenser. Condensed hexane drips down into the thimble containing the grease sample, passes through the filter paper to a space outside the thimble in the extension tube. When filled to the siphon outlet of the tube, the hexane is discharged back down into the flask. This sequence of vaporizing and flushing the hexane through the sample and back to the flask is referred to as a cycle. Extraction in the standard test consists of 20 cycles per hour for 4 hr. After cooling, the flask is removed and placed on a water bath at 85° C for slow distillation. Driving off the hexane solvent leaves the extracted grease in the boiling flask. Grease content is calculated by dividing the increase in weight of the flask by the volume of waste water sample filtered.

Volatile Acids

Measurement of volatile acid concentration in anaerobically digesting sludge is used to monitor the digestion process. The acids, principally acetic, propionic, and butyric, are separated from water solution by column chromatography. A small clarified sample is drawn into a Gooch or fritted-glass crucible packed with granular silicic acid to adsorb the volatile acids. Chloroform-butanol reagent is then added to the column and drawn through to elute selectively the organic acids from the column packing. After purging with a carbon dioxide free gas, the extract is titrated with a standard base to measure acid content. Total volatile acids are expressed in milligrams per liter as acetic acid.

Gas Analysis

Anaerobic decomposition of sludges produces methane, carbon dioxide, and traces of hydrogen sulfide. Digester operation can be monitored by observing the relative amounts of carbon dioxide and methane in the head gases; the ratio is approximately one third and two thirds, respectively. A common laboratory apparatus consists of a water-jacketed buret, catalytic combustion chambers, and a series of glass units containing selective gas absorption reagents for carbon dioxide, oxygen, and carbon monoxide. The catalytic burners are for combustion of methane. A gas sample is drawn into the buret and initial volume recorded. The confined gas can then be passed through a manifold into any one of the absorbing units or through the combustion chambers, and drawn back to the buret again. After several passages through one of the absorbing units, the quantity of gas remaining is measured, and the percentage of that gas constituent in the sample can be calculated.

REFERENCES

1. *Drinking Water Standards and Guidelines*, Water Supply Division, Environmental Protection Agency, 1974.
2. *Standard Methods for the Examination of Water and Wastewater*, 13th

Edition, American Public Health Association, American Water Works Association, and Water Pollution Control Federation, 1971.

3. Sawyer, C. N., and McCarty, P. L., *Chemistry for Sanitary Engineers*, Second Edition, McGraw-Hill Book Co., New York, 1967.

4. *Simplified Procedures for Water Examination*, Manual M12, American Water Works Association, New York, 1964.

5. *Methods for Chemical Analysis of Water and Wastes*, Environmental Protection Agency, Cincinnati, Ohio, 1971.

6. *Laboratory Procedures for Wastewater Treatment Plant Operators*, New York State Department of Health, Health Education Service, Albany, N.Y.

7. Linstromberg, W. W., *Organic Chemistry, A Brief Course*, D. C. Heath and Co., Boston, 1966.

PROBLEMS

2-1 Using atomic weights given in Table 2-1, calculate the molecular and equivalent weights of: (a) $Al_2(SO_4)_3$, (b) calcium carbonate, (c) copperas, and (d) soda ash. [*Answers* (a) 342 and 57.0, others are given in Table 2-3]

2-2 What ions are formed when the following compounds dissolve in water: (a) ammonium sulfate, (b) calcium fluoride, (c) sodium hypochlorite, and (d) sodium carbonate? [*Answers* (a), (b), and (c) ions and radicals are listed in Tables 2-1 and 2-2, and (d) Na^+ and $CO_3^=$ or HCO_3^- depending on pH, Eq. 2-10]

2-3 If a water contains 40 mg/l of Ca^{++} and 10 mg/l of Mg^{++}, what is the hardness expressed in milligrams per liter as $CaCO_3$? (*Answer* 141 mg/l)

2-4 The alkalinity of a water consists of 16 mg/l of $CO_3^=$ and 120 mg/l of HCO_3^-. Compute the alkalinity in milligrams per liter as $CaCO_3$. (*Answer* 125 mg/l)

2-5 Draw a milliequivalent per liter bar graph and list the hypothetical combinations of compounds for the following water analysis:

$$Ca^{++} = 3.0 \text{ meq/l} \qquad HCO_3^- = 2.8 \text{ meq/l}$$
$$Mg^{++} = 1.0 \text{ meq/l} \qquad SO_4^= = 1.7 \text{ meq/l}$$
$$Na^+ = 0.5 \text{ meq/l} \qquad Cl^- = 0.2 \text{ meq/l}$$
$$K^+ = 0.2 \text{ meq/l}$$

2-6 Draw a meq/l diagram for the following water analysis:

$$Ca^{++} = 60 \text{ mg/l} \qquad HCO_3^- = 115 \text{ mg/l as } CaCO_3$$
$$Mg^{++} = 10 \text{ mg/l} \qquad SO_4^= = 96 \text{ mg/l}$$
$$Na^+ = 7 \text{ mg/l} \qquad Cl^- = 11 \text{ mg/l}$$
$$K^+ = 20 \text{ mg/l}$$

(*Answer* sum of the cations = 4.6 meq/l)

2-7 Draw a milliequivalent per liter diagram and list the hypothetical combinations as in Figure 2-1 for the following:

Calcium hardness = 150 mg/l as $CaCO_3$
Magnesium hardness = 65 mg/l as $CaCO_3$
Sodium and potassium = 0.35 meq/l
Bicarbonate alkalinity = 185 mg/l as $CaCO_3$
Sulfate = 29 mg/l
Chloride = 10 mg/l

(*Answer* 3.0 meq/l $Ca(HCO_3)_2$, 0.7 $Mg(HCO_3)_2$, and 0.6 $MgSO_4$)

2-8 Write the chemical reaction between hydrochloric acid with sodium hydroxide. How many mg of NaOH are required to neutralize 1.0 mg of acid? (*Answer* 1.1 mg)

2-9 Write the reaction of lime with water to form lime slurry, $Ca(OH)_2$, and a balanced equation between sulfuric acid and lime slurry. How many pounds of CaO are needed to neutralize one pound of pure acid? (*Answer* 0.57 lb)

2-10 Calcium hydroxide (lime slurry) reacts with calcium bicarbonate to precipitate calcium carbonate. What amount of $CaCO_3$ is formed per pound of $Ca(OH)_2$ applied? (*Answer* 2.7 lb)

2-11 Calcium sulfate reacts with sodium carbonate to precipitate calcium carbonate. Write a balanced equation, and compute the amount of soda ash needed to precipitate 20 mg/l of calcium sulfate. (*Answer* 15.6 mg/l)

2-12 What is the saturation value of dissolved oxygen in pure water at 20° C at sea level? At an altitude of 5000 ft? (*Answers* 9.2 mg/l and 7.7 mg/l)

2-13   A 100-ml sample of water with an initial pH of 7.8 is titrated to pH 4.5 using 8.5 ml of 0.02 N sulfuric acid. Calculate the alkalinity. (*Answer* 85 mg/l as $CaCO_3$)

2-14 In titration for alkalinity, 6.2 ml of 0.02 N sulfuric acid is used to reach the phenolphthalein end point and an additional 9.8 ml to the methyl orange end point for a 100-ml water sample. Calculate the alkalinity values for the two ionic forms present. (*Answer* carbonate: 124 mg/l as $CaCO_3$, bicarbonate: 36 mg/l)

2-15 Name the following compounds: $CH_3CH_2CH_3$, CH_3CH_2OH, CH_3COOH, $CH_3COO^-Na^+$, and CH_3CH_2—O—CH_2CH_3. Write formulas for these chemicals: methane, methyl alcohol, acetone, propionic acid, and sodium propionate.

2-16 Explain why not all of the organic matter in waste water is converted to carbon dioxide and water in biological treatment.

2-17 Convert the following units: (a) 100 mg/l to pounds per mil gal and grains per gallon, (b) 1600 lb/mil gal to milligrams per liter, (c) 2000 mg/l to pounds per million cubic feet, and (d) 6.0 gpg to milligrams per liter. (*Answer* (a) 834 lb/mil gal and 5.84 gpg, (b) 192 mg/l, (c) 125,000 lb/million cubic ft, and (d) 103 mg/l)

2-18 A waste-water treatment plant receives 22 mgd with a COD of 400 mg/l. What is the plant loading in lb COD/day? (*Answer* 73,400 lb COD/day)

2-19 A chlorine dosage of 2.4 mg/l is used in treating an average daily water supply of 3.0 mgd. How many 100-lb chlorine containers are needed per month? (*Answer* 18 containers per month)

chapter 3
Biology

An understanding of the key biological organisms—bacteria, algae, protozoa, crustaceans, and fishes—is essential in sanitary technology. The aquatic food chain, involving all of these organisms in natural waters, can be distressed by water pollution. Bacteria and protozoa comprise the major groups of micro-organisms in the "living" system that is used in secondary treatment of waste waters. A mixed culture of these microorganisms also performs the reaction in the biochemical oxygen demand test to determine waste-water strength. Several waterborne diseases of man are caused by bacteria. Indicator organisms, particularly coliforms, are used to evaluate the sanitary quality of water for drinking and recreation.

3-1 BACTERIA AND FUNGI

Bacteria (*sing.* bacterium) are simple, colorless, one-celled plants that use soluble food and are capable of self-reproduction without sunlight. As decomposers they fill an indispensable ecological role of decaying organic matter both in nature and in stabilizing organic wastes in treatment plants. Bacteria range in size from approximately 0.5 to 5 μ and are, therefore, only visible through a microscope. Figures 3-1a and b are photomicrographs of bacteria magnified 400 times. Individual cells may be spheres, rods, or spiral-shaped and may appear singly, in pairs, packets, or chains.

Bacterial reproduction is by binary fission, that is, a cell divides into two new cells each of which matures and again divides. Fission occurs every 15 to 30 min in ideal surroundings of abundant food, oxygen, and other nutrients. Some species form spores with tough coatings that are resistant to heat, lack of moisture, and loss of food supply as a means of survival.

A binomial system is used to name bacteria. The first word is the genus; the second is the species name. In activated sludge growing on domestic waste, a wide variety of bacteria are found, the majority of which appear to be of the genera *Alcaligenes*, *Flavobacterium*, *Bacillus*, and *Pseudomonas*. Identification of particular types in biological floc is only performed in research studies, since

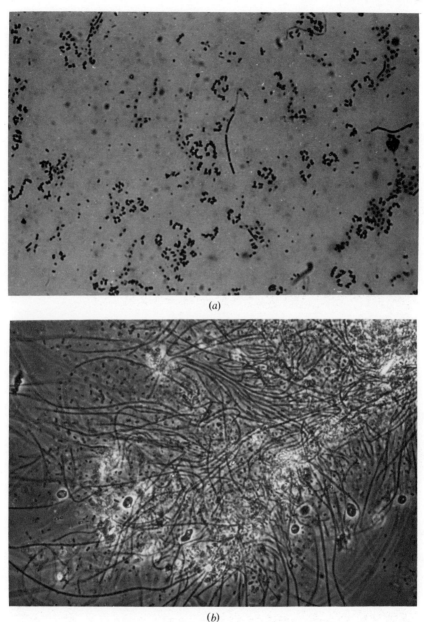

(a)

(b)

Figure 3-1 Photomicrographs of bacterial growths in waste water. (*a*) Dispersed growth of bacteria (400×). (*b*) Strands of *Sphaerotilus* with swarming cells (400×). (*c*) Filamentous activated-sludge floc (100×). (*d*) Normal activated-sludge floc (100×).

it is extremely difficult and of limited value in treatment operations to isolate them. One of the most common problems in aerobic biological treatment is poor settleability of the activated sludge floc related to filamentous growths which produce more buoyant floc (Figures 3-1*c* and *d*). Often this is caused by the bacterium *Sphaerotilus natans* (Figure 3-1*b*) where the cells grow protected in a long sheath; on maturity individual motile cells swarm out of the protective tube seeking new sites for growth. Perhaps the most frequently referred to bacterium in sanitary work is *Escherichia coli*, a common coliform

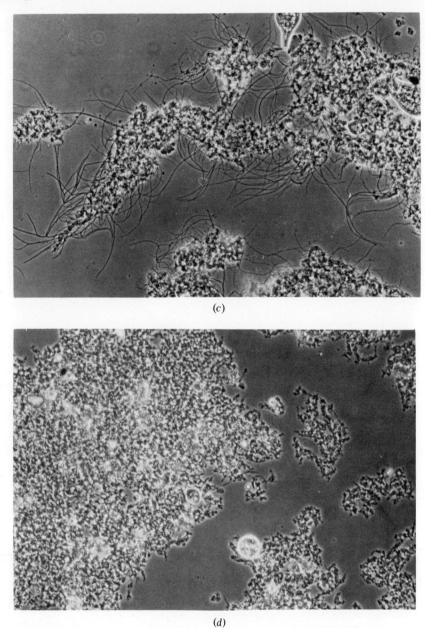

(c)

(d)

Fig. 3-1 (*Continued*)

used as an indicator of the bacteriological quality of water. *E. coli* cells under microscopic examination at 1000 magnification appear as individual short rods.

Bacteria are classified into two major groups as heterotrophic or autotrophic depending on their source of nutrients. Heterotrophs, sometimes referred to as saprophytes, use organic matter as both an energy and carbon source for synthesis. These bacteria are further subdivided into three groups depending on their action toward free oxygen. Aerobes require free dissolved oxygen in decomposing organic matter to gain energy for growth and multiplication, Eq. 3-1. Anaerobes oxidize organics in the complete absence of dissolved oxygen by using oxygen bound in other compounds, such as, nitrate and sulfate. Facultative bacteria comprise a group that uses free dissolved oxygen when available but that can also live in its absence by gaining energy from anaerobic reaction. In waste treatment, aerobic microorganisms are found in activated sludge and trickling filters, but anaerobes predominate in sludge digestion. Facultative bacteria are active in both aerobic and anaerobic treatment units.

Aerobic: $Organics + oxygen \rightarrow CO_2 + H_2O + energy$ (3-1)

Anaerobic: $Organics + NO_3^- \rightarrow CO_2 + N_2 + energy$ (3-2)

$Organics + SO_4^- \rightarrow CO_2 + H_2S + energy$ (3-3)

$Organics \rightarrow organic\ acids + CO_2 + H_2O + energy$ (3-4)

$\hookrightarrow CH_4 + CO_2 + energy$ (3-5)

The primary reason heterotrophic bacteria decompose organics is gain of energy for synthesis of new cells, and for respiration and motility. A small fraction of the energy is lost in the form of heat, Eq. 3-6. Amount of energy

Synthesis: $Energy + organics \longrightarrow new\ cell\ growth$

$\rightarrow respiration\ and\ motility$

$\hookrightarrow lost\ heat$ (3-6)

biologically available from a given quantity of matter depends on the oxygen source used in metabolism. The greatest amount is available when dissolved oxygen is used in oxidation, Eq. 3-1, and the least energy yield is derived from strict anaerobic metabolism, Eq. 3-4. Microorganisms growing in waste water seek the greatest energy yield in order to have maximum synthesis. For example, consider the reactions that would occur in a sample of fresh aerated waste water placed in a closed container. Aerobic and facultative bacteria immediately start to decompose waste organics depleting the dissolved oxygen. Although the strict aerobic organisms could not continue to function, facultative bacteria operating anaerobically can use the bound oxygen in nitrate releasing nitrogen gas, Eq. 3-2. The next most accessible oxygen is available in sulfate (Eq. 3-3) by conversion to hydrogen sulfide (rotten egg odor). Simultaneously other facultative and anaerobic bacteria would partially decompose material to organic acids, alcohols, and other compounds, Eq. 3-4 producing the least amount of energy. If methane-forming bacteria are present, the digestion process is completed by converting the organic acids to gaseous end products of methane and carbon dioxide, Eq. 3-5.

Autotrophic bacteria oxidize inorganic compounds for energy and use carbon dioxide as a carbon source. Nitrifying, sulfur, and iron bacteria are of greatest significance. Nitrifying bacteria oxidize ammonium nitrogen to nitrate in a two-step reaction as follows

$NH_3 + oxygen \xrightarrow{Nitrosomonas} NO_2^- + energy$ (3-7)

$NO_2^- + oxygen \xrightarrow{Nitrobacter} NO_3^- + energy$ (3-8)

Nitrification can occur in biological secondary treatment under the conditions of low organic loading and warm temperature. Although providing a more stable effluent, nitrification is often avoided to reduce oxygen consumption in treatment and to prevent floating sludge on the final clarifier. The latter is caused when sludge solids are buoyed up by nitrogen gas bubbles that are formed as a result of nitrate reduction, Eq. 3-2.

A common sulfur bacterium performs the reaction given in Eq. 3-9, which leads to crown corrosion in sewers. Waste water flowing through sewers

$$H_2S + oxygen \rightarrow H_2SO_4 + energy \qquad (3\text{-}9)$$

often turns septic and releases hydrogen sulfide gas, Eq. 3-3. This occurs most frequently in sanitary sewers constructed on flat grades in warm climates. The hydrogen sulfide is absorbed in the condensation moisture on the side walls and crown of the pipe. Here sulfur bacteria, able to tolerate pH levels of less than 1.0, oxidize the weak acid H_2S to strong sulfuric acid using oxygen from air in the sewer. Sulfuric acid formed reacts with concrete, reducing its structural strength. If sufficiently weakened, corrosion may lead to collapse under heavy overburden loads. Using corrosion resistant pipe material, such as vitrified clay or PVC plastic, is the best protection from corrosion in sanitary sewers. In large collectors where size and economics dictate concrete pipe, crown corrosion can be reduced by ventilation to expel the hydrogen sulfide and to reduce the moisture of condensation, or by chlorinating the waste water as it flows through the sewer to control generation of hydrogen sulfide. Protection of the concrete pipe interior by coatings and linings should be considered.

Iron bacteria are autotrophs that oxidize soluble inorganic ferrous iron to insoluble ferric, Eq. 3-10. The filamentous bacteria *Leptothrix* and

$$Fe^{++}(ferrous) + oxygen \rightarrow Fe^{+++}(ferric) + energy \qquad (3\text{-}10)$$

Crenothrix deposit oxidized iron, $Fe(OH)_3$, in their sheath, forming yellow or reddish-colored slimes. Iron bacteria thrive in water pipes where dissolved iron is available as an energy source and bicarbonates are available as a carbon source. With age the growths die and decompose releasing foul tastes and odors. There is no easy or inexpensive way of controlling bacteria in water distribution systems. The most certain procedure is to eliminate the ferrous iron from solution by water treatment and by controlling internal pipe corrosion. An alternative to removing iron from the water is a continuous ongoing maintenance program of treating and flushing water mains. A section of main to be treated is isolated and the water is then dosed with an excessive concentration of chlorine or other chemical to kill the bacteria. After several hours of contact the pipe section is flushed and returned to service.

Fungi (*sing.* fungus) refer to microscopic nonphotosynthetic plants including yeasts and molds. Yeasts are used for industrial fermentations in baking, distilling, and brewing. Under anaerobic conditions yeast metabolizes sugar, producing alcohol with minimum synthesis of new yeast cells. Under aerobic conditions, alcohol is not produced and the yield of new cells is much greater. Therefore, if the objective is to grow fodder yeast on waste sugar or mollasses, aerobic fermentation is used.

Molds are filamentous fungi that resemble higher plants in structure with branched, threadlike growths. They are nonphotosynthetic, multicellular, heterotrophic, aerobic, and grow best in acid solutions high in sugar content. Growths are frequently observed on the exterior of decaying fruits. Because of their filamentous nature, molds in activated sludge systems can lead to a poor settling floc, preventing gravity separation from the waste-water effluent in the final clarifier. Undesirable fungi growths are most frequently created by low pH conditions often associated with treatment of industrial wastes high in sugar content. Addition of an alkali to increase the pH, and sometimes ammonium nitrogen in high carbohydrate waste water, is necessary to suppress molds and to enhance the growth of bacteria.

3-2 ALGAE

Algae (*sing.* alga) are microscopic photosynthetic plants of the simplest forms, having neither roots, stems, nor leaves. They range in size from tiny single cells, giving water a green color, to branched forms of visible length that often appear as attached green slime. The term diatom is sometimes used in referring to single-celled algae encased in intricately etched silica shells. Figure 3-2 pictures algae filtered from water of an eutrophic lake magnified 100 times. *Anacystis, Anabaena,* and *Aphanizomenon* are blue-green algae associated with polluted water. Long strands of the latter clumped together appear to the naked eye as short grass clippings when suspended in water. *Oocystis* and *Pediastrum* are green algae. There are hundreds of algal species in a wide variety of cell structures in various shades of green and, less commonly, brown and red. An alga is identified by microscopic observation of its essential characteristics. *Standard Methods*[1] contains colored illustrations of common algae associated with taste and odor, filter clogging, polluted water, clean water, and others.

The process of photosynthesis is illustrated by the equation:

$$CO_2 + PO_4 + NH_3 \underset{\substack{\text{dark} \\ \text{reaction}}}{\overset{\substack{\text{energy} \\ \text{from} \\ \text{sunlight}}}{\rightleftarrows}} \frac{\text{new cell}}{\text{growth}} + O_2 \qquad (3\text{-}11)$$

Algae are autotrophic using carbon dioxide, or bicarbonates, as a carbon source, and inorganic nutrients of phosphate and nitrogen as ammonia or nitrate. In addition, certain trace nutrients are required such as: magnesium, boron, cobalt, calcium, and others. Certain species of blue-green algae are able to fix gaseous nitrogen if inorganic nitrogen salts are not available. The products of photosynthesis are new plant growth and oxygen. Energy for photosynthesis is derived from light. Pigments, the most common being chlorophyll which is green, biochemically convert sunlight energy into useful energy for plant growth and reproduction. Although the primary direction for Eq. 3-11 is to the right, in the prolonged absence of sunlight, plants can perform a dark reaction for survival. Algae take in oxygen and degrade stored food to yield energy for essential respiratory reactions. The rate of this survival reaction is significantly slower than photosynthesis.

The purpose of photosynthesis is to produce new plant life thereby increasing the number of algae. Given a suitable environment and proper nutrients algae grow and multiply in abundance. In natural waters the growth of algae may be limited by turbidity blocking sunlight, low temperatures during the winter, or depletion of a key nutrient. Clear, cold, mountain lakes tend to support few algae, while warm water lakes enriched with nitrogen and phosphorus from land runoff, or waste waters, exhibit heavy algal growths

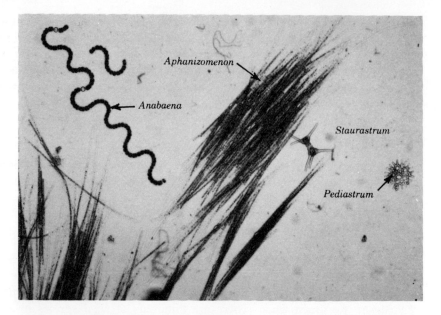

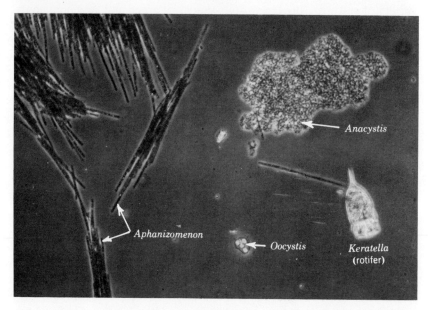

Figure 3-2 Photomicrographs of algae filtered from water of an eutrophic lake (100×).

creating green-colored, turbid water during the plant growing season. Waste-water stabilization ponds support luxurious blooms of algae to the point where the suspension becomes self-shading, that is, surplus nitrogen, phosphorus, and carbon nutrients cannot be synthesized because turbidity caused by the algae limits penetration of sunlight.

Macrophytes are aquatic photosynthetic plants, excluding algae, that occur as floating, submersed and immersed types. Floating plants are not rooted and ride on the water surface; Duckweed, a small three-leafed floating plant about 5 mm in diameter is common, and another is water hyacinth with its flowering tops. All or most of the foliage of submersed plants–pond weed, water weed, and coontail—grow beneath the water surface. These may root at depths greater than 10 ft below the surface, depending on the clarity of the water. Immersed plants, such as pickerelweed and cattail, attached by roots to the bottom mud, extend their principle foliage into the air above the surface. Rock and gravel bottom lakes and nutrient-poor bodies support limited growths of aquatic plants, while eutrophic lakes have extensive weed beds in shallow bays and along shore lines. Disposal of waste water in lakes and streams can promote weeds where other conditions, such as temperature and sunlight, are favorable. Unwanted weeds in waste-water stabilization ponds are controlled by constructing rather steep side slopes and maintaining a minimum depth of liquid of not less than 3 ft to block sunlight penetration to the bottom.

3-3 PROTOZOA AND MULTICELLULAR ANIMALS

Protozoa (or protozoans, *sing.* protozoan) are single-celled aquatic animals that multiply by binary fission. They have complex digestive systems and use solid organic matter as food. Protozoa are aerobic organisms found in activated sludge, trickling filters, and oxidation ponds treating wastes, as well as in natural waters. By ingesting bacteria and algae, they provide a vital link in the aquatic food chain.

Flagellated protozoa (Figure 3-3a) are the smallest type ranging in size from 10 to 50 μ. Long hairlike strands (flagella) provide motility by a whiplike action. While many ingest solid food, some flagellated species take in soluble organics. Amoeba (Figure 3-3b) move and take in food through the action of a mobile protoplasm. Although not as common as other forms of protozoa, amoeba are often found in the slime coating on trickling filter rocks and aeration basin walls. Free-swimming protozoa (Figure 3-3c) have cilia, small hairlike processes, used for propulsion and gathering in organic matter. These are easily

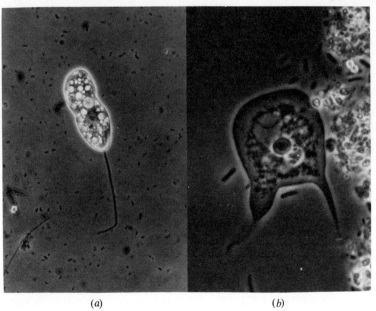

(a)   (b)

Figure 3-3 Photomicrographs of some protozoa and a rotifer found in waste-water aeration basins. (a) Flagellated protozoan (1000×). (b) Amoeba, mobile protoplasm (1000×). (c) Free-swimming protozoa (400×): *Euplotes* (left) and *Blepharisma* (right). (d) Stalked protozoa, *Vorticella* (400×). (e) Rotifer, *Rotaria* (200×).

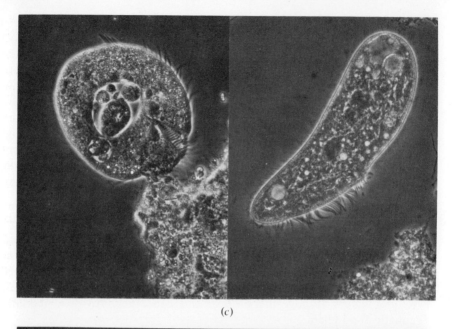

(c)

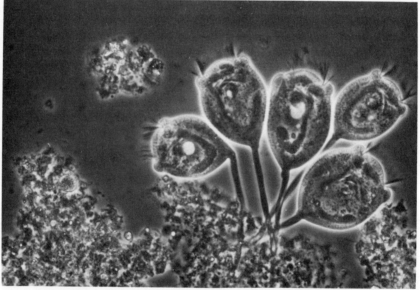

(d)

Fig. 3-3 (*Continued.*)

observed in a wet preparation under a microscope because of their rapid movement and relatively large size, 50 to 300 μ. Stalked forms (Figure 3-3d) attach by a stem to suspended solids and use cilia to propel their head about and bring in food.

Rotifers are simple, multicelled, aerobic animals that metabolize solid food. The rotifer illustrated in Figure 3-3e uses two circular rows of head cilia for catching food. Its head and foot are telescopic and it moves with a leechlike action, attaching the foot to a surface. Figure 3-2 pictures an entirely different type of rotifer with neither appendages nor telescopic head or foot. *Keratella*

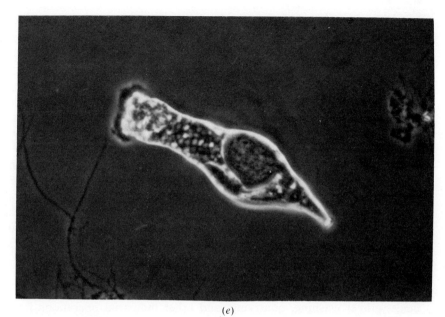

(e)

Fig. 3-3 (*Continued.*)

has a hard protective shell, an anus and, on the front, six spines along with cilia. Rotifers are found in natural waters, stabilization ponds, and extended aeration basins under low organic loading. Crustaceans are multicellular animals, typically 2 mm in size, easily visible with the naked eye. They have branched swimming feet or a shell-like covering with a variety of appendages. In the aquatic food chain they serve as herbivores ingesting algae and, in turn, being eaten by fishes. Most protozoa and higher animals can be identified by microscopic examination using a recognition key as an aid.[2]

3-4 FISHES

Fishes are cold-blooded animals typically with backbones, gills, fins, and usually scales. The internal structure is similar to other animals with flexible backbone and ribs, gullet, stomach, intestine, liver, spleen, and kidney. A two-chambered heart pumps blood to the gills to pick up oxygen before going to other parts of the body. Fishes also have a brain protected by the skull, spinal cord through the backbone, and nerves extending to internal organs and muscles. An air bladder, below the backbone, is used in maintaining buoyancy in water. Like all animals, fishes have a full complement of diseases that arise either internally or as a result of external agents. Illnesses include common organic and degenerative disorders of the internal organs. From outside come viruses, fungi, bacteria, parasitic protozoa, and worms. If they are not killed by disease or disorder, fishes must still survive predators, fishermen, and adverse environment. Often, the latter is caused by water pollution which not only affects adults directly but also their reproductive success.

Fish gills act as lungs. Fine red filaments attached to the gill arch bond contain numerous blood cells that absorb dissolved oxygen from water and give off carbon dioxide. Insufficient oxygen causes death by suffocation. This may occur naturally during the winter in shallow lakes and ponds covered with ice and snow for extended periods of time. However, more frequently it results from bacterial decomposition of waste organics depleting the dissolved oxygen. Municipal waste water disposed of in a stream without adequate treatment, or polluted land drainage from storm runoff, can reduce dissolved oxygen levels and drive fishes away or trap them, resulting in a fish kill. The deep unmixed bottom water in an eutrophic lake may have too little oxygen in late summer, after several months of thermal stratification, thereby limiting fishes to the warmer surface water. Excessive blooms of algae in this upper layer can, under certain conditions, result in oxygen depletion during the night, causing fish deaths.

The body of a fish is covered with scales and/or skin. Scales, formed of bonelike material, are contained in pockets in the skin. A protective slimy substance is produced to help prevent bacteria and other disease-producing organisms from penetrating the skin. Chemical pollutants may attack this protective slime, making the fish more vulnerable to diseases. Fins are used in swimming and to maintain balance. Fishes use an internal ear to detect vibrations and lidless, movable eyes for vision. Predator species, often considered by fisherman as the best game, are seriously impaired by turbid water which limits their ability to catch prey. Silt from erosion of cultivated fields, gravel washing, quarrying, and other industrial operations frequently contributes to unwanted turbidity.

The common reproductive process of fishes is deposition of eggs by the female and fertilization by the male. Some species construct saucer-shaped depressions in which they deposit their eggs and guard their young; others spawn at random and leave the fate of their brood to nature. Sunfishes, crappies, and largemouth bass nest in a circular depression in mud or sand, often among roots of aquatic plants, while others, such as smallmouth bass, prefer a gravel bottom for nesting. Northern pike, walleyes, and carp scatter their eggs over bottom areas in shallow water. Some fishes, for example, trout and salmon, ascend streams to spawn. Success in reproduction can be severely affected by pollution. Siltation covers fish eggs, certain heavy metals and organic chemicals reduce hatch, and stream contamination may inhibit migration to spawning grounds.

Generally, fish can be identified by their external appearance. Size and shape of scales, fins, teeth, head, and body are common criteria. In ichthyology (a branch of zoology dealing with fishes) an elaborate classification has been developed which groups fishes that look alike or have similar characteristics. The most meaningful grouping for a sanitary technologist is by tolerance level. Intolerant species—grayling, white fishes, brook and rainbow trout—require cold water, high dissolved oxygen concentrations, and low turbidity. These are the choice food fishes and, unfortunately, also the most susceptible to adverse environmental changes, whether natural or man-created. Tolerant fishes can withstand wide fluctuations in surroundings, warm water, low dissolved oxygen, high carbon dioxide, and turbidity. Carp, black bullhead, and bowfin are examples. Many moderately tolerant fishes such as walleye and northern pike are both good angling and eating. Their location in warm-water lakes near densely populated areas has promoted them as prized game fishes by sportsmen. However, the waterborne wastes from surrounding agriculture, business, and housing have polluted these warm-water streams and lakes, causing a shift in fish populations to more tolerance species such as bluegill and yellow perch.

3-5 AQUATIC FOOD CHAIN

Aquatic productivity is limited by the supply of raw materials and the biological efficiency of converting them into various life forms. Rivers and lakes well supplied with oxygen, carbon dioxide, nitrogen, phosphorus, and sufficient sunlight are rich in plant and animal life. If any of these nutrients is scarce, if the water is polluted, or does not receive enough sunlight, the production of life is low. Basic forms of life, algae and other green plants, are called primary producers, since they use the energy of sunlight to synthesize inorganic substances into living tissues. Rooted plants, although usually the most conspicuous water vegetation, play a relatively minor role in river or lake productivity. The most numerous plants are algae. Animals, being unable to manufacture their own food, obtain energy and nutrients secondhand by either eating plants or smaller animals. Flow of energy and material from one form of life to another is known as a food chain. The aquatic food chain in Figure 3-4 shows algae as the primary producers. Then come the first-order consumers—animals, such as fly nymphs, copepods, and water fleas. These small herbivores (plant eaters) are in turn eaten by second-order consumers—carnivores (flesh eaters) such as sunfishes. A third-order consumer, such as a bass, may then eat the sunfishes. If the bass dies, it is consumed by scavengers like crayfishes; or the nutrients stored in the dead fish are released by bacterial decomposers and recycled into algae by photosynthesis. Examining a single food chain reveals only part of nature's complex process since hundreds of them exist. Most of the chains are interconnected such that the organization of a community is more accurately described as a food web. For example, rather than dying the bass might be carried away by a heron or captured underwater by a mink. The

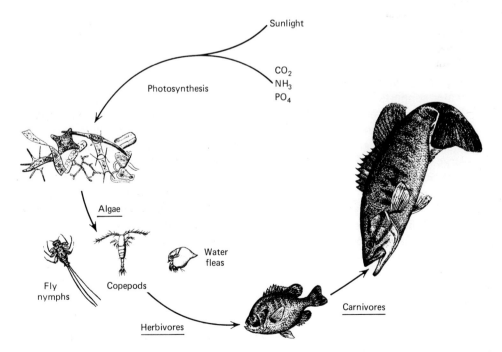

Figure 3-4 Aquatic food chain. Energy and material pass from green plants formed by photosynthesis, through small animals and immature insects that graze on algae, to second- and third-order consumers, fishes.

fly nymph may leave the water environment by emerging as an adult and flying away to be consumed by a swallow.

The complex system of aquatic food relationships can be upset by man-made pollution. For example, primary production is modified by increasing turbidity or overfertilization. Murky water blocks out sunlight slowing plant growth, and by chain reaction reduces the supply of food to all higher forms of life. Overfertilization of water with plant nutrients can be equally disastrous, particularly in lakes. Uninhibited photosynthesis produces more algae than the herbivores can consume. Dense blooms of algae create turbidity, floating scum, and reduction in dissolved oxygen when the mass of algae decompose. More available nutrients also increase aquatic weed growths, which in combination with increased turbidity from algae, upset the predator-prey relationship between carnivores. Prey species, such as bluegills, can effectively hide from bass predators. Thus the total number of smaller fishes increases dramatically compared to larger game species.

Industrial wastes—acids, oils, heavy metals, or other chemicals—have a direct influence on aquatic life. Many wastes can directly kill fishes if they shift the pH of the water sufficiently, cover the surface causing a reduction in dissolved oxygen, or build up to toxic levels. A more subtle effect results if the contaminant dosage is sublethal to the fishes but eliminates their food supply by poisoning the algae or herbivores. Some chemicals are magnified biologically as they pass through the food chain. If traces of mercury, lead, or chlorinated hydrocarbons are in water, their concentration found in algae is much greater. Herbivores consuming the algae accumulate these contaminants in their bodies, which in turn are concentrated to even higher levels in fishes. The seriousness of biological magnification varies with different chemicals. Most frequently the influence is indirect, often resulting in decreased population, either by decreasing reproductive success or by influencing body functions, making the fish a less successful predator. In other instances, the fish flesh may be tainted or unedible.

Biological life in clean water is exemplified by a large number of different species of plants and animals with generally balanced populations. Polluted water supports large populations with little variety. Small polluted lakes may have strong dominance of one type of fish; a pond turbid with silt may support a large population of black bullheads, while a heavily eutrophic lake may dominate in bluegills. A short distance downstream from the introduction of a large quantity of organic matter the dissolved oxygen decreases sharply. Here the most abundant form of life is bacteria, such as *Sphaerotilus*, so-called sewage fungus, rat-tailed maggots, mosquito larvae, and sludge worms. Further downstream protozoa, rotifers, certain fly larvae, polluted-water algae, and tolerant fish species appear. Finally, after the stream has returned to a high dissolved oxygen content, a broad spectrum of life reappears on each level of the food web.

3-6 WATERBORNE DISEASES

Several infectious, enteric (intestinal) diseases of man are transmitted through fecal wastes. Pathogens, disease-producing agents, excreted in the feces of infected persons include all major catagories: bacteria, viruses, protozoa, and parasitic worms. The resultant diseases are most prevalent in those areas where sanitary disposal of human feces is not practiced. Transmission may be by direct contact of infected individuals, via insect vectors such as flies, or through contaminated food and water. Complete control involves instituting an environmental health program that incorporates personal and household hygiene, control of fly species and other insects, monitoring of food processing, proper waste disposal, immunization of the populace where possible, and isolation and treatment of diseased persons.

Three of the most common waterborne bacterial diseases are typhoid fever caused by *Salmonella typhosa*, Asiatic cholera (*Vibrio comma*), and bacillary dysentery (*Shigella dysenteriae*). Typhoid is an acute infectious disease characterized by a continued high fever and infection of the spleen, gastrointential tract, and blood. Cholera symptoms include diarrhea, vomiting, and dehydration. Dysentery causes diarrhea, bloody stools, and sometimes fever. All of these diseases are debilitating and often cause death. Transmission is by direct contact, food, milk, shellfish, and water. Although these diseases are still prevalent in underdeveloped countries and have been responsible for millions of deaths historically, they have been virtually eliminated in the United States by environmental control through pasteurization of milk and chlorination of water supplies.

Viruses of poliomyelitis and infectious hepatitis are excreted in the stools of infected persons, but the modes of transmission are not completely defined. Transfer of polio virus appears to be by direct contact with some evidence of transmission by food and water. Waterborne outbreaks of infectious hepatitis have occurred, although spreading by person-to-person contact appears to be the chief means of transmission. Most authorities believe that outbreaks from municipally treated water supplies are unlikely. The symptoms of hepatitis are loss of appetite, fatigue, nausea, and pain. The most characteristic feature of the disease is a yellow color that appears in the white of the eyes and skin, hence the common name yellow jaundice. Although more severe in older persons, it is seldom fatal.

Amoebic dysentery, the most common enteric protozoal infection, is caused by *Entamoeba histolytica*. It is transmitted by direct contact, by food, and through water in tropical climates but is nontransmittable in temperate climates. Water carriage can be prevented by municipal treatment with heavy chlorination, or by individuals boiling their drinking water.

Schistosomiasis, often called bilharziasis, is a parasitic disease caused by a small, flat worm that infests man's internal organs—heart, lungs, veins, and liver. Eggs of the schistosome worms, living in human abdominal organs, are passed into water with fecal discharges. They hatch into miracidia, enter snails, and develop into sporocysts that produce fork-tailed cercariae. These leave the snail, attach themselves to humans, bore through the skin and enter blood vessels migrating to the internal organs where they grow into adulthood. This debilitating disease, for which there is no immunization, is the world's worst health problem in agricultural communities in Asia, Africa, and South America. The disease does not occur in the continental United States, since the intermediate snail host must be one of several specific species that are not found in this country. Duck schistosomiasis transmitted in a fashion similar to the human disease is found in United States lakes frequented by wild ducks. If infected snails inhabit bathing beaches on these lakes, the cercariae released may come in contact with and penetrate the skin of bathers. The resultant infection, neither a contagious nor a fatal disease, is referred to as swimmer's itch. Red spots appear where the cercariae have penetrated, accompanied by severe irritation which lasts several hours. During the following two days extremely itchy small lumps may appear, sometimes swelling and filling with pus.

Safety surveys of sanitation workers have shown that incidents of water-borne disease is no greater in this group than the population as a whole. Although waste water must be considered potentially pathogenic, the reduced incidence of waterborne disease in this country has diminished the possibility of treatment plant employees becoming infected. Most safety manuals stress that the best defense against infection is the practice of good personal hygiene and prompt medical care for any injury that breaks the skin. The latter appears to be the most critical problem, since infected wounds are far more common than enteric diseases. The most common source of *Clostridium tetani* is human feces; therefore, it is recommended that sanitation workers receive artifical immunity by tetanus toxoid injections.

3-7 INDICATOR ORGANISMS FOR WATER QUALITY

Testing a water for pathogenic bacteria might at first glance be considered a feasible method for determining bacteriological quality. However, on closer examination, this technique has a number of shortcomings that precludes its application. Laboratory analyses for pathogenic bacteria are difficult to perform and generally are not quantitatively reproducible. Furthermore the demonstrated absence of *Salmonella*, for example, does not exclude the possible presence of *Shigella*, *Vibrio*, or disease-producing viruses. Finally, since few pathogens are present in polluted waters, the high frequency of negative results would lead to questioning the validity of the test procedures. It is far more reassuring to an analyst if a few tests yield positive results. For these reasons, the bacteriological quality of water is based on testing for nonpathogenic indicator organsms, principally the coliform group.

Coliform bacteria, as typified by *Escherichia coli*, and fecal streptococci (enterococci) residing in the intestinal tract of man, are excreted in large numbers in feces of man and other warm-blooded animals, averaging about 50 million coliforms per gram. Untreated domestic waste water generally contains more than 3 million coliforms per 100 ml. Pathogenic bacteria and viruses causing enteric diseases in man originate from the same source, namely, fecal discharges of diseased persons. Consequently, water contaminated by fecal pollution is identified as being potentially dangerous by the presence of coliform bacteria.

Drinking water standards generally specify that a water is safe provided that testing in a specified manner does not reveal more than an average of one coliform organism per 100 ml. This criterion is supported by the following arguments. The number of pathogenic bacteria, such as, *Salmonella typhosa*, in domestic waste water is generally less than one per million coliforms, and the average density of enteric viruses has been measured as a virus to coliform ratio of $1:100,000$. The die-off rate of pathogenic bacteria is greater than the death rate of coliforms outside of the intestinal tract of animals; thus, exposure to treatment and residence in surface waters reduces the number of pathogens relative to coliforms. Based on these premises, water that meets a standard of less than one coliform per 100 ml is, statistically speaking, safe for human consumption because of the improbability of ingesting any pathogens. Strictly speaking, this coliform standard as established by the Environmental Protection Agency applies only to processed water where treatment includes chlorination.

Extension of coliform criteria to water quality for purposes other than drinking is poorly defined. Since these bacteria can originate from warm-blooded animals, soil, and cold-blooded animals in addition to feces of man, presence of coliforms in surface waters indicates any one or a combination of three sources: wastes of man, farm animals, or soil erosion. Although a special test can be run to separate fecal coliforms from soil types, there is no way of distinguishing between the human bacteria and those of animals. The significance of coliform testing in pollution surveys then depends on a knowledge of the river basin and the most probable source of the observed coliforms.

Numerical coliform criteria for body-contact water use and general recreation have been established by most states at the upper limits of 200 fecal coliforms per 100 ml and 2000 total coliforms per 100 ml. These values are only considered to be guidelines, since there is no positive epidemiological evidence that bathing beaches with higher coliform counts are associated with transmission of enteric diseases. In fact, these standards appear to be far too conservative from a standpoint of realistic public health risk. A survey of illnesses developed by persons as a result of bathing on public beaches showed that the majority were infections of skin, eye, ear, nose, or throat with only about one tenth related to gastrointestinal disorders. In recent years in the United States, no cases of enteric disease have been linked directly to recreational water use, while several thousand persons have drowned and even greater numbers have been disabled by automobile accidents in traveling to recreational areas. Although substantial risk is involved in recreational water use, the greatest danger is apparently not related to disease transmission. Hence, coliform standards applied to water used for swimming are linked to water-associated diseases of the skin and respiratory passages rather than enteric diseases. This is, of course, completely different than the purpose of the coliform standard for drinking water, which is related to enteric disease transmission. Here tighter restrictions are imperative, since a water distribution system has the potential of mass transmission of pathogens in epidemic proportions.

Streptococci could be used as indicator organisms rather than coliforms. But the use of coliforms was established first, and there does not appear to be any distinct advantages to warrent shifting to a fecal streptococcus system.

3-8 TESTS FOR THE COLIFORM GROUP

Extensive laboratory apparatus are needed to conduct bacteriological tests, including: a hot-air sterilizing oven, autoclave, incubators, sample bottles, fermentation tubes with inverted vials, dilution bottles, petri dishes and dish containers, graduated pipettes and pipette containers, a wire inoculating loop, culture media, and preparation utensils. The hot-air oven is used for sterilizing glassware, empty sample bottles, pipettes, and petri dishes. Heating to a temperature of 160 to 180° C for a period of $1\frac{1}{2}$ hr is adequate to kill microbial cells and spores. An autoclave is used to sterilize culture media and dilution water under steam pressure. Recommended operation is for 15 min at 15 psi steam pressure, which corresponds to a temperature of 121.6° C. Sample bottles with a volume of about 120 ml are usually soft glass with screw-top closures suitable to withstand sterilization temperatures. Fermentation tubes are round-bottomed glass tubular containers designed to accept either a screw cap, slip-on stainless steel closure, or an inserted cotton plug. The inverted vial is a small tube placed upside down in the culture medium in the fermentation tube to collect gas generated by bacterial growth. Screw-cap dilution bottles are to autoclave measured volumes of distilled water for use in diluting water samples prior to testing. Petri dishes are small shallow circular plates with glass covers for holding solid culture media. The dishes are placed in a stainless steel container for sterilization in a hot-air oven. Pipettes used to transfer samples from one container to another are also hot-air treated to prevent sample contamination. A small, thin, wire loop of Chromel or platinum is used to aseptically transfer small quantities of culture. The wire is sterilized by heating red-hot in the flame of a bunsen burner.

Culture media are special formulations of organic and inorganic nutrients to support the growth of microorganisms. Dehydrated culture media (powders) are available commercially. Preparation involves placing a measured amount in distilled water and heating in a double boiler to dissolve without scorching. Lactose broth containing beef extract, peptone (protein derivatives), and lactose (milk sugar) is the primary medium used in testing for coliform bacteria. Some laboratories use lauryl tryptose broth which also contains the critical lactose ingredient. Since occasionally noncoliform aerobic spore-forming bacteria, or groups of bacteria working together, can produce gas in lactose broth, a fermentation tube showing positive results is tested further to confirm the presence of coliforms. The medium used is brilliant green lactose bile broth. The green dye inhibits noncoliform growth while the presence of lactose stimulates the coliforms to overcome the toxic effect of the dye. There are two types of solid inhibitory media—eosin methylene blue (EMB) and Endo agar—that are used in coliform testing. Both contain inhibitory dyes and agar for solidification. These solid media are prepared by dissolving in distilled water, dispensing into bottles for autoclaving, and pouring in the bottom plate of sterile petri dishes to harden. The surface is inoculated by streaking with a wire loop that has been dipped into a culture growing in a fermentation tube. EC medium is a nutrient broth used to test for fecal coliforms.

Examination of Potable Water

Collection of water samples for bacteriological examination must be done carefully to prevent contamination. The sampling fixture is generally a faucet, petcock, or small valve; fire hydrants are poor sampling stations because of the impossibility of sterilizing the long barrel. The fixture nozzle is sterilized by flaming with an alcohol lamp or blowtorch. After draining the tap for several minutes to flush out the service line to the water main, a sterilized bottle is used to catch a sample from the stream of water. Care must be taken not to accidentally contaminate the cap or mouth of the bottle with the fingers. All samples of water collected from chlorinated sources must be dechlorinated at the time of collection. This is achieved by adding a measured amount of prepared sodium thiosulfate solution to the empty sample bottle before sterilization. Its presence in the bottle at the time of collection neutralizes any residual chlorine and prevents continuation of the disinfection action during the time the sample is in transit to the laboratory.

The minimum number of water samples collected from a distribution system and examined each month for coliforms depends on the population that is served by the system. For example, the minimum number required for populations of 1000, 10,000, and 100,000 are 2, 12, and 100, respectively. Compliance with the bacteriological requirements of drinking water standards is determined by the number of positive tests. When 10-ml standard portions are examined, not more than 10 percent in any month should be positive, in other words, the upper limit of coliform density is an average of one per 100 ml.

The coliform group is defined as all aerobic and facultative anaerobic, nonspore forming, Gram-stain negative rods that ferment lactose with gas production within 48 hr of incubation at 35° C. The initial coliform analysis is the presumptive test based on gas production from lactose. Ten milliliter portions of a water sample are transferred using sterile pipettes into prepared fermentation tubes. The tubes contain lactose, or lauryl tryptose, broth, and inverted vials. Inoculated tubes are placed in a warm-air incubator at 35° C ± 0.5° C. Growth with the production of gas, identified by the presence of bubbles in the inverted vial, is a positive test indicating coliform bacteria may be present. A negative reaction, either no growth or growth without gas, excludes the coliform group.

The confirmed test is used to substantiate, or deny, the presence of coliforms in a positive presumptive test. One of two procedures as outlined on the schematic diagram in Figure 3-5 is used. A wire loop is used to transfer growth from a positive lactose tube, either after 24 or 48 hr incubation, to a fermentation tube containing brilliant green lactose bile broth. If growth with gas occurs within 48 hr at 35° C, the presence of coliforms is confirmed. If no gas is produced, the test is negative. The alternate confirmed test is to streak the culture from a positive presumptive test across the surface of Endo or EMB agar in prepared petri dishes by using a sterile wire loop. The inoculated plates are incubated for 24 hr at 35° C. Bacterial colonies growing on the surface of the solid media that exhibit a green sheen, referred to as typical coliform colonies, confirm the presence of coliform bacteria. Atypical growths, light-colored colonies without dark centers or a green sheen, neither confirm nor deny the presence of the coliform group. Absence of bacterial growth in the dishes is a negative test. Figure 3-6 shows fermentation tubes prior to inoculation, positive growth with gas, and a negative test of growth without gas; one of the Endo agar plates pictured exhibits typical coliform colonies.

Normally in examining potable water, coliform testing is terminated with the confirmed test using brilliant green bile broth. Occasionally one may desire to run a completed test as is outlined in Figure 3-5. This involves transferring a colony from an Endo or EMB plate to nutrient agar and into lactose broth. If gas is not produced in the lactose fermentation tube, the colony transferred did not contain coliforms and the test is negative. If gas is produced in the tube, a portion of growth on the nutrient agar is smeared onto a glass slide and prepared for observation under a microscope by using the Gram-stain technique. If the bacteria are short rods, with no spores present, and the Gram-strain is negative (stained cells are pink in color), the coliform group is present and the test is completed. If the culture Gram-stains positive (purple colored), the completed test is proved negative.

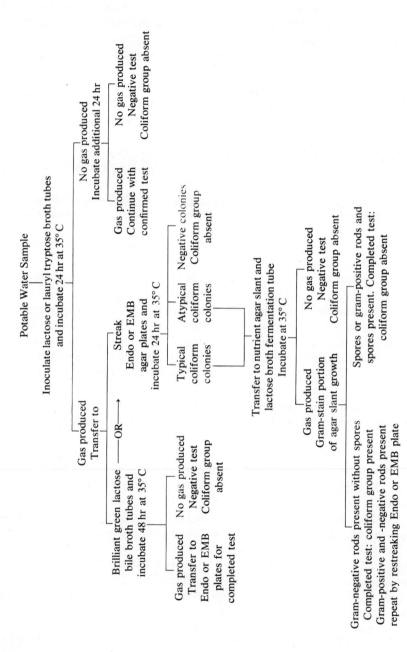

Figure 3-5 Schematic diagram for detection of the coliform group in potable water.

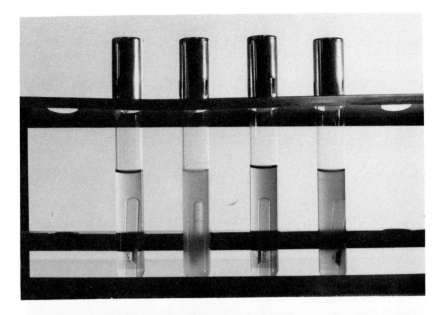

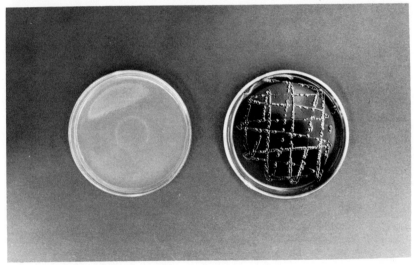

Figure 3-6 Presumptive and confirmed tests for the coliform group. Fermentation tubes shown are, from left to right, sterile lactose broth, growth with gas produced (positive test), sterile brilliant green lactose bile broth, and growth without gas produced (negative test). The petri dish on the left is sterile Endo agar while the plate on the right exhibits typical coliform colonies.

Examination of Surface Water

An elevated temperature coliform test separates microorganisms of the coliform group into those of fecal origin and those of nonfecal sources. This technique is applicable to investigations of stream pollution, raw-water sources, waste-water treatment systems, bathing waters, and general water quality monitoring. The procedure is not recommended as a substitute for the coliform tests used in examination of potable waters. The test procedure, diagramed schematically in Figure 3-7, cannot be used for direct isolation of

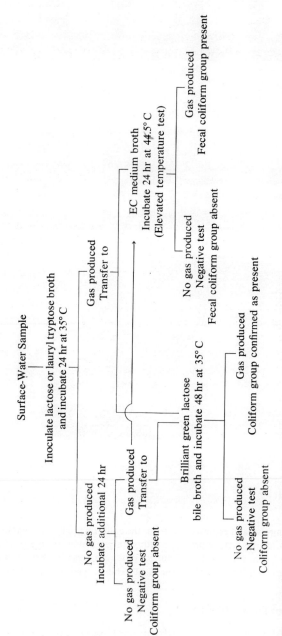

Figure 3-7 Schematic diagram for identification of the coliform group and the fecal-coliform group in water quality monitoring.

coliforms from water but is used as a confirmatory procedure following presumptive testing. The first step is to transfer a measured volume of sample into fermentation tubes containing lactose or lauryl tryptose broth. Those tubes showing growth with gas after 24 or 48 hr are used to inoculate culture tubes containing EC medium broth, which are incubated in a water bath at $44.5 \pm 0.2°\,C$ by immersion to the upper level of the medium. Because of the close temperature tolerance required in the fecal coliform test, the standard warm-air incubators normally used in bacteriological testing are not satisfactory. Growth with gas is positive for the fecal coliform group; no gas demonstrates absence.

In water quality surveys it is necessary to evaluate the number of coliform bacteria presence to determine if a water meets the appropriate standard. The multiple-tube fermentation technique can be used to enumerate positive presumptive, confirmed, and fecal coliform tests. Results of the test are expressed as the most probable number (MPN), since the count is based on statistical analysis of sets of tubes in a series of serial dilutions. MPN is by definition related to a sample volume of 100 ml; hence, a MPN of 10 means 10 coliforms per 100 ml of water.

The common procedure for MPN determination involves the use of sterile pipettes calibrated in 0.1 ml increments, sterile screw-top dilution bottles containing 99 ml of water, and a rack containing six sets of five lactose broth fermentation tubes. Figure 3-8 is a sketch showing test preparation. A sterile pipette is used to transfer 1.0 ml portions of the sample into each of five fermentation tubes, followed by dispensing of 0.1 ml to a second set of five. For the next higher dilution, the third, only 0.01 ml of sample water is required, which it is impossible to pipette accurately. Therefore, 1.0 ml of sample is placed in a dilution bottle containing 99 ml of sterile water and mixed. Now, 1.0 ml portions containing 0.01 ml of the surface water sample can be pipetted into the third set of five tubes. The fourth set receives 0.1 ml from this same

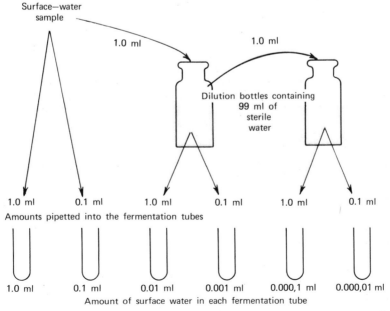

Figure 3-8 Procedure for preparation of the multiple-tube fermentation technique for determining the most probable number (MPN) of coliform bacteria in a surface-water sample.

dilution bottle. The process is then carried one more step by transferring 1.0 ml from the first dilution bottle into 99 ml of water in the second for another 100-fold dilution. Portions from this dilution bottle are pipetted into the fifth and sixth tube sets. After incubation for 48 hr at 35° C, the tubes are examined for gas production, and the number of positive reactions for each of the serial dilutions is recorded.

The MPN index and confidence limits for various combinations of positive and negative multiple-tube fermentation results can be determined from Table 3-1. The first three columns in the table refer to the number of positive tubes out of five containing 10-ml, 1-ml, and 0.1-ml sample portions. If smaller amounts of sample were used in the tubes, which is usually the case, the

Table 3-1. MPN Index and 95 Percent Confidence Limits for Coliform Counts by the Multiple-Tube Fermentation Technique for Various Combinations of Positive and Negative Results When Five 10-ml, Five 1-ml, and Five 0.1-ml Portions Are Used

No. of Tubes Giving Positive Reaction out of			MPN Index per 100 ml	95% Confidence Limits		No. of Tubes Giving Positive Reaction out of			MPN Index per 100 ml	95% Confidence Limits	
5 of 10 ml Each	5 of 1 ml Each	5 of 0.1 ml Each		Lower	Upper	5 of 10 ml Each	5 of 1 ml Each	5 of 0.1 ml Each		Lower	Upper
0	0	0	<2			4	2	1	26	9	78
0	0	1	2	<0.5	7	4	3	0	27	9	80
0	1	0	2	<0.5	7	4	3	1	33	11	93
0	2	0	4	<0.5	11	4	4	0	34	12	93
1	0	0	2	<0.5	7	5	0	0	23	7	70
1	0	1	4	<0.5	11	5	0	1	31	11	89
1	1	0	4	<0.5	11	5	0	2	43	15	110
1	1	1	6	<0.5	15	5	1	0	33	11	93
1	2	0	6	<0.5	15	5	1	1	46	16	120
						5	1	2	63	21	150
2	0	0	5	<0.5	13						
2	0	1	7	1	17	5	2	0	49	17	130
2	1	0	7	1	17	5	2	1	70	23	170
2	1	1	9	2	21	5	2	2	94	28	220
2	2	0	9	2	21	5	3	0	79	25	190
2	3	0	12	3	28	5	3	1	110	31	250
						5	3	2	140	37	340
3	0	0	8	1	19						
3	0	1	11	2	25	5	3	3	180	44	500
3	1	0	11	2	25	5	4	0	130	35	300
3	1	1	14	4	34	5	4	1	170	43	490
3	2	0	14	4	34	5	4	2	220	57	700
3	2	1	17	5	46	5	4	3	280	90	850
3	3	0	17	5	46	5	4	4	350	120	1000
4	0	0	13	3	31	5	5	0	240	68	750
4	0	1	17	5	46	5	5	1	350	120	1000
4	1	0	17	5	46	5	5	2	540	180	1400
4	1	1	21	7	63	5	5	3	920	300	3200
4	1	2	26	9	78	5	5	4	1600	640	5800
4	2	0	22	7	67	5	5	5	$\geqq$2400		

Source. Standard Methods for the Examination of Water and Wastewater.

tabulated answers must be adjusted. If 1.0, 0.1, and 0.01 ml are used, the resultant MPN is 10 times the value given in the table; if the sample quantities are 0.1, 0.01, and 0.001 ml, record 100 times the tabulated value; and so on for other combinations. When more than three dilutions are employed in a decimal series of dilutions, the results from only three of these are used in computing the MPN. The three dilutions selected are the highest dilution giving positive results in all five portions tested (no lower dilution giving any negative results) and the two next succeeding higher dilutions.

The presumptive MPN test can be extended to confirmed testing and fecal coliform determination as is outlined on Figure 3-7. Growth from each of the positive lactose tubes is transferred aseptically using a wire loop to brilliant green bile lactose broth and EC medium broth. These are incubated to confirm the presence of coliforms and to determine whether they are of fecal origin.

EXAMPLE 3-1

The following are results from multiple-tube fermentation analyses of a polluted river water sample for the presumptive, confirmed, and fecal coliform groups. The serial dilutions were set up as is illustrated in Figure 3-7. Determine the MPN for each of the three tests.

Serial Dilution	Sample Portion ml	Number of Positive Reactions out of Five Tubes		
		Lactose Broth	Brilliant Green Bile	EC Medium
0	1.0	5	5	5
1	0.1	5	5	4
2	0.01	5	3	1
3	0.001	1	1	1
4	0.000,1	0	0	0
5	0.000,01	0	0	0

Solution

(a) *Presumptive Coliform MPN.* The three dilutions selected for MPN determination show 5, 1, and 0 positive tubes. The index value from Table 3-1, without regard to size of sample portion, is 33. However, since the sample portions corresponding to the 5, 1, and 0 serial dilutions are 0.01, 0.001, and 0.000,1 ml, the MPN value must be multiplied by 1000. Therefore, the MPN index for the presumptive test series is 33,000, with 95 percent confidence limits of 11,000 and 93,000. (MPN is the most probable number in 100 ml of water, and the confidence limits mean that 95 percent of all MPN analyses run on the same sample should, statistically speaking, fall within this range of limiting values.)

(b) *Confirmed Coliform MPN.* Selected dilutions are 0.1, 0.01, and 0.001 ml showing 5, 3, and 1 positive tubes, respectively. The multiplication factor is 100. From Table 3-1, MPN equals 11,000 with confidence limits of 3100 to 25,000.

(c) *Fecal Coliform MPN.* Selected dilutions are 5, 4, and 1 with a correction factor of 10. MPN equals 1700 with confidence limits of 430 to 4900.

Membrane Filter Technique for Coliform Testing

This method consists of drawing a measured volume of water through a filter membrane fine enough to take out bacteria, and then placing the filter on a growth medium in a petri dish. Each bacterium retained by the filter grows and forms a small colony. The number of coliforms present in a filtered sample is determined by counting the number of colonies and expressing this value in terms of number per 100 ml of water. The membrane filter technique has been widely adopted for use in water quality monitoring studies, since it requires much less laboratory apparatus than the standard multiple-tube technique; and, portable membrane filter apparatus have been developed for conducting coliform tests in the field.

Special apparatus required to conduct membrane filter coliform tests includes: filtration units, filter membranes, absorbent pads, forceps, and culture dishes. The common laboratory filtration unit (Figure 3-9) consists of a funnel that fastens to a receptacle bearing a porous plate to support the filter membrane. The filter-holding assembly, constructed of glass, porcelain, or stainless steel, is sterilized by boiling, autoclaving, or ultraviolet radiation. For filtration, the assembly is mounted on a side-arm filtering flask which is evacuated to draw the sample through the filter. For field use a small hand-sized plunger pump (syringe) is used to draw a sample of water through the small assembly holding the filter membrane.

Filter membranes, available from several commercial manufacturers, are 2-in. diameter disks with pore openings of $0.45 \pm 0.02 \mu$, small enough to retain microbial cells. Filters used in determining bacterial counts have a grid printed on the surface for ease in counting colonies. Filter membranes are sterilized prior to use, either in a glass petri dish or wrapped in heavy paper. Absorbent pads for nutrients are normally furnished with the filters by the manufacturer. After sterilization these pads are placed in culture dishes to absorb the nutrient media on which the membrane filter is placed. During the testing procedure, the

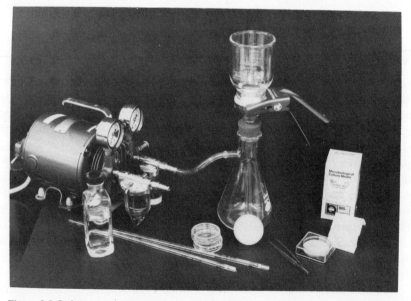

Figure 3-9 Laboratory apparatus for coliform testing by the membrane filter technique.

filters are handled on the outer edges with forceps that are sterilized before use by dipping in alcohol and then ignition.

Either glass or disposable plastic culture dishes may be used. If glass petri dishes are employed, a humid environment must be maintained during incubation to prevent loss of media by evaporation, since these dishes have loose-fitting covers. More popular are disposable plastic dishes with tight-fitting lids reducing the problem of dehydration. Suitable sterile plastic dishes are available commercially; if reuse is necessary, the dishes should be treated by immersion in 70 percent ethanol for 30 min, air dried, and reassembled.

Size of the sample to be filtered is governed by the anticipated bacterial density. An ideal quantity results in growth of about 50 coliform colonies and not more than 200 colonies of all types. Since it may be difficult to anticipate the number of bacteria in a sample, two or three volumes of the same sample should be tested. When the portion being filtered is less than 20 ml, a small amount of sterile dilution water is added to the funnel before filtration to uniformly disperse the bacterial suspension over the entire surface of the filter. Filtration units should be sterilized at the beginning of each filtration series. Rapid decontamination between successive filtrations is accomplished by use of an ultraviolet sterilizer, flowing steam, or boiling water. First, the filter-holding assembly is placed on the suction flask, a sterile filter is placed grid side up over the porous plate of the apparatus, using sterile forceps, and then the funnel is locked in place holding the membrane. Filtration is accomplished by passing the sample through the filter under partial vacuum. A culture dish is prepared by placing a sterile absorbent pad in the upper half of the dish and pipetting enough enrichment media on top to saturate the pad. M-Endo medium is used for the coliform group and M-FC for fecal coliforms. The prepared filter is removed from the filtration apparatus using a forceps and is placed directly on the pad in the dish. The cover is replaced, and the culture is incubated for 24 hr at 35° C. Incubation for fecal coliforms is performed by placing the culture dishes in watertight plastic bags and submerging them in a water bath for incubation at 44.5° C. Typical coliform colonies are pink to dark red with a metallic surface sheen that may vary in size from a small pinhead to complete coverage of the colony surface. Plates with counts within the 20 to 80 coliform colony range are considered to be most valid. Coliform density is calculated in terms of coliforms per 100 ml by multiplying the colonies counted by 100 and dividing this by the milliliters of the sample filtered.

There are two major factors that cause interference with the performance of the membrane filter technique on surface waters. Turbid waters result in suspended solids interfering with filtration and colony development, and water containing a large number of noncoliform bacteria, which develop on the culture media, can overgrow the coliform colonies. In the examination of potable waters, the filter method may be suitable for use in testing certain water supplies. However, when the membrane filter technique is introduced, it is desirable to conduct parallel tests to demonstrate applicability and to familiarize laboratory technicians with the procedures involved.

3-9   BIOCHEMICAL OXYGEN DEMAND

Biochemical oxygen demand (BOD) is the most commonly used parameter to define the strength of a municipal waste water or organic industrial waste.

Widest application is in measuring waste loadings to treatment plants and in evaluating the efficiency of such treatment systems. In addition, the BOD test is used to determine the relative oxygen requirements of treated effluents and polluted waters. However, it is of limited value in measuring the actual oxygen demand of surface waters, and extrapolation of test results to actual stream oxygen demands is highly questionable, since the laboratory environment cannot reproduce the physical, chemical, and biological stream conditions.

BOD is by definition the quantity of oxygen utilized by a mixed population of microorganisms in the aerobic oxidation (of the organic matter in a sample of waste water) at a temperature of 20° C. Measured amounts of a waste water, diluted with prepared water, are placed in 300-ml BOD bottles (Figure 3-10). The dilution water, containing phosphate buffer (pH 7.2), magnesium sulfate, calcium chloride, and ferric chloride, is saturated with dissolved oxygen. Seed microorganisms are supplied to oxidize the waste organics if sufficient microorganisms are not already present in the waste-water sample. The general biological reaction that takes place is Eq. 3-12. The waste water supplies the organic matter (biological food) and the dilution water furnishes the dissolved oxygen. The primary reaction is metabolism of the organic matter and uptake of dissolved oxygen by bacteria, releasing carbon dioxide and producing a substantial increase in bacterial population. The secondary reaction results from the oxygen used by the protozoa consuming bacteria, a predator-prey reaction. Depletion of dissolved oxygen in the test bottle is directly related to the amounts of degradable organic matter. The BOD of a waste water where microorganisms are already present

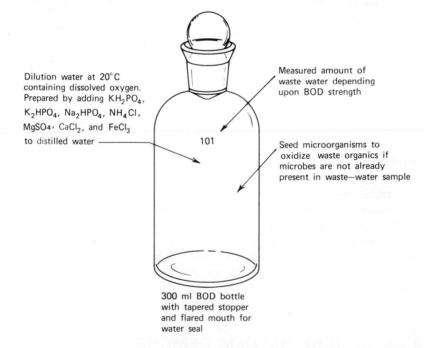

Dilution water at 20° C containing dissolved oxygen. Prepared by adding KH_2PO_4, K_2HPO_4, Na_2HPO_4, NH_4Cl, $MgSO_4$, $CaCl_2$, and $FeCl_3$ to distilled water

Measured amount of waste water depending upon BOD strength

101

Seed microorganisms to oxidize waste organics if microbes are not already present in waste—water sample

300 ml BOD bottle with tapered stopper and flared mouth for water seal

Figure 3-10 Essential constituents in the biochemical oxygen demand (BOD) test are: a measured amount of the waste-water sample being analyzed, prepared dilution water containing dissolved oxygen, and seed microorganisms if not present in the waste water.

in the sample, requiring no outside seed, is calculated using Eq. 3-13. (The standard test has an incubation period of 5 days at 20° C.)

$$\text{Organic matter} \xrightarrow[\text{bacteria}]{\overset{\text{Dissolved oxygen}}{\frown}} CO_2 + \text{Bacterial cells} \xrightarrow[\text{protozoa}]{\overset{\text{Dissolved oxygen}}{\frown}} CO_2 + \text{Protozoal cells} \qquad (3\text{-}12)$$

$$\begin{array}{l}\text{mg/l BOD} \\ \text{(no seed} \\ \text{required)}\end{array} = \dfrac{\dfrac{\text{mg/l initial DO} - \text{mg/l final DO}}{\text{ml of waste water}}}{\text{ml volume of BOD bottle}} \qquad (3\text{-}13)$$

The biochemical oxygen demand of a waste water is in reality not a single point value but time dependent. The curve in Figure 3-11 shows BOD exerted, dissolved oxygen depleted, as the biological reactions progress with time. Carbonaceous oxygen demand, Eq. 3-12, progresses at a decreasing rate with time, since the rate of biological activity decreases as the available food supply diminishes. The shape of the hypothetical curve is best expressed mathematically in the form of Eq. 3-14, where t is time in days and k is a rate constant. Based on this equation for a k equal to 0.1, a common value for domestic waste, 68 percent of the ultimate carbonaceous BOD is exerted after five days.

$$\begin{array}{l}\text{BOD at any} \\ \text{time, } t\end{array} = \begin{array}{l}\text{ultimate} \\ \text{BOD}\end{array} (1 - 10^{-kt}) \qquad (3\text{-}14)$$

Nitrifying bacteria can exert an oxygen demand in the BOD test as in Eqs. 3-7 and 3-8. Fortunately, growth of nitrifying bacteria lags behind that of the microorganisms performing the carbonaceous reaction. Nitrification generally does not occur until several days after the standard 5-day incubation period for BOD tests on untreated waste waters. Treatment plant effluents and stream waters may show early nitrification where the sample has a relatively high population of nitrifying bacteria. No standard method is recommended for

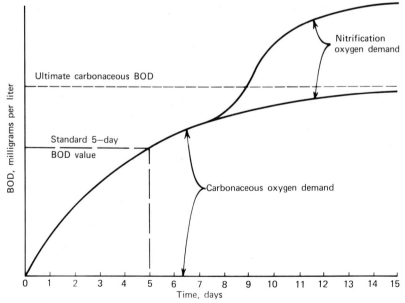

Figure 3-11 Hypothetical biochemical oxygen demand reaction showing the carbonaceous and nitrification demand curves.

preventing nitrification; however, the inhibiting chemicals of thiourea or 2-chloro-6-trichloromethyl pyridine can be used to stop nitrate formation if special laboratory techniques are used.

The degree of reproducibility of the BOD test cannot be defined precisely because of variations that occur in bacterial decomposition of various organic substances. Laboratory technicians can periodically evaluate their procedures by conducting BOD analyses on a standard mixture of glucose and glutamic acid solutions containing 150 mg/l of each component. The mean BOD value should fall in the range of 200 to 215 mg/l, and the k-rate in the bounds of 0.16 to 0.19. This represents a deviation of about ±4 percent from the average value for 68 percent of the tests. However, a series of tests on a real waste water normally shows observations varying from 10 to 20 percent on either side of the mean. The larger variations occur in testing industrial wastes that require seeding, contain substances that inhibit biological activity, and treatment plant effluent samples that are affected by nitrification.

No uniform relationship exists between the COD and BOD of waste waters, except that the COD value must always be greater than the BOD. This is because chemical oxidation decomposes nonbiodegradable organic matter.

BOD of Municipal Waste Water

Following is the description of a typical BOD test on a municipal waste water. The first step is to determine the portion of sample to be placed in each BOD bottle. Either the information in Table 3-2 or Eq. 3-13 may be used to select appropriate amounts. For example, for a waste water with an estimated BOD of 350 mg/l, Table 3-2 indicates sample volumes in the range of 2 to 5 ml per 300-ml bottle would be appropriate; or substituting into Eq. 3-13 a BOD equal to 350 mg/l, bottle volume of 300 ml, and a desired dissolved oxygen decrease of 5 mg/l, the calculated waste-water portion is 4.3 ml. For a valid BOD test, at least, 2 mg/l of dissolved oxygen should be used, but the final DO should not be less than 1 mg/l. Since the initial dissolved oxygen in dilution

Table 3-2. Suggested Waste-Water Portions and Dilutions in Preparing BOD Tests

By Direct Measurement of Waste Water into a 300-ml BOD Bottle		By Mixing Waste Water into Dilution Water $\left[\dfrac{\text{Waste-Water Volume}}{\text{Total Volume of Mixture}}\right]$	
Waste Water (ml)	Range of BOD (mg/l)	Percentage of Mixture	Range of BOD (mg/l)
0.20	3000 to 10,500	0.10	2000 to 7000
0.50	1200 to 4200	0.20	1000 to 3500
1.0	600 to 2100	0.50	400 to 1400
2.0	300 to 1050	1.0	200 to 700
5.0	120 to 420	2.0	100 to 350
10.0	60 to 210	5.0	40 to 140
20.0	30 to 105	10.0	20 to 70
50.0	12 to 42	20.0	10 to 35
100	6 to 21	50.0	4 to 14

Table 3-3. Typical BOD Data from the Analysis of a Municipal Waste Water To Determine the Five-Day BOD Value and Estimate the k-Rate of the Biological Reaction. Results Are Plotted in Figure 3-12. All BOD Bottles Had a Volume of 300 ml

Bottle Number	Waste-Water Portion (ml)	Initial DO (mg/l)	Incubation Period (days)	Final DO (mg/l)	DO Drop (mg/l)	Calculated BOD (mg/l)
			BOD Tests To Determine Average Five-Day Value			
1	2.0	8.3	0			
2	4.0	8.4	0			
3	6.0	8.4	0			
	Average = 8.4					
4	2.0	8.4	5.0	5.9	2.5	375
5	2.0	8.4	5.0	6.0	2.4	360
6	4.0	8.4	5.0	3.8	4.6	345
7	4.0	8.4	5.0	3.5	4.9	365
8	6.0	8.4	5.0	0	(Invalid test)	
9	6.0	8.4	5.0	0	(Invalid test)	
					Average = 360	
			BOD Analyses To Permit Calculation of k-Rate			
10	4.0	8.4	0.5	7.2	1.2	90
11	4.0	8.4	0.5	7.4	1.0	75
12	4.0	8.4	1.0	6.2	2.2	165
13	4.0	8.4	1.0	5.9	2.5	190
14	4.0	8.4	2.0	5.2	3.2	240
15	4.0	8.4	2.0	5.2	3.2	240
16	4.0	8.4	3.0	4.4	4.0	300
17	4.0	8.4	3.0	4.6	3.8	285

water is about 9 mg/l, an average amount of 5 mg/l is available for biological uptake between the minimum desired 2 mg/l and the maximum of 8 mg/l.

The distilled water for dilution is stabilized at 20° C by placing in the BOD incubator. It is then removed and prepared by aeration and addition of the four buffer and nutrient solutions (Figure 3-10). Outside seed microorganisms are not required for municipal waste, since they are already present in the sample. To insure that some of the bottles prepared provide valid test data, three different dilutions of two or three bottles each are set up, plus three bottles for initial dissolved oxygen measurement. Measured portions of the waste water are pipetted directly into empty BOD bottles which are then filled with prepared dilution water by syphoning through a hose. The sample preparation and dissolved oxygen measurements for this illustration are summarized in Table 3-3. Waste-water volumes of 2.0 ml, 4.0 ml, and 6.0 ml were used. The first three bottles were titrated immediately for dissolved oxygen, using the azide modification of the iodometric method. The other nine bottles were incubated for five days at 20° C and then were titrated for remaining dissolved oxygen. The BOD value for each bottle was calculated using Eq. 3-13. The average BOD for four of the six tests was 360 mg/l; two bottles were considered invalid, since the dissolved oxygen was depleted prior to the end of the 5-day incubation period.

Determination of BOD k-Rate

The rate constant k in Eq. 3-14 can be computed from BOD values measured at various times. Bottles numbered 10 to 17 of the analyses in Table 3-3 were analyzed after incubation periods ranging from 0.5 to 3.0 days. These data along with the average 5-day value are plotted in Figure 3-12. The shape of the BOD-time curve is typical of a municipal waste water derived primarily from domestic sources. The upper portion of Figure 3-12 illustrates a graphical method for estimating k-rate. Values of the cubed root of time in days over BOD in milligrams per liter are calculated from the laboratory data and are plotted as ordinates against the corresponding times. The best fit line drawn through these points is used to calculate the k-rate by the following relationship:

$$k = 2.61 \frac{B}{A} \tag{3-15}$$

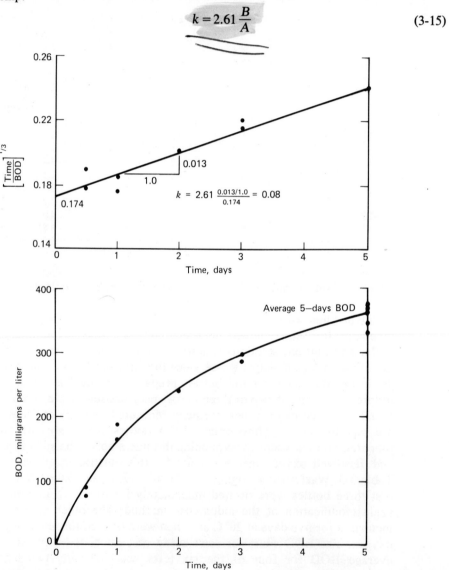

Figure 3-12 BOD-time curve and graphical determination of k-rate for the waste-water data given in Table 3-3.

where

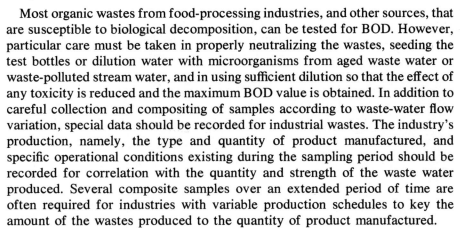

A = intercept of the line on the ordinate axis

B = slope of the line

BOD Analysis of Industrial Wastes

Most organic wastes from food-processing industries, and other sources, that are susceptible to biological decomposition, can be tested for BOD. However, particular care must be taken in properly neutralizing the wastes, seeding the test bottles or dilution water with microorganisms from aged waste water or waste-polluted stream water, and in using sufficient dilution so that the effect of any toxicity is reduced and the maximum BOD value is obtained. In addition to careful collection and compositing of samples according to waste-water flow variation, special data should be recorded for industrial wastes. The industry's production, namely, the type and quantity of product manufactured, and specific operational conditions existing during the sampling period should be recorded for correlation with the quantity and strength of the waste water produced. Several composite samples over an extended period of time are often required for industries with variable production schedules to key the amount of the wastes produced to the quantity of product manufactured.

Initial pretreatment is to neutralize, if necessary, the sample to pH 7.0 with sulfuric acid or sodium hydroxide to remove caustic alkalinity or acidity. The pH of the dilution water should not be changed by addition of the waste water in preparing a BOD test bottle of lowest dilution. Samples containing residual chlorine must be dechlorinated prior to setup. Often the residual dissipates if samples are allowed to stand for one or two hours, but higher chlorine residuals should be destroyed by adding sodium sulfite solution. Industrial wastes containing other toxic substances require special study and pretreatment. In extreme cases where a technique for neutralizing the toxin cannot be developed, BOD testing is abandoned and replaced with COD analyses.

The presence of unknown substances that inhibit biological growth are often detected by carefully conducted BOD tests. Table 3-4 and Figure 3-13 summarize the results from analysis on a food-processing waste water containing an interfering compound. The source of the toxicity was a bactericide used in disinfecting production pipelines and tanks. The first evidence of the potential problem is revealed by comparing the rather wide range of BOD values observed in test bottles at the same dilution—at 0.67 percent the numbers range from 220 mg/l to 485 mg/l. More importantly, the tests show increasing BOD values with increasing dilutions. The concentration of toxin, and consequently the inhibition of biological activity, is greater in lower dilutions (3.0 ml waste water in a 300-ml bottle) than at higher dilutions (1.0 ml/300 ml). The reported BOD should be the highest value obtained in valid tests, this being the average of the highest dilutions providing a minimum uptake of 2.0 mg/l dissolved oxygen.

Very few industrial wastes have sufficient biological populations to perform BOD testing without providing an acclimated seed. The ideal seed is a mixed culture of bacteria and protozoa adapted to decomposing the specific industrial waste organics, with a low number of nitrifying bacteria. Microorganisms for food-processing wastes, and similar organics, can be obtained from aged untreated domestic waste water. The seed material is the supernatant liquid

Table 3-4. BOD Data from the Analysis of a Food-Processing Waste Water Containing a Bactericidal Agent Causing an Increase in BOD Value with Increasing Dilution. Figure 3-13 is a Graphical Presentation of the Results

Waste-Water Portion (ml)	Percentage Dilution of Waste Water	Measured 5-day BOD (mg/l)
3.0	1.00	240
3.0	1.00	205
3.0	1.00	250
3.0	1.00	145
		Average = 210
2.0	0.67	365
2.0	0.67	485
2.0	0.67	315
2.0	0.67	220
		Average = 350
1.0	0.33	520
1.0	0.33	440
1.0	0.33	565
1.0	0.33	520
		Average = 510

from a sample of domestic waste water that has been allowed to age and settle in an open container for about 24 hr at room temperature. Biota for industrial wastes, with biodegradable organic compounds not abundant in municipal waste water, should be obtained from a source having microorganisms acclimated to these wastes. If the industrial discharge is being treated by a biological system, activated sludge from an aeration basin, slime growth from a trickling

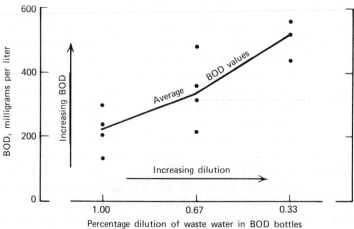

Figure 3-13 Plot of BOD values, given in Table 3-4, showing increasing BOD for the same industrial waste sample set up at increasing dilutions. The sample was a food-processing waste water containing a biologically inhibiting substance.

filter, or water from a stabilization pond contains acclimated microorganisms. When the waste is disposed of in a watercourse, the stream water several miles below the point of discharge may be a good source for seed. An adapted seed culture can be developed by using a small laboratory draw-and-fill activated-sludge unit on a mixed feed of industrial and domestic waste water if a natural source of microbial seed is not readily available, or proves to be unsatisfactory.

The amount of seed material added to the dilution water, or each individual BOD bottle, must be sufficient to provide substantial biological activity without adding too much organic matter. When aged domestic or polluted water is used, the rule of thumb is to add an amount of seed waste water such that 5 to 10 percent of the total BOD exerted by the sample results from oxygen demand of the seed alone. For example, assume that the water being used for seed has an estimated BOD of 150 mg/l. Based on Table 3-2 a 10-ml volume per bottle would be used to run a BOD test on this waste water; however, for seeding purposes, only 5 to 10 percent of this amount is desired. Hence, 0.5 to 1.0 ml should be added in setting up each industrial waste-water test bottle. If the culture is to be placed in the dilution water, rather than directly into each BOD bottle, the percentage-of-mixture data from Table 3-2 is helpful. To conduct a BOD test on a 150-mg/l waste, a 3.5 percent mixture is recommended. Therefore, seeded dilution water should be between 0.17 and 0.35 percent seed waste water, prepared by adding 1.7 to 3.5 ml to each liter of dilution water.

Oxygen demand of the seed is compensated for in computing BOD of a seeded industrial waste test by using Eq. 3-16. The terms D_1 and D_2 are the initial and final dissolved oxygen concentrations in the seeded waste water bottles containing an industrial waste fraction of P. The terms B_1 and B_2 are initial and final dissolved oxygen values from a separate BOD test on the seed material, and f is the ratio of seed volume used in the industrial waste-water test to the amount used in the test on the seed. Hence, $(B_1 - B_2)f$ is the oxygen demand of the seed.

$$\text{mg/l BOD} = \frac{(D_1 - D_2) - (B_1 - B_2)f}{P} \tag{3-16}$$

where

$D_1 = $ DO of diluted seeded waste-water sample about 15 minutes after preparation

$D_2 = $ DO of waste-water sample after incubation

$B_1 = $ DO of diluted seed sample about 15 min after preparation

$B_2 = $ DO of seed sample after incubation

$f = $ ratio of seed volume in seeded waste-water test to seed volume in BOD test on seed

$\quad = \dfrac{\text{percentage or milliliters of seed in } D_1}{\text{percentage or milliliters of seed in } B_1}$

$P = $ decimal fraction of waste-water sample used

$\quad = \dfrac{\text{volume of waste water}}{\text{volume of dilution water plus waste water}}$

A time lag, resulting from insufficient seed, unacclimated microorganisms, or presence of inhibiting substances, often occurs in the oxygen-demand reaction

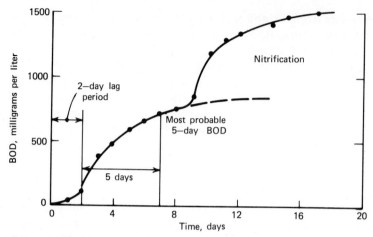

Figure 3-14 Reaction rate curve from a seeded BOD test on a chemical-manufacturing waste water exhibiting a time lag in the biological reaction.

on an industrial waste. To determine BOD accurately, existence of a time lag must be identified during the first few days of incubation. This can be accomplished by either preparing extra bottles for dissolved oxygen testing after $\frac{1}{2}$, 1, 2, and 3 days, or if a dissolved oxygen probe is available, by monitoring the rate of oxygen depletion in two or three bottles each day during incubation. Figure 3-14 illustrates dissolved oxygen testing on a chemical-manufacturing waste water that exhibits a two-day lag period. The most probable BOD can be approximated in a test of this nature by reestablished time zero at the end of the lag period and by measuring the 5-day value based on this new origin. Data throughout the incubation period may also reveal other irregularities influencing selection of a 5-day value. Figure 3-15a is an effluent from an extended aeration plant. Because of the high population of nitrifying bacteria, nitrification started on the third day of incubation and exerted approximately one half of the total oxygen demand by the fifth day of the test. The diphasic curve in Figure 3-15b is a result of rapid oxygen uptake during the first two days caused by a large amount of discharge from a pudding factory into the municipal sewer. Since soluble pudding waste is more easily decomposed by bacteria, it is metabolized first at a very high rate ($k_1 = 0.65$) while the remaining domestic waste organics continue their oxygen demand at a slower rate throughout the duration of the test.

Trade wastes frequently have high strengths that make it difficult, or even impossible, to pipette accurately the small quantity desired for a single test bottle. In this case, the waste water can be diluted by serial dilution to volumes that can be accurately measured. For example, only 0.5 ml of a 3000-mg/l BOD waste is required per BOD bottle. However, if 100 ml of the 3000-mg/l waste is diluted to 1000 ml with distilled water, 5.0 ml of the mixture can be pipetted accurately into each bottle. Waste waters high in suspended solids may be difficult to mix with water; one alternative is to homogenize the sample in a blender to aid dispersion in the dilution water.

The purpose of this discussion is to present some of the problems and pitfalls in BOD testing of industrial wastes. Although difficult to perform, these tests are common in evaluating manufacturing wastes to assess sewer use fees or treatment plant loadings. Improper seeding and unrecognized influence of

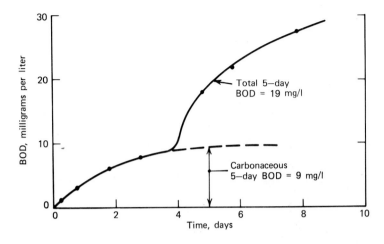

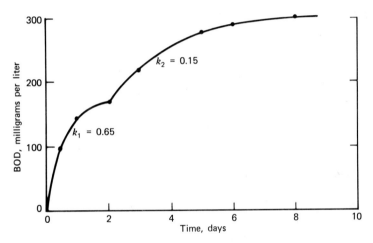

Figure 3-15 BOD analyses showing early nitrification and a diphasic carbonaceous stage. (*a*) BOD curve of a treated waste water from an extended aeration plant showing nitrification starting about the fourth day of incubation. (*b*) BOD curve of a raw municipal waste water exhibiting a diphasic carbonaceous stage resulting from a large amount of soluble pudding-manufacturing waste combined with the domestic.

inhibiting substances cause erroneous results that lead to confusion and consternation. Although occasionally trade wastes defy biological testing, more often they can be tested with reasonable accuracy by following very careful analytical procedures.

Manometric Measurement of BOD

Manometric apparatus, such as the Warburg respirometer, are employed in research studies to measure oxygen uptake from biological cultures. Recently, simplified manometric devices, shown in Figure 3-16, have been marketed for general laboratory use, but not to replace the *Standard Methods* dilution procedure. Measured samples of waste water are placed in each of the brown glass bottles, the amount depending on the anticipated BOD. Buffer and nutrient chemicals are not added to the waste-water portions during routine

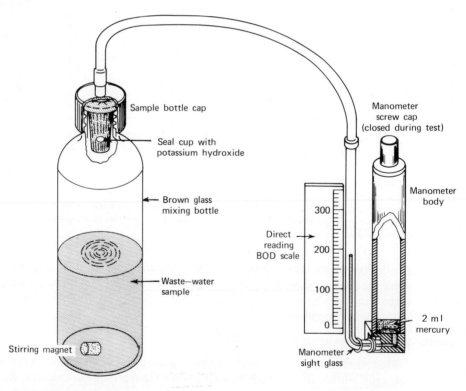

Sample bottle cap

Seal cup with
potassium hydroxide

Brown glass
mixing bottle

Waste—water
sample

Stirring magnet

Manometer
screw cap
(closed during test)

Manometer
body

Direct
reading
BOD scale

300

200

100

0

2 ml
mercury

Manometer
sight glass

Figure 3-16 Diagram and picture of a manometric BOD apparatus. (Courtesy of Hach Chemical Company, Ames, Iowa)

testing, based on the assumption that without dilution the waste water contains sufficient nutrients for biological growth and buffer capacity to prevent a change in pH. A small magnetic stirring bar is placed in each bottle, and a seal cup containing potassium hydroxide is inserted in the bottle cap. The prepared bottles are connected to mercury manometers and are set in position on the stirring apparatus. The samples are mixed continuously by turning the stirring bars driven by a rotating magnet under each bottle. The electric stirring motor is enclosed at one end of the unit. After the initial stirring to reach equilibrium, the bottle closures are tightened, and caps are placed on the manometers to prevent the affect of barometric pressure changes. As the microorganisms take up dissolved oxygen in the liquid, gaseous oxygen is absorbed from air in the closed sample bottle. Carbon dioxide given off by the microorganisms is absorbed and is converted to carbonate ion by the potassium hydroxide solution in the seal cup underneath the bottle cap. Absorption reduces the volume of carbon dioxide gas to zero. Volumetric reduction of the air in a bottle, representing the oxygen demand, is indicated by the manometer sight glass which is calibrated to read directly in milligrams per liter BOD. To maintain 20° C which is required for the standard test, the entire apparatus is placed in an incubator.

Studies have been conducted to compare manometric BOD measurements to those performed by the standard dilution method. For domestic waste water and synthetic glucose-glutamic acid, the two techniques have provided comparable results at 20° C incubation temperature. However, tests on industrial wastes often show significant variance, primarily because the sample being tested in the manometric bottle is undiluted, leading to greater influence of toxic or inhibiting substances. If the manometric method is applied on a regular basis in testing municipal discharges that contain a substantial fraction of industrial wastes, validity of the results should be verified by comparison with the standard dilution technique. The main advantages given in manufacturer's literature are: ease of operation, reduced testing costs, direct reading of BOD without calculations, and continuous readings for monitoring the rate of reaction. Also undiluted tests purportedly are more representative of actual treatment conditions.

EXAMPLE 3-2

Data from an unseeded domestic waste-water BOD test are: 5.0 ml of waste in a 300-ml bottle, initial DO of 7.8 mg/l, and 5-day DO equal to 4.3 mg/l. Compute (a) the BOD, and (b) the ultimate BOD, assuming a k-rate of 0.10 per day.

Solution

(a) From Eq. 3-13

$$BOD = \frac{7.8 - 4.3}{5.0/300} = 210 \text{ mg/l}$$

(b) Using Eq. 3-14

$$\text{Ultimate BOD} = \frac{BOD}{1 - 10^{-kt}} = \frac{210}{1 - 10^{-0.1 \times 5.0}} = \frac{210}{1 - 0.32} = 310 \text{ mg/l}$$

EXAMPLE 3-3

A seeded BOD test is to be conducted on meat-processing waste with an estimated strength of 800 mg/l. The seed is supernatant from aged, settled, domestic waste water with a BOD of about 150 mg/l. (a) What sample portions should be used for setting up the middle dilutions of the waste water and seed tests? (b) Calculate the BOD value for the industrial waste if the initial DO in both seed and sample bottles is 8.5 mg/l, and the five-day DO's are 4.5 mg/l and 3.5 mg/l for the seed test bottle and seeded waste-water sample, respectively.

Solution

(a) Based on data in Table 3-2

Volume required for BOD test on 150-mg/l seed = 10.0 ml pipetted directly into a 300-ml bottle.

Volume of meat-processing waste required, assuming 800 mg/l = 2.0 ml pipetted into a BOD bottle.

Amount of aged waste water for seeding waste sample = $0.10 \times 10.0 = 1.0$ ml pipetted into the BOD bottle, or the addition of 3.33 ml/l of dilution water for a mixture of 0.33 percent.

(b) Substituting into Eq. 3-16

$$BOD = \frac{(8.5 - 3.5) - (8.5 - 4.5)1.0/10.0}{2.0/300} = 690 \text{ mg/l}$$

3-10 BIOLOGICAL TREATMENT SYSTEMS

Biological processing is the most efficient way of removing organic matter from municipal waste waters. These living systems rely on mixed microbial cultures to decompose, and to remove colloidal and dissolved organic substances from solution. The treatment chamber holding the microorganisms provides a controlled environment; for example, activated sludge is supplied with sufficient oxygen to maintain an aerobic condition. Waste water contains the biological food, growth nutrients, and inoculum of microorganisms. Persons who are not familiar with waste-water operations often ask where the "special" biological cultures are obtained. The answer is that the wide variety of bacteria and protozoa present in domestic wastes seed the treatment units. Then by careful control of waste-water flows, recirculation of settled microorganisms, oxygen supply, and other factors, the desirable biological cultures are generated and retained to process the pollutants. The slime layer on the surface of the media in a trickling filter is developed by spreading waste water over the bed. Within a few weeks the filter is operational, removing organic matter from the liquid trickling through the bed. Activated sludge in a mechanical, or diffused-air, system is started by turning on the aerators and feeding the waste water. Initially a high rate of recirculation from the bottom of the final clarifier is necessary to retain sufficient biological culture. However, within a short period of time a settleable biological floc matures that efficiently flocculates the waste organics. An anaerobic digester is the most difficult treatment unit to start up, since the methane-forming bacteria, essential to digestion, are not abundant in raw waste water. Furthermore, these anaerobes grow very slowly and require optimum environmental conditions. Start-up of an anaerobic digester can be hastened considerably by filling the tank with waste water and seeding with a substantial quantity of digesting sludge from a nearby treatment plant. Raw sludge is then fed at a reduced initial rate, and lime is supplied as necessary to hold pH. Even under these conditions, several months may be required to get the process fully operational.

Enzymes are organic catalysts that perform biochemical reactions at temperatures and chemical conditions compatible with biological life. Chemical decomposition of potato or meat requires boiling in a strong acid solution, as performed in the COD test. However, these same foods can be readily digested by microorganisms, or in the stomach of an animal, at a much reduced temperature without strong mineral acids through the action of enzymes. Most enzymes cannot be isolated from living organisms without impairing their functioning capability. Although a discussion of enzymes is beyond the scope of this book, technicians in the field of sanitary science must be aware that enzyme additives sold to enhance biological treatment processes are ineffective. The label on the container generally uses highly scientific terms to convince the purchaser of product worthiness, for example, enzymes for waste water (or for anaerobic digestion, stabilization ponds, septic tanks, etc.), minimum of 10 billion colonies per gram, excellent diastic, proteolytic, amylolytic, and lipolytic activity, a special formulation of enzymes, aerobic and anaerobic bacteria, and the like. In reality, domestic waste water contains an abundant supply of all these enzymes, and to pour in more at a cost of $5 to $10 per pound can be figuratively described as pouring money down the drain.

Factors Affecting Growth

The most important factors affecting biological growth are temperature, availability of nutrients, oxygen supply, pH, presence of toxins and, in the case of photosynthetic plants, sunlight. Bacteria are classified according to their optimum temperature range for growth. Mesophilic bacteria grow in a temperature range of 10 to 40° C, with an optimum of 37° C. Aeration tanks and trickling filters generally operate in the lower half of this range with waste-water temperatures of 20 to 25° C in warm climates and 8 to 10° C during the winter in northern regions. If cold well water serves as a water supply, waste-water temperatures may be lower than 20° C during the summer, and winter operation during extremely cold weather may result in ice formation on the surface of final clarifiers and freezing of stabilization ponds. Anaerobic digestion tanks are normally heated to near the optimum level of 35° C (98° F).

As a general rule, the rate of biological activity doubles for every 10 to 15° C temperature rise within the range of 5 to 35° C (see Figure 3-17). Above 40° C, mesophilic activity drops off sharply and thermophilic growth starts. Thermophilic bacteria have a range of approximately 45 to 75° C, with an optimum near 55° C. This higher temperature range is not used in waste treatment, since it is difficult to maintain that high an operating temperature, and because thermophilic bacteria are more sensitive to small temperature changes. Of particular importance is the dip between mesophilic and thermophilic ranges; operation in this region should be avoided. A technician seeking increased activity in his anaerobic digester may gradually increase the temperature of the digesting sludge and thus realize improved gas production and efficiency. However, if the sludge is heated in excess of the mesophilic optimum, the rate of biological activity will decrease sharply, adversely affecting operation.

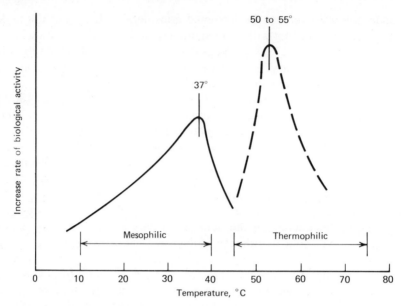

Figure 3-17 Effect of temperature on the rate of biological activity and relative positioning of the mesophilic and thermophilic zones.

Municipal waste waters commonly contain sufficient concentrations of carbon, nitrogen, phosphorus, and trace nutrients to support the growth of a microbial culture. Theoretically, a BOD to nitrogen to phosphorus ratio of 100/5/1 is adequate for aerobic treatment, with small variations depending on the type of system and mode of operation. Average domestic waste water exhibits a surplus of nitrogen and phosphorus with a BOD/N/P ratio of about 100/17/5. If a municipal waste contains a large volume of nutrient-deficient industrial waste, supplemental nitrogen is generally supplied by the addition of anhydrous ammonia (NH_3) or phosphoric acid (H_3PO_4) as is needed.

Diffused and mechanical aeration basins must supply sufficient air to maintain dissolved oxygen for the biota to use in metabolizing the waste organics. Rate of microbial activity is independent of dissolved oxygen concentration above a minimum critical value, below which the rate is reduced by the limitation of oxygen required for respiration. The exact minimum depends on the type of activated sludge process and the characteristics of the waste water being treated. The most common design criterion for critical dissolved oxygen is 2.0 mg/l, but in actual operation values as low as 0.5 mg/l have proved satisfactory. Anaerobic systems must, of course, operate in the complete absence of dissolved oxygen; consequently, digesters are sealed with floating or fixed covers to exclude air.

Hydrogen ion concentration has a direct influence on biological treatment systems which operate best in a neutral environment. The general range of operation of aeration systems is between pH 6.5 and 8.5. Above this range microbial activity is inhibited, and below pH 6.5 fungi are favored over bacteria in the competition for metabolizing the waste organics. Normally the bicarbonate buffer capacity of a waste water is sufficient to prevent acidity and reduced pH; while carbon dioxide production by the microorganisms tends to control the alkalinity of high pH waste waters. Where industrial discharges force the pH of a municipal waste outside the optimum range, addition of a chemical may be required for neutralization. In that case, it is more desirable to have the industry pretreat its waste by equalization and neutralization prior to disposal in the sewer rather than contend with the problem of pH control at the city's disposal plant.

Anaerobic digestion has a small pH tolerance range of 6.7 to 7.4 with optimum operation at pH 7.0 to 7.1. Domestic waste sludge permits operation in this narrow range except during start-up or periods of organic overloads. Limited success in digester pH control has been achieved by careful addition of lime with the raw sludge feed. Unfortunately, the buildup of acidity and reduction of pH may be a symptom of other digestion problems, for example, accumulation of toxic heavy metals which the addition of lime cannot cure.

Biological treatment systems are inhibited by toxic substances. Industrial wastes from metal finishing industries often contain toxic ions, such as nickel and chromium; chemical manufacturing produces a wide variety of organic compounds that can adversely affect microorganisms. Since little can be done to remove or neutralize toxic compounds in municipal treatment, pretreatment should be provided by industries prior to discharging wastes to the city sewer.

Population Dynamics

In biological processing of wastes, the naturally occurring biota are a variety of bacteria growing in mutual association with other microscopic plants and animals. Three of the major factors in population dynamics are: competition for the same food, predator-prey relationship, and symbiotic association. When organic matter is fed to a mixed population of microorganisms, competition arises for this food, and the primary feeders that are most competitive become dominant. Under normal operating conditions, bacteria are the primary feeders in both aerobic and anaerobic operations. Protozoa consuming bacteria is the common predator-prey relationship in activated sludge and trickling filters. In stabilization ponds, protozoa and rotifers graze on both algae and bacteria. Symbiosis is the living together of organisms for mutual benefit such that the association produces more vigorous growth of both species. An excellent example of this is the relationship between bacteria and algae in a stabilization pond.

In an activated sludge process, waste organics serve as food for the bacteria, and the small population of fungi that might be present. Some of the bacteria die and lyse, releasing their contents which are resynthesized by other bacteria. The secondary feeders (protozoa) consume several thousand bacteria for a single reproduction. The benefit of this predator-prey action is twofold: (1) removal of bacteria stimulates further bacterial growth, accelerating metabolism of the organic matter, and (2) settling characteristics of the biological floc is improved by reducing the number of free bacteria in solution. The effluent from the process consists of nonsettleable organic matter and dissolved inorganic salts (Figure 3-18).

Control of the microbial populations is essential for efficient aerobic treatment. If waste water was simply aerated, the liquid detention times would be intolerably long, requiring a time period of about five days at 20°C for 70 percent reduction. However, extraction of organic matter is possible within a few hours of aeration provided that a large number of microorganisms is mixed with the waste water. In practice this is achieved by settling the microorganisms out of solution in a final clarifier and returning them to the aeration tank to

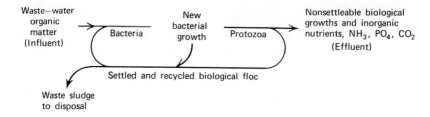

(a)

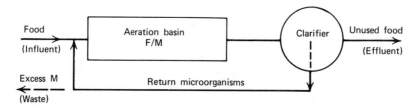

(b)

Figure 3-18 Generalized biological population dynamics in the activated-sludge waste-water treatment process. (*a*) Predator-prey relationship between protozoa and bacteria in the activated sludge process. (*b*) Sedimentation and recirculation maintains the desired food-to-microorganism (F/M) ratio in the aeration basin.

metabolize additional waste organics (Figure 3-18). Good settling characteristics occur when an activated sludge is held in the endogenous (starvation) phase. Furthermore, a large population of underfed biota remove BOD very rapidly from solution. Excess microorganisms are wasted from the process to maintain proper balance between food supply and biological mass in the aeration tank. This balance is referred to as the food-to-microorganism ratio (F/M) which is normally expressed in units of pounds of BOD applied per day per pound of MLSS in the aeration basin (MLSS is the mixed liquor suspended solids). Operation at a high F/M ratio results in incomplete metabolism of the organic matter, poor settling characteristics of the biological floc and, consequently, poor BOD removal efficiency. At a low F/M ratio, the mass of microorganisms are in a near starvation condition that results in a high degree of organic matter removal, good settleability of the activated sludge, and efficient BOD removal.

The relationship that exists between bacteria and algae in a small pond is illustrated in Figure 3-19. Bacteria decompose the organic matter, yielding inorganic nitrogen, phosphates, and carbon dioxide. Algae use these compounds, along with energy from sunlight, in photosynthesis releasing oxygen into solution. This oxygen is in turn taken up by the bacteria, thus closing the cycle. The effluent from a stabilization pond contains suspended algae and excess bacterial decomposition end products. In the summer, BOD reduction in lagoons is very high, commonly in excess of 95 percent. However, during cold temperature operation, microbial activity is reduced, and BOD removal relies to a considerable extent on dilution of the inflowing raw waste water into the large volume of impounded water. Liquid detention in a stabilization pond is rarely less than 90 days and generally is considerably greater. Following a cold winter, particularly if the pond was covered with ice and snow, odorous

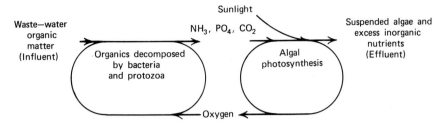

Figure 3-19 Schematic diagram of the mutually beneficial association (symbiotic relationship) between bacteria and algae in a stabilization pond.

conditions can be anticipated during the spring thaw. With increased temperature, the organic matter accumulated during the winter is rapidly decomposed by the bacteria using dissolved oxygen at a faster rate than can be absorbed from the air or supplied by the algae. After a few days to several weeks, depending on climatic conditions and waste load on the lagoon, the algae become reestablished and again supply oxygen to the bacterial cycle. Once this symbiotic relationship is again operational, aerobic conditions are firmly established and odorous emissions cease.

REFERENCES

1. *Standard Methods for the Examination of Water and Wastewater*, 13th Edition, American Public Health Association, American Water Works Association, and Water Pollution Control Federation, 1971.
2. Needham, J. G., and Needham, P. R., *A Guide to the Study of Fresh-water Biology*, Holden-Day, Inc., San Francisco, 1962.
3. McKinney, R. E., *Microbiology for Sanitary Engineers*, McGraw-Hill Book Co., New York, 1962.
4. Hawkes, H. A., *The Ecology of Waste Water Treatment*, Pergamon Press, London, 1963.
5. Clark, J. W., Viessman, W., Jr., and Hammer, M. J., "Biological-Treatment Processes," *Water Supply and Pollution Control*, International Textbook Co., Scranton, 1971, pp. 454–560.

PROBLEMS

3-1 Fresh waste water containing nitrate ions, sulfate ions, and dissolved oxygen is placed in a sealed jar. In what sequence are these oxidized compounds reduced by the bacteria? When would obnoxious odors appear?

3-2 Write the equation for acid production in the moisture of condensation of a sanitary sewer. What preventative measures can be taken to reduce or prevent damage to the crown of pipes?

3-3 The water supply in a small town is untreated well water pumped directly into the distribution system. Bacterial growths on the interior of the mains impart a foul taste to the water. What is promoting the tuberculation and the foul taste? What remedial actions can be taken?

3-4 Write an equation for photosynthesis. Why do plants perform this reaction? What is the dark reaction?

3-5 Describe the aquatic food chain.

3-6 Name the common waterborne bacterial and viral diseases.

3-7 Why can coliform bacteria be used as indicators of quality for drinking water?

3-8 What is the difference between the coliform group and fecal coliforms?

3-9 A multiple-tube fermentation analysis of a river water yielded the following results. What are the MPN and confidence limits? (*Answer* 7000, 2300–17,000)

Serial Dilution	Sample Portion (ml)	Positive Tubes
0	1.0	5 of 5
1	0.1	5 of 5
2	0.01	2 of 5
3	0.001	1 of 5
4	0.000,1	0 of 5

3-10 A bathing beach water is tested for fecal coliforms by the multiple-tube fermentation technique. All five of the 10-ml tubes were positive; none of the 1.0-ml portions showed gas; and two of the 0.1-ml were positive. What is the MPN? (*Answer* 43).

3-11 The membrane filter technique is used to test a polluted water for the coliform group. Three different water sample volumes (5 ml, 50 ml, and 500 ml) were filtered through five filter membranes. The colonies' counts were as follows: 5-ml portions—4, 8, 5, 2, 3; 50-ml portions—24, 38, 25, 32, 30; and 500-ml portions—250, 320, 360, 210, 280. Calculate the coliform count per 100 ml using the most valid data. (*Answer* 60 per 100 ml).

3-12 Calculate the 5-day BOD of a domestic waste water based on the following data: volume of waste water added to 300-ml bottle = 6.0 ml, initial DO = 8.1 mg/l, and 5-day DO = 4.2 mg/l. What is the ultimate BOD assuming a k-rate of 0.1 per day? (*Answer* 195 mg/l and 290 mg/l).

3-13 A seeded BOD test is to be conducted on a poultry-processing waste with an estimated five-day value of 600 mg/l. The seed taken from an existing preaeration tank at the industrial site has an estimated BOD of 200 mg/l. (a) What sample portions should be used for setting up the middle dilutions of the waste water and seed tests? (b) Compute the BOD value for the poultry waste if the initial DO in both the seed and sample bottles is 8.2 mg/l and the 5-day values are 3.5 mg/l and 4.0 mg/l for the seed test bottle and seeded waste-water sample, respectively. Use the volumetric additions from part (a) and assume that the volume of seed used in the waste BOD bottle is 10 percent of that used in the seed test. [*Answers* (a) 2.0 ml poultry waste plus 0.7 ml of seed, 7.0 ml of seed sample (b) 560 mg/l].

3-14 A seeded BOD analysis was conducted on a high strength, food-processing waste water. Twenty milliliter portions were used in setting up the 300-ml bottles of aged, settled, waste-water seed. The seeded sample BOD bottles contained 1.0-ml of industrial waste and 2.0 ml of seed material. The results for this series of test bottles are listed below. Calculate the BOD values. ($B_1 = 8.2$ mg/l, $D_1 = 8.2$ mg/l, $f = 0.10$, $P = 1/300$). Plot a BOD-time curve; what is the five-day value? Determine the k-rate. (*Answers* 5-day BOD = 900 mg/l and $k = 0.09$ per day)

Time (days)	Seed Tests B_2 (mg/l)	Sample Tests D_2 (mg/l)	Time (days)	Seed Tests B_2 (mg/l)	Sample Tests D_2 (mg/l)
0	8.2	8.2	7.9	4.5	4.1
1.0	7.3	7.2	9.8	4.1	3.8
1.9	6.5	6.4	11.8	3.9	3.5
2.9	5.9	5.8	14.0	3.4	2.9
3.8	5.5	5.3	15.6	3.6	2.7
4.7	5.1	5.0	19.0	3.8	2.1
6.0	4.8	4.6			

3-15 Seeded BOD analyses were conducted on a soybean-processing waste water. Three dilutions using 2.0-ml, 3.0-ml, and 4.0-ml portions of waste per 300-ml bottle were incubated and tested over a 7-day period. The BOD test on the aged, settled, domestic waste-water seed was set up applying 15 ml per bottle. The waste samples were seeded by adding 5.0 ml of seed waste-water per liter of dilution water. The initial dissolved oxygen concentration for all bottles, based on the average of several tests, was 8.2 mg/l. Sketch the BOD-time curves for each dilution series, determine the 5-day BOD values, and calculate the k-rates, using the graphical procedure. The dissolved oxygen test results are:

Time (days)	DO in Seed Test (mg/l)	DO Measurements of Waste Portions 2.0 ml (mg/l)	3.0 ml (mg/l)	4.0 ml (mg/l)
0	8.2	8.2	8.2	8.2
1.1	7.9	7.0	6.0	5.7
2.3	7.5	6.1	5.5	4.5
2.9	7.3	6.1	5.2	4.2
3.8	7.0	5.5	4.7	3.6
4.8	6.7	5.4	4.4	3.5
5.9	6.5	5.5	4.4	3.4
6.8	6.4	5.1	4.0	3.4

(*Answers* $f = 0.10$; for $P = 2.0$, BOD = 390 mg/l, $k = 0.21$ per day; $P = 3.0$, BOD = 370, $k = 0.21$; $P = 4.0$, BOD = 350, $k = 0.23$).

3-16 A manometric BOD apparatus is used to conduct a BOD test on a treatment plant effluent. The daily readings are as follows:

Time (days)	BOD (mg/l)	Time (days)	BOD (mg/l)
0	0	5	22
1	2	6	24
2	12	7	30
3	18	8	36
4	20	9	40

Plot the BOD-time curve; identify the lag and nitrification portions. What is the best estimated 5-day value? (*Answer* 24 mg/l).

3-17 List the major factors influencing the rate of biological growth.

3-18 State two reasons why protozoa are significant in activated sludge.

3-19 Why must an activated sludge process be operated in the starvation (low F/M) stage?

3-20 What's wrong with this statement: Algae consume waste organics applied to a stabilization pond; therefore, if the lagoon water is not green-colored, odors result from undecomposed wastes.

3-21 The following loading-temperature-efficiency data were collected during full-scale studies of an extended aeration plant treating domestic waste water. Why is it that the effluent BOD (treatment efficiency) is apparently independent of temperature?

Season of Year	BOD Loading (lb/1000 cu ft)	Effluent BOD (mg/l)	BOD Removal (percent)	Temperature of Mixed Liquor (°C)
Winter	11	17	91	11
Summer	12	16	92	19

chapter 4

Hydraulics and Hydrology

The following sections present the basic principles of hydraulics applicable to water distribution and waste-water collection systems. Low-flow characteristics of streams and stratification of lakes provide background for the study of water pollution.

4-1 WATER PRESSURE

Mass per unit volume is referred to as density of a fluid. The density of water under standard conditions (temperature of 4°C and pressure of one atmosphere) is 1.94 slugs/cu ft in the English system of units. The force exerted by gravity on 1 cu ft (specific weight) of water is 62.4 lb, equal to the density multiplied times the acceleration of gravity [32.2 ft/(sec)(sec)]. In the metric system, the density of water at standard conditions is 1.00 g/cu cm.

Pressure is the force exerted per unit area. Referring to Figure 4-1a, the pressure exerted on the bottom of a 1 cu ft container filled with water is equal to 0.433 psi. In engineering hydraulics, water pressure is frequently expressed in terms of feet of head, as well as psi. The relationship between these units is visually shown in Figure 4-1a where a height of 1 ft of water head exerts a pressure of 0.433 psi. Water pressure increases with depth below the surface linearly such that the pressure in psi is equal to 0.433 times the depth in feet. The sketch in Figure 4-1b shows the pressure acting only horizontally for ease of illustration. In fact, water pressure is exerted equally in all directions. The computation given under the diagram states that a head of 2.31 ft produces a water pressure of 1.0 psi. Thus 1 ft of head is equivalent to 0.433 psi or, alternately, a pressure of 1.0 psi is equivalent to 2.31 ft of head.

A simple piezometer, illustrated in Figure 4-2a, consists of a small tube rising from a container of water under pressure. The height of the water in the piezometer tube denotes the pressure of the confined water. Water piezometers are rarely practical for pressure measurements, since values generally exceed the height of water columns that are convenient to read. A mercury column can be used to measure relatively high pressure values within a limited head range,

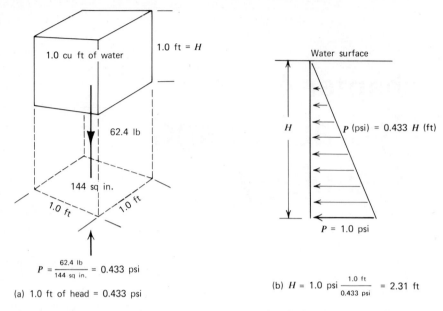

(a) 1.0 ft of head = 0.433 psi

(b) $H = 1.0$ psi $\dfrac{1.0 \text{ ft}}{0.433 \text{ psi}} = 2.31$ ft

Figure 4-1 Basic relationships between water pressure in psi (pounds per square inch) and feet of water head.

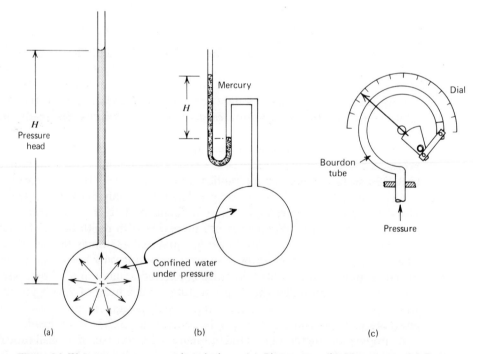

Figure 4-2 Water pressure measuring devices. (*a*) Piezometer. (*b*) Manometer. (*c*) Pressure gage.

since mercury is approximately 13.6 times heavier than water, that is, a 1-ft head of mercury is equivalent to a pressure of 5.9 psi. Figure 4-2b is a sketch of a mercury manometer. The pressure reading is equal to the difference of the fluid pressures in the two legs of the U-shaped tube. The use of manometers is generally restricted to indoor applications where the unit is in a fixed location.

Water pressure is most commonly measured by a Bourdon gage (Figure 4-2c). A hollow metal tube of elliptical cross section is bent in the form of a circle with a pointer attached to the end through a suitable linkage. As pressure inside the tube increases, the elliptical cross section tends to become circular, and the free end of the Bourdon tube moves outward. The dial of the instrument can be calibrated to read gage pressure in psi. Bourdon tubes and manometers actually measure pressure relative to the atmosphere pressure. (Atmospheric or barometric pressure is caused by the weight of the thick mass of air above the earth's surface. Under standard atmospheric conditions, the barometric pressure at sea level is 14.7 psi.) If the pressure measured is greater than atmospheric, this value is sometimes called gage pressure. If the pressure measured is less than atmospheric, it is referred to as a vacuum. Absolute pressure is the term used for a pressure reading that includes atmospheric pressure, that is, the pressure relative to absolute zero.

4-2 PRESSURE-VELOCITY-HEAD RELATIONSHIPS

The association between quantity of water flow, average velocity, and cross-sectional area of flow is given by the equation:

$$Q = VA \qquad (4\text{-}1)$$

where
  Q = quantity, cubic feet per second
  V = velocity, feet per second
  A = cross-sectional area of flow, square feet

This formula is known as the continuity equation. For an incompressible fluid such as water, if the cross-sectional area decreases (the water main becomes smaller), the velocity of flow must increase; conversely, if the area increases, the velocity decreases (Figure 4-3).

The total energy at any point in a hydraulics system is equal to the sum of the elevation head, pressure head, and velocity head.

$$\frac{\text{Total}}{\text{Energy}} = \frac{\text{Elevation}}{\text{Head}} + \frac{\text{Pressure}}{\text{Head}} + \frac{\text{Velocity}}{\text{Head}} \qquad (4\text{-}2)$$

$$E \;=\; Z \;+\; \frac{P}{w} \;+\; \frac{V^2}{2g}$$

where
  E = total energy head, feet
  Z = elevation above datum, feet
  P = pressure, pounds per square foot
  V = velocity of flow, feet per second
  w = unit weight of liquid, pounds per cubic foot
  g = acceleration of gravity = 32.2 ft/(sec)(sec)

The vertical line at point 1 on the left-hand side of Figure 4-4 graphically illustrates the total energy equation. The pressure head is equal to the height to which the water would rise in a piezometer tube inserted in the pipe. The

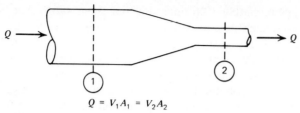

$$Q = V_1 A_1 = V_2 A_2$$

Figure 4-3 Flow equation for incompressible liquids.

elevation head is equal to the vertical distance Z from an assumed datum plane to the pipe. The sum of these two heads is the hydraulic head. The velocity head is equal to the kinetic energy in the water flow, and when added to the hydraulic head yields total energy. If ft-lb-sec units are used in Eq. 4-2, each of the separate head calculations will be in feet.

If the energy is compared at two different points in a piping system (Figure 4-4) Eq. 4-3 results. The term h_L represents the energy losses that occur in any real system. The major loss of energy is due to friction between the moving water and pipe wall; however, energy losses also occur from flow disturbance caused by valves, bends in the pipeline, and changes in diameter. The imaginary line connecting all points of total energy is called the energy gradient or grade line. This line must always slope in the direction of flow showing a decrease in energy, unless external energy is added to the system, for example, by a pump. The hydraulic gradient is defined as the line connecting the level of elevation plus pressure energies, this being defined by the water surfaces in imaginary piezometer tubes inserted in the piping.

$$Z_1 + \frac{P_1}{62.4} + \frac{V_1^2}{64.4} = Z_2 + \frac{P_2}{62.4} + \frac{V_2^2}{64.4} + h_L \tag{4-3}$$

where
Z = elevation, feet
P = pressure, pounds per square foot
V = velocity, feet per second
h_L = head loss, feet
62.4 = unit weight of water
$64.4 = 2\,g$

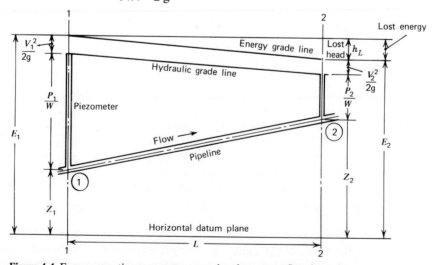

Figure 4-4 Energy equation parameters as related to water flow in a pipe.

Head loss as a result of pipe friction can be computed by using the Darcy Weisbach equation:

$$h_L = f \frac{LV^2}{D2g} \qquad (4\text{-}4)$$

where

h_L = head loss, feet
f = friction factor, Figure 4-5
L = length of pipe, feet
V = velocity of flow, feet per second
D = diameter of pipe, feet

The friction factor, f, is related to the relative roughness of the pipe material and the fluid flow characteristics. For turbulent water flow the f value for Eq. 4-4 can be taken from Figure 4-5 based on pipe diameter and material roughness.

Valves, fittings, and other appurtenances disturb the flow of water, causing losses of head in addition to the friction loss in the pipe. These units in a typical distribution system are at infrequent intervals; consequently, losses due to appurtenances are relatively insignificant in comparison with pipe friction losses. In the case of pumping stations and treatment plant piping, the minor losses in valves and fittings are significant and constitute a major part of the total losses. Unit head losses may be expressed as being equivalent to the loss through a certain length of pipe or by the formula:

$$h_L = \frac{kV^2}{2g} \qquad (4\text{-}5)$$

where

h_L = head loss, feet
V = velocity, feet per second
k = loss coefficient

The equivalent length of pipe that results in the same head loss is expressed in terms of the number of pipe diameters. Table 4-1 gives values for k and equivalent lengths of pipe for various fittings and valves.

Table 4-1. Approximate Minor Head Losses in Fittings and Valves

Fitting or Valve	Loss Coefficient k	Equivalent Length (Diameters of Pipe)
Tee (run)	0.60	20
Tee (branch)	1.80	60
90° bend-	—	—
Short radius	0.90	32
Medium radius	0.75	27
Long radius	0.60	20
45° bend	0.42	15
Gate valve (open)	0.48	17
Swing check valve (open)	3.7	135
Butterfly valve (open)	1.2	40

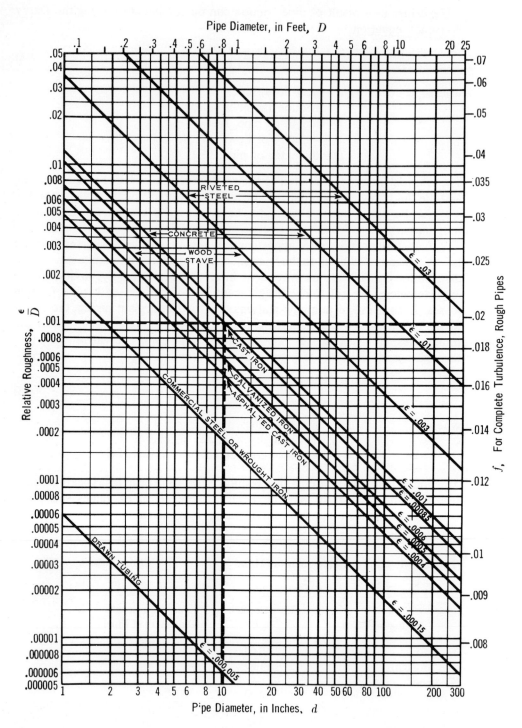

Figure 4-5 Relative roughness of pipe materials and friction factors for turbulent pipe flow. (From *Flow of Fluids*, Crane Co.)

EXAMPLE 4-1

Calculate the head loss in a 24-in. diameter, 5000-ft long, smooth-walled concrete ($\varepsilon = 0.001$) pipeline carrying a water flow of 10 cu ft/sec.

Solution
Using Eq. 4-1,

$$V = \frac{Q}{A} = \frac{10 \text{ cu ft/sec}}{\pi \text{ sq ft}} = 3.2 \text{ ft/sec}$$

From Fig. 4-5 entering with $d = 24$ in. and $\varepsilon = 0.001$, $f = 0.017$. Then substituting into Eq. 4-4,

$$h_L = 0.017 \frac{5000}{2} \frac{(3.2)^2}{2 \times 32.2} = 6.8 \text{ ft}$$

EXAMPLE 4-2

A pump discharge line consists of 200 ft of 12-in. new cast-iron pipe, three 90° medium-radius bends, two gate valves, and one swing check valve. Compute the head loss through the line at a velocity of 3.0 ft/sec.

Solution
Using equivalent lengths for the fittings and valves as given in Table 4-1, the total equivalent pipe length is:

$$200 + 3 \times 27 + 2 \times 17 + 1 \times 135 = 450$$

From Figure 4.5, $f = 0.019$, then using Eq. 4-4,

$$h_L = 0.019 \frac{450}{1.0} \frac{(3.0)^2}{2 \times 32.2} = 1.2 \text{ ft}$$

EXAMPLE 4-3

Calculate the head loss in the pipeline illustrated in Figure 4-4 based on the following: $Z_1 = 15.0$ ft, $P_1 = 40$ psi, $V_1 = 3.0$ ft/sec. $Z_2 = 30.6$ ft, $P_2 = 30$ psi, $V_2 = 3.0$ ft/sec.

Solution

$$P_1 = 40 \frac{\text{lb}}{\text{sq in.}} \times \frac{144 \text{ sq in.}}{\text{sq ft}} = 5760 \frac{\text{lb}}{\text{sq ft}}, \quad P_2 = 4320 \frac{\text{lb}}{\text{sq ft}}$$

Substituting into Eq. 4-3,

$$15.0 + \frac{5760}{62.4} + \frac{(3.0)^2}{64.4} = 30.6 + \frac{4320}{62.4} + \frac{(3.0)^2}{64.4} + h_L$$

$$15.0 + 92.3 + 0.14 = 30.6 + 69.2 + 0.14 + h_L$$

$$h_L = 7.5 \text{ ft} = 3.2 \text{ psi}$$

4-3   FLOW IN PIPES UNDER PRESSURE

The Darcy Weisbach equation for computing head loss is cumbersome and not widely used in water works design and evaluation. A trial-and-error solution is required to determine pipe size for a given flow and head loss, since the friction factor is based on the relative roughness which involves the pipe diameter. Because of this practical shortcoming, exponential equations are commonly used for flow calculations.

The most common pipe flow formula used in the design and evaluation of a water distribution system is the Hazen Williams, Eq. 4-6. This equation relates the velocity of turbulent water flow with hydraulic radius, slope of the hydraulic gradient, and a coefficient of friction depending on roughness of the pipe. Coefficient values for different pipe materials are given in Table 4-2.

$$V = 1.318CR^{0.63}S^{0.54} \tag{4-6}$$

where
V = velocity, feet per second
C = coefficient that depends on the material and age of the conduit, Table 4-2
R = hydraulic radius, feet (cross-sectional area divided by the wetted perimeter)
S = slope of the hydraulic gradient, feet per foot

By combining the Hazen Williams formula with Eq. 4-1 and substituting the value of $D/4$ for R, the formula for quantity of flow in circular pipes flowing full is

$$Q = 0.281CD^{2.63}S^{0.54} \tag{4-7}$$

where
Q = quantity of flow, gallons per minute
C = coefficient, Table 4-2
D = diameter of pipe, inches
S = hydraulic gradient, feet per foot

The nomograph shown in Figure 4-6 solves the above equation for a coefficient equal to 100, represents 15 to 20-year-old cast-iron pipe. Given any two of the

Table 4-2. Values of Coefficient C for the Hazen Williams Formula, Equations 4-6 and 4-7

Pipe Material	C
Asbestos cement	140
Cast iron	
Cement lined	130 to 150
New, unlined	130
5-year-old, unlined	120
20-year-old, unlined	100
Concrete	130
Copper	130 to 140
Plastic	140 to 150
New welded steel	120
New riveted steel	110

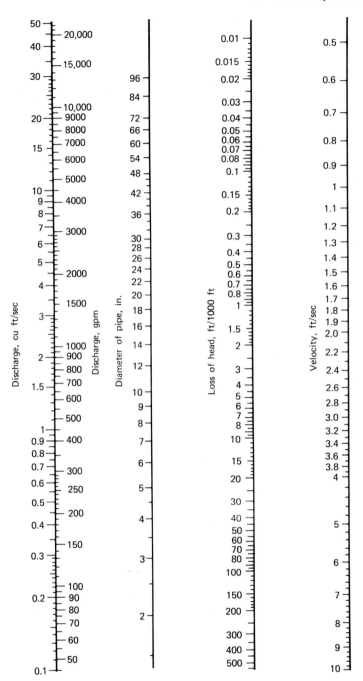

Figure 4-6 Nomograph for Hazen Williams formula, Eq. 4-7, based on $C = 100$.

parameters (discharge, diameter of pipe, loss of head, or velocity) the remaining two can be determined from the intersections along a straight line drawn across the nomograph. For example, the flow of 500 gpm in an 8-in. diameter pipe has a velocity of 3.2 ft/sec with a head loss of 8.5 ft/1000 ft (0.0085 ft/ft). Head losses in pipes with coefficient values other than 100 can be determined by using the correction factors in Table 4-3. For example, if the head loss at

Table 4-3. Correction Factors To Determine Head Losses from Figure 4-6 at Values of C Other Than $C = 100$

\multicolumn{4}{c}{Corrected $h_L = K \times h_L$ at $C = 100$}			
C	K	C	K
80	1.51	120	0.71
100	1.00	130	0.62
110	0.84	140	0.54

$C = 100$ is 8.5/1000 ft, the head loss at $C = 130$ would equal $0.62 \times 8.5 = 5.3$ ft/1000 ft.

EXAMPLE 4-4

If a 10-in. water main is carrying a flow of 950 gpm, what is the velocity of flow and head loss for (a) $C = 100$, and (b) $C = 140$?

Solution

(a) From Figure 4-6, $V = 3.9$ ft/sec and $h_L = 9.0$ ft/1000 ft
(b) $V = 3.9$ ft/sec (same as part a)
Using K from Table 4-3, $h_L = 0.54 \times 9.0 = 4.9$ ft/1000 ft

EXAMPLE 4-5

An extremely simplified water supply system consisting of a reservoir with lift pumps, elevated storage, piping, and load center (withdrawal point) are shown in Figure 4-7.
(a) Based on the following data, sketch the hydraulic gradient for the system:

$Z_A = 0$ ft, $P_A = 80$ psi, $Z_B = 30$ ft, $P_B = 30$ psi,

$Z_C = 40$ ft, $P_C = 100$ ft (water level in tank).

(b) For these conditions compute the flow available at point B from both the supply pumps and elevated storage. Use $C = 100$, and pipe sizes as shown in the diagram.

Solution

(a) Hydraulic head at $A = 0$ ft $+ 80$ psi $\times 2.31 \dfrac{\text{ft}}{\text{psi}} = 185$ ft
At $B = 30 + 30 \times 2.31 = 99$ ft. At $C = 40 + 100 = 140$ ft
The hydraulic gradient is shown as straight lines connecting these hydraulic heads drawn vertically.
(b) h_L between A and $B = 185 - 99 = 86$ ft
h_L per 1000 ft $= 86 \div 5 = 17.2$ ft
Using Figure 4-6, align $h_L = 17.2$ ft/1000 ft and 12-in. diameter, read $Q = 2160$ gpm

h_L per 1000 ft B to $C = \dfrac{140 - 99}{3} = 13.7$ ft

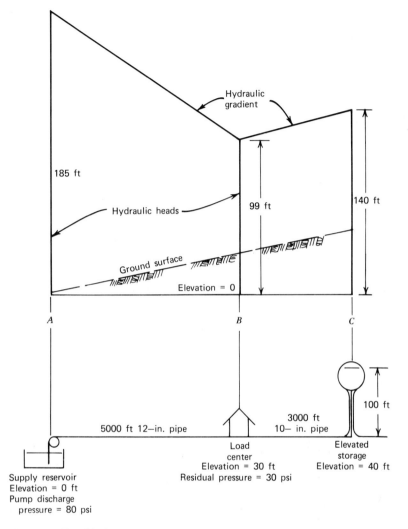

Figure 4-7 Simplified water system for Example 4-5.

For $h_L = 13.7 \text{ ft}/1000 \text{ ft}$ and 10 in., $Q = 1180$ gpm

Total Q available at $B = 2160 + 1180 = 3340$ gpm

4-4 FLOW IN PIPE NETWORKS

Analysis of a water distribution system includes determining quantities of flow and head losses in the various pipelines, and resulting residual pressures. Computations for a large network can often be made easier if a series of pipes of varying diameter are replaced with equivalent pipes. An equivalent pipe is an imaginary conduit that replaces a section of a real system such that the head losses in the two systems are identical for the quantity of flow. For example, pipes of differing diameters connected in series can be replaced by an equivalent pipe of one diameter as follows: assume a quantity of flow and determine the head loss in each section of the line for this flow, then using the sum of the sectional head losses and the assumed flow, enter the nomogram to find the equivalent pipe diameter. For parallel pipe systems, a head loss is assumed, and the quantity of flow through each of the pipes is calculated for that head loss. Then the sum of the flows and the assumed head loss are used to determine the equivalent pipe size. Example 4-6 illustrates equivalent pipe calculations for lines in series and parallel.

EXAMPLE 4-6

Determine an equivalent pipeline 2000 ft in length to replace the pipe system illustrated in Figure 4-8.

Solution

First replace the parallel pipes between B and C with one equivalent line 1000 ft in length. Assume a head loss of 10 ft between B and C. Using the nomograph, the flow in 1000 ft of 8-in. pipe at a head loss of 10 ft/1000 ft = 550 gpm, and the flow in the 6-in. pipe at a loss of 10 ft/800 ft (12.5 ft/1000 ft) = 290 gpm. The equivalent pipe B to C, 1000 ft in length, is then that size pipe which has a 10 ft/1000 ft head loss at a discharge of 550 + 290 = 840 gpm. From the nomograph, diameter = 9.4 in.

Next consider the three pipes in series: 400 ft of 8 in., 1000 ft of 9.4 in., and 600 ft of 10 in. At an assumed flow of 500 gpm the head losses in pipes are: 0.4 × 8.3 = 3.3 ft, 3.8 ft, and 0.6 × 2.7 = 1.6 ft, respectively, for a total of 8.7-ft head loss in 2000 ft. The equivalent pipe A to D is then that diameter which exhibits a loss of 4.4 ft/1000 ft at a flow = 500 gpm. The answer is 9.2 in.

The method of equivalent pipes cannot be applied to complex systems, since crossovers result in pipes being included in more than one loop and because there are normally a number of withdrawal points throughout the system. Network analysis can be performed using the Hardy Cross technique. Assumed flows are assigned to each of the pipelines and then corrections are applied, based on hydraulic calculations, until the computed flows and resultant head losses are in acceptable balance.

The derivation of the expression for flow corrections follows. The Hazen Williams equation (4-7) can be rewritten as

$$Q = KS^{0.54} \tag{4-8}$$

where K is a constant when dealing with a single pipe of specific size and roughness. Rearranging and substituting head loss, h_L, for the hydraulic

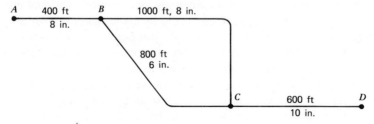

Figure 4-8 Pipe system for Example 4-6.

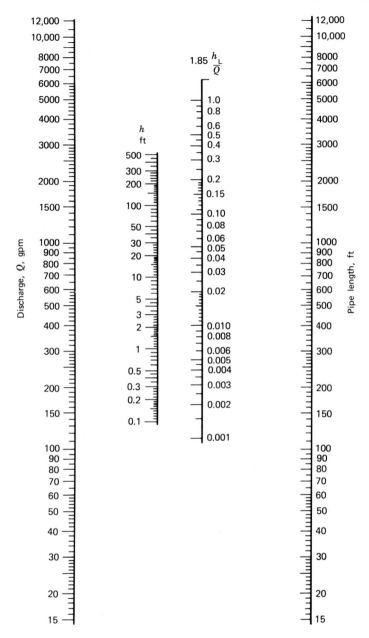

Figure 4-9 Special nomograph for head losses and 1.85 h_L/Q values based on 8-in. pipe with $C = 100$.

gradient, S, results in an equation relating lead loss as a function of flow to the 1.85 power for the Hazen-Williams formula.

$$h_L = KQ^{1.85} \tag{4-9}$$

In balancing a given loop within a network the actual quantity of flow, Q, is determined by applying a correction ΔQ to the assumed flow, Q_1.

$$Q = Q_1 + \Delta Q \tag{4-10}$$

Inserting this term in Eq. 4-9 and expanding the binominal, it becomes

$$KQ^{1.85} = K(Q_1 + \Delta Q)^{1.85} = K(Q_1^{1.85} + 1.85Q_1^{0.85}\Delta Q + \cdots) \qquad (4\text{-}11)$$

Then, the sum of the head losses must equal zero for balanced flow,

$$\sum h_L = \sum KQ^{1.85} = \sum KQ_1^{1.85} + \sum 1.85KQ_1^{0.85}\Delta Q = 0 \qquad (4\text{-}12)$$

And finally, solving for ΔQ and substituting h_L for the $KQ^{1.85}$ from Eq. 4-9, the term for flow correction is Eq. 4-13.

$$\Delta Q = -\frac{\sum h_L}{1.85\sum(h_L/Q)} \qquad (4\text{-}13)$$

The procedure for solving a pipe network using the Hardy Cross method is illustrated in Example 4-7 and involves the following steps:

1. Assume a direction and quantity of flow for each pipe in the system such that the amount entering each pipe junction, or discharge point, equals the quantity flowing out.
2. Select one pipe loop in the system and compute the net head loss for that circuit based on the assumed flows. (The nomograph in Figure 4-6 is normally used for these calculations.)
3. Without regard to sign, calculate the sum of the h_L/Q values.
4. Compute the flow correction using Eq. 4-13 and correct each of the flows in the loop by this amount.
5. Apply this procedure to each pipe loop in the system, repeating earlier circuits as necessary, to arrive at answers within the desired accuracy. (Computational accuracy depends on the precision in reading the nomograph and the use of a slide rule for calculations. In general, an error of 5 to 10 percent in calculating the head losses around a loop is satisfactory.)

Use of the special nomograph for head losses, shown in Figure 4-9, permits a simplified method for determining flows in pipe networks. The lines on the chart are Q, h_L, $1.85h_L/Q$, and L for 8-in. diameter pipe with a $C = 100$. Table 4-4

Table 4-4. Conversion Factors for Various Sized Pipes. Multiply Length of Pipe by Factor Given To Obtain the Equivalent Length of 8-in. Pipe

Diameter (in.)	Factor	Diameter (in.)	Factor
1	24,000	14	0.066
2	840	15	0.047
$2\frac{1}{2}$	286	16	0.034
3	120	18	0.020
4	29	20	0.012
5	9.7	21	0.0093
6	4.06	22	0.0074
8	1.00	24	0.0048
10	0.34	30	0.0016
12	0.14	36	0.00066

gives multiplication factors to convert lengths of various size pipes to equivalent lengths of 8-in. pipe for the same head losses. To use this nomograph in solving for flows in a pipe network, all lines in the system are converted to equivalent lengths of 8-in. pipe with $C = 100$. For example, 1000 ft of 10-in. pipe is equivalent to $1000 \times 0.34 = 340$ ft 8-in. pipe (Table 4-4). If the C value also changes, the equivalent length must be adjusted accordingly, using values given in Table 4-3. Convert 1000 ft 10-in. pipe with $C = 120$ to 8-in. pipe with $C = 100$: $1000 \times 0.34 \times 0.71 = 240$ ft. Figure 4-9 can then be used to find values of h_L and $1.85 h_L/Q$ for a given Q and L of equivalent 8-in. pipe by means of a straight edge aligned across the nomograph. For example, 500 gpm flowing through 1000 ft of 8-in. pipe produces a heat loss of 8.0 ft and the $h_L/Q = 0.03$. When the equivalent pipe is very short, a length 10 times as great can be used and the resulting h_L and h_L/Q values can be divided by 10.

EXAMPLE 4-7

Using the Hardy Cross method, compute the quantities of flow for the pipe network shown in Figure 4-10. Water enters the distribution system from the well supply, 1350 gpm at point A, and storage reservoir, 800 gpm at point J.

Solution

The assumed direction and quantities of flow are shown for each line on Figure 4-10. These estimates are not entirely arbitrary but are based on careful assessment of the network, thus providing the most probable flow pattern. For example, at junction A a significantly greater flow is assumed for the 12-in. diameter pipe AD than the 6-in. line AB, and the quantity entering at J is directed primarily into pipe JI heading to the heavy discharge at point H.

The trial computations are given in Table 4-5. After listing the information on pipe diameter and length for each line in the loop, head losses are determined from the nomograph (Figure 4-6), using the assumed flows. All clockwise flows are positive, and all counterclockwise flows are negative. Head loss resulting from positive flow is considered positive, and negative flows yield negative losses. The head loss is determined with respect to sign for each line, and then is summed to provide total h_L. Values of h_L/Q are computed by dividing the head loss in each line by the assumed flow. The sum of h_L/Q is performed without regard to sign. The flow correction is then calculated using Eq. 4-13. The correction values in this problem are rounded to the nearest 10 gpm, which is consistent with the accuracy of reading the nomograph.

The first trial—loop I resulted in a negligible value for ΔQ indicating a balance of the head losses for the assumed flows. As is shown in trial four, this does not mean that these flows are correct, since two of the lines (DE and EF) are common with other loops. The trial number was placed in parentheses after the assumed flows in loop I, for example, line AB 250 (1).

The second trial—loop II reversed the assumed flow in line GC, and reduced the flow in line DE, common with loop I, from 400 to 300 gpm. Trials three to six were sufficient to provide the required balancing of flows in the network.

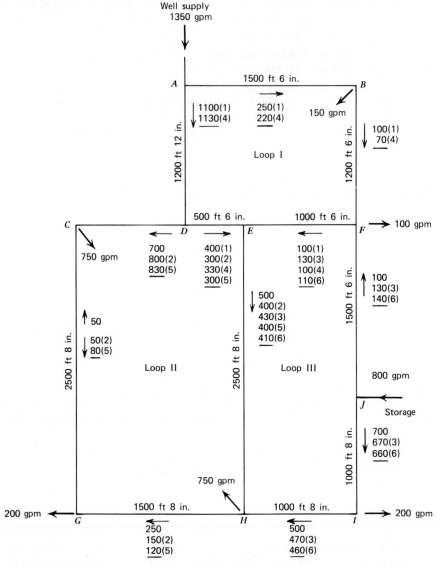

Figure 4-10 Pipe network for Example 4-7. The numbers along each pipeline are quantities of flow from several trial calculations given in Table 4-5. For example, 220(4) on line *AB* means 220 gpm computed during trial 4.

The error in head loss calculations for the sixth trial-loop III was 7 percent. Final quantities for flow for each pipe are underlined.

Alternate Method

The monograph in Figure 4-9 can be used to solve Hardy Cross problems by tabulating values as are shown in Table 4-6. The same trial calculations as in Table 4-5 could be performed in the same sequence by using this format. The final results would be the same. Rather than repeat the entire analysis of the network in Figure 4-10, Table 4-6 loops the outside circuit *ABFJIHGCDA*, using the underlined values as the assumed flows.

Table 4-5. Computations for the Hardy Cross Analysis of the Pipe Network in Figure 4-10, Example 4-7

Line	Diameter (in.)	Length (1000 ft)	Flow (gpm)	Head Loss (ft/1000 ft)	Total h_L (ft)	$\dfrac{h_L}{Q}$	Corrected Flow
			First Trial, Loop I				
AB	6	1.5	+250	+9.2	+13.8	0.055	+250
BF	6	1.2	+100	+1.6	+1.9	0.019	+100
FE	6	1.0	+100	+1.6	+1.6	0.016	+100
ED	6	0.5	−400	−23	−11.5	0.029	−400
DA	12	1.2	−1100	−5.0	−6.0	0.005	−1100
					−0.2	0.124	

$$\Delta Q = -\frac{\sum h_L}{1.85 \sum h_L/Q} = -\frac{-0.2}{1.85 \times 0.124} = \; <1 \text{ gpm (negligible)}$$

Line	Diameter (in.)	Length (1000 ft)	Flow (gpm)	Head Loss (ft/1000 ft)	Total h_L (ft)	$\dfrac{h_L}{Q}$	Corrected Flow
			Second Trial, Loop II				
CD	8	1.0	−700	−16	−16.0	0.024	−800
DE	6	0.5	+400	+23	+11.5	0.029	+300
EH	8	2.5	+500	+8.3	+20.8	0.042	+400
HG	8	1.5	+250	+2.3	+3.4	0.014	+150
GC	8	2.5	+50	+0.1	+0.2	0.004	−50
					+19.9	0.113	

$$\Delta Q = -\frac{+19.9}{1.85 \times 0.113} = -97 \text{ gpm } (-100 \text{ gpm})$$

Line	Diameter (in.)	Length (1000 ft)	Flow (gpm)	Head Loss (ft/1000 ft)	Total h_L (ft)	$\dfrac{h_L}{Q}$	Corrected Flow
			Third Trial, Loop III				
EF	6	1.0	−100	−1.6	−1.6	0.016	−130
FJ	6	1.5	−100	−1.6	−2.4	0.024	−130
JI	8	1.0	+700	+16	+16.0	0.023	+670
IH	8	1.0	+500	+8.3	+8.3	0.017	+470
HE	8	2.5	−400	−5.5	−13.7	0.034	−430
					+6.6	0.114	

$$\Delta Q = -\frac{+6.6}{1.85 \times 0.114} = -30 \text{ gpm}$$

Line	Diameter (in.)	Length (1000 ft)	Flow (gpm)	Head Loss (ft/1000 ft)	Total h_L (ft)	$\dfrac{h_L}{Q}$	Corrected Flow
			Fourth Trial, Loop I				
AB	6	1.5	+250	+9.2	+13.8	0.055	+220
BF	6	1.2	+100	+1.6	+1.9	0.019	+70
FE	6	1.0	+130	+2.7	+2.7	0.021	+100
ED	6	0.5	−300	−13.0	−6.5	0.022	−330
DA	12	1.2	−1100	−5.0	−6.0	0.005	−1130
					+5.9	0.122	

$$\Delta Q = -\frac{+5.9}{1.85 \times 0.122} = -30 \text{ gpm}$$

Line	Diameter (in.)	Length (1000 ft)	Flow (gpm)	Head Loss (ft/1000 ft)	Total h_L (ft)	$\dfrac{h_L}{Q}$	Corrected Flow
			Fifth Trial, Loop II				
CD	8	1.0	−800	−20	−20.0	0.025	−830
DE	6	0.5	+330	+16	+8.0	0.024	+300
EH	8	2.5	+430	+6.5	+16.2	0.038	+400
HG	8	1.5	+150	+0.9	+1.4	0.009	+120
GC	8	2.5	−50	−0.1	−0.2	0.004	−80
					+5.4	0.100	

$$\Delta Q = -\frac{+5.4}{1.85 \times 0.100} = -30 \text{ gpm}$$

(Table continued overleaf)

Table 4-5 *(Continued.)*

Line	Diameter (in.)	Length h (1000 ft)	Flow (gpm)	Head Loss (ft/1000 ft)	Total h_L (ft)	$\dfrac{h_L}{Q}$	Corrected Flow
			Sixth Trial, Loop III				
EF	6	1.0	−100	−1.6	−1.6	0.016	−110
FJ	6	1.5	−130	−2.7	−4.0	0.031	−140
JI	8	1.0	+670	+14.8	+14.8	0.022	+660
IH	8	1.0	+470	+7.4	+7.4	0.016	+460
HE	8	2.5	−400	−5.5	−13.7	0.034	−410
					+2.9	0.119	

$$\Delta Q = -\frac{+2.9}{1.85 \times 0.119} = -10 \text{ gpm}$$

$$\text{Error} = \frac{\text{net } h_L}{\sum h_L} = \frac{2.9}{1.6+4.0+14.8+7.4+13.7} \times 100 = 7 \text{ percent}$$

Table 4-6. Computations for the Hardy Cross Analysis of Circuit *ABF-JKHGCDA*, Figure 4-10, Example 4-7, Using Special Nomograph Figure 4-9 and Table 4-4

Line	Length (ft)	Diameter (in.)	Equivalent 8 in. (ft)	Q (gpm)	h_L (ft)	$\dfrac{1.85h_L}{Q}$	Correct Q
AB	1500	6	6100	+220	+11.0	0.090	
BF	1200	6	4900	+70	+1.0	0.028	Same as
FJ	1500	6	6100	−140	−4.6	0.062	the
JI	1000	8	1000	+660	+13.8	0.038	assumed
IH	1000	8	1000	+460	+7.2	0.028	values
HG	1500	8	1500	+120	+0.8	0.013	of Q
GC	2500	8	2500	−80	−0.6	0.016	
CD	1000	8	1000	−830	−20.5	0.046	
DA	1200	12	170	−1130	−6.5	0.012	
					+1.6	0.333	

$$\Delta Q = -\frac{+1.6}{0.333} = -4.7 \text{ gpm (negligible)}$$

4-5   GRAVITY FLOW IN CIRCULAR PIPES

Sanitary and storm sewers are designed to flow as open channels, not under pressure, although storm sewers may occasionally be overloaded when water rises in the manholds surcharging the sewer. The waste water flows downstream in the pipe moved by the force of gravity. Velocity of flow depends on steepness of the pipe slope and frictional resistance. The Manning formula, Eq. 4-14, is used for uniform, steady, open channel flow. The coefficient of roughness n depends on the condition of the pipe surface, alignment of pipe sections, and method of jointing. The common sewer pipe materials of asbestos cement, vitrified clay, and smooth concrete have n values in the range of 0.011

to 0.015. The lower value is applicable to clear water and smooth joints, while the greater roughness is for waste water and poor joint construction. Corrugated steel pipe used for culverts has considerably greater roughness in the range of 0.021 to 0.026. The common value of n adopted for sewer design is 0.013. Even though certain pipe materials in a new condition may exhibit a lower n, such as plastic or asbestos cement, after they are placed in use pipes accumulate grease and other solids that modify the interior and disturb waste-water flow.

$$Q = \frac{1.49}{n} AR^{2/3} S^{1/2} \tag{4-14}$$

where Q = quantity of flow, cubic feet per second
n = coefficient of roughness depending on material
A = cross-sectional area of flow, square feet
R = hydraulic radius, feet (cross-sectional area divided by the wetted perimeter)
S = slope of the hydraulic gradient, feet per foot

The nomograph shown in Figure 4-11 solves the Manning equation for circular pipes flowing full based on a coefficient of roughness of 0.013. Given any two of the parameters (quantity of flow, diameter of pipe, slope of pipe, or velocity) the remaining two can be determined from the intersections along a straight line drawn across the nomograph. For example, an 8-in. diameter pipe set on a slope of 0.02 ft/ft (2.0%) conveys a quantity of flow of 1.7 cu ft/sec flowing full with an average velocity of 4.9 ft/sec. Quantity of flows at n values other than 0.013 can be calculated by multiplying the nomograph value by 0.013 and dividing the resultant by the desired n factor. The Manning formula can also be solved by using the diagram in Figure 4-12 without using a straightedge. The horizontal and vertical grid lines are quantity of flow and slope of pipe (ft/100 ft), respectively; the slanting grid includes pipe diameter and velocity. For the same example, enter the bottom at 2 ft/100 ft and move vertically to the intersection with the diagonal line representing 8-in. pipe diameter. From this point read 1.7 cu ft/sec along the vertical scale and 4.9 ft/sec on the other diagonal grid line.

Hydraulic problems of circular pipes flowing partly full are solved by using Figure 4-13. To plot these curves, ratios of hydraulic elements were calculated at various depths of flow in a circular pipe using the Manning equation. The symbols q, v, and a relate to the partial flow condition while Q, V, and A are at full flow. Consider flow in a pipe at a depth of 30 percent of the pipe diameter, represented by the horizontal line labeled 0.3. Where this line intersects the curved lines for quantity of flow, area of flow, and velocity, project downward to the horizontal scale and read, respectively, the values 0.2, 0.25, and 0.78; these values mean that partial flow at this depth conveys a quantity of flow equal to 20 percent of the flowing full quantity, the cross-sectional area of flow is 25 percent of the total open area of the pipe, and average velocity of flow is 78 percent of the flowing full velocity. This diagram shows that a pipe flowing at one-half depth conveys one half the flowing-full quantity of flow at a velocity equal to the flowing full velocity. Also the greatest quantity of flow occurs when the pipe is flowing at about 0.93 of the depth, and maximum velocity is at about 0.8 depth. The reason for this is that as the section approaches full flow, the additional frictional resistance caused by the crown of the pipe has a

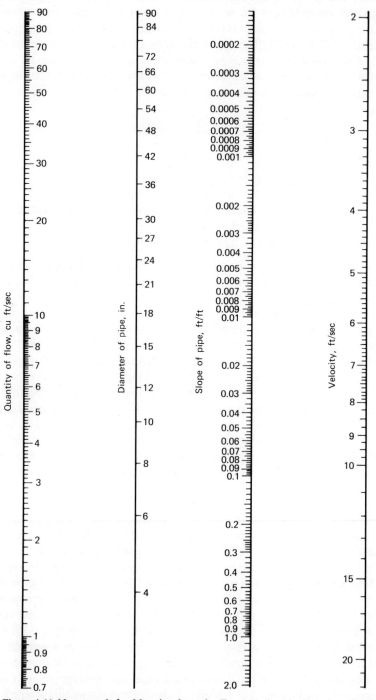

Figure 4-11 Nomograph for Manning formula, Eq. 4-14, for circular pipes flowing full based on $n = 0.013$.

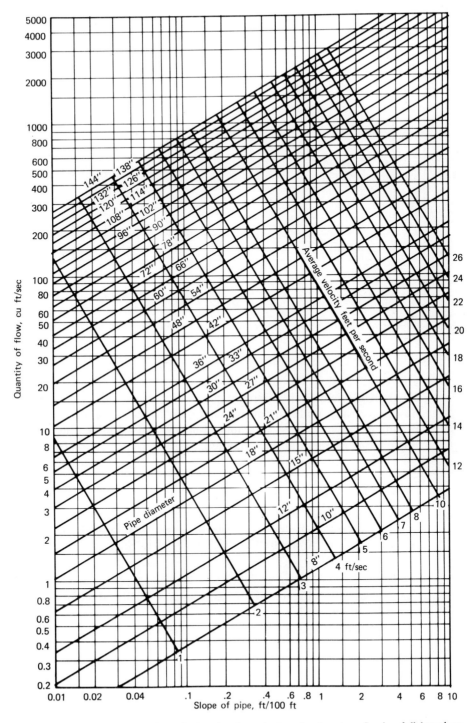

Figure 4-12 Diagram for solution of Manning formula for circular pipes flowing full based on $n = 0.013$. (From *Concrete Pipe Design Manual*, American Concrete Pipe Association)

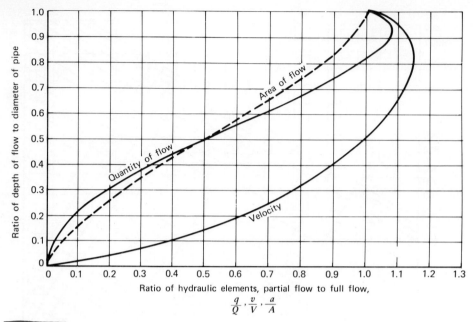

Ratio of depth of flow to diameter of pipe

Ratio of hydraulic elements, partial flow to full flow,

$$\frac{q}{Q}, \frac{v}{V}, \frac{a}{A}$$

Figure 4-13 Relative quantity, velocity, and cross-sectional area of flow in a circular pipe for any depth of flow.

greater affect than the added cross-sectional area. For practical application, Figure 4-13 yields only approximate results, since the theoretical curves are based on steady, uniform flow which does not precisely characterize the nature of actual flow in sewers. Examples 4-8, 4-9, and 4-10 illustrate the use of the nomograms and partial flow diagram.

EXAMPLE 4-8

If a 10-in. sewer is placed on a slope of 0.010, what is the flowing full quantity and velocity for (a) $n = 0.013$, and (b) $n = 0.015$?

Solution

(a) Project across Figure 4-11 through slope $= 0.010$ and diameter $= 10$ in., read $Q = 2.2$ cu ft/sec and $V = 4.0$ ft/sec.

(b) From Eq. 4-14, Q and V are both inversely proportional to n; therefore,

$$Q \text{ at } n = 0.015 = \frac{0.013 \times 2.2}{0.015} = 1.9 \text{ cu ft/sec}$$

$$V \text{ at } n = 0.015 = \frac{0.013 \times 4.0}{0.015} = 3.5 \text{ ft/sec}$$

Alternate solution

(a) On Figure 4-12 locate the intersection of the vertical line for 1.0 ft/100 ft and the diagonal for a 10-in. size, read $V = 4.0$ ft/sec on the opposing diagonal line and project horizontally to $Q = 2.2$ cu ft/sec.

EXAMPLE 4-9

The measured depth of flow in a 48-in. concrete storm sewer on a grade of 0.0015 ft/ft is 30 in. What is the calculated quantity and velocity of flow?

Solution

From either Figure 4-11 or Figure 4-12, the flowing full Q and V are 56 cu ft/sec and 4.4 ft/sec, respectively.
The depth of flow to diameter of pipe ratio is

$$\frac{d}{D} = \frac{30}{48} = 0.62$$

Entering Figure 4-13 with a horizontal line at this ratio, read q/Q and v/V by projecting downward at the intersections with the flow and velocity curves. Read $q/Q = 0.72$ and $v/V = 1.08$.
Then, flow at a depth of 30-in. $= 0.72 \times 56 = 40$ cu ft/sec and $v = 1.08 \times 4.4 = 4.7$ ft/sec.

EXAMPLE 4-10

An 18-in. sewer pipe, $n = 0.013$, is placed on a slope of 0.0025. At what depth of flow does the velocity of flow equal 2.0 ft/sec?

Solution

Velocity flowing full $= 3.0$ ft/sec (Figure 4-12)
Calculate $v/V = 2.0/3.0 = 0.67$
Enter Figure 4-13 vertically at 0.67, intersect the velocity line, and project horizontally to read $d/D = 0.23$.
Depth at a velocity of 2.0 ft/sec $= 0.23 \times 18 = 4.1$ in.

4-6   FLOW MEASUREMENT IN PIPES

The common positive displacement water meter has a measuring chamber of known volume containing a disk that goes through a certain cycle of motion as water passes through (Figure 4-14). The rotation resulting from filling and emptying of the chamber is transmitted to a recording register. The advantages of this meter are: simplicity of construction, high sensitivity and accuracy, small loss of head, and low maintenance costs. Also the accuracy of registration is not materially affected by position. This nutating disk meter is commonly used for small customer services, such as, individual households and apartments.

Several meter manufacturing firms have developed remote registration systems. The simplest is a remotely mounted digital register which is placed so that the person reading the register can do so from outside the house. A more sophisticated system consists of an outdoor receptacle read by plugging in a small portable recording device. The recorder punch card, or magnetic tape, is later used in data processing equipment at the central office for billing and recording in the memory system. Automatic remote meter reading, and billing

Figure 4-14 Cutaway section of a household water meter with a nutating disk geared to a recording register. (Courtesy of Neptune Water Meter Co., Division of Neptune International Corp.)

by computer processing, reduces costs for reading of meters and preparation of water bills.

The measuring device in a current (velocity) meter is a bladed wheel that rotates at a speed in proportion to the quantity of water passing through the blades (Figure 4-15). A recording register is geared to the turbine wheel. The shortcoming of a current meter is poor accuracy at low flow rates when the water is not moving at sufficient velocity to rotate the blades. Consequently, they are of limited value in water systems and are used only in restricted applications, for instance, on sprinkler carts that draw water from hydrants or on water towers. Current meters are unsuitable for customer services because of their inability to register low flow rates.

A general-service compound meter consists of a current meter, a positive displacement meter, and an automatic valve arrangement that directs the water to the current meter during high rates of flow and to the displacement meter at low rates (Figure 4-16). The advantages are high accuracy at all flow rates and relatively wide operating range. The switchover from low- to high-flow operation is controlled by head loss through the disk meter; when it exceeds a certain amount, a valve automatically opens and the current meter in the main line begins to measure the higher flow. Compound meters are used on larger services where widely varying flow rates are typical, such as motels, office buildings, factories, and commercial properties.

Proportional meters are sometimes referred to as "compound meters." The name is derived from the fact that only a portion of the total flow through the

Figure 4-15 Turbine-type water meter used for measuring continuous high rates of flow. (Courtesy of Hersey Products Inc.)

Figure 4-16 Compound water meter designed for services requiring accurate measurements of both large and small flows. (Courtesy of Rockwell International Corporation)

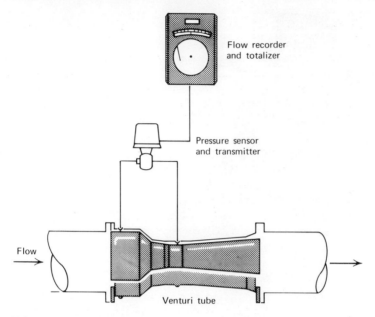

Figure 4-17 Venturi (differential pressure) meter inserted in a water pipeline, with sensor and recorder.

meter passes through a by-pass measuring chamber, which is the actual measuring device. The by-pass meter, either positive displacement or current type, is adjusted to register the total flow through the meter. Different methods of by-passing the water are utilized, such as an orifice in the main throat or a diverging tube. The proportional meter is capable of measuring large volumes of flow at comparatively low head loss and unrestricted flow, as well as accuracy of measurement.

Differential pressure meters, including venturi tubes, orifices, and nozzles, placed in a pipe section are the primary flow measuring devices in water systems. The meter shown in Figure 4-17 consists of a venturi tube with a converging portion, throat, and diverging portion and differential pressure gauge. As water flows through the tube, velocity is increased in the constricted portion and temporarily lowers the static pressure, in accordance with the energy Eq. 4-2. The pressure difference between inlet and throat is measured and correlated to the rate of flow. A flow recorder with a rotating chart and ink pen plots variation of flow with respect to time, and a digital totalizer registers total flow. Venturi meters are shaped to maintain streamline flow for minimum head loss. In addition to mainline flow metering, venturi tubes are used in rate of flow controllers, for example, on the discharge side of water filters. The pressure sensor transmits a controlling signal to a hydraulically, or electrically operated butterfly valve that is inserted in the pipeline ahead of the venturi. By setting in the desired flow rate, the valve automatically opens or closes as necessary to maintain this rate of flow.

4-7 FLOW MEASUREMENT IN OPEN CHANNELS

Waste water contains suspended and floating solids that prohibit the use of enclosed meters. Furthermore, waste water is commonly conveyed by open channel flow rather than in pressure conduits. Therefore, the Parshall flume is the most common device used to measure waste-water flows. A typical flume (Figure 4-18) consists of a converging and dropping open channel section. Flow moving freely through the unit can be calculated by measuring the upstream water level. A stilling well is normally provided to hold a float, bubble tube, or other depth measuring device, which is connected to a transmitter and flow recorder similar to that shown in Figure 4-17. Flumes are commercially available to set in cut-open sections of sewer lines to measure flow from portions of a collection system, or individual industrial waste discharges. The advantages of an open channel flume are low head loss and self-cleansing capacity.

Weirs can also be used for measuring flow of water in open channels. Water flowing over the sharp edge crest must discharge to the atmosphere, that is, air must be allowed to pass freely under the jet. If these conditions are met, the rate of flow can be directly related to the height of water measured behind the weir. Many weirs have been developed and calibrated with the majority having V-notch or rectangular openings.

The most common weir used for measuring waste-water flows is the 90° V-notch weir, illustrated in Figure 4-19. It is particularly well adapted to recording wide variations in flow, and may be used in treatment plants too small to warrant continuous flow recording and the more expensive Parshall flume. V-notch weirs are commonly installed on a temporary basis to make flow measurements associated with industrial waste surveys. Discharge over a 90° V-notch weir can be calculated using the following equation

$$Q = 2.48H^{2.48} \tag{4-15}$$

where Q = free discharge over 90° V-notch weir, cubic feet per second
H = vertical distance (head) from crest of weir to the
free water surface, feet

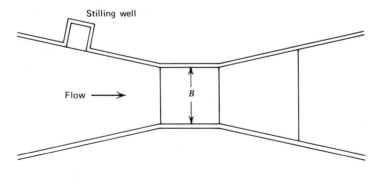

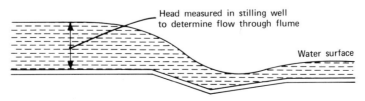

Figure 4-18 Parshall flume for measuring flow in an open channel.

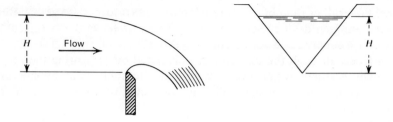

Figure 4-19 Ninety degree V-notch weir used for measuring flow of water in a channel, refer to Eq. 4-15 and Figure 4-20.

Figure 4-20 is a plot of Eq. 4-15 giving the head in inches and discharge in million gallons per day (mgd). Since a weir is commonly placed in a relatively shallow narrow channel for flow measurement, the calculated flow must be adjusted by using correction factors for width and depth as is shown in Figure 4-20. Example 4-11 illustrates the use of this diagram. The height of flow, vertical distance from the weir crest to the free water surface, is measured a short distance behind the weir in a stilling well. In waste-water survey measurements, it is common to place the float of the flow recorder inside a stovepipe held partly submerged below the water surface in the channel some distance behind the weir.

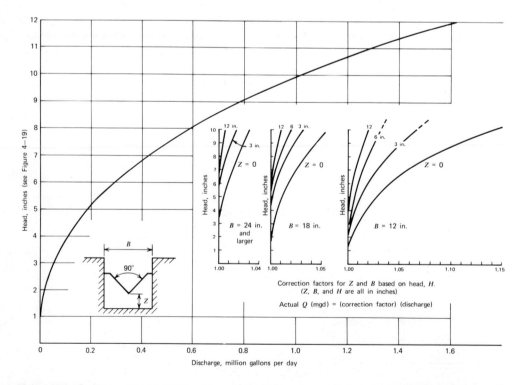

Figure 4-20 Discharge curve for 90° V-notch weir, Eq. 4-15, and correction curves for height of weir crest (Z) and various channel widths.

EXAMPLE 4-11

A 90° V-notch weir is placed in an 18-in. wide channel with the base of the notch 6 in. from the channel bottom. If the height of flow over the weir is 7 in., what is the quantity of flow?

Solution

Using Figure 4-20, the flow from the curve is 0.41 mgd.
Using the correction factor curve for $B = 18$ in., locate the intersection of $H = 7$ in., and $Z = 6$ in. Dropping vertically, read a correction factor of 1.01, which is negligible.
Therefore, actual $Q = 0.41$ mgd.

4-8   CENTRIFUGAL PUMPS

Pumps are used for a variety of functions in water and waste-water systems. Low-lift pumps are used to elevate water from a source, or waste water from a sewer, to the treatment plant; high-service pumps are used to discharge water under pressure to a distribution system or waste water through a force main; booster pumps are used to increase pressure in a water distribution system; recirculation and transfer pumps are used to move water being processed within a treatment plant; well pumps are used to lift water from shallow or deep wells for water supply; still other types are used for chemical feeding, sampling, and fire fighting. Centrifugal pumps are commonly used for low and high service to lift and transport water, reciprocating positive-displacement and screw feed pumps are used to move sludges, vertical turbine pumps are used for well pumping, and pneumatic ejectors are used for small waste-water lift stations. Air-lift, peristaltic, rotary displacement, and turbine pumps are used in special applications. The discussion in this section, however, is restricted to the characteristics of centrifugal pumps; specific types are illustrated and discussed in relationship to their applications in other sections of the book.

Centrifugal pumps are popular because of their simplicity, compactness, low cost, and ability to operate under a wide variety of conditions. The two essential parts are a rotating member with vanes, the impeller, and a surrounding case (Figure 4-21). The impeller driven at a high speed throws water into the volute which channels it through the nozzle to the discharge piping. This action depends partly on centrifugal force, hence, the particular name given to the pump. The function of the pump passages is to develop water pressure by efficient conversion of kinetic energy. A closed impeller is generally used in pumping water for higher efficiency, while an open unit is used for waste water containing solids. The casing may be in the form of a volute (spiral) or may be equipped with diffuser vanes. Several other variations in design involve casing and impeller modifications to provide a wide range of pumps with special operating features.

The head and efficiency characteristics of a centrifugal pump are sketched in Figure 4-22. With the valve in the discharge piping closed, the rotating impeller simply churns in the water, causing the pressure at the outlet to rise to a value referred to as the shutoff head. If the valve is then gradually opened allowing

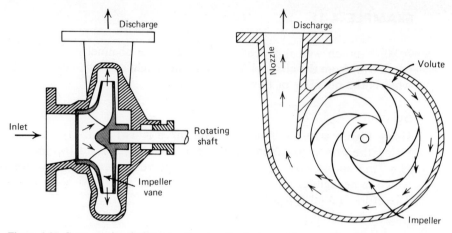

Figure 4-21 Cross-sectional diagrams showing the features of a centrifugal pump.

increasing flow of water, the pump head decreases. With increasing rate of discharge, pump efficiency rises to an optimum value and then slowly falls. The flow rate at peak efficiency is determined by the pump design, and the rate of rotation of the impeller. When a centrifugal pump lifts water from a reservoir into a piping system, the resistance to flow at various rates of discharge is described by a system head curve (Figure 4-22). The two components of discharge resistance are the static head, which is the vertical distance between the water surfaces in the suction and discharge tanks, and friction head loss that increases with pumping rate. A centrifugal pump connected to a piping system operates at a discharge rate defined by the intersection of the pump head and system head curves. This point is predetermined in the design of a pumping system so that the pump selected will operate near optimum efficiency at this point.

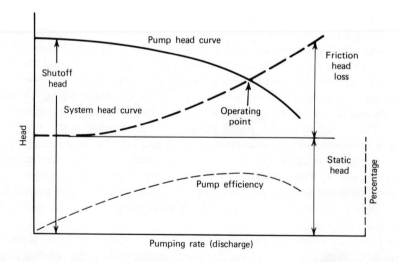

Figure 4-22 Performance curves for a centrifugal pump sketched with a system head curve to illustrate pump operating point.

4-9   AMOUNT OF STORM RUNOFF

The rational method for calculating quantity of runoff for storm sewer design is defined by the relationship

$$Q = CIA \tag{4-16}$$

where Q = maximum rate of runoff, cubic feet per second
   C = coefficient of runoff based on type and character of surface, Table 4-7
   I = average rainfall intensity, for the period of maximum rainfall of a given frequency of occurrence having a duration equal to the time required for the entire drainage area to contribute flow, inches per hour
   A = drainage area, acres

Rainfall may be intercepted by vegetation, retained in surface depressions and evaporate, infiltrate into the soil, or drain away over the surface. The coefficient of runoff is that fraction of rainfall that contributes to surface runoff from a particular drainage area. The coefficients given in Table 4-7 show that the majority of rain falling on paved and built-up areas runs off, while open spaces with grassed surfaces retain the bulk of rain water. The size of the drainage area is determined by field survey, or measurement from a map.

The most complex parameter in the rational formula is the rainfall intensity. The U.S. Weather Bureau maintains recording gauges that automatically chart rainfall rates with respect to time. These data can be compiled and organized statistically into intensity-duration curves like those shown in Figure 4-23. The following illustrates how to read this diagram: for storms with a duration of 30 min, the maximum average rainfall anticipated once every 5 years is 2.9 in./hr, and the 25-year frequency is 3.9 in./hr. Although curves are a popular method of presenting rainfall information, precipitation formulas have been developed for various parts of the United States. In design, 5-year storm frequency is used for residential areas, 10-year frequency for business sections, and 15-year for high-value districts where flooding would result in considerable property damage. The duration of rainfall used to enter Figure 4-23 depends on the time of concentration of the watershed. It is the time required for the maximum runoff rate to develop during a continuous uniform rain or, in other words,

Table 4-7. Coefficients of Runoff for the Rational Method Eq. 4-16 for Various Areas and Types of Surfaces

Description	Coefficient
Business areas depending on density	0.70 to 0.95
Apartment-dwelling areas	0.50 to 0.70
Single-family areas	0.30 to 0.50
Parks, cemeteries, playgrounds	0.10 to 0.25
Paved streets	0.80 to 0.90
Watertight roofs	0.70 to 0.95
Lawns depending on surface slope and character of subsoil	0.10 to 0.25

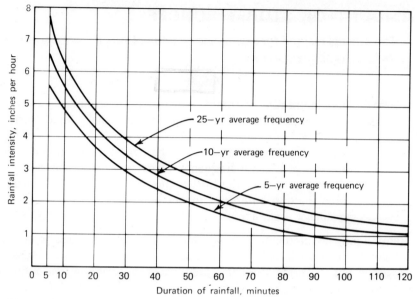

Figure 4-23 Typical rainfall intensity-duration curves used in the rational method for calculating quantity of storm water runoff.

duration of rainfall required for the entire watershed to be contributing runoff. If a portion of the area being considered drains into an inlet and through a storm sewer, the time of concentration equals the inlet time plus the time of flow through the pipe. Inlet times generally range from 5 to 20 min based on the drainage area-amount of lawn area, slope of street gutters, and spacing of street inlets. Example 4-12 illustrates application of the rational formula.

EXAMPLE 4-12

Compute the diameter of the outfall sewer required to drain storm water from the watershed described in Figure 4-24, which gives the lengths of lines, drainage areas, and inlet times. Assume the following: a rainfall coefficient of 0.30 for the entire area, the 5-year frequency curve from Figure 4-23, and a flowing full velocity of 2.5 ft/sec in the sewers.

Solution
Flow time in sewer from manhole 1 to manhole 2

$$= \frac{400 \text{ ft}}{2.5 \text{ ft/sec} \times 60 \text{ sec/min}} = 2.7 \text{ min}$$

From manhold 2 to manhole 3

$$= \frac{600}{2.5 \times 60} = 4.0 \text{ min}$$

Times of concentration from remote points of the three separate areas to manhole 3 are: $5.0 + 2.7 + 4.0 = 11.7$ min for Area 1; $5.0 + 4.0 = 9.0$ min for Area 2; and, 8.0 min (inlet time only) for Area 3. Entering Figure 4-23 with the

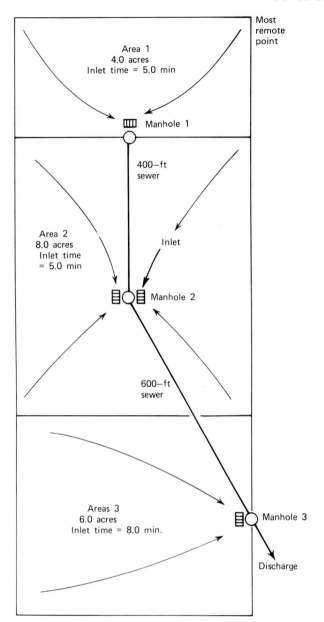

Figure 4-24 Watershed for Example 4-12, illustrating storm-runoff calculations using the rational method.

maximum time of concentration (duration of rainfall) for the watershed of 11.7 min, the rainfall intensity I is 4.6 in./hr for a 5-year frequency.

The sum of the CA values

$$= 0.3 \times 4.0 + 0.3 \times 8.0 + 0.3 \times 6.0 = 5.4$$

Substituting into Eq. 4-16,

$$Q = 4.6 \times 5.4 = 25 \text{ cu ft/sec (acre-inches per hour)}$$

For $Q = 25$ cu ft/sec and $V = 2.5$ ft/sec from Figure 4-12, the required sewer diameter is 42 in. set on a slope of 0.055 ft/100 ft.

4-10　FLOW IN STREAMS AND RIVERS

Permanent gaging stations have been located on streams and rivers throughout the United States by several governmental agencies, including the Corps of Engineers, Bureau of Reclamation, Soil Conservation Service, and Geological Survey. A typical recording gage installation is a stilling well, connected to the river by an intake pipe, with a float-operated water level recorder housed in a shelter above the well. River stage is continuously scribed with a pen on a roll of paper driven by a clock mechanism. In some instances a control weir is constructed across a stream to provide sensitivity at low flows but, more frequently, a rating curve is developed for the natural river channel. Data for a stage-discharge curve are gathered by using current meter measurements for various river discharges. Water levels during the times of survey are plotted against the measured flows. Rating curves must be reexamined periodically to account for shifts that occur as a result of channel filling or scouring. Continuous stage readings recorded on the chart paper are translated into

Table 4-8. Streamflow Records Listing the Lowest Mean Discharge for Seven Consecutive Days for Each Year from 1941 to 1962. The Average Annual Discharge for This Period Was 178 cu ft/sec

Year	Lowest Mean Flow in Cubic Feet per Second for 7 Consecutive Days
1941	19.6
1942	28.6
1943	18.1
1944	34.3
1945	29.3
1946	35.7
1947	35.0
1948	27.0
1949	35.0
1950	36.9
1951	90.3
1952	50.6
1953	35.3
1954	59.4
1955	26.3
1956	30.1
1957	29.4
1958	29.7
1959	30.4
1960	49.6
1961	36.6
1962	59.1

average daily flow by using the rating curve; these are tabulated and published, usually under the title of surface water records, and are distributed to libraries and interested governmental agencies.

Streamflows are of concern in water pollution, since treated waste-water effluents are often disposed of by dilution in rivers and streams; of greatest interest are low flows when the least dilution capacity is provided. The stream classification system of water quality standards establishes maximum allowable concentration of pollutants; for example, ammonia nitrogen in warm-water streams is set at a maximum allowable level of 3.5 mg/l. These standards must be keyed to a specific quantity of flow to determine the amount of pollutant that can be discharged to the watercourse without exceeding the specified concentration. The one-in-ten-year seven-consecutive-day low flow is the value that is commonly adopted. Of course, if the stream has intermittent flow, the critical condition is during the dry period.

The following procedure is used to develop a frequency curve of seven-consecutive-day low flows for determining the one-in-ten-year value. All of the daily discharge records for gaging stations on the stream being evaluated are assembled. By scanning the data for each year, the seven consecutive days of lowest flows are identified and averaged arithmatically. These values are listed by year as shown in Table 4-8. The following statistical method, illustrated in Table 4-9 and Figure 4-25, is used to calculate and plot the frequency curve from the yearly low flow values.

Table 4-9. Streamflow Data from Table 4-8 Organized for Statistical Evaluation as Shown in Figure 4-25

Minimum Flows in Order of Severity (cu ft/sec)	Serial Number, m	$\dfrac{m}{n+1}$
90.3	1	$\dfrac{1}{23} = 0.0435$
59.4	2	$\dfrac{2}{23} = 0.087$
59.1	3	0.130
50.6	4	0.174
49.6	5	0.217
36.9	6	0.261
36.6	7	0.305
35.7	8	0.347
35.3	9	0.390
35.0	10	0.435
35.0	11	0.477
34.3	12	0.520
30.4	13	0.565
30.1	14	0.605
29.7	15	0.650
29.4	16	0.695
29.3	17	0.740
28.6	18	0.782
27.0	19	0.825
26.3	20	0.870
19.6	21	0.912
18.1	$22 = n$	0.955

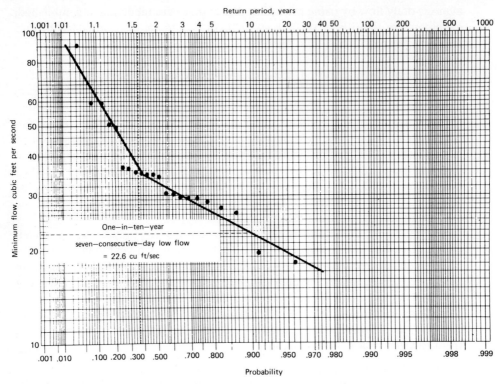

Figure 4-25 Plot of yearly low flows from Table 4-9 on logarithmic probability paper to determine the one-in-ten-year seven-consecutive-day low flow.

1. Arrange the minimum annual flows from historical records in order of severity, that is, from highest to lowest flow.
2. Assign a serial number m to each of the n values: $1, 2, 3, \ldots, n$.
3. Compute the probability plotting position for each serial value as m divided by $n + 1$.
4. Plot flow on the vertical logarithmic scale, and the corresponding probability value along the horizontal axis.
5. Draw the best fit line through the plotted data.

The frequency curve in Figure 4-25 is read by entering the diagram from either the top or bottom, return period or probability, and reading the corresponding low flow on the vertical scale. For a return period of 10 years, or probability of 90 percent, the minimum flow is 22.6 cu ft/sec. In other words, during one 7-day period every 10 years, the lowest average flow for those 7 days is expected to be 22.6 cu ft/sec, no other 7-day period having a lower flow. Ninety percent of the flows during 7-day intervals are expected to be greater than this value. Logarithmic-probability paper is used since the hydrologic extremes, floods and droughts, are skewed and do not follow a normal symmetrical distribution. The break in the line drawn in Figure 4-25 indicates that drought flows did not occur during the 5 years when the low flows were greater than 40 cu ft/sec. The importance of understanding the concept of low flows for waste-water dilution is pointed up by the fact that the average flow in this stream throughout the period of record was 178 cu ft/sec. If the waste assimilative capacity was to be based on average flow rather than the calculated low flow, excessive pollution of the stream would occur a considerable portion of the time. On the other hand, using the lowest daily flow, in this case 18.1 cu ft/sec, leads to extreme conservation in the opinion of many authorities.

4-11   HYDROLOGY OF LAKES AND RESERVOIRS

The parameters used to define the physical characteristics of a lake are: surface area, mean depth, volume, retention time (volume divided by influent flow), color and turbidity of the water, currents, surface waves, thermodynamic relations, and stratification. All of these influence the chemistry and biology of a lake or reservoir and, hence, water quality. Thermal stratification is the most important phenomenon with regard to water supply and eutrophication. Lakes in the temperate zone, or in higher altitudes in subtropical regions, have two circulations each year, in spring and autumn. Thermal stratification is inverse in winter and direct in summer. Lakes of the warmer latitudes in which the water temperature never falls below 4° C at any depth have one circulation each year in winter, and directly stratify during the summer. For example, a lake may stratify from May through September, and may circulate continuously from October to April.

The seasons in a lake are defined graphically in Figure 4-26. In winter the densest water sinks to the bottom, and ice near 0° C covers the surface; the maximum density of fresh water is at 4° C. Shading caused by ice and snow cover inhibits photosynthesis and, if the lake is rich in organic matter, the dissolved oxygen near the bottom gradually decreases. In spring after melting of the ice, the surface waters warm to 4° C and begin to sink, while the less dense bottom waters rise. These convection currents, aided by wind, mix the lake thoroughly for several weeks while the water temperature gradually increases. This is called the spring overturn, or spring circulation. With the approach of summer, the surface waters warm more rapidly and brisk spring winds subside, such that, a lighter surface layer is formed. As the summer season progresses, resistance to mixing between the top and bottom layers of different density becomes greater and thermal stratification is established. The epilimnion, warm surface layer, is continuously mixed by wind and density currents and supports the growth of algae. The hypolimnion, cooler bottom layer, is dark and stagnant. Although the bulk of fish food is found in the epilimnion, many species find the cooler bottom water a more suitable environment. In nutrient-rich bodies of water, the hypolimnion increases in carbon dioxide content and may become devoid of dissolved oxygen after many weeks of stratification. The thermocline is the thin zone of rapid temperature drop between the water layers. The approach of autumn with shorter, cooler days causes the lake to lose heat faster than it is absorbed. When the surface waters are cooled to a higher density than the water in the hypolimnion, vertical currents lead to autumnal circulation. This mixing is supported by wind action until finally the densest water stays on the bottom and the surface freezes.

Thermal stratification in reservoirs and lakes has a direct influence on quality for water supply. In the summer, water drawn from near the surface is warm and may contain algae that cause filter clogging and taste and odor problems. Stagnant, cooler hypolimnion water may be devoid of dissolved oxygen, high in carbon dioxide, and may contain the products of anaerobic conditions, such as

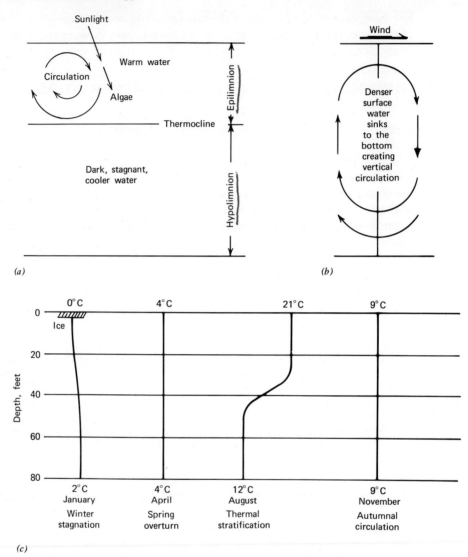

Figure 4-26 Thermal stratification and circulation of a dimictic lake in northern United States. (*a*) Thermal stratification during late summer. (*b*) Spring and autumn overturns. (*c*) Temperature profiles of a lake showing stratification and mixing.

hydrogen sulfide, odorous organic compounds, or reduced iron. Usually the region just below the thermocline provides the most satisfactory water quality during stratification. During winter stagnation, water closer to the surface is likely to be more desirable, since the quality adjacent to the bottom may be poor because of contact with decaying organic matter. The vertical variation that may occur in a body of water illustrates the importance of having a water intake tower with ports at various depths so that the water supply can be drawn from the most advantageous level in the water profile. Spring and autumn circulation mixes the water, spreading any undesirable matter throughout the entire profile. Treatment for taste and odor control may have to be intensified, particularly in the autumn when decaying algae and anaerobic bottom waters are mixed.

PROBLEMS

4-1 (a) What is the hydraulic head equivalent to 20 psi? (b) What is the static pressure equivalent to 75 ft of head? [*Answers* (a) 46.2 ft, (b) 32.5 psi]

4-2 If the pressure in a water main is 60 psi, what is the static pressure at a faucet in a building 40 ft above the main? (*Answer* 42.7 psi)

4-3 Compute the total energy in a pipeline with an elevation head of 30 ft, water pressure of 60 psi, and velocity of flow equal to 4.0 ft/sec; use Eq. 4-3. (*Answer* 170 ft)

4-4 Calculate the head loss in 2000 ft of 14-in. diameter pipe for a flow rate of 1500 gpm. The friction factor (*f*) for the pipe material is 0.025. (*Answer* 6.5 ft)

4-5 Using the nomograph in Figure 4-6, determine the head loss and velocity of flow in a 6-in. pipe carrying 350 gpm. (*Answers* 17.5 ft/1000 ft and 4.0 ft/sec)

4-6 What is the friction head loss, expressed in both psi and feet, in one mile of 12-in. diameter cast-iron pipe carrying 1500 gpm (a) for a $C = 100$, and (b) for a $C = 140$. [*Answers* (a) 46 ft, 20 psi, (b) 25 ft, 11 psi]

4-7 What size pipeline should be used to supply 1500 gpm so that the head loss does not exceed 10 ft/1000 ft assuming a $C = 100$? (*Answer* 12 in.)

4-8 At night, water is pumped from a treatment plant reservoir through distribution piping to an elevated storage tank. Using the energy equation, calculate the pump discharge pressure required to supply a flow of 1000 gpm to the tank. Assume no other withdrawals from the system. The water surface in the supply reservoir is at 10-ft elevation; the lift-pump elevation is 20 ft; the water level in the elevated tank is 140 ft. The pipe network may be considered equivalent to 5000 ft of 10-in. diameter cast-iron pipe, $C = 100$. (*Answer* 74 psi)

4-9 Review Example 4-5. What is the calculated discharge at point *B* in Figure 4-7 if water is neither flowing to, nor from, the elevated storage tank, in other words, the hydraulic gradient between points *B* and *C* is zero. (*Answer* 1500 gpm)

4-10 Draw a hydraulic gradient for the system in Figure 4-27 and compute the quantity of flow available at the load center, point *B*, for the following conditions: lift pumps provide flow at a discharge pressure of 60 psi; residual pressure at point *B* is 20 psi; and, elevated storage and equivalent

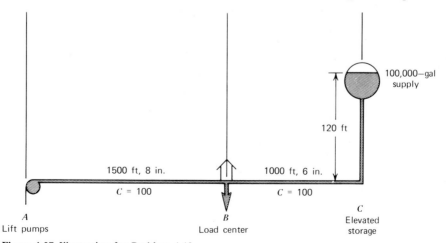

Figure 4-27 Illustration for Problem 4-10.

pipelines are as given in the illustration. Assume that the ground levels at points A, B, and C are at the same elevation. (*Answer* 2200 gpm)

4-11 Determine the diameter of a single, 4100-ft, equivalent pipe between points A and B in Figure 4-28. (*Answer* 7 in.)

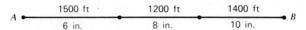

1500 ft	1200 ft	1400 ft
6 in.	8 in.	10 in.

$A \bullet \hspace{3cm} \bullet \hspace{3cm} \bullet \hspace{3cm} \bullet B$

Figure 4-28 Pipes in series for Problem 4-11.

4-12 Find an equivalent, 1200-ft, single pipe to replace the parallel pipes shown in Figure 4-29. (*Answer* 12.5 in.)

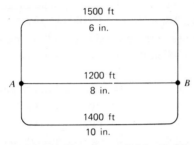

1500 ft
6 in.

1200 ft
8 in.

1400 ft
10 in.

Figure 4-29 Pipes in parallel for Problem 4-12.

4-13 Refer to Example 4-7. For the flow conditions established in Figure 4-10 calculate the required elevation of water in the storage tank at point J if the supply pressure at point A is 60 psi. (*Answer* 131 ft)

4-14 Determine the flows in the pipe network drawn in Figure 4-30 by using the

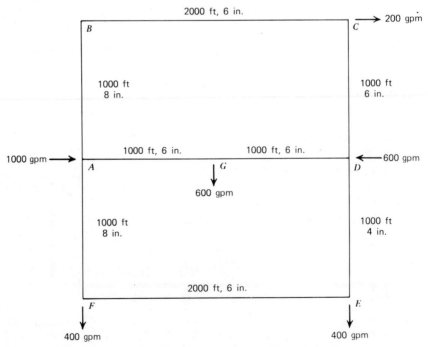

Figure 4-30 Pipe network for Problem 4-14.

Hardy Cross method. (*Answers AB* and *BC* = 62 gpm, *DC* = 138 gpm, *DE* = 150 gpm, *FE* = 250 gpm, *AF* = 650 gpm, *AG* = 288 gpm, and *DG* = 312 gpm)

4-15 Calculate the water pressure at point *I* in the pipe network of Figure 4-31

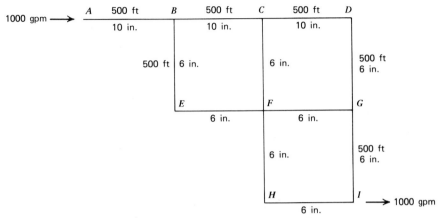

Figure 4-31 Pipe network for Problem 4-15.

if the pressure at point *A* is 80 psi. Assume that all points are at the same elevation. Use the Hardy Cross method to estimate quantities of flow in the pipes; use a *C* = 100. (*Answer* 63 psi, *AB* = 1000 gpm, *BC* = 720, *CD* = *DG* = 360, *GI* = 580, *BE* = *EF* = 280, *FH* = *HI* = 420 gpm)

4-16 Estimate the flows in the system illustrated in Figure 4-32 by using the

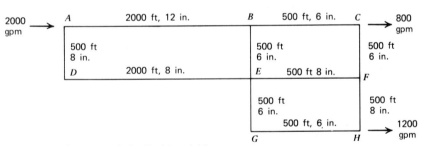

Figure 4-32 Pipe network for Problem 4-16.

Hardy Cross method. (*Answers AB* = 1315 gpm, *BE* = 730, *AD* = *DE* = 685, *BE* = 585, *EF* = 875, *FC* = 70, *EG* = *GH* = 395, *FH* = 805 gpm)

4-17 What is the flowing-full quantity and velocity of flow for a 12-in. sewer on a 0.006 ft/ft slope using (a) *n* = 0.013, and (b) *n* = 0.011? [*Answers* (a) 2.7 cu ft/sec, 3.5 ft/sec, (b) 3.2 cu ft/sec, 4.1 ft/sec]

4-18 What size sewer pipe would you recommend to convey 5.4 cu ft/sec based on the following limitations: maximum allowable velocity of 10 ft/sec and a maximum pipe slope of 3.0 ft/100 ft. (*Answer* 12 in.)

4-19 Compute the design population that can be served by an 8-in. sanitary sewer laid at a slope of 0.40 percent. Assume that the design flow per person is 0.000,62 cu ft/sec (400 gpd). (*Answer* 1200 population)

4-20 A 33-in. sewer pipe, *n* = 0.013, is placed on a slope of 0.40 ft/100 ft. (a) At what depth of flow is the velocity equal 2.0 ft/sec? (b) If the depth of flow is 18 in., what is the discharge? [*Answers* (a) 2.6 in., (b) 20 cu ft/sec]

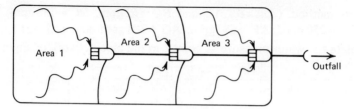

Figure 4-33 Drainage diagram for Problem 4-27.

4-21 A 60-in. diameter storm sewer conveys a flow of 55 cu ft/sec at a depth of flow equal to 40 in. What is the unused capacity of the pipe in cubic feet per second? (*Answer* 14 cu ft/sec)

4-22 Explain why a compound meter is suitable for customer services while a current meter is not.

4-23 Referring to the energy equation, Eq. 4-2, explain the operating principle of a venturi meter (Figure 4-17).

4-24 A 90° V-notch weir is placed in a 12-in. wide channel with the base of the notch 3 in. from the channel bottom. Using Figure 4-20, determine the correction factor and discharge for a height of flow equal to (a) 4.5 in., and (b) 8.0 in. measured in a stilling well behind the weir. [*Answers* (a) 1.01 and 0.14 mgd, (b) 1.05 and 0.62 mgd]

4-25 Explain the relationship between a system head curve and pump head curve (Figure 4-22).

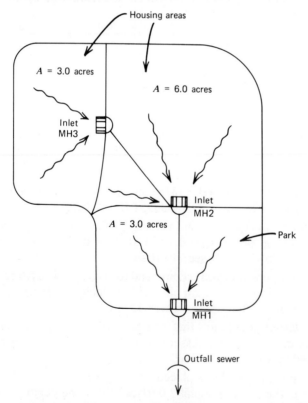

Figure 4-34 Drainage diagram for Problem 4-28.

4-26 If the gate valve on the discharge side of a centrifugal pump is closed for a short period of time, will the pump be damaged?

4-27 Given the drainage area in Figure 4-33, calculate the discharge at the outfall using the rational method. Use the five-year rainfall intensity-duration curve in Figure 4-23. Other data are: for Area 1, $C = 0.50$, area = 1.3 acres, and inlet time = 7 min; for Area 2, $C = 0.40$, area = 2.5 acres, and inlet time 5 min; for Area 3, $C = 0.70$, area = 3.9 acres, and inlet time = 5 min; sewer lines in Areas 2 and 3 are each 500 ft in length; and, the average velocity of flow in the sewers may be assumed to be 3.0 ft/sec. (*Answer* 20 cu ft/sec)

4-28 Determine the size outfall sewer below MH 1 needed to serve the 12-acre drainage area in Figure 4-34. Each area has an inlet time of 10 minutes. The coefficient of runoff for the two housing areas is 0.45 and the park is 0.15. The distance between manholes is 600 ft, and all pipes are set on a slope of 0.0020. The rainfall intensity-duration relationship is: $i = 131$ divided by $t + 19$, where i = inches per hour and t = minutes. (*Answer* 30 in.)

4-29 Determine the one-in-ten-year seven-consecutive-day low flow given the following flow data. (Two-cycle logarithmic probability paper is provided in the appendix.) (*Answer* 85 cu ft/sec)

Year	Cubic Feet per Second	Year	Cubic Feet per Second
1955	120	1963	170
1956	162	1964	82
1957	142	1965	74
1958	137	1966	110
1959	254	1967	184
1960	367	1968	121
1961	145	1969	208
1962	153	1970	145
		1971	198

4-30 Explain why lakes thermally stratify and, in the case of a dimictic lake, circulate twice a year.

chapter 5

Water Quality and Pollution

The many ways in which water promotes the economic and general well-being of society are known as beneficial uses. The major ones, shown in Figure 5-1, are water supply, recreation, and aquatic life. The relative importance of beneficial uses for any particular stream, lake, or estuary depends on the economy of the area and the desires of its people. Many applications are restricted within narrow ranges of water quality, such as public and industrial water supply. Unregulated waste disposal conflicts with use of the water as a municipal source. Therefore, control of quality is required to insure that the best employment of the water is not prevented by indiscriminate use of watercourses for disposition of wastes.

Historically, water quantity has been considered in water law. Riparian rights, derived from British common law, state that the user of water is not entitled to diminish it in quantity. Only recently have laws been developed to protect water quality for downstream users. Past preoccupation with the quantity of water, without regard to quality, has been a handicap in the development of water resources planning.

The Federal Water Pollution Control Act, as amended by the Water Quality Act of 1965, authorizes states to establish water quality standards for interstate waters. Its primary purpose is to protect water character for present and future beneficial uses. After enumerating water uses for each drainage basin, standards with a reasonable margin of safety were set to protect quality for these uses. The standards are chemical, physical, and biological degradation limits beyond which a water may not be polluted for the defined beneficial use.

5-1 TYPES AND SOURCES OF POLLUTION

All domestic, industrial, and agricultural wastes affect in some way the normal life of a river or lake. When the influence is sufficient to render the water unacceptable for its best usage, it is said to be polluted. The following is a discussion of the common types of pollutants. The principle sources of water pollution are shown in Figure 5-2.

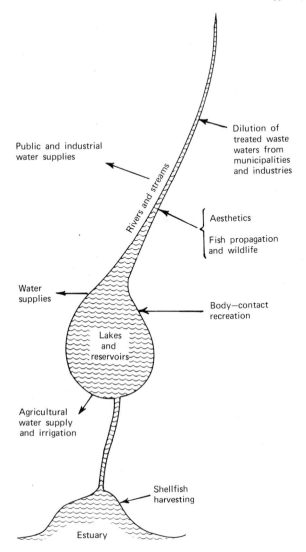

Figure 5-1 Beneficial uses of surface waters.

Oxygen-Consuming Matter

Deoxygenation may be caused by reducing agents that have an immediate oxygen demand, or by biological decomposition of waste organic matter. The latter is a relatively slow reaction, gradually depleting the dissolved oxygen in a river as the water flows downstream. Oxygen is replaced by aeration at the surface and photosynthetic activity of green plants. The maximum oxygen deficit depends on the interrelationship of biological oxygen utilization and reaeration. Fishes and most aquatic life are stiffled by a lack of oxygen, and unpleasant tastes and odors are produced if the content is sufficiently reduced. Settleable organic solids can create sludge deposits that decompose, causing regions of high oxygen demand and intensified odors. Floating solids are unsightly and obstruct passage of light vital to plant growth. Thin films of oil can also reduce the rate of reoxygenation.

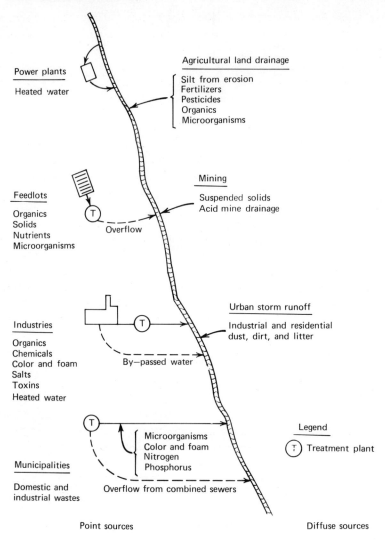

Figure 5-2 Principle sources of water pollution.

Inorganic Solids

Inert suspended solids, such as silt and mine slurries, produce turbidity that reduces light penetration and, therefore, interferes with photosynthesis. Solids that settle out of solution blanket the streambed smothering bottom organisms and hindering the reproduction cycle of fishes.

Poisons

Acids, alkalis, and toxic chemicals adversely affect aquatic life and impair recreational uses of water. Sharp deviations of pH at the point of discharge into a river or lake eliminates less tolerant animal and plant species, and has considerable influence on the toxicity of some poisons. For example, ammonia is much more toxic in alkaline water than acid because free ammonia (NH_3) is more inhibiting than the ionized form (NH_4^+). Certain chemicals are poisonous to man as well as to aquatic life, rendering the water unsuitable for domestic supplies. Heavy metals such as mercury are serious pollutants, since they form stable compounds that persist in nature and are concentrated in the aquatic food chain. The fishing industry has sustained economic losses in recent years because unacceptable levels of mercury, or other heavy metals, were discovered in fishes from contaminated waters, resulting in government condemnation of the affected catches.

Nontoxic Salts

Buildup of salts from domestic wastes and waste brines can interfere with water reuse by municipalities, industries manufacturing textiles, paper, and food products, and agriculture for irrigation water. Salts like sodium chloride and potassium sulfate pass through conventional water and waste-water treatment plants unaffected.

Inorganic phosphorus and nitrogen salts induce the growth of algae and aquatic weeds in surface waters. The majority of phosphates originate from fertilizer washed from agricultural land and phosphate builders used in synthetic detergents. The latter source contributes approximately 60 percent of the phosphorus in domestic waste, and often the majority found in industrial wastes. Ammonia nitrogen is extremely soluble and is readily transported by surface runoff from cultivated farmland. In waste-water treatment, the nitrogen in organic compounds is released as soluble inorganic nitrogen. Removal of nitrogen and phosphorus in conventional biological waste treatment is generally only 30 to 50 percent.

Unesthetic Wastes

Foam-producing matter and color, although often not harmful, lend an undesirable appearance to receiving streams; they are considered indicators of contamination. Taste- and odor-producing compounds interfere with the palatability of the water for drinking purposes. Their source may be an industrial waste discharge, or may result from blooms of algae encouraged by nutrient enrichment from waste disposal. An increase in water temperature often magnifies the offensiveness of polluted water. Discharging heated water, such as cooling water from power plants, accelerates dissolved oxygen depletion, promotes the growth of blue-green algae, intensifies tastes and odors, and may stress fish and other aquatic life.

5-2   WATER QUALITY CHANGES

The water quality changes resulting from municipal use are shown schematically in Figure 5-3. Groundwaters and surface waters drawn for community supplies have a wide range of quality. Although a few well waters may be adequate for public use without treatment, the majority of sources must be processed to improve water quality to the level required for public use. The primary function of municipal water is to carry away unwanted wastes from domestic and industrial operations. Contaminants include human excreta, food-processing scraps, and a wide variety of organic and inorganic trade wastes. Waste waters are treated to improve quality prior to disposal to surface waters. Conventional treatment, including biological secondary, is the lowest level of approved waste processing. Considerable dilution in a natural watercourse is necessary to permit indirect reuse for water supplies. Advanced

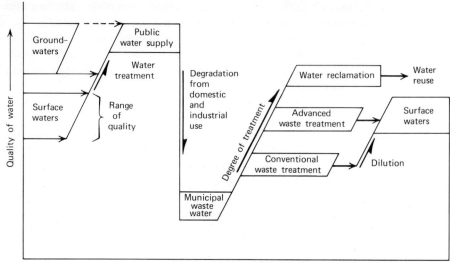

Figure 5-3 Water quality changes resulting from municipal use.

waste-treatment techniques may be applied in certain cases to remove additional organic matter, phosphates, nitrogen compounds, or other pollutants that would degrade surface waters where sufficient dilution is not available. Water reclamation is a combination of conventional and advanced treatment processes employed to return waste water to nearly original quality, thus, reclaiming the water for reuse. Although not used for public water supply, renovated water can be beneficially used for a number of agricultural and industrial applications.

A quality comparison of drinking water with a typical municipal waste water after conventional treatment reveals the following approximate increases of pollutants: 300 mg/l total solids, 200 mg/l volatile solids, 30 mg/l suspended solids, 30 mg/l BOD, 20 mg/l nitrogen, and 7 mg/l phosphorus. The following industrial contaminants are often present in wastes from industrialized cities: phenol, cadmium, chromium, lead, and manganese. Most of these can be removed by additional physical and chemical processes applied in advanced waste treatment and water reclamation. Methods available for removing inorganic nitrogen and phosphorus are both difficult and expensive, and removal of dissolved salts is rarely performed. The latter involves demineralization processes such as, electrodialysis, reverse osmosis, ion exchange, or distillation to separate the salts. Their cost is high and it is not anticipated that the reduction of dissolved solids by any of these methods will find wide application in waste-water treatment. Because of these factors the scheme in Figure 5-3 does not show any waste-water treatment process that returns the water used in a public water supply to drinking water quality.

5-3   WATER QUALITY STANDARDS

Surface Waters

The Water Pollution Control Act of 1972 established a national program to protect the quality of both interstate and intrastate waters. The general aesthetic requirements are that all surface waters should be capable of supporting aquatic life and should be free of substances attributable to waste discharges. When a river or lake is classified in accordance with intended uses, specific physical, chemical, biological, and temperature quality standards are established to ensure that the most beneficial use will not be deterred by pollution. Table 5-1 lists the six most common water use classifications; some states also list navigation and treated waste transportation. In addition to the dissolved oxygen, solids, and coliform criteria listed in Table 5-1, limits are generally specified for pH, temperature, color, tastes and odors, toxic substances, and radioactivity.

The primary purpose for establishing a limit for minimum dissolved oxygen is for protection and propagation of fishes and wildlife, although maintaining adequate oxygen also enhances recreation and reduces the possibility of odor problems that result from the decomposition of waste organics. Cold-water fishes require more stringent limitations, 6 mg/l and 7 mg/l at spawning times, while warm water species, being more tolerant, require a 4- to 5-mg/l limit. Dissolved solids are restricted since high concentrations interfere with most

Table 5-1. Beneficial Uses and Quality Standards for Surface Waters

Water Use	Dissolved Oxygen Minimum Allowable (mg/l)	Solids Allowable Dissolved (mg/l)	Other	Coliforms Maximum Allowable per 100 ml
Public water supply	4.0	500 to 750	No floating solids or settleable solids that form deposits	2000 fecal 10,000 total
Water contact recreation	4.0 to 5.0	None	Same as above	Mean of 1000 (200 fecal) with not more than 10 percent samples exceeding 2000 (400 fecal)
Fish propagation and wildlife	4.0 to 6.0 depending on warm or cold water fishes, fresh water or saltwater	None	Same as above	Mean of 5000
Industrial water supply	3.0 to 5.0 depending on use	750 to 1500	Same as above	Generally none specified
Agricultural water supply	3.0 to 5.0 based on application	750 to 1500 based on use and climate	Same as above	Generally none specified
Shellfish harvesting	4.0 to 6.0 depending on local conditions	None	Same as above	Mean of 70 with no more than 10 percent of samples exceeding 230

water uses and, furthermore, they are difficult to remove by water treatment. Coliforms, as discussed in Section 3-7, are indicators of domestic pollution. The strictest coliform standard applies to shellfish cultivation and harvesting, since these are often eaten without being cooked. The next most stringent is for water contact recreation where persons are likely to ingest water while swimming, wading, or water skiing.

The allowable pH range is normally specified as 6.5 to 8.5 for protection of fish life and to prohibit heavy alkali or acid waste discharges. Temperature standards usually recommend an upper limit of 90° F, with a maximum permissible rise above the naturally existing temperature of 5° F in streams and 3° F in lakes. Trout and salmon waters should not be warmed in order to protect these less tolerant species. In general, discharges of toxic pollutants that cause death, disease, behavioral abnormalities, mutations, or physiological malfunctions in man or other organisms are tightly controlled. Bioassays are the best method for determining safe concentrations of toxicants for aquatic organisms. Test organisms are exposed to various levels of poison concentrations for a specified time span, 96 hr or less, in a laboratory apparatus. The median tolerance limit (TLm) is the concentration that kills 50 percent of the test organisms. The maximum allowable toxicant concentration in surface waters is usually between 0.1 and 0.01 of the TLm value. This factor of safety takes into account long-term exposure, and the fact that other materials already present in the receiving water may create additional stresses on aquatic life.

In addition to classifying watercourses and specifying numerical criteria, surface water standards also provide means for controlling, monitoring, legal enforcement actions, and water resources planning. Effluent standards prescribe the degree of treatment required for particular waste-water discharges. The limit may be a specified amount of pollutant or zero discharge depending on local conditions. Effluent limitations are usually based on the best available control technology, unless more stringent controls are necessary to meet water quality standards. The common interpretation of this rule for municipal waste-water processing is a minimum of secondary treatment plus disinfection. Advanced waste treatment, particularly removal of phosphorus and ammonia nitrogen, may be deemed necessary where discharges enter lakes subject to eutrophication.

A permit system has been established to help maintain a current inventory of waste discharges. Both publicly owned utilities and industries are required to obtain permits to discharge pollutants to surface waters. The purpose is to insure that existing treatment plants comply with effluent limitations and performance standards, and that new plants are designed to provide the greatest degree of purification based on the best available technology and operating methods. The program also includes requirements for monitoring and reporting discharges, and adequate funding and staffing with qualified personnel.

Area-wide planning is essential to urban industrial areas with substantial water pollution problems. The Environmental Protection Agency obligates each state to develop regional plans for water quality management. To receive federal construction grants, publicly owned treatment plants must conform to the area-wide plan. This involves preparation of an environmental impact statement to determine if a proposed facility will have adverse effects on public health and welfare, or on environmental quality. Finally, state standards have an antidegradation clause which recognizes that the existing high quality in certain waters may be better than the assigned classification. This purity is to be maintained unless it can be demonstrated that other uses and different standards are justifiable as a result of necessary economic or social development. Therefore, all proposed new or increased sources of pollution are required to provide the necessary degree of waste treatment to maintain high water quality.

These comments on surface water quality standards constitute an overview for general understanding of actions taken by federal and state agencies to establish controls on water pollution. Personnel involved in water or wastewater systems should study in detail the standards published by their state. Design, operations, sewer ordinances, laboratory testing, and finances are all dictated in some measure by these regulations.

Drinking Water Standards

The Public Health Service first established drinking water standards in 1914 to protect the health of the traveling public and to assist in the enforcement of Interstate Quarantine Regulations. They have been changed and updated several times with the latest revision occurring in 1962. Although the standards are enforceable only on interstate carriers, they have been widely adopted and provide criteria to measure the quality of water supplies. The Safe Drinking Water Act of 1974 gave the Environmental Protection Agency the power to set maximum limits on the level of contaminants permitted in drinking water, and to enforce those standards if the states fail to do so. The bill also directed EPA to develop rules for the operation and maintenance of drinking water systems.

A summary of drinking water standards is given in Table 5-2. Monitoring of bacteriological quality requires testing of treated water at various points in the distribution system for the coliform group. (Sections 3-7 and 3-8 discuss in detail indicator organisms and laboratory testing procedures.) Bacteriological samples should be taken in known problem areas, for instance, reservoirs, dead-end mains, and remote areas of the distribution system. With the approval of the state surveillance agency, chlorine residual tests may be substituted for bacteriological sampling in other parts of a distribution system, provided that an approved minimum chlorine residual is maintained. Chlorine residual testing is better suited to operational control, since it is fast, easy, and provides a performance record signaling abnormal operation. In addition to coliform samples or alternate chlorine residual tests, a proposed standard is to limit the general bacterial population in finished drinking water to a maximum allowable limit of 500 organisms per milliliter.

The physical characteristics of turbidity, color, and odor are limited primarily for esthetic reasons. Drinking water should contain no impurity that offends the sense of sight, taste, or smell. High turbidity in filtered water is often an indication of either inadequate treatment facilities or improper operation and, therefore, the turbidity limit is one standard unit. Most water treatment plants routinely produce water with a turbidity of less than one unit. An amount as high as five units, allowed in the 1962 standards, can interfere with disinfection, maintenance of an effective residual, cause tastes and odors, result in deposits in the distribution system, and cause consumers to question the safety of their water.

Limits of chemical characteristics are presented under the headings of health

Table 5-2. Drinking Water Standards

Bacteriological Characteristics

When 10-ml portions of water are tested by the fermentation tube technique, not more than 10 percent in any month shall show the presence of the coliform group. If portions tested are 100 ml, not more than 60 percent shall be positive. When the membrane filter technique is used, the arithmetic mean coliform density of all standard samples examined per month shall not exceed 1/100 ml.

Physical Characteristics

	Approval Limit
Turbidity	1 unit
Color	15 units
Odor	3 threshold odor number

Chemical Characteristics in Milligrams per Liter

	Approval Limits	
	Esthetics	**Health**
Arsenic (As)		0.05
Barium (Ba)		1.0
Cadmium (Cd)		0.01
Chloride (Cl)	250.0	
Chromium		0.05
Copper (Cu)	1.0	
Carbon Chloroform Extract (CCE)		0.7
Cyanide (CN)		0.2
Fluoride (F)		(Table 5-3)
Iron (Fe)	0.3	
Lead (Pb)		0.05
Manganese (Mn)	0.05	
Mercury (Hg)		0.002
Methylene blue active substances	0.5	
Nitrate Nitrogen (NO_3 as N)		10.0
Selenium (Se)		0.01
Silver (Ag)		0.05
Sulfate (SO_4)	250.0	
Total dissolved solids	(no limits designated)	
Zinc (Zn)	5.0	

Pesticides in Milligrams per Liter

	Approval Limit Health
Aldrin	(pending)
DDT	(pending)
Dieldrin	(pending)
Chlordane	0.003
Endrin	0.0002
Heptachlor	0.0001
Heptachlor epoxide	0.0001
Lindane	0.004
Methoxychlor	0.1

Table 5-2. (*Continued*)

	Pesticides in Milligrams per Liter	Approval Limit Health
Toxaphene		0.005
Organophosphorus insecticides		
Azodrin		0.003
Dichlorvos		0.01
Dimethoate		0.002
Ethion		0.02
Chlorophenoxy herbicides		
2,4-D		0.1
2,4,5-T (2,4,5-TP and Silvex)		0.01

Source. Drinking Water Standards and Guidelines, Water Supply Division, Environmental Protection Agency, 1974.

or esthetics (Table 5-2). The presence of substances in excess of approval limits for human health constitutes grounds for rejection of the water supply. Arsenic, barium, cadmium, chromium, lead, mercury, selenium, and silver are poisons that affect the internal organs of the human body. In arriving at specific limits, the total environmental exposure of man to a specific toxin was considered. The lowest practical level was selected to minimize the amount of toxicant contributed by water particularly when other sources, such as milk, food, or air are known to represent the major exposure of man. The cyanide standard appears to be established for toxicity for fish rather than man. A concentration of 0.2 mg/l, providing a factor of safety of approximately 100, was set because of the rapidly fatal effect of cyanide.

The approval limits for esthetics are based on factors that render a water less desirable for use. These considerations relate to materials that impart objectional tastes and odors to water, render it economically or esthetically inferior, or are toxic to fish or plants. Methylene blue active substances in high concentrations in some detergents may exhibit undesirable taste and foaming. Chloride, sulfate, and dissolved solids have taste and laxative properties, and highly mineralized water affects the quality of coffee and tea. Both sodium sulfate and magnesium sulfate are well-known laxatives with the common names of Glauber salt and Epsom salt, respectively. The laxative effect is commonly noted by travelers or new consumers of waters high in sulfates; however, most persons become acclimated in a relatively short time. Copper is an essential nutritional element and does not constitute a health hazard. The recommended limit is established to prevent a copperous taste. Zinc is also an essential element in the human diet; however, excess amounts act as a gastrointestinal irritant. Carbon chloroform extract includes a wide variety of poorly defined organic residues. The CCE limit is expected to protect against the presence of undetected toxic organic chemicals. Although not dangerous to adults, fatal poisonings in infants can occur following injection of waters that contain high concentrations of nitrate. Most cases of infantile methemoglobinemia have resulted from private well supplies polluted with nitrogen wastes. At present there is no method of economically removing excess

Table 5-3. Recommended and Approval (Health) Limits for Fluoride in Drinking Water in Milligrams per Liter

Annual Average of Maximum Daily Air Temperatures (°F) Based on Temperature Data Obtained for a Minimum of 5 Years	Recommended Limits			Approval
	Lower	Optimum	Upper	Limit
50.0 to 53.7	0.9	1.2	1.7	2.4
53.8 to 58.3	0.8	1.1	1.5	2.2
58.4 to 63.8	0.8	1.0	1.3	2.0
63.9 to 70.6	0.7	0.9	1.2	1.8
70.7 to 79.2	0.7	0.8	1.0	1.6
79.3 to 90.5	0.6	0.7	0.8	1.4

Source. Drinking Water Standards and Guidelines, Water Supply Division, EPA, 1974.

amounts of nitrate from water. Therefore, it is important to warn the population in areas containing high nitrate content water about the potential dangers of using that water for infant feeding. Iron and manganese are objectionable because of brownish-colored stains imparted to laundry and porcelain, and bittersweet taste attributed to iron. The optimum levels of fluoride in drinking water are listed in Table 5-3. Since the amount of water ingested by persons depends on climatic conditions, optimum levels are based on average maximum daily air temperature. Lower than optimum fluoride result in excessive dental caries among the population being served, but concentrations above the approval limits produce objectionable dental fluorosis. The tentative standards for pesticides are listed in Table 5-2.

Surface Water Quality Criteria for Public Water Supplies

Drinking water standards dictate the quality of water that should be achieved in municipal water treatment without reference to desirable raw water quality. Although it is possible to renovate highly polluted surface waters to these standards, the processes required would be both complex and expensive. Even then, some pollutants such as chloride and nitrate would not be removed. Raw water quality criteria have been developed to aid in selection of water sources such that the surface supply chosen can be treated to drinking standards by commonly applied treatment processes that have been proved by demonstration, and that are within reasonable economic limits. Two sets of surface water criteria for public supplies are listed in Table 5-4. Those characteristics in concentrations under the permissible column allow production of a safe, potable water meeting the limits of the drinking water standards after treatment. Desirable criteria represent high quality water that allows treatment at less cost and greater factors of safety than is possible with waters meeting permissible criteria. All of the compounds to which the table footnote applies are refractory to the typical processes used in treatment of surface waters. Although many of these resistant substances may be altered or extracted by more extensive treatment processes, others would pass unchanged through any or all of the systems in current use.

Several words are used in Table 5-4 in lieu of specific values. "Narrative" means that the subject is discussed in the following paragraph. "Absent"

Table 5-4. Surface Water Criteria for Public Water Supplies

Constituent or Characteristic	Permissible Criteria	Desirable Criteria
Physical:		
Color (color units)	75	<10
Odor	Narrative	Virtually absent
Temperature[a]	do	Narrative
Turbidity	do	Virtually absent
Microbiological:		
Coliform organisms	10,000/100 ml	<100/100 ml
Fecal coliforms	2,000/100 ml	<20/100 ml
Inorganic chemicals:	**(mg/l)**	**(mg/l)**
Alkalinity	Narrative	Narrative
Ammonia	0.5 (as N)	<0.01
Arsenic[a]	0.05	Absent
Barium[a]	1.0	do
Boron[a]	1.0	do
Cadmium[a]	0.01	do
Chloride[a]	250	<25
Chromium,[a] hexavalent	0.05	Absent
Copper[a]	1.0	Virtually absent
Dissolved oxygen	≥4 (monthly mean) ≥3 (individual sample)	Near saturation
Fluoride[a]	Narrative	Narrative
Hardness[a]	do	do
Iron (filterable)	0.3	Virtually absent
Lead[a]	0.05	Absent
Manganese[a] (filterable)	0.05	do
Nitrates plus nitrites[a]	10 (as N)	Virtually absent
pH (range)	6.0–8.5	Narrative
Phosphorus[a]	Narrative	do
Selenium[a]	0.01	Absent
Silver[a]	0.05	do
Sulfate[a]	250	<50
Total dissolved solids[a] (filterable residue).	500	<200
Uranyl ion[a]	5	Absent
Zinc[a]	5	Virtually absent
Organic chemicals:		
Carbon chloroform extract[a] (CCE)	0.15	<0.04
Cyanide[a]	0.20	Absent
Methylene blue active substances[a]	0.5	Virtually absent
Oil and grease[a]	Virtually absent	Absent
Pesticides:		
Aldrin[a]	0.017	do
Chlordane[a]	0.003	do
DDT[a]	0.042	do
Dieldrin[a]	0.017	do
Endrin[a]	0.001	do
Heptachlor[a]	0.018	do
Heptachlor epoxide[a]	0.018	do
Lindane[a]	0.056	do

Table 5-4. (*Continued*)

Constituent or Characteristic	Permissible Criteria	Desirable Criteria
Organic chemicals: (*Contd.*)		
Methoxychlor[a]	0.035	do
Organic phosphates plus carbamates[a]	0.1	do
Toxaphene[a]	0.005	do
Herbicides:		
2,4-D plus 2,4,5-T, plus 2,4,5-TP[a]	0.1	do
Phenols[a]	0.001	do
Radioactivity:	(pc/l)	(pc/l)
Gross beta[a]	1,000	<100
Radium-226[a]	3	<1
Strontium-90[a]	10	<2

Source. Report of the Committee on Water Quality Criteria.

[a] Substances that are not significantly affected by the following treatment process: coagulation (less than about 50 mg/l alum, ferric sulfate, or copperas with alkali addition as necessary but without coagulant aids or activated carbon), sedimentation (6 hr or less), rapid sand filtration (3 gpm/sq ft or less) and disinfection with chlorine (without consideration to concentration or form of chlorine residual).

means that the constituent cannot be detected by the most sensitive analytical procedure in *Standard Methods*, and "virtually absent" implies that the substance is present in a very low concentration. The word "do" is an abbreviation for ditto, in this sense meaning, "same as above."

The effectiveness of removing odorous materials from water is highly variable depending on the nature of the source and method of treatment. For this reason, it is not feasible to specify an odor criterion. In general, raw water should be free of objectionable odor, and any present should be removable by conventional means. Surface water temperatures vary with location and climatic conditions; consequently, no fixed criteria are feasible. Turbidity in raw water must be readily removable by chemical treatment. It is desirable not to have frequently changing and varying turbidity characteristics, since such changes cause upset in water treatment plant processing. The standard for a particular water supply should be keyed to the capacity of the treatment works to remove the turbidity adequately and continuously at reasonable cost.

Alkalinity in water should be sufficient to enable chemical coagulation reactions but not so high as to cause physiological stress of consumers. The minimum alkalinity is about 30 mg/l, while the maximum amount should not exceed 400 to 500 mg/l. Fluoride ion is refractory to common water treatment processes, except perhaps lime softening; therefore, the criteria for raw water are the same as for drinking water given in Table 5-3. A singular limit for maximum hardness is not possible, since hardness is a result of natural geological formations and public acceptance of it varies. Excess phosphorus concentrations in raw water are associated with eutrophication problems and coagulation difficulties attributed to complex phosphates. Because of the complicated nature of these topics, phosphorus standards must be established for each particular water supply.

5-4 STREAM POLLUTION

A stream or river that is used for waste-water dilution depends on natural self-purification to assimilate wastes and to restore its own quality. The capacity to recover from a waste discharge is determined by the character of the river, including its climatic setting. A deep meandering river with natural pools has poor reaeration and a slow time of passage. Stream channels that are shallow and steep exhibit good reaeration and high velocities. The most critical condition for waste assimilation normally occurs in the fall of the year, resulting from low flows and high water temperature. For others, the worst circumstances may result in winter under ice cover.

Effect of Dilution

Self-purification of stable chemical wastes is almost entirely dependent on stream flow. In passage down a river, the concentration decreases with greater runoff provided by the increasing drainage area. Many chemicals are reactive and dissipate by adsorption, reaction, or biological decay. Bacteria in domestic effluents, although reduced by dilution, decline in numbers primarily because of unfavorable environmental conditions. Reduced food supply, adverse tempera-ture, and the prey-predator relationship are three of the major determinants in self-purification of microorganisms.

The mixing zone is an area unavoidably polluted for blending of a waste water in a receiving stream. Since this zone may constitute a barrier that blocks migration of fishes and other aquatic organisms, it should be kept as short as possible. Where several discharges are located close together along a river, the mixing regions should lie along the same side to allow continuous passage for aquatic organisms on the opposite side.

The concentration of a pollutant below the mixing zone resulting from a waste discharge can be calculated using Eq. 5-1.

$$C = \frac{C_1 \times Q_1 + C_2 \times Q_2}{Q_1 + Q_2} \tag{5-1}$$

where
C = concentration in combined flows
Q_1 = quantity of flow in stream
C_1 = concentration of constituent in Q_1
Q_2 = quantity of waste-water discharge
C_2 = concentration of constituent in Q_2

Organic Pollution

Many waste discharges from municipalities and industries contain organic compounds that decompose using dissolved oxygen. The rate of biological stabilization is a time-temperature function with deoxygenation increasing as the temperature rises. Oxygen is replenished in the stream water primarily by reaeration from the atmosphere. Thus, quantity of flow, time of passage down the river, water temperature, and reaeration are the four major factors that govern self-purification from organic wastes.

A stream polluted from a substantial point source of organic matter exhibits four fairly well-defined zones (Figure 5-4). The zone of degradation, im-mediately following the sewer outfall, has a progressive reduction of dissolved oxygen used up in satisfying BOD. The zone of active decomposition exhibits

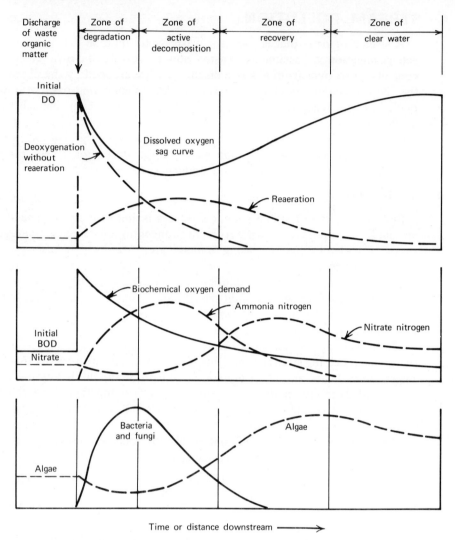

Figure 5-4 Generalized effects of organic pollution on a stream.

the characteristics of significant pollution. Dissolved oxygen is at a minimum level, and often anaerobic decomposition of bottom muds results in offensive odors. Higher forms of life, particularly fishes, find the environment of these polluted zones undesirable. Bacteria and fungi thrive on the decomposition of organics decreasing the BOD and increasing ammonia nitrogen. In the zone of recovery, reaeration exceeds the rate of deoxygenation and the level of dissolved oxygen increases slowly. Ammonia nitrogen is converted biologically to nitrate. Rotifers, crustaceans, and tolerant fish species reappear. Algae thrive on the increase in inorganic nutrients that result from the stabilization of the organic matter. The zone of clean water supports a wide variety of aquatic plants and animals, and more sensitive fishes. Dissolved oxygen returns to its original value, and the BOD has been nearly eliminated. The permanent changes in water quality prior to the waste discharge and the clear water zone include an increase in inorganic compounds, such as nitrate, phosphates, and dissolved salts. These nutrients produce and support higher algal populations if other environmental conditions of sunlight, pH, and temperature are adequate.

Oxygen Sag Equation

The rate of biochemical oxygen demand of organic matter diluted into a stream is proportional to the remaining concentration of unoxidized material. This first-order kinetics relationship is the same as the one used in developing the theoretical equations for the laboratory BOD test. A hypothetical deoxygenation curve without reaeration is shown as a dashed line in Figure 5-4.

Reaeration from the atmosphere is the primary source of oxygen input to a stream. In addition, runoff entering a stream may add oxygen along with dilution water, and algae can produce oxygen by photosynthesis. The latter is neither dependable nor always available, nor can it be expressed in mathematical terms.

The rate of oxygen absorption into water is equal to a coefficient k_2 times the dissolved oxygen deficit (saturation concentration minus the existing DO concentration). This phenomenon is shown diagrammatically in Figure 5-4 by the dashed reaeration curve; oxygen uptake is least when the DO level is high and greatest at the bottom of the sag curve. The combined effect of deoxygenation and reaeration results in a dissolved oxygen sag curve that can be defined mathematically by Eq. 5-2. By rearranging the terms in this formula, Eq. 5-3 can be developed to calculate critical time, that is, time of travel to the minimum DO level of the sag curve.

$$D = \frac{k_1 L_0}{k_2 - k_1}(10^{-k_1 t} - 10^{-k_2 t}) + D_0(10^{-k_2 t}) \qquad (5\text{-}2)$$

where D = dissolved oxygen deficit in time t, milligrams per liter
            D_0 = initial DO deficit at $t = 0$, milligrams per liter
            L_0 = initial ultimate carbonaceous BOD, milligrams per liter
            k_1 = deoxygenation rate, per day
            k_2 = reaeration rate, per day
            t = time of travel, days

$$t_c = \frac{1}{k_2 - k_1} \log \left[\frac{k_2}{k_1} \left(1 - D_0 \frac{k_2 - k_1}{k_1 L_0} \right) \right] \qquad (5\text{-}3)$$

where t_c = critical time, days, time of travel to the
              minimum DO level

Application of these formulas requires prior determination of the deoxygenation and reaeration k-rates. The value of k_1 depends on the rate of oxygen demand of the organic matter in the water. Laboratory BOD test data are generally used to compute the deoxygenation rate even though this is considered to be a questionable procedure; the laboratory environment does not reproduce stream conditions, particularly as they are related to temperature, sunlight, biological populations, and water movement. Depending on the nature of the waste and degree of treatment prior to disposal, k_1 may fall anywhere in the range between 0.05 and 0.2 per day with the most commonly selected value of 0.10 per day at 20° C. A k-rate determined in the laboratory at 20° C can be corrected for river temperature using Eq. 5-4. This equation states that the rate constant increases, or decreases, 4.7 percent for each degree of temperature rise, or fall; and doubles or halves in rate for a 15° C temperature change.

$$k_T = k_{20} \times 1.047^{T-20} \qquad (5\text{-}4)$$

where
k_T = rate constant at temperature T, per day
k_{20} = rate constant at 20° C, per day
T = temperature, degrees centigrade

Although oxygen deficit regulates the rate of reaeration, the actual amount of oxygen absorbed depends on: the physical and hydrological characteristics of the river course, temperature, water depth, surface area, and other factors. Several relationships have been developed for calculating the value of k_2 based on depth and velocity of flow. These formulas usually take on the general form of Eq. 5-5, with specific values for c, n, and m based on characteristics of the river under study. The second part shown in Eq. 5-5 includes some general numerical constants developed from available field and laboratory experimental data. The value of k_2 is also modified by temperature but to a lesser extent than k_1. One degree of temperature rise increases the reaeration coefficient about 1.5 percent; this can be expressed in the form of Eq. 5-4 by substituting 1.015 for 1.047.

$$k_2 = c\,\frac{V^n}{H^m} = 3.3\,\frac{V}{H^{1.33}} \qquad (5\text{-}5)$$

where
k_2 = reaeration rate at 20° C, per day
V = mean velocity of flow, feet per second
H = mean depth of flow, feet

Distance downstream and time of passage are related by the mean velocity of flow. For accurate analysis, the channel must be cross sectioned at regular intervals along its course to calculate volume, surface area, and effective depth. With these data the reaeration rate and travel distance can be calculated for a given quantity of flow in the stream.

There are many abnormalities in self-purification that are not taken into account in the formulation of the sag equation. These include: discharge of a waste that has an immediate oxygen demand, for example, an inorganic chemical that reacts instantaneously with DO; nitrification that results in biological uptake of oxygen; deposits of sludge in pool areas along the river channel that exert abnormal oxygen demand; and, the abnormality of biological extraction where masses of microbial growth attach to the streambed, acting in a manner analogous to a fixed media treatment unit (trickling filter). Further limitations restrict the practical usefulness of the sag equation. The mathematics assumes uniform steady-state conditions all along the river channel below a single point source of organic discharge. Therefore, additional runoff entering from tributaries and k-rate variations resulting from changes in the river channel are not taken into account. Common misuse of the sag equation results from estimating k_2 from limited measurements of channel characteristics, and applying this single value over too long a reach of river. For complex situations, the stream must be subdivided into reaches that take into account changes in time of passage and reaeration rate.

Along rivers containing large numbers of waste sources, changes in channel characteristics, and tributary streams, the simplified DO sag equation may not be able to provide satisfactory results. In that case, deoxygenation and reaeration should be considered separately so that various k_2 values can be applied over relatively short reaches of the river. This rational method involves isolating one section and accounting for oxygen inputs and withdrawals with the net oxygen asset, or deficit, shown as the increase, or decrease, in DO level in the water.

Computed dissolved oxygen profiles must be verified by field observation, regardless of the method used in analysis. Such studies reveal abnormalities or limitations of the mathematical methods applied. The ideal study entails a comprehensive sampling survey under conditions of known waste loads and stable river hydrology. Monitoring must be sufficiently extensive to describe the entire river profile. Intensive sampling over a short time span of known stable conditions is superior to grab samples taken over long periods in which hydrologic conditions or waste loads may vary. Although they are less than desired, these survey data are valuable if passage curves, channel parameters, and waste loadings can verify calculated values. After the assimilative capacity of a river has been defined by verified self-purification factors, expected stream conditions can be forecast for anticipated waste loadings under various hydrologic conditions.

EXAMPLE 5-1

A waste-water effluent of 20 cu ft/sec with a BOD = 50 mg/l, DO = 3.0 mg/l, and temperature of 23° C enters a river where the flow is 100 cu ft/sec with a BOD of 4.0 mg/l, DO = 8.2 mg/l, and temperature of 17° C. From laboratory BOD testing, k_1 of the waste is 0.10 per day at 20° C. The river downstream has an average velocity of 0.60 ft/sec and depth of 4.0 ft. Calculate the minimum dissolved oxygen level and its distance downstream by using the oxygen sag equation.

Solution

The combined flow downstream from the discharge is

$Q_1 + Q_2 = 100 + 20 = 120$ cu ft/sec

Using Eq. 5-1, the BOD, DO, and temperature after mixing of the waste water with the river are calculated as follows:

$$BOD = \frac{100 \text{ cu ft/sec} \times 4.0 \text{ mg/l} + 20 \text{ cu ft/sec} \times 50 \text{ mg/l}}{100 \text{ cu ft/sec} + 20 \text{ cu ft/sec}} = 11.7 \text{ mg/l}$$

$$DO = \frac{100 \times 8.2 \text{ mg/l} + 20 \times 3.0 \text{ mg/l}}{120} = 7.3 \text{ mg/l}$$

$$T = \frac{100 \times 17° \text{ C} + 20 \times 23° \text{ C}}{120} = 18° \text{ C}$$

The results of these computations are illustrated in Figure 5-5.

To apply the oxygen sag equations, k_1, k_2, L_0, and D_0 must be determined. Equation 5-4 is used to adjust $k_1 = 0.10$ at 20° C to 18° C.

$$k_1 \text{ at } 18° \text{ C} = 0.10 \times 1.047^{18-20} = \frac{0.10}{1.047 \times 1.047} = 0.09 \text{ per day}$$

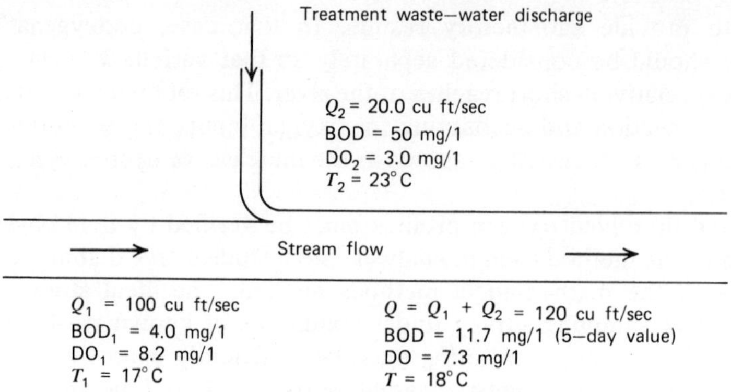

Treatment waste—water discharge

Q_2 = 20.0 cu ft/sec
BOD = 50 mg/1
DO_2 = 3.0 mg/1
T_2 = 23° C

Stream flow

Q_1 = 100 cu ft/sec
BOD_1 = 4.0 mg/1
DO_1 = 8.2 mg/1
T_1 = 17° C

$Q = Q_1 + Q_2$ = 120 cu ft/sec
BOD = 11.7 mg/1 (5—day value)
DO = 7.3 mg/1
T = 18° C

Figure 5-5 Results of flow and concentration calculations for Example 5-1.

From Eq. 5-5

$$k_2 = 3.3 \frac{0.60 \text{ ft/sec}}{(4.0 \text{ ft})^{1.33}} = 0.31 \text{ per day}$$

Correcting for 18° C

$$k_2 \text{ at } 18° C = 0.31 \times 1.015^{18-20} = \frac{0.31}{1.015 \times 1.015} = 0.30 \text{ per day}$$

Rearranging Eq. 3-14,

Ultimate carbonaceous $\text{BOD} = \dfrac{\text{5-day BOD}}{(1 - 10^{-5k})}$

Substituting in the initial 5-day BOD value of 11.7 mg/l and $k_1 = 0.10$

$$L_0 = \frac{11.7}{(1 - 10^{-0.10 \times 5})} = \frac{11.7}{(1 - 10^{-0.50})} = \frac{11.7}{0.68} = 17.1 \text{ mg/l}$$

$(10^{-0.50} = \text{antilog of } -0.50 = \text{antilog of } 0.50 - 1 = 0.32)$

DO saturation at 18° C from Table 2-5 is 9.5; therefore, the initial DO deficit is

$$D_0 = 9.5 - 7.3 = 2.2 \text{ mg/l} \quad \leftarrow$$

Time of travel to minimum DO of sag curve is determined using Eq. 5-3

$$t_c = \frac{1}{0.30 - 0.09} \log \left[\frac{0.30}{0.09} \left(1 - 2.2 \frac{0.30 - 0.09}{0.09 \times 17.1} \right) \right] = 1.8 \text{ days}$$

Distance downstream equals time of travel multiplied by velocity

$$\text{Distance} = \frac{1.8 \text{ day} \times 0.60 \text{ ft/sec} \times 86{,}400 \text{ sec/day}}{5280 \text{ ft/mile}} = 17 \text{ miles}$$

The DO deficit at critical time is calculated using Eq. 5-2

$$D = \frac{0.09 \times 17.1}{0.30 - 0.09} (10^{-0.09 \times 1.8} - 10^{-0.30 \times 1.8}) + 2.2 (10^{-0.30 \times 1.8})$$

$$= 7.3 (10^{-0.16} - 10^{-0.54}) + 2.2 (10^{-0.54}) = 3.6 \text{ mg/l}$$

$(10^{-0.16} = \text{antilog of } -0.16 = \text{antilog of } 0.84 - 1 = 0.69)$

$(10^{-0.54} = \text{antilog of } -0.54 = \text{antilog of } 0.46 - 1 = 0.29)$

Minimum DO = $9.5 - 3.6 = 5.9 \text{ mg/l}$

5-5 EUTROPHICATION

Eutrophication is the process whereby lakes become enriched with nutrients that result in water quality characteristics undesirable for man's use of the water, both for water supplies and recreation. Limnologists categorize lakes according to their biological productivity. Oligotrophic lakes are nutrient poor. Typical examples are a cold-water mountain lake and a sand-bottomed, spring-fed lake characterized by transparent water, very limited plant growth, and low fish production. A slight increase in fertility results in a mesotrophic lake with some aquatic plant growth, greenish water, and moderate production of game fish. Eutrophic lakes are nutrient rich. Plant growth, in the forms of microscopic algae and rooted aquatic weeds, produces a water quality undesirable for body-contact recreation.

The process of eutrophication is directly related to the aquatic food chain (Figure 5-6). Algae use carbon dioxide, inorganic nitrogen, orthophosphate, and trace nutrients for growth and reproduction. These plants serve as food for microscopic animals (zooplankton). Small fishes feed on zooplankton, and large fishes consume small ones. Productivity of the aquatic food chain is keyed to the availability of nitrogen and phosphorus, often in short supply in natural waters. The amount of plant growth and normal balance of the food chain are controlled by the limitation of plant nutrients. Abundant nutrients

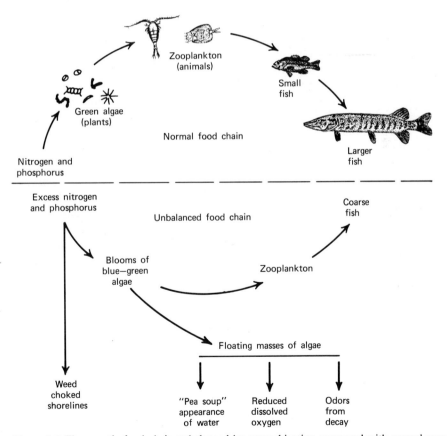

Figure 5-6 The aquatic food chain unbalanced by eutrophication compared with normal succession.

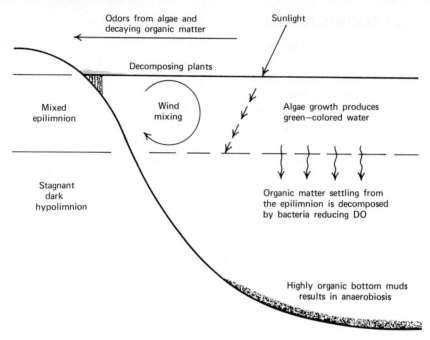

Odors from algae and
decaying organic matter

Sunlight

Decomposing plants

Mixed
epilimnion

Wind
mixing

Algae growth produces
green—colored water

Stagnant
dark
hypolimnion

Organic matter settling from
the epilimnion is decomposed
by bacteria reducing DO

Highly organic bottom muds
results in anaerobiosis

Figure 5-7 Eutrophic lake in the autumn after an extended period of thermal stratification.

unbalance the normal succession and promote blooms of blue-green algae that are not easily utilized as food by zooplankton. Thus the water becomes turbid and under extreme conditions takes on the appearance of "pea soup." Floating masses of algae are windblown to the shore where they decompose producing malodors. Decaying algae also settle to the bottom reducing dissolved oxygen. Shorelines and shallow bays become weed-choked with the prolific growth of rooted aquatics. Preferred food-fishes cannot survive in these unfavorable conditions and, as eutrophication intensifies, are replaced by a succession of increasingly tolerant fishes. Trout are succeeded by warm-water fishes such as perch, walleye, and bass, and these in turn are succeeded by coarse fishes like bullheads and carp.

The sketch in Figure 5-7 shows how overfertilization afflicts a large lake. Even a relatively mild algal bloom can result in accumulation of substantial decaying scum along the windward shoreline because of the lake's vast surface area. Gentle winds passing over the lake can pick up fishy odors from algae blooms. Algal growth developed in the sunlit epilimnion can settle into the hypolimnion, which is dark and stagnant during lake stratification. Bacterial decomposition of these cells and organic bottom muds deplete dissolved oxygen in the bottom zone. Reduced dissolved oxygen has been associated with the reduction of commercial fishing in some eutrophic lakes, and treatment of water supplies is complicated by removal of tastes and odors created by algal blooms in the epilimnion and anaerobic conditions in the hypolimnion. Perhaps the most devastating aspect of eutrophication is that the process appears to be difficult to retard, except in unusual cases. Once a lake has become eutrophic it remains so, at any rate for a very long time, even if nutrients from point sources are reduced. In part, this results from the long turnover time (detention time) which reduces the rate of flushing. Some of the lakes in the United States affected by accelerated eutrophication are: the Great Lakes particularly Lake Erie, Lake Okeechobee, the Madison, Wisconsin lakes, Lake Champlain, and Lake Washington. Some major rivers and estuaries respond to fertilization much as though they were lakes, for instance, the Potomac River and estuary, and San Francisco Bay and its tributaries.

Numerous small lakes and reservoirs used for recreation exhibit symptoms of eutrophication. The source of extra nutrients may be waste waters and, frequently, fertile land drainage. In clear-water impoundments (Figure 5-8a) fertilization leads to abundant plant growth and large populations of tolerant fish species; unfortunately, many of the desired game fish species find the environment unsuitable. Swimming and water skiing may be impaired by weed beds in shallow water. Small bodies of water that are light-limited by soil turbidity support neither heavy aquatic plant growth nor dense algal blooms (Figure 5-8b). Photosynthesis is retarded by the lack of sunlight even though adequate plant nutrients are available. Although limiting productivity may have some aesthetic advantages, the high turbidity is detrimental to reproduction of game fishes.

Macronutrients for plant growth are carbon dioxide, inorganic nitrogen, and phosphate; a variety of trace elements, such as iron, are also needed for growth. The key to controlling rate of lake eutrophication lies in limiting plant

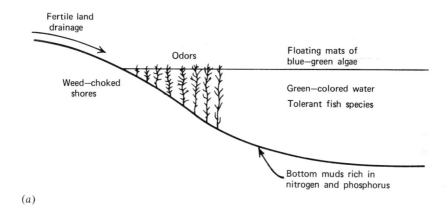

(a)

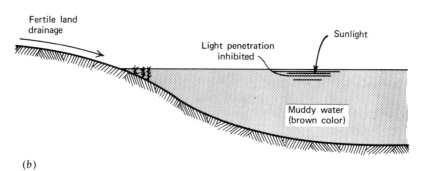

(b)

Figure 5-8 The character of small, eutrophic lakes and reservoirs is often defined by clarity of the water. (a) Fertile, clear-water ponds with low soil turbidity reflect eutrophic conditions. (b) Turbid, light-inhibiting waters prevent excessive plant growth by limiting photosynthesis.

nutrients. Natural waters contain sufficient carbon in the bicarbonate alkalinity system to provide carbon dioxide in excess of growth needs. In the majority of lakes investigated, either phosphorus or nitrogen was the limiting factor. Although not universally accepted, most authorities believe that as a lake becomes more eutrophic the growth-limiting element becomes phosphorus. This concept is reinforced by the fact that blue-green algae can fix atmospheric nitrogen. Because of this, control of fertilization by limiting the nitrogen supply is highly questionable. At present, emphasis is being placed on phosphorus reduction to control the extent of plant growth in lakes.

Only small amounts of nitrogen and phosphorus are transported to surface waters in a simple agrarian economy, principally in runoff from cultivated farmland. However in a complex urban economy, such as the United States, phosphate rock is mined and processed into fertilizers, detergents, animal feeds, and other chemicals. In addition to land runoff, animal excreta, and decaying vegetation, a significant amount of man-generated phosphate ultimately reaches surface waters. Approximately 60 percent of the phosphorus in domestic waste water is from builders used in synthetic detergents. Contributions from agricultural land drainage result from fertilizer additives used to increase crop yields. Our high standard of living has resulted in the use and disposal of large quantities of phosphorus that had previously been underground.

The amounts of inorganic nitrogen and phosphorus needed to produce abundant algae and rooted aquatic weeds are relatively small. The generally accepted upper concentration limits for lakes free of algal nuisances are 0.3 mg N/l of ammonia plus nitrate nitrogen and 0.02 mg P/l of orthophosphate at the time of spring overturn. Lakes with annual mean total nitrogen and phosphorus concentrations greater than 0.8 mg/l and 0.1 mg/l, respectively, exhibit algal blooms and nuisance weed growths during most of the growing season.

The rate of eutrophication of a lake can be retarded by reducing nutrient input. One method used to limit nutrient concentrations has been diversion of nutrient-rich waste waters. Construction of a pipeline to convey the treated waste water from the city of Madison, Wisconsin around a chain of lakes was completed in 1959. Although these lakes still receive nutrients from other uncontrolled sources, such as farmland drainage, the rate of eutrophication has been slowed with no further deterioration since the diversion. Algal blooms are less frequent and less intense because of the appreciable decrease in nitrogen and phosphorus content. Another case history in diversion of waste waters to retard eutrophication involves Lake Washington in Seattle. Between 1950 and 1958, a rapid increase in algae and a corresponding decline in water transparency observed in the lake were correlated to an increase in nitrogen and phosphorus content originating from waste-water discharges. A comprehensive metropolitan sewerage plan was adopted to intercept all waste waters entering Lake Washington and to pipe them to two major waste-water treatment plants on Puget Sound. The multimillion dollar construction work was completed between 1963 and 1968 when the last flow of effluent into the lake was stopped. Significant improvement in water quality has already occurred as a result of the flushing action of the nutrient-poor river water entering the lake; algal growth and water transparency have returned to their original, natural levels. The conventionally treated discharges do not appear to be degrading water quality in Puget Sound where tidal action provides adequate flushing.

Advanced waste treatment can be used to remove nutrients from waste water where diversion is not possible, for example, the Great Lakes. (Conventional biological treatment provides approximately 30 percent phosphorus removal and 40 to 50 percent nitrogen reduction in secondary treatment of domestic waste water. Lower removals may occur depending on nutrient concentrations in the raw waste and methods used in processing of waste sludges.) At present, emphasis is being placed on phosphorus removal, since it is considered the limiting nutrient, and methods are available to precipitate phosphates chemically in waste treatment. Nitrogen compounds are more difficult to eliminate and techniques for nitrogen removal lag behind that of phosphorus. The cost of tertiary treatment for phosphorus removal appears to be in the range of 50 to 100 percent of that of conventional secondary treatment, and processing for nitrogen removal is an even greater expense. Because of the high cost of nutrient removal, diversion should be considered and, of course, the necessity of nutrient reduction to preserve water quality should be verified. For example, if the majority of nitrogen and phosphorus entering a lake or reservoir originates from land drainage, treatment of point sources may be of limited value. Unfortunately, no ready solution exists for the removal of nutrients from land runoff, whether this be from untilled land or cultivated fields. Land management to control soil erosion and loss of fertilizers are related to local weather, for example, rainfall-runoff patterns during thunderstorms which reduce the success of soil and water conservation practices.

Although the only effective means of preventing or reversing eutrophication is nutrient control, several temporary controls have been used to reduce the nuisance effects in enriched lakes and reservoirs including: artificial mixing, harvesting of plants and algae, chemical control, and flushing. Artificial destratification by pumping cold water from the bottom and discharging it at the surface has been effective in improving quality in water supply reservoirs. Mixing adds dissolved oxygen to the hypolimnion and lowers the temperature of the epilimnion. The latter appears to shift algal populations from the less desirable blue-greens, generally associated with nuisance tastes and odors, to green algae, which are less obnoxious. This change appears to be related to cooling of the epilimnion waters producing an environment less desirable for the blue-green algae. In lakes where loss of hypolimnion dissolved oxygen is a serious problem during thermal stratification, deep water diffused aeration can be used in lieu of pumping. One common system is to lay perforated tubing on the bottom to convey and diffuse compressed air.

Power-driven underwater weed cutters with optional barge loading conveyers are available for harvesting aquatic plants. Although costly, weed cutting has been favored for clearing some boating and swimming areas rather than applying herbicides. Harvesting of plants should not be considered a feasible means of removing nutrients from eutrophic lakes. Two tons of aquatic plants by wet weight contain only about one pound of phosphorus and ten pounds of nitrogen. Nutrient removal in fish production is also limited, since fish flesh is only 0.2 percent phosphorus and about 2.5 percent nitrogen. Removal of algae has been suggested as a means for reducing nuisance species and withdrawing nutrients. However, this procedure exists in theory only, since construction of a separator or microstrainer to filter out the algae is prohibitively expensive.

Copper sulfate is commonly used for control of algae in water supply reservoirs. This algicide has several shortcomings as a eutrophication control device. Copper sulfate poisons fish when used in excessive concentrations and has been demonstrated to accumulate in bottom muds of lakes following application over a period of several years. In fertile lakes the copper sulfate must be applied at intervals throughout the growing season to ensure effective algal control; blooms must be anticipated and treated before they occur. This process is very expensive in both man-hours and chemical costs. A large number of herbicides are available to provide relief from aquatic weeds, both preemergent and emergent.

Dissolved salts and nutrients bound in algae are washed out of a lake or estuary in the water passing through the system. The rapid recovery of Lake Washington is attributed to the flushing action of relatively pure mountain runoff and rapid turnover of the lake water because of the short detention time of about five years. Artificial flushing can be performed by introducing a large flow of nutrient-poor water to a small eutrophic lake; however, rarely is such a supply of water available for dilution purposes. In the case of a reservoir with multiple ports at various depths for discharge of water, consideration should be given to releasing that water which contributes most directly to a eutrophication problem.

REFERENCES

1. *Report of the Committee on Water Quality Criteria*, Federal Water Pollution Control Administration, U.S. Department of the Interior (currently Environmental Protection Agency), Washington, D.C., 1968.
2. *Drinking Water Standards and Guidelines*, Water Supply Division, Environmental Protection Agency, 1974.
3. *Manual for Evaluating Public Drinking Water Supplies*, (initially published as PHS Publication No. 1820 in 1969), Water Supply Division, Environmental Protection Agency, Washington, D.C., 1971.
4. Velz, C. J., *Applied Stream Sanitation*, Wiley-Interscience, New York, 1970.

PROBLEMS

5-1 The chloride concentration in a brine waste flow of 100 gpm is 15,000 mg/l. The river flow upstream from the point of discharge is 20.3 cu ft/sec, and the chloride content is 10 mg/l. Calculate the concentration in the stream below the mixing zone. (*Answer* 173 mg/l)

5-2 The limit on increase in streamflow temperature is 4° F. The critical low flow in the river is 40,000 cu ft/sec when the temperature is 74° F. Calculate the maximum allowable temperature of 2000 cu ft/sec of cooling water withdrawn and discharged back to the river, without any loss in quantity. (*Answer* 154° F)

5-3 A 20-mgd domestic waste discharge has a concentration of 22 mg/l of ammonia nitrogen. If the receiving stream contains 1.5 mg/l and the maximum allowable concentration is 3.5 mg/l, compute the dilutional streamflow required in cubic feet per second. (*Answer* 290 cu ft/sec)

5-4 A reactive chemical disposed of by dilution in a river dissipates at a uniform rate of 0.10 mg/l/hr. The waste discharge is 5.0 mgd containing 20 mg/l, while the stream has zero concentration with a flow above the sewer outfall of 105 cu ft/sec. If the mean velocity in the river is 0.5 ft/sec, calculate the distance downstream that the chemical residual persists. (*Answer* 4.7 miles)

5-5 The following treated effluent is discharged to a stream: $Q = 1.0$ cu ft/sec, $DO = 2.0$ mg/l, 5-day $BOD = 40$ mg/l, $k_1 = 0.10$/day, and $T = 20°$ C. Upstream from the outfall the watercourse has the following characteristics: $Q = 9.0$ cu ft/sec, $DO = 8.0$ mg/l, 5-day $BOD = 2.0$ mg/l, and $T = 20°$ C. The stream channel has a $k_2 = 0.30$/day. Calculate the critical dissolved oxygen concentration downstream, and the distance from the outfall to this point assuming a mean velocity of 2.0 ft/sec in the river. (*Answers* 7.0 mg/l, 1.2 days, and 40 miles)

5-6 The characteristics of a river after discharge and mixing of a waste water are: $L_0 = 12.0$ mg/l, $D_0 = 1.0$, $T = 24°$ C, $k_1 = 0.15$/day at $24°$ C, and $Q = 1150$ cu ft/sec. The mean velocity in the river is 0.6 ft/sec and the average depth equals 4.0 ft. Compute the dissolved oxygen level after a time of passage equal to 2.0 days. (*Answer* 5.5 mg/l)

5-7 Solve Example 5-1 using a $k_1 = 0.05$ at $20°$ C. Assume all other parameters remain the same. (*Answer* 6.2 mg/l)

5-8 Based on field studies of a stream during low flow conditions, the 5-day BOD load to critical dissolved oxygen drop was found to be 3.6 mg/l BOD/1.0 mg/l DO. In other words, for each mg/l of BOD in the stream water, the minimum DO level drops 0.28 mg/l. Calculate the allowable BOD concentration in a waste effluent discharged to this stream under the following conditions: maximum DO deficit attributable to this discharge = 2.0 mg/l, quantity of waste water = 13.5 cu ft/sec, and streamflow above sewer outlet = 135 cu ft/sec. (*Answer* 79 mg/l)

chapter 6
Water Distribution Systems

The objectives of a municipal water system are to provide safe, potable water for domestic use, adequate quantity of water at sufficient pressure for fire protection, and industrial water for manufacturing. A typical waterworks consists of source-treatment-pumping and distribution system (Figure 6-1). Sources for municipal supplies are deep wells, shallow wells, rivers, lakes, and reservoirs. About two thirds of the water for public supplies comes from surface-water sources. Large cities generally use major rivers or lakes to meet their high demand whereas the majority of towns use well water if available. Often groundwater is of adequate quality to preclude treatment other than chlorination and fluoridation. Wells can then be located at several points within the municipality and water can be pumped directly into the distribution system. However, where extensive processing is needed, the well pumps, or low-lift pumps from the surface water intake, convey the raw water to the treatment plant site. A large reservoir of treated water (clear-well storage) provides reserves for the high demand periods and the equalizing of pumping rates. The high-lift pumps deliver treated water under high pressure through transmission mains to distribution piping and storage.

The distribution consists of a gridiron pattern of water mains to deliver water for domestic, commercial, industrial, and fire fighting purposes. Elevated storage tanks, or underground reservoirs with booster pumps, reserve water for peak periods of consumption and fire demand. A short lateral line connects each fire hydrant to a distribution main. Shutoff valves are located at strategic points throughout the piping system to provide control of any section or service outlet, including hydrants. These valves are used to isolate units requiring maintenance and to insure that main breaks affect only a small section. A service connection to a residence includes a corporation stop tapped into the water main, a service line to a shutoff valve at the curb, and the owner's line into the dwelling, which incorporates a water meter and a pressure regulator or relief valve if necessary.

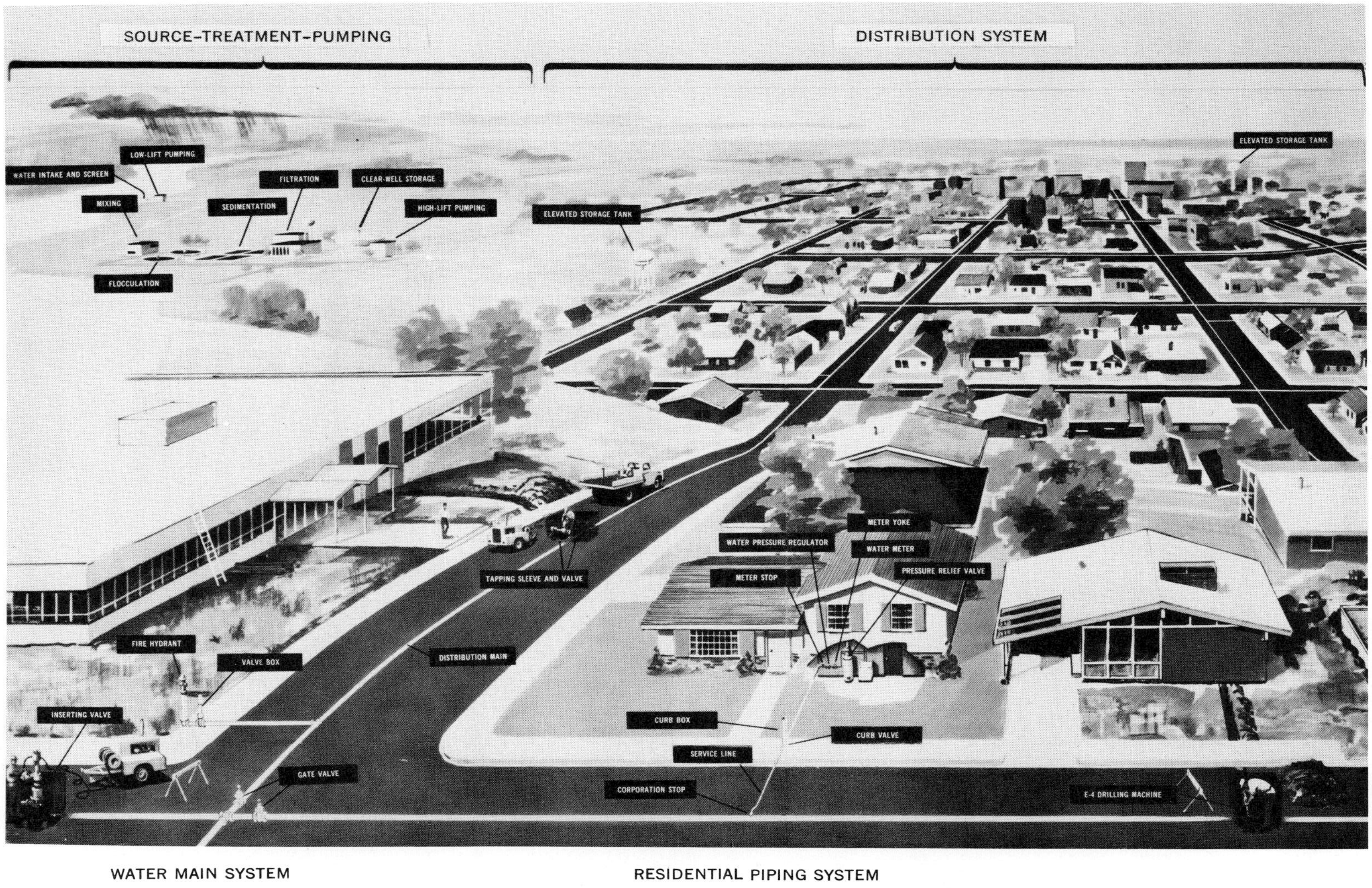

SOURCE–TREATMENT–PUMPING

DISTRIBUTION SYSTEM

ELEVATED STORAGE TANK

LOW-LIFT PUMPING

CLEAR WELL STORAGE

HIGH-LIFT PUMPING

FILTRATION

SEDIMENTATION

MIXING

FLOCCULATION

WATER INTAKE AND SCREEN

ELEVATED STORAGE TANK

METER YOKE

WATER METER

WATER PRESSURE REGULATOR

PRESSURE RELIEF VALVE

METER STOP

CURB VALVE

CURB BOX

SERVICE LINE

CORPORATION STOP

E-4 DRILLING MACHINE

TAPPING SLEEVE AND VALVE

DISTRIBUTION MAIN

GATE VALVE

VALVE BOX

FIRE HYDRANT

INSERTING VALVE

WATER MAIN SYSTEM

RESIDENTIAL PIPING SYSTEM

Figure 6-1 Sketch of a typical waterworks system that includes source, treatment, pumping, storage, and distribution. (Reprinted by permission of Mueller Co.)

6-1 WATER QUANTITY REQUIREMENTS

The amount of water required by a municipality depends on industrial use, climate, and economic considerations. Although industries in the rural countryside frequently maintain private water systems, major plants in urban areas rely on the municipal waterworks. Approximately two thirds of the water withdrawn in the United States is for nonconsumptive industrial uses; more than 90 percent is cooling water returned to the source. Automobile manufacturing, which requires roughly 25,000 gal to make one automobile, derives approximately 60 percent of process water from municipal supplies.

Municipal water use in the United States averages 600 gpd per metered service including residential, commercial, and industrial customers. For residential customers, water consumption in eastern and southern areas is 210 gpd and in central states 280 gpd, while western regions use 460 gpd per household service. Only a small amount of water is sprinkled on lawns where the rainfall exceeds 40 in./yr, while in semiarid climates lawns and gardens are maintained by irrigation. A typical city dweller uses about 90 gpd for personal use.

Lawn sprinkling may have a striking influence on water demand in areas with large residential lots—50 to 75 percent of the total daily volume may be attributed to landscape irrigation. Other climatological factors, for instance, water-chilled air conditioning and swimming pools influence water depletion.

Although water rates increased approximately 30 percent during the past decade, residential water use has continued to rise approximately 1 percent/yr. Most new houses have more water fixtures, modern appliances, spacious lawns, and other conveniences that consume larger volumes of water. A flat-rate residential water charge using a fixed fee per dwelling, rather than a rate based on water consumed, results in wasting of water by residents. For example, water may be allowed to run from a faucet continuously to have an instant cold supply. Metering of individual dwellings and establishing water rates on quantity of flow results in decreased waste and, consequently, in reduced water consumption. Municipalities that are entirely metered use approximately 60 percent of the amount that would be consumed based on flat-rate revenues. Increasing water rates have, in some instances, resulted in a decreasing water use by industry, particularly for cooling. Sewer use fees established in many cities bill industries for the quantity of waste water discharged in municipal sewers. This has resulted in reduced waste-water production and, hence, in reduced water consumption by the industry.

Residential water use varies seasonally, daily, and hourly. Typical daily winter consumption is about 80 percent of the annual daily average, while summer use is 30 percent greater. Variations, from these commonly quoted values, for a particular community may be significantly greater depending on seasonal weather changes. Maximum daily demand can be considered to be 180 percent of the average daily, with values ranging from about 120 to more than 400 percent. Maximum hourly figures have been observed to range from about 1.5 to more than 10 times the average flow in extreme cases; a mean for the maximum hourly rate is 300 percent. Table 6-1 summarizes variations in residential water consumption for domestic and public uses only.

Water flows used in waterworks design depend on the magnitude and variations in municipal water consumption and the reserve needed for fire

Table 6-1. Variations in Residential Water Consumption in Gallons per Capita per Day

	Range	Average
Yearly average consumption	100 to 130	110
Mean winter consumption	50 to 130	100
Mean summer consumption	130 to 260	170
Maximum daily use	160 to 520+	230
Maximum hourly use	200 to 1300+	390

fighting. Quantities of water required for fire demand, as detailed in Section 6-2, are of significant magnitude and frequently govern design of distribution piping, pumping, and storage facilities. Water intakes, wells, treatment plant, pumping, and transmission lines are sized for peak demand, normally maximum daily use where hourly variations are handled by storage. Standby units in the source-treatment-pumping system may be installed for emergency use, convenience of maintenance, or to serve as capacity for future expansion. The required design flow of maximum daily consumption plus fire flow frequently determines the size of distribution mains and results in additional pumping capacity and a need for storage reserves, in addition to that required to equalize pumping rates. If the maximum hourly consumption exceeds the maximum daily plus fire fighting demand it may be the controlling criterion in sizing some units.

6-2 MUNICIPAL FIRE PROTECTION REQUIREMENTS

The Insurance Services Office, Municipal Survey Service,[1,2] has developed a standard schedule for the grading of municipalities with regard to their fire defenses and physical conditions. Fire defenses are weighted for evaluation on the basis of 39 percent for water supply, 39 percent for fire department, 13 percent for fire safety control, and 9 percent for fire service communications. In the evaluation of a municipality, deficiency points are assigned for deviations from the criteria published by the Insurance Services Office.[1] Reliability and adequacy of the following major water supply items are considered in the schedule: water supply source, pumping capacity, power supply, water supply mains, distribution mains, spacing of valves, and location of fire hydrants. These are all essential components for fire fighting facilities of a municipality.

Required Fire Flow

This is the rate of flow needed for fire fighting purposes to confine a major fire to the buildings within a block or other group complex. Determination of this flow depends on size, construction, occupancy, and exposure of buildings within and surrounding the block or group complex. The required fire flow is computed at appropriate locations in each section of the city. The minimum amount is 500 gpm, and the maximum for a single fire is 12,000 gpm. Where local conditions indicate that consideration must be given to simultaneous fires, an additional 2000 to 8000 gpm is required. A municipality will have domestic and commercial water demands at the time fires occur; therefore, an adequate system must be able to deliver the required fire flow for the specified duration with municipal consumption at the maximum daily rate. The maximum daily consumption is defined by the Insurance Services Office as the greatest total amount of water used during any 24-hr period in the past three years. This maximum daily rate, expressed in gallons per minute, is the mean usage during the day of peak delivery. In cases where actual use figures are not available, the maximum consumption is estimated on the basis of use in other cities of similar character and climate. Such estimates are to be, at least, 50 percent greater than the average daily consumption, which is defined as the mean daily usage during a one-year period.

An estimate of fire flow required for a given fire area is calculated by the formula

$$F = 18C(A)^{0.5} \tag{6-1}$$

where F = required fire flow, gallons per minute (answer is rounded off to the nearest 250 gpm)

   C = coefficient related to type of construction: 1.5 for wood frame, 1.0 for ordinary construction, 0.8 for noncombustible, and 0.6 for fire-resistive construction.

   A = total floor area including all stories in the building, but excluding basements, square feet. For fire-resistive buildings, the six largest successive floor areas are used if the vertical openings are unprotected; but where the vertical openings are properly protected, only the three largest successive floor areas are included.

Regardless of the calculated value; the fire flow shall not exceed 8000 gpm for wood-frame or ordinary construction, or 6000 gpm for noncombustible or fire-resistive buildings; except that for a normal one-story building of any type it may not exceed 6000 gpm. The fire flow shall not be less than 500 gpm. For groupings of single-family and small two-family dwellings not exceeding two stories in height, the fire flows in Table 6-2 may be used.

Table 6-2. Required Fire Flows for Single-Family and Two-Family Residential Areas Not Exceeding Two Stories in Height

Distance Between Dwelling Units (ft)	Required Fire Flows[a] (gpm)
Over 100	500
31 to 100	750 to 1000
11 to 30	1000 to 1500
10 or less	1500 to 2000
Continuous buildings	2500

Source: Guide for Determination of Required Fire Flow, Insurance Services Office, December, 1974.

[a] Where wood shingle roofs could contribute to spreading fires add 500 gpm.

The value obtained by Eq. 6-1 may be reduced up to 25 percent for occupancies having a light fire loading, or it may be increased up to 25 percent for high fire loading. Light fire loadings are occupancies of low hazard, such as all forms of housing, churches, hospitals, schools, offices, museums, and other public buildings. However, after credit is applied, the fire flow cannot be less than 500 gpm. High fire hazard loadings encompass all commercial and industrial activities that involve processing, mixing, storage, or dispensing of flammable and combustible materials. Chemical works, explosives, oil refineries, paint shops, and solvent extracting are examples.

Additional adjustments may be applicable to the fire flow from Eq. 6-1 as modified for occupancy. Completely automatic sprinkler protection may reduce the required flow up to 75 percent, but structures within 150 ft of the fire area increase the required fire flow. The magnitude of increase for separation depends on the open distance, number of sides exposed, type of construction, occupancy, and other factors. The charge for one exposed side generally does not exceed a maximum of 25 percent, and the total penalty for all sides shall not exceed 75 percent. After these final corrections, the fire flow shall not exceed 12,000 gpm, or be less than 500 gpm.

The *Guide*[2] was prepared for use by municipal survey and grading personnel of the Insurance Services Office and other fire insurance rating organizations. Although it is available to others as an aid in estimating fire flow requirements, considerable knowledge and experience in fire protection engineering are necessary for detailed application of the guidelines. For example, judgment is used for businesses and industries not specifically mentioned in the *Guide*. A thorough understanding of fire fighting operations is essential when considering the influences of accessibility and configuration of buildings.

Duration

The required duration for fire flow is given in Table 6-3. Major components of a water system, on which the reliability of fire flow depends—such as, pumping capacity, power source, supply mains, and treatment works—must have the ability to deliver the maximum daily consumption rate for several days plus the required fire flow for the number of hours specified at anytime during this interval. The period may be 5, 3, or 2 days depending on the system component under consideration and the anticipated out-of-service time needed for maintenance and repair work.

Pressure

The pressure in a distribution system must be high enough to permit pumpers of the fire department to obtain adequate flows from hydrants. In general, a minimum residual water pressure of 20 psi is required during flow to overcome friction loss in the hydrant and suction hose. Higher pressure is needed where pumpers are not used; a residual pressure of not less than 75 psi permits effective use of streams direct from hydrants that are spaced close enough to allow short hose lines. Sustained high pressures are of value in permitting direct supply to automatic sprinkler systems, and building standpipe and hose systems.

Water Supply Capacity

In evaluating a system, the ability to maintain the maximum daily consumption rate plus fire flow in the municipality, at minimum pressure, is considered

Table 6-3. Required Duration for Fire Flow

Required Fire Flow (gpm)	Required Duration (hr)
10,000 and greater	10
9500	9
9000	9
8500	8
8000	8
7500	7
7000	7
6500	6
6000	6
5500	5
5000	5
4500	4
4000	4
3500	3
3000	3
2500 or less	2

Source: Grading Schedule for Municipal Fire Protection, Insurance Services Office, 1974.

with one or two pumps out of service. To have no insurance grading deficiency, the capacity remaining with the two most important pumps out of service, in conjunction with storage, must provide this flow for the specified duration any time during a five-day maximum consumption period. Some deficiency is charged against a system that can meet the requirement with only one inoperative pump. Where the capacity remaining, alone or with storage, does not equal the maximum daily use rate, only the amount that is available at required pressure may be considered.

Storage is frequently used to equalize pumping rates into the distribution system as well as to provide water for fire fighting. Since the volume of stored water fluctuates, only the normal minimum daily amount maintained is considered available for fire fighting. In determining the fire flow from storage, it is necessary to calculate the rate of delivery during a specified period. Even though the amount available in storage may be great, the flow to a hydrant cannot exceed the carrying capacity of the mains, and the residual pressure at the point of use cannot be less than 20 psi.

Although a gravity system, that is, delivering water without the use of pumps, is desirable from a fire protection standpoint because of reliability, well-designed and properly safeguarded pumping systems can be developed to such a high degree that no distinction is made between the reliability of gravity-feed and pump-feed systems by the Insurance Services Office. Where electrical power is used, the supply should be so arranged that a failure in any power line or repair of a transformer, or other power device, does not prevent delivery of required fire flow. Underground power lines laid directly from a substation of the power utility to the water plant and pumping stations reduce the probability of power failure caused by adverse weather conditions.

Distribution System

Proper layout of supply mains, arteries, and secondary distribution feeders is essential for delivering required fire flows in all built-up parts of the municipality with consumption at the maximum daily rate. Consideration must be given to the greatest effect that a break, joint separation, or other main failure could have on the supply of water to a system. With the most serious failure, no deficiency is considered if the remaining mains from the source of supply and storage can provide the fire flow for the specified duration, during a period of 3 days with consumption at the maximum daily rate.

Supply mains, arteries, and secondary feeders should extend throughout the system properly spaced—about every 3000 ft—and looped for mutual support and reliability of service. The gridiron pattern of small distribution mains supplying residential districts should consist of mains at least 6 in. in diameter. Where long lengths are necessary, exceeding about 600 ft, 8-in. or larger intersecting mains should be used. In new construction 8-in. or larger pipe are used where dead ends and poor gridiron are likely to exist for a considerable time during development, or because of layout of streets and topography. Hydrants for fire protection should never be located on the dead end of 6-in. or smaller mains. In commercial districts, the minimum size main should be 8-in. with intersecting lines in each street, with 12-in. or larger mains used on principal streets and for all long lines that are not connected to other mains at intervals close enough for mutual support.

A distribution system is equipped with a sufficient number of valves located

Table 6-4. Standard Hydrant Distribution

Fire Flow Required (gpm)	Minimum Average Area per Hydrant (sq ft)
1000 or less	160,000
1500	150,000
2000	140,000
2500	130,000
3000	120,000
3500	110,000
4000	100,000
4500	95,000
5000	90,000
5500	85,000
6000	80,000
6500	75,000
7000	70,000
7500	65,000
8000	60,000
8500	57,500
9000	55,000
10,000	50,000
11,000	45,000
12,000	40,000

Source: Grading Schedule for Municipal Fire Protection, Insurance Services Office, 1974.

so that a pipeline break does not affect more than $\frac{1}{4}$ mile of arterial mains, 500 ft of mains in commercial districts, or 800 ft of mains in other districts.

Spacing of fire hydrants is based on required fire flow with the average area served not exceeding that given in Table 6-4. Hydrants should have, at least, two outlets; one must be a pumper outlet and others must be, at least, $2\frac{1}{2}$-in. nominal size. The street connection, not less than 6-in. in diameter, is provided with a gate valve to allow ease of maintenance. The shutoff valve in a hydrant should be designed to remain closed if the barrel is broken.

EXAMPLE 6-1

A three-story wood-frame building with a ground floor area of 7300 sq ft is adjacent to a five-story building of ordinary construction with 9700 sq ft per floor. Determine the fire flow and duration required for each building and the complex assuming the units are connected.

Solution

Using Eq. 6-1, for the three-story building

$$F = 18 \times 1.5(3 \times 7300)^{0.5} = 4000 \text{ gpm}$$

From Table 6-3, the required duration is 4 hr. For the five-story building

$$F = 18 \times 1.0(5 \times 9700)^{0.5} = 4000 \text{ gpm; the duration is 4 hr.}$$

In considering the buildings as communicating, F is calculated by proportioning the $C(A)^{0.5}$ values where A is the total floor area equal to 70,400 sq ft ($3 \times 7300 + 5 \times 9700$). The three-story building is 31 percent of the total area, and the five-story is 69 percent.

$$F = 18[0.31 \times 1.5(70,400)^{0.5} + 0.69 \times 1.0(70,400)^{0.5}] = 5500 \text{ gpm}$$

From Table 6-3, the required duration is 5 hr.

EXAMPLE 6-2

Estimate the fire flow for a 200,000-sq ft, single-story building of ordinary construction.

Solution

$$F = 18 \times 1.0(200,000)^{0.5} = 8000 \text{ gpm}$$

However, since this is a normal single-story building, the maximum required fire flow is 6000 gpm.

6-3 WELL CONSTRUCTION

Water in the voids of underground sand or gravel beds can be tapped for municipal supply by using a drilled well, as is shown in Figure 6-2. The main components of a well are: solid casing, sealed in the upper soil profile; screen set into the aquifer, allowing water to enter the casing while precluding sand and gravel; and turbine pump suspended in the casing on a column pipe.

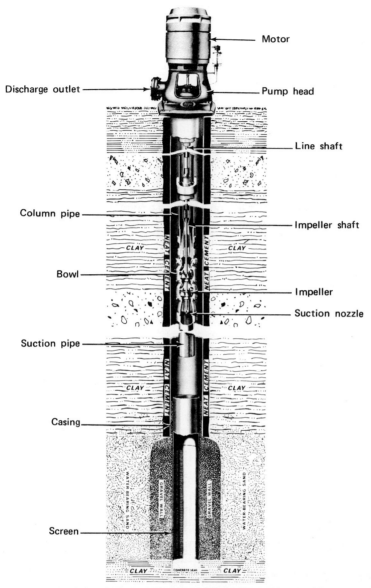

Figure labels:
- Motor
- Discharge outlet
- Pump head
- Line shaft
- Column pipe
- Impeller shaft
- CLAY
- CLAY
- Bowl
- Impeller
- Suction nozzle
- Suction pipe
- CLAY
- CLAY
- Casing
- Screen
- CLAY — CONCRETE SEAL — CLAY

Figure 6-2 Gravel-packed water well in a sand aquifer equipped with a two-stage vertical turbine pump. (Courtesy of Layne & Bowler Division, Singer Water Resources)

Groundwater flows toward the well, enters the casing through screen openings, and is lifted for discharge by the pump.

Naturally developed and artifically gravel packed wells are the two basic types of construction in water-bearing sand and gravel formations. A natural well uses a screen, the same diameter as the well casing, designed with slot openings to permit a predetermined portion of the sand to be removed during development. The finer portion of the formation is drawn through the screen and is removed, leaving the coarser fraction as a natural gravel pack. This highly permeable zone increases the effective diameter of the well and assures freedom from further sand pumping. Construction of an artificially gravel packed well involves placing an envelope of graded gravel around the well screen to provide a zone of high porosity. Grading of pack material is selected relative to the size of the water-bearing sand, so that the completed well delivers sand-free water after development. Artificial packing may be favored by economics or aquifer conditions. Fine uniform sand formations, thick aquifers that are highly stratified, and loosely cemented sandstone formations are best handled by placed gravel packs.

Cable tool percussion and rotary drilling are the two methods used to advance the hole for a well. With cable tool drilling, the pullback method is frequently the simplest and best way to set the well screen. Casing is sunk to the full depth, the screen is lowered inside, and the pipe is then pulled back a sufficient amount to expose the screen to the water-bearing layer. The pullback method of screen installation is also used in rotary drilled wells. A double-casing method can be used in constructing gravel packed wells. After sinking the outer casing to full depth, an inner pipe with well screen is gravel packed inside and the outer casing is lifted, leaving a gravel packed screen in the aquifer. When the pullback method is impractical, a screen is set in an open hole drilled below the casing after it is cemented into position. Gravel can be placed around the telescoped screen, which is then sealed to the bottom of the casing. Figure 6-3 illustrates the steps in this type of construction. Although these methods are common for installing wells, the particular techniques used in any region depend on local conditions.

Several different screens are manufactured for use in municipal wells. Three common types are wound wire (Figure 6-3), shutter (Figure 6-2), and perforated or slotted pipe. The screen openings and grain size of the aquifer media dictate the grading of the gravel pack material. The gravel must be carefully placed, by mechanical or hydraulic means. Merely dumping the material around the casing and screen results in separation of the grain sizes and in bridging, creating poor hydraulic characteristics and possible pumping of sand.

Well development is performed by vigorous hydraulic action through pumping, surging, or jetting. Mechanical surging is done by operating a plunger up and down in the casing like a piston in a cylinder. A heavy baler can be used to produce the surging action, but it is less effective. Air surging is accomplished by lowering an air pipe into the screen section and purging with compressed air. In high-velocity jetting, a tool is slowly rotated inside the screen shooting water out through the openings. Development is intended to correct any damage or clogging of the water-bearing formation that may have occurred as a side effect of drilling: to increase the porosity and permeability of the natural formations surrounding the screen, and to stabilize the aquifer so that the water is free of sand. The net effect is reduction of water drawdown in the well and higher quality water.

The common well pump is a multistage vertical-turbine pump illustrated in Figure 6-2. The four major components are: a motor drive that is mounted at ground level, or is submerged below the pump, a discharge pipe column, pump bowls with enclosed impellers, and suction pipe. The pumping units are bottom-suction centrifugal impellers mounted on a vertical shaft that carry one or more stages submerged below the water level in the well casing.

Pumping a well dewaters the surrounding ground and lowers the static water level forming a cone of depression. Under steady withdrawal, the water table is reduced until equilibrium drawdown is established (Figure 6-4). The cone of depression remains above the well screen under normal conditions of pumping.

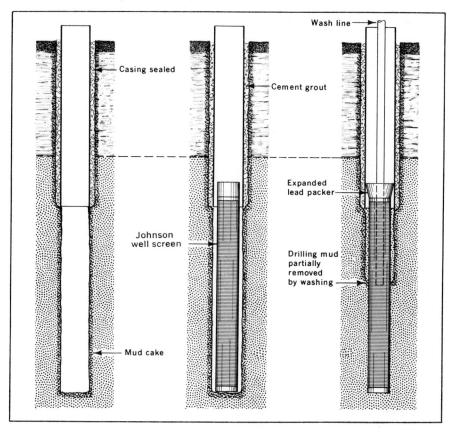

Figure 6-3 Method of placing telescoping screen: (1) after setting casing, hole is extended below by rotary drilling; (2) telescope-size screen is lowered into hole and gravel packed, if necessary; (3) space between top of screen and casing is sealed. The well is then developed to remove mud (clay suspension) used as the drilling fluid to hold the unsupported hole open during construction. (Printed by permission from Catalog 169, Johnson Division, UOP, Copyright 1969, Universal Oil Products Company)

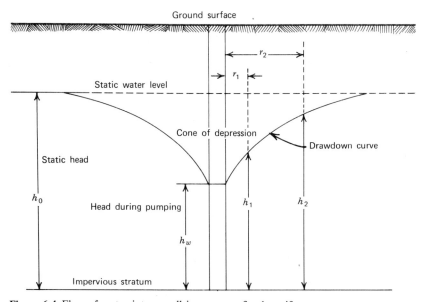

Figure 6-4 Flow of water into a well in an unconfined aquifer.

However, if the discharge exceeds the capacity of the well, the pump suction will draw air.

The factors affecting inflow to an ideal well penetrating a homogeneous aquifer, as is shown in Figure 6-4, are: static head h_0, depth of water in the well while pumping h_w, drawdown at various radii from the well, and a porosity coefficient k of the soil. The mathematical relationship for inflow Q to this ideal well is

$$Q = k \frac{\pi (h_2^2 - h_1^2)}{\log_e (r_2/r_1)} \qquad (6\text{-}2)$$

where
$\qquad Q = $ inflow, gallons per minute

$\qquad h = $ depth in observation wells, feet

$\qquad r = $ distance to observation wells, feet

Pumping tests completed in a homogeneous sand aquifer, extending from the ground surface throughout the depth of the well, have shown that Eq. 6-2 yields reasonable accuracy. In these studies the wells were pumped at a uniform rate, creating a stable cone of depression measured by the drawdown in observation holes. Aquifers that are overlain by impervious strata, artesian aquifers, and nonuniform soil profiles do not relate to this ideal hydraulic model.

6-4 SURFACE-WATER INTAKES

An intake structure is required to withdraw water from a river, lake, or reservoir. In the latter, it is often built as an integral part of the dam. Typical intakes are towers, submerged ports, and shoreline structures. Their primary functions are to supply the highest quality water from the source and to protect piping and pumps from damage or clogging as a result of wave action, ice formation, flooding, or floating and submerged debris. Towers (Figure 6-5) are common for lakes and reservoirs with fluctuating water levels, or variations of water quality with depth. Ports at several depths permit selection of the most desirable water quality any season of the year. For example, in an eutrophic lake during the summer the surface layer is warm and supports abundant

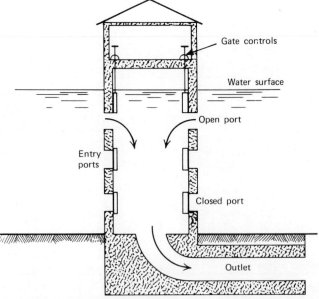

Figure 6-5 Tower water intake for a lake or reservoir supply.

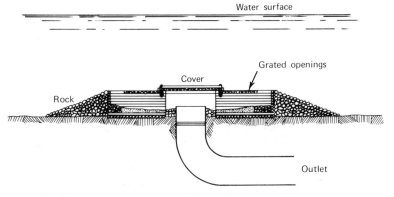

Figure 6-6 Submerged crib intake used for both river and lake sources.

growths of algae, while bottom waters can be devoid of dissolved oxygen causing foul tastes and odors. On the other hand, during the winter, water immediately under an ice cover may be of highest quality. Submerged ports also have the advantage of being free from ice and floating debris. Underflows of sediment laden water have been observed during spring runoff in man-made reservoirs, making bottom waters undesirable and, hence, a tower with upper entry ports is advantageous. Location, height, and selection of port levels must be related to characteristics of the water body. Therefore, a limnological survey should precede design to define degree and depth of stratification, water currents, sediment deposits, undesirable eutrophic conditions, and other factors.

A submerged intake consists of a rock filled crib or concrete block supporting and protecting the end of the withdrawal pipe (Figure 6-6). Because of low cost, underwater units are widely used for small river and lake intakes. Although they have the advantage of being protected from damage by submergence, if repair is needed they are not readily accessible. In reservoirs and lakes the distinct disadvantage of bottom intakes is the lack of alternate withdrawal levels to choose the highest quality available throughout the year.

Shore intakes located adjacent to a river must be sited with consideration for: water currents that might threaten safety of the structure, location of navigation channels, ice floes, formation of sandbars, and potential flooding. In addition, water quality considerations and distances from pumping station and treatment plant are important. Figure 6-7 illustrates the type of screening equipment generally employed in shoreline intakes. A coarse screen of vertical steel bars, having openings of 1 to 3 in. placed in a near vertical position, is used to exclude large objects. It may be equipped with a trash rack rake to remove accumulated debris. Leaves, twigs, small fish, and other material passing through the bar rack are removed by a finer screen having $\frac{3}{8}$-in. openings. A traveling screen consists of wire mesh trays that retain solids as the water passes through. Drive chain and sprockets raise the trays into a head enclosure where the debris is removed by means of water sprays. Heat may be required in winter to prevent ice formation. Travel is intermittent and is controlled by the amount of accumulated material.

Low-lift pumping stations transporting raw water from the source to treatment are located as close as is practical to the intake. In the case of a tower or submerged intake, the station is situated on the near shore of the lake or reservoir. Pumps for a shoreline intake are frequently housed in the same

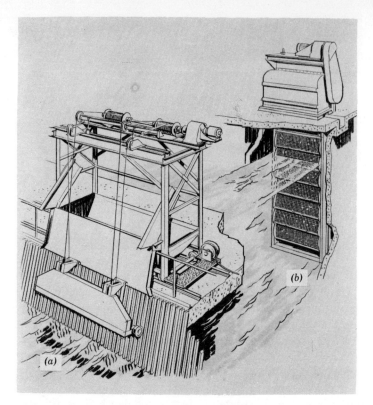

Figure 6-7 Shore water intake screens. (*a*) Coarse screen (bar rack) with power rake and trash hopper for cleaning. (*b*) Traveling water screen with automatic cleaning mechanism to remove debris that passes through bar rack. (Courtesy of Water Quality Control Division, Envirex, a Rexnord Company)

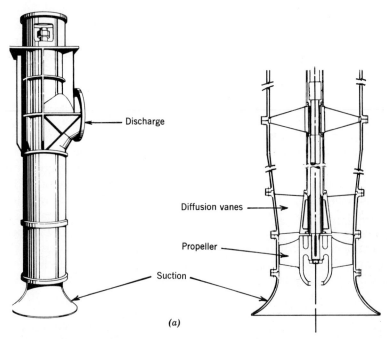

Figure 6-8 Pumps commonly used for low-lift from raw water intake to treatment plants. (*a*) Axial flow (vertical propeller) pump. (*b*) Vertical turbine pump. (Courtesy of Allis-Chalmers)

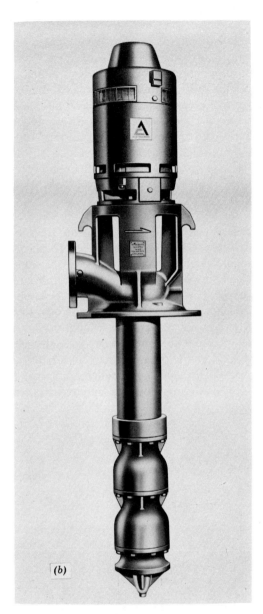

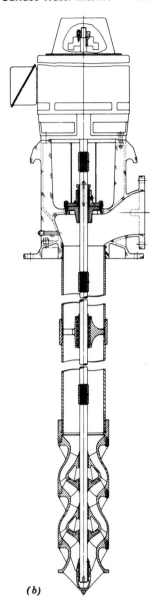

(b)

(b)

Figure 6-8 (*Continued*)

structure as the screens. A typical pumping station consists of a suction well located behind the intake screens, or at the discharge end of the withdrawal pipe, pumping equipment, and associated motors, valves, piping, and control systems. With manual controls, an operator must be available to start and stop pumping equipment, to open and close valves, and to monitor other controls. Highly automated pumping stations can be supervised from a central control panel located in the station, or remotely, in the treatment plant. Standby pumps and duplicate equipment are provided as is necessary to insure uninterrupted water supply during maintenance and repair, and to meet emergency water demands.

Selection of low-lift pumps depends on station capacity, the height to which the water must be lifted, number of pumps required, anticipated method of operation, and costs. The centrifugal pumps common in low-lift pumping at water intakes are vertical turbine pumps and axial flow pumps (vertical propeller) (Figure 6-8). These units have liquid flow parallel to the axis of the pump and are designed to operate efficiently under low-head discharge. In the axial flow pump, head is produced by the vortex and lifting action of the rotating propeller, as well as by centrifugal force. The vertical turbine is a centrifugal pump with one or more impellers located in stages. No priming is needed for these pumps if they are installed with impellers submerged.

6-5   DISTRIBUTION PUMPING AND STORAGE

High-lift pumps move processed water from a basin at the treatment plant into the distribution system. Different sets may be needed to pump against unequal pressures to separate and distinct service areas. If so, some pumps connect directly for main service in lower areas, while booster units are used to reach high elevations in the system. The most common types of pumps in high service are vertical turbine and horizontal split-case centrifugal because of good efficiency and capability to deliver water against high discharge heads. The model illustrated in Figure 6-9 is a double suction design such that water is drawn into both sides of the double volute case, insuring both radial and axial balance for minimum pressure on bearings. The impeller discharges liquid to the ever-increasing spiral casing, gradually reducing the velocity head while

Figure 6-9 Horizontal split-case double-suction centrifugal pump applicable for high-service pumping in waterworks. (Courtesy of Fairbanks Morse Pump Div., Colt Industries)

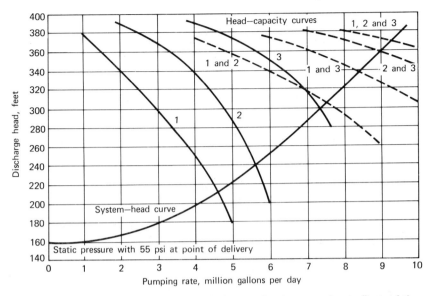

Figure 6-10 Pump capacity-discharge head diagram showing operational effects of three pumps singly and in parallel discharging into a distribution system.

producing an increase in pressure head. This type of pump operates at a range of capacities from design flow to shutoff without excessive loss of discharge pressure or efficiency.

A pump capacity-discharge head diagram of the type shown in Figure 6-10 is developed in the design of pumping stations. The system-head curve is a plot of pump discharge head (pressure against which the pumps must operate) at various pumping rates to deliver water to the distribution system. At higher pumping rates, the pressure resistance from the system increases requiring greater pump discharge heads. Where water levels in distribution system storage tanks vary, or pump suction levels change, this curve is a narrow band rather than a single line. On the same diagram are plotted head-capacity curves for the lift pumps operating individually and in combination. These curves, available from pump manufacturers, show decreasing head with increasing pumping rates. The effect of operating two or more centrifugal pumps in parallel produces an increase in discharge head and, thereby, a decrease in individual pump capacity. Available rates of water delivery into the system are where the pump curves cross the system-head line, for example, from Figure 6-10, operating pump 2 alone supplies 5.6 mgd against the existing system head of 240 ft. During the low-water consumption, pump 1 can supply the demand up to 4.5 mgd while pump 2 can discharge 5.6 mgd. When a pumping rate of 7 to 8 mgd is needed, either pump 3 alone, or pumps 1 and 2 operating simultaneously, can supply the required flow. The selection of operation is determined by which pump operates with highest efficiency at that point on the system-head curve.

Distribution storage may be provided by elevated tanks, standpipes, underground basins, or open reservoirs. Elevated steel tanks are manufactured in a variety of ellipsoidal and spherical shapes and in capacities ranging from 50,000 gal to 3.0 mil gal (Figure 6-11). The advantage of elevated storage is the pressure derived from holding water higher than surrounding terrain. Steel

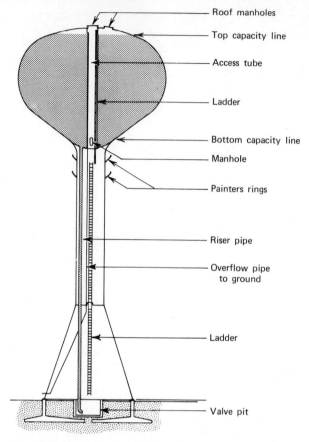

Roof manholes

Top capacity line

Access tube

Ladder

Bottom capacity line

Manhole

Painters rings

Riser pipe

Overflow pipe
to ground

Ladder

Valve pit

Figure 6-11 Elevated spheroid-shaped water storage tank, available in capacities from 0.20 to 1.25 mil gal and specified pedestal height. (Courtesy of Chicago Bridge and Iron Co.)

tanks are normally protected from corrosion by painting the exterior and installing cathodic protection equipment to safeguard interior surfaces. Where gravity water pressure is either not necessary, or is provided by booster pumping, ground-level standpipes or reservoirs are provided. Steel standpipes (Figure 6-12) are usually available in sizes up to 5 mil gal. In general, the term standpipe applies where the height of a tank exceeds its diameter, while one that has a diameter greater than its height is referred to as a reservoir.

Concrete reservoirs may be constructed above or below ground. Recent designs in circular, prestressed concrete reservoirs have increased watertightness and reduced maintenance costs not previously available with plain reinforced concrete basins. Although early reservoirs were constructed without covers, current practice is to enclose storage basins, reducing the possibility of pollution and deterioration of the interior surfaces. Exposure to the atmosphere permits airborne contamination, penetration of sunlight which promotes the growth of algae and, in colder climates, the freezing of the water surface. Most open, paved earth-embankment reservoirs have been abandoned and replaced with other storage facilities, since the cost of covering them is exceedingly high. Clear-well storage at a treatment plant is commonly located under the filter beds that comprise a roof structure. Storage capacity at a treatment plant provides allowance for differences in water production rates and high-lift pump discharge to the distribution system. Additional clear-well volume may be supplied to act as distribution storage.

The choice between elevated and ground storage in water distribution depends on topography, size of community, reliability of water supply, and economics. Ground storage facilities on hills high enough to provide adequate pressures are preferred; however, seldom are hilltop locations in a suitable position. Therefore, elevated tanks, or ground-level reservoirs with booster pumping to provide required water pressure, are needed. The economy and desirability of ground storage with booster pumping, as compared with elevated storage, must be determined in each individual area. In general, elevated tanks are most economical and are recommended for small water systems. Reservoirs and booster pumping facilities are often less expensive in large systems where adequate supervision can be provided. Reliable instrumentation and automatic controls are available for remote operation of automatic booster stations.

The principal function of distribution storage is to permit continuous treatment and uniform pumping rates of water into the distribution system while storing it in advance of actual need at one or more locations. The major advantages of storage are: the demands on source, treatment, transmission, and distribution are more nearly equal, reducing needed sizes and capacities; the system flow pressures are stabilized throughout the service area; and, reserve supplies are available for contingencies, such as fire fighting and power outages. In determining the amount of storage needed, both the volume used to

Figure 6-12 Ornamented steel standpipe for ground-level water storage, designed and built in any capacity. (Courtesy of Chicago Bridge and Iron Co.)

meet variations in demand and the amount related to emergency reserves must be considered. The storage to equalize supply and demand is determined from hourly variations in consumption on the day of maximum water usage. Reserve capacity for fire fighting is computed from required fire flow and duration. Example 6-3 illustrates common techniques in calculating the amount of storage required for a community. As a rule-of-thumb, capacity to balance supply and demand is about 15 to 20 percent of the average daily consumption, and in systems of moderate size, this amount equals 30 to 40 percent of the total needed for both equalization and fire fighting reserves.

The location of distribution storage as well as capacity and elevation are closely associated with water demands and their variation throughout the day in different parts of the system. Normally, it is more advantageous to provide several smaller storage units in different locations than an equivalent capacity at a central site. Smaller distribution pipes are required to serve decentralized storage, and more uniform water pressures can be established throughout the system. In normal operating service, some stored water should be used each day to insure recirculation, and on peak days the drawdown should not consume fire reserves.

EXAMPLE 6-3

Calculate the distribution storage needed for both equalizing demand and fire reserve based on the following information. Hourly demands on the day of maximum water consumption are given in Table 6-5. Listed are: hourly consumption expressed in gallons per minute, gallons consumed each hour of the day, and cumulative consumption starting at 12 midnight. Fire flow requirements are 6000 gpm for a duration of 6 hr for the high-value district.

Solution

Figure 6-13 is a diagram of the consumption rate-time data given in Table 6-5. When the consumption rate is less than the 1860-gpm pumping rate, the reservoir is filling. When it is greater than 1860 gpm, it is emptying. The area under the emptying, or filling, curve is the storage volume needed to equalize demand for the average 24-hr pumping rate of 1860 gpm.

Since the areas on Figure 6-13 are difficult to measure, a mass diagram is commonly used to determine equalizing storage. Figure 6-14 is a plot of the cumulative flow, column 4 in Table 6-5, versus time. A straight line connecting the origin and final point of this mass curve is the cumulative pumpage necessary to meet the consumptive demand; the slope of this line is the constant 24-hr pumping rate. To find the required storage capacity, construct lines that are parallel to the cumulative pumping rate tangent to the mass curve at the high and low point. The vertical distance between these two parallels is the required tank capacity, in this case, 500,000 gal.

However, sometimes it is not expedient to pump water into the distribution system at a constant rate throughout the day and night. For example, a small community may limit treatment plant operations to daylight hours, or to operation of pumps during off-peak periods when power rates are low. Assume in this instance that the lowest power rates may be obtained during the 8-hr period between 12 midnight and 8 A.M. The graphical procedure for determining storage requirements using this 8-hr pumping period is illustrated in Figure 6-14

Table 6-5. Peak Water Consumption Data on Day of Maximum Water Usage for Example 6-3.

Time	Hourly Consumption (gpm)	Hourly Consumption (gal)	Cumulative Consumption (gal)
12 P.M.	0	0	0
1 A.M.	866	52,000	52,000
2	866	52,000	104,000
3	600	36,000	140,000
4	634	38,000	178,000
5	1000	60,000	238,000
6	1330	80,000	318,000
7	1830	110,000	428,000
8	2570	154,000	582,000
9	2500	150,000	732,000
10	2140	128,000	860,000
11	2080	125,000	985,000
12	2170	130,000	1,115,000
1 P.M.	2130	128,000	1,243,000
2	2170	130,000	1,373,000
3	2330	140,000	1,513,000
4	2300	138,000	1,651,000
5	2740	164,000	1,815,000
6	3070	184,000	1,999,000
7	3330	200,000	2,199,000
8	2670	160,000	2,359,000
9	2000	120,000	2,479,000
10	1330	80,000	2,559,000
11	1170	70,000	2,629,000
12	933	56,000	2,685,000
Average = 1860			

using dashed lines. An accumulated pumping line is drawn from the origin, 0 flow and time of 12 P.M., to the end of the pumping period at 8 A.M. on the maximum cumulative consumption for the day. Storage required is then equal to the vertical distance at 8 A.M. between the accumulated demand line and the maximum daily pumpage. (If a starting time for pump operation other than 12 P.M. is to be considered, the data in Table 6-5 must be shifted to the selected time, cumulative consumption values must be recomputed, and Figure 6-14 will have to be redrawn with an origin at the new time.)

Storage required to provide entire fire reserve is equal to flow rate times duration:

$$6000 \frac{gal}{min} \times 60 \frac{min}{hr} \times 6 \, hr = 2,160,000 \, gal$$

The total storage capacity required for equalizing demand for a continuous 24-hr pumping rate plus fire protection is equal to

$$0.50 + 2.16 = 2.66 \, mil \, gal$$

The total considering an 8-hr pumping period plus fire reserve equals

$$2.11 + 2.16 = 4.27 \, mil \, gal$$

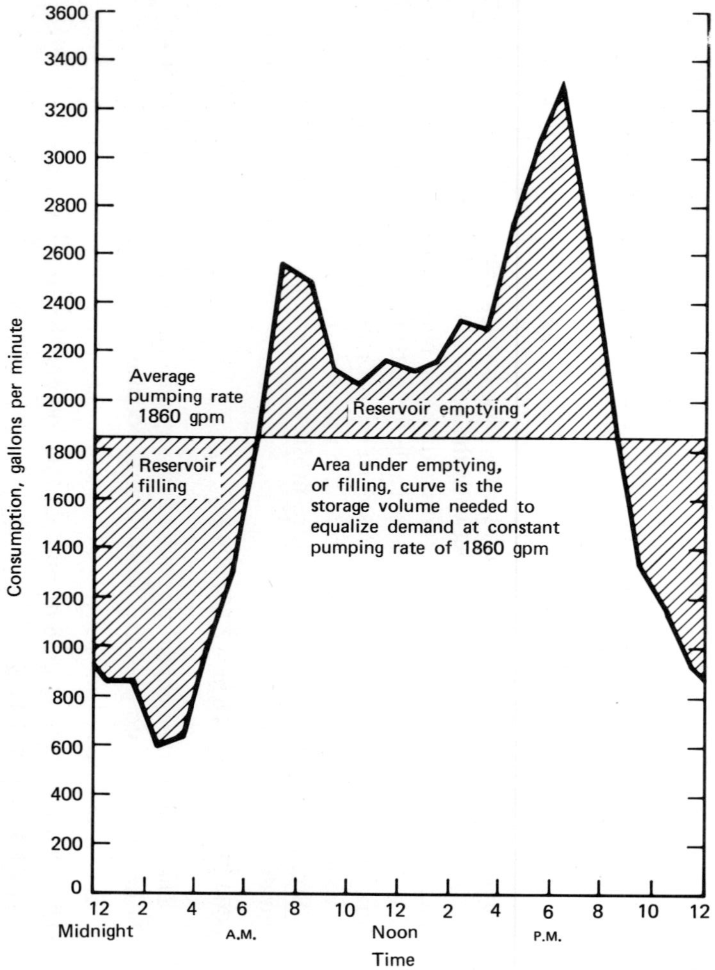

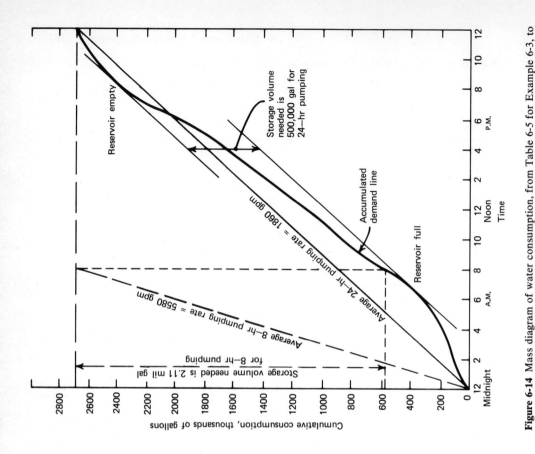

Figure 6-14 Mass diagram of water consumption, from Table 6-5 for Example 6-3, to determine storage needed to equalize demand at constant pumping rates.

Figure 6-13 Plot of hourly water consumption rates, from Table 6-5 for Example 6-3, to determine storage needed to equalize demand at constant pumping rate.

6-6 PIPES

The types of pipe used for distributing water under pressure include: ductile iron, gray cast iron, asbestos cement, concrete, steel, and plastic. Small diameter pipes for house connections are usually either copper or plastic. For use in transmission and distribution systems, pipe materials must have the following characteristics: adequate tensile and bending strength to withstand external loads that result from trench backfill and earth movement caused by freezing, thawing, or unstable soil conditions; high bursting strength to withstand internal water pressures; ability to resist impact loads encountered in transportation, handling, and installation; smooth, noncorrosive interior surface for minimum resistance to water flow; an exterior unaffected by aggressive soils and groundwater; and a pipe material that can be provided with tight joints and is easy to tap for making connections.

Both ductile and gray cast iron are noted for long life, toughness, imperviousness, and ease of tapping as well as the ability to withstand internal pressure and external loads. Ductile iron is produced by introducing a carefully controlled amount of magnesium into a molten iron of low sulfur and phosphorus content, thus, changing the microstructure of the cast iron. This type of pipe is stronger and tougher, and more elastic than gray cast iron. Iron pipes are available in sizes ranging from 2- to 54-in. diameter with several thickness classes in each size (Table 6-6). The selection of a particular thickness class depends on: design internal pressure, including allowance for water hammer; external load due to trench backfill and superimposed wheel loads; allowance for corrosion and manufacturing tolerance; and design factor of safety. The procedure for thickness design of water pipes has been simplified by the use of charts and tables that are published by pipe associations and manufacturers. For example, Figure 6-15 is a diagram that can be used to determine the thickness required for ductile iron pipe under common conditions encountered in construction. The total thickness is dependent on the method of pipe laying, in this case a flat bottom trench with untamped backfill, depth of cover, pipe diameter, and internal working pressure. The value

Table 6-6. Sizes and Thickness Classes of Ductile Iron Pipe, American National Standards Institute

Size (in.)	Thickness Class					
	1	2	3	4	5	6
			(Nominal Thickness in in.)			
3		0.28	0.31	0.34	0.37	0.40
4		0.29	0.32	0.35	0.38	0.41
6		0.31	0.34	0.37	0.40	0.43
8		0.33	0.36	0.39	0.42	0.45
10		0.35	0.38	0.41	0.44	0.47
12		0.37	0.40	0.43	0.46	0.49
14	0.36	0.39	0.42	0.45	0.48	0.51
16	0.37	0.40	0.43	0.46	0.49	0.52
18	0.38	0.41	0.44	0.47	0.50	0.53
20	0.39	0.42	0.45	0.48	0.51	0.54
24	0.41	0.44	0.47	0.50	0.53	0.56

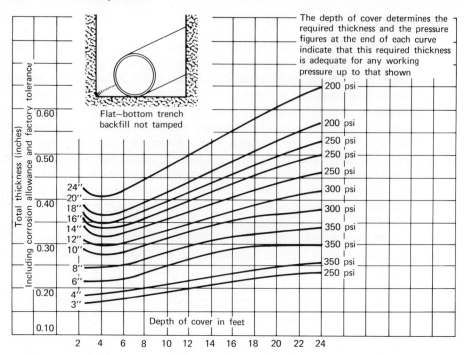

Figure 6-15 Typical chart to determine the thickness design of ductile iron pipe under common field conditions that include: flat-bottomed trench, untamped backfill, factor of safety of 2.5, corrosion allowance of 0.08 in., and a maximum deflection of the pipe of 2 percent. (Courtesy of Clow Corporation)

selected from the chart includes a factor of safety of 2.5, the standard corrosion allowance of 0.08 in., and the specified factory tolerance.

Although cast iron is resistant to corrosion, aggressive waters may cause pitting of the exterior or tuberculation on the interior. Outside protection is usually provided by coating the pipe with a bitumastic tar applied by spray nozzles. The inside is covered by a thin coating of cement mortar about $\frac{1}{8}$-in. thick. This lining, placed while the pipe is rotated at a high speed to compact the mortar, adheres closely to the iron surface such that cutting and tapping the pipe will not cause separation.

Pipe lengths, normally 18 ft, may be joined together by several types of joints; the most common are illustrated in Figure 6-16. One of the earliest was the bell and spigot connection in which the plain end of one pipe was inserted into a flared end of another, and then was sealed with the caulking material such as lead. This development led to the compression type joint, also referred to as push on or slip joint, which is the most popular type used in water distribution piping today. The beveled spigot end pushes into the bell, which has a specially designed recess to accept a rubber ring gasket. When compressed, it produces a tight joint locking the pipes in place against further displacement. The chief advantages of the compression joint are ease of installation, water tightness, and flexibility. The joint permits about 3 percent deflection, making it possible to install pipe on a gradual curve. The mechanical joint utilizes the principal of the stuffing box to provide a fluid tight, flexible connection for distribution piping, however, it is not as popular as the compression union. Flanged joints are made by threading plain-end pipe and

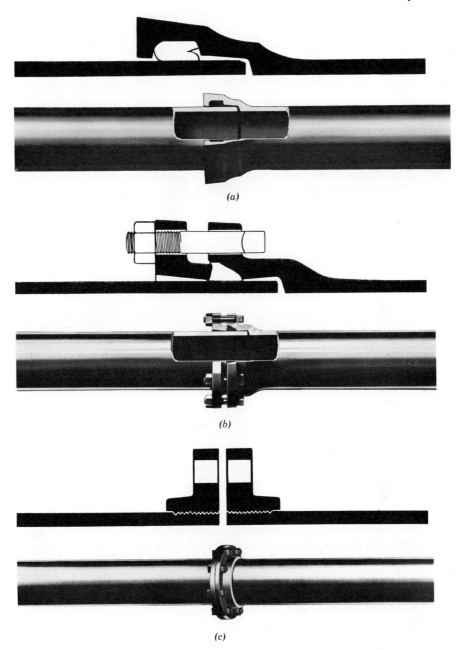

(a)

(b)

(c)

Figure 6-16 Common types of iron pipe joints. (a) Compression-type (slip joint). (b) Mechanical joint. (c) Flanged joint. (Courtesy of Clow Corporation)

screwing on flanges that are then faced and drilled to permit bolting together. Flanged connections are normally used for interior piping in water plants. In addition to these three common joints, particular types are available for special installations. For example, underwater pipelines or other applications requiring a very flexible connection use boltless, or bolted, flexible ball-and-socket joints.

Asbestos cement pipe is composed of a mixture of asbestos fiber, portland cement, and silica sand. A slurry of these ingredients is formed under high

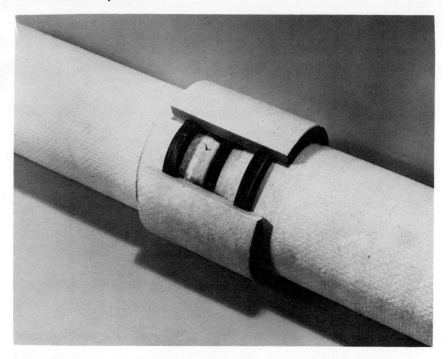

Figure 6-17 Coupling assembly for asbestos cement pipe. Rubber O-rings seal the joint and provide a degree of flexibility. (Courtesy of Certain-teed Products Corporation)

pressure, is heat dried, and then is cured to provide necessary strength and rigidity. Although not as strong as cast-iron pipe, it has the advantages of being a nonconductor immune to electrolysis, resistant to corrosive soil and waters, and smooth, providing minimum surface resistance to hydraulic flow.

Most manufacturers produce asbestos cement water pipe in sizes ranging from 4- to 36-in. diameter in laying lengths of 13 ft. These are provided for three working pressures of 100, 150, and 200 psi, referred to as classes 100, 150, and 200. Pipe lengths are joined by special couplings with rubber sealing rings (Figure 6-17). The 0-rings lock into machined grooves when the coupling and pipe ends are pressed together. This joint provides a watertight seal with some degree of flexibility; pipe ends are separated in the coupling to permit expansion and contraction. The class of pipe chosen for application in a water system depends on internal pressure, external load, and safety. Figure 6-18 is a sample selection curve for various sized asbestos cement pipe based on calculations of the combined loading factors for different pipe-laying conditions, internal pressures, and external loads. Pipe selected from this curve have an overall factor of safety of 2.5 for internal pressure in combination with an external backfill plus a 1000-lb wheel load.

Three types of reinforced concrete pipes are used for pressure conduits: steel cylinder, prestressed with steel cylinder, and noncylinder reinforced that is not prestressed. Concrete pipe has the advantages of durability, watertightness, and low maintenance cost. It is particularly applicable in larger sizes that can be manufactured at or near the construction site by using local labor and materials, insofar as they are available. Nonprestressed steel-cylinder pipe is a welded steel pipe surrounded by a cage of steel reinforcement covered inside

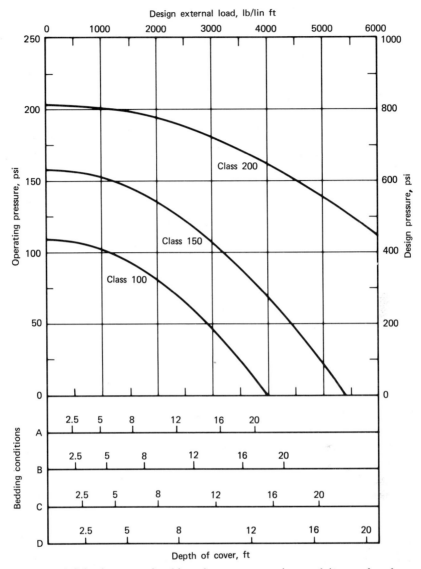

Figure 6-18 Selection curve for 6-in. asbestos cement pipe, applying a safety factor of 4.0 to internal pressures and 2.5 to external loads. Pipe trench bedding conditions: *A*, gravel or sand base, tamped backfill; *B*, same as *A*, backfill not tamped; *C*, pipe barrel resting on flat trench bottom with excavated coupling holes; and *D*, same as *C*, backfill is not tamped. (Reprinted from *Standard Practice for the Selection of Asbestos-cement Water Pipe*, AWWA C401-64, by permission of the Association. Copyrighted 1964 by the American Water Works Association, Inc.)

and out with concrete. The concrete protects the steel from corrosion, provides a smooth interior surface, and contributes high compressive strength to resist stresses from external loading. Applicability is for internal pressures ranging from 40 to 260 psi. For higher pressures, 50 to 350 psi, steel-cylinder pipe is prestressed by winding wire directly on a core consisting of either a steel cylinder lined with concrete, or a pipe of concrete imbedded with a steel cylinder. The exterior of the prestressed core is then finished with a coating of mortar. Reinforced concrete pipe using steel bars or wire fabric, without a steel cylinder, is applicable for low-pressure water transmission requirements, not exceeding 45 psi. Reinforced concrete pipe comes in a variety of diameters ranging from 16 to 144 in. Laying lengths vary with size, the common manufactured lengths being 8, 12, or 16 ft. The usual concrete pipe joint is a modified bell and spigot with steel rings and a sealing element including a rubber seal. A gasket is placed in the outside groove on the steel-faced spigot ring and then is forced into the steel bell ring of another pipe. The rubber gasket provides a watertight seal by filling the groove in a tightly compressed condition. The narrow space between the bell and pipe surfaces is then sealed with a flexible, adhesive plastic gasket and the exposed steel joint faces are painted.

Although not as common as cast iron, asbestos cement, or concrete pipe, steel pipe is used in transmission lines, inplant systems, and sometimes as distribution piping. It exhibits the characteristics of high strength, ability to yield without breaking, and resistance to shock, but careful protection against corrosion is absolutely necessary. A common outside protective coating includes paint primer, coal tar enamel, and wrapping. The particular materials applied depend on the corrosive environment of the pipe and placement, above ground or underground. Interior linings may be either coal tar enamel or cement mortar. The two types of steel pipe fabricated are electrically welded and mill type. Pipe ends are manufactured in a variety of ways for field connection: plain or beveled ends for field welding, flanges for bolting, bumped or lap ends for riveting, and bell and spigot ends with rubber gasket.

Plastic pipe production was initially limited to smaller sizes used for service lines and household plumbing systems. Currently larger sizes are available, and plastic pipe is now being used for water distribution mains and inplant piping systems, as well as service lines. Of the various plastic materials developed for water piping, the three in general use today are PVC (polyvinyl chloride), PE (polyethylene), and ABS (polymers of acrylonitrile, butadiene, and styrene). ABS is primarily for drainage, waste and vent fittings, and piping for interior application. PE and PVC are preferred for water distribution piping because of their greater strength and resistance to internal pressure. All of these materials are chemically inert, are corrosion resistant, and exceptionally smooth, minimizing friction losses in water flow.

All thermoplastic pipe is manufactured by an extrusion process, and fittings are formed in injection molds. The product is slowly cooled and further shaped through sizing devices to insure precise dimensions. By varying the ingredients of the plastic, a range of materials with different physical properties can be produced. For example, there are four types of PVC with design stresses ranging from 1000 to 2000 psi used in the manufacture of pipe with internal pressure rating that range from 50 to 315 psi. Plastic pipe is rated at a standard temperature 73.4° F, since most water temperatures are cooler than this. Pressure resistance of plastic decreases with temperature until a critical point is reached near 150° F. Plastic pipe is available in a wide range of diameters, as follows: ABS, $\frac{1}{2}$ in. to 12 in., PE, $\frac{1}{2}$ in. to 6 in., and PVC $\frac{1}{2}$ in. to 16 in. ABS and PVC, being semirigid products, are normally produced in 20- to 39-ft lengths. Flexible PE pipe is supplied in coils 100 to 500 ft in length.

There are several methods of jointing plastic pipe, varying with size, application, and material. ABS may be connected by screw threaded couplings, solvent weld, or slip couplings. Polyethylene is joined by using insert fittings,

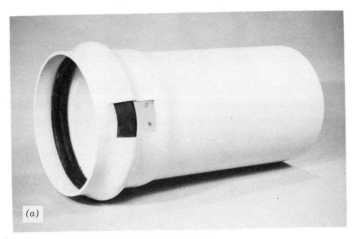

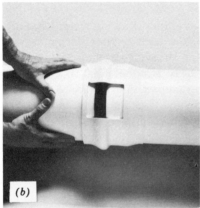

Figure 6-19 PVC bell-and-spigot compression-type joint employing a rubber gasket seal. (Courtesy of Certain-teed Products Corporation)

flaring, or compression jointing. The insert fitting method consists of placing an internal support tube inside the pipe ends and holding the connection by outside clamping. PE pipe can be flared by using special tools with the application of heat to soften the material. This allows joining of plastic service lines directly into conventional curb and corporation stops. The third method uses internal metal stiffener sleeves and 0-rings to form a watertight compression joint. The simplest and most commonly applied joining methods for PVC pipe are the solvent weld system and the belled end coupling. Polyvinyl chloride can be chemically bonded through the use of solvents that dissolve the plastic. The solvated plastic surfaces mix when pressed together and leave a monolithic joint on evaporation of the solvent. Belled end PVC joints may employ a solvent weld or rubber gasket. In the former, the plain pipe end is cemented into a straight bell formed on the opposite end of a second pipe length. The method employs a rubber seal that fits into a recess formed in the belled end (Figure 6-19). The spigot end is beveled for ease of installation, and the joint is made by simply pushing the lubricated spigot into the bell, compressing the rubber gasket for a pressure-tight fit. For distribution system applications, the rubber-gasketed bell and spigot joint is preferred to solvent welding because of the ease and speed of jointing.

6-7 PIPING NETWORKS

A municipal water distribution system includes a network of mains with storage reservoirs, booster pumping stations (if needed), fire hydrants, and service lines. Arterial mains, or feeders, are pipelines of larger size which are connected to the transmission lines that supply the water for distribution. All major water demand areas in a city should be served by a feeder loop, where possible the arterial mains should be laid in duplicate. Two moderately sized lines a few blocks apart are preferred to a single large main. Parallel feeder mains are cross-connected at intervals of about one mile, with valving to permit isolation of sections in case of a main break. Distribution lines tie to each arterial loop, forming a complete gridiron system that services fire hydrants and domestic and commercial consumers.

The gridiron system, illustrated in Figure 6-20, is the best arrangement for distributing water. All of the arterials and secondary mains are looped and interconnected, eliminating dead ends and permitting water circulation such that a heavy discharge from one main allows drawing water from other pipes. When piping repairs are necessary, the area removed from service can be reduced to one block if valves are properly located. Shutoff valves for sectionalizing purposes should be spaced at about 1200-ft intervals, and at all branches from arterial mains. At grid intersections no more than one branch, preferably none, should be without a valve. Feeder mains are usually placed in every second or third street in one direction and every fourth to eighth street in the other. Sizes are selected to furnish the flow for domestic, commercial, and industrial demands, plus fire flow. Distribution piping in intermediate streets should not be less than 6-in. in diameter.

The dead-end system, shown in Figure 6-21, is avoided in new construction and can often be corrected in existing systems by proper looping. Trunk lines placed in the main streets supply submains which are extended at right angles to serve individual streets without interconnections. Consequently, if a pipe break occurs, a substantial portion of the community may be without water. Under some conditions, the water in dead-end lines develops tastes and odors

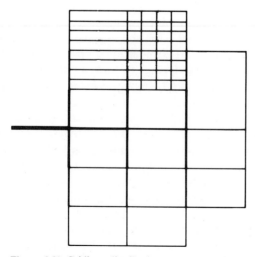

Figure 6-20 Gridiron distribution system consisting of an arterial pipe network with a superimposed system of distribution mains.

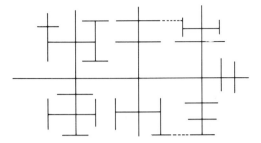

Figure 6-21 Undesirable dead-end distribution system.

from stagnation. To prevent this, dead ends may require frequent flushing where houses are widely separated.

The ideal gridiron system, with duplicating transmission lines, arterial mains, and fully looped distribution piping, may not be feasible in all cases. For example, in communities located on steep hillsides, elevations may vary so greatly that two or more separate systems are needed. Normal pressures in a distribution system, under average conditions of flow, are preferred in the range of 50 to 80 psi. Pressures in excess of 100 psi should be avoided. Since a pressure differential of 50 psi is equivalent to an elevation difference of 115 ft, the desirable elevation change between the high and low points should be limited to this value. Frequently, separate mains serve different elevation zones with either no direct connections or with pressure control valves inserted in the joining lines. Ideally, piping in each zone should be a gridiron system with multiple feeder mains; however, for economic reasons this may not always be feasible. The alternate arrangement is a single arterial main with branches at right angles and subbranches between them such that a grid of interconnected pipes surrounds the central feeder.

6-8 SYSTEM APPURTENANCES

Fire Hydrants

Hydrants provide access to underground water mains for the purposes of extinguishing fires, washing down streets, and flushing out water mains. Figure 6-22 shows the principle parts of a hydrant; the cast-iron barrel is fitted with outlets on top and a shutoff valve at the base is operated by a long valve stem that terminates above the barrel. A typical unit has two $2\frac{1}{2}$ in. diameter hose nozzles and one $4\frac{1}{2}$ in. pumper outlet for a suction line. The barrel and valve stem are designed such that accidental breaking of the barrel at ground level will not unseat the valve, thus preventing water loss. Hydrants are installed along streets behind the curb line a sufficient distance, usually 2 ft, to avoid damage from overhanging vehicles. The pipe connecting a hydrant to a distribution main is normally not less than 6 in. in diameter and includes a gate valve allowing isolation of the hydrant for maintenance purposes. A firm gravel or broken-rock footing is necessary to prevent settling and to permit drainage of water from the barrel after hydrant use. In cold climates the water remaining in a hydrant can freeze and break the barrel. Where groundwater stands at levels above the hydrant drain, generally the drain is plugged at the time of installation and, for service in cold climates, is pumped out after use. The

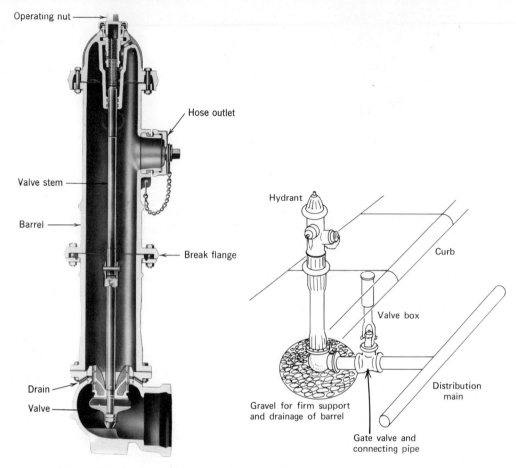

Operating nut

Hose outlet

Valve stem

Barrel

Break flange

Hydrant

Curb

Valve box

Drain

Valve

Gravel for firm support
and drainage of barrel

Distribution
main

Gate valve and
connecting pipe

Figure 6-22 Cutaway section of a fire hydrant and typical installation adjacent to a street. (Reprinted by permission of Mueller Co.)

Insurance Services Office recommends a minimum fire hydrant spacing based on required fire flow (Table 6-4).

Valves

A gate valve consists of a sliding, flat, metal disk that is moved at right angles to the flow direction by a screw-operated stem. When installed in a pipeline, the disk is drawn up into the housing of the case, permitting free flow of the water through the valve opening. The gate is lowered into snugly fitting side channels to block the water flow. The most common valve installed in distribution mains is the double-disk, parallel-seat, cast-iron valve with a nonrising stem (Figure 6-23a). Normally, underground valves in mains are provided with a valve box extending to the street surface. The valve is opened, or closed, by using an extension rod to reach down into the enclosure and turning the nut located on top of the valve stem. Larger gate valves may be placed in vaults or manholes to facilitate operation and maintenance. They are generally equipped with small by-pass valves to reduce pressure differentials on opening and closing (Figure 6-23b) and frequently have spur or beveled gearing to reduce the force required for operation.

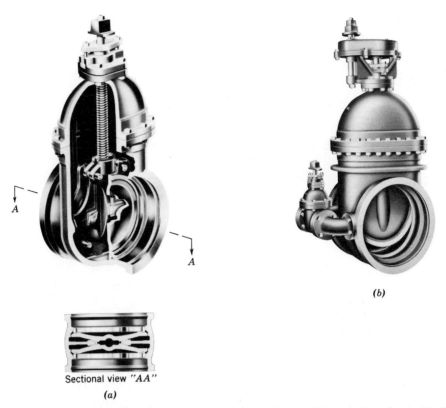

Sectional view "AA"

(a)

Figure 6-23 Typical gate valves used in distribution mains. (a) Cutaway view of a double-disk, parallel-seat gate valve. (b) Larger gate valve equipped with bypass valve and spur gearing for easy operation. (Courtesy of Smith Valve & Hydrant Products, U.S. Pipe and Foundry Co.)

A butterfly valve (Figure 6-24) has a movable disk that rotates on a spindle or axle set in the shell. The circular disk rotates in only one direction from full closed to full open, and seats against a ring in the casing. The main disadvantages of a butterfly valve result from the disk's always being in the flow stream, restricting the use of pipe cleaning tools. On the other hand, the advantages of this valve are: tight shutoff, low head loss, small space requirement, and throttling capabilities. The latter is one of the most popular applications of butterfly valves, for example, use in a rate of flow controller to regulate the rate of discharge from sand filters in a water treatment plant. Recently, and more often in the larger sizes, rubber-seated butterfly valves are being used in distribution systems. Because pressure differences across a butterfly valve disk tend to close the valve, a mechanical operator must be employed to overcome the torque in opening the valve, and to resist this force during closing to prevent slamming. For treatment plant applications, the operator is often a hydraulic cylinder and piston rod assembly used to hold the valve disk in any intermediate position, as well as in opening and closing action. Operators for large valves may also be motor driven. Manual operation for valves in distribution mains is done by using reducing gears or, as is illustrated in Figure 6-24, a threaded stem that operates a lever arm attached to the valve disk.

A check valve is a semiautomatic device designed to permit flow in only one direction. It opens under the influence of pressure and closes automatically

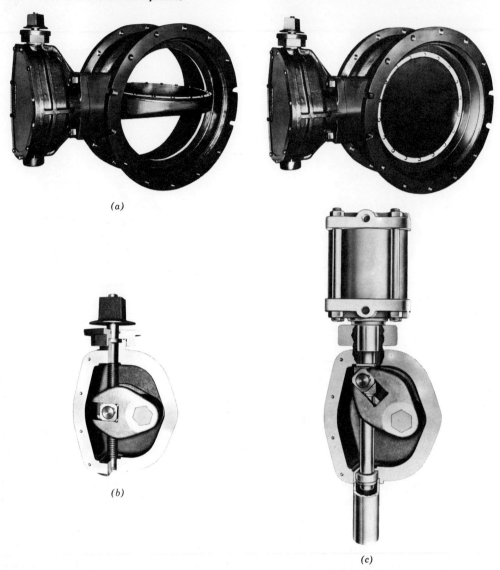

(a)

(b)

(c)

Figure 6-24 Typical butterfly valve used in both water plant and distribution piping. (a) Views of a butterfly valve in open and closed positions. (b) Cutaway view of an operator with a threaded stem for manual operation. (c) Cylinder operator for butterfly valve used in throttling applications to control rate of flow. (Courtesy of Dresser Industries, Inc.)

when flow ceases. These valves are available in a variety of designs. One of the more common is a swing check valve where the valve closure is hinged on top, swinging open during flow, and is spring loaded to close during no flow. Check valves are commonly installed in the discharge piping of centrifugal pumps to prevent backflow when the pump is shut off.

Valving Arrangements for Storage Tanks

Automatically controlled valve systems are used to maintain the desired water level in storage tanks located in a distribution system. The arrangement in Figure 6-25a has separate inflow and outflow pipes. A water-level pressure sensor controls an operator that opens the single-acting sequence valve for

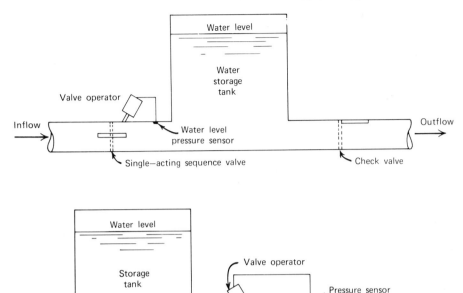

Figure 6-25 Schematics of two typical valving systems used to control flow into and out of storage tanks. (*a*) Single-acting sequence valve opens to permit flow into the tank, while outflow is through a check valve in another line. (*b*) The double-acting sequence valve is a combination butterfly shutoff and swing check. Inflow is controlled by the sequence valve and outflow is through the check valve.

inflow, and closes it when the tank is full. Outflow is through a check valve that opens when pressure on the distribution side is less than the reservoir side, and closes when the pressure differential is reversed.

The double-acting sequence valve in Figure 6-25*b* is a combination shutoff and check valve so that water can enter and leave through the same pipe connection. When distribution pressure reaches a preset amount, the butterfly opens, allowing water to enter the reservoir. After filling, the pressure sensor signals the operator to close the valve. Outflow is managed by differential pressure across a swing check set in the butterfly disk. This allows water to flow into the piping network on demand and, also, prevents water from entering storage when the tank is full and the butterfly valve is closed.

Service Connections

A typical service installation consists of a pipe from the distribution main to a turnoff valve located near the property line (Figure 6-26*a*). The pipe is generally attached to the main by means of a corporation stop that can be inserted by using a special tapping machine while the main is in service under pressure. Occasionally, outlets are provided in the main at the time of installation. Access to the curb stop is through a service box extending from the valve to the ground surface. A number of materials are used for service lines, the most popular being copper, plastic, and cast iron. Copper pipe is generally viewed as the standard. However, recently developed plastic materials have

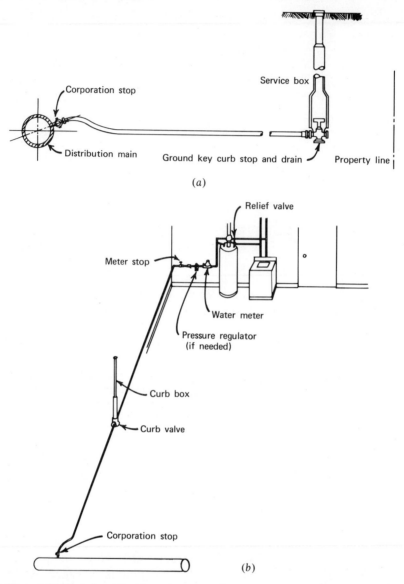

Figure 6-26 Typical residential service connection. (*a*) Service line and meter box installation. (*b*) Complete residential water service.

proved to be just as durable and less expensive. Cast-iron, and sometimes asbestos cement pipe, is used for larger services, generally in a 2-in. size or larger. The water meter, associated shutoff valve, and pressure regulator, if needed, are normally in the basement of the dwelling (Figure 6-26*b*). In some municipalities the meter is placed outside in a box. However, this has the disadvantage of exposing the meter and piping to possible freezing and burial under snow cover. Modern water meters can be equipped with remote readers extended to the outside of dwellings so that water utility personnel do not have to enter buildings to record water usage.

The maximum instantaneous residential consumption rate has increased in recent years because of the larger number of fixture units. The estimated rate for a typical house having two bathrooms, full laundry, kitchen, and one or two hose bibbs, is 15 gpm. With the possible exception of lawn sprinkling a pressure of 15 psi is normally adequate for the operation of any fixture. Allowing 10 psi (23 ft) to overcome static lift from the basement to upper floors, the generally accepted minimum standard of a service is 25 psi at 15 gpm on the customer side of the meter. For a typical residential service consisting of 40 ft of $\frac{3}{4}$ in. copper pipe and $\frac{5}{8}$ in. disk meter, the friction loss for simultaneous fixture use (6 to 12 gpm) is in the range of 5 to 19 psi. Therefore, if 25 psi is to be available to the customer, the pressure in the main must be approximately 40 psi.

6-9 SYSTEM TESTING

Fire Flow Tests

These tests are important in determining the efficiency and adequacy of a distribution system in transmitting water, particularly during days of high demand, and are used to measure the amount of water available from hydrants for fire fighting. The required rate of flow for fire fighting must be available at a specified residual pressure. The minimum is 20 psi if fire department pumpers are used to supply hose streams. If hoses are to be connected directly to hydrants, a residual pressure of 50 to 75 psi is needed, depending on local structural conditions. For automatic sprinkler systems, a pressure of 15 psi is usually specified at the top line of sprinklers.

Flow tests consist of discharging water at a measured rate of flow from one or more fire hydrants and observing the corresponding pressure drop in the mains through another nearby hydrant. Figure 6-27a illustrates a flow test in which hydrants 1, 2, 3, and 4 are discharging while the residual pressure drop is read at hydrant R. The residual hydrant is chosen so that the hydrants flowing water are between it and the larger feeder mains supplying water to the area. The number of hydrants used for discharge, and the rates of flow, should be such that the drop in pressure at the residual hydrant is not less than 10 psi. For single main testing the layouts and hydrant selection are shown in Figure 6-27b.

The typical test procedure is: one of the outlet caps on the residual hydrant is replaced with a pressure gage, the hydrant valve is opened to exhaust air from the barrel, and the initial water pressure reading is recorded. The surrounding hydrants are then opened and discharge is measured using pitot gages. A pitot tube centered in the flow stream from a hydrant nozzle measures the pressure exerted by the velocity of the flowing water. The quantity of discharge from a hydrant nozzle can be calculated from the pitot pressure reading by using Eq. 6-3. Since it is difficult to establish exactly the rate of discharge needed to produce a specific residual pressure, the flow that would be available at a given residual pressure, for example, 20 psi, must be computed from test data by using Eq. 6-4.

$$Q = 29.8C \, d^2(p)^{1/2} \tag{6-3}$$

where Q = discharge, gallons per minute
  C = coefficient, normally 0.90
  d = diameter of outlet, inches
  p = pitot gage reading, pounds per square inch

$$Q_R = Q_F \frac{H_R^{0.54}}{H_F^{0.54}} \cong Q_F \left(\frac{H_R}{H_F}\right)^{1/2} \tag{6-4}$$

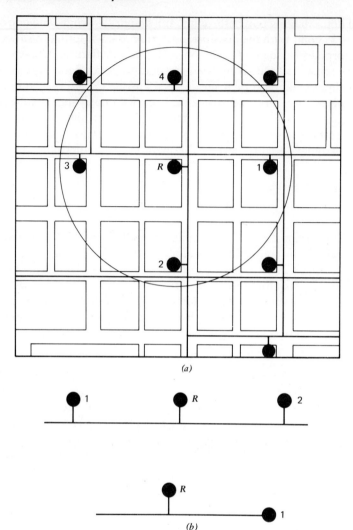

(a)

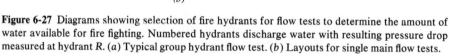

(b)

Figure 6-27 Diagrams showing selection of fire hydrants for flow tests to determine the amount of water available for fire fighting. Numbered hydrants discharge water with resulting pressure drop measured at hydrant R. (a) Typical group hydrant flow test. (b) Layouts for single main flow tests.

where Q_R = computed discharge at the specified residual
          pressure, gallons per minute
    Q_F = total discharge during test, gallons per minute
    H_R = drop in pressure from original value to specified
          residual, pounds per square inch
    H_F = pressure drop during test, pounds per square inch

   Flow test results show the strength of a distribution system but not necessarily the degree of adequacy of the entire waterworks. Consider a system supplied by pumps at one location and having no elevated storage. If the pressure at the pump station decreases during the test, it is an indication that the distribution system is capable of delivering more than the pumps can provide at their normal operating pressure, and the value for the drop in pressure measured during the test must be corrected. It is equal to the actual drop obtained in the field minus the drop in discharge pressure at the pumping station. If sufficient pumping capacity is available at the station and the discharge pressure can be maintained by operating additional pumps, the water system as a whole is able to deliver the computed quantity. If, however, additional pumping units are not available, the distribution system is capable of delivering the computed amount, but the water system as a whole is limited by the pumping.

The corrections for pressure drops in tests on systems with storage are generally estimated from a study of all the pressure drops observed on recording gages at the pumping station. The corrections may be very significant for tests near the pumping station while decreasing to zero for remote tests.

Pressure Tests

These tests are generally performed by attaching a pressure gage to a hose nozzle of a hydrant and opening the valve to fill the barrel with water under residual pressure. Where more than instantaneous readings are desirable, a recording unit may be installed. Operational evaluations normally include monitoring several points continuously for pressure, especially during periods of maximum water consumption.

Hydraulic—Gradient Tests

These tests are conducted to determine the ability of a distribution system to transmit water with adequate residual pressures. Testing is normally performed during periods of peak delivery when the system flows are substantially constant; this often occurs during midmorning or midafternoon. Pressure measurements are taken at various points located along a line from the source of delivery to an outlying section. Pressure observations are plotted vertically along a horizontal distance scale. The curve connecting the plotted values gives a visual picture of the head losses between points as the water is transported through the pipeline. Sections with steep hydraulic gradients represent areas where the greatest friction losses occur.

Coefficient Tests

Occasionally it may be desirable to determine the internal condition of a pipeline with respect to friction loss during water flow. The procedure is to isolate a pipe section to the greatest extent possible, including closing service connections. The hydraulic gradient (pressure drop) is measured between two points a known distance apart while a measured flow is transmitted through the pipe. Assuming that complete isolation of the pipe section being tested is possible, flow can be measured by hydrant discharges as previously described. Otherwise, the quantity of flow is measured by using a modified pitot meter which can be inserted into the pipeline through a 1-in. corporation stop. With these data, the coefficient of roughness can be computed from the Hazen Williams formula. Then, with the proper coefficient known, head losses during other quantities of flow can be calculated.

EXAMPLE 6-4

Calculate the discharge at a residual pressure of 20 psi based on the following flow test. Hydrants 1, 2, 3, and 4 as illustrated in Figure 6-27a were all discharging the same amount of water through $2\frac{1}{2}$ in. diameter hose nozzles. The pitot tube pressure reading at each hydrant was 16 psi. At this flow the residual pressure at hydrant R dropped from 90 psi to 58 psi.

Solution

Using Eq. 6-3

$$Q = (4)(29.8)(0.90)(2.5)^2(16)^{1/2} = 2680 \text{ gpm}$$

From Eq. 6-4,

$$Q_R = 2680 \frac{(90-20)^{0.54}}{(90-58)^{0.54}} = 4090 \text{ gpm}$$

or,

$$Q_R = 2680 \left(\frac{90-20}{90-58}\right)^{1/2} = 3970 \text{ gpm}$$

6-10 SYSTEM EVALUATIONS

Quantity

The supply source plus storage facilities should be capable of yielding enough water to meet both the current daily demands and the anticipated consumption 10 years hence. To quantitatively evaluate water usage, a record of average daily, peak daily, and peak hourly rates of consumption for the past 10 years should be recorded. Projected future needs can be estimated from these values and other factors relating to community growth.

The minimum amount of water available from a supply source should always be sufficient to insure uninterrupted service. Consideration must be given to the probability of a succession of drought years equal to the previous worst drought experience, and the possibility of lowering groundwater levels. For surface supplies the tributary watershed should be able to yield the estimated maximum daily demand 10 years in the future. As a general rule, the storage capacity of an impounding reservoir should be equal to at least 30 days maximum daily demand 5 years into the future. Ideally, for well supplies there should be no mining of water, that is, neither the static groundwater level, nor the specific capacity of the wells (gpm/ft of drawdown), should decrease appreciably as demand increases. Preferably these values should be constant over a period of 5 years except for minor variations that correct themselves within one week.

Intake Capacity

A surface water intake must be large enough to deliver sufficient water to meet municipal consumption and treatment plant needs (e.g., filter backwashing) during any day of peak demand. With respect to fire flow requirements, if no storage is available the intake capacity must be large enough to meet fire demand, maximum hourly flow, and inplant process needs simultaneously. On the other hand, where the quantity of distribution storage is sufficient to meet all fire flow requirements, the intake capacity involves comparing total storage of the system to the maximum amount of water needed, for both present and future demands. Intake facilities should be sized to meet maximum needs projected for at least 5 years into the future.

A water intake system must be reliable; it must be located, protected, or duplicated such that no interruption of service to customers or to fire protection occurs by reason of floods, ice, or other weather conditions, or for reasons of breakdown, equipment repair, or power failure. In other words, intake facilities should be so reliable that no conceivable interruption in any part of the facility could cause curtailment of water supply service to a community.

Pumping Capacity

In a typical surface water supply system, low-lift pumps draw water from the source and transport it to the treatment plant. After processing, high-lift pumps deliver the water from clear-well storage to the distribution system. In the case of a groundwater supply, the well pumps deliver the raw water to treatment. However, if processing is not needed, the wells may discharge water directly to the transmission mains. In large communities, or in areas with widely varying elevations in topography, booster pumping stations may be needed to increase pressure in the distribution network and to extend the system for greater distances from the main pumping station.

With due consideration for the amount of water storage available, a pumping system must have sufficient capacity to provide the amount of water at pressures and flow rates needed to meet both daily and hourly peak demand with required fire flow. Pumping facilities must also be able to meet demands taking into account common system failures and maintenance requirements. For example, the specified maximum pumping capacity should be obtainable with the two largest pumps out of service. In many instances system storage, either elevated or ground level with booster pumping, is a component part of the pumping system to meet peak hourly demands. Example 6-5 is an evaluation of pumping capacity and distribution storage for a simplified system.

Pumping facilities should be sufficiently reliable through duplication of units, standby equipment, and alternate sources of power such that no interruptions of service occur for any reason. In the event of power outage, standby power sources must be capable of sufficiently quick response to prevent exhaustion of water available in distribution storage during peak hourly demands including required fire flow. Where unattended automatic booster stations exist, the control system should report back to a central station both the condition of operation and any departure from normal.

Piping Network

Arterial and secondary feeder mains should be designed to supply water service for 40 or more years after installation. Actual useful service life of mains under normal conditions is 50 to 100 years. Submains should be at least 6-in. in diameter in residential districts and the minimum size in important districts should be 8-in. in diameter with intersecting mains 12-in. Distribution lines are laid out in a gridiron pattern avoiding dead ends by proper looping. An adequate number of valves should be inserted to permit shutoff in case of break so that no more than one block is out of service. The distribution of hydrants is based on Insurance Services standards.

EXAMPLE 6-5

Consider a water supply system serving a city with the following demand characteristics: average daily demand 4.0 mgd (2780 gpm), maximum day

6.0 mgd (4170 gpm), peak hour 9.0 mgd (6250 gpm), and required fire flow 7.2 mgd (5000 gpm) resulting in a maximum 5-hr rate of 13.2 mgd (9170 gpm) maximum daily demand plus fire flow. Assume that the minimum pressure to be maintained in the main district is 50 psi (115 ft) except during fire flow, and that the piping system is equivalent to a 24-in. diameter main with a $C = 100$. Consider the system without storage and with storage beyond the load center.

Solution

Effect of no storage. At each given demand rate the pumping station discharge head must be sufficient to overcome system losses and to maintain a hydraulic gradient with a minimum of 115 ft of head at the load center. Thus, at the average daily demand, the pumping head required is 115 ft plus the head loss in 29,000 ft of 24-in. pipe at 4.0 mgd (using Figure 4-6 for head loss):

$$115 + (0.9 \times 29) = 140 \text{ ft}$$

At maximum daily rate (6.0 mgd):

$$115 + (1.9 \times 29) = 170 \text{ ft}$$

At peak hourly demand (9.0 mgd):

$$115 + (4.0 \times 29) = 230 \text{ ft}$$

And, at maximum daily rate plus fire flow (13.2 mgd):

$$115 + (8.2 \times 29) = 350 \text{ ft}$$

These results are plotted in Figure 6-28.

Storage beyond load center. In the arrangement shown in Figure 6-29, 1.0 mil gal of storage is provided 10,000 ft beyond the load center, 39,000 ft from the pumping station at an elevation of 120 ft. When no water is being taken from storage, the pumping head must be sufficient to pump against the head at the tank and to overcome losses between pumping station and load center. When part of the demand is being supplied from storage, however, the pumping head need only be sufficient to discharge against the head at load center and to overcome losses in the pipeline between the pumping station and load center.

At the average daily demand, the required pumping rate is 4.0 mgd with no water taken from storage. The hydraulic gradient at the load center is thus identical to that at the tank, namely, 120 ft. The pumping head required is equal to the hydraulic gradient at the load center plus the head loss in 29,000 ft of pipe,

$$120 + (0.9 \times 29) = 146 \text{ ft}$$

During the maximum day, the required pumping rate is 6.0 mgd with no water taken from storage. The pumping head required equal to the hydraulic gradient at the load center plus the head loss in 29,000 ft at 6.0 mgd is

$$120 + (1.9 \times 29) = 176 \text{ ft}$$

At the peak hourly demand, consider supplying 3.0 mgd from storage and the remaining 6.0 mgd from pumping. The hydraulic gradient at the load center is that at the tank minus the head loss in 10,000 ft of pipe between the tank and

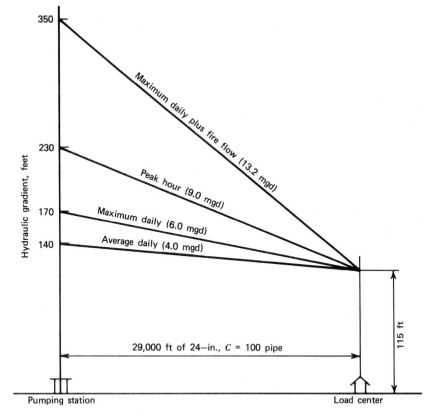

Figure 6-28 Hydraulic gradients with no storage for Example 6-5.

load center at the storage discharge rate of 3.0 mgd:

$120 - (0.5 \times 10) = 115$ ft

The pumping head is then

$115 + (1.9 \times 29) = 171$ ft

Maximum daily plus fire flow. If the 1.0 mil gal in storage is supplied at a uniform rate for the required fire duration of 5 hr, the flow is 3330 gpm (4.8 mgd) with a resulting hydraulic head at the load center of

$120 - (1.2 \times 10) = 108$ ft

The pumping rate needed is $13.2 - 4.8 = 8.4$ mgd, and the pumping head required is

$108 + (3.5 \times 29) = 210$ ft

In this analysis the 1.0 mil gal storage supplied about 35 percent of the fire flow with the pumping station providing the balance. This is realistic, since about one half of the storage is to equalize supply and demand. Considering only 500,000 gal available for fire demand, the head at the load center is 116 ft and the pumping head increases to 275 ft.

A comparison of Figures 6-28 and 6-29 illustrates the benefits of distribution storage. If no storage is provided, significantly greater pumping heads are

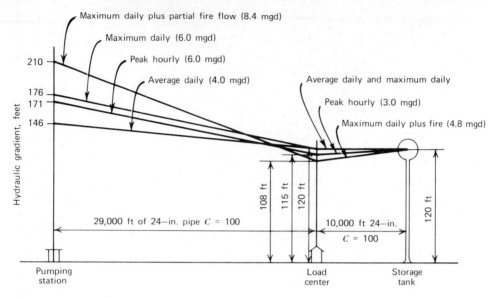

Figure 6-29 Hydraulic gradients with elevated storage beyond load center for Example 6-5.

required to furnish peak demands. The pumping rate at maximum daily plus fire flow more than doubles the power needed. The same is true at peak hourly pumpage: without storage 9.0 mgd at 230 ft compared to 6.0 mgd at 171 ft. During average and maximum daily demands, the pumping heads are approximately the same.

REFERENCES

1. *Grading Schedule for Municipal Fire Protection*, Insurance Services Office, 160 Water Street, New York, N.Y. 10038, 1974.

2. *Guide for Determination of Required Fire Flow*, Insurance Services Office, December, 1974.

PROBLEMS

6-1 How does climate influence water consumption?

6-2 Based on the average water consumption values in Table 6-1, calculate the percentage increase of maximum daily use and maximum hourly use based on mean yearly consumption. (*Answers* 210 percent and 360 percent).

6-3 What is the required fire flow and duration for a 10,000 sq ft building of wood-frame construction? (*Answers* 2700 gpm, 3 hr)

6-4 What is the fire flow needed in a residential area with the houses spaced 50 ft apart, each with 1500 sq ft of floor area? (*Answer* 750 to 1000 gpm)

6-5 Compute the required fire flow and duration for a five-story apartment with 3000 sq ft per floor constructed of ordinary materials with vertical openings protected. (*Answers* 1750 gpm, 2 hr)

6-6 In a densely populated residential area, the required fire flow is 2000 gpm. What is the standard spacing for fire hydrants? (*Answer* At least one hydrant for each 140,000 sq ft)

6-7 Why is a gravel pack placed around the casing and screen of a well?

6-8 What methods can be employed to develop a new well?

6-9 What are the advantages of having entry ports at various depths in a tower intake for a reservoir water supply?

6-10 Explain the purpose of a traveling water screen.

6-11 Refer to Figure 6-10. When the system demand is 8.4 mgd, which pumps are operating and what is the discharge head? (*Answers* 1 and 3, 340 ft)

6-12 What are the functions of water storage in a distribution system?

6-13 The peak water consumption on the day of maximum water usage is as follows.

Time	Gallons per Minute	Time	Gallons per Minute
12 PM	2200	1 PM	6400
1 AM	2100	2	6300
2	1800	3	6400
3	1400	4	6400
4	1300	5	6700
5	1200	6	7400
6	2000	7	9200
7	3500	8	8400
8	5000	9	5000
9	6000	10	3200
10	6400	11	2800
11	7000	12	(2200)
12	6600		

Plot a consumption-time curve like that shown in Figure 6-13. Calculate hourly cumulative consumption values and plot a mass diagram as illustrated in Figure 6-14. What is the constant 24-hr pumping rate and required storage capacity to equalize demand over the 24-hr period. (*Answers* 4780 gpm, 1.47 mil gal)

6-14 What are the advantages of ductile iron pipe?

6-15 An 8-in. ductile-iron water main is buried 14 ft below the ground surface. Based on Figure 6-15, what classification of pipe should be used and what is the maximum allowable working pressure. (*Answers* Thickness Class 2, 350 psi)

6-16 What are the common iron pipe joints for connecting water distribution mains?

6-17 Describe the common jointing methods for asbestos cement and plastic pipe.

6-18 Why is a gridiron system the best arrangement for distributing water?

6-19 Does water gush out of a fire hydrant if the barrel is broken in a vehicular accident?

6-20 Describe the operation and function of a sequence valve connected to an elevated storage tank.

6-21 List the major components of a house water service connection.

6-22 Calculate the discharge at a residual pressure of 20 psi based on the following flow test data. Hydrants 1, 2, 3, and 4 as situated in Figure 6-27a were discharging water with pitot tube pressure readings of 12 psi, 14 psi, 14 psi, and 16 psi, respectively. The hydrant nozzles were all $2\frac{1}{2}$ in. diameter. During these flows, the residual pressure drop at hydrant R was from 70 psi to 45 psi. (*Answers* 3640 gpm or 3540 gpm)

6-23 A fire flow test was conducted on a section of main in a residential district employing three hydrants with 2.5-in. nozzles, as is shown in the upper diagram of Figure 6-27*b*. The pitot tube pressures measured in the discharges from hydrants 1 and 2 were both 10 psi, while the residual pressure drop at hydrant R was from 50 psi to 35 psi. Compute the discharge at 20 psi. (*Answer* 1540 gpm)

6-24 Discuss the relationship between required pumping capacity and amount of water storage provided in the distribution system.

6-25 Review Example 6-5. Applying the techniques explained in Section 4-3, Flow in Pipes Under Pressure, verify the hydraulic gradients under maximum daily flow drawn in Figures 6-28 and 6-29.

6-26 Explain the advantages of placing elevated water storage on the opposite side of the load center from the high-lift pumping station.

chapter 7
Water Processing

The objective of municipal water treatment is to provide a potable supply—one that is chemically and bacteriologically safe for human consumption. For domestic uses, treated water must be aesthetically acceptable—free from apparent turbidity, color, odor, and objectionable taste. Quality requirements for industrial uses are frequently more stringent than domestic supplies. Thus additional treatment may be required by the industry. For an example, boiler feed water must be demineralized to prevent scale deposits.

Common water sources for municipal supplies are deep wells, shallow wells, rivers, natural lakes, and reservoirs. Well supplies normally yield cool, uncontaminated water of uniform quality that is easily processed for municipal use. Processing may be required to remove dissolved gases and undesirable minerals. The simplest treatment illustrated in Figure 7-1a is disinfection and fluoridation. Deep well supplies are chlorinated to provide residual protection against potential contamination in the water distribution system. In the case of shallow wells recharged by surface waters, chlorine both disinfects the groundwater and provides residual protection. Fluoride is added to reduce the incidence of dental caries. Dissolved iron and manganese in well water oxidizes when contacted with air, forming tiny rust particles that discolor the water. Removal is performed by oxidizing the iron and manganese with chlorine or potassium permanganate, and removing the precipitates by filtration (Figure 7-1b). Excessive hardness is commonly removed by precipitation softening, shown schematically in Figure 7-1c. Lime and, if necessary, soda ash are mixed with raw water, and settleable precipitate is removed. Carbon dioxide is applied to stabilize the water prior to final filtration. Aeration is a common first step in treatment of most groundwaters to strip out dissolved gases and add oxygen.

Pollution and eutrophication are major concerns in surface water supplies. Water quality depends on agricultural practices in the watershed, location of municipal and industrial outfall sewers, river development such as dams, season of the year, and climatic conditions. Periods of high rainfall flush silt and organic matter from cultivated fields and forest land, while drought flows may result in higher concentrations of waste-water pollutants from sewer

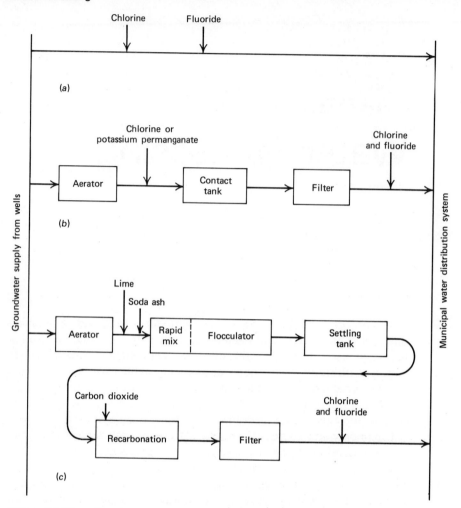

Figure 7-1 Flow diagrams of typical <u>groundwater treatment systems</u>. (*a*) Disinfection and fluoridation. (*b*) Iron and manganese removal. (*c*) Precipitation softening.

discharges. River temperature may vary significantly between summer and winter. The quality of water in a lake or reservoir depends considerably on season of the year. Municipal water quality control actually starts with management of the river basin to protect the raw water supply. Highly polluted waters are both difficult and costly to treat. Although some communities are able to locate groundwater supplies, or alternate less polluted surface sources within feasible pumping distance, the majority of the nation's population draw from nearby surface supplies. The challenge in waterworks operation is to process these waters to a safe, potable product acceptable for domestic use.

The primary process in surface water treatment is chemical clarification by coagulation, sedimentation, and filtration, as illustrated in Figure 7-2. Lake and reservoir water has a more uniform year-round quality and requires a lesser degree of treatment than river water. Natural purification results in reduction of turbidity, coliform bacteria, color, and elimination of day-to-day variations. On the other hand, growths of algae may cause increased turbidity and may produce difficult-to-remove tastes and odors during the summer and fall.

Chlorination is commonly the first and last steps in treatment, providing disinfection of the raw water and establishing a chlorine residual in the treated water. Excess prechlorination and activated carbon are used to remove taste- and odor-producing compounds. The specific chemicals used in coagulation depend on the character of the water and economic considerations. River supplies normally require the most extensive treatment facilities with greatest operational flexibility to handle the day-to-day variations in raw water quality. The preliminary step is often presedimentation to reduce silt and settleable organic matter prior to chemical treatment. As is illustrated in Figure 7-2, many river water treatment plants have two stages of chemical coagulation and sedimentation to provide greater depth and flexibility of treatment. The units may be operated in series, or by split treatment with softening in one stage and coagulation in the other. As many as a dozen different chemicals may be used under varying operating conditions to provide a satisfactory finished water.

The two primary sources of waste from water treatment processes are sludge from the settling tank, resulting from chemical coagulation or softening reactions, and wash water from backwashing filters. These discharges are highly variable in composition, containing concentrated materials removed from the raw water and chemicals added in the treatment process. The wastes are produced continuously, but are discharged intermittently. Historically, the method of waste disposal was to discharge to a watercourse or lake without treatment. This practice was justified from the viewpoint that filter backwash waters and settled solids that were thus returned to the watercourse added no new impurities, but merely returned material that had originally been present in the water. This argument is not now considered valid, since water quality is degraded to the extent that a portion of the water is withdrawn, and chemicals used in processing introduce new pollutants. Therefore, more stringent federal and state pollution control regulations have been enacted requiring treatment of waste discharges from water purification and softening facilities.

The situation of each waterworks is unique and controls to some extent the method of ultimate disposal of plant wastes. For example, they may be piped to

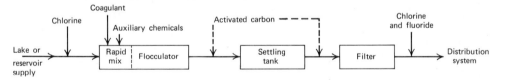

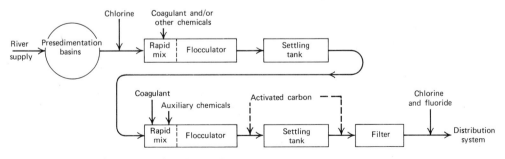

Figure 7-2 Schematic patterns of typical surface-water treatment systems.

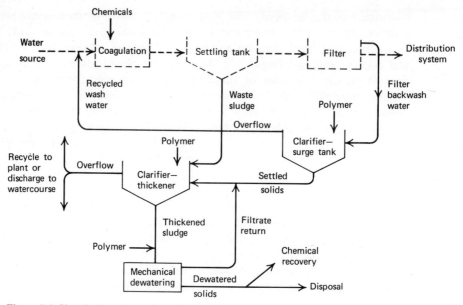

Figure 7-3 Sketch of a dewatering system for alum sludge from a water treatment plant.

a municipal sewer for processing, or may be discharged to lagoons, provided that sufficient land area is available. Ultimate disposal by landfill, or barging to sea, requires thickening for economical handling and hauling. A variety of alternative processing methods are available; however, because of unique characteristics of each plant's waste, no specific process can be universally applied. Figure 7-3 is a typical system for dewatering alum sludge. Filter backwash water is discharged to a clarifier-surge tank. Overflow is recycled to the raw water inlet of the treatment plant while settled solids are discharged, along with waste sludge from the settling tank, to a clarifier-thickener. Supernatant from this unit may be recycled to the head of the plant or may be discharged to a watercourse. Thickened sludge is mechanically dewatered, usually, by centrifugation or filtration. Dewatered solids may be processed for recovery of chemicals or may be discharged to drying beds, land burial, or by barging to sea.

7-1   MIXING AND FLOCCULATION

Chemical reactors in water treatment and biological aeration basins in waste-water processing are designed as either complete mixing or plug flow basins. In an ideal complete mixing unit, the influent is immediately dispersed throughout the volume, and the concentration of reactant in the effluent is equal to that in the mixing liquid (Figure 7-4a), For steady-state conditions the reaction kinetics are given by Eq. 7-1. Applications of complete mixing in water treatment are rapid (flash or quick) mix tanks used to blend chemicals into raw water for coagulation, and the mixing and reaction zone in flocculator-clarifiers. In waste-water treatment, high-rate activated sludge and extended aeration basins are complete mixing.

$$t = \frac{V}{Q} = \frac{1}{K}\left(\frac{C_0}{C_t} - 1\right)$$

(7-1)

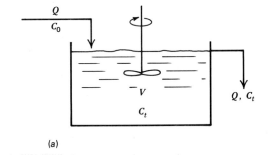

(a)

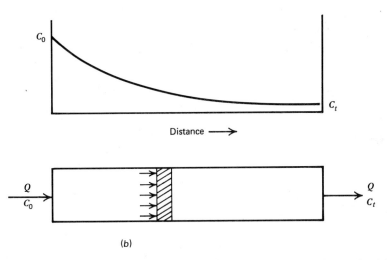

Distance ⟶

(b)

Figure 7-4 Ideal complete-mixing and plug-flow basins. Symbols are defined in Eq. 7-1. (a) Complete mixing. (b) Plug flow.

where
t = detention time
V = volume of basin
Q = quantity of flow
K = rate constant
C_0 = influent reactant concentration
C_t = effluent reactant concentration

In an ideal plug flow system, the water flows through a long chamber at a uniform rate without intermixing. The concentration of reactant decreases along the direction of flow remaining within the imaginary plug of water moving through the basin (Figure 7-4b). For steady-state conditions the relationship between detention time and concentration, applying first-order kinetics, is given in Eq. 7-2. In practice, it is very difficult to achieve ideal plug flow because of short-circuiting and intermixing caused by frictional resistance along the walls, density currents, and turbulent flow. In waste-water treatment, flow through long narrow aeration tanks approaches plug flow. Flocculation basins in chemical water processing are designed as plug flow units using baffles to reduce short-circuiting.

$$t = \frac{V}{Q} = \frac{L}{v} = \frac{1}{K}\left(\log_e \frac{C_0}{C_t}\right)$$

(7-2)

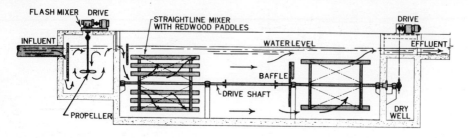

(a)

(b)   *(c)*

Figure 7-5 Straightline rapid mix and flocculation basin used in chemical treatment of water. (*a*) Longitudinal section of flash and horizontal slow mixers. (*b*) Propeller mixer provides rapid blending of chemicals into water. (*c*) Slow, horizontal-paddle mixer allows flocculation. (LINK-BELT Product, Courtesy of Environmental Equipment Division, FMC Corporation)

where $t, V, Q, C_0,$ and C_t are same as Eq. 7-1
                         L = length of rectangular basin
                         v = horizontal velocity of flow
                         K = rate constant for first order kinetics

The in-line mixing and flocculation unit shown in Figure 7-5 has a propeller to provide rapid mixing of chemicals with the raw water, and horizontal paddle units for gentle, slow mixing to build settleable floc. The redwood paddle flocculators (Figure 7-5*c*) are separated by baffles and have variable speed drives so that the peripheral speed of the paddles can be set between 0.6 and 1.5 ft/sec for optimum flocculation.

The 1968 *Recommended Standards for Water Works, Great Lakes-Upper Mississippi River Board of State Sanitary Engineers (GLUMRB)*[1] recommend that quick mixing, for rapid dispersion of chemicals throughout the water, be performed with mechanical mixing devices with a detention time not less than 30 sec.

These recommendations are made for flocculation basins:

a. Inlet and outlet design shall prevent short-circuiting and destruction of floc.
b. Minimum flow-through velocity shall not be less than 0.5 nor greater than 1.5 ft/min with a detention time for floc formation of, at least, 30 min.
c. Agitators shall be driven by variable speed drives with the peripheral speed of paddles ranging from 0.5 to 2.5 ft/sec. Flocculation and sedimentation basins shall be as close together as possible. The velocity of flocculated

water through conduits to settling basins shall not be less than 0.5 nor greater than 1.5 ft/sec. Allowances must be made to reduce turbulence at bends and changes in direction.

d. Baffling (rather than mechanical mixers) may be used to provide flocculation in small plants, but only after consultation with the reviewing authority.

EXAMPLE 7-1

Based on laboratory studies, the rate constant for a chemical coagulation reaction was found to be first-order kinetics with a K equal to 75 per day. Calculate the detention times required in complete mixing and plug flow reactors for a 80 percent reduction, $C_0 = 200$ mg/l and $C_t = 40$ mg/l.

Solution

Using Eq. 7-1 for complete mixing

$$t = \frac{day}{75} \times 1440 \frac{min}{day} \left(\frac{200 \text{ mg/l}}{40 \text{ mg/l}} - 1 \right) = 77 \text{ min}$$

For plug flow, substituting into Eq. 7-2,

$$t = \frac{1440}{75} \left(\log_e \frac{200}{40} \right) = 19.2(2.3 \log 5) = 19.2 \times 2.3 \times 0.7 = 31 \text{ min}$$

7-2 SEDIMENTATION

Sedimentation, or clarification, is the removal of particulate matter, chemical floc, and precipitates from suspension through gravity settling. The common criteria for sizing settling basins are: detention time, overflow rate, weir loading and, with rectangular tanks, horizontal velocity. Detention time, expressed in hours, is calculated by dividing the basin volume by average daily flow, Eq. 7-3.

$$t = \frac{V \times 24}{Q} \tag{7-3}$$

where
$t = $ detention time, hours
$V = $ basin volume, million gallons
$Q = $ average daily flow, million gallons per day
$24 = $ number of hours per day

The overflow rate (surface loading) is equal to the average daily flow divided by total surface area of the settling basin, expressed in units of gallons per day per square foot, Eq. 7-4.

$$V_0 = \frac{Q}{A} \tag{7-4}$$

where
$V_0 = $ overflow rate (surface loading), gallons per day per square foot
$Q = $ average daily flow, gallons per day
$A = $ total surface area of basin, square feet

Most settling basins in water treatment are essentially up-flow clarifiers where the water rises vertically for discharge through effluent channels; hence, the ideal basin shown in Figure 7-6 can be used for explanatory purposes. Water

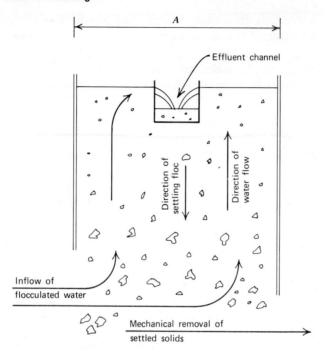

Figure 7-6 Ideal sedimentation basin. Symbols are defined in Eq. 7-4.

entering a settling basin is forced to the bottom behind a baffle wall, and then rises vertically overflowing the weir of a discharge channel at the tank surface. Flocculated particles settle downward, in a direction opposite to the flow of water, and are removed from the bottom by a continuous mechanical sludge removal apparatus. The particles with a settling velocity v greater than the overflow rate Q/A are removed while lighter flocs, with settling velocities less than the overflow rate, are carried out in the basin effluent.

Weir loading is computed by dividing the average daily quantity of flow by the total effluent weir length, and expressing the results in gallons per day per foot.

Sedimentation basins may be rectangular, circular, or square. They are designed for slow uniform water movement with a minimum of short-circuiting. The rectangular tank in Figure 7-7*a* contains partitioning baffles to guide the flow vertically to collecting troughs that extend across, and around the periphery, of the clarifier. Scrapers drawn by endless chains slowly move the settled solids to a sludge hopper at the inlet end. Figure 7-7*b* illustrates a circular clarifier where the influent enters and the effluent leaves from channels around the periphery of the tank. The sludge removal arm, that revolves slowly around a center shaft, is a tapered tube with openings spaced along the length. Instead of a scraping action, the settled solids are picked up by a slight suction action and are conveyed through the arm to be discharged from the bottom of the tank at the center. Figure 7-8 illustrates an in-line flash mixer, flocculator with paddle units set perpendicular to the direction of flow, and a square center-feed clarifier. Water enters the clarifier through a center column and is directed downward by an influent well (circular baffle). The flow radiates out from the center and overflows a peripheral weir. Settled solids are scraped to a hopper for discharge by a center-driven rake arm equipped with corner blades.

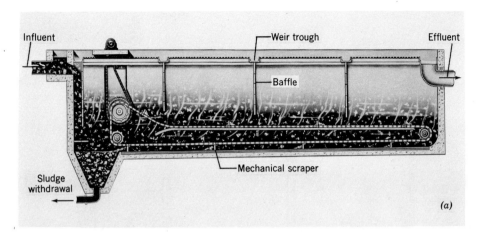

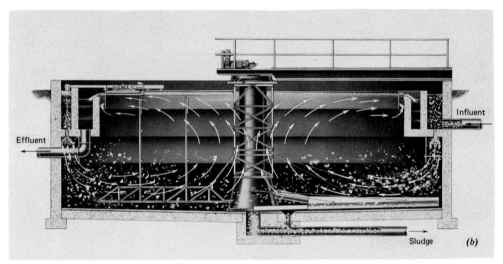

Figure 7-7 Cross-sectional diagrams of rectangular and circular sedimentation basins used in water treatment. (*a*) Rectangular, baffled clarifier with a conventional plow-type scraper mechanism. (*b*) Circular rim-flow clarifier with a rotating, hydraulic sludge pickup arm. (Courtesy of Water Quality Control Division, Envirex, a Rexnord Company)

Presedimentation basins for settling muddy river water are generally circular with hopper bottoms, equipped with scraper-type sludge removal arms. *GLUMRB Standards* recommend a detention time of not less than 3 hr. For basins following chemical flocculation, the *Standards* recommend the following: minimum detention time of 4 hr, maximum horizontal velocity through the settling basin of 0.5 ft/min, and maximum weir loading of 20,000 gpd/ft of weir length. Overflow rates generally fall in the range of 500 to 800 gpd/sq ft.

EXAMPLE 7-2

The in-line system shown in Figure 7-8 has the following sized units: flash mixing chamber with a volume of 1500 gal; a flocculator 15 ft wide, 70 ft long, and 8.0 ft liquid depth; and a settling tank 75 ft square, 12 ft liquid depth, and 300 ft of effluent weir. Calculate the major parameters used in design of these units based on a water flow of 3.0 mgd.

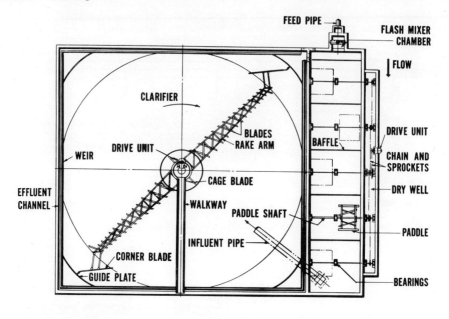

PLAN

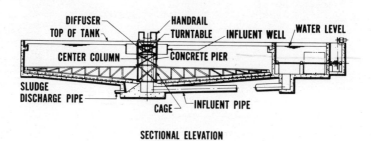

SECTIONAL ELEVATION

Figure 7-8 Flocculator and square sedimentation tank for water clarification. (Courtesy of Dorr-Oliver, Inc.)

Solution

Flow = 3.0 mgd = 2080 gpm = 401,000 cu ft/day = 278 cu ft/min

Detention time in flash mixer

$$t = \frac{1500 \text{ gal}}{2080 \text{ gal/min}} \times 60 \frac{\text{sec}}{\text{min}} = 43 \text{ sec} \qquad (>30 \text{ sec } GLUMRB)$$

Detention time and horizontal velocity in flocculator

$$t = \frac{15 \text{ ft} \times 70 \text{ ft} \times 8.0 \text{ ft}}{278 \text{ cu ft/min}} = 30 \text{ min} \qquad (=30 \text{ min } GLUMRB)$$

$$v = \frac{Q}{A} = \frac{278 \text{ cu ft/min}}{15 \text{ ft} \times 8.0 \text{ ft}} = 2.3 \text{ ft/min} \qquad (>1.5 \text{ ft/min})$$

Settling time, weir loading, and overflow rate in clarifier

$$t = \frac{75 \text{ ft} \times 75 \text{ ft} \times 12 \text{ ft}}{278 \text{ cu ft/min}} \times \frac{\text{hr}}{60 \text{ min}} = 4.0 \text{ hr} \qquad (=4.0 \text{ hr } GLUMRB)$$

$$\text{Weir loading} = \frac{3,000,000 \text{ gal/day}}{300 \text{ ft}} = 10,000 \text{ gpd/ft} \qquad (<20,000 \text{ gpd/ft})$$

$$V_0 = \frac{3,000,000 \text{ gal/day}}{75 \text{ ft} \times 75 \text{ ft}} = 530 \text{ gpd/sq ft}$$

7-3 FLOCCULATOR-CLARIFIERS

Flocculator-clarifiers, also referred to as solids contact units or up-flow tanks, combine the processes of mixing, flocculation, and sedimentation in a single compartmented tank. One such unit, shown in Figure 7-9, introduces coagulants, or softening chemicals, in the influent pipe and mixes the water under a central cone-shaped skirt where a high-floc concentration is maintained. Flow passing under the hood is directed through the sludge blanket at the bottom of

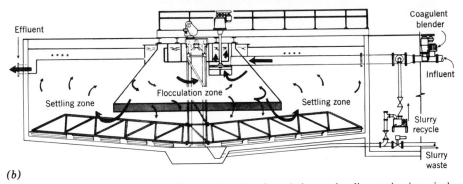

(b)

Figure 7-9 Flocculator-clarifier provides rapid mixing, flocculation, and sedimentation in a single compartmented tank. (a) Photograph of a flocculator-clarifier in a water softening plant. (Metropolitan Utilities District, Omaha, Nebraska) (b) Cross-sectional view of unit pictured above. (Courtesy of Walker Process Division of Chicago Bridge & Iron Company)

the tank, to promote growth of larger conglomerated clusters, where the heavier particles have settled. Overflow rises upward in the peripheral settling zone to radial weir troughs, or circular inboard troughs suspended at the surface. These units are particularly advantageous in lime softening of groundwater, since the precipitated solids help seed the floc, growing larger crystals of precipitate to provide a thicker waste sludge. Recently, flocculator-clarifiers are receiving wider application in the chemical treatment of industrial wastes and surface water supplies. The major advantages promoting their use are reduced space requirements and less costly installation. However, the unitized nature of construction generally results in a sacrifice of operating flexibility.

The *GLUMRB Standards* stress that solids contact units are most acceptable for clarification in combination with softening for the treatment of waters with relatively uniform characteristics and flow rates. They recommend the following in sizing units: flocculation and mixing time of not less than 30 min; minimum detention time of 2 hr for suspended solids contact clarifiers, and 1 hr for suspended solids contact softeners; weir loadings not exceeding 10 gpm/ft for clarifier units, and 20 gpm/ft for softener units; and upflow rates not to exceed 1.0 gpm/sq ft for clarifiers, and 1.75 gpm/sq ft for softeners. The volume of sludge removed from these units should not exceed 3 percent of the water treated for softening units nor 5 percent for clarifiers.

7-4 FILTRATION

The rapid sand, or gravity, filter is the most common type used in water treatment to remove nonsettleable floc remaining after chemical coagulation and sedimentation. A typical sand filter bed (Figure 7-10), is placed in a concrete box with a depth of about 9 ft. The sand filter, about 2 ft deep, is supported by a graded gravel layer containing underdrains. During filtration, water passes downward through the filter bed by a combination of water pressure from above and suction from the bottom. Filters are cleaned by backwashing (reversing the flow) upward through the bed. Wash troughs suspended above the filter surface collect the backwash water and carry it out of the filter box.

The following description of filter operation follows the valve numbering in

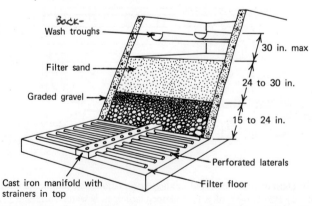

Figure 7-10 Cross section of a typical sand filter. (From *Water Supply and Treatment*, National Lime Association, 1962)

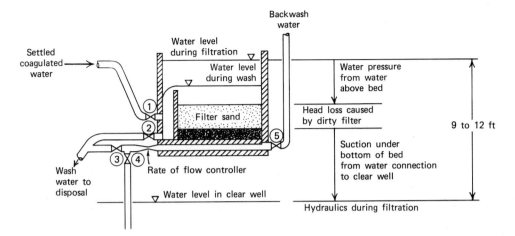

Figure 7-11 Diagrammatic sketch showing operation of a gravity filter.

Figure 7-11. Initially valves 1 and 4 are opened, and 2, 3, and 5 are closed for filtration. Overflow from the settling basin applied to the filter passes through the bed and underdrain system to the clear well underneath. The depth of water above the filter surface is between 3 and 4 ft. The underdrain pipe is trapped in the clear well to provide a liquid connection to the water being filtered, thus, preventing backflow of air into the underdrain. The maximum head available for filtration is equal to the difference between the elevation of the water surface above the filter and the level in the clear well, this is commonly 9 to 12 ft. When the filter is clean, flow through the bed must be regulated to prevent an excessive filtration rate. A rate of flow controller, consisting of a valve controlled by a venturi meter, throttles flow in the discharge pipe restraining the rate of filtration. As the filter collects impurities, the resistance to flow increases and the flow contoller valve opens wider to maintain a preset rate. When the measured head loss through a filter is approximately 8 ft, the bed is cleaned by backwashing. Valves 1 and 4 are closed (3 remaining closed), and 2 and 5 are opened. Clear water flows into the filter underdrain and passes upward through the bed. The sand layer expands hydraulically about 50 percent and the sand grains are scrubbed by rubbing against each other in the turbulent backwash flow. Dirty wash water is collected by troughs and conveyed to disposal. The first few minutes of filtered water at the beginning of the next run is generally wasted to flush the wash water remaining in the bed out through the drain. This is accomplished by opening valve 3 when valve 1 is opened to start filtration (valves 2, 4, and 5 are shut). Opening valve 4 while closing 3 permits filtration to proceed again.

Filter Media

The action that takes place in a filter bed is extremely complex, consisting of straining, flocculation, and sedimentation. Gravity filters do not function properly unless the applied water has been chemically treated, and settled to remove the large floc. Coagulant carryover is essential in removing microscopic particulate matter that would otherwise pass through the bed. If an excessive quantity of large floc overflows the settling basin, a heavy mat forms on the filter surface by straining action and clogs the bed. However, the impurities in improperly coagulated water may penetrate too far into the bed and may be flushed through before being trapped, causing a turbid effluent. Optimum filtration occurs when nonsettleable coagulated floc are held in the pores of the bed and produce "in-depth" filtration. The ideal filter media possesses the following characteristics: it is coarse enough for large pore openings to retain large quantities of floc yet is sufficiently fine to prevent passage of suspended solids; it has adequate depth to allow relatively long filter runs; and it is graded to permit effective cleaning during backwash. The earliest filter medium used was sand of nearly uniform size in the range of 0.4 to 0.5 mm (Figure 7-10). After backwashing, a sand bed becomes hydraulically graded with the largest grains on the bottom and the finest on the top. This forms a bed where most of the removal takes place at or near the surface, which is undesirable. The earliest modifications to this single medium filter was to increase the sand size to the range of 0.6 to 0.8 mm, yielding longer filter runs and greater depth of penetration while reducing the undesirable effect of surface straining. Another problem of early sand filters was poor cleaning during hydraulic backwashing. Mud balls and other heavy impurities, formed during filtration of certain waters, would drop down to the gravel underlayer when the bed was expanded rather than being washed out in the overflow. Higher backwash rates in an attempt to resolve this problem often resulted in disturbance of the graded gravel underdrain, or in carryover of sand in the wash water. Several devices were developed to improve backwashing by increasing the scrubbing action in the expanded bed. Most current installations provide some sort of washing aid, like the surface wash agitator shown in Figure 7-12. This unit consists of a rotating arm, placed just above the sand and beneath the wash troughs, with nozzles directing water jets at right angles to the bed surface. After the sand layer has been expanded by introducing backwash, water under pressure is introduced to the agitator which rotates, increasing turbulence for better washing action. An alternate system consists of fixed nozzles that spray water into an expanded bed.

A coal-sand dual-media filter permits use of a relatively coarse coal medium (1.4 to 1.6 specific gravity) over a finer sand bed (2.6 specific gravity), both generally 12 in. deep. The upper layer of coarser anthracite has voids about 20 percent larger than the sand, and thus a coarse-to-fine grading of media is provided in the direction of flow. After backwashing, the bed stratifies with the heavier sand on the bottom and the lighter, coarser coal medium on top. Larger floc particles are adsorbed and trapped in the surface coal layer, while finer material is held in the sand filter; therefore, the bed filters in greater depth, preventing premature surface plugging (Figure 7-13).

Mixed media beds using coal, sand, and garnet very closely approach the ideal filter (Figure 7-14). Garnet, the finest medium has a specific gravity of about 4.2, which is greater than the sand or coal. The three media are sized so that intermixing of these materials occurs after backwashing with no discrete interface between the three. This eliminates stratification and more closely approximates the idea of a uniform decrease in pore space with increasing filter depth. A typical filter has a particle size gradation decreasing from about 2 mm at the top to 0.2 mm at the bottom. In addition to gravity filters for water treatment, mixed media beds are being used in processing waste water in both open and pressure filters.

Operating rate for filters is determined by quality of raw water, chemical

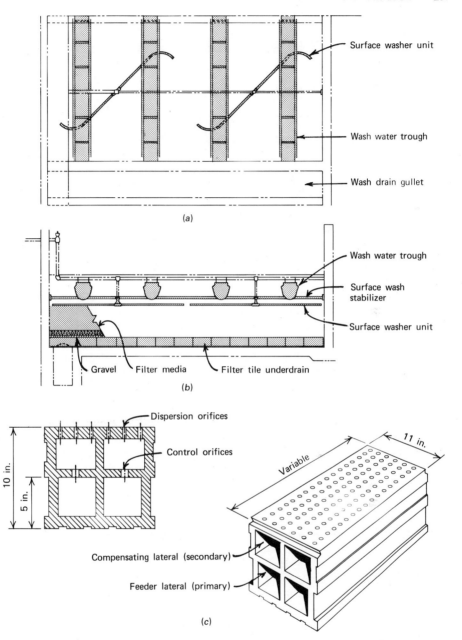

Figure 7-12 A gravity filter showing placement of surface agitators relative to filter bed and wash troughs, and details of underdrain block. (*a*) Plan view of filter bed. (*b*) Sectional view. (*c*) Detail of underdrain filter block. (Courtesy of F. B. Leopold Company, Division of Sybron Corporation)

treatment provided, and filter media. The nominal rate (historically) for a rapid sand filter is 2 gpm/sq ft. Recent systems allow higher rates up to 5 gpm/sq ft, with 8 gpm/sq ft being used in some current designs.

Filter Underdrains

The earliest form of underdrain, still in wide use, is illustrated in Figure 7-10. It consists of a cast-iron manifold with perforated laterals buried in a graded

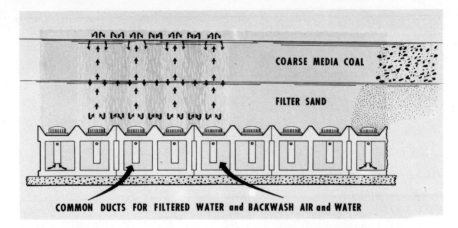

COARSE MEDIA COAL

FILTER SAND

COMMON DUCTS FOR FILTERED WATER and BACKWASH AIR and WATER

Figure 7-13 Sand-coal dual-media filter, with Camp underdrain system, showing air-washing action. (Courtesy of Walker Process Division of Chicago Bridge & Iron Company)

gravel layer up to 2 ft in depth. During filtration the gravel supports the sand and provides large pore openings that permit a flow of water to the collector pipes that lead to the clear well. During backwashing, treated water is flushed up through the bed at a rate of about 15 gpm/sq ft for a period of 5 to 10 min. The gravel underdrain is heavy enough to stay in place while distributing the backwash under the sand bed for uniform hydraulic expansion. A bed is out of operation for about 15 min in the cleaning process, and the amount of water used in backwashing is 2 to 4 percent of the filtered water.

A number of proprietary filter bottoms, including porous plates, filter blocks, synthetic spheres, and nozzles have been developed. The filter bottom in Figure 7-12 is vitrified tile blocks having extruded upper and lower laterals, and punched orifices. An 8-in. gravel layer with a gradation between $\frac{1}{8}$ in. and $\frac{3}{4}$ in. is placed between the block and filter media. Wash water enters the tile

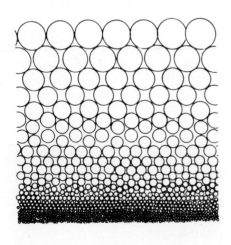

(a)

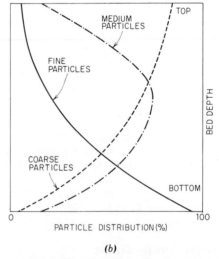

(b)

Figure 7-14 A mixed media filter using coal, sand, and garnet approaches the ideal distribution for a filter bed. (a) Ideal filter bed with decreasing particle size from top to bottom. (b) Particle size distribution in a mixed media filter. (Courtesy of Neptune Microfloc Inc., subsidiary of Neptune International Corp.)

underdrain through the feeder lateral, rises through control orifices (holes) to the compensating lateral and, finally, through the dispersion orifices. The dual lateral feature provides for uniform distribution of water, thus preventing disruption of the gravel layer. Figure 7-12 also shows the plan view of a filter unit with wash troughs spaced approximately 6 ft apart and sloping to a common wash drain gullet.

The underdrain system in Figure 7-13 consists of vitrified clay blocks fitted with thermoplastic resin nozzles. Backwashing of dual media filters have resulted in problems when ordinary backwash and underdrain systems are employed. Nonuniform hydraulic expansion and poor scouring of dual media filters can result in mud balls dropping through the coal medium and lodging on top of the sand layer. A combination air and water backwash system has proved to be effective for cleaning dual filters. Air scour provides thorough agitation of the filter bed breaking up any partially cemented aggregates. The steps used for performing air and water backwashing of the filter in Figure 7-13 are: lower the filter water level below the wash water troughs to about the 20 percent expansion level; introduce air at 4 cu ft/min/sq ft and scour media for 3 min; and then introduce wash water for about 3 to 4 min to purge and stratify the media. Rewash filter if necessary.

Filter Control

The control console for each filter unit has a head loss gauge, flow meter, and rate controller. A filter run is normally terminated when head loss through the filter reaches a prescribed value between 6 and 9 ft, or when effluent turbidity exceeds the amount considered acceptable. As suspended matter collects in the pores of the filter media and head loss increases through the bed, the lower portion of the filter is under suction from the water column connected through the piping to the clear well (Figure 7-11). This partial vacuum permits release of dissolved gases that tend to collect in the filter pores and, as a result, cause air binding and a reduced rate of filtration. In addition to shortening filter runs, the gases accumulated in a bed may tend to discharge violently during the start of the backwash cycle potentially causing disruption of the gravel base.

The rate controller prevents rapid changes in filtration rates and controls the velocity of flow through a clean bed. Continuous-flow photoelectric turbidimeters are used for monitoring filter effluent quality. By detecting turbidity in the filtered water, the operator can terminate a filter run if the quality deteriorates prior to reaching the maximum head loss.

Other Types of Filters

Pressure filters have media and underdrains contained in a steel tank. Media are similar to those used in gravity filters, and filtration rates range from 2 to 4 gpm/sq ft. Water is pumped through the bed under pressure, and the unit is backwashed by reversing the flow, flushing out impurities. Pressure filters are not generally employed in large treatment works because of size limitations; however, they are popular in small municipal water plants that process groundwater for softening and iron removal. Their most extensive application has been treating water for industrial purposes.

Diatomaceous earth filters generally have been limited to swimming pools, portable field units, industrial applications, and installations for small towns. The filter medium is supported on a fine metal screen, porous ceramic material, or synthetic fabric. The filtration cycle consists of precoating the medium, filtering water to a predetermined head loss, and removal of the filter cake. During the precoat operation, a slurry of diatomite in water is used to deposit a filter on the screen. Raw water with a small amount of diatomaceous earth (body feed) is drawn through the filter formed by the initial layer. Body feed continually builds the filter so that surface clogging is minimized. When maximum pressure drop, or minimum filtration rate is reached, the filter medium is washed by reversing the flow discharging the dirty cake to waste. The cycle is then repeated. Common rates of filtration range from 1 to 5 gpm/sq ft.

Microstrainers are mechanical filtering devices with a medium normally consisting of finely woven, stainless steel fabric. The pore openings of the fabric are fine enough to remove microscopic organisms and debris. A common design consists of a rotating drum with the fabric mounted on the periphery; raw water passes from the inside to the outside of the drum. Backwashing is usually continuous using wash water jets. These units have found greatest application in treatment of industrial waters and final polishing filtration of waste-water effluents.

Slow sand filters are of historical interest, although some installations may still be in use. They are limited to low turbidity waters not requiring chemical treatment. Filtering action is a combination of straining, adsorption, and biological extraction. Biological growths form on the surface of the bed and are cleaned off periodically by removing the upper few inches of sand.

EXAMPLE 7-3

The filter unit illustrated in Figure 7-12 is 15 ft by 30 ft. After filtering 2.5 mil gal in a 24-hr period, the filter is backwashed at a rate of 15 gpm/sq ft for 15 min. Compute the average filtration rate, quantity, and percentage of treated water used in washing, and the rate of wash-water flow in each trough.

Solution

$$\text{Filtration rate} = \frac{2{,}500{,}000 \text{ gal/day}}{15 \text{ ft} \times 30 \text{ ft} \times 1440 \text{ min/day}} = 3.9 \frac{\text{gpm}}{\text{sq ft}}$$

$$\text{Quantity of wash water} = \frac{15 \text{ gal}}{\text{min} \times \text{sq ft}} \times 15 \text{ min} \times 15 \text{ ft} \times 30 \text{ ft} = 100{,}000 \text{ gal}$$

$$\frac{\text{Wash water}}{\text{Treated water}} = \frac{100{,}000 \text{ gal}}{2{,}500{,}000 \text{ gal}} \times 100 = 4.0 \text{ percent}$$

$$\text{Wash-water flow in each trough} = \frac{100{,}000 \text{ gal}}{15 \text{ min} \times 4 \text{ troughs}} = 1670 \text{ gpm}$$

7-5   CHEMICAL FEEDERS

A chemical feeder is a mechanical device for measuring quantities of chemical and applying them to a water at a preset rate. Liquid feeders apply chemicals in solutions or suspensions; dry feeders apply them in granular or powdered forms. Some chemicals such as ferric chloride, polyphosphates and sodium silicate must be fed in solution form while others—ferrous sulfate

Figure 7-15 Solution feeder (diaphragm metering pump) for applying chemical solutions and slurries in water treatment. (Courtesy of Wallace & Tiernan Division, Pennwalt Corp.)

and alum—are fed dry. If a chemical does not dissolve readily, it may be applied dry or in suspension provided that the solution is continuously stirred.

Solution feeders are small positive displacement pumps that deliver a specific volume of liquid for each stroke of a piston or rotation of an impeller. Diaphragm and plunger metering pumps are the two general types used for delivery against pressure. The former, illustrated in Figure 7-15, contains a flexible diaphragm driven by a mechanical linkage that converts the rotary motor input to a reciprocating push-rod motion. Inlet and outlet valves are operated by suction and pressure created in the diaphragm chamber. Stroke length, and consequently rate of feed, may be controlled manually at a fixed rate, or with a pneumatic control unit that provides stroking in proportion to water flow or an automatic water quality feedback signal.

The plunger pump has a reciprocating piston that alternately forces solution out of a chamber and then, on its return stroke, refills the chamber by pulling solution from a reservoir. In addition, several types of rotary pumps qualify as positive displacement feeders including gear, swinging vane, sliding vane, oscillating screw, eccentric, and cam pumps.

The criteria used in selecting a solution feeder are: accuracy, durability, capacity, corrosion resistance, and pressure capability. Most manufacturers have separate designs for very corrosive chemicals, milder chemicals, and slurries.

The two types of dry feeders are volumetric and gravimetric, depending on whether the chemical is measured by volume or weight. Volumetric dry feeders are simpler, less expensive, and slightly less accurate than gravimetric units.

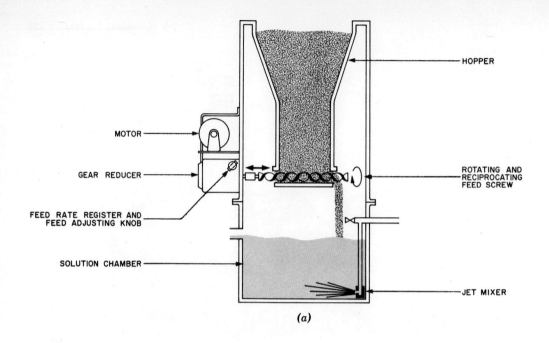

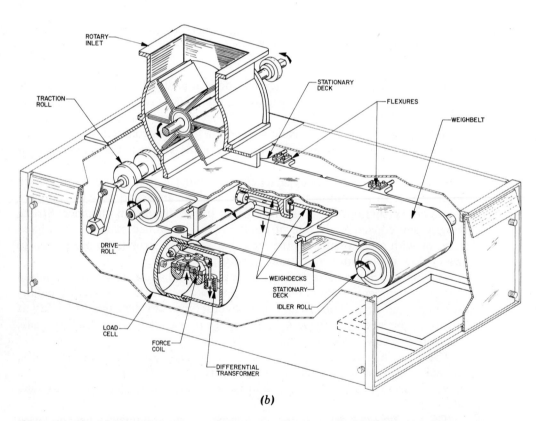

Figure 7-16 Dry feeders, volumetric and gravimetric types, are used for metering powdered chemicals. (*a*) Screw-feed volumetric dry feeder. (*b*) Belt-type gravimetric feeder. (Courtesy of Wallace & Tiernan Division, Pennwalt Corp.)

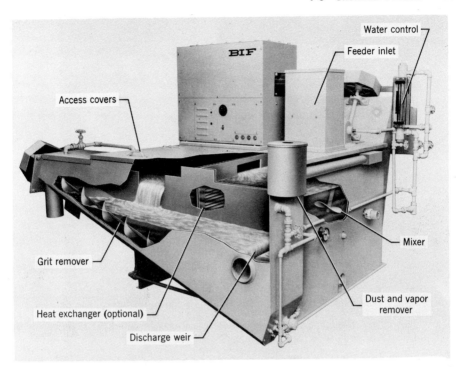

Figure 7-17 Lime slaker mixes powdered lime with water forming a slurry that is fed by a solution pump in treatment of water. (Courtesy of BIF)

Several types are available with a variety of feed mechanisms that include: rotating roller, disk, or screw, star wheel, moving belt, vibratory pan, and oscillating hopper. The unit shown schematically in Figure 7-16a feeds material from a hopper by a rotating screw that moves back and forth. The feed screw is designed so that chemical discharges first from one end of the screw and then the other. Rate of feed is controlled by regulating the speed of the helical screw.

Gravimetric dry feeders are extremely accurate, available in large sizes to deliver high feed rates, readily adaptable to recording and automatic control, and more expensive than volumetric feeders. There are two general types: one weighs the material on a section of moving belt, and the other operates on loss-in-weight of the feeder hopper. The extreme accuracy is accounted for by the fact that material is dispensed by weight, removing the variables of density and consistency. The latter do affect the feed rate of volumetric units. Figure 7-16b is an operational diagram of a belt type feeder. Chemical enters on a weigh belt over a stationary deck. Movement of the belt conveys the material over a weigh deck section which continuously transmits the quantity to a loading beam scale. As the amount of material on the platform varies the scale beam moves, actuating a pneumatic beam balancing cylinder that, in turn, controls belt speed and loading to maintain the preset feed rate.

Applying lime requires a special type of unit that is referred to as a slaker.* The chemical equation for slaking is

$$\text{CaO (lime)} + \text{H}_2\text{O} \rightarrow \text{Ca(OH)}_2 \text{ (lime slurry)} \qquad (7\text{-}3)$$

---

* Slake means: To cause to heat and crumble by treatment with water.

Figure 7-18 Photograph of gravimetric feeders in a water treatment plant. The unit on the right is a lime slaker; the one in the center is feeding alum.

A gravimetric feeder applies powdered lime to a mixing chamber where dilution water is supplied to bring the slurry to desired strength (Figure 7-17). The slaker proportions lime feed to water maintaining the correct concentration; pH controls are available to signal small changes in dosage to compensate for variations in water composition and purity of the lime. The ratio of water to lime is about 5 to 1 and the process time is 30 min. Regulators and alarms protect against excessive temperatures caused by the chemical reaction. An automatic grit separator removes coarse, inert material from the slurry before feeding. The lime is pumped from the holding tank to a slurry feeder that is controlled by an automatic pH and flow control system.

Figure 7-18 shows chemical feeders in a water treatment plant. Each is supplied from a hopper-bottom, chemical storage tank that extends through the ceiling to the floor above. Mechanical vibrators attached to the hoppers keep the chemicals flowing freely out into the feeders. After measuring the dry chemicals into continuously stirred dissolving chambers, the solutions flow either by gravity into an open flume or are transferred by pumps into a pressure main.

7-6 CHLORINATION

Chlorine is used in water and waste-water treatment for disinfection to destroy pathogens and control nuisance microorganisms, and for oxidation. As an oxidant, it is used in iron and manganese removal, destruction of taste and odor compounds, and elimination of ammonia nitrogen.

Chlorine Chemistry

Chlorine is a heavier than air, greenish yellow-colored, toxic gas. One volume of liquid chlorine confined in a container under pressure yields about

450 volumes of gas. It is a strong oxidizing agent reacting with most elements and compounds. Moist chlorine is extremely corrosive; consequently, conduits and feeder parts in contact with chlorine are either silver, special alloys, or nonmetal. The vapor is a respiratory irritant that can cause serious injury if exposure to a high concentration occurs.

Hypochlorites are salts of hypochlorous acid (HOCl). Calcium hypochlorite ($Ca(OCl)_2$) is the predominant dry form used in the United States. High-test calcium hypochlorites, available commercially in granular powdered or tablet forms, readily dissolve in water, and contain about 70 percent available chlorine. Sodium hypochlorite (NaOCl) is commercially available in liquid form at concentrations between 5 and 15 percent available chlorine. Most water treatment plants in the United States use liquid chlorine, since it is less expensive than hypochlorites. The latter are used in swimming pools, small waterworks, and in emergencies.

Chlorine combines with water forming hypochlorous acid which, in turn, can ionize to the hypochlorite ion. Below pH 7 the bulk of the HOCl remains unionized, while above pH 8 the majority is in the form of OCl^-, Eq. 7-4.

$$Cl_2 + H_2O \longrightarrow HCl + HOCl \underset{pH<7}{\overset{pH>8}{\rightleftharpoons}} H^+ + OCl^- \qquad (7\text{-}4)$$

Hypochlorites added to water yield the hypochlorite ion directly, Eq. 7-5.

$$Ca(OCl)_2 + H_2O \rightarrow Ca^{++} + 2OCl^- + H_2O \qquad (7\text{-}5)$$

Chlorine existing in water as hypochlorous acid and hypochlorite ion is defined as free available chlorine.

Chlorine readily reacts with ammonia in water to form chloramines as follows:

$$HOCl + NH_3 \rightarrow H_2O + NH_2Cl \qquad \text{(monochloramine)} \qquad (7\text{-}6)$$

$$HOCl + NH_2Cl \rightarrow H_2O + NHCl_2 \qquad \text{(dichloramine)} \qquad (7\text{-}7)$$

$$HOCl + NHCl_2 \rightarrow H_2O + NCl_3 \qquad \text{(trichloramine)} \qquad (7\text{-}8)$$

The reaction products formed depend on pH, temperature, time, and initial chlorine to ammonia ratio. Monochloramine and dichloramine are formed in the pH range of 4.5 to 8.5. Above pH 8.5 monochloramine generally exists alone, but below pH 4.4 trichloramine is produced. Chlorine existing in chemical combination with ammonia nitrogen or organic nitrogen compounds is defined as combined available chlorine. (The laboratory test procedure for available chlorine is given in Section 2-8.)

When chlorine is added to water containing ammonia, the residuals that develop yield a curve similar to that shown in Figure 7-19. The straight line from the origin is the concentration of chlorine applied, or the residual chlorine if all of that applied appeared as residual. The curved line represents chlorine residuals, corresponding to various dosages, remaining after a specified contact time, such as 20 min. Chlorine demand at a given dosage is measured by the vertical distance between the applied and residual lines. This represents the amount of chlorine reduced in chemical reactions and, therefore, the amount that is no longer available. With molar chlorine to ammonia-nitrogen ratios less than 1 to 1, monochloramine and dichloramine are formed with the relative amounts depending on pH and other factors. Higher dosages of chlorine

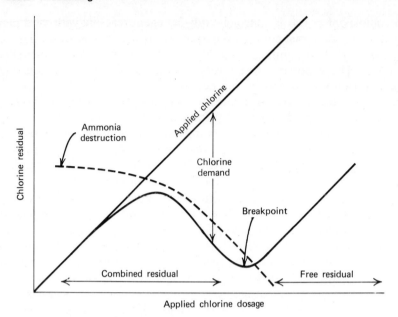

Figure 7-19 Typical breakpoint chlorination curve.

increase the chlorine to nitrogen ratio and result in oxidation of the ammonia and reduction of the chlorine. Theoretically, three moles of chlorine react with two moles of ammonia to drive nitrogen off as gas and to reduce chlorine to the chloride ion,

$$2NH_3 + 3Cl_2 \rightarrow N_2 + 6HCl \tag{7-9}$$

Chloramine residuals decline to a minimum value referred to as the breakpoint. Dosages in excess of this breakpoint produce free chlorine residuals. The breakpoint curve is unique for each water tested, since chlorine demand depends on the concentration of ammonia, presence of other reducing agents, contact time between chlorine application and residual testing, and other factors.

Disinfection

The most common application of chlorination is disinfection of drinking water to destroy microorganisms that cause diseases in man. The bactericidal action of chlorine results from a chemical reaction between HOCl and the bacterial or viral cell structure, inactivating required life processes. Rate of disinfection depends on the concentration and form of available chlorine residual, time of contact, pH, temperature, and other factors. Hypochlorous acid is more effective than hypochlorite ion; therefore, the power of free chlorine residual decreases with increasing pH. Bactericidal action of combined available chlorine is significantly less than that of free chlorine residual.

Information is unavailable to provide specific application rates in chlorination to achieve 100 percent kill of all microorganisms of sanitary significance for the variety of water supplies being treated for domestic use. The water source selected should, of course, be the least polluted available, and regulations should be set up to protect its quality. Current disinfection practice is based on establishing a given kind and amount of chlorine residual during treatment and, then, maintaining an adequate residual to the customer's faucet. Thus a major part of quality control is testing water in the distribution system for chlorine residual. Effectiveness of the disinfection process is determined by testing for the coliform group as indicators of water quality. The sensitivity of bacteria to chlorination is well understood, while the effect on protozoans and viruses has not been clearly delineated. Protozoal cysts and enteric viruses are more resistant to chlorine than are coliforms and other enteric bacteria. On the other hand, very little evidence exists to indicate that current water treatment practices are not adequate. No outbreaks of waterborne viral or protozoal infections have been documented and, furthermore, waterborne diseases attributed to these pathogens are rare in the United States.

Recommended minimum bactericidal chlorine residuals are listed in Table 7-1. These are based on experiments at water temperatures between 20° and 25° C after a 10-min contact for free chlorine and a 60-min contact for combined available chlorine. Minimum residuals for cyst destruction and inactivation of viruses are considerably greater; only undissociated hypochlorous acid is considered an effective antiviral agent. Therefore, even though the chlorine residuals given in Table 7-1 are normally considered adequate, surface waters from polluted rivers and lakes are usually treated with heavier chlorine dosages. For example, breakpoint chlorination as the first step in treatment may be used to establish and retain a free chlorine residual through the treatment plant.

Water chlorination is practiced by using both free and combined residuals. The latter involves application of chlorine to produce chloramines with natural or added ammonia. Anhydrous ammonia is used if insufficient natural ammonia is present in the water. Gaseous ammonia feeding equipment is similar to that used for chlorine. Although combined residual is much less effective as a disinfectant than free chlorine, its most common application is as posttreatment following free residual chlorination to provide initial disinfection. Chloramines provide and maintain a stable residual throughout a distribution system.

Free residual chlorination involves establishing a free residual through the destruction of naturally present ammonia, if necessary, and maintaining it throughout the system. In treating a surface water, this may involve: prechlorination at breakpoint dosages for the first process step, postchlorination to establish a residual leaving the treatment plant, and rechlorination at selected points within the distribution system to retain an adequate free residual. High dosages of chlorine applied during treatment may result in residuals that are

Table 7-1. Recommended Minimum Bactericidal Chlorine Residuals for Disinfection

pH Value	Minimum Free Available Chlorine Residual After 10-Min Contact (mg/l)	Minimum Combined Available Chlorine Residual After 60-Min Contact (mg/l)
6.0	0.2	1.0
7.0	0.2	1.5
8.0	0.4	1.8
9.0	0.8	>3.0
10.0	0.8	>3.0

esthetically objectionable or undesirable for industrial water use. Dechlorination may be performed to reduce the chlorine residual by addition of a reducing agent, often called a dechlor. Sulfur dioxide is the chemical most widely used to dechlorinate in municipal plants; on a smaller scale, sodium bisulfite may be used. Aeration by submerged or spray aerators also releases some residual chlorine.

Oxidation

Chorine is a much stronger oxidizing agent for manganese than is dissolved oxygen. Therefore, one of the treatment processes for iron and manganese removal (Figure 7-1b) uses chlorine to remove these metals from solution. Hydrogen sulfide present in groundwater can be rapidly converted to the sulfate ion using chlorine.

$$H_2S + 4Cl_2 + 4H_2O \rightarrow H_2SO_4 + 8HCl \qquad (7\text{-}10)$$

Breakpoint chlorination in the treatment of surface waters may be used to destroy objectionable tastes and odors and to eliminate bacteria, minimizing biological growths on filters and aftergrowths in the distribution system.

Distribution System Chlorination

A new main after installation should be pressure tested, flushed to remove all dirt and foreign matter, and disinfected by one of the following methods prior to being placed in service. The continuous-feed method involves supplying water to the new main with a chlorine concentration of, at least, 50 mg/l. Either a solution feed chlorinator or a hypochlorite feeder injects chlorine into the water that fills the main. The chlorinated water should remain in the pipe for a minimum of 24 hr while all valves and hydrants along the main are operated to insure their disinfection. At the end of 24 hr, no less than 25 mg/l chlorine residual should be remaining. In the slug method, a continuous flow of water is fed to the main with a chlorine concentration of at least 300 mg/l. The rate of flow is set so that the column of chlorinated water contacts the interior surfaces of the main for a period of, at least, 3 hr. As the slug passes other connections, the valves are operated to insure disinfection of appurtenances. This method is used principally for large diameter mains where the continuous-feed technique is impractical. The tablet method, although commonly used for small diameter mains, is least satisfactory, since it precludes preliminary flushing. Calcium hypochlorite tablets are placed in each section of the pipe, hydrants, and other appurtenances during construction. The new main is then slowly filled with water to dissolve the tablets without flushing them to one end of the pipe. The final solution, with a residual of at least 50 mg/l, should remain in contact for a minimum of 24 hr. Following disinfection by any of these methods, the chlorinated water should be flushed to waste by using potable water. Bacteriological tests should then be conducted before placing the main in service.

A broken main is isolated by closing the nearest valves. The first step in repair involves flushing the broken section to remove contamination while pumping the discharge out of the trench. Minimum disinfection includes swabbing the new pipe sections and fittings with a 5 percent hypochlorite solution before installation, and flushing the main from both directions before returning the system to service. Where conditions permit, the repaired section should be isolated and disinfected by the procedures prescribed for new mains.

Tanks and reservoirs before being placed into service or following inspection and cleaning should be disinfected. One method is to add chlorine directly to the filling water, either using hypochlorite or a portable solution-feed chlorinator. Standard procedure is a 50 mg/l residual maintained for a minimum of 6 hr. An alternate method, where convenient, is to spray the walls and other surfaces with a solution containing about 500 mg/l of available chlorine.

After the construction or repair of wells, disinfection is performed by using a 50- to 100-mg/l free chlorine residual for 12 to 24 hr. A new well should be operated until the water is practically free of turbidity. A quantity of chlorine solution equal to at least, twice the volume of water in the well is added through a pipe or hose. The pump cylinder and drop pipe are then washed with a strong chlorine solution as the assembly is lowered into the casing. After a minimum contact of 12 hr, the chlorinated water is pumped to waste.

Equipment and Feeders

Liquid chlorine is shipped in pressurized steel cylinders. The most popular sizes are 100 or 500 lb, but 1-ton containers may be used in large installations. Storage areas should be cool, well ventilated, and protected from corrosive vapors and continuous dampness. Gas is withdrawn from a container through a valve connection at the top. Liquid chlorine in the cylinder vaporizes to allow continuous gas withdrawal. Design and operation of all facilities should minimize hazards associated with connecting, emptying, and disconnecting containers. The characteristic odor of chlorine provides warning of leaks. Since it reacts with ammonia to form dense white fumes, a leak can be readily detected by holding a cloth swab saturated with strong ammonia water near the suspected area. Calcium hypochlorite is relatively stable under normal conditions; however, reactions may occur with organic substances. Preferably it should be stored in a location segregated from other chemicals and materials.

The most essential unit in a chlorine gas feeder is the variable orifice inserted in the feed line to control rate of flow out of the cylinder. Its operation is somewhat analogous to a water faucet with a constant supply pressure behind it. The amount of water discharged can be governed by opening the faucet, and if supply pressure does not change, a constant rate of flow is maintained for any given setting. The orifice shown in detail in Figure 7-20 consists of a grooved plug sliding in a fitted ring. Feed rate is adjusted by varying the V-shaped opening. However, since chlorine cylinder pressure varies with temperature, the discharge through such a throttling valve does not remain constant without frequent adjustments of the valve setting. Furthermore, the condition on the outlet side may vary as a result of pressure changes at the point of application. To insure that these variable conditions do not affect control, a pressure regulating valve is inserted between the cylinder and the orifice, with a vacuum compensating valve on the discharge side. A safety pressure-relief valve is held closed by vacuum. If vacuum is lost and chlorine under pressure passes the inlet valve, the relief valve opens and the chlorine gas is vented outside the building. The visible flow meter (rotameter), pressure gauges, and feeder rate adjusting knob are located on the front panel of the chlorinator cabinet.

Direct feed of chlorine gas into a pipe or channel has certain limitations; for example, a considerable safety problem would exist if gas piping was extended throughout the treatment plant. To overcome the difficulties of conveyance and direct introduction of chlorine gas, an injector is employed to permit solution

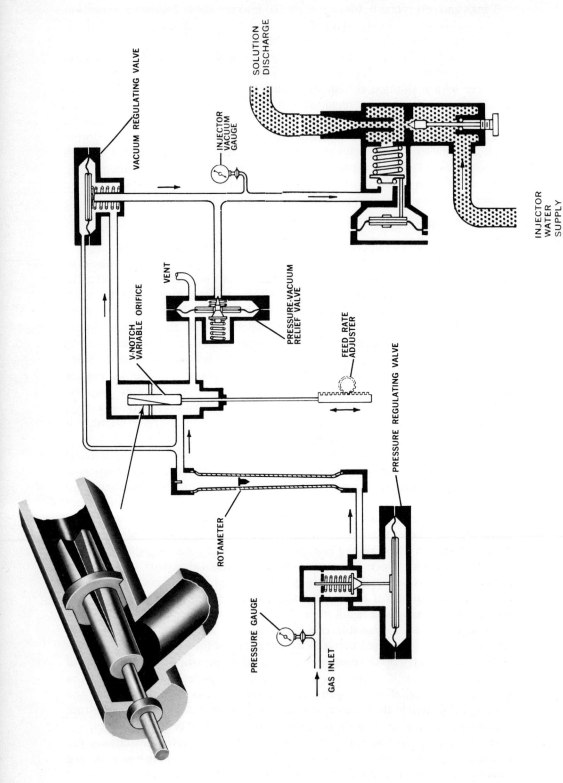

Figure 7-20 Flow diagram of a typical solution feed chlorinator and a detail of the V-notch variable orifice. (Courtesy of Wallace & Tiernan Division, Pennwalt Corp.)

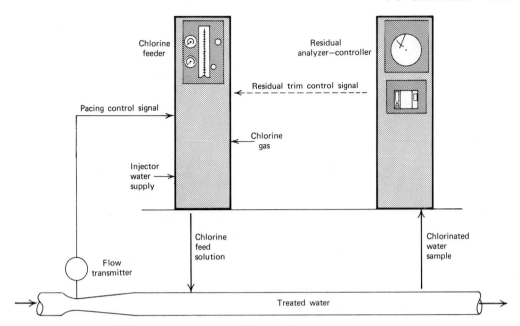

Figure 7-21 Automatic flow and residual control of a chlorine feeder.

feed. Water flowing through the injector creates a vacuum that draws in chlorine gas from the feeder and mixes it with the water supply. This concentrated solution is relatively stable and can be safely piped to various points in the treatment plant for discharge into an open channel, closed pipeline, or suction end of a pump.

Chlorine feeders can be controlled manually or automatically based on flow or chlorine residual or both. Manual adjustment establishes a continuous feed rate. This type of regulator is only satisfactory where chlorine demand and flow are reasonably constant, and where an operator is available to make adjustments as necessary. Automatic proportional control equipment adjusts the feed rate to provide a constant preestablished dosage for all rates of flow. This is accomplished by metering the main flow and using a transmitter to signal the chlorine feeder. Automatic residual control uses an analyzer downstream from the point of application to pace the chlorinator. Combined automatic flow and residual control (see Figure 7-21) maintains a preset chlorine residual in the water independent of demand and flow variations. The feeder is responsive to signals from both the flow meter transmitter and the chlorine residual analyzer. This system is most effective when flow pacing is the primary chlorine feed regulator, and residual monitoring is used to trim the dosage.

Positive displacement diaphragm pumps, mechanically or hydraulically actuated, are used for metering hypochlorite solutions. The hypochlorinator in Figure 7-22 consists of a water-powered pump paced by a positive displacement water meter. The meter register shaft rotates in proportion to the main line flow and controls a cam-operated pilot valve. This, in turn, regulates the flow of water behind the power diaphragm to produce a discharge of hypochlorite proportional to the main flow. Admitting main pressure behind the pumping diaphragm balances the water pressure in the pumping head. The advantage of this is that only a small force is needed for the discharge stroke, placing

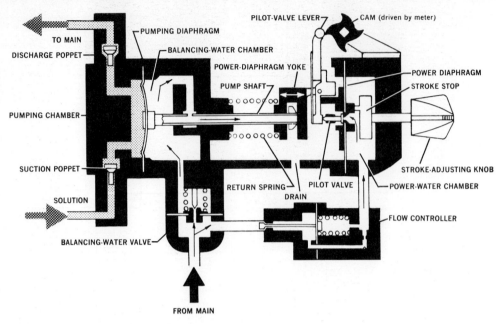

Figure 7-22 Hydraulically actuated diaphragm (water operated) hypochlorinator. (Courtesy of Wallace & Tiernan Division, Pennwalt Corp.)

minimum stress on the diaphragm. Another advantage is that the pump does not require electrical power. Hypochlorite dosage can be manually adjusted by changing the stroke length setting of the pump.

EXAMPLE 7-4

Chlorine usage in the treatment of 5.0 mgd of water is 17.0 lb/day. The residual after 10 min contact is 0.20 mg/l. Compute the dosage in milligrams per liter and chlorine demand of the water.

Solution

$$\text{Dosage} = \frac{17.0 \text{ lb/day}}{5.0 \text{ mil gal/day}} \times \frac{\text{mg/l}}{8.34 \text{ lb/mil gal}} = 0.41 \text{ mg/l}$$

Chlorine demand $= 0.41 - 0.20 = 0.21$ mg/l

EXAMPLE 7-5

A new water main is disinfected using a 50 mg/l chlorine dosage by applying a 2.0 percent hypochlorite solution. (a) How many pounds of dry hypochlorite powder, containing 70 percent available chlorine, must be dissolved in 50 gal of water to make a 2.0 percent (20,000 mg/l) solution? (b) At what rate should this solution be applied to the water entering the main to provide a concentration of 50 mg/l? (c) If 7000 gal of water is used to fill the main at a dosage of 50 mg/l, how many gallons of hypochlorite solution are used?

Solution

(a) $\dfrac{\text{Pounds of hypochlorite powder}}{\text{for 2.0 percent solution}} = \dfrac{50\,\text{gal} \times \dfrac{8.34\,\text{lb}}{\text{gal}} \times \dfrac{2.0}{100}}{0.70} = 11.9\,\text{lb}/50\,\text{gal}$

(b) Feed rate for 50 mg/l $= \dfrac{50\,\text{mg/l}}{20{,}000\,\text{mg/l}} = \dfrac{1\text{ volume of 2.0\% solution}}{400\text{ volumes of water}}$

(c) Solution usage for 7000 gal $= 7000\,\text{gal} \times \dfrac{50\,\text{mg/l}}{20{,}000\,\text{mg/l}} = 17.5\,\text{gal}$

7-7 FLUORIDATION

During the past three decades, hundreds of studies have been undertaken to correlate the concentration of waterborne fluoride and the incidence of dental caries. Examination of children's teeth consuming water with natural fluoride led to the following three distinct relationships: fluoride level in excess of 1.5 mg/l increases the occurrence and severity of dental fluorosis (mottling of teeth) without decreasing the incidence of decaying, missing, or filled teeth; the optimum occurs at a concentration of approximately 1.0 mg/l, resulting in maximum reduction in caries with no esthetically significant mottling; and, below 1.0 mg/l some benefit occurs but decay reduction is not as great and decreasing fluoride levels relate to increasing incidence of caries. Controlled fluoridation in water treatment to bring the natural content up to optimum level achieves the same beneficial results. Recommended limits given in Table 5-3 are based on air temperature, since this influences the amount of water ingested by people. Recent studies have indicated that fluoride also benefits older people in reducing prevalence of osteoporosis and hardening of the arteries. There is no doubt that fluoride in drinking water prevents dental caries and that controlled fluoridation is an acceptable public health measure. Approximately one half of the nation's population consumes water containing near optimum fluoride content with the bulk of these residing in communities that deliberately add a chemical compound to provide fluoride ion.

The three most commonly used fluoride compounds in water treatment are sodium fluoride, sodium silicofluoride, and fluosilicic acid, also known as hydrofluosilicic, hexafluosilicic, or silicofluoric acid. Table 7-2 lists some of the characteristics of these compounds. Sodium fluoride, although one of the most

Table 7-2. Characteristics of the Three Commonly Used Fluoride Compounds

	Sodium Fluoride	Sodium Silicofluoride	Fluosilicic Acid
Formula	NaF	Na_2SiF_6	H_2SiF_6
Fluoride ion, percent	45	61	79
Molecular weight	42	188	144
Commercial purity, percent	90 to 98	98 to 99	22 to 30
Commercial form	Powder or crystal	Powder or fine crystal	Liquid
Dosage in pounds per million gallons Required for 1.0 mg/l at indicated purity	18.8 (98%)	14.0 (98.5%)	35.2 (30%)

expensive for the amount of available F, is one of the most widely used chemicals. The crystalline type is preferred when manual handling is involved, since the absence of fine powder results in a minimum of dust. Fluosilicic acid is a colorless, transparent, fuming, corrosive liquid having a pungent odor and an irritating action on the skin. The major source of acid is a by-product from phosphate fertilizer manufacture. Sodium silicofluoride, the salt of fluosilicic acid, is the most widely used compound primarily because it is the cheapest. The white, odorless, crystalline powder is commercially available in various gradations for optimum application by various feeders.

No one specific type of fluoridation system is applicable to all water treatment plants. Selection is based on size and type of water facility, chemical availability, cost, and type of operating personnel available. For small utilities, some type of solution feed is almost always selected and batches are manually prepared. A simple system consists of a solution tank placed on a platform scale, for the convenience of weighing during preparation and feed, and a solution metering pump with appropriate piping from the tank to the water main for application. If fluosilicic acid is used, it may be diluted with water in the feed tank or may be applied at full strength directly from the shipping drum. When sodium fluoride is used, the feed solution may be prepared to a desired strength or as a saturated solution in a dissolving tank. Sodium fluoride has a maximum solubility of 4.0 percent (18,000 mg F/l), which is essentially independent of water temperature. Saturators are commercially available to prepare a saturated solution by allowing water to trickle through a bed containing a large excess of sodium fluoride. Water used for dissolution should be softened whenever the hardness exceeds 75 mg/l. While NaF is quite soluble, calcium and magnesium fluorides form precipitates that can scale and clog feeders and lines.

Fluoride solution feed must be paced to water flow. If a pump has a fixed delivery rate, the feeder can be tied electrically to turn on and off with pump operation, and the fluoride application can be adjusted to the rate of water discharge. For small water plants, a water meter contactor can be used to pace the feeder. Essentially, a contactor is a switch that is geared to the water meter movement so it makes contact at specific volumetric discharges. A pulse from the meter contactor can be used to energize the feeder in proportion to the meter's response to flow.

Large waterworks use gravimetric dry feeders to apply sodium silicofluoride, or solution feeders to apply fluosilicic acid directly. Automatic control systems use flow meters and recorders to adjust feed rate.

Fluoride must be injected at a point where all the water being treated passes. If there is no single common point, then separate fluoride feeding installations are required for each water facility. In a well pump system, application can be in the discharge line of each pump, or in a common line leading to a storage reservoir. Fluoride applied in a water plant can be introduced in a channel or main coming from the filters, or directly to the clear well. Whenever possible it should be added after filtration to avoid losses that may occur as a result of reactions with other chemicals. Of particular concern are coagulation with heavy alum doses and lime-soda softening. Fluoride injection points should be as far away as possible from any chemical that contains calcium so as to minimize loss by precipitation.

Surveilance of water fluoridation involves testing both the raw and treated water for fluoride. The concentration in treated water should be the amount recommended by drinking water standards. Records of the weight of chemical applied and the volume of water treated should be kept to confirm that the correct amount of fluoride is being added. The amount added to the water should equal the quantity calculated from the increase in concentration and quantity of water treated.

Defluoridation

When water supplies contain excess fluorides, the teeth of most consumers over a period of several years become mottled with a permanent gray to black discoloration of the enamel. Children who have been drinking water containing 5 mg/l develop fluorosis to the extent that the enamel is severely pitted, resulting in loss of teeth.

The two current treatment methods for defluoridation use either activated alumina or bone char. Water is percolated through insoluble, granular media to remove the fluorides. The media are periodically regenerated by chemical treatment after becoming saturated with fluoride ion. Regeneration of bone char consists of backwashing with a 1 percent solution of caustic soda and then rinsing the bed. Reactivation of alumina also involves backwashing with a caustic solution.

Removal of excess fluoride from public water supplies is a sound economic investment when related to the increased cost of dental care and loss of teeth. Despite the obvious need, there are many communities that have not installed defluoridation units, allegedly because of excessive costs of construction and operation. A few communities have been able to seek alternate sources of water to solve their fluoride problem without special treatment.

EXAMPLE 7-6

A liquid feeder applies a 4.0 percent saturated sodium fluoride solution to a water supply increasing the fluoride concentration from the natural level of 0.4 mg F/l to 1.0 mg F/l. The commercial NaF powder contains 45 percent F by weight. (a) How many pounds of NaF are required per million gallons treated? (b) What is the dosage of 4.0 percent NaF solution per million gallons?

Solution

(a) $\dfrac{\text{NaF required}}{\text{Million gallons}} = \dfrac{(1.0 \text{ mg F/l} - 0.4 \text{ mg F/l})}{0.45 \text{ mg F/mg NaF}} \times 8.34 \dfrac{\text{lb/mil gal}}{\text{mg/l}} = 11.1 \text{ lb}$

A 4.0% NaF solution contains 40,000 mg NaF/l, and $0.45 \times 40,000 = 18,000$ mg F/l.

(b) $\dfrac{\text{Solution dosage}}{\text{Million gallons}} = \dfrac{1,000,000 \text{ gal } (1.0 \text{ mg/l} - 0.4 \text{ mg/l})}{18,000 \text{ mg/l}} = 33.3 \text{ gal}$

EXAMPLE 7-7

A dry feeder recorded a weight loss of 25 lb of sodium silicofluoride in the treatment of 1.8 mil gal of water. From Table 7-2, commercial powder is 98 percent pure Na_2SiF_6 and 61 percent of the pure compound is F. Calculate the concentration of fluoride ion added to the treated water.

Solution

$$\text{Fluoride concentration} = \frac{25 \text{ lb} \times 0.98 \times 0.61}{1.8 \text{ mil gal}} \times \frac{1.0 \text{ mg/l}}{8.34 \text{ lb/mil gal}} = 1.0 \text{ mg F/l}$$

7-8   IRON AND MANGANESE REMOVAL

Ferrous iron (Fe^{++}) and manganous manganese (Mn^{++}) are soluble, invisible forms that may exist in well waters or anaerobic reservoir water. When exposed to air, these reduced forms slowly transform to insoluble, visible, oxidized ferric iron (Fe^{+++}) and manganic manganese (Mn^{++++}). The rate of oxidation depends on pH, alkalinity, organic content, and presence of oxidizing agents. If not removed in treatment, the brown-colored oxides of iron and manganese create unaesthetic conditions and may interfere with some water uses.

Preventative measures may sometimes be used with reasonable success. Addition of sodium hexametaphosphate, while not preventing oxidation of the metal ions, may keep them in suspension and, thus, moving through the system without creating accumulations that periodically cause badly discolored water. Success of this treatment is very difficult to predict, since it depends on the concentrations of iron and manganese, the level of chlorine residual established for disinfection, and the time of passage through the distribution system. The latter is established by the extent of the distribution system, pipe sizes in the network, and location and volume of storage reservoirs.

Reduced iron in water promotes the growth of autotrophic bacteria in distribution mains (Section 3-1). Periodic flushing of small distribution pipes may be effective in removing accumulations of rust particles; however, elimination of iron bacteria is generally difficult and expensive. Biological growths are particularly obnoxious when they decompose in the piping system, releasing foul tastes and odors. Heavy chlorination of isolated sections of water mains followed by flushing has been effective in some cases. The only permanent solution to iron and manganese problems is removal by treatment of the water.

Aeration, Sedimentation, and Filtration

The simplest form of iron oxidation in treatment of well water is plain aeration. A typical tray-type aerator has a vertical riser pipe that distributes water on top of a series of trays from which it then drips and splatters down through the stack. The trays frequently contain coke or stone contact beds that develop and support oxide coatings that speed up the oxidation reactions.

$$Fe^{++}(\text{ferrous}) + \text{oxygen} \rightarrow FeO_x(\text{ferric oxides}) \qquad (7\text{-}11)$$
$$\text{Soluble Iron} \qquad\qquad\qquad \text{Insoluble Iron}$$

Manganese cannot be oxidized as easily as iron, and plain aeration is generally not effective. Sometimes, increasing the pH to about 8.5 with lime or soda ash enhances the oxidation of manganese on coke beds coated with manganic oxides; however, high removals are not assured. Inefficient removal of manganese can cause serious problems with post-chlorination for establishing a residual in the distribution system. Manganese not taken out by the aeration-filtration process may be oxidized in the injector mechanism of a solution feed chlorinator and may clog the unit; or in the distribution piping, the oxidizing effect of the chlorine residual may create a staining water.

Aeration, Chemical Oxidation, Sedimentation, and Filtration

This is the sequence of processes, illustrated in Figure 7-1*b*, that is the common method for removing iron and manganese from well water without softening treatment. Preliminary aeration strips out dissolved gases and adds oxygen. Iron and manganese are chemically oxidized by free chlorine residual, Eq. 7-12, or by potassium permanganate, Eqs. 7-13 and 7-14, at rates of oxidation much greater than dissolved oxygen. When chlorine is used, a free available residual is maintained throughout the treatment processes. The specific dosage required depends on the concentration of metal ions, pH, mixing conditions, and other factors. Potassium permanganate is a dark purple crystal or powder available commercially at 97 to 99 percent purity. Theoretically, 1 mg/l of potassium permanganate oxidizes 1.06 mg/l of iron or 0.52 mg/l of manganese. In actual practice, the amount needed is often less than this theoretical requirement. Permanganate oxidation may be advantageous for certain waters, since its rate of reaction is relatively independent of pH.

$$\underset{\text{Soluble Ions}}{Fe^{++} + Mn^{++} + oxygen} \xrightarrow{\underset{\text{residual}}{\text{free chlorine}}} \underset{\text{Insoluble Metal Oxides}}{FeO_x \downarrow + \ MnO_2 \downarrow} \qquad (7\text{-}12)$$

$$\underset{\substack{\text{Ferrous} \\ \text{Bicarbonate}}}{Fe(HCO_3)_2} + \underset{\substack{\text{Potassium} \\ \text{Permanganate}}}{KMnO_4} \longrightarrow \underset{\substack{\text{Ferric} \\ \text{Hydroxide}}}{Fe(OH)_3 \downarrow} + \underset{\substack{\text{Manganese} \\ \text{Dioxide}}}{MnO_2 \downarrow} \qquad (7\text{-}13)$$

$$\underset{\substack{\text{Manganous} \\ \text{Bicarbonate}}}{Mn(HCO_3)_2} + \underset{\substack{\text{Potassium} \\ \text{Permanganate}}}{KMnO_4} \longrightarrow \underset{\substack{\text{Manganese} \\ \text{Dioxide}}}{MnO_2 \downarrow} \qquad (7\text{-}14)$$

Effective filtration following chemical oxidation is essential, since a significant amount of the flocculent metal oxides are not heavy enough to settle by gravity. Iron and manganese, carried over to the filter, coat the media with oxides that enhance filtration removal. Practice has shown that new filters pass manganese until the grains are covered with oxide that develops naturally during filtration of manganese-bearing water.

Manganese Zeolite Process

Manganese zeolite is a natural greensand coated with manganese dioxide that removes soluble iron and manganese from solution. After the zeolite becomes saturated with metal ions, it is regenerated using potassium permanganate. A continuous flow system is illustrated in Figure 7-23. Permanganate solution is applied to the water ahead of a pressure filter that contains a dual-media anthracite and manganese zeolite bed. The iron and manganese oxidized by the permanganate feed is removed by the upper filter layer. Any ions not oxidized are captured by the underlying manganese zeolite layer. If surplus permanganate is inadvertently applied to the water, it passes through the coal medium and regenerates the greensand. When the bed becomes saturated with metal oxides, it is backwashed to remove particulate matter from the surface layer and to regenerate the zeolite with potassium permanganate.

Water Softening

Precipitation softening using lime takes iron and manganese out of solution. If split treatment is employed in softening a groundwater containing iron and

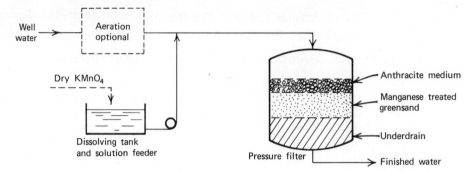

Figure 7-23 Removal of iron and manganese from groundwater by the manganese zeolite process employing a dual-media pressure filter with manganese treated greensand.

manganese, potassium permanganate can be used to oxidize metal ions in the water by-passing the first-stage lime treatment. Lime treatment has been used for the principal purpose of removing organically bound iron and manganese in processing surface waters.

7-9 SOFTENING

Hardness in water is caused by calcium and magnesium ions resulting from water coming in contact with geological formations. Public acceptance of hardness varies, although many customers object to water harder than 150 mg/l. The maximum level considered for public supply is 300 to 500 mg/l. A moderately hard water is generally defined as 60 to 120 mg/l. Hardness interferes with laundering by causing excessive soap consumption, and may produce scale in hot water heaters and pipes. To a considerable extent these disadvantages have been overcome by the use of synthetic detergents and the lining of pipes in small hot-water heaters. Industries generally pretreat boiler water to prevent scaling.

Precipitation softening uses lime (CaO) and soda ash (Na_2CO_3) to remove calcium and magnesium from solution. In addition, lime treatment has the incidental benefits of bactericidal action, removal of iron, and aid in clarification of turbid surface waters. Ion exchange softening uses a resin to remove the bivalent ions and to replace them with the sodium ion; this is the common process used in household water softeners.

Lime-Soda Ash Softening

Lime is sold commercially in the forms of quicklime and hydrated lime. Quicklime, available in granular form, is a minimum of 90 percent CaO with magnesium oxide being the primary impurity. A slaker is used to prepare quicklime for feeding in a slurry containing approximately 5 percent calcium hydroxide. Powdered, hydrated lime contains approximately 68 percent CaO, and may be prepared by fluidizing in a tank containing a turbine mixer. Lime slurry is written as $Ca(OH)_2$ in chemical equations. Soda ash is a grayish-white powder containing, at least, 98 percent sodium carbonate.

Carbon dioxide is a clear, colorless gas used for recarbonation in stabilizing lime-softened water. The gas is produced by burning a fuel, such as coal, coke, oil, or gas. The ratio of fuel to air is carefully regulated in a CO_2 generator to provide complete combustion. Gas under pressure from the combustion chamber is forced through diffusers immersed in a treatment basin. Several manufacturers produce recarbonation systems for generating and feeding carbon dioxide.

The chemical reactions in precipitation softening are:

$$CO_2 + Ca(OH)_2 = CaCO_3 \downarrow + H_2O \qquad (7\text{-}15)$$

$$Ca(HCO_3)_2 + Ca(OH)_2 = 2CaCO_3 \downarrow + 2H_2O \qquad (7\text{-}16)$$

$$Mg(HCO_3)_2 + Ca(OH)_2 = CaCO_3 \downarrow + MgCO_3 + 2H_2O$$

$$MgCO_3 + Ca(OH)_2 = CaCO_3 \downarrow + Mg(OH)_2 \downarrow$$

$$Mg(HCO_3)_2 + 2Ca(OH)_2 = 2CaCO_3 \downarrow + Mg(OH)_2 \downarrow + 2H_2O \qquad (7\text{-}17)$$

$$MgSO_4 + Ca(OH)_2 = Mg(OH)_2 \downarrow + CaSO_4 \qquad (7\text{-}18)$$

$$CaSO_4 + Na_2CO_3 = CaCO_3 \downarrow + Na_2SO_4 \qquad (7\text{-}19)$$

Lime added to water reacts first with any free carbon dioxide forming a calcium carbonate precipitate, Eq. 7-15. Next, the lime reacts with any calcium bicarbonate present, Eq. 7-16. In both of these equations, one equivalent of lime combines with one equivalent of either CO_2 or $Ca(HCO_3)_2$. Since magnesium precipitates as $Mg(OH)_2$ ($MgCO_3$ being soluble) two equivalents of lime are needed to remove one equivalent of magnesium bicarbonate, Eq. 7-17. Noncarbonate hardness (calcium and magnesium sulfates or chlorides) requires addition of soda ash for precipitation. Equation 7-19 shows that one equivalent of soda ash removes one equivalent of calcium sulfate. However, magnesium sulfate needs both lime, Eq. 7-18, and soda ash, Eq. 7-19.

The calcium ion can be effectively reduced by the lime additions defined in the equations that raise the pH of the water to approximately 10.3. But precipitation of the magnesium ion demands a higher pH and the presence of excess lime in the amount of about 35 mg/l CaO (1.25 meq/l) above the stoichiometric requirements. The practical limits of precipitation softening are 30 to 40 mg/l of $CaCO_3$ and 10 mg/l of $Mg(OH)_2$ as $CaCO_3$, while the theoretical solubility of these compounds is considerably less, being approximately one fifth of these limits.

A major advantage of precipitation softening is that the lime added is removed along with the hardness taken out of solution. Thus the total dissolved solids in the water are dramatically reduced. When soda ash is added, Eq. 7-19, the sodium ions remain in the finished water along with the accompanying anion, either sulfate or chloride; however, noncarbonate hardness requiring the addition of soda ash is generally a small portion of the total hardness in a raw water. The chemical reactions of lime-soda softening can also be used to estimate the quantity of solids produced in waste sludge.

Recarbonation is used to stabilize lime treated water, reducing its scale-forming potential. Carbon dioxide neutralizes excess lime precipitating it as calcium carbonate. Further recarbonation converts carbonate to bicarbonate.

$$Ca(OH)_2 + CO_2 = CaCO_3 \downarrow + H_2O \qquad (7\text{-}20)$$

$$CaCO_3 + CO_2 + H_2O = Ca(HCO_3)_2 \qquad (7\text{-}21)$$

Excess lime treatment is used to reduce both calcium and magnesium hardness to the practical limit of about 40 mg/l. Lime and soda ash additions are estimated by using the chemical reactions, plus the excess lime addition needed to remove manganese. After flocculation and sedimentation, the water is scale forming and must be neutralized to a pH of about 9.5 by using carbon dioxide. Excess lime treatment often employs a two-stage water softening scheme that provides a more efficient operation than the single-stage process illustrated in Figure 7-1c. All of the lime is applied to the first stage to precipitate calcium carbonate and magnesium hydroxide. Following mixing and sedimentation, the water is partially recarbonated and soda ash is added. Subsequent mixing may involve recirculating calcium carbonate sludge from the first stage to aid in clarification. Prior to filtration, carbon dioxide is again applied for final stabilization, and approximately 0.5 mg/l of sodium polyphosphate is applied to prevent incrustation of the filter sand.

Selective calcium carbonate removal may be used to soften a water low in magnesium hardness, less than 40 mg/l as $CaCO_3$. The processing scheme is similar to that shown in Figure 7-1c. Enough lime is added to precipitate the calcium hardness; soda ash may or may not be required depending on the amount of noncarbonate hardness. Sludge recirculation may be used to aid in precipitation, and recarbonation is usually practiced to produce a stable effluent. Sodium polyphosphate is applied to inhibit incrustation of the filter sand and scale formation in the distribution piping.

Split treatment consists of dividing the raw water into two portions for softening in a two-stage system illustrated by the flow diagram in Figure 7-24. The larger portion is given excess lime treatment in the first stage by using a flocculator-clarifier, or in-line mixing and sedimentation basins. Soda ash is added to the second stage where the split flow is blended with the treated water. Excess lime used to force precipitation of magnesium in the first stage now reacts with the calcium hardness that was by-passed around the lime treatment. Thus the excess lime is used in the softening process instead of being wasted at the expense of carbon dioxide neutralization. Recarbonation is not customarily required, but it may be desirable in the treatment of some waters for stabilization. Since hardness levels of 80 to 100 mg/l are generally considered acceptable, split treatment can result in considerable chemical savings. Lime and recarbonation costs are lower than excessive lime treatment, while the advantage of reducing magnesium hardness to less than 40 mg/l is still possible. A concentration of magnesium hardness greater than 40 mg/l is not recommended because of the possible formation of hard magnesium silicate scale in high temperature (180° F) services. The amount of magnesium in the finished water can be calculated by multiplying the by-pass flow times the magnesium concentration in the raw water plus the lime-treated flow times 10 mg/l divided by the total raw water flow. The quantity of flow split around the first stage is often determined by the level of magnesium desired in the softened water.

Softening surface waters may not permit by-passing flow without treatment to remove turbidity. In this case, the split leg is frequently treated by using a coagulant, such as alum. Additionally, coagulant aids or activated carbon or both may be applied to second-stage blending.

Ion Exchange

In this process ions of a given species are displaced from an insoluble exchange material by ions of different species from solution. In practice, the

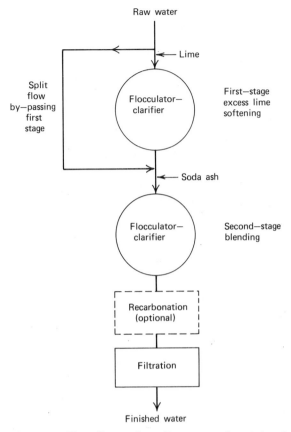

Figure 7-24 Flow diagram for split-treatment lime-soda ash softening process.

water being treated is passed through a filter bed of ion exchange material. When the capacity for exchanging ions has been depleted, a regenerating solution with a high concentration of the original ions is pumped through the bed. This process displaces the contaminant ions and rejuvenates the exchanger.

Three common cation exchangers used in water softening are greensand, siliceous gel-type zeolite, and polystyrene resins. Natural greensand (glauconite) is mined, processed, and chemically treated to produce a durable greenish-black mineral that is relatively inert except for ion exchange properties. Synthetic zeolites are prepared by mixing and reacting chemicals to form a gel that is dried and crushed to form a particulate column packing. Currently these two materials have only limited use, since polystyrene resins now dominate the softening field. Synthetic resins are manufactured by chemical processes in the form of small spheres that are approximately 0.5 mm in diameter.

Schematic Eq. 7-22, in which R represents the anionic component of the ion exchanger, indicates the reaction that takes place in cation exchange softening. When the exhausted exchanger is regenerated, the reverse action takes place. Backwashing the bed with a brine solution displaces the bivalent hardness ions and regenerates the resin with sodium, Eq. 7-23. Operating characteristics of polystyrene ion-exchange resin are: operating exchange capacity 20,000 to 35,000 grains per cubic foot, softening flow rate 2 to 6 gpm/cu ft, backwash flow rate 5 to 6 gpm/sq ft, regenerating salt dosage 5 to 20 lb/cu ft, and brine contact time 25 to 45 min.

$$\begin{matrix} Ca^{++} \\ Mg^{++} \end{matrix} + Na_2R \longrightarrow \begin{matrix} CaR \\ MgR \end{matrix} + Na^+ \qquad (7\text{-}22)$$

$$\begin{matrix} CaR \\ MgR \end{matrix} + NaCl \xrightarrow[\text{salt}]{\text{excess}} Na_2R + \begin{matrix} Ca^{++} \\ Mg^{++} \end{matrix} \qquad (7\text{-}23)$$

Demineralization

This process employs both cation and anion exchangers in single or dual process tanks to completely remove all metal and nonmetal ions from solution. A strong-acid resin replaces cations with hydrogen ions, and a weak-base resin exchanges hydroxide ions for the removed anions. Deionization is not applied in municipal water treatment because of high cost; however, it is extensively used in industry to produce water of extremely high quality.

EXAMPLE 7-8

Water defined by the following analysis is to be softened by excess lime treatment. Assume that the practical limit of hardness removal for $CaCO_3$ is 30 mg/l, and that of $Mg(OH)_2$ is 10 mg/l as $CaCO_3$.

$CO_2 = 8.8$ mg/l $Alk(HCO_3^-) = 135$ mg/l as $CaCO_3$
$Ca^{++} = 40.0$ mg/l $SO_4^- = 29.0$ mg/l
$Mg^{++} = 14.7$ mg/l $Cl^- = 17.8$ mg/l
$Na^+ = 13.7$ mg/l

(a) Sketch a meq/l bar graph and list the hypothetical combinations of chemical compounds in solution (Section 2-2).
(b) Calculate the softening chemicals required, expressing lime dosage as CaO and soda ash as Na_2CO_3.
(c) Draw a bar graph for the softened water before and after recarbonation. Assume that one half the alkalinity in the finished water is in the bicarbonate form.
(d) Sketch a bar graph of the treated water after it is softened to zero hardness by the cation exchange process.

Solution

Component	mg/l	Equivalent Weight	meq/l
CO_2	8.8	22.0	0.40
Ca^{++}	40.0	20.0	2.00
Mg^{++}	14.7	12.2	1.21
Na^+	13.7	23.0	0.60
Alk	135	50.0	2.70
SO_4^-	29.0	48.0	0.60
Cl^-	17.8	35.0	0.51

(a) From the meq/l bar graph drawn in Figure 7-25a, the hypothetical combinations are: $Ca(HCO_3)_2$, $Mg(HCO_3)_2$, $MgSO_4$, Na_2SO_4, and NaCl. Calcium

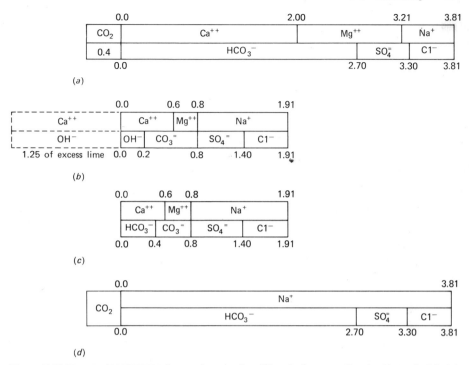

(a)

(b)

(c)

(d)

Figure 7-25 Bar graphs of raw and treated waters in milliequivalents per liter for Example 7-8. (a) Raw water prior to any treatment. (b) Water after excess lime softening, but before recarbonation and filtration. (c) Lime-soda ash softened water following recarbonation and filtration. (d) Water after softening by ion exchange.

hardness $= 2.00 \times 50 = 100$ mg/l as $CaCO_3$ and magnesium hardness $= 1.21 \times 50 = 60.5$ mg/l.

Component	meq/l	Applicable Equation	Lime meq/l	Soda Ash meq/l
CO_2	0.40	7-15	0.40	0
$Ca(HCO_3)_2$	2.00	7-16	2.00	0
$Mg(HCO_3)_2$	0.70	7-17	1.40	0
$MgSO_4$	0.51	7-18 & 7-19	0.51	0.51
			4.31	0.51

(b) Required lime dosage equals the amount needed for the softening reactions plus 35 mg/l CaO of excess lime to precipitate the magnesium.

Dosage $= 4.31 \times 28 + 35 = 156$ mg/l CaO $= 1300$ lb/mil gal

Dosage of soda ash $= 0.51 \times 53 = 27$ mg/l $Na_2CO_3 = 225$ lb/mil gal

(c) A hypothetical bar graph for the water after addition of softening chemicals is shown in Figure 7-25b. The dashed box to the left of zero is the excess lime (35 mg/l CaO $= 1.25$ meq/l) added to increase the pH high enough to precipitate the $Mg(OH)_2$. The 0.6 meq/l of Ca^{++} (30 mg/l as $CaCO_3$) and 0.2 meq/l of Mg^{++} (10 mg/l as $CaCO_3$) are the practical limits of hardness reduction. The 1.11 meq/l of Na^+ is the sum of the sodium originally in the water (0.60 meq/l) plus the amount increased by soda ash addition (0.51 meq/l). Alkalinity consists of 0.20 meq/l of OH^- associated with $Mg(OH)_2$, and the remainder is 0.60 meq/l carbonate ion. The $SO_4^=$ and Cl^- meqs/l are unchanged by the softening process.

Recarbonation converts the excess lime to calcium carbonate precipitate, Eq. 7-20, which is removed by second-stage sedimentation and filtration. Further carbon dioxide addition converts $CO_3^=$ to HCO_3^-, Eq. 7-21. Figure 7-25c illustrates the composition of the stabilized, finished water with a total hardness of 40 mg/l.

(d) The bar graph for ion exchange softened water is given in Figure 7-25d. All of the calcium and magnesium ions in the raw water are replaced by sodium ions. A serious disadvantage of ion exchange softening is the lack of reduction in total dissolved solids; for example, the sum of ions in the raw water is 280 mg/l while the weight of ions in Figure 7-25d is 300 mg/l. Lime-soda ash softening reduced the dissolved salts to 145 mg/l in the finished water (Figure 7-25c). The ion exchange process does not remove carbon dioxide, therefore, blending or lime addition is needed for corrosion control.

EXAMPLE 7-9

Consider precipitation softening by split treatment of the raw water described in Example 7-8. Assume that 40 percent of the flow is by-passed, and that 60 percent is given excess lime treatment in the first stage. Compute the chemicals required and the hardness of the finished water.

Solution

Flow passing through the first stage is processed in the same fashion as the excess lime treatment described in the solution to Example 7-8. However, the chemical additions are reduced to 60 percent, since only that fraction of the raw water is being treated.

Lime dosage $= 0.60 \times 1300 = 780$ lb CaO/mil gal

Soda ash $= 0.60 \times 217 = 130$ lb Na_2CO_3/mil gal

Figure 7-25a is the bar graph of the by-passed water, and Figure 7-25b is the analysis of the first-stage effluent. These two are blended in the second stage where the excess lime reacts with the untreated water. The amount of excess hydroxide in the mixed flow streams is equal to

$$0.60(1.25 + 0.20) = 0.87 \text{ meq/l}$$

The other components of interest in the blended water are

$$CO_2 = 0.4 \times 0.4 = 0.16 \text{ meq/l}$$
$$Ca(HCO_3)_2 = 0.4 \times 2.00 = 0.80 \text{ meq/l}$$

First, the carbon dioxide is eliminated by the excess lime

$$0.87 \text{ meq/l} - 0.16 \text{ meq/l} = 0.71 \text{ meq/l}$$

The balance of the hydroxide ion reacts with calcium bicarbonate reducing it to

$$0.80 \text{ meq/l} - 0.71 \text{ meq/l} = 0.09 \text{ meq/l}$$

Final calcium hardness equals this remainder plus the limit of calcium carbonate removal,

$$(0.09 \text{ meq/l} + 0.60 \text{ meq/l})50 = 35 \text{ mg/l as CaCO}_3$$

Magnesium hardness in the finished water is

$$\frac{0.40 \times 60.5 + 0.6 \times 10}{1.00} = 30 \text{ mg/l}$$

Total hardness in the finished water is then 65 mg/l.

7-10   TURBIDITY REMOVAL

Surface waters generally contain suspended and colloidal solids from land erosion, decaying vegetation, microorganisms, and color-producing compounds. Coarser materials such as sand and silt can be eliminated to a considerable extent by plain sedimentation, but finer particles must be chemically coagulated to produce larger floc which is removable in subsequent settling and filtration. Destabilization of colloidal suspensions is discussed in Section 2-5, and typical flow schemes for surface water treatment plants are diagramed in Figure 7-2.

Coagulation and flocculation are sensitive to many variables, for instance, the nature of the turbidity-producing substances, type and dosage of coagulant, pH of the water, and the like. Of the many variables that can be controlled in plant operation, pH adjustment appears to be most important. Generally, the type of coagulants and aids available are defined by the plant process scheme; of course, dosages of these substances can be regulated to meet changes in raw water quality. Also, mechanical mixing can be adjusted by varying the speed of the flocculator paddles.

Jar tests are widely used to determine optimum chemical dosages for treatment. This laboratory test (Section 2-8) attempts to simulate the full-scale coagulation-flocculation process and can be conducted for a wide range of conditions. The interpretation of test results involves visual and chemical testing of the clarified water. Ordinarily, a treatment plant gives better results than the jar test for the same chemical dosages. Of course, to confirm optimum chemical treatment, the water should be tested at various processing stages including the finished effluent. One common monitoring technique is analysis of the filtered water for turbidity; another is to record the length of time between filter runs.

Coagulants

Commonly used metal coagulants in water treatment are: (1) those based on aluminum, such as aluminum sulfate, sodium aluminate, potash alum, and ammonia alum; and (2) those based on iron, such as ferric sulfate, ferrous sulfate, chlorinated copperas, and ferric chloride. The description that follows gives some of the relevant properties and chemical reactions of these substances in the coagulation of water. The latter are presented in the form of hypothetical equations with the understanding that they do not represent exactly what happens in water. Studies have shown that the hydrolysis of iron and aluminum salts is far more complicated than these formulas would indicate; however, they are useful in approximating the reaction products and quantitative relationships.

Aluminum sulfate, $Al_2(SO_4)_3 \cdot 14.3H_2O$, is by far the most widely used coagulant; the commercial product is commonly known as alum, filter alum, or alumina sulfate. Filter alum is a grayish-white crystallized solid containing approximately 17 percent water-soluble Al_2O_3 and is available in lump, ground, or powdered forms as well as concentrated solution. Ground alum is commonly measured by a gravimetric type feeder into a solution tank from which it is transmitted to the point of application by pumping. Amber-colored liquid aluminum sulfate contains about 8 percent available Al_2O_3.

The hydrolysis of aluminum ion in solution is complex and is not fully defined. In pure water at low pH, the bulk of aluminum appears as Al^{+++} while in alkaline solution complex species such as $Al(OH)_4^-$ and $Al(OH)_5^=$ have been shown to exist. In the hypothetical coagulation equations, aluminum floc is written as $Al(OH)_3$. This is the predominant form found in a dilute solution near neutral pH in the absence of complexing anions other than hydroxide. The reaction between aluminum and natural alkalinity is given in Eq. 7-24. If lime or soda ash is added to the water with the coagulant, the theoretical reactions are as shown in Eqs. 7-25 and 7-26.

$$Al_2(SO_4)_3 \cdot 14.3H_2O + 3Ca(HCO_3)_2 \rightarrow 2Al(OH)_3 \downarrow + 3CaSO_4 + 14.3H_2O$$
$$+ 6CO_2 \quad (7\text{-}24)$$

$$Al_2(SO_4)_3 \cdot 14.3H_2O + 3Ca(OH)_2 \rightarrow 2Al(OH)_3 \downarrow + 3CaSO_4 + 14.3H_2O$$
$$(7\text{-}25)$$

$$Al_2(SO_4)_3 \cdot 14.3H_2O + 3Na_2CO_3 + 3H_2O \rightarrow 2Al(OH)_3 \downarrow + 3Na_2SO_4 + 3CO_2$$
$$+ 14.3H_2O \quad (7\text{-}26)$$

Based on these reactions, 1.0 mg/l of alum with a molecular weight of 600 reacts with: 0.50 mg/l natural alkalinity, expressed as $CaCO_3$; 0.39 mg/l of 95 percent hydrated lime as $Ca(OH)_2$, or 0.33 mg/l 85 percent quicklime as CaO; or, 0.53 mg/l soda ash as Na_2CO_3. When lime or soda ash are reacted with the aluminum sulfate, the natural alkalinity of the water is unchanged. Sulfate ions added with the alum remain in the finished water. In the case of natural alkalinity and soda ash, carbon dioxide is produced. The dosages of alum used in water treatment are in the range of 5 to 50 mg/l, with the higher concentrations needed to clarify turbid surface waters. Alum coagulation is generally effective within the pH limits of 5.5 to 8.0.

Sodium aluminate, $NaAlO_2$, may be used as a coagulant in special cases. The commercial grade has a purity of approximately 88 percent and may be purchased either as a solid or solution. Because of its high cost, sodium aluminate is generally used to aid a coagulation reaction instead of being the primary coagulant. It has been found effective for secondary coagulation of highly colored surface waters, and as a coagulant in the lime-soda ash softening process to improve settleability of the precipitate.

Ferrous sulfate, $FeSO_4 \cdot 7H_2O$ (also commonly known as copperas), is a greenish-white crystalline solid that is obtained as a by-product of other chemical processes, principally from the pickling of steel. Although available in liquid form from processed spent pickle liquor, the common commercial preparations are granular. Ferrous iron added to water precipitates in the oxidized form of ferric hydroxide; therefore, the addition of lime or chlorine is generally needed to provide effective coagulation. Ferrous sulfate and lime coagulation, shown in Eq. 7-27, is effective for clarification of turbid water and other reactions conducted at high pH values, for example, in lime softening.

$$2FeSO_4 \cdot 7H_2O + 2Ca(OH)_2 + \tfrac{1}{2}O_2 \rightarrow 2Fe(OH)_3 \downarrow + 2CaSO_4 + 13H_2O \quad (7\text{-}27)$$

Chlorinated copperas is prepared by adding chlorine to oxidize the ferrous sulfate. The advantage of this method, relative to lime addition, is that coagulation can be obtained over a wide range of pH values, 4.8 to 11.0. Theoretically, according to Eq. 7-28, each mg/l of ferrous sulfate requires 0.13 mg/l of chlorine, although additional chlorine is generally added to insure complete reaction.

$$3FeSO_4 \cdot 7H_2O + 1.5Cl_2 \rightarrow Fe_2(SO_4)_3 + FeCl_3 + 21H_2O \qquad (7\text{-}28)$$

Ferric sulfate, $Fe_2(SO_4)_3$, is available as a commercial coagulant in the form of a reddish-brown granular material that is readily soluble in water. It reacts with natural alkalinity of water according to Eq. 7-29, or with added alkaline materials such as lime or soda ash, Eq. 7-30.

$$Fe_2(SO_4)_3 + 3Ca(HCO_3)_2 \rightarrow 2Fe(OH)_3 \downarrow + 3CaSO_4 + 6CO_2 \qquad (7\text{-}29)$$

$$Fe_2(SO_4)_3 + 3Ca(OH)_2 \rightarrow 2Fe(OH)_3 \downarrow + 3CaSO_4 \qquad (7\text{-}30)$$

In general, ferric coagulants are effective over a wide pH range. Ferric sulfate is particularly successful when used for color removal at low pH values; at high pH, it may be used for iron and manganese removal and as a coagulant in precipitation softening.

Ferric chloride, $FeCl_3 \cdot 6H_2O$, is used primarily in the coagulation of waste water and industrial wastes, and it finds only limited use in water treatment. Normally, it is produced by chlorinating scrap iron, and is available commercially in solid and liquid forms. Being highly corrosive, the liquid must be stored and handled in corrosion-resistant tanks and feeders. The reactions of ferric chloride with natural and added alkalinity are similar to those of ferric sulfate.

Coagulant Aids

Difficulties with coagulation often occur because of slow-settling precipitates, or fragile flocs that are easily fragmented under hydraulic forces in basins and sand filters. Coagulant aids benefit flocculation by improving settling and toughness of flocs. The most widely used materials are polyelectrolytes, activated silica, adsorbent-weighting agents, and oxidants.

Synthetic polymers are long-chain, high-molecular-weight, organic chemicals commercially available under a wide variety of trade names. Polyelectrolytes are classified according to the type of charge on the polymer chain. Those possessing negative charges are called anionic, those positively charged are called cationic; and those carrying no electric charge are nonionic. Anionic or nonionic are often used with metal coagulants to provide bridging between colloids to develop larger and tougher floc growth (Figure 2-4c). The dosage required as a flocculent aid is generally in the order of 0.1 to 1.0 mg/l. In the coagulation of some waters, polymers can promote satisfactory flocculation at significantly reduced alum dosages. The potential advantages of polymer substitution are in reducing the quantity of waste sludge produced in alum coagulation, and in changing the character of the sludge such that it can be more easily dewatered.

Cationic polymers have been used successfully in some waters as primary coagulants for clarification. Although the unit cost of cationic polymers is about 10 to 15 times higher than the cost of alum, the reduced dosages required may nearly offset the increased cost of chemicals. Furthermore, unlike the gelatinous and voluminous aluminum hydroxide sludges, polymer sludges are relatively dense and easier to dewater for subsequent handling and disposal. Sometimes cationic and nonionic polymers may be used together to provide an adequate floc, the former being the primary coagulant and the latter a coagulant aid. Although significant strides have been made in the application of polyelectrolytes in water treatment, their main application is still as an aid rather than as a primary coagulant. Many waters cannot be treated by using polymers alone but require aluminum or iron salts. Jar tests and actual plant operation must be used to determine the effectiveness of a particular proprietary polyelectrolyte in flocculation of a given water.

Activated silica is sodium silicate that has been treated with sulfuric acid, aluminum sulfate, carbon dioxide, or chlorine. A dilute solution is partially neutralized by an acid and then is allowed to age for a period of time up to 2 hr; additional aging runs the risk of solidification of the entire solution by gelation. The aged silicate solution is generally further diluted before being applied to the water. As a coagulant aid, it offers the advantages of: increased rate of chemical reaction, reduced coagulant dose, extended optimum pH range, and production of a faster settling, tougher floc. One of the disadvantages of activated silica, relative to polyelectrolytes, is the precise control required in preparation and feeding. Activated silica is normally used with aluminum coagulants at a dosage between 7 and 11 percent of the alum dose, expressed as milligrams per liter of SiO_2.

Bentonitic clays may be used in treating waters containing high color, low turbidity, and low mineral content. Iron or aluminum floc produced under these conditions is frequently too light to settle readily. Addition of clay results in a weighting action that improves settleability. Clay particles may also adsorb organic compounds, improving treatment. Although exact dosage must be determined by testing, 10 to 15 mg/l frequently results in formation of a good floc. Other weighting agents that have been used besides clay are powdered silica, limestone, and activated carbon; the latter possesses the additional advantage of high adsorptive capacity.

Problems in clarification of surface waters and color removal can often be minimized by applying oxidants. The most common practice is breakpoint chlorination of raw water. In this procedure, sufficient chlorine is added to oxidize interfering organic compounds. Other, less widely used oxidants are potassium permanganate, ozone, and chlorine dioxide.

EXAMPLE 7-10

A dose of 50 mg/l of alum is used in coagulating a turbid surface water. (a) How much natural alkalinity is consumed? (b) What changes take place in the ionic character of the water? (c) How many mg/l of aluminum hydroxide are produced?

Solution

(a) From the text, 1.0 mg/l of alum reacts with 0.5 mg/l of natural alkalinity, therefore,

$$50 \text{ mg/l (alum)} \frac{0.50 \text{ mg/l (alk)}}{1.0 \text{ mg/l (alum)}} = 25 \text{ mg/l of alkalinity as } CaCO_3$$

(b) 50 mg/l of alum is equivalent to 0.50 meq/l; therefore, based on Eq. 7-24, 0.50 meq/l of sulfate ion is added to the water. The aluminum ions precipitate out of solution, the calcium content is not affected, and 0.50 meq/l of bicarbonate, is converted to carbon dioxide.

(c) From Eq. 7-24, one mole of alum (600 g) reacts to produce two moles of aluminum hydroxide ($2 \times 78 = 156$ g). Therefore,

$$\frac{156}{600} = \frac{\text{mg/l of Al(OH)}_3 \text{ produced}}{50 \text{ mg/l alum dose}}, \qquad \text{or Al(OH)}_3 = 13 \text{ mg/l}$$

EXAMPLE 7-11

A surface water is coagulated by adding 2.5 grains per gallon of ferrous sulfate and an equivalent dosage of lime. How many pounds of coagulant are used per million gallons of water processed? How many pounds of lime are required at a purity of 80 percent CaO?

Solution

Consumption of ferrous sulfate equals

$$2.5 \text{gpg} \times 17.1 \frac{\text{mg/l}}{\text{gpg}} \times 8.34 \frac{\text{lb/mil gal}}{\text{mg/l}} = 356 \text{ lb/mil gal}$$

One equivalent weight of ferrous sulfate (139) reacts with one equivalent of 80 percent CaO ($28 \div 0.80 = 35$). Therefore,

$$\text{Lime dosage} = 356 \frac{\text{lb}}{\text{mil gal}} \times \frac{35}{139} = 90 \text{ lb/mil gal}$$

7-11   TASTE AND ODOR CONTROL

One of the objectives in water treatment is to produce a palatable water that is aesthetically pleasing. Repeated presence of objectionable tastes, odors, or colors in a water supply may cause the public to question its safety for consumption. Flavor may be affected by inorganic salts or metal ions, a variety of organic chemicals found in nature or resulting from industrial wastes, or by products of biological growths. Algae are the most frequent cause of taste and odor problems in surface supplies. Their metabolic activities impart odorous compounds that produce fishy, grassy, musty, or foul odors. Actinomycetes are moldlike bacteria that create an earthy odor.

Problems related to the palatability of water are generally unique in each system, and they must be individually studied to determine the best approach for prevention and cure. In groundwater treatment, aeration is frequently effective, since the odor compounds are often dissolved gases that can be stripped from solution. However, aeration is rarely effective in processing surface waters where the odor-producing substances are generally nonvolatile.

First consideration should be given to preventative measures in surface water supplies. If the interference can be traced to an industrial waste discharge, the source may be removed. In an eutrophic reservoir or lake, regular copper sulfate applications to the impounded water is effective in suppressing algal blooms that cause taste and odor, and filter clogging.

Oxidation by breakpoint chlorination is the technique most common for destruction of odor. This process as a first step in treatment provides both odor control and disinfection. Occasionally, potassium permanganate is more effective than chlorine as an oxidizing agent, and it may be used in conjunction with chlorination to destroy tastes and odors. Although not common in the United States, ozone is also a strong oxidant.

Activated carbon may be applied in conjunction with coagulation, sedimentation, and filtration processes for taste and odor control. Carbons used in the water industry are prepared principally from paper char, hardwood charcoal, or lignite, and are activated by controlled combustion to develop adsorptive characteristics. Each particle is honeycombed with thousands of molecular-sized pores that adsorb odorous substances. Activated carbon is available in powdered and granular forms. Granular carbon adsorption beds have not been used extensively in municipal water processing because of economic considerations; however, they are used extensively by bottling industries in purification of product water. Carbon in municipal treatment is a finely ground, insoluble black powder that can be applied either through dry feed machines or as a slurry. It can be introduced in any stage of processing before filtration where adequate mixing is available to disperse the carbon, and where the contact time is 15 min or more before sedimentation or filtration. The optimum point of application is generally determined by trial and error, and previous experience. Carbon adsorbs chlorine and, therefore, these two chemicals should not be applied simultaneously, or in sequence without an appropriate time interval.

7-12 REMOVAL OF DISSOLVED SALTS

Demineralization processes of distillation, electrodialysis, ion exchange, and reverse osmosis can separate salts from water. Distillation is a vapor-liquid transfer in which water vapor is driven off by heating in a still followed by condensation. This is one of the methods used in desalting sea water. Electrodialysis employes an induced electrical current to isolate the positive and negative ions in solution by means of selective membranes that permit ions to pass through the material from the diluted solution on one side to the concentrated solution on the other. Problems associated with dialysis in water renovation include chemical precipitation of salts with low solubility and membrane clogging as a result of colloidal matter. To reduce membrane fouling, surface waters must be given pretreatment by chemical precipitation and activated carbon to remove organic molecules and colloids. Demineralization by ion exchange is discussed in Section 7-9. None of these processes are expected to find widespread application in water treatment because of their high cost.

Reverse osmosis involves the forced passage of water through a membrane against the natural osmotic pressure to accomplish separation of water and ions. The process of osmosis is illustrated in Figure 7-26a where a thin

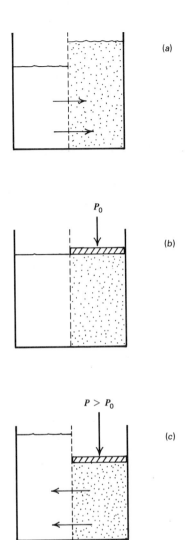

Figure 7-26 Illustrations describing the process of reverse osmosis which is used in removal of dissolved salts from water. (a) Normal osmotic flow of water through a semipermeable membrane is from the low to high concentration salt solution. (b) The pressure at which no net flow occurs is referred to as the osmotic pressure, P_0. (c) In reverse osmosis, water is forced by high pressure from the salty solution through the membrane that blocks the passage of ions.

membrane of cellulose acetate (plastic) separates two salt solutions. Water from the side of lower salt concentration flows through the membrane to the solution of higher concentration attempting to equalize the salt content, while the membrane allowing water flow blocks passage of salt ions. If pressure is applied to the side of higher salt content, flow of water can be prevented; this pressure, at no net flow, is termed the osmotic pressure. If force is increased, the water flow is reversed passing from the salt water to the fresh water, thus separating the salts from solution. This process is called reverse osmosis. The rate of water transfer is dependent primarily on the difference in salt concentration between the solutions, characteristics of the membrane, and magnitude of the pressure applied. Commercial reverse-osmosis devices are available in a variety of designs, each having particular advantages. The module of 24 units

Figure 7-27 A module of 24 reverse-osmosis permeator units used to desalt a municipal ground-water supply. (Manufactured by E. I. DuPont de Nemours and Co., Inc.)

shown in Figure 7-27 are hollow fiber permeators each having the following characteristics: 5.5 in. diameter and 47 in. shell length, 400 psi operating pressure, 1500 sq ft of fiber surface area, membrane rate of 1.8 gal/sq ft/day, and a nominal product water capacity of 200 gal/day.

In addition to the permeators, a basic reverse osmosis system consists of pretreatment, pumps to provide the operating pressure, tanks and appurtenances for cleaning and flushing, and a disposal system for waste brine. Pretreatment may require suspended solids removal by filtration or carbon adsorption beds to prevent membrane fouling and/or pH adjustment and addition of sodium hexametaphosphate to reduce salt precipitation. Posttreatment may be necessary to stabilize the finished water because carbon dioxide is able to pass through the membranes with the product water. Aerators or vacuum degasifiers may be used to remove dissolved gases, while lime or soda ash additions provide final pH adjustment. Periodic cleaning of the membrane surface is required to maintain high water transfer efficiency. Units are flushed with acid rinses and cleaning agents to remove any buildup of metal ions, salt precipitates, or organic matter.

Disposal of concentrated reject water from reverse osmosis can be handled in several ways, such as deep well injection or evaporation ponds. A typical reverse osmosis plant that is operating on a feed of 1000 mg/l total dissolved solids can economically achieve about 75 percent conversion, that is, 100 gal of feed produces 75 gal of water with a brine rejection of 25 gal containing about 4000 mg/l of dissolved solids. The costs involved in disposing of this large volume of reject water is a critical economic and environmental problem associated with the reverse osmosis. On the other hand, several installations have been constructed for processing water for small municipalities in locations where more suitable water supplies are unavailable.

7-13   CORROSION CONTROL AND STABILIZATION

Ferrous metal when placed in contact with water or moist soil results in an electric current caused by the reaction between the metal surfaces and existing chemicals in the earth or water. The basic electrode reactions are represented by Eqs. 7-31 and 7-32. At the anode, iron goes into solution as the ferrous ion, and at the cathode, when dissolved oxygen is present, hydroxide ion is formed.

$$\text{Anode:}\quad Fe \rightarrow Fe^{++} + 2\,\text{electrons} \tag{7-31}$$

$$\text{Cathode:}\quad 2\,\text{electrons} + H_2O + \tfrac{1}{2}O_2 \rightarrow OH^- \tag{7-32}$$

Ferrous (Fe^{++}) iron is further oxidized to the ferric (Fe^{+++}) state in the presence of oxygen and precipitates as insoluble ferric hydroxide, or rust, Eq. 7-33.

$$2Fe^{++} + 5H_2O + \tfrac{1}{2}O_2 \rightarrow 2Fe(OH)_3 \downarrow + 4H^+ \tag{7-33}$$

The rate of corrosion is controlled by the concentration and diffusion rate of dissolved oxygen to the surface of the iron. Diffusion of oxygen to the anode is limited somewhat by the physical barrier of rust forming on the surface. A much denser barrier of calcium carbonate scale is formed at the cathode under alkaline conditions if sufficient calcium and bicarbonate ions are available, Eq. 7-34. The net result of pipe corrosion is: pitting and scaling of the surface referred to as tuberculation, discoloration of the water, and eventual failure of the pipe.

$$Ca^{++} + HCO_3^- + OH^- \rightarrow CaCO_3 \downarrow + H_2O \tag{7-34}$$

Scaling

The easiest way to protect an iron pipe against corrosion is by coating or lining. Cast-iron pipe is normally coated with coal tar or enamel on the exterior, and is lined with a thin layer of cement mortar to protect the interior. Asbestos cement and plastic pipe are corrosion resistant and immune to tuberculation. In addition to cement lining, a calcium carbonate film (scale) on the pipe interior provides protection. The method most successful in corrosion control is upward adjustment of pH, using either lime or soda ash, and addition of metaphosphates. Bringing the water pH above its calcium carbonate saturation value preserves a thin protective coating on the pipe interior. Metaphosphates sequester the excess calcium and carbonate ions preventing formation of further calcium carbonate scale. Therefore, an equilibrium is reached whereby the $CaCO_3$ coating does not dissolve because the water is saturated with calcium ions; yet, these ions cannot precipitate to form scale because of the sequestering action of the sodium hexametaphosphate. The latter, also known as glassy phosphate, is available in both powdered and granular form for application in municipal and industrial water treatment.

Cathodic protection is a means of counteracting corrosion by preventing the dissolution of iron shown in Eq. 7-31. Either a sacrificial galvanic anode, composed of a metal higher in the electromotive series, or an electrolytic anode energized by an external source of direct current may be used to protect ferrous structures. Magnesium or zinc galvanic anodes provide a battery action with iron such that they corrode and are sacrificed while the iron structure they are connected to is protected from dissolution. This type of system is common to small hot-water heaters.

Large steel waterworks structures are given cathodic protection using electrolytic anodes that are made anionic by application of a direct current.

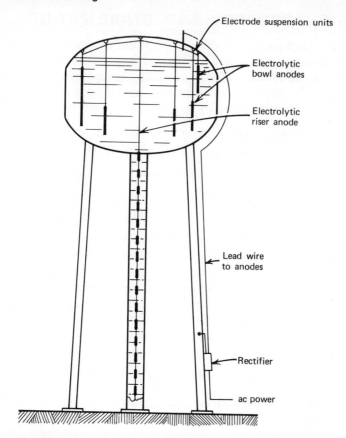

Electrode suspension units

Electrolytic
bowl anodes

Electrolytic
riser anode

Lead wire
to anodes

Rectifier

ac power

Figure 7-28 Cathodic protection can be used to prevent corrosion of the interior surfaces of a water storage tank.

Anodes may be of many different metals, for instance, graphite, carbon, platinum, aluminum, iron, or steel alloys. They are energized by connecting to the positive terminal of a direct current power supply, usually rectifier, while the structure being protected is connected to the negative terminal. The impressed current attracts electrons to the iron preventing ionization and, hence, corrosion. Figure 7-28 sketches the application of cathodic protection to the interior of an elevated storage tank. Under some conditions, galvanic anodes are employed in lieu of, or in combination with, the rectifier depending on the tank condition and chemical characteristics of the water. Buried tanks are protected from exterior corrosion by placing the anodes in the ground surrounding the tank. Except in unusual cases, cathodic systems are not used to protect water distribution piping because of the high cost.

7-14   SOURCES OF WASTES IN WATER TREATMENT

Wastes originating from water treatment in approximate order of abundance are: residues from chemical coagulation, precipitates from softening, filter backwash water, settled solids from presedimentation, oxides from iron and manganese removal, and spent brines from regeneration of ion exchange units. These wastes vary widely in composition containing the concentrated materials removed from raw water and chemicals added in treatment. They are produced continuously and discharged intermittently. Settled floc is allowed to accumulate in clarifiers over relatively long periods of time, while backwashing of filters produces a high flow of waste water for a few minutes, usually once a day for each filter.

Coagulation Wastes

The chief constituent in coagulation sludge is either hydrated aluminum from alum or iron oxides from iron coagulants. Small quantities of activated carbon and coagulant aids, such as polyelectrolytes and activated silica, may be included. Particulate matter entrained in the floc is mostly inorganic in nature, being principally silt and clay. Since the organic fraction is small, the sludge does not undergo active biological decomposition. Aluminum hydroxide sludges are gelatinous in consistency, which makes them difficult to dewater. Settled sludges have low solids concentrations, usually between 0.2 and 2.0 percent. Iron precipitates are slightly denser than alum sludges. The total solids produced in alum coagulation of a surface water can be estimated by using the following relationship, as is illustrated in Example 7-12.

$$\frac{\text{Sludge solids,}}{\text{lb/mil gal}} = 8.34\left(\frac{\text{alum dosage, mg/l}}{4} + \frac{\text{raw water}}{\text{turbidity, Jtu}}\right) \quad (7\text{-}35)$$

Softening Sludge

Lime-soda ash softening of groundwaters yields a sludge free of extraneous inorganic and organic matter consisting of calcium carbonate, magnesium hydroxide, and unreacted lime. This residue is generally stable, dense, inert, and relatively pure, allowing recovery of the lime by recalcination where economically feasible. The settled solids concentration in softening sludges is in the range of 2 to 15 percent depending on the ratio of calcium to magnesium precipitates. Calcium carbonate provides a compact sludge easy to handle while magnesium hydroxide, like aluminum hydroxide, is gelatinous and does not consolidate well by gravity settling. Highly turbid, hard surface waters are often treated by both precipitation softening and coagulation. The character and settleability of these combined sludges varies considerably depending on factors such as nature of turbidity in the raw water, ratio of calcium to magnesium in the softening precipitate, type and dosage of metal coagulant, and filter aids used. In general, lime and iron sludges have solids concentrations greater than 10 percent and often greater than 15 percent, while alum-lime sludges have settled solids densities of less than 10 percent. The quantity of sludge solids produced in softening can be estimated, as is shown in Example 7-13, by using the precipitation softening Eqs. 7-15 to 7-19.

Other Residues

The amount of sludge attributable to iron and manganese removal is relatively small and negligible compared to coagulation or softening process wastes. Where the sole treatment is iron and manganese removal from groundwater, 50 to 90 percent of the hydrated ferric and manganic oxides are trapped in the filters and appear in dilute backwash water. Since the concentration of these metals in raw water is generally low, the amount of solids produced and volume of waste are correspondingly small.

Where turbid surface waters are given presedimentation, the accumulated solids consist of fine sands, silts, clays, and organic debris. Sometimes chlorine and other chemicals are applied to the raw water to improve settleability; yet, they rarely contribute a measurable amount of solids to the sludge. Presedimentation residue may be hauled to landfill or, in some instances discharged back to the watercourse.

Spent brine solutions from regeneration of ion exchange softeners contain regenerate sodium chloride plus calcium and magnesium ions displaced from the exchange resin. The character and volume of the waste liquid depends on the amount of hardness removed and operation of the exchange unit; the volume of spent brine is usually 3 to 10 percent of the treated water.

Filter Wash Water

Backwashing of filters produces a relatively large volume of waste water with low solids concentration in the range of 0.01 to 0.1 percent (100 to 1000 mg/l). The total solids content depends on efficiency of prior coagulation and sedimentation and may be a substantial fraction, say 30 percent, of the residue resulting from treatment. Two to three percent of all water processed is used for filter washing; the exact amount is contingent on the type of treatment system and the filter backwashing technique. Wash water may be discharged to a recovery basin and recycled for processing with the raw water. In the case of a lime softening plant that is treating groundwater, the backwash may be collected, mixed, and returned to the inlet of the plant without solids removal. However, in surface water plants this often creates a buildup of undesirable solids, for example, algae, that keep cycling through the system. Here, the suspended solids are allowed to settle, often after the addition of a polyelectrolyte to improve flocculation, and only the overflow is returned for reprocessing, as is illustrated in Figure 7-3. Settled sludge is discharged from the bottom of the clarifier-surge tank to a sludge thickener, or dewatering unit, or directly to disposal. Sometimes backwash water is released to a sanitary sewer for final disposition at the waste-water treatment plant; a surge tank may still be required to even out the flow so that the sewer is not hydraulically overloaded. Where lagoons are used to dispose of sludge solids, the wash water may be directed to these ponds and, occasionally, the pond water is decanted from near the surface and recycled to the plant.

EXAMPLE 7-12

A surface water treatment plant coagulates a raw water having a turbidity of 9 Jackson candle units by applying an alum dosage of 30 mg/l. Estimate the total sludge solids production in pounds per million gallons of water processed. Compute the volume of sludge from the settling basin and filter backwash water using 1.0 percent solids concentration in the sludge and 500 mg/l of solids in the washwater. Assume that 30 percent of the total solids are removed in the filter.

Solution

Applying Eq. 7-35,

$$\text{Total Sludge Solids} = 8.34\left(\frac{30}{4}+9\right) = 137 \text{ lb/mil gal}$$

$$\text{Solids in Sludge} = 0.70 \times 137 = 96 \text{ lb}$$

Solids in Backwash Water $= 0.30 \times 137 = 41$ lb

$$\text{Volume} = \frac{\text{sludge solids (lb)}}{\text{solids fraction} \times 8.34 \text{ (lb/gal)}}$$

$$\text{Sludge volume} = \frac{96}{\dfrac{1.0}{100} \times 8.34} = 1150 \text{ gal/mil gal}$$

$$\text{Wash-water volume} = \frac{41}{\dfrac{500}{1,000,000} \times 8.34} = 9830 \text{ gal/mil gal}$$

EXAMPLE 7-13

Calculate the residue produced in the excess lime softening of the water defined in Example 7-8.

Solution

Component in Water		Applicable Equation	Precipitate Produced	
Formula	meq/l		$CaCO_3$ meq/l	$Mg(OH)_2$ meq/l
CO_2	0.40	7-15	0.40	0
$Ca(HCO_3)_2$	2.00	7-16	4.00	0
$Mg(HCO_3)_2$	0.70	7-17	1.40	0.70
$MgSO_4$	0.51	7-18 and 7-19	0.51	0.51
Excess lime	1.25	7-20	1.25	0
			7.56	1.21
Minus the practical limits (solubility)			−0.60	−0.20
			6.96	1.01

$$\text{Residue} = CaCO_3 + Mg(OH)_2$$
$$= 6.96 \times 50 + 1.01 \times 29.2 = 378 \text{ mg/l} = 3150 \text{ lb/mil gal}$$

7-15 DEWATERING AND DISPOSAL OF WASTES FROM WATER TREATMENT PLANTS

Prior to regulations governing the disposal of processing wastes, settled sludges and backwash waters were discharged to rivers. Lagoons, or sludge drying beds, were used where a watercourse was not conveniently available for waste discharge. In some municipalities, sanitary sewers received the waste flows and, in rare cases, softening plants used recalcination to process calcium carbonate precipitate back to lime for reuse. Since residues from each water plant are unique, no specific treatment process for dewatering and disposal will yield the same results. In fact, although a variety of alternative methods are available, only one or two techniques may be applicable to a particular location.

Lagoons

Ponding is a popular and acceptable method for dewatering, thickening, and the temporary storage of waste sludge. Lagoons are constructed by excavation or enclosing of an area by dikes. Drainage may be improved by blanketing the bottom with sand and installing underdrains. An overflow device is normally constructed to decant clear supernatant, particularly if filter wash water is discharged to the lagoon. Although ponding is an inefficient process, where there is suitable land area available that is inexpensive, any other disposal method may be difficult to justify. The amount of land needed depends on factors such as: type, character, and solids content of the waste sludge; design features including underdrains and decanters; climate; and method of operation.

Sludges from water softening plants are readily dewatered in lagoons. Slow consolidation of the settled precipitates yields concentrations of about 50 percent solids. This material can be removed by a scraper or dragline; however, lime sludge is considered a poor landfill and final disposition of lagoon residue can be a serious problem where burial sites are scarce. Alum sludge is more difficult to dewater and, even after a long period of time, consolidates to only 10 to 15 percent solids density. If the surface of settled alum sludge is exposed, evaporation may provide a hard crust but the remaining depth has the character of a gel that turns into a viscous liquid on agitation. This slurry may be removed by a dragline and dumped on the banks to air dry prior to hauling. As with lime sludge, it does not make a good landfill. In cold climates, freezing can enhance the dewatering character of alum sludge. On thawing, the moisture content decreases and the solids are in the form of small granular particles that dry to a brown powder.

Drying Beds

Sand drying beds are generally applicable to small water plants in areas where land is readily available. A bed consists of 6 to 12 in. of coarse sand underlain by layers of graded gravel. The earth bottom is graded slightly to tile underdrains placed in trenches. A surrounding wall is used to contain the liquid sludge when applied to the bed; filling depths usually range from 2 to 5 ft. Dewatering action is essentially gravity drainage aided by air drying, although the operation may include decanting supernatant. After repeated sludge applications over a period of several months, the sludge cake formed on the surface is removed either by hand shoveling or mechanical means.

Gravity Thickening

Separate thickeners may be used as the first step in sludge processing for volume reduction when consolidation of sludge in settling basins used to clarify coagulated water is inadequate. Polymers or other coagulant aids may be applied to improve the settleability of the sludge solids. Figure 7-29 illustrates a typical gravity thickener. Compared to a conventional settling basin, the tank is deeper to accomodate a greater volume of sludge and has a heavier raking mechanism. Pickets or palings attached to the collector arms stir through the sludge providing cavities for the release of trapped water. Flow enters from behind an inlet well in the center of the tank and is directed downward. Supernatant overflows a peripheral weir, while underflow of thickened sludge is drawn from a bottom sump in the tank. In waste-water treatment, thickening is normally considered a continuous-flow process, but in handling water treatment plant or industrial waste sludges flow to the thickener is often intermittent. While performing clarification, tanks fed intermittently can be

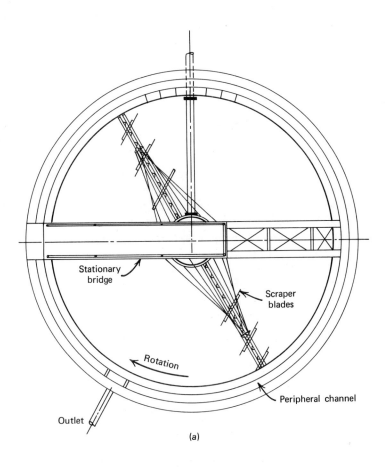

Stationary
bridge

Scraper
blades

Rotation

Peripheral channel

Outlet

(a)

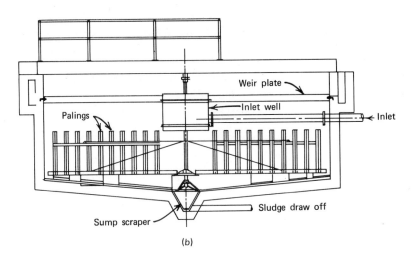

Weir plate

Inlet well

Palings

Inlet

Sump scraper

Sludge draw off

(b)

Figure 7-29 (*a*) Plan and (*b*) cross-sectional view of a gravity sludge thickener. (Courtesy of Westinghouse–Infilco)

designed to allow for storage of accumulated sludge. This holding tank function is often essential to efficient subsequent dewatering by mechanical units, such as, filters. or centrifuges. Gravity thickeners are often sized on the basis of solids loading per square foot of tank area; values in the range of 5 to 10 lb/sq ft/day are common. However, individual sludges vary widely in their consolidating characteristics, and thickeners should be selected on the basis of local conditions.

Centrifugation

The first centrifuge developed was a basket type that is sketched in Figure 7-30a. It consists of a spinning cylinder creating high centrifugal forces to push solids against the drum wall. Feed slurry enters in the center at the bottom and clarified liquid discharges over a lip ring at the top. When cake depth in the bowl reaches a preset thickness, the centrifugation process is interrupted to scrape out collected sediment using a skimmer or knife plow. After cake discharge, feed is again started.

The most popular centrifuge in dewatering water and waste-water sludges is the solid bowl scroll type illustrated in Figure 7-30b. The unit consists of a rotating solid bowl in the shape of a cylinder with a cone section on one end, and an interior rotating screw conveyer. Feed slurry, entering from the center, is held against the bowl wall by centrifugal force. Settled solids are moved by the conveyer to one end of the bowl while clarified effluent discharges at the other end. Additional details are shown in the cutaway view in Figure 7-31. The conical bowl shape at the solids discharge end enables the conveyer to move the solids out of the liquid for drainage before being discharged.

A major advantage of the solid bowl centrifuge is operational flexibility. Machine variables include pool volume, bowl speed, and conveyer speed. The depth of liquid held against the bowl wall can be controlled by an adjustable plate dam at the discharge end of the bowl. Pool volume adjustment changes the drainage deck surface area of the solids discharge section. Bowl speed affects gravimetric forces on the settling particles, and conveyer speed controls the solids retention time. The driest cake product results when bowl speed is increased, pool depth is at the minimum allowed, and differential speed between the bowl and conveyer is the maximum possible. Flexibility of operation allows a wide-range moisture content in the solid discharge, varying from a dry cake to a thickened liquid sludge. Feed rates, solids content, and prior chemical conditioning are also variables that influence performance. Removal of solids can be enhanced by adding polyelectrolytes or other coagulants with the slurry feed. The scroll centrifuge also operates with very little surveillance.

Centrifuges are being used to dewater treatment plant sludges drawn from settling tanks or gravity thickeners. Lime sludges compact readily producing a cake with 60 to 70 percent solids (toothpaste consistency). In recalcination plants, centrifuges commonly dewater lime softening sludges because of their ability to selectively thicken calcium carbonate precipitate. A machine can be operated such that the bulk of undesirable magnesium hydroxide appears in the liquid discharge, thus providing a purer sludge cake for recalcining. Although aluminum hydroxide sludges do not dewater as readily as lime precipitates, centrifugation with polymer addition to aid flocculation provides a viscous, liquid, solids discharge suitable for either further processing or disposal. Sludge

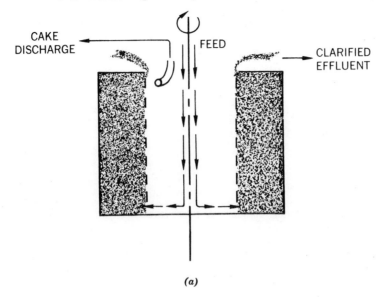

CAKE DISCHARGE

FEED

CLARIFIED EFFLUENT

(a)

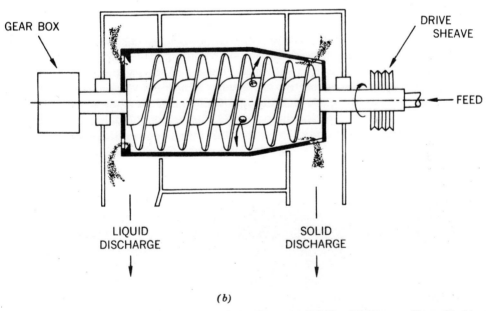

GEAR BOX

DRIVE SHEAVE

FEED

LIQUID DISCHARGE

SOLID DISCHARGE

(b)

Figure 7-30 Schematic diagrams for two types of centrifuges. (a) Solid bowl basket centrifuge with skimmer. (b) Solid bowl scroll centrifuge, also illustrated in Figure 7-31, is the common type used in dewatering waste sludges from water and waste-water treatment. (Courtesy of Pennwalt Chemicals Corp., Sharples Equipment Division)

containing about one-half aluminum hydroxide can be thickened to 10 to 15 percent solids, while approximately one-quarter hydrate slurry can be thickened to 20 to 25 percent. Removal of solids from the feed ranges from 50 to 95 percent depending on operating conditions and polymer dose. Holding all other variables constant, the percentage of solids recovery and density of the thickened discharge are directly related to polyelectrolyte dosage.

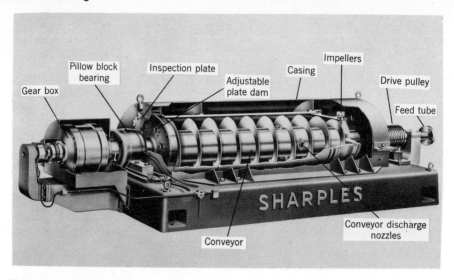

Figure 7-31 Cutaway view of a horizontal, solid bowl scroll centrifuge. (Courtesy of Pennwalt Chemicals Corp., Sharples Equipment Division)

Pressure Filtration

The common type of pressure filter adopted for dewatering waste sludges is the plate and frame type like the one shown in Figure 7-32. A view of the plate section of this type unit is shown in Figure 7-33. Steps in operation include: chemical conditioning of the raw sludge if necessary, precoating the filter media, pressure sludge dewatering, and discharge of the sludge cakes. The filter media is precoated by feeding a water suspension of diatomaceous earth. Precoat prevents binding of the filter cloth with sludge particles and facilitates discharge of cake at the end of the filter cycle. Conditioned sludge enters the center feed port immediately after the precoat slurry and is distributed

Figure 7-32 Pressure filter for sludge dewatering. (Courtesy of Passavant Corporation)

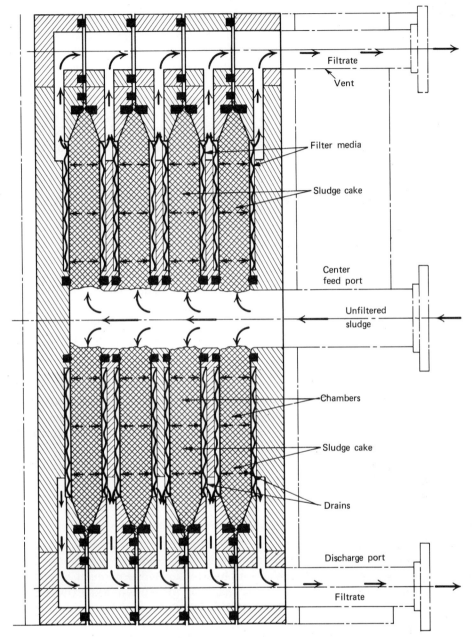

Figure 7-33 Cross-sectional view and flow diagram of the plate section of a pressure filter. (Courtesy of Passavant Corporation)

throughout the chambers located between the individual plates. Water is forced through the filter cloth and out through drains between the frames. When the chambers have filled with accumulated solids, a high-pressure cycle consolidates the cakes. Compressed air at about 200 psi is applied to the sludge inlet and maintained until the desired density is obtained. After the filter cycle, a core flow valve opens and compressed air blows the feed sludge remaining in the center feed port back to a holding tank. The filter plates are then separated and the dewatered sludge cakes drop into a hopper equipped with a conveyer mechanism. The filter cycle, taking from 1 to 2 hr, is completely automated including precoat, filtration, and discharge. Some of the advantages attributed to pressure filtration are high solids concentration in the filter cake, clear filtrate with low suspended solids, and minimum operator supervision.

Pressure filtration appears to be particularly advantageous for thickening alum sludges where a high solids content is most desirable. With lime or polymer conditioning, aluminum hydroxide sludges can be pressed to a solids content of 30 to 40 percent, which is adequate for hauling by truck.

Vacuum Filtration

Vacuum filters are not used extensively in dewatering water plant sludges even though they are extremely popular in processing waste-water treatment sludges; refer to Section 11-4. Where applied to pure lime sludges, belt filters with synthetic cloth have worked rather successfully while coil filters were plagued with encrustation of the coils. Vacuum filtration of aluminum hydroxide precipitates appears to be both economically and technologically questionable. Even with polymers and precoating it is difficult to consistently achieve a suitably dry cake. Based on preliminary studies, it appears that vacuum filtration will not receive broad application in the dewatering of chemical sludges, since other processes seem to be more economical and to achieve better results.

Recovery of Chemicals

Lime can be recovered from softening sludges by recalcination. The basic steps are: gravity thickening; dewatering by centrifugation, or filtering, with classification of solids to remove magnesium hydroxide and other impurities; for some processes, flash drying by injecting the dewatered sludge into a stream of hot off-gases from the recalciner; and, recalcining which is accomplished by heating the dried solids in a rotary kiln or fluidized-bed incinerator to produce lime and carbon dioxide, Eq. 7-36. An impurity that must be removed

$$CaCO_3 \xrightarrow{\text{heat}} CaO + CO_2 \qquad (7\text{-}36)$$

to produce a high quality lime is the magnesium hydroxide precipitate. At present this is wasted prior to recalcination, although recovery of magnesium carbonate from the magnesium hydroxide sludge is under consideration if it should prove to be a favorable coagulant for reuse in water treatment. Another problem associated with recalcining is interference from clay turbidity where the softening plant treats a surface water. Clay mixed with the calcium carbonate sludge results in buildup of insolubles in the finished lime; currently, there is no feasible means of separating colloidal impurities like clay. Froth flotation is one technique being considered.

Recovery of alum is rarely practiced because of low-process efficiency and poor quality of the regenerated alum. The key problem is separation of aluminum hydroxide from other impurities in the waste sludge. Where practiced, the sludge is mixed with sulfuric acid and discharged to a gravity thickener for separation of alum solution from settleable insoluble residue. The decanted supernatant contains the bulk of the alum, but separation is incomplete. Carryover of colloidal matter, including organic materials charred by the sulfuric acid, contaminates the recovered alum solution. Reclaimed alum is held in tanks and metered to points of application in the treatment plant. Underflow from the thickener is wasted by treating with lime, dewatered, and hauled away for land burial.

REFERENCES

1. *Recommended Standards for Water Works*, Great Lakes–Upper Mississippi River Board of State Sanitary Engineers, 1971 Edition (Health Education Service, Albany, N.Y.).
2. *Water Chlorination Principles and Practices*, American Water Works Association, Manual M20, New York, 1973.
3. Bellack, E., *Fluoridation Engineering Manual*, Environmental Protection Agency, Water Supply Programs Division, 1972.
4. *Water Quality and Treatment*, American Water Works Association, Third Edition, McGraw-Hill Book Co., 1971.

PROBLEMS

7-1 What are the most common impurities removed in treating groundwater for municipal supply?

7-2 Why do plants for processing river water generally have the most extensive and flexible treatment systems of any water supply source?

7-3 Laboratory tests were conducted on the chemical treatment of a water using both a long, narrow flocculation tank (plug flow) and a complete-mixing unit. The observed rate constants (K) under steady-state conditions were first order kinetics equal to "6 per day" for plug flow and "18 per day" for complete mixing. If the desired degree of removal is 90 percent for an influent concentration of 60 mg/l, which process requires the shortest retention time? (*Answers* plug flow at 9.1 hr, complete-mixing 12 hr)

7-4 Refer to the top diagram in Figure 7-2. For a water flow rate equal to 1000 gpm (1,440,000 gpd) what minimum size units (volumes) would you recommend for the rapid mix, flocculator, and settling tank. (*Answers* minimum volumes of 500 gal, 30,000 gal, and 240,000 gal, respectively)

7-5 A water treatment plant designed for a flow of 12 mgd has units as illustrated in Figure 7-8. The flocculator is 25 ft wide, 125 ft long, and 12 ft liquid depth; the clarifier is 125 ft by 125 ft, and 16 ft deep. Calculate settling time in hr, flocculation time in min., and horizontal velocity through the flocculator in feet per minute. (*Answers* 3.7 hr, 34 min, 3.7 fpm)

7-6 A flocculator designed to treat 20 mgd is 100 ft long, 40 ft wide and 15 ft deep. Compute the detention time and horizontal flow-through velocity. Do these values satisfy the *GLUMRB Standards*? (*Answers* $t = 32$ min > 30, yes; $v = 3.1$ fpm > 1.5, no)

7-7 Two rectangular clarifiers each 90 ft long, 16 ft wide, and 12 ft deep settle 1.5 mgd. Total effluent weir length is 162 ft. Calculate the detention period, overflow rate, and weir loading. (*Answers* 4.1 hr, 520 gpd/sq ft, and 9300 gpd/ft)

7-8 What diameter circular clarifier and side water depth are needed for a 4.0 mgd flow based on a maximum overflow rate of 400 gpd/sq ft and detention time of 4.0 hr? (*Answers* dia = 113 ft, SWD = 9.0 ft)

7-9 If the overflow rate of a flocculator-clarifier is 1.0 gpm/sq ft, what is the overflow rate expressed in gpd/sq ft? If the depth is 10 ft, what is the detention time? (*Answers* 1440 gpd/sq ft, 1.2 hr)

7-10 What is meant by the expression "in depth" filtration?

7-11 Why is the end of the outlet pipe from a rapid sand filter immersed in the clear well?

7-12 What is the function of the rate-of-flow controller in filter operation?

7-13 What are the advantages and disadvantages of a gravimetric feeder compared with a volumetric type?

7-14 Why is a slaker used in applying powdered lime to water?

7-15 Why is anhydrous ammonia sometimes added to water after chlorination?

7-16 Results of a chlorine demand test on a raw water are as follows:

Sample Number	Chlorine Dosage (mg/l)	Residual Chlorine After 10-min Contact (mg/l)
1	0.20	0.19
2	0.40	0.36
3	0.60	0.50
4	0.80	0.48
5	1.00	0.20
6	1.20	0.40
7	1.40	0.60
8	1.60	0.80

Sketch a chlorine demand curve as is shown in Figure 7-19. What is the breakpoint dosage, and what is the chlorine demand at a dosage of 1.2 mg/l. (*Answers* 1.0 mg/l, 0.8 mg/l)

7-17 What is the recommended minimum free chlorine residual to insure disinfection of a water at pH 8? Minimum combined chlorine residual? (*Answers* 0.4 mg/l, 1.8 mg/l)

7-18 Why are pressure regulating and vacuum compensating valves needed on a solution feed chlorinator?

7-19 A dosage of 0.60 mg/l liquid chlorine is needed to maintain a residual of 0.20 mg/l in the distribution system. What dosage rate in pounds per day is needed to treat 6.0 mgd? (*Answer* 30 lb/day)

7-20 How many pounds of available chlorine are contained in 1 gal of 15 percent sodium hypochlorite? How many gallons would be required for a dosage of 0.6 mg/l to 6.0 mgd? (*Answers* 1.25 lb/gal, 24 gal/day)

7-21 How many pounds of dry hypochlorite powder with 70 percent available chlorine must be added to 100 gal of water to make a 1.0 percent solution? (*Answer* 11.9 lb)

7-22 In treating 25 mil gal of water, 167 lb of liquid chlorine are applied. Compute the dosage in milligrams per liter. (*Answer* 0.80 mg/l)

7-23 A sodium fluoride solution is prepared by dissolving 8.0 lb of 98 percent NaF in 50 gal of water. The solution tank mounted on scales has a measured weight loss of 250 lb in the treatment of 300,000 gal of water. Calculate the dosage of fluoride in milligrams per liter. (*Answer* 0.83 mg/l)

7-24 A liquid feeder applies 12 gal of 4.0 percent saturated sodium fluoride solution to 400,000 gal of water. What is the concentration of fluoride added? (*Answer* 0.54 mg/l)

7-25 How many pounds of commercial fluosilicic acid with a purity of 30 percent should be added per million gallons of water to increase the fluoride ion concentration from 0.4 mg/l to 1.0 mg/l? (*Answer* 21.1 lb/mil gal)

7-26 Calculate the theoretical dosage of potassium permanganate required to oxidize 1.2 mg/l of iron and 0.8 mg/l of manganese. (*Answer* 2.7 mg/l)

7-27 The analysis of a raw water is: calcium 2.0 meq/l, magnesium 0.8 meq/l, sodium 0.7 meq/l, potassium 0.2 meq/l, bicarbonate 2.5 meq/l, sulfate 1.0 meq/l, and chloride 0.2 meq/l. Calculate the lime and soda ash additions required for excess lime softening. Sketch bar graphs for the raw water, and lime-soda ash softened water following recarbonation and filtration. (*Answers* 127 mg/l CaO and 16 mg/l Na_2CO_3)

7-28 Raw water analysis: carbon dioxide 26 mg/l, calcium hardness 150 mg/l as $CaCO_3$, magnesium 65 mg/l as $CaCO_3$, sodium and potassium 0.29 meq/l, bicarbonate alkalinity 185 mg/l as $CaCO_3$, sulfate 29 mg/l, and chloride 10 mg/l. Draw a meq/l bar graph, and compute the necessary dosages in pounds per million gallons for precipitation softening. Assume that the soda ash is pure sodium carbonate and that the lime is 75 percent CaO. (*Answers* 2310 lb/mil gal of 75 percent CaO, and 265 lb/mil gal Na_2CO_3)

7-29 Consider precipitation softening by split treatment of the raw water described in Problem 7-27. Assume that one half of the flow is by-passed, and the other one half is given excess lime treatment in the first stage. Compute the chemicals required and hardness of the finished water. (*Answers* 64 mg/l CaO, 8 mg/l Na_2CO_3, and 69 mg/l)

7-30 Alum is applied at a dosage of 30 mg/l in coagulation of a surface water. How many mg/l of alkalinity are consumed in the reaction? (*Answer* 15 mg/l)

7-31 A water is coagulated by 3.0 grains per gallon of pure ferric chloride with an equivalent amount of lime. What is the lime dosage in mg/l as CaO? (*Answer* 27 mg/l)

7-32 How much lime with 70 percent CaO is needed to react with a copperas dose of 20 mg/l? (*Answer* 5.8 mg/l)

7-33 List the most common chemicals used for destruction or removal of tastes and odors in water treatment.

7-34 Explain how a protective calcium carbonate film can be maintained on the interior of piping in a distribution system.

7-35 Estimate the sludge solids produced in coagulating a surface water having a turbidity of 12 Jackson candle units with an alum dosage of 40 mg/l? What is the sludge volume if the settled waste sludge and filter backwash water are concentrated to 1000 mg/l of solids in a clarifier-thickener? (*Answers* 180 lb/mil gal, 22,000 gal/mil gal)

7-36 Calculate the residue produced in the excess lime softening of the water defined in Problem 7-27. (*Answer* 2600 lb/mil gal)

7-37 What are the relative advantages of dewatering sludges by centrifugation and pressure filtration?

7-38 Why does sludge processing for the recovery of alum lag behind recalcination?

chapter 8

Operation of Waterworks

Management of a water utility requires extensive knowledge in distribution systems and water processing; refer to Chapters 6 and 7, respectively. Water quality (Chapter 5) and testing for both chemical and biological characteristics are also prerequisite to systems control. This chapter provides an insight to operation of waterworks by discussing a groundwater plant, surface water plant, distribution system, and water rates.

8-1 GROUNDWATER TREATMENT PLANTS

This presentation is based on the operation of a 60-mgd groundwater plant placed in operation in 1968. Water is supplied from 37 wells, each pumping from 400 to 2000 gpm, located in a sand and gravel aquifer adjacent to the Platte River. The raw water has the following average characteristics: pH of 7.7, total hardness of 214 mg/l as $CaCO_3$ (61 mg/l calcium and 15 mg/l magnesium), 0.01 mg/l iron, 0.13 mg/l manganese, and 0.48 mg/l fluoride. The mean water temperature is 56° F with average yearly variations from 42° F during March to a maximum average of 70° F in late summer. Treatment objectives are iron and manganese removal, hardness reduction, fluoridation, and chlorination to provide disinfection and establish a protective residual. The physical treatment units, shown schematically in Figure 8-1, are six flocculator-clarifiers, eight dual-media filter beds, and two clear-well storage reservoirs. High service pumps deliver finished water to the city through a 60-in. diameter transmission main. Two high-service units are powered by synchronous electric motors and two by natural gas engines. Standby capacity is also provided in the well field where enough pumps have alternate gas engines to enable the plant to operate at 50 percent load in the event of electrical power failure.

Manual operation of a treatment plant of this complexity is not feasible. Therefore, a control network links pumping, chemical dosing, and filter operation to a control room that contains a panel and operator's console (Figure 8-2). Well pumps can be started or stopped from this point at any time, and running lights on the panel indicate which pumps are in service. The

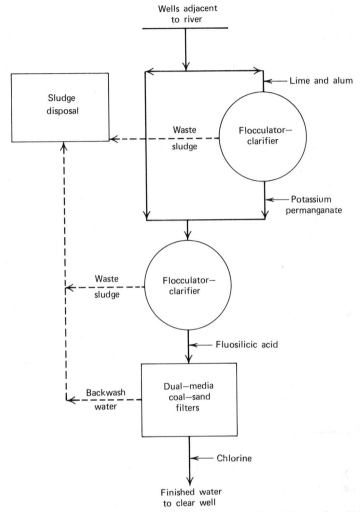

Figure 8-1 Schematic of the Platte River Water Treatment Plant, Metropolitan Utilities District, Omaha, Nebraska. Processes are hardness reduction by split treatment, iron and manganese removal, fluoridation, and disinfection.

capacity of each well is marked on the board, enabling the operator to select the proper combination for any desired pumpage. Flow of raw water into the treatment basins is controlled from the console by rate of flow and level controllers. Each inlet line has a butterfly valve regulated by signals from a flow meter and a bubbler tube for level control. Since there is always some imbalance in flow because of signal oscillation, two or more basins are set for approximate flows and one for level control. The latter takes minor surges as a result of pressure fluctuations and valve cycling. If all of the basins are placed on automatic flow control, there is a tendency for valves to "hunt" or oscillate as each, in turn, corrects for the slight variation that another makes. Flow passing into each unit is recorded automatically at the control panel, which receives a propeller meter pulse from each influent line.

Water forced into the plant by the well pumps discharges to selected flocculator-clarifiers (Figure 7-9). Lime slurry is applied by gravity feed to the mixing zones of the basins. Slakers and feeders (Figure 7-18) are capable of

Figure 8-2 Control panel at the Platte River plant. Displayed, from left to right, are: well pump operating lights; six flocculator-clarifiers and eight filters, with flow and level controllers mounted underneath; warning alarm panel; and chlorine dosage and residual indicators. The edge of the operators console is in the foreground of the picture.

automatic pacing at preset feed rates using the same propeller meter pulse that transmits the rate of flow signals to the control panel. Units are also provided for adding a coagulant, potassium permanganate, and fluosilicic acid. Applying a small amount of either alum or ferric sulfate coagulant along with the lime improves clarification. Feeder operation is displayed on the control panel by running lights, and is tied to the alarm section. Warning lights and a horn announce interruption of feed caused by blockage in the hopper or other malfunction. Although it is possible to treat the entire plant flow by the sequence of lime softening followed by recarbonation, split treatment has been adopted as a more economical alternative. This process involves lime softening about 50 percent of the flow, with the remaining raw water by-passed for blending with the treated portion in a second-stage clarifier (Figure 8-1). Mixing of the two flows stabilizes the effluent. Lime applied to a pH of 10.3 to 10.6 in the first stage reduces the calcium in the finished water to about 40 mg/l, while magnesium precipitation occurs only at a pH above 11.0. While pH of the blended water is generally high enough to remove iron and manganese, a small amount of potassium permanganate added either to the blending basin, or to the raw water by-pass, assists in manganese oxidation. Stoichiometric quantities of permanganate have not been found necessary in operation. Contact of precipi-tated manganese dioxide with the raw water in the flocculator-clarifiers is, at least, partially responsible for manganese removal. Yet, the chief factor appears to be oxidation of the metals at high pH with the small amount of permanganate catalyzing the reaction.

The chemically treated clarified water is filtered by using dual-media beds containing 14 in. of filter sand topped by 14 in. of anthracite. The media rests directly on filter blocks with plastic nozzles, as is illustrated in Figure 7-13. The common filter rate is 3 to 4 gpm/sq ft with continuous filter runs up to 100 hr.

For this performance, however, influent turbidity should not exceed 3 to 4 units, and floc carryover must be held to a minimum. Fine particles of manganese dioxide occasionally tend to penetrate the filter media, sometimes shortening filter cycles. At temperatures below about 45° F, the precipitated manganese may not grow sufficiently to be removed at high filtration rates. Chemical characteristics of the finished water are: pH of 9.1, total hardness of 147 mg/l, zero iron, less than 0.01 mg/l manganese, and a fluoride level of 1.0 mg/l. The latter is increased from the natural level by addition of fluosilicic acid prior to filtration.

Filter backwash is initiated by the operator either when the head loss increases to 8 ft or after 100 hr of operation. After manual starting of the cycle, the air and water process continues on an automatic program. The bed is scoured by air agitation at a rate of about 1.5 cfm/sq ft at approximately 5-lb pressure for 2 to 3 min. The bed is then backwashed at a rate of 14 to 18 gpm/sq ft for a duration of 6 to 12 min. Timers for both air and water washes are adjustable at an auxiliary control panel. A volume of 100,000 to 160,000 gal is used in washing one 35 ft by 40 ft bed. After cleaning, the filter is held out of service for, at least, 2 hr to improve the initial filtrate quality. An alternate procedure is to waste the first portion of filtered water, which is likely to show slightly higher turbidity.

Filtered water is routed to clear-well storage or may be pumped directly into the distribution system. Chlorine is applied to filter influent and/or to high-service pump discharge. Chlorine rates and residuals can be read at the main control panel. Treated water is constantly sampled by means of a closed loop, and residuals are adjusted to a preset level. Chlorinators and evaporators are placed in a separate room which is equipped with a detector and automatic blower system for ventilation should an accident occur. The detection unit is also connected to the alarm and warning horn in the control room. Gas masks and compressed air packs are stored just outside the chlorine room for emergency use.

Fluosilicic acid, found to be the most convenient and economical chemical for fluoridation, is metered into the water prior to filtration by using a solution feeder from a tank mounted on a platform scale. A recorder graphs loss of weight of the tank as a function of time. This arrangement allows the operator to constantly confirm feed pump accuracy against the quantity of acid applied. Regular laboratory testing of both raw and finished supplies is used both to establish treatment rate and confirm that the proper dosage is being applied.

Clear-well storage capacity is approximately 10 percent of the plant design flow. High-service pumps are controlled from the main panel, and running lights indicate their status to the operator. Both pumpage and discharge pressure are charted on recorders. While electrically powered pumps require little attention, the natural gas engines need periodic operator observation. An hourly checklist includes running speed, cooling water and lubrication oil temperatures, water jacket temperature and pressure, crankcase vacuum, manifold pressures, and other items. High-service engines require prelubrication before starting and warm-up prior to applying a pumping load.

The present method of sludge and backwash water disposal is by dilution in the Missouri River. A possible alternative is dewatering in lagoons and disposal by hauling to land burial. Backwash water and sludge from the flocculator-clarifiers can be discharged to several earthen basins near the plant. A pilot-plant study showed that a 4-ft depth of almost pure calcium carbonate sludge air dries to approximately 20 percent moisture during a 1-yr cycle. Sludge cake removed from the ponds could be used for agricultural liming, or disposed of in landfill. The estimated sludge yield with the plant operating at design flow is 22 to 30 tons of dry solids per day depending on raw water hardness.

Recalcining to process lime for reuse was considered as an alternative to wasting the sludge. One of the technical problems involved would be separation of magnesium hydroxide from the calcium carbonate thereby preventing buildup of magnesium in the recycled lime. Being more soluble than calcium carbonate, the magnesium could be separated by recarbonation and centrifuging prior to dewatering. However, this still leaves a magnesium waste stream which requires disposal.

A recalcining plant would represent a substantial investment in dewatering and drying equipment; would require approximately the same number of employees as the treatment plant itself; and would need a substantial supply of fuel. For example, about 9000 cu ft of gas would be required for drying and burning one ton of recalcined lime. Total costs for recalcining operations were calculated at $12 to $15 per ton of lime produced, which was substantially more than lagoon treatment and land disposal.

This water treatment plant with modern equipment and automation permits operation with minimum manpower, usually a two-man operating shift. This savings in operating personnel is partially offset by the need for a highly skilled maintenance crew to perform constant inspection and service functions. During the evening and night shifts, operating personnel generally take over some of the maintenance chores, such as machine cleanup, lubrication, and minor adjustments. The alarm panel may summon the shift operator 20 to 30 times in an 8-hr period to handle irregularities in normal treatment and pumping operation. Chemical feeder and chlorinator stoppages most frequently demand the operator's attention. Other problems can be contributed by air compressors, steam boilers, sludge pumps, air handling units, and flow controllers. Pneumatic chemical unloading systems now require operator time since truck shipments, especially, may arrive at any time of the day or night.

8-2 RIVER WATER TREATMENT PLANT

Omaha's Missouri River plant, shown in Figure 8-3, was first placed in service in 1889 using open settling basins. In 1923 sand filtration was added, and in the mid-1940s baffled mixing basins were placed ahead of the open reservoirs. Final stages of expansion added flocculator-clarifiers and pre-sedimentation basins as are shown in Figure 8-3. In addition to minor modifications, the most notable recent change has been to construct a cover over the clear well. As with many water plants in older municipalities, the orientation of the treatment units and physical facilities installed are a result of development and growth over a long period of time, in this case about 85 years.

The process flow diagram for the Missouri River plant is given in Figure 8-4. Operational flexibility is necessary to handle the highly variable quality of the river water. Under normal operation, chemical processing as shown on the right side of the flow diagram is practiced. Additional chemical treatment, during spring runoff when the water quality in the river deteriorates, is needed

Figure 8-3 Omaha's Missouri River Water Treatment Plant was first placed in service in 1889 using open settling basins. Several stages of expansion and modification led to the existing facility as shown in this 1970 photograph. (Courtesy of the Metropolitan Utilities District, Omaha, Neb.)

as indicated on the diagram in italic type. In addition to flexibility, the plant has great depth of treatment with four stages in series provided with chemical treatment facilities. Liquid retention time in the plant at the maximum design flow of 140 mgd is greater than 16 hr from intake to clear well. In contrast, the flow time from the wells to clear-well storage in the groundwater plant (Figure 8-1) is 3.0 hr at design flow.

After screening the raw river water is held for desilting in mechanically-cleaned presedimentation basins. During certain times of the year, polyelectrolytes may be added to enhance clarification. Next, breakpoint chlorination is used to oxidize tastes and odors, and for disinfection. The chlorinated water is held in a contact basin and then is fed to flocculator-clarifiers for split treatment. One leg is lime softened while the other is coagulated with alum. These two streams are blended for stabilization and final settling. The water is then filtered through gravity sand beds and chemicals are applied for chlorine residual and pH adjustment.

The operating challenge of this plant arises from the variable characteristics of the raw water that can drastically change from season to season and day to day. In contrast to the minor treatment problems encountered in a groundwater plant, the superintendent of a river water facility may have a few weeks during the year when his problems seem unsurmountable. During critical seasons a laboratory technician is on duty around the clock to run jar tests, and to perform quality surveillance. A delay in modifying chemical treatment for even a few hours, during a period of rapid water quality deterioration, can lead to several days of taste and odor problems in the finished water. Problems of

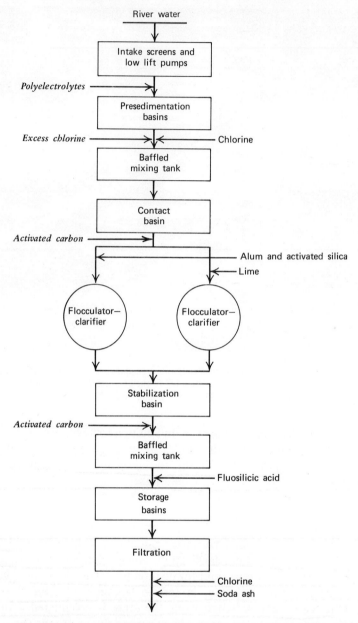

Figure 8-4 Process flow diagram for the Missouri River Water Treatment Plant pictured in Figure 8-3. Normal chemical applications are shown on the right side. Those listed on the left in italic type are required during spring runoff.

coagulation, taste and odor removal, and color reduction are intensified when the water quality deteriorates during spring runoff and chemical reactions proceed more slowly at water temperatures close to freezing. Furthermore, disinfection efficiency is likely to suffer because of the higher turbidity in the raw water. Breakpoint chlorination usually aids in eliminating tastes and odors, but it may also accentuate the problem. For example, decaying vegetation releases aromatic compounds that on chlorination produce disagreeable chlorophenol tastes. Either another oxidizing agent or large quantities of

Table 8-1. Typical Chemical Applications and Costs

	Pounds per Million Gallons	Dollars per Million Gallons
Surface Water Treatment Plant on the Missouri River		
Lime	729	9.67
Alum	69	2.19
Polymer	Variable	0.77 (approx)
Sodium silicate	20	0.51
Activated carbon	Variable	1.85 (approx)
Chlorine	46	2.05
Fluosilicic acid (23%)	23	0.75
Soda ash	37	0.61
		18.40
Groundwater Treatment Plant on the Platte River		
Lime	682	9.04
Alum	17	0.71
Chlorine	18	0.95
Fluosilicic acid (23%)	25	0.90
Potassium permanganate	0.5	0.17
		11.77

activated carbon are then necessary to produce a palatable water. Major rivers are subject to pollution from oil spills, industrial accidents that contribute exotic contaminants, and agricultural drainage which increases the organic content as well as producing a variety of pesticide residuals. Therefore, plant personnel can never relax their vigilance in water quality monitoring.

The difference in water quality between a well supply and surface source is reflected in chemical consumption and treatment costs. If a surface plant has a good quality raw water, for example, from a clear lake or clean stream, the cost of treatment may be only a few dollars per million gallons more than groundwater processing. But on major rivers, the difference in costs may be significant. Table 8-1 lists the average chemical additions and dollar values for treatment of Missouri River water and groundwater from the Platte Valley, both produced for the same distribution system. Although the mineral content between the two waters is not significantly different, the river contains an average of 480 mg/l suspended solids, 2.3 mg/l BOD, 22 mg/l COD, and 10 color units, which are considerably higher than the levels found in raw well water.

8-3   WATER QUALITY CONTROL

A quality control program is essential to the operation of any water treatment plant. For surface supplies it is mandatory but, even if processing well water is limited to iron and manganese removal, adequate laboratory control is beneficial to overall operation. Some small plants routinely carry out comprehensive sampling and testing programs, while others rely completely on state or local health agencies for surveillance. But, even most small waterworks now find it within their ability to train operators in many of the common chemical and bacteriological tests by using proprietary kits and reagents. A well planned program need not result in excessive cost. For example, total operating costs for the quality control programs at the Platte River groundwater plant and the Missouri River surface waterworks have been approximately $2 per million gallons and $4 per million gallons, respectively.

Availability of complete laboratory reports are valuable for public relations and in combating unwarranted criticism of system operations. Industries served by a water system are often interested in the chemical characteristics of the treated water and request monthly reports. A typical analysis includes results of all common physical and chemical tests, and concentrations of major cations and anions. Periodic trace metal analyses may be conducted for arsenic, barium, cadmium, and other heavy metals. With increasing concern for organic contaminants, particularly chlorinated hydrocarbons, comprehensive testing also includes examination for the common pesticides.

Almost all states now require analysis of finished water for coliform bacteria; the number of tests required is based on the population served. Fecal coliform counts, while not generally required by regulatory agencies, are not difficult and can give additional information regarding sources of contamination. Chemical standards such as chlorine residual, turbidity, dissolved solids, nitrate, and color are sometimes specified. Measurement of chlorine residuals in the distribution system is mandatory to determine if the chlorination practice is adequate. Other laboratory testing is related to control of chemical treatment, trouble-shooting problems in the distribution system, and handling customer complaints on quality. Shipments of treatment chemicals, if purchased under specifications, should be routinely tested, and penalties should be assessed for deficiencies. For example, lime is generally purchased on a basis of 88 to 90 percent CaO, alum as 17 percent available Al_2O_3, and activated carbon on a phenol value specification. A penalty clause in the contract supported by an approved laboratory analysis can protect the plant against inferior material. Recent water pollution control legislation now requires laboratory monitoring of waste discharge streams which represents another addition to the duties of the quality control section.

The smallest plant laboratory may contain only a turbidimeter, chlorine residual comparator, pH meter, and glassware for hardness and alkalinity determinations. A completely equipped laboratory for a metropolitan area includes infrared, ultraviolet, and atomic absorption spectrophotometers, a gas chromatograph, amperometric titrator, and conductance meter, plus the common ovens and glassware. Incubators, autoclave, special glassware, and culture media are needed for bacteriological work. If a nuclear power plant is located upstream, some radiochemical surveillance is necessary and a proportional counter and scintillation counter may be needed. Laboratories serving large water systems are usually supervised by a graduate chemist or chemical engineer. Additional professionals and two or three laboratory technicians may also be employed, depending on the extent of the testing program. Laboratory personnel may also serve part-time or full-time in plant supervision. Shift operators can be trained to do routine tests like those for turbidity, alkalinity, and chlorine residual. Standard coliform counts can also be handled by operators who have had the proper training. The advantage of laboratory-trained operators is that operating shifts cover the entire 24-hr day while laboratory personnel typically work only an 8-hr day. Capability for around-the-clock surveillance is important in detecting sudden changes in raw water quality.

Laboratory jar tests and past experience are both used to determine optimum chemical dosages for coagulation and lime softening processes. Problems may be anticipated by careful testing of the raw water so that variations in the treatment process can be made promptly. Past treatment records are helpful, for example, efficient lime softening can be constantly evaluated by maintaining a record of lime feed to hardness removed. The ratio of CaO applied to hardness precipitated may range from 0.8 to 1.2 depending on water temperatures and retention time, but it should be fairly uniform during any one season of the year.

Taste and odor problems in surface water supplies can easily be the primary concern in water quality control for several months of the year. The specific causative compounds are often dissolved organics that are exceedingly difficult, if not impossible, to identify. The standard chlorine demand test gives an excellent indication of overall water quality. Other useful tests include those of potassium permanganate demand and the standard 4-AAP phenols procedure. One of the more promising new methods involves gas chromatographic analysis to establish an organic profile. An adverse change in water quality is identified by a change in the profile. The purpose of any of these tests is to shift the treatment scheme and to apply the best chemical for dealing with each taste and odor problem. Past experience also dictates whether chlorine or potassium permanganate oxidation will be effective, or whether heavy doses of activated carbon are needed.

Determination of threshold odor number is another means of evaluating water quality. The test is conducted by a panel of observers who attempt to detect odor in successively diluted samples. Odor number is calculated from the number of dilutions necessary for the last traces of odor to disappear. Although a useful tool, the test lacks reproducibility, and many instances have been reported in which treated water with a low threshold odor number has been responsible for numerous customer complaints. Odor dilution, it seems, is not always linear, and rather offensive samples may occasionally respond well to only one or two dilutions.

8-4   WATER DISTRIBUTION MAINTENANCE AND SURVEILLANCE

Operation of a water distribution system relies first on proper design and installation of facilities including pumps, mains, storage capacity, valving, and fire hydrants. A systematic inspection program is then essential to insure proper functioning. Valves and hydrants are customarily operated twice a year to detect any defects. Hydrant inspection includes flushing the mains to remove deposits and identify any serious corrosion problems. Frequency of main flushing depends on the character of the water, in most cases, once or twice a year is sufficient. Dead-end lines are particularly susceptible to deterioration of water quality. Leak detection surveys help to keep losses at an acceptable level, and large systems find it profitable to have one full-time crew performing this service. Pressure tests during periods of maximum consumption are the best way to reveal hydraulic deficiencies in the distribution system.

Maps and card files document the exact location of mains, valves, and service connections. Record-keeping is often neglected in smaller water systems where the person in charge is familiar with the location of appurtenances but never bothers to catalog the information for his successors. Valves that cannot be rapidly located in case of main breaks or struck hydrants are of little value to the system. A large main break can drain storage reservoirs and leave major areas without service. Equipment for emergency repair should always be available, and most utilities have an inventory of the basic equipment necessary including excavating equipment, pipe cutters, portable pumps and chlorinators, and generating and welding equipment.

Unlined cast-iron pipe may become heavily encrusted and result in reduced pipe capacity because of tuberculation and red-colored water from iron oxides. The common solution is hydraulic pressure cleaning using an internal pipe scraper, commonly called a pig. The cutting tool consists of a series of scraping and polishing blades of high carbon spring steel that are mounted in series on a rod. Another cleaning tool manufactured is a bullet-shaped, plastic ball with crisscross strips to scrape the interior pipe surface. The first step in the cleaning process is to cut out a few feet of pipe at each end of the section to be cleaned. The cutting tool is inserted in one end, and the section of pipe replaced so that water can be supplied from the main behind the tool. This pushes it through the main, by-passing some water through the cutter to wash the scrapings out of the pipe. At the discharge end, a pipe rising to the surface is placed over the cut end. As is shown in Figure 8-5, this snorkel pipe conveys wash water out onto the street and allows convenient recovery of the cleaning tool.

Unless some type of protective coating is applied to prevent future corrosion, cleaning provides only a temporary solution. Several techniques and proprietary processes are used to line mains with cement or bitumen without removing the pipe. One process applies cement mortar centrifugally to the pipe wall with a lining machine, which is followed by rotating steel trowels to smooth the surface. Before and after cleaning and lining, the Hazen Williams coefficient C should be recorded.

Management of small distribution systems is generally performed from the treatment plant where pumpage and pressures can be controlled. As the area served increases, however, system pressures can no longer be accurately controlled, or even determined, from one point. Operations must then be extended into the pipe network to monitor pumps and reservoirs to insure adequate service to all parts of the system. The following is a brief description of the surveillance techniques employed by the Metropolitan Utilities District, which has more than 100,000 water customers. The system's control center is a data gathering and computing facility for supervising water operations. A control panel (Figure 8-6) allows the operator to enter data into the computer, interrogate information stored in its memory, control data logging, and set system alarm limits. The overall function of the system is to establish water production rates from the two treatment plants, and to operate nine booster pumping stations and three reservoirs.

Data from pumping stations, reservoirs, and pressure points are telemetered continuously to the center. Flows for each pumping station and consumption in the two pressure zones are calculated by the computer, which also stores the information and hourly prints out the rate and quantity of flow for each pumping station and pressure zone. The controller is warned of low-pressure

Figure 8-5 Photograph showing hydraulic pipe cleaning. The cutting tool is forced through the pipe by water pressure. A pipe riser, or snorkel, on the discharge end is used to recover the scraper and to convey wash water onto the street away from the excavation. (Courtesy of the Metropolitan Utilities District, Omaha, Neb.)

conditions, depleted water storage, chlorine leaks, and equipment malfunctions by an audible alarm and printout on the teletype. Discharge pressure from three major booster pumping stations is recorded on continuous strip charts, and pump speed for the variable speed units is displayed on an indicator. Several reservoir valves can be remotely controlled to regulate flow into, or out of, storage. Some valves can be positioned to control rate of flow while others are

Figure 8-6 Operators control the supervisory console for the water system of the Metropolitan Utilities District, Omaha, Neb. (Courtesy of Metropolitan Utilities District)

simple open-or-close valves. Pumps at all booster stations are started and stopped by remote control. Pump check valves are hydraulic or pneumatic operated ball valves. On the electric motor-driven pumps, these valves open automatically when the unit is operated. Panel lights indicate valve positions and pump operation.

Central management of operations and control of the water system provides uniform and accurate data at one location. This allows the supervisor to make the best possible operational decisions. Also, critical conditions can be anticipated and emergencies can be handled rapidly.

8-5 WATER RATES*

A water utility must receive sufficient revenue to cover the costs of adequate service. Basic revenue requirements are for: operation and maintenance expenses, debt service including interest, principle, and stipulated reserves, utility extensions and improvements, and plant replacement for perpetuation of the system. The total revenue collected should reflect not only recent cost experience but also should recognize anticipated future costs during the nominal period for which rates are being established, for example, five years.

Contrary to usual lay opinion, a schedule of water rates can be developed to recover the costs of serving different classes of customers while maintaining reasonable equity. Proper rate design does not take quantity discount into consideration nor does it advocate lower rates simply for water sold in larger amounts, as has been the popular past practice. Modern water rates recognize the costs of supplying the amount of water consumed, the rate of use or demand for use, and the expense involved in maintaining the customer account. Recognition of different classes of customers is reasonable in measuring costs to serve these diverse groups. Charges should be commensurate with the service rendered. For example, a customer with a high peak rate of use in comparison with his average consumption would require larger capacity pumps, pipes, and other facilities than a customer who has a comparable total consumption but uses water continuously at essentially a constant rate. Accordingly, cost-allocation procedures should recognize the particular service requirements of a user for not only total volume but also peak rates, and other factors.

Development of water rates involves the following major areas of study: determination of the amount of revenue required for the water utility; allocation of this annual revenue requirement to basic cost functions, which in turn are attributed to customer classes in accordance with respective class requirements for service; and, design of water rates that recover from each customer class the respective costs of providing service. The latter, of course, requires an evaluation of water use for each class of customer based on local conditions that are influenced by climate, industrial demand, and other factors.

One technique for allocating the cost of service to cost functions is referred to as the commodity-demand method. This separates the three primary cost functions of demand, consumption, and customer costs. Demand costs are associated with providing facilities to meet peak rates of use. They include capital charges and operating costs on the part of the treatment works designed

---

* *Water Rates*, AWWA Manual M1, Second Edition, American Water Works Association, 1972.

to meet peak requirements. Consumption costs include power, chemicals, and other incremental expenses that vary with the quantity of water produced. Customer expenses are associated with serving consumers irrespective of the amount of water used, such as meter reading, billing, accounting, and maintenance on meters and services.

The next step is to ascertain customer classes on the basis of hourly or daily demand characteristics, although some special categories may be based on unusual annual requirements. Typically the three principal classes are residential, commercial, and industrial. In any given system, there may be users with unusual water-use characteristics that should be given separate consideration, for example, large institutions such as a university.

System costs for demand, consumption, and customer services are now related to the chosen customer classes to establish cost responsibility. Each customer's demand expense is based on his fair share of the total system demand. Consumption costs are allocated according to total annual use, and customer costs are allocated on the basis of number of metered services. Finally, a rate schedule is fixed to collect sufficient revenue to meet expenses of the waterworks.

chapter 9

Waste-Water Flows and Characteristics

Domestic or sanitary waste water refers to liquid discharge from residences, business buildings, and institutions. Industrial waste is discharge from manufacturing plants. Municipal waste water is the general term applied to the liquid collected in sanitary sewers and treated in a municipal plant. In addition, interceptor sewers direct dry weather flow from combined sewers to treatment, and unwanted infiltration and inflow enters the collector pipes. A schematic of the system is given in Figure 9-1.

Storm runoff water in most communities is collected in a separate storm sewer system, with no known domestic or industrial connections, and is conveyed to the nearest watercourse for discharge without treatment. Rain water washes contaminants from roofs, streets, and other areas. Although the pollutional load of the first flush may be significant, the total amount from separated storm-water systems is relatively minor compared with other waste discharges. Several large cities have a combined sewer system where both storm water and sanitary wastes are collected in the same piping. Dry weather flow in the combined sewers is intercepted and conveyed to the treatment plant for processing but, during storms, flow in excess of plant capacity is by-passed directly to the receiving watercourse. This can constitute significant pollution and a health hazard in cases where the receiving body is used for a drinking water supply. One solution would be to replace the combined sewers with separate pipes, but the cost in large cities would be prohibitive, although this technique can be applied where only a few combined sewers exist in a municipal system.

9-1 DOMESTIC WASTE WATER

The volume of waste water from residential areas varies from 50 to 100 gal per capita per day (gpcd) depending on the type of dwellings. Largest flows come from single-family houses that have several bathrooms, automatic washing machines, and other water-using appliances. BOD contributed to the waste water per person is approximately 0.17 lb/day. Disposal of household

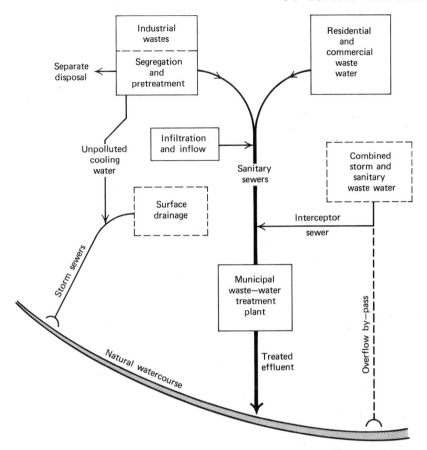

Figure 9-1 Sources of municipal waste water in relation to collector sewers and treatment.

kitchen wastes through garbage grinders increases the per capita BOD by 30 to 50 percent. Table 9-1 lists these values, plus estimated flows for other establishments. Mobile homes and hotels generate less waste water than residences, since they have fewer appliances. The quantity and strength of waste from schools, offices, factories, and other commercial establishments depend on hours of operation and available eating facilities. Although cafeterias do not provide a great deal of flow, the waste strength is increased materially by food preparation and cleanup.

The commonly accepted value for sanitary waste water is 100 gpcd; this includes residential and commercial wastes plus reasonable infiltration, but excludes industrial wastes. Characteristics of this waste water prior to treatment, after settling, and following conventional biological processing are given in Table 9-2. Total solids, residue on evaporation, include both dissolved salts and organic matter; the latter is represented by the volatile fraction. BOD is a measure of the waste-water strength (Section 3-9). Sedimentation of a typical domestic waste water diminishes BOD approximately 35 percent and suspended solids 50 percent. Processing, including secondary biological treatment, reduces the suspended solids and BOD content more than 85 percent, volatile solids 50 percent, total nitrogen about 40 percent, and phosphorus only 30 percent.

Table 9-1. Approximate Waste-Water Flows for Various Types of Establishments and Services

Type	Gallons per Person per Day	Pounds of BOD per Person per Day
Domestic waste from residential areas		
Large single-family houses	100	0.20
Typical single-family houses	75	0.17
Multiple-family dwellings (apartments)	60 to 75	0.17
Small dwellings or cottages	50	0.17
(If garbage grinders are installed multiply BOD by a factor of 1.5)		
Domestic waste from camps and motels		
Luxury resorts	100 to 150	0.20
Mobile home parks	50	0.17
Tourist camps or trailer parks	35	0.15
Hotels and motels	50	0.10
Schools		
Boarding schools	75	0.17
Day schools with cafeterias	20	0.06
Day schools without cafeterias	15	0.04
Restaurants		
Each employee	30	0.10
Each patron	7 to 10	0.04
Each meal served	4	0.03
Transportation terminals		
Each employee	15	0.05
Each passenger	5	0.02
Hospitals	150 to 300	0.30
Offices	15	0.05
Drive-in theaters, per stall	5	0.02
Movie theaters, per seat	3 to 5	0.02
Factories, exclusive of industrial and cafeteria wastes	15 to 30	0.05

Table 9-2. Approximate Composition of Average Sanitary Waste Water (mg/l) Based on 100 gpcd

Parameter	Raw	After Settling	Biologically Treated
Total solids	800	680	530
Total volatile solids	440	340	220
Suspended solids	240	120	30
Volatile suspended solids	180	100	20
Biochemical oxygen demand	200	130	30
Inorganic nitrogen as N	15	15	20
Total nitrogen as N	35	25	20
Soluble phosphorus as P	7	7	7
Total phosphorus as P	10	8	7

The surplus of nutrients in the treated effluent indicates that sanitary waste water has nitrogen and phosphorus in excess of biological needs. The generally accepted BOD/N/P weight ratio required for biological treatment is 100/5/1 (100 mg/l BOD to 5 mg/l nitrogen to 1 mg/l phosphorus). Raw sanitary waste with a ratio of 100/17/5 and after settling 100/19/6, thus containing abundant nitrogen and phosphorus for microbial growth. (The exact BOD/N/P ratio needed for biological treatment depends on the process method and availability of the N and P for growth; 100/6/1.5 is often related to unsettled sanitary waste water, while 100/3/0.7 is used for wastes where the nitrogen and phosphorus are in soluble forms.) Another important characteristic of sanitary waste is that not all of the organic matter is biodegradable. Although a substantial portion of the carbohydrates, fats, and proteins are converted to carbon dioxide by microbial action, a waste sludge equivalent to 20 to 40 percent of the applied BOD is generated in biological treatment.

Loadings on waste treatment units are often expressed in terms of pounds of BOD per day or pounds of solids per day, as well as quantity of flow per day. The relationship between the parameters of concentration and flow is based on the following conversion factors: 1.0 mg/l, which is the same as 1.0 part per million parts by weight, equals 8.34 lb/mil gal, since 1 gal of water weighs 8.34 lb; and, used less frequently, the value 62.4 lb/mil cu ft, since 1 cu ft of water weighs 62.4 lb. These relationships are defined by the following equations:

$$\text{Pounds of } C = \text{concentration of } C(\text{mg/l}) \times Q(\text{mil gal}) \times 8.34 \qquad (9\text{-}1)$$

or

$$\text{Pounds of } C = \text{concentration of } C(\text{mg/l}) \times Q(\text{mil cu ft}) \times 62.4 \qquad (9\text{-}2)$$

where C = BOD, SS, or other constituent, milligrams per liter

Q = volume of waste water, million gallons or million cubic feet

$$8.34 = \frac{\text{lb/mil gal}}{\text{mg/l}}$$

$$62.4 = \frac{\text{lb/mil cu ft}}{\text{mg/l}}$$

Calculations in Example 9-1 show that 100 gal of the sanitary waste water as described in Table 9-2 contains 0.17 lb of BOD and 0.20 lb of suspended solids; Examples 9-2 and 9-3 illustrate applications of Eqs. 9-1 and 9-2.

EXAMPLE 9-1

Sanitary waste water from a residential community is 100 gpcd containing 200 mg/l BOD and 240 mg/l suspended solids. Compute the pounds of BOD per capita and pounds of SS per capita.

Solution
Using Eq. 9-1

$$BOD = 200 \text{ mg/l} \times 0.000,100 \text{ mil gal} \times 8.34 \frac{\text{lb}}{\text{mil gal} \times \text{mg/l}} = 0.17 \text{ lb}$$

$$SS = 240 \text{ mg/l} \times 0.000,100 \text{ mil gal} \times 8.34 \frac{\text{lb}}{\text{mil gal} \times \text{mg/l}} = 0.20 \text{ lb}$$

EXAMPLE 9-2

Industrial wastes (Table 9-4) have a total flow of 2,930,000 gpd, BOD of 21,600 lb/day, and suspended solids of 13,400 lb/day. Calculate the BOD and suspended solids concentrations.

Solution

From the relationship in Eq. 9-1

$$\text{BOD concentration} = \frac{21{,}600 \text{ lb/day}}{2.93 \text{ mil gal/day} \times 8.34} = 880 \text{ mg/l}$$

$$\text{SS concentration} = \frac{13{,}400 \text{ lb/day}}{2.93 \text{ mgd} \times 8.34} = 550 \text{ mg/l}$$

EXAMPLE 9-3

An aeration basin with a volume of 70,000 cu ft contains a mixed liquor (aerating activated sludge) with a suspended solids concentration of 2000 mg/l. How many pounds of mixed liquor suspended solids are in the tank?

Solution

Applying Eq. 9-2

$$\text{MLSS} = 2000 \text{ mg/l} \times 0.070 \text{ mil cu ft} \times 62.4 \frac{\text{lb}}{\text{mil cu ft} \times \text{mg/l}} = 8700 \text{ lb}$$

9-2 INDUSTRIAL WASTES

Industries within municipal limits ordinarily discharge their waste water to the city's sewer system after pretreatment. In joint processing of waste water, the municipality accepts responsibility of final treatment and disposal. The majority of manufacturing wastes are more amenable to biological treatment after dilution with domestic waste water; however, large volumes of high-strength wastes must be considered in sizing of a municipal treatment plant. Uncontaminated cooling water is directed to the storm sewer.

A sewer code, user fees, and separate contracts between an industry and city can provide adequate control and sound financial planning while they accomodate industry by joint waste treatment. Pretreatment at the industrial site must be considered for wastes having strengths or characteristics significantly different from sanitary waste water. Consideration should be given to modifications in industrial processes, segregation of wastes, flow equalization, and waste strength reduction. Process changes, equipment modifications, by-product recovery, and in-plant waste-water reuse can result in cost savings for both water supply and waste treatment. Modern industrial plant design dictates segregation of separate waste streams for individual pretreatment, controlled mixing, or separate disposal. The latter applies to both uncontaminated cooling water that can be discharged directly to surface watercourses, and toxic wastes that cannot be adequately processed by the municipal plant and must be handled by the industry. Manufacturing plants using a diversity of operations may be required to equalize wastes by holding them in a basin for stabilization prior to their discharge to the sewer. Unequalized waste flows may have dramatic fluctuations in quality that would upset the efficiency of a biological treatment system. Certain industrial discharges such as dairy wastes can be more easily reduced in strength by treatment in their concentrated form at the industrial site. Others, like metal-plating wastes, often require pretreatment for the removal of toxic metal ions. If reuse of the municipal waste water is planned, rather stringent controls on industrial discharges are needed, since many of the substances in manufacturing wastes are refractory to conventional treatment and will interfere with water reuse, for example, dissolved salts.

The characteristics of four selected industrial waste waters are listed in Table 9-3 for comparison with sanitary waste water in Table 9-2. BOD concentrations range from 5 to 20 times greater than domestic waste. Total solids are also greater but vary in character from colloidal and dissolved organics, abundant in food processing wastes, to predominantly inorganic salts, such as the chlorophenolic waste. Suspended solids concentration relative to BOD is important when considering conventional primary sedimentation and secondary biological treatment. Settling of the synthetic textile waste with a suspended solids to BOD ratio of 2000 mg/l to 1500 mg/l would be as effective as clarifying a sanitary waste water with a ratio of 240/200, but settling a milk-processing waste with a suspended solids to BOD ratio of 300 mg/l to 1000 mg/l would remove very little organic matter. In addition to high strength and settleability, particular consideration must be given to nutrient content, grease, and toxicity. Food-processing wastes generally contain sufficient nitrogen and phosphorus for biological treatment while discharges from chemical and materials industries are deficient in growth nutrients. Handling animal fats, plant oils, and petroleum products may result in a waste too high in grease content for admission to a municipal system without pretreatment. The chlorophenolic waste in Table 9-3 could not be discharged to a sewer without extensive reduction in phenol; the limit applied by sewer ordinances is in the range of 0.5 to 1.0 mg/l. Metal finishing wastes are pretreated to remove oil, cyanide, chromium, and other heavy metals such that the pretreated discharge has fewer contaminants than domestic waste water. Each municipality should have an inventory of industrial wastes discharged to the sanitary sewer system as is illustrated in Table 9-4. In this city the major waste contributors are

Table 9-3. Average Characteristics of Selected Industrial Waste Waters

	Milk Processing	Meat Packing	Synthetic Textile	Chlorophenolic Manufacture
BOD, mg/l	1,000	1,400	1,500	4,300
COD, mg/l	1,900	2,100	3,300	5,400
Total solids, mg/l	1,600	3,300	8,000	53,000
Suspended solids, mg/l	300	1,000	2,000	1,200
Nitrogen, mg N/l	50	150	30	0
Phosphorus, mg P/l	12	16	0	0
pH	7	7	5	7
Temperature, °C	29	28	—	17
Grease, mg/l	—	500	—	—
Chloride, mg/l	—	—	—	27,000
Phenols, mg/l	—	—	—	140

Table 9-4. Results from a Municipal Industrial Waste Survey Listing Discharges to the Sanitary Sewer in a City with a Population of 145,000

	Flow (gpd)	BOD (mg/l)	BOD (lb/day)	Suspended Solids (mg/l)	Suspended Solids (lb/day)	COD (mg/l)	Grease (mg/l)
Meat processing	1,200,000	1,300	13,000	960	9,600	2,500	460
Soybean oil extraction	478,000	220	880	140	560	440	—
Rubber products	189,000	200	310	250	390	300	—
Ice cream	138,000	910	1,050	260	300	1,830	—
Cheese	110,000	3,160	2,900	970	890	5,600	—
Metal plating	108,000	8	7	27	24	36	—
Carpet mill	103,000	140	120	60	51	490	—
Candy	97,700	1,560	1,270	260	210	2,960	200
Motor scooters	93,500	30	23	26	20	70	—
Potato chips	90,400	600	450	680	510	1,260	—
Flour	83,100	330	230	330	250	570	—
Milk processing	65,100	1,400	760	310	170	3,290	—
Industrial laundry	50,000	700	290	450	190	2,400	520
Pharmaceuticals	40,700	270	91	150	50	390	160
Chicken hatchery	35,300	200	59	310	90	450	—
Luncheon meats	20,900	270	47	60	10	420	—
Soft drinks	16,000	480	64	480	64	1,000	—
Milk bottling	12,700	230	24	110	12	420	—
Totals	2,930,000		21,600		13,400		

food-processing industries. The manufacturing wastes from rubber products, metal working, and carpet weaving have strengths comparable to, or less than, domestic waste water.

Industrial wastes expressed in terms of quantity of flow and pounds of BOD are relatively meaningless to the general public. Therefore, the quantity and strength of waste flows may be related to the number of persons that would be required to contribute an equivalent quantity of waste water. Hydraulic and BOD population equivalents, based on average sanitary waste water, are 100 gpcd and 0.17 lb BOD per person per day, respectively. In addition to equivalent populations, it is desirable to express the quantity of waste produced per unit of raw material processed or finished product manufactured. Examples 9-4 and 9-5 illustrate waste production and equivalent population calculations.

EXAMPLE 9-4

A dairy processing about 250,000 lb of milk daily produces an average of 65,100 gpd of waste water with a BOD of 1400 mg/l. The principal operations are bottling of milk and making ice cream, with limited production of cottage cheese. Compute the waste-water flow and BOD per 1000 lb of milk received, and the equivalent populations of the daily waste discharge.

Solution

Flow per 1000 lb of milk $= \dfrac{1000 \text{ lb}}{250,000 \text{ lb/day}} \times 65,100 \text{ gpd} = 260 \text{ gal}$

$$\text{BOD per 1000 lb of milk} = \frac{0.0651 \text{ mil gal/day} \times 1400 \text{ mg/l} \times 8.34}{250 \text{ thousands of lb/day}} = 3.0 \text{ lb}$$

$$\frac{\text{BOD equivalent}}{\text{population}} = \frac{0.0651 \text{ mil gal/day} \times 1400 \text{ mg/l} \times 8.34}{0.17 \text{ lb BOD/person/day}} = 4500 \text{ persons}$$

$$\frac{\text{Hydraulic equivalent}}{\text{population}} = \frac{65,000 \text{ gal/day}}{100 \text{ gal/person/day}} = 650 \text{ persons}$$

EXAMPLE 9-5

A meat-processing plant slaughters an average 1,100,000 lb of live beef per day. The majority is shipped as dressed halves with some production of packaged meats. Blood is recovered for a salable by-product, paunch manure (undigested stomach contents) is removed by screening and hauled to land burial, and process waste water is settled and skimmed to recover heavy solids and some grease for inedible rendering with other meat trimmings. After this pretreatment, the waste discharged to the municipal sewer is 1,200,000 gpd containing 1300 mg/l BOD. Calculate the BOD waste per 1000 lb LWK (live weight kill) and the equivalent populations of the daily waste-water flow.

Solution

$$\text{BOD per 1000 lb LWK} = \frac{1.2 \text{ mil gal/day} \times 1300 \text{ mg/l} \times 8.34}{1100 \text{ thousands of lb/day}} = 11.8 \text{ lb}$$

$$\frac{\text{BOD equivalent}}{\text{population}} = \frac{1.20 \text{ mil gal/day} \times 1300 \text{ mg/l} \times 8.34}{0.17 \text{ lb BOD/person/day}} = 76,500 \text{ persons}$$

$$\frac{\text{Hydraulic equivalent}}{\text{population}} = \frac{1,200,000 \text{ gal/day}}{100 \text{ gal/person/day}} = 12,000 \text{ persons}$$

9-3 INFILTRATION AND INFLOW

Infiltration is groundwater entering sewers and building connections through defective joints, and broken or cracked pipe and manholes. Inflow is water discharged into sewer pipes or service connections from sources such as foundation drains, roof leaders, cellar and yard area drains, cooling water from air conditioners, and other clean-water discharges from commercial and industrial establishments. In comparison to storm sewers, sanitary lines are small, being sized to handle only domestic and industrial wastes plus reasonable infiltration. Excessive infiltration and inflow may create several serious problems including: surcharging of sewer lines with backup of sanitray wastes into house basements, flooding of street and road areas, overloading of treatment facilities, and by-passing of pumping stations and treatment works.

The quantity of infiltration water entering a sewer depends on condition of pipe and pipe joints, groundwater levels, and permeability of the soil. Seepage into new lines is controlled by proper design, selection of sewer pipe, close supervision of construction, and limiting infiltration allowances. Construction specifications usually permit a maximum infiltration rate of 500 gpd/mile/in. of pipe diameter. The quantity of this seepage flow is equal to 3 to 5 percent of the peak hourly domestic flow rate, or approximately 10 percent of the average flow. With development of better pipe jointing materials and tighter control of construction methods, infiltration allowances as low as 200 gpd/mile/in. of pipe diameter are being considered. Correction of infiltration conditions in existing sewer systems involves evaluation and interpretation of waste-water flow conditions in determining the source and rate of excessive infiltration, followed by consideration of corrective measures. Present techniques to reduce infiltration are grouting or sealing of soils surrounding the sewer pipe, pipe relining, and sewer replacement; all of them are costly.

Inflow is the result of deliberately planned, or expediently devised, connections of extraneous water sources to sanitary sewer systems. Although unwanted storm water or drainage should be disposed of in storm sewers, the sanitary system is often a more convenient conduit because of greater depth of burial and more convenient location. Excess inflow can be prevented by establishing and enforcing a sewer use regulation that excludes storm and surface waters from separate sanitary collectors. The ordinance should be explicit in directing surface runoff from roofs and other areas, foundation drainage, unpolluted water from air conditioning systems, industrial cooling operations, swimming pools, and the like, to storm lines leading to natural drainage outlets. A few ordinances allow cellar drainage into sanitary sewers, however, this is no longer considered proper under present-day conditions; this permit was probably derived from the days when basements were built with stone walls and unpaved floors. Where inflow problems already exist, surveys can be conducted to locate connections and to institute corrective measures.

EXAMPLE 9-6

Calculate the infiltration and compare this quantity to the average daily and peak hourly domestic waste-water flows for the following:

Sewered population = 24,000 persons

Average domestic flow = 80 gpcd

Peak hourly domestic flow = 240 gpcd

Infiltration rate = 500 gpd/mile/in. of pipe diameter

Sanitary sewer system

4-in. building sewers = 36 miles

8-in. street laterals = 24 miles

10-in. submains = 6 miles

12-in. trunk sewers = 6 miles

Solution

$$\text{Infiltration (gpd)} = \text{rate}\left(\frac{\text{gal}}{\text{day} \times \text{miles} \times \text{in.}}\right) \times \text{dia (in.)} \times \text{length (miles)}$$

$$= 500(4 \times 36 + 8 \times 24 + 10 \times 6 + 12 \times 6) = 234{,}000 \text{ gpd}$$

Average domestic flow = $24{,}000 \times 80 = 1{,}920{,}000$ gpd

$$\frac{\text{Infiltration}}{\text{Average domestic flow}} = \frac{234{,}000}{1{,}920{,}000} \times 100 = 12.2 \text{ percent}$$

Peak hourly domestic flow = $24{,}000 \times 240 = 5{,}760{,}000$ gpd

$$\frac{\text{Infiltration}}{\text{Peak hourly flow}} = \frac{234{,}000}{5{,}760{,}000} \times 100 = 4.1 \text{ percent}$$

9-4 MUNICIPAL WASTE WATER

As is shown in Figure 9-1, waste water in sanitary sewers is a composite of domestic and industrial waste, infiltration and inflow, and intercepted flow from combined sewers. Collector sewers must have hydraulic capacities to handle maximum hourly flow including domestic and infiltration, plus any additional discharge from industrial plants. New sewer systems are usually designed on the basis of an average daily per capita flow of 100 gal which includes normal infiltration. However, pipes must be sized to carry peak flows that are often assumed to be: 400 gpcd for laterals and submains when flowing full, 250 gpcd for main, trunk, and outflow sewers; and in the case of interceptors, collecting from combined sewer systems, 350 percent of the average dry weather flow. Peak hourly discharges in main and trunk sewers are less than the maximum

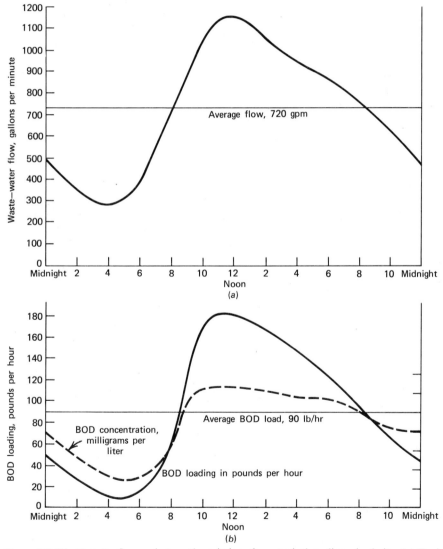

Figure 9-2 Waste-water flow and strength variations for a typical medium-sized city. (*a*) Typical municipal waste-water flow pattern. (*b*) Variation in concentration of BOD in waste water and resultant BOD loading pattern.

flows in laterals and submains, since hydraulic peaks tend to level out as the waste water flows through a pipe network picking up an increasing number of connections.

A typical discharge pattern from a separate sanitary sewer system is illustrated in Figure 9-2a. Hourly flow rates range from a minimum to a maximum of 20 to 250 percent of the average daily rate for small communities, and from 50 to 200 percent for larger cities. Lowest flows occur in early morning about 5 A.M. and peak discharge takes place near midday. The BOD concentration in waste water also varies with time of day in a path that follows the flow variation (Figure 9-2b). Waste strength is greatest during the workday when household and industrial activities are contributing a large amount of organic matter, and is reduced during the night when entering flow is less contaminated and slow velocities in pipes permit settling of solids. If both flow and BOD concentration variations are known, the time-BOD loading on a treatment plant can be calculated and plotted as shown in Figure 9-2b. Knowledge of influent hydraulic and BOD loadings are essential in evaluating the operation of a treatment plant.

The quantity and characteristics of waste water fluctuate with season of the year and between weekdays and holidays. Summer discharges frequently exceed winter flows by 10 to 20 percent, and industrial contributions are reduced on Sundays. Hourly fluctuations in large cities are modified in comparison with small towns because of the diversity of activities and operations that take place throughout the 24-hr day. Large volumes of high-strength industrial waste contributions can distort typical flow and BOD patterns by accentuating the peak hydraulic and BOD loadings during operational hours. Excessive infiltration and inflow, while diluting waste-water strength, can have considerable impact on a treatment facility by increasing both the average and peak flows during periods of high rainfall. All of these factors must be considered in assessing the waste-water flow and strength variations for a particular community.

EXAMPLE 9-7

The sanitary and industrial waste from a community consists of: domestic waste water from a sewered population of 7500 persons; potato processing waste of 30,000 gpd containing 550 lb BOD; and creamery waste flow of 120,000 gpd with a BOD concentration of 1000 mg/l. Estimate the combined waste-water flow in gallons per day and BOD concentration in milligrams per liter.

Solution

Source	Flow in Gallons per Day	BOD in Pounds per Day
Domestic	$7,500 \times 100 = 750,000$	$0.17 \times 7,500 = 1,270$
Potato	$= 30,000$	$= 550$
Creamery	$= 120,000$	$0.120 \times 1,000 \times 8.34 = 1,000$
Total	900,000	2,820

$$\text{BOD Concentration} = \frac{2820 \text{ lb/day}}{0.90 \text{ mil gal/day} \times 8.34} = 380 \text{ mg/l}$$

EXAMPLE 9-8

A city with a sewered population of 145,000 has an average waste-water flow of 18.9 mgd with a mean BOD of 320 mg/l. An inventory of the industrial wastes entering the sanitary sewer is given in Table 9-4. (a) Compute the equivalent populations for the composite municipal waste water, industrial wastes, and remainder attributed to residential plus commercial flows. (b) Determine the per capita contribution of domestic flow and BOD based on the city's population exclusive of industrial wastes.

Solution

(a) Using 100 gpcd and 0.17 lb BOD/person/day:

	Hydraulic Equivalent Population	BOD Equivalent Population
Municipal waste water (domestic + industrial)	189,000	$\dfrac{18.9 \times 320 \times 8.34}{0.17} = 296{,}000$
Industrial wastes (from Table 9-4)	−29,000	$\dfrac{21{,}600}{0.17} = -127{,}000$
Residential plus commercial	160,000	169,000

(b) Domestic flow $= \dfrac{18{,}900{,}000 - 2{,}930{,}000}{145{,}000} = 110$ gpcd

Domestic BOD $= \dfrac{18.9 \times 320 \times 8.34 - 21{,}600}{145{,}000} = 0.20$ lb/person/day

9-5 EVALUATION OF WASTE WATER

Proper sampling techniques are vital for accurate testing in evaluation studies. To be representative of the entire flow, samples should be taken where the waste water is well mixed. An instantaneous grab sample represents conditions at the time of sampling only, and cannot be considered to represent a longer time period, since the character of a waste-water discharge is not stable. A composite sample is a mixture of individual grabs proportioned according to the waste-water flow pattern. Compositing is commonly accomplished by collecting individual samples at regular time intervals, for example, every hour on the hour, and by storing them in a refrigerator or ice chest; coincident flow rates are read from an installed flow meter, or are determined from some other flow recording device. A representative sample is then integrated by mixing together portions of individual samples relative to flow rates at sampling times. Example 9-9 illustrates this procedure.

Composite samples representing specified time periods are tested to appraise plant performance and loadings. Weekday specimens collected over a 24-hr period are most common. Average daily BOD and suspended solids data are used to calculate plant loadings while mean influent and effluent concentrations yield treatment efficiencies. Integrated samples during the period of peak flow, usually 8 to 12 hr depending on influent variation, allow determination of maximum loadings on treatment units.

The most common laboratory analyses for defining the characteristics of a municipal waste water are BOD and suspended solids. BOD and flow data are basic to the design of biological treatment units, while the concentration of suspended solids relative to BOD indicates to what degree organic matter is removable by primary settling. Occasionally comprehensive testing to yield the type of information given in Table 9-2 should be performed. Treatability as well as waste-water characteristics can be defined from these data. In addition, it may be advantageous to record waste-water temperature, pH, COD, alkalinity, color, grease, and presence of specific heavy metals. Selection of these tests depends on the industrial wastes contributed to the sewer system. Importance of comprehensive BOD testing, as described in Section 3-9, is extremely vital, since the BOD test relates more closely than any other to the operaton of a biological treatment system. Besides the five-day value and k-rate, the ratios of BOD to COD and BOD to volatile solids indicate biodegradability of waste organics. Presence of inhibiting substances or toxins, resulting from industrial wastes, are often indicated by increasing BOD values with increasing dilution, lag periods at the beginning of properly seeded tests, and erratic test results. To be significant, laboratory analyses must be correlated with plant operations that prevailed at the time of sampling. Waste-water flow, day of the week, weather and rainfall, and abnormal waste-water discharges, for example, breakdown of pretreatment facilities at a major industry, are essential for future interpretation of recorded testing data. Also, it may be advantageous to record sludge production, equipment performance, chemical and power usage, operational problems, and information on industrial waste-water flows.

Occasionally a biological process treating municipal waste water is afflicted by unknown substances that reduce efficiency by inhibiting microbial activity, for example, bulking of activated sludge in an aeration system. If the plant operates satisfactorily under the same loadings at other times, the problem is most likely related to an industrial waste being discharged to the sewer system. Often laboratory testing of the industrial waste, or comprehensive tests on the municipal waste, can pinpoint the difficulty. However, if results do not prove adequate, a treatability study using a small, laboratory, biological treatment unit is warranted.

Waste-water treatability studies are particularly appropriate for evaluating proposed industrial waste discharges from new industries that plan to use the municipal plant for handling their wastes. Although most trade wastes are more amenable to biological treatment after dilution with domestic waste water, they should be subjected to laboratory analyses prior to permitting their discharge to a municipal sewer. This is especially true if the projected waste volume is substantial in comparison to the sanitary waste flow, and if the type of waste involved has a history of creating treatment problems. Management of the proposed industry should be required to furnish precise information as to the quantity of waste-water flow, including anticipated flow patterns both daily and weekly, waste-water characteristics with anticipated hourly and daily variations from the norm, and proposed pretreatment facilities.

A laboratory apparatus for studying aerobic treatability is pictured in Figure 9-3. Waste is pumped from a refrigerated supply bottle into the aeration chamber; air is supplied through a porous stone diffuser at the bottom. Aerated mixed liquor flows through a connecting tube to the clarifier for gravity separation. Clear supernatant is the effluent while the settled solids are returned

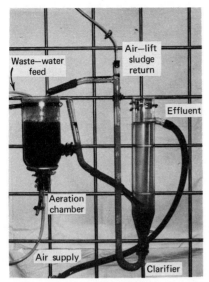

Figure 9-3 Laboratory apparatus for aerobic treatability studies on industrial and municipal waste waters.

to the aeration cylinder by air lifting. Aeration period and BOD loading should be similar to those used in the full-scale treatment system being simulated. Treatability of the applied waste is most easily measured in terms of BOD or COD removal efficiency, settleability of the mixed liquor, and microscopic examination of the activated sludge biota. The unit should be operated both on a feed of pure industrial waste water and in a mixture with domestic waste water. When treated alone, the industrial waste may require neutralization and/or addition of inorganic nitrogen and phosphate for nutrient balance. Joint treatment should be conducted at several different ratios of industrial waste to domestic waste water to include the dilution anticipated in the actual municipal system.

EXAMPLE 9-9

Hourly samples were taken of the waste water entering a treatment plant. The recorded flow pattern is given in Figure 9-2a. Tabulate the portions to be used from the hourly grabs to provide composite samples for the 24-hr duration and during the period of maximum 8-hr loading, between 9 A.M. and 5 P.M. The composite sample volumes needed for laboratory testing are approximately 2500 ml.

Solution

The following relationship determines the portion per unit of flow needed from individual samples to provide a desired composite volume.

$$\frac{\text{Portion of sample needed}}{\text{per unit of flow}} = \frac{\text{total volume of sample desired}}{\text{average flow rate} \times \text{number of portions}} \qquad (9\text{-}3)$$

$$\frac{\text{Portion for the}}{\text{24-hr period}} = \frac{2500 \text{ ml}}{720 \text{ gpm} \times 24} = 0.15 \text{ ml/gpm}$$

$$\frac{\text{Portion for the}}{\text{8-hr period}} = \frac{2500 \text{ ml}}{1000 \text{ gpm} \times 8} = 0.3 \text{ ml/gpm}$$

Calculations for the portions of hourly samples to be used in compositing are tabulated as follows:

Time	Flow (gpm)	Portions of Hourly Samples in Milliliters for:	
		24-hr Composite	8-hr Composite
Midnight	490	$0.15 \times 490 = 74$	
1 A.M.	420	$0.15 \times 420 = 63$	
2 A.M.	360	$0.15 \times 360 = 54$	
3 A.M.	310	$0.15 \times 310 = 47$	
4 A.M.	290	$0.15 \times 290 = 43$	
5 A.M.	310	$0.15 \times 310 = 46$	
6 A.M.	390	$0.15 \times 390 = 58$	
7 A.M.	560	$0.15 \times 560 = 84$	
8 A.M.	620	$0.15 \times 620 = 93$	
9 A.M.	900	$0.15 \times 900 = 135$	$0.3 \times 900 = 270$
10 A.M.	1040	$0.15 \times 1040 = 156$	$0.3 \times 1040 = 310$
11 A.M.	1130	$0.15 \times 1130 = 170$	$0.3 \times 1130 = 340$
Noon	1160	$0.15 \times 1160 = 174$	$0.3 \times 1160 = 350$
1 P.M.	1120	$0.15 \times 1120 = 168$	$0.3 \times 1120 = 340$
2 P.M.	1060	$0.15 \times 1060 = 159$	$0.3 \times 1060 = 320$
3 P.M.	1000	$0.15 \times 1000 = 150$	$0.3 \times 1000 = 300$
4 P.M.	950	$0.15 \times 950 = 143$	$0.3 \times 950 = 290$
5 P.M.	910	$0.15 \times 910 = 136$	
6 P.M.	870	$0.15 \times 870 = 130$	
7 P.M.	810	$0.15 \times 810 = 121$	
8 P.M.	760	$0.15 \times 760 = 114$	
9 P.M.	690	$0.15 \times 690 = 103$	
10 P.M.	630	$0.15 \times 630 = 94$	
11 P.M.	540	$0.15 \times 540 = 81$	
Total composite sample volumes		2596 ml	2520 ml

PROBLEMS

9-1 Using values from Table 9-1, estimate the daily quantity of waste water and pounds of BOD produced by an airport facility, based on the following: 20 airport employees, 4000 passengers handled each day, a motel with 50 double rooms, and a restaurant that serves 1000 patrons daily with a staff of 10 employees. Calculate the average BOD concentration in the waste water by Eq. 9-1. (*Answers* 35,600 gpd, 132 lb BOD/day, and 445 mg/l)

9-2 Calculate the removal efficiencies in biologically treated sanitary waste for total solids, BOD, and total phosphorus from values listed in Table 9-2. (*Answers* 34, 85, and 30 percent, respectively)

9-3 Data given in Table 9-4 can be used to practice converting pollutant concentrations in milligrams per liter to pounds per day, and vice versa, employing Eq. 9-1. For example, calculate the pounds per day of BOD and SS in 1.2 mgd of meat-processing waste with 1300 mg/l BOD and 960 mg/l SS. Or, compute the concentration of BOD in a soybean oil extraction waste of 478,000 gal containing 880 lb of BOD.

9-4 Compute the hydraulic and BOD equivalent populations for the cheese processing waste listed in Table 9-4. (*Answers* 1100 and 17,000 persons, respectively)

9-5 The daily waste-water flows from the luncheon meats, soft drinks, and milk-bottling industries, listed at the bottom of Table 9-4, are blended together for joint treatment. What are the calculated average BOD, SS, and COD concentrations in milligrams per liter? Compute the hydraulic and BOD equivalent populations for the combined daily waste flow. (*Answers* 327 mg/l BOD, 208 mg/l SS, 608 mg/l COD; 500 and 800 persons, respectively)

9-6 The combined waste-water in a municipality with a sewered population of 2000 persons includes wastes from a dairy and a poultry plant. The milk waste is 20,000 gpd with a BOD concentration of 900 mg/l. The poultry-dressing industry processes 5000 chickens per day discharging 16,000 gpd containing 150 lb BOD. Estimate the total combined waste-water flow from the community in gallons per day and the average BOD concentration in milligrams per liter. Compute the BOD population equivalents for the dairy waste, poultry plant discharge, and combined municipal waste water. (*Answers* 236,000 gpd, 325 mg/l BOD; 880, 880, and 3760 persons, respectively)

9-7 What is the BOD/N/P ratio of the synthetic textile waste described in Table 9-3? If a BOD/N/P ratio of 100/3.0/0.7 is considered the minimum nutrient level for complete-mixing activated-sludge treatment, calculate the additions of NH_4OH and H_3PO_4 required in pounds per million gallons of textile waste water. (*Answers* 100/2.0/0, 280 lb H_3PO_4, and 310 lb NH_4OH)

9-8 The waste-water flow applied to a biological aeration process is a composite of 75 percent domestic waste with 130 mg/l BOD, 25 mg/l nitrogen, and 8 mg/l phosphorus, and 25 percent industrial waste with 1200 mg/l BOD, 40 mg/l nitrogen, and no phosphorus. Calculate the BOD/N/P ratio of the waste water. Does it contain adequate nutrients? (*Answers* 100/7.2/1.5, yes)

9-9 Samples were taken at one-half hour intervals of the industrial waste-water discharge plotted in Figure 10-11. Tabulate the portions to be used from half-hour grabs to provide a composite sample for the 12-hr period from 7 A.M. to 7 P.M. The sample volume needed for laboratory testing is approximately 2500 ml. (*Answers* sample portion 5.0 ml/gpm, composite sample volume. 2270 ml)

chapter 10
Waste-Water Collection Systems

Sewers are underground, watertight conduits for conveying waste waters by gravity flow from urban areas to points of disposal. The earliest drainage systems, constructed in the 16th- and 17th-century cities, were to carry storm runoff from built-up areas protecting them against inundation. Privies and cesspools were used for disposal of human excreta, and household wastes were often thrown on the streets. Although this created deplorable sanitary conditions, cities such as London and Philadelphia, prohibited discharge of household wastes to storm drains as late as 1850. The development of steam-driven pumps and cast-iron pipe for pressure water distribution, however, lead to indoor plumbing and flush toilets. Soon cesspools were banned and waterborne wastes were piped to the storm drains, converting them to combined sewers. While this water carriage system improved conditions within the city, untreated wastes were now being directed to surface watercourses.

Combined sewers were constructed in many cities of the United States prior to 1900 without recognizing the need for segregation and treatment of domestic and industrial wastes. Although these systems still exist in older municipalities, separate sewers have dominated construction during the 20th century. Storm sewers carry only surface runoff and other uncontaminated waters to natural channels directly, while sanitary sewers convey domestic and industrial waste waters to treatment works for processing prior to their disposal. Where combined drain pipes exist, intercepting sewers have been constructed to collect the dry-weather flow from a number of transverse outfalls and to transport it to a treatment plant; wet-weather flow in excess of treatment plant capacity is still routed directly to disposal in most cases.

10-1 STORM SEWER SYSTEM

Surface waters enter a storm drainage system through inlets located in street gutters or depressed areas that collect natural drainage. Cooling water from industries and groundwater seepage that enters footing drains are pumped to the storm sewer, since the pipes are usually set too shallow for gravity flow;

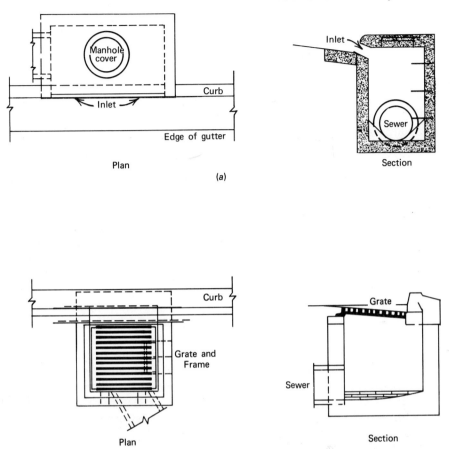

Figure 10-1 Storm-water inlets for street drainage. (*a*) Curb inlet. (*b*) Gutter inlet.

furthermore, direct connections would be subject to backflow when the pipe surcharges. Figure 10-1 illustrates two common types of storm-water inlets for streets. The curb inlet has a vertical opening to catch gutter flow. Although the gutter may be depressed slightly in front of the inlet, this type offers no obstruction to traffic. The gutter inlet is an opening covered by a grate through which the drainage falls. The disadvantage is that debris collecting on the grate may result in plugging of the gutter inlet. Combination inlets composed of both curb and gutter openings are also common. Street grade, curb design, and gutter depression define the best type of inlet to select; nevertheless, minimizing traffic interference and eliminating plugging often take precedence over hydraulic efficiency.

Catch basins under street inlets are connected to the main storm sewer located in the street right-of-way, often along the center line, by short pipelines. Manholes are placed at curb inlets, intersections of sewer lines, and at regular intervals to facilitate inspection and cleaning. Pipeline gradients follow the general slope of the ground surface such that water entering can flow downhill to a convenient point for discharge. Sewer pipes are set as shallow as possible to minimize excavation while providing 2 to 4 ft of cover above the pipe to reduce the effect of wheel loadings. Sewer outlets that terminate in natural channels subject to tides or high water levels are equipped with flap gates to prevent back-flooding into the sewer system. Backwater gates are also used on combined sewer outfalls and effluent lines from treatment plants where needed.

The rational method, described in Section 4-9, is used to calculate the quantity of runoff for sizing storm sewers. Climatic conditions are incorporated by using local rainfall intensity-duration formulas or curves. In dry regions, sewers may be placed only in high-value districts, while streets and roadside ditches serve as surface drains in sparsely populated areas. On the other hand, in regions of the country having intense thunderstorm weather, lined open channels are often found to be more economical compared with large buried conduits; sewers leading from small drainage areas terminate in grassed or concrete-lined ditches that discharge to surface watercourses.

Flowing full velocities used in design of storm sewers are a minimum of 3.0 ft/sec and a maximum of about 10 ft/sec. The lower limit is set so that the lines are self-cleansing to avoid deposition of solids, and the upper limit is fixed to prevent erosion of the pipe by grit transported in the water.

A major difference in design philosophy between sanitary and storm sewers is that the latter are assumed to surcharge and overflow periodically. For example, a storm drain sized on the basis of a 10-yr rainfall frequency presumes that one storm every 10 years will exceed the capacity of the sewer. Sanitary sewers are designed and constructed to prevent surcharging. Where backup of sanitary sewers does occur, it is more frequently attributable to excess infiltration of groundwater through open pipe joints and unauthorized drain connections. A second easily recognizable difference between sanitary and storm sewers is the pipe sizes that are needed to serve a given area. As is illustrated in Example 10-1, storm drains are many times larger than the pipes collecting domestic waste water. Consequently, only a small amount of infiltrating rain water results in overloading domestic sewers.

Circular concrete pipe is commonly used for storm sewers. Nonreinforced concrete pipe is available in sizes up to 24-in. diameter in lengths of 3 or 4 ft. Reinforced pipe has steel imbedded in the concrete to give added structural strength. Circular pipe is manufactured in diameters from 12 to 144 in., and in laying lengths that range from 4 to 12 ft. Elliptically shaped and arch-type concrete pipe are also manufactured for special applications. Various types of joints are available for connecting pipe sections. The one selected depends on construction conditions, pipe size, manufacturer, and whether the pipe is reinforced or nonreinforced. The single-rubber-ring gasket joint has largely superseded all other types for use in concrete sewer lines because of economy, ease of construction, and satisfactory performance. In manufacture, the pipe ends are carefully cast with a space left for the rubber ring. With the gasket in place and lubricated, the spigot end is pushed into the bell to seat the joint properly.

EXAMPLE 10-1

(a) What is the maximum population that can be served by an 8-in. sanitary sewer laid at minimum grade using a design flow of 400 gpcd and a flowing full velocity of 2.0 ft/sec? (b) Compute the diameter of storm drain to serve the same population based on: population density = 30 persons per acre, coefficient of runoff $= 0.40$, 10-yr rainfall frequency curve in Figure 4-23, a duration (time of concentration) $= 20$ min, and a velocity of flow $= 5.0$ ft/sec.

Solution

(a) Based on Figure 4-12, Q flowing full at a velocity of 2.0 ft/sec for an 8-in. diameter pipe is 0.77 cu ft/sec. The maximum population that can be served is

$$= \frac{0.77 \text{ cu ft/sec} \times 7.48 \text{ gal/cu ft} \times 86,400 \text{ sec/day}}{400 \text{ gal/day}} = 1200 \text{ persons}$$

(b) Drainage area $= \dfrac{1200 \text{ persons}}{30 \text{ persons per acre}} = 40$ acres

From Figure 4-23 for duration $= 20$ min., $I = 4.2$ in./hr
Using Eq. 4-16, $Q = 0.40 \times 4.2 \times 40 = 67$ cu ft/sec
Based on Figure 4-12, for $Q = 67$ cu ft/sec and $V = 5.0$ ft/sec, pipe diameter $= 48$ in.
Therefore, a 48-in. diameter storm sewer is needed to drain the housing area of 1200 persons that can be served by an 8-in. diameter sanitary sewer.

10-2 SANITARY SEWER SYSTEM

Sanitary sewers transport domestic and industrial wastes by gravity flow to treatment facilities. A lateral sewer collects discharges from houses and carries it to another branch sewer, and has no tributary sewer lines. Branch or submain lines receive waste water from laterals and convey it to large mains. A main sewer, also called trunk or outfall sewer, carries the discharge from large areas to the treatment plant. A force main is a sewer through which waste water is pumped under pressure rather than by gravity flow.

Design flows for sewer systems are based on population served, using the following per capita quantities: laterals and submains 400 gpcd, main and trunk 250 gpcd, and interceptors 350 percent of the estimated average dry-weather flow. These figures include normal infiltration and are based on flowing full capacity. Excluded are industrial wastes and excessive infiltration. Sewer slopes should be sufficient to maintain self-cleansing velocities; this is normally interpreted to be 2.0 ft/sec when flowing full. Table 10-1 lists sewer size, minimum slope for 2 ft/sec, and the corresponding quantity of flow. Slopes slightly less than those listed may be permitted in lines where the design average flow provides a depth of flow greater than one third the diameter of the pipe. Where velocities are greater than 15 ft/sec, special provision must be made to protect the pipe and manholes against displacement by erosion and shock hydraulic loadings.

Sanitary sewers are placed at sufficient depth to prevent freezing and to receive waste water from basements. As a general rule, laterals placed in the street right-of-way are set at a depth of not less than 11 ft below the top of the house foundation. To provide economical access to the sewer after street construction, service connections are generally extended from laterals to outside the curb line at the time of sewer placement. An alternative is to place the sanitary sewer behind the curb on one side of the street, making it readily accessible for service connections on that side. Pipe connections from the opposite side of the street are accomplished by excavating working pits on each

Table 10-1. Minimum Slopes for Various Sized Sewers at a Flowing Full Velocity of 2.0 ft/sec and Corresponding Discharges, Based on Manning's Formula with $n = 0.013$

Sewer Diameter (in.)	Minimum Slope (ft/100 ft)	Flowing Full Discharge	
		(cu ft/sec)	(gpm)
8	0.33	0.7	310
10	0.25	1.1	490
12	0.19	1.6	700
15	0.14	2.4	1080
18	0.11	3.5	1570
21	0.092	4.8	2160
24	0.077	6.3	2820
27	0.066	8.0	3570
30	0.057	9.8	4410
36	0.045	14.1	6330

side and by jacking the house sewer into position for connection to the lateral on the opposite side. In jacking, pipe sections are pushed beneath the roadway by hydraulic jacks. For hard soils, boring machines are used to cut an opening through which the pipeline is pushed. The cutter head operates in front of the first pipe section and an auger pulls the excavated material out through the pipe to the jacking pit.

Ease of maintenance dictates many of the design criteria for waste-water collection systems. The minimum recommended size for laterals is an 8-in. diameter. Manholes located at regular intervals allow access to the pipe for inspection and cleaning. Pipes laid on too flat a grade require periodic flushing and cleaning to remove deposited solids and to prevent pipe plugging. Sewers less than 24 in. should be laid on a straight line between manholes. Although in recent years curves have been permitted on smaller lines, this is a questionable practice that may interfere with sewer cleaning. For example, a cable riding on the inside wall of the pipe around a curve may damage the pipe interior. Because of the high cost and problems associated with maintenance of pumping stations in the collection system, they are used only when it is impracticable to continue the sewer by gravity flow.

House Connections

House sewers are laid on a straight line and grade using 4-in. or 6-in. diameter pipe. The preferred minimum slope is 2 percent, or $\frac{1}{4}$ in./ft, although slopes as shallow as $\frac{1}{8}$ in./ft are occasionally used. In some housing developments, the setback of dwellings from the street dictates the slope of the connection. Pipe trenches for laying the service line should be straight and should be excavated to the required slope. Where the soil is suitable for pipe support, the natural floor of the trench can be shaped to support the barrel of the pipe. Crushed stone or coarse sand may be used for bedding when fill is required to provide uniform support. High-quality pipe, watertight joints, and good workmanship are required to minimize infiltration and root penetration.

The connection to the street sewer should be made with a wye or tee branch; the latter is preferred because a wye may be broken if the service line is rodded for cleaning. The tee may be installed, as is shown in Figure 10-2, with the branch turned about 45 degrees from the horizontal so that back-flooding does not occur when the collecting sewer is flowing full. For a deep sewer, the vertical pipe riser is encased in concrete to prevent damage during backfilling. All possible provisions should be made for future connections in the original construction by extending service lines from the street lateral to the curb line. The free end is then closed with a carefully fitted stopper until needed for a building connection. When joining to an existing sewer, without a tee provided, the connection must be carefully made to be watertight, and so that jointing materials do not extend into the sewer pipe. A section of the main sewer may be broken out and replaced with a tee branch. By removing half of the bell on the fitting, it can be slid into position, replacing the section of straight pipe that was removed.

Building vents attached to sewer drains provide a connection between the air in sewer pipes and the atmosphere. Plumbing traps on toilet and sink drains prevent backup of sewer gases into the building interior, while the fall of waste water down the stack draws air into the pipe. Oxygen in the sewer atmosphere reduces production of hydrogen sulfide, and ventilation carries off volatile gases that may originate from illegal disposal of flammable liquids. In some instances where natural aeration is not sufficient because of long sewer lines with few service connections, forced ventilation is installed to draw air out of the sewer, exhausting it to a high stack or some deodorizing process.

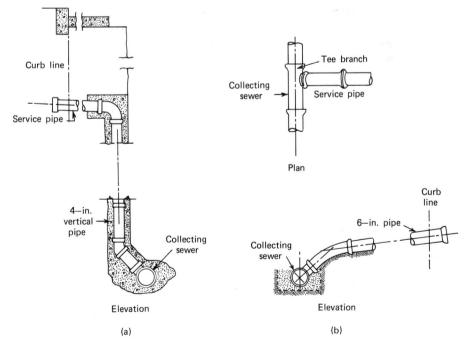

Figure 10-2 Typical sanitary sewer service connections. (*a*) Connection to a deep sewer. (*b*) Connection to a shallow sewer.

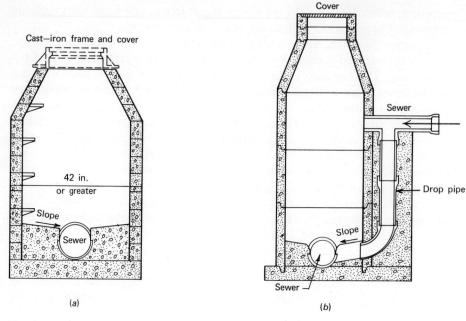

Figure 10-3 Sewer manholes. (*a*) Typical sewer manhole. (*b*) Drop manhole.

Manholes

Most manholes are circular in shape with an inside diameter of 4 ft, which is considered sufficient to perform sewer inspection and cleaning (Figure 10-3). For small diameter pipes, the manhole is usually constructed directly over the center line of the sewer. For very large sewers, access may be provided on one side with a landing platform for the convenience of introducing cleaning equipment. Manhole frames and covers are usually cast iron with a minimum clear opening of 21 in. Solid covers are used on sanitary sewers, while open-type covers are common on storm sewers. Steps or ladder rungs are placed for access. Walls may be constructed of precast concrete rings, concrete block, brick, or poured concrete.

Waste-water flow is conveyed through the manhole in a smooth U-shaped channel formed in the concrete base. Where more than one sewer enters a manhole, the flowing-through channels should be curved to merge the flow streams. If a sewer changes direction in a manhole without change of size, a drop of 0.05 to 0.10 ft is provided in the manhole channel to account for head loss. When a smaller sewer joins a larger one, the bottom of the larger pipe should be lowered sufficiently to maintain uniform flow transition. An approximate method for achieving this is to place the 0.8 depth point of both sewers at the same elevation; an alternate technique is to have the pipe crowns at the same height.

A drop manhole is used when it is necessary to lower the elevation of a sewer in a manhole more than 24 in. (Figure 10-3). This construction is needed to protect a man entering the structure, and to eliminate nuisance created by solids splashed onto the walls. The tee pipe section extending past the drop into the manhole allows access to the sewer line for cleaning tools.

Manholes should be placed: at all changes in sewer grade, pipe size, or

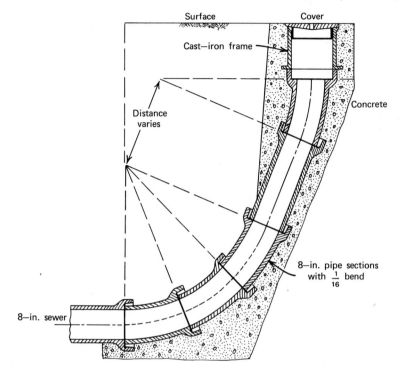

Figure 10-4 Terminal cleanout for ends of short branch or lateral sewers.

alignment; at all intersections; at the end of each line; and, at distances not greater than 400 ft for sewers 15 in. or less, and 500 ft for sewers 18 to 30 in. In some municipalities, a spacing of 300 ft is considered the maximum for smaller sized sewers. Distances of greater than 500 ft may be used where the pipe is large enough to permit walking in the sewer. A terminal cleanout sometimes substitutes for a manhole at the end of a short lateral sewer. The cleanout structure shown in Figure 10-4 is an upturned pipe coming to the surface of the ground for flushing or inserting cleaning tools. This type of unit should not be installed at the end of a lateral greater than 150 ft from a manhole.

Inverted Siphons

A siphon is a depressed sewer that drops below the hydraulic gradient to avoid an obstruction, such as a stream, railway cut, or depressed highway. The design incorporates provisions for maintenance-free operation and minimum hydraulic head loss. Since a depressed sewer acts as a trap, the velocity of flow in the pipes should be greater than 3 ft/sec to prevent deposition of solids. This is accomplished by constructing an inlet splitter box that directs flow to two or more siphon pipes placed in parallel. For example, in Figure 10-5, flow in the 18-in. diameter sewer is carried by two depressed pipes. During low flow in the main sewer, all of the waste water is directed to the 8-in. pipe. When the depth of flow in the main sewer exceeds 6 in., the waste-water stream is split in the inlet box and is directed to both siphon pipes; with the main sewer flowing half full, the velocity in both is greater than 3 ft/sec. In addition to control of flow, the inlet and outlet structures provide access for cleaning the sewer lines.

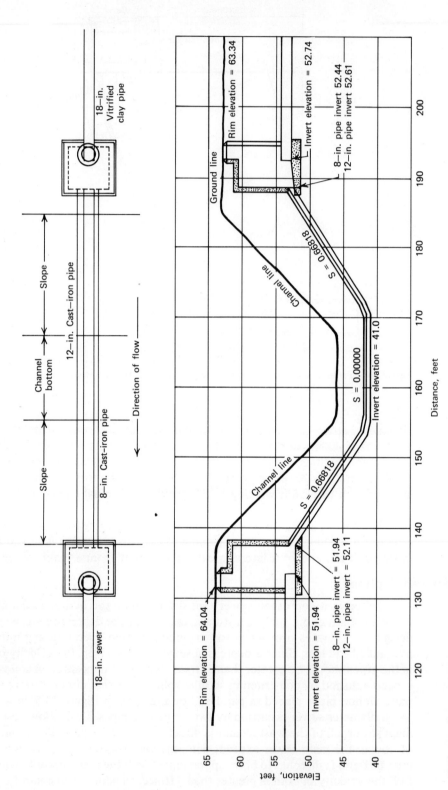

Figure 10-5 Plan and profile of an inverted siphon.

320

Sewer Profile

The drawing for a section of sanitary submain is given in Figure 10-6. To illustrate more detail, the vertical scale is ten times greater than the horizontal. The profile shows the ground surface, sewer line giving slope and diameter, manholes, connecting sewers, all other underground utilities, elevation of the sewer invert at critical points, and special features, such as an inverted siphon. In design drawings the profile is always accompanied by a plan view diagraming the location of the sewer in the street right-of-way.

Layout of new sewer lines and water mains must insure adequate separation to prevent any possible passage of polluted water into the potable supply. Generally sewers are laid at least 10 ft horizontally from a water main. If local conditions prevent this lateral separation, a sewer may be laid closer if the crown of the sewer is, at least, 18 in. below the bottom of the water main, and the pipes are placed in separate trenches. The minimum vertical separation for sewers crossing under water mains is 18 in. between the bottom of the water main and top of the sewer. When this separation cannot be provided the water main should be relocated, or the sewer should be constructed of materials and with joints that are equivalent to water main standards of construction and are tested to insure watertightness. If a sewer must pass over a water main, the main should be constructed with compression-type or mechanical joint pipe and the sewer should be made with mechanical joint cast-iron pipe; both services are then pressure tested for leakage. Sewer plan and profile drawings must show existing or proposed water mains and special sewer pipe construction where minimum distances between facilities cannot be maintained.

Measuring and Sampling of Flow in Sewers

Monitoring sewer use requires measuring of flow and composite sampling at critical points in the collection system. Evaluation of peak flow periods defines the unused capacity of pipes and indicates infiltration-inflow problems. Sampling stations on industrial waste discharges are necessary to regulate sewer use and to determine flow and strength data to establish user fees.

The estimated quantity of flow in a sewer can be computed from the measured depth of flow using the Manning formula if the slope of the pipe is known (Section 4-5). Another approximate method is to multiply the measured

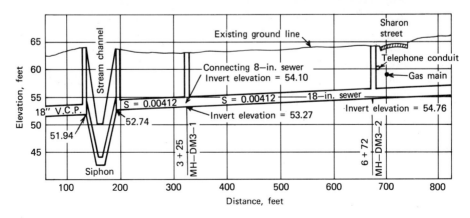

Figure 10-6 Profile of a sanitary sewer.

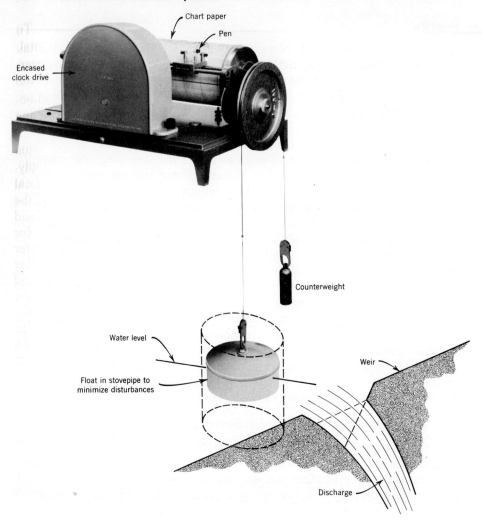

Figure 10-7 Continuous water level recorder for plotting depth of flow over a weir with respect to time. (Courtesy of Leupold & Stevens, Inc.)

wetted cross-sectional area of flow times the velocity obtained by a current meter or the use of dyes or floats.

Weirs are employed where accurate flow measurement is desired. For installation in a manhole, a watertight bulkhead, containing the triangular or rectangular weir opening on the top, is constructed at right angles to the waste-water flow. Water elevation in the stilling pool behind the bulkhead relates to the depth of flow over the weir for calculating discharge. Periodic observations may be performed with a vertical rule or hook gauge. However, for flow variation and total quantity of discharge, a continuous recording device is convenient and more accurate. The unit shown in Figure 10-7 has a float linked mechanically to a clock driven, drum type, head recording device. Rise and fall of the float with changing water levels turns the recording drum proportionally as the clock-controlled pen moves across the chart at constant speed. The resulting graph traced on removable chart paper is a record of water level with respect to time (Figure 10-10). With the proper weir formula, this

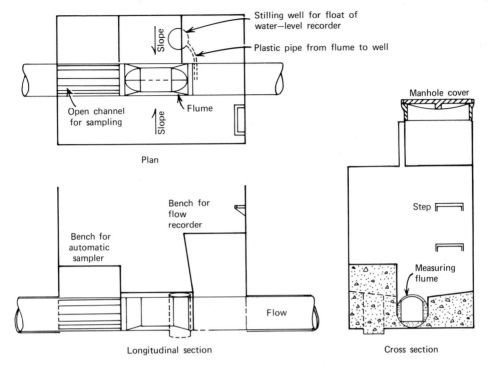

Figure 10-8 Sampling chamber to monitor industrial waste-water discharge to a municipal sewer where only intermittent analyses are required. Measuring flume is permanently installed but portable water level recorder and automatic sampler are needed to perform composite sampling.

head-time plot can be converted into a flow diagram (Figure 10-11) as is demonstrated in Example 10-2.

Industrial-waste sampling stations vary from a manhole to a separate building that is large enough to house flow recording and automatic sampling equipment. A manhold on the service sewer to a commercial business is generally adequate, provided that neither flow measurements nor composite sampling are necessary. Frequently for small enterprises, the quantity of waste flow can be determined from water consumption, and grab sampling is adequate for monitoring waste-water quality.

For moderate-sized industries and commercial businesses, a sampling station as shown in Figure 10-8 is appropriate. The sewer passes through a reinforced concrete chamber of sufficient size to conveniently set up sampling equipment. The flow measuring device is a flume inserted in the sewer line; these devices are available commercially for various sized pipes. A slightly elevated bench and stilling well are furnished to accomodate a water-level recorder of the type shown in Figure 10-7. A portable sampler (Figure 10-9) is used to collect waste-water samples automatically at set time intervals. The intake hoses are placed in the flow downstream from the flume, and the sampler is set on the bench above. Preparation involves evacuating the sample bottles by using a vacuum pump. A clock mechanism periodically trips hose clamps so that waste-water portions are drawn up into the bottles. Composite samples can then be integrated from the individual portions based on the recorded flow pattern.

For large industries, completely automated systems are essential for

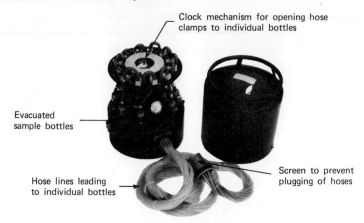

Clock mechanism for opening hose
clamps to individual bottles

Evacuated
sample bottles

Screen to prevent
plugging of hoses

Hose lines leading
to individual bottles

Figure 10-9 Automatic waste-water sampler that collects portions of flow at regular intervals of time. (Manufactured by North Hants Engineering Co. Ltd., Four Marks, Alton, Hampshire, England)

monitoring waste discharges economically. The flow measuring system consists of a Parshall flume with flow recorder and totalizer, while the sampler performs continuous, automatic compositing of waste-water samples. This may be done by connecting a positive displacement sampling pump electrically to the flow recorder such that the amount pumped to a refrigerated collection bottle is proportional to the waste flow. Although costly, this type of station eliminates the time-consuming task of installing and removing portable equipment, and of compositing samples manually.

EXAMPLE 10-2

Waste-water flow entering a municipal sewer from an industry was monitored by using a continuous flow recorder installed in a stilling chamber behind a 90° V-notch weir (Figure 10-7). The recorded water-level variation with time is given in Figure 10-10; flow was negligible prior to 7 A.M. Plot the flow pattern in gallons per minute versus time, and compute the total discharge between 7 A.M. and 7 P.M.

Solution

Height at zero flow was located by marking the pen position on the chart paper when the waste-water level in the stilling chamber reached the weir crest. After removing the paper from the recorder drum at 7 P.M., the horizontal time scale and vertical discharge-head scale were written on the chart.

To develop a flow pattern from the water-level graph, the heights of flow at various breaks in the curve in Figure 10-10 were converted to flow using the conversion table in the Appendix. For example, at 8:15 A.M. $H = 2$ in. which corresponds to a flow of 12.4 gpm, while at 9:00 A.M. $H = 3\frac{3}{8}$ in. for a discharge of 46.0 gpm, interpolating between 41.8 gpm and 50.3 gpm.

Figure 10-11 is a plot of the waste-water flow diagram from the water level chart in Figure 10-10. The total quantity of flow from 7 A.M. to 7 P.M. was 15,000 gal and the average flow rate was 20.8 gpm. Total discharge is determined graphically by counting the square spaces under the curve, or by measuring the area using a planimeter. Average flow equals total discharge divided by time.

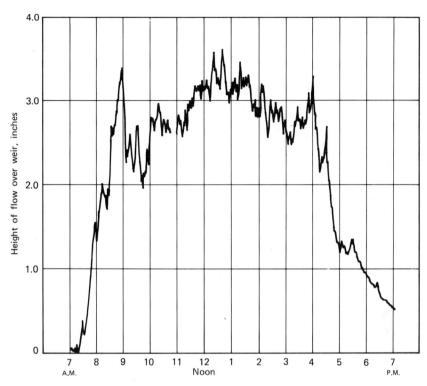

Figure 10-10 Water level chart of an industrial waste-water discharge for Example 10-2. Flow recorder and 90° V-notch weir were arranged as illustrated in Figure 10-7.

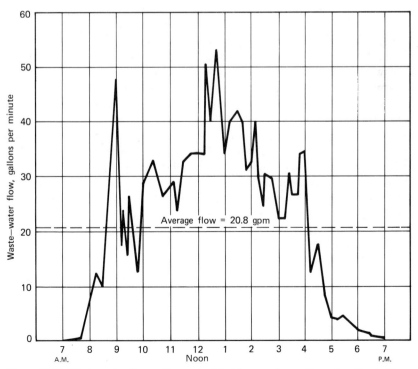

Figure 10-11 Waste-water flow pattern calculated from weir overflow recording in Figure 10-10 as the solution to Example 10-2.

EXAMPLE 10-3

A survey of the sewer line shown in Figure 10-6 was conducted to determine the approximate unused capacity. The maximum depth of flow in the pipe entering manhole DM3-1 was 12 in. Calculate the quantity of flow at this depth, and the unused capacity of the sewer based on the Manning formula (Eq. 4-14) using $n = 0.013$.

Solution

From Figure 4-11, for an 18-in. diameter and slope of 0.00412 ft/ft, flowing full $Q = 6.7$ cu ft/sec.
Entering Figure 4-13 with the depth to diameter ratio of $12/18 = 0.67$, the partial flow to full flow $q/Q = 0.80$.
Therefore,
Quantity of flow at 12 in. $= 0.80 \times 6.7 = 5.3$ cu ft/sec
Unused capacity $= 6.7 - 5.3 = 1.4$ cu ft/sec

10-3   SEWER PIPES AND JOINTING

Circular sewer pipe is manufactured with inside diameters from 4 to 144 in., at intervals of 2 in. from 4 to 12 in., increases of 3 in. between 12 and 36, and 6-in. diameter intervals from 36 to 144. Maximum sizes vary with materials, for example, the maximum clay pipe is 36-in. diameter. The physical characteristics essential for sewer pipe are: durability for long life; abrasion-resistant interior to withstand scouring action of waste water carrying gritty materials; impervious walls to prevent leakage of water; and, adequate strength to resist failure or deformation under backfill and traffic loads. Joints should be durable, easy to install, and watertight to prevent leakage or entrance of roots.

The most important chemical characteristics of a pipe material are resistance to dissolution in water, and corrosion. Pipe surfaces must be able to withstand both electrochemical and chemical reactions from the surrounding soil and waste water conveyed in the pipe. Figure 10-12 illustrates the process of crown corrosion in sanitary sewers. Bacterial activity in anaerobic waste water produces hydrogen sulfide gas; particularly in warm climates when sewers are laid on flat grades. Hydrogen sulfide absorbed in the water condensed on the crown is converted to sulfuric acid by aerobic bacterial action. If the pipe is not chemically resistant, the acid deteriorates it and eventually results in collapse of the crown. The most effective preventative measure is to select a pipe material resistant to corrosion, such as vitrified clay or plastic. When reinforced concrete pipe is needed for larger sizes, an interior protective coating of coal tar, vinyl, or epoxy should be considered. Generation of hydrogen sulfide in a sewer can be reduced by placing the pipe on as steep a gradient as possible, and by ventilating if necessary. Corrosion of the pipe bottom is caused by disposal of acidic industrial waste waters. This problem is best solved by limiting the discharge of acid wastes to the municipal sewer system. In concrete pipe, corrosion resistant liners, such as vitrified clay plates, may be placed in the invert of the sewer for protection.

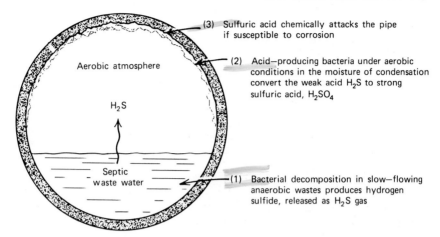

Figure 10-12 Environmental conditions leading to crown corrosion occur in sanitary sewers as a result of flat grades and warm waste waters.

Vitrified clay is the most common material used in sanitary sewer pipe. It is manufactured of clay, shale, or combinations of these materials that are pulverized and mixed with a small amount of water. The moistened clay is extruded through dies under high pressure to form the barrel and socket, dried, and then fired in a kiln for vitrification. Vitrified clay pipe (VCP) is manufactured in both standard and extra strength in diameters from 4 to 36 in., and in laying lengths of 2 to 7 ft depending on diameter. Curves, elbows, and branching fittings, including single and double tee and wye shapes, are available in most pipe sizes.

Plain bell and spigot pipe may be connected by hot-poured or precast asphaltic or bituminous joints. Although these joints are flexible and overcome the objection of cement mortar joints, they require care and skill in application to insure watertightness. More popular is the factory-made compression joint. One type consists of a cast plastic ring on the spigot and a plastic liner in the bell of the pipe. When they are joined, these preformed ends lock together with the rings compressed. A second type is composed of plastic material applied to the bell and spigot ends and a separate compression ring. The mating ends are designed so that when the pipes are joined, the O-ring compresses in the annular space between the bell and spigot to form a closing seal as is detailed in Figure 10-13b.

Asbestos cement pipe comes in three types: for service connections, gravity sewers, and pressure pipe for force mains. The pipe material is composed of cement, silica, and asbestos fibers. It is formed under high pressure and is heat treated to develop strength. Asbestos cement is durable and has a smooth surface without lining, but epoxy-lined pipe is available for special applications. Joints are made with an outside coupling and flexible rubber rings (see Figure 6-17). Nonpressure sewer pipe is available in the size range from 4 to 42-in. diameter in 13-ft lengths; shorter lengths are available in smaller sizes, and $6\frac{1}{2}$-ft sections are available in 10- to 21-in. sizes. Pipe is available in class designations based on supporting strength.

Plastic pipe used most frequently in sewer systems is of PVC (polyvinyl chloride) and PE (polyethylene). PVC pipe is produced in two strength classifications in sizes from 4 to 12 in. A few manufacturers make pipe up to

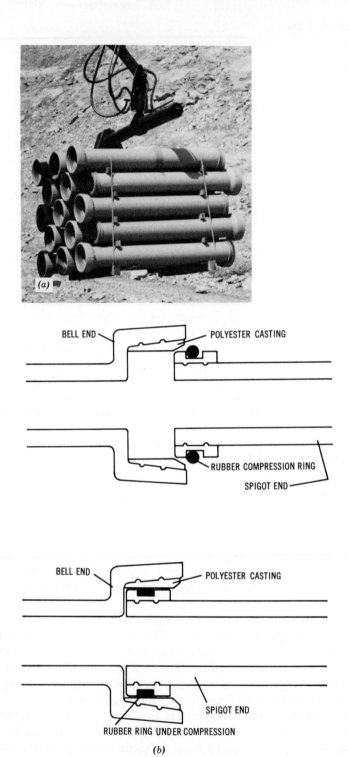

BELL END — POLYESTER CASTING

RUBBER COMPRESSION RING

SPIGOT END

BELL END — POLYESTER CASTING

SPIGOT END

RUBBER RING UNDER COMPRESSION

(b)

Figure 10-13 Picture of vitrified clay sewer pipe and details of an O-ring compression joint. (*a*) Vitrified clay sewer pipes packaged for delivery to job site. (*b*) Details of an O-ring compression joint. Both bell and spigot ends have factory-made joining surfaces with **an** annular space for sealing with a rubber ring. (Courtesy of the Logan Clay Products Co.)

30-in. diameter. The standard length is 20 ft with others available. PVC pipe sections have a deep-socket, bell end to accomodate a chemical weld joint. Solvent is spread both on the inside of the bell and on the plain end before they are pushed together. PE pipe is joined by softening the aligned faces of the ends in a suitable apparatus and pressing them together under controlled pressure. PVC pipe is used for building connections and branch sewers, while the popular application of PE pipe has been for long pipelines, often laid under adverse conditions, for example, in swamp or underwater crossings.

Precast concrete sewer pipe may be obtained in many sizes with several types of joints. Choice of a particular type of pipe and joint depends on application, location, and conditions for installation. Nonreinforced concrete pipe is available in 4- to 24-in. diameters in 3- or 4-ft lengths. It is manufactured in standard strength and extra strength grades. The bell and spigot ends are normally joined by using a rubber ring gasket. Circular reinforced concrete pipe, produced in a size range from 12 to 108 in., is available in five classes based on strength. Precast conduits are also manufactured in elliptical and arch-type shapes. Tongue and groove joints, which are common in reinforced pipe, use a mastic compound or rubber gasket to form a watertight seal.

Concrete pipe is used extensively in storm sewer systems for its abrasion resistance, availability in large sizes, high crushing strength and, generally, lower cost relative to other types of pipe. Since concrete is an alkaline material subject to attack by acids, it should not be used for small-sized sanitary sewers where industrial wastes or hydrogen sulfide production is likely to cause internal corrosion. However, reinforced concrete pipe is installed for sanitary trunk sewers where the diameter exceeds the available sizes of vitrified clay pipe. Here the installation must be protected by control of acidic, high-temperature, or high-sulfate wastes by adding chemicals to control biological growth, maintaining flushing velocities and adequate ventilation, and installation of pipe lining if necessary. Epoxy or plastic lining may be cast into the concrete pipe during manufacture; or bitumastic or coal-tar epoxy may be painted on the concrete pipe surfaces after installation.

Cast-iron or ductile-iron pipe is used for force mains, inverted siphons, in pumping stations and treatment plants, when proper separation from water mains cannot be maintained, and where poor sewer foundation conditions exist. Other pipe materials for special applications in waste-water collection systems are smooth wall and corrugated steel pipe, bituminized fiber, and reinforced resin pipe.

10-4 LOADS ON BURIED PIPES

Sewer design requires prior knowledge of soil and site conditions to determine overburden loads that will be placed on the buried pipe. Crushing strength of the pipe and type of bedding define the load that can be safely supported by a sewer. The paragraphs that follow are an overview of considerations in sewer design.

Backfill load on a pipe depends on trench width, depth of burial, unit weight of the fill material, and frictional characteristics of the backfill. These parameters are formulated in Eq. 10-1 and are defined in Figure 10-14. Soil exploration data, size of pipe, depth of burial, and trench excavation technique by the contractor are needed to compute backfill load. Unit weight of the fill and

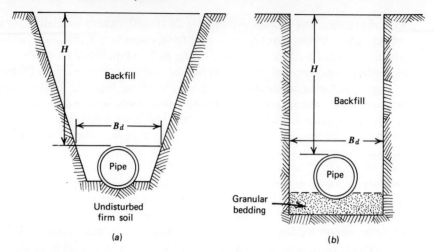

Figure 10-14 Common trench cuts used for sewer installations that define the dimensions B_d and H as applied in Eq. 10-1 and Figure 10-15. The minimum trench width for working space to install pipe is commonly taken as 4/3 the pipe diameter plus 8 in. (a) Trench with sloping sides in cohesionless soil. (b) Vertical trench walls in stiff clayey soil.

trench width are applied directly in Eq. 10-1. Coefficient C_d depends on the ratio of trench depth to width and type of backfill material as graphed in Figure 10-15.

$$W_d = C_d w B_d^2 \qquad (10\text{-}1)$$

where W_d = load on buried pipe as a result of backfill, pounds per linear foot
   C_d = coefficient based on type of backfill and ratio of trench depth to width, Figure 10-15
   w = unit weight of backfill, pounds per cubic foot
   B_d = width of trench at top of pipe, feet

Design manuals published by the American Concrete Pipe Association, National Clay Pipe Institute, and others provide tables listing backfill loads for various conditions based on Eq. 10-1. An example is Table 10-2 for 8-in.

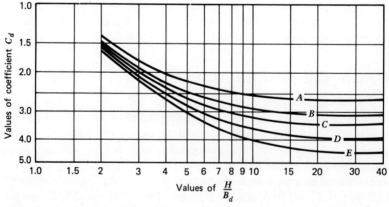

Figure 10-15 Diagram for the coefficient C_d in Eq. 10-1 based on H/B_d and type of trench backfill. A, granular material without cohesion. B, sand and gravel. C, maximum for saturated top soil. D, ordinary clay. E, maximum for saturated clay.

Table 10-2. Backfill Loads in Pounds per Linear Foot on 8-in. Circular Pipe in a Trench Installation Based on 100 lb/cu ft Ordinary Clay Fill. The Transition Width Column Gives the Trench Width at Which the Backfill Load on the Pipe Is a Maximum and Remains Constant Regardless of Increase in Trench Width. The Bold Printed Figures are the Maximum Load at the Transition Width for any Given Height of Backfill.

Height of Backfill H Above Top of Pipe, Feet	Trench Width at Top of Pipe										Transition Width
	1'-6"	1'-9"	2'-0"	2'-3"	2'-6"	2'-9"	3'-0"	3'-3"	3'-6"	4'-0"	
5	501	**603**									1'- 9"
6	559	694	**724**								1'-10"
7	608	761	**847**								1'-11"
8	649	819	**969**								2'- 0"
9	683	868	**1088**								2'- 0"
10	712	911	1119	**1213**							2'- 1"
11	736	948	1170	**1332**							2'- 2"
12	757	979	1215	**1458**							2'- 3"
13	774	1007	1254	1513	**1575**						2'- 4"
14	788	1030	1289	1560	**1698**						2'- 4"
15	801	1051	1319	1603	**1818**						2'- 5"
16	811	1068	1346	1640	**1942**						2'- 6"
17	819	1083	1369	1674	1993	**2065**					2'- 7"
18	827	1096	1390	1703	2034	**2182**					2'- 7"
19	833	1107	1408	1730	2070	**2308**					2'- 8"
20	838	1117	1424	1754	2103	**2429**					2'- 9"
21	842	1125	1438	1775	2133	**2553**					2'- 9"
22	846	1133	1450	1793	2159	2545	**2673**				2'-10"
23	849	1139	1461	1810	2184	2578	**2788**				2'-11"
24	851	1144	1470	1825	2205	2607	**2910**				2'-11"
25	854	1149	1478	1838	2225	2635	**3041**				3'- 0"
26	855	1153	1486	1850	2242	2658	**3154**				3'- 0"
27	857	1156	1492	1861	2258	2682	3128	**3278**			3'- 1"
28	858	1159	1498	1870	2273	2702	3155	**3398**			3'- 2"
29	859	1162	1502	1878	2286	2721	3181	**3514**			3'- 2"
30	860	1164	1507	1886	2297	2738	3204	**3646**			3'- 3"
31	861	1166	1511	1892	2308	2753	3225	**3776**			3'- 3"
32	862	1167	1514	1898	2317	2767	3245	3748	**3880**		3'- 4"
33	862	1169	1517	1904	2326	2780	3263	3772	**4004**		3'- 4"
34	862	1170	1519	1908	2333	2791	3279	3794	**4124**		3'- 5"
35	863	1171	1522	1913	2340	2802	3294	3815	**4241**		3'- 5"
36	863	1172	1524	1916	2346	2811	3308	3834	**4384**		3'- 6"
37	863	1173	1525	1920	2352	2820	3321	3851	**4495**		3'- 6"
38	864	1173	1527	1922	2357	2828	3333	3868	4431	**4603**	3'- 7"
39	864	1174	1528	1925	2362	2835	3343	3883	4451	**4740**	3'- 7"
40	864	1174	1529	1927	2366	2842	3353	3896	4470	**4877**	3'- 8"

Source. Concrete Pipe Design Manual, American Concrete Pipe Association.

circular pipe and ordinary clay fill with an unit weight of 100 lb/cu ft. Tables of this type have been developed for other pipe diameters and backfill materials. From the proper table, height of fill above the pipe and trench width are used to select backfill load. This number is then adjusted for the proper unit weight; for backfill weighing 110 lb/cu ft increase loads 10 percent, for 120 lb/cu ft increase 20 percent, and so forth.

Wheel loads from trucks and other vehicles transmit live loads to buried sewer lines. As is illustrated in Figure 10-16, the vertical pressure produced in underlying soil dissipates laterally with depth, such that only a portion of the concentrated tire pressure is realized by the underground pipe. Live loads on the surface rarely influence design of sanitary sewers because of their great depth and small size. On the other hand, supporting strength of storm sewers placed at shallower depths must consider loads transmitted from heavy vehicles. The equations for computing live loads are complex and, therefore, designers often use precalculated data that are organized as illustrated in Table 10-3. Knowing pipe diameter and height of fill, the load transmitted by a standard wheel load can be selected. If a pipe has less than 3-ft cover, the traffic load is multiplied by an impact factor as recommended in Table 10-4.

The three-edge bearing test is the standard for crushing strength of sewer pipe. The laboratory apparatus (see Figure 10-17) loads full-size pipe sections until failure occurs. Specifications published by the American Society for Testing and Materials list the minimum crushing strength requirements for

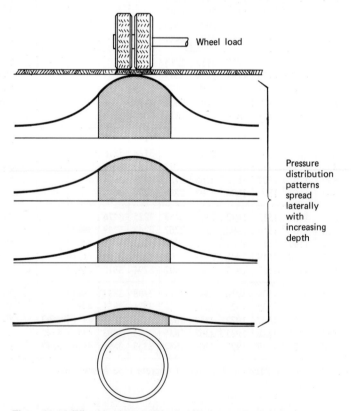

Figure 10-16 Wheel loads and other live loads are distributed with depth such that only a portion of the surface pressure reaches a buried pipe.

Table 10-3. Highway Truck Loads Transmitted To Buried Circular Pipe in Pounds per Linear Foot. Values Based on a 16,000-lb Dual-Tire Wheel Load (32,000-lb Axle Load) and 80-psi Tire Pressure.

Pipe Diameter, inches	Height of Fill H Above Pipe (ft)										
	0.5	1.0	1.5	2.0	2.5	3.0	3.5	4.0	5.0	6.0	
12	4849	2730	1542	954	574	361	316	237	168	46	12
15	**5915**	3322	1872	1163	700	441	358	289	205	56	15
18		3926	2196	1369	825	520	454	341	242	66	18
21		**4498**	2525	1580	950	599	522	392	279	77	21
24			2856	1788	1077	679	592	444	316	87	24
27			**3096**	1998	1201	757	660	496	353	97	27
30				2204	1328	837	731	548	390	107	30
33				**2370**	1452	915	798	599	426	117	33
36					1580	996	869	652	463	127	36
42					**1679**	1154	1007	755	537	147	42
48						**1221**	1146	859	611	168	48
54							**1185**	962	684	188	54
60								**998**	758	208	60
66									832	228	66
72									**857**	249	72
78										269	78
84										**275**	84

Source. Concrete Pipe Design Manual, American Concrete Pipe Association.

various types and sizes of pipe. These values can then be reliably used in design when the pipe manufacturer subscribes to ASTM specifications. Crushing strength for clay pipe based on three-edge bearing are given in Table 10-5. Similar tables are available for asbestos cement and nonreinforced concrete pipe.

Strength requirements for reinforced concrete sewer pipe are given in Table 10-6. Since numerous pipe sizes are available, three-edge bearing strengths are classified by D loads to produce both a 0.01-in. crack and ultimate failure of the section. The D-load concept provides a strength classification independent of

Table 10-4. Recommended Impact Factors for Calculating Loads on Pipe with Less Than 3-ft Cover Subjected to Highway Truck Loads

Height of Cover H	Impact Factor
0 to 1 ft-0 in.	1.3
1 ft-1 in. to 2 ft-0 in.	1.2
2 ft-1 in. to 2 ft-11 in.	1.1
3 ft-0 in. and greater	1.0

Source. AASHO Standard Specifications for Highway Bridges.

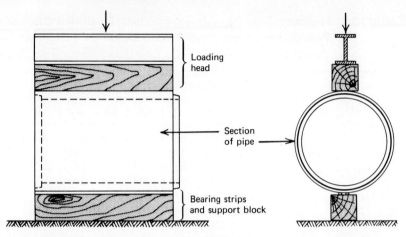

Figure 10-17 The three-edge bearing test is the standard laboratory procedure for determining crushing strength of sewer pipe.

pipe diameter; consequently, strength is expressed as pounds per linear foot per foot of inside diameter. Three-edge bearing test load in pounds per linear foot for a specific size pipe is calculated by multiplying the D load for the particular class pipe times inside diameter in feet.

If sewers were simply laid by placing the pipe barrels on the flat trench bottom, as is illustrated in Figure 10-18a, the pipe would not be able to support a load significantly greater than the three-edge bearing load. However, if bedding bears, at least, the lower quadrant of the barrel and backfill material is carefully placed and tamped around the sides of the pipe, the supporting

Table 10-5. Crushing Strength Requirements for Vitrified Clay Sewer Pipe Based on the Three-Edge Bearing Test, Pounds per Linear Foot

Nominal Size (inches)	Standard Strength	Extra Strength
4	1200	2000
6	1200	2000
8	1400	2200
10	1600	2400
12	1800	2600
15	2000	2900
18	2200	3300
21	2400	3850
24	2600	4400
27	2800	4700
30	3300	5000
33	3600	5500
36	4000	6000

Source. ASTM Specification C700 for Standard and Extra Strength Clay Pipe.

Table 10-6. Strength Requirements for Reinforced Concrete Sewer Pipe Based on the Three-Edge Bearing Test, Pounds per Linear Foot per Foot of Inside Pipe Diameter

Classification	D Load To Produce 0.01-in. Crack	D Load at Failure	Pipe Size (diameter in inches)
Class I	800	1200	
Concrete strength 4000 psi			60 to 96
Concrete strength 5000 psi			102 to 108
Class II	1000	1500	
Concrete strength 4000 psi			12 to 96
Concrete strength 5000 psi			102 to 108
Class III	1350	2000	
Concrete strength 4000 psi			12 to 72
Concrete strength 5000 psi			78 to 108
Class IV	2000	3000	
Concrete strength 4000 psi			12 to 66
Concrete strength 5000 psi			60 to 84
Class V	3000	3750	
Concrete strength 6000 psi			12 to 72

Source. ASTM Specification C76-66T.

strength is increased significantly. Load factor expresses this increase numerically as the ratio of the supporting strength in the trench, as determined by bedding conditions, to the three-edge bearing test strength. Supporting strength is 50 percent greater when the trench bottom is carefully shaped to carry the lower quadrant of the pipe (Figure 10-18b). Carefully placed granular bedding in the trench bottom and tamped under the pipe haunches up to the center line raises the load factor to 1.9. Concrete bedding or encasement provides even higher load factors (Figure 10-18d).

A factor of safety is applied to decrease the calculated supporting strength to an allowable load on the pipe as expressed in Eq. 10-2. In effect, this gives a margin of safety against variations in soil conditions, quality of construction, and other deviations not anticipated during design. The factor of safety normally used for clay or plain concrete pipe is 1.2 to 1.5, while for reinforced concrete pipe a factor of 1.0 based on 0.01-in. crack load is generally considered sufficient.

$$\text{Allowable load on buried pipe,} = \frac{\text{supporting strength, lb/lin ft}}{\text{factor of safety}} \qquad (10\text{-}2)$$
$$\text{lb/lin ft}$$

where Allowable load must be greater than, or equal to, backfill load (W_d) plus live load (W_l)

For clay or plain concrete pipe:

Supporting strength = Crushing strength × load factor

Factor of safety = 1.2 to 1.5

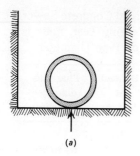

(a)

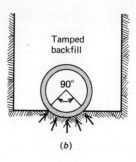

Tamped
backfill

90°

(b)

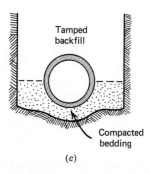

Tamped
backfill

Compacted
bedding

(c)

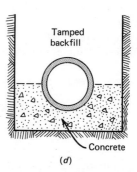

Tamped
backfill

Concrete

(d)

Figure 10-18 Common types of bedding for sewer pipe installations and corresponding load factors. (a) Unprepared bedding, improper support (not to be used). (b) Ordinary bedding, shaped trench bottom (load factor = 1.5). (c) First-class bedding, compacted backfill (load factor = 1.9). (d) Concrete cradle bedding (load factor = 2.3 or 3.2).

For reinforced concrete pipe:

$$\text{Supporting strength} = \text{D load} \times \text{diameter} \times \text{load factor}$$

$$\text{Factor of safety} = 1.0 \text{ based on the 0.01-in. crack strength}$$

EXAMPLE 10-4

An 8-in. diameter, vitrified clay sewer is being placed in a 2.0-ft-wide trench at a depth of 19.2 ft to the pipe invert. The backfill is ordinary clay with a unit weight of 120 lb/cu ft. Determine the strength of pipe and type of bedding that should be used for a factor of safety of 1.5.

Solution

$H = \text{depth to bottom} - \text{pipe diameter}$

$H = 19.2 - 0.7 = 18.5 \text{ ft}$

$\dfrac{H}{B_d} = \dfrac{18.5}{2.0} = 9.2$

From Figure 10-15, for $H/B_d = 9.2$ and curve D for ordinary clay,

$C_d = 3.5$

By Eq. 10-1,

$W_d = 3.5 \times 120 \times 2.0 \times 2.0 = 1680$ lb/lin ft

Alternate solution is from Table 10-2, under the 2'-0" column interpolating at $H = 18.5$, load $= 1400$ lb/lin ft for 100 lb/cu ft backfill.
Therefore, $W_d = 1400 \times 1.2 = 1680$ lb/lin ft for 120 lb/cu ft
Consider standard strength pipe with a crushing strength of 1400 lb/lin ft (Table 10-5) and ordinary bedding (Figure 10-18b), substituting into Eq. 10-2

Allowable load $= \dfrac{1400 \times 1.5}{1.5} = 1400$ lb/lin ft < 1680 (No Good)

Try first-class bedding (Figure 10-18c)

Allowable load $= \dfrac{1400 \times 1.9}{1.5} = 1770$ lb/lin ft > 1680 (Adequate)

Consider extra strength pipe and ordinary bedding

Allowable load $= \dfrac{2200 \times 1.5}{1.5} = 2200$ lb/lin ft > 1680 (Adequate)

Either standard strength VCP with first-class bedding, or extra strength with ordinary bedding, is satisfactory.

EXAMPLE 10-5

What class of reinforced concrete pipe should be placed for a storm sewer based on the following: pipe diameter $= 33$ in., $H = 6.0$ ft, $B_d = 4.5$ ft, sand and gravel backfill with a unit weight of 110 lb/cu ft, ordinary bedding, highway truck loads, and a factor of safety $= 1.0$ based on the 0.01-in. cracking load.

Solution

From Figure 10-15, for sand and gravel backfill, and

$\dfrac{H}{B_d} = \dfrac{6.0}{4.5} = 1.3$, $C_d = 1.1$

Using Eq. 10-1

$W_d = 1.1 \times 110 \times 4.5 \times 4.5 = 2450$ lb/lin ft

From Table 10-3 for $H = 6.0$ ft and diameter $= 33$ in., $W_1 = 117$ lb/lin ft
Therefore, total load $= 2450 + 117 = 2570$ lb/lin ft
Substituting into Eq. 10-2 to find required D load

$2570 = \dfrac{\text{D load} \times 33 \times 1.5}{12 \times 1.0}$

D load $= 620$ lb/lin ft/ft of diameter
Based on D loads in Table 10-6, Class I at 800 lb/lin ft/ft of diameter would be satisfactory, but 33-in. pipe is not precast in this classification. Therefore, use a

Class II pipe and 4000-psi concrete with a specified D load of 1000 lb/lin ft/ft of diameter.

10-5   SEWER INSTALLATION

Trenches can be excavated with vertical walls in stiff clayey soils, but sloping sides are often necessary in less cohesive soils. With careful control of trench depth, the pipe can be placed on undisturbed soil by trimming the bottom to fit the pipe shape. It may be less costly in some cases to overexcavate and backfill with granular bedding. In unstable soils a special bedding, such as a reinforced concrete cradle is needed. Digging below the water table in permeable soils requires a well point system for dewatering to keep the water level below the trench bottom. A backhoe is popular for excavating, since it can also serve for backfilling and lifting pipe sections. Most trench excavations require bracing to protect workmen during pipe placement.

In pipe placement, an offset line paralleling the proposed sewer is located with transit and tape. The trench location is then staked out on the ground surface by measuring from the offset line. After excavation at the proper width, grade bars (batter boards) are placed across the trench at 25- or 50-ft intervals and are supported by stakes as illustrated in Figure 10-19. The center line of the

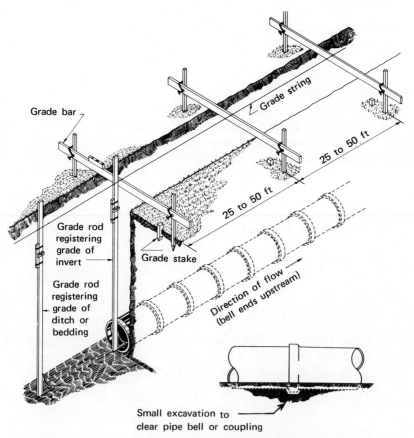

Figure 10-19 Diagram showing the common method of laying sewer pipe to line and grade. The grade bars carry both vertical and horizontal levels. (*Clay Pipe Engineering Manual*, National Clay Pipe Institute)

sewer is located on the grade bars by a nail or edge of an upright cleat. Elevations are then run and the grade bars are set, or a mark is placed on each cleat, at some even foot distance above the invert of the sewer. A cord is stretched from grade bar to grade bar along the center line. Sewer alignment is transferred from this stringline to the trench bottom by means of a plumb bob. Sewer slope is transferred by using a wooden rod, marked in even foot marks, that has a short piece fastened at right angles to the lower end. After bedding and joining each pipe section, grade is checked by placing the foot of the grade rod on the sewer pipe invert and noting whether the proper foot mark touches the cord (Figure 10-19).

Special laser systems have been designed for aligning sewers. Such a unit eliminates the stringline and the need to transfer line and grade into the trench every time a pipe section is laid. It also serves as a guide in the excavation of the trench to proper slope. The laser shown in Figure 10-20a projects a pencil-thin ray of red light through the pipe onto a target for position reference. After the instrument is leveled, desired grade is set by a dial.

A laser installed in a manhole is illustrated in Figure 10-20b. The base plate is set over the starting point by using a center hole for location, and a vertical pole, with a bullseye level on top, is placed in position on the base and held plumb with extension braces. The laser is then attached to the vertical pole at the proper height to put the beam over the pipe center line. The beam is aligned in the proper horizontal direction by using an above ground tilting telescope assembly, or transit, to transfer a distant reference point into the trench just

Figure 10-20 A laser beam for laying sewer pipe eliminates the need for grade bars and a stringline. (a) Laser with a dial for setting the desired sewer grade. (b) Laser installed in a manhole. (c) A transit is used to set the laser beam on line. (Courtesy of Spectra-Physics)

ahead of the laser. The procedure is to swing the transit so that the vertical hairline centers on the next manhole, and then to dip it to focus on the trench bottom in front of the laser. Next, the laser beam is swung into view and is set so that it coincides with the vertical hairline on the transit. The laser instrument is then leveled, and the proper grade is set on the dial. Each pipe section is placed with a removable target set in the bell so that the pipe may be shifted into position by bringing the laser spot on target center.

New sewers are tested for watertightness by monitoring infiltration, measuring exfiltration with the pipeline full of water, and by air testing. Infiltration testing consists simply of measuring, through use of a weir, the amount of flow in a sewer before any building connections are made. This technique, of course, is not applicable if the sewer line is above the water table. Flooding of the trench to produce a high groundwater condition rarely simulates actual immersion of a sewer line. Even when sewers are laid below the groundwater level, interpretation of results are questionable, since the head of water above the pipe greatly influences the quantity entering through pipe cracks and joint defects. Another problem with the infiltration procedure is that long sections of sewers must be tested to get measurable flow rates, for example, infiltration at a rate of 500 gal/in. of sewer diameter per mile per day would yield a flow rate of only 0.2 gpm for a 400-ft manhole-to-manhole section of 8-in. pipe. Testing of long sewer lines, while having the advantage of higher flow yields, may not identify the precise location of broken pipes and poor joints.

Exfiltration testing, the reverse of measuring actual infiltration, is used mainly in dry areas where the groundwater table is below the pipe crown. Filling the pipe with water and observing the loss during a specified time period is a positive method, since it subjects the sewer and manholes to a known water pressure. Excessive pressures can produce destructive results in lower sections of a sewer; however, testing of sections between manholes has few hazards. The maximum hydrostatic head applied is usually 10 ft, and the water is

Table 10-7. Conversion of Infiltration or Exfiltration Rates from Gallons per Inch of Diameter per Mile per Day to Gallons per Hour per 100 ft for Various Sized Sewer Pipes

Diameter of Sewer inches	Filtration Rates in gal/hr/100 ft for the Following Rates in gal/in. diameter/mile/day				
	100	200	300	400	500
8	0.63	1.3	1.9	2.5	3.2
10	0.79	1.6	2.4	3.2	4.0
12	0.95	1.9	2.8	3.8	4.7
15	1.2	2.4	3.5	4.7	5.9
18	1.4	2.8	4.3	5.7	7.1
21	1.7	3.3	5.0	6.6	8.3
24	1.9	3.8	5.7	7.6	9.5
27	2.1	4.3	6.4	8.5	10.7
30	2.4	4.7	7.1	9.5	11.8
36	2.8	5.7	8.5	11.4	14.2
42	3.3	6.6	10.0	13.3	16.6
48	3.8	7.6	11.4	15.2	19.0

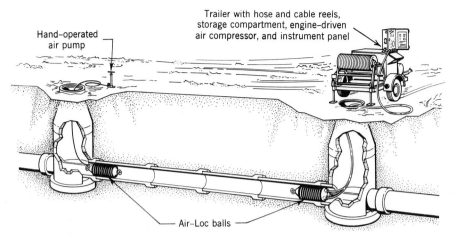

Hand-operated air pump

Trailer with hose and cable reels, storage compartment, engine–driven air compressor, and instrument panel

Air–Loc balls

Figure 10-21 Low pressure air testing to determine the tightness of newly constructed sewer lines. (Courtesy of Cherne Industrial, Inc.)

allowed to stand for at least 4 hr before measuring exfiltration. This allows the pipe and joint material to become saturated with water and permits the removal of entrapped air. The maximum allowable exfiltration specified varies from 100 to 500 gal/in. of diameter/mile/day. For example, a typical allowable limit is 290 gal/in. of diameter/mile/day under a 10-ft head of water, while another specifies a maximum rate of 200 gal/in. of diameter/mile/day plus 10 percent increase for each 2 ft of head over the initial 2 ft. Table 10-7 gives infiltration and exfiltration rates for various sized sewer pipes.

Low-pressure air testing for tightness is a quick and easy method for evaluating new sewer sections (Figure 10-21). Both ends of the line at adjacent manholes are plugged and all service connections tightly capped. Air pressure applied and raised to 4 psi is held for, at least, 2 min while all plugs are checked for leakage. The air supply is then disconnected and the time required for the

Table 10-8. Allowances in Low-pressure Air Testing of Sewers Based on an Air Loss of 0.003 cu ft/min/sq ft of Internal Pipe Surface Under a Test Pressure of 3.0 psi Above Groundwater Pressure

Pipe Diameter (inches)	Minimum Time Lapse for an Air Pressure Drop of 1.0 psi (minutes)
6	2.7
8	3.7
10	4.7
12	5.7
15	7.1
18	8.5
21	9.9
24	11.3

pressure to drop from 3.5 to 2.5 psi is measured using a stopwatch. If the pipe being tested is submerged in groundwater, a pipe probe is inserted in the backfill to the depth of the pipe center line, and the probe pressure is determined by slowly passing air through it. All gauge pressures in the test are increased by the amount of this back pressure as a result of groundwater submergence. Although acceptable leakage rates are expressed in different ways, the recommended specification is based on an allowable air loss of 0.003 cu ft/min/sq ft of interior pipe area at a test pressure of 3.0 psi above groundwater pressure. Allowances based on this criterion for various diameter sewers are given in Table 10-8.

10-6   PUMPING STATIONS

Lift stations are required to elevate and transport waste water in collection systems when continuation of gravity flow is no longer feasible. In flat terrain, sewers en route to a treatment plant may increase in depth to the point where it is impractical to continue gravity flow. Here, a pumping station can be installed to lift the waste water to an intercepting sewer at a higher level. Or, the pumps can discharge to a force main conveying waste water under pressure to the treatment plant. New subdivisions or industries locating on the edge of an existing sewer system may find that the pipe is too shallow to receive sewer connections from basement level. In this case, a small lift station can pump wastes from the area to the main sewer.

Pumping stations are expensive to install, require periodic inspection and maintenance, and strain public relations if they malfunction and allow backup of domestic wastes into residences or businesses. For these reasons, they are avoided wherever possible by recognizing sewer planning and treatment plant location in municipal land use zoning and comprehensive planning.

Pneumatic ejectors are commercially available for pumping flows not exceeding about 100 gpm, for example, for several homes or a commercial building. They have the important advantage of being able to handle un-screened domestic waste water without clogging. (Although low capacity pumps can be employed, the smallest centrifugal pump designed to pass 3-in. solids and operate with reasonable freedom from clogging has a rated capacity greater than 100 gpm.) The essential components of a pneumatic ejector are an air compresser, waste-water receiving chamber, piping with appropriate valves, an automatic system for operation. Waste water flows by gravity into an enclosed receiving chamber displacing air through a vent. When the receiver fills, an electrical control system shuts off the vent connection and supplies compressed air to the chamber. The pressure closes the inlet check valve and forces the waste water through a discharge check valve in the outlet pipe. After it is emptied, the chamber is again vented, allowing waste water to flow in while the outlet check valve shuts, preventing the backflow of ejected waste water. Except in the smallest installations, it is recommended that at least two ejector units be provided, each having sufficient capacity to handle the expected maximum flow.

Wet-well pumping stations are often installed for peak flows between 100 and 600 gpm. They can be mounted directly over an enlarged manhole providing a compact low-cost station. The pumps may be vacuum primed or submerged-impeller centrifugal pumps. The pump control devices hanging in the wet well

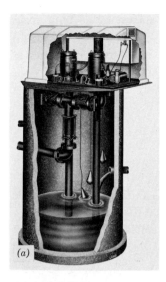
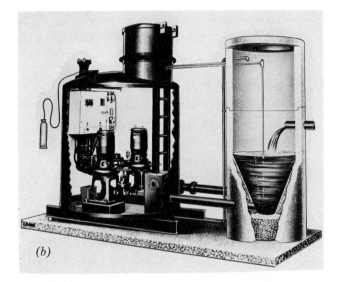

Figure 10-22 Typical sewer-system pumping stations for lifting waste water from a deep sewer to an intercepting sewer, or transporting to a treatment plant by a force main. (*a*) Wet well-mounted lift station for moderate waste-water flows. (*b*) Standard pump station with separate wet and dry well structures. (Courtesy of Smith & Loveless Division, Ecodyne Corp.)

Figure 10-22*a* are mercury switches enclosed in teardrop-shaped plastic casings. These units tip when floated and activate the enclosed switch.

Pumping stations with separate wet and dry wells are preferred to wet well-mounted lift stations. The unit illustrated in Figure 10-22*b* is a factory-built pumping station that is available for maximum flows in the range of 100 to 1500 gpm. The pumps are automatically controlled using an air-bubbler pipe to sense the depth of waste water in the wet well. Back pressure on the air supply to the bubbler tube actuates a mercury pressure switch that controls the pump starter. Waste water flows from the wet well through a suction pipe, gate valve, pump housing, check valve, and discharge gate valve to the force main. The dry well is ventilated and dehumidified to protect the pump control equipment and to make the chamber safe for entry by workmen.

REFERENCES

1. *Concrete Pipe Design Manual*, American Concrete Pipe Association, Arlington, Va, 1970.

2. *Clay Pipe Engineering Manual*, National Clay Pipe Institute, Washington, D.C., 1974.

3. Symons, G. E., *Wastewater Systems Pipes and Piping*, The Reuben H. Donnelley Corp., New York, N.Y., 1967.

4. *Design and Construction of Concrete Sewers*, Portland Cement Association, Skokie, Ill., 1968.

PROBLEMS

10-1 If a storm sewer surcharges during a heavy rainstorm was it poorly designed? Explain.

10-2 What is the maximum population that can be served by an 18-in. main laid on minimum slope? (*Answer* 9000)

10-3 For a peak flow of 1000 gpm what size sewer is needed for construction at minimum slope? At a slope of 0.40 ft/100 ft? (*Answers* 15 in., 12 in.)

10-4 What is a potential maintenance problem if small sewers are placed on curves rather than straight lines between manholes?

10-5 How are lateral sewers ventilated?

10-6 Where are drop manholes used and why?

10-7 How are inverted siphons designed to reduce plugging by deposition of solids in the depressed pipes?

10-8 The depth of flow in a 24-in. diameter sewer placed on a 0.0030 ft/ft grade is 16 in. What is the quantity of flow? (*Answer* 9.6 cu ft/sec)

10-9 A 90° V-notch weir was installed in a manhole to measure flow in a sewer line. If the maximum height of flow over the weir crest was 6.5 in., what was the maximum flow? (*Answer* 0.35 mgd)

10-10 What is the cause of crown corrosion in sanitary sewers? Invert corrosion?

10-11 Where is cast-iron pipe employed in a waste-water collection system?

10-12 In Example 10-4, what strength of pipe and type of bedding should be used if the backfill was saturated clay with a unit weight of 130 lb/cu ft? All other conditions remaining the same. (*Answers* standard strength with concrete cradle bedding, or extra strength with ordinary bedding)

10-13 Vitrified clay pipe with an 8-in. diameter is being placed as illustrated in Figure 10-14b with: first-class bedding, 2 ft-3 in. trench width, and ordinary clay backfill with a unit weight of 120 lb/cu ft. What is the maximum allowable depth of burial for standard strength pipe maintaining a factor of safety of 1.5? For extra strength pipe? (*Answers* $H = 12$ ft, and $H > 40$ ft)

10-14 Standard strength 12-in. diameter VCP is placed as shown in Figure 10-14a with: $H = 5$ ft, $B_d = 2.0$ ft, ordinary bedding, 110 lb/cu ft sand and gravel backfill, and subjected to highway truckloads. What is the calculated factor of safety for the pipe installation? (*Answer* 2.9)

10-15 What is the supporting strength of a 33-in. diameter, class II, reinforced concrete pipe based on the 0.01-in. crack strength and load factor of $= 1.5$? (*Answer* 4100 lb/lin ft)

10-16 How is a grade string set for sewer placement?

10-17 When a laser is used in a sewer pipe installation, how is the laser beam aligned in the proper direction and set on the desired slope?

10-18 What is the maximum allowable infiltration flow from a 400-ft line of 10-in. diameter sewer if the specification permits 400 gal/in. diameter/mile/day? (*Answer* 12.8 gal/hr)

10-19 During exfiltration testing of 350 ft of 18-in. sewer, the water loss was 8.0 gal/hr. Does this exceed 200 gal/in. diameter/mile/day? (*Answer* 8.0 < 9.8 gal/hr, no)

chapter 11
Waste-Water Processing

Conventional waste-water treatment is a combination of physical and biological processes designed to remove organic matter from solution. The earliest method was plain sedimentation in septic tanks. Imhoff tanks used by municipalities were two-story septic tanks that separated the upper sedimentation zone from the lower sludge digestion chamber by a sloping bottom with a slot opening. Solids settling in the upper portion of the tank passed through the slot into the bottom compartment from which the digested sludge was periodically withdrawn for disposal. The final step in development of primary treatment was complete separation of sedimentation and sludge processing units. Currently, raw sludge is handled independently by mechanical dewatering processes or biological digestion.

Primary sedimentation of municipal waste water has limited effectiveness, since less than one half of the waste organics are settleable. The initial attempt at secondary treatment involved chemical coagulation to improve settleability of the wastes. Although this provided considerable improvement, the heavy chemical dosages resulted in high cost and dissolved organics were still not removed. The first major breakthrough in secondary treatment occurred when it was observed that the slow movement of waste water through a gravel bed resulted in rapid reduction of organic matter. This process, referred to as trickling filtration, was developed for municipal installations starting in about 1910. A more accurate term for a trickling filter is biological bed, since the process is microbial oxidation of organic matter by slimes attached to the stone, rather than a straining action.

A second major advancement in biological treatment took place when it was observed that biological solids, developed in polluted water, flocculated organic colloids. These masses of microorganisms, referred to as activated sludge, rapidly metabolized pollutants from solution and could be subsequently removed by gravity settling. In the 1920s, the first continuous-flow treatment plants were constructed using activated sludge to remove BOD from waste waters. (Biological systems are also discussed in Section 3-10.)

The pictorial diagram in Figure 11-1 summarizes processes applied in

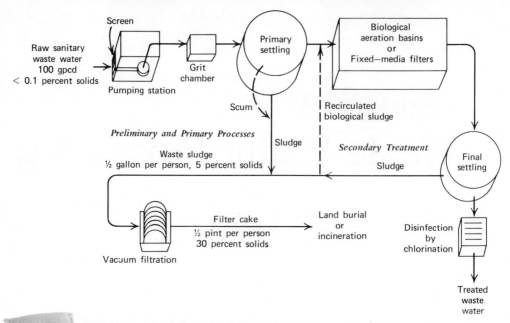

Figure 11-1 Schematic of a conventional, municipal, waste-water treatment plant. Floating, settleable, and biologically flocculated waste solids are removed from the water and thickened for ease of disposal.

conventional municipal waste-water treatment. Preliminary steps include: screening to remove large solids, grit removal to protect mechanical equipment against abrasive wear, flow measuring, and pumping to lift the waste water above ground. Primary treatment is to remove settleable organic matter, amounting to 30 to 50 percent of the suspended solids, and scum that floats to the surface. Secondary treatment is by aeration in open basins with return biological solids, or fixed-media (trickling) filters, followed by final settling. Excess microbial growth settled out in the final clarifier is wasted while the clear supernatant is disinfected with chlorine prior to discharge to a receiving watercourse. Waste sludges from primary settling and secondary biological flocculation are thickened and dewatered in preparation for disposal. Anaerobic bacterial digestion may be used to stabilize the sludge prior to dewatering, but in large plants vacuum filtration is often used to extract water directly from raw sludge after chemical conditioning. Ultimate disposal of dewatered solids may be by landfill or incineration.

The overall process of conventional waste treatment can be viewed as thickening; pollutants removed from solution are concentrated in a small volume convenient for ultimate disposal. The contribution of raw sanitary waste water is about 100 gal/person with a total solids content of less than 0.1 percent, 240 mg/l suspended solids, and 200 mg/l BOD. Liquid waste sludge withdrawn from primary and secondary processing amounts to approximately $\frac{1}{2}$ gal/person with a solids content of 5 percent by weight (Figure 11-1). This is further concentrated to a handleable material by mechanical dewatering; the extracted water is returned for reprocessing. Cake from a vacuum filter amounts to about $\frac{1}{2}$ pt/person with a 30 percent solids concentration. This type of physical-biological scheme is effective in reducing the organic content of waste water, and accomplishes the major objective of BOD and suspended solids removal. Dissolved salts and other refractory pollutants are removed to a lesser extent. Referring to Table 9-2, 50 percent of total volatile solids, 60 percent of total nitrogen, and 70 percent of total phosphorus remain in the effluent after secondary biological treatment. Advanced waste treatment processes are needed to remove these refractory contaminants. Orthophosphate can be precipitated by chemical coagulation, nitrogen may be reduced by biological nitrification and denitrification, and activated carbon takes out refractory soluble organics.

Complete mixing aeration without primary sedimentation (Figure 11-2a) is popular for treatment of small waste-water flows, for example, from subdivisions, villages, and towns. The size of aeration basins ranges from factory-built metal tanks with diffused aeration to accomodate the waste flow from a few hundred persons to mechanically aerated concrete-lined basins, such as, an oxidation ditch, to serve a town with a population of several thousand. Elimination of primary settling dramatically affects the character of waste sludge. Instead of a septic sludge with relatively high solids content, the waste is aerobic and much more voluminous with solids in the range of 0.5 to 2 percent. Therefore, the solids handling system for the flow scheme in Figure 11-2a frequently consists of aerobic digesters (aerated sludge holding tanks) and hauling of the stabilized liquid sludge by tank wagon to land burial or surface spreading on farmland. Larger plants may reduce the volume of sludge by gravity thickening in hopper-bottomed holding tanks, or mechanical flotation thickening, to consolidate the waste solids.

Hundreds of villages and commercial establishments in rural areas use stabilization ponds for waste treatment (Figure 11-2b). The organic loading

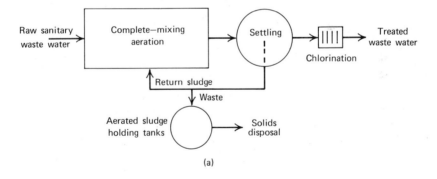

(a)

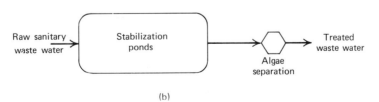

(b)

Figure 11-2 Processing diagrams of systems for treatment of small waste-water flows. (a) Biological processing without primary sedimentation. (b) Natural biological stabilization in lagoons with removal of suspended solids from the effluent.

applied to lagoons is very low and the liquid retention time is long, normally more than 90 days. The natural biological processes that stabilize the waste are sketched in Figure 3-19. In dry climates the evaporation rate may equal or exceed the liquid loading so that the ponds provide complete retention of the waste water. In other cases, stabilized waste water is used for irrigation. Where lagoons overflow to streams, the effluent may require separation of algae and disinfection to achieve adequate quality in terms of suspended solids and fecal coliform bacteria.

11-1   TREATMENT PLANT DESIGN STANDARDS

Municipal facilities are conventionally designed on the basis of quantity of flow and organic content of the raw waste water. Required degree of treatment is designated by effluent standards including BOD and suspended solids concentrations, fecal coliform counts, and pH.

Design Loading

The waste-water quantity for sizing basins is the average weekday flow during that season of the year when discharge is greatest. For example, in warm climates summer flows are often 10 to 20 percent greater than the annual daily average. Saturdays and Sundays generally have lower flows, since industries are not in operation. If sewer pipes in the collection system are in poor condition, the selected design flow includes normal infiltration that occurs during the wet season. Storm runoff collected in combined sewers draining to the plant must also be considered. Where few data are available, the engineer may increase the quantity of flow from areas with older sewers by about 25 percent and refers to this value as the wet-weather design flow.

Maximum and minimum hourly rates, in addition to the average daily, are applicable for sizing certain units, for example, lift pumps. Normally low flows range from 20 to 50 percent of the average daily, while maximum rates are 200 to 250 percent. Sometimes these daily extremes are confusing, since they may be expressed in units of million gallons per day while referring to flows that occur only over a 1-hr period. For example, a specification may state that an aeration system is planned for an average daily flow of 5 mgd and a maximum (or peak) flow of 10 mgd. This means that the unit has a capability of treating 5 mil gal during one day with a flow variation such that the peak hourly rate does not exceed 10 mgd, and should not be interpreted that the system is able to process 10 mil gal in a 24-hr time span.

Organic content of a municipal waste water is defined by the concentrations of BOD and suspended solids. Design organic loadings on a treatment plant are expressed in terms of average pounds per weekday. Values given in the units of milligrams per liter can be converted to pounds per day, using Eq. 9-1, p. 299.

Effluent Quality

Rules and regulations of the Environmental Protection Agency describe the minimum level of effluent quality that must be attained by secondary treatment. Acceptable secondary effluent is defined in terms of BOD, suspended solids, fecal coliform bacteria, and pH as follows:

BIOCHEMICAL OXYGEN DEMAND (5-DAY)

"1. The arithmetic mean of the values for effluent samples collected in a period of 30 consecutive days shall not exceed 30 milligrams per liter."

"2. The arithmetic mean of the values for effluent samples collected in a period of seven consecutive days shall not exceed 45 milligrams per liter."

"3. The arithmetic mean of the values for effluent samples collected in a period of 30 consecutive days shall not exceed 15 percent of the arithmetic mean of the values for influent samples collected at approximately the same times during the same period (85 percent removal)."

SUSPENDED SOLIDS

"1. The arithmetic mean of the values for effluent samples collected in a period of 30 consecutive days shall not exceed 30 milligrams per liter."

"2. The arithmetic mean of the values for effluent samples collected in a period of seven consecutive days shall not exceed 45 milligrams per liter."

"3. The arithmetic mean of the values for effluent samples collected in a period of 30 consecutive days shall not exceed 15 percent of the arithmetic mean of the values for influent samples collected at approximately the same times during the same period (85 percent removal)."

FECAL COLIFORM BACTERIA

"1. The geometric mean of the value for effluent samples collected in a period of 30 consecutive days shall not exceed 200 per 100 milliliters."

"2. The geometric mean of the values for effluent samples collected in a period of seven consecutive days shall not exceed 400 per 100 milliliters."

pH

"The effluent values for pH shall remain within the limits of 6.0 to 9.0."*

Design Parameters

Standards for design of treatment units are expressed by a variety of terms. The following common definitions are inserted here as examples. Sedimentation basins are sized on overflow rate, detention time, and tank depth. Overflow rate is the average daily effluent divided by the surface area of the tank, expressed in units of million gallons per day per square feet. Detention time in hours is calculated by dividing the tank volume by influent flow. Tank depth is the water depth at the side of the tank measured from the bottom to top of the overflow weir.

Organic loadings on biological treatment units are stated in terms of pounds of applied 5-day BOD. Loading on an aeration basin is commonly expressed as pounds BOD applied per 1000 cu ft of tank volume per day. A biological filter loading uses the same units except the volume refers to the quantity of media rather than liquid volume. Aeration period in an activated sludge process is equal to volume of the aeration basin divided by flow of raw waste water. Since biological filters do not contain a liquid volume, hydraulic loading is presented

* *Source.* Federal Register, Vol. 38, No. 159, August 17, 1973.

as the amount of waste water applied per unit of surface area, for example, million gallons per acre per day (mgad).

The specific numerical values of design parameters used by an engineer in sizing treatment units are frequently recommended in published standards. For example, the 1971 *Recommended Standards for Sewage Works, Great Lakes Upper Mississippi River Board of State Sanitary Engineers (GLUMRB)* are used by regulatory agencies in the 10 states surrounding the Great Lakes in reviewing plans for sanitary facilities. However, many design specifications are not listed and must be selected by the engineer based on experience, or on recommendations of equipment manufacturers. These may not be readily available in published literature, and for a particular treatment plant the criteria applied must be obtained from the designing engineer or equipment supplier.

Construction drawings show plant layout, flow schemes, and tank dimensions, but do not list design loadings or show proprietary equipment. Contract specifications describe quality of workmanship, materials, and equipment to be installed and include performance requirements for mechanical units, such as pumps and aerators. The preliminary study report, written prior to preparation of plans and specifications, does discuss conditions relative to plant design and includes loadings, flows, and alternate treatment processes. However, values in this document or amendments to it must be confirmed by comparing the proposed scheme of the system selected against the as-built drawings, since designs may be modified between the time of preliminary report writing and plant construction. In recent years, the designing engineer has been required to write operational manuals that should include design loadings and effluent quality requirements. The latter should be confirmed by referring to the most recent state surface water quality standards. These are subject to change, and current effluent limitations may be more stringent than those applied during design.

11-2   PRELIMINARY TREATMENT

Screening, pumping, flow measuring, and grit removal are normally the first steps in processing a municipal waste water. Flotation is sometimes included to handle greasy industrial wastes, but these are usually more effectively handled by pretreatment at the industrial site. Chemical coagulation is sometimes incorporated to increase removal in primary settling. However, this is costly and generally is applied only when a treatment plant is overloaded. Chlorination of raw waste water may be employed for odor control and to improve settling characteristics of the waste. Arrangement of preliminary units varies but the following general rules always apply. Screens protect pumps and prevent solids from fouling subsequent units; therefore, they are always placed first. A Parshall flume is located ahead of constant speed lift pumps because the turning on and off of the pumps produces a pulsing flow that cannot be graphed by a flow recorder. With variable speed pumps, the flume may be placed on either side, since the pumping rate is paced to be identical to the influent flow. Grit removal reduces abrasive wear on mechanical equipment and prevents accumulation of sand in tanks and piping. Although ideally it should be taken out ahead of the lift pumps, grit chambers located above ground are far more economical and offset the cost of pump maintenance. Two possible arrangements for preliminary units are illustrated in Figure 11-3. The upper scheme is

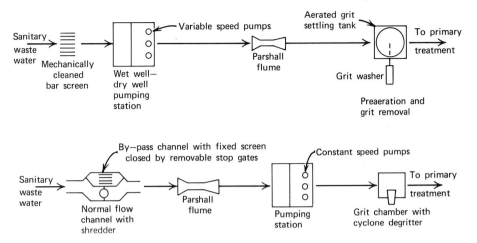

Figure 11-3 Typical arrangements of preliminary treatment units in municipal waste-water processing. The lower sequence is common to smaller plants.

typical of a large plant while the lower sequence is common to small municipalities.

Screens and Shredders

Mechanically cleaned screens have clear bar openings of approximately 1 in. As is illustrated in Figure 11-4, collected solids are removed from the bars by a traveling rake that lifts them to the top of the unit. Here the screenings may be fed to a shredder and returned to the waste-water flow, or they may be deposited in a portable receptacle for hauling to land burial. Inclined bar screens with front rakes are also manufactured. Mechanically cleaned screens are preceded by hand-cleaned bar racks for protection in large combined sewer systems.

A shredder or comminutor cuts solids in the waste water passing through the device to about one-quarter inch size. They are installed directly in a flow channel and are provided with a by-pass so that the section containing the shredder can be isolated and drained for machine maintenance. A comminutor is not often used with a mechanically cleaned bar screen; nevertheless, if installed, it is placed at some point after the screen. In a small municipal plant, a shredder can be placed in the flow stream. For this type of arrangement (Figure 11-3), the by-pass channel contains a hand-cleaned bar screen for emergency use. The channels are equipped with stop gates so that waste water can be directed through the fixed screen while maintenance is performed on the shredder.

Two common types of shredding devices are pictured in Figure 11-5. The comminutor consists of a vertical revolving slotted drum screen submerged in a flow channel such that the waste water passes into the drum through slots and is discharged from the bottom of the unit. Solids too large to enter the slots are cut into pieces small enough to pass through while stringy material that enters partway is cut by a shear bar. The *Barminutor*, combination screening and comminuting machine, has a high-speed rotating cutter that travels up and down on the front of the bar screen. Accumulated solids are shredded and flushed through with the waste water. Cutter action can be automatically

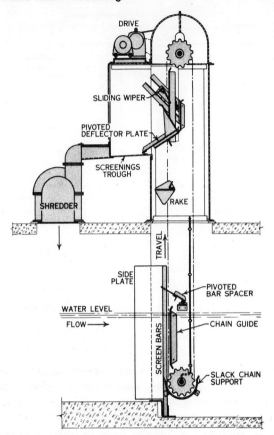

Figure 11-4 Waste-water screen with vertical bars cleaned by a traveling rake. (LINK-BELT Product, Courtesy of Environmental Equipment Division, FMC Corporation)

initiated when solids have accumulated on the screen face and can be stopped when they are shredded and passed through.

Pumping Stations

Municipal lift stations have both wet and dry chambers separated by a common wall as is shown in Figure 11-6. Raw waste water after screening, and sometimes flow measurement, enters the hopper-bottomed wet well. Automatic controls governing pump operation maintain the water surface between preset levels. At the high mark all pumps are running, while lowering to the minimum level shuts off the last pump before it can suck in air. Effective capacity of a wet well, volume between low and high water levels, provides a holding period of less than 10 min to prevent septic conditions. In addition to the influent sewer, drain and recirculation lines often return in-plant wastes to the wet well. These include supernatant from digesters, filtrate from sludge dewatering, underflow from final clarifiers in filtration plants, and waste activated sludge in aeration systems.

Centrifugal waste-water pumps located in the dry well can be mounted with either horizontal or vertical drive shafts. Placing motors in the housing above the dry well has the advantages of accessibility for maintenance and protection in case the dry well is accidently flooded. A separate sump pump is installed in a dry well but its capacity is limited to removal of leakage or drainage that

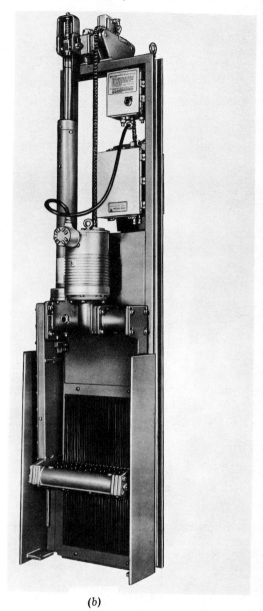

(a)   (b)

Figure 11-5 Two common types of shredders. (a) The revolving slotted drum screen of the comminuter has cutting teeth that pass through a stationary comb mounted on the main casing. As waste water passes through the drum screen, large solids are captured and shredded to pass through. (b) The *barminutor* has a high-speed rotating cutter that travels up and down the vertical bar screen shredding accumulated solids. (CHICAGO PUMP Product, Courtesy of Environmental Equipment Division, FMC Corporation)

enters the chamber. Pump housings are set below the low-water level in the wet well for automatic priming. Suitable valves are placed on the suction and discharge lines of each pump so that it can be removed for maintenance. A check valve is essential in the discharge line, between shutoff valve and pump, to prevent waste water from flowing back through the pump when not in operation. Independent mechanical ventilation is required for the dry and wet

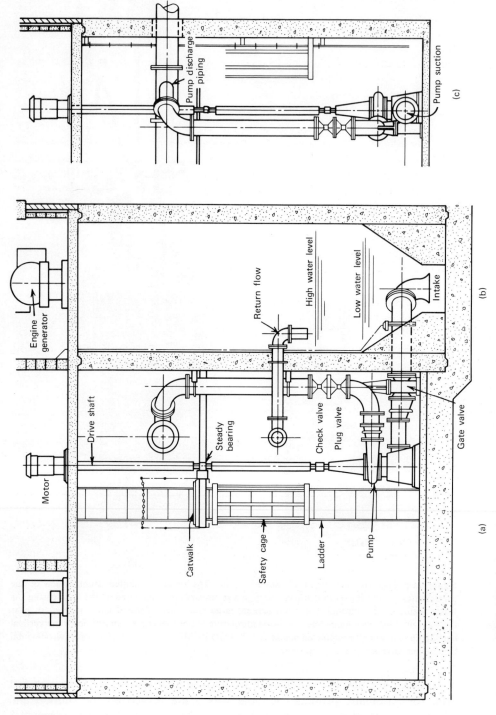

Figure 11-6 Cross-section of a typical municipal waste-water pumping station. (*a*) Dry well. (*b*) Wet well. (*c*) Side view of end pump in dry well.

chambers, even if the latter is not covered. Switches for operation are marked and located conveniently and, if the ventilation system is intermittent rather than continuous, the electrical switches are interconnected with the pit lighting system. A safe means of access by either a stairway or caged ladder, with rest landings at least every 10 ft, is needed for dry wells, and wet wells containing mechanical equipment. Alarm systems are installed to notify plant personnel in cases of power or pump failure. Provision of an emergency power supply is accomplished by using two incoming lines with automatic switching equipment to transfer the load from one utility system to a standby source in the event of failure of the former. The ideal situation is two independent public utilities; if they cannot be provided, standby engine-driven generator sets should be installed.

Centrifugal pumps for lifting raw waste water are designed for ease of cleaning or repair, and are provided with open impellers to reduce the risk of clogging. Most horizontal pumps are made with a split casing so that one half can be removed for access to the interior. For removing obstructions, the casing may also be provided with a hand-sized cleanout. Freedom from clogging depends on type of impeller, shape of casing, and operating clearance between the two. Impellers are generally made with only two or three vanes, set between two parallel plates, to that any object entering the pump passes through. Pumps preceded by a bar screen with clear openings less than $2\frac{1}{2}$ in. should be capable of passing spheres of, at least, 3-in. diameter. Nonclog sludge pumps may have an open impeller where the vanes are attached to a single plate. The characteristics of centrifugal pumps are discussed in Section 4-8, and pump capacity discharge head diagrams are explained in Section 6-5.

Grit Chambers

Grit includes sand and other heavy particulate matter such as seeds and coffee grounds that settle from waste water when the velocity of flow is reduced. If not removed in preliminary treatment, grit in primary settling tanks can cause abnormal abrasive wear on mechanical equipment and sludge pumps, can clog pipes by deposition, and can accumulate in sludge holding tanks and digesters. Grit chambers are designed to remove particles equivalent to a fine sand, defined as 0.2 mm diameter particles with a specific gravity of 2.7, with a minimum of organic material included. A variety of systems are employed depending on the quantity of grit in the waste water, the size of treatment plant, and the expenditures allocated to installation and operation. Standard chambers include: channel-shaped settling tanks, aerated units with hopper bottoms, and clarifiers with mechanical scraper arms. Separated grit may be further processed in a screw-type grit washer or cyclone separator.

The earliest grit chamber consisted of two or more long narrow channels constructed in parallel with space for grit storage. While one was in service, the idle channel could be cleaned by shoveling out the accumulated grit. A proportional weir placed at the discharge end controlled the flow such that the horizontal velocity was maintained at about 1.0 ft/sec independent of the quantity of flow. This allowed the grit to settle while providing a scouring velocity to flush through organic suspended solids. Currently, channel-type grit removal units are equipped with mechanical grit collectors. Buckets on a continuous chain scrape material from the bottom into a receptacle, thus avoiding hand cleaning and the need for more than one channel.

The most popular type of grit chamber in smaller plants is a hopper-bottomed tank with the influent pipe entering on one side and an effluent weir on the opposite. The chamber is small with a detention time of approximately 1 min at peak hourly flow, and is often mixed by diffused aeration to keep the organics in suspension while grit settles out. Solids are removed from the hopper bottom by an airlift pump, screw conveyer, bucket elevator, or gravity flow, if possible. Settled solids from this type of unit are sometimes relatively high in organic content thus creating a nuisance. A cyclone separator or grit washer may be used to wash and dewater the grit slurry returning the overflow to the raw waste water. A common application for a cyclone separator is in processing grit slurry drawn from the hopper of a grit chamber. Feed entering tangentially creates a spiral flow thereby throwing heavy solids towards the inner wall for discharge from the apex while the liquid vortex moves in the opposite direction for discharge.

Grit clarifiers, sometimes called detritus tanks, are generally square with influent and effluent weirs on opposite sides. A centrally driven scraper arm pushes the grit into a hopper on the bottom for removal by a mechanical

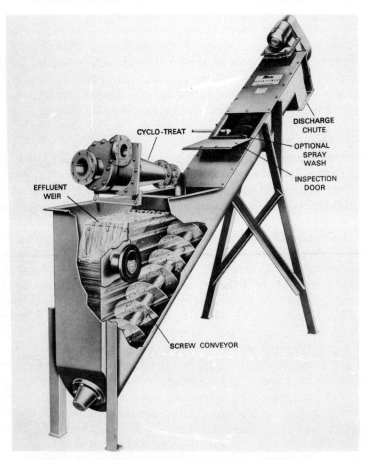

Figure 11-7 The slurry containing grit enters the washer in a liquid chamber above the screw conveyor. Heavy settled solids are moved up the inclined conveyor by the screw while a water spray washes the grit before discharge. (Courtesy of Water Quality Control Division, Envirex, a Rexnord Company)

conveyer or gravity flow. In most installations, the slurry withdrawn is washed and dewatered by a unit like Figure 11-7. Here the slurry is diluted with clean wash water to flush out suspended matter. Grit is moved out of the wash water and up an inclined screw conveyer for drainage prior to discharge. The clarifier may be a shallow tank with a short detention time or a deeper aerated chamber to improve grit separation while freshening the raw waste water.

Preaeration of raw wastes prior to primary settling is practiced to increase dissolved oxygen and scrub out entrained gases for the purpose of improving settleability. The detention period in a combined grit removal preaeration basin is generally 15 to 20 min.

Flow Measuring

All treatment plants should have a means of monitoring waste-water flow. The best system is a Parshall flume that is equipped with an automatic flow recorder and totalizer (refer to Section 4-7). Advantages of a flume are low head loss and smooth hydraulic flow preventing deposition of solids. Some existing treatment plants have venturi meters in the discharge piping from the wet well, but these are not recommended because of potential operation and maintenance problems.

Small disposal plants without installed flow recording equipment usually have provisions for placing a weir in either the plant influent or effluent channel. Recording of flow then involves inserting the weir and installing a portable water level recorder as is illustrated in Figure 10-7. Measuring flow in this fashion is time consuming and is not recommended for plants where regular waste-water compositing is an anticipated requirement. The measured flow for several study periods can be correlated with pump operating time. Then, the quantity of influent in the future can be estimated by reading the total hours of pump operation from timers attached to the electrical controls of the pumps.

11-3 SEDIMENTATION

Clarification is performed in rectangular or circular basins where the waste water is held quiescent to permit particulate solids to settle out of suspension. To prevent short-circuiting and hydraulic disturbances in the basin, flow enters behind a baffle to dissipate inlet velocity. Overflow weirs, placed near the effluent channel, are arranged to provide a uniform effluent flow. Floating materials are prevented from discharge with the liquid overflow by placing a baffle in front of the weir. A mechanical skimmer collects and deposits the scum in a pit outside of the basin. Settled sludge is slowly moved toward a hopper in the tank bottom by a collector arm. Clarifiers following activated sludge may be equipped with hydraulic pickup pipes for rapid sludge return.

Criteria for sizing settling basins are: overflow rate (surface settling rate), tank depth at the side wall, and detention time. Surface settling rate is defined as the average daily overflow divided by the surface area of the tank, expressed in terms of gallons per day per square foot, Eq. 11-1. Area is calculated by using inside tank dimensions, disregarding the central stilling well or inboard weir troughs. The quantity of overflow from a primary clarifier is equal to the waste-water influent, since the volume of sludge withdrawn from the tank bottom is negligible. However, secondary settling tanks may have recirculation lines drawing liquid from the tank bottom, for example, recirculation of activated sludge, in which case the influent flow is equal to the effluent plus returned flow. For these, the flow used for design is the effluent, or overflow, and not the influent which includes recirculation.

$$V_0 = \frac{Q}{A} \tag{11-1}$$

where V_0 = overflow rate (surface settling rate), gallons per day per square foot

Q = average daily flow, gallons per day

A = total surface area of basin, square feet

Detention time is computed by dividing tank volume by influent flow expressed in hours, Eq. 11-2. Numerically, it is the time that would be required to fill the tank at a uniform rate equivalent to the design average daily flow. Depth of a tank is taken as the water depth at the side wall measuring from the tank bottom to the top of the overflow weir; this excludes the additional depth resulting from the slightly sloping bottom that is provided in both circular and rectangular clarifiers. Effluent weir loading is equal to the average daily quantity of overflow divided by the total weir length, expressed in gallons per day per linear foot.

$$t = 24\frac{V}{Q} \tag{11-2}$$

where t = detention time, hours

V = basin volume, million gallons

Q = average daily flow, million gallons per day

24 = number of hours per day

Primary Clarifiers

Settling basins that receive raw waste water prior to biological treatment are called primary tanks. Figure 11-8 is a sectional view of a rectangular tank. Raw waste water enters through a series of ports near the surface along one end of the tank. A short baffle dissipates the influent velocity directing the flow downward. Water moves through at a very slow rate and discharges from the opposite end by flowing over multiple effluent weirs. Settled solids are scraped to a sludge hopper at the inlet end by redwood flights that are attached to endless chains riding on sprocket wheels. Sludge is withdrawn periodically from the sludge hopper for disposal. The upper run of flights protrude through the water surface pushing floating matter to a manual skimmer placed in front of the effluent weir. The scum trough is a cylindrical tube with a slit opening along the top. When manually rotated, scum collected on the surface flows through the slot into the tube that slopes toward a scum pit. Length to width ratio of rectangular tanks varies from about 3 : 1 to 5 : 1 with liquid depths of 7 or 8 ft. The bottom has a gentle slope toward the sludge hopper.

Views of a circular primary clarifier are shown in Figure 11-9. Raw waste water enters through ports in the top of a central vertical pipe, and flows radially to a peripheral effluent weir. The influent well directs flow downward to reduce short-circuiting across the top. A very slowly rotating collector arm

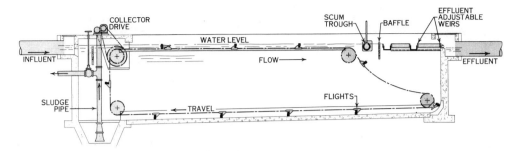

Figure 11-8 Longitudinal section of a rectangular, primary settling tank. Sludge collector scrapes settled solids to hopper at inlet end. The upper run of flights pushes floating matter to a scum trough. (LINK-BELT Product, Courtesy of Environmental Equipment Division, FMC Corporation)

plows settled solids to the sludge drawoff at the center of the tank. Floating solids migrating toward the edge of the tank are prevented from discharge by a baffle set in front of the weir. A skimmer attached to the arm collects scum from the surface and drops it into a scum box that drains outside the tank wall. Circular tanks are from 30 to 150 ft in diameter, although some are as large as 200 ft. Side water depths range from 7 to 12 ft, and bottom slopes are about 8 percent.

Circular basins are generally preferred to rectangular tanks in new construction because of lower installation and maintenance costs. Bridge or pier supported, center-driven collector arms have fewer moving parts than the chain-and-sprocket scraper mechanisms in rectangular tanks. Although inlet turbulence is greater behind the small influent well of a circular clarifier, as the flow radiates toward the effluent weir the waste-water movement slows, thus reducing the exit velocity. Greater weir lengths can be more easily achieved

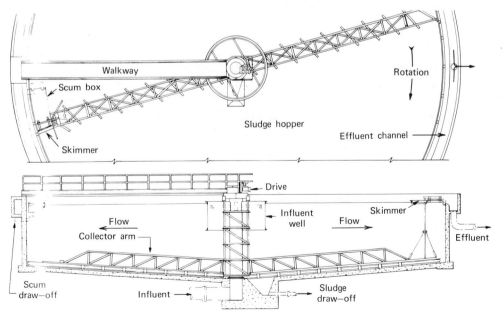

Figure 11-9 Circular primary clarifier with a pier-supported center drive and peripheral effluent weir. (Courtesy of Peabody Welles)

around the periphery of a circular tank than across the end of a rectangular one. A typical installation of two primary settling basins located on opposite sides of a sludge pumping station is pictured in Figure 11-10. Pumps housed inside draw sludge from the bottom of the clarifiers at preset time intervals and discharge it to holding tanks for further processing. (Where holding tanks are not employed, sludge is allowed to accumulate in the clarifiers and is pumped directly to digesters or dewatering equipment.) Scum boxes drain to pits adjacent to the pump building so that the scum can be disposed of with the waste sludge. On the right side of the picture are two aerated grit chambers with grit washers housed in an attached building.

Overflow rates for sizing primary clarifiers fall in a range between 400 and 800 gpd/sq ft, with 600 being a common design value. These hydraulic loadings realize 30 to 40 percent BOD removal in settling raw domestic waste water. Effectiveness of plain sedimentation, of course, depends largely on the character of the waste water. If a municipal waste contains a large amount of soluble organic matter, BOD removal may drop to less than 20 percent while, on the other hand, industrial wastes that contribute settleable solids may increase BOD removal as high as 60 percent. Hydraulic loading also influences the density of the accumulated sludge. With overflow rates less than 600 gpd/sq ft, the settled solids tend to thicken in the bottom of the tank as the collector arm slowly moves through the accumulated sludge. Rates in excess of 800 gpd/sq ft create hydraulic movements in the tank that inhibit consolidation of the sludge. In addition to a more dilute waste sludge, a hydraulically overloaded clarifier is identified by increased turbidity in the effluent during the period of maximum waste-water flow. Storing sludge in a basin for too long a time can also upset clarification, particularly if waste activated sludge is returned to the head of the plant for settling with the raw waste water. Microorganisms

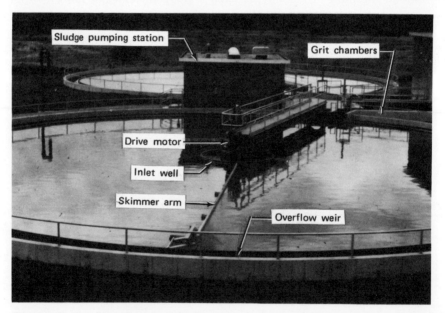

Figure 11-10 Two 90-ft diameter, primary settling basins located on opposite sides of a sludge pumping station. Waste sludge drawn from the clarifiers is pumped to sludge holding tanks at preset time intervals. Floating solids are collected by skimmers and are deposited in scum boxes that drain to a pit for disposal with the waste sludge. (Grand Island, Neb.)

decomposing the waste organics produce gas that makes the solids more buoyant, thus, expanding the sludge blanket and reducing the solids concentration. A severe case of detrimental biological activity is characterized by foul odors, floating sludge, and a darkening of the waste-water color.

The *Recommended Standards for Sewage Works, Great Lakes-Upper Mississippi River Board of State Sanitary Engineers (GLUMRB)* suggest that the liquid depth of mechanically cleaned primary tanks be as shallow as practicable but not less than 7 ft. Designers sometimes increase the side water depth to 7.5 or 8 ft to provide additional volume for accumulated sludge. Detention time is generally not a specified criterion for sizing primary clarifiers, since it is already defined by overflow rate and depth as related in Eq. 11-3. For example, an overflow rate of 600 gpd/sq ft and depth of 7 ft yields a detention time of 2.1 hr.

$$t = \frac{180 \times H}{V_0} \qquad (11\text{-}3)$$

where

t = detention time, hours

H = depth of water in tank, feet

V_0 = overflow rate, gallons per day per square foot

$180 = 7.48$ gal/cu ft $\times 24$ hr/day

Weir loading is the hydraulic flow over an effluent weir. For primary tanks, weir loadings are not to exceed 10,000 gpd/lin ft for plants of 1 mgd or smaller, and preferably not more than 15,000 gpd/ft for design flows above 1 mgd. These values limit the water velocity approaching the effluent weir to minimize carryover of suspended solids.

EXAMPLE 11-1

Two primary settling basins are 90 ft in diameter with a 7-ft side water depth. Single effluent weirs are located on the peripheries of the tanks. For a waste-water flow of 7.0 mgd, calculate the overflow rate, detention time, and weir loading.

Solution

Calculating surface area and volume

Surface area $= 2\pi r^2 = 2 \times 3.14(45)^2 = 12{,}700$ sq ft

Volume $= 12{,}700 \times 7 = 89{,}000$ cu ft $= 0.665$ mil gal

From Eq. 11-1, $V_0 = 7{,}000{,}000/12{,}700 = 550$ gpd/sq ft

By Eq. 11-2, $t = (24 \times 0.665)/7.0 = 2.3$ hr

Alternate calculation for t using Eq. 11-3

$$t = \frac{180 \times 7}{550} = 2.3 \text{ hr}$$

Weir length $=$ circumference of both tanks $= 2\pi d$

Weir loading $= \dfrac{7{,}000{,}000}{2 \times 3.14 \times 90} = 12{,}400$ gpd/ft

Intermediate Clarifiers

Sedimentation tanks between trickling filters, or between a filter and subsequent biological aeration, in two-stage secondary treatment are called intermediate clarifiers. *GLUMRB Standards* recommend the following criteria for sizing intermediate settling tanks: overflow rate should not surpass 1000 gpd/sq ft, minimum side water depth of 7 ft, and weir loadings less than 10,000 gpd/lin ft for plants of 1 mgd or smaller and should not be over 15,000 for larger plants.

Final Clarifiers

Settling tanks following biological filters are similar to the one illustrated in Figure 11-9. Sometimes the effluent channel is an inboard weir trough that allows overflow to enter the channel from both sides to reduce weir loading. Common criteria for final clarifiers of trickling filter plants are: overflow rate not exceeding 800 gpd/sq ft, minimum side water depth of 7 ft, and maximum weir loadings the same as for intermediate tanks, although lower values are preferred.

The purpose of gravity settling following filtration is to collect biological growth, or humus, flushed from filter media. These sloughed solids are generally well-oxidized particles that settle readily. Therefore, a collector arm that slowly scrapes the accumulated solids toward a hopper for continuous or periodic discharge gives satisfactory performance. Gravity separation of biological growths suspended in the mixed liquor of aeration systems is far more difficult. Greater viability of activated sludge results in lighter, more buoyant flocs with reduced settling velocities. In part, this is the result of microbial production of gas bubbles that buoy up the tiny biological clusters. Depth of accumulated sludge in a trickling filter final is normally a few inches if recirculation flow is drawn from the tank bottom. Even if sludge is drained only twice a day, the blanket of settled solids rarely exceeds 1 ft. In contrast, the accumulated microbial floc in a final basin for separating activated sludge may be 1 to 2 ft thick in a well-operating plant. During peak loading periods, the sludge blanket may expand further to incorporate one third to one half of the tank volume; this is particularly true in high-rate aeration systems.

The clarifier in Figure 11-11 is specially designed for an activated sludge secondary. The liquid flow pattern is the same as that of other circular clarifiers but the sludge collection system is unique. Uptake tubes are attached to and spaced along a V-plow-scraper mechanism rotated by a turn-table above the liquid surface. Discharge elevation of suction pipe headers in the sight well is lower than the water surface in the tank so that sludge is forced up and out of the uptake pipes by hydraulic action. Sludge flows by gravity from the sight wells to a manifold encircling the bottom of the center-drive cage and out under the basin. Rate of discharge is controlled by an adjustable weir outside the tank. Cleanout ports are provided to permit easy rodding in the event of nozzle plugging. While the unit illustrated is referred to as a sight well clarifier, other manufacturers produce similar units using descriptive titles such as rapid sludge removal clarifier, suction collector, or manifold draw collector. Each has advantageous features that may be of distinctive importance in the design of a specific biological aeration system. For example, one proprietary unit separates sludge collected by the uptake pipes from the heavier solids that are plowed to

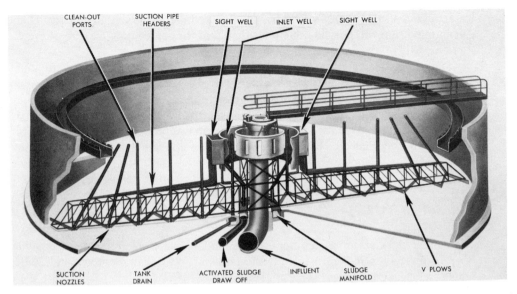

Figure 11-11 Final clarifier designed for use with biological aeration. Activated sludge is drawn through uptake pipes that are located along the collector arm for rapid return to the aeration basin. Sludge flowing from each pipe can be observed in the sight well. (Courtesy of Walker Process Division of Chicago Bridge & Iron Company)

a sludge hopper. The former is returned as activated sludge while the solids scraped from the bottom are wasted.

Rapid uniform withdrawal of sludge across the entire bottom of an activated sludge final clarifier has two distinct advantages. The retention time of solids that settle near the tank's periphery is not greater than those that land near the center; thus, aging of the biological floc and subsequent floating solids due to gas production is eliminated. With a scraper-type collector, the residence time of a settled solids depends directly on the radial distance from the sludge hopper. The second advantage is that the direction of activated sludge return flow is essentially perpendicular to the tank bottom, rather than horizontally toward a centrally located sludge hopper. Downward flow through a sludge

Table 11-1. Recommended Design Parameters for Final Settling Tanks Following Activated Sludge Processes

Type of Process	Average Design Flow (mgd)	Minimum Detention Time (hr)	Maximum Overflow Rate (gpd/sq ft)
Conventional, high	to 0.5	3.0	600
rate, and step	0.5 to 1.5	2.5	700
aeration	1.5 up	2.0	800
Contact	to 0.5	3.6	500
stabilization	0.5 to 1.5	3.0	600
	1.5 up	2.5	700
Extended	to 0.05	4.0	300
aeration	0.05 to 0.15	3.6	300
	0.15 up	3.0	600

blanket enhances gravity settling of the floc and increases sludge density. This is an important factor when one considers that the return flow may be as great as one half of the influent flow.

Suggested design parameters for final settling tanks following activated sludge processes are listed in Table 11-1. The specific overflow rate and detention time selected for basin sizing depend on type of process and average design flow. Weir loadings are generally lower than the suggested maximum for primary clarifiers to reduce the velocity of water approaching the weir. This is commonly accomplished by installing an inboard effluent channel with weirs on both sides as is shown in Figure 11-11.

EXAMPLE 11-2

Two final clarifiers of a step aeration secondary were sized using the design parameters given in Table 11-1. If the tanks are 25.2 ft in diameter and 10 ft deep, what is the design flow for the treatment plant?

Solution

Design criteria for final settling tanks following step aeration are listed on the first three lines of Table 11-1 according to size of plant. To determine which one is applicable, each of the overflow rates are multiplied by the surface area of the tanks and are compared with the corresponding average design flow limitation.

Surface area $= 2 \times 3.14(12.6)^2 = 1000$ sq ft

For 600 gpd/sq ft,

$$\frac{600 \frac{\text{gpd}}{\text{sq ft}} \times 1000 \text{ sq ft}}{1,000,000 \frac{\text{gal}}{\text{mil gal}}} = 0.60 \text{ mgd} > 0.5 \text{ mgd} \qquad \text{(No Good)}$$

For 700 gpd/sq ft,

$$\frac{700 \times 1000}{1,000,000} = 0.70 \text{ mgd}, \qquad 0.5 \text{ mgd} < 0.7 \text{ mgd} < 1.5 \text{ mgd} \qquad \text{(OK)}$$

For 800 gpd/sq ft,

$$\frac{800 \times 1000}{1,000,000} = 0.80 \text{ mgd} < 1.5 \text{ mgd} \qquad \text{(No Good)}$$

Since the overflow rate of 700 gpd/sq ft yields a design flow of 0.70 mgd that falls within the flow range of 0.5 to 1.5 mgd for a 700 gpd/sq ft rate, the design flow of the plant must be 0.70 mgd. This value can be verified by calculating the detention time, which according to Table 11-1, should be 2.5 hr. By Eq. 11-3

$$t = \frac{10 \times 180}{700} = 2.6 \text{ hr} \qquad \text{(OK)}$$

11-4   BIOLOGICAL FILTRATION

Fixed-growth biological systems are those that contact waste water with microbial growths attached to the surfaces of supporting media. Where the waste water is sprayed over a bed of crushed rock, the unit is commonly referred to as a trickling filter. Unfortunately, this is a misnomer and a better term would be biological bed, because the process is one of biological extraction rather than filtration. With the development of synthetic media to replace the use of stone, the term biological tower was introduced, since these installations are often about 20 ft in depth rather than the traditional 6-ft rock-filled filter. Another type of fixed-growth system is the biological disk where a series of circular plates on a common shaft are slowly rotated while partly submerged in a trough of waste water. Microbes attached to the disks extract waste organics. Although the physical structures differ, the biological process is essentially the same in all of these fixed-growth systems.

Biological Process

Domestic waste water sprinkled over fixed media produces biological slimes that coat the surface. The films consist primarily of bacteria, protozoa, and fungi that feed on waste organics. Sludge worms, fly larvae, rotifers, and other biota are also found, and during warm weather sunlight promotes algal growth on the surface of a filter bed. The schematic in Figure 11-12 describes the biological activity. As the waste water flows over the slime layer, organic matter and dissolved oxygen are extracted, and metabolic end products such as carbon dioxide are released. Dissolved oxygen in the liquid is replenished by absorption from the air in the voids surrounding the filter media. Although very thin, the biological layer is anaerobic at the bottom. Therefore, although biological filtration is commonly referred to as aerobic treatment, it is in fact a facultative system incorporating both aerobic and anaerobic activity.

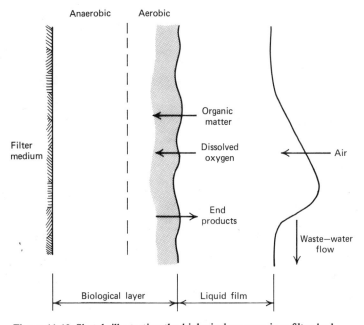

Figure 11-12 Sketch illustrating the biological process in a filter bed.

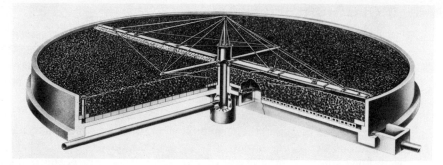

Figure 11-13 Cutaway view of a rock-filled trickling filter. (Courtesy of Dorr-Oliver, Inc.)

Organisms attached to the media in the upper layer of a bed grow rapidly, feeding on the abundant food supply. As the waste water trickles downward, the organic content decreases to the point where microorganisms in the lower zone are in a state of starvation. Thus, the majority of BOD is extracted in the upper 2 or 3 ft of a 6-ft filter. Excess microbial growth sloughing off of the media is removed from the filter effluent by a final clarifier. Purging of a bed is necessary to maintain voids for passage of waste water and air. Organic overload of a stone-filled filter, in combination with insufficient hydraulic flow, can result in plugging of passages with biological growth causing ponding of waste water on the bed, reduced treatment efficiency, and foul odors from anaerobic conditions.

Rock-filled Trickling Filters

A cutaway view of a trickling filter is shown in Figure 11-13. The major components are a rotary distributor, underdrain system, and filter media.

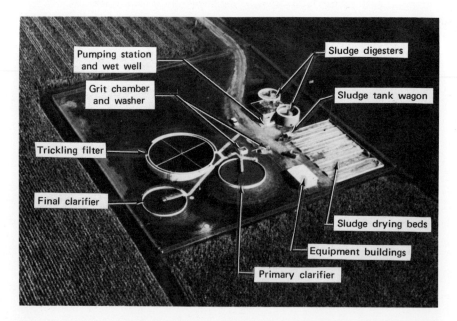

Figure 11-14 Aerial view of a single-stage high-rate trickling filter plant. (Shenandoah, Ia.)

Influent waste water is pumped up a vertical riser to a rotary distributor for spreading uniformly over the filter surface. Rotary arms are driven by reaction of the waste water flowing out of the distributor nozzles. Bed underdrains carry away the effluent and permit circulation of air. Ventilation risers and the effluent channel are designed to permit free passage of air. In some installations, the underdrain block empty into a channel between double exterior walls to allow improved aeration and access for flushing of underdrains.

The most common media in existing filters are crushed rock, slag, or field stone that are durable, insoluble, and resistant to spalling. The size range preferred for stone media is 3 to 5-in. diameter. Although smaller stone provides greater surface area for biological growth, the voids tend to plug and limit passage of liquid and air. Bed depths range from 5 to 7 ft; greater depths do not materially improve BOD removal efficiency. Rock-filled filters in the treatment of municipal wastes are always preceded by primary settling to remove larger suspended solids.

An aerial view of a single-stage trickling filter plant is shown in Figure 11-14. The sequence of treatment units is a hopper-bottomed grit chamber with separate grit washer, primary settling tank, trickling filter, final clarifier with a gravity-flow recirculation line back to the wet well, waste sludge treatment by anaerobic digestion, and digested sludge disposal by drying beds or spreading on adjacent farmland using a tank wagon. The following waste flows return to the raw waste-water wet well: overflow from grit washer, drainage from drying beds, supernatant from digesters, and underflow recirculation (humus return) from the final clarifier. Waste sludge is pumped from the bottom of the primary clarifier into floating-cover digestion tanks.

BOD load on a trickling filter is calculated using the raw BOD in the primary effluent applied to the filter, without regard to any BOD contribution in the recirculated flow from the final clarifier, Eq. 11-4. Hydraulic loading is the amount of liquid applied to the filter surface including both untreated waste water and recirculation flows, Eq. 11-5. This surface loading is commonly expressed in units of million gallons per acre of surface area per day (mgad) or gallons per minute per square foot. Recirculation ratio, calculated by Eq. 11-6, is the ratio of recirculated flow to raw waste water entering the treatment plant.

$$\text{BOD load} = \frac{Q \times \text{settled waste-water BOD} \times 8.34}{\text{volume of filter media}} \qquad (11\text{-}4)$$

where   BOD load = lb BOD applied per 1000 cu ft per day

Q = raw waste-water flow, million gallons per day

Settled BOD = raw waste-water BOD remaining after primary settling, milligrams per liter

Volume of media = volume of rock or slag in beds, thousands of cubic feet

8.34 = conversion factor, milligrams per liter to pounds per million gallons

$$\text{Hydraulic load} = \frac{Q + Q_R}{A} \qquad (11\text{-}5)$$

where

Hydraulic load = million gallons per acre per day (or gpm/sq ft)

Q = raw waste-water flow, million gallons per day (or gpm)

Q_R = recirculation flow, million gallons per day (or gpm)

A = surface area of filters, acres (or sq ft)

$$R = \frac{Q_R}{Q} \qquad (11\text{-}6)$$

where $\qquad\qquad\qquad\qquad$ R = recirculation ratio

Q_R and Q = same as above

Typical loadings for trickling filter secondarys are listed in Table 11-2. Low-rate filter plants return only sufficient flow from the final clarifier hopper to the wet well for removal of accumulated settled solids. Consequently, the flow through the trickling filter is essentially only the raw waste water. To prevent stalling of the distributor arm during low waste-water flow at night, a dosing siphon is installed between the primary clarifier and trickling filter. It accumulates and periodically discharges primary effluent at a rate sufficient to turn the distributor arm and, hence, the operation of the filter is intermittent. Low-rate filters are no longer considered feasible for new construction, since the design BOD loading is only 15 lb BOD/1000 cu ft/day. Most plants originally constructed as low-rate systems have been converted to high-rate filters.

High-rate filtration systems use in-plant recirculation of waste water to increase liquid flow through the filter bed. The primary purpose is to allow greater organic loading without filling the bed voids with biological growths that would inhibit aeration. Experience has shown that BOD loads in excess of about 25 lb/1000 cu ft/day require a minimum hydraulic flushing of 10 mgad to keep a rock-filled bed open. In addition, BOD removal efficiency is enhanced by passing waste water through a filter more than once. The common design loading for a high-rate filter is 45 lb/1000 cu ft/day, which means that the volume of filter media required is only one third that of a low-rate unit.

Numerous recirculation patterns are applied in existing high-rate filter plants. The most popular in new design is: gravity return of underflow from the final clarifier to the wet well during periods of low waste-water flow (Figure 11-15); and, direct recirculation by pumping filter discharge back to the influent as

Table 11-2. Typical Loadings for Trickling Filters with a 5- to 7-ft Depth of Rock or Slag Media

	Low Rate	High Rate	Two Stage
BOD loading			
lb/1000 cu ft/day	5 to 25	30 to 90	45 to 70
lb/acre-ft/day	200 to 1100	1300 to 3900	2000 to 3000
Hydraulic loading			
mil gal/acre/day	2 to 5	10 to 30	10 to 30
gpm/sq ft	0.03 to 0.06	0.16 to 0.48	0.16 to 0.48
Recirculation ratio	0	0.5 to 3.0	0.5 to 4.0

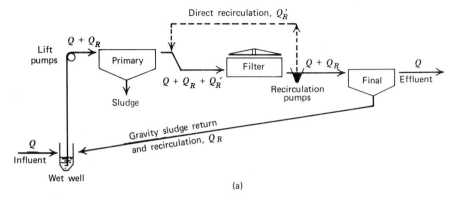

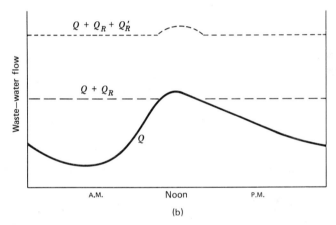

(b)

Figure 11-15 Profile of a high-rate filter plant and related waste-water flow diagrams including in-plant recirculation. (*a*) Flow diagram for a single-stage high-rate trickling filter plant. (*b*) General flow patterns: Q = waste-water influent and final clarifier overflow, $Q + Q_R$ = flow through primary tank and applied to the final based on a gravity return of about $0.5Q$, and $Q + Q_R + Q'_R$ = flow through the filter with both indirect and direct recirculation.

shown by the dotted line in Figure 11-15*a*. Recycling of final clarifier underflow is generally limited to a recirculation ratio of about 0.5, since this is adequate to return settled solids for removal in the primary clarifier and to maintain adequate flow for turning the distributor arm. Yet, by limiting this return flow, the peak overflow rate of the primary clarifier is not increased. Direct recirculation has the advantage of influencing neither the primary nor secondary settling tank; the disadvantage is that a separate pumping station is required.

A two-stage trickling filter consists of two filter-clarifier units in series (Figure 11-16); sometimes the intermediate settling tank is omitted. This type of system is needed to achieve an effluent BOD of 30 mg/l when processing a waste that is stronger than average domestic waste water. Both filters are normally constructed the same size for economy and optimum operation. Generally, several options for recirculation are incorporated in design. For example, in Figure 11-16, underflow from the intermediate and final clarifiers is returned to the wet well for solids disposal, while direct recirculation around

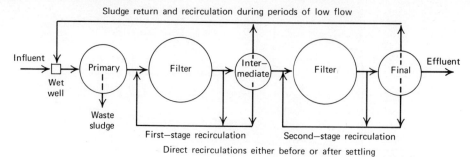

Figure 11-16 Typical flow diagram for a two-stage trickling filter plant.

each filter stage is possible. Direct recirculation may be pumped from the bottom of a settling tank, or directly from the influent well (filter discharge).

BOD Removal Efficiency

Several mathematical equations have been developed for calculating BOD removal efficiency of biological filters based on factors such as depth of bed, type of media, temperature, recirculation, and organic loading. The exactness of these formulas depends on the filter media having a uniform biological layer, and an evenly distributed hydraulic load. These conditions rarely occur in rock filters that may develop unequal biological growths resulting in short-circuiting of waste water through the bed. Therefore, general practice has been to use empirical relationships based on operational data collected from existing treatment plants. One of the most popular formulations evolved from National Research Council (NRC) data that were collected from filter plants at military installations in the United States during the early 1940s. The results, graphed in Figure 11-17, are considered applicable to single-stage rock-filled trickling filters followed by a final settling tank and treating settled domestic waste water with a temperature of 20° C. Efficiency is read along the bottom scale below the point where the horizontal load line intersects the curved line with the proper recirculation ratio. For example, at a load of 40 lb BOD/1000 cu ft/day and a recirculation ratio of 0 the efficiency is 74 percent, with the same BOD load and $R = 3$, efficiency is 81 percent.

A second-stage filter is less efficient than the first stage because of the decreased treatability of the waste fraction applied to the second bed. In other words, the most available biological food is taken out first, passing the more difficult to remove organics through to the second-stage filter. Based on NRC observations, this effect can be incorporated by increasing the actual load to the second-stage filter as is shown in the following relationship:

$$\frac{\text{Second-stage BOD load}}{\text{adjusted for treatability}} = \frac{\text{Actual second-stage BOD load}}{\left[\dfrac{(100 - \text{percentage of first-stage efficiency})}{100}\right]^2}$$

$$(11\text{-}7)$$

The second-stage BOD load is calculated by Eq. 11-4 where the settled waste water is taken as the overflow from the intermediate clarifier. After adjusting for treatability by Eq. 11-7, efficiency of the second stage can be determined from Figure 11-18.

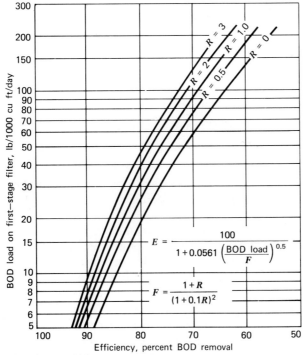

Figure 11-17 Efficiency curves for single-stage rock-filled trickling filters treating domestic waste water at 20° C, based on the National Research Council data.

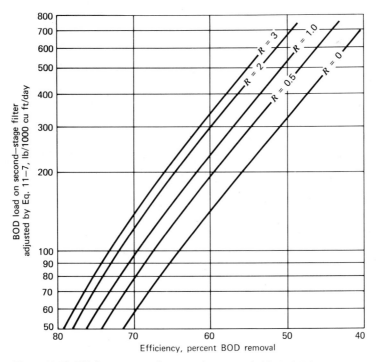

Figure 11-18 Efficiency curves for second-stage rock-filled trickling filters treating domestic waste water at 20° C based on the National Research Council data. Adjusted BOD load for entering diagram is calculated using Eq. 11-7.

Overall treatment plant efficiency of a two-stage filter system can be calculated by Eq. 11-8.

$$E = 100 - 100\left[\left(1 - \frac{35}{100}\right)\left(1 - \frac{E_1}{100}\right)\left(1 - \frac{E_2}{100}\right)\right] \qquad (11\text{-}8)$$

where E = treatment plant efficiency, percent

35 = percentage of BOD removed in primary settling

E_1 = BOD efficiency of first-stage filter and intermediate clarifier corrected for temperature, percent

E_2 = BOD efficiency of second-stage filter and final clarifier corrected for temperature, percent

BOD removal in biological filtration is influenced significantly by waste-water temperature. Filters in northern climates operate at efficiencies about 5 percent or more below the yearly average during winter months. The plot in Figure 11-19 can be used to adjust efficiencies from Figures 11-17 and 11-18 for temperatures above or below 20° C. Filter covers may be required in cold climates to achieve the effluent standard of 30 mg/l BOD or less. An installation of the type shown in Figure 11-20 protects the filters from wind and snow, thus preventing excess cooling of waste water sprayed onto the beds. Positive ventilation is provided to maintain passage of air through the bed and to dissipate corrosive gases, namely, hydrogen sulfide.

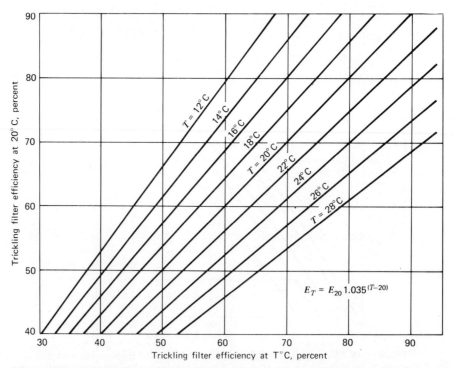

Figure 11-19 Diagram for correcting BOD removal efficiency from NRC data at 20° C to efficiency at other temperatures between 12° C and 28° C.

Figure 11-20 Trickling filters that are domed with fiberglas covers to reduce cooling of the waste water and prevent loss of treatment efficiency during winter months. (Kearney, Neb.)

EXAMPLE 11-3

A trickling filter plant as illustrated in Figure 11-14 has the following: a primary clarifier with a 55-ft diameter, 7.0-ft side water depth, and single peripheral weir; an 85-ft diameter trickling filter with a 7.0-ft deep rock-filled bed; and a final settling tank with a 50-ft diameter, 7.0-ft side water depth, and single peripheral weir. The normal operating recirculation ratio is 0.5 with return to the wet well from the bottom of the final during periods of low influent flow (Figure 11-15). The daily waste-water flow is 1.38 mgd with an average BOD of 180 mg/l, essentially all domestic waste. Calculate the loadings on all of the units, and the anticipated effluent BOD at 20° C and 16° C.

Solution

Primary clarifier

$$A = \frac{3.14 \times (55)^2}{4} = 2370 \text{ sq ft}$$

$$V = 2370 \times 7 = 16{,}600 \text{ cu ft} = 0.124 \text{ mil gal}$$

From Eq. 11-1

$$V_0 \text{ without recirculation flow} = \frac{1{,}380{,}000 \text{ gpd}}{2370 \text{ sq ft}} = 580 \text{ gpd/sq ft}$$

Using Eq. 11-2

$$t = 24 \frac{\text{hr}}{\text{day}} \frac{0.124 \text{ mil gal}}{1.38 \text{ mgd}} = 2.2 \text{ hr}$$

Weir loading $= 1{,}380{,}000/(3.14 \times 55) = 8000 \text{ gpd/lin ft}$

Trickling filter

$$A = \frac{3.14 \times (85)^2}{4} = 5660 \text{ sq ft}, \qquad A = \frac{5660 \text{ sq ft}}{43,560 \frac{\text{sq ft}}{\text{acre}}} = 0.130 \text{ acres}$$

$$V = 5660 \times 7 = 39,600 \text{ cu ft}$$

Substituting Eq. 11-6 in Eq. 11-5

$$\text{Hydraulic load} = \frac{Q + R \times Q}{A} = \frac{1.38 \text{ mgd} + 0.5 \times 1.38 \text{ mgd}}{0.130 \text{ acres}} = 15.9 \text{ mgad}$$

Since the waste water is primarily domestic and the primary overflow rate is reasonable, assume a BOD removal of 35 percent by settling in the primary clarifier. Therefore, settled waste-water BOD $= 0.65 \times 180 = 117$ mg/l. By Eq. 11-4,

$$\text{BOD load} = \frac{1.38 \text{ mgd} \times 117 \text{ mg/l} \times 8.34}{39.6 \text{ thou cu ft}} = 34 \text{ lb BOD/1000 cu ft/day}$$

Final settling tank

$$A = \frac{3.14 \times (50)^2}{4} = 1960 \text{ sq ft}, \qquad V = \frac{1960 \times 7 \times 7.48}{1,000,000} = 0.103 \text{ mil gal}$$

$$V_0 = \frac{1,380,000}{1960} = 705 \text{ gpd/sq ft}, \qquad t = 24 \frac{0.103}{1.38} = 1.8 \text{ hr}$$

$$\text{Weir loading} = \frac{1,380,000}{3.14 \times 50} = 8800 \text{ gpd/lin ft}$$

Treatment plant efficiency and effluent BOD

The trickling filter loading is in the normal range and the final settling tank overflow rate is reasonable. Therefore, Figure 11-17 can be used to estimate filter efficiency. Entering with a BOD load of 34 lb/1000 cu ft/day and $R = 0.5$, efficiency is 78 percent at a waste-water temperature of 20° C.

$$\text{Effluent BOD} = \frac{100 - 78}{100} \, 117 \text{ mg/l} = 26 \text{ mg/l}$$

The trickling filter efficiency at 16° C, entering Figure 11-19 with 78 percent at 20° C, is 68 percent.

$$\text{Effluent BOD at } 16° \text{C} = \frac{100 - 68}{100} \, 117 \text{ mg/l} = 37 \text{ mg/l}$$

Overall treatment plant efficiencies are:

$$\text{at } 20° \text{C}, \; E = \frac{180 - 26}{180} \, 100 = 86 \text{ percent}$$

$$\text{at } 16° \text{C}, \; E = \frac{180 - 37}{180} \, 100 = 80 \text{ percent}$$

EXAMPLE 11-4

The design flow for a two-stage trickling filter plant is 1.2 mgd with an average BOD concentration of 400 mg/l. Calculate the unit loadings, and treatment plant efficiency at a waste-water temperature of 17° C.

Primary tank surface area = 2400 sq ft

First-stage filter: area = 4880 sq ft = 0.112 acres

volume = 29,300 cu ft

Intermediate clarifier area = 1500 sq ft

Second-stage filter is identical to first stage

Final clarifier area = 1500 sq ft

Recirculation pattern is as shown in Figure 11-16: return to wet well is 0.3 mgd underflow from each clarifier for total Q_R = 0.6 mgd; and direct recirculation around each filter Q_R' is 0.8 mgd.

Solution

Primary settling tank

V_0 without recirculation flow = 1,200,000/2400 = 500 gpd/sq ft

V_0 with Q_R of 0.6 mgd = 1,800,000/2400 = 750 gpd/sq ft

Both of these values are satisfactory; therefore, assume a BOD removal of 35 percent.

BOD of settled waste water = 0.65×400 = 260 mg/l

First-stage filter and intermediate clarifier

$$\text{BOD load} = \frac{1.2 \text{ mgd} \times 260 \text{ mg/l} \times 8.34}{29.3 \text{ thou cu ft}} = 89 \text{ lb/1000 cu ft/day}$$

$$\text{Hydraulic load} = \frac{1.2 + 0.6 + 0.8}{0.112} = 23 \text{ mgad (OK)}$$

From Eq. 11-6, $R = (0.6 + 0.8)/1.2 = 1.2$

Efficiency from Figure 11-17 at 89 lb/1000 cu ft/day and $R = 1.2$ is 71 percent at 20° C.

Efficiency corrected for 17° C from Figure 11-19 is 64 percent.

V_0 of intermediate clarifier = (1,200,000 + 300,000)/1500 = 1000 gpd/sq ft (OK)

Effluent BOD from intermediate clarifier = 0.36×260 = 94 mg/l

Second-stage filter and final clarifier

$$\text{BOD load} = \frac{1.2 \text{ mgd} \times 94 \text{ mg/l} \times 8.34}{29.3 \text{ thou cu ft}} = 32 \text{ lb/1000 cu ft/day}$$

From Eq. 11-7,

$$\text{Adjusted BOD load} = \frac{32}{(1 - 0.64)^2} = 250 \text{ lb/1000 cu ft/day}$$

$$\text{Hydraulic load} = \frac{1.2 + 0.3 + 0.8}{0.112} = 21 \text{ mgad (OK)}$$

$$R = \frac{0.3 + 0.8}{1.2} = 0.9 \qquad (\textit{Note.} \text{ 0.3 mgd was returned to wet well in first stage.})$$

Efficiency from Figure 11-18 at 250 lb/1000 cu ft/day and $R = 0.9$ is 59 percent. Efficiency corrected for 17° C from Figure 11-19 is 53 percent

$$V_0 \text{ of final clarifier} = \frac{1,200,000}{1500} = 800 \text{ gpd/sq ft (OK)}$$

Plant effluent BOD $= 0.47 \times 94 = 44$ mg/l

Treatment plant efficiency at 17° C

$$E = \frac{(400 - 44)}{400} 100 = 89 \text{ percent}$$

EXAMPLE 11-5

Estimate the design flow and BOD load for a high-rate rock-filled trickling filter plant with the following sized units:

	Diameter (ft)	Area (sq ft)	Depth (ft)
Primary clarifier	60	2820	8.0
Trickling filter	90	6360	7.0
Final clarifier	48	1810	7.0

The flow pattern is shown in Figure 11-15.

Solution

Assuming a final clarifier overflow rate of 800 gpd/sq ft, the design flow is

$$Q = 1810 \text{ sq ft} \times 800 \frac{\text{gpd}}{\text{sq ft}} = 1,440,000 \text{ gpd} = 1.44 \text{ mgd}$$

Check the overflow rate of the primary settling tank at this flow (1) assuming no recirculation flow, and (2) assuming a recirculation ratio of 0.5.

$$1. \ V_0 = \frac{1,440,000 \text{ gpd}}{2820 \text{ sq ft}} = 510 \frac{\text{gpd}}{\text{sq ft}} (<800 \text{ is approved})$$

$$2. \ V_0 = \frac{1.5 \times 1,400,000 \text{ gpd}}{2820} = 766 \frac{\text{gpd}}{\text{sq ft}} (<800 \text{ is approved})$$

Assuming an allowable BOD loading of 45 lb/1000 cu ft/day on the trickling filter and 35 percent BOD reduction in the primary, the raw waste-water BOD load is

$$\text{BOD} = \frac{1}{0.65} \times 6360 \text{ sq ft} \times 7.0 \text{ ft} \times \frac{45\text{-lb BOD}}{1000 \text{ cu ft/day}} = 3080 \text{ lb/day}$$

$$\text{BOD concentration} = \frac{3080 \text{ lb/day}}{1.44 \text{ mgd} \times 8.34} = 256 \text{ mg/l}$$

Biological Towers

In recent years, several forms of manufactured media have been marketed for trickling filters. The main advantage, relative to crushed rock, is the high specific surface (sq ft/cu ft) with a corresponding high percentage of void volume that permits substantial biological slime growth without inhibiting passage of air supplying oxygen. Other advantages include: a uniform media for better liquid distribution, light weight facilitating construction of deeper beds, chemical resistance, and the ability to handle high-strength and unsettled waste waters. Several companies produce plastic packing sold under trade names, for example, *Flocor* which is illustrated in Figure 11-21a. Another type is *Del-Pak Bio-Media* (Figure 11-21b) consisting of redwood slats with a specific surface of 14 sq ft/cu ft. The corrugated surface of plastic packing and the rough-sawn texture of the redwood facilitates retention of biological films.

Introduction of synthetic media has broadened the application of biological filtration in treating both industrial and domestic waste waters. High-strength soluble food-processing wastes, while generally not suited to rock-filled filters, can be handled by multiple-stage biological towers. Flow diagrams in Figure 11-22 illustrate possible applications of manufactured media in municipal treatment. Rock or slag in existing filters may be replaced to improve operation or degree of treatment. However, this type of installation with a rotary distributor and bed depth of only 5 to 7 ft does not provide optimum use of the media. Better results are achieved, and greater organic loadings can be applied, when the beds are 20 ft or more in depth. These towers allow a greater contact time and the liquid can be applied continuously by fixed distributors, instead of by rotating arms. In special cases, existing treatment plants may install a biological tower preceding primary settling. This type of unit, referred to as a roughing filter, improves overall plant efficiency by: reducing influent BOD, enhancing settleability by preaeration, and leveling out (absorbing) shock loads of high-strength industrial wastes discharged to the municipal system.

Biological towers in new construction of municipal plants are generally installed as illustrated in Figures 11-22b and 11-22c. Normal high-rate filtration plants use flow patterns similar to rock-filled filter systems. Direct recirculation is preferred for increased BOD removal, and gravity flow from the final clarifier returns settled solids to the head of the plant for removal in the primary. For strong municipal waste waters, two towers may be set in series with or without an intermediate clarifier. Increased BOD removal is possible by returning settled solids from the final clarifier to the tower influent. If the waste is sufficiently strong, this recirculation results in a buildup of activated sludge in the flow passing through the tower. Thus the system responds as both a filter, with fixed biological growth, and a mechanical aeration system with suspended microbial floc. Another alternative for processing strong municipal wastes is shown in Figure 11-22d. The tower serves as a roughing filter, with direct recirculation to absorb shock loads, prior to activated sludge treatment in a second-stage aeration basin.

Construction features of a biological tower are illustrated in Figure 11-23. Modules of self-supporting media are stacked so that they interlock for structural stability. The underdrain system consists of supporting block with drain troughs leading to an effluent channel. Waste water is pumped to the top of the bed and is spread by fixed distributors. Piping is supported by horizontal beams that rest on vertical columns. The siding may be corrugated metal, plastic, wood, or block; all of these permit architectural design to enhance the appearance of a treatment plant.

Allowable organic loadings on deep beds range from 25- to 150-lb BOD/1000 cu ft/day with hydraulic loadings up to 2 gpm/sq ft. BOD removal efficiencies

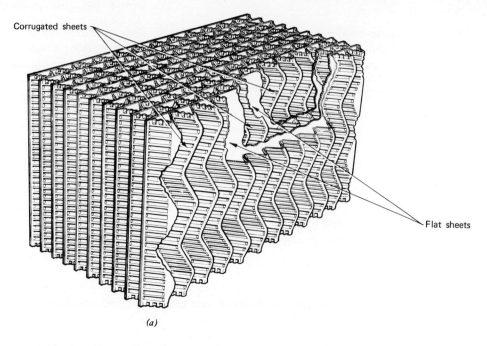

Corrugated sheets

Flat sheets

(a)

(b)

Figure 11-21 Illustrations of plastic and redwood media for biological towers. (*a*) *Flocor* polyvinyl chloride packing with 29 sq ft of surface area per cubic foot consists of alternate flat and corrugated sheets bonded together in modules 2 ft wide, 4 ft long, and 2 ft deep. (Courtesy of Imperial Chemical Industries Limited, Pollution Control Systems) (*b*) *Del-Pak Bio-Media* consists of rough-sawn horizontal redwood slats ($\frac{3}{8}$ in. × $1\frac{1}{2}$ in.) stapled to rails ($1\frac{3}{4}$ in. deep) to form racks 4 ft by 4 ft. For installation, the racks are stacked with slats and rails aligned vertically. (Courtesy of Neptune Microfloc, Inc., subsidiary of Neptune International Corp.)

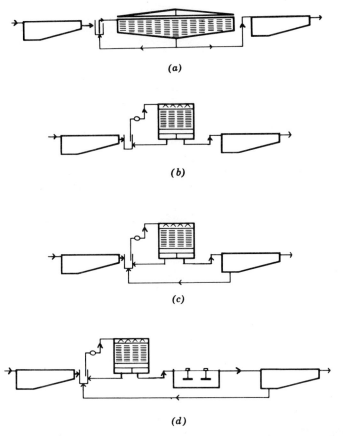

(a)

(b)

(c)

(d)

Figure 11-22 Flow diagrams illustrating a variety of possible applications for biological towers (Bio-Cells). (*a*) Synthetic packing used to replace rock media in existing filters. (*b*) Biological tower as a high-rate trickling filter. (*c*) Biological tower as an activated biological filtration system; microbial solids settled in the final clarifier are returned to develop an activated sludge that is mixed with the waste water entering the tower. (*d*) Activated biological filter with supplemental aeration and clarification to maintain operating stability in plants that frequently encounter organic, or flow, shock loads. (Courtesy of Neptune Microfloc, Incorporated)

rely on volumetric organic loading, and are generally independent of bed depth if it is greater than about 10 ft; normally towers are constructed with 20 ft depth of media. The exact design loading for a biological tower is based on: placement of the unit within the treatment scheme; waste-water recirculation pattern and ratio; type of synthetic media employed; and strength, biodegradability, and temperature of the waste water. Design parameters for a particular synthetic media and application are available from the manufacturer.

Operational Problems

Two major problems of rock-filled trickling filters are effluent quality and odors; both are associated with organic loading, industrial wastes, and cold weather operation. The average BOD removal efficiency of a high-rate single-stage filter plant is about 85 percent. Therefore, to achieve an effluent BOD of 30 mg/l, the raw waste water must be essentially a domestic waste with a BOD not greater than 200 mg/l. If a municipal waste water contains significant industrial waste contributions, a two-stage filtration system is necessary to

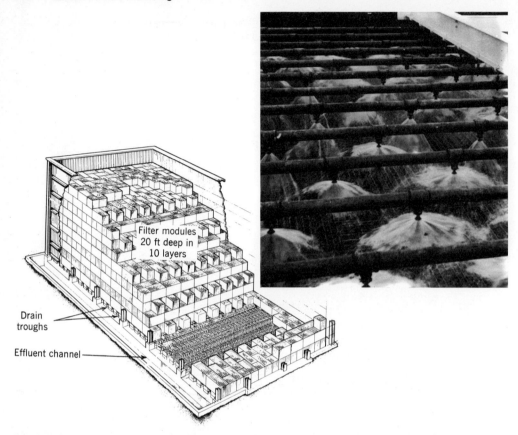

Filter modules
20 ft deep in
10 layers

Drain
troughs

Effluent channel

Figure 11-23 Cutaway view showing construction of a rectangular biological tower, and a photograph of fixed distributors. Waste water is piped to the top of the tower, spread over the packing, and trickles down through the bed. The underflow is conveyed by drains to an effluent channel. (Courtesy of Imperial Chemical Industries Limited, Pollution Control Systems)

meet the required effluent standards. In northern climates, the temperature of waste water passing through the bed may be considerably lower in the winter and may adversely influence BOD removal. Covers may be placed over trickling filters to help maintain temperature during cold weather.

The microbial zone immediately adjacent to the surface of the media is anaerobic and capable of producing metabolic end products that have offensive odors. Reduced compounds formed in treating domestic wastes, such as hydrogen sulfide, appear to be oxidized as they move through the aerobic zone with adequate aeration. However, if voids fill with excess biological growth, foul odors can be emitted during spring and fall when air temperatures reduce natural air circulation through the bed. Industrial wastes, particularly from food-processing industries, have characteristic odors that are not easily oxidized in a trickling filter, and they can cause problems even when design loadings are not exceeded. Filter covers have been used in some locations to reduce the spread of nuisance odors, but the installation must be carefully planned. For example, forced air ventilation with an air scrubbing tower to remove the odors from exhausted air may be required to maintain adequate air passage through the bed and to prevent a corrosive atmosphere under the dome.

Biological towers using synthetic media do not appear to be as susceptible to operational difficulties of quality control and odors as are rock-filled beds. This is primarily attributed to improved aeration and hydraulic distribution of the waste water. Nevertheless, potential odor problems and the influence of cold weather must be considered in design and operation of treatment systems that employ biological towers.

Filter flies, *Psychoda*, are a nuisance problem near filters during warm weather. They breed in sheltered zones of the media, and on the inside surfaces of the retaining walls. Although wind can carry these small flies considerable distances, their greatest irritation is to operating personnel. Periodic spraying of the peripheral area and walls of the filter with an insecticide is a common method of fly control.

Biological Disks

The *Bio-Surf* disk illustrated in Figure 11-24 is a 12-ft diameter corrugated plastic plate. A series of these, mounted on a horizontal shaft, are placed in a contour-bottomed tank and immersed approximately 40 percent. The disks are spaced so that during submergence waste water can enter the separation between the corrugated surfaces. When rotated out of the tank, the liquid trickles out of the voids between plates and is replaced by air. A fixed-film biological growth, similar to that on a trickling filter medium, adheres to the rotating surfaces. Alternating exposure to organics in the waste water and oxygen in the air during rotation is like the dosing of a trickling filter with a rotating distributor. Excess biomass sloughs from the disks and is carried out in the process effluent for gravity separation.

The flow diagram of a biological-disk treatment system is given in Figure 11-25. After primary settling the waste water is applied to disk chambers separated by baffles, with each containing several stages of media. Organics are

Figure 11-24 *Bio-Surf*, biological disks of corrugated plastic 11 ft 10 in. in diameter used as biological growth media as shown in Figure 11-25. (Bio-systems Division, Autotrol Corp.)

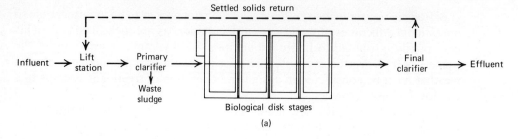

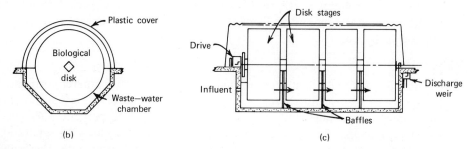

Figure 11-25 Waste-water flow pattern and details of a biological disk treatment system. (*a*) Plant flow diagram. (*b*) Cross section. (*c*) Longitudinal section. (Courtesy of Bio-systems Division, Autotrol Corp.)

extracted by the biological growths as waste water slowly passes through the disk stages. Solids sloughed from the media are collected in a final clarifier and are recycled to the head of the plant for removal in the primary clarifier. The bio-disk process is a once-through treatment so that in-plant recycling is not practiced. In northern climates, the disks must be covered for protection from freezing temperatures. A molded plastic cover with thermal insulation is available for enclosing disk stages. Otherwise, they may be placed in a suitable building that has been provided with adequate ventilation.

Three of the design variables in a biological disk system are detention time of the waste water in the chambers, rotational velocity of the media, and arrangement of the disk stages. Normally, waste water passes through four stages separated by baffles to simulate plug flow. The shafts of the disk units may be set either parallel, as in Figure 11-25, or perpendicular to the direction of flow. Advantages of bio-disk treatment, relative to other biological systems, include: ease of operation, high degree of BOD removal achievable, and good settleability of solids flushed from the disk surfaces.

11-5   BIOLOGICAL AERATION

Understanding of aeration processes in waste-water treatment requires a greater knowledge of biology than that required for biological filtration. Therefore, the reader is encouraged to review the discussions on bacteria and protozoa presented in Sections 3-1 and 3-3, and population dynamics in Section 3-10.

The generalized biological process that takes place in an aeration system is sketched in Figure 11-26. Raw waste water flowing into the aeration basin contains organic matter (BOD) as a food supply. Bacteria metabolize the waste solids, producing new growth while taking in dissolved oxygen and releasing

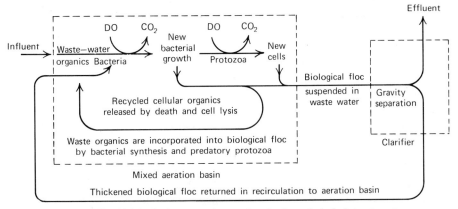

Figure 11-26 Generalized biological process in aeration (activated sludge) treatment.

carbon dioxide. Protozoa graze on bacteria for energy to reproduce. Some of the new microbial growth dies releasing cell contents to solution for resynthesis. After the addition of a large population of microorganisms, aerating raw waste water for a few hours removes organic matter from solution by synthesis into microbial cells. Mixed liquor is continuously transferred to a clarifier for gravity separation of the biological floc and discharge of the clarified effluent. Settled floc is returned continuously to the aeration basin for mixing with entering raw waste.

The liquid suspension of microorganisms in an aeration basin is generally referred to as a mixed liquor, and the biological growths are called mixed liquor suspended solids (MLSS). The name activated sludge was originated in referring to the return biological suspension, since these masses of microorganisms were observed to be very "active" in removing soluble organic matter from solution. This extraction process is a metabolic response of bacteria in a state of endogenous respiration, or starvation.

Activated sludge systems are truly aerobic, since the microbial floc is suspended in mixed liquor containing oxygen. Even though the dissolved oxygen may reduce to zero in the final clarifier, this rarely produces offensive odors since the organic matter has been throughly oxidized during aeration. Transfer to the waste water is a two-step process (Figure 11-27). Air bubbles are created by compressed air forced through a submerged diffuser, or by mechanical aeration where turbulent mixing entrains air in the liquid. The relationship for transfer from air to dissolved oxygen is expressed in Eq. 11-9. The K-factor depends on waste-water characteristics and, more importantly, on physical features of the aeration system, such as type of diffuser, depth of aerator, mixing turbulence, and basin configuration.

$$R = K(\beta Cs - Ct) \tag{11-9}$$

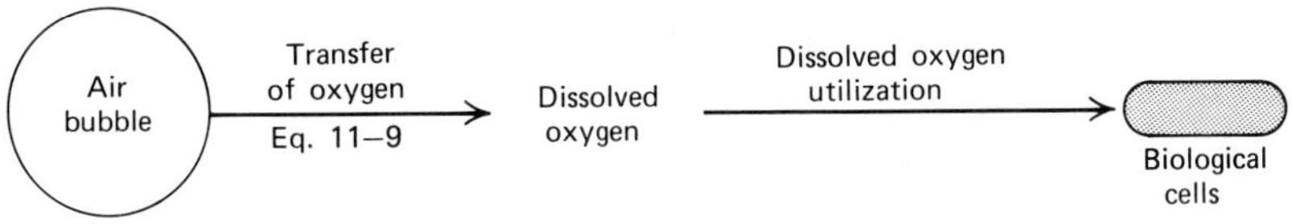

Figure 11-27 Two-step oxygen tranfer in activated sludge aeration. First, gaseous oxygen is converted to dissolved oxygen and, then, is taken up by the biological floc.

where R = rate of oxygen transfer from air to dissolved oxygen, milligrams per liter per hour

K = transfer coefficient depending on aeration equipment and waste-water characteristics, per hour

β = oxygen saturation coefficient of the waste water, usually 0.8 to 0.9

Cs = dissolved oxygen concentration at saturation for pure water, milligrams per liter (Table 2-5)

Ct = dissolved oxygen concentration existing in the mixed liquor, milligrams per liter

$(\beta Cs - Ct)$ = dissolved oxygen deficit, milligrams per liter

The rate of dissolved oxygen utilization is essentially a function of the food-to-microorganism ratio (BOD loading and aeration period) and temperature. This biological uptake is generally less than 10 mg/l/hr for extended aeration processes, about 30 mg/l/hr for conventional, and as great as 100 mg/l/hr for high rate aeration. Aerobic biological activity is independent of dissolved oxygen above a minimum critical value. Below this concentration, the metabolism of microorganisms is limited by reduced oxygen supply. Critical concentrations reported for various systems range from 0.2 to 2.0 mg/l, the most common being 0.5 mg/l.

Knowledge of basic concepts of oxygen transfer and uptake which are sketched in Figure 11-27, is helpful in understanding operational problems generally associated with aeration processes. It is possible to have a deficiency in an aerating basin if the rate of biological utilization exceeds the capability of the equipment. For example, organic overloading of an extended aeration system that is equipped with coarse bubble diffusers set at a shallow depth can result in a dissolved oxygen level less than 0.5 mg/l even though the tank contents are being vigorously mixed by air bubbles emitting from diffusers. Perhaps a situation that occurs more frequently in practice is uneconomical operation from overaeration producing a dissolved oxygen level greater than is necessary in the mixed liquor. Since biological activity is just as great at low levels and the transfer rate from air to dissolved oxygen increases with decreasing concentration, it is logical to operate a system as close to critical minimum dissolved oxygen as possible. It may be feasible to operate the air compressors at reduced capacity, or even turn off one blower on weekends, to conserve electrical energy with no adverse effects to the biological process. The best method for determining suitable operation is to make oxygen measurements at various times, particularly during periods of maximum loading, and then to adjust the air supply accordingly.

GLUMRB Standards provide the following guidelines as minimum design air requirements for diffused air systems: conventional, step aeration, and contact stabilization 1500 cu ft of air applied per lb BOD aeration tank load; modified or high-rate 400 to 1500 cu ft per lb BOD load; and extended aeration 2000 cu ft per pound of BOD load. They further recommend that the aeration system produce thorough mixing of the basin contents, and be capable of maintaining a minimum of 2.0 mg/l of dissolved oxygen at all times. For mechanical aeration units, as well as diffused air, the equipment should be capable of transferring at least 1.0 lb of oxygen into the mixed liquor per lb of BOD aeration tank loading.

EXAMPLE 11-6

An aeration system has a transfer coefficient K of 3.0 per hour for a waste water at 20° C with a β of 0.9. What is the rate of oxygen transfer when the dissolved oxygen is 4.0 mg/l? When the DO is 1.0 mg/l?

Solution

From Table 2-5, $Cs = 9.2$ mg/l.
Using Eq. 11-9

R at DO of 4.0 mg/l $= 3.0(0.9 \times 9.2 - 4.0) = 13$ mg/l/hr

R at DO of 1.0 mg/l $= 3.0(0.9 \times 9.2 - 1.0) = 22$ mg/l/hr

These calculations illustrate that the rate of oxygen transfer is greater, and the aeration system more efficient, at an increased dissolved oxygen deficit.

EXAMPLE 11-7

A conventional diffused aeration process supplies 1000 cu ft of air per pound of BOD applied to the tank. The installed equipment is capable of transferring 1.0 lb of atmospheric oxygen to dissolved oxygen per pound of BOD applied, as specified by *GLUMRB*. Calculate the oxygen transfer efficiency of the aeration equipment. (One cubic foot of air at standard temperature and pressure, 20° C and 760 mm, contains 0.0174 lb oxygen.)

Solution

$$\frac{\text{Oxygen supplied}}{\text{lb BOD applied}} = 1000 \text{ cu ft} \times 0.0174 \frac{\text{lb O}_2}{\text{cu ft}} = 17.4 \text{ lb}$$

$$\frac{\text{Oxygen transferred}}{\text{lb BOD applied}} = 1.0 \text{ lb (specified rate)}$$

$$\text{Oxygen transfer efficiency} = \frac{1.0}{17.4} \times 100 = 6 \text{ percent}$$

This means that about 1 percent of the air diffused into the mixed liquor is absorbed, since air is only 21 percent oxygen by volume.

Aeration Basin Loadings

Aeration period, BOD loading per unit volume, and food-to-microorganism ratio define an activated sludge process. Aeration period is calculated in the same manner as detention time. Using Eq. 11-2, V is the volume of the aeration tank and Q the quantity of waste water entering, without regard to recirculation flow. BOD load is usually expressed in terms of lb BOD applied per day per 1000 cu ft of liquid volume in the aeration basin. This can be calculated by Eq. 11-4 where V is the liquid volume of the aeration basins, rather than volume of the trickling filter media. The food-to-microorganism ratio (F/M) is a way of expressing BOD loading with regard to the microbial mass in the system. Equation 11-10 calculates the F/M value as lb BOD applied/day/lb of MLSS in the aeration tank.

$$\frac{F}{M} = \frac{Q \times BOD}{V \times MLSS}$$ (11-10)

where   F/M = food-to-microorganism ratio, pounds of BOD per day per pound of MLSS

Q = raw waste-water flow, million gallons per day

BOD = raw waste-water BOD, milligrams per liter

V = liquid volume of aeration basin, million gallons

MLSS = mixed liquor suspended solids in the aeration basin, milligrams per liter

Some authors express the food-to-microorganism ratio in terms of lb BOD/day/MLVSS (mixed liquor volatile suspended solids).

BOD loading per unit volume and aeration period are interrelated parameters dependent on the concentration of BOD in the waste water entering and the volume of the aeration basin. (For example, if a 200 mg/l waste water flows into a basin with an aeration period of 24 hr, the resulting BOD load is 12.5 lb/1000 cu ft/day. If the aeration period is reduced to 8 hr, by increasing the waste flow, the BOD load would be tripled or 37.5 lb/1000 cu ft/day.) But, the F/M ratio as an expression of BOD loading relates to the metabolic state of the biological system rather than to the volume of aeration basin. The advantage of this expression is that lb BOD/day/lb MLSS defines an activated sludge process without reference to aeration period or strength of applied waste water. Two systems quite different may operate at the same F/M ratio. For example, an extended aeration process with a 24-hr aeration period and MLSS concentration of 600 mg/l would have an F/M of one third for an influent waste water with a BOD of 200 mg/l. A conventional activated sludge system can treat this same waste water at the same F/M ratio by operating with a higher MLSS of 1800 mg/l to compensate for reducing the aeration period from 24 to 8 hr.

A summary of volumetric BOD loadings, F/M ratios, aeration periods, return sludge rates, and typical BOD removal efficiencies are listed in Table 11-3.

Table 11-3. Summary of Loading and Operational Parameters for Aeration Processes

Process	BOD Loading ($\frac{lb\ BOD/day}{1000\ cu\ ft}$)	F/M Ratio ($\frac{lb\ BOD/day}{lb\ MLSS}$)	Aeration Period (hr)	Return Sludge Rates (percent)	BOD Efficiency (percent)
Extended aeration	10 to 30	0.05 to 0.2	20 to 30	100	85 to 95
Conventional	30 to 40	0.2 to 0.5	6.0 to 7.5	30	90 to 95
Step aeration	30 to 50	0.2 to 0.5	5.0 to 7.0	50	85 to 95
Contact stabilization	30 to 50	0.2 to 0.5	6.0 to 9.0	100	85 to 90
High rate	80 up	0.5 to 1.0	2.5 to 3.5	100	80 to 85
High purity oxygen	120 up	0.6 to 1.5	1.0 to 3.0	50	90 to 95

Descriptive titles for the various aeration processes are defined by illustrations and flow diagrams in subsequent paragraphs. The rate of return sludge from the final clarifier to the aeration basin is expressed as percentage of the raw waste-water influent. For example, if the return activated sludge rate is 30 percent and the raw waste-water flow into the plant is 10 mgd, the recirculated flow equals 3.0 mgd. BOD efficiency is calculated by dividing the quantity of BOD removed in aeration and subsequent settling by the raw BOD entering.

Sludge Settleability

Degree of treatment achieved in an aeration process depends directly on settleability of the activated sludge in the final clarifier. A biological floc that agglomerates and settles by gravity leaves a clear supernatant for discharge. Conversely, poorly flocculated particles (pin floc), or buoyant filamentous growths, that do not separate by gravity contribute to BOD and suspended solids in the process effluent. Excessive carryover of floc resulting in inefficient operation is referred to as sludge bulking. This may be caused by adverse environmental conditions which are created by insufficient aeration, lack of nutrients, presence of toxic substances, or overloading.

Settleability of a biological sludge, under a normal operating environment, depends on the food-to-microorganism ratio. Figure 11-28 graphically relates F/M values to sludge settleability. Extended aeration systems with long aeration periods and relatively high MLSS concentrations operate in the endogenous growth phase. This yields high BOD efficiency since starving microorganisms effectively scavenge the organic matter, and flocculate readily under quiescent conditions. At the opposite extreme, high-rate aeration employs a high F/M ratio which allows a greater volumetric BOD loading and

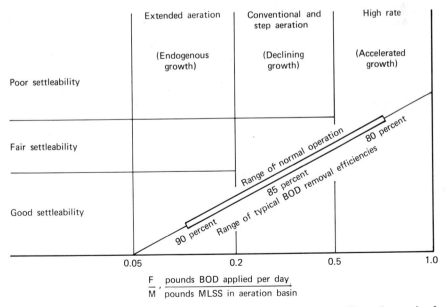

Figure 11-28 Approximate relationship between activated sludge settleability and operating food-to-microorganism ratio.

reduced aeration period. The resulting activated sludge has reduced settleability which can be partly overcome by using a rapid sludge return clarifier that helps to thicken the biological floc hydraulically (Figure 11-11). Nevertheless, high-rate systems are not as efficient in BOD removal primarily because of carry-over of microscopic floc, for instance, groups of bacteria or protozoa not captured by the settleable matter. This can result in suspended solids greater than 30 mg/l in the effluent. The F/M range between 0.5 and 0.2 appears to be optimum for achieving the efficiency needed for treating municipal waste water. Volumetric loadings in this range are great enough to allow aeration periods of 5 to 7 hr, thus permitting economical construction and operating costs.

Another aspect of aeration systems that relates to loadings and the F/M ratio is sludge age. While liquid retention times (aeration periods) vary from 3 to 30 hr, retention of biological solids in a system is much greater and is measured in terms of days. In other words, while waste water passes through aeration only once and rather quickly, the extracted waste organics and resultant biological growths are repeatedly recycled from the final clarifier back to the aeration basin. Several formulations have been developed to compute sludge age; two of the most common are given in Eqs. 11-11 and 11-12. Activated sludge processes operate with a sludge age of several days as calculated by either of these formulas.

$$\frac{\text{Suspended solids}}{\text{sludge age (days)}} = \frac{\text{lb MLSS in aeration basin}}{\text{lb SS in effluent and waste sludge per day}} \quad (11\text{-}11)$$

$$\frac{\text{BOD sludge}}{\text{age (days)}} = \frac{M}{F} = \frac{\text{lb MLSS in aeration basin}}{\text{lb BOD applied to basin per day}} \quad (11\text{-}12)$$

Mathematical Relationships

One of the most common tests for monitoring operation of an aeration system is the sludge volume index (SVI). The procedure involves determining the mixed liquor suspended solids concentration (MLSS), and sludge settleability using a standard laboratory graduated cylinder with a capacity of 1 liter. Mixed liquor for testing is drawn from the aeration basin near the discharge end. Sludge volume is measured by filling the graduated cylinder to the 1-liter mark, allowing undisturbed settling for 30 min, and then reading the volume occupied by the settled solids in milliliters. A second portion of the mixed liquor, or contents of the cylinder after remixing, is tested for MLSS by the procedure for suspended solids presented in Section 2-8. Sludge volume index is then calculated by Eq. 11-13. SVI is the volume in milliliters occupied

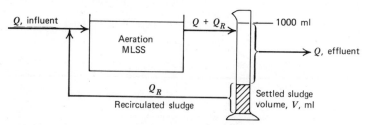

Figure 11-29 Hypothetical relationship between the settled sludge volume from the SVI test and required recirculation flow of activated sludge in an aeration process, Eqs. 11-13 and 11-15.

by 1 gram of settled suspended solids. In general, the range of 50 t
indicates a good settling sludge. The MLSS in conventional and ste
basins is nórmally held between 1500 and 2500 mg/l; high-rate proce
about 4000 mg/l.

$$\text{SVI} = \frac{V \times 1000}{\text{MLSS}} \qquad (11\text{-}13)$$

where   SVI = sludge volume index, milliliters per gram

   V = volume of settled solids in a 1-liter graduated
      cylinder after 30 min, milliliters per liter

   MLSS = mixed liquor suspended solids, milligrams per liter

   1000 = milligrams per gram

Sludge volume index can be hypothetically related to the quantity and solids
concentration in return activated sludge as is depicted in Figure 11-29. In the
following mathematical relationships, the final clarifier of the treatment system
is assumed to respond identically to the graduated cylinder used in the SVI test.
This assumption appears reasonable with respect to extended aeration and
conventional processes, while in high-rate systems there may be considerable
deviation in sludge settleability between an actual clarifier and that measured in
a laboratory container. In either case, the illustration gives a pictorial representation of process operation. Quantity of return sludge flow is keyed to settled
sludge volume by the relationship in Eq. 11-14. These ratios were derived by
noting in Figure 11-29 that recirculated sludge flow is to settled sludge volume
as the flow entering the clarifier relates to clarifier volume. This assumes that
sludge recirculation is exactly at the required rate. A greater quantity returns
clarified waste water unnecessarily, while a lesser flow leaves settled solids in
the clarifier and results in eventual loss in the process effluent. By rearranging
Eq. 11-14 to the form in Eq. 11-15, the theoretical required flow of recirculated
activated sludge can be calculated based on settled sludge volume. Concentration of suspended solids in the recirculated sludge can be computed from the
SVI by Eq. 11-16. This formula is useful in estimating the solids concentration
in excess activated sludge that is usually wasted from the return sludge line.

$$\frac{Q_R}{Q + Q_R} = \frac{V}{1000} \qquad (11\text{-}14)$$

$$Q_R = \frac{V \times Q}{1000 - V} \qquad (11\text{-}15)$$

where Q_R = flow of recirculated activated sludge,
      million gallons per day (or gpm)

   Q = average flow of waste water to aeration basin,
      million gallons per day (or gpm)

   V = volume of settled solids in 1-liter graduated cylinder,
      milliliters per liter

   1000 = milliliters per liter

$$\frac{\text{SS in recirculated}}{\text{activated sludge}} = \frac{1,000,000}{\text{SVI}} \qquad (11\text{-}16)$$

where $\qquad$ SS = suspended solids, milligrams per liter

$\qquad\qquad$ SVI = sludge volume index, milliliters per gram

EXAMPLE 11-8

The mixed liquor suspended solids concentration in the basin of a step aeration system is 2400 mg/l, and the sludge volume after settling in a 1-liter graduated cylinder is 220 ml. Compute the SVI. Is this value within the proper range? Based on the hypothetical relationships in Figure 11-29, what is the required return sludge ratio and suspended solids concentration in the recirculated sludge?

Solution

Applying Eq. 11-13,

$$SVI = \frac{220 \text{ ml/l} \times 1000}{2400 \text{ mg/l}} = 92 \text{ ml/g} \ (<150 \text{ OK})$$

By Eq. 11-15,

$$\frac{Q_R}{Q} = \frac{220}{1000-220} = 0.28 \qquad \text{or 28 percent}$$

From Eq. 11-16,

$$SS = \frac{1,000,000}{92} = 11,000 \text{ mg/l} = 1.1 \text{ percent}$$

Extended Aeration

The most popular application of this process is in treating small flows from schools, subdivisions, trailer parks, and villages. Cross-sectional diagrams of two common designs are given in Figure 11-30. Aeration basins may be cast-in-place concrete or steel tanks fabricated in a factory. Continuous complete mixing is either by diffused air or mechanical aerators, and aeration periods are 24 to 36 hr. Because of these conditions, as well as low BOD loading, the biological process is very stable and can accept intermittent loads without upset. For example, a unit serving a school may receive waste water during a 10-hr period each day, for only five days a week.

Clarifiers for small plants are conservatively sized with low overflow rates, ranging from 200 to 600 gpd/sq ft, and long detention times. Sludge may be returned to the aeration chamber through a slot opening as in Figure 11-30a or by employing an air-lift pump. A slot return requires periodic cleaning to prevent plugging by settled solids. Although satisfactory performance can be achieved, returning settled solids by pumping provides a more positive process control. Sludge that floats to the surface of the sedimentation chamber is returned to the aeration tank by either hydraulic action or through a skimming device attached to the air-lift pump return.

There is usually no provision for wasting of excess activated sludge from small extended aeration plants. Instead, the mixed liquor is allowed to increase in solids concentration over a period of several months and then is discharged directly from the aeration basin. This is performed by allowing the suspended

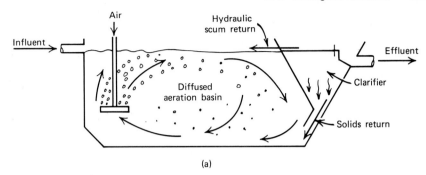

(a)

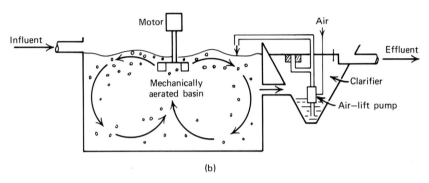

(b)

Figure 11-30 Cross-sectional diagrams illustrating the operation of typical small extended aeration plants. (*a*) Diffused aeration with a slotted-bottom clarifier for return of settled solids to the aeration basin. (*b*) Mechanical aeration with an air-lift pump for return of settled solids and scum to the aeration basin.

solids to settle in the tank with the aerators off, and then pumping the concentrated sludge from the bottom into a vehicle for hauling away. The MLSS operating range varies from a minimum of 1000 mg/l to a maximum of about 10,000 mg/l. In treating domestic waste water under normal loading, the mixed liquor concentration increases at the rate of approximately 50 mg/l suspended solids per day.

Larger extended aeration plants consisting of an aeration basin, clarifier, and aerobic digester (Figure 11-2*a*) are used by small municipalities. The basin may be a concrete tank with diffused aeration, a lined earth basin with mechanical aerators, or a race-track-shaped oxidation ditch. Final clarifiers are generally separate circular concrete tanks with mechanical sludge collectors. Aerobic digesters and sludge holding tanks are discussed in Section 11-13. In most cases, the BOD loading of these systems is at the upper limit of 30 lb BOD/1000 cu ft/day for extended aeration, and the aeration period is as low as 12 hr. The treatment plants, in addition to lighter loadings, are distinguished from conventional and step aeration processes by applying unsettled waste water directly to aeration without primary settling.

EXAMPLE 11-9

A small extended aeration plant without sludge wasting facilities is loaded at a rate of 10.5 lb BOD/1000 cu ft/day with an aeration period of 24 hr. The

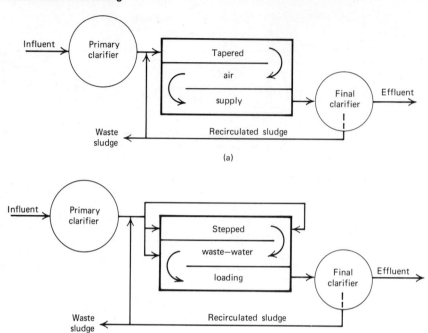

Figure 11-31 Flow schemes for conventional and step-aeration activated sludge processes. (*a*) Conventional tapered aeration. (*b*) Step aeration.

measured suspended solids buildup rate in the aeration tank is 30 mg/l/day. What percentage of the raw influent BOD is converted and retained as MLSS? If the MLSS concentration is allowed to increase from 1000 mg/l to 7000 mg/l before wasting solids, how long would this buildup take?

Solution

Since the aeration period is 24 hr,

$$\frac{\text{BOD load/day}}{\text{liter of tank}} = 10.5 \frac{\text{lb BOD/day}}{1000 \text{ cu ft}} \times \frac{\text{mg/l}}{62.4 \text{ lb/1,000,000 cu ft}} = 168 \text{ mg/l}$$

$$\frac{\text{MLSS buildup}}{\text{BOD applied}} = \frac{30 \text{ mg/l/day}}{168 \text{ mg/l/day}} \times 100 = 18 \text{ percent}$$

$$\text{Buildup time} = \frac{7000 \text{ mg/l} - 1000 \text{ mg/l}}{30 \text{ mg/l}} = 200 \text{ days}$$

Conventional and Step Aeration

These processes are similar to the earliest activated sludge systems that were constructed for secondary treatment of municipal waste water. As is diagramed in Figure 11-31, the aeration basin is a long rectangular tank with air diffusers along one side for oxygenation and spiral flow mixing. In a conventional basin, the air supply is tapered along the length of the tank to provide greatest aeration at the head end where raw waste water and return activated sludge are introduced. Air is provided uniformly in step aeration while waste water is introduced at intervals, or steps, along the first portion of the tank. Fine-bubble air diffusers are set at a depth of 8 ft or more to provide deep mixing and

Figure 11-32 Picture of a conventional diffused aeration tank. Air headers on the right side with diffusers attached are retracted for inspection.

adequate oxygen transfer. The air header is connected to a jointed arm so that diffusers can be swung out of the tank for cleaning and maintenance (Figure 11-32). A variety of diffusers are marketed; two popular types are synthetic cloth tubes, easily removed for laundering, and diffuser nozzles that can be detached from the air header pipe.

Plug-flow pattern of long rectangular tanks produces an oscillating biological growth pattern. The relatively high food-to-microorganism ratio at the head of the tank decreases as mixed liquor flows through the aeration basin. Since the aeration period is 6 to 8 hr, and can be considerably greater during low flow, the microorganisms move into the endogenous growth phase before their return to the head of the aeration basin. This weak starving microbial population must quickly adapt to a renewed supply of waste organics. The process has few problems of instability where waste-water flows are greater than 0.5 mgd; however, because of wide hourly variations in waste loads from small cities, the conventional plug-flow system can experience serious problems of biological instability. This phenomenon was a major factor contributing to the development of complete mixing aeration for handling small flows.

EXAMPLE 11-10

The following are average operating data from a conventional activated sludge secondary:

Waste-water flow = 7.7 mgd

Volume of aeration basins = 300,000 cu ft = 2.24 mil gal

Influent total solids = 599 mg/l

Influent suspended solids = 80 mg/l

Influent BOD = 173 mg/l

Effluent total solids = 497 mg/l

Effluent suspended solids = 12 mg/l

Effluent BOD = 10 mg/l

Mixed liquor suspended solids = 2500 mg/l

Recirculated sludge flow = 2.7 mgd

Waste sludge quantity = 54,000 gal/day

Suspended solids in waste sludge = 9800 mg/l

Based on these values calculate the following: aeration period; BOD load in lb BOD/1000 cu ft/day; F/M ratio in lb BOD/day/lb MLSS; total solids, suspended solids, and BOD removal efficiencies; BOD sludge age; suspended solids sludge age; and return sludge rate.

Solution

$$t = \frac{2.24 \text{ mil gal}}{7.7 \text{ mgd}} \times 24 = 7.0 \text{ hr}$$

$$\text{BOD load} = \frac{7.7 \text{ mgd} \times 173 \text{ mg/l} \times 8.34}{300 \text{ thou cu ft}} = 37.1 \frac{\text{lb BOD/day}}{1000 \text{ cu ft}}$$

$$\frac{F}{M} = \frac{7.7 \text{ mgd} \times 173 \text{ mg/l} \times 8.34}{2.24 \text{ mil gal} \times 2500 \text{ mg/l} \times 8.34} = 0.24 \frac{\text{lb BOD/day}}{\text{lb MLSS}}$$

$$\text{Total solids efficiency} = \frac{599 - 497}{599} \times 100 = 17 \text{ percent}$$

$$\text{Suspended solids efficiency} = \frac{80 - 12}{80} \times 100 = 85 \text{ percent}$$

$$\text{BOD efficiency} = \frac{173 - 10}{173} \times 100 = 94 \text{ percent}$$

$$\text{BOD sludge age} = \frac{M}{F} = \frac{1}{0.24} = 4.2 \text{ days}$$

Suspended solids in effluent = 7.7 mgd × 12 mg/l × 8.34 = 770 lb/day

Suspended solids in waste sludge = 0.054 mgd × 9800 mg/l × 8.34 = 4400 lb/day

$$\text{Suspended solids sludge age} = \frac{2.24 \text{ mil gal} \times 2500 \text{ mg/l} \times 8.34}{770 \text{ lb/day} + 4400 \text{ lb/day}} = 9.0 \text{ days}$$

$$\text{Return sludge rate} = \frac{2.7 \text{ mgd}}{7.7 \text{ mgd}} \times 100 = 35 \text{ percent}$$

EXAMPLE 11-11

Estimate the design flow and BOD loading for a step aeration activated-sludge plant based on the following data:

Primary clarifiers: A = 30,000 sq ft, depth = 8.0 ft

Aeration basins: V = 4.50 mil gal = 602,000 cu ft

Aeration capacity: V = 25,000 cu ft air per minute + standby compressor

Final clarifiers: A = 22,500 sq ft, depth = 8.0 ft

Solution

Final clarifier overflow rate from Table 11-1 is 800 gpd/sq ft

$$\text{Design flow} = 22{,}500 \text{ sq ft} \times 800 \, \frac{\text{gpd}}{\text{sq ft}} = 18{,}000{,}000 \text{ gpd} \; (> 1.5 \text{ mgd OK})$$

Check overflow of primary settling tanks

$$V_0 = \frac{18{,}000{,}000 \text{ gpd}}{30{,}000 \text{ sq ft}} = 600 \text{ gpd/sq ft} \; (< 800 \text{ is approved})$$

Check aeration period

$$t = \frac{4.50 \text{ mil gal}}{18.0 \text{ mgd}} \times 24 = 6.0 \text{ hr} \; (6.0 \text{ to } 7.5 \text{ is common})$$

Assume that the design criterion for aeration capacity was 1500 cu ft/lb of BOD applied to the aeration basin, and a BOD removal of 35 percent in primary settling. Then,

$$\text{Design BOD loading} = \frac{1}{0.65} \, 25{,}000 \, \frac{\text{cu ft}}{\text{min}} \times 1440 \, \frac{\text{min}}{\text{day}} \times \frac{\text{lb BOD}}{1500 \text{ cu ft}}$$

$$= 37{,}000 \text{ lb BOD/day}$$

Check BOD load on aeration basins

$$\text{BOD load} = \frac{0.65 \times 37{,}000 \text{ lb/day}}{602 \text{ thou cu ft}} = 40 \, \frac{\text{lb BOD/day}}{1000 \text{ cu ft}} \; (30 \text{ to } 40 \text{ is common})$$

$$\text{BOD concentration} = \frac{37{,}000 \text{ lb/day}}{18.0 \text{ mgd} \times 8.34} = 247 \text{ mg/l}$$

Contact Stabilization

The sequence of operations in this process are: aeration of raw waste water with return activated sludge, sedimentation to yield a clarified effluent, and reaeration of the clarifier underflow with a portion wasted to an aerobic digester. Supernatant drawn from the digester is returned to the process influent. The raw waste-water aeration chamber, also referred to as the contact zone, is approximately one third of the total aeration volume. Therefore, the aeration chamber has a detention time of 2 to 3 hr while reaeration is a 4- to 6-hr period. Normal operating sludge recirculation is 100 percent. Aerated compartments may be rectangular to simulate plug flow, or complete mixing.

The sequence of aeration-clarification-reaeration in processing large waste flows is neither as economical nor as efficient in BOD reduction as conventional or step aeration. Hence, current use is largely limited to factory-built field-erected plants capable of handling 0.05 to 0.5 mgd. A typical unit (Figure 11-33) consists of two concentric circular tanks about 14 ft deep with the inner shell 15 to 30 ft in diameter and the outer tank 30 to 70 ft across. The doughnut-shaped space between the two walls is segmented into three chambers for aeration, reaeration, and aerobic digestion. The central tank serves as a clarifier.

High Rate Aeration

The primary motivation in developing the high rate process was to reduce the cost of construction by increasing BOD load per unit volume of aeration tank,

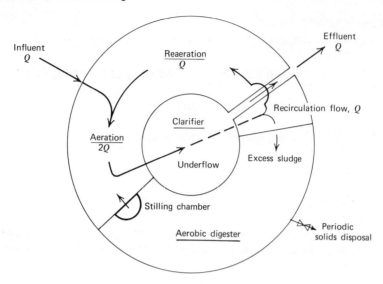

Figure 11-33 Diagram of a typical factory-built field-erected contact stabilization plant.

with a corresponding decrease in aeration period. This was achieved by operating at a higher food-to-microorganism ratio and reduced sludge age, while increasing the mixed liquor suspended solids carried in the basin to 4000 or 5000 mg/l. Problems in oxygen transfer and sludge settleability were rectified by modifying processing equipment. Complete mixing was necessary to distribute shock loads throughout the tank contents and maintain a homogeneous mixed liquor for efficient oxygen transfer. The unit illustrated in Figure 11-34 uses a combination of compressed air aeration and mechanical mixing. Air is introduced through holes in a sparge ring located at the base of the mixer

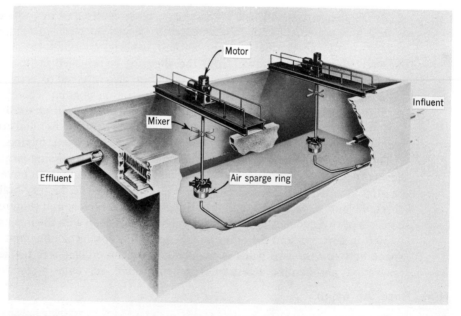

Figure 11-34 High-rate aeration basin with a large-bubble air sparge ring that is mounted under dual mixing impellers. (Courtesy of Dorr-Oliver, Inc.)

shaft, and turbines churn the tank contents shearing the coarse bubbles. Hydraulic thickening action of a rapid sludge return clarifier is mandatory to offset decreased settleability of the biological floc. If a high-rate process is not pushed to excessive BOD loads, exceeding 80 to 100 lb BOD/1000 cu ft/day, or extremely shortened aeration periods, the features of complete mixing aeration and rapid sludge return clarification provide a stable efficient treatment process. Overloading of a high-rate system is generally reflected by insufficient aeration capacity to maintain adequate dissolved oxygen during peak loading periods and carryover of suspended floc in the clarifier effluent.

High Purity Oxygen

The major components of a system using high purity oxygen in lieu of air are: a gas generator, a specially compartmented aeration tank, final clarifier, pumps for recirculating activated sludge, and sludge disposal facilities. Oxygen supply is generated either by manufacturing liquid oxygen or by producing high purity gas by adsorption separation from air. Standard cryogenic air separation, involving liquefaction of air followed by fractional distillation to separate the major components of nitrogen and oxygen, is employed at large installations. For the majority of treatment plants, the simpler pressure-swing adsorption system is more efficient. Air is compressed in a vessel filled with a granular adsorbent that takes in carbon dioxide, water, and nitrogen gas under high pressure, leaving a gas relatively high in oxygen. The adsorber is regenerated by depressurizing and purging. Three vessels are installed for continuous flow of high purity oxygen, with two units alternating between operation and regeneration and the third as a standby or backup unit.

The aeration tank is divided into stages by means of baffles and covered with a gas-tight enclosure; thus, the liquid and gas phases flow concurrently through the sections (Figure 11-35). Raw waste water, return activated sludge, and oxygen gas under a slight pressure are introduced to the first stage. Oxygen may be mixed with the tank contents by recirculation through a hollow shaft to a rotating sparger device as illustrated in Figure 11-35. Or, a surface aerator may be installed at the top of the mixer turbine shaft to contact oxygen gas with the mixed liquor. Successive aeration chambers are connected to each other so that liquid flows through submerged ports, and head gases pass freely from

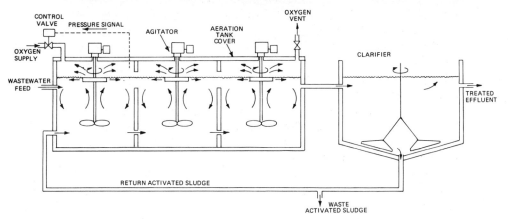

Figure 11-35 Schematic longitudinal section of a high-purity oxygen aeration system. (Courtesy of UNOX System, Union Carbide Corp.)

stage to stage with only a slight pressure drop. Exhausted waste gas is a mixture of carbon dioxide, nitrogen, and about 10 to 20 percent of the oxygen applied. Effluent mixed liquor is settled in either a scraper type or rapid sludge return clarifier, and activated sludge is recirculated to the aeration tank.

Several advantages are attributed to high-purity oxygen aeration compared with air systems. As is shown in Table 11-3, high efficiency is possible at increased BOD loads and reduced aeration periods. Disposal of waste sludge is easier because of the higher solids content and smaller quantity produced; the settled sludge generally contains 2 to 3 percent solids. Effective odor control is easier to achieve with oxygen aeration, covered tanks, and the reduced volume of exhaust gases.

Operation and Control

Aeration processes are regulated by the quantity of air supplied, rate of activated sludge recirculation, and the amount of sludge wasted which, in turn, controls the MLSS and F/M ratio. A properly designed system, treating domestic waste water, experiences few problems if the system is operated in a steady-state condition. Namely, the quantity of waste water applied each day is approximately the same, aeration mixing and dissolved oxygen concentration are relatively steady, and excess sludge is wasted continuously in quantities necessary to maintain a proper food-to-microorganism ratio. The latter may require reduced sludge wasting on weekends when the organic load on the plant is less.

Perhaps the most common problem associated with domestic waste treatment is overaeration resulting in a bulking activated sludge. When a plant is operated considerably under the design BOD load, nitrifying bacteria can convert ammonia to nitrate in the aeration basin. During subsequent detention in the final clarifier, the nitrate can be used as an oxygen source under anaerobic conditions, releasing nitrogen gas that buoys up biological floc. The best solution is to increase sludge wasting to deplete the nitrifying bacterial populations, and to decrease the air supply to reduce the dissolved oxygen concentration, provided that these control measures do not reduce the carbonaceous BOD removal efficiency.

Laboratory tests for monitoring activated sludge treatment are dissolved oxygen, MLSS concentration, SVI, and effluent BOD and suspended solids. Influent BOD and flow are needed to calculate organic loading, F/M ratio, and aeration period. Concentration of solids in the return activated sludge, effluent quality from the final clarifier, probing to determine depth of the sludge blanket in the clarifier, and SVI data yield information to establish the proper recirculation rate for optimum process efficiency and maximum solids concentration in the waste sludge.

Industrial wastes are frequently the source of operational instability of aeration processes. Shock loads of high strength waste water can deplete dissolved oxygen and unbalance the biological system. Toxic wastes in sufficient amounts interfere with microbial metabolism, and high carbohydrate wastes may cause nutient deficiency. All of these result in loss of MLSS in the clarifier overflow, thus decreasing treatment efficiency and shifting the F/M ratio by this unintentional wasting of microorganisms. Section 9-5 discusses techniques to evaluate the treatability of waste water. Hydraulic shock loads resulting from excessive inflow and infiltration to the sewer system can be just as detrimental as toxins or organic overload. High rates of overflow from a final clarifier, even for short time periods, can carry over a substantial portion of the viable activated sludge needed in the system. It may take several days to rebuild the density of mixed liquor and to return to a proper operating F/M ratio.

11-6   STABILIZATION PONDS

Stabilization ponds, also called lagoons or oxidation ponds, are generally employed as secondary treatment in rural areas. Although they serve only about 7 percent of the population, there are approximately 3500 lagoon installations in the United States with 90 percent in communities of less than 10,000 population. Ponds are classified as facultative, tertiary, aerated, and anaerobic according to the type of biological activity that takes place in them. Each of these kinds of ponds are discussed in the following paragraphs.

Facultative Ponds

These are the most common lagoons employed for stabilizing municipal waste water. The bacterial reactions include both aerobic and anaerobic decomposition and, hence, the term facultative pond. Figure 11-36 illustrates the basic biological activity. Waste organics in suspension are broken down by bacteria releasing nitrogen and phosphorus nutrients, and carbon dioxide. Algae use these inorganic compounds for growth, along with energy from sunlight, releasing oxygen to solution. Dissolved oxygen is in turn taken up by the bacteria, thus closing the symbiotic cycle. Oxygen is also introduced by reaeration through wind action. Settleable solids decomposed under anaerobic conditions on the bottom yield inorganic nutrients and odorous compounds, for instance, hydrogen sulfide and organic acids. The latter are generally oxidized in the aerobic surface water thus preventing their emission to the atmosphere.

Bacterial decomposition and algal growth are both severely retarded by cold temperature. During winter when pond water is only a few degrees above freezing, the entering waste organics accumulate in the frigid water. Microbial activity is further reduced by ice and snow cover that prevents sunlight penetration and wind reaeration. Under this environment, the water can become anaerobic causing odorous conditions during the spring thaw, until algae become reestablished. This may take several weeks depending on climatic conditions and the amount of waste organics accumulated during the cold weather.

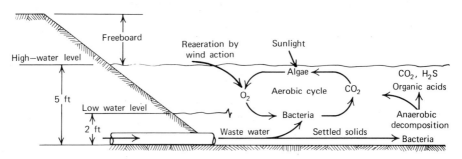

Figure 11-36 Schematic of a facultative stabilization pond showing the basic biological reactions of bacteria and algae. (Also refer to Figure 3-19.)

Operating water depths range from 2 to 5 ft, with 3 ft of dike freeboard above the high water level (Figure 11-36). The minimum 2-ft depth is needed to prevent growth of rooted aquatic weeds, but exceeding a depth of 5 ft may create excessive odors because of anaerobiosis on the bottom.

Typical construction is two or more shallow pools with flat bottoms enclosed by earth dikes (Figure 11-37). Waste water enters through an inlet division box, flows between cells by valved cross-connecting lines, and overflows through an outlet structure. Inlet lines, controlled by stop gates in the division box, discharge near the pond centers. Operating water depth is managed by a valving arrangement in the discharge structure. Connections between cells permit either parallel or series operation. Dikes are constructed with relatively flat side slopes to facilitate grass mowing and reduce slumping of the earth wall into the lagoon. Often the inside slopes along the waterline are protected by stone riprap to prevent erosion by wave action. If the soil is pervious, the pond bottoms should be sealed with bentonite clay, or lined with plastic, to prevent groundwater pollution. The area should be fenced to keep out livestock and discourage trespassing.

BOD loadings on stabilization ponds are expressed in terms of lb BOD applied per day per acre of water surface area, or sometimes as BOD equivalent population per acre. The maximum allowable loading is about 20 lb BOD per acre per day in northern states to minimize odor nuisance in the spring of the year. In those climates where ice coverage does not prevail, higher organic loadings may be used, for example, in the south and southwest a loading of 50 lb BOD per day per acre is practical. These loadings are 0.9 to 1.3 lb BOD per 1000 cu ft per day for a 5-ft water depth, which are substantially less than the volumetric loads applied to aeration and filtration units. Retention time of the waste water in lagoons is 3 to 6 months depending on applied load, depth of waste water, evaporation rate, and loss by seepage.

Lagoon treatment is regulated by the sequence of pond operation and water-level control. Series operation prevents short-circuiting and increases

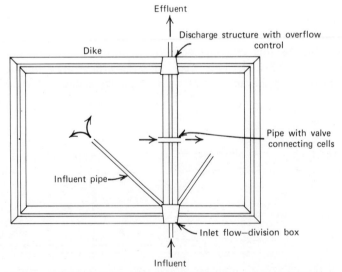

Figure 11-37 Typical plan view of two stabilization ponds that can be operated in either parallel or series. The latter flow pattern is shown by the arrows.

BOD reduction, but the load on the first cell is increased and may cause odor problems. Functioning in parallel distributes the raw BOD load but allows short-circuiting. Regulating lagoon discharge to a receiving stream can minimize water pollution. Pond water level may be slowly lowered in the fall and early winter when there is adequate dilution flow in the receiving stream. Discharge in the winter is then minimized or stopped, and influent waste water is stored until spring. Overflow is then permitted after the weather warms and the biological processes reduce the BOD concentration in the impounded water to an acceptable effluent level.

Facultative ponds treating only domestic waste water normally operate odor free except for a short period of time in the spring of the year. On the other hand, lagoons treating municipal waste water that include industrial wastes can produce persistent obnoxious odors. Often this is the result of organic overload from food-processing industries or a result of the odorous nature of the industrial waste itself or both. The best solution is to require pretreatment of the offending waste waters prior to discharge to the sewer system.

Meeting the effluent standard of 30 mg/l of suspended solids is a serious problem in lagoon treatment, since algae suspended in the water generally contribute 50 to 70 mg/l. In some instances, the effluent standard can be met by series operation and careful control of effluent discharge. Alternatives are land disposal or installation of an additional treatment unit to remove the suspended solids prior to disposal. Colloidal and particulate solids may be removed by a gravity filter similar to that used in water treatment when this is preceded by chemical flocculation, or it can be taken out by an upflow filter. The latter passes water up through the filter media at a controlled rate to capture a portion of the suspended solids. The most common land disposal system is spray irrigation (Section 14-6). In addition to solids separation, pond overflow may require chlorination to meet the effluent quality for fecal coliform bacteria.

In summary, facultative ponds are best suited for small towns that do not anticipate industrial expansion, and where extensive land area is available for construction and effluent disposal. The advantages of low initial cost and ease of operation, as compared to a mechanical plant, can be offset by operational difficulties. The key problems are poor assimilative capacity for industrial wastes, odorous emission, and meeting the minimum effluent standards for disposal in surface waters.

EXAMPLE 11-12

Stabilization ponds for a town of 3000 population are constructed as is shown in Figure 11-37 with the larger cell having 14 acres and the smaller 7 acres. The average daily waste-water flow is 0.24 mgd containing 450 lb of BOD; this is equivalent to 80 gpcd and 0.15 lb BOD per person per day, or 225 mg/l BOD. (a) For series operation, calculate the BOD loadings based on both the total area and the first cell only. (b) Estimate the number of days of winter storage available between the 2-ft and 5-ft water levels assuming an evaporation and seepage loss of 0.10 in. of water per day.

Solution

(a) BOD load on total pond area = 450/21 = 21.4 lb per day per acre

Equivalent population load = 21.4/0.17 = 125 persons per acre

BOD load on first cell $= 450/14 = 32.1$ lb per day per acre

(b) Storage volume

$= (5 \text{ ft} - 2 \text{ ft}) \times 21 \text{ ac} \times 43,560 \text{ sq ft per acre} = 2,740,000 \text{ cu ft}$

$= 20.5 \text{ mil gal}$

$$\text{Water loss} = \frac{0.10 \text{ in./day}}{12 \text{ in./ft}} \times 21 \text{ ac} \times 43,560 \frac{\text{sq ft}}{\text{acre}} \times \frac{7.48 \text{ gal}}{\text{cu ft}}$$

$$= 57,000 \text{ gpd} = 0.057 \text{ mgd}$$

$$\frac{\text{Storage time}}{\text{available}} = \frac{20.5 \text{ mil gal}}{0.24 \text{ mgd} - 0.057 \text{ mgd}} = 112 \text{ days}$$

Tertiary Ponds

These units, also referred to as maturation or polishing ponds, serve as third-stage processing of effluent from activated sludge or trickling filter secondary treatment. Stabilization by retention and surface aeration reduces suspended solids, BOD, fecal microorganisms, and ammonia. The water depth is generally limited to 2 or 3 ft for mixing and sunlight penetration. BOD loads are less than 15 lb BOD per acre per day, and detention times are relatively short at 10 to 15 days.

Aerated Lagoons

Complete-mixing aerated ponds, usually followed by facultative ponds, are used for first-stage treatment of municipal waste waters and for pretreatment of industrial wastes. The basins are 10 to 12 ft deep and are aerated with pier-mounted or floating mechanical units (Figure 11-38). The aerators are designed to provide mixing for suspension of microbial floc and to supply dissolved oxygen. The biological process does not include algae, and organic stabilization depends on the mixed liquor that develops within the basin, since there is no provision for settling and returning activated sludge. BOD removal is a function of aeration period, temperature, and nature of the waste water. Design aeration periods are normally in the range of three to eight days depending on the degree of treatment desired and waste-water temperature during the cold season of the year. For example, aerating a typical municipal waste water for five days at 20° C provides about 85 percent BOD reduction while lowering the temperature to 10° C reduces the efficiency to approximately 65 percent.

Problems of odors and low efficiency result when aerated lagoons are improperly designed or poorly operated. Thorough mixing and adequate dissolved oxygen insure odor-free operation. If the aeration equipment is inadequate, deposition of solids and reduced oxygenation can result in anaerobic decomposition that leads to foul odors. Pretreatment and control of industrial wastes are required, since large inputs of either biodegradable or toxic wastes can cause process upset. Infiltration entering the sewer collection system during wet weather can have a detrimental effect on an aerated lagoon by reducing the aeration period and flushing microbial floc out of the basin. Where infiltration is a problem, it is best to divert a portion of wet-weather flow around the aerated lagoon to the second-stage facultative ponds, thus preventing undesirable hydraulic loading on the aerated basins. In the winter, the

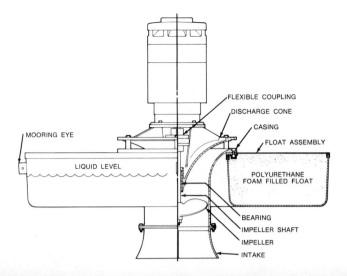

Figure 11-38 Cutaway section of a floating aerator and picture of a unit in operation. (Courtesy of Peabody Welles)

aerators should be adjusted and windbreaks should be set up to reduce cooling of the lagoon water.

A second type of aerated stabilization pond is illustrated in Figure 11-39. Compressed air is introduced through rows of plastic tubing strung across the pond bottom. Streams of bubbles rising to the surface provide vertical mixing and distribution of dissolved oxygen. The cells normally have 10 ft of liquid depth and operate in series with a total detention time of 25 to 35 days. Tube aeration systems have been most successful in locations where the pond surfaces are frozen for several months in the winter and an external air supply

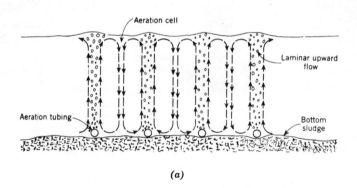

Aeration cell

Laminar upward flow

Aeration tubing

Bottom sludge

(a)

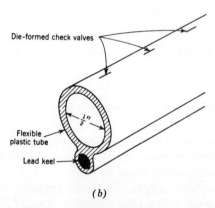

Die-formed check valves

Flexible plastic tube

Lead keel

$\frac{1}{2}''$

(b)

Figure 11-39 Stabilization pond aerated and mixed by compressed air emitting from plastic tubing laid on the bottom of the lagoon. (*a*) Schematic of aerating and mixing action. (*b*) Detail of aeration tube. (Courtesy of Hinde Engineering Co.)

is needed to maintain aerobic conditions. In this climate, the effluent of unaerated facultative ponds is generally unsatisfactory and surface aerators are impaired by ice formation.

Anaerobic Lagoons

Bacteria anaerobically decompose organic matter to gaseous end products of carbon dioxide and methane. In addition, intermediate odorous compounds, such as organic acids and hydrogen sulfide, are formed. The two primary advantages of anaerobic treatment compared with an aerobic process are the low production of waste biological sludge and no need for aeration equipment. The disadvantages are that incomplete stabilization requires a second-stage aerobic process and the relatively high temperature required for anaerobic decomposition. The important characteristics for a waste water to be amenable to anaerobic treatment are: high organic strength particularly in proteins and fats, relatively high temperature, freedom from toxic materials, and sufficient biological nutrients. Typical meat-processing waste water with a BOD of 1400 mg/l, grease content of 500 mg/l, temperature of 82° F, and neutral pH exhibits these characteristics. A schematic diagram of a first-stage anaerobic lagoon for treatment of slaughterhouse waste water in rural locations is given in Figure 11-40. Minimum pretreatment of the raw waste water includes: blood recovery for a salable by-product, screening to remove coarse solids (paunch manure), and skimming to reduce the grease content. The basins are constructed with steep side walls and a depth of 15 ft to minimize the surface area relative to total volume. This construction allows several inches of grease accumulation to form a natural cover for retaining heat, suppressing odors, and maintaining anaerobic conditions. Influent waste water enters near the bottom so that it mixes with the active microbial solids in the sludge blanket. The discharge pipe is located on the opposite end and is submerged below the grease cover. Upward flow of the discharge allows settling of the bacterial floc so that the anaerobic mixed liquor is retained in the lagoon. Sludge recirculation is not necessary, since gasification and the inlet-outlet flow pattern provides adequate mixing. Series operation is not recommended because it is difficult to maintain an adequate grease cover on a second-stage lagoon.

The normal operating standards to achieve a BOD removal efficiency of, at least, 75 percent are a loading of 20 lb BOD per 1000 cu ft per day, a minimum detention time of four days, and a minimum operating temperature of 75° F. The most common operating problems result from reduced temperature in the liquid caused by an insufficient cover of grease for thermal insulation and protection from wind mixing. Too low a BOD loading and overly efficient pretreatment can result in inadequate grease input to build up and maintain a

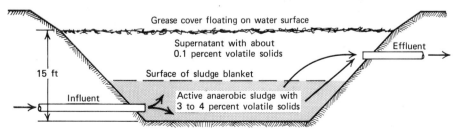

Figure 11-40 Schematic of an anaerobic lagoon for treating meat-processing waste water.

cover layer. Anaerobic lagoons do not create serious odor problems when operating properly, that is, when there is complete anaerobiosis and adequate grease cover. One exception is when the water supply to the industry is high in sulfate ion which is reduced in the anaerobic environment and emitted as hydrogen sulfide.

Success of anaerobic lagoons in handling meat-processing wastes should not be translated to municipal applications. Domestic waste water with its relatively low BOD and grease content and cool temperature is simply not suitable to develop the system in Figure 11-40. Even in instances where municipal waste water has a high industrial waste content, application of first-stage anaerobic treatment has led to serious odor problems. This can generally be attributed to the fact that the lagoons are facultative cells instead of being strictly anaerobic.

11-7   DISINFECTION

Although other bactericidal agents may be used to disinfect waste water, chlorine is the only one that has widespread application. The purpose for chlorinating waste-water effluents is to protect public health by inactivating pathogenic organisms including enteric bacteria, viruses, and protozoan cysts. Prior to establishing rigid bacteriological requirements, disinfection of waste water was defined simply as the addition of sufficient chlorine so that a residual of 0.5 mg/l existed after 15 min. Now, satisfactory biological quality of a secondary effluent is defined by an average fecal coliform count of less than 200/100 ml. The number of coliform organisms in effluents from trickling filters, activated sludge units, and stabilization ponds are in the range of 100,000 to 10,000,000/100 ml. Safety of effluent disposal by dilution in surface waters after chlorination is based on the argument that reduction of fecal coliforms from 10^6/100 ml to 200/100 ml eliminates the great majority of bacterial pathogens and inactivates large numbers of enteric viruses.

A properly designed system provides rapid initial mixing of chlorine solution with the waste water, contact time in a plug flow basin for, at least, 30 min at peak flow, and automatic chlorine residual control. Chlorine solution should be applied either in a pressure conduit under conditions of highly turbulent flow, or in a channel immediately upstream from a mechanical mixer. Adding the solution to an open channel results in very little dispersion, and in poor chlorination efficiency because the flow is usually stratified. Following initial blending, the most efficient basis for disinfection is a baffled contact chamber that approaches plug flow conditions. Circular and rectangular tanks that allow backmixing are not as effective, and the design criterion of detention time has little meaning.

Precise measurements and control of chlorine residual are extremely important for efficient and economical operation. An automatic residual monitoring and feedback control is essential to prevent low concentrations and inadequate disinfection, as well as excessive chlorination resulting in discharge of an effluent toxic to aquatic life. Existing installations may require structural modifications to insure proper disinfection without discharging waste water with harmful chlorine residuals. Some regulatory agencies have specified maximum chlorine residuals in undiluted effluent of 0.1 to 0.5 mg/l to prevent potential toxicity in the receiving stream. If it is poorly designed, dechlorination may be required to detoxify the discharge after disinfection has been achieved. The least expensive and most effective method is by adding sulfur dioxide; for small flows, sodium metabisulfite may be considered as an alternative.

Chlorine dosage needed for disinfection depends on waste-water pH, presence of interfering substances, temperature, and contact time. Applications of 8 to 15 mg/l provide adequate disinfection in well-designed units with a minimum contact time of 20 to 30 min. (Chlorine chemistry and chlorination equipment are discussed in Section 7-6.)

11-8 INDIVIDUAL HOUSEHOLD DISPOSAL SYSTEMS

Approximately one fourth of the homes in the United States, comprising about 70,000,000 people, are located in unsewered areas and must rely on separate treatment. The quantity of waste water is from 60 to 100 gpcd with BOD concentrations varying from 200 to 400 mg/l. The number of bathrooms, automatic washing machines, and garbage grinders, as well as the number of residents, define the magnitude and character of the waste.

The most popular system is the septic tank and absorption field because of their low cost and the desirable characteristic of underground disposal of effluent (Figure 11-41). A septic tank is an underground concrete box sized for a detention time of approximately two days. With garbage grinders and automatic washers, the recommended minimum capacity is 750 gal for a two-bedroom house, 900 gal for three bedrooms, 1000 for four bedrooms, and 250 for each additional bedroom. Inspection and cleaning ports must be accessible for maintenance, usually by removing about 1 ft of earth cover. Inlet and outlet pipe tees prevent clogging of the drains with scum that accumulates on the liquid surface. The functions of a septic tank are settling of solids, flotation of grease, anaerobic decomposition of accumulated organic matter, and storage of sludge. Retaining large solids is essential to prevent plugging of the percolation field.

An absorption field, where the majority of the biological stabilization takes place, consists of looped or lateral trenches 18- to 24-in. wide and, at least, 18 in. deep. Drain tile or perforated pipe in an envelope of gravel are used to distribute the waste water uniformly over the trench bottom. Organics are decomposed in the aerobic-facultative environment of the bed, and water seeps down into the soil profile. Air enters the drain tile through the backfill covering

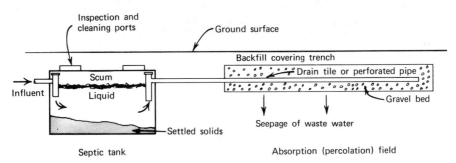

Figure 11-41 A typical septic tank-absorption field system for disposal of household waste water. The tank provides for sedimentation, scum accumulation, anaerobic destruction of organic matter, and sludge storage. The absorption field permits aerobic-facultative decomposition of organics and seepage of the waste water.

of the trenches and by ventilation through the house plumbing stack. The percolation area required depends directly on soil permeability, and for a four-bedroom dwelling the area needed ranges from 300 to 1300 sq ft. Trench area for a particular location is determined by subsurface soil exploration and percolation tests. The standard design guide for septic tank systems is *The Manual of Septic Tank Practice*,* although most state environmental control agencies and county health departments have guidelines for installations which are based on local conditions.

The most frequent complaints in operation of septic tank-absorption field systems concern plumbing stoppages and odorous seepage through the ground surface. Tanks fill with nonbiodegradable solids and must be pumped out every few years. When cleaning them, it is best to leave a small amount of the black-colored digesting sludge in the tank to insure adequate bacterial seeding to continue solids digestion. Introduction of chemical or enzymatic conditioners have not been shown to be of any significant value in reviving or increasing bacterial activity in a septic tank. Attempting to delay cleaning by adding a large quantity of caustic chemical can be harmful to the disposal system. It disrupts bacterial decomposition in the tank and "cleans" the unit by discharging suspended solids into the drain tile, which promotes clogging of the percolation field.

Proper functioning of the absorption field relies on leaching of the waste water into the soil profile and adequate aeration of the bed. Obstruction of percolation may result for any of the following reasons: construction in clayey soils of low permeability that are simply inadequate for placement of a seepage system, high water table conditions that saturate the soil profile during the wet weather season, inadequate design resulting in hydraulic and organic overloading, and improper operation of the septic tank. In attempting to solve the problem of leaching fields in unsuitable soils, small aeration treatment units have been developed in recent years.

Individual household aeration units are manufactured in many different shapes and sizes using concrete, steel, or fiber glass tanks. The cross-section of a typical compartmented aeration tank is shown in Figure 11-42a. Some designs have a presedimentation chamber that performs similarly to a septic tank. The aeration portion is sized for a 24- to 48-hr retention period, and is mixed with either compressed air or some type of mechanical aerator. A final settling chamber is for separation and return of activated sludge. Although these systems have the appearance of an extended aeration process, in reality, they seem to function more like an aerated stabilization pond. The main reason is that gravity return of solids is not effective in maintaining an adequate mixed liquor solids concentration, particularly with the surging influent caused by high rates of discharge from household appliances. Consequently, the aeration unit is not efficient in solids capture or retention, which leads to inefficient treatment and a poor quality effluent.

Several methods have been proposed for the disposal of aeration effluent (Figure 11-42b). An underground absorption field is acceptable, but then there is no purpose in installing a more expensive aeration tank when an ordinary septic tank produces an equally suitable effluent for leaching. Evaporation-transpiration beds may be satisfactory with proper climatic conditions;

* Public Health Service Publication 526, U.S. Department of Health, Education, and Welfare.

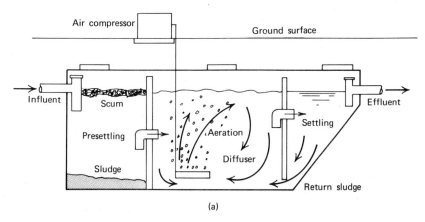

(a)

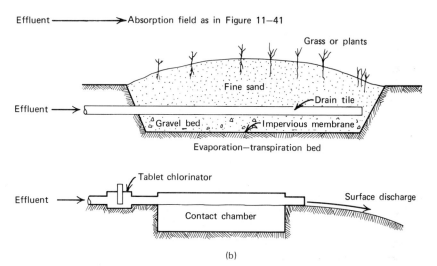

(b)

Figure 11-42 Schematics of an aeration system for treating waste water from an individual household and alternatives suggested for disposal of the process effluent. (a) Compartmented aeration tank. (b) Alternative methods proposed for disposal of effluent from the aeration unit above.

however, the bed must not be subjected to freezing. Furthermore, evapotranspiration must be predictable, and there is always the potential problem of salt buildup.

The main motivation for development of a miniature aeration system was an attempt to permit surface effluent discharge and, thus, to avoid the problems associated with absorption fields. But this raises the question of disinfection to meet the required bacteriological quality. The simplest system would be a chlorine contact chamber equipped with a tablet-type chlorinator. One such unit consists of a slotted tube containing solid tablets of calcium hypochlorite suspended in the waste-water flow such that a suitable amount of chlorine is dissolved for disinfection. Control problems include potential overchlorination during low flow periods and underchlorinating at peak flows, and the possibility that homeowners may not be conscientious in replacing hypochlorite tablets because of the inconvenience and cost. Also, a poorly operating aeration tank discharges excessive suspended solids that prevent effective disinfection. Little evidence exists to substantiate the idea that effluent standards of 30 mg/l BOD and 30 mg/l suspended solids can be met by small compartmented aeration tanks.

In summary, septic tanks are economical and practical for the disposal of household wastes in areas with acceptable soils for percolation. Where leaching fields are impossible, rather extensive treatment is required to meet established quality standards for surface discharge. Few of the proposed aeration-chlorination systems appear to provide realistic performance. The labor and cost of operating, maintaining, and monitoring to insure satisfactory performance may approach a level of economic infeasibility.

11-9   CHARACTERISTICS AND QUANTITIES OF WASTE SLUDGES

The purpose of waste-water settling and biological aeration is to remove organic matter and to concentrate it in a much smaller volume for ease of handling and disposal. As is illustrated in Figure 11-1, the solids in 100 gal of waste water are consolidated into approximately 2 qt of liquid sludge; this amounts to about 5000 gal of slurry per 1.0 mil gal of water treated. The cost of facilities for stabilizing, dewatering, and disposal of this concentrate is about one third of the total investment in a treatment plant. Operating expenses in sludge handling may amount to an even larger fraction of the total plant operating costs depending on the system employed. For these reasons, it is essential to have a properly designed and efficiently run sludge disposal system.

The quantity and nature of sludge generated relates to the character of the raw waste water and processing units employed. Primary settling produces an anaerobic sludge of raw organics that are being actively decomposed by bacteria. Therefore, these solids must be handled properly to prevent emission of obnoxious odors. In comparison with secondary biological waste, primary sludges thicken and dewater readily because of their fiberous and coarse nature. The following formula can be used to estimate the raw solids that are removed by plain sedimentation.

$$W_p = f \times SS \times Q \times 8.34 \qquad (11\text{-}17)$$

where W_p = raw primary sludge solids, pounds of dry weight per day

f = fraction of suspended solids removed in primary settling (f is about 0.5 for domestic waste water)

SS = suspended solids in unsettled waste water, milligrams per liter

Q = daily waste-water flow, million gallons per day

8.34 = pounds per million gallons per milligram per liter

Waste from aeration is flocculated microbial growths with entrained non-biodegradable suspended and colloidal solids. It is relatively odor-free because of biological oxidation, but the finely divided and dispersed particles make it difficult to dewater. Excess activated-sludge solids from aeration processes and humus from biological filtration can be estimated by Eq. 11-18, which relates solids production to BOD load. Coefficient K depends on the process food-to-microorganism ratio shown in Figure 11-43. Although this formulation is

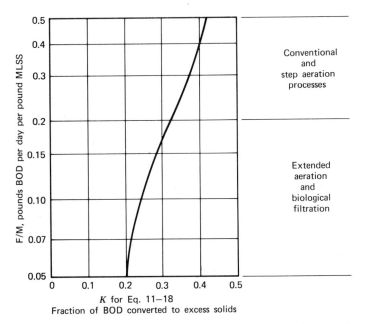

Figure 11-43 Hypothetical relationship between the food-to-microorganism ratio and coefficient K in Eq. 11-18 for calculating excess activated sludge production in biological aeration of waste water, based on an effluent BOD of approximately 30 mg/l.

reasonable for domestic waste, calculated values may differ considerably from real sludge yields when treating municipal discharge that contains a substantial portion of industrial waste.

$$W_s = K \times BOD \times Q \times 8.34 \qquad (11\text{-}18)$$

where W_s = biological sludge solids, pounds of dry weight per day

K = fraction of applied BOD that appears as excess biological solids from Figure 11-43 assuming about 30 mg/l BOD remaining in the process effluent

BOD = concentration in applied waste water, milligrams per liter

Q = daily waste-water flow, million gallons per day

8.34 = pounds per million gallons per milligram per liter

Design and operation of a sludge disposal system are based on volume of the wet sludge as well as the dry solids content. Once the dry weight of solids has been determined from Eqs. 11-17 and 11-18, the volume of wet sludge can be calculated using Eq. 11-19 by knowing the percentage of solids, or water content. This formula assumes a specific gravity for the wet sludge of 1.0, which is sufficiently accurate for normal computations. For example, a slurry with 10 percent organics has a specific gravity of about 1.02.

$$V = \frac{W}{\left(\dfrac{s}{100}\right)8.34} = \frac{W}{\left(\dfrac{100-p}{100}\right)8.34} \qquad (11\text{-}19)$$

where V = volume of wet sludge, gallons

W = weight of dry solids, pounds

s = solids content, percent

p = water content, percent

Typical solids concentrations in waste sludges are: raw primary sludge only, 6 to 8 percent; primary plus humus from secondary filtration, 4 to 6 percent; primary plus waste activated sludge from secondary aeration, 3 to 4 percent; and excess activated sludge only, 0.5 to 2.0 percent. In all of these wastes, the percentage of volatile solids is about 70 percent of the total dry solids. For a given quantity of dry solids, the percentage concentration dramatically influences the sludge volume. Doubling the solids concentration reduces the wet volume by one half, conversely, the amount would double if the slurry were diluted to one half the original percentage of solids. For example, using Eq. 11-19, the volume of wet sludge containing 10 lb of dry solids at 2 percent concentration is 60 gal. If thickened to 4 percent solids, the volume becomes 30 gal, and a further concentration to 8 percent reduces the quantity to 15 gal. With this concept in mind, it is easier to understand why dilute secondary sludges are more difficult to process and handle than primary wastes.

EXAMPLE 11-13

A domestic waste water with 200 mg/l BOD and 220 mg/l of suspended solids is processed in a trickling filter plant. (a) Using Eqs. 11-17 and 11-18, estimate the quantities of sludge solids from primary settling and secondary biological filtration per million gallons of waste water treated.
(b) How many pounds of dry solids are produced relative to the BOD equivalent population load on the plant?
(c) What is the volume of wet sludge per million gallons of waste water if underflow from the final clarifiers returns filter humus to the plant influent, and the combined sludge drawn from the primaries has a solids content of 5.0 percent.

Solution

(a) Assuming 50 percent suspended solids reduction from settling the raw waste water, and substituting into Eq. 11-17

$$W_p = 0.50 \times 220 \times 1.0 \times 8.34 = 916 \frac{\text{lb dry solids}}{\text{mil gal}}$$

From Eq. 11-18, assuming 35 percent BOD removal in the primary and an estimated K of 0.24 from Figure 11-43,

$$W_s = 0.24 \times 0.65 \times 200 \times 1.0 \times 8.34 = 260 \frac{\text{lb dry solids}}{\text{mil gal}}$$

(b) For 1.0 mgd with a BOD of 200 mg/l,

$$\frac{\text{BOD equivalent}}{\text{population}} = \frac{1.0 \times 200 \times 8.34}{0.17} = 9800 \text{ persons}$$

$$\frac{\text{Sludge production}}{\text{per capita}} = \frac{W_p + W_s}{\text{population}} = \frac{916 + 260}{9800} = 0.12 \frac{\text{lb dry solids}}{\text{person}}$$

(*Note.* The value of 0.20 lb of dry solids per person per day has often been used as a conservative estimate of solids yield in trickling filter plants treating domestic waste water.)

(c) Substituting into Eq. 11-19 with $W = W_p + W_s$

$$V = \frac{916 + 260}{\left(\frac{5.0}{100}\right)8.34} = 2800 \frac{\text{gal of sludge}}{\text{mil gal of waste water}}$$

11-10 SELECTION AND ARRANGEMENT OF PROCESSES

Techniques for processing waste sludges depend on the type, size, and location of the waste-water plant, unit operations employed in treatment, and the method of ultimate solids disposal. The system selected must be able to receive the sludge produced and economically to convert it to a product that is environmentally acceptable for disposal. Popular methods are listed in Table 11-4.

Cost of chemicals for sludge conditioning and optimum use of biological digesters is directly related to the concentration of solids in waste sludge. As a general rule, the solids content must be, at least, 4 percent for feasible dewatering. Secondary biological sludge from aeration processes is thin and dilutes the more concentrated primary waste, and blending the two is likely to produce a sludge with less than 4 percent solids. Therefore, gravity thickening is often applied to mixtures of primary and secondary discharges, and dissolved air flotation may be employed for concentrating waste activated sludge.

ImporTanT

Table 11-4. Common Methods for Handling, Processing, and Disposing of Waste Sludge

Storage Prior to Processing
    In the primary clarifiers
    Separate holding tanks
Thickening Prior to Dewatering or Digestion
    Gravity settling
    Dissolved air flotation
Conditioning Prior to Dewatering
    Chemical treatment
    Stabilization by anaerobic digestion
    Stabilization by aeration
Dewatering
    Vacuum filtration
    Pressure filtration
    Centrifugation
    Drying beds or lagoons
Solids Disposal
    Burial in landfill
    Incineration
    Spreading on farmland
    Production of soil conditioner

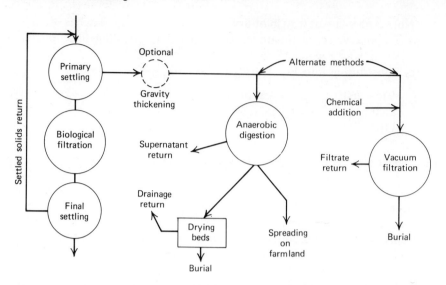

Figure 11-44 Typical unit operations for processing waste sludge from filtration plants.

Typical unit operations selected for processing sludge from filtration plants are diagramed in Figure 11-44. Humus from intermediate or final clarifiers is returned with recirculated flow to the head of the plant for separation in the primary. Underflow from the clarifiers may be gravity thickened, although, this is not common, since the solids concentration averages an acceptable 5 percent. Anaerobic digestion is a popular method for stabilizing raw sludge. Liquid digested sludge may be applied to farmland or may be dried on sand beds from which the solids are later removed for burial. In larger installations, raw solids are trucked to landfill after chemical treatment and dewatering by vacuum filtration.

The scheme for filter plants is not optimum for activated sludge systems. Most aeration plants designed for returning secondary waste sludge to the primary tank have operating problems that relate both to upset of primary clarification as a result of biological activity in the settled sludge, and high chemical costs resulting from dewatering thin sludge. Modern design blends the two wastes in a holding tank separate from the primary clarifier and concentrates the slurry prior to processing (Figure 11-45). Waste activated sludge may be thickened separately by air flotation, or the slurry of primary and secondary wastes may be gravity thickened after mixing.

EXAMPLE 11-14

A conventional aeration plant treats 7.7 mgd of municipal waste water with a BOD of 240 mg/l and suspended solids of 200 mg/l. The sludge flow pattern is as shown in Figure 11-45 with a flotation thickener to concentrate the excess biological floc, and a holding tank to blend primary and secondary wastes. Estimate the quantities and solids contents of the primary, secondary, and mixed sludges. Assume the following: 50 percent SS removal and 35 percent BOD reduction in the primary, a raw sludge with 94.0 percent water content, an operating F/M ratio of one third in the aeration basin, a solids concentration of

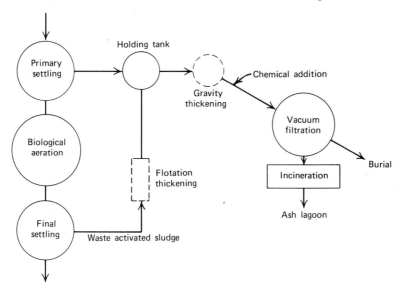

Figure 11-45 Possible alternatives to handling waste sludges from activated sludge treatment.

15,000 mg/l (1.5%) in the waste activated, and 3.5 percent solids in flotation thickener discharge.

Solution

Applying Eqs. 11-17 and 11-19 for the primary sludge

$$W_p = 0.50 \times 200 \text{ mg/l} \times 7.7 \text{ mgd} \times 8.34 = 6410 \text{ lb/day}$$

$$V_p = \frac{6410}{\left(\dfrac{100-94}{100}\right)8.34} = 12,800 \text{ gal/day}$$

From Figure 11-43, for a F/M ratio of 1/3, the K is 0.38. Substituting in Eqs. 11-18 and 11-19

$$W_s = 0.38 \times 0.65 \times 240 \text{ mg/l} \times 7.7 \text{ mgd} \times 8.34 = 3800 \text{ lb/day}$$

$$V_s = \frac{3800}{(1.5/100)8.34} = 30,400 \text{ gal/day}$$

After flotation thickening, assuming 100 percent solids capture, the waste activated sludge volume is

$$V_s = \frac{3800}{(3.5/100)8.34} = 13,000 \text{ gal/day}$$

The blended sludge, after mixing primary and thickened secondary, has the following characteristics.

$$W = W_p + W_s = 6410 + 3800 = 10,200 \text{ lb/day}$$

$$V = V_p + V_s = 12,800 + 13,000 = 25,800 \text{ gal/day}$$

$$s = \frac{100 \times W}{8.34 \times V} = \frac{100 \times 10,200}{8.34 \times 25,800} = 4.7 \text{ percent}$$

11-11   THICKENING OF WASTE SLUDGES

Mechanical thickening is performed in circular settling tanks that are equipped with scraper arms having vertical pickets. The process is described on page 272 and a typical unit is illustrated in Figure 7-29. Waste slurry withdrawn from primary clarifiers or sludge holding tanks is applied to the gravity thickener through a central inlet well. Overflow containing the non-settleable fraction is returned to the wet well for reprocessing, while the concentrate is drawn from the tank bottom for dewatering and ultimate disposal. Unit loadings are normally in the range of 6 to 12 lb of solids per square foot of tank bottom per day, or are expressed as overflow rates in the range of 400 to 800 gpd/sq ft. In handling domestic wastes, the underflow solids concentration is often double that of the applied sludge. For example, expected underflow solids content would be 6 percent for a mixture of primary and activated sludge, and 8 percent for primary plus filter humus. Precise performance of a thickener is difficult to predict because of the varying character of waste sludges. Coagulant chemicals may be used to improve capture of suspended solids and density of the thickened waste.

Dissolved air flotation is achieved by releasing fine air bubbles that attach to sludge particles and cause them to float. Figure 11-46 is a flow diagram for a typical unit. Waste activated sludge enters the bottom of the flotation tank where it is merged with recirculated flow that contains compressed air. The latter is a portion of the clarified effluent that has been held in a separate retention tank under an air pressure of approximately 60 psi. On pressure release, air dissolved in the recirculated-pressurized flow forms fine bubbles that agglomerate with the suspended solids. Process underflow is returned to waste-water treatment, and the overflow, discharged by a mechanical skimming device, is the thickened sludge. Flotation thickeners are used primarily with waste activated sludge and normally produce a concentrate of approximately 4 percent solids with a total solids recovery of 85 percent. Polyelectrolytes or other flocculents may be added to increase capture of solids to about 95 percent (Figure 11-47). Chemical treatment, however, may not increase the concentration of the thickened sludge. Practical loading rates for waste-activated sludge appear to be in the range of 2 to 4 lb solids per square foot of flotation area per hour. Since the characteristics of biological floc are not consistent, even in the same treatment plant, the response to chemical flotation aids and solids loadings must be determined in each individual case.

11-12   ANAEROBIC DIGESTION

The purpose of sludge digestion is to convert bulky, odorous, raw sludge to a relatively inert material that can be rapidly dewatered in the absence of obnoxious odors. The bacterial process, as summarized in Eq. 11-20, consists of two successive processes that occur simultaneously in digesting sludge. The first stage consists of breaking down large organic compounds and converting them to organic acids along with gaseous by-products of carbon dioxide, methane, and trace amounts of hydrogen sulfide. This step is performed by a variety of facultative bacteria operating in an environment devoid of oxygen. If the process was to stop there, the accumulated acids would lower the pH and would inhibit further decomposition by "pickling" the remaining raw wastes. In

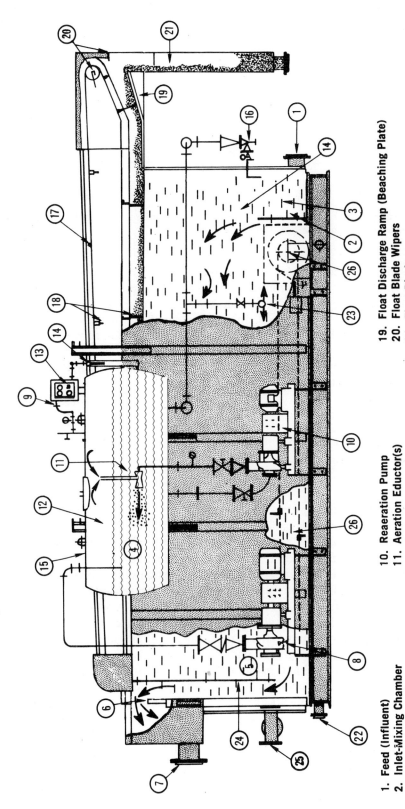

1. Feed (Influent)
2. Inlet-Mixing Chamber
3. Inlet Flow Zone
4. Recycle Being Activated
5. Discharge Flow Zone
6. Adjustable Discharge Weir
7. Unit Effluent
8. Recirculation Pump
9. Air Inlet
10. Reaeration Pump
11. Aeration Eductor(s)
12. Air Cushion
13. Automatic Air Feed Controller
14. Liquid Level Sensor Tube
15. Retention Tank
16. Pressure Reducing Control Valve
17. Skimmer Chain
18. Float Skimmer Blades
19. Float Discharge Ramp (Beaching Plate)
20. Float Blade Wipers
21. Float Discharge
22. Unit Drain
23. Back Pressure Distributors
24. Baffle
25. Auxiliary Water Connection
26. Bottom Collector & Drive (Optional)

Figure 11-46 Cross-sectional view of a dissolved air flotation unit employed to thicken waste activated sludge. (Courtesy of Komline-Sanderson, Peapack, New Jersey)

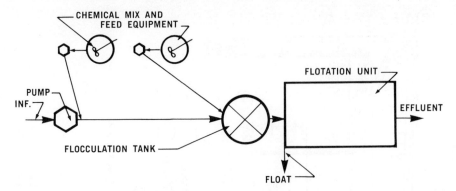

Figure 11-47 Flow diagram for chemical pretreatment of waste sludge prior to flotation thickening. (Courtesy of Komline-Sanderson, Peapack, New Jersey)

order for digestion to occur, second-stage gasification is needed to convert the organic acids to methane and carbon dioxide. Acid-splitting methane-forming bacteria are strict anaerobes and are very sensitive to environmental conditions of temperature, pH, and anaerobiosis. In addition, methane bacteria have a slower growth rate than the acid formers, and are very specific in food supply requirements. For example, each species is restricted to the metabolism of only a few compounds, mainly alcohols and organic acids, while carbohydrates, fats, and proteins are not available as energy sources.

$$\underset{\substack{\text{Acid-forming}\\\text{bacteria}}}{\text{Organic}\atop\text{Matter}} \xrightarrow{\hspace{1cm}} \overset{\substack{CO_2,\,CH_4\\ \to H_2S}}{} \underset{\substack{\text{Acid-splitting}\\\text{methane-forming}\\\text{bacteria}}}{\text{Organic}\atop\text{Acids}} \xrightarrow{\hspace{1cm}} {CH_4 \atop \text{and} \atop CO_2} \qquad (11\text{-}20)$$

Stability of the digestion process relies on proper balance of the two biological stages. Buildup of organic acids may result from either a sudden increase in organic loading, or a sharp rise in operating temperature. In either case, the supply of organic acids exceeds the assimilative capacity of the methane-forming bacteria. This unbalance results in decreased gas production and eventual drop of pH, unless the organic loading is reduced to allow recovery of the second-stage reaction. Accumulation of toxic substances from industrial wastes, such as heavy metals, may also inhibit the digestion reaction. It is often difficult to determine the exact cause of digester problems. Monitoring volatile solids loading, total gas production, volatile acids concentration in the digesting sludge, and percentage of carbon dioxide in the head gases are the methods most frequently employed to give advance warning of pending failure. These measurements can also indicate the most probable cause of difficulties. Gas production should vary in proportion to organic loading. Volatile acids content is normally stable at a given loading rate and operating temperature. The percentage of carbon dioxide should also remain relatively constant. Monitoring digestion by pH measurements is not recommended, since a drop in pH does not herald failure but announces that it has occurred. Table 11-5 lists the general operating and loading conditions for anaerobic digestion.

Table 11-5. General Operating and Loading Conditions for Anaerobic Sludge Digestion

Temperature	
Optimum	98° F (35° C)
General operating range	85° F to 95° F
pH	
Optimum	7.0 to 7.1
General limits	6.7 to 7.4
Gas Production	
Per pound of volatile solids added	6 to 8 cu ft
Per pound of volatile solids destroyed	16 to 18 cu ft
Gas Composition	
Methane	65 to 69 percent
Carbon dioxide	31 to 35 percent
Hydrogen sulfide	trace
Volatile Acids Concentration	
General operating range	200 to 800 mg/l
Alkalinity Concentration	
Normal operation	2000 to 3500 mg/l
Volatile Solids Loading	
Conventional single stage	0.02 to 0.05 lb VS/cu ft/day
First-stage high rate	0.05 to 0.15 lb VS/cu ft/day
Volatile Solids Reduction	
Conventional single stage	50 to 70 percent
First-stage high rate	50 percent
Solids Retention Time	
Conventional single stage	30 to 90 days
First-stage high rate	10 to 15 days
Digester Capacity Based on Design Equivalent Population	
Conventional single stage	4 to 6 cu ft/PE
First-Stage high rate	0.7 to 1.5 cu ft/PE

Single-Stage Digestion

The cross section of a floating-cover digester is illustrated in Figure 11-48. Raw sludge is pumped into the tank through feed pipes that terminate both near the center of the tank and in the gas dome. Contents stratify with a scum layer on top, a middle zone of supernatant (water of separation) underlain by actively digesting sludge, and a bottom layer of digested concentrate. A limited amount of mixing is provided in the upper half of the tank by withdrawing digesting sludge, passing it through a sludge heater, and returning it through the inlet piping. Supernatant is withdrawn from any one of a series of pipes extended through the tank wall. Digested sludge is taken from the tank bottom.

The digester cover floats on the sludge surface, and liquid extending up the sides provides a seal between the tank wall and the side of the cover. Gas rising out of the digesting sludge is collected in the gas dome and is burned as a fuel in the sludge heater, with the excess wasted to a gas burner. The cover can rise vertically from the landing brackets to near the top of the tank wall guided by rollers around the circumference to keep it from binding. The volume between the landing brackets and the fully raised cover position is the amount of storage available for digested sludge; this is approximately one third of the total volume.

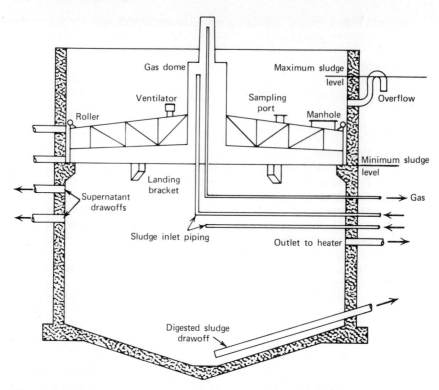

Figure 11-48 Diagram of a single-stage floating-cover anaerobic digester.

Digestion in a single-stage floating-cover tank performs the functions of volatile solids digestion, gravity thickening, and storage of digested sludge. When sludge is pumped into the digester from the primary settling tanks, the floating cover rises making room for the sludge. Unmixed operation permits daily drainage of supernatant equal to approximately two thirds of the raw sludge feed. Being high in both BOD and suspended solids, the withdrawn water is returned to the inlet of the treatment plant. Periodically digested sludge is removed for dewatering and disposal. In large plants, digested sludge may be dewatered mechanically; however, in small installations it is frequently spread in liquid form on farmland or is dried on sand beds and hauled to land burial. Weather often dictates the schedule for land disposal and, consequently, substantial digester storage volume is required in northern climates. Typical operation lowers the cover to the landing brackets in the fall of the year to provide maximum storage volume for the winter.

A drying bed consists of a 12-in. sand layer underlain by graded gravel that envelopes tile or perforated pipe underdrains. Large beds are partitioned by concrete walls into sections 25 ft wide by 100 to 200 ft long. A pipe header, with gated opening to each cell, is used to apply digested sludge to the bed. Seepage collected in the underdrains is returned to the treatment plant wet well. The total area of open drying beds at a treatment plant is usually 1 to 2 sq ft/BOD design population equivalent.

Digested sludge applied to a depth of 8 or 10 in. dries to a fibrous layer of 3 or 4 in. in a period of a few weeks. With sufficient bed area and digester storage volume, this process of dewatering can be compatible for most climatic regions. However, the laborious process of removing the digested cake is a major problem for operating personnel. Although some plants have mechanical equipment, the time-honored method is manual removal using a shovel-like fork. Attempting to employ front-end loaders leads to disturbance of the bed and excessive loss of sand. In most instances, an operator will seek an alternate technique, such as spreading the liquid sludge on crop or pastureland or lagooning in shallow trenches that permit the use of a tractor for scraping up dried cake for hauling to landfill.

Two-Stage Digestion

In this process, two digesters in series separate the functions of biological stabilization from gravity thickening and storage. The first-stage high-rate unit is complete mixing and heated for optimum bacterial decomposition. Mixing is accomplished either mechanically by an impeller suspended from the cover, or by recirculation of compressed digestion gases. The latter is done by one of the following three methods: gas withdrawn from the dome is injected back into the tank through a series of small diameter pipes hanging into the digesting sludge; compressed gas is discharged into a draft tube located in the center of the tank to lift recirculating sludge from the bottom and to spill it out on top; or pressurized gas is forced through diffusers mounted on the bottom of the tank. These systems are available for installation in either fixed or floating cover tanks. By using a floating cover, digested sludge does not have to be displaced simultaneously with raw sludge feed as is required with a fixed-cover tank. In either case, however, the sludge cannot be thickened in a high-rate process because continuous mixing does not permit formation of supernatant. Actually, the discharged sludge has a lower solids concentration than the raw feed because of the conversion of volatile solids to gaseous end products.

The second-stage digester must be either a floating-cover or open tank with provisions for withdrawing supernatant. The unit is often unheated, depending on the local climate and degree of stabilization accomplished in the first stage. By minimizing hydraulic disturbances in the tank, density of the digested sludge and clarity of the supernatant are both increased.

Two-stage digestion may be advantageous in some plants while conventional operation may be better in others. The determining factors include size of treatment plant, flexibility of sludge handling processes, method of ultimate solids disposal, storage capacity needed, and the interrelated element of climatic conditions. For large plants with a number of digesters, series operation often provides better utilization of digester capacity, but for small plants with limited supervision the conventional operation is frequently more feasible.

Sizing of Digesters

Historically, conventional single-stage tanks have been sized on the basis of population equivalent load on the treatment plant. Heated digester capacity for a trickling filter plant processing domestic waste water was established at 4 cu ft per capita of design load. For primary plus secondary activated sludge, the total tank volume requirement was increased to 6 cu ft per capita. These values are still used as guidelines for sizing conventional digesters for small treatment works.

Total digestion capacity can be calculated for conventional single-stage operation using Eq. 11-21. Application of this formula requires knowing the characteristics of both the raw and digested sludges.

$$V = \frac{V_1 + V_2}{2} T_1 + V_2 \times T_2 \qquad (11\text{-}21)$$

where V = total digester capacity, gallons

V_1 = volume of daily raw sludge applied, gallons per day

V_2 = volume of accumulated digested sludge, gallons per day

T_1 = period required for digestion, days (approximately 30 days at a temperature of 85° to 95° F)

T_2 = period of digested sludge storage, days

Volume needed for the high-rate unit in a two-stage digestion system is based on a maximum volatile solids loading and minimum detention time. For new design the generally adopted maximum allowable loading is 0.08 lb VS/cu ft/day and the minimum liquid detention time is 10 days. At these loadings and a temperature of 95° F, volatile solids reduction should be 50 percent or greater. There are no specific design criteria established for second-stage tanks in high-rate systems, since thickening and digested sludge storage requirements depend on local sludge disposal procedures.

Start-up of Digesters

Anaerobic digestion is a difficult process to start because of the slow growth rate and sensitivity of acid-splitting methane-forming bacteria. Furthermore, the numbers of these microorganisms is very low in raw sludge compared with acid-forming bacteria. The normal procedure for start-up is to fill the tank with waste water and to apply raw sludge feed at about one tenth of design rate. If several thousand gallons of digesting sludge from an operating digester are used as seed, the new process can be operational in a few weeks. However, if only raw sludge is available, developing the biological process may take months. Careful additions of lime added with raw sludge are helpful in maintaining the pH near 7.0, but erratic dosage can result in sharp pH changes detrimental to the bacteria. After gas production and volatile acids concentration have stabilized, the feed rate is gradually increased by small increments to full loading. Daily monitoring of this process involves plotting the daily gas production per unit of raw sludge fed, percentage of carbon dioxide in the head gases, and concentration of volatile acids in the digesting sludge.

EXAMPLE 11-15

An anaerobic sludge digester operating at a feed rate of 0.06 lb VS/cu ft/day has good gas production and volatile solids reduction. The external sludge heater fails and is not repaired for about 6 weeks. During heater shutdown, the same feed rate is maintained and there is a noticeable decrease in gas production as the temperature of the digesting sludge drops from 95° F to 75° F. When the heater is returned to service, the digester temperature is rapidly elevated back to 95° F with a sudden initial increase in gasification. However, after a few days of renewed operation, gas production decreases sharply, volatile acids are measured at about 2000 mg/l, and pH of the digesting sludge begins to drop. Describe what has happened to the anaerobic biological process.

Solution

Initially, acid-forming and methane-forming populations are in balance and, because they are operating near optimum temperature, the majority of the volatile solids are being converted to gaseous end products. When the heater fails and temperature of the digesting sludge drops, a new equilibrium of populations is established at reduced gasification efficiency. Therefore, raw organic matter accumulates in the digesting sludge. These dormant volatile solids suddenly become available to the bacteria when the temperature is raised back to 95° F. The sharp increase in gas production is related primarily to carbon dioxide produced when first-stage bacteria convert volatile solids to organic acids. The acid-splitting methane formers attempt to respond to the increased organic acid supply. However, their populations have been reduced by the lack of food during low temperature operation. Furthermore, the methane bacteria being more sensitive are inhibited by the acid conditions and, hence, total gas yield decreases. The solution is to drop the temperature of the digesting sludge back to 75° F and to increase it slowly, perhaps one degree per week, to allow the methane formers to adjust gradually to the increasing production of organic acids.

EXAMPLE 11-16

Calculate the capacity required per population equivalent for single-stage floating-cover digester based on the following: 0.20 lb solids contributed per person, 4.0 percent solids content in raw sludge, 7.0 percent solids concentration in digested sludge, 40 percent total solids reduction during digestion, 30-day digestion period, and 90-day storage period for digested sludge.

Solution

Using Eq. 11-19, the daily volumes of raw sludge produced and digested sludge accumulated are

$$V_1 = \frac{0.20 \text{ lb/person/day}}{(4.0/100)8.34} = 0.60 \text{ gal/person/day}$$

$$V_2 = \frac{0.60 \times 0.20 \text{ lb/person/day}}{(7.0/100)8.34} = 0.21 \text{ gal/person/day}$$

Then, the digester capacity by Eq. 11-21 is

$$V = \frac{0.60 + 0.21}{2} \times 30 + 0.21 \times 90 = 31 \text{ gal} = 4.1 \text{ cu ft}$$

(*Note.* This calculation verifies the single-stage digester capacity based on design equivalent population listed in Table 11-5.)

EXAMPLE 11-17

A trickling filter plant has two floating-cover digesters each with a total capacity of 30,000 cu ft with 20,000 cu ft below the landing brackets. The remaining 10,000 cu ft in each tank is the volume between the lowered and fully raised cover positions (storage volume). The sludge piping is arranged so that they may be operated in parallel as conventional digesters or in series as a two-stage system. One digester is equipped with gas mixing to serve as a first-stage high-rate process. Daily raw sludge production is 6400 gal containing 2800 lb of dry solids that are 70 percent volatile. Digested sludge is 8.0 percent solids, and the process converts 60 percent of the volatile solids to gas.

(a) For conventional single-stage operation with one half of the waste applied to each tank, calculate the volatile solids loading, supernatant produced, and number of days of digested sludge storage available.

(b) For two-stage operation, compute the volatile solids loading on the first tank, solids content of the digested sludge leaving the high-rate digester, and available sludge storage in the system.

Solution

(a) For conventional single-stage operation

VS applied to each tank $= (0.70 \times 2800)/2 = 980$ lb/day

$$\frac{\text{Loading with}}{\text{covers lowered}} = \frac{980 \text{ lb VS/day}}{20,000 \text{ cu ft}} = 0.049 \text{ lb VS/cu ft/day}$$

$$\frac{\text{Loading with}}{\text{covers raised}} = \frac{980 \text{ lb VS/day}}{30,000 \text{ cu ft}} = 0.033 \text{ lb VS/cu ft/day}$$

$$\frac{\text{Supernatant}}{\text{return per day}} = \frac{\text{raw sludge volume}}{\text{applied per day } (V_1)} - \frac{\text{digested sludge volume}}{\text{accumulated/day } (V_2)}$$

$V_1 = 6400$ gal

$$V_2 = \frac{\text{inert solids} + \text{remaining VS}}{(\text{solids content}/100)8.34}$$

$$= \frac{0.30 \times 2800 + 0.40(0.70 \times 2800)}{(8.0/100)8.34} = 2400 \text{ gal}$$

Supernatant return $= 6400 - 2400 = 4000$ gal/day

$$\frac{\text{Digested sludge}}{\text{storage available}} = \frac{2 \times 10,000 \text{ cu ft} \times 7.48 \text{ gal/cu ft}}{2400 \text{ gal/day}} = 62 \text{ days}$$

(b) For two-stage operation

VS applied to high-rate tank $= 0.7 \times 2800 = 1960$ lb/day

$$\frac{\text{Loading with}}{\text{cover lowered}} = \frac{1960}{20,000} = 0.098 \text{ lb VS/cu ft/day}$$

$$\frac{\text{Detention time}}{\text{with cover lowered}} = \frac{20,000 \times 7.48}{6400} = 23 \text{ days}$$

With cover raised the volume is 30,000 cu ft,

Loading $= 0.065$ lb VS/cu ft/day and detention time $= 35$ days

Solids content in digested sludge leaving the first-stage high-rate digester (Eq. 11-19)

$$s = \frac{0.30 \times 2800 + 0.40(0.70 \times 2800)}{6400 \times 8.34/100} = 3.0 \text{ percent}$$

(*Note.* The raw sludge is 5.2 percent solids. The percentage of solids reduces to 3.0 percent in high-rate digestion, since volatile solids are converted to gas while no supernatant is withdrawn to thicken the digested sludge.)

Digested sludge storage, available only in the second-stage tank, is equal to 31 days. This reduced storage capacity is a disadvantage of high-rate digestion compared with parallel operation that permits 62 days storage.

11-13   AEROBIC DIGESTION

Waste sludge can be stabilized by long-term aeration that biologically destroys volatile solids. The aerobic digestion process was developed specifically to handle excess activated sludge from aerobic treatment plants without primary clarifiers (Figure 11-2a). Early attempts to anaerobically digest this sludge failed because of the low solids concentration and aerobic nature of the waste. High water content in the range of 98 to 99 percent also prevented economical dewatering by mechanical means without prior thickening. Furthermore, since most of the aeration plants served small communities, large investments for mechanical equipment could not be justified for waste sludge processing. Consequently, aerated holding tanks, or digesters, were introduced as shown in Figure 11-33 to stabilize and store waste biological floc from the aeration process.

Design standards for aerobic digestion vary with character of the waste sludge and method of ultimate disposal. Small systems often have 2 to 3 cu ft per design population equivalent of the treatment plant. This provides a loading in the range of 0.01 to 0.02 lb VS/cu ft/day, although substantially greater volatile solids loadings in the neighborhood of 0.07 could be stabilized. The minimum aeration period is 10 to 20 days depending on temperature. Digested sludge is relatively free of obnoxious odors and has good drainability so that it can be disposed of either by spreading on farmland, lagooning, or drying on sand beds.

Problems associated with aerobic digestion relate to the fact that the digested sludge does not thicken by gravity without chemical treatment. In fact, if the suspended solids concentration exceeds about 5000 mg/l, a clear supernatant cannot be decanted even after prolonged quiescent holding. Therefore, supernatant separation chambers that are generally installed in aerobic digesters are rarely effective, and overflow carries waste suspended solids back to the aeration chamber of the treatment plant. To produce a high quality plant effluent, design must take care of storage and elimination of the entire volume of waste activated sludge independent of supernatant return. Although this is not difficult in a small package plant, the problem of voluminous sludge disposal can be a serious operational problem in larger plants. If sand drying beds are used, they must be specially designed for dewatering dilute sludge. Several feet of sludge must be applied to achieve a dried cake of a few inches. Design standards for drying beds that handle anaerobically digested sludge are not appropriate.

11-14   VACUUM FILTRATION

Rotary-drum vacuum filtration is the most common mechanical method to dewater sludge. The cylindrical drum covered with a filter medium rotates partially submerged in a vat of chemically treated waste. The filter medium may be a belt of synthetic cloth, woven metal, or two layers of stainless steel coil springs arranged in corduroy fashion around the drum (Figure 11-49). As the

Figure 11-49 Picture of a vacuum filter with coil-spring media dewatering waste sludge. (Courtesy of Komline-Sanderson, Peapack, New Jersey)

drum slowly rotates, vacuum is applied immediately under the filter medium to attract sludge solids as the drum dips into the vat. Suction continues to draw water from the accumulated solids as the filter slowly rotates. Radial vacuum pipes convey filtrate from collecting channels in the drum surface behind the media. In the discharge sector, suction is broken and the belt or coil springs are drawn over a small diameter roller for removal of the dried cake. Fork tines help release the solids layer. The filter medium is washed by water sprays, located between wash and return rollers, before it is reapplied to the drum surface for another cycle.

Many small plants draw settled sludge directly from the primary tanks without thickening. Although this is often satisfactory for trickling filter installations, vacuum filtration of unthickened sludges from aeration plants often leads to unsatisfactory performance. The generally accepted minimum solids concentration for economical filtration is 4 percent.

Mixtures of raw primary and secondary wastes are gravity thickened, while waste activated may be increased in solid content by flotation. Digested sludge seldom is thickened separately because of its high density when drawn from digesters. However, elutriation may be practiced to remove alkalinity and fine solids to improve filter yields and reduce chemical dosage. This process consists of countercurrent washing of the digested sludge with plant effluent.

Components of a vacuum filter installation are illustrated in Figure 11-50. Positive displacement pumps meter sludge to a conditioning tank. Here chemicals are mixed with the liquid to flocculate the solids. Chemical treatment is necessary to capture fines that would otherwise be drawn through the medium, and to prevent cracking of the filter cake during dewatering, which would result in excess air entering the vacuum system. The mixture of filtrate and air drawn by the vacuum pump is separated in a receiving vessel. Air is exhausted through a silencer to reduce noise, and the water is pumped back to the inlet of the treatment plant. Cake peeled from the filter drops onto a belt conveyer for transporting to a truck for hauling to landfill, or directly to an incinerator. Filtrate return accounts for the majority of the process discharge. For example, dewatering 1000 gal of sludge with 5 percent solids to a 30 percent cake results in 830 gal of extracted water and about 140 lb of wet cake; this amounts to a wet sludge weight reduction of 83 percent.

Common chemicals in conditioning are ferric chloride and lime, or polyelectrolytes. Ferric chloride is dissolved in water and is applied by a solution feed pump, while lime is fed in a slurry. Polymers available in powder, pellet, or liquid form are dissolved to a stock solution of 0.5 to 5 percent. Most manufacturers recommend further dilution to 0.01 to 0.05 percent before application. This permits uncurling of the long chain molecules. Chemical dosages for conditioning vary considerably among different sludges. Solids concentration and type of sludge are the most critical factors. With rare exception, digested sludges require considerably greater chemical additions than do fresh raw sludges. Typical inorganic chemical requirements are 2 to 4 percent $FeCl_3$ and 5 to 10 percent CaO, or polyelectrolyte amounts of less than 1 percent. These chemical consumptions are expressed as percentages of the dry solids filtered, for example, 3 percent conditioner means 3 lb of chemical per 100 lb of dry solids in the filter cake. The choice of conditioning chemicals depends on economics, desired filter yield, and method of ultimate disposal. It is advantageous to have a flexible installation so that any of the inorganic metal

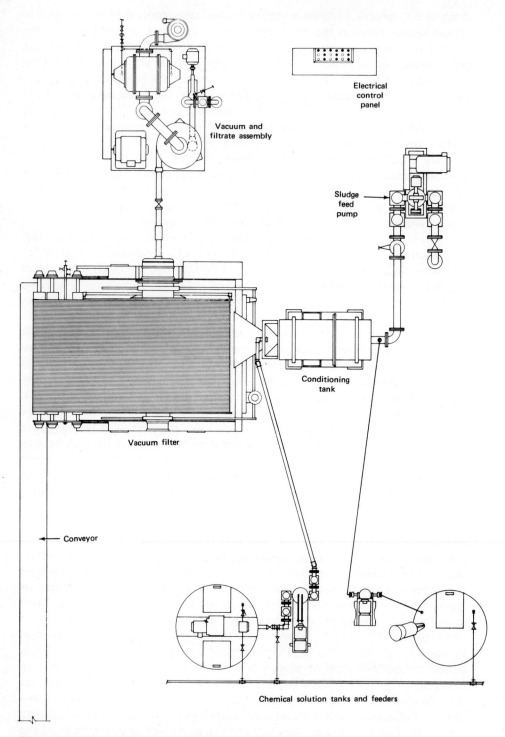

Figure 11-50 Schematic layout showing the major components of a vacuum filtration system. (Courtesy of Komline-Sanderson, Peapack, New Jersey)

coagulants, lime, or polyelectrolytes can be handled by the system. The advantages of ferric chloride and lime are that they provide disinfection and stabilization of the waste sludge thus reducing health hazards and odors. The disadvantage is that they are more difficult to handle and feed, result in increased filter maintenance, and add weight to the filter cake. Polymers do not provide disinfection, but they are easier to feed and often more economical. The most suitable chemicals for a particular situation can only be determined by laboratory and plant-scale investigations. Final filter cake disposal dictates the extent of dewatering required. Concentration to 20 to 25 percent solids is generally satisfactory for hauling to landfill. Cake with a higher moisture content is troublesome to handle and does not discharge readily from a dump truck. If long distance trucking is involved, it may be more economical to dewater to a lower moisture content and thus reduce the total weight of cake transported. When disposed of by incineration, it is desirable to dewater to as low a moisture content as possible, usually to 30 percent solids or greater. This allows burning without auxiliary fuel.

Rate of vacuum filtration is expressed as yield in pounds of dry solids produced per square foot of filter area per hour. Table 11-6 lists the usual filter yields expected in dewatering various types of domestic waste sludges. These values reflect both typical characteristics and solids concentration of the different sludges. As a general rule, the expected filter yield of any chemically conditioned sludge is about 1 lb per square foot per hour for each percentage of solids. In addition to yield, solids in the filtrate and final moisture content of the cake are important in judging filter performance. Figure 11-51 shows typical variations of these parameters relative to chemical conditioning. Yield increases continuously with increasing chemical dosage. Filtrate solids are very high when conditioning is not sufficient to flocculate properly the finer particles. It is essential to operate in the region of low filtrate solids to prevent returning an excessive amount of solids back to the treatment plant. Final cake moisture decreases with increasing chemical addition; however, it is more dramatically influenced by solids content of the sludge feed. Where a dry sludge is necessary for incineration, thickening prior to filtration results in chemical savings that usually offset the cost of the thickening operation.

Size of vacuum filters for new installations is based on estimated dry solids production, a selected filter yield depending on thickness and character of the sludge, and operating time. For small plants, 35 hr per week is often selected for design purposes. This allows a 7-hr day for five days with 1 hr for start-up

Table 11-6. Common Vacuum Filter Yields for Chemically Conditioned Waste Sludge

	Yield, Pounds per Square Foot per Hour	
Source of Sludge	Raw	Digested
Primary only	8	7
Primary plus trickling filter humus	7	6
Primary plus waste activated	6	5
Thickened waste activated only	3 to 5	

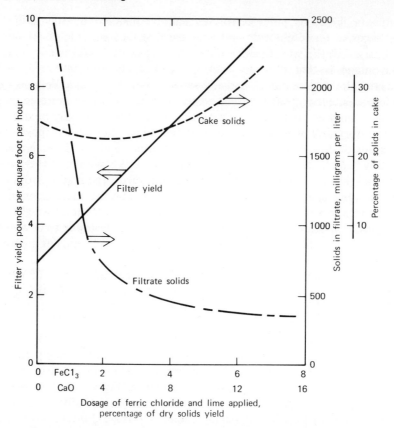

Figure 11-51 Typical variations of filter yield, filtrate solids, and cake moisture relative to dosage of conditioning chemicals.

and wash-down. It is often more economical in larger plants, particularly where incineration is employed, to filter for a period of 16 to 20 hr per day.

Where sludge samples are available prior to design and installation, it is advantageous to test sludge filterability in the laboratory. The Buchner funnel method consists of pouring a chemically conditioned sludge sample into a filter-paper-lined funnel and dewatering under vacuum. With this procedure, it is possible to predict suitability of a sludge for filtration and applicability of certain chemicals, but it is of little value in sizing or predicting plant-scale operation. A second test procedure uses a filter leaf with 0.10 sq ft of effective area attached to a vacuum apparatus. A medium with characteristics similar to a full-scale filter is placed over the face of the leaf. After applying suction, the unit is inserted upside down in a representative slurry to simulate cake formation, and then is withdrawn and dewatered for a time period related to drum rotation speed. Cake removed from the leaf face and filtrate drawn into the vacuum flask can then be tested for solids content. If carefully conducted, results can be correlated to indicate expected full-scale vacuum filter operation.

Good filter operation relies on installation of the proper sludge handling equipment, careful monitoring of the process, and a willingness to experiment to determine optimum performance. Uniform sludge and proper chemical dosage are the most essential factors. If the character of the sludge varies, as may be the case if it is drawn directly from primary clarifiers, it is difficult to maintain consistent operation. For example, hour to hour fluctuations of solids concentration may lead to poor cake and filtrate quality, or more likely, chemical overdosing by the controller to prevent such problems. Laboratory monitoring should include analyses to determine filtrate quality, yield, and cake moisture. The operator should observe sludge level in the vat to insure proper submergence, clarity of the filtrate, cake thickness, which is normally about $\frac{1}{8}$ in., and drum speed. Slow rotation produces a thicker drier cake with less yield, while selection of a faster speed produces moister cake and a greater yield. Finally, because of the variety of chemicals marketed for sludge conditioning, particularly polyelectrolytes, special full-scale studies must be conducted periodically to insure that the selected coagulant is most economical for the performance desired.

EXAMPLE 11-18

The smallest size vacuum filter produced by one of the leading manufacturers has a nominal surface area of 60 sq ft; the drum is 7 ft in diameter and 2 ft 10 in. in width. Estimate the equivalent population that can be served by this unit. Assume 0.20 lb dry solids per capita per day, a filter yield of 5.0 lb per square foot per hour (5 percent solids content in sludge), and 7 hr of operation per day.

Solution

$$\text{Total yield per day} = \frac{5.0 \text{ lb}}{\text{sq ft} \times \text{hr}} \times 7.0 \text{ hr} \times 60 \text{ sq ft} = 2000 \text{ lb}$$

$$\text{Equivalent population} = \frac{2000 \text{ lb/day}}{0.20 \text{ lb/capita/day}} = 10{,}000 \text{ persons}$$

EXAMPLE 11-19

A conventional aeration plant produces a daily waste sludge of 25,800 gal containing 10,200 lb of dry solids which is a 4.7 percent solids concentration. (These values are from Example 11-14.) The installed vacuum filter has a nominal area of 300 sq ft, and the filtration characteristics of the sludge are given in Figure 11-51. For a filter yield of 6.0 lb per square foot per hour, calculate the following: daily consumption of chemicals, amount of cake produced, quantity of filtrate, and hours of filter operation per day.

Solution

From Figure 11-51 for a yield of 6.0 lb per square foot per day, chemical dosages are 3.1 percent ferric chloride and 6.2 percent lime, the filtrate contains 500 mg/l of suspended solids, and the percentage of solids in the cake is 23 percent.
Chemical dosages are:

$$\text{FeCl}_3 \text{ applied per day} = 0.031 \times 10{,}200 = 320 \text{ lb}$$

$$\text{CaO applied per day} = 0.062 \times 10{,}200 = 630 \text{ lb}$$

After conditioning, the quantity of sludge is increased by the weight and volume of chemical additions. Assume that the FeCl_3 and CaO solutions amount to a total of 700 gal/day. Then,

Conditioned sludge solids $= 10,200 + 320 + 630 = 11,200$ lb/day

Conditioned sludge volume $= 25,800 + 700 = 26,500$ gal/day

Weight of filter cake, computed from cake solids content of 23 percent, is

$$= \frac{11,200 \text{ lb dry solids}}{0.23 \dfrac{\text{lb dry solids}}{\text{lb wet sludge}}} = 48,700 \text{ lb/day}$$

And, the volume of cake using a specific gravity of 1.06 is

$$= \frac{48,700}{1.06 \times 8.34} = 5500 \text{ gal/day}$$

Quantity of filtrate, volume of conditioned sludge minus the cake volume,

$$= 26,500 - 5500 = 21,000 \text{ gal/day}$$

Filtrate solids $= 500 \text{ mg/l} \times 0.021 \text{ mgd} \times 8.34 = 88$ lb/day

Filter operation required per day

$$= \frac{10,200 \text{ lb/day}}{300 \text{ sq ft} \times 6 \text{ lb/sq ft/hr}} = 5.7 \text{ hr/day}$$

11-15   CENTRIFUGATION AND PRESSURE FILTRATION

Centrifuges are employed for dewatering raw, digested, and waste activated sludges, although they are not as popular as vacuum filters. A discussion of centrifugation is presented on page 274, and the common types employed are illustrated in Figures 7-30 and 7-31. Typical operation produces a cake with 15 to 30 percent solids concentration, depending on the character of the sludge. Without chemical conditioning the solids capture is in the range of 50 to 80 percent, while proper chemical pretreatment can increase solids recovery to 80 to 95 percent. Disposal of a centrate with high suspended, nonsettleable solids as a result of sludge characteristics and inadequate chemical conditioning can cause problems. For example, high solids in the return to the head of the treatment plant can result in a large load of fine solids recirculating around and around through the centrifuge and the waste-water processing units. Both polyelectrolytes, and ferric chloride and lime may be used as chemical conditioners. The choice between centrifugation and vacuum filtration is based on performance and economics. The latter involves both initial cost of installation and operating costs including labor, chemicals, and power.

Pressure filtration is discussed on page 276 and a typical unit is illustrated in Figure 7-32. Although not widely adopted in the United States, the suitability of pressure filtration in dewatering biological sludges has been verified by systems employed in England and Europe. Two advantages of pressure filtration are low filtrate suspended solids concentration, ordinarily less than 100 mg/l, and a drier cake with 40 to 50 percent solids. Denser cake is easier to incinerate, lighter for hauling, and relatively odorless if ferric chloride and lime are used for conditioning. Highly automated filter systems are easy to manage and operate unattended, except during initial and final stages of the filter cycle. Selection of pressure filtration, as opposed to anaerobic digestion or vacuum filtration, bears on economics and other considerations including type of sludge, size of treatment plant, and method of ultimate cake disposal.

11-16   INCINERATION AND DRYING

Mechanical dewatering is used as pretreatment prior to drying or burning of waste solids. Sludge drying involves reducing water content by vaporizing the water to air. Auxiliary fuel is burned to heat the ambient air and to provide the latent heat of evaporation. Incineration is an extension of the drying process and converts solids into an inert ash which can be disposed of easily. If dewatered to approximately 30 percent solids, the process is usually self-sustaining without supplemental fuel, except for initial warm-up and heat control. It is generally preferable to burn raw rather than digested sludge because of its higher heat value.

The multiple hearth furnace illustrated in Figure 11-52 has proved to be successful for both sludge drying and incineration. The furnace consists of a circular steel shell containing several hearths arranged in a vertical stack, and a central rotating shaft with rabble arms. Sludge cake is fed onto the top hearth and is raked slowly in a spiral path to the center. Here, it drops to the second level where it is pushed to the periphery and drops, in turn, to the third hearth where it is again raked to the center. The two upper levels allow for evaporation of moisture, the middle hearths burn the solids producing temperatures exceeding 1400° F, and the bottom zone cools the ash prior to discharge.

The hollow central shaft is cooled by forced air vented out the top. A portion of this preheated air from the shaft is piped to the lowest hearth and is further heated by the hot ash and combustion as it passes up into the furnace. The air then cools as it gives up its heat to dry the incoming sludge. Countercurrent flow of air and sludge solids optimizes combustion efficiency. After the air passes through the furnace twice, it is discharged through a wet scrubber to remove fly ash. This furnace can also be designed as a dryer only. Hot gases from an external furnace flow parallel with the sludge downward through the multiple hearths to dry the solids without scorching.

The main air pollution of concern in operating a sludge incinerator is emission of particulate matter. The latest performance standard specifies that the gas discharged into the atmosphere cannot contain particulate matter in excess of 70 mg/cu meter, nor exhibit 10 percent or greater opacity (shadiness), except for 2 min in any 1 hr. The presence of uncombined water is the only reason for failure to meet the latter requirement. Performance tests are conducted with the incinerator operating at design load and burning representative sludge. Both the cyclonic water jet scrubber and the perforated plate-type jet scrubber are able to meet the particulate air quality standard, as well as the acceptable stack emissions of nitrogen oxides, sulfur oxides, and odors.

Two other heat-treating processes applied in sludge incineration are the flash-drying incineration system and the fluidized bed reactor. Flash drying involves pulverizing wet sludge cake with recycled dried solids in a cage mill. Hot gases from the incinerating furnace suspend the dispersed sludge particles up into a pipe duct where they are dried. A cyclone separator removes the dried solids from the moisture-laden hot gas which is returned to the furnace. Part of the dried sludge returns to the mixer for blending with incoming wet cake. The remainder is either withdrawn for fertilizer or incinerated.

A fluidized bed reactor uses a sand bed which is kept in fluid condition by an upflow of air, as a heat reservoir to promote uniform combustion of sludge solids. The bed is preheated to approximately 1200° F by using fuel oil or gas.

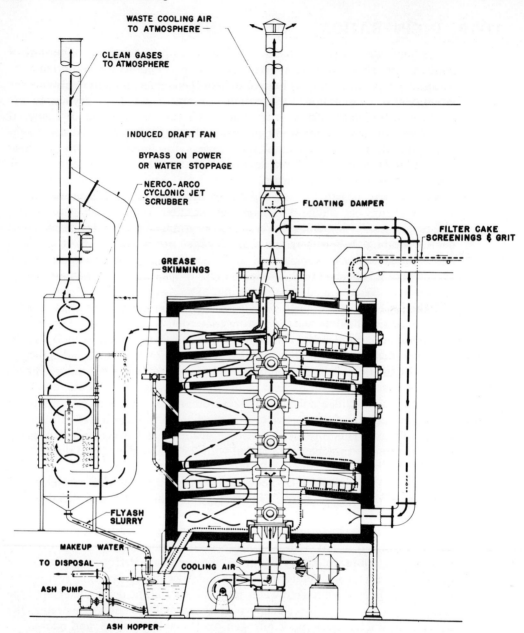

WASTE COOLING AIR TO ATMOSPHERE —

CLEAN GASES TO ATMOSPHERE

INDUCED DRAFT FAN

BYPASS ON POWER OR WATER STOPPAGE

NERCO-ARCO CYCLONIC JET SCRUBBER

FLOATING DAMPER

FILTER CAKE SCREENINGS & GRIT

GREASE SKIMMINGS

FLYASH SLURRY

MAKEUP WATER

TO DISPOSAL

ASH PUMP

COOLING AIR

ASH HOPPER —

Figure 11-52 Multiple hearth furnace for burning dewatered waste sludge. (Courtesy of Nichols Engineering and Research Corp., subsidiary of Neptune International Corp.)

When sludge cake is introduced, rapid combustion occurs maintaining an operating temperature of about 1500° F. Ash and water vapor are carried out with the combustion gases. A cyclonic wet scrubber is used to remove ash from the exhaust gases. Finally, the ash is separated from the scrubber water in a cyclone separator.

11-17 LAND DISPOSAL

Lagoon disposal of digested sludge is often economical where inexpensive land is available at or close to the treatment plant site. The advantages are simplicity of operation, flexibility, and suitability for plants of any size. Lagoon drying can be a regular operational procedure, or can be used temporarily during peak load periods. Disadvantages are the possibility of groundwater pollution and odor nuisance if the sludge is not completely digested.

Liquid disposal of digested sludge on agricultural land is popular where acceptable sites are located within convenient hauling distance. This process is most applicable to small plants that are located in rural areas and are equipped with anaerobic digesters to biologically condition the sludge before spreading. Even though hauling costs may be high, the alternatives of solid-liquid separation on drying beds or by mechanical dewatering are often more costly. Furthermore, digested solids are useful as a soil conditioner and fertilizer.

Landfills are commonly used as final disposal sites for incinerator ash, dried digested sludge, and dewatered raw sludge cake. The latter poses both nuisance and health problems unless the fill sites are designed to prevent ground and surface water pollution, and are covered daily with earth. Many cities bury waste sludge in their sanitary landfill along with municipal refuse. The basic operations include spreading, compacting, and covering the solid wastes with excavated soil daily.

Ocean disposal is practiced by several large seacoast communities. Liquid sludge may be pumped by pipeline into the ocean, or dewatered sludge may be transported to open water in barges. Although this method of disposal is relatively inexpensive, it may not be acceptable in the future. Regulatory agencies have restricted this practice for environmental and esthetic reasons.

PROBLEMS

11-1 Describe the EPA effluent quality standards required for municipal waste-water treatment.

11-2 Where is a Parshall flume placed relative to lift pumps in preliminary processing of waste water?

11-3 What is the detention time in an aerated grit clarifier 10 ft by 10 ft with 8 ft liquid depth based on a flow of 0.55 mgd? (*Answer* 15.7 min)

11-4 Compute the overflow rate and detention time in an 80-ft diameter, 7.5-ft deep primary clarifier for a flow of 2.75 mgd. (*Answers* 550 gpd/sq ft, 2.5 hr)

11-5 Two rapid sludge return final clarifiers following high rate aeration are 60-ft in diameter with a 9-ft side water depth. The effluent weir is an inboard channel set on a diameter of 55 ft. For a total flow of 3.4 mgd, calculate the overflow rate, detention time, and weir loading. (*Answers* 600 gpd/sq ft, 2.7 hr, 4900 gpd/ft)

11-6 Estimate the design loading for an extended aeration plant with final clarifier that has a 30-ft diameter and 10-ft side water depth. (*Answer* 0.42 mgd)

11-7 What is the BOD removal efficiency of a single-stage rock-filled trickling filter secondary at 16° C, $R = 0.5$, and a BOD loading of 45 lb/1000 cu ft/day. (*Answer* 66%)

11-8 Calculate the BOD efficiency of a two-stage trickling filter secondary at 16° C with the first filter loaded at 90 lb/1000 cu ft/day and $R = 0.5$. The second filter is the same size as the first and operates at $R = 0.5$. (*Answer* 80%)

11-9 Compute the hydraulic and BOD loadings on a high-rate trickling filter,

40 ft in diameter with 7-ft depth of media, for a raw waste-water flow of 0.40 mgd with 130 mg/l BOD and a recirculation flow of 0.20 mgd? (*Answers* 49 lb/1000 cu ft/day, 21 mgad)

11-10 A high-rate trickling filter with a 70-ft diameter and 7-ft depth of media is operated as is illustrated in Figure 11-15. Settled waste-water flow is 0.80 mgd with 130 mg/l BOD. Indirect recirculation during low-flow periods is 0.40 mgd, and direct recirculation is 600 gpm. Calculate the BOD and hydraulic loadings, recirculation ratio, BOD removal efficiency, and effluent BOD at a temperature of 16° C. (*Answers* 32 lb/1000 cu ft/day, 23 mgad, 1.6, 71%, and 38 mg/l BOD)

11-11 Solve Example 11-4 for a raw waste-water BOD equal to 280 mg/l. (*Answers* $E_1 = 67\%$, $E_2 = 56\%$, $E = 90\%$)

11-12 A conventional activated-sludge basin is 24-ft wide, 98-ft long, and has 13-ft liquid depth. The influent flow is 0.91 mgd containing 985 lb BOD. Compute the BOD loading and aeration period. The operating MLSS is 2200 mg/l, and the settled sludge volume in the SVI test is 230 ml/l. Compute the F/M ratio, SVI, recommended sludge recirculation rate, and solids concentration in the return sludge. (*Answers* 32 lb BOD/1000 cu ft/day, 6.0 hr, 0.23 lb BOD/day/lb MLSS, 105 ml/g, 0.27 mgd, and 9500 mg/l)

11-13 A step-aeration activated-sludge secondary is operating under the following conditions: influent waste-water flow = 5.50 mgd, average influent BOD = 126 mg/l, volume of aeration basins = 150,000 cu ft, and MLSS concentration 2500 mg/l. Determine the following: BOD loading, aeration period, F/M ratio, and estimated effluent BOD assuming good operation. (*Answers* 38.5 lb BOD/1000 cu ft/day, 4.9 hr, 0.25 lb BOD/day/lb MLSS, 15 to 20 mg/l BOD)

11-14 A 6000-cu ft complete mixing aeration basin treats 0.35 mgd with an average BOD of 200 mg/l. The operating MLSS is 4000 mg/l, the effluent suspended solids concentration equals 40 mg/l, and waste sludge solids average 180 lb per day. Compute the BOD loading, aeration period, F/M, BOD, and SS sludge ages. (*Answers* 97 lb BOD/1000 cu ft/day, 3.1 hr, 0.39 lb BOD/day/lb MLSS, 2.6 days and 5.0 days)

11-15 An aeration tank and clarifier system, without primary settling, is used to treat a dairy waste water of 0.30 mgd at 1000 mg/l BOD. For an aeration tank having a volume of 62,500 cu ft, compute the BOD loading and aeration period. If the MLSS in the aeration basin is 2000 mg/l, calculate the F/M ratio. If the settled sludge volume in the SVI test is 200 ml, what is the SVI value? What is the suggested sludge recirculation rate? (*Answers* 40 lb BOD/1000 cu ft/day, 37 hr, 0.32 lb BOD/day/lb MLSS, 100 ml/g, 25% = 0.075 mgd)

11-16 Estimate the design flow and BOD loading for an extended aeration system with floating mechanical aerators. The units sizes are: volume of aeration basins = 2.50 mil gal, oxygen transfer capability of aerators = 340 lb/hr at 2.0 mg/l DO, and 10-ft deep final clarifiers with a surface area of 33,400 sq ft. (*Answers* 20 mgd and 24 lb BOD/1000 cu ft/day)

11-17 Facultative ponds with a total surface area of 14.8 acres serve a community with a waste-water flow of 140,000 gpd at a BOD of 320 mg/l. Calculate the BOD loading, and winter storage available between the 2-ft and 5-ft depths assuming a water loss of 40 in./yr. (*Answers* 25 lb BOD/day/acre, water loss = 44,000 gpd, 150 days)

11-18 Based on a loading of 25 lb BOD/day/acre, what stabilization pond area is needed for an average daily flow of 0.84 mgd with 250 mg/l BOD? Calculate the minimum water depth for a retention time of 90 days assuming no water loss by evaporation and seepage. (*Answers* 70 acre, 3.3 ft)

11-19 Estimate the quantity of sludge produced by a trickling filter plant treating 3.0 mgd with a suspended solids concentration of 240 mg/l and BOD of 260 mg/l. Assume primary removals of 50 percent suspended solids and 35 percent BOD. The water content of the sludge is 95 percent. (*Answers* $W_p = 3000$ lb/day, $W_s = 1000$ lb/day, $V = 9600$ gal)

11-20 What is the estimated waste activated sludge from a conventional aeration process treating 7.7 mgd with 173 mg/l BOD operating at a F/M of 0.24 lb BOD/day/lb MLSS. Assume a suspended solids of 9800 mg/l in the waste sludge. (*Answers* 3900 lb/day, 47,000 gpd)

11-21 A gravity thickener processes 66,000 gpd of waste sludge increasing the solids content from 3.0 to 7.0 percent with 90 percent solids recovery. Calculate the quantity of thickened sludge. (*Answer* 25,000 gpd)

11-22 A trickling filter plant treats an average domestic waste water of 0.40 mgd (4000 persons at 0.20 lb dry solids/capita). Estimate the capacity of conventional digesters needed based on equivalent population, and compute the volume required assuming: raw sludge water content = 96 percent, volatile solids in raw sludge solids = 70 percent, solids in digested sludge = 6.0 percent, volatile solids destruction = 50 percent, and a storage period of 90 days. (*Answers* 20,000 cu ft (5 cu ft/capita), 19,400 cu ft)

11-23 Raw sludge applied to a high-rate digester, with a volume of 4800 cu ft, is 2400 gpd with 800 lb of solids that are 70 percent volatile. Compute the VS loading and detention time. (*Answers* 0.12 lb VS/cu ft/day, 15 days)

11-24 A single-stage anaerobic digester has a capacity of 13,800 cu ft of which 10,600 cu ft is below the landing brackets. The following conditions prevail: average raw sludge solids = 580 lb/day, raw sludge solids content = 4.0 percent, digestion period = 30 days, total solids reduction during digestion = 45 percent, digested sludge moisture content = 94 percent, and digested sludge storage required = 90 days. Calculate the digester loading with the cover in the lowered position, assuming 70 percent of the solids are volatile. Determine the digester capacity based on Eq. 11-21. (*Answers* 0.038 lb VS/day/cu ft, 12,400 cu ft)

11-25 A contact stabilization plant has an aerobic digester with a volume of 5400 cu ft. The design population is 2000, and the anticipated waste volatile solids production is 0.045 lb VS/capita. Compute the volume provided per person and the VS loading. (*Answers* 2.7 cu ft/person, 0.017 lb VS/day/cu ft)

11-26 A treatment plant produces a raw waste sludge with a solids content of 5.0 percent containing 2760 lb of dry solids daily. What area of vacuum filter surface would you recommend assuming a 7-hr operating period per day. Estimate the weight of cake produced assuming chemical additions of 7 percent lime and 3 percent ferric chloride, and a moisture content of 75 percent. (*Answers* 80 sq ft, 12,000 lb/day)

11-27 A vacuum filter with an area of 250 sq ft dewaters 16,000 gal of sludge contains 6.5 percent solids in 6 hr of operation. Compute filter yield, and weight of cake using 20 percent solids content. Estimate the volume of filtrate assuming a specific gravity of 1.3 for dry solids. Since a low dosage of polyelectrolyte is used in conditioning, the weight and volume of chemical feed may be ignored. (*Answers* 5.8 lb/sq ft/hr, 43,400 lb/day, 11,100 gal)

chapter 12

Operation of Waste-Water Systems

Sewer maintenance requires a thorough knowledge of the layout and appurtenances used in collection systems covered in Chapter 10. Waste-water flows, infiltration, and inflow are discussed in Chapter 9. Prerequisite information regarding the function of various unit operations and how they relate to each other is presented in Chapter 11. The following sections are restricted to cleaning and inspection of sewer systems, infiltration and inflow surveys, regulation of sewer use, and performance evaluation of treatment plants.

12-1 SEWER MAINTENANCE AND CLEANING

Serious and expensive sewer problems can result from improper design or poor construction. Adequate slopes to maintain self-cleaning velocities are essential to minimizing maintenance. Selection of a suitable pipe joint is vital to prevent penetration of roots and excessive infiltration. Cutting of tree roots from sewer lines can be an expensive and reoccurring cleaning process. Groundwater entering joints carries with it soil from around the pipe which ultimately causes structural failure. In addition to review of new design and supervision of construction, building permits should require careful inspection of all service connections before backfilling. It is important to make certain that unused service lines are properly capped when buildings are demolished.

A successful maintenance program operates on a planned schedule and requires keeping effective records. Maps are used to show location of manholes, flushing inlets, service connections, and other appurtenances. Records should be kept on maintenance performed with particular stress on troublesome lines that are known to require more frequent inspection or cleaning. While large sewers on adequate slopes may never require flushing or cleaning, others must be placed on a regular schedule that may range from every month to once a year. The number of emergency sewer blockages can be materially reduced by such preventative maintenance.

Sewer stoppages are caused chiefly by sand, greasy materials, sticks, stones, and tree roots. The latter are most troublesome. Common cleaning techniques are flushing with water, scraping with mechanical tools, hydraulic scouring with high pressure jets, and addition of chemicals. Periodic flushing helps to keep lines clear and is often performed in association with inspection. The usual procedure for developing scouring velocity is to insert a fire hose into the sewer through a manhole. This is most advantageous in cleaning lines in residential sections that do not have sufficient connections to provide cleansing flows of waste water. Flushing has limitations, since it merely moves debris from one section of a sewer into another; it is assumed that flows in downstream pipes are sufficient to suspend the solids and keep them moving.

Operation of a power rodding machine is shown schematically in Figure 12-1. Sections of flexible steel rod are coupled together to hold the cleaning tool. The trailer unit, with a reel for holding several hundred feet of rod, has an engine-driven mechanism with variable speed control. The rods, fed in through a hose guide braced in the manhole, may be pushed or pulled with the reel rotating or stationary. A number of different tools are used depending on the

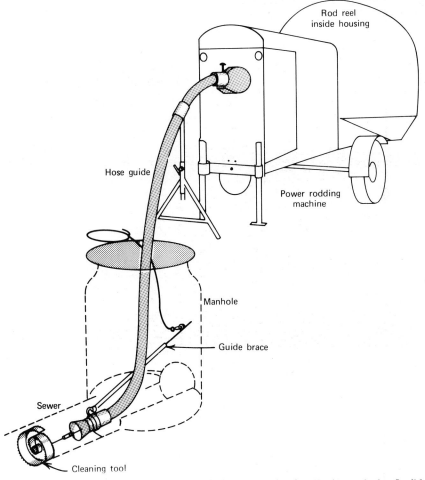

Figure 12-1 Power rodding machine for mechanical sewer cleaning. Tools attached to flexible steel rods can be rotated while being pushed or pulled through the pipeline.

reason for the stoppage. A rotating blade cutter is pulled through a sewer to cut grease and roots, or a root saw is pushed through the pipe to accomplish the same thing. An auger or corkscrew may be used to remove large objects. In addition, there are porcupine scrapers, spearheads, hole openers, and other appurtenances for specific applications.

Normally, when a stoppage is reported, manholes are inspected to determine the location. With few exceptions, it is best to rod up the line from the first dry manhole downstream. This procedure allows a clear manhole from which to work and the flushing effect of the waste-water flow in clearing the line. However, when roots are suspected as the major cause, operating from the upstream manhole may have considerable advantage, since roots grow in a downstream direction. The large tap root is therefore more easily reached from the upstream side, eliminating the necessity of working through all the fine feeder roots.

A bucket machine cleans grit from sewer lines. Two power winches are centered over adjacent manholes with cables connected to both ends of a cleaning bucket (Figure 12-2). The cable is initially threaded through the pipeline by using a power rodder, or floating cone. Connecting cables fastened to the bucket ride over surface-hung manhole rollers so that it can be pulled in either direction by the machine at the appropriate end. The bucket is pulled through the sewer until it is loaded with debris. That winch is then taken out of gear, and the opposing unit retrieves the bucket that automatically closes when the reverse pull starts. A chute at the discharge manhole is used to load the grit directly onto a truck. A power winch and cable can also pull a porcupine scraper or sewer ball through a pipe for clearing a stoppage.

A hydraulically propelled sewer cleaning tool depends on water pressure to provide cleaning force and propulsion. The unit may be attached to the end of a fire hose or a truck-mounted hose reel that can supply water under high pressure. Action of the backward-pointing jets moves the nozzle forward pulling the hose behind it. Debris is flushed back to the inlet manhole where it can be removed hydraulically.

Routine use of copper sulfate is successful in control of tree roots. Periodic

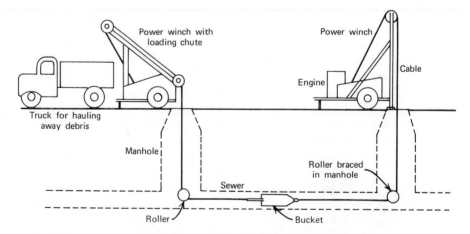

Figure 12-2 A bucket machine consists of two power winches set at adjacent manholes. The bucket is pulled through the sewer collecting grit and depositing it in a truck.

dosages may be added at manholes or through residential water closets. The amount required is determined by trial and error, and experience.

Several techniques are applied for inspecting sewer lines. Low pressure smoke detection can help locate openings in a pipeline, service connections left unplugged from houses that have been removed, cross-connections with the storm sewer, and unauthorized drainage connections. Blowing smoke into a sewer is a good way to locate holes accessible to rodents. Rats inhabiting sewer lines need access to a dry area to build their nests. Dyes or other tracer chemicals may also be used to locate extraneous infiltration or exfiltration points. While large sewers allow visual inspection, the inside of small-diameter pipes require photographing or closed-circuit television inspection. Skid-mounted cameras can be drawn through lines to observe existing conditions. Most small cities find it more economical to contract for services to perform this specialized inspection.

12-2 INFILTRATION AND INFLOW SURVEYS

Entrance of extraneous waters into sewer systems is of concern for several reasons. These include sewer surcharging during periods of intensive rainfall resulting in flooded basements, overloading of treatment plant facilities, overtaxing of pumping stations, excessive costs in processing diluted waste-water flows, and health hazards from discharging raw waste water. In the past, problems of excessive infiltration and inflow were often handled by construction of relief sewers than by-passed treatment works and flowed directly to surface watercourses. This practice allowed adequate waste treatment 90 to 95 percent of the time and appeared at the time to be the best economical design choice. However, the objective of new federal legislation is to eliminate these pollutional discharges by requiring treatment of all waste flows even during peak periods.

Infiltration results from groundwater entering sewer lines through poor joints and cracks in manholes and sewer pipe; the sources are widespread, and the flow is relatively steady during times of high groundwater levels. Inflow comes from direct connections such as roof drains and result in sudden high rates of flow of short duration. The quantity of infiltration-inflow is considered to be the maximum waste-water flow minus the peak domestic and industrial discharge; in other words, it is the difference between peak flow measurements during wet weather and dry weather.

A systematic approach to sewer system evaluation includes identifying the quantity and nature of infiltration-inflow, isolating problem areas, and then evaluating the most economical corrective measures. The basic alternatives are either to rehabilitate the sewer system to reduce extraneous flows, or to extend treatment facilities to handle peak wet-weather flows.

The diagram in Figure 12-3 outlines the general approach applied in infiltration-inflow surveys. The first step is to analyze and identify the magnitude of the problem. The sewer system is divided into areas that drain to key points, such as a manhole or pumping station. Flow tests are then conducted to determine peak discharges by the methods outlined on page 321. Obviously measurements must be taken at the proper times during both dry weather, and high groundwater and rainfall situations. Evaluation of these data reveals whether infiltration and inflow are reasonable. Detailed field investigations are

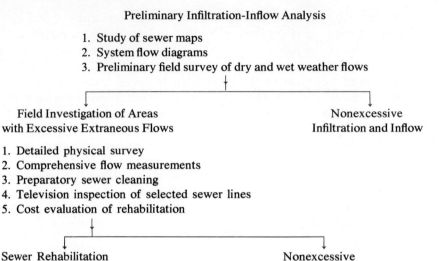

Preliminary Infiltration-Inflow Analysis

1. Study of sewer maps
2. System flow diagrams
3. Preliminary field survey of dry and wet weather flows

Field Investigation of Areas
with Excessive Extraneous Flows

Nonexcessive
Infiltration and Inflow

1. Detailed physical survey
2. Comprehensive flow measurements
3. Preparatory sewer cleaning
4. Television inspection of selected sewer lines
5. Cost evaluation of rehabilitation

Sewer Rehabilitation

Nonexcessive
Based on Economics

1. Sewer replacement
2. Repair of defects
3. Sealing of pipe

Expansion of Treatment works

Figure 12-3 Diagram outlining the general steps that are included in infiltration and inflow evaluations of sanitary sewer collection systems.

then conducted in those areas where flows are considered excessive. Field investigations start with a survey of physical conditions which requires descending every manhole in the study area and visually inspecting the sewer lines. This often uncovers deficiencies that are not obvious by merely looking into manholes from the street level. Cross-connections from storm sewers and unauthorized drains attached to sanitary sewers may be detected by using tracer dyes in water or low pressure smoke. Often elimination of inflows that are observed during the physical survey materially reduces peak discharges. For this reason, additional flow measurements may be required. Finally, those sewer lines identified as the source of excessive infiltration and inflow are inspected by using closed circuit television. Preparatory cleaning is required prior to televising.

The final report of the field survey presents data from investigations, recommendations for a sewer rehabilitation program and, most important, an economic evaluation of the problem. Expenses of rehabilitation are compared to costs of expanding treatment works to process extraneous water. Decisions are made regarding the feasibility of repairing different sections of the collection system. Where it is more economical to treat than to repair, the extraneous water flows are considered to be nonexcessive, that is, in an economic sense. Excessive flows are those where the treatment costs dictate rehabilitation. Sewer repair may involve replacing pipelines, repairing structural deterioration or defects, or sealing openings by external grouting or pipe relining. The most economic and best techniques must be determined from local conditions and experience. For example, the success of grouting with cement or polymers depends to a considerable extent on the type of soil. Care must be taken in external grouting to insure that pipes are not fractured by grout injection. The economics of alternate repair methods are also considered; it may cost more to test and seal every joint in a manhole reach than for complete sewer replacement. With manually accessible lines, hand-mortaring coatings and guniting are primary repair methods. Internal sealing of small diameter sewers commonly involves forcing sealants through pipe openings to gel in the surrounding soil, thus sealing the leaks.

12-3 REGULATION OF SEWER USE

The key purposes of a sewer ordinance are to control discharges to the sanitary sewer, to insure that water quality standards of the receiving watercourse can be achieved, and to establish equitable customer charges for waste-water service. A comprehensive code contains: regulations requiring use of public sewers where available, control of private waste disposal in the absence of public sewers, construction of service connections, control of the quantity and character of waste waters admissible to municipal sewers, procedures for waste-water sampling and analyses, provisions for the powers and authority of inspectors, an enforcement (penalty) clause, and other legal clauses and signatures to validate the document. A model ordinance in the *Regulation of Sewer Use*[3] can be used as a guide in developing a regulation in a municipality where none has previously existed. Although several items included are common to all ordinances, each must be modified to incorporate local concerns of water pollution control and existing city documents, for example, the plumbing code.

Proper use and separation of sanitary and storm sewers are essential in protecting public health and reducing the quantity of inflow. Privies and septic tanks should not be permitted where sewer service is available. In the absence of public sewers, regulations should be established regarding construction, installation, and monitoring of private household systems to insure that surface water and groundwater are not contaminated. Unpolluted waters must be excluded from the sanitary collection system and must be directed to storm sewers or a natural drainage outlet where feasible. These include inflow from downspouts, footing drains, surface water inlets, swimming pools, and cooling water from air conditioners and industrial refrigeration units.

Certain wastes cannot be admitted to sanitary sewers, while others must be carefully controlled by specifying discharge limitations. These can be considered in the following four categories: (1) wastes that create a fire or explosion hazard; (2) substances that impair hydraulic capacity, (3) contaminants that create a hazard to people, the physical sewer system, or the biological treatment process, and (4) the refractory wastes that pass through treatment and result in degradation of the receiving watercourse. Examples of flammable liquids are gasoline, fuel oil, and cleaning solvents. Solids and viscous liquids that create sewer stoppages include ashes, sand, metal shavings, paunch manure, unshredded garbage, grease, and oil. The most common sewer stoppage however is related to root growth in sewer lines. This can be minimized by discouraging the planting of certain trees like elm, poplar, willow, sycamore, and soft maple near sewer lines. An alternate preventative measure is to use special jointing methods and materials when sewers are installed in areas of root influence.

Uncontrolled industrial wastes may contain corrosive or toxic compounds. For example, sulfur compounds and high temperature wastes can promote bacterial formation of sulfuric acid and lead to sewer crown corrosion. Acid wastes cause invert corrosion and, if not sufficiently diluted, may interfere with treatment processes. Toxic metal ions such as chromium and zinc and organic chemicals in rather small concentrations can lead to inhibition of the biological activity in secondary aeration and anaerobic digestion. Dissolved salts, color, and odor-producing substances are only partially removed by conventional treatment. For these, protection of the receiving watercourse is best achieved by separate disposal at industrial sites rather than by discharge to the sewer system. Examples are spent brine solutions, dye wastes, and phenols. Where industrial wastes are highly variable, it may be necessary to install equalizing tanks to prevent shock loads on the treatment works. In addition to controlling quality variations by neutralization and dilution, pretreatment by equalization can also provide discharge control to smooth out flow and to prevent shock hydraulic loads.

Flow measuring and sampling of waste waters entering the sewer system are essential for enforcing a sewer ordinance and establishing equitable fees for sewer use. Each industry is required to install and maintain a suitable sampling station which may range from a simple manhole for grab samples to a structure that includes a flow recorder and automatic sampler; these are discussed in Section 10-2, page 322. Inspectors must have the authority to enter all properties for the purposes of observation, measurement, sampling, and testing pertinent to waste discharge. Facilities for sampling and flow measurement are absolutely essential for instituting a sewer ordinance and, therefore, their construction must be given primary consideration. For a small enterprise, the cost may involve only the installation of a manhole for sampling. Additional water meters may be needed to determine waste-water discharge if the waste water is assumed to equal the water supplied minus the consumption for lawn watering, or discharge to the storm sewer. Automated sampling stations for large industries, particularly with more than one outlet sewer, may be a substantial investment.

Violation of an ordinance is declared a misdemeanor, or sometimes disorderly conduct. Penalties usually involve a fine not exceeding $500 for each violation plus costs of prosecution. If a user fails to correct an unauthorized discharge, the city is commonly authorized to discontinue sewer or water service or both. Repeated violations may result in revoking such an industry's discharge permit. Any of these actions require written notice prior to institution. A hearing board may be defined in an ordinance to arbitrate differences between the city and an institution aggrieved by a penalty.

An ordinance also includes a section on charges for waste-water service. Sometimes, this is put in a separate document apart from the regulations governing sewer use so that the two subjects can be handled independently at public hearings and during enactment. This approach is particularly practical when instituting a new code, since fees cannot be collected on the basis of waste-water flow or excessive strength until after installation of sampling stations.

Annual revenues of a waste-water system include: costs associated with operation and maintenance of the collection system, treatment plant, and other facilities; principle and interest payments of providing debt financing for major capital improvements, as well as required reserve payments for general obligation or revenue bonds; and, costs of capital additions not debt-financed, such as interim replacements, betterments, and minor extensions to physical facilities that are paid for from current revenues. The basis for allocating revenue requirements must give consideration to all costs of providing service. These vary with each municipality. Payment for service from each customer should be in proportion to use and benefits received. Arriving at a fair, proper, and practical system for raising annual revenue from users according to their responsibility for costs is a complex subject (see *Financing and Charges for Wastewater Systems*).[4]

Financial support may be obtained by ad valorem property taxes; however, these do not collect payments in proportion to customers' uses or benefits received. For example, a disproportionately large charge may be placed on industries that produce little or no industrial waste. Waste-water service charges based on volume and strength reflect the actual physical use of the system to a considerable degree. But raising all revenues on the basis of service does not provide for payment by undeveloped property, for having facilities available whether they are used or not, or for the general community benefit that results from having adequate waste-water facilities. Yet, a service charge system yields a relatively stable source of funds that allows for orderly planning, upgrading, and expansion. A major advantage in handling industrial wastes is that volume and strength charges encourage reduction of waste-water loads by in-plant modifications, improved housekeeping, by-product recovery, and water reuse. For most cities, the fairest system for raising revenues is a combination of property taxes and service charges.

Sewer collection systems are designed for peak hourly flows, while treatment facilities are sized on average daily flow. Although ideally desirable, it is impractical to base volume charges on discharge rates. Therefore, user charges are keyed to volume of flow without regard to time pattern of release, as long as excessive hydraulic shock loads are avoided. An exception is seasonal industries that should be assessed some charges during the nonuse season. The most equitable charge for flow is a uniform rate for all users regardless of quantity, with no tapered schedule of charges. In many cities, household production is considered to be the winter water consumption. This assumes that the water drawn is disposed of through the house drains and that none is applied for lawn watering. Water meters can also be used to determine waste-water flows from small industries and commercial establishments. Where uncontaminated water is discharged to a storm sewer, the flow is metered and this volume is subtracted from the metered water supply to yield the quantity of sanitary waste water. Large industries install flow recorders that directly measure the quantity discharged to the municipal sewer.

The second component of a service charge is related to waste strength. This is generally applied as a surcharge for concentrations of pollutants that are greater than those found in average domestic waste water. The strength parameters most frequently employed are BOD and suspended solids. These must be determined by analyzing composite samples of industrial discharges in accordance with guidelines that are specified in the ordinance. Equation 12-1 is a typical formula for calculating service charges. The surcharge portion is for BOD in excess of 250 mg/l and suspended solids greater than 300 mg/l. Example 12-1 illustrates application of this formula.

$$\frac{\text{Waste-water}}{\text{service charge}} = \text{charge for volume} + \text{surcharge for strength}$$

$$= VR_V + V[(\text{BOD} - 250)R_B + (\text{SS} - 300)R_S]8.34 \qquad (12\text{-}1)$$

where
V = waste-water volume, million gallons
BOD = average BOD, milligrams per liter
SS = average suspended solids, milligrams per liter
R_V = charge rate for volume, dollars per million gallon
R_B = charge rate for BOD, dollars per pound
R_S = charge rate for suspended solids, dollars per pound

A sewer fee schedule should stipulate a minimum monthly charge for each property. The amount is often set on a graduated scale that increases with the size of water service to the property. One reason for a minimum bill is to cover the costs for maintaining capacity in the system even though it is not used. This is referred to occasionally as the right-to-service factor. Cities may also have a fee for making service connections to the municipal system. The charge may be small, only offsetting the costs of inspection, or much larger to pay a share of the cost involved in extending the sewer system.

EXAMPLE 12-1

Calculate the service charge for a dairy waste by using Eq. 12-1 based on the following: daily flow = 150,000 gal, average BOD = 910 mg/l, average suspended solids 320 mg/l, service charge for flow = $250.00/mil gal, and surcharges of 1.38¢/lb excess BOD and 1.03¢/lb excess SS.

Solution
Substituting into Eq. 12-1,

$SC = 0.15 \times 250 + 0.15[(910 - 250)0.0138 + (320 - 300)0.0103]8.34$

$= 37.50 + 11.68 = \$49.18 \text{ per day}$

12-4 PERFORMANCE EVALUATION OF TREATMENT PLANTS

Comprehensive studies are needed to determine treatment efficiency and economical operation in waste-water processing. As-built drawings of all treatment units provide dimensions of tanks and interconnecting piping. From these, a process flow diagram can be sketched showing normal plant operation. Changes in plant operation to meet unusual flow and load conditions should be noted. Physical facilities for flow-measuring and sampling at various points are essential. The influent Parshall flume should be checked for accuracy, since significant error in raw waste-water flows precludes satisfactory results. Flow measuring at various points within the plant may be accomplished by calibrating pump discharges, or by installing temporary weirs in flow channels. Sampling points must be carefully selected to insure collection of representative portions for compositing. Often, a lack of adequate flow-measuring facilities and accessibility for sampling in-plant flows jeopardizes or prevents the study of individual unit operations. Physical modifications may be needed in some plants to permit evaluation. Finally, a laboratory facility is needed to perform at least routine tests, such as total and suspended solids, BOD, pH, fecal coliforms, and chlorine residual. Metropolitan plants require additional equipment for analyses of COD, grease, alkalinity, phosphates, various forms of nitrogen, sulfides, volatile acids, gas analysis, sludge filterability, and biological oxygen uptake. In addition it may be desirable to have testing facilities for heavy metals and total organic carbon.

The person supervising a plant evaluation must be thoroughly familiar with each unit operation and how it fits into the overall plant process. Characteristics of the raw waste water must be completely defined by flow patterns, waste strength parameters, types of industrial wastes, and infiltration-inflow quantities. Gathering these data relies on a year-round testing and sampling program of influent waste water and monitoring of industrial wastes discharged to the sewer system. To insure that the degree of treatment is satisfactory, the superintendent should be familiar with local, state, and federal water quality and effluent standards. A complete study also includes a review of maintenance procedures and accurate records of operating costs.

The process diagram for a typical treatment plant in Figure 12-4 indicates the minimum testing program for evaluation. Effluent standards require daily monitoring of average BOD and suspended solids concentrations, pH, and fecal coliform counts. In some locations, tests for chlorine residual, presence of heavy metals, phosphates, and ammonium nitrogen content may also be specified by regulatory agencies. Percentage of organic matter removal is traditionally calculated by comparing influent and effluent BOD and suspended solids. The relationship between BOD and suspended solids, or volatile suspended solids, in the raw waste water is an important parameter in defining the effectiveness of primary sedimentation. Plain settling is expected to be effective if the suspended solids and BOD concentrations have an approximate ratio of 1 to 1. However, if the soluble organic content is several magnitudes greater than the suspended organic matter, in other words at high BOD to volatile suspended solids ratio, the majority of organics will be removed in secondary aeration rather than in the primary tank. Sludge solids discharged from a treatment plant also define the effectiveness of operations. Economics of solids disposal relate directly to moisture content. Excess water in filter cake can result in increased costs of hauling or, in the case of incineration, can require auxiliary fuel for burning.

Examination of individual unit operations within a treatment plant requires testing of all influent and effluent flow streams. Also, loading parameters should be calculated for each unit process to determine whether the system is being stressed beyond its intended design capacity. Consider first primary sedimentation where measurements include flow, BOD and suspended solids of the influent, BOD and suspended solids of the effluent, volume of sludge withdrawn, and its total solids content. Overflow rate, weir loading, and detention time can be calculated from the waste-water flow, and these values can be compared with design parameters to indicate the loading status of basins. Although efficiency of primary settling is most often related to BOD and suspended solids removal, the quantity and solids content of sludge withdrawn is equally important as a measure of satisfactory performance. Poor sludge thickening in a sedimentation basin may be a result of hydraulic overload, or

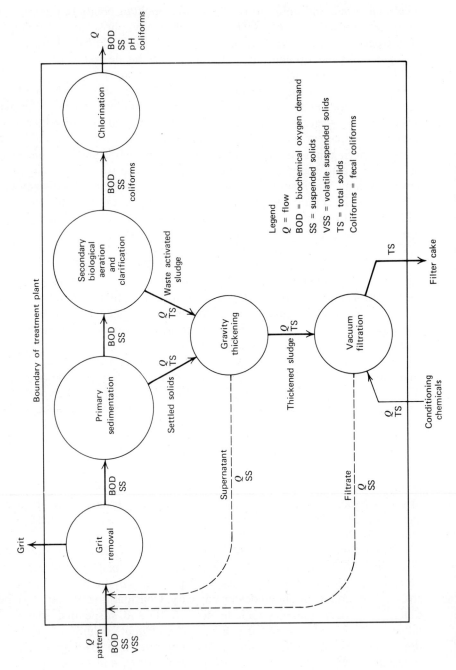

Figure 12-4 Typical process diagram for secondary waste-water treatment showing interrelated unit processes.

poor operating procedure. For example, a voluminous amount of thin sludge can be the result of high pumping rates that cause withdrawal of water from above the sludge blanket.

Secondary biological aeration should be examined on the bases of both organic matter removal and characteristics of the waste activated sludge. To view the aeration-clarification function only in terms of BOD and suspended solids removals, with no regard for concentration of solids in the waste sludge, is ignoring the thickening function of biological flocculation. Air supply, rate of activated sludge return, and sludge wasting should be adjusted to provide the thickest waste possible and yet still maintain a clear waste-water effluent. Operation at various dissolved oxygen levels, mixed liquor suspended solids concentrations, and food-to-microorganism ratios with careful surveillance testing can define optimum performance. The final processing step in secondary treatment is chlorination of the effluent. Careful studies are needed to insure efficient use of chlorine, since poorly operated disinfection units can result in excessive chlorine residuals. Chlorine residuals at different dosages should be measured under varying flow conditions and should be compared with laboratory tests on waste-water samples held for the same contact time.

Very careful consideration should be given to all unit operations that discharge flow back to the head of the plant. Sludge processing may return supernatant from anaerobic or aerobic digesters, overflow from gravity thickeners, underflow from flotation thickeners, centrate from centrifuges, or filtrate from vacuum filters. Excessive return of suspended solids can result in recycling of fines within the treatment plant. For example, if insufficient conditioning chemical is applied for sludge dewatering, the filtrate may carry a large amount of solids drawn through the filter medium back to the plant influent. Being primarily colloidal in nature, these solids pass through primary sedimentation for capture in biological aeration. They are then returned in the waste activated sludge for dewatering again. Cycling solids can lead to overloading and upset of all systems. However, their presence is normally first noticed in thinning of the primary sludge and increased oxygen demand in the aerobic secondary. Study of gravity thickening includes measuring the influent and effluent sludge flows and their solids content. If supernatant suspended solids are too high, greater capture in thickening can be achieved by chemical addition. Review of vacuum filter operation normally requires extensive study since it involves economic considerations related to operation time, chemical dosage, and disposal of filter cake. Since the quantity of chemical used is directly related to the thickness of the sludge feed, satisfactory filtration is keyed to the previous unit operations of waste-water treatment and sludge thickening. Solids processing, whether it be filtration or digestion, should be emphasized in plant evaluations; yet, it is frequently ignored. (*Note.* Normally, sludge analysis is performed by residue on evaporation which is essentially the same as a suspended solids measurement. This simplifying procedure of performing total solids testing, rather than suspended solids determinations, on sludge does not create significant error, since the dissolved solids content usually amounts to less than 5 percent of the total solids. On the other hand, dissolved solids may be a major portion of the total solids in supernatant or filtrate because of the low suspended solids concentration.)

Complete and accurate records of all phases of plant operation and maintenance are essential. However, too often, extensive testing and flow records are filed daily and are not reviewed until the plant is having serious operational problems or requires expansion. A rigid testing program without a directed purpose wastes manpower and, worst yet, does not achieve its objectives. An effective sampling and testing program is one of continuous evaluation. Data collected may be applied to calculate existing hydraulic and organic loadings on all unit processes and to pinpoint problem areas in operations that are related to industrial wastes or excessive infiltration and inflow. Errors or omissions in laboratory testing will be recognized in the process of analyzing accumulated records. One procedure of value in compiling data is to trace suspended solids through the treatment processes, namely, removal from suspension in primary and secondary units, thickening, dewatering, and ultimate disposal. In-plant data on sludge production, thickening efficiency, vacuum filter operation, and others are extremely valuable to an engineer in modifying or expanding an existing facility. Effluent quality is of value as a measure of overall plant performance.

REFERENCES

1. *Sewer Maintenance*, WPCF Manual of Practice No. 7, Water Pollution Control Federation, Washington, D.C., 1966.

2. *Prevention and Correction of Excessive Infiltration and Inflow into Sewer Systems*, Water Pollution Control Research Series, 11022 EFF 01/71, Environmental Protection Agency, Water Quality Office, 1971.

3. *Regulation of Sewer Use*, WPCF Manual of Practice No. 3, Water Pollution Control Federation, Washington, D.C., 1974.

4. *Financing and Charges for Wastewater Systems*, A Joint Committee Report, American Public Works Association, American Society of Civil Engineers, and Water Pollution Control Federation, 1973.

chapter 13

Advanced Waste Treatment

Historically, the practice of discharging waste into rivers was directed toward utilizing the dilutional and natural treatment capacities of moving waters to maintain public health and satisfactory environment for fish propagation. Treatment plants were designed to remove biodegradable organics in order to maintain a specified minimum dissolved oxygen level in receiving waters. Later, chlorination of effluents was adopted to reduce the potential contamination of natural waters with pathogens. As the dilutional capacity of watercourses reached saturation and increased water consumption required greater indirect reuse, more stress was placed on waste-water treatment to curb pollution. It has been deemed necessary in some instances to provide waste processing beyond conventional secondary, for example, to remove phosphates that induce the growth of algae. Nutrient salts, foam, color, and other refractory pollutants can be significantly reduced only by special waste treatment techniques.

Advanced waste treatment refers to methods and processes that remove more contaminants from waste water then are usually taken out by the present conventional treatment. The term advanced waste treatment may be applied to any system that follows secondary processing, or that modifies or replaces a step in conventional handling. The expression tertiary treatment is often used as a synonym, but the two meanings are not precisely ther same. Tertiary suggests a third step which is applied after primary and secondary processing.

The title water reclamation implies that the combination of conventional and advanced waste treatment processes employed returns the waste water to its original quality, reclaiming the water. Renovation and reuse of waste water is becoming more important as the relatively fixed natural water supply is strained by the demands of an expanding population. Envisioned reuse ranges from agricultural and industrial process water to potable water, with the degree of purification required varying according to the specific use. Chapter 14 discusses several case histories of water reuse and presents land disposal techniques.

13-1   SUSPENDED SOLIDS REMOVAL

Inability of gravity sedimentation in final clarifiers to remove small particles is a major limitation of BOD and suspended solids removal by conventional waste-water treatment. Where needed a tertiary liquid-solids separation step, either microstraining or filtration, may be added to upgrade performance. Microscreening is a physical straining process to remove particulate matter as small as 20 to 50 μ. Feed water enters a rotating drum covered with a fine screen, that deposits solids on the inner surface. These are dislodged by pressure jets of effluent water and are captured in a waste hopper for return to the head of the treatment plant.

Unsettled suspended solids in effluents can also be removed by filtration through beds similar to those employed in water treatment (Section 7-4). Design must take into account, however, the higher suspended solids content and fluctuating flow rates of waste-water discharges not common to water processing. The preferred beds are sand-coal dual media or mixed media that contains anthracite coal, garnet, sand, and gravel. A sand medium alone cannot accomodate the increased quantity of coarse solids that create the problem of surface plugging. Multimedia beds allow in-depth filtration and greater solids holding capacity, resulting in longer filter runs. Efficient backwashing requires auxiliary scour, since hydraulic suspension of the media, by upward flow of water through the bed, does not provide adequate cleaning. Either air scrubbing or a backwashing device such as a rotating agitator improves scouring action. Filter beds may be either gravity or pressure units depending on available water head. Gravity filtration can generally be performed satisfactorily with a head loss of 8 to 10 ft, while pressure units use higher head losses up to 20 ft.

Proper design of filter systems depends on providing sufficient bed area. Although removal efficiency is relatively independent of hydraulic loading rate, too small a surface area causes short filter cycles. Excess backwash water cycled to the head of the plant, and filter downtime results in inefficient operation. A minimum filtration time of 24 hr between backwashes is desirable under normal conditions of waste-water flow and suspended solids concentration. Effective suspended solids removal relies on installation design, and characteristics of the applied waste water. In general, the best effluent quality achievable by plain filtration is about 10 mg/l suspended solids and 10 mg/l BOD. If further reduction is desired, chemical coagulation must precede filtration to flocculate the colloidal solids.

13-2   PHOSPHORUS REMOVAL

Thirty to 50 percent of the phosphorus in domestic waste water is from sanitary wastes, while the remaining 50 to 70 percent is attributable to phosphate builders used in household detergents. Total phosphorus contribution is about 3.5 lb per capita per year resulting in an average concentration of 10 mg/l in domestic waste. The most common forms of phosphorus are organic compounds, orthophosphates, and polyphosphates. The latter, such as sodium hexametaphosphate, gradually hydrolyze in aqueous solution to the ortho form, and bacterial decomposition of organic compounds releases orthophosphates. The majority of phosphorus compounds in waste water, being soluble, are removed only sparingly by plain sedimentation. Secondary biological treatment removes phosphorus by biological uptake; however, there is surplus relative to the quantities of nitrogen and carbon necessary for biological synthesis. Consequently, primary and activated sludge secondary remove a maximum of only 20 to 30 percent, leaving an average effluent of 7 mg P/l.

The rate of fertilization of lakes can be retarded by reducing nutrient input. At present, emphasis is being placed on phosphorus extraction, since it is generally considered the limiting nutrient for controlling eutrophication. However, perhaps an equally important reason is the fact that nitrogen compounds are much more difficult to remove. Several states have adopted waste-water effluent standards for phosphorus. These standards have taken the form of an effluent concentration limit, and a requirement for a specified percentage reduction during treatment. Effluent limits range from 0.1 to 2.0 mg/l as P, with many established at 1.0 mg/l, while treatment efficiency requirements range from 80 to 95 percent.

One of the earliest concepts of phosphorus control was to make a substitute for the phosphate builders used in laundry detergents. At that time, this appeared to be a reasonable approach, since the majority of phosphorus in domestic wastes was derived from this source. Unfortunately, no suitable substance was found. Caustic additives did not clean as well, were irritating to skin, and certain brands were sufficiently strong to damage eyes and mucus membranes if inhaled or eaten. NTA, sodium nitrilotriacetate, which appeared to be the best phosphate substitute, was viewed as a hazard to human health. The Surgeon General of the United States suggested that for the time-being housewives should use phosphate detergents because of their safety. Another factor that was revealed during the phosphate detergent controversy was that eutrophication is not a nationwide problem. Waste waters from approximately 55 percent of the population are discharged into the ocean, or into major river systems that flow to the ocean. Another 30 percent of the population is in rural unsewered communities. That leaves only about 15 percent contributing waste discharges to inland lakes in danger of eutrophication. These include all of the Great Lakes, the Potomac River and estuary, San Francisco Bay and its tributaries, Lake Tahoe, and numerous other large and small lakes and reservoirs. This argument presumes, of course, that phosphates are not a significant problem in river systems. Evidence to date supports this view, since observed concentrations as high as 2 to 3 mg P/l in flowing waters have not resulted in serious water degradation.

Phosphate and inorganic nitrogen are taken out of solution by algal synthesis. However, growing and harvesting algae to remove nutrients from waste water has not proved to be feasible. Biological problems of proper balance of carbon to nitrogen to phosphorus ratio, adequate sunlight intensity, proper pH, and temperature control, physical limitations of the large land area required for adequate detention time, and costly mechanical harvesting techniques have prevented application of photosynthesis as a practical means of nutrient extraction.

Chemical precipitation using aluminum and iron coagulants and lime is effective in phosphate removal. Although coagulation reactions are complex and only partially understood, the primary action appears to be a combining of orthophosphate with metal cations. Polyphosphates and organic phosphorus compounds are probably removed by being entrapped, or adsorbed, on floc particles. Aluminum ions combine with phosphate ions as follows:

$$Al_2(SO_4)_3 \cdot 14.3H_2O + 2PO_4^= \rightarrow 2AlPO_4\downarrow + 3SO_4^= + 14.3H_2O \qquad (13\text{-}1)$$

The molar ratio for Al to P is 1 to 1, and the weight ratio of commercial alum to phosphorus is 9.7 to 1. Coagulation studies have shown that greater than stoichoimetric quantities of alum are necessary to precipitate phosphorus from waste water. One of the competing reactions, that accounts in part for the excess alum requirement, is with natural alkalinity. Phosphorus reductions of 75 percent, 85 percent, and 95 percent require alum to phosphorus weight ratios of about 13 to 1, 16 to 1, and 22 to 1, respectively. To achieve 85 percent phosphorus removal from a waste water containing 10 mg/l of P, the alum dosage needed would be $16 \times 10 = 160$ mg/l, which is substantially greater than the amount used in municipal water supply clarification.

Iron coagulants also precipitate phosphates. Ferric ions combine to form $FePO_4$ at a molar ratio of 1 to 1. Just as with aluminum, a larger amount of iron is required in actual coagulation than the chemical reaction predicts. Since the reaction of ferric chloride with natural alkalinity is relatively slow, lime or some other alkali is also applied to raise the pH and to increase hydroxyl ion concentration. The chemical reaction between ferrous ions and phosphate is not as well understood. Although ferrous sulfate could form a phosphate precipitate with an Fe to P molar ratio of 3 to 2, experimental results indicate that the molar ratio is essentially the same as when ferric salts are used. Commercially available iron salts are ferric sulfate, ferric chloride, ferrous sulfate, and waste pickle liquor from the steel industry. Provided that the waste water has sufficient natural alkalinity, ferric salts applied without coagulant aids result in phosphorus removal at Fe to P dosages of 1.8 to 1 or greater. This is equivalent to an application of approximately 100 mg/l as $FeCl_3$. Pickle liquor is variable in composition, depending on the metal treatment process. Ferrous sulfate produced from pickling with sulfuric acid is the most common form, but ferrous chloride from hydrochloric acid pickling is also available. Waste liquors have an iron content from 5 to 10 percent and free acid ranges from a low of 0.5 percent to a high of 15 percent. Their use necessitates addition of lime or sodium hydroxide for good results. For example, typical dosages in primary sedimentation are about 40 mg/l iron, 70 mg/l lime, and 0.5 mg/l polymer for removal of approximately 80 percent of the total phosphorus and 60 percent of the BOD.

Lime action differs from the hydrolyzing metal coagulants. When added to waste water it increases pH and reacts with the carbonate alkalinity to precipitate calcium carbonate. Calcium ion also combines with orthophosphate in the presence of hydroxyl ion to form gelatinous calcium hydroxyapatite as is shown in Eq. 13-2.

$$5Ca^{++} + 4OH^- + 3HPO_4^= \rightarrow Ca_5(OH)(PO_4)_3\downarrow + 3H_2O \qquad (13\text{-}2)$$

If sufficient lime is added, precipitation softening continues with the formation of magnesium hydroxide. A pH in the range of 9.5 to 11.5 is required to remove the major fraction of phosphorus. Lime dosages of 150 to 300 mg/l as CaO remove 80 to 90 percent of phosphate in a typical domestic waste. The actual amount required depends primarily on the alkalinity, phosphorus concentration, and degree of removal desired.

Field-scale studies have shown that alum and ferric salts can be used for chemical coagulation in combination with activated sludge aeration. In this process, the coagulant is added to the aerating mixed liquor or effluent of the aeration tank prior to final settling. Although the chemical-biological sludge has fewer protozoa, BOD removal efficiency is not adversely influenced. Chemicals added increase the volume of waste sludge by about 50 percent. Where dictated by water quality standards, the effluent may be filtered to remove carry-over floc prior to discharge.

Coagulation of raw waste water precipitates phosphates along with settleable organic matter in primary clarification. Where waste pickle liquor and lime are used, this is the common point of addition, and phosphorus removal is normally in the range of 70 to 80 percent. Excess lime addition in the primary can be used to extract up to 80 percent of the phosphorus without adversely affecting secondary biological treatment, if the pH is not raised above about 9.5. Microbial production of carbon dioxide in the activated sludge secondary is sufficient to maintain a neutral pH in the aeration compartment. If the process scheme is physical-chemical treatment, as is illustrated in Figure 13-1, the dosage of lime may be increased and subsequent recarbonation can be used to lower the pH. Although a single-stage process may be satisfactory, two-stage lime treatment is recommended to insure a consistent phosphorus removal of 90 percent. Lime is added in the first stage to a pH of 11.5 and is followed by recarbonation to a pH of about 9.5. Flocculation chemicals are then added to enhance settling in a second-stage clarifier. Lime treatment and recarbonation are followed by filtration to remove nonsettleable solids.

Chemical precipitation can also be applied as tertiary treatment following secondary biological processing (Figure 14-4). The flow scheme is similar to the water treatment processes of rapid mix, flocculation, sedimentation, and filtration. The primary advantage of tertiary treatment is isolation of the chemical and biological sludges which may allow more economical waste disposal. At Lake Tahoe, reclaiming of lime by recalcining in a multiple hearth furnace was considered feasible only for tertiary lime sludge without significant organic content.

13-3   NITRIFICATION AND NITROGEN REMOVAL

Nitrogen in municipal waste water results from human excreta, ground garbage, and industrial wastes, particularly from food processing. Approximately 40 percent is in the form of ammonia and 60 percent is organic, with negligible nitrate. Total content in municipal wastes is in the range of 8 to 12 lb of nitrogen per capita per year. Conventional primary and secondary treatment removes 40 percent or less.

Common forms of nitrogen are organic, ammonia, nitrate, nitrite, and gaseous nitrogen. Decomposition of nitrogenous organic matter releases ammonia to solution, Eq. 13-3. Under aerobic conditions, bacteria perform Eq. 13-4, oxidizing ammonia to nitrite and subsequently to nitrate. Nitrifying bacteria are autotrophic, using energy from ammonia oxidation and carbon from carbon dioxide for synthesis. Bacterial denitrification, Eq. 13-5, occurs

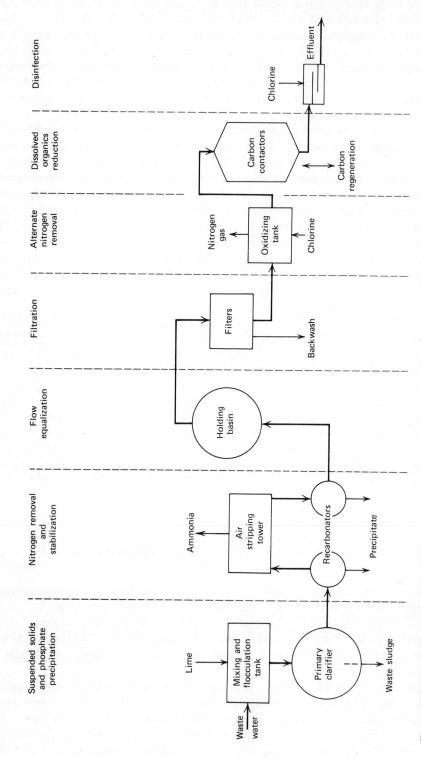

Figure 13-1 Schematic of physical-chemical treatment including various unit operations for different degrees of treatment. Primary lime application precipitates phosphorus and removes the majority of suspended solids and BOD. Air stripping, with backup breakpoint chlorination, extracts nitrogen. Flow equalization allows uniform operation of filters and contactors. Filtration removes nonsettleable suspended solids, and activated carbon columns reduces dissolved organics. Chlorination provides final disinfection.

under anaerobic conditions when organic matter (AH$_2$) is oxidized and nitrate is used as a hydrogen acceptor releasing nitrogen gas.

$$\text{Organic Nitrogen Compounds} \xrightarrow[\text{decomposition}]{\text{bacterial}} NH_3 \text{ (ammonia)} \qquad (13\text{-}3)$$

$$NH_3 + O_2 \xrightarrow[\text{nitrification}]{\text{bacterial}} NO_3^- \text{ (nitrate)} \qquad (13\text{-}4)$$

$$NO_3^- + AH_2 \xrightarrow[\text{denitrification}]{\text{anaerobic}} A + H_2O + N_2 \uparrow \qquad (13\text{-}5)$$

Pollution problems related to nitrogen are: decrease of dissolved oxygen in streams and lakes resulting from oxidation of ammonia nitrogen, toxic effect of ammonia on fishes, limitation of nitrate nitrogen in drinking water to protect public health, and nitrogen as a plant nutrient in eutrophication of lakes and estuaries.

Biological Nitrification-Denitrification

This process involves oxidation of ammonia to nitrate followed by reduction to nitrogen gas (Figure 13-2). The important parameters in bacterial nitrification kinetics are temperature, pH, and dissolved oxygen concentration. Reaction rate is decreased markedly at reduced temperatures with about 8° C being the minimum reasonable value. Optimum pH is near 8.4, and the dissolved oxygen level should be greater than 1.0 mg/l. Continuous flow aeration systems require a long sludge retention time to prevent excessive loss of viable nitrifying bacteria. In other words, the growth rate of nitrifying microorganisms must be rapid enough to replace bacteria lost through sludge wasting and washout in the plant effluent. Although it may be possible to perform nitrification along with organic matter removal in a single-stage extended aeration unit in southern climates, two-step treatment is necessary at reduced operating temperatures. The first stage is for removing BOD, without oxidation of ammonia nitrogen, to produce a suitable effluent for second-stage nitrification. A high ammonia nitrogen to BOD ratio provides greater growth potential to the nitrifiers, relative to the heterotrophs, allowing operation at an increased sludge age in the nitrification step to compensate for lower operating temperatures. Because the rate of oxidation of ammonia is nearly linear, the tank configuration should be plug flow to minimize short-circuiting. Since biological nitrification destroys alkalinity, lime may be needed to raise the pH to the optimum level in the nitrification tank.

Nitrate is reduced to nitrogen gas by a variety of facultative bacteria in an anaerobic environment. An organic carbon source (AH$_2$ in Eq. 13-5) is needed to act as a hydrogen donor and to supply carbon for biological synthesis. Numerous reduced organic substances have been successfully tested as a carbon source, including acetic acid, acetone, ethanol, methanol, and sugar. Methanol has been preferred in most applications because it is the least expensive synthetic compound available that can be applied without leaving a residual BOD in the process effluent. However, this is not to imply that methanol treatment is cheap; it is estimated to contribute about one half of the total costs of denitrification. Methanol demand for a typical domestic waste water is about 60 mg/l. The recommended denitrification system consists of a plug-flow basin with underwater mixers followed by a clarifier for sludge

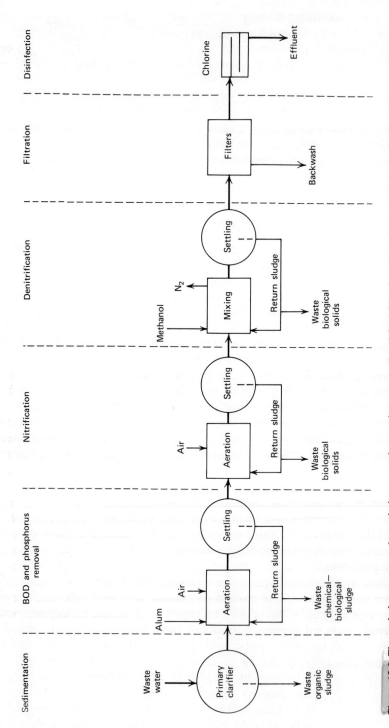

Figure 13-2 Three-sludge system schematic incorporating nutrient removal and final filtration. Organic matter is removed by primary settling and secondary activated sludge. Phosphorus is precipitated in the aeration secondary using a metal coagulant. Second-stage aeration converts ammonia to nitrate, and denitrification liberates nitrogen gas. Filtration removes nonsettleable solids and chlorination disinfects the plant effluent.

separation and return. The level of agitation must keep the microbial floc in suspension, but controlled to prevent undue aeration. Denitrification can also be accomplished in a submerged (anaerobic) filter; however, there are few data on field-scale studies as a basis for design.

Nitrification-denitrification in a two-stage system, preceded by secondary biological treatment, should achieve 90 percent inorganic nitrogen reduction and 80 to 95 percent total nitrogen removal at design flows. Advantages of biological nitrogen removal are that the nitrification process can be built to meet a current requirement for ammonia removal, adding denitrification in the future if required. Also, the system is adaptable as an addition to existing secondary treatment.

Air Stripping

Ammonia can be air stripped from solution at an alkaline pH of about 11, Eq. 13-6. After lime treatment for precipitation of phosphorus, a waste water can be pumped to the top of a cooling tower and distributed over the column packing (Figure 13-1). Forced air is drawn through the media to extract ammonia from the water droplets. The simplicity of the process makes it the least expensive method of denitrification where pretreatment with lime is employed for phosphorus removal. At Lake Tahoe during warm weather, air stripping can achieve 95 percent removal of ammonia nitrogen at a pH of about 11.5 by using 400 cu ft of air per gallon of waste water.

$$NH_4^+ + OH^- \underset{\text{acidic}}{\overset{\text{basic}}{\rightleftarrows}} NH_4OH \xrightarrow[\text{stripping}]{\text{air}} H_2O + NH_3 \uparrow \qquad (13\text{-}6)$$

Several technical problems must be resolved, however, before air stripping of ammonia can receive wide application. Experience with packed cooling towers has revealed the following opertional problems: scale formation on the tower packing that must be removed frequently by acid washing or mechanical scrubbing; ice formation in the tower during the winter season; reduced removal efficiency because of increased ammonia solubility at low temperatures, possibly requiring tower heating during the winter; and, the fact that nitrate nitrogen inadvertently produced in the biological pretreatment is not removed by air stripping. Attempts are being made to resolve the scaling problem by experimenting with media less susceptible to accumulation of calcium carbonate scale, such as smooth plastic packing or switching to fine water spray to eliminate tower packing entirely. Inability to operate at ambient air temperatures below 32° F may be solved during short periods of time by providing an alternate method, for example, breakpoint chlorination. Where weather is not severe, high pH holding ponds with long detention times and surface agitation are being studied. Climate may also limit the application of air stripping, since ammonia released to the atmosphere can be returned by rainfall.

Breakpoint Chlorination

The chemistry of breakpoint reactions between ammonia and chlorine are not particularly well defined. Equation 13-7 is an unbalanced chemical reaction to illustrate the possible products formed in the oxidation of ammonia; in order of importance, these are nitrogen gas, nitrous oxide, and nitrite-nitrate nitrogen. Weight ratios of chlorine to ammonia nitrogen required for breakpoint chlorination of waste waters have been reported as ranging from

8 to 1 to 10 to 1 as Cl_2 to N; the lower value being applicable for the most highly pretreated waste water. Tests have shown that breakpoint chlorination in the pH range of 6.5 to 7.5 can yield 95 percent ammonia removal, and for initial ammonia nitrogen concentrations of 8 to

$$NH_3 + HOCl \rightarrow N_2\uparrow + N_2O\uparrow + NO_2^- + NO_3^- + Cl^- \qquad (13\text{-}7)$$

15 mg/l, the nitrate and nitrogen trichloride residuals have never exceeded 0.5 mg/l. Chlorination is adaptable to physical-chemical treatment, and the process is relatively inexpensive and easy to operate and control. One disadvantage to heavy chlorination is that essentially all of the chlorine added is reduced to chloride ion, thus contributing to dissolved solids concentration in the treated water. For example, at an 8 to 1 weight ratio, the oxidation of 20 mg/l of ammonia nitrogen contributes 160 mg/l of chloride ion. In many instances, something less than complete ammonia removal may be sufficient to meet water quality objectives. However, at sub-breakpoint levels the production of chloramines may be too high and may present problems if they are discharged directly to receiving waters. Activated carbon is effective in destroying both free and combined residuals of chlorine; therefore, one solution is to pass the treated water through carbon columns.

Ion Exchange

Ammonium and nitrate ions are present in low concentrations relative to other ions in waste water, and their behavior does not differ significantly. Consequently, they are difficult to selectively extract by ion exchange. In order for denitrogenation by ion exchange to be an economical process, materials of relatively high specificity for the inorganic nitrogen are required, since the cost of removal of all ions by demineralization is not feasible for municipal treatment. Currently, a specific ion exchange resin for the nitrate ion is not available, but for the ammonium ion, clinoptilolite has unusual selectivity. It is a natural inorganic zeolite material currently available in small quantities. Research is underway in an attempt to develop a synthetic material.

Pretreatment prior to cation exchange involves clarification by chemical coagulation and filtration. Adjustment to a pH 6.5 is necessary to convert ammonia to ammonium ion, since free ammonia does not absorb or enter into ion exchange reactions with clinoptilolite. Spent exchange material is regenerated with a lime slurry, which is subsequently air stripped discharging ammonia to the atmosphere. Nevertheless, this process of ammonium ion exchange is still experimental, and it is difficult to speculate as to its future application in removal of nitrogen from waste water.

13-4   REDUCTION OF DISSOLVED SOLIDS

Activated carbon is effective in extracting dissolved organics in residual BOD, COD, color, and taste-and-odor producing compounds not completely removed by conventional biological treatment. Carbon removes organics through the action of adsorption and biodegradation. Molecules in solution are captured in the porous surface of granulated carbon, while other materials are retained through precipitation and biological assimilation. Theoretically, adsorption is the principle mechanism by which dissolved organics are removed, while biological activity regenerates sites by reopening the surfaces of the carbon. Although adsorption probably does dominate when a carbon column is first put into service, it appears that the biological contribution to removal capacity is quite significant. Consequently, toxic substances that inhibit microbial activity can reduce the carbon's removal capacity. High pH waste waters, resulting from prior chemical clarification, must be neutralized before carbon filtration. Since understanding of carbon treatment is incomplete, performance on a particular waste water should be based on testing, preferably a pilot-scale evaluation.

Carbon contactors can be used as a tertiary step to upgrade biological or physical-chemical treatment. Pretreatment usually includes filtration. Water is passed through a vessel filled with either carbon granules, or a carbon slurry, for a contact time ranging from 15 to 40 min. The columns, being subject to plugging, need backwashing to clean the bed. When the carbon adsorptive capacity is exhausted, the spent carbon is regenerated in a furnace at about 1700° F. Reactivated carbon along with a small amount of fresh material, to make up for loss in regeneration, is replaced in the contactor.

Demineralization processes can be used to remove dissolved salts from water. The systems most frequently employed in water treatment are discussed in Section 7-12. Because of high costs, it is not anticipated that reduction of dissolved salts by any of these methods will find wide application in waste treatment.

REFERENCES

1. *Physical-Chemical Wastewater Treatment Plant Design*, Environmental Protection Agency, Technology Transfer, August 1973.

2. *Nitrification and Denitrification Facilities, Wastewater Treatment*, Environmental Protection Agency, Technology Transfer, August 1973.

3. Culp, R. L., and Culp, G. L., *Advanced Wastewater Treatment*, Van Nostrand Reinhold Company, New York, 1971.

chapter 14

Water Reuse and Land Disposal

Indirect reuse of waste waters occurs when effluents that have been discharged into a river are drawn as a water supply downstream. In some instances, treated wastes discharged to streams comprise a significant portion of drought flows. The practice of discharging to surface waters and withdrawal for reuse provides dilution and separation in time and space, thus, allowing the processes of natural purification to take place. In contrast to this, planned direct reuse of reclaimed waters is being practiced for several applications, for instance, industrial processing, irrigation, and recreation. In a few cases, reclaimed waters are used for groundwater recharge and, in isolated examples, for reuse in municipal water systems.

14-1 WATER QUALITY MANAGEMENT

Water quality management is essential where indirect reuse occurs or if direct application is being considered. Long-range (50-yr) regional studies can provide a framework for day-to-day development of integrated water and waste-water systems. The major objectives of comprehensive planning are: to establish a water quality monitoring system; to identify all waste discharges; to evaluate water and waste-water treatment plants; to conduct special studies of unique problems; and to maintain updated water quality standards. The latter are fundamental to quality control. Figure 14-1 shows how various established standards relate to water use and treatment. Surface water standards define the quality acceptable for beneficial uses, for example, for public water supplies. Waste-water effluent standards regulate discharges of industries and municipalities to insure their compliance with surface water criteria; industries located in cities are controlled by sewer use ordinances. Finally, drinking water standards apply to potable public supplies.

The first standards were for drinking water, and were initially established without regard to waste treatment and disposal practices. However, with

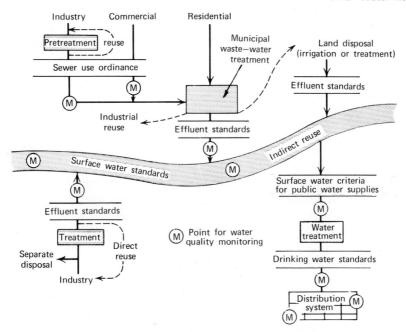

Figure 14-1 This illustration diagrams the various water quality standards used to manage an integrated water and waste-water system.

population growth and dramatic industrial expansion, untreated waste-water discharges exceeded the self-purification capacity of surface waters, and new emphasis on pollution control was needed. Deterioration of surface-water sources for public supplies was of particular concern to the waterworks industry. Certain substances are not significantly affected by conventional chemical water processing (Table 5-4). These refractory compounds were often identified with municipal waste waters or with industrial wastes discharged directly to watercourses. A public awareness of water pollution forced new legislation in the early 1960s. Surface water standards were established first, including criteria for public water supplies. Next effluent standards and sewer-use ordinances were instituted to limit discharges, particularly elements dangerous to human health and aquatic life. Refractory toxicants entering surface waters may pass through water treatment plants to the consumer. Therefore, the best point of control is pretreatment or separate disposal at industrial sites. One of the recent trends in waste management is land disposal. Although the intent is to protect the water environment, control of trade wastes in the municipal waste water is needed to safeguard the soil-plant filter and, of course, return drains and surface runoff are subject to effluent standards.

14-2   WATER REUSE

Many areas of the United States are confronted with increasing demands for water supply. To insure adequate quantities for both domestic and industrial use, several municipal and regional water agencies have either undertaken water reuse projects or are developing comprehensive studies of water resources. The objectives are to plan for necessary treatment and distribution piping, to promote and monitor water quality, to investigate potential new water sources, and to provide for proper financing for the entire system. Depending on local conditions, new sources may include: surface waters, natural groundwater or underground well recharge water, and processed waste water. Following are historical cases of water reuse and two examples of metropolitan areas that have undertaken extensive water use studies.

Chanute, Kansas[1]

During the years 1953 to 1957, a record drought occurred in the tributary area of the Neosho River; flow in the river decreased and nearly ceased by early 1956. The city's water supply reservoir on the river was forecast to go dry unless additional water was discharged to the impoundment. Since there was no feasible alternative water source, the city officials were faced with the option of either shutting down industries and severely curtailing water consumption or reusing the municipal waste water. The decision was to impound treated effluent from the trickling filter plant and to pump it back into the water supply reservoir. The mean cycle time through reservoir, water treatment plant, distribution system, waste treatment plant, and back to the reservoir was approximately 20 days. The recycling system operated for 5 months during with Chanute reused the same water approximately seven times.

Although the waste treatment plant removed about 85 percent of the BOD and bacteriological quality of the tap water was apparently maintained by extensive chlorination, the following operational problems occurred in water treatment: extremely high chlorine demand of the raw water reduced effectiveness of chlorination for taste and odor control, color was not adequately reduced, frothing occurred in the recarbonation basins and during filter backwashing chemical coagulation was impaired, and carry-over of solids from the sedimentation tanks resulted in plugging and coating of the filter sand. The treated water gradually developed a pale yellow color and an unpleasant musty taste and odor. It contained undesirable quantities of dissolved minerals and organic substances, and foamed readily when agitated. Chloride ion increased dramatically from about 40 to 670 mg/l. Near the end of the 5 months recycling, the impounded water took on the character of a treated waste water with about 17 mg/l of combined nitrogen, 6 mg/l of ABS, 13 mg/l of BOD, and 1000 to 1200 mg/l total solids. Although the practice at Chanute far exceeded the limit of reasonable reuse, it did point out dramatically the shortcomings of wastewater reuse with conventional treatment processes and insufficient dilution with fresh water. Where high concentration of refractory compounds are involved, special treatment operations must be employed to insure quality control during reuse.

Santee, California[2,3]

This community of approximately 15,000 persons is located in a semiarid region about 20 miles northeast of San Diego. With no local water supplies available, the Santee district depends on imported water from the Colorado River. The rising cost of this water has encouraged continued reuse of waste water starting in 1959 for recreational lakes (Figure 14-2). The initial Santee water reclamation project consisted of land-spreading of secondary wastewater effluent after polishing treatment in a large oxidation pond. The percolation area is a canyon floor consisting of a 10- to 12-ft layer of sand and gravel

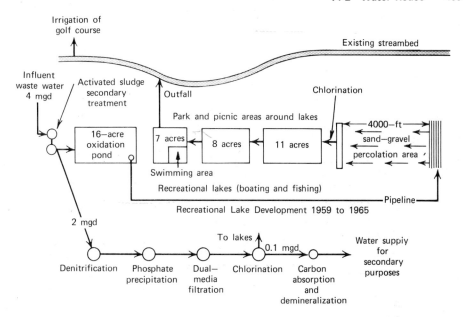

Figure 14-2 Water reclamation and reuse at Santee, California.

underlain by a clay stratum. After percolating through this natural filter for a distance of about 4000 ft, the underground flow is intercepted by a well extending across the valley floor, chlorinated, and is pumped into the first of four recreational lakes constructed in series. The second and third lakes are used for fishing and boating, and the final one includes swimming. The latter is maintained in accordance with the established state standards for artificial pools, including disinfection of the entire lake volume. Recent plans include improving the waste-water reclamation system to reuse water for secondary purposes, construction of a separate water distribution system for delivery of reclaimed water, facilities for recharging the underground aquifer for the purpose of storing and treating water prior to reuse, and withdrawal of brackish water from the aquifer for municipal supply after demineralization.

The first phase in 1965 consisted of storing surplus reclaimed waste water in a gravel pit for irrigation of a golf course. This was done in lieu of pumping excess waste water for disposal in the ocean about 20 miles away. Between 1967 and 1970, a distribution system for reclaimed water was extended to customers for irrigation of parks and institutional landscaping. A tertiary treatment facility was established to augment the soil percolation system. The treatment processes include removal of nitrates by biological nitrification-denitrification, reduction of phosphates by chemical flocculation and sedimentation using alum and lime, and final polishing by filtration through dual media filters. The finished water is chlorinated and transported to the recreational lakes. Demonstration pilot plants consisting of carbon filters, ion exchangers, and electrodialysis units are being tested for the removal of refractory organics and reduction of total dissolved solids to a level below drinking water standards. However, the district has no intention of returning this water directly to the potable system.

Windhoek, Southwest Africa[4]

The city of Windhoek, capital of Southwest Africa, is situated in a very arid region where surface water resources are scarce and expensive to develop. Under these circumstances in 1968, a waste-water reclamation plant was put into operation that provided approximately one third of the city's total domestic water supply. Windhoek is the first city in the world to practice large-scale waste-water reuse for drinking purposes.

The scheme for waste-water reclamation as is shown in Figure 14-3 includes both conventional and advanced waste treatment processes. After primary settling and secondary treatment using trickling filters, the waste water flows to a series of three stabilization ponds with a total retention time of about 20 days. Algal growth in these ponds reduces the inorganic nitrogen and phosphate levels. These maturation ponds also provide time separation reducing other unwanted contaminants. Pond effluent is recarbonated, lowering the pH from 9.0 to 7.5, and is dosed with 150 mg/l alum for flotation separation of the algae. Skimmer arms collect the algal scum for disposal, and the underflow is next treated by foam fractionation. Compressed air applied near the bottom of the tank agitates the water to produce a foaming action. The foam is then skimmed off the top of the water and is collapsed by a water spray for ease of disposal. Breakpoint chlorination is applied to the effluent from the foam fractionater to oxidize and drive off the majority of the remaining inorganic nitrogen and to establish a free chlorine residual. A small dosage of lime, about 30 mg/l, is

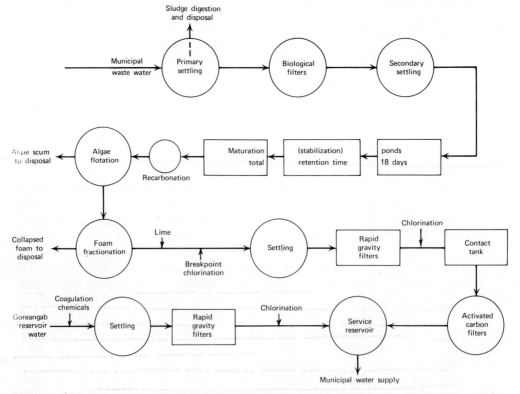

Figure 14-3 Schematic of the water reclamation plant and water supply system for Windhoek, Southwest Africa.

supplied with the chlorine to improve suspended solids settleability. The clarified water is filtered through rapid sand filters followed by granular activated carbon columns. Taste, color, and odor are improved by removal of residual dissolved materials in carbon adsorption. The columns are back-washed to remove suspended materials and the carbon is replaced when necessary. Spent carbon is stored for future regeneration.

Effluent from the activated carbon columns is rechlorinated and blended with treated Goreangab Dam water before being pumped to the distribution system. Reservoir water is given conventional chemical coagulation, filtration, and chlorination treatment. Reclaimed water has a higher total dissolved solids concentration than is acceptable for drinking water, but blending with the less saline fresh water produces a combined water within the World Health Organization standards. The reclamation plant contributed an average of 13.4 percent of the total city water consumption during the 2-yr period of 1969 to 1970, with monthly proportions ranging from 0 to 27.7 percent.

The reclamation plant has had operational difficulties. Mechanical failure of gear boxes driving the scraper arms in algal flotation and foam fractionation units have caused temporary stoppages. Recharge of the activated carbon filters with new carbon have caused further shutdowns. However, the most significant problem involves winter operation of the stabilization ponds. Reduced bacterial and algal activity results in increased levels of ammonia in the pond effluent requiring chemical oxidation using chlorine; this is considered to be uneconomical. No solution to this problem is in sight without resorting to an alternate treatment method, such as excess lime dosing and ammonia stripping. Process losses from the reclamation plant are higher than anticipated. Losses in algae flotation and foam fractionation are 7.5 percent of the total water produced, and wash water from the sand and activated carbon filters amounts to 3.0 percent.

Lake Tahoe, California[5]

The South Tahoe Public Utility District created in 1951 was empowered to provide long-range planning for preservation of water quality, with particular emphasis directed toward Lake Tahoe which is a deep oligotrophic alpine lake. After a detailed investigation, it was found that effluent disposal by spray irrigation on land could not accommodate the rapid population growth, and advanced waste treatment was recommended.

The present water reclamation plant (Figure 14-4) with a design capacity of 7.5 mgd consists of conventional secondary treatment followed by tertiary chemical-physical treatment. Primary and secondary treatment are performed by an activated sludge system, with waste sludge dewatering and incineration. The phosphorus and nitrogen removal processes consist of lime precipitation and air stripping of the secondary effluent. A lime dosage of about 400 mg/l as CaO is necessary for maximum phosphate precipitation. At the resulting high pH, the waste water is pumped through a countercurrent ammonia stripping tower for nitrogen removal. Recarbonation is then applied to reduce the pH to 7.5 prior to filtration through mixed-media pressure units. Activated carbon columns take out refractory soluble organics not removed by lime coagulation, and the final step is post-chlorination. The lime sludge is recalcined for reuse in the treatment process.

The most significant operational problem has been associated with the

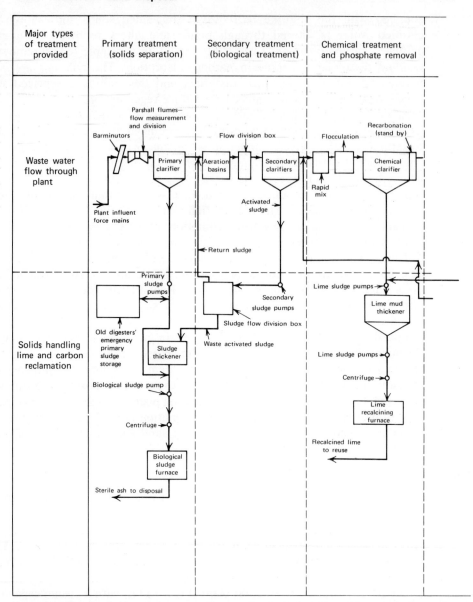

Figure 14-4 Schematic flow and process diagram for the Lake Tahoe Water Reclamation Plant. issue of *Water and Wastes Engineering.*)

ammonia stripping process. Calcium carbonate scale accumulated on the tower packing of rough-cut hemlock slats. The inaccessible packing scaled to the extent of interfering with water droplet formation and air flow, thus resulting in loss in tower efficiency. Experience also has shown that it is not practical to operate the tower at ambient air temperatures below 32° F. Ice formation and reduced ammonia removal efficiency, less than 30 percent, made tower operation of little value at low temperatures. A modified ammonia stripping system consisting of high pH ponds, air spraying with forced ventilation, and breakpoint chlorination is being investigated. Clarified lime-treated waste water would flow to ponds equipped with air-spraying systems for a detention of less than one day. Sprays of recirculated pond water could provide partial

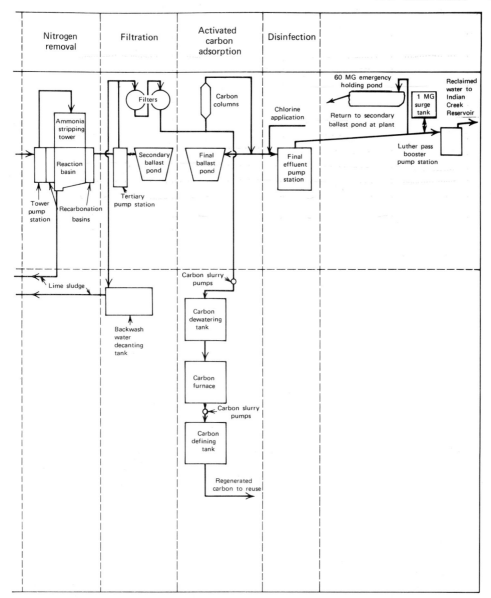

| Nitrogen removal | Filtration | Activated carbon adsorption | Disinfection |

release of ammonia to the atmosphere. Pumping pond effluent through a stripping tower, without packing, using water spray nozzles, and forced air ventilation could complete the ammonia stripping. Breakpoint chlorination prior to filtration would serve as a standby system.

Although advanced treatment reclaims the waste water to nearly drinking water quality, a small quantity of phosphorus and trace elements still exist in the effluent, along with significant quantities of inorganic nitrogen. Therefore, the decision was to transport the effluent entirely out of the natural drainage area of Lake Tahoe. An export pipeline (Figure 14-5) conveys the water over Luther Pass to the reservoir 26 miles away where it is used for irrigation and recreation. Water impounded in the reservoir is approximately 70 percent

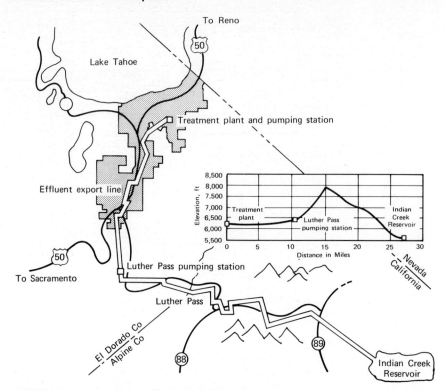

Figure 14-5 Export pipeline for reclaimed water from the Lake Tahoe Water reclamation Plant. (Courtesy of South Tahoe Public Utility District)

reclaimed water and 30 percent natural runoff mostly from the snow melt. The surface area of the lake varies from about 100 to 160 acres between conservation pool (300 million gallons) and full reservoir (1 billion gallons). Chemical and biological characteristics of the impounded water are similar to those of a fertile lake which is suitable for recreation and fish propagation. Field studies have shown that the reservoir is not experiencing excessive blooms of algae, apparently because of the limited supply of available phosphorus in the water. However, during periods of lake stratification the hypolimnion experiences significant reductions in dissolved oxygen concentration and a buildup of ammonia nitrogen. Although there is a loss of about 80 percent of the influent ammonia nitrogen from the reservoir, 15 mg/l to 4 mg/l apparently the result of natural nitrification-denitrification, the residual was toxic to trout stocked in the reservoir. The solution to this problem was providing aeration to destratify the reservoir and to supply dissolved oxygen to the hypolimnion.

Pleasanton, California[6]

One million gallons per day of secondary effluent from the municipal plant, serving a population of 10,000, is applied to irrigated pasture (Figure 14-6). Each day 1 ft of water is sprinkled over 2.5 acres of the 90-acre area providing an irrigation cycle of 30 to 35 days. This application rate, of about 11 ft/yr, is the maximum volume that can be handled in addition to the 20 in. of winter rainfall without resulting in overland runoff.

The soil profile is a moderately permeable loam containing lenses of sand and

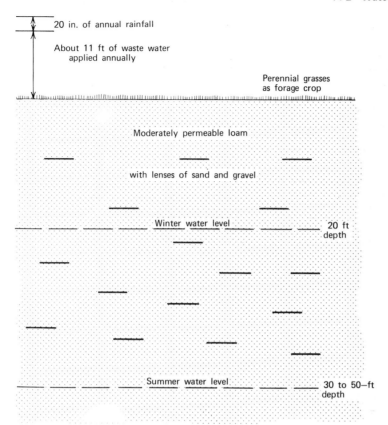

20 in. of annual rainfall

About 11 ft of waste water
applied annually

Perennial grasses
as forage crop

Moderately permeable loam

with lenses of sand and gravel

Winter water level

20 ft
depth

Summer water level

30 to 50—ft
depth

Figure 14-6 Spray irrigation land disposal at Pleasanton, California.

gravel. The water table is about 20 ft below the surface in winter, dropping to a depth of 30 to 50 ft in summer. Liquid loading on the pasture is approximately four times greater than the irrigation requirement of the perennial grasses grown as a forage crop. The quantity of nutrients also exceeds plant needs. Effluent applied to the soil each year contains an estimated 400 lb of nitrogen per acre, 200 lb of phosphorus, and 500 lb/acre of potassium.

Observations have led to the following hypotheses regarding operation of the living filter at Pleasanton. Most fecal coliforms and biodegradable material are removed during the first 5 to 10 ft percolation. Nitrogen, phosphorus, and potassium stimulate plant growth and under optimum conditions the pasture grasses utilize about 300 lb of nitrogen per acre per year. Cyclic application permits alternate aerobic and anaerobic conditions in the soil profile leading to loss of nitrogen by biological nitrification and denitrification. Most of the phosphate in excess of plant needs precipitates in the soil as insoluble inorganic salts. After 20 years of operation, periodic analyses of groundwater beneath the irrigated field indicate that the soil continues to fix and hold insoluble inorganic salts. Salts that do not precipitate have not created a problem in the root zone of the soil. Apparently the sodium salts are leached from the profile and are adequately diluted in the groundwater. Tests have revealed no dangerous amounts of nitrate, phosphorus, sodium chloride, or other pollutants. The land treatment as practiced is considered to be a successful operation for waste-water disposal.

Denver, Colorado[7]

Current water sources for the Denver area are the South Platte River system and two transmountain diversion systems (Figure 14-7).One transports water from the upper reaches of the Fraser River through the Moffat Tunnel for 6 miles en route to Denver. This diversion was completed in 1937, and the Blue River Project was added in 1964. Water from the Blue River Drainage Basin is impounded in Dillon reservoir and diverted to Denver through a 23-mile tunnel.

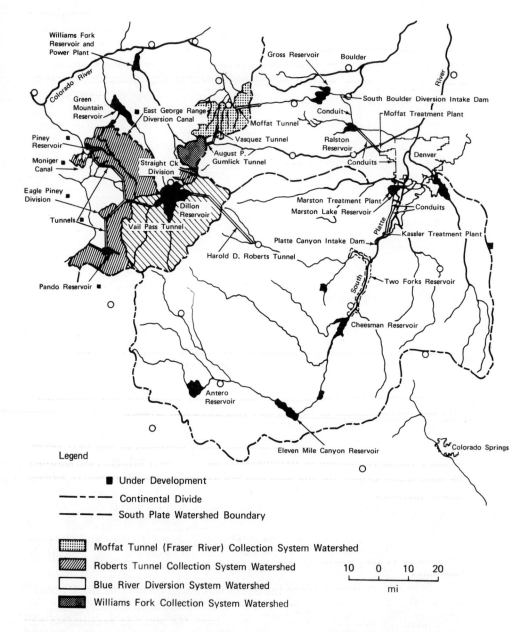

Figure 14-7 Water supply system for Denver, Colorado. (From K. D. Linstedt, K. J. Miller, and E. R. Bennett, "Metropolitan Successive Use of Available Water," *Journal American Water Works Association*, Vol. 63, October 1971. Copyright 1971 by the American Water Works Association, Inc.)

Potential future water sources include reapportionment of waters presently used for agriculture to municipal and industrial uses, importation of additional western slope water, and reuse of treated waste water. Present projections indicate that additional supplies delivered through the existing Moffat and Blue River systems will be sufficient until the year 2010. The prediction of additional needs within the next 40 years has prompted considerable interest in water reuse. The following study program has been outlined for the next decade: review of water reclamation processes, cataloging of beneficial uses, investigating quality requirements for various uses, examining necessary alterations to the distribution system, legal considerations, and evaluation of public acceptance. Initial reuses being considered are cooling water for power plants and industry, plus landscape irrigation on public parks, golf courses, and similar areas. A beneficial use study is identifying potential water reuse customers and their locations within the service area. This information is prerequisite to locating water reclamation plants and distribution piping. Quality needs for various applications are required to determine the degree of treatment necessary to reclaim waste water. The question of public acceptance of renovated waste water is related to the proposed uses and the information that reaches the public. Surveys are planned every three or four years to determine the levels of public information and attitudes toward domestic reuse. This measurement of public opinion can provide direction for an informational and educational program regarding reuse of waste water.

Los Angeles, California[8]

Waste-water reclamation is most desirable and necessary in more arid regions of the country. For example, California is presently reclaiming about 50 billion gallons of waste water annually at more than 100 reclamation operations. Because the bulk of its water must be imported, pilot and operational studies have been made by the City of Los Angeles since 1927. Initial water supplies were provided by damming the major stream channels draining mountain areas and deep-well pumping. By the early 1900s local water supplies could no longer sustain the city's developing growth and aqueduct systems were developed. The Los Angeles Owens River Aqueduct was completed in 1913 and was extended in 1970. The Colorado River Aqueduct was completed in 1941, and the California Aqueduct was completed in 1972. The latter two sources were planned to meet increasing water demands of the next several decades, but development of individual reclaimed water projects are adding to the total water supply for the area and are extending the adequacy of existing import facilities.

Because rainfall occurs only a few winter months in storms of high intensity but short duration, separate systems were constructed for storm-water drainage and waste-water collection. The Hyperion regional treatment plant was designed to collect wastes by gravity flow from a 328,000-acre area in the San Fernando Valley with a contributing population of three million. The sea level location of the plant adjacent to the ocean was based on the traditional concept of collecting waste water by gravity flow and discharge to the ocean. This site has limited the economic feasibility of reusing the 300 mgd reclaimable flow (Figure 14-8). Present investigations consider about 100 mgd for the purposes of industrial reuse and well injection to prevent seawater intrusion. Distance from other users makes economical reclamation more difficult and in some

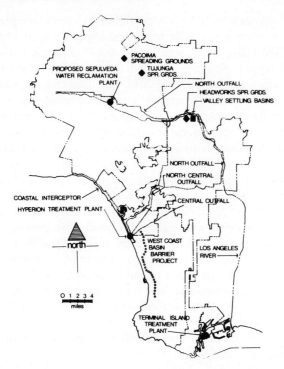

Figure 14-8 Los Angeles water reclamation plan showing the locations of the regional (Hyperion) waste-water treatment plant, aquifer injection barrier project to control seawater intrusion, spreading areas for groundwater recharge, and proposed locations of water reclamation plants. (From R. D. Bargman, G. W. Adrian, and D. C. Tillman, "Water Reclamation in Los Angeles," *Journal of the Environmental Engineering Division, American Society of Civil Engineers*, Vol. 99, No. EE6, December 1973)

cases impossible. For example, to use Hyperion effluent to recharge the San Fernando Valley aquifer would require 30 to 60 miles of pipeline through congested city areas and an increase of elevation of about 1000 ft. The proposed locations of new water reclamation plants are being considered relative to reuse sites. The city's master plan includes waste-water collection, treatment, and water reuse in terms of sewer location, treatment plant sites, quality of imported water, and quality available for reuse. The uses of reclaimed water evaluated are: groundwater recharge by spreading, aquifer injection for the control of seawater intrusion, industrial supplies for processing waters and cooling, landscape irrigation, and recreational lakes.

Table 14-1 gives the cost of potable water sources, including delivery to the distribution system, and estimated costs for reclaimed water use. The latter values include necessary tertiary treatment, pumping, and piping to the point of application. These costs consider the feasible uses of waste water from the Hyperion plant and renovated water from proposed reclamation plants at sites 10 to 20 miles inland from the ocean. Indications are that where it is technologically possible to produce a reclaimed water of acceptable quality, several specific reuses appear economically competitive with fresh-water supplies. However, if extremely high standards for discharge to the ocean and for reuse are set by state and federal agencies, the economics of reuse could change dramatically.

Table 14-1. Potable Water Costs and Estimate Reuse
Costs for the City of Los Angeles

Potable Source	Dollars per Acre-Foot
Los Angeles aqueduct	32
Colorado River	61
Local wells	21
California aqueduct	62
Reclaimed Water Use	
Coastal industrial supply	15 to 25
Recreational lakes in Los Angeles River	15 to 25
Groundwater recharge by spreading:	
Secondary effluent	20 to 30
After tertiary treatment	70 to 80
Irrigation	
Nearby parks	10 to 40
Freeway landscaping	750 to 1000
Aquifer injection for seawater barrier	50
Injection wells in San Fernando Valley	75 to 85

Source. R. D. Bargman, G. W. Adrian, and D. C. Tillman, "Water Reclamation in Los Angeles," *Journal of the Environmental Engineering Div., American Society of Civil Engineers,* Vol. 99, No. EE6, December 1973.

14-3 PROBLEMS ASSOCIATED WITH MUNICIPAL REUSE

The studies at Denver and Los Angeles point out the importance of regional planning in development of water reuse systems. Waste-water collection piping and treatment facilities must be located near reuse projects. Otherwise the cost of transporting renovated water to the reuse site may deem the project unfeasible. Treatment facilities should be designed to match reuse needs, both in terms of quality and quantity. This requires continuing studies of reclaimed water markets and availability of water sources. Technological developments and costs of advanced waste treatment for different degrees of water renovation must be accessible to planners. Finally, approval must be obtained from both citizenry and controlling governmental organizations for reuse projects to reach fruition. If careful, continuous, and comprehensive planning is not practiced, the full potential of water reuse cannot be realized.

Conventional secondary waste-water treatment plants can remove a majority of the organic matter and, to a considerable extent, the heavy metals, although these are best controlled at the originating industrial sites. Chemical treatment methods are now available for the removal of phosphates, and both chemical and biological systems are being developed for the removal of nitrogen; but dissolved solids are only sparingly reduced by these processes. Since domestic use adds approximately 300 mg/l of total dissolved solids, waste waters may require demineralization prior to reuse. Current technology provides techniques such as distillation, ion exchange, reverse osmosis, and dialysis for removal of dissolved solids, but they are very costly and create a problem of brine disposal. If demineralization of a waste water becomes necessary prior to reuse, this process cost may be greater than the cost differential of using conventional sources, thus making reuse economically unfeasible.

Bacteriological quality is of major importance to domestic reuse. The American Water Works Association's policy statement regarding the use of reclaimed waste waters as a public water supply source is as follows: "The Association is of the opinion, however, that current scientific knowledge and technology in the field of waste-water treatment is not advanced sufficiently to permit direct use of treated waste waters as a source of public water supply, and it notes with concern current proposals to increase significantly both indirect and direct use of treated waste waters for such purposes." [9] However, the AWWA does encourage increased use of reclaimed waste waters for beneficial purposes such as industrial cooling and processing, irrigation of crops, recreation and, within the limits of historical practice, groundwater recharge. It urges that steps be taken through intensive research and development to identify: the full range of possible contaminants present in treated waste waters, the degree to which these contaminants are removed by various treatment processes, long-range physiological effects of continued use of reclaimed waste waters, definition of testing procedures, methodology, and monitoring systems to be employed with respect to reuse of reclaimed waste waters, development of greater capability and reliability in treatment processes, and, improved capabilities of operating personnel. The Association concludes its policy statement with: "The Association believes that the use of reclaimed waste water for public water purposes should be deferred until research and development demonstrates that such use will not be detrimental to the health of the public and will not effect adversely the wholesomeness and potability of water supplies for domestic use."

14-4 LAND DISPOSAL OF WASTE WATERS

Both irrigation of waste water in dry climates and soil systems for reclamation of wastes to reduce treatment costs have been used for several decades, although on a limited basis. Applications of effluent for irrigation and groundwater recharge are not synonymous with the previous practice of spreading of untreated waste which is no longer acceptable. Before waste water can be recycled it must be appropriately treated to be compatible with the soil system and environmental concerns. Land disposal projects must consider the following factors prior to implementation: quality and quantity of waste water available; method of application, including rate and total annual quantities; characteristics of the strata in the soil profile and groundwater levels; elapsed time between application and withdrawal based on rate of groundwater flow; distance between point of application and drainage collectors; and, anticipated quality of the reclaimed water. There are three types of land disposal systems: spray irrigation, rapid infiltration, and overland runoff.

In spray irrigation, the choice of equipment depends on site topography and crop. For large, permanent facilities the pipe network is often buried with only risers and spray nozzles appearing above the ground surface. Irrigation pipe may be placed on the ground with attached risers and spray nozzles for temporary use. On level sites, rotating spray booms supported by a central tower, or riding on a circumferential track, are often used (Figure 14-9). Choice of spray equipment and general site layout are also dependent on health and

Figure 14-9 Large rotating, twin boom applicator that is designed for spray application of waste-water effluents. The low-pressure nozzles on the spray booms are directed downward, providing for low application rate on the cover crop with a minimum of aerosols. (Courtesy of McDowell Manufacturing Company, DuBois, Pa.)

hygiene considerations. High-pressure, high trajectory, fine droplet sprays are desirable to avoid erosion and maximize coverage, but this type of spray also increases the risk of aerosol loss that could carry pathogenic organisms. Application rates range from 0.2 in./week to a maximum 6 in./week when air temperatures are sufficiently high to prevent freezing. The allowable application rate to handle the liquid loading depends on soil type, topography, infiltration capacity, and weather conditions. A conservative estimate appears to be an average of 2 in./week (54,300 gal/acre) at a rate of 0.25 in./hr for an 8-hr period when site and climatic conditions are suitable for spray irrigation. Ideally, the surface soil should be of a silt loam texture to a depth of about 5 ft. Since substantial land area is generally required, several soil types might be encountered on a disposal site. A flexible spray irrigation system can be designed to apply waste water at loading rates compatible with the varying soil characteristics of the site.

Rapid infiltration is performed by spreading or flooding waste water onto land capable of a percolation rate of several feet per week. The function of most infiltration areas located in the Southwest and California is groundwater recharge. Generally water is applied to a series of ponds for 10 to 14 days followed by a drying period of 10 to 20 days, depending on season of the year. Percolation rates during the application period ranged from 1 to 4 ft/day with a maximum of about 330 ft/year. The drying cycle is necessary for oxidation of organic matter to restore permeability, since aerobic conditions that occur during flooding can result in clogging of the soil. The bottom of the basins may be barren or grassed. Grasses are beneficial to prevent clogging of the soil pores and to retain high infiltration rates; Bermuda is one species of grass that is able to withstand periods of shallow inundation. Ideal site conditions would be a shallow layer of sandy loam to support grass growth, underlain by gravels and sands with few or no fines, and a groundwater table at a depth of 10 to 20 ft. The grass bottom and surface soils contribute to renovation capacity, while the deeper clean coarse sands and gravels have, by comparison, insignificant reclamation ability. Specific data on renovation versus depth at rapid infiltration sites are limited. Implications are that poor soils under high loadings remove very limited amounts of dissolved solids. Although analyses of groundwater under infiltration basins have shown reduction of such mobile ions as chloride and nitrate, this is more likely because of dilution of the percolated water with fresh groundwater than renovation by the soil column.

Overland runoff is applicable only for sloping sites with relatively impervious soils, such as clays, clayey sands, or silts. Waste water is discharged by spraying or spreading to produce sheet flow down the slope, rather than infiltration, as is sketched in Figure 14-10. Renovation is achieved by movement of the water through grass and decaying debris on the soil surface. A continuous vegetative cover is essential for filtering action. Sites must be leveled and graded to provide slopes between 2 and 6 percent. For each slope length of 200 to 300 ft, a cutoff trench is constructed to intercept the flow in the soil mantle. Application points and collector ditches are located to provide the necessary contact time for treatment. Buried or above ground pipe laterals equipped with spray nozzles are generally used for waste water distribution. Application rates vary from 1.5 to 3 in./week during a daily application period of 4 to 6 hr. The actual wetted coverage is about one half of the area. Sites should have, at least, 6 to 8 in. of good topsoil to support grass growth. Grasses that have been tested include: Kentucky blue, Bermuda, Reed canary, seaside bent, red top, and fescue. Reed canary grass seems to be preferred primarily because of its market value as a high quality hay.

The two major restrictions to overland runoff are difficulty of maintaining consistent quality in the renovated water and site preparation cost. Since water flowing over the ground surface is exposed to weather, biological activity and, consequently, degree of renovation can be adversely affected by lack of sunshine and cold temperatures. Careful management and control of operations are necessary to insure a satisfactory final water quality. Under adverse conditions, reductions of nitrogen, phosphorus, and metal cations may drop to as low as 50 percent, while 85 percent or more are removed in an optimum

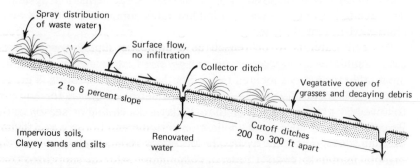

Figure 14-10 Sketch of an overland runoff land disposal system.

environment. Although the degree of renovation from a spray irrigation system for a particular soil type and depth can be predicted, this is not true with overland runoff.

14-5 APPLICATION OF RAPID INFILTRATION

The great advantage of a high infiltration system is the large quantity of waste water that can be discharged to a small land area. The Flushing Meadows Project,[10] located in the Salt River bed west of Phoenix, Arizona, was installed in 1967 to determine the feasibility of renovating secondary waste-water effluent by groundwater recharge using infiltration basins. The test site included six basins each 20 by 700 ft with a total bottom area of slightly less than 2 acres. The annual hydraulic loading was about 300 ft (0.27 mgd/acre, or a hydraulic population equivalent of 2700 persons per acre) in contrast to the recommended average spray irrigation rate of 9 ft/yr. Soil beneath the basins consisted of 3 ft of fine loamy sand (approximately 85 percent sand, 10 percent silt, and 5 percent clay) underlain by coarse sand and gravel layers to a depth of 240 ft, where a clay layer begins. The groundwater table was at a depth of about 10 ft. The optimum cycle was determined to be 2 weeks of inundation at depths of 6 to 12 in. of standing water, followed by a 10-day dry-up period in the summer and 20 days in the winter. Year-round operation was possible in Arizona. Observation wells were installed under the basin area for obtaining samples of the groundwater.

Nitrogen content of the secondary waste water applied averaged about 30 mg/l; this yielded a nitrogen loading of about 24,000 lb per acre per yr at an annual infiltration rate of 300 ft. The estimated overall nitrogen removal, based on average nitrogen concentration in the renovated water under the basins was about 30 percent. Since only a few hundred pounds of nitrogen could be removed by harvesting the grasses, denitrification was probably the main mechanism for the nitrogen reduction. Essentially all the nitrogen applied was in the form of ammonia. During infiltration, ammonium ions were absorbed by the cation exchange complex in the soil until saturation, and then ammonia was carried into the groundwater. During the drying period, ammonia complexed in the soil was converted to nitrate under aerobic conditions. On reflooding, a portion of the nitrate was converted to nitrogen gas while the majority was flushed down into the underlying water, since nitrate ions are relatively unaffected by the ion exchange capacity of a soil. After three years of operation, the ammonia nitrogen content of the renovated water averaged about 10 mg/l. Nitrate nitrogen concentrations varied from near 0 to greater than 20 mg/l during inundation. These data illustrate the limitation of high infiltration in removing inorganic nitrogen. Serious problems can occur if the percolated water is to be reused as a domestic supply, since the drinking water standard is a maximum of 10 mg/l of nitrate nitrogen.

Other water renovation tests in coarse aquifers have indicated the reclamation limitations of rapid infiltration operations. Although multivalent ions such as orthophosphate appear to be extracted rather easily within travel distances of approximately 200 ft, others such as chloride and boron travel significantly long distances. The high hydraulic loads and reduced physical and chemical filtration actions of coarser soils, in combination with the short travel distances between basin bottom and groundwater table, limit water renovation capacity.

While a grass cover effectively removes suspended solids and organic matter, only a small portion of nutrient salts are utilized. There is significant risk of pathogen transmission to underground water, particularly viruses. The same compounds that are refractory to waste water and water treatment systems appear to be only sparingly removed by high-rate percolation. Thus it appears that the chief limitation of rapid infiltration is potential of groundwater pollution.

The rapid infiltration system proposed at Phoenix, Arizona, based on the Flushing Meadows Pilot Project, is illustrated in Figure 14-11. Secondary effluent from the present activated-sludge treatment plant is to be applied to basins located along both sides of the Salt River bed. Renovated water will be pumped from wells in the center of the riverbed and will be returned to surface flow. The design is intended to avoid loss of renovated water into the aquifer outside the river bed; to insure a minimum underground travel time and distance of several weeks and several hundred feet, respectively; and, to avoid a water table rise higher than about 5 ft below the bottom of the basins during infiltration. Data from the completed pilot project show that a system of this kind can produce renovated water of sufficient quality to permit unrestricted irrigation and recreational use. The Arizona standards for tertiary treatment require an aesthetically acceptable effluent with BOD and suspended solids concentrations of less than 10 mg/l, and a fecal coliform density below 200 per 100 ml.

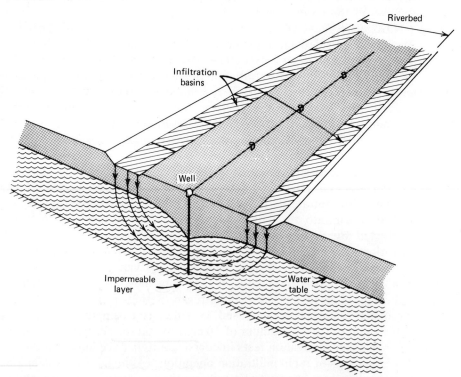

Figure 14-11 Proposed rapid infiltration system at Phoenix, Arizona, which includes a system of infiltration basins on both sides of the Salt River bed with wells in the center for pumping renovated water to the surface. (Herman Bouwer, "Renovating Municipal Wastewater by High-rate Infiltration for Ground-Water Recharge," *Journal American Water Works Association*, Vol. 66, March 1974)

14-6 APPLICATION OF SPRAY IRRIGATION

Spray irrigation is the best of the three land disposal systems in terms of predictability of the renovated water quality, and ease of operation and maintenance. Suitable soil types are available in most areas of the United States. The biggest restriction to waste-water application is climate; although spraying may be conducted at below 32° F, the ground must not be frozen. Spray irrigation is based on the premise that the distributed waste water percolates into the ground with no surface runoff. In warm southern climates operations can by conducted year-round; however, in the northern states spraying must be suspended during the winter months and reduced water renovation efficiency should be anticipated outside of the normal crop growing season. In the Midwest, spray irrigation could be conducted for a maximum of about 250 days per year.

The most recent crop irrigation and soil infiltration system for tertiary treatment of municipal waste water is in Muskegon, Michigan. The system has six components: collection and transportation piping, aerated stabilization ponds, storage basins, irrigation land and facilities for crop production, a natural filter of sandy soil, and a drainage network (Figure 14-12). Raw waste waters are collected from households and industries and are conveyed by gravity flow and force maining to the reclamation site 11 miles east of Muskegon. After biological treatment in three aerated lagoons, the flow enters storage basins that are sealed on the bottom to prevent exfiltration. A large storage volume is needed to accumulate the municipal discharge for at least four months of the year when spray irrigation is not possible because of frozen ground. Effluent leaving the outlet, no. 5 on Figure 14-12, is chlorinated and then pumped through a piping system to 55 center-pivot spray irrigation units. A network of underground pipes collects the infiltration, which is pumped to surface waters for disposal. A series of berms around the lowland edges of the

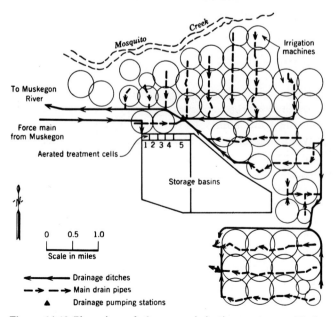

Figure 14-12 Plan view of the spray irrigation system at Muskegon, Michigan. (From *Civil Engineering*, ASCE, May 1973)

irrigation circles are to contain storm runoff water until it can be handled by the drainage system. The anticipated degree of treatment under design loadings is 99 percent reduction of BOD and suspended solids, 90 percent removal of phosphorus, and 76 percent or more removal of nitrogen.

Spray Irrigated Crops

The choice of vegetation depends upon climate, soil conditions, potential use of the harvested crop, and rules covering waste-water irrigation. State regulations generally prohibit the use of raw waste water to irrigate crops. Settled waste is sometimes permitted on industrial grain and fodder crops, and on vegetables grown for seed purposes; but not on plants grown for human consumption, such as vegetables, berries, low growing fruit, and vineyards and orchards during fruit growth. Complete treatment is preferred for feed and pasture crops that are used for animal consumption. Well-oxidized, disinfected waste-water effluent that conforms to the bacterial requirements of drinking water standards is generally allowed on all crops.

Many different crops have been successfully used in land disposal operations. These include wheat, corn, oats, silage corn, red clover, alfalfa; and, Reed canary, red top, tall fescue, Bermuda, and Sudan grasses. Most of these are intended as animal forage. Corn and some of the grasses grown as hay crops seem to have more favorable renovating capacity, since both demonstrate significant nutrient uptake and have potential market value after harvest. Corn may have a higher market value, but is also requires more effort necessitating annual plowing and planting. Perennial grasses require only harvest and have a further advantage in that their root system is fully established at the start of the spraying season and, therefore, can provide immediate response. Annual crops such as corn and other grains may allow significant nitrogen losses during the early growth period. Reed canary grass has been used successfully at a variety of sites in the United States and seems to be a desirable species.

Soil Types and Waste-Water Renovation Effectiveness

Soil factors that determine effectiveness of waste-water renovation are: clay mineral content, organic matter, soil permeability, pH and depth to groundwater. In addition, the soil's ability to renovate waste water depends on the design of the spray application system hydraulic loading rate and maintenance of the site. Spray irrigation is best on soil with fair to poor drainage characteristics, namely, silty sands and gravels, inorganic silts, very fine sands, or clayey sands. Based on studies of Pleasanton, California,[6] the following recommendations were proposed for disposal of secondary waste-water effluent on pasture. The surface soil should be, at least, 5 ft in depth with a surface texture ranging from sandy loam through silty clayey loam. Subsurface permeability should be moderate with a rate of 0.5 to 1.5 in./hr. Average water-holding capacity in the top 5 ft of soil should be 8 to 12 in., and the soil should possess little or no salinity or alkalinity. The water table should remain, at least, 20 ft below the ground surface. If these criteria are followed, satisfactory renovation is anticipated if the irrigation dosage does not exceed an average of 2 in./week on a year-round basis that is applied on a biweekly irrigation cycle using sprinklers.

Pollutional Concerns of Spray Irrigation

Municipal waste water is a mixture of domestic and industrial discharges, plus infiltration. Domestic waste is a potential source of viral and bacterial pathogens, and industries may discharge toxic organic compounds or heavy metals. Both add to organic content. Because of the potential hazard of transmitting diseases, altering physical, chemical, and biological characteristics of soils, and chemical pollution of water supplies, it is not acceptable practice to spread untreated waste waters. In general, secondary biological treatment plus chlorination is adequate to satisfy environmental concerns and to provide an irrigation water compatible with the soil system. Although there are materials that are resistant to conventional treatment processes, many of the compounds that would interfere with land disposal are eliminated by secondary biological treatment, for example, organic matter and pathogenic bacteria. Activated sludge cannot tolerate excessive quantities of acids, alkalies, and toxic chemicals. Therefore, such a system functioning properly is an indication that the levels of these pollutants are low in the raw waste water, and in resulting treated effluent. Inorganic solids, foam-producing matter, and color are not readily removed.

Removal of dissolved and suspended organic solids in spray irrigation is largely due to the high filtering effectiveness of the cover crop and soil mantle. Although treated waste-water nitrogen is essentially all in the form of ammonia, spray irrigation readily converts it to the nitrate form. In a properly operated system, the majority of nitrate is removed from the soil profile by synthesis into grass. On rewetting with the next irrigation application, any remaining nitrate may be denitrified to gaseous nitrogen and lost to the atmosphere, or the nitrate ions may percolate through the soil profile and eventually appear in the groundwater. Spray irrigation should remove about 80 percent or more of the nitrogen. However, the reclaimed water under the site may still contain measurable amounts of the applied nitrogen, primarily as nitrate. The amount of phosphorus removed in harvested grass crop is about one fifth that of nitrogen. Most municipal waste waters have phosphorus in excess of this amount but, since orthophosphate is readily complexed by multivalent cations, the remainder is removed in the soil mantle by adsorption or ion exchange or both. Thus overall phosphorus removal is expected to be 90 to 99 percent.

Retention of heavy metals by soil depends on adsorption by clay minerals and organic matter, and formation of insoluble compounds at alkaline pH. Although information is limited from field situations, under proper conditions the retention of the surface soil prevents significant movement of heavy metals to lower horizons, at least, for many years. Based on current information, it seems reasonable to assume that heavy metals will be retained by soils under conditions suitable for spray irrigation operations. Uptake of trace quantities by plants tends to extend the renovation capacity of the soil, but at some point in time retention capacity of the surface layers may be fully utilized. When final equilibrium is reached, drainage through the surface layer into subsoil will contain essentially the same concentration of metals found in the applied liquid.

In trace quantities many inorganic materials are essential growth factors for plants, but higher concentrations of these same materials can cause toxicity. Boron is a typical example. Many cereals and species of grass are sensitive to high boron levels, while some boron may be taken up by these plants. The balance may be retained in the soil, or may pass through into underlying strata. The sodium content of waste water is also significant. Salinity is important relative to the amount of calcium and magnesium. A high sodium to multivalent cation ratio adversely affects both soil and plants. It is difficult for plants to obtain water from a saline solution, and if the sodium adsorption ratio of the water is too high, the soil structure loses porosity. Salinity is more critical for irrigation in arid climates where rapid evaporation tends to increase the concentration of salts. In more humid climates of northern United States, it is unlikely that salt accumulations will be critical to the forage crops grown. The concentration of the dissolved minerals in the water can also be of significance, particularly, if some direct reuse of the product water is intended. The most common soluble salts are sodium, potassium, magnesium, and calcium sulfates and chlorides. Although some of these are retained in the soil by ion exchange, the total dissolved solids in the underflow may be nearly the same as the applied waste water. Boron, selenium, and nitrate are not retained significantly by soils and accompany the flow of water through the soil profile once the plant and microbial zones are passed.

Bacteria and viruses are removed by filtration, flocculation, and chemical interaction with the soil. Field observations indicate that bacteria and viruses are primarily captured in the soil mantle, although viruses may be transported to greater depths in some instances. Unless fissures or dissolution channels are present for organism transport, percolation through even a coarse soil provides some reduction. Since the soil environment is not conducive to survival, pathogens generally do not persist for long periods of time. The longevity of an extremely small percentage of residual pathogens is possible, but even in cold climates survival in the soil profile would not usually be expected to exceed one month. For liquid loadings and soil profiles associated with spray irrigation, fecal microorganisms are commonly trapped during the first 5 to 10 ft of percolation. Infiltration through porous media, or piping in the soil structure, may transport bacteria and viruses for several hundred feet. Furthermore, pathogens can be transported in the air by spray irrigation systems, which appears to be their greatest health hazard. Consequently, it is recommended that waste waters be chlorinated prior to land application.

The economics of spray irrigation is related to large area requirements. For example, for a city of 100,000 population 1290 acres, or about 2 square miles, is required. In addition, storage basins are needed to hold water when climatic conditions prohibit application. Apparent benefits include: recycling waste water for reuse may be an advantage; land disposal may be cheaper than other alternate tertiary treatment methods; pastures for effluent disposal can maintain open space and create green belts near urban areas; the system is consistent with the effort to expand and improve the used of irrigated pastures; and, when considered strictly as a source of irrigation water, effluent use may compare favorably with deep-well irrigation and, hence, provide a reduction in cost.

REFERENCES

1. Metzler et al., "Emergency Use of Reclaimed Water for Potable Supply at Chanute, Kansas," *Journal American Water Works Association.*, Vol. 51, No. 8, August, 1958, pp. 1021–60.

2. Merrell, J., and Stoyer, R., "Reclaimed Sewage Becomes a Community Asset," *American City*, April 1964, pp. 97–101.

3. Houser, E. W., "Santee Project Continues to Show the Way," *Water and Wastes Engineering*, May 1970, pp. 40–44.

4. Clayton, A. J., and Pybus, P. J., "Windhoek Reclaiming Sewage for Drinking Water," *Civil Engineering—ASCE*, September 1972, pp. 103–106.

5. Culp, R. L., and Culp, G. L., *Advanced Wastewater Treatment*, Van Nostrand Reinhold Co., New York, 1971.

6. Bouwer, H., "Returning Wastes to the Land, A New Role for Agriculture," *Journal of Soil and Water Conservation*, Vol. 23, No. 5, 1968.

7. Linstedt, K. D., Miller, K. J., and Bennett, E. R., "Metropolitan Successive Use of Available Water," *Journal American Water Works Association*, Vol. 63, No. 10, 1971.

8. Bargman, R. D., Adrian, G. W., and Tillman, D. C., "Water Reclamation in Los Angeles," *Journal of the Environmental Engineering Division, Proceedings of the Am. Society of Civil Eng.*, Vol. 99, No. EE6, December 1973.

9. "AWWA Policy Statement on Use of Reclaimed Wastewaters as a Public Water-Supply Source," *Journal American Water Works Association*, Vol. 63, No. 10, 1971, p. 609.

10. Bouwer, Herman, "Renovating Municipal Wastewater by High-rate Infiltration for Ground-water Recharge," *Journal American Water Works Association*, Vol. 66, No. 3, 1974, p. 159.

Appendix

COMMON ABBREVIATIONS USED IN TEXT

ac	acres	l	liters
BOD	biochemical oxygen demand	lb	pounds
°C	degrees Celsius	m	meters
cfm	cubic feet per minute	meq	milliequivalents
cm	centimeters	meq/l	milliequivalents per liter
cu ft	cubic feet	mg	milligrams
DO	dissolved oxygen	mg/l	milligrams per liter
°F	degrees Fahrenheit	mgad	million gallons per acre per day
F/M	food-to-microorganism ratio	mgd	million gallons per day
fps	feet per second	mi	miles
ft	feet	mil gal	million gallons
g	grams	min	minutes
gal	gallons	mm	millimeters
gpcd	gallons per capita per day	MPN	most probable number
gpg	grains per gallon	ppm	parts per million
gpm	gallons per minute	psi	pounds per square inch
gr	grains	sec	seconds
hr	hour	sq ft	square feet
in.	inches	yr	year

COMMON CONVERSION FACTORS

Multiple	By	To Obtain
acres	43,560	square feet
acres	0.001,56	square miles
acre-feet	43,560	cubic feet
acre-feet	325,850	gallons
atmospheres	76.0	centimeters of mercury
atmospheres	29.92	inches of mercury
atmospheres	33.90	feet of water
atmospheres	14.7	pounds per square inch
centimeters	0.3937	inches
cubic centimeter	0.061,02	cubic inches
cubic feet	1728	cubic inches

Multiple	By	To Obtain
cubic feet	7.481	gallons
cubic feet	28.32	liters
cubic feet per second	0.6463	million gallons per day
cubic feet per second	448.8	gallons per minute
cubic meters	35.31	cubic feet
cubic meters	264.2	gallons
cubic meters	1000	liters
cubic meters per second	2,120	cubic feet per minute
day	1440	minutes
day	86,400	seconds
feet	30.48	centimeters
feet	12	inches
feet	0.3048	meters
feet of water	0.4335	pounds per square inch
gallons	3785	cubic centimeters
gallons	0.1337	cubic feet
gallons	231	cubic inches
gallons	3.785	liters
gallons of water	8.345	pounds of water
gallons per minute	0.002,23	cubic feet per second
gallons per minute	8.021	cubic feet per hour
grains per gallon	17.12	milligrams per liter
grains per gallon	142.9	pounds per million gallons
gram	0.035,27	ounces
gram	0.002,21	pounds
inches	2.540	centimeters
kilograms	2.205	pounds
kilograms per cubic meter	0.062,43	pounds per cubic foot
kilogram per square centimeter	14.22	pounds per square inch
kilometers	0.6214	miles
liters	1000	cubic centimeters
liters	0.035,31	cubic feet
liters	0.2642	gallons
liters per second	15.85	gallons per minute
meters	100	centimeters
meters	3.281	feet
meters	39.37	inches
meters per second	3.281	feet per second
microns	0.001	millimeters
milligrams per liter	1	parts per million
milligrams per liter	0.0584	grains per gallon
milligrams per liter	8.345	pounds per million gallons
pounds	7000	grains
pounds	453.6	grams
pounds per square foot	0.006,95	pounds per square inch
square centimeter	0.1550	square inches
square meters	10.76	square feet
square miles	640	acres
temperature		
°C	9/5	+32 = °F
°F − 32	5/9	°C
tons	2000	pounds

FOUR-PLACE LOGARITHMS TO BASE 10

The logarithms to base 10 of numbers between 1 and 10, correct to four places, are given in the tables on the following pages.

If the decimal point in the number is moved n places to the right (or left), the value of n (or $-n$) is added to the logarithm, thus:

$$\log \quad 3.14 \quad = 0.4969$$
$$\log 314. \qquad = 0.4969 + 2 \text{ or } 2.4969$$
$$\log \quad 0.0314 = 0.4969 - 2, \text{ which may be written } \bar{2}.4969$$
$$\text{or } 8.4969 - 10$$

If the given number has more than four significant figures, it should be reduced to four figures, since those beyond four figures will not affect the result in four-place computations.

The logarithm of a number having four significant figures must be interpolated by adding to the logarithm of the three figure number, the amount under the fourth figure, as read in the proportional parts section of the table.

Thus, the logarithm of 3.1416 is found as follows:

a. Reduce the number to four significant figures: 3.142
b. The log of 3.14 is 0.4969
c. The value of the proportional part under 2 (the fourth figure) is 3
d. Then, the log $3.142 = 0.4969 + 0.0003$ or 0.4972

Natural Logarithms

Many calculations make use of natural logarithms (Base $e = 2.7183$). To convert base 10 (common) logarithms to natural logarithms, multiply the value for the former by 2.30258.

Natural logarithms are also called Hyperbolic or Naperian logarithms.

$$\log ab = \log a + \log b \qquad \log a^n = n \log a$$

$$\log \frac{a}{b} = \log a - \log b \qquad \log \sqrt[n]{a} = \frac{\log a}{n}$$

Four-Place Logarithms to Base 10

N	0	1	2	3	4	5	6	7	8	9	Proportional Parts 1 2 3 4 5 6 7 8 9
1.0	0000	0043	0086	0128	0170	0212	0253	0294	0334	0374	4 8 12 17 21 25 29 33 37
1.1	0414	0453	0492	0531	0569	0607	0645	0682	0719	0755	4 8 11 15 19 23 26 30 34
1.2	0792	0828	0864	0899	0934	0969	1004	1038	1072	1106	3 7 10 14 17 21 24 28 31
1.3	1139	1173	1206	1239	1271	1303	1335	1367	1399	1430	3 6 10 13 16 19 23 26 29
1.4	1461	1492	1523	1553	1584	1614	1644	1673	1703	1732	3 6 9 12 15 18 21 24 27
1.5	1761	1790	1818	1847	1875	1903	1931	1959	1987	2014	3 6 8 11 14 17 20 22 25
1.6	2041	2068	2095	2122	2148	2175	2201	2227	2253	2279	3 5 8 11 13 16 18 21 24
1.7	2304	2330	2355	2380	2405	2430	2455	2480	2504	2529	2 5 7 10 12 15 17 20 22
1.8	2553	2577	2601	2625	2648	2672	2695	2718	2742	2765	2 5 7 9 12 14 16 19 21
1.9	2788	2810	2833	2856	2878	2900	2923	2945	2967	2989	2 4 7 9 11 13 16 18 20
2.0	3010	3032	3054	3075	3096	3118	3139	3160	3181	3201	2 4 6 8 11 13 15 17 19
2.1	3222	3243	3263	3284	3304	3324	3345	3365	3385	3404	2 4 6 8 10 12 14 16 18
2.2	3424	3444	3464	3483	3502	3522	3541	3560	3579	3598	2 4 6 8 10 12 14 15 17
2.3	3617	3636	3655	3674	3692	3711	3729	3747	3766	3784	2 4 6 7 9 11 13 15 17
2.4	3802	3820	3838	3856	3874	3892	3909	3927	3945	3962	2 4 5 7 9 11 12 14 16
2.5	3979	3997	4014	4031	4048	4065	4082	4099	4116	4133	2 3 5 7 9 10 12 14 15
2.6	4150	4166	4183	4200	4216	4232	4249	4265	4281	4298	2 3 5 7 8 10 11 13 15
2.7	4314	4330	4346	4362	4378	4393	4409	4425	4440	4456	2 3 5 6 8 9 11 13 14
2.8	4472	4487	4502	4518	4533	4548	4564	4579	4594	4609	2 3 5 6 8 9 11 12 14
2.9	4624	4639	4654	4669	4683	4698	4713	4728	4742	4757	1 3 4 6 7 9 10 12 13
3.0	4771	4786	4800	4814	4829	4843	4857	4871	4886	4900	1 3 4 6 7 9 10 11 13
3.1	4914	4928	4942	4955	4969	4983	4997	5011	5024	5038	1 3 4 6 7 8 10 11 12
3.2	5051	5065	5079	5092	5105	5119	5132	5145	5159	5172	1 3 4 5 7 8 9 11 12
3.3	5185	5198	5211	5224	5237	5250	5263	5276	5289	5302	1 3 4 5 6 8 9 10 12
3.4	5315	5328	5340	5353	5366	5378	5391	5403	5416	5428	1 3 4 5 6 8 9 10 11
3.5	5441	5453	5465	5478	5490	5502	5514	5527	5539	5551	1 2 4 5 6 7 9 10 11
3.6	5563	5575	5587	5599	5611	5623	5635	5647	5658	5670	1 2 4 5 6 7 8 10 11
3.7	5682	5694	5705	5717	5729	5740	5752	5763	5775	5786	1 2 3 5 6 7 8 9 10
3.8	5798	5809	5821	5832	5843	5855	5866	5877	5888	5899	1 2 3 5 6 7 8 9 10
3.9	5911	5922	5933	5944	5955	5966	5977	5988	5999	6010	1 2 3 4 5 7 8 9 10
4.0	6021	6031	6042	6053	6064	6075	6085	6096	6107	6117	1 2 3 4 5 6 8 9 10
4.1	6128	6138	6149	6160	6170	6180	6191	6201	6212	6222	1 2 3 4 5 6 7 8 9
4.2	6232	6243	6253	6263	6274	6284	6294	6304	6314	6325	1 2 3 4 5 6 7 8 9
4.3	6335	6345	6355	6365	6375	6385	6395	6405	6415	6425	1 2 3 4 5 6 7 8 9
4.4	6435	6444	6454	6464	6474	6484	6493	6503	6513	6522	1 2 3 4 5 6 7 8 9
4.5	6532	6542	6551	6561	6571	6580	6590	6599	6609	6618	1 2 3 4 5 6 7 8 9
4.6	6628	6637	6646	6656	6665	6675	6684	6693	6702	6712	1 2 3 4 5 6 7 7 8
4.7	6721	6730	6739	6749	6758	6767	6776	6785	6794	6803	1 2 3 4 5 5 6 7 8
4.8	6812	6821	6830	6839	6848	6857	6866	6875	6884	6893	1 2 3 4 4 5 6 7 8
4.9	6902	6911	6920	6928	6937	6946	6955	6964	6972	6981	1 2 3 4 4 5 6 7 8
N	0	1	2	3	4	5	6	7	8	9	1 2 3 4 5 6 7 8 9

Four-Place Logarithms to Base 10 — continued

N	0	1	2	3	4	5	6	7	8	9	Proportional Parts								
											1	2	3	4	5	6	7	8	9
5.0	6990	6998	7007	7016	7024	7033	7042	7050	7059	7067	1	2	3	3	4	5	6	7	8
5.1	7076	7084	7093	7101	7110	7118	7126	7135	7143	7152	1	2	3	3	4	5	6	7	8
5.2	7160	7168	7177	7185	7193	7202	7210	7218	7226	7235	1	2	2	3	4	5	6	7	7
5.3	7243	7251	7259	7267	7275	7284	7292	7300	7308	7316	1	2	2	3	4	5	6	6	7
5.4	7324	7332	7340	7348	7356	7364	7372	7380	7388	7396	1	2	2	3	4	5	6	6	7
5.5	7404	7412	7419	7427	7435	7443	7451	7459	7466	7474	1	2	2	3	4	5	5	6	7
5.6	7482	7490	7497	7505	7513	7520	7528	7536	7543	7551	1	2	2	3	4	5	5	6	7
5.7	7559	7566	7574	7582	7589	7597	7604	7612	7619	7627	1	2	2	3	4	5	5	6	7
5.8	7634	7642	7649	7657	7664	7672	7679	7686	7694	7701	1	1	2	3	4	4	5	6	7
5.9	7709	7716	7723	7731	7738	7745	7752	7760	7767	7774	1	1	2	3	4	4	5	6	7
6.0	7782	7789	7796	7803	7810	7818	7825	7832	7839	7846	1	1	2	3	4	4	5	6	6
6.1	7853	7860	7868	7875	7882	7889	7896	7903	7910	7917	1	1	2	3	4	4	5	6	6
6.2	7924	7931	7938	7945	7952	7959	7966	7973	7980	7987	1	1	2	3	3	4	5	6	6
6.3	7993	8000	8007	8014	8021	8028	8035	8041	8048	8055	1	1	2	3	3	4	5	5	6
6.4	8062	8069	8075	8082	8089	8096	8102	8109	8116	8122	1	1	2	3	3	4	5	5	6
6.5	8129	8136	8142	8149	8156	8162	8169	8176	8182	8189	1	1	2	3	3	4	5	5	6
6.6	8195	8202	8209	8215	8222	8228	8235	8241	8248	8254	1	1	2	3	3	4	5	5	6
6.7	8261	8267	8274	8280	8287	8293	8299	8306	8312	8319	1	1	2	3	3	4	5	5	6
6.8	8325	8331	8338	8344	8351	8357	8363	8370	8376	8382	1	1	2	3	3	4	4	5	6
6.9	8388	8395	8401	8407	8414	8420	8426	8432	8439	8445	1	1	2	2	3	4	4	5	6
7.0	8451	8457	8463	8470	8476	8482	8488	8494	8500	8506	1	1	2	2	3	4	4	5	6
7.1	8513	8519	8525	8531	8537	8543	8549	8555	8561	8567	1	1	2	2	3	4	4	5	5
7.2	8573	8579	8585	8591	8597	8603	8609	8615	8621	8627	1	1	2	2	3	4	4	5	5
7.3	8633	8639	8645	8651	8657	8663	8669	8675	8681	8686	1	1	2	2	3	4	4	5	5
7.4	8692	8698	8704	8710	8716	8722	8727	8733	8739	8745	1	1	2	2	3	4	4	5	5
7.5	8751	8756	8762	8768	8774	8779	8785	8791	8797	8802	1	1	2	2	3	3	4	5	5
7.6	8808	8814	8820	8825	8831	8837	8842	8848	8854	8859	1	1	2	2	3	3	4	5	5
7.7	8865	8871	8876	8882	8887	8893	8899	8904	8910	8915	1	1	2	2	3	3	4	4	5
7.8	8921	8927	8932	8938	8943	8949	8954	8960	8965	8971	1	1	2	2	3	3	4	4	5
7.9	8976	8982	8987	8993	8998	9004	9009	9015	9020	9025	1	1	2	2	3	3	4	4	5
8.0	9031	9036	9042	9047	9053	9058	9063	9069	9074	9079	1	1	2	2	3	3	4	4	5
8.1	9085	9090	9096	9101	9106	9112	9117	9122	9128	9133	1	1	2	2	3	3	4	4	5
8.2	9138	9143	9149	9154	9159	9165	9170	9175	9180	9186	1	1	2	2	3	3	4	4	5
8.3	9191	9196	9201	9206	9212	9217	9222	9227	9232	9238	1	1	2	2	3	3	4	4	5
8.4	9243	9248	9253	9258	9263	9269	9274	9279	9284	9289	1	1	2	2	3	3	4	4	5
8.5	9294	9299	9304	9309	9315	9320	9325	9330	9335	9340	1	1	2	2	3	3	4	4	5
8.6	9345	9350	9355	9360	9365	9370	9375	9380	9385	9390	1	1	2	2	3	3	4	4	5
8.7	9395	9400	9405	9410	9415	9420	9425	9430	9435	9440	0	1	1	2	2	3	3	4	4
8.8	9445	9450	9455	9460	9465	9469	9474	9479	9484	9489	0	1	1	2	2	3	3	4	4
8.9	9494	9499	9504	9509	9513	9518	9523	9528	9533	9538	0	1	1	2	2	3	3	4	4
9.0	9542	9547	9552	9557	9562	9566	9571	9576	9581	9586	0	1	1	2	2	3	3	4	4
9.1	9590	9595	9600	9605	9609	9614	9619	9624	9628	9633	0	1	1	2	2	3	3	4	4
9.2	9638	9643	9647	9652	9657	9661	9666	9671	9675	9680	0	1	1	2	2	3	3	4	4
9.3	9685	9689	9694	9699	9703	9708	9713	9717	9722	9727	0	1	1	2	2	3	3	4	4
9.4	9731	9736	9741	9745	9750	9754	9759	9763	9768	9773	0	1	1	2	2	3	3	4	4
9.5	9777	9782	9786	9791	9795	9800	9805	9809	9814	9818	0	1	1	2	2	3	3	4	4
9.6	9823	9827	9832	9836	9841	9845	9850	9854	9859	9863	0	1	1	2	2	3	3	4	4
9.7	9868	9872	9877	9881	9886	9890	9894	9899	9903	9908	0	1	1	2	2	3	3	4	4
9.8	9912	9917	9921	9926	9930	9934	9939	9943	9948	9952	0	1	1	2	2	3	3	4	4
9.9	9956	9961	9965	9969	9974	9978	9983	9987	9991	9996	0	1	1	2	2	3	3	3	4
N	0	1	2	3	4	5	6	7	8	9	1	2	3	4	5	6	7	8	9

Source. Flow of Fluids, Crane Company.

GEOMETRIC FORMULAS FOR BASINS OF VARIOUS SHAPES

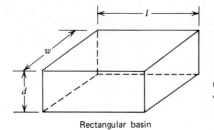

Surface area = $w \times l$
Circumference = $2w + 2l$
Volume = $w \times l \times d$

Rectangular basin

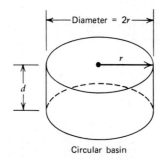

Surface area = $\pi \times r^2 = 3.14 \times r \times r$
Circumference = $2 \times \pi \times r = 6.28 \times r$
Volume = $\pi \times r^2 \times d$

Circular basin

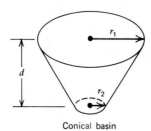

$Area_1 = \pi \times r_1$
$Volume = \frac{d}{3} \times (A_1 + A_2 + \sqrt{A_1 \times A_2})$
$Area_2 = \pi \times r_2^2$

Conical basin

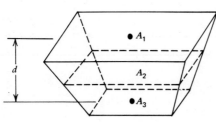

$Volume = \frac{d}{6} \times (A_1 + 4A_2 + A_3)$

where A_1 = surface area

A_2 = area of midsection

A_3 = bottom area

d = depth

Prismoidal basin

DISCHARGE FROM TRIANGULAR NOTCH WEIRS WITH END CONTRACTIONS

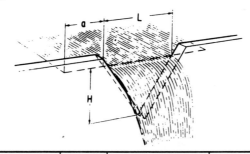

Head (H) in Inches	Flow in Gallons Per Min.		Head (H) in Inches	Flow in Gallons Per Min.		Head (H) in Inches	Flow in Gallons Per Min.	
	90° Notch	60° Notch		90° Notch	60° Notch		90° Notch	60°° Notch
¼	0.07	0.04	6¼	214	124	14½	1,756	1,014
½	0.42	0.24	6½	236	136	15	1,912	1,104
¾	1.10	0.62	6¾	260	150	15½	2,073	1,197
1	2.19	1.27	7	284	164	16	2,246	1,297
1¼	3.83	2.21	7¼	310	179	16½	2,426	1,401
1½	6.05	3.49	7½	338	195	17	2,614	1,509
1¾	8.89	5.13	7¾	367	212	17½	2,810	1,623
2	12.4	7.16	8	397	229	18	3,016	1,741
2¼	16.7	9.62	8¼	429	248	18½	3,229	1,864
2½	21.7	12.5	8½	462	267	19	3,452	1,993
2¾	27.5	15.9	8¾	498	287	19½	3,684	2,127
3	34.2	19.7	9	533	308	20	3,924	2,266
3¼	41.8	24.1	9¼	571	330	20½	4,174	2,410
3½	50.3	29.0	9½	610	352	21	4,433	2,560
3¾	59.7	34.5	9¾	651	376	21½	4,702	2,715
4	70.2	40.5	10	694	401	22	4,980	2,875
4¼	81.7	47.2	10½	784	452	22½	5,268	3,041
4½	94.2	54.4	11	880	508	23	4,565	3,213
4¾	108	62.3	11½	984	568	23½	5,873	3,391
5	123	70.8	12	1,094	632	24	6,190	3,574
5¼	139	80.0	12½	1,212	700	24½	6,518	3,763
5½	156	89.9	13	1,337	772	25	6,855	3,958
5¾	174	100	13½	1,469	848			
6	193	112	14	1,609	929			

Based on formula:
$$Q = (C)\ (4/15)\ (L)\ (H)\ \sqrt{2gH}$$
in which Q = flow of water in cu. ft. per sec.

        L = width of notch in ft. at H distance above apex.

        H = head of water above apex of notch in ft.

        C = constant varying with conditions, .57 being used for this table.

        a = should be not less than ¾ L.

For 90° notch the formula becomes
$$Q = 2.536H^{5/2}$$
For 60° notch the formula becomes
$$Q = 1.408H^{5/2}$$

Source. Clay Pipe Engineering Manual, National Clay Pipe Institute.

DISCHARGE FROM RECTANGULAR WEIR WITH END CONTRACTIONS

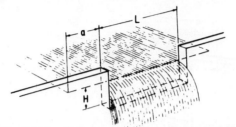

Figures in Table
are in Gallons
Per Minute

Head (H) in Inches	Length (L) of Weir in Feet				Head (H) in Inches	Length (L) of Weir in Feet		
	1	3	5	Additional g.p.m. for each ft. over 5 ft.		3	5	Additional g.p.m. for each ft. over 5 ft.
¼	4.5	13.4	22.4	4.5	7¾	2,238	3,785	774
½	12.8	38.2	63.8	12.8	8	2,338	3,956	814
¾	23.4	70.2	117.0	23.4	8¼	2,442	4,140	850
1	35.4	108	180	36.1	8½	2,540	4,312	890
1¼	49.5	150	250	50.4	8¾	2,656	4,511	929
1½	64.9	197	330	66.2	9	2,765	4,699	970
1¾	81.0	248	415	83.5	9¼	2,876	4,899	1,011
2	98.5	302	506	102	9½	2,985	5,098	1,051
2¼	117	361	605	122	9¾	3,101	5,288	1,091
2½	136	422	706	143	10	3,216	,5,490	1,136
2¾	157	485	815	165	10½	3,480	5,940	1,230
3	178	552	926	187	11	3,716	6,355	1,320
3¼	200	624	1,047	211	11½	3,960	6,780	1,410
3½	222	695	1,167	236	12	4,185	7,165	1,495
3¾	245	769	1,292	261	12½	4,430	7,595	1,575
4	269	846	1,424	288	13	4,660	8,010	1,660
4¼	294	925	1,559	316	13½	4,950	8,510	1,780
4½	318	1,006	1,696	345	14	5,215	8,980	1,885
4¾	344	1,091	1,835	374	14½	5,475	9,440	1,985
5	370	1,175	1,985	405	15	5,740	9,920	2,090
5¼	396	1,262	2,130	434	15½	6,015	10,400	2,165
5½	422	1,352	2,282	465	16	6,290	10,900	2,300
5¾	449	1,442	2,440	495	16½	6,565	11,380	2,410
6	477	1,535	2,600	538	17	6,925	11.970	2,520
6¼		1,632	2,760	560	17½	7,140	12,410	2,640
6½		1,742	2,920	596	18	7,410	12,900	2,745
6¾		1,826	3,094	630	18½	7,695	13,410	2,855
7		1,928	3,260	668	19	7,980	13,940	2,970
7¼		2,029	3,436	702	19½	8,280	14,460	3,090
7½		2,130	3,609	736				

$$Q = 3.33 \ (L - 0.2H) \ H^{1.5}$$

in which

Q = cu. ft. of water flowing per second.

L = length of weir opening in feet (should be 4 to 8 times H).

H = head on weir in feet (to be measured at least 6 ft. back of weir opening).

a = should be at least $3H$.

Source. Clay Pipe Engineering Manual, National Clay Pipe Institute.

TABLE OF CHEMICAL ELEMENTS

Name	Symbol	Atomic Number	Atomic Weight	Name	Symbol	Atomic Number	Atomic Weight
Actinium	Ac	89	—	Mercury	Hg	80	200.59
Aluminum	Al	13	26.9815	Molybdenum	Mo	42	95.94
Americium	Am	95	—	Neodymium	Nd	60	144.24
Antimony	Sb	51	121.75	Neon	Ne	10	20.183
Argon	Ar	18	39.948	Neptunium	Np	93	—
Arsenic	As	33	74.9216	Nickel	Ni	28	58.71
Astatine	At	85	—	Niobium	Nb	41	92.906
Barium	Ba	56	137.34	Nitrogen	N	7	14.0067
Berkelium	Bk	97	—	Nobelium	No	102	
Beryllium	Be	4	9.0122	Osmium	Os	76	190.2
Bismuth	Bi	83	208.980	Oxygen	O	8	15.9994
Boron	B	5	10.811	Palladium	Pd	46	106.4
Bromine	Br	35	79.904	Phosphorus	P	15	30.9738
Cadmium	Cd	48	112.40	Platinum	Pt	78	195.09
Calcium	Ca	20	40.08	Plutonium	Pu	94	—
Californium	Cf	98	—	Polonium	Po	84	—
Carbon	C	6	12.01115	Potassium	K	19	39.102
Cerium	Ce	58	140.12	Praseodymium	Pr	59	140.907
Cesium	Cs	55	132.905	Promethium	Pm	61	—
Chlorine	Cl	17	35.453	Protactinium	Pa	91	—
Chromium	Cr	24	51.996	Radium	Ra	88	—
Cobalt	Co	27	58.9332	Radon	Rn	86	—
Copper	Cu	29	63.546	Rhenium	Re	75	186.2
Curium	Cm	96	—	Rhodium	Rh	45	102.905
Dysprosium	Dy	66	162.50	Rubidium	Rb	37	85.47
Einsteinium	Es	99	—	Ruthenium	Ru	44	101.07
Erbium	Er	68	167.26	Samarium	Sm	62	150.35
Europium	Eu	63	151.96	Scandium	Sc	21	44.956
Fermium	Fm	100	—	Selenium	Se	34	78.96
Fluorine	F	9	18.9984	Silicon	Si	14	28.086
Francium	Fr	87	—	Silver	Ag	47	107.868
Gadolinium	Gd	64	157.25	Sodium	Na	11	22.9898
Gallium	Ga	31	69.72	Strontium	Sr	38	87.62
Germanium	Ge	32	72.59	Sulfur	S	16	32.064
Gold	Au	79	196.967	Tantalum	Ta	73	189.948
Hafnium	Hf	72	178.49	Technetium	Tc	43	—
Helium	He	2	4.0026	Tellurium	Te	52	127.60
Holmium	Ho	67	164.930	Terbium	Tb	65	158.924
Hydrogen	H	1	1.00797	Thallium	Tl	81	204.37
Indium	In	49	114.82	Thorium	Th	90	232.038
Iodine	I	53	126.9044	Thulium	Tm	69	168.934
Iridium	Ir	77	192.2	Tin	Sn	50	118.69
Iron	Fe	26	55.847	Titanium	Ti	22	47.90
Krypton	Kr	36	83.80	Tungsten	W	74	183.85
Lanthanum	La	57	138.91	Uranium	U	92	238.03
Lead	Pb	82	207.19	Vanadium	V	23	50.942
Lithium	Li	3	6.939	Xenon	Xe	54	131.30
Lutetium	Lu	71	174.97	Ytterbium	Yb	70	173.04
Magnesium	Mg	12	24.312	Yttrium	Y	39	88.905
Manganese	Mn	25	54.9380	Zinc	Zn	30	65.37
Mendelevium	Md	101	—	Zirconium	Zr	40	91.22

TWO-CYCLE LOGARITHMIC PROBABILITY PAPER FOR USE IN SOLVING PROBLEM 4-29

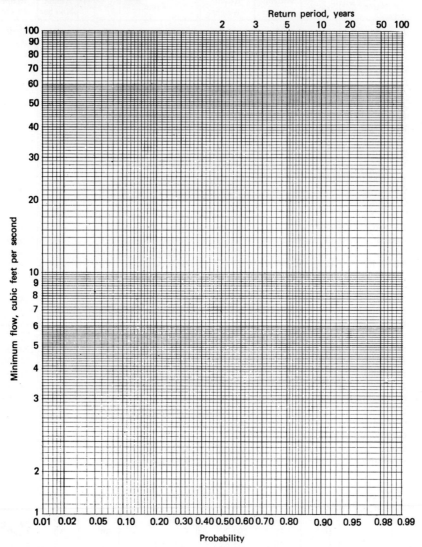

Return period, years

Minimum flow, cubic feet per second

Probability

Index

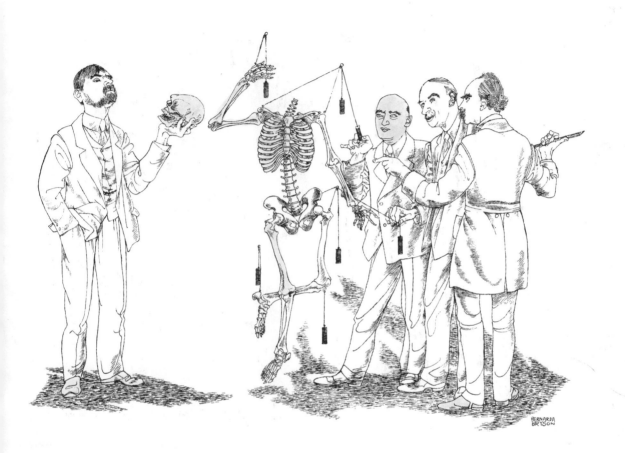

Carrying on the Anatomical Studies

Left to right: Thorstein Veblen, Joseph Schumpeter, John Maynard Keynes, and Alfred Marshall.

Ex Libris

David M. Livingston

Robert L. Heilbroner

Lester C. Thurow

Englewood Cliffs, New Jersey

The Economic Problem

fourth edition

PRENTICE-HALL, INC.

Library of Congress Cataloging in Publication Data

Heilbroner, Robert L
    The economic problem.

    Includes bibliographical references.
        1. Economics.   I.   Thurow, Lester C., joint author.
II.   Title.
HB171.5.H39   1975   330.9'04   74-11284
ISBN 0-13-233320-1

THE ECONOMIC PROBLEM, fourth edition
Robert L. Heilbroner and Lester C. Thurow
© 1975 by Prentice-Hall, Inc., Englewood Cliffs, N.J.
© 1972, 1970, 1968 by Robert L. Heilbroner
10 9 8 7 6 5 4 3 2 1

Prentice-Hall International, Inc., *London*
Prentice-Hall of Australia, Pty. Ltd., *Sydney*
Prentice-Hall of Canada, Ltd., *Toronto*
Prentice-Hall of India Private Limited, *New Delhi*
Prentice-Hall of Japan, Inc., *Tokyo*

Contents

From Growth to Fluctuations,   316

The Demand for Output,   329

Saving and Investment,   349

Consumption Demand,   363

Investment Demand,   378

The Problem of Inflation,   496   *32*

The Problem of
Unemployment,   515   *33*

Problems of   *34*
    Economic Growth,   531

5   INTERNATIONAL
    ECONOMICS

The Gains from Trade,   551   *35*

The Mechanism of   *36*
International Transactions,   570

The International   *37*
Monetary Problem,   584

The Multinational   *38*
Corporation,   601

*SOCIOECONOMIC
SYSTEMS*
6

The Underdeveloped   *39*
World,   617

From Market to Planning, 638

*QUANTITATIVE
METHODS*

An Introduction to Statistics
   and Econometrics, 669

Is Capitalism the Problem? 653

A guide to the instructor

How can one best use *The Economic Problem?* The answer depends on the length of the course, the special interests of the teacher, and the inclinations and training of the students. No single prescription will serve all purposes, and we offer our suggestions with diffidence. Nonetheless, we recognize that the text will not be read from cover to cover in the great majority of cases—indeed, we have deliberately planned a text that can be adapted to various situations. Herewith, then, our suggestions; first rather general, then chapter by chapter.

Part 1   The Economy in Overview

Part 1 is a quick, overall view of economics—a "vision" of the economic problem, as Schumpeter might have put it. It is not necessary reading, although we hope it is interesting reading. Instructors who are pressed for time can either omit it entirely or assign it as background material.

Part 2   Fundamental Concepts

Here are four chapters that also give an overview of economics, but from the worm's-eye rather than the bird's-eye perspective. Chapters 4 and 5, "The Market Mechanism" and "The Tools of Economic Analysis," are really essential for all students. Chapter 3, "Economic Behavior and Nature," and 6, "Some Basic Problems," will appeal to students who really want to go into the subject; but 3 and 6 can be omitted from a short course.

Part 3   Microeconomics~Anatomy of a Market System

First a word to those who want to begin with macroeconomics. There is no reason why you cannot. The macro section was written before the micro, to make sure that it would stand on its own feet.

Most of the micro section follows the established tradition, but pointing out some differences may help you decide what to assign and what to

skip. Chapters 7, 8, and 9—"Introduction to the Microeconomy," "Prices and Allocation," and "The Market in Movement," are pretty essential, although one can leave out the treatment of utility at the end of Chapter 8 and the full discussion of elasticity and market stability in Chapter 9. Chapter 10 brings our analysis to bear on agriculture. Some instructors may linger over this chapter; others may assign it as supplementary reading or omit it and push on to the next four chapters on the firm.

Chapters 11, 12, and 13 cover the basic elements of the competitive firm and market imperfections. It's difficult to see how these could be passed over. Chapter 14 is, however, a discussion of big business, which we again single out. Some teachers will focus on this chapter; others may leave it to the student to read by himself. We have made our next chapter, "Economics and the Environment," a mixture of theoretical and institutional analysis. The theory is a bit demanding; the institutional discussion is not. We hope everyone will find some part of this chapter to his or her taste.

Chapters 16, 17, and 18 offer the greatest range of choice. Chapter 16, "The Market for Factors of Production," is a straightforward presentation of marginal productivity theory, and most instructors will probably cover it carefully. Chapter 17 is a critique of marginal productivity theory from two points of view: a review of the many institutional imperfections in the labor market and a critical look at the underlying validity of the marginal productivity concept itself. Teachers who are interested in institutional problems will probably emphasize the first set of problems in this chapter: analytically oriented instructors or students may prefer work on the second part; some teachers may assign the whole chapter for reading rather than classroom work. Last and most controversial is Chapter 18, "Changing the Distribution of Income." We know that some instructors will spend time on this, especially the last section on value judgments; others may decide it is not right for their course. We hope it will be used for reading, even if not for classwork.

Part 4 Prosperity and Recession~ the Economics of the Macro System

It is difficult for us to give much advice about the macro section, where each chapter fits into the whole. Be sure that Chapter 4," The Market Mechanism," and Chapter 5, "The Tools of Economic Analysis," are covered. Thereafter, it is possible to assign Chapters 20, "The Growth of Output," and 21, "From Growth to Fluctuations," for background reading, getting right down to business with Chapter 22, "The Demand for Output." Certainly the supplementary sections—on input-output, national income and product accounts, and the military subeconomy—are to be relegated to a supporting role, if time is short. The remaining chapters seem crucial to us, culminating in Chapters 32 and 33, "The Problem of Inflation" and "The Problem of Unemployment." Chapter 34, "Problems of Economic Growth," is another place where the instructor's taste (and schedule) will be decisive; pp. 531–36 can be omitted if need be; the remainder should provoke a spirited discussion.

Part 5 International Economics

Not all introductory courses cover international economics, but a textbook would be lacking if it failed to pay attention to the subject. Chapters 35 and 36, "The Gains from Trade" and "The Mechanism of International Transactions," are the core of the analytical material. Chapters 37 and 38, "The International Monetary Problem" and "The Multinational Corporation," may be more interesting to students. If time is short and you intend to touch on international problems, we suggest assigning the first two for classwork and the second two for background reading.

Part 6 Socio-economic Systems

Here is a section deliberately planned to be of use to institutional-minded teachers and students. The problems of this section are perhaps among the most important in political economy, but they are less demanding in terms of standard analysis than are the micro/macro foreign trade sections, and they can therefore be used as "readings" if one wishes.

Part 7 Quantitative Methods

The longish Chapter 42 is a very brief introduction to a few central ideas of statistics and econometrics. We imagine that very few users of our book will assign these chapters, but instructors may refer students to various parts of it when words such as *correlation* come up in class.

Workbook and Reader

Before turning to a more detailed guide, permit us to add a word about the combined student workbook and reader that has been designed to accompany the text. We know that not all users of the text will want to use the companion volume, but we should like to point out that the second book offers a reexposition of certain points that may present difficulties, a detailed set of "do-it-yourself" exercises, and a battery of true-false and multiple-choice questions (with answers), as well as a considerable number of readings. We hope that students will find this a good way to master the text. In addition, there are tests for each major section, but answers for these are provided only in the *Instructor's Manual*. These tests are designed to help you see how well the class is following along.*

Now for a chapter-by-chapter analysis. In the following table we have tried to classify each chapter as "basic," "advanced," or "institutional." The basic chapters are those that seem to us essential for a one-year course in introductory economics. The advanced chapters can be omitted or assigned as supplementary reading. The institutional chapters are those for which the students need little or no assistance with regard to the formal apparatus of economics. Need we add that these chapters, which seem "easy," include some of the most difficult areas of economic thought? (You will note that we have included several chapters under more than one heading.)

*Additional test material is included in the *Instructor's Manual*.

A CHAPTER GUIDE TO

THE ECONOMIC PROBLEM

ACKNOWLEDGMENTS

Among those we wish to thank, we can name only a few:

Herbert Werner, University of Missouri;

John Murgo, Massasoit Community College;

John P. David, West Virginia Institute of Technology;

William Casey, Babson College;

James W. Foley, University of Miami;

L. H. Zincone, Jr., East Carolina University;

Mostafa I. Shaaban, West Virginia Institute of Technology;

Steven Call, University of Wisconsin at Milwaukee;

Donald R. Wentworth, Pacific Lutheran University;

Peter Danner, Marquette University;

Albert Domenico, Rhode Island Junior College;

John M. Murphy, North Shore Community College;

Peter Wogart, University of Miami; and

Kurt Adams, University of Missouri at Rolla.

The
Economic
Problem

1

*THE
ECONOMY
IN
OVERVIEW*

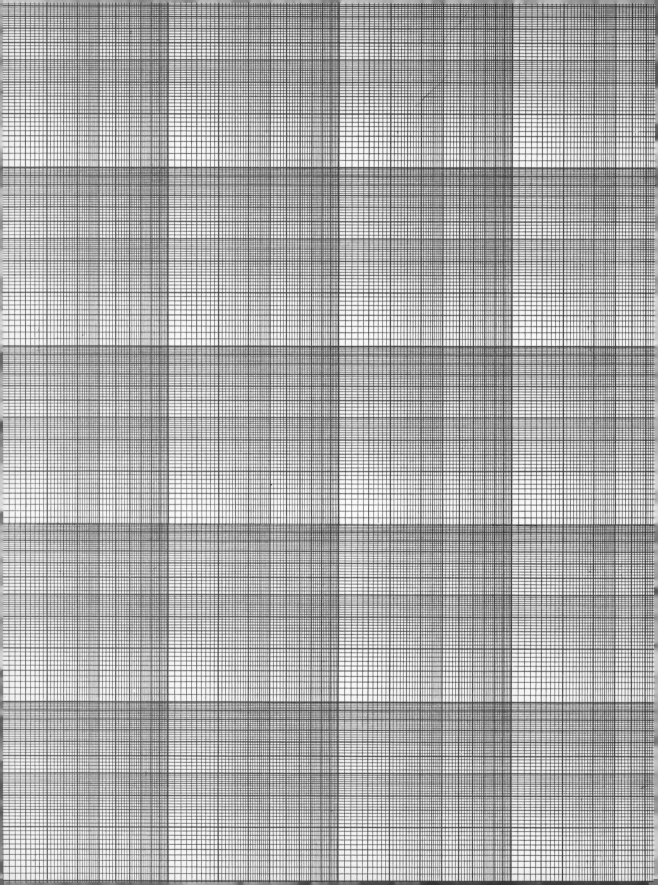

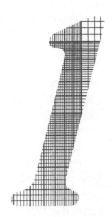

The Economic Problem

ONE NEED HARDLY TELL A STUDENT who is about to plunge into a multihundred-page text that the subject is complicated. A riffle through the pages of this book, with its incomprehensible graphs and unfamiliar vocabulary, makes it very clear that economics *is* complicated. Indeed, the reader may feel that he is embarking on a long and hazardous expedition in which he will end up hopelessly lost, and from which he will be lucky to emerge at all.

Fortunately, the trail through this forbidding territory is a good deal less difficult than a first impression may convey. This is particularly the case if the student has a general idea of what the country is like, where the great mountain ranges lie, and where the journey is headed.

Hence this first section, to give the reader his bearings before we begin to cover the ground in a more systematic and thorough way. Unlike later chapters, which move along slowly and carefully on the ground, these initial chapters will take us on an aerial reconnaissance, a quick overflight to give us a first impression of that *terra incognita* we call economics. Moreover, unlike later chapters, many of which will no doubt require underlinings and notes in the margins, this first short section is just to be read, possibly even enjoyed. The hard work will come later.

THE PROVISIONING PROBLEM

Everyone is generally aware of some of the things that economics is "about": inflation and poverty, unemployment and monopoly, the mysteries of money, to name but a few. What is needed to begin with, however, is a feeling for the subject as a whole, for what we might call The Economic Problem. What underlying theme runs through these chapters that touch on so many problems?

The answer is surprisingly simple. *Economics is the study of how mankind copes with the problem of provisioning itself.* However complex or sophisticated the individual aspects of the subject may be, at bottom the unifying question is how man earns his daily bread or how he divides it among his brethren.

This hardly seems like a particularly demanding subject: most of us take for granted the matter of sustenance. But if we look back over history or across the oceans to the realities of life in Africa, Asia, or Latin America, the problem takes on an unexpected urgency. For man has by no means solved the economic problem satisfactorily. Over most of his tenure of occupancy on earth, life has been as Thomas Hobbes described it: "poor, nasty, brutish, and short." Even today, for the peoples of the underdeveloped world, Hobbes's description is all too true. According to the 1967 report of the President's Science Advisory Panel on World Food Supply, malnutrition affects some 600 million inhabitants of the backward world, 200 million of whom suffer from undernutrition or actual slow starvation. Not many years ago an Indian demographer made a chilling calculation that of one hundred Asian and one hundred American newborn infants, more of the Americans would be alive at age sixty-five than the Indians at age *five*.

In our own world, of course, such considerations seem very remote. Yet even in America, the problem of provisioning asserts its importunate demands. In most American cities a few blocks separate a world of air-conditioned affluence from one of mean squalor, and within that squalor one can still find populations afflicted by malnutrition and touched by undernutrition. Moreover, even in affluent America there remains, however unnoticed, a reminder of the underlying problem of brute survival. *This is our helplessness as economic individuals.*

THE INDIVIDUAL AND SOCIETY

For it is a curious fact that as we leave the most impoverished peoples of the world, where the human being scratches out for himself a meager existence, we find the economic insecurity of the individual many times multiplied. The solitary Eskimo, Bushman, Indonesian, African, left to his own devices, will survive a considerable time. Living close to the soil or to his animal prey, such an individual can sustain his own life, at least for a while, singlehanded. With a community numbering only a few hundred, he can live indefinitely. Indeed, a very large percentage of the human race today lives in precisely such fashion—in small, virtually self-contained peasant communities that provide for their own survival with a minimum of contact with the outside world. This large portion of mankind suffers great poverty, but it also knows a certain economic independence. If it did not, it would have been wiped out centuries ago.

When we turn to the New Yorker or the Chicagoan, on the other hand, we are struck by exactly the opposite condition, by a prevailing ease of material life coupled with an extreme *dependence* on others. We can no longer envisage the solitary individual or the small community surviving unaided in the great metropolitan areas where most Americans live, unless they loot warehouses or stores for food and necessities. The overwhelming majority of Americans have never grown food, caught game, raised meat, ground grain into flour, or even fashioned flour into bread. Faced with the challenge of clothing themselves or building their own homes, they would be hopelessly untrained and unprepared. Even to make minor repairs in the machines that surround them, they must call on other members of the community whose business it is to fix cars or repair plumbing or whatever. Paradoxically, perhaps, the richer the nation, the more apparent is this inability of its average inhabitant to survive unaided and alone.

DIVISION OF LABOR

There is, of course, an answer to the paradox. We survive in rich nations because the tasks we cannot do ourselves are done for us by an army of others on whom we can call for help. If we cannot grow food, we can buy it; if we cannot provide for our needs ourselves, we can hire the services of someone who can. This enormous *division of labor* enhances our capacity a thousandfold, for it enables us to benefit from other men's skills as well as our own. In our next chapter it will play a central role.

Along with this invaluable gain, however, comes a certain risk. It is a sobering thought, for example, that we depend on the services of less than 138,000 men, out of a national labor force of 90 million, to provide us with that basic commodity, coal. An even smaller number of workers are responsible for running the locomotives that haul all the nation's rail freight. A much smaller number—roughly 40,000—comprises our total airline pilot crew. Failure of any one of these very small groups to perform its functions would cripple us: in the case of airplane pilots, slightly; in the case of locomotive engineers, badly; in the case of coal miners, perhaps disastrously. As we know, when from time to time we face a bad strike, our entire economic machine may falter because a strategic group ceases to perform its accustomed tasks.

Thus along with the abundance of material existence as we know it goes a hidden vulnerability: our abundance is assured only insofar as the organized cooperation of huge armies of people is to be counted upon. Indeed, our continuing existence as a rich nation hinges on the tacit precondition that the mechanism of social organization will continue to function effectively. *We are rich, not as individuals, but as members of a rich society, and our easy assumption of material sufficiency is actually only as reliable as the bonds that forge us into a social whole.*

ECONOMICS AND SCARCITY

Strangely enough, then, we find that man, not nature, is the source of most of our economic problems, at least above the level of subsistance. To be sure, the economic problem itself—that is, the need to struggle for existence—derives ultimately from the *scarcity* of nature. If there were no scarcity, goods would be as free as air, and economics—at least in one sense of the word—would cease to exist as a social preoccupation.

And yet if the scarcity of nature sets the stage for the economic problem, it does not impose the only strictures against which men must struggle. For scarcity, as a felt condition, is not solely the fault of nature. If Americans today, for instance, were content to live at the level of Mexican peasants, all our material wants could be fully satisfied with but an hour or two of daily labor. We would experience little or no scarcity, and our economic problems would virtually disappear. Instead, we find in America—and indeed in all industrial societies—that as the ability to increase nature's yield has risen, so has the reach of human wants. In fact, in societies such as ours, where relative social status is importantly connected with the possession of material goods, we often find that "scarcity" as a psychological experience and goad becomes more pronounced as we grow wealthier: our desires to possess the fruits of nature race out ahead of our mounting ability to produce goods.

Thus the "wants" that nature must satisfy are by no means fixed. But, for that matter, nature's yield itself is not a constant. It varies over a wide range, depending on the social application of human energy and skill.

Scarcity is therefore not attributable to nature alone but to "human nature" as well; and economics is ultimately concerned not merely with the stinginess of the physical environment, but equally with the appetite of the human being and the productive capability of the community.

THE TASKS OF ECONOMIC SOCIETY

Hence we must begin a systematic analysis of economics by singling out the functions that social organization must perform to bring human nature into social harness. And when we turn our attention to this fundamental problem, we can quickly see that it involves the solution of two related and yet separate elemental tasks a society must:

1. *organize a system to assure the production of enough goods and services for its own survival, and*

2. *arrange the distribution of the fruits of its production so that more production can take place.*

These two tasks of economic continuity are, at first look, very simple.* But it is a deceptive simplicity. Much of economic history is concerned with the manner in which various societies have sought to cope with these elementary problems; and what strikes us in surveying their attempts is that most of them were partial failures. (They could not have been *total* failures, or society would not have survived.) So we had better look more carefully into the two main economic tasks, to see what hidden difficulties they may conceal.

*In later chapters we will return to the Economic Problem in a somewhat more technical perspective. There we will note that the production problem is itself a twofold task, involving both a choice of different *ends* to which effort may be put and also of different *methods* by which those ends may be sought. Here, where the challenge of social organization is emphasized, we combine these two tasks into a single concern.

Production and Distribution Problems

MOBILIZING EFFORT

What obstacles does a society encounter in organizing a system to produce the goods and services it needs?

Since nature is usually stingy, it seems that the production problem must be essentially one of applying engineering or technical skills to the resources at hand, of avoiding waste and utilizing social effort as efficaciously as possible.

This is indeed an important task for any society, and a great deal of formal economic thought, as the word itself suggests, is devoted to economizing. Yet this is not the core of the production problem. Long before a society can even concern itself about using its energies "economically," it must first marshal the energies to carry out the productive process itself. That is, *the basic problem of production is to devise social institutions that will mobilize human energy for productive purposes.*

This basic requirement is not always so easily accomplished. For example, in the United States in 1933, the energies of nearly one-quarter of our work force were not directed into the production process at all. Although millions of unemployed men and women were eager to work, although empty factories were available for them to work in, despite the existence of pressing wants, somehow a terrible and mystifying breakdown short-circuited the production process, with the result that an entire third of our previous annual output of goods and services simply disappeared.

We are by no means the only nation that has, on occasion, failed to find work for willing workers. In the very poorest nations, where production is most desperately needed, we frequently find that unemployment is a chronic condition. The streets of Asian cities are thronged with people who cannot find work. But this, too, is not a condition imposed by the scarcity of nature. There is, after all, an endless amount of work to be done, if only in cleaning the filthy streets or patching up the homes of the poor, building roads, or digging ditches. What is lacking is a social mechanism to put the unemployed to work.

Both these examples point out to us that the production problem is not solely a physical and technical struggle with nature. On these "scarcity" aspects of the problem will depend the ease with which a nation may forge ahead and the level of well-being it can reach with a given effort. But the original mobilization of productive effort itself is a challenge to its *social organization,* and on the success or failure of that social organization will depend the volume of the human effort that can be directed to nature.

ALLOCATING EFFORT

But putting men to work is only the first step in the solution of the production problem. Men must not only be put to work; they must be put to work *in the right places.* They must produce the goods and services that society needs. Thus, *in addition to assuring a large enough quantity of social effort, the economic institutions of society must also assure a viable allocation of that social effort.*

In a nation such as India or Bolivia, where the great majority of the population is born in peasant villages and grows up to be peasant cultivators, the solution to this problem offers little to vex our understanding. The basic needs of society—food and fiber—are precisely the goods that its peasant population "naturally" produces. But in an industrial society, the proper allocation of effort becomes an enormously complicated task. People in the United States demand much

more than bread and cotton. They need, for instance, such things as automobiles. Yet no one "naturally" produces an automobile. On the contrary, in order to produce one, an extraordinary spectrum of special tasks must be performed. Some people must make steel; others must make rubber. Still others must coordinate the assembly process itself. And this is but a tiny sampling of the far from "natural" tasks that must be performed if an automobile is to be produced.

As with the mobilization of its total production effort, society does not always succeed in the proper allocation of its effort. It may, for instance, turn out too many cars or too few. Of greater importance, it may devote its energies to the production of luxuries while large numbers of its people are starving. Or it may even court disaster by an inability to channel its productive effort into areas of critical importance.

Such allocative failures may affect the production problem quite as seriously as a failure to mobilize an adequate quantity of effort, for a viable society must produce not only goods, but the *right* goods. And the allocative question alerts us to a still broader conclusion. It shows us that the act of production, in and of itself, does not fully answer the requirements for survival. Having produced enough of the right goods, society must now *distribute* those goods so that the production process can go on.

DISTRIBUTING OUTPUT

Once again, in the case of the peasant who feeds himself and his family from his own crop, this requirement of adequate distribution may seem simple enough. But when we go beyond the most primitive society, the problem is not always so readily solved. In many of the poorest nations of the East and South, urban workers have often been unable to deliver their daily horsepower-hour of

TRADITION IN ACTION

Tradition not only provides a solution to the production problem of society, but also regulates the distribution problem. Take the Bushmen of the Kalahari Desert in South Africa, who depend for their livelihood on their hunting prowess. Elizabeth Marshall Thomas, a sensitive observer of these peoples, reports on the manner in which tradition solves the problem of distributing their kill.

The gemsbok has vanished ... Gai owned two hind legs and a front leg. Tsetchwe had meat from the back, Ukwane had the other front leg, his wife had one of the feet and the stomach, the young boys had lengths of intestine. Twikwe had received the head and Dasina the udder.

It seems very unequal when you watch Bushmen divide the kill, yet it is their system, and in the end no person eats more than any other. That day Ukwane gave Gai still another piece because Gai was his relation, Gai gave meat to Dasina because she was his wife's mother ... No one, of course, contested Gai's large share, because he had been the hunter and by their law that much belonged to him. No one doubted that he would

work because they have not been given enough of society's output to run their human engines to capacity. Worse yet, they have often languished on the job while granaries bulged with grain and the well-to-do complained of the ineradicable laziness of the masses. At the other side of the picture, the distribution mechanism may fail because the rewards it hands out do not succeed in persuading people to perform their necessary tasks. Shortly after the Russian Revolution, some factories were organized into communes in which managers and janitors pooled their pay, and from which all drew equal allotments. The result was a rash of absenteeism among the previously better-paid workers and a threatened breakdown in industrial production. Not until the old unequal wage payments were reinstituted did production resume its former course.

As was the case with failures in the production process, distributive failures need not entail a total economic collapse. Societies can exist—and most do exist—with badly distorted productive and distributive efforts. Only rarely, as in the instances above, does maldistribution actively interfere with the actual ability of a society to staff its production posts. More frequently, an inadequate solution to the distribution problem reveals itself in social and political unrest or even in revolution.

Yet this, too, is an aspect of the total economic problem. For if society is to insure its steady material replenishment, it must parcel out its production in a fashion that will maintain not only the capacity but the willingness to go on working. And thus again we find the focus of economic inquiry directed to the study of human institutions. For a viable economic society, we can now see, must not only overcome the stringencies of nature, but also contain and control the intransigence of human nature.

The Three Solutions to the Economic Problem

Thus to the economist, society presents itself in an unaccustomed aspect. Underneath the problems of inflation or monopoly or money, he sees a process at work that he must understand before he can turn his attention to the issues of the day, no matter how pressing. That process is society's basic *mechanism for survival,* a mechanism for accomplishing the complicated tasks of its production and distribution necessary for its own continuity.

But the economist sees something else as

well, something that at first seems quite astonishing. Looking not only over the diversity of contemporary societies, but back over the sweep of all history, he sees that man has succeeded in solving the production and distribution problems in but three ways. That is, within the enormous diversity of the actual social institutions that guide and shape the economic process, the economist divines but three overarching *types* of systems that separately or in combination enable humankind to solve its economic challenge. These great systemic types can be called economies run by *Tradition*, economies run by *Command*, and economies run by the *Market*. Let us briefly see what is characteristic of each.

TRADITION

Perhaps the oldest and, until a very few years ago, by far the most generally prevalent way of solving the economic challenge has been that of tradition. It has been a mode of social organization in which both production and distribution were based on procedures devised in the distant past, rigidified by a long process of historic trial and error, and maintained by heavy sanctions of law, custom, and belief.

Societies based on tradition solve the economic problems very manageably. First, they deal with the production problem— the problem of assuring that the needful tasks will be done—by assigning the jobs of fathers to their sons. Thus a hereditary chain assures that skills will be passed along and jobs will be staffed from generation to generation. In ancient Egypt, wrote Adam Smith, the first great economist, "every man was bound by a principle of religion to follow the occupation of his father and was supposed to commit the most horrible sacrilege if he changed it for another." And it was not merely in antiquity that tradition preserved a productive orderliness within society. In our own Western culture, until the fifteenth or sixteenth century, the hereditary allocation of tasks was also the main stabilizing force within society. Although there was some movement from country to town and from occupation to occupation, birth usually determined one's role in life. Born to the soil or to a trade and on the soil or within the trade, one followed in the footsteps of one's forebears.

Thus tradition has been the stabilizing and impelling force behind a great repetitive cycle of society, assuring that society's work would be done each day very much as it had been done in the past. Even today, among

9

the less industrialized nations of the world, tradition continues to play this immense organizing role. In India, for example, until recent years, one was born to a caste that had its own occupation. "Better thine own work is, though done with fault," preached the Bhagavad-Gita, the great philosophic moral poem of India, "than doing other's work, even excellently."

THE COST OF TRADITION

Traditional solutions to the economic problems of production and distribution are most commonly encountered in primitive agrarian or nonindustrial societies where, in addition to serving an economic function, the unquestioning acceptance of the past provides the necessary perseverance and endurance to confront harsh destinies. Yet even in our own society, tradition continues to play a part in solving the economic problem. It plays its smallest role in determining the distribution of our own social output, although the persistence of traditional payments such as tips to waiters, allowances to minors, or bonuses based on length of service are all vestiges of old traditional ways of distributing goods, as is the differential between men's and women's pay for equal work.

More important is the continued reliance on tradition, even in America, as a means of solving the production problem—that is, in allocating the performance of tasks. Much of the actual process of selecting an employment in our society is heavily influenced by tradition. We are all familiar with families in which sons follow their fathers into a profession or a business. On a somewhat broader scale, tradition also dissuades us from certain employments. Sons of American middle-class families, for example, do not usually seek factory work, even though factory jobs may pay better than office jobs, because blue-collar employment is not in the middle-class tradition.

Thus, even in our society—clearly not a "traditional" one—custom provides an important mechanism for solving the economic problem. But now we must note one very important consequence of the mechanism of tradition. *Its solution to the problems of production and distribution is a static one.* A society that follows the path of tradition in its regulation of economic affairs does so at the expense of large-scale rapid social and economic change.

The economy of a Bedouin tribe or a Burmese village has not changed essentially over the past hundred or even thousand years. The bulk of the peoples living in tradition-bound societies repeat, in the daily patterns of their economic life, much of the routine that characterized them in the distant past. Such societies may rise and fall, wax and wane; but external events—war, climate, political adventures and misadventures—are mainly responsible for their changing fortunes. Internal, self-generated economic change is but a small factor in the history of most tradition-bound states. *Tradition solves the economic problem, but it does so at the cost of economic progress.*

COMMAND

A second manner of solving the problem of economic continuity also displays an ancient lineage. This is the method of imposed authority, of economic command. It is a solution based not so much on the perpetuation of a viable system by the changeless reproduction of its ways, as on the organization of a system according to the orders of an economic commander-in-chief.

The mode of authoritarian economic organization was by no means confined to ancient Egypt. We encounter it in the despotisms of medieval and classical China which produced, among other things, the colossal Great Wall, or in the slave labor by which many of the great public works of ancient

COMMAND IN ACTION

Not infrequently we find the authoritarian method of economic control superimposed upon a traditional social base. Thus the Pharaohs of Egypt exerted their economic dictates above the timeless cycle of traditional agricultural practice on which the Egyptian economy was based. By their orders, the supreme rulers of Egypt brought into being the enormous economic effort that built the pyramids, the temples, the roads. Herodotus, the Greek historian, tells us how the Pharaoh Cheops organized the task.

[He] ordered all Egyptians to work for himself. Some, accordingly, were appointed to draw stones from the quarries in the Arabian mountains down to the Nile, others he ordered to receive the stones when transported in vessels across the river.... And they worked to the number of a hundred thousand men at a time, each party during three months. The time during which the people were thus harassed by toil lasted ten years on the road which they constructed, and along which they drew the stones; a work, in my opinion, not much less than the Pyramid.*

*Histories, trans. Cary (London: 1901), Book II, p. 124.

Rome were built, or for that matter, in any slave economy, including that of pre-Civil War U.S.A. Of course, we find it today in the dictates of the communist economic authorities. In less drastic form we find it also in our own society; for example, in the form of taxes—that is, in the preemption of part of our income by the public authorities for public purposes.

Economic command, like tradition, offers solutions to the twin problems of production and distribution. In times of crises, such as war or famine, it may be the only way in which a society can organize its manpower or distribute its goods effectively. Even in America, we commonly declare martial law when an area has been devastated by a great natural disaster. Then we may press people into service, requisition homes, impose curbs on the use of private property such as cars, or even limit the amount of goods a family may consume.

Quite aside from its obvious utility in meeting emergencies, command has a further usefulness in solving the economic problem. Unlike tradition, the exercise of command has no inherent effect of slowing down economic change. Indeed, the exercise of authority is the most powerful instrument society

has for *enforcing economic change.* Authority in modern China or Russia, for example, has effected radical alterations in the systems of production and distribution. So, too, even in our own society, it is sometimes necessary for economic authority to intervene in the normal flow of economic life, to speed up or bring about change. The government may utilize its tax receipts to lay down a network of roads that will bring a backwater community into the flux of active economic life. It may undertake an irrigation system that will dramatically change the economic life of a vast region. It may deliberately alter the distribution of income among social classes.

THE IMPACT OF COMMAND

Economic command that is exercised within the framework of a democratic political process is very different from that exercised by a dictatorship: there is an immense social distance between a tax system controlled by Congress and outright expropriation or labor impressment by a supreme and unchallengeable ruler. Yet whilst the means may be much milder, the *mechanism* is the same. In both cases, command diverts economic effort toward goals chosen by a higher authority. In both cases it interferes with the existing order of production and distribution, to create a new order ordained from "above."

This does not in itself serve to commend or condemn the exercise of command. The new order imposed by the authorities may offend or please our sense of social justice, just as it may improve or lessen the economic efficiency of society. Clearly, command can be an instrument of a democratic as well as of a totalitarian will. There is no implicit moral judgment to be passed on this second of the great mechanisms of economic control. Rather, it is important to note that no society—certainly no modern society—is without its elements of command, just as none is devoid of the influence of tradition. *If tradition is the great brake on social and economic change, so economic command can be the great spur to change.* As mechanisms for assuring the successful solution to the economic problem, both serve their purposes, both have their uses and their drawbacks. Between them, tradition and command have accounted for most of the long history of man's economic efforts to cope with his environment and with himself. The fact that human society has survived is testimony to their effectiveness.

THE MARKET

But there is also a third solution to the economic problem, a third way of maintaining socially viable patterns of production and distribution. This is the *market organization of society*—an organization that, in truly remarkable fashion, allows society to insure its own provisioning with a minimum of recourse either to tradition or command.

Because we live in a market-run society, we are apt to take for granted the puzzling—indeed, almost paradoxical—nature of the market solution to the economic problem. But assume for a moment that we could act as economic advisers to a society that had not yet decided on its mode of economic organization.

We could imagine the leaders of such a nation saying, "We have always experienced a highly tradition-bound way of life. Our men hunt and cultivate the fields and perform their tasks as they are brought up to do by the force of example and the instruction of their elders. We know, too, something of what can be done by economic command. We are prepared, if necessary, to sign an edict making it compulsory for many of our men to work on community projects for our national development. Tell us, is there any other way we can organize our society so that it will function successfully—or better yet, more successfully?"

Suppose we answered, "Yes, there is another way. Organize your society along the lines of a market economy."

"Very well," say the leaders. "What do we then tell people to do? How do we assign them to their various tasks?"

"That's the very point," we would answer. "In a market economy, no one is assigned to any task. In fact, the main idea of a market society is that each person is allowed to decide for himself what to do."

There is consternation among the leaders. "You mean there is no assignment of some men to mining and others to cattle raising? No manner of designating some for transportation and others for weaving? You leave this to people to decide for themselves? But what happens if they do not decide correctly? What happens if no one volunteers to go into the mines, or if no one offers himself as a railway engineer?"

"You may rest assured," we tell the leaders, "none of that will happen. In a market society, all the jobs will be filled because it will be to people's advantage to fill them."

Our respondents accept this with uncertain expressions. "Now look," one of them finally says, "let us suppose that we take your advice and allow our people to do as they please. Let's talk about something specific, like cloth production. Just how do we fix the right level of cloth output in this 'market society' of yours?"

"But you don't," we reply.

"We don't! Then how do we know there will be enough cloth produced?"

"There will be," we tell him. "The market will see to that."

"Then how do we know there won't be *too much* cloth produced?" he asks triumphantly.

"Ah, but the market will see to that too!"

"But what is this market that will do these wonderful things? Who runs it?"

"Oh, nobody runs the market," we answer. "It runs itself. In fact there really isn't any such *thing* as 'the market.' It's just a word we use to describe the way people behave."

"But I thought people behaved the way they wanted to!"

"And so they do," we say. "But never fear. They will want to behave the way you want them to behave."

"I am afraid," says the chief of the delegation, "that we are wasting our time. We thought you had in mind a serious proposal. What you suggest is inconceivable. Good day, sir."

Could we seriously suggest to such an emergent nation that it entrust itself to a market solution of the economic problem? That will be a problem to which we shall return toward the end of our book. But the perplexity that the market idea would rouse in the mind of someone unacquainted with it may serve to increase our own wonderment at this most sophisticated and interesting of all economic mechanisms. How does the market system assure us that our mines will find miners, our factories workers? How does it take care of cloth production? How does it happen that in a market-run nation each person can indeed do as he wishes and, withal, fulfill needs that society as a whole presents?

ECONOMICS AND THE MARKET SYSTEM

Economics, as we commonly conceive it and as we shall study it in much of this book, is primarily concerned with these very problems. Societies that rely primarily on tradition to solve their economic problems are of less interest to the professional economist than to the cultural anthropologist or the sociologist. Societies that solve their economic problems primarily by the exercise of command present interesting economic questions, but here the study of economics is necessarily subservient to the study of politics and the exercise of power.

It is a society that solves its economic problems by the market process that presents an aspect especially interesting to the economist. Clearly, many (although not all) of the problems we encounter in America today have to do with the workings or misworkings of the market system. And precisely *because* our contemporary problems often arise from the operations of the market, we study economics itself. Unlike the case with tradition and command, where we quickly grasp the nature of the economic mechanism of society, when we turn to a market society we are lost without a knowledge of economics. For in a market society it is not at all clear that even the simplest problems of production and distribution will be solved by the free interplay of individuals without guidance from tradition or command; nor is it clear how and to what extent the market mechanism is to be blamed for society's ills—after all, we can find poverty and neglect and pollution and armaments in nonmarket economies too.

In subsequent parts of this book we shall

analyze these puzzling questions in more detail. But the task of our initial exploration must now be clear. As our imaginary interview with the leaders of an emergent nation has suggested, the market solution appears very strange to someone brought up in the ways of tradition or command. Hence the question arises: how did the market solution itself come into being? Was it imposed, full-blown, on our society at some earlier date? Or did it arise spontaneously and without forethought? This is the focusing question of economic history to which we now turn, as we retrace the evolution of our own market system out of the tradition- and authority-dominated societies of the past.

KEY WORDS

Provisioning
wants

Scarcity

Production

Distribution

Division
of labor

Tradition

Command

Market

CENTRAL CONCEPTS

1. Economics is the study of how man assures his material sufficiency, of how societies arrange for their *material provisioning*.

2. Economic problems arise because the wants of most societies exceed the gifts of nature, giving rise to the general condition of *scarcity*.

3. Scarcity, in turn (whether it arises from nature's stinginess or man's appetites) imposes two severe tasks on society: it must
   a) mobilize its energies for *production*—producing not only enough goods, but the right goods, and
   b) resolve the problem of *distribution*, arranging a satisfactory solution to the problem of Who Gets What?

4. These problems afflict all societies, but they are especially difficult to solve in advanced societies in which there exists a far-reaching *division of labor*. Men in wealthy societies are far more socially interdependent than men in simple societies.

5. Over the course of history, three types of solutions to the two great economic problems have evolved: *Tradition, Command,* and the *Market System.*

6. Tradition solves the problems of production and distribution by enforcing a continuity of tasks and rewards through social institutions such as the caste system. *Typically, the economic solution imposed by Tradition is a static one,* in which little change occurs over long periods of time.

7. Command solves the economic problem by imposing allocations of effort or reward by *governing authority*. Command can be a means for achieving rapid and far-reaching economic *change*. It can take extreme totalitarian or mild democratic forms.

8. The market system is a complex mode of organizing society, allowing order and efficiency to emerge spontaneously from a seemingly uncontrolled society. We shall investigate the market system in great detail in the chapters to come.

QUESTIONS

1. If everyone could produce all the food he needed in his own backyard, and if technology were so advanced that we could all make anything we wanted in our basements, would an "economic problem" exist?

2. Suppose that everyone were completely versatile—able to do everyone else's work just as well as his or her own. Would a division of labor still be useful in society? Why?

3. Modern economic society is sometimes described as depending on "organization men" who allow their lives to be directed by the large corporations for which they work. Assuming that this description has some glimmer of truth, would you think that modern society should be described as one of Tradition, Command, or the Market?

4. In what way do your own plans for the future coincide with, or depart from, the occupations of your parents? Do you think that the so-called generational split is observable in all modern societies?

5. Economics is often called the science of scarcity. How can this label be applied to a society of considerable affluence such as our own?

6. What elements of Tradition and Command do you think are indispensable in a modern industrial society? Do you think that modern society could exist without any dependence on Tradition or without any exercise of Command?

7. Much of production and distribution involves the creation or the handling of *things*. Why are production and distribution *social* problems rather than engineering or physical problems?

8. Do you consider man's wants to be insatiable? Does this imply that scarcity must always exist?

The Emergence of Economic Society ~ and Economics

WE TEND TO THINK THAT MARKETS ARE THE normal form of social organization, that they must have always existed. And in a manner of speaking, they have. Men traded with one another at least as far back as the last Ice Age—the mammoth-hunters of the Russian steppes obtained shells from the Mediterranean region, as did the hunters of the central valleys of France. In fact, on the moors of Pomerania in northeastern Germany, archeologists have come across an oaken box, replete with the remains of its original leather shoulder strap, in which were a dagger, a sickle head, and a needle, all of Bronze Age manufacture. According to the conjectures of experts, this was very likely the sample kit of a prototype of the traveling salesman, an itinerant representative who collected orders for the specialized production of his community.[1]

Thus it seems as if we could discover evidences of a market organization of society deep in the past. But these surprising notes of modernity must be interpreted with caution. If markets, buying and selling—even highly organized trading bodies—were well-nigh ubiquitous features of ancient Greece or Rome or of feudalism, they should not be confused with the presence of a market *society*. Trade has existed as an important adjunct to society from earliest times, but the fundamental tasks of production or distribution were largely divorced from the market

[1] *Cambridge Economic History of Europe* (Cambridge: Cambridge University Press, 1952), II, 4.

process. Over most of mankind's history, markets were *not* the means used by men to solve the economic problem.

The reason for this is easy to grasp. In the marketplaces of antiquity, men traded goods or bought and sold services. *But they did not buy or sell land, labor, or capital—the so-called "factors of production"—that entered into the production of these goods and services.* Nor, for that matter, did they entrust the distribution of the output of those factors of production mainly to the marketplace. Production and Distribution were largely organized by the rules of tradition or by the imperatives of command—not by a vast trading network in which human labor or the use of land or capital were directed by the invisible push and pull of market forces. This point is so important to an understanding of the economic problem in our own day that we must take a moment to explain it further.

Land, Labor, and Capital

How did pre-modern society combine its factors of production? The answer comes as something of a shock: *there were no factors of production.* Of course, labor has always existed, for men have always had to expend effort to remain alive; land has always played its essential role as the great source of sustenance, together with the sea; and capital is as old as the first hunting implements or digging sticks.

Yet, *labor, land, or capital were never considered to be commodities for sale.* Labor was performed as part of the social duties of a peasant cultivator, a serf, or a slave, but in none of these roles was the laborer *paid* for his work. The peasant or serf raised his crop, handed over to the landlord his often onerous rent—usually by surrendering a portion of the actual crop—and kept body and soul together with what was left. The slave was

the property of his owner; he was given his subsistence but certainly not remunerated by anything resembling a wage. Here and there, labor was performed in exchange for money— for example, the work of skilled artisans, jewelers, armorers, and the like—but probably something like 70 to 80 percent of all the productive activity that sustained economic life in the pre-market economy was totally unconnected with anything that resembled a "market."

The same was true for land and capital. Land was the basis of power and status, and perhaps the most important form of wealth, but it was rarely, and only under great duress, bought and sold. A medieval lord, for example, would no more have thought of selling a portion of his ancestral estates to a neighboring lord than the governor of Connecticut would think of selling off a few counties to the governor of Rhode Island. Capital, too, was an agency of the productive process for which nothing like a market existed. A few merchants dealt in gold and silver, and the owner of a ship might sell his capital equipment to another. But the idea that one's productive property was an "asset," worth a certain price and producing a certain income, was as foreign to pre-market society as the idea, today, that shares of common stock should be considered a family heirloom to be handed down from generation to generation.

THE ECONOMIC REVOLUTION

How did wageless labor, unrentable land, and private treasures become "factors of production"—that is, abstract commodities to be bought and sold like so many yards of cloth or bushels of wheat? How did the peasant, the serf, or the slave become the modern worker who earns his income by selling his labor—as a farmer, lathe operator, engineer,

PROPERTY IN MEN

Property is a fundamental concept of economics, and its complexity is nowhere more clearly shown than in the rights that society permits men to exercise over one another.

The most extreme form of property in men is *slavery*, widespread in antiquity. A slave is literally the physical property of another man, who can sell him, abuse him, and—in some societies— even kill him. Usually there are limits to the "use" one can make of human slaves; but some societies, including ancient Greece, literally worked them to death.

A less extreme form of property, *serfdom*, appears in many forms in European history, sometimes approaching slavery, sometimes quite different. In Western Europe the distinctive feature of serfdom was its structure of *mutual obligations*. The serf owed his lord days of labor or various dues and tolls. The lord owed his slave protection against marauding invaders and a minimum of subsistence from his stores if crops failed. Also, although the serf was "tied" to the strips of land he cultivated and could not leave his homestead for another (even under the jurisdiction of the same lord) without permission or payment, the lord was stayed by "the customs of the manor" from displacing the serf from "his" land. The serf was thus the property of the lord, but in a very different sense from that in which a slave might be his property.

Finally, a modern worker has a *property in his own labor*, which he is entitled to sell for as much as he can get. No slave or serf has this property in himself. At the same time, the worker, who is no man's property, is also no man's obligation. The employer buys his employees' labor, not their livelihood. All responsibility for the laborer ends when he leaves the owner's factory, which is *the owner's* property.

or advertising man—for whatever price is offered for that kind of work? How did lords of manors become real estate developers; how did petty merchants or guildmasters become capitalists?

The answer is that a vast revolution undermined the world of tradition and command and brought into being the market relationships of the modern world. Beginning roughly in the sixteenth century—although with roots that can be traced much further back—a gradual process of change tore apart the bonds and customs of the medieval world and ushered in the market society we know.

THE ENCLOSURE MOVEMENT

We cannot trace here the long, tortuous, often bloody history of that revolution, a revolution more profoundly disturbing and upsetting than the French or the American or even the later Russian revolutions. But a few examples may make that dimly glimpsed change come alive. Take, for example, the matter of labor performed for wages.

Returning from a triumphal tour of her kingdom at the end of the sixteenth century, Queen Elizabeth wrote: "Paupers are everywhere!" How could this be, when the boast of England had been its sturdy class of yeomen, tilling their own land as peasant proprietors?

The reasons are many, but one of the most important and surely the most dramatic was the forcible expulsion of the "cottager" from his land. Beginning in Elizabethan times and proceeding irregularly until the middle of the nineteenth century, *enclosure* became an enormous movement of expropriation. As the word suggests, it enclosed the open land, usually with hedges and fences surrounding a veritable army of grazing sheep, the source of the wool that rapidly became England's chief export.

The reason for the enclosure movement was the desire of landed proprietors to raise the increasingly profitable crop of wool. But we are interested here in the effects of the movement on the peasant proprietor. Those effects can only be called devastating. By taking away from him, in exchange for an often trifling sum, his right to use the "common land" to graze his goat or cow, it robbed the small cultivator of access to the land he needed to maintain his economic independence. Sheep flourished, peasants starved. As one John Hales wrote in 1549: "... where XL persons had their lyvings, now one man and his shepherd hath all ... Yes, these shepe is the cause of all theise mischiefs, for they have driven husbandrie out of the countries, by the which was encreased before all kynde of victuall, and now altogether shepe, shepe."

It is difficult to imagine the scope and force of the enclosure movement, which for two centuries gradually pushed the peasant off the land. Already by the middle of the sixteenth century, peasants were rioting in protest; in one such uprising 3,500 persons were killed. And yet the process continued. As late as 1820, for example, the Dutchess of Sutherland dispossessed 15,000 tenants from 794,000 acres of land, replaced them with 131,000 sheep, and rented her evicted families an average of 2 acres of submarginal land

THE FACTORS OF PRODUCTION

What is important for us to note, however, is the fate of the dispossessed small proprietor. Deprived by enclosure of his ability to support himself, he became a new economic individual—a landless worker, an agricultural proletarian. As such he was forced to hire himself out for whatever work was to be had; and when none was available, he turned to vagabondage or brigandage or descended into the paupery that so alarmed Elizabeth. Grad-

ually he drifted toward the towns, where work might be had in small shops and "manufactories"; but he came not as a yeoman, with the security of a self-supporting farm, or as a guild apprentice, protected by the network of medieval rules and regulations. He came as a new personage in economic history—as the factor of production *labor,* selling his manpower for whatever the market offered.

No less painful was the process of changing land from a unit of administrative and political power to real estate. Much of this took place as the consequence of a gradual "monetization" of life. Beginning in the sixteenth century, precious metals started to flow into Europe from the merchant-adventurers of Spain and Portugal and England, who had begun to open up America and the fabled treasures of the East. At the same time, the flow of merchandise—silks and spices and expensive armor—gained in volume as trade routes across the face of Europe became more secure.

Hence the lords of the manor, eager to buy these new goods, began to convert their traditional peasant dues from days of work or ells of cloth to actual cash. But this soon put the lords into a squeeze between a fixed income and rising prices. For like everything in feudal life, once these cash dues were set, they were fixed and unchangeable. But as gold and silver poured into Europe, the prices of merchandise steadily rose.

Hence we begin to find another new economic individual, the impoverished aristocrat. In the year 1530, for instance, in the Gevaudan district of France, 121 lords had an aggregate income of 21,400 livres, but one of these seigneurs accounted for 5,000 livres of the sum, another for 2,000, and the rest averaged a mean 121 livres each.[2] Meanwhile, the richest town merchants had in-

[2]*Cambridge Economic History of Europe,* I, 557–58.

comes up to 65,000 livres. Thus the balance of power turned against the landed aristocracy, reduced some to shabby gentility, and elevated the upstart merchants, who lost no time in acquiring lands that they soon came to regard as property, not as an ancestral estate.

Thus a long, wrenching process of economic change impelled by many background forces—the monetization of life, the enclosure movement, the rise of national states out of feudal principalities, and still other causes—gradually imposed a market system on a world that had known markets but had never imagined that land, labor, and capital could be bought and sold, hired and fired, to carry out the daily process of social provisioning.

What we must carry away from this brief glance at economic history is one very important realization. It is that the factors of production, with which modern economic inquiry is concerned, are not eternal attributes of a natural order. *They are the creations of a process of historic change,* a change that divorced "labor" from social life and made it an abstract quantity of effort offered for sale to bidders for labor-power; a change that separated the value of "land" from its ancient prerogatives of status and power; a change that brought the idea of "capital" to a society which had always known wealth, but had never conceived of it as something whose form and shape were of no consequence, but whose "yield" was all-important.

THE PROFIT MOTIVE

The emergence of a market society, however, with its new factors of production, was not the only creation of the historic forces we have examined in this chapter. Along with the new relationships of man to man in the marketplace, there arose a new form of *social control* to take over the guidance of the economy from the former aegis of tradition and command.

What was this new form of control? Essentially, it was a pattern of social behavior, of normal, everyday action that the new market environment imposed on society. In the language of the economist, this pattern of behavior was the drive to *increase one's income* by concluding the best possible bargains in the marketplace. In ordinary language, it was the drive to buy cheap and sell dear; in business terminology, the *profit motive.*

The market society had not, of course, invented this motive. Perhaps it did not even intensify it. But it did make it a ubiquitous and necessitous aspect of social behavior. Although men may have *felt* acquisitive during the Middle Ages or antiquity, they did not enter en masse into market transactions for the basic economic activities of their livelihoods. Even when, for instance, a peasant sold his few eggs at the town market, rarely was the transaction a matter of overriding importance for his continued existence. Market transactions in a fundamentally nonmarket society were a subsidiary activity, a means of supplementing a livelihood which, however sparse, was largely independent of buying or selling.

With the monetization of labor, land, and capital, however, transactions became universal, critical activities. Now everything was for sale, and the terms of transactions were anything but subsidiary to existence itself. To a man who sold his labor in a society that no longer assumed any responsibility for his upkeep, the price at which he concluded his bargain was all-important. So it was with the landlord and the budding capitalist. For each of these a good bargain could spell riches; a bad one, ruin. Thus the pattern of economic maximization was generalized throughout society and given an inherent urgency that made it a powerful force for shaping human behavior.

The Invention of Economics

The new market society did more than merely bring about an environment in which men were forced to follow their self-interest. It also brought a puzzle of great importance and considerable difficulty. The puzzle was to understand the workings of a world in which profit-seeking individuals were no longer constrained to follow the ways of their forefathers or to shape their economic activities according to the dictates of a ruling lord or king.

THE "PHILOSOPHY" OF TRADE

The new order needed a "philosophy"—a reasoned explanation of how such a society would hang together, would "work." And such a philosophy was by no means self-evident. In many ways the new world of profit-seeking individuals appeared as perplexing and fraught with dangers to its contemporaries as it did to the imaginary village elders to whom we described it in our last chapter.

Hence it is not surprising that the philosophers of trade disagreed. In England, a group of pamphleteers and merchants, the so-called Mercantilists, put forward an explanation of economic society that stressed the importance of gold and extolled the role of the merchant whose activities were most likely to bring "treasure" into the state by selling goods to foreigners. The Mercantilists also urged that the wages of labor be kept as low as possible, for they feared that men would not work at all unless they were literally driven to it by hunger: "To make society happy," wrote one spokesman, "it is necessary that large numbers be wretched as well as poor."

In France, a school of thinkers we call the Physiocrats held quite different ideas. They extolled the virtues of the farmer, not the merchant. All wealth ultimately came from nature's bounty, they argued, dismiss-ing merchants and even manufacturers as belonging to a "sterile" class that added nothing to the wealth produced by the farmer. Labor was assumed to be poor, although not necessarily "wretched."

With such diverse views, it is obvious that nothing like unanimity prevailed concerning proper economic policy. Should competition be regulated or left alone? Should the export of gold be prohibited or should "treasure" be permitted to enter or leave the kingdom as the currents of trade dictated? Should the agricultural producer be taxed because he was the ultimate source of all wealth, or should taxes fall on the prosperous merchant class?

There is no purpose in pursuing further the various answers that we can find to these questions. More interesting, in retrospect, is the gradual emergence of a common set of questions. As the eighteenth century progressed, the practitioners of "moral philosophy" began to apply scientific methods to the social universe, just as practitioners of "natural philosophy" applied them to the natural universe. And within the field of moral philosophy emerged a special study that struggled to make sense of those peculiar goings-on in the social universe involving the production and distribution of wealth. We call this field economics—or to give it the name of its day, political economy.

ADAM SMITH (1723–1790)

The roots of economics can be traced far back to antiquity, when Aristotle worried about problems such as the "value" of objects. We have glanced at its initial formulations in the works of the Mercantilists and the Physiocrats, but its full-fledged expression awaited the advent of Adam Smith, patron saint of our discipline and a figure of towering intellectual stature.

Born in the tiny town of Kirkaldy, Scotland, in 1723, Smith soon showed remarkable

A PORTRAIT OF ADAM SMITH

"I am a beau in nothing but my books," Adam Smith once said of himself. He was indeed a homely man, with heavy-lidded eyes and a rather protrusive lower lip, a man burdened with curious personal traits including a stumbling manner of speech and a gait a friend described as "vermicular." He was, moreover, given to famous fits of reverie and absent-mindedness. On one occasion, absorbed in a discussion with a friend, he fell into a tanning pit.

Few adventures befell Smith in the course of his rather uneventful life. Indeed, perhaps the high point was reached at age four, when he was abducted from his home by a band of passing gypsies. His captors held him for only a few hours. Perhaps they sensed what a biographer would later write: "He would have made, I fear, a poor gypsy."

But if his life was bland and his manner distracted, his prose was sharp as a knife. Not least among the pleasures of the *Wealth of Nations* are the phrases and insights with which it abounds. It was Smith who wrote: "By nature a philosopher is not in genius and disposition half so different from a street porter, as a mastiff is from a greyhound"; who called England "a nation of shopkeepers"; who castigated "the mean rapacity, the monopolizing spirit" of the manufacturers whose economic role he celebrated; and who observed, "People of the same trade seldom meet together but the conversation ends in a conspiracy against the public, or in some diversion to raise prices." Anyone who has the idea that Smith is a "conservative" philosopher has a surprise coming to him when he reads the *Wealth of Nations*.

abilities, and at age 17 was sent to Oxford to complete his education. Oxford was not then a great center of learning. Students rarely bothered to attend the equally rarely held classes. The professoriat was less concerned with the intellectual training of its charges, than with their moral rectitude: Smith was nearly expelled for owning a copy of a dangerous tract, David Hume's *A Treatise of Human Nature,* a book we now regard as one of the masterpieces of eighteenth century philosophy.

FROM HISTORY AND PHILOSOPHY TO ECONOMICS

Of necessity, Smith spent most of his time educating himself. What is astonishing is that by the time he returned to Scotland, as a young man in his twenties, he had worked out in his mind a great scheme of historical evolution that included the first detailed exposition of a market system.

The scheme of history is interesting enough to warrant a moment. Smith believed that we could trace a great evolutionary pattern in social history. It began with man in his "rude" state; passed through his organization as a society of agricultural nomads, such as that of Ghenghis Khan; then entered a period of settled manorial agriculture (which we would call the medieval period), until it emerged into the stage of commerce and manufacture, which Smith saw burgeoning around him.

Each period had its appropriate form of law and social organization, a conception of history very much akin to that later developed by Karl Marx. But what interested Smith were the particular laws and forms of organization demanded by the stage of commerce and manufacture. Smith called this necessary framework the system of *perfect liberty,* meaning a system in which the restrictions and regulations of feudal life, or of the "mercantilist" period that superseded it, were replaced by the widest possible liberty accorded to every person to determine for himself his economic scope and place. Later, Sir Henry Maine, the distinguished British historian, would call this a transition from a society of status to one of contract. We would call it the development of the legal underpinnings of a society of *private enterprise,* or of *capitalism.*

THE THEORY OF MORAL SENTIMENTS

In the lectures Smith gave at Glasgow and at Edinburgh, following his return to Scotland, he sketched only general outlines of the broad concept of historic evolution. Then in 1759 he published *The Theory of Moral Sentiments,* a brilliant, delightfully written book in which he posed a curious philosophical question: how does it happen that man, a creature of self-interest, is able to form detached and impersonal moral judgments about things? Smith answered by describing an "impartial Spectator," a "man within" who resided, so to speak, inside our selfish selves, whispering dispassionate advice into our ears. (The idea is not quite so naïve as it sounds if we substitute the Freudian idea of a "super-ego" for the "man within" as the source of our moral promptings.)

The book made Smith's name overnight and brought him a tempting offer to tutor a young English lord at a salary considerably higher than his professorial fees, with a pension promised thereafter. Smith embarked with his pupil on a two-year European tour that brought him in touch with not only Voltaire and Rousseau, but with Francois Quesnay, a physician at the court of Louis XIV, who was also the central figure in the Physiocratic school.

During these years Smith began to develop in his mind the outlines of a major treatise on political economy; and after his return to Scotland in 1766, he made it his central preoccupation, one that would take him ten years to complete. When it was done,

published in the very year of the American revolution, Smith had written a masterpiece. In *The Wealth of Nations* he presented to society the conception it most needed, a clear and convincing picture of the way it worked.

THE GROWING WEALTH OF NATIONS

The world that Smith described was very different from our own. It was a world of very small enterprises: Smith's famous description of a pin factory describes a manufacturing establishment that employs ten people. It was still hampered by medieval guild restrictions: in Smith's time no master hatter in England could employ more than two apprentices; in the famous Sheffield silver trade, no master cutler could employ more than one. Still more important, it was a world in which government-protected monopolies were accorded to certain fields of commerce, such as the trade with the East Indies. Yet, for all the differences from modern economic society, the basic vision that Smith gave to his time can still elucidate the tasks of economics in our own time.

Two main problems occupied Smith's attention. The first is implicit in the title of the book. This is Smith's theory of the most important tendency of a society of "perfect liberty"—its *tendency to grow.*

Economic growth—that is, the steady increase in the output of goods and services enjoyed by a society—was hardly a concern for philosophers of tradition-bound societies; or even of societies ruled by imperial-minded emperors. But what Smith discerned amid the seeming turmoil of a market society was a hidden mechanism that would operate to enlarge the "wealth of nations"—at any rate, those nations that enjoyed a system of perfect liberty and did not tamper with it.

What was it that drove society to increase its riches? Basically it was the tendency of such a society to encourage a steady rise in the *productivity* of its labor, so that over time, the same number of working people could turn out a steadily larger output.

And what lay behind the rise in productivity? The answer, according to Smith, was the gain in productiveness that was to be had by achieving an ever finer *division of labor*. Here Smith's famous pin factory serves as an example:

One man draws out the wire, another straits it, a third cuts it, a fourth points it, a fifth grinds it at the top for receiving the head; to make the head requires two or three distinct operations; to put it on is a peculiar business; to whiten it is another; it is even a trade by itself to put them into paper ... I have seen a small manufactory of this kind where ten men only were employed and where some of them consequently performed two or three distinct operations. But though they were poor, and therefore but indifferently accomodated with the necessary machinery, they could, when they exerted themselves, make among them about twelve pounds of pins in a day. There are in a pound upwards of four thousand pins of middling size. Those ten persons, therefore, could make among them upwards of forty-eight thousand pins in a day.... But if they had all wrought separately and independently ... they could certainly not each of them make twenty, perhaps not one pin in a day.

ADAM SMITH'S GROWTH MODEL

This begins to unravel the reasons why a society of free enterprise tends to grow. But it does not fully explain the phenomenon. For what is it that drives such a society to a division of labor? And how do we know that the tendency to growth will not peter out, for one reason or another?

This leads us to the larger picture that Smith had in mind. We would call it a growth model, although Smith used no such modern term himself. What we mean by this is that Smith shows us both a *propulsive force* that will put society on an upward growth path

and a *self-correcting mechanism* that will keep it there.

First the driving force. One of the fundamental building blocks of Smith's conception of human nature was what he called the "desire for betterment"—what we have already described as the profit motive. And what does the desire for betterment have to do with growth? The answer is very important. *It impels every manufacturer to expand his business in order to increase his profits.*

And how does this business expansion result in a higher division of labor? The answer is very neat. The main road to profit consists in equipping workmen with the necessary machinery that Smith mentions in his description of the pin factory, for it is this machinery that will increase their productivity. Thus the path to growth lies in what Smith called *accumulation,* or in more modern terminology, in the process of *capital investment.* As capitalists seek money, they invest in machines and equipment. As a result of the machines and equipment, their men can produce more. Because they produce more, society's output grows.

THE DYNAMICS OF THE SYSTEM

This answers the first part of our query. But there is still the question of how we know that society will continue to grow, that its trajectory will not flatten out. Here we come to the cleverest part of Smith's model. For at first look, it might seem as if the drive to increase capital investments would be self-defeating. The reason is that the steady increase in the demand for workmen to run the new machines would drive up their wages; and as wages rose, they would cut into the manufacturer's profits. In turn, as profits were eaten away, the very source of new investment would dry up and the growth curve would soon level off.

Not so, according to Smith. To be sure,

the rising demand for workmen *would* tend to drive up wages. But this was only half the picture. The same upward tendency of wages would also tend to increase the supply of workingmen. The reason is not implausible. In Smith's day, infant mortality was shockingly high: "It is not uncommon," Smith remarked, "... in the Highlands of Scotland for a mother who has borne twenty children not to have two alive." But as wages rose, infant and child mortality would tend to diminish, and therefore more of the population would arrive at working age (ten or younger in Smith's day).

The outcome must already be clear. Along with an increase in the demand for workingmen (and working children) comes an increase in their supply. This increase in the number of available workers meant that the competition for jobs would increase. Therefore the price of labor would *not* rise, at least not enough to choke off further growth. Like a vast self-regulating machine, the mechanism of capital accumulation would provide the very thing it needed to continue unhampered: a force to prevent wages from eating up profits. And so the growth process could go on undisturbed.

We will not concern ourselves here with the full details of Smith's growth model. What interests us is a first view of a problem that will absorb our attention in Part Four— the problem of the dynamics of growth of a market system. Of course, Smith's "model" is not directly applicable to the modern world, where (at least in industrialized nations) most children do not die before they reach working age and where his "safety valve" therefore has no relevance. But none-

theless in Smith's model we get a sense of the imaginative reach and capacity for enlightenment that economic analysis can bring.*

THE MARKET MECHANISM

The wealth (we would say the *output*) of nations was not, however, the only major problem on which Smith's treatise threw a clarifying light. There was also the question of how a market system held together, of how it provided an orderly solution to the problems of production and distribution.

This brings us to Smith's description and explanation of what we would call the market mechanism. Here Smith begins by elucidating a perplexing problem. The actors in Smith's drama, as we know, are driven by the desire for self-betterment and guided mainly by their self-interest. "It is not from the benevolence of the butcher, the brewer or the baker that we expect our dinner," writes Smith, "but from their regard to their self-interest. We address ourselves not to their humanity, but to their self-love, and never talk to them of our necessities, but of their advantages."

The problem here is obvious. How does a market society prevent self-interested, profit-hungry men from holding up their fellow citizens for ransom? How does a socially workable arrangement emerge from such a socially dangerous set of motivations?

The answer introduces us to a central mechanism of a market society, the mechanism of *competition*. For each man, out to do the best for himself, with no thought of others, is faced with a host of similarly moti-

*It seems necessary to add a word to the student who gets sufficiently interested in Smith's model to look into the *Wealth* itself. He will look in vain, in this vast, discursive book for a clear-cut exposition of the interactions we have just described. The model is implicit in Smith's exposition, but it lies around the text like a disassembled machine, requiring us to put it together in

our minds. Nonetheless it is there, if one fits together the pieces. For a full exposition, see A. Lowe "The Classical Model of Economic Growth" *Social Research,* Vol. 21 (1954); and Joseph Spengler's "Adam Smith's Theory of Economic Growth," *Southern Economic Journal,* Vols. 26 and 27 (1959).

THE DIVISION OF LABOR AT WORK

"Observe the accommodation of the most common artificer or day labourer in a civilized and thriving country, and you will perceive that the number of people of whose industry a part, though but a small part, has been employed in procuring him this accommodation, exceeds all computation. The woollen coat, for example, which covers the day-labourer, as coarse and rough as it may seem, is the produce of the joint labour of a great multitude of workmen. The shepherd, the sorter of the wool, the woolcomber or carder, the dyer, the scribbler, the spinner, the weaver, the fuller, the dresser, with many others, must all join their different arts in order to complete even this homely production. How many merchants and carriers besides, must have been employed ... how much commerce and navigation ... how many ship-builders, sailors, sail-makers, rope makers....

"Were we to examine, in the same manner, all the different parts of his dress and household furniture, the coarse linen shirt which he wears next to his skin, the shoes which cover his feet, the bed which he lies on ... the kitchen-grate at which he prepares his victuals, the coals which he makes use of for that purpose, dug from the

vated individuals who are in exactly the same position. Each is only too eager to take advantage of his competitor's greed if it urges him to raise his price above the level "set" by the market. If a pin manufacturer tried to charge more than his competitors, they would take away his trade; if a workman asked for more than the going wage, he would not be able to find work; if a landlord sought to exact a rent steeper than another with land of the same quality, he would get no tenants.

THE MARKET AND ALLOCATION

But the market mechanism does more than impose a competitive safeguard on the price of products. It also arranges for the production of the right *quantities* of the goods that society desires. Suppose that consumers want more pins than are being turned out, and fewer shoes. The public will buy out the existing supply of pins, while business in the shoe stores will be dull. Pin prices will tend to rise as the public scrambles for shrinking supplies, and prices of shoes will tend to fall as merchants try to get rid of their burdensome stocks.

And now, once again, a restorative force comes into play. As pin prices rise, so will the profits of the pin business, and as shoe prices sag, so will profits in shoemaking. Again, self-interest and the desire for betterment go to work. Pin manufacturers will expand their output to take advantage of higher prices; shoe factories will curtail production to cut their losses. Employers in the pin business will seek to hire more factors of production—more workers, more space, more capital equipment; and employers in the shoe business will reduce their use of the factors of production, letting workers go, giving up leases on land, cutting down on their capital investment.

Hence pin output will rise and shoe output will fall. *But this is exactly what the public wanted in the first place!* Through what Smith called, in a famous phrase, "an invisible hand," the selfish motives of men are transmuted by the market mechanism to yield the most unexpected of results: social well-being.

THE SELF-REGULATING SYSTEM

Thus Smith showed that a market system, far from being chaotic and disorderly, was in fact the means by which a solution of the strictest discipline and order was provided for the economic problem.

bowels of the earth, and brought to him perhaps by a long sea and a long land carriage, all the other utensils of his kitchen, all the furniture of his table, the knives and forks, the earthen or pewter plates upon which he serves up and divides his victuals, the different hands employed in preparing his bread and his beer, the glass window which lets in the heat and the light and keeps out the wind and the rain, with all the knowledge and art requisite for preparing that beautiful and happy invention ...; if we examine, I say, all those things ... we shall be sensible that without the assistance and cooperation of many thousands, the very meanest person in a civilized country could not be provided, even according to what we very falsely imagine, the easy and simple manner in which he is commonly accommodated. Compared indeed with the more extravagant luxury of the great, his accommodation must no doubt appear extremely simple and easy; and yet it may be true, perhaps, that the accommodation of a European prince does not always so much exceed that of an industrious and frugal peasant, as the accommodation of the latter exceeds that of many an African king, the absolute master of the lives and liberties of ten thousand naked savages."

From *Wealth of Nations*, Modern Library ed. p. 11, 12.

First, he has explained how the motive of self-interest provides the necessary impetus to set the mechanism to work. Next, he has shown how competition prevents any individual from exacting a price higher than that set by the marketplace. Third, he has made clear how the changing desires of society lead producers to increase production of wanted goods and to diminish the production of goods that are no longer as highly desired.

Not least, he has shown that the market system is a *self-regulating process*. For the beautiful consequence of a competitive market is that it is its own guardian. If prices or profits or wages stray away from the levels determined by the market system, forces exist to drive them back into line. Thus a curious paradox emerges; the competitive market, which is the acme of individual economic freedom, is at the same time the strictest of economic taskmasters. One may appeal the ruling of a planning board or win the dispensation of a minister, but there is no appeal, no dispensation, from the anonymous pressures of the competitive marketplace. Economic freedom is thus more illusory than it appears. You may do as you please, but if you please to do that which the market disapproves of, the price of freedom is economic ruin.

MARKETS AND MICROECONOMICS

We have by no means fully explained or criticized Smith's analysis of the market mechanism. We have left unexamined for example, the prices that the market "sets" or the operation of a market system when competition is impeded by monopolies of capital or labor. But we have gone far enough to grasp another main area of economics, one that we shall study in detail in Part Three, as "Microeconomics."

When we do study microeconomics, we will do so under assumptions very different from those of Smith. The problem of giant firms, of powerful industrial unions, of environmental danger—all problems unknown in Smith's day—create issues far removed from those in the *Wealth of Nations*. But the essential puzzle to which Smith addressed himself will still be there: will a system driven by a hunger for profit yield socially acceptable results? Will a society powered by a million motors of self-interest manage to coordinate the interests of its constituents through institutions of the market system?

A FINAL OVERVIEW

All that in good time. Our purpose in taking this first look into economic philosophy was not to polish the finer points of economic analysis or to bring us directly to a confrontation with today's problems. It was to get an initial sense of what economics is about, of what the *terra incognita* looks like from an aerial overview. So perhaps we should conclude with a few very general, but important, points.

First, *we have discovered that economics is principally the study of how market societies operate.* Tradition and command have solved the economic problem over most of history, and they continue to hold sway over vast areas of the globe, from the underdeveloped lands to the highly industrialized Soviet Union. Economists are interested in the operation of these societies, especially planned (command) societies, but they are primarily concerned with the mechanisms of the market system that we find in the so-called Western world, mainly North America, Europe, Japan, and Oceania.

Second, *we have learned that the market system is a historical development,* and a fairly recent historical development at that. The "factors of production" that are its moving parts are not to be found in societies of tradi-

tion or command. They are the special products of a society in which men *sell* their labor, *rent* their land, *invest* their capital, with as little interference from tradition or command as possible. Thus there is a historical setting, both for our economic system and for the discipline called economics, which analyzes that system. It is well to keep this in mind, for we live in a period in which that historical setting is rapidly changing, often in ways that conflict with the operation of a market economy.

Third, *Adam Smith's* Wealth of Nations *reveals two major problems with which economics is concerned.* The first is the large-scale tendencies of the entire system, a problem that we call *macroeconomics.* The second is the internal mechanism of the system, a problem we study under the name of *microeconomics.*

A brief introductory section cannot describe everything that economics is about, but the main outlines of the subject should now begin to take shape. We have an idea of the problems that lie ahead. We are ready to begin the expedition on foot.

KEY WORDS

Market society

Factors of production

Land, labor, and capital

CENTRAL CONCEPTS

1. We must differentiate between markets, which have a very ancient pedigree, and market societies, which do not. *In a market society, the economic problem itself—both production and distribution—is solved by means of a vast exchange between buyers and sellers.* Many ancient societies had markets, but these markets did not organize the fundamental activities of these societies.

2. The advent of free laborers, capitalists, and landlords, each selling his services on the market for land and capital and labor, made it possible to speak of the *"factors of production."* By this, two things were implied: the *physical categories* of land, labor, and capital as distinguishable agents in the production process, and the *social categories* of laborers, landowners, and capitalists as distinct groups or classes entering the marketplace.

3. The factors of production were brought into being by a profound and long-lasting *economic revolution* that wrested marketable labor from work, rentable real estate from land, investible capital from wealth. Central processes in this change were the *enclosures* that forced peasants off the land and the gradual *monetization* that eroded the power of the feudal class and enhanced the power of the merchant class.

Monetization

4. As part of this process of change we find the emergence of the *profit motive* at all levels of society, not as an acquisitive drive (which may have existed for centuries), but as the pervasive necessity for all individuals to strive for higher

Profit motive

incomes for economic survival.

5. Along with the new economic society came a new interest in the mechanism of a market society. This new direction of philosophy was called *political economy*,

Political economy

or later, *economics*.

6. The greatest of the early economists was *Adam Smith*, author of the *Wealth of Nations*. Essentially a philosopher, Smith turned his powerful and far-ranging

Adam Smith

inquiry to the problem of how a society of "perfect liberty" (a society of competitive free enterprise) operated.

7. In the *Wealth*, he described two attributes of such a society. The first was its

Productivity

tendency to grow. Smith showed how growth resulted from the increase in labor *productivity* that came from the ever-finer *division of labor*. This enhancement in productivity was brought about by capitalists' *investment* in *capital*

Division of labor

equipment as a means to higher profit.

8. Smith actually described a "*growth model*" for a market system. The model showed that growth would continue, even though the demand for labor bid up

Growth model

wages, imperiling profits. In Smith's model, the rise in wages also increased the supply of labor, thereby preventing a sharp increase in wages from undermining the process of capital investment.

Market mechanism

9. Smith also described the *market mechanism*. In this mechanism, *competition* played a key role in preventing individuals from exacting whatever price they pleased from buyers.

10. The *market mechanism* also revealed how changing demands for goods would

Competition

change the production of goods, to match that demand. Thus the capstone of Smith's treatise was the demonstration of the *self-regulating* nature of a competitive market, in which an "invisible hand" brought socially useful ends from

"Invisible hand"

selfish and private means.

QUESTIONS

1. We tend to speak of certain economic attributes of our economy as if they were part of "human nature." Do you think that they *are*? For example, are all societies and cultures oriented toward profit? Are they all competitive? Do they all esteem individual economic freedom? How do you explain this?

2. Why was the economic revolution so disruptive? Can you imagine a change in the economic organization of society today that would be equally disruptive? Suppose, for example, that ecological problems make it imperative for society to cut its production of man-made energy by half. What new forms of economic organization might be needed? What resistance to such a reorganization would you expect? How would you feel about living in a society in which each family's consumption of electricity or gasoline were rationed—and cut in half?

3. What is meant by a "factor of production"? Dis-

tinguish carefully between the physical services rendered by labor, land, and capital, and the social representation of those services.

4. What sort of "economics" would be useful in describing the economic problem as it is solved by the Kalahari Bushmen? By the Cubans or the Chinese?

5. Describe what Smith meant by the "invisible hand." What is the mechanism by which selfish interests are made compatible with—indeed, made the agent for—successful social provisioning?

6. Can you see a relation between Smith's micro and macro models? Could the macro model work if the micro forces did not operate?

7. Can you describe what probably would be the outcome of Smith's growth model if the supply of labor did not increase? Can you guess what might be the outcome of his micro model if competition were inoperative?

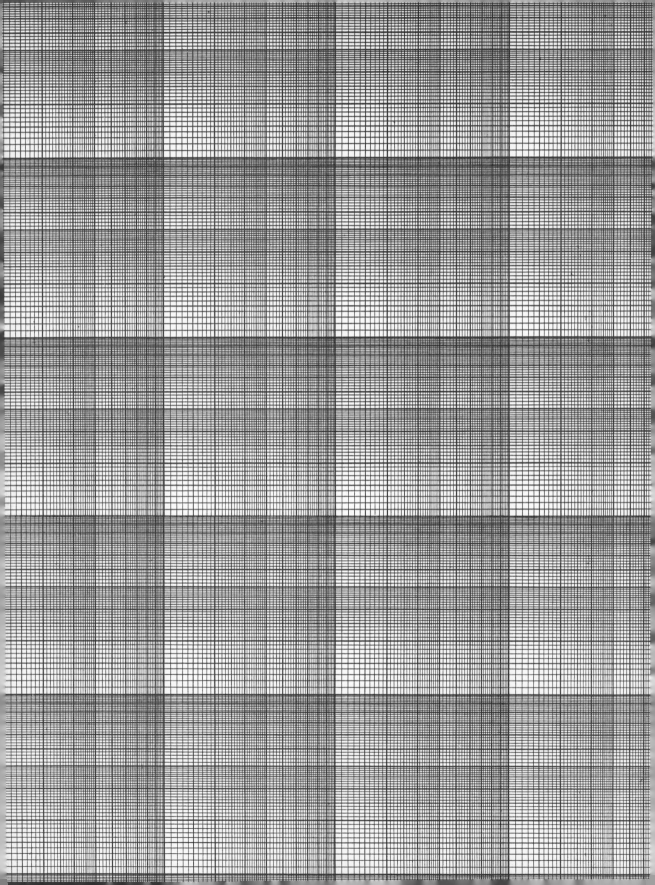

2

FUNDAMENTAL CONCEPTS

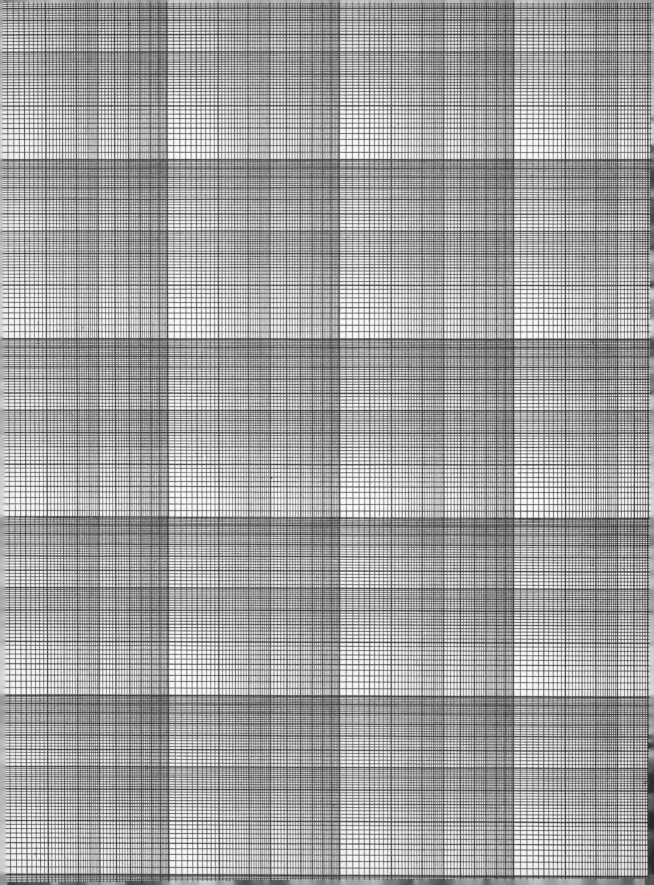

Economic Behavior and Nature

HAVING LOOKED OVER THE TERRITORY, we may be eager to strike out for the lands of Micro- or Macroeconomics. The going would be slow, however, unless we were better prepared for the task. We now have a rough map of the area, but we still have no familiarity with the equipment of the expedition—the intellectual equivalent of the surveyor's theodolite and his knowledge of trigonometry. Therefore we must pause in this section to learn something about the fundamental working notions of economics, before we can explore the great issues that economics presents.

A word before we set off. There are a great many ideas presented in this chapter, most of them probably unfamiliar to you. Let us therefore take some of the worry out of the next pages by stressing that *you need not learn these ideas "cold" at this point*. We will meet them again many times as we proceed; before long you will be very much at home with them.

Our purpose here is only to give you a first acquaintance with some basic economic concepts. That should make these concepts much simpler to master when you meet them in subsequent chapters.

THE TWO FUNDAMENTAL ASSUMPTIONS

How does economics "work"? If we think back over the matters of our initial chapters, we may be struck by a thought that puts our subject into sharp focus. Essentially, when economists build models and theorize about the problems of society, they depend on two fundamental assumptions:

One of these is a set of hypotheses about behavior—our activities when we work with nature to produce goods or when we distribute our production among the members of society. The other is a set of assumptions about nature, the phys-

ical "resistance" that the material world offers when we seek to use it for the purposes of production.

Behavior

MAN THE MAXIMIZER

Let us begin with the fundamental notions about behavior. They can be summed up in a single sentence—although, as we shall see, the sentence has deeper implications and raises more profound questions than might at first appear. The sentence states that *man is a maximizer.* By this we mean that economic theory begins with the assumption that men, whether they are engaged in producing or in distributing the fruits of production, seek to gain as much wealth or pleasure as they can. We call this wealth or pleasure "utility."

We will look later into the ambiguities and difficulties of this basic assumption, but it will help us if we begin by trying to understand it sympathetically, rather than examining it critically. When we say that man is a maximizer, we mean something very close to Adam Smith's contention that man is engaged in "bettering his condition."

Immediately this alerts us to the fact that *economics is concerned with the kind of behavior that we find in a market society.* In a traditional society, men may be perfectly content with their condition, whatever it is, and entertain no thought of "bettering it," especially by the pursuit of wealth. Perhaps in a socialist society of the future a different hypothesis about behavior would have to serve as our starting point; men may then be driven by the desire to better the condition of others rather than themselves.

But in a market system, men and women are presumed to be motivated by self-interest rather than altruism and to be driven to fulfilling that self-interest by actively pursuing whatever things they desire, whether they be gold, land, consumption goods, or simply leisure. The human being is thus conceived as an acquisitive animal, not necessarily by an inborn instinct (as Adam Smith would have maintained), but simply because a market society encourages and cultivates this frame of mind.

ECONOMIC MAN

This leads us to inquire further into the view of man that underlies economic theory. Economists do not believe that men and women are only and solely creatures of acquisition. They are fully aware that a hundred motivations impel them—rational and irrational, esthetic, political, religious, or whatever. If they concentrate on the acquisitive element in man, it is because they believe it to be decisive for his economic behavior—that is, for the explanation of his productive and distributive activities.

Economic theory is therefore a study of the effects of one attribute of man as it motivates him to undertake his worldly activities. Very often, as economists well know, other attributes will override or blunt the acquisitive, maximizing orientation. To the extent that this is so, economic theory loses its clarity or may even suggest outcomes different from those that we find in fact. But economists think that maximizing as a description of human behavior is universal and strong enough to serve as a good working hypothesis—a useful "first approximation" on which to build their complicated theories.

RATIONALITY

Economic man is conceived not only as a maximizer but as a *rational* maximizer. By this, economists mean that men in a market milieu, wishing to "better their condition," stop to consider the various ways open to them and to calculate in some fashion the means that will best suit their maximizing aims. There may be, for example, two different ways of producing a good: as a rational maximizer, man will choose the method that will yield him the good for the smallest cost (we will soon see exactly what we mean by *cost*).

This concept of "rational" maximizing does not mean that human beings may not wish, on some occasions, to go to more "trouble" than necessary. After all, men could worship God in very simple buildings or out-of-doors, but they go to extraordinary lengths in erecting magnificent churches and in decorating them with sculpture and paintings. It is meaningless to apply the word *rational* to pursuits such as these, which may be among the most important things that man does in his life.

But when men are engaged in producing the goods and services of ordinary life, seeking to "better their condition" by achieving the largest possible incomes or the most satisfaction-yielding patterns of consumption, the economist assumes that men *will* stop to add up the differing ways of attaining a given end and then choose the way that is least costly.

Rational economic man is not altogether a pretty concept. The picture of man that it suggests is not attractive. Selfishness and calculation figure large in this image; love and generosity are overlooked. Hence let us emphasize once more that economists do not set forth this picture of man as an ideal. It is not their conception of how men should behave in their relations to one another, or even their description of how men always do behave to one another. It is, rather, their generalization about *how men tend to behave in the pursuit of their economic aims*.

COMPETITION

Two more behavioral attributes should be added to our notion of maximizing, rational economic man. One is that he is *competitive*. This is a quite separate assumption, since he might be a maximizer unwilling to better his condition at the expense of another man. By the assumption of competition we mean that he is indeed willing to do so—that he will outbid his neighbor for a good if he has the ability to do so and if he wants it enough; or that to maximize his income he will, if necessary, "steal" an order or a job from a competitor by shaving his price. Economic man is presumed to be highly individualistic.

The assumption about competitiveness is also a statement about the kind of society to which economic theory turns its analytic powers. Just as maximizing man is partly a product of a market system, so is competitive man. In a traditional society, a man may not be inclined to outbid or undersell another; in a command society, he may not be allowed to do so. In a market system, he is not only "supposed" to act competitively, but he is often *forced* to do so by the pressures exerted on himself by other maximizing, competitive individuals.

UTILITIES AND DISUTILITIES

The second behavioral assumption, in addition to competitiveness, is that *man dislikes and tries to avoid work*. Indeed, when we spoke before of men trying to maximize their utilities, whether by the pursuit of wealth or in any other way, what was implied was really a twofold process. On the one hand, economists assume that men want as much income as possible. On the other, they assume that they want to gain this income by the least possible expenditure of effort. This expenditure of effort is called *disutility*. What rational economic man tries to maximize are therefore not his gross utilities but his net utilities—that is, the utilities he gains from his income minus the disutilities he expends in the pursuit of it.

It need hardly be said that this, too, is a statement about men in a certain kind of society, not men in all kinds of societies. The assumption that work is unpleasant may be true enough as a generalization, but it is certainly not true that all men dislike all work or that they dislike the same work to the same degree. Later we will inquire into some of the consequences of the differing disutilities, or even positive utilities, of work insofar as they help us explain economic behavior. But once again, in setting up our basic model of economic behavior, we begin with the assumption—not too unplausible—that work of the kind and amount normally experienced in our society is deemed unpleasant and counted as a loss to be measured against the gains that work makes possible.

INSATIABLE WANTS

If we think about the conception of economic man pursuing his acquisitive aims, we are forced to recognize another set of assumptions with which economics begins. It is that man's wants are *insatiable,* that they can never be fulfilled. If this were not so, men would sooner or later reach a condition of satiation, after which it would be pointless to continue to pursue disutility-yielding work for economic rewards that brought no further utilities.

Are mens' wants in fact insatiable? Is man shackled by his nature to a treadmill of striving that can never bring him to a point of contentment? The question is difficult to answer from one point of view, simple from another. If we include among man's aims leisure as well as goods, more time to enjoy

himself as well as more income to be enjoyed, it seems plausible that something very much like a condition of insatiability afflicts most people, at least in the kind of society that encourages striving for status and success and that sets high store on consumption and recreation.

Perhaps in a different kind of society, that chronic dissatisfaction would disappear; we do not know. But as a general description of the state of mind of most individuals in today's market society, it is probably accurate enough. We *are* driven to acquire goods, to gain higher income, to increase our enjoyment of leisure. Surveys regularly show that men and women at all economic levels express a desire for more income (usually about 10 percent more than they actually have), and *this drive for "more" does not seem to diminish as we move up the economic scale.* If it did, we would be hard put to explain why businessmen, who are generally in the upper echelons of the distribution of wealth and income, work just as hard as, or even harder than, those on the lower rungs of the economic ladder.

SATIABLE WANTS

But the assumption that wants are insatiable applies only to wants as a whole. Just as important for economic theory is a second set of assumptions about wants, insofar as they apply to any particular good or service. Indeed, one of the cardinal notions about man's wants is that they can be lessened by having more and more of *any one thing,* including leisure itself, within a given period of time. An hour off from work may be sheer heaven; a second hour off, mere earthly paradise; a third hour off, a bore. One car is "indispensable," a second a convenience, a third a luxury, a fourth a nuisance.

DIMINISHING MARGINAL UTILITY

The idea that within a stated period of time, successive additions of any one good or service yield successively less utility to the possessor leads to a basic notion of economics that is called the principle of *diminishing marginal utility*.

This principle has very important implications for behavior. If successive units of a good or service yield us less and less enjoyments, *we will typically pay less for these additional units*. A thirsty man will pay a great deal (albeit with complaints about being "held up") if we offered to sell him a glass of water; but he would pay less for the second, probably very little for the third, and would throw the fourth in our face.

DEMAND CURVES

The idea of diminishing marginal utility leads to a very widely used intellectual tool of economics called a *demand curve*. A demand curve simply describes the changing prices we

are willing to pay for additional units of a good (within some specified period of time). Typically, such a curve slopes downward, as in Fig. 3·1.

In the bar chart on the left, we show the ever smaller amounts of money we are willing to pay for additional units of some good or service, simply because each additional unit gives us less utilities than its predecessor. In the graph on the right we draw a *demand curve* to generalize this basic relation between the quantity of a good we are interested in acquiring and the price we are willing to pay for it.

THE LAWS OF BEHAVIOR

We will be dealing with demand curves much more fully in our next chapter, when we will discuss how they enter into the operation of the market system. Here let us stop to consider them as an example of the *laws of behavior* on which economic science builds.

What do we mean by "laws" of be-

FIGURE 3 · 1
Diminishing marginal utility

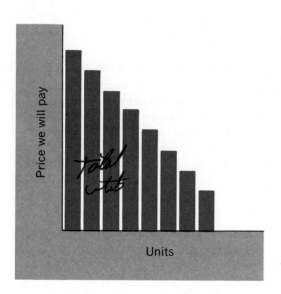

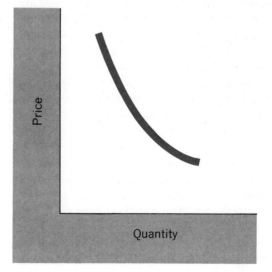

UTILITIES AND DEMAND

Does diminishing marginal utility really determine how much we buy? The idea seems far removed from common sense. But is it? Suppose we decide to buy a cake of fancy soap. In commonsense language, we'll do so only "if it's not too expensive." In the language of the economist this means we'll only do so *if the utilities we expect from the fancy soap are greater than the utilities we derive from the money we have to spend to get the soap.*

But if we buy one or two cakes, doesn't this demonstrate that the pleasure of the soap is greater than the pleasure of holding onto the money or spending it for something else? In that case, why don't we buy a year's supply of the soap? The commonsense answer is that we don't want *that much* soap. It would be a nuisance; we wouldn't use it all for months and months, etc. *In the language of the economist, the utilities of the cakes of soap after the first few, would be less than the utilities of the money they would cost.*

In the diagram (right) we show these

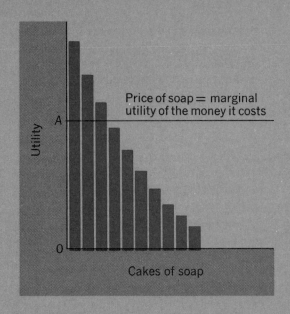

Price of soap = marginal utility of the money it costs

diminishing marginal utilities of successive cakes. The price of soap represents the utility of the money we have to pay. As you can see, if soap costs OA, we'll buy three cakes, no more.

havior? The answer is not easy. Certainly such laws are not like the laws of nature, for no one would claim that man's drives and propensities operate with the unchanging exactness of the laws of gravity. In addition, as we have been careful to point out, the "laws" describe the way men behave in market economies, and therefore at most they are descriptions that apply to specific kinds of societies and not to "human nature."

Nonetheless, with all these qualifications, *economics does take as one of its fundamental assumptions that behavior is sufficiently regular to let us predict how men will behave in market situations.* The maximizing impulse, the competitive drive, the rational choice of the course calculated to give the highest net utilities, the phenomenon of diminishing marginal utility—these are all

parts of the behavioral premises on which economic theory is ultimately built. To put it differently, if we did not think that man's behavior were generally describable by such "laws," we would be unable to make *any* predictions about his economic actions. We could not then say that men would generally buy in the cheapest market and sell in the dearest or that men would take their cues from price changes in the market and alter their course in predictable ways, as our next chapter will show. The laws of behavior may be less binding than the laws of nature, but they are no less essential for economics.

Nature

We have said that economic theory is based on two sets of assumptions about the world. One set involves how man behaves. It leads, as we have seen, to the conception of economic man rationally pursuing his acquisitive ends. But man does not maximize in a vacuum or—to speak in more economic terms—in a world where all goods are free, available effortlessly and in infinite amounts. Instead, man exerts his maximizing efforts in a world where nature "opposes" those efforts, in the sense that goods and services are not free but must be won by working with the elements of the physical world—its land, its resources, its manmade artifacts inherited from past generations.

THE LAWS OF PRODUCTION

If man is assumed by economic theory to be a maximizer, we might say that economic theory assumes that nature is a "constrainer." We put the word in quotes because nature does not willfully withhold its potential treasures from man. It is simply a fact that man encounters problems and obstacles as he works with nature. What is more interesting and important is that these problems and obstacles can be described in systematic ways that we can call *laws of production.*

We can distinguish three general tendencies or laws as man applies his energies to the physical world.

1. The law of diminishing returns

Why can we not grow all the world's food in a flowerpot? The absurd question makes us think about a property of nature that is of utmost importance for economics. As we add more and more of one kind of input (seed, in this case) to a fixed quantity of other inputs (the soil in the flowerpot), the amount of additional output will, after a time, diminish. It may even fall to zero or become a negative quantity.

We can visualize this relationship in Fig. 3·2. On the vertical axis, we measure additional quantities of output *as we add more*

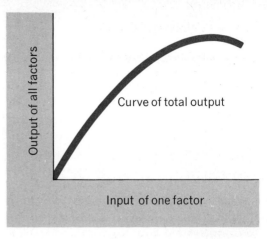

FIGURE 3 · 2
Law of diminishing returns

and more units of one kind of input to a fixed supply of others. The curve of production at first mounts, but then begins to taper off and may even decline. This declining curve describes the phenomenon we call the *law of diminishing returns.*

We get the same kind of curve (although not, of course, always a curve of exactly the same shape) if we add additional units of, say, labor to a fixed supply of land and capital. Suppose that we start with a factory equipped with machinery, and that we measure output as we add more and more units of labor. The factory output will at first increase sharply; but after a certain point, as the plant gets more and more crowded, the equipment and space will no longer yield as much additional output as did the initial inputs of labor; and at a certain point, output will level off or even decline.

Later we will study the law of diminishing returns in much greater detail, but the idea of this basic constraint should now be clear. Note that it is a constraint imposed by nature, not by human behavior. Our productive powers depend in large part on our skills and technology—maybe some day a miracle seed *will* enable us to feed the world from a flowerpot. But even then we can expect additional inputs of seed eventually to yield diminishing returns of food. This has to do with the germinative properties of seeds and soil, not with the behavior of men.

2. The law of increasing cost

A second aspect of the "constraining" tendencies of nature resembles the law of diminishing returns but is differently caused. In the phenomenon of diminishing returns we watch what happens to the output of a product as we add more and more of *one kind* of input, holding the others constant. In the law of increasing cost we watch what happens to the output of a product as we combine more and more of *all kinds of inputs to make that product*. Once again we eventually experience a decreasing increment of output as we increase inputs.

Suppose we have a community that divides its land, labor, and capital between two occupations: dairy-farming and grain-raising. If the community should decide to use all of its inputs to produce milk, it must convert the land, labor, and capital formerly used for grain to the production of milk. At first the result may be very favorable. But unless there is no difference between one piece of land and another, between the skills of dairymen and grain farmers, or between the equipment used for milk and grain, productivity is bound to fall as we move from using resources best suited to milk production to those least suited to it. Finally we will be forced to take land, labor, and capital that is ill-suited to producing milk and put it into pasturage, where its yield will be very low. As a result, the successive increments to milk output will decline as we concentrate more and more of all our resources and efforts on this one output alone.

The same phenomenon will take place in reverse if we now switch back from milk to grain. At first we will experience large increases in grain output as we take land ill-adapted to milk-farming and put it into seed. But as we put more and more land, labor, and capital into grain, we are bound after a while, to experience a fall in the increments

that each additional unit of land-and-labor-and capital will bring.

THE MEANING OF COST

We show this law of increasing cost in Fig. 3·3. Suppose we were producing both grain and milk, as shown by the lines OA and OA'. If we move a given quantity of resources AB out of grain and into milk, grain production will fall by the length of the line AB; milk production will rise by $A'B'$. Now imagine that we continue to concentrate on milk at the expense of grain, until we are producing only OC worth of grain and OC' worth of milk. Once again we move the *same amount* of resources from grain to milk ($CD = AB$). Look how much smaller is the gain in milk production: $C'D'$ compared with $A'B'$.

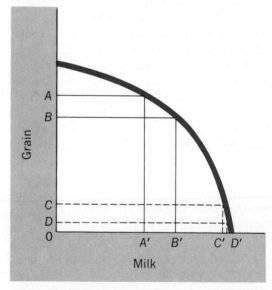

FIGURE 3 · 3
Production possibility curve

In a later chapter we will study this kind of *production possibility* curve much more carefully. But note that this curve gives us a new understanding of the word *cost*. The cost of a larger quantity of milk is the amount of grain we have to give up to get that milk.

The cost of producing a quantity of milk is the amount of grain we would have to give up to get that much more milk. In other words, we must trade off gains in one commodity against losses in another.

Economists call this trade-off *opportunity cost.* The term drills home a very important realization: *there is no economic activity that does not have a cost.* That cost is the measure of other actions we might have taken but could not, because we were engaged in the course we chose. As the saying goes, economics teaches us that there is no such thing as a "free lunch." Everything that utilizes labor or resources has a cost, whether a charge is levied or not. The cost is the alternative benefits that could have been enjoyed by using those resources for some other purpose.

3. Returns to scale

A third attribute of nature that enters into economic analysis is somewhat different from the previous two. It sets nature in the role of an efficiency engineer as well as a constrainer. This attribute is called *scale* or *returns to scale.* It refers to a simple but profoundly important fact about production, namely that *size matters.*

For example, can we produce 10 automobiles per year in a small garage as cheaply per car as we can produce 1 million in a large assembly plant? Obviously not. There are certain attributes of automobile making that cannot be efficiently reduced in size or scale. We cannot produce a car efficiently in a garage because there would be no way to arrange a flow of production through such a small area. We cannot produce steel cheaply by the pound because the equipment that it takes to make steel cannot today be miniaturized.

Most products or processes exhibit these so-called *economies of scale.* The attainment of a certain size enables us to use more efficient technologies or to attain a finer subdivision of labor, as in the case of Adam

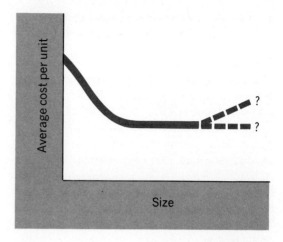

FIGURE 3 · 4
Economies of scale

Smith's pin factory. Eventually, as the scale of production becomes very large, costs per unit may again increase because of the sheer difficulties of running a huge operation. This is a matter we will turn to again when we look into the economics of big business. But leaving that question in abeyance, we can see a "profile" of average costs of production per unit in Fig. 3·4.

CONSTRAINTS AND SUPPLY

All these properties of nature set the stage for the maximizing behavior of man. Men seek utilities through the production and exchange of goods and services, but they do not maximize in a world where goods can be limitlessly and effortlessly obtained. Nature offers us its services easily or reluctantly, depending on whether we are trying to maximize output by adding more and more of one kind of input (when we encounter diminishing returns); or whether we are trying to increase the output of one good or service at the expense of others (when the law of increasing cost comes into play); or whether we are seeking to organize our production in accordance with the technological characteris-

IS THERE SUCH A THING AS A FREE LUNCH?

Two farmer-hunters live side by side in an isolated community. One is a very adept husbandman; and by virtue of four hours' work every morning, he raises enough wheat for his wife to bake large quantities of bread each day, which he and his children eat for breakfast, stuffing themselves like Strasbourg geese. In the afternoon, the farmer, whom we will call Adams, hunts for four hours. He is, however, only an indifferent hunter, and he brings home so little game that the family usually has slim pickings of partridge or venison for dinner. They don't eat anything at all between breakfast and dinner.

Next to Adams lives another farmer, Smith, whose talents lie just in the opposite direction. Smith is a rather poor farmer, and each member of his family accordingly has only a thin slice of bread for breakfast. But he is an ace hunter, and he regularly brings home coveys of birds and heavy deer, so that the family gorges at dinner on an all-meat diet. The Smiths also have no midday meal.

Now Adams and Smith chance to meet and discuss their respective situations; and having read *The Wealth of Nations*, they simultaneously see a solution to their problems. "We will specialize our labor!" cries Smith. "Right on!" says Adams. "I'll put in eight hours a day of farming, and you hunt for eight hours, and we'll split the proceeds."

Putting their plan to work, the Adamses and Smiths soon discover that they have two large ovens full of bread and two great catches of meat and fowl to divide among them. The result is that enormous breakfasts and huge dinners leave the Smiths and the Adamses so stupefied that their efficiency both in the field and on the farm diminishes alarmingly. They meet to resolve their new dilemma, and this time it is Adams who has the key. "We will eat a little less for breakfast and dinner," he declares, "since those huge meals

---

tics of the agencies of production (economies of scale).

These constraints introduce us to another fundamental concept, to which we will return in our next chapter. It is the idea of a *supply curve*, the counterpart of the demand curve whose acquaintance we made a few pages back.

Demand curves, we recall, told us something about the prices we would be *willing* to pay to acquire more of a good. Supply curves tell us the opposite—the amount of money we will *have to pay* to acquire more of a good, for the constraints of nature make it clear that in many cases it will become increasingly difficult to produce larger quantities of output. If we want to acquire them, we will have to give up more resources to get them.

But how about economies of scale? Is this not a property of nature that promises us decreasing costs as we increase the volume of output? Indeed this is the case—up to a point. Therefore we must learn to think of two kinds of supply curves. In the short run, before we have time to enlarge the scale of output, we will almost surely encounter rising costs as we expand production. In the longer run, economies of scale may well give us lower costs, at least for a while. Meanwhile the law of increasing costs may be working for us or against us, depending on whether we are starting from one end or the other of our production possibility curve.

SUPPLY CURVES

Therefore we can think of supply curves as typically forcing higher and higher costs on us in the short run, when the scale of operation is given; and as offering constant costs or falling costs or possibly rising costs in the long run, depending on the obstacles or opportunities afforded by nature. We show these two kinds of supply curves in Fig. 3·5. In the chapters to follow we will learn much more about them.

are actually counterproductive in terms of our output."

"And what will we do with the leftovers?" asks Smith.

"The answer," declares Adams triumphantly, "is that we will invent a new meal to be eaten at midday, which we will call *lunch*."

"But won't that cost us something?" asks Smith, remembering his freshman course in economics.

"Not a thing," answers Adams. "Will you work any harder?"

"No," admits Smith. "Same hours, and more agreeable work, too."

"Will you consume any less?"

"Well," says Smith, "breakfast and dinner will be smaller; but since I eat more than I want at those meals, my total utility from the three meals will be greater than that from the previous two. I can't see any cost there."

"In other words," says Smith, "you will be getting a free...."

Question: Is it a free lunch?

Answer: In specializing their labor, Smith and Adams increase their total output.

Therefore they could have enjoyed the same diet as before with a smaller expenditure of working time. Had Smith and Adams both worked at their specialized occupations only, say, six hours a day, and thereafter divided their produce, each family could have had the same meals as before, and each family would have enjoyed two hours of leisure. The cost of the lunch was therefore the value of these two hours of "unnecessary" labor. The cost of the lunch was the opportunity cost of the foregone trout fishing or whatever else the Adamses and Smiths sacrificed when they decided to put in a full eight-hour day to enjoy a higher total production, instead of a six-hour day with the same level of food consumption as before. And then, too, there is the extra labor of Mrs. Smith and Mrs. Adams in fixing that new third meal.

SOCIAL CONSTRAINTS

Perhaps we can already see the makings of a powerful analytic device in the interplay of maximizing and constraining portrayed in demand and supply curves. In our next chapter we will put those curves to work. But before we move on, it is important to recognize that nature is not the only constraint on the maximizing force of behavior.

Equally important are constraints imposed by the society in which we exert our

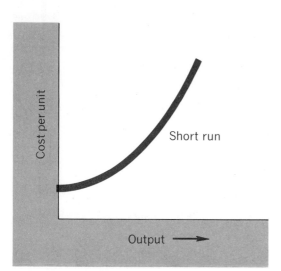

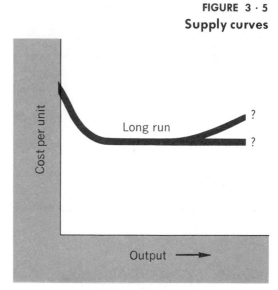

FIGURE 3 · 5
Supply curves

MARXIAN ECONOMICS

By and large, economics takes most social institutions for granted, especially that underlying institution, the market system. There is one branch of economics, however, that tries to answer the question of how social institutions come into being and of how they continue to change. This is Marxian economics, founded on the writings of that extraordinary social thinker, Karl Marx (1818–1883).

Marxian economics differs in a number of ways from the economics we teach in this text. It is primarily interested in the evolution of the market system as a socioeconomic form called capitalism, and not so interested in the problems of economic policy within capitalism, with which this text is concerned. It is focused on questions such as the changing balance of power among the social classes of society, and is therefore impatient with the nonpolitical manner in which conventional economics treats these classes as "factors of production."

Marxian economics is a vast subject that we must leave largely unexamined in this book. In some ways it is more perceptive than conventional economics; in many ways it is less useful. One surprising point is worth mentioning, however. If there is one economist whose historical and political vision embraced as large a panorama as did that of Marx, it was Adam Smith! There is surely food for thought in the fact that the aim of the greatest "conservative" economist was not so different from that of the greatest "radical" economist. Both sought to embrace philosophy, history, and political economy in a single unified system of thought. If nothing else, this should induce "conservative" students to look into Marx, and "radical" students to look into Smith.

acquisitive impulses. One such constraint is our *budget*. We are constantly forced to temper our maximizing impulses because we own only limited amounts of resources, including those all-important "resources," time and money. Other constraints are introduced by social institutions. The law is a major constraining factor on our acquisitive propensities. Competition itself, a part of our maximizing behavior, becomes a constraint because it limits our freedom of action, preventing us from maximizing our incomes by charging more for goods or services than other maximizing individuals will allow us to get away with. The banking system, big business, the market mechanism itself with its legal underpinnings of private property are all institutions that operate like the constraints of nature in curbing the unhampered exercise of our maximizing behavior. We will spend a lot of time later investigating how these institutional constraints affect economic behavior.

THE BASIC ASSUMPTIONS AGAIN

All that, however, in good time. Here let us briefly review the basic propositions that we have covered in this first look into economic analysis. They can be summed up very simply.

Economics is able to theorize about, and to predict, the operations of a market society because it depends on generalizations with regard to human behavior and the "behavior" of nature. If we were not able to make such generalizations—if we could not begin with the plausible hypothesis that man is a maximizer and that nature (and social institutions) constrain his behavior in clearly defined ways—we could not hazard the simplest predictive statement about economic society. We could not explain why a store that wants to sell more goods marks its prices down rather than up, or why copper costs will probably rise if we try to double copper production in a short period of time.

Perhaps these simple generalizations about behavior and nature do not seem like an impressive foundation for a social science. But ask yourself whether we can match these economic generalizations when we think in political or sociological terms. Are there political or social "laws of behavior" that we can count on with the same degree of certainty we find in those of economics? Are there constraints of nature, comparable to the laws of production, discoverable in the political and social areas of life? Alas, there are not, which is why our ability to predict

political or sociological events is so much less than our ability to predict economic events.

We will soon discover that economic prediction has its sharp limitations. Nonetheless, in the presence of an underlying structure of behavioral and natural laws lies a unique strength of economics, whose capabilities we must now explore. And the place to begin must be obvious from our look into Adam Smith and our first examination of supply and demand. It is the market mechanism.

KEY WORDS

Laws of behavior

Laws of nature

Attributes of economic man

Maximizing

Diminishing marginal utility

Demand

Laws of production

Diminishing returns

Increasing cost

Opportunity cost

Economies of scale

Short-run and long-run supply curves

CENTRAL CONCEPTS

1. Economic analysis is built on two set of propositions: (1) that we can depend on certain "*laws*" of behavior, and (2) that we can also depend on *certain laws of nature*. Without these, economic theory would be impossible.

2. The laws of behavior describe the way man generally behaves in a market system. He seeks to *maximize his utilities*. He is *competitive*. He is *rational*. His *total wants are insatiable*. He *dislikes work*.

3. Although man seeks to maximize his total utilities, he does not seek to do so by maximizing the possession of any one good or service. *Typically, the more of any particular good a man possesses, the less utility the marginal addition yields.*

4. This phenomenon leads to the generalization that *demand* is normally characterized by unwillingness to buy more of any good unless its price declines.

5. The assumptions about nature cast it in the role of a constrainer. We describe its constraints in terms of three *laws of production*.

   ●The law of *increasing cost* constrains output by yielding less and less output of one kind of input to a fixed supply of other inputs.

   ● The law of *increasing cost* constrains output by yielding less and less output of any one kind, as we transfer resources into the production of any one good.

   ● *Economies of scale* usually work in the opposite direction, *giving us lower costs as we increase the scale of output*, at least up to some very large size.

6. The law of increasing cost teaches us the real meaning of cost. The cost of any economic activity is the need to forego some other activity which we might otherwise have carried on. We call this unavoidable "trade-off" opportunity cost.

7. Natural and technological characteristics of production make it possible to attain *economies of scale* in most production processes. This results from the fact that most processes become more efficient beyond a certain size.

8. These constraints of nature affect costs. Generally speaking, we have to pay more to produce more in the short run. In the long run, costs may rise or fall as

Social constraints

output increases. Thus we speak of rising *short-run supply curves* and rising, constant, or falling *long-run supply curves*.

9. In addition to the constraints of nature, *society imposes important constraints* on our maximizing behavior. Our budget is one such constraint. Institutional limitations of many kinds, such as laws, are another. Time is often a crucial constraint.

Importance of the laws of behavior and nature

10. *The capacity of economics to predict the outcome of the economic process depends on the presence of these laws of behavior and nature. If there were no such laws, economics could not make the simplest prediction. Although these laws are far from perfect, their presence gives economics a unique analytical capability.*

QUESTIONS

1. Do you feel like a "maximizer"? Are you content with your income? If you are not, do you expect that some day you will be?

2. Do you feel that you act "rationally" when you spend money? Do you consciously try to weigh the various advantages or utilities of buying this versus that and to spend your money for the item that will have the greatest marginal utility? Consciously or not, do you think you generally act as a "rational maximizer"?

3. Do you think men are naturally competitive? Do you think that work is distasteful to most people? To you? All kinds of work?

4. Have you ever experienced the feeling of diminishing marginal utility? How would you describe the feeling that comes from having eaten too much of a good thing?

5. How valid do you think the "laws" of behavior are? If you think they are *not* valid, why does economic society "work" and not collapse? If they *are* valid, why can't economists predict better?

6. Suppose that you had a very large flowerpot and extraordinary chemicals and seeds. Could you conceivably grow all the world's food in it? Why would you still get diminishing returns?

7. Discuss the law of increasing cost in a society that ordinarily devoted half its efforts to farming, half to hunting. What would be the cost of increasing the output of farm products? Of game? Show this on a production possibility curve.

8. Describe the economies of scale that might be anticipated if you were opening a department store. What economies might be expected as the store grew larger? Do you think you would eventually reach a ceiling on these economies?

9. What is the meaning of *oppostunity cost*? What is the opportunity cost of sleep? Of an hour's work? Of a restaurant meal?

10. What is the cost of a "free lunch," such as the pretzels a bar may serve?

11. In what way is "competition" an institution? To what degree is it fostered by laws? Would you have competition in a society that denied spatial or social mobility to labor, as under feudalism?

12. Can you think of any "laws" that describe political activity comparable to economic maximizing or nature's constraints? Is there a law of political power? Are there constraints of national size? Do you think it might be possible to devise an "economics of politics"?

The Market Mechanism

WE HAVE ALREADY GAINED A GENERAL UN-DERSTANDING of the market system from our quick look at Adam Smith's model of the economy. Now we must learn much more about that system, as the next step in building our understanding of the fundamental concepts of economics.

What impresses us first when we study the market as a solution to the economic problem? The striking fact is that the market uses only one means of persuasion to induce men to engage in production or to undertake the tasks of distribution. It is not time-honored tradition nor the edict of any authority that tells the members of a market society what to do. *It is price.*

PRICES AND BEHAVIOR

Thus the first attribute of a market system that we must examine is how price become the guide to behavior, taking the place of tradition or command.

The answer lies in our postulate of maximization. *Through prices, acquisitive individuals learn what course of action will maximize their incomes or minimize their expenditures.* This means that in the word *price* we include the very important prices of labor or capital or land—prices that we call wages or profits and interest or rent. Of course, within the term *prices* we also include those ordinary prices that we pay for the goods and services we consume or the materials we purchase in order to build a home, or to operate a store or factory or whatever. In each case, the only way that we can tell how to maximize our receipts and how to minimize our costs is by "reading" the signals of price that the market gives us.

Therefore, if we are to understand how the market works as a mechanism—that is, how it acts as a guide for the solution to the economic problem—we must first understand how the market "sets" prices. But perhaps we will remember from our discussion with the village elders that we said there was no such thing as "the market." When we say "the market" we mean the activity of buying and selling or, in more precise economic language, *demand and supply.* Let us discover how demand and supply interact to establish prices.

GOODS AND SERVICES

What is a good? What is a service? It is very easy to answer the first question; not so easy to answer the second.

A *good* is any object that we can legally possess. In a slave society, a slave is a good. In our kind of society, goods range from ordinary objects of daily use, which we call *consumption goods*, to machines, factories, or a plant, which we call *capital goods*. Actually, the same good can be either a consumption good or a capital good, depending on whether we use it for our personal enjoyment or in the process of production: oil is a consumption good in a home, a capital good in a factory.

Services are enjoyments or useful activities that we buy without purchasing the agent that gives rise to them. Movies are a service, since we buy the right to enter the theater, not the theater and the projection equipment. Education, legal advice, fire protection, haircuts are all services. About a third of all our expenditures for personal use (consumption) are services, as we will see in a later chapter.

The difficulty lies in the fact that *all* economic activity is reducible to services. Even the food we literally consume is useful because it yields the services of nutrition. Clothes give us the services of warmth and adornment. Machines and land are also valuable to us only because they give rise to a flow of useful services—physical and chemical effects—that enter into production. In other words, *all economic activity ultimately consists in the production and utilization of services.* The term *goods* refers to the strictly legal or institutional fact that we privately own some service-yielding goods and not others.

DEMAND

When you "enter" the market for goods and services (almost every time you walk along a shopping street), two factors determine whether or not you will actually become a buyer and not just a window-shopper. The first factor is your *taste* for the good. It is your taste that determines in large degree whether a good offers you utility, and how much utility. The windows of shops are crammed with things you could afford to buy, but which you simply do not wish to own, because they do not offer you sufficient utility. Perhaps if they were cheaper, you might wish to own them; but there are other goods you would not want, even if they were free. For such goods, for which your tastes are too weak to motivate you, your demand is zero.

On the other hand, taste is by no means the only component of "demand." The shop windows are also full of goods that you might very much like to own but cannot afford to buy; your demand for Rolls Royces is also apt to be zero. Thus, demand hinges on the *ability to buy*—the possession of sufficient wealth or income—as well as the *willingness* of the buyer. If it did not hinge on ability as well as willingness to buy, the poor, whose wants are always very large, would constitute a great source of demand.

BUDGET CONSTRAINTS

Note that your demand for goods depends on your willingness and ability to buy goods or services *at their going price*. From this it follows that the amount of goods you demand will change as their prices change, just as it also follows that the amounts you will demand change as your wealth or income changes. There is no difficulty in understanding why changing prices should change our ability to buy: our wealth simply stretches further or less far. In economic language, our budget constraint is loosened when prices fall and tightened when they rise.

DIMINISHING MARGINAL UTILITY

But why should our willingness to buy be related to price? The answer returns us again to the behavioral hypotheses of our last chapter. Man is a maximizing creature, but that does not mean that he wants ever more of the *same* commodity. On the contrary, as we saw, economists take as a plausible generalization that additional increments of the same good or service, within some stated period of time, will yield smaller and smaller increments of utility. As we know, these increments of utility are called *marginal utility,* and the general tendency of marginal utility to diminish is called the *law of diminishing marginal utility.* Remember: diminishing marginal utility refers strictly to behavior and not to nature; the units of goods we continue to buy are not smaller—only the pleasure associated with each additional unit.

This gives us a more complete understanding of the downward sloping demand curve we encountered in Fig. 3·1. Each glass of water affords us less marginal utility, so we are not willing to pay as much for the next glass as for the one we just bought. This does not mean that the *total utility* we derive from 3 or 4 glasses is less than that derived from the first. Far from it. But the addition to our utility from the last glass is much lower than the addition of the first or second.

The notion of diminishing marginal utility also clears up another puzzle of economic life. This is why we are willing to pay so little for bread, which is a necessity for life, and so much for diamonds, which are not. The answer is that we have so much bread that the marginal utility of any loaf we are thinking of buying is very little, whereas we have so few diamonds that each carat has a very high marginal utility. If we were locked inside Tiffany's over a long holiday, the prices we would pay for bread and diamonds, after a few days, would be very different from those we would have paid when we entered.

SUPPLY

What about the supply side? Here, too, willingness and ability enter into the seller's actions. But as we would expect, they bring about reactions different from those in the case of demand.

At high prices sellers are much more *willing* to supply goods and services because they will take in more money. They will also be much more easily *able* to offer more goods because higher prices will enable less efficient suppliers to enter the market or will cover the higher costs of production that may result from the effect of diminishing returns.

Therefore, as we saw in our previous chapter, we depict supply curves as rising in the short run. These rising curves contrast the eager response of sellers to high prices with the negative response of buyers.

SUPPLY AND DEMAND

The idea that buyers welcome low prices and sellers welcome high prices is hardly apt to come as a surprise. What is surprising is that the meaning of words *supply* and *demand* differs from the one we ordinarily carry about in our heads. It is very important to understand that when, as economists, we speak of demand, we do not refer to a single purchase at a given price. *Demand in its proper economic sense refers to the various quantities of goods or services that we are willing and able to buy at different prices at a given time.* That is the relationship that is shown by our demand curve.

The same relationship between price and quantity enters into the word *supply.* When we say *supply,* we do not mean the amount a seller puts on the market at a given price. We mean the various amounts offered at different prices. Thus our supply curves, like our demand curves, portray the relationship between willingness and ability to enter into transactions at different prices.

INDIVIDUAL AND COLLECTIVE SUPPLY AND DEMAND

We must add one last word before we investigate the market at work. Thus far we have considered only the factors that make a single individual more willing and able to buy as prices fall, or less willing and able to sell. But generally when we speak of supply and demand we refer to markets composed of many suppliers and demanders. That gives us an additional reason for relating price and behavior. If we assume that most individuals have somewhat different willingnesses and abilities to buy, because their incomes and their tastes are different or that they have unequal willingnesses or abilities to sell, then we can see that *a change in price will bring into the market new buyers or sellers.* As price falls, it will tempt or permit one person after another to buy, thereby adding to the quantity of the good that will be purchased at that price; and conversely, as prices rise, the number of sellers drawn into the market will increase, and the quantity of goods they offer will rise accordingly.

We can see this graphically in Figure 4·1. Here we show three individuals' demand curves. At the going market price of $2, A is either not willing or not able to buy any of the commodity. B is both willing and able to buy 1 unit; C buys 3 units. If we add up their demands we get a *collective or market demand curve.* At the indicated market price of $2, the quantity demanded is 4 units. What would it be (approximately) for each unit, and for the group, at a price of $1?

The same, of course, applies to supply. In Figure 4·2 we show individual supply curves and a collective or market supply curve that is 7 units at $2 market supply. What would total supply be at a price of $1? What would seller A's supply be at $1?

BALANCING SUPPLY AND DEMAND

We are now ready to see how the market mechanism works. Undoubtedly you have al-

FIGURE 4 · 1
Individual and market demand curves

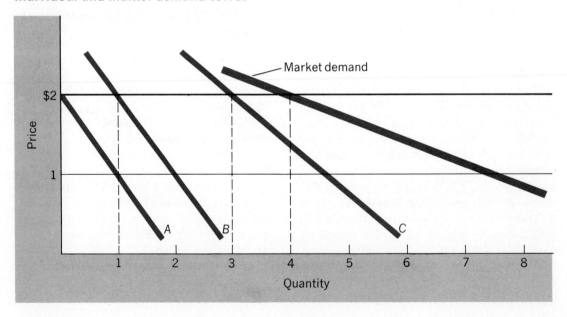

ready grasped the crucial point on which the mechanism depends. This is the *opposing behavior* that a change in prices brings about for buyers and sellers. Rising prices will be matched by an increase in the willingness and ability of sellers to offer goods, but in a decrease in the willingness and ability of buyers to take goods.

It is through these opposing reactions that the market mechanism works. Let us examine the process in an imaginary market for shoes in a small city. In Table 4·1 we show the price-quantity relationships of buyers and of sellers: how many thousand pairs will be offered for sale or sought for purchase at a range of prices from $50 to $5. We call such an array of price-quality relationships a *schedule* of supply and demand.

As before, the schedules tell us that buyers and sellers react differently to prices. At high prices, buyers are either not willing or unable to purchase more than small quantities of shoes, whereas sellers would be only too willing and able to flood the city with

TABLE 4·1
Demand and Supply Schedules

Price	Quantity demanded (1,000 prs.)	Quantity supplied (1,000 prs.)
$50	1	125
$45	5	90
$40	10	70
$35	20	50
$30	25	35
$25	**30**	**30**
$20	40	20
$15	50	10
$10	75	5
$ 5	100	0

them; at very low prices, the quantity of shoes demanded would be very great, but few shoe manufacturers would be willing or able to gratify buyers at such low prices.

But if we now look at *both* schedules at *each* price level, we discover an interesting thing. *There is one price*—$25 in our exam-

FIGURE 4 · 2
Individual and market supply curves

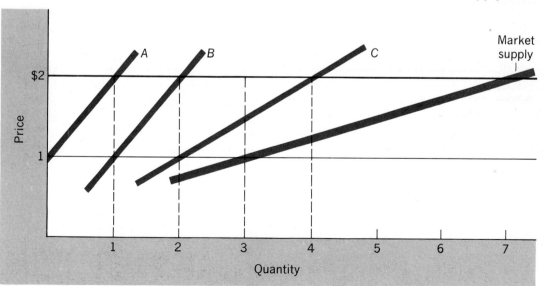

ple—*at which the quantity demanded is exactly the same as the quantity supplied.* At every other price, either one schedule or the other is larger, but at $25 the amounts in both columns are the same: 30,000 pairs of shoes. We call this balancing price the *equilibrium price,* and we shall soon see that it *is* the price that emerges spontaneously in an actual market where supply and demand contend.*

Emergence of the Equilibrium Price

How do we know that an equilibrium price will be brought about by the interaction of supply and demand? The process is one of the most important in all of economics, so we should understand it very clearly.

Suppose in our example above that for some reason or other the shoe retailers put a price tag on their shoes not of $25 but of $45. What would happen? Our schedules show us that at this price shoe manufacturers will be pouring out shoes at the rate of 90,000 pairs a year, whereas customers would be buying them at the rate of only 5,000 pairs a year. Shortly, the shoe factories would be clogged with unsold merchandise. It is plain what the outcome of this situation must be. In order to realize some revenue, shoe manufacturers will begin to unload their unsold stocks of shoes at lower prices. *They do so because this is the rational course for competitive maximizers to pursue.*

As they reduce the price, the situation will begin to improve. At $40, demand picks up from 5,000 pairs to 10,000, while at the same time the slightly lower price discourages some producers, so that output falls from

*Of course we have made up our schedules so that the quantities demanded and supplied would be equal at $25. The price that actually brought about such a balancing of supply and demand might be some odd number such as $24.98.

90,000 pairs to 70,000. Shoe manufacturers are still turning out more shoes than the market can absorb at the going prices, but the difference between the quantities supplied and the quantities demanded is smaller than it was before.

Let us suppose that the competitive pressure continues to reduce prices so that shoes soon sell at $30. Now a much more satisfactory state of affairs exists. Producers will be turning out 35,000 pairs of shoes, and consumers will be buying them at a rate of 25,000 a year. But still there is an imbalance, and some shoes will still be piling up, unsold, at the factory. Prices will therefore continue to fall. Eventually they reach $25. At this point, the quantity of shoes supplied by the manufacturers—30,000 pairs—is exactly that demanded by customers. There is no longer a surplus of unsold shoes hanging over the market and acting to press prices down.

THE MARKET CLEARS

Now let us quickly trace the interplay of supply and demand from the other direction. Suppose that prices were originally $5. Our schedules tell us that customers would be standing in line at the shoe stores, but that producers would be largely shut down, unwilling or unable to make shoes at those prices. We can easily imagine that customers, many of whom would gladly pay more than $5, let it be known that they would welcome a supply of shoes at $10 or even more; they too are trying to maximize their utilities. If enough customers bid $10, a trickle of shoe output begins. But the quantity of shoes demanded at $10 far exceeds the available supply. Customers snap up the few pairs around, and tell shoe stores they would gladly pay $20 a pair. Prices rise accordingly. Now we are getting closer to a balance of quantities offered and bid for. At $20 there will be a demand for 40,000 pairs of shoes, and output will have risen to 20,000 pairs. But still the pressure of unsatisfied demand raises prices further. Finally a price of $25 is tried. Now, once again, the quantities supplied and demanded are exactly in balance. There is no further pressure from unsatisfied customers to force the price up further, because at $25 no customer who can afford the going price will remain unsatisfied. The market "clears."

FUNCTION OF EQUILIBRIUM PRICES

Thus we can see how *the interaction of supply and demand brings about the establishment of a price at which both suppliers and demanders are willing and able to sell or buy the same quantity of goods.* We can visualize the equilibrating process more easily if we now transfer our supply and demand schedules to graph paper. Figure 4·3 is the representation of the shoe market we have been dealing with.

The graph shows us at a glance the situation we have analyzed in detail. At the price of $25, the quantities demanded and supplied are equal—30,000 pairs of shoes.

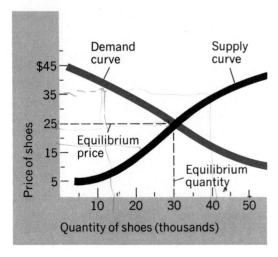

FIGURE 4 · 3
Determination of an equilibrium price

But the graph also shows more vividly than the schedules why this is an *equilibrium* price.

Suppose that the price were temporarily lifted above $25. If you will draw a horizontal pencil line from any point on the vertical axis above the $25 mark to represent this price, you will find that it intersects the demand curve before it reaches the supply curve. In other words, *the quantity demanded is less than the quantity supplied at any price above the equilibrium price, and the excess of the quantity supplied means that there will be a downward pressure on prices, back toward the equilibrium point.*

The situation is exactly reversed if prices should fall below the equilibrium point. Now the quantity demanded is greater than that supplied, and the pressure of buyers will push the price up to the equilibrium point.

Thus equilibrium prices have two important characteristics:

1. **They are the prices that will spontaneously establish themselves through the free play of the forces of supply and demand.**
2. **Once established, they will persist unless the forces of supply and demand themselves change.**

DOES "DEMAND EQUAL SUPPLY"?

There is one last thing carefully to be noted about equilibrium prices. They are the prices that bring about an equality in the *quantities demanded* and the *quantities supplied*. They are not the prices that bring about an equality of "supply and demand." Probably the most common beginning mistake in economics is to say that supply and demand are equal when prices are in equilibrium. But if we remember that both supply and demand mean the *relationships* between quantities and prices, we can see that an equality of

SUPPLY AND DEMAND, AGAIN

Here is one of the oldest "puzzles" in economics.
Suppose that the price of A.T.& T. stock rises. Because the price rises, the demand for the stock falls. Therefore the price of A.T.& T. must decline. It follows that the price of A.T.& T. should never vary, or at least should quickly return to the starting point.

Tell that to your broker. Better, tell it to your instructor, and show him—and yourself—with a graph of supply and demand, where the fallacy of this puzzle lies. Hint: When the price rises, does the demand for A.T.& T. stock fall—or does the quantity demanded fall? Will the price fall again?

supply and demand would mean that the demand schedule and the supply schedule for a commodity were alike, so that the curves would lie one on top of the other. In turn, this would mean that at a price of $50, buyers of shoes would be willing and able to buy the same number of shoes that suppliers would be willing to offer at that price, and the same for buyers at $5. If such were the case, prices would be wholly indeterminate and could race high and low with no tension of opposing interests to bring them to a stable resting place.

Hence we must take care to use the words *supply* or *demand* to refer only to relationships or schedules, and to use the longer phrase *quantity demanded* or *quantity supplied* when we want to speak of the effect of a particular price on our willingness or ability either to buy or sell.

THE ROLE OF COMPETITION

We have seen how stable, lasting prices spontaneously emerge from the flux of the marketplace, but we have silently passed over a basic condition for the formation of these prices. This is the role played by competition in the operation of the market mechanism.

In our discussion of the economists' view of behavior, competition appeared as a somewhat unpleasant attribute of economic man. Now, however, we can see that it is an attribute that is indispensable if we are to have socially acceptable outcomes for a market process.

For competition is the regulator that "supervises" the orderly working of the market. Because economic competition (unlike the competition for prizes outside economic life) is not a single contest but a *continuing process,* it monitors a race that no one ever wins, but a race where all must go on endlessly trying to stay in front, to avoid the economic penalties of falling behind.

Second, unlike the contests of ordinary life, economic *competition involves not just a single struggle among rivals, but two struggles,* one of them between the two sides of the markets and the other among the marketers on each side. For the competitive marketplace is not only where the clash of interest between buyer and seller is worked out by the opposition of supply and demand, but also where buyers contend against buyers and sellers against sellers.

It is this double aspect of the competitive process that accounts for its usefulness. A market in which buyers and sellers had no conflict of interest would not be competitive, for prices could then be arranged at some level convenient for both sides, instead of representing a compromise between the divergent interests of the two. And a market that was no more than a place where opposing forces contended would be only a tug of war, a bargaining contest with an unpredictable outcome, unless we knew the respective strengths and cunning of the two sides.

Competition drives buyers and sellers to a meeting point because each side of the price contest is also contesting against itself. Vying takes place not merely *between* those who want high prices and those who want low ones, but on each side of this divide *among* marketers whose self-interest urges them to meet the demands of the other side. If some unsatisfied shoe buyers, although preferring low prices to high ones, did not want shoes enough to offer a little higher price than the prevailing one, and if some unsatisfied sellers, although hoping for high prices, were not driven by self-interest to offer a price a little below that of their rivals, the price would not move to that balancing point where the two sides arrived at the best possible settlement.

Thus, whereas buyers as a group want low prices, each individual buyer has to pay as high a price as he can to get "into" the market; and whereas sellers as a group want high prices, each individual seller has to trim his prices if he is to be able to "meet" the competition.

MAXIMIZING SUBJECT TO CONSTRAINTS

Does the extraordinary market mechanism bear a relation to the general notion of maximizing and constraints? Indeed it does. Both buyers and sellers are driven by maximizing impulses and constrained by forces of nature and social institutions, as well as by their limited incomes or wealth.

For example, both buyers and sellers are *willing* to respond to price signals because they wish to maximize their incomes or utilities. But neither can maximize at will. Buyers are constrained by their budgets and sellers are constrained by their costs. Thus the *ability* of buyers or sellers to respond to price signals is limited by obstacles of budgets or cost.

In addition, buyers and sellers are both constrained by the operation of the market in which their own maximizing (and competitive!) behavior places them. A seller might like to sell his goods above the market price and a buyer might like to buy goods below the market price, but the presence of competitors means that a seller who quotes a price above the market will be unable to find a buyer, and that a buyer who makes a bid below the market, will be unable to find a seller.

Thus the market mechanism is a very important example of what economists call "maximizing subject to constraints." Furthermore, we can see that it is the very interaction of the maximizing drives and the constraining obstacles that leads the market to the establishment of equilibrium prices. We can also see that if we could know these maximizing forces and constraints beforehand, we would know the supply and demand curves of a market and could actually predict what its equilibrium price would be! In actual fact, our knowledge falls far short of such omniscience, but the imaginary example nonetheless begins to open up for us the analytical possibilities of economics.

KEY WORDS

CENTRAL CONCEPTS

Price signals

Supply and
demand

Willingness and
ability

Demand:
budget con-
straint and
diminishing
marginal
utility

Marginal vs.
total utility

Supply:
maximizing and
higher costs

Individual and
collective demand
and supply

Supply and
demand as
opposing
behavior

Quantities
demanded and
supplied

Unsatisfied buyers
and sellers

Equilibrium price
"demand" ≠
"supply"

Competition:
● across the
   market
● on each side of
   the market

Constraints

1. In a market system, prices give us *signals* that tell us how to maximize our receipts or utilities. Our responses to these signals are called *supply* and *demand*.

2. Demand and supply both hinge on two factors: *willingness* and *ability*. Willingness refers to our motivation; ability to our constraints.

3. *In the case of demand, we are able to buy more when prices fall, because our budget constraint is loosened. We are willing to buy more because the lower price offsets the fall in marginal utility as we buy more of a good.*

4. Diminishing marginal utility refers to the utility of only the last item we buy, not to the total utility of all the items we may possess. The total utility of water is very great—far greater than gold—but under ordinary conditions the marginal utility of another ounce of water is worth much less to us than that of another ounce of gold.

5. Supply also involves willingness and ability. *We are willing to sell more as prices rise because this maximizes our incomes. We are able to sell more because higher prices cover the higher costs we typically encounter in the short run when we expand output.*

6. *Demand and supply curves have their typical downward and upward slopes for two reasons.* (1) As prices fall, individuals are able and willing to buy more, or supply less; and vice versa when prices rise. (2) When prices fall, additional persons may enter the market, adding their demand to existing demand; and when prices rise, new suppliers may enter the market, adding their output to that of former suppliers.

7. It is important to remember that the words *supply* and *demand* refer to behavior *as prices change.* They describe the relationship between different market prices and our actions as buyers and sellers. The market mechanism "works" because buyers and sellers exhibit *opposing behavior* when prices change. As prices rise, buyers are less willing and able to sell; sellers are more willing and able to sell.

8. *In a competitive market, prices will be determined by the interactions of buyers and sellers, each side trying to maximize its incomes or utilities.* At high prices, there will be a larger quantity supplied than demanded, and many sellers will be unable to find buyers. Competition among these unsatisfied sellers will cause prices to fall. At low prices there will be a larger quantity demanded than supplied. Many buyers will not be able to find goods, and competition among them will drive prices up.

9. *At one price the quantities offered and demanded will be equal. This is the equilibrium price.* It will emerge spontaneously from the market process, and it will persist unless the willingness and ability of either buyers or sellers should change. *At equilibrium prices "demand" does not equal "supply";* quantity demanded equals quantity supplied.

10. Competition provides the control over the operation of the market. In part, this control is exercised by the contest between buyers and sellers as groups with opposing interests. But it is also exercised by the contest among buyers and among sellers, each of whom must try to meet the demands of the other side if he is not to be left out of the market entirely.

11. The market mechanism is an example of *maximizing subject to constraints.* In the market, men seek to maximize against the constraints of budgets, costs, and competition, which lead them in the direction of equilibrium prices.

QUESTIONS

1. Why does the same price signal mean different things to buyers and sellers?

2. What is meant by saying that the quantity we demand depends on price? Does this mean that we will buy more of a good when its price goes up? What are the motivational roots of "demand"? Why is not our ability to buy—our budget constraint—enough to give us a demand schedule?

3. Why would the marginal utility of bread and diamonds change if you were locked in Tiffany's over the weekend? Why are so many necessities of life relatively inexpensive, when they are indispensable?

4. Draw up a hypothetical demand schedule for yourself for one year's purchases of books, assuming that the average price of books changed from $10 through $1.

5. Draw up a supply schedule for your local bookstore, assuming that book prices ranged from $1 to $10.

6. What is the equilibrium price of books in your examples above? Exactly what do you mean by equilibrium price? What is special about the quantities demanded and supplied at this price?

7. What is the difference between "supply and demand" and "quantity supplied" and "quantity demanded"?

8. What do we mean when we say that there are two aspects to competition? What are they? Why is any one not enough?

9. What are the maximizing and the constraining forces in a competitive market?

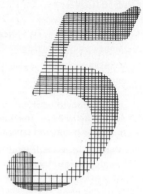

The Tools of
Economic Analysis

WE HAVE TAKEN A QUICK LOOK AT THE BASIC assumptions underlying economic theory, and made a more detailed examination of the workings of the market mechanism. So we are almost ready to begin a full-fledged exploration of the insights of micro and macro analysis. But first there is a necessary chore: to learn something more about the working techniques of economic analysis itself.

In this chapter we will take up a series of concepts that are important to understand in order to think clearly about economics. Some of the concepts seem very simple but are more subtle than they appear at first look; others may look demanding at first but are actually very simple. There are seven of these intellectual tools in all. Try to master them all, for we will be using them continuously from here on.

1. Ceteris paribus

The first concept is one that sounds simple but is actually very complex. It is the assumption that when we examine the relationship between any two economic activities, we can disregard the effects of everything but the particular elements under examination.

This assumption of holding "other things equal" is called by its Latin name ceteris paribus. It is extremely easy to apply in theory—and extremely difficult to apply in practice. For example, in our examination of the demand curve, we assume that the *income* and *tastes* of the person (or of the collection of persons) are unchanged, while we examine the influence of price on the quantities of shoes they are willing and able to buy. The reason is obvious. If we allowed their incomes or tastes to change, both their willingness *and* their ability would also change. If prices doubled, but a fad for shoes developed; or if price tripled, but income quadrupled, we would not find that "demand" decreased as prices rose.

Ceteris paribus is applied every time we speak of supply and demand, and on many other occasions as well. Since we know that in reality prices, tastes, incomes, population size, technology, "moods," and many other elements of society are continually changing, we can see why this is a heroic assumption and one that is almost impossible to trace in actual life or to correct for fully by special statistical techniques.

Yet we can also see that unless we apply ceteris paribus, at least in our minds, we cannot isolate the particular interactions and causal sequences that we want to investigate. The economic world then becomes a vast Chinese puzzle. Every piece interlocks with every other, and no one can tell what the effect of any one thing is on any other. But if economics is to be useful, it must be able to tell us something about the effect of changing *just* price or *just* income or *just* taste or any *one* of a number of other things. We can do so only by assuming that other things are "equal" and by holding them unchanged in our minds while we perform the intellectual experiment in whose outcome we are interested.

2. Functional relationships

Economics, it is already very clear, is about relationships—relationships of man and nature, and relationships of man and man. The laws of diminishing marginal utility or diminishing returns or supply and demand are all statements of those relationships, which we can use to explain or predict economic matters.

We call these relationships, the effect of one thing on another, functional relationships. For example, functional relationships may relate the effect of price on the quantities offered or bought, or the effect of successive inputs of the same factor on outputs of a given product, or the effect of population growth on economic growth, or whatever.

One important point. Functional relationships are not "logical" relationships of the kind we find in geometry or arithmetic, such as that the square of the hypotenuse of a triangle is equal to the sum of the squares of the other two sides, or that the number six is the product of two times three or three times two. Functional relationships cannot be discovered by deductive reasoning. They are descriptions of real events that we can discover only by empirical investigation. We then search for ways of expressing these relationships in graphs or mathematical terms. In economics, the technique used for discovering these relationships is called *econometrics.* You can get a first taste of this subject, if you wish, in Chapter 42, pp. 685ff.

3. Identities

Before going on, we must clarify an important distinction between functional relationships and another kind of relationship called an *identity.* We need this distinction because both relationships use the word *equals,* although the word has different meanings in the two cases.

A few pages ahead we will meet the expression

$$Q_d = f(P)$$

which we read "Quantity demanded (Q_d) equals $f(P)$" or "*is* a function of price [$f(P)$]." This refers to the kind of relationship we have been talking about. But we will also find another kind of "equals," typified by the statement that $P \equiv S$ or purchases equals sales. $P \equiv S$ is *not* a functional relationship, because purchases do not "depend" on sales. They are *the same thing* as sales, viewed from the vantage point of the buyer instead of the seller. P and S are identities: Q_d and P are not.

Identities are true by definition. They cannot be "proved" true or false, because there is nothing to be proved. On the other hand, when we say that the quantity purchased will depend on price, there is a great deal to be proved. Empirical investigation may disclose that the suggested relationship is not true. Or

THE IMPORTANCE OF TIME

Of all the sources of difficulty that creep into economic analysis, none is more vexing than *time*. The reason is that time changes all manner of things and makes it virtually impossible to apply *ceteris paribus*. That is why, for example, we always specify "within a fixed period of time" when we speak of something like diminishing marginal utility. There is no reason for the marginal utility of a meal tomorrow to be less than one today, but good reason to think that a second lunch on top of the first will bring a sharp decline in utilities.

So, too, supply and demand curves presumably describe activities that take place within a short period of time, ideally within an instant. The longer the time period covered, the less is *ceteris* apt to be *paribus*.

This poses many difficult problems for economic analysis, because it means that we must use a "static" (or timeless) set of theoretical ideas to solve "dynamic" (or time-consuming) questions. The method we will use to cope with this problem is called comparative statics. We compare an economic situation at one period with an economic situation at a later period, without investigating in much detail the path we travel from the first situation to the second. To inquire into the path requires calculus and advanced economic analysis. We'll leave that for another course.

it may show that a relationship exists but that the nature of the relationship is not always the same. Identities are changeless as well as true. They are logical statements that require no investigations of human action.

Sometimes identities and behavioral equations are written in the same manner with an equal sign (=). Technically, identities should be written with an identity sign (≡). Unfortunately for generations of students, that is also read "equals." Since it is important to know the difference between definitions, which do not need proof, and hypotheses, which *always* need demonstration or proof, we shall carefully differentiate between the equal sign (=) and the identity sign (≡). *Whenever you see an equal sign, you know that a behavioral relationship is being hypothesized. When you see the identity sign, you will know that a definition is being offered, not a statement about behavior.*

The fact that identities are always true does not make them unimportant. Definitions are very important. They are the way we establish a precise working language. Learning this language, with its special vocabulary, is essential to being able to speak economics accurately.

4. Tautologies

Tautologies are statements that resemble identities but are not quite the same thing. *They are statements that cannot be proven false because there is no empirical or operational way to examine them.*

Tautologies play an important part in economic thought. Take, for example, our conception of economic man. We have said that his aim is to maximize his utilities. Now this is a very elusive statement. If we had said that his aim was to maximize his *money* income, we might put the statement to the test of observation. We might discover that it was untrue. A businessman, for example, may refuse to cheat, although cheating might make his money income larger.

When we say that individuals seek to maximize their *utilities*, however, we are dealing with a concept that defies empirical testing. No matter what a person does—cheating or not cheating, working hard or not working hard, seeking a million dollars or leading a life of ease—we can always claim that he is maximizing his utilities, and there is no way of disproving the contention.

We will look further into this problem in our next chapter. Here we must realize that their untestability does not mean that tautologies are useless. We have said that maximizing refers to an acquisitive propensity, to an effort aimed at "bettering our condition," that has a ring of experiential truth. The problem is that there is no way of *measuring* this "betterment," which includes not alone money but our whole conduct of life. Thus the information conveyed by a tautology may be real and useful, but it cannot be objectively accepted or rejected in the way that functional relationships can be. Nor can tautologies be accepted as definitions, the way identities can. They occupy a special position, giving us statements about reality that reflect our beliefs or feelings, but not statements that can be subjected to empirical scrutiny.

5. Schedules

We are familiar with the next item in our kit of intellectual tools. It is one of the techniques used to establish functional relationships: the technique of drawing up *schedules* or lists of the different values of elements.

We have already met such schedules in our lists of the quantities of shoes supplied or demanded at various prices. *Schedules are thus the empirical or hypothetical data whose*

functional interconnection we wish to investigate. In our next chapter we will look into some of the problems of drawing up such schedules in real life. But we must understand that we use them in economic analysis as "examples" of the raw material of behavior scrutinized by economic theory.

6. Graphs

The depiction of functional relationships through schedules is simple enough, but economists usually prefer to represent these relationships by graphs or equations. This is so because schedules only show the relationship between *specific* quantities and prices, or specific data of any kind. Graphs and equations show the *generalized* relationship, the relationship that covers all quantities and prices or all values of any two things we are interested in.

The simplest and most intuitively obvious method of showing a functional relationship in its general form is through a graph. Everyone is familiar with graphs of one kind or another, but not all graphs show functional relationships. A graph of stock prices over time, as in Fig. 5·1, shows us the level of prices in different periods but

FIGURE 5 · 1
Stock market prices

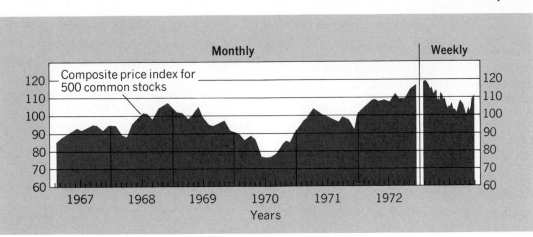

While most demand curves slope downward, in three interesting cases they don't. The first concerns certain *luxury goods* in which the price itself becomes part of the "utility" of the good. The perfume Joy is extensively advertized as "the world's most expensive perfume." Do you think its sales would increase if the price were lowered and the advertisement changed to read "the world's second-most expensive perfume"?

The other case affects just the opposite kind of good: certain basic staples. Here the classic case is potatoes. In 19th-century Ireland, potatoes formed the main diet for very poor farmers. As potato prices rose, Irish peasants were forced to cut back on their purchases of other foods, to devote more of their incomes to buying this necessity of life. More potatoes were purchased, even

---

does not show a behavioral connection between a date and a price. Such a graph merely describes and summarizes history. No one would maintain that such and such a date *caused* stock market prices to take such and such a level.

On the other hand, a graph that related the price of a stock and the quantities that we are willing and able to buy *at that price, ceteris paribus,* is indeed a graphic depiction of a functional relation. If we look at the hypothetical graph below, we can note the dots that show us the particular price/quantity relationships. Now we can tell the quantity that would be demanded at any price, simply by going up the "price" axis, over to the demand curve, and down to the "quantity" axis. In the graph, Fig. 5·2 (left) for example, at a price of $50, the quantity demanded is 5,000 shares per day.*

7. Equations

A third way of representing functional relationships is often used for its simplicity and brevity. *Equations are very convenient means of expressing functional relationships, since they allow us to consider the impact of more than one factor at a time.* A typical equation for demand might look like this:

$$Q_d = f(P)$$

Most of us are familiar with equations but may have forgotten their vocabulary. There are three terms in the equation above: Q_d, f, and P; and each has a name. Two of these names are simple. We are interested in seeing

---

FIGURE 5·2

Price/quantity relationship of a hypothetical stock

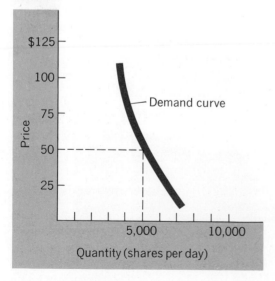

*Technically we would need a schedule of survey results showing the quantities demanded for every conceivable price, in order to draw a graph. In fact, we obtain results for a variety of prices and assume that the relationship between the measured points is like that of the measured points. The process of sketching in unmeasured points is called *interpolation*.

though their prices were rising, because potatoes were the cheapest thing to eat.

Such goods, with upward sloping demand curves, are called *inferior goods*. Up to a point, the higher the price, the more you (are forced to) buy. Of course, when potatoes reach price levels that compete with, say wheat, any further price rises will result in a fall in the quantity demanded, since buyers will shift into wheat.

Finally, there is a very important upward sloping curve that relates quantities demanded and *incomes*: the higher our incomes, the more we buy. We will use this special demand relationship a great deal in macroeconomics. However, in microeconomics, the functional relation is mainly between *price* and quantities demanded, not income and quantities demanded. Still, it is useful to remember that the functional relationships involving demand do not all slope in the same direction.

how our quantity demanded (Q_d) is affected by changes in prices (P). In other words, our "demand" is dependent on changes in price. Therefore the term Q_d is called the *dependent variable:* "variable" because it changes; "dependent" because it is the result of changes in P. As we would imagine, the name for P is the *independent variable*.

Now for the term f. The definition of f is simply "function" or "function of," so that we read $Q_d = f(P)$ as "quantity demanded is a function of price." If we knew that the quantity demanded was a function of both price *and* income (Y) we would write $Q_d = f(P, Y)$. Such equations tell us what independent variables affect what dependent variables, but they do not tell us *how* Q_d changes with changes in P or Y.

The "how" depends on our actual analysis of actual market behavior. Let us take a very simple case for illustrative purposes. Suppose that a survey of consumer purchasing intentions tells us that when the price of a particular product goes up by $1, the quantity demanded goes down by 2 units. In this case, the functional relationship would be represented by this equation:

$$Q_d = -2(P)$$

The function, f, has been found to be -2. Suppose we also find out that consumers would take 100 units of this product if its price were zero—that is, if it were given away free—and that they would buy one-half unit less each time the price went up by $1. The demand equation would then be:

$$Q_d = 100 - .5(P)$$

Thus, if price were $10, buyers would take $100 - .5 \times 10$, or 95 units.

We should stop to note one important property of ordinary price/quantity demand or supply functions. It is that they have opposite "signs." A normal demand function is negative, showing that quantities demanded *fall* as prices rise. A supply function is usually positive, showing that quantities supplied *rise* as prices rise. A survey of producers, for example, might tell us that the quantity supplied would go up by 2 units for every $1 increase in price, or

$$Q_s = 2(P)$$

Note that the sign of the function 2 is positive, whereas the sign of the demand function was negative, $-.5$.

ECONOMIC TECHNIQUES REVIEWED

The seven items in our kit of intellectual tools do not tell us more about the basic assumptions that economics makes with respect to the facts of economic society. We

EQUILIBRIUM IN EQUATIONS

It is very easy to see the equilibrium point when we have a supply curve and a demand curve that cross. But since equations are only another way of representing the information that curves show, we must be able to demonstrate equilibrium in equations. Here is a simple example:

Suppose the demand function, as before, is:

$Q_d = 100 - .5(P)$, and that the supply function is:

$Q_s = 2(P)$

The question is, then, what value for P will make Q_d equal to Q_s? The answer follows:

If $Q_d = Q_s$, then
$100 - .5(P) = 2(P)$.
Putting all the P's on one side
$2(P) + .5(P) = 100$, or $2.5P = 100$. Solving, $P = 40$.

Substituting a price of 40 into the demand equation we get a quantity of 80. In the supply equation we also get 80. Thus 40 must be the equilibrium price.

---

have seen that these facts can be summed up in two sets of general propositions or laws—laws about behavior and laws about production. What we have been learning in this chapter are the *techniques* of economic analysis, rather than its basic premises.

These techniques, as we have seen, revolve around the central idea of functional relationships. Because behavior or production is sufficiently "regular," functions enable us to explain or predict economic activity. Their relationships are presented in the form of graphs or equations derived from the underlying schedules of data.

As we have seen, the ability to establish functional relationships depends critically on the *ceteris paribus* assumption. Unless we hold other things equal, either by econometric means or simply "in our heads," we cannot isolate the effect of one variable on another.

Finally, economic analysis also relies on identities and tautologies. Identities are definitional—they are merely two names for one thing. Tautologies are descriptions that cannot be subjected to empirical investigation. Both identities and tautologies can be useful in clarifying our thought, but they must not be confused with functional relationships.

ECONOMIC FALLACIES

No chapter on the mode of economic thought would be complete without reference to economic fallacies. Actually there is no special class of fallacies that are called economic. The mistakes we find in economic thought are only examples of a larger class of mistaken ways of thinking that we call fallacies. But they are important enough to warrant a warning in general and some attention to one fallacy in particular.

The general warning can do no more than ask us to be on guard against the sloppy thinking that can make fools of us in any area. It is easy to fall into errors of false syllogisms,* of trying to "prove" an argument "post hoc, ergo propter hoc" ("after the fact, therefore the cause of the fact"); for example, "proving" that government spending must be inflationary by pointing out that the government spent large sums during periods when inflation was present, ignoring other factors that may have been at work.

The gallery of such mistaken conclusions is all too large in all fields. But there is one fallacy that has a special relevance to the

---

*See the questions at the end of this chapter for examples.

kinds of problems that economics considers. It is called the *fallacy of composition.* Suppose we had an island community in which all farmers sold their produce to one another. Suppose further that one farmer was able to get rich by cheating—selling his produce at the same price as everyone else, but putting fewer vegetables into his bushel baskets. Does it not follow that all farmers could get rich if all cheated?

We can see that there is a fallacy here. Where does it arise? In the first example, when our cheating farmer got rich, we ignored a small side effect of his action. The side effect was that a loss in real income was inflicted on the community. To ignore that side effect was proper so long as our focus of attention was what happened to the one farmer. But when we now broaden our inquiry to the entire community, the loss of income becomes important. Everyone loses as much by being shortchanged as he gains by shortchanging. The side effects have become central effects; what was true for one turns out not to be true for all. Later on, in macroeconomics, we will find a very important example of exactly such a fallacy when we encounter what is called the Paradox of Thrift.

KEY WORDS

Techniques vs. theory

Ceteris paribus

Functional relationships

Identities

Tautologies

Schedules

Graphs

Equations

● *Independent and dependent variables*

● *Function*

Fallacy of composition

CENTRAL CONCEPTS

1. **This is a chapter about the techniques of economic analysis, not about the basic assumptions underlying economic theory. We should become familiar with a few useful ideas, or tools.**

2. **Ceteris paribus is the assumption that everything other than the two variables whose relationship is being investigated is "kept equal."** *Without ceteris paribus we cannot discern functional relationships.*

3. *Functional relationships*—**relationships that show that x "depends" on y—lie at the very center of economic analysis. They are not logical or deductive relationships, but relationships that we discover by** *empirical investigation.*

4. *Identities* **are purely definitional and therefore** *not subject to proof or to empirical investigation.* **Such definitions can, however, be very important.**

5. *Tautologies* **are statements that may convey feelings or beliefs but are** *not capable of empirical measurement* **and therefore cannot be proved or disproved.**

6. **Functional relationships use three techniques:**

   schedules, or lists of data
   graphs, or visual presentations
   equations

   We should learn the meaning of three equational terms: the *independent variable,* the "causative" element that interests us; the *dependent variable,* the element whose behavior is affected by the independent variable; and the *function,* a mathematical statement of the relation between the two. We should learn to read the sentence $x = f(y)$ as "x is a function of y." Here, x is the dependent variable; y is the independent variable.

7. **We must be on guard against** *economic fallacies,* **especially against the** *fallacy of composition.*

QUESTIONS

1. Suppose we discover that the quantity of food you buy per week increases by 10 percent every time the price of food goes down. If we have no more information than that, can we derive a functional relationship?

2. What other information would you need to know? Would changes in your income have to be taken into account? Changes in your tastes? Changes in the prices of other goods? Explain.

3. Can you write a hypothetical function that might relate your demand for food and the price of food, assuming *ceteris paribus*?

4. "The quantity of food bought equals the quantity sold." Is this statement a functional relationship? If not, why not? Is it a tautology? An identity?

5. Which of the following are tautologies:

   Everything always turns out for the best in the long run.

   Men always act in their own self-interest.

   Men always behave to maximize their incomes.

   The effort to maximize incomes always results in economic success.

   HINT: to which of these statements could you apply some kind of empirical test?

6. Here is a schedule of supply and demand:

Price	Units supplied	Units demanded
$1	0	50
2	5	40
3	10	30
4	20	25
5	30	20
6	50	10

   Does the schedule show an equilibrium price? Can you draw a graph and approximate the equilibrium price? What is it?

7. How do we read aloud the following?

   $C = f(Y)$ where C = consumption and Y = income.

   Which is the independent variable? The dependent?

8. Which of the following statements is a fallacy?

   All X is Y
   Z is Y
   Therefore Z is X

   All X is Y
   Z is X
   Therefore Z is Y

Try substituting classes of objects for the X's and Y, and individual objects for the Z's. Example: All planets (X's) are heavenly bodies (Y). The sun (Z) is a heavenly body (Y). Therefore the sun (Z) is a planet (X). Clearly, a false syllogism. But: the sun is a heavenly body (Z is Y) is correct.

Other fallacies:

If I can move to the head of the line, all individuals can move to the head of the line.

If I can save more by spending less, all individuals should be able to save more by spending less. (Hint: if all spend less, what will happen to our incomes?)

The fact that Lenin called inflation a major weapon that could destroy the bourgeoisie indicates that inflations are part of the Communist strategy for the overthrow of capitalism.

Some Basic Problems

WE ARE ALMOST READY TO PROCEED to the main task at hand. But before we leave this introductory section, we must stop to take a critical look at what we have learned. For economics is by no means a "finished" discipline. Unlike elementary geometry or high-school chemistry, its relations or facts are not logically beyond question or empirically secure. Rather, like all the social sciences, economics is built on foundations that we *know* to be unfinished, inadequate, perhaps even wrong.

Thus economics is constantly in the process of self-examination, and the analytical explanations it offers are constantly in need of reevaluation. This is the case partly because the society that economics examines is changing, and partly because our understanding of that society is also changing. Therefore part of learning economics is learning what we do not know, becoming aware of our ignorance as well as of our knowledge. Learning what economists don't know and are trying to find out may make the subject a little disturbing to the student who would like his economics served up like the multiplication table, but at least it promises that the subject matter will not be as boring as the multiplication table.

DATA PROBLEMS

Economics tries to generalize about the results that emerge as men expend their energies in the acts of production and distribution. But how does it gain its knowledge about these results? The answer, obviously, is from the collection of facts about economic activity that we call *economic data.*

The question is: how accurate are our data on production and distribution? The answer, unhappily, is that we know much less than the smooth curves of our diagrams indicate.

There are many reasons why our knowledge is limited. One reason is that it is almost impossible to carry out experiments that will give us the actual "shapes" of curves such as diminishing returns or supply and demand or even changes in scale. The difficulty arises in part because society is not a machine that can be tinkered with to establish functional relationships in a laboratory-like manner. But it is also due to *ceteris paribus,* rearing its ugly head, making it very difficult to isolate the particular facts we seek from the raw data of experience. Even if we seek something so simple as a schedule of prices of automobiles, for example, we en-

CAUTIONS ABOUT DATA

Nothing is "harder" than data; nothing is "softer." The experienced economic statistician learns to look for certain pitfalls when he tries to collect facts about the economic world. Here are some of them.

1. Unseasoned data.

If you examine many figures, even in the most official statistical sources, you will discover that they change from one edition to the next. In the 1970 edition of the *U.S. Statistical Abstract*, the figure for U.S. gross national product in 1969 is $932.3. In the edition published two years later, the 1969 figures are $929.1. What accounts for the change? Largely the fact that many data take considerable time to collect. Most current data are revised, often substantially, in a matter of a few years. Moral: do not base your findings rashly on small numerical differences that may disappear by the time your paper is published.

2. Definitions.

The Statistical Abstract (1972) shows that 10.5 million households were below official poverty counter vexing problems. Do our prices reflect the discounts that dealers may give? Do they take into account trade-ins? Have we eliminated price changes due to changes in the style or quality of cars? Thus the raw data of automobile prices comes to us in a fashion that is often misleading if we are to use it, unexamined, to prove a case.

AGGREGATION PROBLEMS

A second difficulty with data has to do with the problem of *aggregation*. Suppose that we want to know about the output of steel, perhaps in order to establish functional relationships between steel output and inputs of labor or capital. If we go out into the marketplace, we discover a curious fact: there is no such thing as "steel." There are steel bars and steel sheets, steel ingots and structural beams, but "steel" as such is a nonexistent commodity. (Even ingots vary in the quality of steel they contain.)

Aggregates make it difficult to establish functional relationships, because we are caught in the problem of comparing apples and pears. Data that indicate a rising output of "steel" may conceal the fact that the kinds or qualities of steel products are changing, so that the figure for "total output" may lead us to conclusions that a finer analysis would not support. And if even products such as steel or wheat or coal are actually aggregates, how much more deceptive are data such as "clothing" or "food" or that largest aggregate of all, gross national product!

What can we do in the face of this difficulty? In part we can "correct" for the distortions of aggregates by the use of econometric techniques. But to a large degree we are left with a problem that is insoluble. Some degree of aggregation is present in nearly all economic data, and therefore some degree of deception is inherent in all economic data. There is no escape from this dilemma of economic statistics. We cannot reason about economic phenomena without the use of data. But their inherent problems teach us to be cautious and careful in jumping to conclusions on the basis of "facts" that may turn out to be a great deal less solid than they appear.

ABSTRACTION

Problems with data lead to a further, inherent problem in economic reasoning: the *level of abstraction* necessary for carrying on our investigations.

Like all sciences, economics struggles with

levels in 1970. Does this tell us how many *families* were below this level? Answer: no, it does not. A household is not necessarily a family (it can be a single individual). Actually, in 1970, only about 5.5 million families were below the poverty line. Moral: carefully read all table headings and footnotes.

3. Statistical deceptions.

The endless figures on rising thefts, assaults, etc., attesting to the existence of "crime waves" are perhaps correct. But before we can show that a crime wave exists, we would have to know circumstances such as whether population rose (normally bringing a larger *number* of crimes but not necessarily a larger proportion of crime *per capita*); whether crime reporting methods changed (affecting the number of crimes we *know about*, but not necessarily the number that really occurred); and such seemingly unconnected things as changes in prices (causing more thefts to involve sums that made them, by law, major rather than minor offenses). Moral: caution first, last, and always in dealing with data.*

*There is more on this problem in "Quantitative Methods," Part 7.

the hopeless complexity of real life by *abstracting* crucial aspects of complex processes. It was only by abstraction that Adam Smith was able to construct a "model" of economic growth or of the workings of the market mechanism. It is only by abstraction that we arrive at concepts such as that of economic man, whose living counterpart is not to be discovered in real life, although some of his attributes are certainly discernible.

Abstraction is both the central strength and one of the most vulnerable weaknesses of economics. Economic "reality" is at least as complex as the newspapers, which are themselves drastic simplifications of the real events of economic life. But unless we abstract from the confusion of newspaper events we cannot even begin to describe the basic functional relationships that are one of the main objectives of economic inquiry. Think of trying to construct a supply and demand diagram from the financial pages! *Thus it is wrong to think of abstraction as a defect of economics. It is an indispensable prerequisite of all scientific thinking.*

What is the weakness of abstraction, then? It lies in the fact that there is no single valid "level" of abstraction, no absolutely sure rule that will focus our attention on the invisible forces we wish to uncover. In abstracting from the richness of life to arrive at the concept of economic man, for example, we ignore impulses and feelings that lead man away from maximizing. If we are incorrect in ignoring those other forces, which may make men behave in illogical ways or may make generosity and compassion central constituents of behavior, we may end up with an abstraction that leads to unsubstantiated economic conclusions. Indeed, the errors that economics makes—and they are many—are rarely the result of a faulty train of reasoning. *Almost always they stem from mistaken premises, ill-founded initial assumptions about behavior or possibly about nature.* These mistakes, in turn, are often traceable to abstractions that fail to do justice to important attributes of behavior or the physical world.

THE IMPORTANCE OF INSTITUTIONS

In the inescapable process of abstraction, an important element of error involves the importance that we assign institutions. To take an example that is central to our inquiry, we study the economy as if it were a market system. But is it? The market certainly describes very well the organization of economic life in many areas, as in the determination of prices for used cars or perhaps (although somewhat less well) in the determination of prices and outputs of new cars.

But does it describe a process that is tacitly taken for granted—even though it is fundamental to the operation of the automotive markets—the decision to build highways or mass transit? Or to take another area, does the market mechanism explain how resources are allocated to peace or war, space medicine or home medicine, public housing or private homes?

Few economists would claim that these basic decisions can be explained by economic analysis. Yet they are determinative for our society. *Do we then misrepresent the basic problem of our social system when we examine it as a market mechanism and not as a power mechanism with markets playing an important, but subordinate role?*

There is no clear answer to this problem, but the examples we have given should add a cautionary note to our use of economic analysis. By setting the level of abstraction so high that institutional realities drop out of focus, crucial aspects of contemporary society can be easily removed from the picture. On the other hand, if we change the focus to highlight institutional problems, we tend to end up with a newspaper account of economic life without any discernible patterns or tendencies.

The aim of economics, and of all social science, is to discover a middle level of abstraction, where the most important facts are clearly present and where large movements are also discernible. This is an objective that is a great deal easier to specify on paper than to attain in fact.

TAUTOLOGIES AGAIN

A related difficulty of economic theory is that, as we know, some of its underlying concepts are tautologies—statements that cannot be verified by observation or testing in the real world. Take the central idea of equilibrium. Is equilibrium a tautology? We know that markets bring about equilibrium prices at which quantitities supplied and demanded are equal. But how do we know, looking at the real world, whether or not a price is an equilibrium price?

The question requires that we return to our distinction between behavioral equations and identities. When we look at the actual world, we see markets in which purchases equal sales at all sorts of prices. But just because purchases *always* equal sales, how do we know whether a *particular* level of purchases-and-sales represents the level that corresponds to an equilibrium level? The answer is that we don't. The notion of an equilibrium price presupposes that we know the schedules of demand and supply. If we do not know these schedules, we cannot tell whether a given price is in fact an equilibrium price—and certainly the fact that purchases $\equiv$ sales is no such indication.

Since we do not know these schedules in the great majority of cases, it is very difficult to say whether a given price is or is not an equilibrium price. Indeed, if we had some system of ascertaining the elusive supply and demand schedules, we would probably find that most prices were *not* in equilibrium. They should be in the process of moving *toward* equilibrium, but it is possible that they are moving away. Even when they are moving toward equilibrium, they seldom reach it, because the equilibrium point itself is constantly changing as *ceteris paribus* conditions change—tastes or incomes or the prices of other goods.

THE PROBLEM OF PREDICTION

Basic to economic reasoning is its power of prediction. Indeed, what is so remarkable about the market mechanism is that it works in predictable ways. That is why we can say

A PREDICTIVE SCOREBOARD

Despite very sophisticated economic aids, the predictive records of the best-informed "insiders" in the stock market has been dismal. Peter L. Bernstein, a well-known Wall Street economist, has analyzed the performance of the major pension funds, which employ the very best economic advice and the most sophisticated econometric techniques in their attempt to "out-perform" the market.

Between 1962 and 1971, the stock market (as measured by Standard and Poor's index of 500 stocks) earned an average return of 7.1 percent per year in dividends plus growth. The 300 biggest pension funds earned only 6.2 percent, *with 75 percent of the funds doing less well than the market.* Bernstein also analyzed the showing of 35 leading "growth-oriented" mutual funds in mid 1972. At that time, 7 were still below their values

of 1961, although the market was over 60 percent higher; an additional 21 funds were below their 1965 levels, although the market was up by about 25 percent from that date; and *only 8 funds had outperformed the market over the 10-year period as a whole.*

Further, Bernstein points out that the predictions of the best informed "insiders" have been more often wrong than right. For 10 years, leading investment bankers have been polled each November for their forecasts on the level of the market for the following June. Their record appears in the table below.

As we can see, the investment bankers are certainly not good predictors. There was a famous cartoon in *The New Yorker* many years ago, showing a suburban lady instructing her broker: "Just before the market goes down, I want you to sell my stocks, and then buy them back just before it goes up." If she had gone to the investment bankers, she'd be broke.

Year of prediction (November)	Dow-Jones Stock Average on that date	Prediction for June	Actual Dow-Jones following June
1961	730	86% above 730*	561
1962	640	92% under 700	707
1963	750	91% under 800	832
1964	870	85% between 800–950	868
1965	940	94% above 850	870
1966	800	95% under 900	860
1967	875	69% between 800–950	898
1968	975	81% above 900	875
1969	800	94% above 750	690
1970	800	82% between 750–900	890
1971	800	47% under 900	929

*That is, 86 percent of predictions picked a Dow-Jones average of 730 or higher for the following June.

that if prices rise, *ceteris paribus,* quantities demanded will fall; or that if prices fall, *ceteris paribus,* the quantity supplied will fall.

No other social science has this seemingly simple but enormously valuable predictive capacity. That is, no other social science can lean with confidence on functional relationships of the kind that underpin economics.

But with how much confidence can economics lean on those relationships? The answer is disconcerting. The actual predictive record of economics is spotty, even poor. In certain fields, such as the trend of stock market prices, economics (and economists!) have shown no predictive success whatever. In other, more important areas such as the prediction of national income, the batting average is far from good. Thus despite the clarity of the schedules, graphs, and equations of our previous chapter, the predictive power of economics is a great deal more impressive in theory than in fact. Why?

IMPERFECT KNOWLEDGE

The first reason for inadequacy is probably self-evident from our foregoing discussion. An ability to predict the outcome of the market process hinges on our foreknowledge of the shapes and positions of supply and demand curves, or the functions we can insert into equations of supply and demand.

It can hardly come as a surprise that this foreknowledge is both small and shaky. We know very little about the actual willingness and ability of masses of individuals to buy and sell various goods. We do not have very good information about the costs of production of a great many commodities. Worse, unless we possess complete knowledge, we cannot act as the rational maximizers we are

supposed to be. How would a businessman maximize his profits unless he knew the price of every possible alternative source of supply or the state of the marketplace in every nook and corner of the world?

In fact, of course, no businessman can hope to encompass all the knowledge that would enable him to make a perfect calculation of his most profitable course. As for consumers, they do not begin to have the encyclopedic knowledge that would enable them to behave as our market model supposes. Moreover, the *costs of ascertaining the knowledge* that would enable one to maximize may be so great that we can never become more than pale approximations of the economic creature we are supposed to be. How much time can we spend comparison shopping? How many catalogs must one consult before buying a chair?

Information is particularly scanty when we concern ourselves with the future. Buyers and sellers often have no inkling of what the future will bring, and no way of finding out. Think of the effect of weather on crops. The task of acting as a rational maximizer in this situation requires that one have a mathematician's knowledge of probabilities. In fact, the difficulties of being a rational maximizer are so great that we act on hunches, guesses, rumor, and faith.

THE PROBLEM OF EXPECTATIONS

But there is still another even more vexing problem. It is that we often cannot depend on economic theory as a predictive tool, because the basic supply and demand functions are capable of bewildering and totally unforeseeable shifts—not just in their numerical values but in their *signs.*

Take, for example, the normal case of a downward sloping demand curve. The curve shows us that the quantities we demand will increase as prices fall and decrease as prices rise. In the main, that is no doubt true. But assume that prices rise and that buyers interpret this rise as indicative of a further rise in the future. Suppose the prices of machine tools go up and that the general talk is that they will be going higher, perhaps because of an impending wage settlement that will send up costs. Now buyers may well rush in to buy *more* machine tools at their higher price, hoping to get their needs filled before things get worse.

The same "perverse" reaction can affect the sign of the supply function. Ordinarily, lower prices will result in less goods being offered for sale; higher prices, in more. But again, suppose that suppliers interpret a price fall as indicative of a worse fall coming—rumors of a coming depression or of vast imports in transit might create such a frame of mind. In that case, suppliers will react to lower prices by trying to sell *more* at the lowered price, before prices go lower still. Conversely, if prices are going up and are expected to go still higher, sellers will not increase their offerings but will hold back, waiting for still better opportunities.

Thus the "signal" of higher or lower prices does not always result in the behavior that is described by any given supply or demand function. *Behavior depends on the state of mind, the expectations of buyers and sellers.* Ordinarily we have no way of knowing this state of mind. Moreover, expectations can change overnight. That is another reason why prediction is much more difficult in fact than in our initial presentation.

It is instructive in this regard to compare the problems of economic prediction with those of meteorology. Both have some obvious difficulties in common, mainly the enormous mass of data needed for accurate prediction. But economics has one problem that

weather forecasting does not. The weatherman does not have to contend with winds or clouds "changing their minds." Economists, on the other hand, must deal with not only the often unmanageable (or unknown) data of the economic weather, but also with the intentions, expectations, moods, and feelings of the data he is observing. To put it differently, the economist is concerned with data that "behave" in a meaning of that word that never troubles the natural scientist.

INCREASING INDETERMINACY

There is a last, but very important reason why prediction is difficult, and why the reliability of economic reasoning may suffer accordingly. This reason has to do with the increasing indeterminacy of behavior in an affluent economic setting.

Here it may help us to go back to the tautology of "maximizing utilities." Suppose we suggested that maximizing be defined in a more testable way, such as the effort to gain as much income or profit as possible (within normal moral and legal limits) or to spend as little money as possible for the purchase of goods of a given kind. Now imagine that we were to go to a very poor country—say England in Adam Smith's day—to see if we could verify this hypothesis about behavior. It seems plausible that we would find that market activities roughly approximated what we would expect. Workers and capitalists both, in that setting, were more or less forced to strive after every last penny. Most consumers were forced to squeeze every sixpence when they went shopping.

When we turn to behavior in our own much more affluent society, however, the hypothesis of income maximization (or expense minimization) is much less likely to be valid. We can by no means be sure, for instance, that we will always choose a higher paying over a lower paying job. Just because we are better off than members of a poor society, we can weigh aspects other than the differential in pay. We noted in our first chapter that most middle-class people prefer white-collar to blue-collar work, even though blue-collar work may pay more. So, too, we do not always buy the cheapest item available to us; we may feel that its very cheapness means that it may not be as good as a competitive item, even though the two are presumably the same. How else would we explain consumers' preferences for brand-name aspirin over nonbrand name, although most consumers "know" that all aspirin is chemically the same?

Nor is maximization a word that gives us much predictive insight into the behavior of big firms. When we speak of a small profit-maximizing firm, such as Adam Smith's pin factory, we can foretell fairly well what measures its manager will have to take, as we will see when we study the competitive firm in Chapters 11 and 12. But this clearcut predictive capability becomes increasingly fuzzy when we examine the behavior of large firms who are not maximizing in order to survive from day to day, but to grow over the long term future.

Ford and General Motors for example, may choose very different strategies, according to their differing interpretations of the outlook for the coming years. One may embark on a heavy program of capital investment; the other may retrench, preparing for an expected recession. Yet, officials of both companies would declare that they were seeking to maximize their firm's sales or profits or asset values over the period in question. Thus, *maximization* becomes a word that loses its clear predictive content, once we leave the world of "Dickensian" behavior and pin factories, and enter the world of affluence and giant firms.

THE USES OF ECONOMICS: CONCEPTUAL CLARITY

In the face of all this criticism, how do we justify economics? What is left of our fundamental elements? What good is a study of the market mechanism?

There are three answers to this extremely important question. First, economics gives us a *set of concepts* to dispel the fog as we cope with the flux of economic reality. The kind of puzzle we posed in our box, "Supply and Demand, Again," in Chapter 4, is the sort of tangled thinking that we are rescued from by the ideas of supply and demand and equilibrium prices, even though we have great difficulty in finding real-world counterparts to these ideas.

This situation is not unique to economics. Physics deals with ideas like "force" or "matter," although it has great difficulty in giving empirical meaning to these terms. As in physics, the ideas of economics must be judged by their *usefulness,* not necessarily by their representation of unambiguous "things."

ORDINARY APPLICABILITY

Second, if we except periods of rapid change, or shifting expectations, *economic theory provides a roughly accurate guide to the outcomes that we can expect from changes in the market.* Perhaps, given the immense complexity of the structure of society, this is no more than we can reasonably expect. In this chapter we have stressed the shortcomings and inadequacies of economic theory, but it might be well to redress the balance somewhat by being grateful for what we have.

This underlying reliability is aided because it is much easier to predict and explain *average* behavior than *individual* behavior. Many individuals act in ways that do not accord with the assumptions of economic theory, but to a great extent these "abnormal" actions are mutually offsetting. One person may invest in the stock market because his horoscope says to do so, rather than because he has rationally calculated how to maximize his gains; but he is apt to be cancelled by another who will not invest in the market because *his* horoscope tells him *not* to.

In the aggregate, our market decisions should be sufficiently linked to the stimuli of economics to enable us to make generalizations about *average* economic behavior, without consulting the stars. Our predictional system breaks down only when large numbers of people behave "irrationally" in those recurrent "moods" or epidemic-like shifts in expectations that sweep over us.

NORMATIVE USES OF ECONOMICS

Third, *price theory can be useful* as *an instrument to help us gain desired goals.* That is, we can use the theory of the market mechanism to tell us what kind of behavior or what sorts of expectations we will *require* for a smoothly functioning price system.

Price theory can then help us design the policies that will enable the market to solve the economic problem in the way we want it to. When we think of it, business schools are premised on the assumption that it is possible to teach people to be rational profit maximizers.

The use of economic theory as a means to an end is *normative,* in contrast to *positive,* or predictive, use. Economics today is caught halfway between these two uses of theory.

Positive theory is still indispensable for many everyday situations: the managers of the A. & P. when they wish to increase their sales, do not advertise that their prices are *higher* than those of their competitors. Normative theory is increasingly important in a world where affluence makes it ever more difficult to foretell the behavior of consumers or firms, and where we are therefore constantly seeking to *alter* their behavior—by Presidential exhortations, by controls, by advertising, or by laws.

ECONOMIC REASONING AND VALUES

We have spent a good deal of time looking into the question of the reliability of economic reasoning. But before we leave this chapter, in which we have learned something of the fundamental problems of economics, we must look into one last question. This is the relation between economic theories and values—*values* meaning our beliefs, moral standards, or preferences.

All social systems ultimately rest on a series of value judgments. Societies are the expression of values, the very incorporation of morality. These values emanate from the distilled experience of its citizenry and from the special wisdom of its great men—philosophers, poets, priests. In this process of the *establishment* of values, economists have no special claim to expertise. The economist takes his place along with other trained members of society to whom the community may look for advice, but he has no greater competence than anyone else to say what is "better" or what is "worse."

Thus economists speak of their discipline as if it were, or at least should try to be, "value-free." The task of the economist, they maintain, is only to analyze the economic behavior of a society, a task that requires him to have a thorough knowledge of the society's values, but that removes him from passing judgment on the behavior he analyzes. In theory at least, an economist should be able to apply his objective analytic reasoning equally well to a hierarchical society or an egalitarian one, a society dedicated to plunder or to peace, to making money or to cultivating its gardens.

In fact, however, it is not so easy to divorce values from analysis and not as simple as it sounds to apply analytic techniques to all kinds of societies. Most contemporary Western economists are mainly concerned with investigating the problems and predicting the outcome of a market system. To understand such a system he must understand the value judgments that make a market society operate—value judgments such as "more is better" (or that maximizing is a commendable social objective), or that individuals in a market society are supposed to make their own decisions on how to maximize their utilities, rather than abiding by a course laid out for them by tradition or command. An economist need not personally subscribe to these values, but he could not hope to understand the operations of a market system unless he understood them.

It is not so easy to understand the kinds of value systems that underlie economic behavior in a very different setting—say the value systems of a slave society, of feudalism, of communist China. Hence it is difficult to develop analytic apparatus for explaining or predicting their economic behavior. An economist applies his knowledge with less assurance in these areas than in the market system. Worse, since nonmarket systems may reject the values with which he normally works (and lives!), he is apt to reject *their* values as inferior to his. As a result, despite their protestations of "value neutrality,"

economists sometimes give the impression that market systems are the only way or the best way to solve economic problems, which may not be the case at all.

Still more difficult is the problem of *economic policy*. Economists do not just analyze; they advise. Economists are constantly being called on to suggest how an economic system should reach a certain goal, such as full employment or a low rate of inflation. In these policy tasks the economist is put in a very difficult position. Presumably he should operate as a completely detached observer, a technician devoid of values, presenting the various alternatives open to society and analyzing the consequences of utilizing each of these alternatives.

In fact, the economist is hopelessly caught between his dedication to value-free analysis and his moral position as a member of the society for which he is prescribing. He may be asked to present only an "objective" range of possibilities to the designated decision-makers of society—its political leaders—who will make a final choice. But even then, the values of the economist will insinuate themselves into this choice by determining what alternatives are to be considered in the first place. An economist asked to design a program for improving American productivity is not likely to recommend the adoption of forced labor, even though that *might* actually raise productivity more than any of the measures he proposes.

There is no easy escape from this intermingling of subjective values and objective analysis in the determination of policy. Perhaps there should not be; the economist is a member of society, and to suppress all his feelings and values would be to distort his perception of the situation he must analyze. But that being the case, the economist should *declare his values,* insofar as he is aware of them, so that he does not deceive his audience into thinking that his counsel has some kind of awesome "scientific" finality.

Most economic arguments and disagreements, like most ordinary arguments and disagreements, are about value differences, not about analytic techniques; for value judgments of one sort or another lie at the basis of *all* the premises of economic reasoning. Learning to distinguish where these value judgments stop and economic reasoning begins is one of the most difficult tasks for a student in economics. He should be consoled to learn, however, that it is no easier for his instructors.

KEY WORDS	CENTRAL CONCEPTS
Data	1. Economics is an unfinished study because it deals with a social reality that is in constant change.
Aggregation problems	2. Certain problems are inherent to economic theory. But a separate source of difficulty arises from the difficulties of *economic facts or data*. Our "data base" is far from complete and is made complex by the problems of ceteris paribus that blur the data. In addition, problems of *aggregation* are inherently difficult to cope with and impossible wholly to overcome.
Level of abstraction	3. The difficulties of abstraction, inherent to all scientific inquiry, may lead us into serious error in our initial premises. One of the problems of *the level of abstraction* is the degree of importance we assign to institutions. For example, how much does power set the basic framework for the operation of the market mechanism? There is no wholly satisfactory answer to this question. Too high or too low a level of abstraction both pose dangers. We seek a hard-to-define middle ground.
	4. Equilibrium, a prime concept in economic theory, may also be a tautology, which lessens its usefulness.
Prediction	5. Economics hopes, through its functional relationships, to be able to make *reliable predictions* about the effects of changes in the independent variables of the market mechanism. Its record is not good in this regard, however.
Imperfect knowledge	6. The failure of prediction is attributed to several causes. One is inadequate or *imperfect knowledge*, especially regarding the future. The *cost of ascertaining perfect knowledge may be prohibitive*.
Expectations	7. A separate problem of prediction relates to the importance of *expectations* in determining the reactions of buyers and sellers. "Perverse" reactions are perfectly rational if buyers expect prices to rise further, or sellers expect them to fall further.
Conceptual clarity	8. What defenses can economic theory muster? They are three:
	• It gives us *conceptual clarity*.
Normal reliability	• We can make *fairly accurate predictions about normal economic events*, even if we cannot predict unusual events.
Normative vs. positive theory	• We can use economic theory *normatively*, to help us attain a goal, as well as *positively*, or predictively.

Value free

9. **We must learn to distinguish between** *economic analysis and economic value judgments.* **Presumably, economic analysis is** *value free,* **although values may get smuggled in when we sometimes (erroneously) assert that the mechanism of a market system is always "better" than some other. On the other hand, in eco-**

Economic policy

nomic policy it is impossible to keep values out of analysis. What the economist must do is to declare his values, not conceal them.

QUESTIONS

1. If you were asked to compile an index for the price of "shoes," how would you go about doing so? (HINT: look into Chapter 42 to see the process of "weighting" averages.) Would any index that you constructed be wholly accurate if the "mix" of shoe styles changed?

2. Make a list of various levels of abstraction. What is "the real world"? What is a workable level of abstraction—the newspapers? Which newspapers? What about magazines? Books? Theories? Do you think such a systematic ladder of abstractions can be made? Do you think we can deal with the world without abstractions? How do we know if our abstractions are correct?

3. How much research would you have to do to act as a utility-maximizing consumer if you want to buy a car? Some clothes? A TV? Potatoes? Do you think there is some relationship between our degree of imperfection of knowledge and the role we play in the economic system; i.e., consumer or producer? Is the *kind of market* pertinent?

4. Do you think that power is an institution? Is it an abstraction? Can we deal with power theoretically? (There are no "answers" to these questions, but they should be thought about carefully.)

5. Describe carefully the effect of a *change in expectations* on your rational maximizing behavior as a trader in stocks. Show how your buying and selling behavior would both be altered.

6. Can you give an operational meaning to "maximum income-seeking" for a small businessman in a country like India? Can you make the same predictive statement for, say, the actions that A.T.&T. is likely to take? What accounts for the difference? Do you think consumer behavior should be more predictable in India than in the United States? Why?

7. Can you think of a test that *might* show whether a price was at equilibrium? Suppose that inventories were piling up. Would that be suggestive of the fact that prices were higher or lower than equilibrium? Might such a pile-up arise from reasons other than a price that was different from equilibrium—for instance, the expectations of businessmen?

8. What is the difference between predicting and prescribing? If you prescribed a certain kind of economic behavior as a means to an end, is that a normative use of economics? What kinds of actions or pronouncements might restore "perverse" markets to normal operation?

9. Which of the following statements is "value laden"?

   • The market is a rationing mechanism.

   • The market is an efficient rationing mechanism because it requires no external supervision.

   • The market is an equitable rationing mechanism because it allows individuals to act as they please.

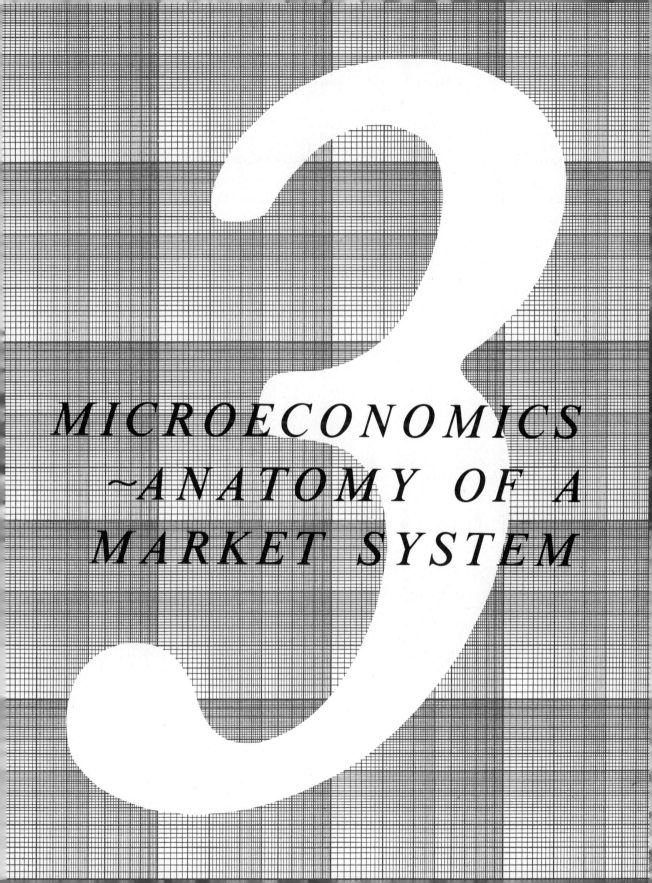

3

MICROECONOMICS
~ANATOMY OF A MARKET SYSTEM

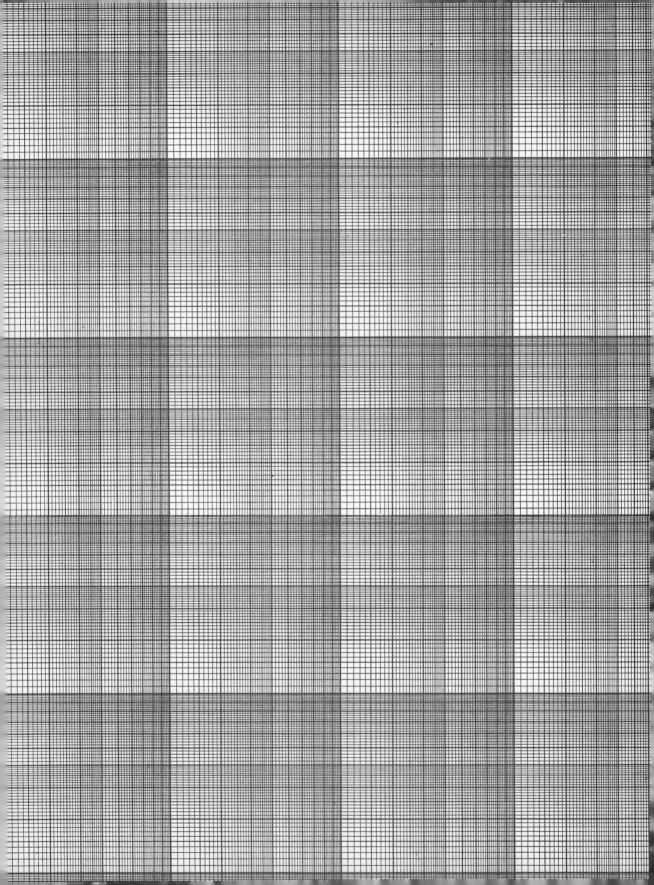

Introduction to the Microeconomy

7

WE HAVE COMPLETED OUR PREPARATORY WORK and are ready to venture into the territory of microeconomics. Actually, we know a good deal about that territory from our first two sections. We glimpsed the essential problem of microeconomics when we looked into Adam Smith's *Wealth of Nations* and discovered the market mechanism as the means by which a market society generated an orderly solution to the economic problem. In Part Two, we learned about how the market works and about the concepts we use to analyze its operation.

Now we are going to fortify our comprehension by going more carefully into the dynamics of the system. We still have much to find out about demand and a lot to learn about supply—in particular, how firms operate in a market setting. A little further along we will be investigating the special problems raised by the existence of giant firms that clearly differ in important ways from the pin factory of Smith's competitive model. The vitally important question of the dis-

tribution of income—perhaps the most "politically" significant task of the market—must be looked into with special care. But all that in good time.

MICRO- AND MACROECONOMICS

There is a certain amount of preliminary work that we must do before we can venture into these major questions. And the place to begin is with a careful look at the overall field itself. What is microeconomics, as contrasted with macroeconomics? Why do we study it separately, instead of joining the two in a unified study of the operation of the economy as a whole?

The answer is that we study economics from two perspectives because we need two different approaches to the economic world to understand two different kinds of problems we find in that world. Both micro- and macroeconomics are relevant for the study of how we produce and how we distribute wealth. The difference is that one approach, as we saw in our discussion of Adam Smith, emphasizes the *total* production or distribution of the system, while the other—as Smith's treatise also made clear—concerns the *composition of output* and the *determination of individual incomes within that total flow.*

The fact that we need two approaches to illumine the movement of the economy as a whole and the currents within the economy does not mean that we will not constantly be using the same concepts, such as supply and demand or the laws of production, in both inquiries. Rather, it reflects the fact that the problems characteristic of micro- or macro-economics become blurred when we use the approach of one to examine the other.

The fallacy of composition is also involved here: some truths at the micro level are untruths at the macro level, and vice versa.

Take, for example, the problem of inflation. Obviously, inflation has to do with the relation between the total demand for goods and the total supply of goods. Inflation has often been described as "too much money chasing too few goods." This problem lends itself naturally to an analysis that begins with an aggregative approach to demand, such as that implicit in *national* income, or an aggregative conception of supply, suggested by gross *national* output. But it is very hard to grasp the problem of inflation if we begin from the other end of the scale, where we concentrate on the production of particular goods or the demands of particular individuals or groups of individuals. To put it differently, microeconomics will help us understand why prices rise in, say, the market for beef or wheat, cars or houses, but it is less helpful in explaining why prices rise in *all* markets simultaneously.

THE NEED FOR TWO APPROACHES

Economists thus use two lenses to study the economy: one to see the actions and in-

teractions of its individual participants: one to see the economy in a bigger scale. *Neither perspective by itself is sufficient.* In particular, a view through the macro lens does not tell us much about the process of allocating resources to various uses or distributing income within an economic society. Macroeconomics, as we will see, is interested in the size of total output, but it throws almost no analytical light on whether that output consists of Cadillacs or Chevrolets, of bread or cake. So, too, macroeconomics is interested in the size of total incomes, but it does not explain why income is divided in such unequal shares.

At the same time, microeconomics by itself is also inadequate. It tells us a great deal about Chevrolets and Cadillacs, bread and cake, but it does not tell us whether total income and output will be large or small. Yet, there is clearly some relation between the proportions of Cadillacs to Chevrolets or cake to bread and the level of wealth that a society enjoys. Thus we cannot fully analyze the motives of microeconomic behavior without taking into account the macroeconomic setting for that behavior.

THE CIRCULAR FLOW

In this section we will be looking through the microeconomic lens. Our first task, therefore, is to set that lens at the proper focal length to highlight the activities we want to investigate.

What are those activities? Essentially, we want a perspective that will show us the microeconomy as a *system*. That is, we want to focus on the real goings-on in the economic world—men and women shopping, working, offering their services or their assets—in such a way that the transactions of innumerable markets can all be seen as part of an integrated whole, a great chain of activities that linked into one integrated network.

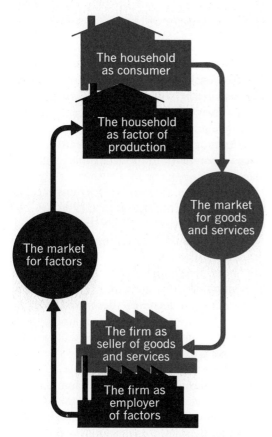

FIGURE 7·1
Circular flow in microview

We can find exactly such a focal length if we look at the microeconomy as a *circular flow,* as in Fig. 7·1. The arrows show that purchasing power streams from households into the market for goods and services, whence it flows into the firms who supply commodities to that market. They also show how firms spend money in the market for factors as they hire land, labor, and capital, and how this money in turn flows into households who supply these factor services. *Thus the first insight of the circular flow is that money travels in a great loop from household to firm and back from firm to household.*

Note that the flow also allows us to organize the complexity of the marketplace into a few very simple overall elements. In particular, *we distinguish between two basic institutions in the market: households and firms.* (We omit government institutions here, or group them mentally with "firms," because this makes it simpler to grasp the basic flow and does not fundamentally distort the relationships out model seeks to bring to the fore.)

We also note two main markets—a market for goods and services and a market for factors. In the market for goods and services, buyers and sellers are meeting in immense numbers of submarkets, and the prices of the enormous range of commodities produced by the system are being formed by their encounters. In the market for factors, the services of land, labor, and capital are being bought and sold in innumerable other encounters whose outcome will be the wages and rents and interest that reward the factors of production.

THE TWO MARKETS

Our model gives us a further insight. *Households and firms both participate in each of the two basic markets, but on different "sides" of each market.* In the market for goods, the household is a buyer; in the market for factors, it is a seller, as its members offer their services for hire. In the market for goods, the firm is a seller; in the market for factors, a buyer. Thus we can redraw our first model of the circular flow with supply and demand curves that show the twofold participation of each of the basic participants in the two markets.

Notice in Fig. 7·2 how the household is the source of the demand curve in the market for goods and of the supply curve in that for factors; while the firm lies behind the supply curve in the goods market and the demand curve in the factor market.

Thus, far from being a chaos of buying and selling, the market system is a "seamless web"—a network of transactions with

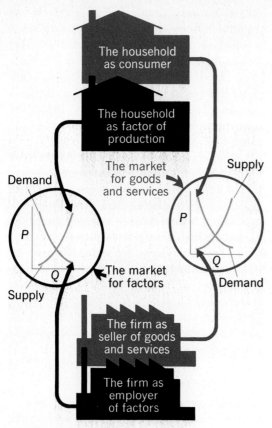

The household as consumer

The household as factor of production

The market for goods and services

Supply

Demand

Supply

The market for factors

Demand

The firm as seller of goods and services

The firm as employer of factors

FIGURE 7 · 2

Demand and supply curve in the circular flow

demand on one market reflected in supply in another, and supply in one market reflected in demand on another. As we will see, this circular flow—this linkage of demand and supply—will be one of the main keys to understanding how the economy works as a whole.

PRODUCTION AND DISTRIBUTION

Last, our model gives us a way of connecting the market mechanism with the economic problem itself; that is, with the production and distribution of goods and services. The market for goods is clearly the central locus where the production decisions of society will be determined. The market for factors is the place where distribution of incomes will take place.

Moreover, the two main institutions, households and firms, also locate key *decision-making points*. Within the household there will be crucial determinations of taste or of the willingness to buy. We have identified these determinations as major elements in demand. Within the firm there will occur very important technical decisions affecting supply. We are generally acquainted with some of the laws of production that will affect those decisions.

Like many models, ours operates at a very high level of abstraction, lumping together the richest investor and the poorest laborer in the market for "factors," and jumbling the sale of caviar and hospital care into one undifferentiated market for "goods and services." Nevertheless, this abstract conception begins to untangle the flux of market activities in the world around us. Moreover, it clarifies our task in the chapters ahead. One by one we must study the two different markets and the two different institutions to find out the actions and motivations characteristic of each. We will begin in our very next chapter by taking a searching look at households and their demand for goods and services.

KEY WORDS

Microeconomics

Macroeconomics

Deficiencies of one
approach alone

Circular flow

Households

Firms

Market for goods
and services

Market for factors

CENTRAL CONCEPTS

1. *Micro- and macroeconomics are two separate approaches to the analysis of economic society. Microeconomics begins its analysis with the motives and actions of individuals and firms; macroeconomics begins with the large-scale movements of the system as a whole.*

2. *Neither a micro nor macro perspective by itself gives a complete picture of the system. A macro approach does not shed much light on the composition of output or the distribution of income. A micro approach does not reveal at what level of income firms and individuals have to carry on their microeconomic pursuits.*

3. *We can picture the operations of the market mechanism as a circular flow. In this flow there are two main institutions—households and firms—and two main markets—a market for goods and services and a market for factors.*

4. *Each institution enters both markets, but on different sides. The household is a buyer in the goods market, a seller in the factor market. The firm enters the goods market from the supply side, the factor market on the demand side. The two markets interact, since the market for factors will determine the income of households, and the market for goods will determine the income of firms.*

5. *Finally, the circular flow shows us where the market system exerts its influences on the economic problem. The market for goods is where a market society determines what goods it will produce. The market for factors is where its incomes are determined. The firm is where laws of production will have to be reckoned with. The household is the center of decision-making about the demand for goods and services.*

QUESTIONS

1. Do you think there is an analog to micro- and macroeconomics in other social sciences? Take politics. What is micropolitics? What is macro-politics? (How about voting behavior versus revolution?)

2. Which of the following pairs do you think would best be analyzed by a microeconomic approach: inflation or high prices of meat; national unemployment or black unemployment; automobile output or small-car output; the total supply of housing or the supply of cooperatives? What is char-acteristic of your choices? Can you add some more examples?

3. What is the difference between the market for goods and for factors? Can a household be a buyer in the market for factors? (Suppose it hires a maid!) Can it be a seller in the market for goods? (Suppose it sells vegetables on its front porch?) *Is a household then really a firm?*

4. What is the interaction between the market for goods and services and the market for factors?

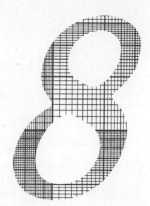

Prices and Allocation

WE HAVE CLEARED UP A LARGE AREA of the mystery about how prices are formed in a market system. Although we have not looked very fully into supply curves (and we cannot until we probe the operations of the firm) we understand in general that prices in the marketplace for goods reflect the interplay of the demand schedules of consumers and the supply schedules of producers. In our next chapter we will see how changes in demand affect prices and how various characteristics of demand exert different influences on the price structure.

But before we turn to the dynamics of supply and demand, there is a further illumination that our understanding of the price mechanism can shed on the problems of microeconomics. It begins to explain to us how the market system solves the problems of allocation. In particular, it clears up the puzzle of how the market system *rations* goods among all claimants and uses for the good.

RATIONING

In one form or another, rationing—or the allocation of goods among claimants—is a disagreeable but inescapable task that every economic system must carry out; for in all societies, the prevailing reality of life has been the inadequacy of output to fill the needs of the people. In traditional economies, we will remember, rationing is performed by a general adherence to rigidly established rules that determine the rights of various individuals to share in the economic product, whether by caste or class or family position or whatever. In command societies, the division of the social product is carried out in a more explicitly directed fashion, as the authorities—lords, priests, kings, commissars—determine the rights of various groups or persons to share in the fruits of society.

A market society, as we know, dispenses with the heavy hand of tradition or the authoritative one of command, but it too must impose some system of rationing to prevent what would otherwise be an impossibly destructive struggle among its citizens. This critical allocative task is also accomplished by the price mechanism. *For one of the prime functions of a market is to determine who shall be allowed to acquire goods and who shall not.*

TABLE 8 · 1

Price	$11	$10	$9	$8	$7	$6	$5	$4	$3	$2	$1
Number willing and able, at above price, to											
buy one unit	0	1	2	3	4	5	6	7	8	9	10
sell one unit	10	9	8	7	6	5	4	3	2	1	0

HOW THE MARKET RATIONS

Imagine a market with ten buyers—each willing and able to buy one unit of a commodity, but each having a different maximum price that is agreeable to him—and ten suppliers, each also willing and able to put one unit of supply on the market, again each at a different price. Such a market might look like Table 8·1.

As we can see, the equilibrium price will lie at $6, for at this price there will be five suppliers of one unit each and five purchasers of one each. Now let us make a graph and let each bar stand for a single individual. The height of the bar tells us the maximum each individual will be willing to pay for the unit of the commodity or the minimum he would sell it for. If we line up our marketers in order of their demand and supply capabilities, our market will look like Fig. 8·1.

What we have drawn is in fact nothing but a standard supply and demand diagram.

FIGURE 8 · 1
How the market rations

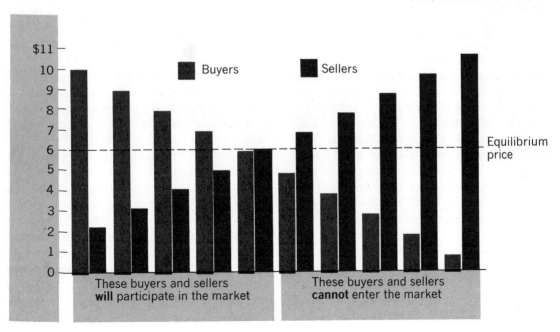

But look what it shows us. All the buyers who can afford and are willing to pay the equilibrium price (or more) will get the goods they want. All those who cannot, will not. So, too, all the sellers who are willing and able to supply the commodity at its equilibrium price or less will be able to consummate sales. All those who cannot, will not.

Thus the market, in establishing an equilibrium price, has in effect allocated the goods among some buyers and withheld it from others, and permitted some sellers to do business and denied that privilege to others. In our previous case in Chapter 4, anyone who could pay $25 or more got a pair of shoes, and all those who could not were unable to get shoes; while all producers who could turn out shoes for $25 or less were able to do business, and those who could not meet that price were unable to make any sales at all.

Note that the market is in this way a means of excluding certain people from economic activity—namely, customers with too little money or with too weak desires, or suppliers unwilling or unable to operate at a certain price.

PRICE VS. NONPRICE RATIONING

The rationing system of the market is both its triumph and its trouble. At the outset of our book we briefly surveyed the problems of nonmarket control mechanisms. In the case of tradition, we remember, the problem is the profound inertia that comes from a static arrangement of economic duties and rewards. In the case of command economies, the problem lies in the difficulty of administering a system without resort to bureaucratic inefficiency on the one hand or dictatorial intervention on the other.

Against these very grave difficulties of other systems, the price system has two great advantages: (1) *it is highly dynamic,* and (2) *it is self-enforcing.* That is, on the one hand it provides an easy avenue for change

to enter the system, while on the other, it permits economic activity to take place without anyone "overseeing" the system.

The second (self-enforcing) attribute of the market is especially useful with regard to the rationing function. In place of ration tickets with their almost inevitable black markets or cumbersome inspectorates or queues of customers trying to be first in line, *the price system operates without any kind of visible administration apparatus or side effect.* The energies that must go into planning or the frictions that come out of it are alike rendered unnecessary by the self-policing market mechanism.

MARKET PROBLEMS

On the other hand, the system has the defects of its virtues. If it is efficient and dynamic, it is also devoid of values: it recognizes no priorities of claim to the goods and services of society except those of wealth and income. If all shared alike or all incomes were distributed in accordance with some universally approved principle, this neutrality of the market would be perfectly acceptable, for then each would enter the market on equal terms or at least with advantages and disadvantages that bore the stamp of social approval. But when inheritance still perpetuates large fortunes made in the past, and unemployment or old age can bring extreme deprivation, the rationing results of the market often affront our sense of dignity.

Therefore every market society interferes to some extent with the "natural" outcome of the price rationing system. In times of emergency, it issues special permits that take precedence over money and thereby prevents the richer members of society from buying up all the supplies of scarce and costly items. In depressed areas, it may distribute basic food or clothing to those who have no money to buy them. And to an ever-increasing extent it uses its taxes and transfer payments to

RATIONING BABIES

Because the market is such an efficient distributive mechanism, it has been proposed as a means to achieve Zero Population Growth, assuming that this were the declared national policy. Since a sizeable minority (probably about 15 percent) of all families voluntarily choose to have no children or only one, a country can achieve ZPG even if some families have more than two children. The question is how to decide which families should be allowed to have the extra children? Professor Kenneth Boulding has ventured an answer that leans heavily on the market mechanism. He proposes that each girl and boy at adolescence be given 110 green stamps, of which 100 are required if a woman is to have a legal child. (The penalty for having an illegal child would be very severe, possibly even sterilization.) Unwanted or surplus stamps would then be sold in a market organized for that purpose. It can be seen that the total number of stamps would permit the population as a whole to have 2.2 children per family— the ZPG rate. The market would therefore serve to ration the extra stamps, making them available to those with higher incomes or a greater desire for children. "As an incidental benefit," writes Boulding, tongue in cheek, "the rich will have loads of children and become poor, and the poor will have few children and become rich."

When this scheme was first published, it provoked a storm of criticism. Commenting on its reception, Boulding observes: "This modest and humane proposal, so much more humane than that of Swift, who proposed that we eat the surplus babies, has been received with so many cries of anguish and horror that it illustrates the extraordinary difficulty of applying rational principles to processes involving human generation."* What do you think of the idea?

*Kenneth E. Boulding, *Economics as a Science* (New York: McGraw-Hill, 1970), p. 39.

redistribute the ration tickets of money in accordance with the prevailing sense of justice.

SHORTAGES

Our view of the price system as a rationing mechanism helps to clarify the meaning of two words we often hear as a result of intervention into the market-rationing process: *shortage* and *surplus*.

What do we mean when we say there is a *shortage* of housing for low income groups? The everyday meaning is that people with low incomes cannot find enough housing. Yet in every market there are always some buyers who are unsatisfied. We have previously noted, for instance, that in our shoe market, all buyers who could not or would not pay $25 had to go without shoes. Does this mean there was a shoe "shortage"?

Certainly no one uses that word to describe the outcome of a normal market, even though there are always buyers and sellers who are excluded from that market because they cannot meet the going price. Then what does a "shortage" mean? *We can see now that it usually refers to a situation in which the price has been fixed by some nonmarket agency, such as the government, below the equilibrium price.*

Figure 8·2 shows us such a situation. Note that at the price established by the government, the quantity demanded is much greater than the quantity supplied. If this were a free market, the price would soon rise to the equilibrium point, and we would hear no more about a shortage. But so long as the price is fixed at the ceiling, this equilibrating process cannot take place. Thus the quantity demanded will remain larger than the quantity supplied, and some buyers will go unsatisfied *even though they are willing and able to pay the current price.*

This bears directly on the problem of

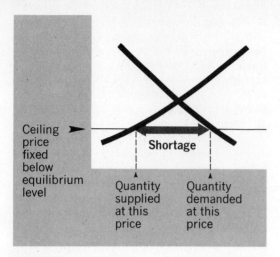

FIGURE 8 · 2
Shortages

price controls that we will discuss in Chapter 32. The problem with such controls is that they tend to fix prices that are below the level that would be established in a free (inflationary) market. As a result, some buyers who would ordinarily have been "priced out" of the market remain *in* the market, although there are not enough goods offered to satisfy their demands. The result tends to be queues in stores to buy things before they are gone, under-the-counter deals to get on a "preferred list," or black or gray markets selling goods illegally at higher prices than are officially sanctioned.

SURPLUSES

The opposite takes place with a surplus. Here, in Fig. 8·3, we see a price floor fixed above the equilibrium price, as when the government supports a crop above its free market price.

In this situation, the quantity supplied is greater than that demanded (note that we should *not* say that "supply" is greater than "demand"). In a free market, the price would

fall until the two quantities were equal. But if the government continues to support the commodity, then the quantity bought by private industries will not be as large as the quantity offered by farmers. The unsold amounts will be a "surplus," bought by government.

Thus the words shortage *and* surplus *mean situations in which there are sellers and buyers who are willing and able to enter the market at the going price but who remain active and unsatisfied because the price mechanism has not eliminated them.* This is very different from a free market where there are unsatisfied buyers and sellers *who cannot meet the going price* and who are therefore not taken into account. Poor people have no demand for fresh caviar at $60 per pound and therefore do not complain of a caviar shortage, but if the price of fresh caviar were set by government decree at $1 a pound, there would soon be a colossal "shortage."

What about the situation with low cost housing? Essentially what we mean when we talk of a shortage of inexpensive housing is that we view the outcome of this particular

FIGURE 8 · 3
Surpluses

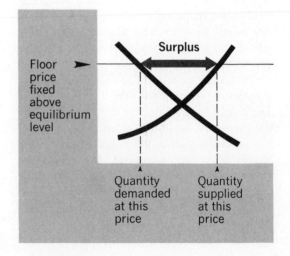

market situation with noneconomic eyes and pronounce the result distasteful. By the standards of the market, the poor who cannot afford to buy housing are simply buyers at the extreme lower right end of the demand curve, and their elimination from the market for housing is only one more example of the rationing process that takes place in *every* market. When we single out certain goods or services (such as doctor's care or higher education) as being in "short supply," we imply that we do not approve of the price mechanism as the appropriate means of allocating scarce resources in these particular instances. This is not because the market is not as efficient a distributor as ever. It is because in these instances we feel that the underlying distribution (or maldistribution) of income causes the outcome of the market rationing process to clash with other standards of the public interest that we value even more highly than the standard of efficiency.

RESERVATION PRICES

Now take another look at Fig. 8·1, which we reproduce here for convenience.

Let us call the *maximum amount* that a buyer is willing and able to spend for a unit of a good his *reservation price*. The first consumer on the left has a reservation price of $10; the last one on the right has a reservation price of $1. We can also call the *minimum amounts* that sellers demand their reservation prices: the first seller on the left has a reservation price of $2; the last one on the right has one of $11.

Now look at the difference between these reservation prices and the actual market price. The consumer on the left, who would have been willing to pay $10 for the commodity, gets it for only $6; he has a *consumer's surplus* of $4. The first seller on the left, who would have been willing to sell at $2 and who actually sells at $6, has a *producer's surplus*

FIGURE 8·1 (repeated)
How the market rations

RATIONING WITHOUT TEARS

Although we now understand that the price system is a rationing system, when we say "rationing" we usually mean a system of coupons or publicly determined priorities. For example, if there were a permanent shortage of gasoline—meaning that at *going* prices, the quantity of gas sought would be larger than the quantity offered—we might ration by allowing each car owner an equal amount or by assuring that certain vehicles, such as ambulances, always had first crack at supplies.

No sooner do we begin to think about rationing by coupon or by priority than we begin to see the complexity of the problem. Clearly, the purpose of rationing is to prevent rich people from riding about in Cadillacs while poor people can't afford the gas to ride to work in their Volkswagens. But imagine that you were in charge of nonprice rationing. Suppose that the number of gallons of gas expected to be available were 100 billion. Would you now determine the basic ration by dividing this number by the population, giving each person an equal allotment? That would enormously benefit a family with one car and many children, and penalize a single person who might desperately depend on his car. And what would a family do if it got its coupons but

of $4. Notice also that every consumer who can afford to buy gets some consumer's surplus, except for the marginal (last) consumer, who is forced to pay his full reservation price. The same is true for sellers. All get producers' surpluses, except for the marginal seller who can just meet the market clearing price of $6.

MAXIMIZING CONSUMERS' AND PRODUCERS' SURPLUSES

Now we can see two things the market has accomplished, in addition to rationing the good and establishing an equilibrium price for it. First, *it has rationed the good to those consumers who get the largest consumer surplus from it.* As we can see from our diagram, those consumers to whom the good is not "worth" $6 are not given any of the good; and only those whose estimation of the good is equal to or above its market price will get it.

There is a very important warning that we must quickly and forcefully interject here.

Although our market maximizes consumers' surpluses, we cannot say that it establishes the "best" allocation of goods, because the ability to buy the good varies according to the income of each consumer. With a different initial distribution of income, a different line-up of consumers with different reservation prices would almost certainly occur. Hence consumers' surplus is maximized, *given the initial distribution of income.* But this is not to say that the initial distribution is a good one. In Chapter 18, we will come back to the crucial question of what that initial distribution might be.

Second, note that *the market has also maximized producers' surpluses,* because those producers who enjoy a surplus are those whose costs are below $6. The market, therefore, permits only the most efficient sellers or producers to supply goods. It thereby serves as an agency for assuring the *efficiency* of production, since it caters to sellers who are able to provide the good with the least cost, which usually means the smallest use of society's resources.

did not own a car? Or would you perhaps ration supplies per car owner, rather than per person? Here, of course, the trouble is that you would be giving the same allotment to all car owners, without knowing their respective needs. Some owners, such as Hertz and Avis, would be desperate for supplies. Other owners, who hardly used their cars, would not need all their coupons.

Or might these very difficulties prompt you to follow a scheme that resembles Boulding's proposal for rationing babies? Suppose you issued to each adult a book of coupons entitling him to his basic allotment of gallons, and *suppose that you allowed individuals to buy or sell these coupons!* To be sure, rich citizens would now be in a posi-

tion to buy up coupon books, but no poor citizen would have to sell his book. If he needed his basic allotment, he would keep his coupons. If he did not need his allotment, he could supplement his income by selling it.

The point of such a plan is to use the market as a means by which individuals can determine their own economic activities according to their marginal utilities, and to combine that use with the overall fairness that a market may not attain. The ration books would insure a basically democratic sharing of one part of the national wealth, but they would permit individuals to maximize their surpluses in a way that rationing alone would not.

RESERVATION PRICES AND MARGINAL UTILITIES

Aren't reservations very much like marginal utilities? Indeed they are, but there is an important difference. When we discussed utilities, we more or less evaded the question of how we would *measure* those utilities. The obvious way would be in money: the utility of the first bottle of wine being equal to $10, the second to $9, etc. Note that in our previous examples, however, we didn't use prices. The reason is that the *marginal utility of money also changes as we spend it and therefore have less of it,* so that trying to measure the utilities of goods with money is like trying to measure something with a ruler whose length is changing.

By speaking of reservation prices, we get around that problem. Reservation prices are prices that we would actually pay for the first, second, or *n*th unit of a good. Behind these prices lurks the psychological fact of utility, but we manage to evade problems of measuring utility by talking in terms of

prices. We can think about *motivations* in terms of utilities; we describe *actions* in terms of reservation prices.

MAXIMIZING INDIVIDUAL CONSUMERS' SURPLUS

We have shown how a market maximizes consumers' surpluses in a market composed of different individuals with different reservation prices (or differing marginal utilities) competing for one good. Now let us see how the market mechanism also maximizes the consumer's surplus (or total utilities) of one individual who shops in many markets for many goods.

An intuitive example may help us begin. Suppose that you had to spend your weekly income each Monday, but that you had to make up your shopping list, once and for all, before leaving your house. If you had enough price catalogs, that would not be impossible to do, although you might debate the merits of this item versus that one. *But suppose you had to make up the list without knowing what prices were!*

Two problems would present themselves. First, you would not know how many goods you could buy, *in toto,* because you would not know whether your income would suffice to buy few goods or many. Second, you would have no way of "ranking" the priority of your purchases. Knowing the prices of bread and cake, you can decide how much you want to spend on each. But not knowing these prices, how could you make a rational decision whether to buy many units of bread and no cake, or fifty-fifty, or some other combination?

You might think, perhaps, that a "rational" man would buy bread first, then cake. But suppose after he had made his irrevocable decision he found that bread was very expensive and cake very cheap. He might then regret having decided to buy so much bread and dearly wish he had chosen cake instead.

THE OPTIMUM ALLOCATION OF INCOME

This seemingly trivial example contains more than may at first meet the eye, for it shows

us how the existence of prices enables us to behave as rational maximizers in disposing of our incomes. Therefore let us pursue our line of reasoning a little further.

In Fig. 8·4 we show our reservation prices for three commodities. In each case, our reservation price for another unit of the same good diminishes because the good gives us less marginal utility. At the same time, as the diagram makes clear, the schedule of reservation prices is very different for each good. Good A is very important to us, so our initial reservation price is very high; Good B less so; Good C still less. (We have drawn our reservation prices in steplike fashion and overlaid a generalized schedule of reservation prices, which is, of course, our familiar demand curve.)

MARGINS VS. TOTALS

The question we want to elucidate is this: *how much of each good will we buy to get the largest possible satisfaction from our income?*

Suppose we had an unlimited income. This is the same thing as supposing that the goods were free, that their prices were zero. How much of Goods A, B, and C would we then acquire? An unlimited amount? Cer-

FIGURE 8 · 4
Allocating income

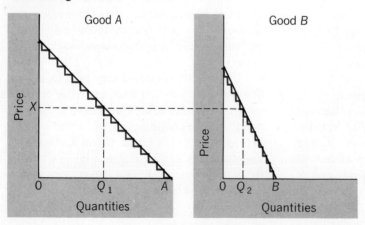 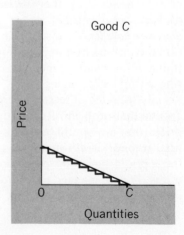

tainly not. As our diagram shows, we don't want unlimited quantities of A, B, and C. Beyond a certain point, their marginal utilities are negative. They are nuisances. We could even have negative reservation prices: we would pay someone to take the stuff away. Thus, with no budget constraint we consume quantities $OA + OB + OC$.

Now notice something interesting about this unlimited consumption. The three demand curves reflect differing marginal utilities of the three goods. Looking at these curves, we see that the *total* utility we get from A will be greater than that we get from B or C, and that the *total* utility of B will be greater than that of C. Why then don't we take more of A and B, since their total utility is so large? The answer, also apparent from the graph, is that after we have acquired quantity OA of Good A *we don't get any further utility from it*. The same is true of B after we have OB of it, and of course of C, after OC.

This begins to clarify a very important point. To get the maximum amount of enjoyment from our income—even from an *unlimited* income—we need pay no attention to the total utilities we get from various commodities. Their marginal utilities are all we need to know. We will reach a maximum of satisfaction from our total expenditure when we get as much utility from the marginal unit of one good as from another. *Indeed, the rule for maximizing our total satisfaction is to acquire goods until their marginal utilities per dollar of expenditure are equal.*

THE EQUIMARGINAL RULE

This equimarginal rule has many applications in economics, for it has an astonishing property. It means that we don't have to stop to compare totals when we maximize values. *We need compare only margins.* Later we will see that this applies to entrepreneurs trying

to maximize total revenues and to minimize total costs; they too will need to look at only marginal costs and incomes. For ourselves as consumers, it means that we do not have to try to compare whether we get more "total" satisfaction out of bread or out of cake. All we have to do is worry about whether we want one more loaf of bread or one more piece of cake. If we buy whichever we want more of at the moment, we will automatically be maximizing our total well-being. When we equally desire another unit of each, we have spent our income as efficiently as possible.

BUDGET CONSTRAINTS

But we have so far imagined that we had no budget constraint. Of course we do have such constraints. Our incomes are limited, and prices are not zero. Then how does the equimarginal principal apply?

If you will turn back to Fig. 8·4 you will see an X on the price axis of Good A. If you draw a line parallel to the quantity axis all the way across the diagram, we can imagine that this is the price of Goods A, B, and C. We picture them having the same price: e.g., $5 for a basket of fruit (Good A), a necktie (Good B), a movie ticket (Good C). (We could draw different prices for each good, but that would only complicate the diagram without changing the principle.)

Now how much of A, B, and C do we buy? If you will drop a line from the point where price intersects the demand curve, you will see that we buy (approximately) OQ_1 of Good A, and OQ_2 of Good B. We buy none of Good C. Why? Because the price is higher than our top reservation price. We don't want to go to the movies at $5, given our budget constraint.

Now look at Goods A and B. You obviously have much more consumer surplus from A than from B. Why then, don't you buy more of A and less of B? The answer is that you are getting as much satisfaction *at the margin* for Good A as for Good B. If you bought another unit of Good A, and one less unit of Good B, you would be giving up more consumer's surplus than you would be getting. *Thus budget constraints limit the amount of goods we can buy, but we still maximize our well-being by seeking equal marginal utilities from those we buy.*

EQUALIZING MARGINAL UTILITIES

Now let us take one last step. We have just seen that we maximize our personal well-being by equalizing the marginal utilities of goods, not their total utilities. *We can then see that we will spend our income optimally when we get the same satisfaction from a dollar spent on each good.* If the marginal utility of a dollar's worth of bread is equal to that of a dollar's worth of cake, we have obviously achieved our aim.

But when we speak of "a dollar's worth" of bread, we are speaking of its price. Therefore we can set up a formula that will describe the way we allocate our incomes to maximize our satisfactions.

$$\frac{\text{Marginal Utility of Good A}}{\text{Price of Good A}} = \frac{\text{Marginal Utility of Good B}}{\text{Price of Good B}}$$

or in slightly more abstract terms:

$$\frac{MU_1}{P_1} = \frac{MU_2}{P_2} = \frac{MU_n}{P_n}$$

where MU stands for the marginal utilities of Goods 1, 2 ... n, and P stands for their respective prices.

Here is the equimarginal principle at work. We are maximizing our well-being by equating the *marginal* utilities of different goods in proportion to their prices, so that each dollar of expenditure for each good

FIGURE 8 · 5
The equimarginal principle

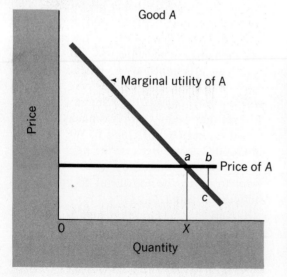

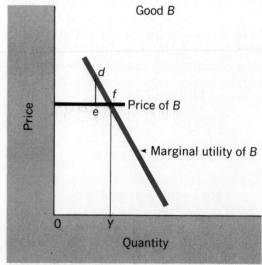

gives us the same enjoyment. We may still get a much larger amount of enjoyment from one kind of good than from another, but we will only decrease our total welfare if we lose sight of the equimarginal principle.

For one last demonstration of this, examine Fig. 8-5. It shows us two goods, A and B. A is very cheap and gives us much total pleasure. B is dearer and provides less total satisfaction but more marginal satisfaction per unit. At their respective prices, we buy *OX* of Good A and *OY* of Good B. Should we buy more A and less B? If we did, we would incur the consumer's loss (negative surplus) shown by triangle *abc* and forego the potential consumer's surplus shown by triangle *def*. We are obviously best off when we follow the equimarginal rule.

A TAUTOLOGY?

Do we *really* use prices to maximize our consumers' surpluses? Do we really follow the equimarginal principle when we push a grocery cart through the supermarket aisles?

In point of fact, we are faced here with a tautology—a statement that cannot be demonstrated empirically. We cannot measure marginal utilities, so we cannot divide them by prices to see if equalization is occurring. Therefore, there is really no way of proving that we follow the marginal rule, nor any way of teaching us how to do so. Nonethe-

less, we get a certain intellectual insight in discovering that we tend to be rational maximizers, dutifully trying to follow the equimarginal rule.

For who would deny that we experience something that the words *consumer surplus* describe very well? Who has not gone to buy something that he expected to pay $10 for, found it on sale, and felt at firsthand the pleasures of that little triangle *def?** Who has not bought something on impulse "because it was cheap" and later regretted it—a clear case of *abc?* However imperfectly, it seems that we do try to spend our limited incomes so that we will get about as much pleasure from the last dollar spent for food as for clothes or entertainment or whatever. Indeed, when we go to spend our incomes, what other guide can we use than the marginal utilities that we will get from this versus that, compared with their prices?

Thus the idea of maximizing our individual well-being by using prices to allocate our incomes according to the equimarginal principle seems plausible. Although it remains a tautology, the hypothesis is a fruitful one, and economics has built on it a powerful theory of how we behave as consumers and why.

---

*In this case, we expected to buy *OY* of Good B at a price *d*, rather than *e*.

KEY WORDS

Allocatory
function

Rationing

Self-enforcing

Shortage

Surplus

Reservation prices

Consumers'
surplus

Producers'
surplus

Maximizing
surpluses

Marginal utility
vs. reservation
price

Equimarginal rule

Budget constraint

The equimarginal
formula

CENTRAL CONCEPTS

1. Prices perform an *allocatory function*, *rationing goods* among those buyers and sellers who are willing and able to enter the market at the going price. This rationing system is *highly dynamic* and is *self-enforcing*. At the same time, it recognizes no distributive principles or values except those of wealth.

2. *Shortages and surpluses* refer to situations in which prices have been imposed on a market and are below or above the equilibrium price. We do not count as victims of shortage or surplus the unsatisfied buyers or sellers we encounter in an equilibrium market. When we do label a situation as "shortage" or "surplus," it means that we do not approve of the market system as an appropriate rationing device for that good or service.

3. Reservation prices are the *maximum prices* a consumer will pay for another unit of a good, or the *minimum price* a producer (seller) will charge for a good.

4. *Consumers' surplus* refers to the difference between the top price a consumer would pay for a good and the price he finds on the market. *Producers' surplus* is the difference between a producer's reservation price and the selling price established by the market.

5. The market allocates goods in such a way that *consumers' surpluses are maximized.* This does not mean that social welfare is maximized, because a different *distribution of income* would establish different reservation prices. *The market also maximizes producers' surpluses by "assigning" production to low-cost producers.*

6. Reservation prices are actual prices that we will pay (or accept) to enter a market. They reflect marginal utilities (on the part of the demanders). Marginal utilities cannot be measured in the way we can measure reservation prices, because they refer to "psychic" well-being. This well-being has no standard of measure, even money, because the marginal utility of money changes as we spend it.

7. Prices enable us to allocate our incomes in order to maximize the total satisfaction of the goods we buy. We maximize our total satisfaction by buying goods only up to the point where we get equal satisfaction from the last unit of each good. By following the *equimarginal rule*, we maximize our total well-being.

8. Even if we had unlimited incomes, we would acquire goods only up to the point at which their equimarginal gains were the same. In that case, the last unit would yield zero marginal (additional) utility. Given a *budget constraint*, we buy fewer goods, but we still maximize our total utilities by allocating our purchases so that the marginal utilities of the goods we buy are the same.

9. We can generalize the equimarginal rule in terms of the formula:

$$MU_1/P_1 = MU_2/P_2 = MU_x/P_x$$

10. The concept of consumer's surplus, like that of utility, is a tautology, a statement not susceptible of empirical demonstration. But it seems to accord with our introspective understanding of how we actually behave.

QUESTIONS

1. Why is rationing an inescapable problem in our economic society? Is it inescapable in all societies? Traditional ones? How is it solved there?

2. Explain how the market rations a commodity like automobiles. What other means of allocating autos could we devise? What are the advantages and disadvantages of a market rationing process?

3. Under what circumstances is the market *not* regarded as a satisfactory rationing mechanism?

4. If income distribution were determined by majority vote, do you think there would ever be a public demand for nonmarket rationing?

5. What do we mean when we say there is a shortage of low-cost housing? Is there a shortage of high-cost housing?

6. Do you have reservation prices for, say, gasoline? What would be your reservation price for 10 gallons per month? Twenty? One hundred? Is this the same as your demand curve for gas?

7. If your reservation price for a certain movie that would be shown only once in your neighborhood were $5 and the admission price were $2, what would be your consumer's surplus? Does that mean you would see the movie *twice*?

   HINT: Your reservation price for the second showing would probably be a lot lower than for the first.

8. Explain how a market maximizes consumer surpluses. How does income distribution enter into this picture? Does income distribution enter into the determination of producers' surpluses?

9. How can *total* consumer surplus be so much larger for a commodity such as water than for one such as wine? Why do we pay attention to the relation only at the margin, and not to total consumer surplus?

The Market in Movement

WE HAVE LEARNED HOW EQUILIBRIUM PRICES emerge from the wholly unsupervised interaction of competing buyers and sellers; and we have seen how those prices, once formed, silently and efficiently perform the necessary social task of allocating goods among buyers and sellers. Yet our analysis is still too static to resemble the actual play of the marketplace, for one of the attributes of an equilibrium price, we remember, is its lasting quality, its persistence. But the prices around us in the real world are often in movement. How can we introduce this element of change into our analysis of microeconomic relations?

The answer is that the word equilibrium does not imply changelessness. Equilibrium prices last only as long as the forces that produce them do not change. To put it differently, if we want to explain why any price changes, we must always look for changes in the forces of supply and demand that produced the price in the first place.

Shifts in Demand and Supply

But what makes supply and demand change? If we recall the definition of those words, we are asking: What might change our willingness or ability to buy or sell something at any given price? Having asked the question, it is not difficult to answer it. If our incomes rise or fall, that will clearly alter our *ability* to buy. Similarly, a change in our tastes will change our *willingness* to buy. On the sellers' side things are a bit more complicated. If we are talking about factors of production, once again changes in incomes or tastes will change our ability and willingness, as owners of labor-power, or land, or capital, to offer these services on the market. If we are talking about firms, changes in *cost* will be the main determinant. We will study these changes when we turn to the firm in later chapters.

SHIFTS IN CURVES VS. SHIFTS ALONG CURVES

Thus changes in taste or attitudes or in income or wealth will shift our whole demand schedule, and the same changes, plus any change in costs, will shift our whole supply schedule.

Note that this is very different from a change in the quantity we buy or sell when *prices* change. In the first case, as our willingness and ability to buy or sell is increased or diminished, *the whole demand and supply schedule (or curve) shifts bodily*. In the second place, when our basic willingness and ability is unchanged, but prices change, our schedule

FIGURE 9 · 1

Changes in quantities demanded or supplied vs. changes in demand or supply

CHANGES IN QUANTITIES
DEMANDED OR SUPPLIED vs . . . . . . CHANGES IN DEMAND OR SUPPLY

A change in price alone changes the QUANTITY we demand or supply

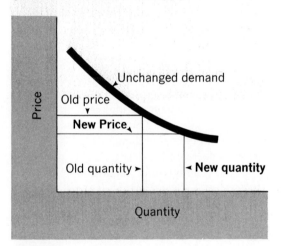

A change in our willingness or ability changes our whole DEMAND SCHEDULE

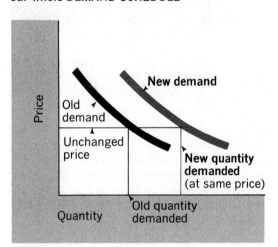

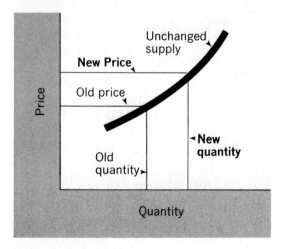

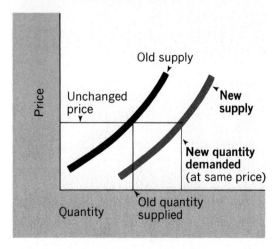

(or curve) is unchanged, but *we move back or forth along it*.

Here are the two cases to be studied carefully in Fig. 9·1. Note that when our demand schedule shifts, we will buy a *different amount at the same price*. If our willingness and ability to buy is enhanced, we will buy a larger amount; if they are diminished, a smaller amount. Similarly, the quantity a seller will offer will vary as his willingness and ability are altered. Thus demand and supply curves can shift about, rightward and leftward, up and down, as the economic circumstances they represent change. In reality, these schedules are continuously in change, since tastes and incomes and attitudes and technical capabilities (which affect costs and therefore sellers' actions) are also continuously in flux.

PRICE CHANGES

How do changes in supply and demand affect prices? We have already seen the underlying process at work in the case of shoes. Changes in supply or demand will alter the *quantities* that will be sought or offered on the market at a given price—an increase in demand, for instance, will raise the quantity sought. Since there are not enough goods offered to match this quantity, prices will be bid up by unsatisfied buyers to a new level at which quantities offered and sought again balance. Similarly, if supply shifts, there will be too much or too little put on the market in relation to the existing quantity of demand, and competition among sellers will push prices up or down to a new level at which quantities sought and offered again clear.

In Fig. 9·2, we show what happens to the equilibrium price in two cases—first, when demand increases (perhaps due to a sudden craze for the good in question); second, when demand decreases (when the craze is over). Quite obviously, a rise in demand—other things being equal—will cause prices to rise; a fall will cause them to fall.

We can depict the same process from the supply side. In Fig. 9·3, we show the impact on price of a sudden rise in supply and of a fall. Again the diagram makes clear what is intuitively obvious: an increased supply (given an unchanging demand) leads to lower prices; a decreased supply to higher prices.

And if supply and demand *both* change?

FIGURE 9 · 2
Shifts in demand change equilibrium prices

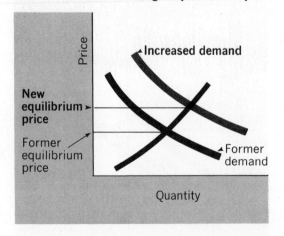

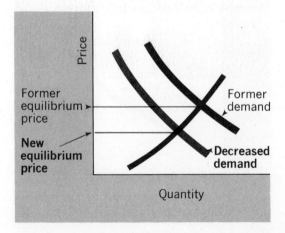

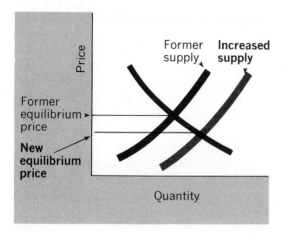

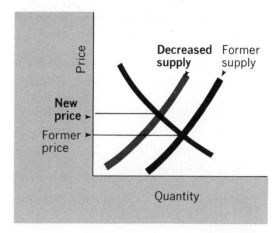

FIGURE 9 · 3
Shifts in supply change equilibrium prices

FIGURE 9 · 4
How shifts in both supply and demand affect prices

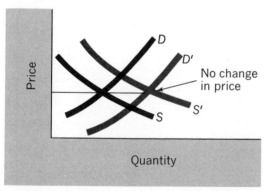

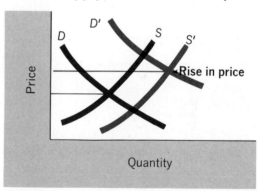

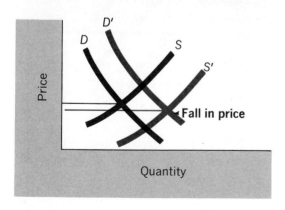

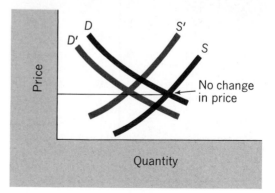

Then the result will be higher or lower prices, depending on the shapes and new positions of the two curves—that is, depending on the relative changes in the willingness and ability of both sides. Figure 9·4 shows a few possibilities, where *S* and *D* are the original supply and demand curves, and *S'* and *D'* the new curves.

LONG AND SHORT RUN

There is one point we should add to conclude our discussion of supply and demand. Students often wonder which "really" sets the price—supply or demand. Alfred Marshall, the great late-nineteenth-century economist, gave the right answer: *both do,* just as both blades of a scissors do the cutting.

Yet, whereas prices are always determined by the intersection of supply and demand schedules, we can differentiate be-

tween the *short run* when demand tends to be the more dynamic force, and the *long run,* when supply is the more important force. In Fig. 9·5 we see (on the left) short-run fixed supply, as in the instance of fishermen bringing a catch to a dock. Since the size of the catch cannot be changed, the supply curve is fixed in place, and the demand curve is the only possible dynamic influence. The broken lines show that changes in demand alone will set the price.

But now let us shift to the long run and draw a horizontal supply curve representing the average cost of production of fish (and thus the supply price of fish) in "the long run." Fluctuations in demand now have no effect on price, whereas a change in fishing costs that would raise or lower the supply curve would immediately affect the price.

In both cases, do not forget, *both* demand and supply enter into the formation of

FIGURE 9 · 5
Short- and long-run supply curves

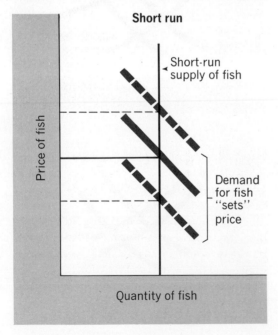

Short run

Short-run supply of fish

Price of fish

Demand for fish "sets" price

Quantity of fish

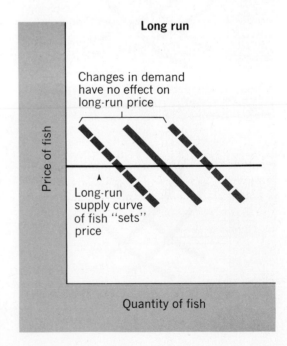

Long run

Changes in demand have no effect on long-run price

Price of fish

Long-run supply curve of fish "sets" price

Quantity of fish

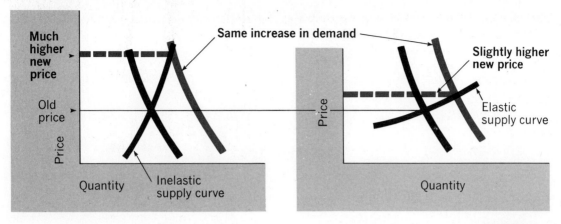

FIGURE 9 · 6
How elasticity of supply curve affects price

FIGURE 9 · 7
How elasticity of demand curve affects price

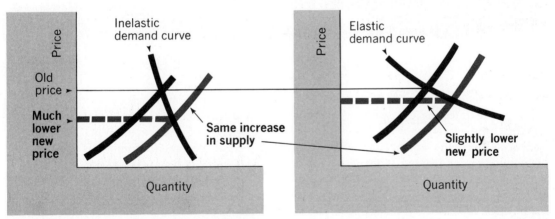

price. But in the short run, as a rule, changes in demand are more likely to affect changes in prices, whereas in the long run, changes in supply are apt to be the predominant cause of changes in price.

Elasticity

We have seen how shifts in demand or supply affect price. But *how much* do they affect price? Suppose, for example, that demand schedules have increased by 10 percent. Do we know how large an effect this change will have on price?

These questions lead us to a still deeper scrutiny of the nature of supply and demand, by way of a new concept called *elasticity* or, more properly, *price elasticity*. Elasticities describe the shapes of supply and demand curves and thereby tell us a good deal about whether a given change in demand or supply

INCOME ELASTICITIES

We should notice that we can use another term—*income elasticity*—to describe how our willingness or ability to buy or sell responds to a change in *income*, rather than price. With many commodities, income elasticities of both demand and supply are more significant than price elasticities in actual economic life.

The idea of income elasticity is exactly the same as price elasticity. Sales of an income-elastic good or service rise proportionately *faster* than income. Sales of an income-inelastic commodity rise *less than proportionately* with income. These relationships are graphed in the accompanying figure.

Do not be fooled into thinking that these are supply curves because they slope upwards. "Income" demand curves show a functional relation different from that of "price" demand curves. In the curves shown, we assume that prices are unchanged, otherwise we would not have *ceteris paribus*.

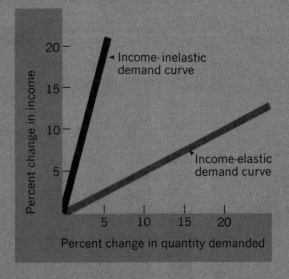

As an exercise, try drawing an income-elastic and an income-inelastic supply curve.

will have a small or large effect on price. Figure 9·6 illustrates the case with two supply curves. Our diagrams show two commodities selling at the same equilibrium prices and facing identical demand schedules. Note, however, that the two commodities have very different supply curves. In both cases, demand now increases by the same amount. Notice how much greater is the price increase for the good with the inelastic (steep) supply curve.

Similarly, the price change that would be associated with a change in supply will be greater for a commodity with an inelastic demand curve than for one with an elastic (gently sloping) demand curve. Figure 9·7 shows two identical supply curves matched against very different demand curves. Notice how the commodity with inelastic demand suffers a much greater fall in price.

Elasticities are powerful factors in ex-

plaining price movements, because the word "elasticity" refers to our sensitivity of response to price changes. What we mean by an elastic demand (or supply) is that, as buyers or sellers, our willingness or ability to buy or sell is strongly affected by changes in price, whereas when our schedules are inelastic, the effect is small. In more precise terms, *an elastic demand (or supply) is one in which a given percentage change in prices brings about a larger percentage change in the quantity demanded (or supplied).* An inelastic schedule or curve is one in which the response in the quantities we are willing and able to buy or sell is proportionally less than the change in price.

It helps if we see what elasticities of different kinds look like. Figure 9·8 is a family of supply and demand curves that illustrates the range of buying and selling responses associated with a change in prices.

FIGURE 9 · 8

A family of supply and demand curves

TOTALLY INELASTIC DEMAND OR SUPPLY. The quantity offered or sought is unchanged despite a change in price. Examples: Within normal price ranges there is probably no change at all in the quantity of table salt bought. Similarly, a fisherman landing a catch of fish will have to sell it all at any price within reason.

INELASTIC DEMAND OR SUPPLY. Quantity offered or sought changes proportionately less than price. Examples: We probably do not double bread purchases if the price of bread halves. On the supply side, the price of wheat may double, but farmers are unable (at least for a long time) to offer twice as much wheat for sale.

UNIT ELASTICITY. This is a special case in which quantities demanded or supplied respond in exact proportion to price changes. (Note the shape of the demand curve, a rectangular hyperbola.) Examples: Many goods may fit this description, but it is impossible flatly to state that any one good does so.

ELASTIC DEMAND OR SUPPLY. Price changes induce proportionally larger changes in quantity. Examples: Many luxury goods increase dramatically in sales volume when their price is lowered. On the other side, elastic supply usually affects items that are easy to produce, so that a small price rise induces a rush for expanded output.

TOTALLY ELASTIC DEMAND OR SUPPLY. The quantity supplied or demanded at the going price is "infinite." Examples: This seemingly odd case turns out to be of great importance in describing the market outlook of the typical small competitive firm. Merely as a hint: For an individual farmer, the demand curve for his output at the going price looks horizontal because he can sell all the grain he can possibly grow at that price. A grain dealer can also buy all he wants at that price.

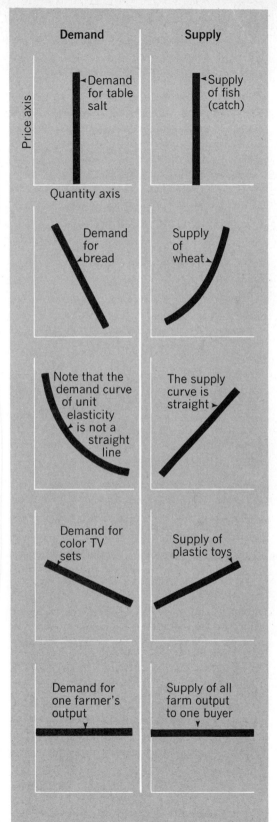

CAUTIONS ABOUT CALIBRATION

There are two important cautions about the family of curves in Fig. 9·8. The first is that we have omitted "calibrating" the scales with actual units. We have done so purposely, to alert you to a problem. A curve can "look" elastic or inelastic, but can in fact not be what it looks like if the calibration is tricky.

Here is an example. Figure 9·9 "looks" like a very elastic demand curve.

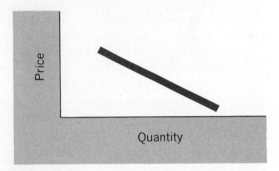

FIGURE 9 · 9
Noncalibrated curves

But suppose that the calibration shows the units to be those in Fig. 9·10.

Now we can see that at a price of $2, the quantity demanded is 10 units, and at a price

FIGURE 9 · 10
Tricky calibrations I

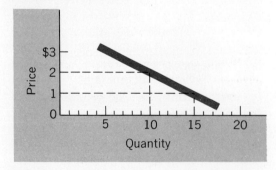

of $1 it is 15 units. In other words, a 50 percent fall in price has been accompanied by the *same percentage increase* in quantity demanded. That is unit elasticity of demand, not "elastic" demand. Further, if you sketch in with a pencil the result of dropping the price from $1 to 50¢, you will see that a further fall of 50 percent in price brings an increase in quantity demanded from 15 units to something less than 20. This is a *smaller proportionate price increase* and therefore reveals an inelastic demand.

For practice, look at Fig. 9·11. The demand curve looks highly inelastic. Is it?

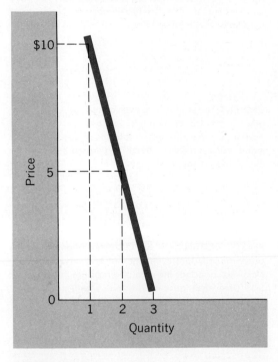

FIGURE 9 · 11
Tricky calibrations II

ARC AND POINT ELASTICITY

The second caution is actually closely connected with the first. It is *that the elasticity of*

MORE ON ELASTICITIES

For rough and ready purposes, we speak of steeply sloping demand curves as *inelastic* and of gradually sloping ones as *elastic*, and similarly for supply curves.

But as economic analysts, we should note that all sloping demand curves can be divided into elastic and inelastic portions. Take the curve below. We show that above the point of unitary elasticity it is an elastic demand curve; below the unitary point, an inelastic one. You should prove this by figuring out the elasticity for, say, a fall in price from $9 to $8; and again from $2 to $1.

If you have made these calculations, you

should be able to see why the following formula is used for elasticity:

$$n = -\frac{\Delta Q/Q}{\Delta P/P}$$

In this formula, n stands for "elasticity"; ΔP for the *change* in price and ΔQ for the change in quantity. Thus the formula shows us the *percentage* change in quantity, in relation to a *percentage* change in price. (The Greek letter Δ, *delta*, stands for "change in." Note that we use the initials P and Q as denominators in our formula.)

Why does the formula have a minus sign in front of it? The reason is that a *decrease* in price is associated with an *increase* in quantity demanded. Dividing the positive quantity change by the negative price change will give us a "minus" answer. For convenience, we therefore put a minus sign in front of the equation, to make the answers positive instead of negative numbers. Any number larger than 1 therefore stands for an elastic price response; any number less than 1, for an inelastic price response. For example, an elasticity coefficient of -2 means that a 1 percent price rise would cause a 2 percent decrease in quantity demanded.

Final note. Elasticity works just as well going up as coming down. We have used illustrations showing the change in Q_d associated with a fall in price. The same principles and the same formula apply to a rise in price and a fall in quantity. Try working your previous examples "backward." Notice that you get slightly different elasticities, because the proportionate change from, e.g., $9 to $8 is not the same as that from $8 to $9!

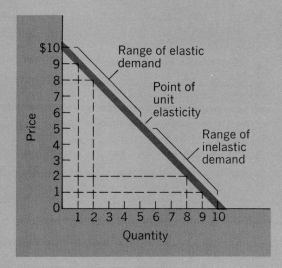

most demand curves changes from point to point! Actually there are only three normally shaped curves whose elasticity does not change: (1) totally inelastic (vertical) curves; (2) totally elastic (horizontal) curves; and (3) curves of unitary elasticity of demand or supply. Note in Fig. 9·11 that a fall in price of 50 percent, from $10 to $5, brings an increase in demand of 100 percent, from 1 unit to 2. But a second fall of 50 percent, from $5 to $2.50 would clearly not result in another 100

percent increase in purchases. As we go down the curve, the proportionate response of quantity to price steadily diminishes. Our curve becomes less and less elastic.

What, then, do we mean by "elastic" or "inelastic" demand, when elasticity changes as we move up or down the curve? There are two answers:

1. We can measure the elasticity of a given schedule or a *portion of the curve;* for ex-

QUANTITIES DEMANDED			
Price	Inelastic demand	Unit elasticity	Elastic demand
$10	100	100	100
9	101	$111\frac{1}{9}$	120
8	102	125	150
7	103	143	200
6	104	$166\frac{2}{5}$	300
5	105	200	450
4	106	250	650
3	107	$333\frac{1}{3}$	900
2	108	500	1,400
1	109	1,000	3,000

TABLE 9 · 1
*Demand
Schedules
for Three
Goods*

ample, the portion that lies between the price of $10 and $5 or any other two points. The answer gives us the "average" elasticity for that part of the curve, and for that part only. We call this *arc elasticity*.

2. We can also measure (by using the calculus) the elasticity at any point on the curve. This is *"point elasticity";* and this measure of elasticity will change from point to point. In theoretical economics, we often use point elasticity for accuracy, but in practical economics we usually use arc elasticity because we are interested in large-scale effects.

ELASTICITIES, EXPENDITURES, AND RECEIPTS

Elasticities not only affect the determination of market prices, they also have a very great effect on the fortunes of buyers and sellers in the marketplace. That is, it makes a great deal of difference to a buyer whether the supply curve of a commodity he wants is elastic or not, for that will affect very drastically the amount he will have to spend on that particular commodity if its price changes; and it makes an equal amount of difference to a

seller whether the demand curve for his output is elastic or not, for that will determine what happens to his total revenues as prices change.

Here is an instance in point. Table 9·1 shows three demand schedules: elastic, inelastic, and of unit elasticity. Let us see how these three differently constituted schedules would affect the fortunes of a seller who had to cater to the demand represented by each.

Notice that a very interesting result follows from these different schedules. *The total amount spent for each commodity (and thus the total amount received by a firm) will be very different over the indicated range of prices.* In Table 9·2 are the amounts spent (price times quantity).

To a seller of goods, it makes a lot of difference whether or not the demand he faces is elastic. *If demand is elastic and he cuts his price, he will take in more revenue.* If his demand is inelastic and he cuts his price, he will take in *less* revenue. (This has typically been the case with farm products, as we will see in our next chapter.)

Conversely, a businessman who raises his price will be lucky if the demand for his product is inelastic, for then his receipts will actually increase. Compare the fortunes of

Price	Good with inelastic demand schedule	Good with unit elasticity demand schedule	Good with elastic demand schedule
$10	1,000	1,000	1,000
9	909	1,000	1,080
8	816	1,000	1,200
7	717	1,000	1,400
6	612	1,000	1,800
5	525	1,000	2,250
4	424	1,000	2,600
3	321	1,000	2,700
2	216	1,000	2,800
1	109	1,000	3,000

TABLE 9·2
Total Expenditures (or receipts)

the two businessmen depicted in Fig. 9·12. Note that by blocking in the change in price times the change in quantity, we can show the change in receipts. (Because we have ignored changes in costs, we cannot show changes in profits.)

Our figure shows something else. If we reverse the direction of the price change, our businessmen's fortunes take a sharp change. A demand curve that is elastic spells bad news for a businessman who seeks to raise his price, but the same demand curve brings good fortune to a businessman who intends to cut prices. Just the opposite is the case with an inelastic demand curve: now the condition of demand is favorable for a price rise, since the seller will hold most of his sales even at the higher price; but inelastic demand is bad for someone who cuts prices, since he will gain few additional customers (or his old ones will increase their purchases only slightly) when prices fall.

FIGURE 9 · 12
How elasticities affect receipts when prices change

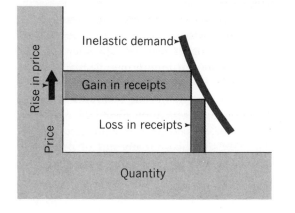

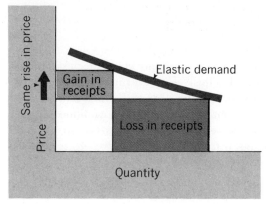

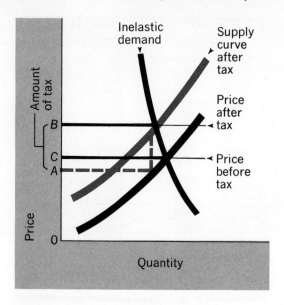 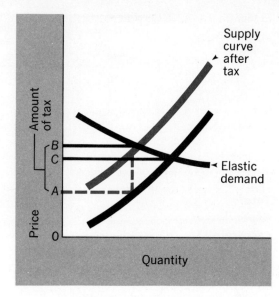

FIGURE 9·13
The incidence of taxation

Obviously, what every businessman would like to have is a demand for his product that was inelastic in an upward direction and elastic at lower than existing prices, so that he stood to gain whether he raised or lowered prices. As we shall see when we study pricing under oligopoly, just the opposite is apt to be the case.

THE INCIDENCE OF TAXATION

We can see the importance of elasticities in another illustration. In Fig. 9·13, we see the effect of adding a tax on a firm's output, such as an excise tax per bottle of liquor or per pack of cigarettes or a sales tax on the final sales price. The supply curve shifts upward by the amount of the tax (*AB*).

Who pays such a tax? The answer hinges significantly on the elasticity of demand for the product. In the lefthand portion of the graph we show a product with a highly inelastic demand curve. Note that the equilibrium price will rise from *OC* to *OB*, nearly as much as the tax. The consumer will pay virtually all of the tax.

In the figure on the right, demand is very elastic. Now note that the effect of the same tax is mainly to decrease the quantity demanded. The new equilibrium price *OB* will be slightly higher than the old price *OC*, but the consumer will not pay nearly the full amount of the tax (*AB*) in higher prices. Who will pay most of it? The producer will. The tax will force out of business all producers who cannot meet the new equilibrium price. The tax will come largely out of producers' surpluses (or profits) and result in the reduction of total payments to factors of production.

THE EFFECT OF WAGE INCREASES

Can you also see that exactly the same analysis would apply if we substituted a wage increase for a tax increase? If the product faces an inelastic curve, the employer can "pass along" the wage increase by raising his prices enough to cover the new wage cost. Consumers will pick up most of the bill, and only a few marginal producers will be forced to leave the industry. On the other hand, if the demand is highly elastic, a wage increase cannot be passed along. Or rather, if it *is* passed along, sales will diminish, and inefficient producers will have to leave the industry. Eventually a new equilibrium price will be set, only slightly above the old one (as in the right-hand diagram).

BEHIND ELASTICITIES OF DEMAND

Because elasticities are so important in accounting for the behavior of prices, we must press our investigation further. However, we must leave the supply side of elasticity, to be studied when we look into the behavior of factors and firms. Here we will ask why are demand curves shaped the way they are? Why is our price (or income) sensitivity for some commodities so great and for others so slight?

If we think of a good or service for which our demand might be very inelastic— say eyeglasses (assuming we need them)— and compare it with another for which our demand is apt to be highly elastic—say, a trip to Europe—the difference is not difficult to grasp. One thing is a necessity; the other is a luxury. But what do we mean by *necessity* and *luxury?*

One attribute of a necessity is that it is not easily replaced by a substitute. If we need eyeglasses, we will spend a great deal of money, if we must, to acquire a pair. Hence such a necessity has a very inelastic demand curve.

MARGINAL UTILITY AGAIN

Necessities are never absolute in the sense that nothing can be substituted for the commodity in question. High enough prices will drive buyers to *some* substitute, however imperfect.* Just when will the buyer be driven to the "next-best thing"? As we know, economists say that the decision will be made by a comparison of the marginal utility derived from a dollar's worth of the high-priced item with that derived from the lower-priced substitute. As the price of champagne goes up and up, there comes a point when we would rather spend our next dollar for a substantial amount of beer, rather than for a sip of champagne.

NECESSITIES AND ELASTICITY

We have seen that necessities have inelastic demand curves, so that we stick to them as

---

*What *would* be the substitute for eyeglasses? For a very nearsighted person, the demand for one pair of glasses would be absolutely inelastic over a considerable price range. But when glasses got to be, say, $500 a pair, substitutes would begin to appear. At those prices, one could hire someone to guide him around or to read aloud. Admittedly this is less satisfactory than having glasses; but if the choice is between spending a very large amount on glasses and on personal help, the latter might seem preferable. Of course, there are some goods without any substitutes—air, for example. Such goods are "free goods," because no one owns them. If a *good* such as air could be owned, it would have to be subject to stringent public control, to prevent its owners from exacting a horrendous price for it.

THE SEARCH FOR SUBSTITUTES

The search for substitutes is a complicated process that can lead to equally complicated supply-and-demand reactions. Just after a taxi strike in New York City, when fares went up 50 percent, people switched to substitutes (they rode buses, subways, or walked), and the taxi business suffered severely. Then after the shock of the increase wore off, business revived. People got used to higher fares; that is, they discovered that the marginal utility of a high-priced taxi ride was still greater than the marginal utility of the money they saved with other transportation, plus the marginal utility of the time they wasted or the business they lost because they weren't taking taxis. In other words, the substitutes weren't satisfactory. Gradually, people began taking more cab rides, and taxi receipts were higher than before the fare hike. The quantity of taxi service consumed was down somewhat, but not by as much as the original drop. In this case, demand proved to be more *inelastic* over time than it was in the very short run.

prices rise. But what about when they fall? Won't we rush to buy necessities, just because they *are* necessities? Won't that make their demand curves elastic?

Surprisingly, the answer is that we do not rush to buy necessities when their prices fall. Why? The answer is that necessities are the things we buy *first,* just because they are necessities. Having bought what we needed before the fall in price, we are not tempted to buy much more, if any more, after the fall. Bread, as we commented before, is a great deal more valuable for life than diamonds are, but we ordinarily have enough bread, so that the marginal utility of another loaf is no greater than that of an equivalent expenditure on any other good. Thus, as the price of bread drops, the quantity we seek expands only slightly. So, too, with eyeglasses.

Compare the case with a luxury, such as a trip to Europe. There are many substitutes for such a trip—trips out West, trips South, or some other kind of vacation. As a result, if the price of a European trip goes up, we are easily persuaded to switch to some alternative plan. Conversely, when the price of a European trip gets cheaper, we are quick to substitute *it* for other possible vacation alternatives, and our demand accordingly displays its elastic properties.

Do not make the mistake, however, of thinking that elasticity is purely a function of whether items are "expensive" or not. Studies have shown that the demand for subway transportation in New York City is price-elastic, which hardly means that riding in the subway is the prerogative of millionaires. The point, rather, is that the demand for subway rides is closely affected by the comparative prices of substitutes—bus fares and taxis. Thus *it is the ease or difficulty of substitution that always lies behind the various elasticities of demand schedules.*

Other Influences on Demand

IMPORTANCE OF TIME AND OTHER PRICES

Time also plays an important role in shaping our demand curves. Suppose, for example, that the price of orange juice suddenly soared, owing to a crop failure. Would the demand for orange juice be elastic or inelastic?

In the short run, it would generally be more inelastic than in the longer run. Lovers of orange juice would likely be willing to pay a higher price for their favorite juice because (they would believe) there was really no other juice quite as good. But as the weeks went by, they might be tempted to try other breakfast juices, and no doubt some of these experiments would "take." Substitutes would be found, after all.

The point is that it takes time and information for patterns of demand to change. Thus demand curves generally become more elastic as time goes on, and the range of discovered substitutes becomes larger.

Because substitutes form a vast chain of alternatives for buyers, changes in the prices of substitutes change the positions of demand curves. Here is a new idea to be thought about carefully. Our existing demand curve for bread or diamonds has the shape (elasticity) it does because substitutes exist at various prices. But when the prices of those substitutes *change,* the original commodity suddenly looks "cheaper" or "more expensive." If the price of subway rides rises from 35 cents to 50 cents, while the price of taxi rides remains the same, we will be tempted to switch part of our transportation from subways to taxis. If subway rides went to $1, there would be a mass exodus to taxis. Thus we should add changes in the prices of substitutes to changes in taste and in income when we consider the possible causes of a shift in demand. If the price of a substitute commodity rises, the demand for the original commodity will rise; and as the price of substitutes falls, demand for the original commodity will fall. This may, of course, bring changes in the price of the original commodity.

BEHAVIOR AND NATURE, AGAIN

There is a last point we should make before we leave the subject of substitution. We have

seen that the substitutability of one product for another is the underlying cause of elasticity. Indeed, more and more we are led to see "products" themselves as bundles of utilities surrounded with other competing bundles that offer a whole range of alternatives for a buyer's satisfaction.

What is it that ultimately determines how close the substitutes come to the commodity in question? As with all questions in economics that are pursued to the end, the answer lies in two aspects of reality before which economic inquiry comes to a halt. One of these is human behavior, with its tastes and drives and wants. One man's substitute will not be another's.

The other ultimate basing point is the technical and physical nature of the world that forces certain constraints upon us. Cotton may be a substitute for wool because they both have the properties of fibers, but diamonds are not a substitute for the same end-use, because they lack the requisite physical properties. Diamonds, as finery, may be a substitute for clothes made out of cotton; but until we learn how to spin diamonds, they will not be a substitute for the cloth itself.

COMPLEMENTS

There is another connection between commodities besides that of substitution—the relationship of *complementarity*. Complementarity means that some commodities are technically linked, so that you cannot very well use one without using the other, even though they are sold separately. Automobiles and gasoline are examples of such complementary goods, as are cameras and film.

Here is another instance of change in the price of one good actually affecting the position of the demand curve for the other. If the price of film goes up, it becomes more expensive to operate cameras. Hence the demand for cameras is apt to drop. Note that the price of cameras has not changed in the first instance. Rather, when the price of the complementary good—the film—goes up, the whole demand curve for cameras shifts to the left. Thereafter, the price of cameras is apt to fall as well.

The Market as a Self-Correcting Mechanism

We have begun to see how the market for goods and services operates as a dynamic, constantly altering—and yet self-adjusting—mechanism. From the interplay of supply and demand schedules for goods emerge equilibrium prices of those goods. However, these prices are rarely at rest. As the incomes and tastes of the throngs of goods-seekers change —or as the conditions of supply change—the prices of goods are continuously seeking new levels that will equate, if only for a moment, the quantities offered and sought. Moreover, as the prices of goods change, new ripples of disturbance are set into motion, for each good is a substitute for (or perhaps a complement of) others; and thus as each price changes, it will induce shifts in the demand curves that affect other prices. Meanwhile, the ever-shifting pattern of prices is silently carrying out its task of allocating the products on the market among the throng seeking them, distributing goods and permitting sales to those who meet the test of going prices, and quietly but uncompromisingly refusing them to those who do not.

How is order maintained in this extraordinary flux of activity? We have learned about some of the basic principles that keep the process smoothly and continuously working. One of these, we have seen, is the maximizing motive that drives buyer and seller to seek their respective economic advantages—

the buyer searching for the cheapest market for the good or service he wants, the seller looking for the highest price for the commodity he wants to sell. The other is the continuous constraint of competition that serves to bring buyer and seller together at a point acceptable to both.

Do self-interest and competition alone guarantee the orderly working of the market? As we shall see, they do not. But to understand why not, we must investigate what we mean by an "orderly" market, in the first place.

STABILITY AND INSTABILITY

We have seen one meaning for the word *orderly* in the extraordinary ability of a market situation to produce equilibrium prices that serve to bring together buyers and sellers in a stable situation of exchange, despite the different directions in which their self-interests propel them.

FIGURE 9 · 14
Short-run change in equilibrium price

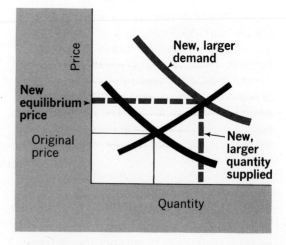

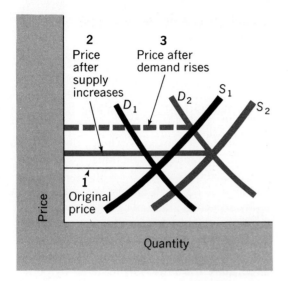

FIGURE 9 · 15
Possible long-run change
in equilibrium price

2
Price
after
supply
increases

3
Price after
demand rises

D_1 D_2 S_1 S_2

1
Original
price

Price

Quantity

adjustment is apt to be more complicated.* At higher shoe prices, enterpreneurs in other lines are apt to move into the industry, with the result that the whole supply curve of shoes shifts to the right. As a result, shoe prices will again fall (although we cannot tell by how much unless we know more about the costs of the industry). This solution would look like Fig. 9·15.

What would be an orderly outcome for a *fall* in demand? In the simplest short-run case, a new lower price and a smaller output. A longer-run solution might entail an exodus of suppliers and capital equipment from the industry, so that the supply curve moved left-ward and prices again recovered from their initial drop. Whichever the solution, in the end we would again have stable prices and outputs.

But we cannot confine orderliness to such a static solution. What happens when things change? Suppose, for example, that the demand curve for shoes shifts because the number of people in town has increased. The first effect of this, we know, will be a new, higher equilibrium price for shoes, as Fig. 9·14 shows. Notice also that at the new higher price, the quantity of shoes supplied will also be larger than it was before.

For the short run, this provides an orderly solution to the problem of change. There is now a new level of prices as stable as before, and a new equally stable level of output. (To be sure, there have been shifts in patterns of spending and in incomes that may exert pressures elsewhere on the economy, but we will ignore that for the moment. As far as the market for shoes is concerned, the problem of change has been met in an orderly way.)

In the long run, however, the process of

UNSTABLE SOLUTIONS

Now consider a less reassuring case. Suppose that the initial shift in demand does not simply result in a new higher, stable price, or in a rightward shift in the supply schedule, giving us the orderly solution we have just investigated. Suppose, instead, that it gives rise to the kind of perverse reaction we looked into in Chapter 6, when rising prices led to the expectation of still higher prices. In our present case, an increase in demand could lead suppliers to the expectation of a still higher demand in the future—an expectation that would lead them to hold back supplies in the hope of reaping larger profits subsequently.

In Fig. 9·16 we show this "perverse"

*The "long run" is not just a vague figure of speech. It means, in cases like this, the time necessary for producers to build new plant and equipment. Conversely, in the "short run," producers are limited to adjusting their supplies as best as they can from their given plant and equipment.

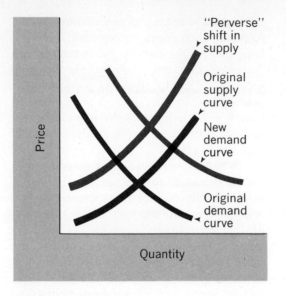

"Perverse" shift in supply

Original supply curve

New demand curve

Original demand curve

Price

Quantity

FIGURE 9 · 16
Perverse supply shifts

reaction—perverse in its effect on an orderly solution to the problem, but perfectly rational from the viewpoint of the supplier. Notice that the rightward shift in the demand curve gives rise to a *leftward* shift in the supply curve, causing prices to rise still faster. If both buyers and sellers *expect* this rise to continue, things can get worse, not better. Buyers will not reduce their quantities demanded as prices rise, but will shift their demand schedules still further to the right; and sellers may be sufficiently encouraged by rising prices to contract their offerings still further.

The *problem here is that the rise in price itself acts to move the schedules, rather than to change the quantities demanded or supplied along the former schedules.* Instead of leading to orderly markets, such situations lead to disruptive and disorderly markets: prices zoom—or collapse if expectations point in the wrong direction.

PREDICTIVE AND NORMATIVE PRICE THEORY

How are such instabilities handled? What effect do they have on the market? In the nature of things, usually they are short-lived affairs that result in wild splurges and busts that disrupt particular markets but do not derange the system as a whole. Sometimes, though, destabilizing reactions can become generalized throughout the whole goods market, as they do when buyers and sellers generally expect an inflation or a depression, and rush in or hold back in a way that aggravates the very thing they dread. At such times, destabilizing reactions can be of the gravest importance and must be countered by public policies or public pronouncements aimed at changing peoples' anticipations of the future, so that self-interest will again move them in the direction that brings order rather than disorder.

The narrow line that separates order from disorder, and stability from instability, brings us again to consider the relation of economic analysis to economic policy—a problem we broached at the end of Chapter 6.

The cases we have been describing can be *analyzed,* but they cannot be very well *predicted,* by economic reasoning. We do not know, in the ordinary course of events, when a shift in demand will result in a perfectly ordinary and orderly change in prices, or when it will result in an unstable and disorderly solution of runaway prices. Economists cannot foretell when a rise in land prices will become a "land boom," when a fall in the price of a stock will become a panicky rout, or when a gently inflating economy will suddenly give evidences of generalized "perverse" behavior that threatens to turn "orderly" inflation into frightening inflation.

Yet, even in these events, microeconomics has a special relevance. As we have seen, we can then use the insights of theory in a *normative* way, as a guide to restore normality. Microeconomic theory tells us what kind of behavior we require to restore calm in a badly functioning economy, and it may therefore help our policymakers determine the announcements or regulations that will bring about the expectations on which a well-behaved economic system rests.*

*For a full discussion of these problems, see Adolph Lowe, *On Economic Knowledge* (New York: Harper, 1966).

KEY WORDS

Demand and
supply schedules

Shifts of curves
and shifts along
curves

Elastic

Inelastic

Arc and point
elasticity

Substitutes

CENTRAL CONCEPTS

1. Equilibrium prices change when *supply or demand schedules change.* In turn, these schedules change when our willingness or ability to buy or sell is altered.

2. *Changes in taste or income* or other prices lie behind shifts in demand schedules; behind changes in supply lie *changes in attitudes or costs.*

3. A change in demand (or supply) means a shift of the whole *schedule.* This must be contrasted with changes in the *quantity* demanded (or supplied), which refers to movements *along given curves.*

4. Shifts in demand or supply mean that *different quantities* will be sought or offered *at the same price.*

5. Price elasticities measure the proportionate change in quantities offered or sought when prices change. *Elastic demand* (or supply) means that the *percentage change in quantities demanded* (or supplied) *will be larger than the percentage change in price;* inelastic demand (or supply) means that the percentage quantity change is smaller than the percentage price change.*

6. Elasticities can be measured along segments of demand (or supply) curves or at each point along these curves. The former measures the "average" responsiveness of demand or supply over a given schedule and is called *arc elasticity.* The latter is named *point elasticity.*

7. Price elasticities very greatly affect price changes. Price changes will be larger when demand and supply curves are inelastic than when they are elastic. *Elasticities affect the receipts of the seller or the expenditures of the buyer.*

8. Elasticities reflect the ease or difficulty of *substitution.* Hence, elasticities typically increase over time, as new substitutes are found or as information about them spreads. The *ease of substitution* is an important concept that helps us define what a "commodity" is.

9. Substitutes are commodities for which the demand rises when the price of the original commodity rises, and for which demand falls when the price of the original commodity falls.

*The formula for elasticity is

$$n = -\frac{\Delta Q}{Q} \bigg/ \frac{\Delta P}{P} \text{ (See box, p. 109.)}$$

Complements

10. **Complements** are commodities that are technically linked to one another. The demand for one of them therefore falls when the price of the linked commodity rises, and the demand rises when the price of the linked commodity falls. This situation is just the opposite from that of substitutes.

Orderly markets

11. The market is ordinarily a self-correcting and orderly mechanism. By *orderly*, we mean that *changes in supply and demand produce new, stable prices and outputs*.

Expectations

12. In some instances, the market does not produce an orderly outcome. This is the case *when expectations cause shifts in demand and supply schedules, rather than movements along those schedules.*

QUESTIONS

1. What changes in your economic condition would increase your demand for clothes? Draw a diagram to illustrate such a change. Show on it whether you would buy more or less clothes at the prices you formerly paid. If you wanted to buy the same quantity as before, would you be willing and able to pay prices different from those you paid earlier?

2. Suppose that you are a seller of costume jewelry. What changes in your economic condition would decrease your supply curve? Suppose that costs dropped. If demand were unchanged, what would happen to the price in a competitive market?

3. Draw the following: an elastic demand curve and an inelastic supply curve; an inelastic demand curve and an elastic supply curve; a demand curve of infinite elasticity and a totally inelastic supply curve. Now give examples of commodities that each one of these curves might represent. Caution: remember that elasticities change in most curves.

4. Show on a diagram why elasticity is so important in determining price changes. (Refer back to the diagrams on p. 107 to be sure that you are right.)

5. Draw a diagram that shows what we mean by an increase in the quantity supplied; another diagram to show what is meant by an increase in supply. Now do the same for a decrease in quantity supplied and in supply. (Warning: it is very easy to get these wrong. Check yourself by seeing if the decreased supply curve shows the seller offering less goods at the same prices.) Now do the same exercise for demand.

6. How does substitution affect elasticity? If there are many substitutes for a product, is demand for it elastic or inelastic? Why?

7. Show on a diagram (or with figures) why you would rather be the seller of a good for which demand was elastic, if you were in a market with falling prices. Suppose prices were rising—would you still be glad about the elasticity of demand?

8. If you were a legislator choosing a product on which to levy an excise tax, would you choose a necessity or a luxury? Which would yield the larger revenue? Show how your answer hinges on the different elasticities of "luxuries" and "necessities."

9. By and large, are luxuries apt to enjoy elastic or inelastic demands? Has this anything to do with their price? Can high-priced goods have inelastic demands?

10. Why is demand more apt to become elastic over time?

11. The price of pipe tobacco rises. What is apt to be the effect on the demand for pipes? On the demand for cigars?

12. What is meant by an orderly market? If prices fall, how should you behave as a buyer to insure an orderly market? As a seller? What might make you behave differently?

The Microeconomics of Farming

10

Has our analysis of supply and demand, elasticities, and all the rest seemed very abstract, remote from the problems of real life? Let us bring the tools of economic analysis to bear on a sector of the economy of great importance in a world of runaway populations and limited resources—our farm sector.

A BRIEF BACKGROUND

Traditionally, farming has been a trouble-ridden occupation. All through the 1920s, the farmer was the "sick man" of the American economy. Each year saw more farmers going into tenantry, until by 1929 four out of ten farmers in the nation were no longer independent operators. Each year the farmer seemed to fall further behind the city dweller in terms of relative well-being. In 1910 the income per worker on the farm had been not quite 40 percent of the nonfarm worker; by 1930, it was just under 30 percent.

Part of this trouble on the farm, without question, stemmed from the difficult heritage of the past. Beset now by drought, now by the exploitation of powerful railroad and storage combines, now by his own penchant for land speculation, the farmer was proverbially an ailing member of the economy. In addition, the American farmers had been traditionally careless of the earth, indifferent to the technology of agriculture. Looking at the average individual farmer, one would have said that he was poor because he was unproductive. Between 1910 and 1920, for instance, while nonfarm output per worker rose by nearly 20 percent, output per farm worker actually fell. Between 1920 and 1930, farm productivity improved somewhat, but not nearly so fast as productivity off the farm. For the great majority of the nation's agricultural producers, the trouble appeared to be that they could not grow or raise enough to make a decent living.

If we had looked at farming as a whole,

however, a very different answer would have suggested itself. Suppose that farm productivity *had* kept pace with that of the nation. Would farm income as a whole have risen? The answer is disconcerting. The *demand* for farm products was quite unlike that for manufactured products generally. In the manufacturing sector, when productivity rose and costs accordingly fell, the cheaper prices of manufactured goods attracted vast new markets, as with the Ford car. Not so with farm products, however. When food prices fell, people did not tend to increase their actual consumption very greatly. Increases in over-all farm output resulted in much lower prices but not in larger cash receipts for the farmer. Faced with an *inelastic demand,* a flood of output only leaves sellers *worse* off than before, as we saw in our last chapter.

That is very much what happened during the 1920s. From 1915 to 1920, the farmer prospered because World War I greatly increased the demand for his product. Prices for farm output rose, and his cash receipts rose as well; in fact, they more than doubled. But when European farms resumed production following the war, the American farmers' crops simply glutted the market. Although prices fell precipitously (40 percent in the single year 1920–21), the purchases of farm products did not respond in anything like equal measure. As a result, the cash receipts of the farmer toppled almost as fast as prices. In turn, an ailing farm sector contributed to a general economic weakness that would culminate in the Great Depression.

THE NEW DEAL

At its core, the trouble with the farm sector was that the market mechanism did not yield a satisfactory result for farmers. Two causes were evident. One was the inelastic demand for food. The second was the inability of a vast, highly competitive industry like agriculture to limit its own output, so that it would not constantly "break the market" every time a bumper crop was harvested.

This chronic condition of agriculture was one of the first problems attended to by Franklin Roosevelt's New Deal administration. The New Deal could not alter the first cause, the inelasticity of demand, for that arose from the nature of the consumer's desire for food. But it could change the condition of supply, which hurled itself, self-destructively, against an unyielding demand. One of the earliest pieces of New Deal legislation—the Agricultural Adjustment Act—sought to establish machinery to be used by farmers, as a group, to accomplish what they could not do as competitive individuals: curtail output.

The curtailment was sought by offering payments to farmers who agreed to cut back their acreage or in other ways hold down their output. In the first year of the act, there was no time to cut back acreage, so that every fourth row of growing cotton had to be plowed under, and six million pigs were slaughtered. In a nation hungry and ill-clad, such a spectacle of waste aroused sardonic and bitter comment. Yet, if the program reflected an appalling inability of a society to handle its distribution problem, its attack on overproduction was not without results. In both 1934 and 1935 more than thirty million acres were taken out of production in return for government payments of $1.1 billion. Farm prices rose as a result. Wheat, having slumped to 38¢ a bushel in 1932, rose to $1.02 in 1936. Cotton doubled in price, hog prices tripled, and the net income of the American farmer climbed from the fearful low of $2.5 billion in 1932 to $5 billion in 1936.

Later in this chapter we will trace the heritage of the New Deal idea and explore the turns of recent farm policy. But already we may have seen that microeconomics is not

THE FARM SECTOR

What is a farm? According to the Bureau of the Census it is an agricultural establishment of 10 acres or more whose annual sales amount to at least $50, or a plot of less than 10 acres whose annual sales come to $250 or more. One of the authors of this text grows tomatoes in his back yard. If he could sell $250 worth—quite a lot of tomatoes—he would be a ''farmer'' and might qualify for income support.

In 1972, there were 2.8 million qualified ''farms.'' About half produced less than $5,000 worth of output; *in toto*, they accounted for only 5 percent of the nation's farm production. Thus half the ''farms'' in the nation could disappear with virtually no effect on food supply. Moreover, the majority of these very small farmers earned substantial income from ''off-the-farm'' jobs. Farming at the bottom of the scale is often a means of supplementing an income earned in trade or manufacturing.

How about big farms? In 1972, farms with sales of more than $40,000 comprised only 8 percent of farming establishments, but earned 56 percent of cash receipts in agriculture. Corporations account for about 14 percent of all farms, but most of these are small family corporations: 95 percent of all farms are still family-owned. The 5 percent that are not family owned are agri-businesses, often run by big food packers. This agribusiness sector produced about 35 percent of all farm output.

Two important conclusions. First, there are still at least 1.4 million farmers who could be forced from the land in coming decades without much loss in output. Because most of them have other employments, the effect on unemployment would not be as large as the sheer numbers of such submarginal farmers indicate, but the social dislocation implicit in the figures is substantial.

Second, the notion that farming is somehow ''less advanced'' than industry is a notion that should be forgotten once and for all. The successful family farm and the big agribusiness is the match for, or the superior of, most businesses, in technology, capital requirements, and managerial skills. Just to illustrate: the capital invested per worker in manufacturing is $28,500. In farming, it is $68,000!

without its cutting edge when applied to the facts of real life. Let us now cut a little deeper, by using our newly acquired analytic techniques to trace the extraordinary events of 1972 and 1973.

THE RUSSIAN WHEAT DEAL OF 1972

In midsummer 1972, the Nixon administration startled the world with the announcement that it had concluded the largest agricultural transaction in history, selling 20 percent of the American wheat crop to the Soviet government. Prices for wheat began to soar. In mid-July, when the announcement was made, wheat was selling at $1.70 per bushel; five months later, it was $2.40; in 13 months, it reached $5.00.

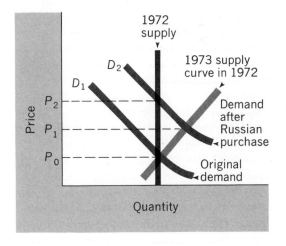

FIGURE 10 · 1
The wheat sale in diagram

123

It is easy to understand why. When the Russian sale was made in July, 1972, the grain crop had been completely planted and to a substantial extent completely harvested. Thus the 1972 supply curve of grain was totally inelastic—vertical. Just as we would expect, the addition of Russian demand to American demand resulted in a jump in price from P_0 to P_2, as Fig. 10·1 shows.

Notice that if the Russians had contracted to buy grain for 1973 delivery rather than for 1972 delivery, the situation would have been different. The supply schedule would have been much more elastic. Farmers would have had time to expand their production. As a result, P_1, the equilibrium 1973 price, given a year's advance warning, would have been much lower than P_2, the equilibrium 1972 price, given no advance warning. Prices would still have gone up, but not by as much.

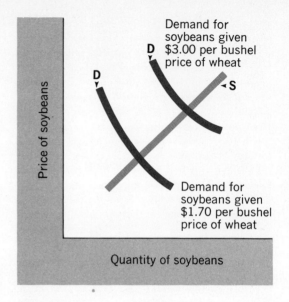

FIGURE 10 · 2
Induced shift in demand

THE SHIFT TO SUBSTITUTES

But we are not through tracing out the impacts of the Russian grain deal. We have seen that the substitutability of products for one another is the underlying cause of elasticity. Grain is no exception to the rule that all products have near if not perfect substitutes. If the price of wheat rises, individuals will substitute other grain crops such as oats, soybeans, or rye, wherever possible. Therefore an increase in the price of wheat will cause an outward shift in the demand schedule for other grain products, as Fig. 10·2 shows. Obviously, the more perfect a substitute a commodity is, the greater the induced shift in the demand curve. With an outward shift in demand, *other* grain crops will rise in price. Thus, although the Russian purchases may have initially affected only wheat prices, they soon induced price increases among all substitutes for wheat. In 13 months, the price of oats rose by 150 percent; soybeans, 257 percent; rye, 282 percent.

FURTHER COMPLICATIONS

It was not the Russian sale alone that accounted for the jump in farm prices. For reasons unknown, the Humboldt current shifted its position off the coast of Peru, so that the normal supplies of anchovies ran very short. Since millions of pounds of anchovies are ground up as fish meal and fed to animals as a close substitute for feed grains, the reduction in the supply of anchovies led to an increase in the price of fish meal, and thus to a *further* induced shift in the demand for soybeans and other American feed grains.

In addition, the U.S. dollar was devalued during 1972 and 1973. Devaluation is a subject that we will cover in detail later, but it is easy to see its impact on agricultural prices. Let us assume that yen and dollars have been trading at 400 yen to the dollar. A Japanese importer is thinking about buying soybeans, the cost of which is 800 yen or $2 per bushel. Now let us assume that the exchange rate between yen and dollars falls

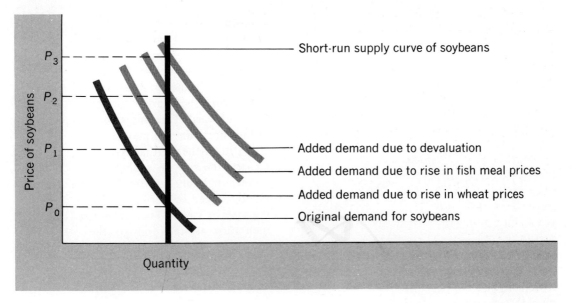

Short-run supply curve of soybeans

Added demand due to devaluation

Added demand due to rise in fish meal prices

Added demand due to rise in wheat prices

Original demand for soybeans

FIGURE 10 · 3
Further induced demands

to 300 yen to the dollar. The price of soybeans in America remains unchanged at $2 per bushel; but from the viewpoint of Japan, they have declined from 800 yen to 600 yen. Given their lower price in yen, the Japanese importer decides to buy more American soybeans. This leads to another outward shift in the demand curve for soybeans. Because the short-run supply curve of soybeans will be essentially vertical, large price increases result.

Figure 10·3 shows how these successive additions to demand, stemming from different causes, brought about a rise in soybean prices, from P to P_3, actually from $3.50 in July to a peak of $12.00 in late summer of 1973.

EFFECTS ON BEEF SUPPLY

Soybeans, however, are wanted not only directly as human food, but indirectly as feed for animals. As a result, an increase in the price of feed grains has reverberations in the supply and demand for meat. A sudden increase in the price of animal feeds means that the cost of producing meat has taken a corresponding jump upward. For a given price, each farmer is now willing to supply less meat.

But now one of the peculiarities of the agricultural market arises. When a manufacturer wants to cut back his production, he simply fires his workers and closes his factories. The rancher does something else. His breeding herd of cattle is part of his equipment, but it is also the same commodity that he sells on the market. To cut his production he must cut back on his breeding herd, but *to cut back on his breeding herd, he must sell cattle that he would otherwise use for breeding.* In order to get a long-term decrease in his breeding herd, he cannot avoid a short-term *increase* in the supply of meat brought to market. Thus the price of beef will first go down; and then, after breeding herds have

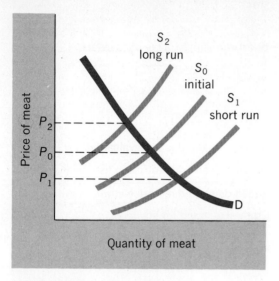

FIGURE 10 · 4
Supply curves for beef

been reduced, will rise sharply. We see this in Fig. 10·4.

Omaha steers were, in fact, selling at about $37.50 in midsummer 1972. A few months later they had fallen in price to roughly $33 ($P_1$ in our diagram). Thereafter prices rose to over $50 before they finally peaked.

CORN-HOG CYCLE

This sequence of events can lead to still further problems. The long-term movement of prices is up, but the short-term movement is down. If ranchers interpret this short-term movement as a long-term tendency, they will plan to reduce their breeding herds even further, bringing *more* cattle or hogs to market and pushing prices still lower!

Meanwhile, the fall in breeding herds will begin to bring another result. It will greatly diminish the demand for feed. As a consequence, corn prices (or cattle- and hog-feed prices generally) will fall. As feed prices fall, prospects for making money improve. Ranchers will stop reducing breeding stock

and think about enlarging their capacity for *future* delivery of meat.

Now we encounter problems on the other side. To increase the size of his breeding herd, a rancher must withhold pigs or cattle from the market, moving the supply curve leftward. This raises beef or hog prices in the short run and may lead the cattleman into still further cutbacks in sales if he interprets the rise as leading to a future rise. Sooner or later, however, his increased demand for feed will bring up the price of raising sows or steers. The "corn-hog" cycle will then turn around and go the other way. Prices may rise and fall and never reach equilibrium.

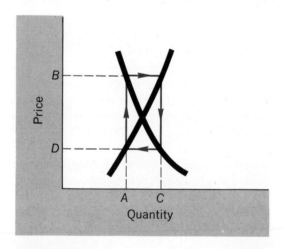

FIGURE 10 · 5
The cobweb

COBWEBS

Agricultural markets are also associated with instability because of the long time-lags between decisions involving production and the ultimate sale of that output. Take the case of Christmas trees.

In Fig. 10·5, we show the supply and demand curves for Christmas trees and imag-

NUMBERS GAME

We can see the difficulties of the farm problem with a simple numerical example. Assume that a farmer has a breeding herd of 100 head of cattle, and that each cow has a calf every year. He sells these calves when they reach two years of age. This means that he can sell 100 head per year, which are then replaced by 100 newborn calves.

Now let's assume that he wants to increase his breeding herd by 50 percent. A cow cannot calve until it is about 3 years old. To add fifty breeding cows to his stock he must therefore withhold fifty two-year-old cattle that he would otherwise sell. His sales will fall from 100 to 50.

The following year, however, he will have a breeding herd of 150, producing 150 calves, although they will not be two-year old calves and will not be ready for market. He will have only 100 two-year-old calves—the crop of his original herd, now come of age. The *following* year, he can market 150 calves, if he wishes, and maintain his herd steady, as the table below shows.

Year	Cows	1-yr.-olds	2-yr.-olds	Sales
1	100	100	100	100
2	100	100	100	50*
3	150	150	100	100
4	150	150	150	150

*Holding back 2-yr.-olds to become cows.

ine that the quantity supplied is initially indicated by point *A* on the supply schedule.

We can see that quantity *A* will sell at price *B*. Figuring that this will be *next year's* price, tree growers now plant the amount they are willing and able to offer at price *B*—quantity *C*. Alas, when the harvest comes, it is found that quantity *C* will fetch only price *D*. Now the process goes into re-

verse. Growers will figure that next year's price will be *D*, and they plant amount *A*, since at price *D* the quantity they wish to supply is no more than that. Thereupon, next harvest time, the price goes back to *B*—and around we go. If the supply and demand schedules were differently sloped, we *could* have a so-called *cobweb* that converged toward equilibrium, as we show on the left of

FIGURE 10 · 6
Stabilizing and explosive cobwebs

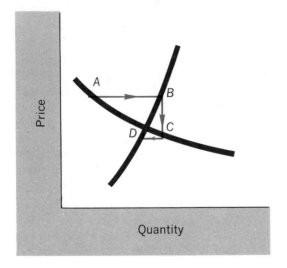

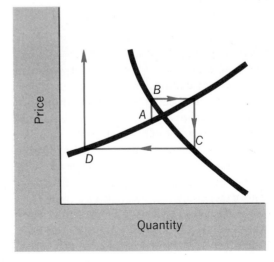

Fig. 10·6; and we could also have one that "exploded"—as we can see on the right.

Cobwebs are peculiar to agricultural production for two reasons. First, agriculture is a competitive industry where each producer takes the price as a "given" that he cannot affect. Second, the production processes in agriculture are very long and regular, compared to most industries. Think of how long it takes to produce a bushel of wheat or a pound of beefsteak, as opposed to the time necessary to produce a car. Nature also synchronizes those production decisions so that they all must be made at approximately the same time. Thus it is not possible to wait to see what others are going to do, and then adjust your production in light of your neighbors' production decisions.

Above all, the cobweb illustrates an endemic weakness of agricultural markets. Farm prices, like all prices, serve to allocate existing supplies. But unlike prices in markets where production decisions can be quickly changed, farm prices do not always accurately depict the direction to be taken by *future* equilibrium price-quantity relations. Thus they can lead to erroneous decisions, such as the corn-hog cycle or the famous cobweb.

STABILIZING THE MARKET

What can help farmers avoid these problems? One possibility is to have the U.S. Department of Agriculture predict future agricultural prices and tell the farmers what they believe these prices will be. Ideally, they would predict future equilibrium market prices, and farmers could adjust their production plans to these prices rather than to current prices. Farmers might not, however, believe that the prices will actually come into being. If they do not, they will not plan their production in accordance with these equilibrium prices.

PRICE SUPPORTS

One way out of this situation is for the government to announce what it believes to be the equilibrium market price and then stand ready to buy crops at this announced price. In this way, it can give credibility to its own forecast. If the government is an accurate forecaster, it will not have to buy farm products, since the market will purchase what is offered at the announced price.

While a price support system run in this manner could stabilize agricultural prices, the actual support system inaugurated in the New Deal sought to *raise farm incomes* and not necessarily to stabilize farm production. This means that support prices were set at a level *higher* than equilibrium market prices. Given these support prices, farmers could confidently plan their future production, since they knew their output would be bought. But because prices were above equilibrium levels, they chose to grow more than the consumer was willing to consume at the support price. In other words, surpluses emerged, as in Fig. 10·7.

FIGURE 10 · 7
Price-support surplus

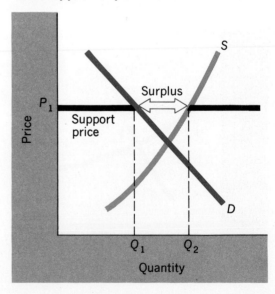

THE IMPORTANCE OF MULES

Professor Eldon Weeks points out that technology had been catching up with—and creating problems for—the farmer for a long time. One of the reasons for the overproduction of the 1920s was that we were steadily cutting back on the acreage needed to sustain horses and mules, as tractors came into general use. Before World War I, we used to devote over a quarter of our cropland to sustaining draft animals. After 1940, this fell to just over 10 percent. Much of the land not needed for animals went into production for the market, thereby adding its load of straw to the camel's back.

At support price P_1 consumers want to buy quantity Q_1. Farmers, however, produce quantity Q_2. The difference has to be bought and stored by the government. In 1960, government warehouses bulged with unsold crops worth $6 billion.

To avoid these surpluses, the government limited the acreage that farmers could plant if they wanted to qualify for support payments. In this way, as Fig. 10·8 shows, they hoped to push back the supply curve so that the market generated an equilibrium at quantity Q_1.

The strategy might have worked, were it not for the extraordinary increase in agricultural productivity, resulting from new technologies. Between 1940 and the late 1960s, harvested acreage declined by 15 percent, but the yield per acre increased by over *70 percent*. The result was a flood of output. Huge quantities had to be purchased and stored by the government. Only the massive distribution of these supplies to the underdeveloped lands during the late 1960s prevented the surplus problem from becoming a permanent national embarrassment.

The result was a transfer of income from the consumer to the farmer (see Fig. 10·9).

FIGURE 10 · 8
The effort to limit supply

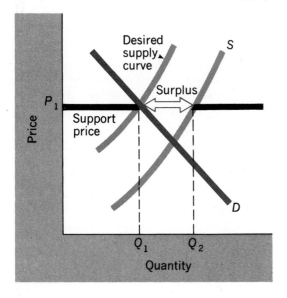

FIGURE 10 · 9
Loss of consumers' surplus

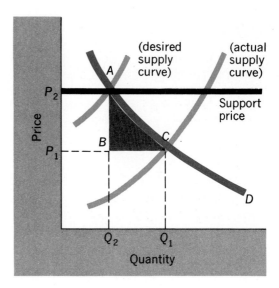

FUTURES MARKETS

Another means of avoiding the extreme risks inherent in the long production process of agriculture is the use of *futures markets* to "hedge" against uncertainty.

All basic agricultural products are sold on markets called commodities exchanges, the most famous being the Chicago Board of Trade. In this marketplace two kinds of transactions take place. Actual commodities, sitting in warehouses or freight cars, are bought and sold, just as other goods are traded, in any marketplace. But in the commodity markets you can also buy "futures"— future deliveries of grain, or cotton, or meat.

This seems mysterious. How can anyone buy or sell something that does not exist? The answer is that one can buy or sell *contracts* to deliver fixed amounts of an agricultural commodity at some fixed date in the future. The person who buys a *futures contract* buys the promise that a given quantity of product will be delivered to him on a future given date for a price agreed upon now. If you are a grain miller seeking to plan your own production, you may well want to buy futures contracts so that you can know *now* how much

grain will cost you over the course of the next several months, when the grain will be delivered to you.

If you were a farmer, you could *sell* a futures contract. Knowing the futures price of wheat a year from now, you could determine how much you wanted to produce at that price, and sign a contract to deliver that production at a given price a year into the future.

Most sellers of futures contracts are not farmers, however, but speculators. When a speculator sells a futures contract he is betting that a year from now he can go into the market for immediate, or "spot," delivery and buy the required amount of grain for less than he is now selling it for. If he can sell grain for delivery a year hence at $3.00 per bushel, and a year from now meet his obligations by buying it for $2.00 per bushel, he obviously has made a good profit. On the other hand, if grain actually sells for $3.50 a year from now, he will lose money, since he sold grain at $3.00 per bushel and must fulfill the contract by buying it at $3.50 per bushel.

Thus the speculator is trying to guess what next year's equilibrium market price will be. If he thinks that next year's actual price is going to be lower than the current "futures price" for delivery next year, he will sell futures. If he thinks

Instead of paying price P_1 and buying quantity Q_1, the consumer was forced by the market to pay price P_2 and to buy quantity Q_2. As a result, the consumer lost the consumer's surplus represented by the shaded triangle ABC, and he found his real income reduced by $P_2 - P_1$ times Q_2.

INCOME SUPPORT

To avoid these unwanted and politically unwelcome effects, the Nixon administration in 1973 finally adopted a plan that had been proposed almost a quarter-century earlier by Charles Brannan, Secretary of Agriculture

under President Truman. The so-called Brannan Plan supports farm *incomes,* not prices. A "target price" is established by law for various crops, but this target price does not apply to the actual selling prices of the crops. The free play of forces on the market allows prices to reach whatever levels supply and demand dictate, thereby getting rid of the ABC triangle, because (given our supply and demand schedule), crops will sell at a price P_1 and not at the target price P_2.

The farmer is fully protected, nonetheless. If actual market prices are below target prices, the government will send a farmer a check for the difference that results from sell-

that next year's price is going to be higher than the current futures price, he will buy futures.

Do speculators cause prices to fluctuate more than they otherwise would? Occasionally speculators can "corner" a market by buying up a large amount of, say, September wheat (wheat to be delivered in September) and thereby forcing up the price. Such corners are illegal, and the speculators are liable to prosecution. Usually, however, speculators fulfill a very useful function of taking the risk out of certain industrial operations.

Suppose you are a meatpacker and that each month you buy 1,000 tons of beef. If you wait, month to month, and buy on the current or spot market, prices may be high one month and low the next. Since you sell at a price that is not likely to fluctuate with spot prices, you find yourself rich one month, when beef supplies are cheap, but possibly broke the next, when you are forced to buy beef at prices too high to give you a profit when you sell the processed meat.

To avoid that problem, you do two things. For each month, you buy a *futures contract and also sell a futures contract!* Suppose you buy beef for delivery a month hence at, say, $50 per hundredweight and also sell it at the same price per pound. When the time for delivery comes around, beef for immediate delivery might have risen to $60. Your futures purchase contract, bringing you an actual delivery of beef at $50, would therefore give you a profit. Alas, you will have to forego that profit by buying spot beef at $60, to honor your disadvantageous sale contract for $50 beef. Your only profit will be the manufacturing profit you make.

Why then did you sell beef? The answer is that beef might have fallen in price, and you would be accepting beef at your contract price of $50, although spot beef was only, say, $40. A competitor who bought in the spot market would have an edge in his cost of production. But now your futures sale pays off. You have sold your futures at $50; you can "cover" the sale at $40. You have therefore recouped in the selling contract the loss you incurred in the buying contract.

This process of simultaneously buying and selling is called *hedging.* Its purpose is to take out the risk that would otherwise be inescapable. *Having hedged, you can neither win nor lose, as spot prices change.* Your beef will come to you without gain or loss arising from changes in beef prices. You make your money as a manufacturer, not as a speculator. But you are able to do so only because speculators create a market for futures, which enables you to buy and sell for future delivery.

ing his crop below the target level. (Moreover, there is a limit of $20,000 per farmer in these support income payments, whereas there was no limit under previous plans that sent enormous sums to some very large farm operators). Two results follow:

1. When production is high and prices fall, consumers get the benefit of cheaper food prices, although as taxpayers they will still have to give up a certain amount of income to be transferred to farmers.

2. Because target prices are fairly high, farmers can plan with assurance for high outputs, without worrying about "breaking the market." In an era of world food shortages and high domestic food prices, this helps assure a high level of farm output.

Controls in a Competitive Industry

Now let's suppose that as a result of various circumstances, food prices take a sharp upward turn, as they did during the spring of 1973. The government decides that to control the rate of inflation it is going to freeze all retail food prices.

If all prices were at or above their equilibrium when controls were imposed and if these equilibrium levels were not changing, price freezes would have no immediate impact, since they would be the same prices that the market would have sooner or later set. But let's assume that this is not the case and that the controls are going to hold some prices below their equilibrium level, as in Fig. 10·10.

PROBLEMS OF PRICE CONTROLS

As we have seen, whenever prices are held below their equilibrium level, shortages will occur—that is, people will want to buy more of the commodity than suppliers are willing to supply. Since price is not being used to ration the existing supplies, some other technique must be found. The only alternatives are to distribute goods and services on a first-come first-served basis or to establish formal rationing. Note that if the government control procedures break down and black markets are established, *we are right back to rationing by prices*. The only difference is that in the black market, prices are illegal. Purchasers or sellers at these illegal prices can be thrown in jail.

FIGURE 10 · 10
Price controls

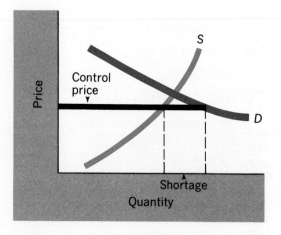

If formal rationing is used, governments seek to reduce the demand curve for the commodity by insisting that you must have both money *and* ration coupons to buy a pound of beefsteak. By limiting the amount of beef covered by ration coupons to the known supply of beef, it becomes possible to push the demand curve back to the point where it just crosses the supply curve at the desired price. The problem with this procedure is that the farmer now has no incentive to expand his production. The only way he will produce more is if he is able to get a higher price. Thus the controls that reduce the price of food retard the expansion of the food supply.

Suppose, instead, that we ration on a first-come, first-served basis. Now the problem is the opportunity cost of shopping time. The price of purchasing a good is now the money you must pay plus the time you must wait in line to get what you want. For most people, beefsteak that sells for 50¢ per pound but requires a 2-hour wait in line is not cheap beefsteak.

Now let's assume that the farmer has the option of selling his production to foreigners as well as to Americans. A freeze on retail prices is not a freeze on prices on the farm. Foreign buyers can therefore offer farmers more than American retailers can. Accordingly, farmers sell their crops to foreign buyers. This not only leads to greater U.S. shortages, but it can also create other problems. After the freeze is established, suppose soybean prices rise because of foreign demand. With higher soybean prices, U.S. chicken growers may find that chickens sell below the cost of feeding them. If this occurs, they will stop raising chickens and further aggravate the shortage of meat. Another possibility is that producers may simply hang onto their production, waiting for the freeze to be lifted. If cattle are withheld from the market, for example, shortages are once again exacerbated.

As a result, there are a host of adverse consequences stemming from the effort to control the increase in retail food prices. If retail controls are to be effective in holding prices below their equilibrium levels, they require formal rationing as a complement to price ceilings, plus some nonprice effort to increase the supply of foodstuffs. Eliminating acreage controls would be one nonprice action to increase production in the face of a price freeze.

THE FARM PICTURE IN PERSPECTIVE

As we have seen from our quick tour of the agricultural markets, the farmer leads anything but a serene life. He must make difficult guesses about the course of prices years into the future. Prices take large swings, seldom settle into stable patterns, and may not be a very good guide to the future.

Moreover, efforts to correct these difficulties have not been marked by great success. The original price support program was costly, inefficient, and anything but equitable. The new income-support program hopes to avoid many of these problems, but it may be expensive if market prices fall and target prices are set too high, under political pressure.

Yet, in all this, it is important to gain a sense of long-term perspective. We have seen that the farmer, unable to control economic swings, has traditionally been their victim. What we have tried to do, in the last four decades, is to intervene in the market process, to make it yield results more in accord with our conceptions of social justice. And despite all the difficulties we have discussed, the results are far from unimpressive. Agriculture, as an income-producing activity has benefited substantially—especially for the million-odd successful farmers who produce 90 percent of our marketed farm products. Between 1940 and 1964, farm operator families enjoying the use of electricity increased from 33 percent to more than 95 percent; telephones increased from 25 percent to 76 percent; refrigerators from 15 percent to more than 90 percent. In the West, Midwest, and Northeast, the independent farm operator is today, as he never was, at a close parity with the urban middle class in living standards. In the South, traditionally a laggard area, farm incomes are rising faster than the national average. The important lesson is that it is possible to intervene in the market process to bring about desired social ends, although as our chapter has made abundantly clear, the process of intervention is far from simple and full of unexpected pitfalls.

KEY WORDS	CENTRAL CONCEPTS

KEY WORDS

Inelastic demand

New Deal

Substitutes

Corn-hog cycle

CENTRAL CONCEPTS

1. **The farm sector has been a traditional source of trouble in the American economy. Its problems have mainly been the result of *inelastic demand* and *the inability to control supply*.**

2. **The New Deal marked the first of a series of efforts to intervene in the agricultural sector and *minimize the fluctuations* that penalize farmers.**

3. **When agricultural demand increases (as with the Russian wheat deal of 1972), prices jump, and we find a *shift to substitutes*.**

4. **Agriculture suffers from the problem that everyone must make his production decisions at the same time. Farm output is also peculiar in that it can serve both as *capital input* and as a *final good for sale*. Therefore, in order to increase fu-**

ture production, the farmer must cut back on current production. These complications lead to a famous agricultural dilemma called the corn-hog cycle.

Cobweb

5. Farm output is also subject to the *cobweb* when farmers plan future production on the basis of current prices, leading to sharp fluctuations in prices and quantities sold.

Support prices

6. In order to avoid these fluctuations, the government has long sought means of *stabilizing prices*. One means has been *support prices*. Because support prices have usually been set above equilibrium prices, they have usually led to surpluses. To avoid these surpluses, the government has limited acreage. A problem up to the mid-1960s was that yields grew faster than acreage restrictions, leading to surpluses bought by the government. In effect, this transferred income from consumers to farmers.

Surpluses

7. To avoid the surplus effect, current farm legislation calls for establishing *target prices* but allows the market to set actual prices. *The farmer is given the differential if the actual price is below the target price.*

Target prices

8. Controls on food prices at the retail level will be ineffective if retail prices are below equilibrium prices. *Retail price controls must be accompanied by rationing in order to be effective.*

Rationing

9. Despite all of the difficulties, farm programs have succeeded in raising farm incomes.

QUESTIONS

1. Explain why the farmer has been traditionally regarded as the sick man of the economy.

2. Suppose that there is a sudden increase in the demand for corn. You are a farmer whose land is now used equally for wheat or oats. Would you expect the price of corn to affect the price of wheat and oats differently? Why? (HINT: wheat is not generally used to feed animals.)

3. Suppose U.S. wheat production were run by one firm. Anticipating a world wheat shortage, it decides to increase wheat output. Can it maintain its current volume of wheat sales? HINT: wheat is also a capital input and a final output.

4. Is there any industrial product that is both the means of its own production and the final product itself? (HINT: think about machine tools.)

5. Draw a supply and demand diagram that will yield an exploding cobweb.

6. Draw another supply and demand diagram that will eventually reach equilibrium prices.

7. India has established price ceilings to hold down prices. Why has this aggravated the Indian food problem?

8. What considerations would you take into account if you were given the job of establishing target

prices? (HINT: farm income is the major item at stake.)

9. [optional] You manufacture textiles and need 1,000 bales of cotton per month. Suppose cotton costs $100 per bale this month. As a manufacturer, you do not want to speculate on the future price of cotton. Accordingly, you buy cotton for next month's delivery at $110, and you also sell a contract for next month's delivery at $110. When next month arrives, the spot price of cotton is $120. Have you gained or lost money on the contract that you bought or on the contract that you sold? Suppose the price of cotton had fallen to $90 rather than risen to $120. What would your gains and losses be as a buyer and seller? Can you see that the contracts have taken the risks out of changes in cotton prices? Can you also see that the speculators with whom you have dealt have made it possible for you not to speculate?

10. Suppose that the price of beef rises, and consumers complain. What would be the most effective way of keeping beef prices from rising higher?

11. Why would it be easier to make a fair distribution of meat coupons then of gas coupons? Or don't you think it would be?

Operating a Competitive Firm

11

MOST OF US HAVE FIRSTHAND KNOWLEDGE of the household, whose role is crucial in determining the demand for goods. We know a lot less about the firm, whose function is to supply us with goods. In this chapter, we enter the factory gates of a competitive firm to find out what goes on inside this all-important institution in a market system.

THE RATIONAL, MAXIMIZING, COMPETITIVE FIRM

Note *competitive* firm. We will leave the operation of the giant corporation until later. There are several good reasons for this initial limitation of our investigation.

First, the kind of small, highly competitive firm we will describe exists as an important reality on the business scene, and we must understand how such firms do in fact survive and what role they play. Second, even the biggest and most monopolistic firms bear certain family resemblances to small com-

petitive ones, so that our study will lay the groundwork for a later analysis of very big business. And last, until we have grasped how the market mechanism works with small competitive firms, we will not be in a position to understand and to measure the difference that large firms and imperfect competition add to the system.

In addition to limiting our scrutiny to markets made up of small, competitive firms, we will add two assumptions about the behavior of these firms—one of them realistic and one perhaps not. The unrealistic assumption is that our firm acts *intelligently and rationally* in the pursuit of its goals. Since the world is littered with the bankruptcies that result from mistaken calculations and foolish decisions, this assumption may strain our credulity, but at least it will do no violence to understanding the *principles* of the market system. We recall learning about this assumption of rationality in Chapter 3.

Second, as we also remember from that chapter, we will assume that our firm is motivated by a desire to *maximize its profits.* Later, when we study the operations of very large firms, we shall have to ask whether this maximizing assumption makes much sense in their case. But when we deal with the "atomistic" competition of small firms, the assumption stands up very well because, as we shall see, *it is only by trying to maximize short-term profits that small firms manage to survive at all.* This gives to microtheory both a positive and a normative function. It tells us how small firms usually do behave—and thereby enables us to make predictions about the way the market system works. It also serves as a goal-oriented description of how small business firms *should* conduct their affairs if they hope to survive.

Economics of the Firm

The first person we encounter inside the factory gates is an economic personage we have not previously studied. It is the boss of the works, the organizer of the firm, the *entrepreneur.*

ENTREPRENEURSHIP

Our entrepreneur is not necessarily the capitalist—that is, the person who has supplied capital to the business. The capitalist may be a group of people who have lent money to the business but never visited the premises; or it might be a bank; or it might be the entrepreneur himself, acting as the risker of capital. But the entrepreneur provides a service that is essentially different from that of putting up capital. His contribution is *organizational.* Indeed, some economists have suggested that it is proper to think of four factors of production: labor, land, capital, and entrepreneurship, instead of the traditional first three alone.

ECONOMIC PROFIT

As a "fourth" factor of production, the entrepreneur is paid a wage, *the wage of management.* It can, of course, be very high, since entrepreneurship is a valuable skill—the skill of maximizing a firm's *economic profit.*

This is not the profit of everyday usage. In ordinary usage we call "profit" any sum left over after a firm has paid its wages and salaries, rents, costs of materials, taxes, etc. Included in that ordinary profit is an amount that an economist excludes from his definition of a true economic profit. This is the interest that is owed to the capitalist for the use of his capital. In other words, if a firm has a plant and equipment worth $1 million and makes a profit of $50,000, an economist would first ask whether the firm had taken into account the interest owing to it on this capital, before declaring the $50,000 to be a true economic profit. If interest rates were 5 percent, and no such allowance had been made (and it usually is not in ordinary accounting practice when the firm owns its own capital), an economist would say that no real economic profits were earned.

Economic profit, in other words, refers to the residual—if there is one. Indeed, our analysis shows us how an entrepreneur tries to create economic profits after appropriately remunerating all the factors, including capitalists, and how the operation of the market constantly tends to make this economic profit disappear, despite his best efforts. (That will become clear in our next chapter, rather than in this one.)

One last point. Who gets the residual? It goes to the owner of the business, who is legally entitled to any profits it enjoys. That owner, as we have said, may or may not be the entrepreneur. In a cooperatively owned factory it might be the work force. Usually it is a proprietor, a group of partners, or shareholders. But in all cases, this residual economic profit is over and above any recom-

THE SMALL BUSINESS SCENE

There are roughly 12 million "small" businesses in America, counting every one from the newsstand, through the farm, up to enterprises that employ several hundred people and count their dollar value in tens of millions. Among these 12 million business units, corporations number only about 1.7 million, but these 1.7 million corporations do nearly *five times* as much business as all the proprietorships and partnerships. In turn, however, most of the 1.7 million corporations are small, even when measured by a small-business yardstick. Over 60 percent do less than $100,000 a year in sales, with the result that the whole group of little corporations accounts for only 1.3 percent of all corporate sales.

Just to get an idea of scale: the 500th largest American industrial firm in 1973 (in terms of sales) was Avery Products (San Marino, Calif.) with sales of $243 million. The 500th biggest in terms of assets was National Beef Packing, worth $33 million. At the other end of the list, the largest industrial corporation in terms of sales was General Motors, with a sales volume of $35.8 *billion*. Biggest in terms of assets was Exxon with assets of almost $25 billion. This does not count A.T.&T., which is not classified as an "industrial" company, but which is the largest enterprise in the nation, by far, in terms of assets—($67 billion in 1973).

Thus, in terms of sales (or assets), little business is little indeed. Yet, by virtue of numbers, the small businesses of America are by no means unimportant in the national economic picture. Small business collectively employs roughly 40 percent of the American labor force. (This compares with 35 percent of the labor force that works for the "not-for-profit" sector—state, local, and federal government, hospitals, social service agencies, clubs, etc.,—and with 25 percent that works for big business.) Millions of small entrepreneurs constitute the very core of the American "middle class," and thus give a characteristic small-business view to much of American political and social life.

pense for the services these factors supply, including the services of making capital available.

THE ENTREPRENEUR AT WORK

Exactly what does the entrepreneur do to earn his wages of management? The answer is essentially simple: he does three things; he

- buys factors in the factor market
- combines factors in the enterprise, to produce output as inexpensively as possible
- sells the finished product in the goods market

These three tasks may be enormously complicated in real life. Buying factors in the factor market is likely to entail bargaining with unions, establishing credit connections with banks, and arranging complex real estate deals. Producing output for the goods market will surely need technical skills, including those of a designer to make the output of one entrepreneur distinguishable from that of another. Selling goods and services will require the special skills of advertising personnel and a sales force.

As a result, entrepreneurship is often revealed by the ability of one firm to gain an edge over another in hiring labor and talented designers, obtaining credit or land, and devising selling techniques. But in the competitive firm whose factory we are visiting, none of these real-life situations will be presumed to exist. The entrepreneur will be hiring labor, capital, and land on terms exactly equal with his competitors. Moreover, he is such a small part of the industry that he cannot influence the markets for factors or goods, whatever he does. Within his plant, furthermore, we will assume that he produces some product that cannot be "differentiated" from that of his competitors. A farmer's wheat, a metal manufacturer's nails or screws, the stampings of a small plastics factory are examples of these kinds of goods. Last, because his output is exactly the same as that of his competitors, he will have no use for the wiles of advertising or elaborate selling costs. What good would it do for a farmer to advertise "Buy wheat!"?

In subsequent chapters, we will look more carefully into these conditions of atomistic competition, as well as into the problems of entrepreneurship in situations where these special conditions do not apply. But the very severity of the setting for our competitive firm highlights the essential economic function of entrepreneurship. For if we think of the entrepreneur buying his factors of production and selling his output in terms of absolute equality with his competitors, and producing a product that is indistinguishable from theirs, only one task remains. *He must buy the right amounts of factors' services* and combine these efficiently, *in order to produce output as cheaply as possible.*

THE PROBLEM OF SCALE

When we now enter the factory, we begin to understand the economic problem that the entrepreneur faces. Around us we see machines, men, space. Why are they combined in the proportions we find? Why did the entrepreneur hire this many workers, that much floor space, so much capital?

The answer lies partly in engineering and partly in economics. We will learn about these matters bit by bit in this chapter and the next one, so that you should not expect an instant answer. But a good place to begin is with a fact of technology or engineering called *indivisibility*. It is an aspect of something we first discussed in Chapter 3; namely, that *size matters*. In this case it matters because with each factor there is a certain *minimum scale* that must be attained if we are to generate production at less than very high costs.

From one industry to another, this minimum size changes, largely because of technical or engineering considerations. Suppose, for example, that we were considering opening a bookstore. A bookstore may be very small, but it must occupy *some* space; it must stock a reasonable number of books and employ at least one person to sell them. If we were in agriculture, we would need a farm of at least a certain area depending on our crop, and a basic amount of capital in the form of buildings, equipment, fertilizer, seed. If we were in manufacturing, there would be a minimum size of plant or machinery essential for our operation. If we were in a mass-production business, such as steel, the smallest efficient plant might run into an investment of millions of dollars and a work force of thousands. But that would take us well out of the world of atomistic competition to which we are still devoting our attention.

In other words, the first decision about hiring factors involves the physical (or engineering) fact that there is a certain amount of each factor that we must hire for technical reasons. This is what we mean by minimum scale. If problems of scale did not exist we could produce television sets as efficiently in a garage as in a vast plant or raise cattle as cheaply in our back yard as on a range.

THE FACTOR MIX

Indivisibilities of land, labor, or (mostly) capital, help explain the initial decisions that an entrepreneur must make. There is a certain critical size or scale he must have, to produce at all, although this size differs from one activity to another. He must have the capital to buy the factors needed to operate at that scale, or he cannot start up in business in the first place.

But suppose that he does have the capital, and that he hires the smallest possible amount of labor, land, and capital. Is there any reason to believe that this is the *combination* of factors that will enable him to produce as cheaply as possible? The question brings us to the next problem of entrepreneurship—*deciding what mix of factors will be most profitable.* How does an entrepreneur decide how much labor, how much land, how much capital to hire?

Law of Variable Proportions

As we shall see, this is by no means a simple question to answer, for many considerations bear on the decision to hire a factor or not. But in the center of any businessman's calculations lies one extremely important fact to which we must now pay close attention. *This is the changing physical productivity that results from combining different amounts of one factor with fixed amounts of the others.* Here the entrepreneur encounters the laws of production that we first encountered in Chapter 3.

PRODUCTIVITY CURVE: INCREASING RETURNS

Let us begin with a case that is very simple to imagine. Suppose we have a farmer who has a farm of 100 acres, a certain amount of equipment, and no labor at all. Now let us observe what happens as he hires one man, then a second man of the same abilities, then

a third, and so on. Obviously, the output of the farm will grow. What we want to find out, however, is whether it will grow in some clearly defined pattern that we can attribute to the changing amounts of the factor that is being added.

What would such a curve of productivity look like? Assume that one man, working the 100 acres alone as best he can, produces 1,000 bushels of grain. A second man, helping the first, should be enormously valuable, because two men can begin to specialize and divide the work, each doing the jobs he is better at and saving the time formerly wasted by moving from one job to the next. As a consequence of this division of labor, output may jump to 3,000 bushels.

MARGINAL PRODUCTIVITY

Since the *difference* in output is 2,000 bushels, we speak of the *marginal productivity* of labor, when two men are working, as 2,000 bushels. Note that we should not (although in carelessness we sometimes do) speak of the marginal productivity of the second *man*. Alone, his efforts are no more productive than those of the first man: if we fired the first man, worker number 2 would produce only 1,000 bushels. What makes the difference is the jump in the combined productivity of the *two* men, once specialization can be introduced. Hence we should speak of the changing marginal productivity of *labor,* not of the individual.

It is not difficult to imagine an increasing specialization taking place with the third, fourth, and fifth man, so that the addition of another unit of labor input in each case brings about an output larger than was realized by the average of all the previous men. Remember that this does not mean the successive factor units themselves are more productive. *It means that as we add units of one factor, the total mix of these units plus* *the fixed amounts of other factors, forms an increasingly efficient technical combination.**

We call the range of factor inputs, over which average productivity rises, a range of *increasing average returns*. It is, of course, a stage of production that is highly favorable for the producer. Every time he adds a factor, efficiency rises. (As a result, as we shall see in our next chapter, costs per unit of output fall.) The rate of increase will not be the same, for the initial large marginal leaps in productivity will give way to smaller ones. But the overall trend of productivity, whether we measure it by looking at *total* output or at *average* output per man, will still be up. And all this keeps on happening, of course, because the factor we are adding has not yet reached its point of maximum technical efficiency with the given amount of other factors.

DIMINISHING RETURNS

Then our farmer notices a disconcerting phenomenon. At a certain point, marginal output no longer rises when he adds another man. Total output will still be rising, but a quick calculation reveals that the last man on the team has added less to output than his predecessor.

What has happened is that we have overshot the point of maximum technical efficiency for the factor we are adding. Labor is now beginning to "crowd" land or equipment. Opportunities for further specializa-

---

*With each additional man, the proportions of land, labor, and capital are altered, so that the change in the level of output should rightfully be ascribed to new levels of efficiency resulting from the interaction of *all three factors*. But since labor is the factor whose input we are varying, it has become customary to call the change in output the result of a change in "labor productivity." If we were altering land or capital alone, we would call the change the result of changes in their productivities, even though, as with labor, the real cause is the changing efficiency of *all* factors in different mixes.

tion have become nonexistent. *We call this condition of falling marginal performance a condition of decreasing or diminishing returns.* As the words suggest, we are getting back less and less as we add the critical factor—not only from the "marginal" man, but from the combined labor of all men.

If we now go on adding labor, we will soon reach a point at which the contribution of the marginal man will be so small that average output per man will also fall. Now, of course, costs will be rising per unit. If we went on foolishly adding more and more men, eventually the addition of another worker would add nothing to total output. In fact, the next worker might so disrupt the factor mix that *total* output would actually fall and we would be in a condition of negative returns.

AVERAGE AND MARGINAL PRODUCTIVITY

This changing profile of physical productivity is one of the most important generalizations that economics makes about the real world. It will help us to think it through if we now study the relationships of marginal and average productivity and of total output in Table 11·1.

TOTAL, AVERAGE, AND MARGINAL PRODUCT

All three columns are integrally related to one another, and it is important to understand the exact nature of their relationships.

1. *The column for total output is related to the column for marginal output, because the rise in total output results from the successive marginal increments. For instance, the reason total output goes from 7,800 bushels with 4 men to 9,800 with 5 men is that the marginal output as-*

TABLE 11 · 1

Number of men	Total output	Marginal productivity (change in output)		Average productivity (total output ÷ no. of men)	
1	1,000	1,000 ⎫	Increasing	1,000 ⎫	
2	3,000	2,000 ⎬	marginal	1,500 ⎪	Increasing
3	5,500	2,500 ⎭	productivity	1,833 ⎬	average
4	7,800	2,300 ⎫		1,950 ⎪	productivity
5	9,800	2,000 ⎪	Decreasing	1,960 ⎭	
6	11,600	1,800 ⎬	marginal	1,930 ⎫	Decreasing
7	13,100	1,500 ⎪	productivity	1,871 ⎬	average
8	14,300	1,200 ⎭		1,790 ⎭	productivity

sociated with the fifth man is 2,000 bushels. Thus, if we know the schedule of total outputs, we can always figure the schedule of marginal outputs simply by observing how much total output rises with each additional unit of factor input.

2. It stands to reason, therefore, that if we know the schedule of marginal outputs, it is simple to figure total output: we just add up the marginal increments.

3. Finally, the meaning of average productivity is also apparent. It is simply total output divided by the number of men (or of any factor unit in which we are interested).

One thing must be carefully studied in this example. Note that marginal productivity begins to diminish with the fourth man, who adds only 2,300 bushels to output, and not 2,500 as did his predecessor. Average productivity, however, rises until we hire the sixth man, because the fifth man, although producing less than the fourth, is still more

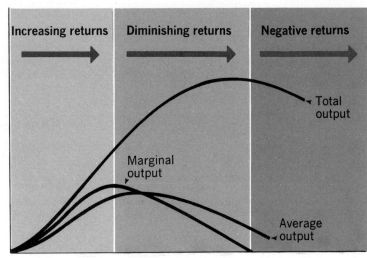

**FIGURE 11 · 1
The law of
variable proportions**

Increasing returns Diminishing returns Negative returns

Output of all factors

Total output

Marginal output

Average output

Input of one factor

productive than the average output of all four men. *Thus marginal productivity can be falling while average productivity is still rising.*

The three curves in Fig. 11·1 all show the same phenomenon, only in a graphic way. The top curve shows us that as we add men to our farm, output at first rises very rapidly, then slowly, then actually declines. This is the curve we met in Chapter 3. The marginal productivity curve shows us *why* this is happening to total output: as we add men, the contribution they can make to output changes markedly, at first each man adding so much that average output grows rapidly; thereafter marginal output falling although average output still rises; finally each man adding so little that he actually pulls down the average that obtained before his hiring. And the average curve, as we have just indicated, merely sums up the overall output in an arithmetical way by showing us what the average person contributes to it.

Put into the form of a generalization, we can say that *as we add successive units of one factor to fixed amounts of others, the marginal output of the units of the variable factor will at first rise and then decline. We call this the law of variable proportions or the law of diminishing returns, or we can simply talk about it as the physical productivity curve.*

THE LAW REVIEWED

The generalizations of the law of variable proportions constitute one of the key insights that microeconomics gives us into the workings of the real world. Let us be certain that we understand exactly what the law says and implies.

1. The law of variable proportions describes what happens to physical productivity when we add units of one factor and *hold the others constant.* As we added labor in the example above, we did not also add land or capital. Had we done so, there would have been no way of ascribing changes in output to the addition of one factor rather than the other.

2. The law applies to adding *any* factor to fixed quantities of the others. Suppose that we had started with a fixed amount of labor and capital and had added successive acres of land. The first acre would not have been very productive, for we would have had to squeeze too much labor and capital into its area. The second acre would have permitted a better utilization of all three factors, and so *its* marginal productivity would have been much higher. But in time the addition of successive units of land would pass the point of optimum mix, until another acre would add so little yield that the average production of all acres would be pulled down. And the same pattern of increase, diminution, and final decrease would of course attend the addition of doses of capital—say successive bags of fertilizer or additional tractors—to a fixed amount of land and labor.

3. Unlike many other "laws" in economics, the law of variable proportions has nothing to do with behavior. The actions of men on the marketplace or the impulses or restraints of utility and disutility play no role in diminishing returns. *Essentially, the law expresses a constraint imposed by the laws of nature.* If there were no such constraint, we could indeed grow all the world's food on a single acre or even in a flowerpot, simply by adding more and more labor and capital.

Marginal Revenue and Marginal Cost

But now we must get back to our point of interest—the firm seeking to hire factors to its own best advantage. Is a knowledge of

NATURE'S CONSTRAINTS AGAIN

Here is a good place to review the different constraints of nature that we first encountered in Chapter 3. We are now familiar with the law of diminishing returns. The key element in this law is the effect on productivity of adding one factor only, *while holding the others constant.*

The law of increasing cost differs in an important regard. It describes the effect on output of adding *all factors,* not just one. Suppose we wanted to increase the national production of wheat and began to take land, labor, and capital from other uses and put them into wheat farming. At first we might encounter no change in productivity per combined "unit" of land + labor + capital (say, per hundred men, hundred acres, and $10,000 worth of equipment). But soon we would be forced to use factors that were less and

less appropriate to wheat output. Imagine trying to grow wheat in Nevada, using casino croupiers as workers and hotel limousines hitched to plows! *The law of increasing cost therefore refers to the decline in output that results from using factors that are not specialized in the output of the good in question.* In the law of diminishing returns this problem of specialization does not enter: all our units of added factor input are presumed to be homogenous.

Last, we have constraints having to do with scale. We discuss the problem in our next chapter. As we shall see, its constraints have to do with technological considerations that make operations more efficient over a certain size. Changes in scale are one way we avoid diminishing returns, which presuppose that the scale of an operation is fixed because the input of the two complementary factors does not change.

factor productivity all a businessman needs? To revert to our first illustration, suppose we knew that a single salesperson in our bookstore had a productivity of (i.e., could sell) 5 books a day. Would that alone tell us if we should hire him?

The question answers itself. Before we can hire the clerk or any other factor, we have to know two other things: (1) *what the unit of the factor will cost,* and (2) *how much revenue our firm will get as a result of hiring that unit of the factor.*

If the price of a salesperson is $6,000 and if we think he will sell 5 books a day at an average markup of $5 per book, then if he works for 250 days he will bring in revenues of $6,250 (5 × $5 × 250). Obviously it will pay to hire him. On the other hand, if his productivity were less—if he sold only 4 books per day—then the revenues from hiring him would be only $5,000 (4 × $5 × 250),

or $1,000 less than his wage. He would be a dead loss.

MARGINAL REVENUE PRODUCT

Here we can add a simple term to our vocabulary. We call the marginal revenue we get from adding a unit of a factor its *marginal revenue product.* It is simply the *physical* increase in output from the additional factor, multiplied by the price at which we sell that output.

The hiring rule is then very simple. If the marginal revenue product of a factor is greater than its cost, hire it. If it isn't, don't. (Furthermore, it will raise our profit if we fire any factors whose marginal revenue products are less than their cost.)

TABLE 11 · 2

Number of men	Marginal cost per man @ $4,500	Marginal physical output per man (from Table 11·1)	Marginal revenue product per man (output × $2.50)	Marginal profit or loss
1	$4,500	1,000	$2,500	−2,000
2	4,500	2,000	5,000	500
3	4,500	2,500	6,250	1,750
4	4,500	2,300	5,750	1,250
5	4,500	2,000	5,000	500
6	4,500	1,800	4,500	0
7	4,500	1,500	3,750	− 750
8	4,500	1.200	3,000	−1,500

Let us see how this actually works in practice. In the schedules below we go back to the farm, this time armed with two new pieces of information. We now know that labor costs $4,500 per man and that a bushel of wheat sells for $2.50. (Later we will look into *how* we know these things, but here we can take them for granted.) Our farm schedule of marginal costs and marginal revenue products therefore looks like Table 11·2.

What does our table tell us? Our first man seems to be very unprofitable, for he costs us $4,500 and brings in only $2,500. We suspect, however, that he is so unprofitable because he is trying to spread his one unit of labor over the whole farm. The addition of a second man confirms our suspicions. He also costs us $4,500, but brings in $5,000. (Remember it is not the second man himself who does so, but the two men working together who give rise to an increase in revenues of that amount.) The third and fourth and fifth men also show profits, when we compare their marginal revenue products to their marginal costs, but when we reach the sixth man, the law of diminishing returns brings its decisive force to bear. The sixth man is unable to increase the revenues of the

team by more than $4,500, which is just his hire. It is not worthwhile to engage him.

TOTAL COSTS AND TOTAL REVENUES

Are we certain that hiring five men will really maximize the profits of the farm? We can find out by adding up our *total* costs and our *total* revenues and figuring our profit at each level of operation. Table 11·3 does just that.

TABLE 11 · 3

Number of men	Total cost of men	Total revenue	Profit (total revenue less total cost)
1	$ 4,500	$ 2,500	−$2,000
2	9,000	7,500	− 1,500
3	13,500	13,750	250
4	18,000	19,500	1,500
5	22,500	24,500	2,000
6	27,000	29,000	2,000
7	31,500	32,750	1,250
8	36,000	35,750	− 250

We are really fooling ourselves when we "check" on our former calculations about marginal changes by looking at the totals. For just as with marginal and total output, the totals are themselves nothing but the sum of the marginal changes! As long as each man brings *some* addition to revenue, large or small (that is, so long as marginal revenue is larger than marginal cost when that man is hired), then the total of all revenues must be growing larger too. When we add up the marginal contributions, we measure each different-sized contribution. We should not consider it surprising that the whole is the sum of what we put into it.

As Table 11·3 shows, our best profit comes with the hiring of 5 men. The addition of a sixth does us no good. A seventh lowers our net income—not because of his lack of skill or effort, but because with seven men the mix of labor and the fixed amounts of other factors is no longer so efficient as before. An eighth involves us in an actual loss.

We can also see now how neatly the physical productivity curve ties together marginal cost and marginal revenue product. For three things would entice us to hire the sixth or even the seventh man.

1. *A fall in cost.* If wages dropped to any figure under $4,500, our sixth man will immediately pay his way. By how much would they have to drop to make it worthwhile to hire the seventh man? Marginal revenue product when seven men are working is only $3,750. The wage level would have to drop below that point to bring a profit from seven men.

2. *A rise in the price of output.* If the demand for grain increased, and the price of grain went to $3, our calculations would change again. Now the sixth man is certainly profitable: adding him now brings a marginal revenue product of 1,800 bushels × $3, or $5,400, far above his wage. Is

the seventh man profitable at his wage of $4,500? His physical productivity is 1,500 bushels. At $3 per bushel he is not quite worth hiring. At $3.01 he would be.

3. *An increase in productivity.* If a change in skills or techniques raises the physical output of each man, this will also change the margin of profitable factor use. Any small increase will lead to the employment of the sixth man.

CHOICE AMONG FACTORS

We have talked, so far, as if an entrepreneur had only one "scarce" factor and as if his only task were to decide how much of that factor to add. But that is not quite the problem faced by the businessman. He has to make up his mind not only *whether* to add to his output at all, but *which* factor to hire in order to do so.

How does a businessman make such a choice? How would we choose between adding a salesclerk or inventory to our bookstore, or how would the farmer decide between hiring labor or spending the same sum on capital or land? Suppose the farmer has already hired four men to work for him and has also rented a certain amount of land and used a certain amount of capital. Now when he thinks about expanding output, he needs to know not only how much the addition of another man will yield him, but *what the alternatives are.*

They might look like this.

1. He can hire a fifth worker for $4,500, who will, as we know, bring about an increase of 2,000 bushels in production.

2. He could spend the same amount on land, renting an acre that would increase his output by, let us say, 2,100 bushels.

3. For the same outlay he could engage the services of a tractor that would add 2,200 bushels a year to his output.

The next step is easy. The entrepreneur compares *the cost* of hiring a unit of each factor with the *marginal revenue product* of the factor. Thus, in the example above:

Factor	Marginal physical product	Selling price of goods	Marginal revenue product
Labor	2,000 bu.	$2.50	$5,000
Land	2,100 bu.	2.50	5,250
Capital	2,200 bu.	2.50	5,500

Now, comparing marginal revenue products and cost:

Factor	Marginal revenue product	Cost	Profit
Labor	$5,000	$4,500	$ 500
Land	$5,250	$4,500	750
Capital	$5,500	$4,500	1,000

Obviously, the tractor is the best buy, dollar for dollar, and the entrepreneur should spend his money on tractors, as long as the relative prices and productivities of the factors are unchanged.*

BIDDING FOR FACTORS

But will factor prices remain unchanged? We have not yet inquired into the factor market, where wages, rents, and interest are priced. But to an entrepreneur in a competitive firm, that process doesn't matter. Prices "exist" for

land, labor, and capital, and he has no alternative but to accept them.

Nonetheless, we can now see how an entrepreneur's reactions to the prices of factors can affect those prices. You will remember we assumed, at the outset of our discussion, that the entrepreneur farmer or manufacturer—was able to bid for additional factors without thereby affecting their prices.

This assumption followed from the premise of atomistic firms from which we started. The amount of land or labor or capital that such a small firm can add to its operations is so insignificant a portion of the total supply of that factor that the individual firm's demand does not affect the price of the factor appreciably, if at all. If 10,000 young women are looking for sales work in New York City, the addition of a few salespeople in any single business will not change the going price for salespeople at all. Neither will a small farmer's demand, by itself, alter the rent of land, nor will one firm's demand for capital change the rate of interest.

Therefore, for *one small firm, the supply curve of any factor looks like a horizontal line.* It is infinitely elastic, because a firm can engage all of the factor it requires at the going price without affecting the price of that factor at all.

But this is not the case when many small firms all begin to demand the same factor. If all stores are looking for sales help, the salaries of salespeople will rise. If many farmers seek land or capital, rentals or interest (or the prices of capital goods) will go up. As a result, each firm will find that the going price for the factor in general demand has a mysterious tendency to rise, as Fig. 11·2 shows.*

---

*Ideally, an economist would wish to compare the marginal productivities of much smaller amounts of these three factors; i.e., an hour's worth of labor or a very small parcel of land or a day's use of a tractor. We have dealt in big chunks because factors often do come in indivisible units, and because this is the way the problem usually looks in the business world.

*The diagram does not emphasize a very important difference in *scale*. One inch along the horizontal axis of the industry diagram on the left may represent 100,000 units. The same inch along the firm's horizontal axis would then stand for only a few units.

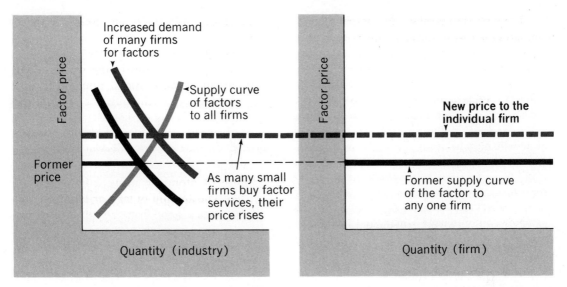

Increased demand of many firms for factors

◄Supply curve of factors to all firms

Factor price

Former price

As many small firms buy factor services, their price rises

Quantity (industry)

Factor price

New price to the individual firm

Former supply curve of the factor to any one firm

Quantity (firm)

FIGURE 11 · 2

Supply curve to all firms vs. supply curve to one firm

Supply and demand sets the prices of factors, just as they do when consumers buy goods. But now the demand is exercised by a cluster of firms or an industry. The individual firm has no impact on the market and no choice but to accept its price as given.

FACTOR PRICING

We can now finish the first part of our analysis. We remember that in our last example it was profitable for farmers to buy tractors, rather than land or labor. Now we must suppose that many farmers, finding themselves in the same situation, all bid for tractors. The result, of course, is that tractor prices will rise. The consequence of this, plainly, is that tractors become less profitable to the farmer, compared with land or labor, than they were originally. Suppose that the price of tractor

hire goes up to $5,000, so that for an outlay of $4,500 a farmer can now afford a tractor only four days out of the week instead of five (or if he buys a tractor, he has to get a smaller model). Since the expenditure of $4,500 will now buy fewer tractor inputs, the tractor's addition to output will also decrease. Let us imagine that tractor output falls by almost one-fifth with the cut in tractor hours. Then $4,500 of expenditure will bring an increase in output of 1,800 bushels instead of 2,200. At $2.50 per bushel, the marginal revenue from using a tractor now amounts to only $4,500. There is no net profit in using tractors at all!

EQUIMARGINAL RETURNS

Actually things will work out in less dramatic fashion than this. As tractor prices rise (and as the marginal productivity of tractors falls, owing to their more intensive use), the profit advantage between tractors and land will narrow. At a certain point, land will be just as attractive to farmers as tractors. Thereafter, since both land and tractors are still better buys than labor, the price of *both* land and tractors will continue to rise. Again as a result, the profit difference between them and labor will narrow until the same profit will be derived from an equal expenditure on any of the three factors. Now, finally, the prices of *all three* factors may rise if there is still a profit obtainable from using them. Only when rising prices and falling marginal productivities make it no longer profitable to seek any more factors at all will the bidding stop and factor prices stabilize.

Here is (have you recognized it?) the *equimarginal* rule again! Our entrepreneur maximizes his profits first by seeking *equal marginal revenues,* just as the consumer seeks equal marginal utilities. Thereafter, he hires factors until there is no marginal profit to be had from any of them, just as the consumer buys commodities until there is no marginal consumers' surplus available from any of them.

Now perhaps we can stand back and review the process.

1. An entrepreneur who makes a profit in a competitive market will seek to expand his output to increase that profit. This will require hiring additional factors.

2. When deciding which factor to hire, an entrepreneur will compare the dollar return he will get from an equal dollar outlay on different factors. The size of these relative returns will depend on the relative costs of the various factors and on their different marginal revenue products.

3. Entrepreneurs in a given industry will normally discover that one factor is more profitable than any other, at a given time, and will concentrate their demand on that factor. As a result, the price of that factor will rise; and as more of it is used,

its marginal productivity will decline. Hence its profit advantage over the other factors declines.

4. As the result of continuously bidding for the most profitable factor, the profitability of all factors will be equalized. There will then be no advantage to spending a given sum on land instead of labor or on capital instead of land, and so on. Instead, all factors will now be equally sought, *according to the equimarginal rule of maximizing.*

5. Finally, all factors will be bid up in price until there is no longer any profit to be had from using any more of any of them. At that point the expansion of the industry stops, and the prices of factors remain steady.

THE REMUNERATION OF FACTORS

At the conclusion of our analysis, we see *that in a competitive market all factors will be paid their marginal revenue products.* Any factor that was not paid its marginal revenue product would still yield a profit to the entrepreneur. That factor will be added until its marginal physical product had dropped and there was no longer a profit to be had from its employment; or it would be bid for by many entrepreneurs, in which case its cost would rise. In either case, each factor will be remunerated to the full extent of its marginal revenue product.

This conclusion threatens to take us away from the factory into the market for factors; but we must not allow it to, because we have not completed our investigation into the tasks of the entrepreneur. So let us leave in abeyance the relation between wages and interest and rents and marginal revenue products. We will complete that part of our microeconomic analysis when we study the factor market carefully as part of our inquiry into the distribution of income.

First, however, there is an important task to be done. We have learned how an enterpriser decides which factors to hire.

But we have not really tied his hiring rule into the task that we defined as his main objective: producing goods as cheaply as possible. Let us therefore turn to the competitive firm in action, to watch the entrepreneur minimizing costs—and to see his profit-making efforts frustrated by the actions of the market itself.

KEY WORDS

Maximization

Entrepreneurship

Economic profit

Entrepreneurial task

Minimum scale

Increasing and decreasing returns

Law of diminishing returns or law of variable proportions

Marginal productivity and marginal revenue product

Marginal revenue product vs. cost

CENTRAL CONCEPTS

1. The study of the firm concentrates initially on the small enterprise engaged in "atomistic" competition. It assumes that such enterprises act intelligently and rationally and that they try to maximize their profits in the short run.

2. The firm introduces us to a new "factor of production," the *entrepreneur*. (It is a matter of convention whether or not we choose to call him a fourth factor of production.) He is paid a *management wage*, and his function is to maximize economic profit.

3. *Economic profit* does not include the normal return to capital (interest). It is *the residual after a firm pays all factors their marginal revenue product and any other costs.*

4. The entrepreneur of a competitive firm has three tasks. *He buys factors in the factor market, sells output in the market for goods and services, and combines the factors in the enterprise.* No single entrepreneur has any influence over the market for factors or the market for goods because his operation is too small to affect these markets. He buys and sells on terms of complete equality with his competitors. Therefore *his task is to combine the factors in the right proportions to produce output as cheaply as possible.*

5. For all factors there is some minimum amount that is necessary for normally efficient production. This is determined by the smallest "indivisible" unit of land, labor, and capital that can be employed in a particular activity. It determines the *minimum scale* of the enterprise.

6. As an entrepreneur adds factors (over that minimum initial amount) he is faced with the problem of their *changing productivity*. As he adds more and more of any *one* factor to a fixed amount of the others, he finds that marginal output initially rises, and subsequently, declines. This sequence of rising and then declining marginal productivity is reflected in *rising and then declining average productivity*. The phenomenon is called the *law of diminishing returns*, or *the law of variable proportions*.

7. The law of diminishing returns is entirely a *law of production*, imposed by nature. It is not a law of behavior, such as maximizing. It applies to all factors including entrepreneurship, always with the proviso that we hold all other factors fixed.

8. Marginal productivity is a critical element in determining *the marginal revenue product* of a factor. This marginal revenue product is its physical marginal product multiplied by the price of its output.

9. The crucial determination in whether or not to hire a factor *is made by comparing its marginal revenue product with its cost.* Unless the marginal revenue product is greater than cost, we will not hire a factor, for it would entail a loss.

10. The entrepreneur will compare the returns to be had from all factors yielding the *highest net return per dollar of cost.* He will buy the most profitable input.

*Rising
factor prices*

11. Any individual entrepreneur can add as much of any factor as he wishes without raising its price. But when *many entrepreneurs bid for the same factor, they do raise its price.*

12. The result of the competitive bidding of entrepreneurs for factors will have several results. *First, it will equalize the profitability of a dollar of outlay for all factors. Second, it will drive up the price of factors until their remuneration equals their marginal revenue product.*

*Maximizing
by the
equimarginal
rule*

13. The entrepreneur will follow the *equimarginal rule*, seeking to maximize profit by (1) obtaining equal marginal profits from all factors, and (2) continuing to hire factors until their marginal profits are zero.

QUESTIONS

1. Suppose that you were about to open a small business—say a drugstore. What do you think would be the factors critical in determining the scale of your operation? Suppose it were a farm? A factory?

2. Once you had started the business, what consideration would be uppermost in your mind when you were deciding how much of the factors to hire? What is the cardinal rule you would have to bear in mind in deciding if a unit of a factor would or would not pay its way?

3. One thing that would affect your decision to hire or not to hire a factor would be the amount of physical increase in output it would yield. What is the generalization we make about the change in output associated with combining more and more of one factor with a fixed combination of others? Is this generalization based on behavior? State the law of variable proportions as carefully as you can.

4. What is meant by marginal productivity? What is its relation to average productivity? Suppose you were considering the increase in your drug sales that would result from adding square feet of space. Draw up a schedule showing that the addition of square footage (in units of 100 sq ft) would at first yield increasing returns (dollars of sales) and then diminishing returns.

5. Suppose that a manufacturer had the following information for a given plant and number of men:

Number of machines	1	2	3	4	5	6	7	8
Total output (units)	100	250	450	600	710	750	775	780

What would be the marginal productivity of each successive machine? The average productivity from using additional machines? When would diminishing *marginal* productivity set in? Diminishing average productivity?

6. Why must we hold the other factors constant to derive the law of variable proportions?

7. Suppose that each machine in Question 5 cost $1,000 and that each unit of output sold for $10. How many machines would it be most profitable to hire? (Figure out the marginal revenue, product and marginal cost for each machine added.)

8. What would be the most profitable number to have if the cost of the machine rose to $1,500? If the price per unit of sales fell to $9?

9. Suppose the manufacturer found he had the following alternatives. He could

   · spend $1,000 on a machine that would add 115 units to sales (each unit selling at $10)

   · spend $5,000 to hire a new man who would increase output by 510 units

   · rent new space for $10,000 that would make possible an increase in output of 1,100 units

   How would he know which was the best factor to hire? Show that he would begin by asking what his dollar return would be per dollar of cost for each factor. What is this in the case of the machine? The new man? The land?

10. If one manufacturer in a competitive market adds to his factor inputs, will that affect their price? What happens when all manufacturers bid for the same factor? In the example above, which factor will be bid for? What will happen to its price? To its marginal productivity? Which factor will then become the "best buy"? What will happen to *its* price? What will be the final outcome of the bidding?

11. Explain how the equimarginal rule for maximizing profit resembles that for maximizing a consumer's utility. Is profit-maximizing a tautology?

The Competitive Firm in Action

LET US ONCE AGAIN put ourselves in the shoes of an imaginary entrepreneur. Since we have already become familiar with the firm's calculations in regard to buying factor services, let us extend our knowledge into a full appreciation of what the cost problem looks like to the entrepreneur.

FIXED AND VARIABLE COSTS

We know that a firm's total costs must rise as it hires additional factors. Yet if we put ourselves in a businessman's position, we can see that our total costs will not rise proportionally as fast as our additional factor costs, because there are some costs of production that will not be affected by an increase in factor input. Real estate taxes, for example, will remain unchanged if we hire one man or 100—so long as we do not acquire additional land. We can assume that the depreciation cost of machinery will not be affected by additions to land or labor. Rent will be unchanged, unless the premises are expanded.

The cost of electric light will not vary appreciably despite additions to labor or machinery. Neither will the salary of the president. *Thus some costs, determined by legal contract or by usage or by the unchanging use of one factor, do not vary with output. We call these fixed costs.*

In sharp contrast with fixed costs is another kind of cost that does vary directly with output. Here are many factor costs, for generally we vary inputs of labor and capital (and sometimes land) every time we seek a new level of production. To increase output almost always requires the payment of more wages and the employment of more capital (if only in the form of inventories or goods in process) and sometimes the rental of more space. *We call all costs that vary with output variable costs.*

UNIT COSTS

This important dichotomy between fixed and variable costs also requires us to shift our view a little within the business enterprise. We have been mainly concerned with calculating the costs of *inputs* in relation to their marginal productivities. Now we must turn around and begin to think about costs in relation to *output*.

In particular, we have to learn to think in terms of *cost per unit of output*. As we have seen, when a manufacturer (or a farmer or a storekeeper) expands his output, his total costs usually rise because his variable costs are going up, but total costs do not rise proportionally as fast as variable costs, because fixed costs are set. But it is not easy to work with this upward rising curve of total costs. Hence businessmen and economists usually convert the figures for total cost into *unit costs* by dividing the total cost by the number of units of goods produced. This results, of course, in a figure for the *average cost per unit of output*. We shall see that this gives us a very useful way of figuring what happens to costs as output expands.

FIXED AND VARIABLE COSTS PER UNIT

There is certainly no difficulty in picturing what happens to fixed costs per unit of output as output rises. By definition, they must fall. Suppose a manufacturer has fixed costs (rent, certain indirect taxes, depreciation, and overhead) of $10,000 a year. If he produces 5,000 units of his product per year, each unit will have to bear $2 of fixed costs as its share. If output rises to 10,000 units, the unit share of fixed costs will shrink to $1. At 100,000 units it would be a dime. Thus a curve of fixed costs per unit of output would look like Fig. 12·1.

What about variable costs per unit? Here the situation is more complex, for it depends directly on the analysis of the productivity curve we discussed in our last chapter. Hence, let us first set up a hypothetical schedule of output for our manufacturer (Table 12·1).

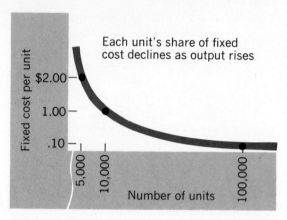

Each unit's share of fixed cost declines as output rises

FIGURE 12 · 1
Profile of fixed costs per unit

As in the case of our farm, once again we see the law of variable proportions at work. The total numbers of units he produces will rise at first rapidly, then more slowly, as he adds labor input to his plant.

TABLE 12 · 1

Number of men	Total output (units)	Marginal product
1	5,000	5,000
2	13,000	8,000
3	23,000	10,000
4	32,000	9,000
5	39,000	7,000
6	44,000	5,000
7	47,000	3,000
8	49,000	2,000

To convert this schedule of physical productivity into a unit cost figure, we must do two things:

1. We must calculate total variable cost for each level of output.

2. We must then divide the total variable cost by the number of units, to get average variable cost per unit of output.

Table 12·2 shows the figures (assuming that the going wage is $5,000).

Notice that average variable costs per unit decline at first and thereafter rise. The reason is by now clear enough. Variable cost increases by a set amount—$5,000 per man—as factors are added. Output, however, obeys the law of variable proportions, increasing rapidly at first and then displaying diminish-

ing returns. It stands to reason, then, that the variable cost *per unit* of output will be falling as long as output is growing faster than costs, and that it will begin to rise as soon as additions to output start to get smaller.

If we graph the typical variable cost curve per unit of output, it will be the dish-shaped or U-shaped profile that Fig. 12·2 shows.

TOTAL COST PER UNIT

We can now set up a complete cost schedule for our enterprise by combining fixed and variable costs, as in Table 12·3. Notice how

Number of men	Total variable cost @ $5,000 per man	Total output (units)	Average variable cost per unit of output (cost ÷ output)
1	$ 5,000	5,000	$1.00
2	10,000	13,000	.77
3	15,000	23,000	.65
4	20,000	32,000	.63
5	25,000	39,000	.64
6	30,000	44,000	.68
7	35,000	47,000	.74
8	40,000	49,000	.82

TABLE 12 · 2

FIGURE 12 · 2
Profile of changing variable costs per unit

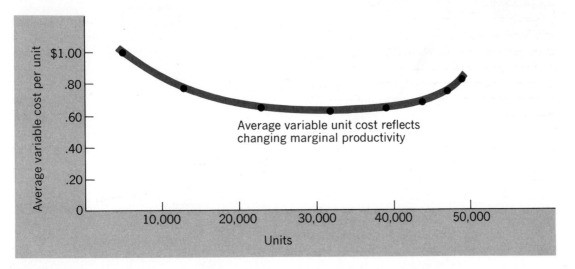

Average variable unit cost reflects changing marginal productivity

TABLE 12·3

			COST PER UNIT OF OUTPUT			
Number of men	Total cost ($10,000 fixed cost + $5,000 per man)	Output (units)	Average (total cost ÷ output)		Marginal (change in cost ÷ change in output)*	
1	$15,000	5,000	$3.00 ⎞		$— ⎞	Falling
2	20,000	13,000	1.54 ⎟ Falling		.63 ⎬ marginal
3	25,000	23,000	1.09 ⎬ avg.		.50 ⎠ cost
4	30,000	32,000	.94 ⎠ cost		.55 ⎞
5	35,000	39,000	.90		.71 ⎟ Rising
6	40,000	44,000	.91 ⎞ Rising		1.00 ⎬ marginal
7	45,000	47,000	.96 ⎬ avg.		1.67 ⎠ cost
8	50,000	49,000	1.02 ⎠ cost		2.50

*Ideally, we should like to show how marginal cost changes with *each* additional unit of output. Here our data show the change in costs associated with considerable jumps in output as we add each man. Hence we estimate the marginal cost per unit by taking the *change in total costs* and dividing this by the *change in total output*. The result is really an "average" marginal cost, since each individual item costs actually a tiny fraction less, or more, than its predecessor. We have shown the data this way since it is much closer to the way businessmen figure.

marginal costs begin to turn up *before* average costs.

If we graph the last two columns of figures—average and marginal cost per unit—we get the very important diagram in Fig. 12·3.

FIGURE 12 · 3
Average and marginal cost per unit

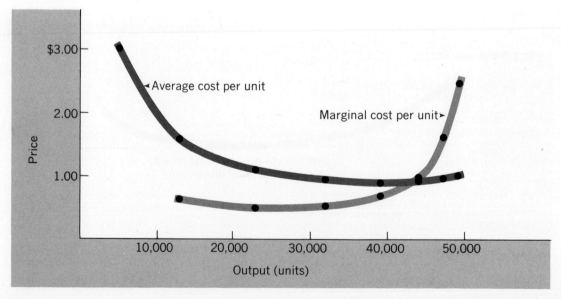

THE COST PROFILE

We have reached the end of our cost calculations, and it will help to take stock of what we have done. Actually, despite all the figures and diagrams, the procedure has been quite simple.

1. We began by seeing what would happen to our *fixed costs* per unit as we expanded output. Since fixed costs, by their nature, do not increase as production increases, the amount of fixed cost that had to be charged to each unit of output fell sharply as output rose.

2. Next we calculated the *variable costs* that would have to be borne by each unit as output increased. Here the critical process at work was the law of variable proportions. As the marginal productivity of factors increased, variable cost per unit fell. But when the inevitable stage of diminishing returns set in, variable costs per unit had to rise.

3. Adding together fixed and variable costs, we obtained the *total unit cost* of output. Like the variable cost curve, average total unit costs are dish-shaped, reflecting the changing marginal productivity of factors as output grows.

4. Finally, we show the changing *marginal cost per unit*—the increase in total costs divided by the increase in output. As before, it is the changing marginal costs that the entrepreneur actually experiences when he alters output. It is the increase at the margin that alters his total cost and therefore determines his average cost.

AVERAGE AND MARGINAL COSTS

Actually, the cost profile that we have worked out would be known by any businessman, whether or not he had ever studied microeconomics. Whenever a firm starts producing, its average cost per unit of output is very high. A General Motors plant turning out only a few hundred cars a year would have astronomical costs per automobile.

But as output increases, unit costs come down steadily, partly because overhead (fixed costs) is now spread over more units, partly because the factors are used at much greater efficiency. Finally, after some point of maximum factor efficiency, average unit costs begin to mount. Even though overhead continues to decline, it is now so small a fraction of cost per unit that its further decline does not count for much, while the rising inefficiency of factors steadily pushes up variable cost per unit. If General Motors tries to jam through more cars than a plant is designed to produce, the cost per auto will again begin to soar.

So much for the average cost per unit. By directing our attention to the *changes* that occur in total cost and total output every time we alter the number of factors we engage, the marginal cost curve per unit simply tells us why all this is happening. In other words, as our plant first moves into high gear, the cars we add to the line (the marginal output) will cost considerably less than the average of all cars processed previously; later, when diminishing returns begin to work against us, we would expect the added (marginal) cars to be high-cost cars, higher in cost than the average of all cars built so far.

Since the cost of marginal output always "leads" the cost of average output in this way, we can understand an important relationship that all marginal and average cost curves bear to each other. *The marginal cost curve always cuts the average cost curve at the lowest point of the latter.*

Why? Because as long as the additional cars are cheaper than the average of all cars, their production must be *reducing* average cost—that is, as long as the marginal cost curve is lower than the average cost curve, the average cost curve must be falling. Conversely, as soon as additional output is more expensive than the average for all previous output, that additional production must *raise* average costs—again, (look at the previous diagram) as soon as marginal cost

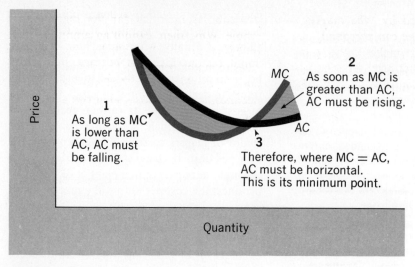

FIGURE 12 · 4
Relation of marginal and average cost per unit

is above average cost, average cost must begin to rise. Hence it follows that the *MC* (marginal cost) curve must cross the *AC* (average cost) curve at the minimum point of the latter. This relationship has nothing to do with economics, as such, but with simple logic, as Fig. 12·4 may elucidate.*

FROM COST TO REVENUE

The cost profile gives us a clear picture of what happens to unit costs as our firm hires additional factors. But that is only half the information we need for understanding how a firm operates with one foot in the factor market and the other in the market for goods. Now we need a comparable profile of

---

*It follows that the marginal productivity curve always crosses the average productivity curve at its peak. Look at Fig. 11·1. As long as marginal productivity is *higher* than average productivity, the average curve must be *rising*. As soon as additional (marginal) output is *less than* the preceding average output, the lower marginal output must *diminish* the average. The relation is exactly that of Fig. 12·4, only upside down.

what happens to revenues as the firm sells the output its factors have made for it.

FROM SUPPLY TO DEMAND

This brings us over from supply to demand—from dealing with factors who are selling their services, back to householders who are buying goods. What the entrepreneur wants to know is whether "the market" will buy his goods and if so, at just what price it will buy them. In other words, he wants to know the demand curve for his particular output. If he knows that, he can easily figure what his revenues will be.

What does the demand curve look like for a small competitive firm? Let us take the case of the manufacturer with whose costs we are now familiar, and assume that the "units" he is making are simple metal stampings selling at the rate of several million each year. The manufacturer knows two things about the market for those stampings. First, he knows that a "going" price for stampings (say $1.50) is established by "the market." Second, he knows that he can personally sell all the stampings he can make at the going price without altering that price by so much as a penny—that "the market" will not be affected whether he closes down his shop entirely or whether he sells every last stamping he can afford to make at the price the market offers.

BETWEEN TWO HORIZONTAL CURVES

What our manufacturer knows in his bones, we can translate into economics. The price of any commodity is set in the goods market by the interplay of supply and demand. Our firm is one of the many suppliers whose willingness and ability to sell at different prices (largely determined by their costs) makes up the supply curve. The demand curve for the commodity is familiar to us as the expression of the consumers' willingness and ability to buy.

But now we can also see that as far as the output of any *one* small firm is concerned, the demand for *its* output is a horizontal line—that there is an "infinite" willingness to buy its product, provided it can be supplied at the going price. The output of any one seller, in other words, is too small to affect the equilibrium price for the market as a whole. Why, then, cannot an ambitious firm make an "infinite" profit by expanding its sales to match demand? The shape of its cost curve gives the answer. As factor productivity declines, marginal costs soon rise above selling price.

Thus the competitive firm operates between two horizontal curves. On the supply side it faces a perfectly elastic supply of factors, meaning that it can hire all the factors it wishes without changing prices an iota in the factor market. On the demand side it also faces a perfectly elastic curve, meaning that it can sell as much as it can produce without any perceptible price effect here, either. As a result, the firm is squeezed between two forces that it is powerless to change. It must therefore devote all its energies to those parts of the market process that are in its control—the efficient combining of factors to minimize its costs and the selection of the most profitable scale and line of output.

AVERAGE AND MARGINAL REVENUE

Facing a known demand curve, the manufacturer can now calculate his revenues. He

FIGURE 12 · 5
Average and marginal costs under competition

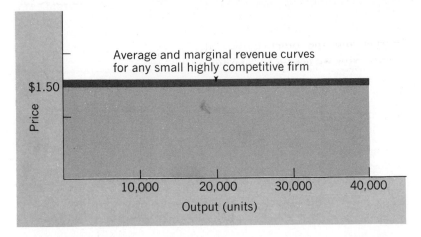

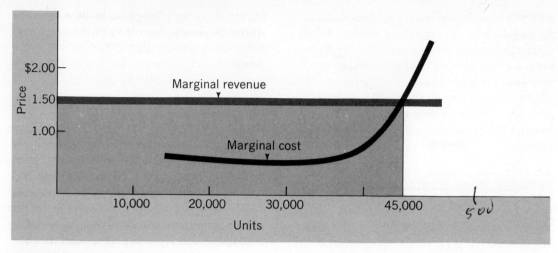

FIGURE 12 · 6
The point of optimum output

will take in an amount determined by his total unit output multiplied by the price of each unit. And since, with a horizontal demand curve, the price of each unit will be exactly the same price as the previous one, the marginal revenue of each unit sold—that is, the additional amount it will bring it—will be unchanged no matter how much is sold by the firm. If the selling price is $1.50, then the marginal revenue per unit will be $1.50. As a result, average revenue per unit will also be $1.50. The schedules of revenue will look like Table 12·4. We can see that a graph of the average and marginal costs curves for this (or any) small, highly competitive firm would be horizontal, as in Fig. 12·5.

MARGINAL REVENUE
AND
MARGINAL COST

Now we have all the information we want. We have a cost profile that tells us what happens to unit costs as we hire or fire factors. We have a revenue profile that tells us what happens to unit revenues as we do the same. It remains only to put the two together to discover just how much output the firm should make to maximize its profits.

We can do this very simply by superimposing the revenue diagram on top of the cost diagram. *The point where the marginal*

TABLE 12 · 4

Output (units)	Price per unit	Marginal revenue per unit	Total revenue	Average revenue per unit
5,000	$1.50	$1.50	$ 7,500	$1.50
10,000	1.50	1.50	15,000	1.50
20,000	1.50	1.50	30,000	1.50
40,000	1.50	1.50	60,000	1.50

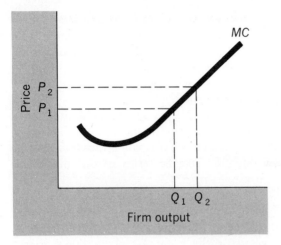

FIGURE 12 · 7
The firm's supply curve

revenue and the marginal cost curves meet should indicate exactly what the most profitable output will be. As we can see in Fig. 12·6, it is just about at 45,000 units.

Why is this the point of best possible output? Because another unit of output—as we can see by looking at the two curves—will cost more than it brings in, while a unit less of output would deprive us of the additional net revenue that another unit of output could bring.

FIRM AND MARKET SUPPLY CURVES

Our discussion of the relation between the marginal cost and marginal revenue curves leads to a further insight. *The marginal cost curve is the firm's supply curve.* Suppose price rises from P_1 to P_2 in Fig. 12·7. As far as the firm is concerned, it does not matter where its average cost is. What counts is whether additional production will be profitable. Further, it will be profitable only if the marginal cost of that additional production does not exceed marginal revenue. Therefore, when price rises, the firm's output increases from Q_1 to Q_2, a point determined by the intersection of the *MC* curve and the new price.

It follows that the supply curve of a competitive industry is the horizontal summation of the marginal cost curves of its firms, just as the market demand for goods is the summation of the demands of indi-

FIGURE 12 · 8
The industry's supply curve

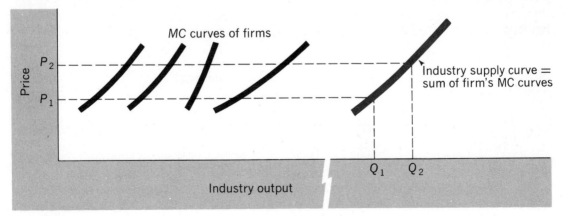

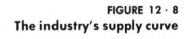

viduals. We show such a market supply curve in Fig. 12·8, for simplicity's sake drawing only the rising portion of the *MC* curves.

The point we must bear in mind is more than just a geometrical demonstration. *It is that marginal costs, not average costs, determine most production decisions.* When prices rise or fall, the change in quantity will reflect the ease or difficulty of adding to, or diminishing, production, as that ease or difficulty is reflected in the shape of its MC schedule.

PROFITS

Now let us return to the firm producing the quantity of stampings that just equates marginal cost and marginal revenue. What is the total amount of economic profit at the firm's best level of output? This is very difficult to tell from diagrams that show only marginal costs and marginal revenues. As we have just seen, these curves tell us how large *output* will be. But we must add another curve to enable us quickly to see what the *profit* will be at each level of output.

This is our familiar average cost curve. Average costs, we know, are nothing but total costs reduced to a per-unit basis. Average revenues are also on a per-unit basis. Hence, *if we compare the average unit revenue and cost curves at any point, they will tell us at a glance what total revenues and costs look like at that point.*

Figure 12·9 reveals what our situation is at the point of optimum output. (This time we generalize the diagram rather than putting it into the specific terms of our illustrative firm.)

The diagram shows several things. First, as before, it indicates our most profitable output as the amount *OA* — the output indicated by the point *X,* where the marginal revenue and marginal cost curves meet. Remember: *We use marginal costs and marginal revenues to determine the point of optimum output.*

Second, it shows us that at output *OA,* our *average cost* is *OC* (= *AB*) and our *average revenue* is *OD* (= *AX*), the same as our marginal revenue, since the demand curve for the firm is horizontal. Our profit on the *average* unit of output must therefore be *CD* (= *BX*), the difference between average costs and average revenues at this point. The *total* profit is therefore the rectangle *CDXB,* which

The firm in equilibrium with profits

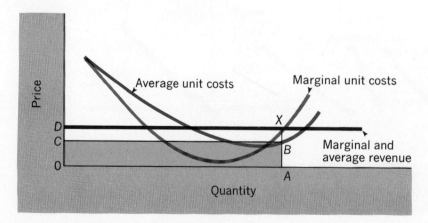

is the average profit per unit (*CD*) times the number of units. Remember again: *we use average costs and average revenues to calculate profits.*

WORKING OUT AN EXAMPLE

We can translate this in terms of our firm. At the point where *MC = MR,* it is making about 45,000 stampings, as Fig. 12·6 shows, at an average cost that we will estimate at 92¢ per unit. (Table 12·3 does not show us the exact cost at 45,000 units, but we will assume it is 1¢ more than the 91¢ cost of making 44,000 units). Since the selling price is $1.50, we are now taking in a total of $1.50 × 45,000 units, or $67,500, while our total cost is 92¢ × 45,000 units or $41,400. Our profit is the difference between total revenues and total costs, or $26,100.

Note that at an output of 44,000 units our revenues would have been $1.50 × 44,000 or $66,000, and our total costs 91¢ × 44,000 or $40,040, with the result that our profit would have been slightly smaller ($25,960). If we sold a larger quantity, 46,000 units, at a cost of, say, 94¢ per unit, our revenues would have come to $69,000 and our costs to $43,-

240, with a profit of the difference, or $25,760, also less than the $26,100 we made at an output of 45,000 units. Thus, given our cost figures and our selling price, an output of 45,000 is the optimal level for our firm. It gives us the maximum economic profit obtainable.

However satisfactory from the point of view of the firm, this is not yet a satisfactory stopping point from the point of view of the system as a whole. If our firm is typical of the metal stamping industry, then small firms throughout this line of business are making profits comparable to ours. Unhappily for them, there are numerous businesses in other lines of endeavor that do not make $26,100 of economic profit. *Hence entrepreneurs in these lines will now begin to move into our profitable industry.*

ENTRY AND EXIT

Perhaps we can anticipate what will now happen. Our firm is going to experience the same "mysterious" change in prices that we have already witnessed in the factor market, when many firms altered their demands for

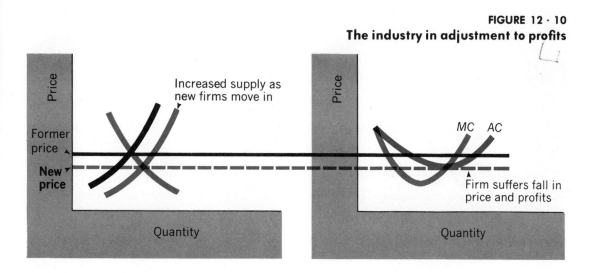

FIGURE 12 · 10
The industry in adjustment to profits

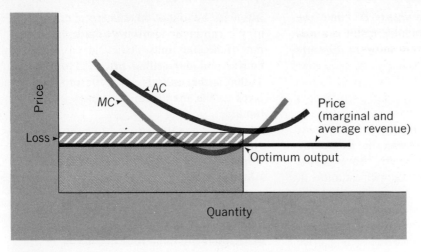

FIGURE 12 · 11
The firm suffering a loss

land or labor or capital and the prices of these factor services changed accordingly. Only this time, it is the price of goods, not of factors, that will change, for the influx of entrepreneurs from other areas will move the industry supply curve to the right and thereby reduce the going price. As the going price falls, our own business will be powerless to stop a fall in the price for its goods. We can see the process in Fig. 12·10.

How long will this influx of firms continue? Suppose that it continues until price falls *below* the average cost curve of our representative firm. Now its position looks like Fig. 12·11. Output will still be set where $MC = MR$ (it always is), but now the average

FIGURE 12 · 12
Industry adjustment to losses

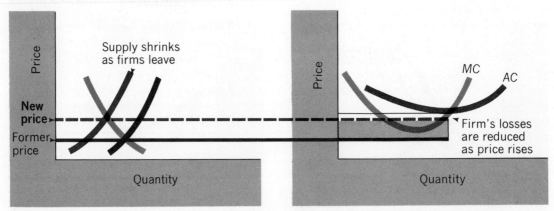

cost curve is above the average revenue curve at this point. The unavoidable result is a loss for the firm, as the diagram shows.

What will happen? Clearly, we need a reverse adjustment process—an exodus of firms into greener pastures, so that the supply curve for our industry can move to the left, bringing higher prices for all producers. This may not be a rapid process, but eventually the withdrawal of producers should bring about the necessary adjustment, shown in Fig. 12·12.

MINIMIZING LOSSES

The process of minimizing losses (which is as close as an unfortunate entrepreneur can get to maximizing profits) is worth a careful look because it again illustrates the importance of marginal, rather than average, costs. Figure 12·13 reproduces the relevant aspects of our previous diagram, showing a firm whose selling price is below average cost. It produces quantity Q_1, where $MC = P_1$, and it clearly suffers losses because AC is higher than price.

Why does the firm not quit entirely? The answer is that a decision to quit would bring larger losses, not smaller ones, because fixed costs—depreciation, interest, certain taxes, etc.—must be paid even if production is cut to zero. Therefore, *as long as price is as high as marginal cost, a firm will continue to produce, although on a reduced scale, since each unit of output contributes something over and above variable cost to fixed cost.*

Only if price falls below MC (P_2 in Fig. 12·13) will the firm quit entirely. Now production does not even cover the direct variable costs to which it gives rise, much less contribute anything to fixed cost. Therefore zero production is the way to minimize losses.

How long can a firm go on incurring losses? The answer depends on how rapidly it can terminate its fixed costs, such as getting rid of its buildings, machinery, etc., or how long it can incur losses without going bankrupt. The firm may limp along for an extended period, continuing to add production to the market and thereby delaying the leftward shift of the industry's supply curve.

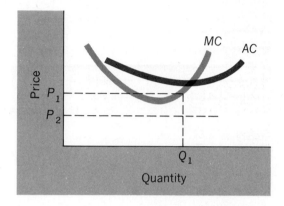

FIGURE 12 · 13
Adjusting to losses

LONG-RUN EQUILIBRIUM

Sooner or later, whether through the entry of new firms or the gradual withdrawal or disappearance of old ones, we reach a point of equilibrium both for the firm and the industry. It looks like Fig. 12·14.

Note that this position of equilibrium has two characteristics.

1. **Marginal cost equals marginal revenue, so there is no incentive for the individual entrepreneur to alter his own output.**
2. **Average cost equals average revenue (or price), so there is no incentive for firms to enter or leave the industry.**

Thus we can state the condition for the equilibrium resting point of our firm and industry as being a four-way equality:

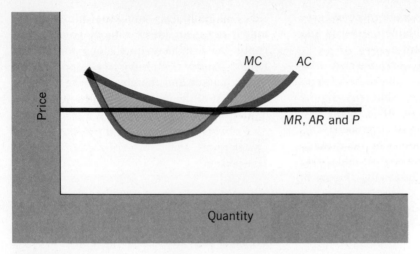

FIGURE 12 · 14
The marginal firm in equilibrium with no profit

$$MC \quad = \quad MR \quad = \quad AC \quad = \quad P$$

| Marginal cost | = | Marginal revenue | = | Average cost | = | Price (average revenue) |

PROFITS AND EQUILIBRIUM

We have reached an equilibrium both for the firm and for the industry, but it is certainly an uncomfortable one for ourselves as typical manufacturers. For in the final resting point of the firm, it is clear that *profits have been totally eliminated.* Is this a realistic assumption?

The question forces us again to confront the slippery question of what "profits" are. By definition, they are not returns to factors, for these payments have already been made by the firm—including all payments made to capitalists and entrepreneurs for the full value of their contribution to output.

To put it differently, we do not include in the term *profit* any revenues the firm *must* have to stay in business. An accountant, examining the books of a marginal firm in an industry, might find that there was a small bookkeeping profit. But an economist, looking at these revenues, would not call this sum a true *economic profit* if it were necessary to maintain the firm (or its entrepreneur) in operation. We should note as well that profits are usually figured as a return on the capital invested in a firm, not as a return on each unit sold. It is simpler for our purposes, however (and it does no violence to the argument) to talk of profits in relation to output (= sales) rather than as a return on investment.

What are economic profits, then? There have been numerous attempts to define them as the return for risk or for innovation, and so on. But however we describe them, we are driven to the conclusion that in a "perfect" competitive market, the forces of competition would indeed press toward zero the returns of the *marginal* firms in all industries, so that the cost and revenue profile of the last firms able to remain alive in each industry would look like our diagram.

Note, however, that these are marginal

firms. Here is a clue to how profits can exist even in a highly competitive situation. In Fig. 12·15, we show the supply curve of an industry broken into individual supply curves of its constituent firms. Some of these firms, by virtue of superior location or access to supplies or managerial skills will be lower-cost producers than others. When the industry price is finally established, they will be the beneficiaries of the difference between the going price (which reduces the profits of the marginal firm to zero) and the lower unit costs attributable to their superior efficiency.

These intramarginal profits are called *quasi rents,** a term we will encounter again when we discuss factor incomes. They result from scarcities—of location or managerial talents or whatever—that earn a high return. If, through a fall in price, any one of these firms were suddenly put at the margin of production, it would continue to produce, fully covering its costs, but earning no profits. Even without a fall in price, we would expect

*Note that economists use the terms *quasi rent, economic rent, producers' surplus,* and *economic profit* pretty much interchangeably.

intramarginal rents to diminish over time, owing to factor mobility. Badly located firms will pick up and move, or newcomers will locate favorably and thereby displace a firm at the margin. Managerial skills will be learned elsewhere or hired. Thus in the long run we must expect the *tendency* in a fully competitive market to be a constant pressure toward lower prices and toward the elimination of quasi rents.

Long Run and Short Run

We have seen how the market makes it necessary for firms to produce goods at the lowest points along their average unit cost curves and to sell those goods at prices that will yield no profit to the least efficient firm in the industry. Yet there is one last adjustment process to consider. In all our investigations into the firm's operations, we have hitherto taken for granted that the *scale of output* would remain unchanged. As a result, all of our adjustments have involved us in moving back and forth along a cost curve that was

FIGURE 12 · 15
Intramarginal profits

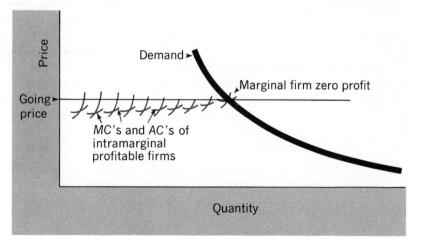

A THEORY OF CONCENTRATION?

Suppose that intramarginal profits are *not* quickly eliminated. Then what? We would expect that a firm enjoying these profits would use them to buy capital equipment or land in order to expand its scale of output. If economies of scale (see below) exist, this will lower average costs even further, enabling the firm to grow still larger.

The result could then be the gradual evolution of a highly "atomistic" industry into a "concentrated" one in which a few firms dominated the market. Something very much like this process characterized many American industries in their early days. By 1900, for example, the number of textile mills, although still large, had dropped by one-third from the 1880s; over the same period, the number of manufacturers of agricultural implements had fallen by 60 percent, the number of leather manufacturers by two-thirds. In all these and many other cases, the change in market structure was the result of the growth of especially dynamic firms whom we can picture as lying to the left along our cluster of competitive firms in Fig. 12·15. In our next chapter we shall be concerned with the problems faced by industries in which the process of concentration has progressed until these industries can no longer be considered competitive, but fall into a new category of oligopoly.

basically set in place by one or more limiting factors.

This may be accurate enough in the short run, when most firms are circumscribed by a given size of plant, but it certainly does not describe the long run. *For in the long run, all costs are variable.* A firm can usually enlarge its scale of plant; and as the scale increases, it is often possible to realize additional savings in cost, as a result of still finer specialization of the production process. If our expanding firm confines its growth to a single plant, the cost curves of these successive scales of output are apt to look like Fig. 12·16.

ECONOMIES OF SCALE

The decision to move from a plant of a given scale to the next largest size is a complex one.

FIGURE 12 · 16
Long-run cost curve

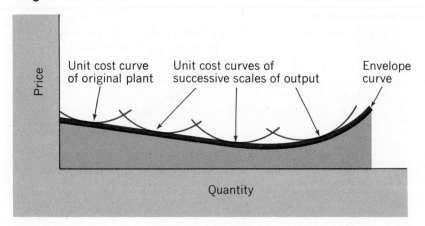

But if we imagine a firm choosing among a very large number of "blueprints," we could picture its long-run average cost profile in terms of the envelope curve we have indicated in Fig. 12·16. We have explained that this long-run average unit cost curve (or envelope curve) initially slopes downward, owing to *economies of scale.*

Our diagram shows, however, that economies of scale do not go on forever. At some point—again determined by technology —the limits of efficient plant operation are reached. A sprawling enterprise begins to stretch too thin the coordinating powers of management. *Diseconomies of scale* enter, and the long-run unit cost curve again begins to mount.

Note, however, that these long-run cost curves apply to individual *plants.* What about a firm that pushes each plant to its optimum size and then adds a new plant, or a firm that diversifies its efforts among many different kinds of businesses, like the conglomerates? Does it also face a long-run, upward sloping cost curve, owing perhaps to eventual diseconomies of management? We really do not know the answer. To hazard an unsubstantiated guess, it may well be that the new technology of information retrieval has so increased the efficiency of management that the economically effective size of multiplants or diversified plants is today extremely large. For all practical purposes, the long-run cost curve is probably horizontal or perhaps even falling.

INCREASING OR DECREASING LONG-RUN COSTS

There is still, however, another situation that can alter our firm's long-run costs. Together with all of its competitors it may be subject to *increasing or decreasing costs for the industry as a whole.*

The source of these changes in cost do not lie within the firm, in the relative efficiency of various factor mixes. Rather, they are changes thrust upon the firm—for better or worse—by the interaction of the growing industry of which it is a part and the economy as a whole. A new industry, for example, by its very expansion may bring into being satellite firms that provide some of its necessary inputs; and as the main new industry grows, the satellites also expand and thereby realize economies of scale that will benefit the main industry itself. The automobile industry was surely an instance of such long-run falling costs (for a long period, at least) resulting from the economies of scale enjoyed by makers of tires, batteries, and other equipment. In turn, the rise of low-cost trucking brought other such economies to many other industries.

Industries may also experience long-run rising costs if their expansion pushes them up against factor scarcity of a stubbornly inelastic kind. Extractive industries, for example, may be forced to use progressively less-accessible mineral deposits; or agricultural industries may be forced to use progressively less-fertile or less-conveniently-located land. Such industries would experience a gradual rise in unit costs as their output increased.

Are most industries the beneficiaries of decreasing cost or the victims of increasing cost? Empirical studies seem to suggest that save for youthful, growing industries, and for the special case of extractive ones, most industries enjoy a middle position of roughly constant long-run unit costs, at least over a considerable period of time.

The Competitive Environment

Now we must ask a question that has surely occurred to the reader. Does the world really behave as microtheory has portrayed it? Do firms really equate marginal costs and revenues or balance the advantages of one factor versus another with the fine precision that our model has indicated? That is not a question we can fully answer until our next chapter. But we already know one condition that must be fulfilled if firms are to behave as we have pictured them. This is the condition of "pure" competition, the environment we take for granted during our theoretical microeconomic investigations. Now is the time to look carefully into exactly what the term means.

PURE COMPETITION DEFINED

In general, when economists use the term *pure competition,* they imply three necessary attributes of the market situation:

1. *Large numbers of marketers*

Unless there are numerous buyers and sellers facing one another across the market and jostling one another on each side of it, the competitive process will not fully work itself out. When the number of marketers is few (whether as buyers or sellers), the vying among them that gives competition its resistless force is apt to be muted or even lacking entirely. As the extreme case of this we have outright *collusion* (a few buyers or sellers agree to bid at only one low price or to offer at only one high price). But even when collusion is absent, fewness of buyers or sellers will lead, as we shall see in our next chapter, to results that are considerably at variance with those of the competitive process we have assumed.

How many buyers or sellers does it take to make a fully competitive market? There is no clear-cut answer. The critical number is reached when no firm, by varying its scale of output, is able to affect the price of the factors it buys or the product it sells. We have pictured this condition in terms of the horizontal factor supply and market demand curves that present the purely competitive firm with the unalterable data of factor and goods prices to which it must accommodate itself. *Thus under conditions of pure competition, the only thing the firm can control is its scale of output and the mix of factors it uses; all prices are beyond its power to influence.*

2. *Ease of entry into, and exit out of, industries*

A second prerequisite for so-called pure competition is a condition that we have already relied upon frequently in discussing the operation of the market. *This is the ability of firms and factors to move freely and easily from one industry to another in search of the highest possible return.* Only in this way can supply and demand schedules move rightwards and leftwards, bringing about the needed adjustments of quantities and prices.

We have seen as well that this is by no means an easy set of conditions to achieve. With firms, as previously with factors, it rules out all legal barriers to interindustry movement, such as restrictive patents. But beyond this it implies that if the initial scale of manufacture is very large, that industry cannot be considered as meeting the requirements for pure competition. In automobile or steel manufacturing, for example, the minimum size of plant entails an investment of millions of dollars. The degree of competitive pressure from "outside" entrepreneurs, therefore, is obviously much less than in the stationery store business, where a newcomer can enter for an investment of a few thousand dollars.

Ease of exit is a no less necessary and equally demanding requirement. The competitive process will not shift about supply curves if some producers cannot withdraw their investments in land or capital and move them to alternative uses. Yet, as we have seen, this inability to move out can indeed retard adjustments of supply, particularly in industries with large fixed investments that are highly "specific" in their use. Such industries may go on producing even if they cannot cover their full costs, as long as their revenues bring in enough to nibble away at fixed expenses (see p. 163). Thus the problems of securing easy entry and exit further restrict the environment of pure competition to industries requiring no large or technically specific investments.

3. Nondifferentiated goods

But even these strict conditions still do not define the state of competition we have implicitly assumed. It is possible for a market to consist of many small buyers and sellers, each operating with relatively simple equipment, and yet these firms may not compete fully against one another. This is the case when each firm sells a product that is *differentiated* (or distinguishable) from that of its competitors. For if there is a difference, however slight, between one man's product and another's, the demand curve for the product of each will be sloping rather than horizontal, even if the slope is very small. As a result, product differentiation will enable a seller to hold onto *some* of his trade even if his price is a trifle higher than his competitor's, whereas in a purely competitive market where goods and services are indistinguishable from one another, no marketer can depart in the slightest degree from the prevailing price.

In some markets, perfectly anonymous undifferentiated commodities are sold. In the market for grain or for coal, for example, no seller can ask even a penny more than the going price for his product. In the great bulk of retail and wholesale markets, however, such totally undifferentiated products are the exception rather than the rule.

Why must commodities be exactly alike for a state of pure competition to exist? The answer is that *only identical commodities compete solely on the basis of price.* Much of what we call "competition" in the impure markets of the real world consists in differentiating products through style, design, services, etc., so that they will *not* have to compete just on price. We shall look into this very common case of "imperfect competition" in our next chapter. But we must rule it out as a permissible form of competition to bring about the exact results of our market analysis thus far. The essential rule for a purely competitive market is that the word "competition" must mean *price competition only.*

COMPETITION IN THE REAL WORLD

Obviously, pure competition is an extremely demanding state of affairs. It requires numerous small firms selling identical products in a highly mobile and fluid environment. Do such markets in fact exist?

Farming is often treated as if it were a perfect example of pure competition, since its units are very numerous and its main products undifferentiated. While this is true, farm markets do not adjust rapidly, because of the difficulties of entry and exit. Efficient farms require large investments that few can make. Since land is a major cost of farming, and usually a fixed cost, marginal costs are far below average costs. As a result, large losses must be sustained before production actually halts. In addition, to move off the farm there must be something better to move to. Of the roughly 2.8 million farms in the United States, half are, by any definition, uneconomical operations. Yet submarginal farmers are reluctant to leave an occupation that gives them minimal security for the frightening insecurity (and perhaps even worse economic luck) of the city.

The world of retailing qualifies much more readily in terms of easy entry and exit, and it is characterized by many small firms. But retailing loses out on the question of differentiation. The very essence of most retail establishments, even if they sell exactly the same wares as the competitor down the street (in fact, especially if they do), is to try to be "different."

Hence the search for perfect examples of pure competition is apt to end with very few cases. Why, then, do we spend so much time analyzing it?

The answer is twofold. In part it lies in the fact that as much as 40 or 50 percent of the output of the nation comes from sectors that *resemble*—even though they do not exactly qualify for—pure competition. The service trades, the wholesale markets, much retailing, some raw material production are near enough to being "pure" in their competitive structures, so that we can apply the reasoning of price theory very closely in understanding the market results we see in those industries.

But there is a second reason as well. Even in those sectors or industries where pure competition obviously does not apply—in the monopolistic or oligopolistic situations we will examine in our next chapter—the mechanism of the competitive firm will still be applicable. Supply and demand, factor productivities, marginal revenue and marginal cost will continue to be the prevailing guidelines. Hence we must understand the basic workings of the small firm, because most of them will still apply. And unless we know to what results these workings lead us in the ideal environment of pure competition, we will hardly be in a position to know what a difference monopoly or oligopoly makes to the workings of the market system.

KEY WORDS

CENTRAL CONCEPTS

Fixed and variable costs

1. We divide costs within the firm into *fixed and variable* costs. Variable costs change with the addition or discharge of factors. Fixed costs are contractual costs or costs that are associated with the unchanged use of one factor. They do not vary with output.

2. Both fixed and variable (and total) costs are usually calculated *per unit of output*. Fixed costs decline per unit of output as total output increases.

U-shaped unit cost curve

3. *Variable costs first decline and then rise*, reflecting the increasing and then diminishing marginal productivity of the factors. The typical shape of the variable unit cost curve is *U-shaped* or *dish-shaped*.

4. Adding together fixed and variable costs, we get a *dish-shaped unit cost curve* for the firm.

Marginal and average unit cost curve

5. The marginal cost curve shows us the actual operative element at work. The relation of marginal and average figures to one another is such that the *marginal unit cost curve always cuts the average unit cost curve at the lowest point of the latter*.

The competitive firm between two horizontal curves

6. *The competitive firm operates between two horizontal curves: The demand curve for its product is infinitely elastic. The supply curve of factors to itself is also infinitely elastic. That is, it can sell all it can profitably make without dis-*

turbing the market price, and it can hire or fire all the factors it wishes without disturbing their price. The only process under its control is the combination of factors to secure maximum efficiency.

Optimum output: MC = MR

7. Profits are at a *maximum where marginal revenue just equals marginal cost.* Thence, the rule for finding optimum *output* is always to equate *MC* to *MR*.

Profit: AC and AR

8. The rule of *MC = MR* establishes optimum output. But the amount of profits depends on the relation of *AC* to *AR*, or *average* cost to price.

Minimizing losses

9. If a competitive firm enjoys profits when *MC = MR*, it will experience *an influx of firms* from other fields. This will cause industry supply to increase and prices to fall. Conversely, if the firm experiences a loss, there will be an exodus of firms until industry supply falls and prices rise. However, *a firm will continue to produce as long as marginal revenue is greater than marginal cost,* even though marginal revenue is below average cost. This will *minimize its losses,* because it will be able to cover some of its fixed costs. It will quit producing only when marginal revenue fails to cover *MC* or when it can liquidate its fixed costs or when it simply goes bankrupt.

Equilibrium = MC = MR = AC = P

10. The *equilibrium point for the competitive industry is reached when marginal revenue equals marginal cost, and average cost equals price.* At this point there is no incentive for the entrepreneur to alter his scale of output or for firms to leave or enter the industry.

Intramarginal firm

11. In a competitive market, the *marginal* firm enjoys no economic profit. Intramarginal firms may have quasi rents.

Economies of scale

Envelope curve

12. In addition to changes in cost from moving along a fixed unit cost curve, firms can enjoy *economies of scale,* provided that the technology of the industry leads to these. At some point, economies of scale cease and the long-run cost curve or envelope curve again turns up.

Long-run costs

13. In addition, *economies or diseconomies* in other industries can affect the cost curves of all firms within an industry. For most industries a condition of roughly *constant* long-term costs seems to prevail.

Pure competition

14. The environment of the firm is assumed at first to be that of pure competition. By this we mean a market that includes:

• Large numbers

• *Large numbers of marketers.* As a result, each firm is powerless to affect prices of factors and goods. Only the factor mix is under the control of the competitive entrepreneur.

• Exit and entry

• *Ease of exit and entry.* This requires the absence of all barriers to mobility, including that of size and technical specificity.

• Nondifferentiated products

• *Nondifferentiated products.* This insures the restriction of competition to price alone.

QUESTIONS

1. If you were a retail grocer, what kinds of costs would be fixed for you? If you were a manufacturer who owned a large computer, would its maintenance be a fixed cost? If you *rented* the computer, would it be?

2. Assume that your fixed costs are $500 a week and that your output can vary from 100 to 1,000 units, given the scale of your enterprise. Graph what happens to fixed costs per unit.

3. Assume that your plant hires 6 men successively, and that output changes as follows:

Number of men	1	2	3	4	5	6
Total units of output per week	100	300	550	700	750	800

   What is the marginal product of each man? If each worker costs you $100 per week, what is the variable cost per unit as you add men?

4. If you add fixed costs at $500 per week to the variable cost you have just ascertained, what is the average cost per unit? What is the marginal cost per unit? (Remember, this is figured by dividing the *change in total cost by the change in total output.*)

5. Graph the curve of average total unit costs and marginal unit costs. Why does the marginal unit cost curve cross the average unit cost curve at its lowest point?

6. What does average revenue mean and what is its relation to price? What is meant by marginal revenue? Why is marginal revenue the same as average revenue for a competitive firm?

7. Explain carefully why a competitive firm operates between two horizontal curves, one on the demand side and one on the factor supply side.

8. Suppose (in the example above) you sell the output of your firm at $1.35 per unit. Draw in such a marginal revenue curve. Now very carefully indicate where the MR and MC curves meet. Show on the diagram the output corresponding to this point.

What is the approximate average cost at this output? Is there a profit here? Indicate by letters the rectangle that shows the profit per unit of output and the number of units.

9. What will be the result, in a competitive industry, of such a profit? Draw a diagram showing how an influx of firms can change the ruling market price. Will it be higher or lower?

10. Draw a diagram showing how price could drop below the lowest point on the average total unit cost curve, and indicate the loss the firm would suffer. Explain, by means of a diagram, why a manufacturer may remain in business even though he cannot sell his output for the full cost of producing it. What will determine whether or not it is worth his while to quit entirely?

11. Carefully draw a diagram showing the equilibrium position for the firm. Explain how MR and MC are all the firm is concerned with. How do AR and AC enter the picture? Why do MR and MC, by themselves, fail to give an equilibrium price in a competitive industry?

12. Suppose that you are a druggist and you know that the least efficient druggist in town makes virtually no profit at all. Assuming that you sell in the same market as he does, at the same prices, and that you hire factors at the same prices also, what causes could bring about a profit to your enterprise? What would you expect to be the trend of these profits?

13. Would you expect economies of scale from greatly enlarging your drugstore? Why or why not?

14. Do you think as the entire drugstore business expands it enjoys external economies or suffers from external diseconomies? How about the goldmining business?

15. Is the drugstore business an example of pure competition? Explain carefully in what ways it might qualify and in what ways it might not.

13

Monopolies and Oligopolies

MONOPOLY (AND NOWADAYS OLIGOPOLY) are bad words to most people, just as competition is a good word. But not everyone can specify exactly what is good or bad about them. Often we get the impression that the aims of the monopolist are evil and grasping, while those of the competitor are wholesome and altruistic, and therefore the essential difference between a world of pure competition and one of very impure competition is one of motives and drives—of well-meaning competitors and ill-intentioned monopolists.

The truth is that *exactly the same motives drive the monopoly and the competitive firm.* Both seek to maximize their profits. Indeed, the competitive firm, faced with the necessity of watching costs and revenues in order to survive, is apt to be, if anything, *more* penny-pinching and more intensely profit-oriented than the monopolist who (as we shall see) can afford to take a less hungry attitude toward profits. The lesson to be learned—and it is an important one—is that motives have nothing to do with the problem of less-than-pure competition. *The difference between a monopoly, an oligopoly, and a situation of pure competition is entirely one of market structure—that is, of the number of firms, ease of entry or exit, and the degree of differentiation among their goods.*

Monopoly

PRICE TAKERS VS. PRICE SEARCHERS

We have noted a very precise distinction between the competitive situation (numerous firms, undifferentiated goods) and markets with few sellers or highly differentiated goods. In the competitive case, as we have seen, each firm caters to so small a section of the market that the demand curve for its product is, for all intents and purposes, horizontal. By way of contrast, in a monopolistic or oligopolistic market structure there are so few firms that each one faces a downward sloping demand curve. That means that each firm, by varying its output, can affect the price of its product.

HOW MONOPOLISTIC
IS A MONOPOLY?

How do we recognize a monopoly? Because of the problem of substitutes, there is no very clear sign, except in a few cases such as the "natural" monopolies discussed above. Usually when we speak of monopolies we mean (a) very large businesses with (b) much higher than competitive profits and (c) relatively little direct product competition. Not many firms satisfy all three conditions. Exxon or

GM is a monopoly by criteria *a* and *b*, but not by *c*; Polaroid by *b* and *c* but not *a*. Note, furthermore, that even small businesses can be monopolies, as "the only gambling house in town" famed in all Westerns.

An interesting case in point is the telephone company. Most people assume that A.T.&T. is a monopoly, if ever there was one. Yet in fact, although the Bell Telephone system provides 82 percent of all telephone service in the country, there are actually 1800 telephone companies, quite independent of the Bell System.

Another way of describing this difference is to call purely competitive firms, who have no control over their price, *price-takers* and to label monopolies or oligopolies or any firm that can affect the price of its product, *price-searchers*.

"PURE" MONOPOLIES

By examining the economic problems faced by a "pure" monopoly, let us see how such a price-searcher operates. Why do we place the word "pure" in quotes? Because monopoly is not as easy to define as one might think. Essentially, the word means that there is only *one* seller of a particular good or service. The trouble comes in defining the "particular" good or service. In a sense, any seller of a differentiated good is a monopolist, for no one else dispenses *quite* the same utilities as he does: each shoe-shine boy has his "own" customers, some of whom would probably continue to patronize his stand even if he charged slightly more than his competition.

Thus at one end of the difficulty is the fact that there is an element of monopoly in many seemingly competitive goods, a complication we shall come back to later. At the

other end of the problem there are so-called "natural" monopolies, where economies of scale lead to one seller supplying the whole market, such as a local utility company. Yet even here there are substitutes. If power rates become exhorbitant, we *could* switch from electric light to candlelight. Hence, before we can draw conclusions from the mere fact that a company provides the "only" service of its kind, we need to know how easy or difficult it would be to find substitutes, however imperfect, for its output.

FIGURE 13 · 1
Demand curve for a monopoly

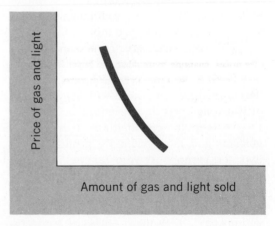

LIMITS OF MONOPOLY

Evidently the problem of defining a "pure" monopoly is not easily resolved. Let us, however, agree to call the local power company a monopoly, because no one else sells gas and electricity to the community. In Fig. 13·1 we show what the demand curve of such a monopoly looks like.

One point is immediately clear. *The monopolistic firm faces the same kind of demand curve that the competitive industry faces.* That is because both cater to *all* the demand for that particular product. A corollary follows. The monopolist faces a fundamental limitation on his power to control the market imposed by the demand curve itself. Suppose a monopolist is selling quantity *OX* at price *OA* as shown in Fig. 13·2. He would like to sell quantity *OY* at price *OA*, but *he cannot*. He has no way of forcing the market to take a larger quantity of his product—unless he lowers the price to *OB*.*

The situation is very similar (on the seller's side) to a *union*. A union can raise the price of labor, since it controls the supply of labor, but it cannot force employers to hire more labor than they want. Hence the question "Can unions raise wages?" must be answered "Yes," insofar as those who continue to be hired are concerned. But until we know

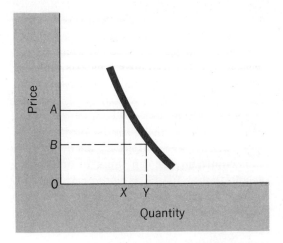

FIGURE 13 · 2
Demand curve as a constraint on the monopolist

the elasticity of the demand for labor, we cannot say if unions can raise the total amount of labor's revenues. We will study this question again in Chapter 18.

There is one thing a monopolist can do, however, that neither a union nor a purely

FIGURE 13 · 3
Advertising and demand

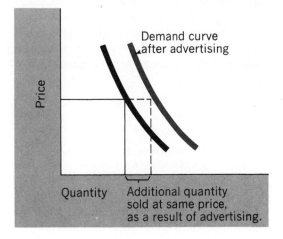

Demand curve after advertising

Additional quantity sold at same price, as a result of advertising.

*What would be the *most* profitable course for the monopolist to follow? It would be to sell his goods at *varying* prices, charging more when the buyer is willing and able to pay a high price. (One could imagine a monopolist doling out his product this way at an auction.) In fact, it amounts to a transfer of *all* consumers' surplus to the producer!

Why does the monopolist not follow it? (1) In some industries it is illegal. (2) It is extremely difficult to carry out. This is because buyers can trade among themselves, setting up a "secondhand" market that will compete with the monopolist. Nevertheless, discriminatory pricing is not uncommon in certain fields: antiques, used cars, pawnshops—and perhaps some professional fees.

competitive firm can. *He can advertise and thereby seek to move to the right, or change the slope of, the demand curve for his product.* Advertising does not "pay" in a purely competitive firm selling undifferentiated goods, for such a firm has no way of being sure that *its* goods—and not a competitor's—will benefit. But advertising *can* be profitable for a monopolist who will get all the demand he can conjure up. We can think of advertising as an attempt to sell larger quantities of a good or service without reducing prices, by shifting the demand curve itself. Figure 13·3 shows us this important effect, and we will talk about it further in a moment.

COST CURVES FOR THE MONOPOLIST

We have seen in what way the shape of the demand curves faced by monopolists differ from those faced by competitive firms. Are cost curves similarly different?

In general, they are not. We can take the cost profile of a monopoly as being exactly the same as that of a competitive firm. The monopoly, like the competitive firm, buys factors and exerts no control over their prices. A.T.&T. does not affect the level of wages or the price of land or capital by its decisions to expand production or not.* The monopolist, like the competitive entrepreneur, experiences the effects of changing productivity as he hires additional factors, and again like the competitive firm, he shops for

---

*There is a special situation with only one *buyer*—a "monopsony." A large employer who is the only substantial buyer of labor in a small town may be a monopsonist, and his decisions to hire labor or land or capital *will* affect their prices. For the monopsonist, the marginal cost curve is affected not only by the changing productivity of a factor but by its rising price, as more and more of the factor is hired. Because this is a situation infrequently found in the marketplace, we shall not analyze it further here. The principles involved are in no way different from those of monopoly.

the best buy in the factor markets. Thus the same U-shaped average cost curve and the same more steeply sloped marginal cost curve will describe the cost changes experienced by a monopolist quite as well as those of a competitive firm.

SELLING COSTS

There is, however, one item of cost for a monopolist (or as we shall later see, for any seller in an imperfectly competitive market) that does not arise in a milieu of pure competition. This is the need to incur *selling cost,* in an effort to affect the position or the shape of the demand curve. Advertising, a sales and service force, or the cost of product design are all important elements in the total cost picture of many businesses that sell differentiated products.

Selling costs are partly fixed and partly variable. Some selling expenses, such as the hire of designers to style a product, are little affected by changes in output. Other selling costs vary more or less directly with output, the size of a company's sales force being an example. Advertising is probably the most difficult selling cost to generalize about. A monopoly (or any firm selling in an imperfectly competitive market) may decide to increase its advertising if its sales fall; however, this is generally a short-run phenomenon. If the advertising effort fails, the firm is apt to suffer serious losses and may even go out of business. If the advertising works, the firm's sales rise, and advertising then becomes, in most cases, some more or less regular fraction of revenues.

Selling costs are very important in the *strategy* of firms. But they do not change the basic configuration of *MR* and *MC* curves, nor the imperative rule that the road to profit maximization is to bring *MC* equal to *MR*.

From Cost to Revenue

MONOPOLY REVENUES

It is when we come to the revenue side of the picture that we meet the critical distinction of monopoly. Unlike a competitive firm, *a monopoly has a marginal revenue curve that is different from its average revenue curve.* The difference arises because each time a monopolist sells more output, he must reduce his price, not on just the last unit sold, but on *all* units, whereas a competitive firm sells its larger output at the same price. Therefore, as the monopolist's sales increase, his *marginal revenues will fall.*

A table may make this clear. Let us suppose we have a monopoly that is faced with an average revenue or price schedule as in Table 13·1.

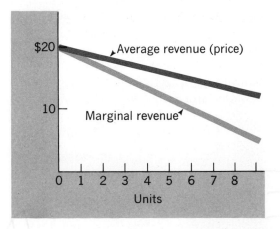

FIGURE 13 · 4
Average and marginal revenue for a monopolist

TABLE 13 · 1

Quantity sold	Price	Total revenue	Marginal revenue
1	$20	$20	$20
2	19	38	18
3	18	54	16
4	17	68	14
5	16	80	12
6	15	90	10

The graph of such a marginal revenue curve looks like Fig. 13·4. Note that at an output of 6 units, *AR* (price) = $15; *MR* = $10.

What determines the shape of the marginal revenue curve? Obviously, the change in quantity demanded that will be brought about by a drop in price. In turn, this reflects—as we remember from our discussion in Chapter 9—the elasticity of demand

which, in turn, hinges on our tastes and the availability of substitutes. Note especially that the *MR* curve lies below the *AR* curve because the monopolist must drop his prices on *all* units sold, not on just the marginal unit. Thus each additional item "drags down" the revenue of all output.

EQUILIBRIUM FOR THE MONOPOLY

The next step is obvious. We must superimpose the cost and the revenue profiles to determine the equilibrium position for the monopolist. We can see it in Fig. 13·5, p. 178.

What will be the equilibrium position? *The monopolist who seeks to maximize profit is guided by exactly the same rule as the competitive firm: he adds factors so long as the marginal revenue they bring in is greater than their marginal cost.* Hence we look for the intersection of the *MC* and *MR* curves on Fig. 13·5 to discover his optimum output, as the graph shows.

And what will the profit be? As before, profit reflects the spread of *average* cost

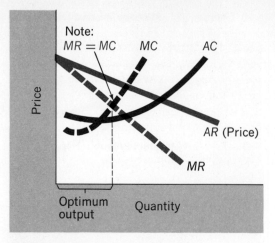

FIGURE 13 · 5
Equilibrium for the monopolist

and *average* revenue. The intersection of the *MC* and *MR* curves determines what our output will be; and knowing that, we can tell what our average cost and price (or average revenue) will be. Hence we can easily block in the profit of the enterprise. We do this in Fig. 13·6.

MONOPOLY VS. COMPETITIVE PRICES

What is the difference between this price and that of a purely competitive market? We remember the formula for the equilibrium price

FIGURE 13 · 6
Monopoly profit

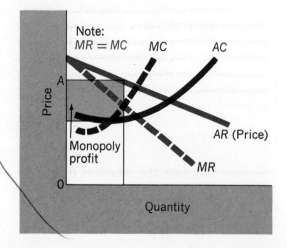

of such a market: $MC = MR = AC$ = Price. In the monopoly situation, MC still equals MR (this is always the profit-maximizing guide), but price certainly does not equal AC. Whereas the competitive firm is forced to price its goods at the lowest point on its cost curve, the monopolist sells at a price far above cost. Then, too, there is no pressure from "outside" forcing the monopolist to reduce costs. Hence his entire cost curve may well lie above that of a competitive industry producing the same product. It is interesting to note that when hard pressed, some big auto firms (not even monopolies) have reduced overhead expenses by as much as a third. But most of the time, monopolies are not hard pressed.

If this were the case in a competitive market, we know what the remedy would be. An influx of firms would move the supply curve to the right; and as a result, prices would fall until excess profits had been wiped out. *But in a monopoly situation, by the very definition of a monopoly, there is no entry into the market.* Hence the monopolist is able to restrict his output to the amount that will bring in the high profit he enjoys.

THE COST OF MONOPOLY

Monopoly thus imposes two related burdens on society. It sells wares at a *higher price* than that of the competitive firm, and its *output* is *smaller* than would be the case under competitive conditions. The consumer gets less and pays more for it. In more technical language, some consumer's surplus gets transferred to the monopolist as Fig. 13·7 shows.

This is not a full accounting of the "social cost" of monopoly. It is a description of the *economic* difference between the market solutions of a competitive and a monopolistic firm. It ignores entirely such matters as power or influence, which may make monopoly politically undesirable, and it omits as well any consideration of the possible use-

$MR \neq DC$

fulness of some monopolies (like the phone company) in providing a unified service or considerable economies of large-scale production.

Finally, we should note that most "natural" monopolies, such as the utility companies, are regulated by public authorities. By imposing price ceilings on these monopolies (usually calculated to allow the com-

panies to earn a "fair return" on their invested capital), the regulating commissions seek to approximate the results of a competitive environment. As Fig. 13·8 shows, its *MR* will be horizontal along the price ceiling, just like that of a competitive firm, and therefore its output will expand beyond the level of an unregulated monopoly.

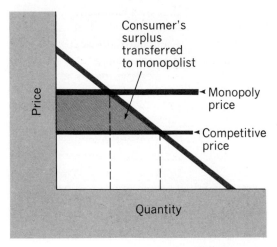

FIGURE 13 · 7
Loss of consumers' surplus under monopoly

FIGURE 13 · 8
Regulation of monopoly

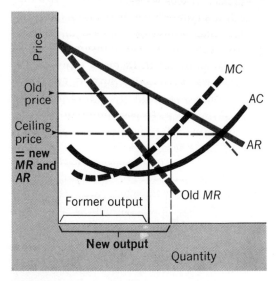

Oligopoly

Monopoly in its "pure" form is a rarity, and unregulated monopoly is rarer still. Most big corporations operate in a market structure of oligopoly, rather than pure monopoly. In an oligopolistic market situation, a *few* sellers divide the market and typically compete with one another by means of advertising, product differentiation, service, etc., rather than by the classic competitive means of price. The reason is that competition by advertising seems to give the oligopolist a chance to move his demand curve to the right, at the expense of his competitors. Competition by price, however, affects his own revenues as well as his competitors'. It therefore seems a much riskier tactic and is shunned by oligopolists as "self-destructive."

What does a typical oligopoly look like under the lens of price theory? On the cost side, it is much the same as a monopolist or a perfect competitor. There is, however, an essential difference between the demand curve of a monopolist and that of an oligopolist. The demand curve for a monopolist, since it comprises the entire demand for the commodity, is the familiar downward sloping curve. But the demand curve for an oligopolist, although it is also downward sloping, has a shape that is new to us and quite unusual.

THE "KINKED" DEMAND CURVE

Suppose you were the president of a large company that, along with three other very

THE CURIOUS DISCONTINUOUS MARGINAL REVENUE CURVE

It's easy to understand the kinked demand curve, but not so easy to see why the marginal revenue curve is discontinuous. The diagram at right may explain things.

Notice that our oligopolist has two demand curves, one above and one below the kink. Call them AR_1 and AR_2, and their respective marginal revenue curves, MR_1 and MR_2. Suppose our firm is selling a quantity OX just to the left of the kink. It will be working on AR_1 and will enjoy the marginal revenue (BX) from this output. Now suppose it shifts to an output just to the right of the kink and sells output OZ. It has shifted from AR_1 to AR_2, and its marginal revenue curve is now MR_2. Notice that this marginal revenue is ZC.

You can see that at the point of the kink there will be a sudden shift from MR_1 to MR_2 with a discontinuous drop. What this means is that if our oligopolist went below the kink (which would

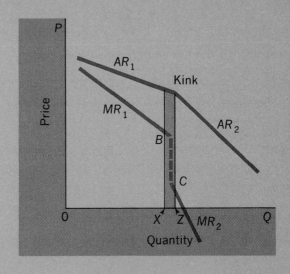

mean that he dropped his price and that all his competitors followed suit), his marginal revenue would no longer sink slowly, but would suddenly plummet to a new lower level.

similar companies, sold roughly 80 percent of a certain commodity. Suppose also that a price had been established for your commodity. It yielded you and your competitors a "reasonable" profit, but you and your fellow officers were considering how that profit might be increased.

One possibility that would certainly be discussed would be to raise the price of your product and hope that your customers would continue to be loyal to you. But your company economists might point out that their analyses showed a very elastic demand for your product *if you raised your price, but your competitors did not*. That is, at the higher price, many of your "loyal" customers would switch to a competitive brand, so that your revenues would fall sharply and your profits decline.

Suppose, then, you took the other tack and gambled on that very elasticity of de-

mand by cutting your prices. Would not other firms' customers switch to you and thereby raise your revenues and profits? This time your advisors might point out that if you cut your price, your competitors would almost certainly do the same, to prevent you from taking a portion of "their" market. As a result, with prices cut all around, you would probably find demand very much less elastic.

As Fig. 13·9 shows, you are facing a *kinked* demand curve. In this situation, you might well be tempted to sit tight and do nothing, for a very interesting thing happens to the marginal revenue curve that is derived from a kinked demand curve.

As Fig. 13·10 shows, we now get two marginal revenue curves: one applicable to the upper, elastic section of the demand curve; the other applicable to the lower, inelastic section. At the point of the kink, the marginal revenue curve is discontinuous, dropping vertically from the end of one slope to the beginning of the next. *As a result, there is no single point of intersection of the marginal revenue and marginal cost curves.* This means that an oligopolist's costs can change

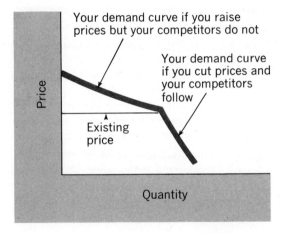

Your demand curve if you raise prices but your competitors do not

Your demand curve if you cut prices and your competitors follow

Existing price

Quantity

FIGURE 13 · 9
The kinked demand curve

FIGURE 13 · 10
The discontinuous MR curve

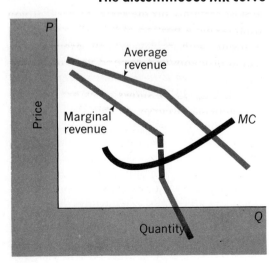

Average revenue

Marginal revenue

MC

Quantity

considerably before he is forced to alter his optimum volume of output or selling price.

The kinked demand curve helps explain why oligopoly prices are often so unvarying even without collusion among firms. But it does not really explain how the existing price is arrived at; for once cost or demand conditions have changed enough to overcome oligopolistic inertia, a new "kink" will again appear around the changed price. The kinked curve thus shows what forces affect *changes* in the oligopolistic situation, rather than how supply and demand originally determine the going price.

Price theory, as such, does not shed as much light on how the price of steel or cars or tires or tin cans is arrived at, as it does in a fully competitive industry. But why should this be so? We know in general what the demand curves look like for oligopolistic commodities, and we have a general idea of their supply behavior. Does it not follow therefore that price will be set by the standard supply-and-demand interplay?

THE MAXIMIZING ASSUMPTION

The answer—or rather, the lack of a clear-cut answer—causes us to examine another underpinning of classical price theory. We have already looked into the assumptions of pure competition—indeed, in this chapter we are tracing the consequences of failing to meet those assumptions. But the oligopoly problem forces us to confront a second fundamental premise of microeconomic theory. This is the assumption that firms maximize short-run profits.

Under pure competition, as we have seen, such profit maximization has a powerful motive working for it, quite above that of an entrepreneur's desire to get rich. *This is the motive of self-preservation.* Firms that do not constantly equate marginal cost and marginal revenue—the iron rule of maximizing—will simply fail to survive the competitive test. Hence, in the classical model, we can accept short-run profit maximizing as a fair description of reality, as well as a valid guideline for a well-working market system.

Not so, however, with oligopoly or monopoly. To be sure, a great corporation that failed to make any profits for many years would eventually fail, but there is little danger that a large company will go under if it does not make the last dollar—or even the last ten million dollars—of possible profits in any one year or even over two or three years running.

OLIGOPOLISTIC INDETERMINACY

Do the great oligopolistic concerns try to equate marginal cost and marginal revenue, and thus maximize their income?

The question is difficult to answer, by virtue of the very indeterminacy of the oligopolistic equilibrium resting point. We have just seen that a kinked demand curve makes likely a very high degree of oligopolistic inertia. The classic route to higher profits—beating one's competitor on price—tends to be shunned. What we find instead of the single-minded pursuit of gain is a strategic pursuit of long-run growth and strength that may very well lead to short-run decisions *not* to make as much money as possible.

Two factors in particular make the oligopolist's day-by-day behavior difficult to predict in theory. The first is his limited but not insignificant ability to influence the demand curve for his own product. Oligopolies are typically the largest advertisers, for one way to engage in competition without disturbing the price structure is to induce customers to switch brands at the same price. Thus a considerable degree of oligopolistic success depends not on pursuing the traditional course of profit-maximizing, but on pursuing a successful strategy of demand management.

Second, the *long-term* considerations that guide the behavior of oligopolies introduce a new element into the profit-maximizing picture that is missing in the competitive case. The entrepreneur of the small firm lives in a world where his costs and his demand curve are both subject to change without notice. He must do the best he can, at each moment, to obtain the highest income for himself. The oligopolist, on the other hand, has a considerable degree of control over his future. Since his overhead costs are usually high, he is able to make stringent economies if the need arises. And since he lives in a market situation where he is reasonably sure of not having to face price competition, he plans for the future in terms of improving or changing his product or his advertising or both.

This necessarily introduces into his plans a *time dimension* of much greater depth than in the case of classic competition. And in turn, the ability—not to say necessity—to plan several years in advance gives greater latitude to whatever decisions may be in the best immediate interests of the firm. *Thus profit-maximizing, however valid as a description of the state of mind or aim of the oligopolist, no longer serves, as it does in the competitive case, to describe the exact price and output decisions that we can expect firms to undertake in the face of market pressures.* Oligopolies are, within broad limits, free to follow whatever course—aggressive or defensive—their managements decide is best, and competing oligopolies have often followed *different* courses. Hence microtheory can say relatively little about how the market will impinge on the oligopolistic firm, since its economic pressures are much less imperious than they are in pure competition.

Imperfect Competition

We shall return once more to an assessment of the market under the conditions of oligopoly, but we must first finish our review of the marketplace in reality, rather than theory. For oligopoly, although perhaps the most striking departure from the ideal of pure competition, is not the most common departure. Once we pass from the manufacturing to the retail or service sectors where competition is still intense and characterized by numerous small units, we encounter a new kind of market situation, equally strange to the pages of a text on pure competition. This is a situation in which there are many firms, with relatively easy entrance and exit, but where *each firm sells a product slightly differentiated from that of every other*. Here is the world of the average store or the small competitive manufacturer of a brand-name product—indeed, of every seller who can "identify" his product to the public and who must face the competition of many other makers of similar but not exactly identical products.

Economists call this market situation tinged by monopoly *imperfect competition* or *monopolistic competition*. How does it differ from pure competition? Once again, there is no difference on the cost side. That is the same for both a perfectly competitive and an imperfectly competitive firm, except for the presence of selling costs in imperfect competition. The difference, again, comes in the nature of the demand curve.

We recall that the special attribute of the demand curve facing a firm in a purely competitive situation is its horizontal character. By way of contrast, *in a market of imperfect competition, the demand curve facing each seller slopes gently downward* because his good or service is not exactly like that of his competitors, and because he therefore has some ability to raise price without losing all his business.

EQUILIBRIUM IN MONOPOLISTIC COMPETITION

What is the equilibrium position of such an imperfectly competitive firm—say a dress manufacturer? In Fig. 13 · 11, on the left, an imperfect competitor is obviously making substantial profits. Note that his best position where $MR = MC$ is *exactly* like that of any firm, monopolies included.

But our firm is not a monopolist, and its profits are therefore not immune to erasure by entry into its field. In Fig. 13 · 11, we show the same firm after *other entrepreneurs have moved into the industry* (with additional, similar, although not identical, products) and thereby have taken away some of our firm's market and *moved its demand curve to the left*.

Note that our final position for the marginal dress firm has no more profit than that of a purely competitive seller because $AC = AR$. On the other hand, because his demand curve slopes, the equilibrium point cannot be at the lowest point on the average cost curve, nor will output have reached optimum size. (Of course, intramarginal firms can be more profitable than the marginal case we have graphed.)

This outcome clearly dissipates economic well-being. The fact that firms are forced to operate to the left of the optimums on their cost curves means that *they have not been able to combine factors to yield their greatest efficiency*—a failure that penalizes

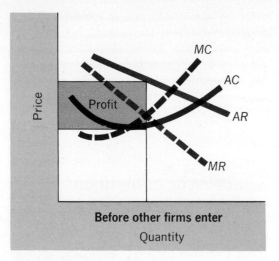

Before other firms enter

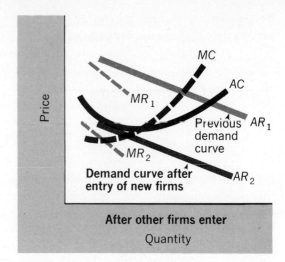

After other firms enter

FIGURE 13 · 11
Monopolistic competition

factors, once when they are paid too little be-cause their potential marginal productivity has not been reached, and again as consumers when they are forced to pay too much for products that have not been produced at lowest possible cost. In addition, wastage is incurred because the attempt to differentiate products leads in many instances to too many small units—for example, four gas stations at one intersection.

Inefficient though it may be, just as in a purely competitive industry, monopolistic competition yields the marginal firm in an industry no profit. The entrepreneur therefore feels fully as hard-pressed as would the producer of an undifferentiated commodity. The difference is that a monopolistic competitive businessman has the possibility of *further differentiating* his product, hoping thereby to tilt his demand curve in a more inelastic position. In turn, this might allow him to raise his prices slightly and to squeeze out a tiny "pure monopoly" profit. The result is that monopolistic competition fosters a tre-mendous variety of goods—the ladies garment industry being a prime example.*

Strategies in Economics:
The Case of Oil

As our discussion has moved from the firm in pure competition through various market situations of imperfect competition, a critical dividing line has become apparent. We have already distinguished between price-takers and price-searchers, to differentiate the case of pure competition from that of monopoly (or any other seller in an imperfectly com-

---

*Oligopolists as well as monopolistic competitors march to the drums of competition through product differentia-tion. In the fields of cigarettes, soaps, and breakfast cereals, the market is pretty much made up of three firms, and thirty or forty years ago most grocery stores stocked only about three or four brands of each. But today the shelves holding these products have been ex-panded again and again as each of the three firms came forward with a new differentiation or one to match a new, highly successful brand of one of its competitors.

STRATEGY IN ACTION

The problems of strategy have led to the development of game theory, a new approach to economic analysis. Suppose that we have two arch-rival firms who dominate an industry. The board of directors of Firm A meets one day to decide on price policy and is told by the president that he has been informed ("by trustworthy sources I am unable to reveal") that if Firm A raises its prices, Firm B, its main competitor, will raise its prices. Both firms will thereupon net $10 million more in profits. The president asks the directors for approval of a price rise, but he is interrupted by the company treasurer, who is a graduate of the Harvard Business School.

"Gentlemen," says the treasurer, "I must recommend against this proposal on two counts. First, Firm B may well be baiting a trap for us. If we raise our prices, they will *not* raise theirs. We will thereupon lose a vast amount of business to B. I estimate that this will cause us a loss of $10 million, and that Firm B will make a profit of $20 million.

"Second, suppose that Firm B *does* raise prices. Then, as profit-maximizing directors, we should certainly not raise ours; for in that case, by keeping our prices low, we will steal away a great deal of business from B, and we will make a profit of $20 million, while they lose $10 million."

"Well reasoned," says the chairman of the board. "It is not only clear that we should not raise prices, but I suggest that we inform Company B—discreetly, of course—that we *will* raise prices, as they suggest, in order to tempt them to do so. That will make our strategy foolproof."

In due course, this information is transmitted to Firm B, where a similar discussion takes place. Firm B resolves to "accept" Firm A's offer, but in fact not to raise prices at all. Result: neither firm raises prices, because neither trusts the other. Instead of each gaining $10 million, which would have been the result of a price rise by both firms, the two firms "stand pat" and accept the much smaller profits that accrue from *minimizing risk*. Perhaps you can see that this is an example of the Prisoners' Dilemma we discuss in Chapter 27, pp. 429–30.

---

petitive market). Now we can see that another dividing line lies between firms that can have "strategies" and those that cannot.

STRATEGIES

A competitive firm has no strategy, because it has no options. It is forced to maximize its short-run profits, or it will shortly be forced out of business. But the introduction of market imperfections opens the way for numerous strategies. We have noted that large companies often choose different expansion programs, according to their estimates of demand in the future. So, too, imperfectly competitive firms have options with respect to product differentiation. One popular product may be designed to look very much like another one, to attract sales because of this very resemblance. But another entrepreneur with the "same" product may decide to differentiate his good sharply, to gain the patronage of people who don't like the looks of the well-known product.

Strategies cover a gamut of behavior, from the selection of advertising campaigns to outright violations of the law. We cannot begin to cover their range of variety, but a look into the behavior of the oil industry may suggest some of the problems that strategies pose for economic analysis.

THE MARKET FOR OIL

The oil industry is one of the biggest and most important in the United States. Among the fifteen largest industrial companies, 5 are oil companies. Collectively these 5 big companies sell about 35 percent of all the gasoline in the United States.

But an oil company is much more than a seller of gasoline. The big oil companies are "integrated" vertically. This means that they own oil-producing properties; oil storage properties; oil transportation equipment (tankers and pipelines); oil refineries, and gas service stations. They make, in addition to gasoline, a long line of heating and industrial oils, as well as petrochemicals. They usually operate in many countries, not only as producers of crude but as refiners and final sellers.

This gives oil companies a great many options. They can decide whether to take more oil from one country or another. They can decide how to price the oil they load into their tankers—whether to price it high, thereby increasing the profits of their foreign producing subsidiaries, or low, thereby giving their refineries "cheap" raw materials and increasing profits in the refining end of the business. They can try to price their various products—heating oil, industrial oil, ordinary gasoline, aviation gasoline, etc.— according to the local availability of substitutes for each, thereby acting as discriminatory monopolists (see fn. p. 175).

EXCESS CAPACITY

Thus each oil company has many options and can follow many profit-maximizing strategies. At the same time, each company finds itself locked in a game of uneasy competition with its opposite numbers. For example, a chronic problem of the oil industry in the past has been excess capacity. Most integrated companies produce more gasoline than they can sell through their own retail chains. Because of the kinds of problems we saw in analyzing the kinked demand curve, none of the companies wants to cut retail

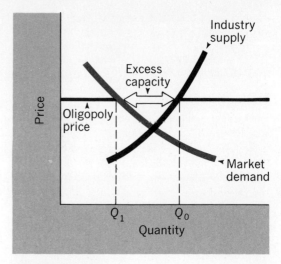

FIGURE 13 · 12
Excess capacity

gas prices—that would only start a "price war." *All* companies would cut prices, and the original price cutter would benefit little, if at all.* Therefore the big refiners have typically made special deals with so-called independents—gasoline distributors who do not have refineries—to sell their excess production under another name. Often this excess production has been sold at retail at cut rates, so that in fact the big companies were forced to compete against themselves.

Excess capacity is a chronic complaint of many oligopolistic industries. It arises because the price established by the industry is higher than a competitive market equilibrium price would be. Despite the fact that the industry "knows" that the price is higher than it would be under purely competitive conditions, each seller within the industry is tempted to enlarge his production capacity to the size that the administered price indicates. The result, as Fig. 13 · 12 shows, is excess capacity—that is, the ability to supply

*Price wars do break out from time to time, nonetheless, throwing the industry into a panicky search for "stability."

more output (Q_2) than the market will take at that price (Q_1).

INDUSTRY COORDINATION

Sales to independents, usually at special rates, have been a form of secret competition among the big companies. At the same time, like most oligopolies, the oil corporations have tried to concert their plans so that they could act together as if they were a monopoly. Under the Sherman Antitrust Act and other federal statutes, companies are not allowed to meet together for purposes of "restraining trade." Although the oil companies are excluded from this act for many purposes, such as dealing with oil-producing countries, they are still legally enjoined from "conspiring" with regard to their selling prices. This has left the oil companies with three options in seeking to coordinate their price policies.

1. Collusion

They could meet together illegally to plan their joint price policies. We do not know if such collusive meetings have taken place. In other oligopolistic industries there is evidence that they do; in the famous "General Electric-Westinghouse" case, the government caught the major heavy electrical companies conspiring in hotel rooms to rig their bids on equipment and to take turns as the "low bidder" on big contracts. Fines and prison sentences were imposed.

2. Price leadership

A very common practice for an oligopolistic industry is the recognition of a "price leader," who is tacitly charged with setting a monopolistic price for other members of the industry to match. When economic conditions change, the price leader changes his price, and the industry follows along. The oil industry has frequently used this system of price setting. When charged with using the price-leader system as a means of avoiding competition, the industry responds that since all companies are selling at the same price, there *is* competition!

This is a specious reply. There will normally be only one price in any market, competitive or "administered." What distinguishes competitive markets from administered ones is the presence of much higher than competitive profits in the latter. In 1971 profits in the petroleum industry averaged 6.3 percent on sales, compared with 3.8 percent in all manufacturing. However, this is not necessarily a true measure of the industry's monopoly power, because of their options in "transfer pricing" (deciding what price to put on the oil sold by their producing companies to their refining subsidiaries).

3. Informal coordination

A third method of coordination is to use an informal association or trade journal as a mechanism for coordination. An editorial in a trade paper calling for higher or lower prices can be read as a suggestion on what price strategy seems best for the industry to follow. The "official" journal for the oil industry is the *Oil and Gas Journal.*

THE GREAT GAS SQUEEZE OF 1973

Now let us consider the shortages of gasoline that first took the nation by surprise in the summer of 1973—*before* the Arab-Israel war that led to an Arab oil embargo. For many weeks, cars queued up in front of gas stations in some eastern and midwestern states, waiting for the few gallons that station owners were rationing to each customer.

Why the shortage? To begin with, the demand curve for gasoline had been moving gradually to the right, year by year, as the American automobile armada has grown in size. The 1973 increase, however, was much in line with past increases, not a sudden jump, as in the Russian wheat deal. Yet,

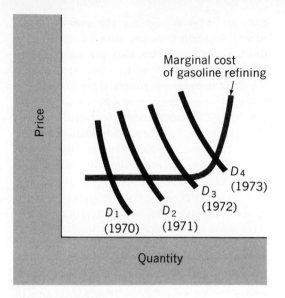

FIGURE 13 · 13
Marginal cost of gasoline refining

gasoline supply capacity failed to keep pace with demand. Despite the steady rightward shift in the demand curve, *no refineries were built in the United States for five years prior to 1973*. With gradually increasing demand and a fixed supply, the time came when refining capacity was fully utilized, and additional output could be had only at very high marginal cost. The situation was like Fig. 13·13.

In addition to raising prices for their own customers, the full utilization of refining capacity allowed the oil companies to stop selling oil products to independent distributors.* This contributed to the profits

*Why, a student may ask, was there a shortage that showed up in queues of cars? Why didn't the price of gas rise until the quantity demanded equaled the quantity supplied? There were two answers. First, many gas station owners feared raising their prices to the very high levels that might have cleared the market. They preferred accepting smaller profits to the wrath of the motorist. Second, the government froze prices in the spring of 1973, so that gas stations were prohibited from raising prices. The result was rationing by first-come, first-served methods, instead of by price.

of the big companies, but it raised a more fundamental question: did the gas shortage arise from a deliberate strategy of the major producers?

The evidence was ambiguous. On the one hand, there was every reason to anticipate the growing demand for gasoline, so that additional refineries could be expected to earn substantial profits. Moreover, although the oil companies had built no additional U.S. capacity, they had built refineries in Europe and elsewhere, and they had also invested millions of dollars in retail gas stations in the United States. Therefore the failure to build refineries here could be interpreted as a deliberate strategem designed to *create* a gas shortage and to enable the big companies to increase the sales of gasoline through their own service stations, at the expense of the independents.

On the other side, the oil companies argued that sound economic reasons had prevented them from increasing refinery capacity. Environmental groups were protesting the construction of refineries because of the pollution they caused. The uncertain future of the American oil quota policy, limiting imports of oil from abroad, made them wonder where additional supplies would be coming from. The future of the Alaskan oil pipeline was still clouded in uncertainty. Fearing that a refinery built in the wrong place would become a white elephant, they chose to follow the course of least risk and to build no refineries at all.

AN UNCERTAIN FUTURE

The resolution of this matter will not be known for many years. The Justice Department has initiated a major antitrust suit against the oil companies, but it is likely to remain in the courts for a very long time. Eventually, we may see a new structure enjoined by law on the oil industry. But no one is yet clear what that structure will or should be. Should the major companies be allowed to integrate vertically, or should producing companies be divorced from refining companies and these, in turn, from distributing companies? If vertical integration is allowed, should competition be judged by the number of companies operating within the United States or by the number operating *in each region?*

THE STRUGGLE FOR OIL

There are no easy answers to these questions. But while they are being thrashed out, a second struggle is also coming to a head. This is the contest between the oil consuming nations and the oil producing nations. Oil is produced in many nations and many areas of the world, but the bulk of the world's reserves lies in the nations of the Middle East. These nations have formed an organization called OPEC (Organization of Petroleum Exporting Countries) and have tried, with considerable success, to coordinate their policies as sellers of oil. They have done so by demanding (and obtaining) considerably larger shares in the profits of foreign-owned oil producing companies, and in some cases by expropriating the property of these companies.

No one knows exactly the marginal costs of producing oil in the great Middle Eastern fields, but it is probably much less than half, perhaps even as little as a tenth, of the marginal cost of oil coming from Texas or the North Sea fields or the Alaskan slopes. Because the demand for oil far exceeds the production of the Middle Eastern fields, OPEC producers can sell their oil at a price well above cost.

Thus the OPEC producers enjoy large intramarginal rents. But the situation is more complicated than the mere existence of intramarginal rents determined by geography and

geology. This is because the major buyers of oil—the refiners—are also the major producers of oil. In effect, they are selling oil to themselves, as well as to other companies. There is therefore a complex tug of war between oil-producing companies and the oil-producing countries, whose interests sometimes coincide and sometimes conflict.

BARGAINING POWER

It is impossible to predict the outcome of this tangled situation. Powerful oil producing countries—political allies who are often at each other's throats—clash with powerful companies—economic allies who are also often bitter enemies. Perhaps a solution may arise in the formation of an organization of oil importing countries for bargaining as a group, with the oil exporting countries. We would call such a market, having one buyer and one seller, a *bilateral monopoly*. As Fig. 13·14 shows, there is an indeterminate area between the reservation prices of seller and buyer that sets upper and lower limits to price. Within that area, price will be actually determined by the *bargaining power* of each side. Bargaining power is often determined by staying power—how long buyer or seller

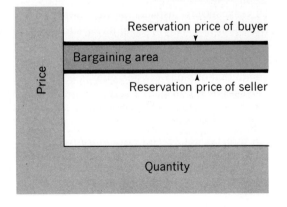

FIGURE 13 · 14
Bilateral bargaining

can hold out. In the contest for oil, it is difficult to know which side has the greater staying power—the consuming countries that must have oil to maintain their economies or the producing countries who must sell oil to maintain *their* economies. In the short run, perhaps, the oil producers have the upper hand; but in the longer run, as the oil consuming countries develop new energy sources, bargaining power is apt to swing in their favor.

THE ENERGY CRISIS

Meanwhile, there is no question that we and all other industrialized nations face an energy crisis, a realization brought home with sudden urgency when the OPEC countries sharply curtailed their deliveries to Japan, Europe, and the United States in the late fall of 1973. Although that was a political crisis, it was premonitory of a deeper-lying problem. The use of energy has been growing for many years at a rate that doubles in less than 20 years. Known reserves of petroleum are also still growing as new efforts and new technologies of discovery and extraction are applied, but no one doubts that sooner or later we will run out of oil. Then what?

In Chapter 34 we will return to the problem of resource depletion as a threat to the growth of industrial systems. But there are microeconomic aspects of the energy crisis as well as macroeconomic ones, and this is a good place to look into them.

Suppose we do run out of oil. Does that mean that one day the gas pumps go dry and the economy comes to a screeching halt? Our analysis allows us to see that this is not the way a depleting resource exerts its effect. As oil reserves get scarcer, the marginal cost of producing oil rises, shifting supply curves leftward. (Remember that the marginal cost curve *is* the supply curve.) Meanwhile as the

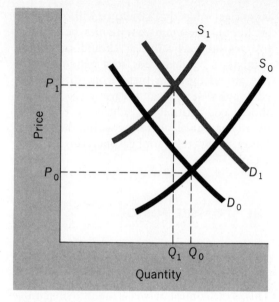

FIGURE 13·15
A resource "squeeze"

demand for oil continues to grow, demand curves shift to the right.

The result is, of course, a rise in the price of gasoline (or of any other scarce resource). At higher prices we will consume less, partly by personal economies, partly because there will be strong incentives to redesign automobiles to use less gas. As Fig. 13·15 shows, our demand curve for gas—the price we are able and willing to pay—may be rising, even though the amount we actually purchase is diminishing. Note that Q_1 is less than Q_0. How high can the price of gasoline go? The answer, of course, depends on the availability of substitutes. But this in turn hinges on the urgency to find substitutes. At 40¢ a gallon there may be "no alternative" to gasoline. At $4.00 per gallon there will be a tremendous incentive to find alternatives: public transportation, bicycles, electric vehicles, and the like. Indeed, at very high prices, gasoline-driven vehicles are likely to become as rare as electric vehicles (once fairly popular) are today. To put it differently: at very high prices, the demand curve for gas will become increasingly elastic, as Fig. 13·16 shows. After we reach price P_1 further decreases in supply have no further effect on price but simply result in decreased use.

As price rises from P_0 to P_1, we may, for a time, be acutely aware of a "shortage" of gas. Once consumption shrinks to Q_1 we are likely to have adjusted to other energy sources

(including our own feet) and may be almost unaware of the further contraction of consumption if the supply curve continues its leftward shift.

STRATEGY, POWER, POLICY

The economics of oil is itself virtually a field of economics, and we have barely scratched the surface of the problems it contains. But excursion into the strategies and dilemmas of oil has been undertaken not merely to give you a glimpse of an area that you may wish later to pursue. Rather, it has led us gradually away from the study of monopoly and oligopoly, approached mainly from the viewpoint of the monopolist and the oligopolist, to a consideration of imperfect competition, from the viewpoint of the public.

Alternative strategies faced by powerful sellers imply alternative outcomes for the consumer or the worker. The impersonal operation of the market mechanism has been replaced by the highly personal operation of option-choosing firms, even though their options are far from limitless. This raises a problem that we must consider in our next chapter. We have learned a good deal about the economics of big business. Now we must learn something about the problems of economic policy as it concerns big business.

FIGURE 13 · 16
The effect of very high prices

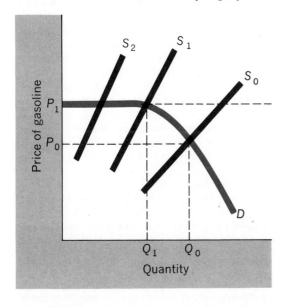

KEY WORDS

Monopoly

Price-searcher
Price-taker

Selling costs

KEY CONCEPTS

1. The motives of monopoly and oligopoly are exactly those of competition. What is different is the *structure of the market* in which these motives are turned into action.

2. The monopoly or oligopoly faces a downward sloping rather than a horizontal demand curve. It is a *price-searcher rather than a price-taker.*

3. A monopoly is difficult to define narrowly because of the existence of *substitutes* for all commodities. The monopolistic *firm* enjoys the same demand curve as does the competitive *industry,* and it is forced to obey the limitations imposed by its demand curve. It can, however, attempt to *shift the curve to the right or change its shape by advertising.* This results in *selling costs* as a kind of cost peculiar to imperfect competition.

Marginal revenue

4. The cost curves of monopolists and oligopolists are shaped the same as those of competitive firms. However, monopolists face *marginal revenue curves that are not identical with their average revenue curves*. The difference arises because the monopolist must lower his price to increase his sales. Therefore, each additional unit brings in a smaller addition to total revenues than the previous unit brought.

Equilibrium for monopoly

5. Equilibrium for the monopoly (as for the competitive firm) is that output where $MR = MC$. This is the point of maximum profit. AC will not equal P, however, for there is no influx of firms to push prices down to this point.

Oligopoly

6. The result of monopoly is twofold: prices will be higher than in competitive firms, and output will be smaller.

7. Oligopoly is a more common market structure than monopoly. It is characterized by a few sellers, rather than one.

Kinked demand curve

8. The oligopolist's *demand curve is typically kinked*. This shape reflects the difficulty of holding his market share if he raises prices alone, and the difficulty of maintaining his price advantage if he cuts price, which will be quickly followed by his fellow oligopolists.

Monopolistic competition

9. Oligopolies often need not obey the dictates of short-run maximizing. On the contrary, they can plan considerably far into the future. As a result, *profit-maximizing*, however accurate a description of their goal, does not lead to the exactly predictable behavior that it does in the competitive firm.

Differentiation

10. Imperfect or monopolistic competition is used to describe markets where many sellers offer *differentiated* products. The point of difference is the existence of a *slightly sloping demand curve*.

Strategies

11. Imperfect competition leads to two main results. First, output *does not reach the point of lowest cost and greatest efficiency*, thereby penalizing both consumers and factors. Second, there is a tendency for *differentiation to increase* in the hope of gaining a small profit.

Excess capacity

12. Sellers in imperfectly competitive markets have options (alternative courses of behavior) that are denied to competitive firms. This makes *strategy* important for their success.

13. Like many oligopolies, the oil industry tries to act as a monopoly. Their action results in *excess capacity*, since the oligopoly price is above the equilibrium price. The attempt to sell their excess production leads to secret competition. Most oligopolies try to avoid this unwanted competition by *collusion, price leadership*, or *informal coordination* through a journal or trade association.

Collusion Price leadership

OPEC nations Transfer pricing

14. The gasoline shortage of 1973, primarily the consequence of a lack of refinery capacity, raises a basic question: was the failure to build capacities a deliberate strategy of big, *vertically integrated* companies? The answer is not clear, and perhaps will not be known for many years, until the present antitrust suit is resolved.

15. A complex struggle for oil is taking place between the OPEC nations and the oil consuming nations. The oil companies, being both producers and consumers, are placed in a cross-current of self-interest involving *transfer pricing* as well as their relation to the producing countries and their contest for markets. This *bargaining* situation resembles that of *bilateral monopoly*, in which the staying power of contestants is apt to determine outcomes. It is difficult to predict whose staying power is greater in oil, since a unified buying group does not exist.

Bilateral monopoly

Energy crisis and substitution

16. The oil problem is part of a larger *energy crisis*. However important this crisis, we must learn to analyze it, as economists, in terms of a *search for substitutes* and a gradual leftward shift of demand curves that may become horizontal at very high prices.

QUESTIONS

1. In what way, if any, do the motives of a monopolist differ from those of a perfect competitor? In what way does the time span of his decision-making differ? Can this affect his behavior (as contrasted to his motivation)?

2. How would you define a monopoly? Are monopolies necessarily large? What constraints does a demand curve put on the behavior of a monopoly?

3. Suppose that you were the only seller of a certain kind of machinery in the nation. Suppose further that you discovered that your demand curve looked like this:

Price	$100	$90	$80	$70
Quantity of machines sold		1	2	3	4

What is the average revenue at each price? What is the marginal revenue at each price? Draw a diagram showing the marginal and average revenue.

4. Now superimpose on this diagram a hypothetical cost profile for your business. Where is the point of equilibrium for the monopolist? Is this the same, in terms of MC and MR, as the point for the competitive firm? Now show the equilibrium output and price.

5. Does the equilibrium output of the monopolist yield a profit? What are the relevant costs for figuring profit, average costs per unit, or marginal costs? Show on your diagram the difference between average cost and selling price.

6. Why will a monopolist's selling price not be pushed to the lowest point on the cost curve?

7. What is the difference between monopoly and oligopoly? Between oligopoly and pure competition? Between pure competition and monopolistic (or imperfect) competition? Between the latter and oligopoly?

8. What is meant by the kinked demand curve of an oligopolist? How do you explain the kink?

9. Why does the profit-maximizing assumption lead to such clear-cut results when we speak of pure competition, and to such indeterminate ones for oligopolies? Does the answer have something to do with the time over which the consequences of actions can be calculated?

10. What is a differentiated commodity? Give examples. Draw the demand curve for a farmer selling wheat and that for a toy manufacturer selling his own brand of dolls. What will happen if the doll manufacturer makes a large profit? What will his final point of equilibrium look like if he has many competitors?

11. Compare on two diagrams the equilibrium of the purely competitive firm and of the imperfectly competitive one. Show, by a dotted line, what the demand curve would look like for the imperfectly competitive firm if its product differentiation were removed. Where would its output now be located? What would be its selling price? Would the consumer gain from this?

12. Make a list of all the institutional changes you can think of that would be needed if we were to institute a system of perfect competition.

13. How would you advise the Department of Justice to determine whether or not oil refiners deliberately refrained from building refineries in the United States?

14. What is the difference between a kinked demand curve and the curve shown in Fig. 13·16, where demand becomes completely elastic at some point because of substitution?

15. In economic terms, what does it mean to say that there is a long-run scarcity of oil?

The Problem of Big Business

IN A FAMOUS STUDY published in 1932, Adolf A. Berle, a law professor at Columbia University, and Gardiner Means, an economic statistician, pointed out that between 1909 and 1928 the 200 largest nonfinancial corporations increased their gross assets over 40 percent more rapidly than all nonfinancial corporations. Looking into the future, Berle and Means concluded:

Just what does this rapid growth of the big companies promise for the future? Let us project the trend of the growth of recent years. If the wealth of the large corporations and that of all corporations should each continue to increase for the next twenty years at its average annual rate for the twenty years from 1909 to 1929, 70 percent of all corporate activity would be carried on by two hundred corporations in 1950. If the more rapid rates of growth from 1924 to 1929 were maintained for the next twenty years 85 percent of corporate wealth would be held by two hundred huge units. . . . If the indicated growth of the large corporations and of the national wealth were to be effective from now until 1950, half of the national

wealth would be under the control of big companies at the end of that period.[1]

Indeed, warned the authors, if the trend of the past continued unchecked, it was predictable that in 360 years all the corporate wealth in the nation would have become fused into one gigantic enterprise having an expected life span equal to that of the Roman Empire!

CORPORATE POWER TODAY

What has happened to this trend toward concentration in recent years? We might begin by looking at Table 14·1, showing us the strategic position of the giant corporation within various divisions of the economy.

The table speaks for itself. Note that the giant corporation is much more dominant in some sectors, such as transportation and

[1] *The Modern Corporation and Private Property* (New York: Macmillan, 1948), p. 36, 40–41.

TABLE 14 · 1
Giant Corporations

RELATIVE SHARES OF LARGE CORPORATIONS IN VARIOUS SECTORS, 1967

Sector	All corporations		Corporations with assets of $250 million or more	
	Number	*Total assets ($ billion)*	*Number*	*Percent of assets in that sector*
Mining	14,441	$ 18.1	12	31%
Manufacturing	197,023	448.0	243	58
Transportation, communication, utilities	66,045	221.1	95	79
Finance, insurance, real estate	388,115	1,097.3	517	55
Wholesale and retail trade	465,841	144.1	29	20
All industrial divisions[a]	**1,534,360**	**2,010.7**	**958**	**53**

Source: *Statistical Abstract*, 1970, p. 475. (Data calculated from tax returns.)
[a]Including divisions not shown in table.

communication, than in others, such as retail and wholesale trade. In the latter sector, where we find nearly a half-million corporate enterprises (note that we are not even counting the millions of proprietorships and partnerships), 29 giant companies, such as Sears Roebuck, A&P, Safeway, and others, nonetheless control almost 20 percent of the total assets.

CONCENTRATION RATIOS: WHAT IS AN INDUSTRY?

This overall appraisal, however, conceals wide variations from industry to industry and from region to region. Economists generally measure the degree of concentration by comparing the ratio of total sales or total assets of the top 4 or the top 8 companies to the total sales or assets of their industry. Immediately we encounter the difficult problem of defining *industry*. For example, the top 4 companies make 81 percent of a commodity for which there is really no adequate substitute: salt. Yet if we put salt into an industrial classification called "chemical preparations not elsewhere specified," the share of the top 4 companies falls to a paltry 23 percent of the output of that larger group of products.

Take another example. A housewife shopping for salad dressing is a consumer in an "industry" in which the top 4 producers sell 57 percent of the product. But if she thinks of herself as a consumer browsing in

THE CORPORATION

Much has been written, quite rightly, about the abuses of corporations, but it is important to recognize how valuable is this ingenious legal innovation in encouraging the accumulation of capital and in creating the organizational means to supervise and direct that capital into production. The corporation is a marvelously adaptive legal form of organizing production and an important spur to the growth of the economy. The reason is that, unlike the personal proprietorship or partnership, *the corporation exists quite independently of its owners*, survives their deaths, and can enter into binding contracts in "its" own name. Further, by limiting the liability of its owners to the value of the stock they have bought, it protects a capitalist against losing his entire fortune, and thereby encourages him to invest.

an industry called "pickles and sauces," the top 4 salad dressing makers' share of output is only 29 percent. So, too, a farmer, looking for a tractor is buying in a "market" in which the top 4 companies make 72 percent of the total output; but if we think of a tractor as belonging to a larger industry called "farm machinery and equipment," the share of the top 4 is but 44 percent.[2]

CROSS-ELASTICITIES OF SUBSTITUTION

How *should* we draw the lines around products? Is carpeting, for example, an industry unto itself or only part of a larger industry called "floor coverings" that includes linoleum and tiles? These questions, bringing us again to the problem of substitutes, are of enormous importance in political economics, for they enter directly into the considerations that bear on antitrust suits. Economists and lawyers differ about the best way to draw the lines. One way is through the so-called *cross elasticity of substitution;* that is, by measuring how much the sales of product A increase if the price of product B rises. If A and B are close substitutes, a rise in the price of B will induce many consumers to switch to A. But

that still leaves undecided how large the coefficient of cross-elasticity must be to define an "industry" in a way that will satisfy everyone.

REGIONAL CONCENTRATION

In addition, the problem of measuring or even defining oligopoly is complicated by the question of the regional division of sales. For example, the top 8 beer and ale companies sell only 28 percent of the nation's total consumption of beer; but because these drinks are usually produced and marketed within a limited region, the concentration of sales among a few brewers is much greater in any one area, such as New York or Chicago, than it is in the nation as a whole.

Concentration ratios are generally much higher when we consider regions or localities than when we measure sales in the nation's market as a whole. For instance, the ice cream and frozen dessert industry is only mildly concentrated if we measure the percentage of business going to the top 4 firms on a national basis—37 percent—but when we descend to the level of localities, we find that the average proportion of a local market claimed by the top 4 firms in *that* locality rises to 70 percent. (Of course, the top 4 firms are not the same in every locality.)

[2] John M. Blair, *Economic Concentration* (New York: Harcourt, 1972), p. 9.

RISING CONCENTRATION RATIOS

The figures in Table 14·1 do not give us a sense of the *movement* toward corporate concentration that is still going on, not only in the U.S., but in all market systems.* During the decade of the 1950s and 1960s, a burst of merger activity dwarfed the previous great merger boom of the late nineteenth century. Between 1951 and 1960, *one-fifth* of the top 1,000 corporations disappeared—absorbed within the remaining four-fifths. As a result of this and other growth, by 1970 the *100* largest manufacturing corporations owned 49 percent of the assets of all manufacturing corporations—*a larger percentage than the top 200 corporations had held in 1948!*

Moreover, the pace of this centralizing activity has been steadily growing. Between 1963 and 1966 the value of assets acquired by the big mining and manufacturing companies averaged $4 billion to $5 billion a year. This rate rose to $10 billion in 1967, then to $15 billion in 1968, and reached an all-time peak *rate* of $20 billion in the first quarter of 1969. The subsequent break in the stock market thereupon brought mergers to an abrupt halt, but the total for 1969 was nonetheless almost at the 1968 level.[3]

CONCENTRATION AND COMPETITION

What does this disquieting trend in concentration imply for the competitive market structure of the economy? There is no ques-

*By and large, concentration ratios—and the trend to concentration—are greater in Europe and Japan than in the United States. In those areas, governments tend to *encourage* mergers because they offer economies of scale. Deliberate efforts to increase concentration have been a means for improving the international competitive strength of these nations. More on this in Chapter 38 on the multinational corporation.

[3] *Statistical Abstract* 1970, pp. 476, 483. See also Federal Trade Commission, *Staff Report on Corporate Mergers,* October 1969, esp. pp. 184–98.

tion that a massive concentration of corporate power exists in America and that the degree of concentration within business as a whole is increasing. At the same time, rather surprisingly, the degree of concentration in the *marketplace*—an area of critical importance for the control mechanism of the capitalist society—does not seem to be getting significantly worse.

How can that seemingly contradictory state of affairs exist? How can corporations be getting bigger and not be increasingly monopolistic? The answer is that the merger wave that has so dramatically boosted the figures for the national concentration of business wealth has taken place largely by the rise of so-called *conglomerates*—corporations that have grown by merging with other corporations not *within* a given market but in a *different* market.

We do not yet know what may be the long-term consequences of the rise of such giant, diversified companies. Many of them have been put together more with an eye to realizing the profits that could be had from exchanges of shares than with any careful consideration of operating efficiencies. Some came unglued in the 1969–1970 and 1973 stock market declines. Others may indeed prove to be efficient combinations of diverse activities that will enjoy the advantages of access to a central pool of capital and a top-flight supermanagement. Still others may run afoul of antitrust laws and may be forced to dissolve.

STABILITY OF MARKET SHARES

But as matters now stand, it seems unlikely that the conglomerates will be adding to their size by buying up *competitors*. Hence it is likely that the structure of the individual markets in which they operate will show no more change than they have in the past. And

PORTRAIT OF A CONGLOMERATE

Consider the case of International Telephone and Telegraph, originally a company wholly engaged in running foreign communications systems. In 1961 ITT determined to embark on a major acquisition and diversification program. During the next 7 years it acquired 52 domestic and 55 foreign companies with combined assets of $1.5 billion. In 1969 alone, the directors approved an additional 22 domestic and 11 foreign acquisitions. As a result it has jumped from 34th to 9th in industrial size, with sales and assets of $9 billion each and 428,000 employees—the 4th largest private employer in the world. More important, whereas it had once been a "one-product" company, selling and operating telecommunications systems, by 1972 ITT rented cars (Avis), operated motels and hotels (Sheraton), built homes (Levitt), baked bread (Continental), sold insurance, produced glass, made consumer loans, managed a mutual fund and processed data—among other things.

here, in the critical area of the marketplace, we discover a truly surprising long-run stability. For more than half a century, now, there has been no substantial increase in "monopoly" in the nation's markets, considered as a whole. Going back to 1901, we find some industries—tobacco, chemicals, stone, clay and glass, transportation equipment—where industrial concentration has risen; but in others, no less important—food, textiles, pulp and paper, petroleum and coal products, rubber, machinery—concentration has fallen since 1901.

The same conclusion holds for more recent years. If we take the four largest companies in any industry between 1947 and 1967 (the latest figures available) and compare the value of their total shipments to the value of all shipments in the industry, we find a similar mixed trend. In a few instances, such as the automobile industry, where some early postwar competitors were shaken out, the concentration ratio has increased appreciably: in autos, from 56 percent in 1947 to 79 percent in 1967. In the majority of industries, however, the movement toward concentration was insubstantial; and in a considerable number, concentration actually declined.

Perhaps most convincing of all is a government study that shows the degree of concentration within 213 industrial markets to be virtually unchanged between 1947 and 1966. In 1947, the average share of all markets going to the top 4 firms was 41.2 percent; in 1966 it was 41.9 percent. Moreover, the number of highly concentrated industries—where the top 4 firms did three-quarters or more of all sales—dropped sharply during the period.

CAUSES OF STABILIZATION: ANTITRUST

What has produced this stabilization of concentration within industry? Two explanations are plausible.

The first is the *restraining effect of antitrust legislation.* Beginning in the 1930s under Franklin Roosevelt, a much stronger enforcement of the various antitrust laws began to gain favor. A vigorous campaign to block the drift toward concentration resulted in a number of suits brought against major corporations for restraint of trade. No less important was a change in the prevailing judicial view, which now construed much less narrowly the powers of the Constitution to impose social controls over business enterprise. In more recent years, new amendments to the antitrust acts and a growing stringency of Justice Department rulings have made the marriage of competitive firms increasingly difficult.

One result of these obstacles in the way of direct "concentration-affecting" mergers is that corporations have been forced to seek acquisitions in fields considerably removed from their original base of operations. This is one source of that trend toward conglomerates we noted above. What interests us here is that this process of diversification has put many large corporations into fields they do not dominate. One study shows that in a thousand different product markets, the 100 biggest firms are not even among the 4 biggest sellers in almost half these markets.[4] By way of illustration, in 1965 the Radio Corporation of America acquired Random House, a book publisher. In its manufacturing activities, RCA is a large producer in a moderately concentrated field (the top 4 companies make just under half of all radios); in its broadcasting activities, as the owner of NBC, it is one of three great broadcasting networks; but as a book publisher it sails in a highly competitive race.

THE NEW CORPORATE EXECUTIVE

A second reason for the stabilization within markets is more diffuse, but no less significant. It lies in a decisive *change in the character of business management.*

Throughout the first century of rapid corporate expansion, the men who ran the corporations were themselves the men who had put up the capital or who owned large blocks of stock in the enterprises. The Carnegie Steel Company, the Standard Oil Company, the Ford Motor Company were all extensions of the personalities of Carnegie, Rockefeller, and Ford. This personal direction of affairs was the case in the overwhelming majority of other large and small enterprises of the day.

But with time, a significant change set in. As the original founders of the great businesses died, their stock was inherited by heirs who often did not have business ability and who receded into the background. In addition, the widening dispersion of stock ownership among more and more investors made it unnecessary for any group to own an actual majority of the stock to exercise control over a company. In 1928, for example, the board of directors of U.S. Steel (which included two of the largest stockholders in the company) held only 1.4 percent of the company's stock. In that year, the biggest stockholder in A.T.&T. held but seven-tenths of 1 percent of the company's total stock.

The result was that the active direction of the big firms passed from owner-capitalists to a new group of "managers" who ruled the corporation by virtue of their *expertise* rather than because they owned it. The new management was different in many ways from its predecessors. In 1900, for example, half the top executives of the biggest corporations had followed paths to the top that could be described as "entrepreneurial" or "capitalist"—that is, half had built their own businesses or had risked their own capital as the means to business preeminence. By 1925 only a third of the top corporate executives had followed this path, and in 1960 *less than 3 percent* had done so. More and more, the route to success lay through professional skills, whether in law or engineering or science or in the patient ascent of the corporate hierarchical ladder. Significantly there was a marked change in the educational background of the top corporate officials. As recently as the 1920s, a majority of the topmost

[4] A. D. H. Kaplan, *Big Enterprise in a Competitive System* (Washington, D.C.: Brookings, 1964), p. 286.

corporate leaders had not gone to college; to-day over 90 percent have college degrees and a third hold graduate degrees.[5]

BUREAUCRATIZATION

The new management brought important changes in its wake. The affairs of big companies increasingly became matters to be handled in systematized, "professional" ways, rather than in the often highly personal, often idiosyncratic mode of the founders. A certain *bureaucratization* thus crept into the conduct of business life at the very time that the business community was waxing most vocal about the dangers of bureaucracy in government. More important, the new management now adopted a new strategy for corporate growth—or rather, abandoned an older one. Advances in technology, changes in product design, vigorous advertising, the wooing of businesses to be acquired in other fields—all these provided ample outlets for the managerial impulse toward expansion. But one mode of growth—the mode that the founders of the great enterprises had never hesitated to use—was now ruled out: growth was no longer to be sought by the direct head-on competition of one firm against another in terms of *price*.

OLIGOPOLY AND MARKET BEHAVIOR

Thus, the evolution of a more "statesman-like" attitude in business affairs brought a significant change in business tactics. But the change cannot be ascribed solely to a new outlook on the part of big-business men. Behind that new outlook was *a change in the market environment*.

[5] Mabel Newcomer, *The Big Business Executive* (New York: Columbia University Press, 1955), pp. 61–63; and Jay Gould, *The Technical Elite* (New York: Augustus Kelley, 1966), pp. 160–71.

In many of the most important industrial markets, as we have seen, it is oligopoly rather than pure competition that is the order of things today. In industry after industry, economies of large-scale production have brought about a situation in which a few large producers divide the market among themselves. As we saw in the case of oil, one very large firm often serves as price leader, raising or lowering its prices as general economic conditions warrant, and being followed up or down by everyone else in the field. U.S. Steel and General Motors have more or less consistently "led" their industries in this fashion. By and large, as we would expect, these prices are considerably higher than the prices that a pure competitive market would enforce. General Motors, for instance, "targets" its prices to attain a 15 to 20 percent return after taxes, *calculating its costs on the assumption that it will use only 60 to 70 percent of its total plant capacity.* U.S. Steel sets its prices high enough so that it can earn a small profit *even if it operates only two days out of five.* In fact "target pricing" has come to be the established procedure for many leading manufacturing firms.

Price leadership does not mean that in each of these industries firms do not vie with one another. On the contrary, if you ask a General Motors or a U.S. Steel executive, he will tell you of vigorous competition. He may show you that Ford has edged out General Motors in such-and-such a line, or that Bethlehem Steel has captured some of U.S. Steel's business. The point, however, is that the *competition among oligopolists typically uses every means except price cutting.* The tactics of lowering price to secure a rival's business is not considered "fair play," although it may take place in covert ways, as we saw in the oil industry's special deals with independents. But in the main, "competition" among oligopolists means winning business away from each other by every tactic—advertising, customer service, product design—except that of "chiseling" on price. This leaves everyone the gainer, except the person the market mechanism was supposed to serve—the consumer.

The Cost of Market Imperfection

What does the rise of a highly oligopolized market structure cost the consumer? In theory the answer is very clear. In a purely competitive market, the consumer is king; indeed, the rationale of such a market is often described as *consumer sovereignty.*

The term means two things. First, in a purely competitive market *the consumer determines the allocation of resources by virtue of his demand.* Second, *the consumer enjoys goods that are sold as cheaply and produced as abundantly as possible.* As we have seen, in a purely competitive market there exist no profits (except transitory intramarginal rents). Each firm is producing the goods that consumers want, in the largest quantity and at the lowest cost possible, given its cost curves.

In an oligopolistic market the consumer loses much of this sovereignty. Firms have *strategies,* including the strategy of influencing consumer demand. Profits are not competed away, so consumers' surplus is transferred to producers. Output is not maximized but is reduced by whatever amount results from higher-than-competitive prices.

CONSUMER SOVEREIGNTY TODAY: ADVERTISING

No one contests these general conclusions. How great are they, however, in actuality? Here the problem becomes muddier.

Take the question of consumer demand. In 1867 we spent an estimated $50 million to persuade consumers to buy products. In 1900 advertising expenditures were $500 million.

In 1971 they were $21 *billion*—roughly two-thirds as much as we spend on primary and secondary education. Indeed, advertising expenditures can be considered as a vast campaign to educate individuals to be good consumers.

THE EFFECT OF ADVERTISING: INFORMATION VS. MANIPULATION

To what extent does advertising infringe on consumer sovereignty? The question is perplexing. For one thing, it is no longer possible to think of consumers as having "natural" tastes, once we go beyond a subsistence economy. For that reason, much advertising has a genuine informational purpose—people do have to be made aware that it is possible (and imaginable) for, say, a factory worker to take a vacation by airplane rather than in the family car.

Moreover, numerous efforts to create tastes have failed. In the mid-1950s the Ford Motor Company poured a quarter of a billion dollars into a new car, the Edsel, and performed prodigies of advertising to make the American public like it. The public did not, and the car was quietly discontinued. So, too, consumers spontaneously decided to buy small sports cars, beginning in the 1950s; and after valiant efforts to turn the tide, the major American manufacturers capitulated and admitted that the American car buyers *did* want small cars.

Yet, it is obvious that all advertising is not informational and that consumer's tastes are manipulated to a considerable (although not clearly measureable) degree. We are mainly creatures of brand preference as a result of advertising exposure, not because we have sampled all the choices and made up our minds. It is difficult to contemplate the battles of aspirins, soaps (up to 10 percent of the price of soap is selling expense), cars, and cigarettes, without recognizing that much of this represents a waste of resources, including the very scarce resource of talent largely devoted to annulling the talent in a different advertising agency.

IS PRODUCT DIFFERENTIATION A GOOD THING?

Product differentiation is also an ambiguous case. Few would deny that the proliferation of "models" is often carried to the point of absurdity—and more important, to the point of substantial economic waste. It has been calculated that annual model changes added as much as $700 (in 1962) to the price of a car.

Yet, as with advertising, the question is where to draw the line. Where product differentiation results in variations in the actual product, and not merely in its "image," one must ask whether affluent society should aim to produce the largest possible quantity of a standardized product at the least possible cost or to offer an array of differing products that please our palates, admittedly at somewhat higher costs. Few consumers in a rich society would prefer an inexpensive uniform to more expensive but individualized clothes. From this point of view, even the wasteful parade of car styles has a certain rationale.

Thus, as with advertising, *some* production differentiation plays a useful and utility-increasing function. The question is how much? It is difficult to form a purely objective judgment, for even if the amount of "useless" product differentiation is relatively small, its impact on the public taste may be disproportionately large. The problem is perhaps particularly acute insofar as much of our "taste" for style seems to be the product of the deliberate advertising efforts of manufacturers. No doubt there is a real aesthetic pleasure in variety, but one doubts that it would take the form of a yearning for "this year's model" without a good deal of external stimulation. Product differentiation thus becomes in part

an effort to maximize the public's utilities; but it is also in part an effort to create those "utilities" in order to maximize the producers' profits.

MONOPOLY AND INEFFICIENCY

What about the second main attribute of consumer sovereignty—the ability to buy goods as cheaply as possible? To what extent does oligopoly introduce inefficiency into the system or transfer consumers' surplus to producers?

Once again the evidence in fact is murkier than in theory. For one thing, we tend to leap to the conclusion that a competitive firm, which has managed to combine its factors as profitably as possible, has also reached the frontiers of technological efficiency. But is this so? Suppose that the competitive firm cannot afford the equipment that might lead to economies of large-scale production. Suppose it cannot afford large expenditure on research and development. Suppose that its workers suffer from low morale and therefore do not produce as much as they might.

These are not wild suppositions. There is good evidence that many large firms are more efficient, in terms of productivity per man-hour, than small firms, although of course some large, monopolistic firms tolerate highly inefficient practices simply because of the lack of competition. Moreover, estimates for the increase in production that would take place as a result of a demolition of all monopolistic restraints amount to only about $5 billion for 1954—an amount that would only lower prices and increase total output insignificantly.

Once again, however, we must consider the other side. Profits in monopolistic industries as a whole are 50 to 100 percent higher than those in competitive industries: in 1962, for example, the 10 largest manufacturing corporations enjoyed profits of 8.7¢ per dollar of sales, compared with 3.5¢ per dollar for corporations doing less than $50,000 worth of annual sales. In certain fields, such as prescription medicines, there is evidence that consumers are mercilessly exploited. Brand name aspirins, for example, sell for up to three times the cost of unbranded versions of the same product. Certain medicines, such as antibiotics and the like, have enjoyed enormous profits—which is to say, have forced consumers to pay far more than they would have had to pay were the rate of profit a "competitive" one.

Another complication is introduced by virtue of the fact that oligopolies have often, although not always, provided more agreeable working conditions, more handsome offices, and safer plants than have small competitive firms. Thus some of the loss of consumers' surplus is regained in the form of lessened disutilities of work. Needless to say, this is not solely the result of a "kindlier" attitude on the part of big producers but reflects their sheltered position against the harsh pressures of competition. Nonetheless, the gains in work conditions and morale are real and must be counted in the balance.

Economic Power and Its Control

Thus the economic arguments regarding consumer sovereignty are not easy to resolve. The gains and losses are difficult to measure and cannot easily be weighed on the same scale. One thing is clear, however. The basic change in the market structure has brought a new element with which society must cope, however it may elude the net of economic analysis. This is the *problem of economic power,* not diffused among a host of consumers but concentrated in a relatively small number of giant producers whose influence—political and social as well as economic—is

surely a major concern of contemporary economic society.*

POWER AND RESPONSIBILITY

What can be done about the fortresses of power that have emerged in modern capitalism? Is there a way of imposing public responsibility on the labor union, the government office, the big corporation? Let us explore this question by examining some frequently encountered proposals for increasing the social responsibility of the giant corporation.

1. Profit as social responsibility

The first suggestion is most prominently associated with the name of Milton Friedman. Professor Friedman is a philosophic conservative whose response to the question of what a corporation should do to discharge its social responsibility is very simple: *make money.*

The function of a business organization in society, argues Friedman, is to serve as an efficient agent of production, not as a locus of social improvement. It serves that productive function best by striving after profit—conforming, while doing so, to the basic rules and legal norms of society. It is not up to business to "do good"; it is up to government to prevent it from doing bad.

Moreover, as soon as a businessman tries to apply any rules other than money-making, he takes into his own hands powers that rightfully belong to other parts of society, such as its political authorities. Friedman would even forbid corporations to give

---

*Power is actually always present in the institutional framework that legitimates private property itself. Capitalism, like socialism or precapitalist societies, is in part a system of legitimated economic power. But formal economics does not discuss these institutions; it discusses their *consequences.* Radical economics is interested in bringing the institutions themselves back into the center of focus.

money to charities or universities. Their business, their responsibility to society, he insists, is *production*. Let the dividend-receivers give away the money the corporations pay them, but do not let corporations become the active social welfare agencies of society.[6]

The counterarguments to Friedman's position are not difficult to frame. They are two:

1. Friedman assumes that the stockholders' moral claim to earnings of the vast semimonopolies they "own" is superior to that of the consumers or workers from whose pockets these profits are plucked. This is at least a debatable point: since the stockholders are *not* active entrepreneurs, they make little or no contribution to the profits of the corporation. Why, then, should their claim to corporate earnings be given a top priority?

2. Friedman assumes that the government, whose purpose is to set the rules and oversee the operations of business, acts *independently* of the corporations they regulate. But many studies show that the so-called regulatory agencies of the government usually act on *behalf of the big corporations* they "regulate," rather than on behalf of the consumer. For example, the Food and Drug Administration banned cyclamates as a dangerous food additive in 1969, *nineteen years* after the first warnings of their dangerous effects had been brought to its attention! In the long interim, it failed to act, largely because of its reluctance to incur the wrath (and political counterattack) of the industry it was supposedly "regulating."[7] We shall come back to this point again.

2. The corporation as social arbiter

Quite a different approach to the problem of social responsibility has been widely espoused by many concerned corporate executives. This view recognizes that the corporation, by virtue of its immense size and strength, has power thrust upon it, whether it wishes to have it or not. The solution to this problem, as these men see it, is for corporate executives to act "professionally" as the arbiters of this power, doing their best to adjudicate equitably among the claims of the many constituencies to whom they are responsible: labor, stockholders, customers, and the public at large. The executive of one of the largest enterprises in the nation said, two decades ago, "The manager is becoming a professional in the sense that like all professional men he has a responsibility to society as a whole."[8]

There is no doubt that many top corporate executives think of themselves as the referees among contending groups, and no doubt many of them use caution and forethought in exercising the power of decision. But the weaknesses of this argument are also

[6] Friedman, *Capitalism and Freedom*. The late Professor Frank Knight, a distinguished predecessor of Friedman at the University of Chicago, agreed with the proposal and carried it even further: force corporations, he urged, to pay out to stockholders *all* their earnings every year! That would deprive them of the market power that stems from the very large "retained earnings" they do not distribute as dividends. Under Knight's proposal, a corporation would be under constant economic scrutiny from its stockholders, who would have to be persuaded to reinvest their earnings with the company if it were to grow.

[7] See James S. Turner, "The Chemical Feast," in *The Report on the Food and Drug Administration*, eds. Ralph Nader and Summer Study Group (New York: Grossman, 1970), pp. 5–30. See also John C. Esposito, "Vanishing Air," in *The Ralph Nader Study Group Report on Air Pollution* (New York: Grossman, 1970); and Grant McConnell, *Private Power and American Democracy* (New York: Knopf, 1966).

[8] R. W. Davenport and *Fortune* editors, *U.S.A.: The Permanent Revolution* (Englewood Cliffs, N.J.: Prentice-Hall, 1951), p. 79. (The executive is Frank Abrams of Standard Oil, N.J.)

not difficult to see. Unlike other professions, there are neither criteria for "qualifying" as a corporate executive nor penalties for failing to accept social responsibilities. The executive of a corporation who fails to act responsibly may incur the opprobrium of the public, but the public has no way of removing him from office or reducing his salary.

Nor is there any clear guideline, even for the most scrupulous executive, defining the manner in which he is *supposed* to exercise his responsibility. Is his concern for the prevention of pollution to take precedence over his concern for turning in a good profit statement at the end of the year or giving wage increases or reducing the price of his product? Is the contribution of "his" company to charity or education supposed to represent *his* preferences or those of his customers or workers? Has Xerox a right to help the cause of public broadcasting; Exxon to help finance Harlem Academy (a private school aimed at assisting Harlem youths to go on to college); the makers of firearms, to help support the National Rifle Association?

These questions begin to indicate the complexity of the issue of "social responsibility" and the problems implicit in allowing these extremely important *social* decisions to be made by private individuals who are in no way publicly accountable for their actions.

3. Dissolution of monopoly

A third approach to the problem of responsibility takes yet another tack. It suggests that the power of big business be curbed by dividing large corporations into several much smaller units. A number of studies have shown that the largest *plant* size needed for industrial efficiency is far smaller (in terms of financial assets) than the giant firms typical of the *Fortune* list of the top 500 industrial corporations (or for that matter of the next 500). Hence a number of economists have suggested that a very strict application

of antitrust legislation should be applied, not only to prohibit mergers but to separate a huge enterprise such as General Motors into its natural constituent units: a Buick Company, an Oldsmobile Company, a Chevrolet Company, and so on.

One major problem stands in the way of this frontal attack on corporate power. It is that size and social responsibility are by no means clearly correlated. Professor J. K. Galbraith has wittily remarked that: "The showpieces [of the economy] are, with rare exceptions, the industries which are dominated by a handful of large firms. The foreign visitor, brought to the United States ... visits the same firms as do the attorneys of the Department of Justice in their search for monopoly."[9]

The other side of that coin, as we have already noted, is that small competitive industry is typically beset by low research and development programs, antilabor practices, and a general absence of the kinds of amenities we associate with "big business."

Moreover, there is no reason to believe that smaller firms would be more pollution-conscious (indeed, owing to competitive pressures, they might be less inclined to minimize pollution) or that they would be more conscientious in advertising, racial nondiscrimination, and other practices. In other words, the loss of *political* power, which might well accompany the fractioning of the largest firms, is apt to be offset by a rise in certain forms of economic ruthlessness or even antisocial behavior.

Competition, it has been remarked more than once, is a social condition paid homage to by all parties in an enterprise economy, but taken seriously only by economists. Business and labor both spend much of their energies trying to avoid competition or to minimize it, and the attempt to intensify

---

[9] *American Capitalism* (Boston: Houghton-Mifflin, 1952), p. 96.

competition by breaking up large firms into smaller ones might bring about worse problems than it would alleviate.

4. Regulation

Regulation has been a long-standing American answer to the problem of corporate power. Regulation has sought to prevent corporations from carrying on practices that were destructive to *other* groups (such as the discriminatory pricing of the railroads in the late nineteenth century); regulation has also tried to prevent industries from injuring *themselves* through cutthroat competition (for example, when that competition threatened to bring about bankruptcies).

The problem with regulation is that typically the regulatory commission has become the captive and servant of the very industry it was set up to control. The Interstate Commerce Commission, established in 1887 to regulate the railroads, is a prime example of this reversal of roles. When the ICC was established, the railroads were a monopoly that badly needed public supervision. Autos and trucks had not yet come into existence, so that there were few alternative means of bulk transportation in many areas.

By the end of the first quarter of the twentieth century, however, the railway industry was no longer without effective substitutes. Cars, trucks, busses, planes, pipelines—all provided effective competition. At this point, the ICC became interested in protecting the railroads against competition, rather than in curbing abuses. One by one, these alternative modes of transport fell under its aegis (or under that of other regulatory agencies), and quasi-monopoly prices were set, as little "empires" were established for each form of transportation.

An example turned up by the Senate Select Committee on Small Business concerned a small trucking firm that wished to extend trucking service to two Alabama towns not directly served by any large carrier. After *4½ years* of proceedings, the ICC granted the applicant limited approval to serve one of the towns, but not the other. In its report, the commission stated that these towns had "only limited transportation needs" and that additional service was therefore not warranted. In effect, the commission *prevented* the second town from enjoying trucking service.

In similar actions, the ICC has prevented private truckers from choosing new routes that would greatly shorten trucking hauls, has forbade trucks that carried goods one way from picking up goods for a return load, and has limited the products that certain carriers might legally haul—ruling, for example, that a live chicken is an "agricultural commodity" but a dead chicken is not; that nuts in the shell qualify as agricultural goods but not shelled nuts, etc. These regulations follow from a law that states that motor vehicles carrying agricultural commodities are exempt from ICC regulation. It was therefore in the "interest" of the ICC to define an agricultural commodity as narrowly as possible, to minimize the competition that would otherwise arise.[10]

Another example is the difference in the air fare from Washington to Boston and from Los Angeles to San Francisco. The distances are the same, but the California flight is not regulated by the Civil Aviation Board because it is not in interstate commerce. The unregulated fare in 1973 was $18; the regulated fare, $40. So, too, the Texas Railroad Commission, which regulates oil prices and production in the Texas oil fields, the Federal Power Commission, the many state power commissions, the Securities and Exchange Commission, and numerous other agencies have gradually become "defenders" of their industries, rather than guardians of the public interest.

---

[10] Blair, *Economic Concentrations*, pp. 398–99.

Recently, partly as a result of effective publicity from reformers such as Ralph Nader and others, there has been a tightening-up of some regulatory bodies. Yet there remains a deep skepticism about the ability of commissions to retain their initial purposes. When commissions are established to regulate industries there is a need to find personnel who are familiar with the industry's problems. Quite naturally, these experts tend to come from within the industry itself. Moreover, the long contact between the industry and the commission tends to bring a gradual "understanding" of the industry's problems that becomes, bit by bit, indistinguishable from the position of the industry itself. Indeed, some critics of the system have maintained that regulatory commissions and agencies are in fact the means by which the dominant members of the industry suppress the competition of fringe members, thereby achieving through public regulation the "orderly"—that is oligopolistic—solution to the problems that they were unable to attain without government help.[11]

5. Nationalization

Then why not nationalize the large firms? The thought comes as rank heresy to a nation that has been accustomed to equating nationalization with socialism. Yet Germany, France, England, Sweden, Italy, and a host of other capitalist nations have nationalized industries ranging from oil refineries to airlines, from automobile production to the output of coal and electricity. Hence, Professor Galbraith has suggested that we should nationalize corporations charged with the public interest, such as the giant armaments producers who are wholly dependent on the Pentagon, in order to bring such firms under public control.

[11] See Gabriel Kolko, *The Triumph of Conservatism* (New York: Free Press, 1963).

But would nationalization achieve its purpose of assuring social responsibility? In 1971, the Pentagon arranged special contracts and "loans" to save Lockheed Aircraft, one of its "ward" companies, from bankruptcy—the fate of an ordinary inefficient firm. Outright nationalization would only cement this union of political and economic power, by making Lockheed a part of the Pentagon and thus making it even more difficult to put pressure on it to perform efficiently.

Or take the Tennessee Valley Authority, perhaps the most famous American public enterprise: it is currently being sued for the environmental devastation it has wrought by its strip-mining operations. So, too, the Atomic Energy Commission, which operates "nationalized" plants, has been severely criticized for its careless supervision of radioactive processes, and the Post Office has been converted into a semi-autonomous agency, having proved a bureaucratic disaster as a "nationalized" industry.

The problem is that nationalization not only removes the affected enterprise *entirely* from the pressures of the market, but almost inevitably brings it under the political shelter of the government, further removing the venture from any effective criticism.

OTHER POSSIBILITIES

All these difficulties make it clear that the problem of social responsibility will not be easy to solve (or for that matter, even to *define*), no matter what step we choose, from Professor Friedman's laissez faire to Galbraith's nationalization. And for each of these problems with regard to the corporation, we could easily construct counterparts that have to do with the control over labor unions or over the government itself.

What, then, is to be done? A number of other lines of action suggest themselves. One is the widening of the *legal responsibility* of the corporation to include areas of activity for which it now has little or no accountability. Environmental damage is one of these: consumer protection is another. Ralph Nader has further suggested that a top corporate official be legally charged with seeing to it that his company complies with the law in full, and that the top officers of noncomplying companies be suspended, as is the case in certain violations of SEC regulations. He also strongly advocates the federal, rather than state, incorporation of big companies.

A second step would be a widening of *public accountability through disclosure*—the so-called fishbowl method of regulation. Corporations could be required to report to public agencies their expenditures for pollution control, for political lobbying, and so on. Corporate tax returns could be opened to public scrutiny. Unions and corporations both could be required to make public disclosure of their race practices, with regard to hiring or admission, advancement, and rates of pay. Public responsibility for advertising, with formal proofs submitted to back all claims, is yet another means of securing better accountability to the public.

Still another course of action would be to appoint *public members* to boards of directors of large companies or to executive organs of large unions and to charge these members with protecting the consumers' interest and with reporting behavior that seemed contrary to the public interest. Worker-members of boards of directors might also serve such a useful purpose (there are such members in Germany). The mobilization of the votes of concerned stockholders (nonprofit institutions own tens of billions of corporate securities) is still another way of bringing social pressure to bear.

Finally, there is the corrective action of dedicated private individuals such as Ralph

Nader, who rose to fame on his exposé of the safety practices of the auto industry, and who has since turned his guns on pollution, other irresponsibilities of big business, and on poor performance in the federal bureaucracy. Such public pressure is necessarily sporadic and usually short-lived, but it has been a powerful source of social change.

POWER: THE UNRESOLVED PROBLEM

It would be a mistake to conclude this recital with the implication that corporate (or union or government) power can be easily brought under control through a few legal remedies or by the power of public opinion. Certainly, many abuses can be curbed, and much better levels of social performance achieved.

But it is a sober fact that mass organizations seem an inescapable concomitant of our age of high technology and increasing social interdependence. Here we should note that, depending on our interests, we stress different aspects of this universal phenomenon. To some, who fear the continued growth of very large-scale business, the most significant aspect is that we have not managed to control business power. To others, concerned over the emergence of large labor unions, it is labor power that most dangerously eludes effective control. And to still others who are most worried by the growth of big government, it is the growth of public power that is the main problem.

What is common among these concerns is the awareness that very large, only half-controlled organizations have come to dominate much of the market system. But the problem is bigger than that. For if we look to the nonmarket systems, we see the growth of socialist ministries of production and administration that display much the same bureaucratic indifference and mixed political and economic power as do our corporations, unions, and government agencies, together with the same uncertainty about how to reconcile their power with their conception of the public good.

Thus the question of economic power remains, at best, only partially resolved. As Adolph Berle has written: "Some of these corporations can be thought of only in somewhat the same way we have heretofore thought of nations." Unlike nations, however, their power has not been rationalized in law, fully tested in practice, or well defined in philosophy. Unquestionably, the political and social influence and the economic power of the great centers of production pose problems with which capitalism—indeed, all industrialized societies—will have to cope for many years to come.

KEY WORDS

Concentration

Conglomerates

Concentration ratios

Stability of market shares

Antitrust

Bureaucratization

Consumer sovereignty

Advertising:
    informational
    manipulative

Waste

Losses and gains from monopoly

Controls over corporate power

CENTRAL CONCEPTS

1. The trend toward *economic concentration* is long-standing in the United States (and in other capitalist nations).

2. In recent years, concentration has dramatically increased on a national basis in the United States, largely as a result of the rise of *conglomerates*—huge diversified corporations.

3. *Concentration ratios* vary greatly from industry to industry. They are difficult to define accurately, owing to the arbitrary definition of an *industry*, and the much greater degree of *regional* (over *national*) concentration.

4. Despite growing concentration, the degree of oligopolization of most markets has remained roughly unchanged. This static condition can be traced to several causes:
   - More effective *antitrust* legislation
   - Diffusion of conglomerate growth into *new markets*
   - *Bureaucratization* of management

5. Market imperfection calls into question the idea of *consumer sovereignty*. Consumer sovereignty assumes that consumers alone are the source of the allocation of resources.

6. Market imperfection has eroded consumer sovereignty, although to an extent difficult to measure. Consumers' tastes are certainly manipulated by advertising. Advertising nonetheless has a legitimate *informational* component. Advertising can also lead to considerable waste of resources, as can product differentiation.

7. Monopolistic profits are not very large, on a national basis, but may be considerable in certain industries. To some extent the losses in consumers' surplus are probably offset by the gains from monopoly—more efficient production, more research and development, better working conditions.

8. A number of proposals have been advanced to curb corporate power. These include (1) strict attention to *profit-making* only; (2) "*professional*" standards of business conduct and self-conscious attention to social needs on the part of corporations; (3) the *break-up of big business* into smaller, more competitive units; (4) *regulation*; (5) *nationalization*; (6) and *publicity*.

Problems of control

9. Each of these proposals has weaknesses. (1) The abdication of social responsibility by business assumes that stockholders have a "right" to profits and that business power does not exist. (2) The professionalization of management ignores the absence of any standards of professionalism and overlooks the essential arbitrariness of corporate decisions in social areas. (3) The break-up of business assumes that smaller business units will be more socially responsible, which is doubtful. (4) Regulation ignores the record of "capture" of the regulatory agency by the industry it is supposed to regulate. (5) Nationalization ignores the sorry public record of many (not all) state-owned enterprises; and (6) publicity is apt to be sporadic and often short-lived.

New legal requirements

10. Other means of improving corporate behavior may be a redefinition of their *legal responsibilities*, new areas of *public accountability*, *public or labor representation* on their boards, and heightened scrutiny by outside investigators.

Problem of power

11. The overall *problem of power* remains recalcitrant, not only in market societies but in all industrialized nations.

QUESTIONS

1. Can you name the chief corporate executives of the top ten industrial firms in the U.S. (General Motors, Exxon, Ford, General Electric, Socony, U.S. Steel, Chrysler, Texaco, Gulf, Western Electric)? How many names of leading businessmen do you know? What does this suggest as to the character of business leadership today contrasted with the 1890s?

2. Suppose Congress decided to foster a return to classical competition in the United States. What changes would have to be wrought in the American business scene? Do you think this is a practical possibility?

3. Do you consider yourself to be a "sovereign" consumer? Do you think you are influenced by advertising? If you could imagine a nonadvertising society, do you think you would change your life style?

4. What do you consider to be the most desirable characteristic of bigness in business? The most undesirable?

5. Suppose you wanted to measure concentration in an industry. What attributes of the firms in that industry would interest you: their respective sales? their assets? their number of employees? Might different measures give different concentration ratios?

6. What are the important differences between "pure" competition and oligopoly regarding the position of the consumer? The producer? On net balance which do you think is preferable? Why?

7. Do you think businessmen should be more socially responsible? How would you go about achieving a higher level of social responsibility in the following areas: (1) truth in advertising, (2) absence of political interference with government, (3) colorblind hiring and promotion, (4) high levels of antipollution performance? What measures would you propose for numbers 2 and 3 with regard to labor unions? How would you bring government agencies to a higher level of social responsibility?

8. Do you think corporations have a right to back right-wing groups? Left-wing groups? Center-wing groups? Modern art? Old-fashioned art? Universities? Sports groups? Political parties? How do you justify your answers?

9. Can you find areas ripe for deregulation other than the transportation industry? How would you set up regulatory agencies so that they cannot be captured by producers?

10. Do you think it would help to nationalize the big arms companies? Could you imagine a situation in which you would favor the nationalization of any company? How about the airlines, if they

were about to go bankrupt and leave the nation stranded? How about GM if it defied a government regulation with regard to producing a pollution-free vehicle? How about a textile company that admitted to racial discrimination in its advancement policies?

11. What relation do you see between the growth of power in all areas of the society and the development of modern technology? Discuss with relation to the automobile, the airplane, nuclear power, the computer.

Economics and the Environment

15

THERE IS MORE AND MORE AWARENESS of the fact that we are living in a period of environmental strain. We have already looked briefly into the question of resource depletion, a problem we will again review in Chapter 34, but in recent years a new, hitherto almost overlooked problem has come to the fore. This is the problem of pollution, the frightening rate at which we are narrowing the margin of safety in the air we breathe, the water we drink, the wastes we discharge into the environment.

EXTERNALITIES

What is the cause of this pollution problem? Much of it is often blamed on the selfish attitudes of our corporations or on our carelessness as individuals. This is saying only that pollution occurs because corporations and individuals defile the environment. That does not answer the question in a way that satisfies an economist. He seeks reasons for

the economic behavior that leads to pollution, so that he can suggest policies that will effectively lessen it.

This brings us to examine another aspect of production—its output of things we do *not* want, along with those we do. In the current vocabulary, dirt, wastes, noise, congestion are "bads" that are integrally connected with the production of "goods." We call these "bads" *externalities*. They are "joint products" with goods, joint products that have the peculiar and distasteful (or dangerous) effect of lowering the quality of the environment for those who have the bads thrust upon them, sometimes as a part of purchasing goods, sometimes as the result of a wholly unconnected process.

EXTERNALITIES AND THE MARKET

Why do externalities exist? The basic answer is, of course, technological—we do not know

POLLUTION PROBLEMS

Annual throwaway list

214 million tons of carbon monoxide, sulfur oxides, hydrocarbons, particulates, and nitrous oxides (over a ton per American)

55 billion tin cans (275 for each of us)

20 billion bottles (100 apiece)

65 billion metal and plastic bottle caps (325 each)

7 million cars (our annual output has to go *somewhere* after 100,000 miles or so)

10 million tons of steel and iron scrap (an average of 1,000 pounds each)

150 million tons of garbage and trash (4.1 pounds per person daily)

3 billion tons of tailings, mine debris, and waste (4 tons for each of us each day)

ITEM: Inhabitants of Los Angeles and St. Louis have been warned on more than one occasion against indulging in any kind of exercise, including jogging and golf, because deep breathing is dangerous.

ITEM: Nitrates and phosphates in fertilizers and detergents cause water pollution problems that are beyond the control of present technology. The use of these chemicals is estimated to increase by tenfold before the year 2000. Professor Barry Commoner, ecologist, predicts a major agricultural crisis in the United States from this source alone within 50 years.

ITEM: It is estimated that by the year 2000, there will be, at any time, some 3,000 six-ton trucks in transit carrying dangerous radioactive wastes to burial sites.

how to produce goods "cleanly"; i.e., without wastes and noxious by-products such as smoke. But the economic answer calls our attention to another aspect of the problem. *Externalities refer to the fact that the output of "bads" does not pass through the market system.* A factory may produce smoke, etc., but it does not have to pay anyone for producing these harmful goods. So, too, certain inputs used by firms—air, water, even space in an esthetic rather than economic sense—are available without charge, so that there is no constraint to urge a firm to use air or water sparingly or to build a handsome rather than an ugly factory or building.*

In other words, pollution exists because it is the cheapest way to do many things, some having to do with producing goods, some with consuming them. It is cheaper to

*Some externalities are not "bads," but "goods." For example, a new office building may increase the property value of a neighborhood. Here is a positive externality. The benefit gained by others results from the new building, but is not paid to the owners of that building.

litter than to buy waste cans (and less trouble, too); cheaper to pour wastes into a river than to clean them up. That is, it is cheaper for the individual or the firm, but it may not be cheaper for the community. A firm may dump its wastes "for free" into a river, but people living downstream from the firm will suffer the costs of having to cope with polluted water.

MARGINAL PRIVATE AND SOCIAL COSTS

The point is so important it is worth repeating. In back of the conception of the marketplace as an *efficient* allocator of goods and incomes was the assumption that all the inputs going into the process were owned by some individual and that all the outputs were bought by some person or firm or agency. Presumably, then, the price at which commodities were offered would include all the costs that were incurred in the process; and presumably also, the price that would be offered would reflect all the prospective benefits accruing to the buyer. In other words, prices established in the marketplace were supposed to take into account all the disutilities involved in the process of production, such as the fatigue or unpleasantness of work, and all the utilities ultimately gained by the final consumer of those goods.

What the environmental crisis has brought home is that the *market is not a means for effectively registering a great many of these costs and benefits.* The examples of the damage wrought by smoke that is not charged against the factory or of a neighborhood nuisance such as a bar or a hideous advertising sign are instances of economic activities in which *private costs are less than social costs.* These "social" costs are, of course, private costs incurred by other people. Contrariwise, when a person spends money to educate himself, he benefits not only himself but also the community, partly because he becomes a more productive citizen and partly because he presumably becomes a more responsible one. Thus the *social benefits* of some expenditures may be greater than their *private benefits.*

SOCIAL AND PRIVATE MARGINAL COSTS AND BENEFITS

But suppose that the social and private gains and losses do not coincide. How does that reflect on the efficiency of the market process?

The answer is very plain. If we are interested ultimately in maximizing everyone's well-being, we want the price system to act as a guide to firms and factors and consumers, urging them to buy more or produce more of socially useful goods and to make or buy fewer socially deleterious ones. We want an economy that produces more goods and less "bads." But under a pricing system that takes

no account of the externalities wrought by consumption and production, the economy will produce too many external bads and too few goods. Because no charge for smoke is levied on the factory owner, there is no economic incentive to economize on smoke, perhaps by switching to more expensive but less smoky fuel. If there is no reduction in the price of education, it will be bought only to the point at which the marginal dollar spent on education by a consumer is equal to the marginal *private* gain he hopes to get from that dollar; whereas if the community's benefit were calculated in the price, education would be subsidized so that the consumer would be tempted to buy more of it for himself, at the same time creating external benefits for the society.*

Controlling Externalities

How can we bring the process of pollution under social control? Basically, we can attack the problem in three ways. We can

1. **Regulate** *the activity that creates it*

2. **Tax** *the activity that creates it*

3. **Subsidize** *the* *polluter,* *to* *stop (or lessen) his activity*

REGULATION

Faced with the ugly view of smoke belching from a factory chimney, sludge pouring from a mill into a lake, automobiles choking a city, or persons being injured by contaminants, most ecologically concerned persons cry for regulation: "Pass a law to forbid smoky chimneys or sulfurous coal. Pass a law to make mills dispose of their wastes elsewhere or purify them. Pass a law against auto-

---

*It is precisely because society recognizes that education does bring beneficial externalities that we have public education.

mobiles in the central city. Pass a law against DDT."

What are the economic effects of regulation? Essentially the idea behind "passing laws" is *to internalize a previous externality.* That is, a regulation seeks to impose a cost on an activity that was previously "free" for the individual or firm, although not free, as we have seen, for society. This means that individuals or firms must stop the polluting activity entirely or bear the cost of whatever penalty is imposed by law, or else find ways of carrying out their activities without giving rise to pollution.

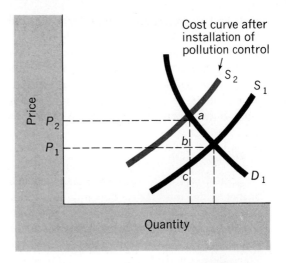

FIGURE 15 · 1
The effects of regulation

THE COSTS OF REGULATION

Let us take the case of a firm that pollutes the environment as a joint product of producing goods or services. Suppose a regulation is passed, enjoining that firm to install antipollution devices—smoke scrubbers or waste-treatment facilities. Who bears this cost?

The answer seems obvious at first look: the firm must bear it. And in fact it would, in the form of reduced profits, if it were a monopoly or an oligopoly prevented from raising prices (say by price ceilings).

But if the firm is competitive or if it does pass its higher costs along in higher selling prices, we arrive at a different answer. Examine Fig. 15·1. Our firm's original marginal cost (or supply) curve is S_1. The need to install new anti-pollution equipment raises it by an amount ac to S_2. Now a little economic analysis will show us that the cost is in fact borne by three groups, not just by the firm. First, the firm will bear some of the cost because at the higher price, it will sell less output. How much less depends on the elasticity of demand for its product. But unless demand is totally inelastic (a vertical line), its sales and income must contract.

But two other groups also bear part of the cost. One group is the factors of produc-

tion. Fewer factors will be employed because output has fallen. Their loss of income is therefore also a part of the economic cost of antipollution regulation. Last, of course, is the consumer. Prices will rise from P_1 to P_2. Note that the rise in price (*ab*) is less than the full rise in costs (*ac*), so that the consumer will not bear all the costs (unless, once again, we have vertical demand curves).

THE GAINS FROM REGULATION

Offsetting all these costs is the fact that each of these three groups and the general public now have a better environment. There is no reason, however, why each of these three groups, singly or collectively, should think that *its* benefit outweighs *its* costs. Most of the benefit is likely to go to the general public, rather than to the individuals actually involved in the production or consumption of the polluting good or service.

Thus a regulation forcing car manufacturers to make cleaner engines will cost the manufacturers some lost sales, will cost the consumer added expense for a car, and will cost lost income for whatever land, labor, and capital is no longer employed at higher production costs. As part of "the public," all three groups will benefit from cleaner air, but each is likely to feel its specific loss more keenly than its general gain.

IS REGULATION USEFUL?

Regulation can be a good or a bad way to control pollution. This depends on two things: the quality of the regulation itself and the efficiency with which it is carried out. For example, regulations may be poorly phrased in law, so that many loopholes exist for firms or individuals, or they may fail to discriminate among different cases of pollution. An antismoke regulation may be sensible in a city but foolish if applied to a plant in the country where air currents easily disperse the smoke.

So, too, regulations are good or bad depending on their ease of enforcement. Compare the effectiveness of speed limits, which attempt to lessen the externality of accidents, and of regulations against littering. It is difficult enough to enforce speed laws, but it is almost impossible to enforce anti-littering laws. On the other hand, regulation of the disposal of radioactive wastes is simpler to enforce because the polluters are few and easily supervised.

This in turn is largely a matter of cost. If we were prepared to have traffic policemen posted on every mile of highway or every city block, regulation could be just as effective for speed violations or littering as for radioactive disposal. But the cost would be horrendous, and so would most people's reaction to being overpoliced.

TAXATION

A second way to cope with pollution is to tax it. When a government decides to tax pollution (often called effluent charges), it is essentially creating a price system for disposal processes. Ideally, the tax would be set high enough so that the city or state would have enough tax revenue to install the devices for cleaning up whatever pollution remains. If an individual company found that it could clean up its own pollutants more cheaply than paying the tax, it would do so—thereby avoiding the tax. If the company could not clean up its own pollutants more cheaply than the tax cost (which is often the case because of economies of scale in pollution control), it would pay the necessary tax and look to the state to clean up the environment.

The effluent charge looks like, but is not, a "license to pollute." It is a license that allows you to give some of your pollutants to the state, *for a price.*

As a result of effluent charges, an activity that was formerly costless, is no longer so. Thus, in terms of their economic impacts, these charges are just like government regulations. In fact, they are a type of government regulation. They raise the supply curve for the good in question, with all of the corresponding ramifications. The difference is that each producer can decide for himself whether it pays to install clean-up equipment and not pay the tax, or to pollute and pay whatever tax costs are imposed.

ANTIPOLLUTION TAXES VS. REGULATIONS

Which is better, regulation or taxation? As we have seen, regulation affects all polluters alike, and this is both its strength and its weakness. Taxation enables each polluter to determine for himself what course of action is best. Some polluters will achieve low pollution targets more cheaply by installing antipollution equipment, thereby avoiding taxes on their effluents, while other polluters will find it more profitable to pay the tax.

Here practical considerations are likely to be all-important. For example, taxation on effluents discharged into streams is likely to be more practical than taxation on smoke coming from chimneys. The state can install a sewage treatment plant, but it cannot clean up air that is contaminated by producers who find it cheaper to pay a pollution tax than to install smoke-suppressing equipment. Moreover, to be effective, a pollution tax should vary with the amount of pollution—a paper mill or a utility plant paying more taxes if it increases its output of waste or smoke. One of the problems with taxation is that of installing monitoring equipment. It is difficult to make accurate measurements of pollution or to allow for differences in environmental harm caused by the same amount of smoke coming from two factories located in different areas.

SUBSIDIES

The third way of dealing with pollution is to *subsidize polluters to stop polluting;* that is, the government pays polluters to install the necessary equipment to clean up their effluents.

As we might expect, subsidies have impacts quite different from those of regulation or taxation. Because the government pays the costs of the antipollution equipment, the private firm incurs no costs. Its supply curves do not shift. No fewer factors are employed. Prices to the consumer remain unchanged. One curious effect is that the total amount of resources devoted to pollution control will therefore be larger under subsidy than under taxation or regulation. The reason is obvious: there will be no reduction in output, as in the case of the other two techniques.

ARE SUBSIDIES USEFUL?

Economists typically object to subsidies because they camouflage the true economic costs of producing goods and services cleanly. When regulations or taxes increase the price of paper or steel, the individual or firm becomes aware that the environment is not free and that there may be heavy costs in producing goods in a way that will not damage the environment. The increased price will lead him to demand less of these goods. But when he gets clean environment through the allocation of a portion of his taxes, he has no "price signal" to show him the cost of pollution associated with particular commodities.

Nevertheless, there are cases when subsidies may be the easiest way to avoid pollution. For example, it might be more effective to pay homeowners to turn in old cans and bottles than to try to regulate their garbage disposal habits or to tax them for each bottle or can thrown away. Subsidies may therefore sometimes be expedient means of achieving a desired end, even if they may not be the most desireable means from other points of view.

Costs and Benefits

Whether we seek to lessen pollution through regulation, taxation, or subsidy, there are costs involved—often high costs. How much should we be prepared to pay to gain cleaner air, quieter airfields, less congested streets?

MARGINAL SOCIAL COSTS OF POLLUTION CONTROL

To begin with, we must recognize that cleanup costs are not only the direct expense of antipollution equipment. It is also the reduction in the output of goods and services (or the reduction in the incomes of taxpayers) that results from "clean" production. A regulation requiring that effluents be cut below a certain level may force a plant out of business because it cannot afford to install the necessary equipment. For those who will lose their jobs, this is a terrific cost. Or regulations or taxes requiring the installation of pollution control equipment (such as cleaner auto engines) may price a car out of the pocketbook of a large number of would-be consumers. For them, too, this is a severe cost.

Thus the social costs of controlling pollution are difficult to measure. But in general we can probably describe the shape of the marginal social cost curve as rising from the origin, like most cost curves. Moreover, the *marginal* cost of reducing pollution is apt to rise as we reduce it more and more. The second 10 percent reduction is apt to cost more than the first 10 percent, in out-of-pocket expenses or in terms of indirect costs imposed on workers or consumers. Reduction by 100 percent may be prohibitive. You might construct such a cost curve in your mind by imagining how much it would cost to clean up all beaches and parks from their present state to a condition of "perfection."

DIMINISHING MARGINAL SOCIAL BENEFITS

There is also a marginal benefit curve (or a demand curve for benefits). As part of the general public, we would benefit a lot from a 10 percent reduction in smoke, wastes, litter, and noise. Our ability and willingness to pay for this effort may set a high reservation price on initial antipollution efforts. Improving the environment another 10 percent yields lower marginal benefits—diminishing marginal utility sets in. Attaining absolute purity after we have reached "90 percent purity" may be of very little importance to us. Construct an imaginary demand schedule for the amounts the public might be willing to pay for successive clean-ups of beaches and parks.

SOCIAL VS. PRIVATE BENEFITS

At this point, however, the environmental analysis diverges from the analysis of conventional economic goods and services, because *my* expenditures on environmental control help clean up not only *my* environment but also *your* environment. Similarly your expenditures on environment control help clean up my environment. As a result, there are *positive externalities* from our clean-up expenditures, just as there were negative externalities from our disposal processes.

This means that to determine the *social* demand or benefit curve for a clean environment we must add up individual benefit curves *vertically* rather than horizontally, as we did in calculating the market demand for conventional goods and services where there were no externalities.

FIGURE 15 · 2
Social benefit curves

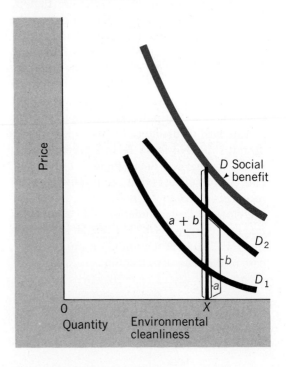

In Fig. 15·2 citizen 1 will pay sum *a* for *OX* "amount" of clean environment; and citizen 2 will pay sum *b*. *But because each has benefited the other, OX amount of cleanliness is "worth" a + b to the two individuals together.*

PRIVATE VS. COLLECTIVE DECISIONS

Thus in talking about pollution problems it is important to understand that we are not talking about private markets in which each person makes his or her individual decision largely unaffected by the decisions of others. The aim of antipollution policy is to spend as much on antipollution efforts as is needed to equate marginal *social* benefits and marginal *social* cost, as Fig. 15·3 shows.

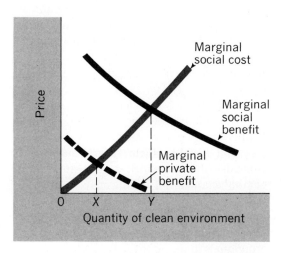

FIGURE 15 · 3
Social costs and benefits

But this result will not come about by pure market forces alone. From society's point of view the social benefit or demand curve lies much further to the right than the private

benefit or demand curve. This position is due to the small amount of antipollution expenditure each person is willing to make *by himself*, knowing that most of his efforts will benefit others.

As a result, only *collective* decision making will approximate an expenditure that equates marginal social benefits and costs. If there is no means of exerting a collective demand that reflects, however roughly, social benefits, the amount of private environmental cleaning-up *OX* will fall far short of the amount that would be socially optimal, *OY*.

PRACTICAL PROBLEMS

This leads us to a gap between theory and practice. Through the political process, we reveal our preferences for collective action, but it is very difficult to obtain or devise a schedule of costs and benefits from which rational judgments can be made. How are we to assess the full costs of cleaning up the environment, such as the loss of jobs for workers? How are we to assess the benefits of clean air? Smoke over a city not only adds to cleaning bills and causes ill health, but defaces buildings, delays air traffic, and causes people to move to the suburbs (with secondary costs that we will examine below).

Putting dollars and cents on these kinds of effects is inevitably a somewhat arbitrary procedure, often little more than an informed guess. At best, cost-benefit analysis forces us to consider the complexities—including the hidden costs and benefits—of pollution control, so that we can try to frame the issue as thoughtfully as possible for the public's decision. It does not lead to open and shut answers. In the end, the effort we put into cleaning the environment depends at least as much on our values and preferences as on the outcome of any "rational" economic calculus.

Part of the urban environmental problem results from the sheer interaction of people in crowded environments. William Baumol has explained the reason for this in an important article in the June 1967 *American Economic Review*. Consider the two circles, representing two equal-sized cities. The dots are people (in thousands or millions), and the crosses are (smoky) chimneys, also in some suitable multiple. Suppose that the number of chimneys is primarily determined by the number of people, say as the number of private houses grows. Then, as the different shadings of the two circles suggest, the absolute amount of smoke pollution per unit/area (say, per block) within the two cities will be determined directly by the numbers of their respective populations.

But the matter does not stop here. Take the city on the right, in which population and the number of chimneys has increased from two to six. The absolute amount of smoke over the city has now tripled—*but this tripled density of smoke is*

The Urban Environment

All our environment is not natural. Most of us (68 percent) live in cities where the environment is affected by *social* factors such as congestion and crime as much as it is by purely physical ones. Many of these problems can be approached through economic analysis. Here we will consider two special problems of the urban environment: economic segregation and commuter transportation. Once again, what we want to discover is what light, if any, an analysis of economic motives and interactions throws on these problems and what hints it offers for alleviating them.

ECONOMIC SEGREGATION

If we go back before World War II, most of the population of any large metropolitan area lived in the central city. There was no choice, since one had to live in the central city if one were to work in the city. With the development of commuter transportation, particularly the automobile, our large metropolitan areas have developed into a maze of suburbs, each with its own local government surrounding a central city. People have been liberated from having to live close to their place of work.

Where *do* they want to live? Obviously, in the place that has the most pleasant environment. But the environment is created by two basic forces. Part of one's environment is purchased privately in the form of a home and yard. But part of the environment is purchased *collectively* in the form of good schools, parks, clean streets, police and fire protection.

The privately purchased portion of the environment is not an economic problem. People simply purchase what they can afford. (If it is a social problem, it is because of income distribution, which we will study in the coming chapters.) The collectively purchased part, however, presents a different series of questions. Since local governments basically depend upon property taxes, the amount that individuals have to pay as a share of the necessary collective expenditures depends not only upon the value of *their* homes, but also upon the value of the homes around them.

Imagine purchasing a certain amount of collective environment—schools, police, etc.—that costs $2,000 per year per person. *The amount that you personally are going to have now inflicted on three times as many people. Thus the total social impact of pollution has gone up not three times, but nine!*

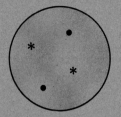 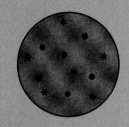

Smoke pollution is not the only form of environmental damage that goes up by the square of the population. Noise pollution is another environmental hazard that may be directly produced by population density and that feeds back in turn across the whole population. Perhaps the "damage" caused by the sheer ugliness of crowded neighborhoods is another. Certainly the number of manhours spent fuming in traffic jams is on the list, a fact that may add understanding, if not consolation, the next time you are trapped on a clogged freeway.

to pay depends largely upon the value of your neighbors' houses. The more expensive their houses, the less that you will have to pay. Conversely, the less expensive their houses, the more you will have to pay. As a result, you will want to live in a neighborhood with homes at least as expensive as yours, and you will want to keep out all homes that are less expensive than yours. If you allow cheaper homes to be built, your costs will go up, since the number of people in the community will rise, but their tax payments will not go up equivalently. They will be paying less than the average tax.

THE ECONOMICS OF ZONING

If residents of a town restrict homebuilding to houses that cost more than the average existing homes, their own taxes will fall. If residents allow less-than-average cost homes to be built, their taxes will rise. As a result, zoning rules are used to maintain a minimum standard for housing in each area.

This process leads to a condition of economic segregation, because the central city and its suburbs have people of very different income levels and very different per capita tax bases. In poor suburbs it becomes expensive to buy good schools, because the effective property tax rate would have to be very high. Conversely, in rich suburbs it becomes cheap to buy good schools because the effective property tax rate can be very low. As a result, poor neighborhoods buy fewer schools than rich neighborhoods, even if their desires for schools are exactly the same.*

As a result of economic segregation, we also find wide disparities in the distribution of public services. The poor man with a poor private environment has a poor public environment, not because *he* is poor but because he lives with other poor people. If we think of governmental expenditures as being one of the prime means of providing equal opportunity through good education, recreational, and cultural facilities, then we see that *local governments are unable to fulfill this role because of the economic segregation that has occurred.*

*Recent Supreme Court decisions have challenged the constitutionality of these unequal taxes' bases, and may bring about a more equitable distribution of educational expenditure.

THE NEIGHBORHOOD EFFECT

But let's return to the central city. There, another economic process exacerbates the pull of economically homogeneous suburbs. The value of a home depends partly upon the condition of the home itself and partly upon the condition of the neighborhood. The *same* house in different areas can bring very different prices. A poor neighborhood can bring the value of a good house down by as much as 60 to 70 percent.

The value of your home therefore depends mostly upon the actions of others. What your neighbors do to maintain their own homes and what the city does to provide public services will dominate the price at which you can sell your house. *As a result, you have a rational economic incentive to undermaintain your home.* To do so cuts your costs and has little impact on the price of your home, as long as the *other* homes are well maintained and good public services are provided. If other homes are not maintained or public services are bad, good maintenance on your home will be wasted. It will not show up in the price you can get for your home.

As a result, economic incentives lead to neighborhood deterioration once anything happens to disrupt the stable character of the neighborhood. Movement to the suburbs starts; stability is broken. The quality of public services may start to deteriorate as high income and high tax paying citizens move out. This tempts the economic man in all of us to quit maintaining our property and move to the suburbs to get the quality housing and public services we want. If we yield to this temptation, we make the collective problem worse and accelerate the process.

Once again we have each individual acting in his own economic self-interest, with results that may *not* be socially optimal—the very obverse of Adam Smith's "invisible hand"—and once again there is no market correction for the situation. No one is being *economically* irrational. The only solution is therefore to change the structure of the market. Metropolitan-wide tax raising authorities would be one solution, since they could establish uniform tax rates that would make it impossible to lower your own tax bill by moving into areas without poor people (and hopefully with people wealthier than you).

COMMUTER TRANSPORTATION

In most cities, the possibility of living long distances from one's place of work did not exist until the automobile became an instrument of mass transportation. Curiously, however, as transportation has become more efficient, there is no evidence that there has been any reduction in the average commuting time. What has happened is that the average commuter now lives many more miles away from his work than he previously did. Therefore the urban transportation system exists not as an instrument to cut commuting times but as a means to allow us to live farther away from our jobs and to purchase the housing we want in the neighborhood that we like. In all probability, further improvements in urban transportation will have the same effect. This is not to say that improved transportation would be worthless. Being able to purchase a desired house with desired neighbors may be of great value to an individual.

In the two decades after World War II, the automobile was overwhelmingly the preferred instrument of mass transportation. Immense investments were made in cars and roads. Since the late 1960s, however, it has become difficult, if not impossible, to improve further the efficiency of the automobile transportation system in most urban areas. Pollution is one reason. Most urban areas are reaching a critical point where the air is barely able to digest the automobile fumes emitted. The volume of cars must be curtailed, or the characteristics of the automobile engine must be modified.

Land utilization problems present another problem. Automobiles are such a land-intensive (land-using) method of transportation, that they require great quantities of land to move a limited number of people. Only 1.1 persons occupy the average commuter car. This land-intensive aspect of auto use comes into conflict with a stubborn fact—the limited land area in central cities. The question is where to find land to locate roads to allow more people to enter the city or to move around in it. We soon begin to reach the point where the only solution is to tear down the city to build roads. Alas, this also destroys the city as a place where one can find jobs.

MASS TRANSPORTATION

The "obvious" solution is more mass transportation. But the existence of the automobile makes it difficult, if not impossible, to finance and utilize other types of urban mass transit. The automobile has great private advantages. No alternative mode of transportation allows you to walk out the front door and find your mode of transportation waiting for you. Busses and trains will never be quite as convenient. Bicycles cannot give you the same air-conditioned, poshly tailored, private compartment with your own radio and tape deck. Most public transportation systems require many people to transfer from one vehicle to another. This takes time and is inconvenient.

The peculiar character of automobile cost curves also makes it difficult for public transportation to compete with the automobile. The automobile has high average

costs, but very low marginal costs. Volkswagen advertises "three pennies a mile." That's the marginal cost. But according to *Consumer's Report,* a New Yorker pays almost $2,000 per year—in depreciation, insurance, taxes, etc.—to *own* a car, before he ever drives a mile. That enters the average cost but is not included in the 3¢ per mile.

PUBLIC VS. PRIVATE TRANSPORTATION COSTS

Thus the *average* cost per auto commuter trip may be several dollars, although the marginal cost may be just a few pennies. Unfortunately, when the commuter is judging whether public transportation is cheaper than private transportation, he will compare not the average costs of his automobile with the public transit fare, but the marginal costs. Public transit, however, if it is to be self-financed must charge enough to cover its average costs. These average costs can be well below the average costs of the automobile but still well above the auto's marginal costs. *As a result, public transit can attract automobile riders only if it charges a fare that is below the auto's marginal costs. But this means running a huge deficit in the transit authority.* The perceived quality of the automobile being what it is, even a free public transportation system might not attract the desired number of riders.*

Any deficit must be picked up by the taxpayers, a group that may not benefit at all. Certainly the central city taxpayers will not benefit from a transit system that extends far into the suburbs. Yet they may be called upon to pay the deficit caused by distant

*Alternatively, the marginal cost of auto travel could be raised by heavy tolls. This is hardly likely to be a popular solution.

commuters. The taxpayer does not want to pay the bill for something he does not use, and there is no way to charge the commuter a price that will cover costs.

As a result, the economic system leads to incentives that encourage automobile driving, despite attendant problems of pollution and congestion, and make it difficult to change that mode of transportation. Public opinion polls illustrate the problem. Many automobile commuters are in favor of building an urban transportation system—not because they would use it, but because they think that others might use it, enabling them to ride their autos along uncongested streets. This attitude can be seen in a "scientific sample" taken from the Los Angeles *Times* Reader Panel. People were asked, "If by some miracle a rapid transit system were in effect tomorrow, would you ride to work on it?" In response, 50 percent said they "wouldn't, definitely, flatly." Another 10.8 percent said they probably would not; 2.6 percent said they probably would; and only 4.7 percent said they definitely would. On the other hand, an overwhelming number (86.6 percent) said that they believed that Los Angeles needed a new rapid transit system!

Public transportation is therefore viable only if the public is willing to take steps to curtail its own utilization of the automobile. Its willingness to do so will perhaps depend upon the seriousness of the health hazard created by automobile pollutants, or by the mounting problems of an energy crisis.

Once again, *collective decision making, through political processes, is the only way to solve a problem that the market mechanism cannot solve.* Whether the political process can convince people that their long-term state of well-being would be improved by setting aside their immediate individual preferences is a possibility that remains to be seen. It need hardly be said that it is one of the most important decisions the next decade will bring.

KEY WORDS

CENTRAL CONCEPTS

Externalities

1. By *externalities* we mean *effects of production or consumption that are not reflected in the price mechanism.* One such effect is the damage of smoke imposed upon the consumer but not included in the costs of the smoke producer. Externalities may also be positive instead of negative, as they are when a handsome building creates economic and esthetic values for a neighborhood.

Social vs. private costs and benefits

2. Externalities are important because they are instances in which the market price of a good does not accurately reflect all the disutilities or utilities involved in producing or consuming that good. That is, the *private costs of the good may be greater than or less than its social costs,* and the social benefits of consuming the good may also be different from its private benefits.

Marginal costs and benefits

3. As a result of this divergence of social and private marginal cost and benefit, society will tend to produce too many "bads" (goods with less social value than private value) and too few "goods" (whose social benefits are greater than their private ones).

Internalizing externalities

4. As a ubiquitous feature of all economic systems, externalities pose a severe problem for the effective operation of a market economy. *To the extent that the costs or benefits of externalities can be "internalized"—that is, picked up by taxes or regulation, so that they are included within the price of the commodity—the market itself can be used as an administrative device for controlling the effects of externalities.*

Regulation

5. *Regulation is a commonly advocated means of internalizing externalities.* The costs of regulation are shared among three groups: the firm whose production is cut, the factors whose services are required in smaller amounts, and the consumer who must pay more for the product. The efficiency of regulation depends on the quality of the laws or administrative arrangements, and on the ability to enforce them.

Taxation

6. *Taxation resembles regulation insofar as it also raises the cost curves of firms.* Taxation, however, enables a firm to choose between installing antipollution devices and avoiding taxes on its effluents, or emitting effluents as before but paying taxes on them. Practical rather than economic reasons determine whether regulation or taxation is a better means of controlling pollution.

Subsidy

7. A third method of lowering pollution is by *subsidy*. *Subsidy does not affect the price that consumers pay,* thereby failing to give a market signal as to the true price of the product. In some cases subsidies may, however, be the most convenient means of dealing with pollution.

Cost-benefit analysis

8. *Cost-benefit analysis* provides a theoretical guide to how many resources should be devoted to pollution control. *Ideally, we would equate the marginal social cost of control with the marginal social benefit from control.* This is easier said than done. Social benefit curves are cumulated vertically rather than horizontally. Social costs are very diffuse and difficult to assess.

Collective decision making

9. *The market, in and of itself, will not lead to solving pollution problems. Only collective decision making can bring the desired results.* This process is difficult to guide because there is no economic means for the public to reveal its preferences for collective action.

Economic segregation

10. Urban environments are also deteriorating and can be subject to economic analysis. *Economic segregation* is an instance of a social effect that follows from private efforts at maximizing one's well-being. Each individual seeks to enjoy maximum collective environment by preventing the influx of neighbors who have less wealth than he has. This leads to *zoning*.

Zoning

Disparities of local service

11. Economic segregation increases provision of collective consumption (such as schools and swimming pools) for residents in high-tax yielding, rich communities and decreases provision in poor communities. In rich areas, local governments can provide services that governments in poor areas cannot afford.

Neighborhood effects

12. In the central city we find *neighborhood effects,* the importance of the neighborhood in determining the value of an individual's home. If a neighborhood begins to deteriorate, there is a strong incentive for a homeowner to allow his house to deteriorate, since its value will depend more on the neighborhood than on its intrinsic condition. This incentive aggravates the deterioration of the neighborhood.

Commuter transportation

13. *Commuter transportation* presents difficult economic problems. The automobile is a highly land-intensive means of transportation and is threatening to destroy the city. But the switch to public transportation is difficult because of conveniences that cars offer. Among these is the fact that the *marginal cost of an automobile ride is "negligible,"* even though the average cost may be high.

Marginal and average costs of auto travel

Mass transit deficits

14. *Mass transit must therefore match these low marginal costs if it is to attract passengers from cars.* But to charge very low (or zero!) mass transit fares means that large deficits will be incurred. Taxpayers may not be willing to pick up this bill, since they may not personally benefit from mass transit. Once again, only public intervention—collective decision making—can undo a situation in which private decisions do not lead to socially satisfactory *results*.

QUESTIONS

1. Can you think of an externality imposed by a producer on a consumer? (That's easy.) Of one imposed by a producer on another producer? (How about the effect on work efficiency caused by the noise of erecting a building?) Of an externality imposed by a consumer on another consumer? Of one imposed by consumers on producers?

2. How would you go about systematically trying to reduce the many externalities imposed by traffic in a big city? How would you figure marginal costs and benefits? Could you arrange a schedule of fees or taxes or tolls that might reduce the net burden of congestion? Suppose you decided to ban traffic from downtown during certain hours. How would you estimate the costs and benefits involved?

3. Explain carefully why a perfectly competitive market will turn out an assortment of goods in which the sum of the private costs and benefits imposed by, or enjoyed from, those goods will be different from the sum of the social costs and benefits associated with those goods.

4. What do you think would be the best way to control the following externalities? (1) a smoke-producing utility company, (2) roadside littering, (3) overfishing in a stream, (4) noise from a jetport, (5) radiation hazards in a hospital, (6) billboards, (7) pornography, (8) overcutting forests, (9) noise from diesel trucks and motorcycles. In each case discuss the relative merits of taxation, subsidy, or outright regulation.

5. Does it matter if we tax producers of (say) paper or users of paper? HINT: work it out with supply and demand curves.

6. Can you think of any ways to get people to reveal their preferences about the value of a cleaner environment?

7. Would you represent the demand for education as a social demand or benefit curve or as a market demand curve?

8. Explain why market forces will lead to economic segregation.

9. If the marginal costs of public transportation were also far below the average costs, would this solve the problem of trying to compete with the low marginal cost automobiles?

10. What measures can you think of to prevent adverse "neighborhood effects" from taking place? Can private incentives do the trick or will effective action probably require government intervention? Of what kind?

11. Could tax policy change the average and marginal costs of cars vs. mass transportation? Explain how the effect of a tax on cars differs from a tax on gasoline.

The Market for
Factors of Production

16

WITH THIS CHAPTER WE ENTER the last and perhaps most important major area of microeconomics. We have explored in some detail the dynamics of the demand for goods and the calculations of firms—both competitive and monopolistic—in the provision of goods. But we have left unexamined until now a crucial area of the economy. This is the market for factors where wages, rent, and interest are determined.

This area is crucial for two reasons. First, it leads us to complete the analysis of the circular flow, which has provided the basic pattern of our analysis. When we have understood the factor market we will have knit together the household and the firm in their second major interaction—this time with the household acting as supplier rather than demander, and with the firm providing demand rather than supply.

Second, our analysis is crucial because it takes us to the consideration of a problem of major political and social importance. This is the *distribution of income*—the main theme

that will occupy us in this and the next two chapters to come.

FACTORS AND FACTOR SERVICES

We must begin with a simple but important fact that distinguishes the market for factors from that for goods. When we buy or sell goods, we take possession of, or deliver, an actual commodity. But when we speak of buying or selling factors, we mean only that we are buying or selling a stream of services that a factor produces. When we buy or sell "labor," we do not buy or sell the human being who produces that labor, but only the value of his work efforts. So, too, when we buy or sell "capital," we are not talking about purchasing or selling a sum of money or a capital asset, but about hiring or offering the use of that money or equipment. Land, too, enters the market for factors as an agency of production whose services we rent but need not actually purchase as so much real estate.

Obviously we can buy land as real estate, and we can buy capital goods and various forms of capital, such as stocks and bonds. We cannot buy human labor as an entity, because slavery is illegal: in former times one *could* buy labor outright. It stands to reason, then, that there should be a relationship between the price of factors, considered as assets (actual capital goods or real estate) and the price of the stream of services that factors produce—the *interest* earned on capital, the *rent* of land. There is such a relationship, which we will study later under the heading of "capitalization." But here we must recognize that the market for factors is not a market in which assets are sold, but in which the productive services—the earnings—of these assets are priced.

As in every market, these prices will be determined by the interplay of supply and demand. Therefore to study the factor market we must first look into the forces that determine the demand for factors, and then into the forces that determine their supply.

DIRECT DEMAND FOR FACTORS

Who buys factor services? Part of the answer is very simple. A portion of the services of labor, land, and capital is demanded directly by consumers for their own personal enjoyment, exactly as with any good or service. This kind of demand for factors of production takes the guise of the demand for lawyers and barbers and servants or the demand for plots of land for personal dwellings or the demand for cars and washing machines or other personal capital goods or the demand for loans for consumption purposes. To the extent that factors are demanded directly for these utility-yielding purposes, there is nothing in analyzing the demand for them that differs from the demand curves we have previously studied.

DERIVED DEMAND

But most factors of production do not earn their incomes by selling their services directly to final consumers. They sell them, instead, to firms. In turn, firms want the services of these factors not for the firm's personal enjoyment, but to put them to profitable use.

Thus we speak of the firm's demand for factors as *derived demand*. Its demand is derived from the consumers' demand for the output the firm makes. We recall that each firm will hire labor (or any other factor) until the factor's marginal revenue product (its physical product multiplied by the selling price of the product) just equals the cost to the firm of hiring that factor. We see this in Fig. 16·1, where the derived demand of Firm 1 for a factor leads it to employ quantity *OA,* that of Firm 2 results in the hire of *OB,* and the summed demand of both firms gives us a market demand of *OC.*

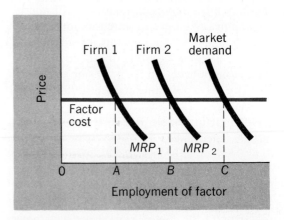

FIGURE 16 · 1
Derived demand for factors

This however leaves us with the question of factor price only half explained. We understand the elements that enter into the demand curve for the factor. But we cannot understand how the price of the factor is established until we know something about the supply curves in the factor market. That is the missing element we need to complete the full chain of the circular flow.

The Supply Curve of Factors

What do we know about the willingness and ability of the owners of land, capital, and labor to offer their services on the marketplace at varying prices for these services?

LABOR

By far the most important supply curve in the factor market is that of labor—meaning, let us remember, labor of all grades from the least skilled workman to the most highly trained scientist or the most effective entrepreneur.

As Fig. 16·2 shows, the supply curve for the labor of most individuals has a curious shape. Up to wage level *OA,* we have no trouble explaining things. As we know, economists assume that labor involves *disutility.* Moreover, just as we assume that increasing amounts of a utility-yielding good give us

diminishing marginal utility, so we assume that increasing amounts of labor involve *increasing marginal disutility.* Therefore the curve rises up to level *OA* because we will not be willing to work longer hours (i.e., to offer a larger quantity of labor services within a given time period), unless we are paid more per hour.

How then do we explain the "backward-bending" portion of the rising curve above wage level *OA?* The answer lies in adding to the rising marginal disutility of labor the falling marginal utility of *income* itself, on the assumption that an extra dollar of income to a man who is making $10,000 is worth less than the utility of an additional dollar when he was making only $5,000. Above a certain income level he prefers leisure to more income.

Together, these two forces explain very clearly why the supply curve of labor bends backward above a certain level. Take a man who has been tempted to work 70 hours a week by wage raises that have finally reached $5 an hour. Now suppose that wages go up another 10 percent. It is possible of course, that the marginal utility of the additional income may outweigh the marginal disutility of these long hours, so that our man stays on the job or even works longer hours. If, however, his marginal utility of income has reached a low enough point and his marginal disutility of work a high enough point, the raise may bring a new possibility: he may work *fewer* hours and enjoy the same (or a somewhat higher) income as well as additional leisure. For example, as his pay goes up 10 percent, he may reduce his workweek by 5 percent.

Backward bending supply curves help explain the long secular trend toward reducing the workweek. Over the last century, weekly hours have decreased by about 40 percent. Although many factors have converged to bring about this result, one of them is certainly the desire of individual men

FIGURE 16 · 2
The backward-bending supply curve of labor

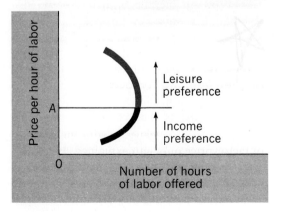

and women to exchange the marginal utility of potential income for that of increased leisure.

INDIVIDUAL VS. COLLECTIVE SUPPLY

A cautionary note is useful here. We can speak with some degree of confidence about the backward bending supply curve of individual labor, especially when labor is paid by the hour or by the piece. But we must distinguish the supply curve of the individual from that of the labor force as a whole. As the price of labor rises, some persons will be tempted to enter the labor market because the opportunity cost of remaining "leisured" is now too great. This accounts for the entrance of many housewives into the market as the price for part-time office work goes up. Later we will study this problem of the "participation rates" of the population, when we look into the macroeconomic problem of employment. *Here it is important to understand that the collective supply curve of labor is probably upward sloping rather than backward bending.* (It would be backward bending if wages rose so high that married women, for example, dropped out of the labor force as their husbands' earnings rose, but that does not seem to be the case.)

PSYCHIC INCOME

The supply curve of labor is further complicated because work brings not only disutilities but positive enjoyments. Jobs bring friendships, relieve boredom, may lead to power or prestige. Many people derive deep satisfactions from their work and would not change jobs even if they could improve their incomes by doing so. Indeed, it is very likely that individuals seek to maximize their "psychic incomes" rather than their money incomes, combining the utilities derived from their earnings and the quite separate utilities

from their work, and balancing these gains against the disutilities that work also involves.

The difficulty in speaking about psychic incomes is that it involves us in an unmeasureable concept—a tautology. Therefore when we speak about the supply curve of labor, we generally make the assumption that individuals behave roughly as money maximizers, an assumption that has prima facie evidence in the long run exodus from low-paying to higher-paying occupations, and from low-wage regions to high-wage regions.

MOBILITY OF LABOR

More than a million American families change addresses in a typical year, so that over a decade the normal mobility of the labor force may transport 15 million to 20 million people (including wives and children) from one part of the country to another. Without this potential influx of labor, we would expect wages to shoot up steeply whenever an industry in a particular locality expanded, with the result that further profitable expansion might then become impossible. Isn't this like the labor supply mechanism in Adam Smith's model (p. 24)!

We also speak of mobility of labor in a vertical sense, referring to the movement from occupation to occupation. Here the barriers to mobility are not usually geographical but institutional (for instance, trade union restrictions on membership) or social (discrimination against the upward mobility of blacks) or economic (the lack of sufficient income to gain a needed amount of education). Despite these obstacles, occupational mobility is also very impressive from generation to generation, as the astounding changes in the structure of the U.S. labor force have demonstrated (see box p. 232).

Here again the upward streaming of the population in response to the inducement of better incomes makes the long-run elasticities

THE PUZZLE OF LABOR DISUTILITY

Do people really work less as they get richer, trading leisure for income? The question is more puzzling than at first appears. The table below shows us what appears to be a steady decline in hours worked per year, for an average employee.

Year	Yearly hours worked
1890	2789
1900	2766
1910	2705
1920	2564
1930	2477
1940	2278
1950	2131
1960	1985
1970	1990

Yet, on second look the table is deceptive. For the "average" employee is an average of the hours worked by men and women. As more and more women have entered the labor force, *most of them working part time*, of course the average working hours per employee has dropped. If we look at men and women separately, we find quite another conclusion. Both men and women were working *longer* hours (about 7 percent more) over the last decade, even though their real incomes were up.

Has the individual backward-bending curve changed its time-honored shape? Many answers have been suggested to explain the countertrend to longer hours *and* higher real incomes. One explanation is that leisure is now more expensive (campers instead of canoes), so that we work more hours to earn a shorter but more expensive vacation. Another explanation is that we work longer hours because the time spent in unpaid work—household chores—has dropped, so that we have more time "available" for paid work. Again, it has been suggested that we work longer hours because we have switched to less "productive" but more enjoyable work, leaving the assembly line for the office.

Which of these answers is correct? *We do not know.* The upward twist in the individual labor supply curve remains an ill-understood fact. It should lead us, however, to be skeptical of claims that Americans are about to perish of boredom. Evidently work is more popular than the textbooks make it out to be.

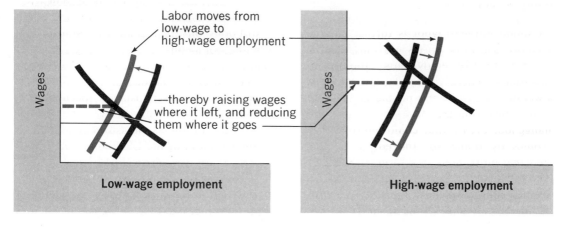

FIGURE 16 · 3
Effect of labor mobility on relative wages

UNITED STATES' WORK PROFILE

What kinds of work do people do in the United States? The table below gives us a picture of the shifting occupational profile of the nation's work force over the last seventy-odd years, together with projections of the U.S. Department of Labor for 1980.

Note particularly the dramatic change from 1900 to 1968. This striking growth in professional and managerial work, together with the drift from blue- to white-collar jobs, lies behind most discussion of "postindustrial" society as the direction in which we may be headed.

	Percent of labor force		
	1900	1968	1980
Managerial and professional			
Professional and technical workers	4.1%	14.4%	16.3%
Managers, officials, and proprietors (nonfarm)	5.9	10.5	10.0
White-collar			
Clerical workers	3.1	17.9	18.2
Sales workers	4.8	6.0	6.0

	Percent of labor force		
	1900	1968	1980
Blue-collar			
Skilled workers and foremen	10.3	12.7	12.8
Semiskilled workers	12.8	17.6	16.2
Unskilled workers	12.4	4.6	3.7
Household and other service workers	8.9	12.4	13.8
Farm			
Farmers and farm managers	20.0	2.2	2.7
Farm laborers	17.6	1.9	

of supply of favored professions much greater than they are in the short run. This is a force tending to reduce the differences between income extremes, since the mobility of labor will not only shift the supply curve to the right in the favored places it moves to (thereby exerting a downward pressure on incomes), but will move the supply curve to the left in those industries it leaves, bringing an upward impetus to incomes. Figure 16·3 shows how this process works.

CAPITAL

What about capital? First, let us be clear that we understand what we mean by the supply curve of capital. We mean the willingness and ability of individuals (or businesses) to offer their *savings* on the market as the price of savings (the interest rate) changes.

For it is savings that constitute the supply of capital. The assets I already own—the stocks or bonds in my portfolio or the equipment I use as an entrepreneur—constitute my *stock* of capital. At any moment, this stock of bonds or machines is fixed: the supply curve of total capital is a perfectly inelastic curve. But we are constantly adding to this stock, or possibly using it up. Thus, when we speak of the supply curve of this factor, we focus on the change in the stock of capital.

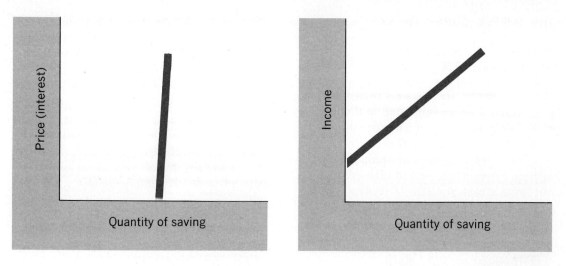

FIGURE 16 · 4
The supply curve of savings

FIGURE 16 · 5
The reallocation of savings

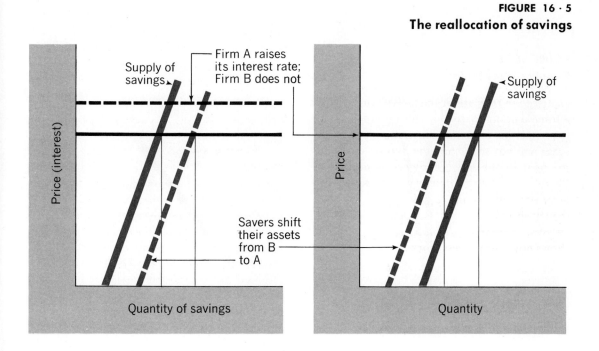

THE SUPPLY CURVE OF SAVINGS

Do individuals or businesses supply more savings, out of a given income, if interest rates rise? The answer, by and large, seems to be no. Savings are relatively price inelastic, particularly if we are interested in the flow of total saving. Income is much more important than price in determining the supply of saving. When we turn to macroeconomics, the relation between income and savings will turn out to be a central focus of study. Thus we should become familiar with two conceptions of the supply curve of capital: a highly *price-inelastic* supply and a highly *income-elastic* supply. We see these in Fig. 16·4.

At the same time, the *use* of savings is very responsive to price. We frequently reallocate the way we dispose of our savings, shifting our "portfolios" among checking accounts, savings accounts, insurance, or stocks and bonds, as the various returns offered by these assets change (or as our assessment of their risk changes). For example, the rise in interest rates in the early 1970s resulted in a tremendous movement from noninterest-bearing checking accounts into interest-bearing savings accounts.

Thus, when we picture the supply curve for saving, we should also bear in mind what this curve looks like to any particular demander. From the point of view of an enterprise in the market for capital, savings are highly price elastic; that is, a firm can attract savings by offering a higher return than a competitive enterprise offers. As a result, savers will shift their capital from one enterprise to the other, as in Fig. 16·5. This bears an obvious relation to the mobility of labor, and we can indeed speak of it as the *mobility of capital.*

LAND

Finally, let us consider the supply curve of land. Here we face considerations of still another kind. At any moment, the total supply of land, like capital, is fixed. But there really is no behavioral counterpart to saving.

FIGURE 16·6
Supply curves for land

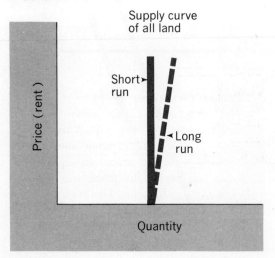

Supply curve of all land

Price (rent)

Short run

Long run

Quantity

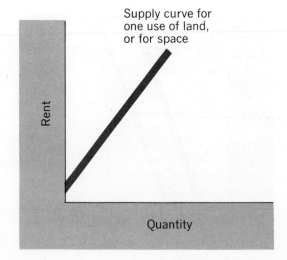

Supply curve for one use of land, or for space

Rent

Quantity

We can add to land by dredging, clearing forests, reclaiming deserts, etc., but this takes inputs of labor and capital. What then does the supply curve of land look like?

As with capital, there are several ways of looking at the question. The first is that the supply curve of all land is highly inelastic, even moreso than that of capital. The only way we can add to this stock is to rent land that we own but have not previously put on the market. This amount is probably very small.

As with capital, however, it is possible to speak of the *mobility of land.* This seems strange, since land is obviously not moveable. But land can be used for very different purposes, depending on the returns to be had from them. Thus if we picture the supply curve of land for, say, shopping centers or orange groves or industrial sites, we can picture an upward sloping curve, just like the supply curve of capital for any one use. Land can be—and constantly is—moved from use to use, as various enterprises bid for it.

In addition, we must differentiate land from *space.* We can create more space in the same way that we can create more land—by adding capital to land. Every time we build a highrise building where there was previously a low one, we have converted a fixed amount of land into a larger amount of space. The supply curve of space will therefore also be upward sloping, with more space offered as its price rises. In Fig. 16·6 we see these supply curves for land and space.

Rents and Incomes

The importance of time in bringing about increases in land or space alerts us to a very important reason for the existence of very large incomes (and very large disparities of incomes) in the short run. This is the phenomenon of *quasi rent* (also called *economic rent*).

QUASI RENTS

Quasi rents, or economic rents, are not the same as the rent on land, and the unfortunate fact that the terminology is so close has justifiably aggravated a good many generations of students. Hence let us begin by making a clear distinction between the two. *Rent is the payment we make to induce the owner of land to offer its services on the market.* If we cease to pay rent or pay less rent, the amount of land offered on the market will fall. If we pay more, it will rise. Thus rent is both a payment made to a factor of production to compensate its owner for its services and an element of cost that must enter into the calculation of selling prices. If a farmer must pay $100 rent to get an additional field he needs, that $100 will clearly be part of the cost of producing his new crop. Quasi rents or economic rents are different from this.

First, quasi rent is not a return earned by the factor, in that the payment has nothing to do with inducing the factor to enter the market.

Second, quasi rents are not a cost that helps to determine selling price; they are earnings determined by selling price.

Last, quasi rents apply to all factors, not just to land.

Let us see what these differences mean.

QUASI RENTS AND SCARCITY

An illustration may clarify the problem. Figure 16·7 shows the supply curve for "first class" office space in New York City. Notice that over a considerable range there is an unchanging price for space: up to amount *OX,* you can get all the space you want at rent *OA* (per square foot). This is real rent, in the sense of being a necessary payment for a factor service. If there were no rent paid, no space would be forthcoming.

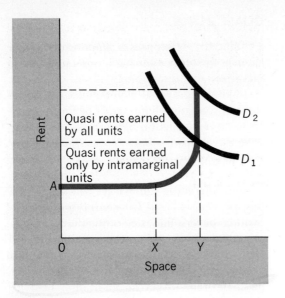

FIGURE 16 · 7
Rents and quasi rents

Next, look at the situation after we have used up *OX* amount of space. We can rent additional space up to *OY,* but only at rising prices (rents). Perhaps expenditures are needed to induce landlords to upgrade "second-class" space. Each new area that is tempted onto the market at a higher price earns real rent, in that it would not appear unless a higher price were paid. But notice that all the offices previously offered at lower prices are now in a position to ask higher rentals because the price of the marginal unit has increased. Thus here we have a mixture of real rents and quasi rents. *The marginal landlord is receiving real rent. All the other (intramarginal) landlords who can now get higher rentals are the beneficiaries of a scarcity price that gives them a bonus over the real rent they originally received. This bonus is quasi rent.*

Finally we reach the point *OY,* at which there is no more space to be had at any price. Suppose the demand for space continues to

rise from D_1 to D_2. The price of office space will rise as well, and *all* landlords will be receiving quasi rents on top of their real rents.

QUASI RENTS AND PRICES

Here is where we make the distinction between price-determined and price-determining. True rent is a cost that must be paid to bring land into production. Quasi rent is not such a cost. It is wholly the result of scarcity and plays no role in determining the real cost to society of producing goods or services. If demand fell from D_2 to D_1, a great deal of quasi rent would disappear, although not a foot of office space would be withdrawn.

Thus true rent helps determine price, whereas quasi rent is determined by price. We can see that rent must be paid if office space *OX* is to appear, but that quasi rent is not a necessary cost of production for space thereafter. This conclusion, however, applies only to our analysis of costs from a social point of view. From the point of view of the individual producer, the distinction disappears. A renter must pay his landlord whatever it costs to rent space, be it rent or quasi rent. Thus quasi rents enter into *his* cost and help determine the price he must ask for his goods.

QUASI RENTS AND ALLOCATION

From this point of view, quasi rent is an economic waste. If we could eliminate it, say by taxing it away, it would not diminish production at all, only the incomes of the owners of scarce resources. Quasi rents are therefore wholly "unearned" incomes.

Nonetheless, they have an important and useful function for society. They allocate a scarce commodity—in this case, office space—among various claimants for that commodity. The fact that quasi rents are a "monopoly return" is neither here nor there, so far as their rationing function is concerned. If there were no quasi rents, office space would be leased at prices that failed to equate quantities demanded and supplied. There would be a "shortage" of space, and rationing would take place on some other basis—first come, first served or political influence or having an "in" with the landlord, instead of through the price mechanism.

QUASI RENTS AND INCOMES

Finally, *we must make careful note of the fact that quasi rents (or economic rents) are not returns limited to land.* Take car rentals, for instance. Suppose we can rent all the car transportation we need at the going price. Suddenly there is a jump in demand, and the rentable fleet is too small. Rentals will rise. If no additional cars come onto the market, the fleet owners will simply enjoy economic rents (quasi rents) on their cars.

In the same way, the earnings of actors or authors or of anyone possessing scarce talent or skills are likely to be partly economic rents. An actress might be perfectly happy to offer her services for a good movie role at $50,000, but she may be able to get $100,000 because of her name. The first $50,000, without which she would not work, is her wage; the rest is an economic rent. So, too, a plumber who would be willing to work for standard wages, but who gets double because he is the only plumber in town, earns economic rents. And as we have already seen, economic rent explains a part of business profits, as returns going to intramarginal firms.

CAPITALIZATION

Economic rents lead us to an important process called *capitalization.* As we have mentioned, buyers of land and capital often have the option of buying the actual real

estate or machinery they use, instead of just paying for its services. Capitalization is the means by which we place a value on a factor as an asset.

Suppose that an office building makes $100,000 a year in rentals and that we decide to buy the building. How much would it be worth? The answer depends, of course, on the riskiness of the investment: some buildings, like some machines or bonds or businesses, are safer buys than others. But for each class of risk there is a rate of return established by other buyers and sellers of similar assets. Suppose that the rate of return applicable to our building is 10 percent. Then we will capitalize it at $1 million. This is because $1 million at 10 percent gives us a return of $100,000.

We can capitalize any factor (or any business) by dividing its current earnings by the appropriate interest rate. The appropriate interest rate tells us the opportunity cost of our money: how much we could get for it if we bought some other factor or asset of similar risk. Notice that as the interest rate falls, so does the opportunity cost: we have to be content with a smaller return on our money. Thus if we are buying land that rents for $1,000, and the appropriate interest rate is 5 percent, the land will sell for $20,000 ($20,000 × .05 = $1,000). If the interest rate *falls* to 4 percent, the value of the land will *rise* to $25,000.

CAPITALIZATION AND QUASI RENTS

Capitalization is important in determining the asset prices of factors. It is also important because it gets rid of private quasi rents! Suppose the $100,000 rental income were virtually all quasi rent and that the building were sold for $1 million. The new owners of the building will still be making $100,000 a year, *but it will no longer be a quasi rent.* It will be a normal market return on the

THE SINGLE TAX

In 1879 Henry George, an *enfant terrible* of economics, stunned the economic world with a savage attack on rent. Struck by the land millionaires and the land speculation around him, George wrote in his *Progress and Poverty*:

Take now ... some hard headed businessman, who has no theories, but knows how to make money. Say to him: "Here is a little village; in ten years it will be a great city—in ten years the railroad will have taken the place of the stagecoach, the electric light of the candle; it will abound with all the machinery and improvements that so enormously multiply the effective power of labor. Well, in ten years, will interest be any higher?"

He will tell you, "No!"

"Will the wages of common labor be any higher...?"

He will tell you, "No, the wages of common labor will not be any higher...."

"What then will be higher?"

"Rent, the value of land. Go, get yourself a piece of ground, and hold possession."

And if, under such circumstance, you take his advice, you need do nothing more. You may sit down and smoke your pipe; you may lie around

capital they expended. They will have paid for their quasi rents in the opportunity cost of their capital.

The gainer in the transaction will be the original owner of the building. Perhaps the building cost him very little, and the $1 million sale price came as a great capital gain. He will have capitalized his quasi rents into a large sum of capital; but on the sum, in the future, he can expect to make no more than the normal rate of interest.

THE MARKET PRICE FOR FACTORS

We are finally in a position to assemble the pieces of the puzzle. We have traced the forces that give rise to the supply curves of the different factors, and we understand the explanation for the demand curves as well. Hence we understand how factors are priced in the market, which is to say we understand the mechanism for determining *factor incomes.*

Can we generalize about the result? One would think, in the light of all the varieties of supply curves we have discussed, that this would be impossible. If we think about the nature of the demand curve for factors, however, we see that this is not so. The shape and position of the supply curve for any factor will determine how much of it appears on the market at any given price. *But whatever the amount of the factor, it will be paid a return equal to the marginal revenue product of that factor.* Remember that the entrepreneur will go on hiring all factors as long as their marginal revenue products are greater than their marginal costs. Thus, whatever the shape or position of the supply curve, the earnings of the factor will always be equal to the marginal revenue it brings to the buyer, as Fig. 16·8 shows.

We could draw other curves, showing the collective supply of labor, the supply of savings to a given user, or the total quantity of land, but the conclusion would be the same. In each case, the market rate of wages, interest, and rent would be equal to the marginal revenue product of that factor.

like the lazzaroni of Naples or the leperos of Mexico; you may go up in a balloon or down in a hole in the ground; and without doing one stroke of work, without adding one iota of wealth to the community, in ten years you will be rich! In the new city you may have a luxurious mansion, but among its public buildings will be an almshouse.

George advocated . . . 100 percent tax on rent —the so-called single tax—to soak up this wealth. Was he justified in doing so? He was correct in claiming that the landowner was the recipient of value that resulted from other peoples' activities, although he overlooked the risk involved in buying land in a "little village" that did *not* become a thriving city. He also over-

looked one important fact. A tax that absorbed 100 percent of rent (or even 100 percent of quasi rent) would remove any incentive to put this land on the market. However profitable the land might be, its net return to its owner, after payment of the 100 percent tax, would be zero. Therefore there would be no reason for owners of high-productivity lands to sell them and no reason for entrepreneurs to buy them. There would therefore be no mechanism to allocate land to its most productive use. This would not be true if the "single tax" were scaled down to 50 percent or even 90 percent on rental incomes. The trick would be to leave sufficient income to the land-owner to induce him to act as a maximizer and not to lie around as a lazzarone, doing nothing.

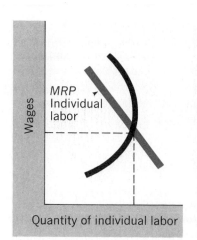 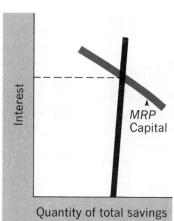 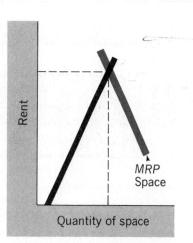

FIGURE 16 · 8
Marginal revenue products and earnings

THE MARGINAL PRODUCTIVITY THEORY OF DISTRIBUTION

We call this generalization about factor prices the *marginal productivity theory of distribution.* What it tells us is very simple. The

income of any factor will be determined by the contribution each factor makes to the revenue of the enterprise. Its income will be higher or lower, depending on the willingness and ability of suppliers of factor services to enter the market at different prices, but at all

SURPLUS VALUE

What would a Marxist economist say to the argument that a perfect market system removes all exploitation?

He would reply that we are confusing the idea of "productivity" with that of the legitimacy of reward. A Marxian economist would not deny that land and capital vastly augment production. He would object to the *owners* of land and capital receiving incomes because of the contribution that "their" factors made to social output. According to Marxian analysis, all the output from land and capital and labor belongs to society as a whole, not to the individuals who are permitted by society to exercise legal claims over some portion of the output. Exploitation is thus a consequence of *ownership*, not of labor being paid less than its marginal revenue product. Exploitation—so-called surplus value—consists of the marginal revenue products of land and capital.

This does not mean that a Marxian economist would claim that labor is therefore rightfully due the sum of the marginal revenue products of all the factors. The Marxist differentiates between living labor and "dead" labor, embodied in the accumulated wealth of society. The labor of the past aids the labor of the present, and thus "society" as well as present labor has a claim on output. This claim is exercised partly in withholding from labor that part of output needed to renew the depleted wealth of society, and partly in accumulating new wealth for the future.

$$\frac{MP_A}{MP_B} = \frac{P_A}{P_B}$$

prices, factors will earn amounts equal to the marginal revenue they produce.

NO EXPLOITATION

Two conclusions follow from this theory. *The first is that there can be no exploitation of any factor.* Each factor will receive an amount exactly equal to the revenue it produces for the firm.* There can be no unpaid labor or unrewarded land or capital. A worker or capitalist or landlord may not be willing to offer his services at going rates of pay, but none can claim that his earnings are less than the revenues he brings into the firm; for as we have seen, every profit-maximizing entrepreneur must keep on hiring factors until their marginal revenue products equal their marginal cost.

Moreover, no factor can claim that some other factor is paid too much. To be sure, some factors will earn quasi rents, a waste

*When a factor is bought directly by a consumer, the factor will receive the money value of the marginal utility it produces for the consumer.

from society's point of view, as we have seen. But we have also seen that quasi rents are a temporary phenomenon, slowly eliminated by mobility of factors into an area of scarcity. In the long run, the returns to land, capital, and labor will reflect the actual contributions that each makes to output.

PRICES AND PRODUCTIVITIES

The second conclusion follows from the first. *It is that all factors will be paid in proportion to their productivity.*

Suppose that an entrepreneur can hire a unit of Factor A, say an acre of land, which will produce a marginal revenue product of $20, or an extra workman whose marginal revenue product will be $40. We know that the entrepreneur will have to pay $20 to hire the acre of land and cannot pay more than that, and that he will have to pay $40 for the man and cannot pay less than that.

But what accounts for the difference between the earnings of a unit of land and a unit of labor? Their marginal revenue products. But these marginal revenue products, we recall, are nothing but the *physical* marginal products of the factors multiplied by the market value of the output to which they contribute. The value of the output must be the same, since land and labor will both be adding to the output of the same commodity. The difference in their marginal *revenue* products is therefore solely the result of the fact that a unit of land creates fewer units of output than a unit of labor.

It follows, therefore, that factor prices must be proportional to their physical productivities, or that:

$$\frac{\text{Price of Factor A}}{\text{Marginal Productivity of Factor A}} = \frac{\text{Price of Factor B}}{\text{Marginal Productivity of Factor B}}$$

MARGINAL PRODUCTIVITY AND "JUSTICE"

This is a very remarkable solution to the problem of distribution. What it says is that in a competitive system, all factors will be rewarded in proportion to their contribution to output. If an acre of land is only half as productive as a unit of labor, it will be paid only half as much; if it is twice as productive, it will be paid double. If skilled labor produces three times as much as unskilled, its wage rate will be three times that of unskilled, and so on. The resulting pattern of income distribution thus seems both "just" and "efficient." It seems just because everyone is getting all the income he or she produces, and because no one is getting any income he or she has not produced. It seems efficient because entrepreneurs will use factors in a way that maximizes their contribution to output, thereby not only giving the factors their largest possible reward but giving society the greatest overall output to be had from them.

Is this conclusion valid? Do the earnings of land, labor, and capital reflect their contributions to output? The question takes us from a consideration of how the factor market works "in theory"—that is, under conditions of perfect competition—to a consideration of how it works in fact. It brings us also to look carefully into the question of whether or not marginal productivity establishes a pattern of rewards that can rest its case on some definition of "justice." We will look into these extremely important questions in our next chapters.

KEY WORDS

CENTRAL CONCEPTS

Market for factors

1. The market for factors is important for two reasons. It is the last link in the *circular flow,* joining households as suppliers and firms as demanders of productive services. And it is crucial in explaining *income distribution.*

Factor services

2. We must distinguish between factors and *factor services.* It is services that are priced in the market for factors—wages, rent, and interest. The price of the factors is obtained by capitalizing their earnings.

Direct demand

3. There are two sources of demand for factor services. One is a *direct demand* for factor services as consumer goods. The other is *derived demand* from firms who hire factors for profit. Firms hire factors until their marginal revenue products (MRPs) equal their marginal costs.

Derived demand

Backward bend-ing supply curve	4. The individual labor supply curve is *backward bending*. This reflects a rising preference for leisure over work—the result of the *increasing marginal disutility of work and the decreasing marginal utility of income.*
Collective supply curve	5. The *collective* supply curve of labor is positively sloping, owing to the increasing entrance of previous nonworkers into the labor force as wages rise.
Psychic income	6. The supply curve of labor is rendered complex because individuals seek to maximize *psychic income* as well as (or instead of) money income. We assume that labor is generally impelled by income maximizing—enough to account for the widespread phenomenon of labor *mobility* toward higher-paying work.
Mobility of labor	
Supply curve of savings	7. The supply curve of capital is *savings*, and the price of savings is interest. The supply curve of savings is price inelastic but income elastic. Although total saving is price inelastic, *the supply of saving to an individual user is responsive to price.* That is, savings are also mobile.
Mobility of savings	8. The supply of land is fixed in the short run, and the supply curve is therefore vertical. But the supply to any particular user is not fixed; it varies with rent. Land is therefore mobile, as is capital. In addition, *space* can be created by adding capital to land. Space is price-responsive.
Mobility of land	
Land vs. space	
Quasi rents	9. *Quasi rents* or *economic rents* are an important component of all earnings in which mobility (or elasticity) is impaired. Quasi rents must be distinguished from real rents in that they play no role in inducing an increase in supply. *Rent is a real cost that helps determine price. Quasi rents are wholly determined by the price paid for the marginal unit.* However, once established, quasi rents enter into costs of production and help determine prices for individual producers. Quasi rents also play a useful role in allocating scarce goods.
Economic rents	
Price-determined vs. price-determining	
Capitalization	10. The relation between the price of factor *services* and the price of factors as *assets* is determined by the rate at which we *capitalize* factor services. We capitalize any asset by dividing its yield by the rate of interest or by some other appropriate opportunity cost. Capitalization converts quasi rents or profits into costs for the new owner and gives the original owner of these rents or profits a capital sum on which he will thereafter make only normal returns, *ceteris paribus.*
Opportunity cost	
Marginal productivity theory of distribution	11. In a competitive market, *factors will be priced according to their marginal revenue products. This is called the marginal productivity theory of distribution.* In theory, this removes all exploitation and prices all factors in proportion to their contribution to output. This appears to be both a just and efficient solution to the problem of distribution.

QUESTIONS

1. When a suburban homeowner buys real estate, what services is he actually buying? When he hires domestic help, what services is he buying? When he borrows money from a bank?

2. What is meant by the *derived* demand for labor? What is its relationship to the marginal revenue product of labor? To clarify your understanding, do the following: (1) run over in your mind the law of diminishing returns (2) be sure you understand how this law affects marginal *physical* productivity; (3) explain how we go from marginal physical product to marginal *revenue* product; and (4) explain how marginal revenue product influences the willingness of the employer to hire a factor.

3. Suppose you had $10,000, that you kept half in cash, and invested the rest at 6 percent. If the rate of interest went up to 7 percent, would you be tempted to invest more? Or the same? Or might you think that it would be wise to invest a little less, since your income was now higher? Are there reasonable arguments for all three?

4. If the rent of a piece of land is $500 and the rate of interest is 5 percent, what is the value of the land? Suppose the rate falls to 2½ percent; does the value of the land rise or fall? If the rental increases to $1,000 and interest is unchanged, what happens to the value of the land?

5. What do you think the supply curve of executive labor looks like? Is is backward bending? Would you expect it to be more or less backward bending than the supply curve of common labor? Why?

6. What do you think are the main impediments to factor mobility in the labor market? Location? Education? Discrimination? Wealth? How would you lessen these barriers?

7. Exactly what is rent? How does it differ from quasi rent? Is there any similarity between rent and interest, or rent and wages? What is the role of ownership in rent, profits, *and* wages?

8. What is meant by saying that rent is price determining? What does it mean to say that quasi rent is price-determined? Show on a diagram.

9. Explain carefully how factor returns are proportional to their marginal productivities. What conditions would be necessary to make this theoretical conclusion true in the real world?

Problems in the Distribution of Income

17

IN OUR LAST CHAPTER WE TOOK A FIRST LOOK into the factor market, partly to learn something about the behavioral elements that give us the various supply curves of factors, partly to gain a general understanding of how incomes are determined in a market system. Now we want to put our first grasp of the subject to the test of a searching criticism. At the end of the last chapter we reached the conclusion that the prices of factors could be explained by their marginal productivities. Does this mean that we now have the key to understanding the actual distribution of income in our society? Can we explain riches and poverty, high-paid professions and low-paid ones, in terms of marginal productivity?

THE FACTS OF INCOME DISTRIBUTION

Before we look into problems, let us look at facts. How is income actually distributed in the United States? There are a number of ways of answering this question. We can inquire into the *share* of national income going to the poorest 20 percent, the richest 20 percent, and other groups. We can inquire into the actual number of dollars received by households in these different groups. We can sort out income distribution by families and single individuals, by race, and by region. We can look at the distribution accruing to wages, rents, and profits.

Each of these approaches highlights certain aspects of income distribution, but we would have a textbook cluttered with figures if we tried to present all the possible arrangements of the data. Let us therefore begin with a table that presents a few central facts that will serve as a useful jumping-off point:

Our first impression from this table is that *income is very unequally distributed*. Notice that the upper 20 percent of the nation's households, with incomes over $16,920, garner almost 42 percent of all income. By way of contrast, note that the lowest fifth, with incomes below $3,667, get only 5.5 percent of the nation's income.

Family units income rank*	Percent of total income	Mean income of that fifth	Lowest income of that fifth
Lowest fifth	5.5%	$ 3,185	(negative)
Second fifth	11.9	6,892	$ 3,667
Third fifth	17.4	10,077	8,636
Fourth fifth	23.7	13,726	11,814
Highest fifth	41.6	24,093	16,920

TABLE 17 · 1
Share of Money Income by Family Units 1971

*A family unit is two or more people living in the same dwelling unit and related to each other by blood, marriage, or adoption. A single person unrelated to other occupants in the dwelling unit or living alone is a separate family unit.

Second, we should know (although the table does not show it), that *this basic income distribution pattern has been very stable over long periods of time.* In 1960, for example, the top 20 percent received 43 percent of all income; the bottom 20 percent received 4 percent; and statistics for the 1940s and 1950s do not show marked changes, although the share of the top groups was probably somewhat higher in those earlier years.

THE CAUSES OF INEQUALITY

Can we explain the actual pattern of incomes in the United States by marginal productivity theory? That is, can we account for the difference between the incomes of the families in the top and the bottom fifths by claiming that the marginal revenue product of a household making $20,000 or more was at least 7 or 8 times as high as that of a family in the bottom group?

No economist would make such a claim. Marginal productivity does not explain many kinds of low income, such as that arising from unemployment (or zero productivity). Nor does it explain great wealth. Marginal productivity theory may tell us something about the contribution made by capital to production, but it offers no explanation of why capital is owned by so few individuals, instead of being spread more equally among the population. To explain riches and poverty we must look to the institutional arrangements and failures of the social system. At best, *marginal productivity may shed some light on the distribution of earnings among the employed population.*

How much light does the theory shed on the variation of earnings? In Table 17·2 we show the range of incomes of broad classes of occupations for a recent year.

Here is where marginal productivity theory should be most applicable. Does it in fact explain the difference between the earnings of a surgeon and those of a day laborer?

TABLE 17 · 2
Median Earnings of Males, 1971*

Physicians and surgeons (self-employed)	over $25,000
Engineers, salaried	14,509
Sales workers	10,650
Primary and secondary school teachers	9,913
Managers, self-employed	9,280
Clerical workers	9,124
Manufacturing operatives	7,966
Service workers	7,111
Laborers, nonfarm	6,866

*Full-time, year-around.

THE RICH

Income distribution statistics do not tell us much about two groups of great importance—the rich and the poor. Here are a few salient facts about the top of the income pyramid. On the facing page are a few about the bottom.

How much income does it take to be "rich"? There is no official designation, whereas there *is* an officially drawn "poverty line." One problem is that you can be rich in assets but have a small income (tax loopholes enable some millionaires to claim a net "loss" for their year's income), or you can have a large income but no assets (many a prize fighter has died broke). For a full account of what it means to be rich, read Herman P. Miller's *Rich Man, Poor Man* or Ferdinand Lundberg's *The Rich and the Super-Rich.* Meanwhile, here are a few highlights:

1. To be classified as a "top wealthholder" in the *Statistical Abstract* you must, on your death, leave a gross estate of at least $60,000. In 1962 (last available data), there were just over 4 million such top wealthholders. With inflation, there must be many more today. As the size of the estate rises, the number of wealthholders falls precipitously. In 1962 less than 900,000 individuals had estates over $200,000, just over 70,000 had estates of $1 million, and only 2,000 had estates of $10 million or more.

2. An income of $50,000 is probably enough to classify someone as rich. How many have such incomes? That depends on whether you include unrealized capital gains and undistributed corporate earnings. If you count as income only monies actually received, some 200,000 families were rich in 1968, enjoying 2 percent of all U.S. incomes. If we include undistributed earnings and gains, the number of families rises to 900,000 and their share of total income to 11 percent.

3. The very rich *are* very rich. In 1962 the top ½ of 1 percent of the adult population owned about 25 percent of the $752 billion of stocks, bonds, savings accounts, real estate, and other personal wealth in the nation. Some years ago it was estimated that the top ⅕ of 1 percent of all households owned roughly two-thirds of all corporate stock, worth, in toto, $325 billion in 1962. Finally, we might note that income of the top 4,000 families in 1968 was just under $4 billion.

Market Imperfections

This brings us to the subject of market imperfections—a range of problems that prevents the market from bringing about the results that theory would predict. Market imperfections play the same role in the factor market that noncompetitive market structures play in the market for goods and services. They result in prices higher or lower than the long-run equilibrium prices that a competitive supply and demand setting would produce. Let us look into some of these imperfections.

IGNORANCE AND LUCK

Markets often fail to bring about a level of earnings corresponding to relative marginal productivities, because the marketers do not have all the relevant information. A skilled mason in Connecticut may not know that there is a brisk demand for his skill in Arizona—or even Rhode Island. A high-school graduate looking for his first job may not know that the possibilities for high wages are much greater in printing than in retailing. Moreover, the costs of acquiring the information needed to maximize one's income may be prohibitive. It becomes perfectly rational, in that case, to operate in partial ignorance—as we all do. This ignorance, however, leads to results other than those we would expect from a perfect market. Earnings may be higher or lower than they would be if everyone knew everything: some lucky persons have high abilities; others

THE POOR

How many people are poor in the United States? Official statistics count as "poor" anyone whose income falls below a designated "povery line." This line varies from $4,725 for a city-living family of four in 1972 to $1,774 for a single individual living in the country. In 1972 there were 24.5 million persons in these "low income" brackets. This is a decline in absolute numbers from 39.5 million in 1959 (with the poverty levels adjusted for changes in the cost of living). It is also a decline from 22 percent of all persons at the earlier date to just under 12 percent at the later date.

What characteristics distinguish poor families? One is color. About 9 percent of the white population is poor; about a third of the black population. A second characteristic is family structure. Only 17 percent of all families with a male head-of-household are poor; over 37 percent of all families headed by women are poor. A third characteristic is age. Ten percent of the total U.S. population is 65 or older, but almost double that percent poor. (On the other hand, over 40 percent of the poor are under age 18, mainly the children of poor households.) A fourth characteristic is place of residence. Eight percent of metropolitan dwellers fall into low income groups; 12 percent of nonmetropolitan groups. The incidence of poverty, of course, is especially severe in slums, urban or rural. Example: average income in the Kentucky regions of Appalachia is about *one-half* of average income in the United States.

Many of the poor depend on welfare for existence. Public assistance provides about 8 percent of their total income. Note that this is *only* 8 percent. The rest of the income of the poor comes from Social Security, other retirement income, and from the direct earnings of the poor. The problem is that these earnings are not enough to lift them out of poverty. In 1972 one out of 33 families was in poverty, even though it was headed by a person who worked full time all year.

Perhaps it should be added that in 1972 it would have required additional public assistance of about $12 billion to lift all poor families "above poverty." This was a little more than 1 percent of gross national product, or about as much as we spent for drinking and smoking.

will enter overcrowded markets. _Luck therefore becomes an important element in the distribution of income._

TIME LAGS

Of equal importance is the problem of time lags. As equilibrium prices for factors change, substantial periods of time may be needed to bring actual earnings into line with long-run equilibrium earnings.

Think about Medicare and Medicaid and the consequent outward shift in the demand for medical care and doctors. Although there is a long-run upward sloping demand curve for doctors, the supply is more or less fixed at the moment. Thus, as Fig. 17·1 shows, when demand shifts from D_o to D_1, the incomes of doctors (or price of doctors' services) will at first rise from P_1 to P_2, and only gradually fall to P_3. Supposing that it takes 5 to 10 years to train doctors, their earnings will not be at their long-run equilibrium for a 5 to 10 year period. If we add the time necessary to expand medical schools, the lags can easily be much longer. Moreover, by the time the schools are built, the long-run equilibrium price for medical care has probably moved again.

The opposite situation arises when demand curves fall. Because it is difficult to move into new occupations as a person grows older, it may take a long period of time for the existing supply of doctors (or teachers or skilled workers) to shrink to the new long-run equilibrium. For substantial periods of

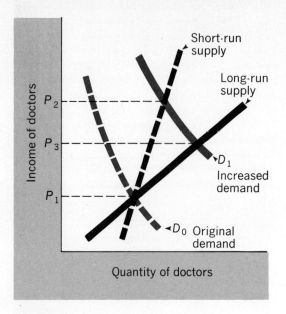

FIGURE 17 · 1
Time lags and earnings

time, the wages paid to such occupations can therefore be below the long-run equilibrium level.

In both cases, the actual distribution of income accords with the marginal productivity of doctors, or whomever. But this is a short-run rather than a long-run measure of their marginal productivity. In one case, doctors will be receiving incomes *above* their *long-run* marginal revenue product; in the other, they will be receiving *less* than their long-run marginal revenue product.

MONOPOLIES

Ignorance and time lags are involuntary market imperfections. More striking are market imperfections introduced by institutions that *deliberately seek to set factor prices above or below equilibrium prices.* This is the case whenever there is an element of monopolization in the factor market; for instance, in a

small town where one landowner controls virtually all the real estate or where a single bank controls the availability of local capital. In these cases we would expect the level of rents or interest to be higher than the equilibrium level. In the same way, if one company dominates the labor market, we may find that wages are *below* their equilibrium levels—that labor is paid less than its marginal revenue product because it has nowhere else to look for work.

One of the most important institutions creating factor "monopolies" is the labor union. Essentially, a union tries to establish a floor for wages above the equilibrium rate that the market would establish. In this way, the economic effect of a union is exactly like that of a minimum wage law.

In dealing with monopolies of all kinds in the factor market, we have to distinguish between two questions. The first has to do with the earnings of the factors. No one doubts that factor earnings can be depressed below their competitive equilibrium levels in a one-company town or can be raised above their competitive levels by a powerful union. The more interesting question is the effect of such changes on the total earnings of the factor. Can a union, by raising wages, increase total payrolls, or will the effect of higher wages be to shrink payrolls? We will come back to this question in our next chapter when we talk about policies designed to alter income distribution.

RENTS AND DISCRIMINATION

We have already seen how quasi rents create unearned incomes. Anything that inhibits the mobility of factors—anything that impedes their movement from lower-paid to higher-paid occupations—creates or perpetuates economic rents and enters into the explanation of income differences. Barriers of race and wealth, of patents and initiation fees, of

ECONOMIC RENT IN HIGH PLACES

Much of the very high incomes of corporate managers is also probably economic rent. In 1962, according to Robert Averitt (*The Dual Economy*, Norton, 1968, p. 178) the salaries and bonuses paid to the fifty-six officers and directors of General Motors exceeded the combined remuneration received by the President of the United States, the Vice-President, 100 senators and 435 representatives, 9 Supreme Court Justices, 10 cabinet members, and the governors of 50 states.

How could we ascertain whether the incomes of the General Motors executives (or for that matter, of the officials of government) contained economic rent? The answer is simple: we would have to reduce their incomes and observe whether they reduced their output of work. Presumably that is what the income tax tries to do, and studies indicate that the payment of a portion of income to the government does not seem to affect the supply curve of labor for executive skill. The presumption, then, is that a good part of their income *is* an economic rent, which the income tax siphons off in part.

geography and social custom—all give rise to shelters behind which economic rents flourish. If blacks, for instance, are systematically excluded from managerial positions, the supply of managers will be smaller than otherwise. Existing (white) managers will therefore enjoy economic rents.

The other side of this story is that barriers can constitute sources of *discrimination that lower the earnings of those who are discriminated against.*

Economic discrimination can take many forms. Wage discrimination occurs when two identical people are paid different wages for the same work. Employment discrimination exists when unemployment is concentrated among a preselected group. Occupational discrimination leads to limits on entry into good jobs and a corresponding increase in the supply of labor to less desirable jobs. Human capital discrimination occurs when individuals are not allowed to acquire certain types of human capital, such as education or training. All of these forms of discrimination will lead to lower incomes for the person discriminated against and higher incomes for someone else.

There are essentially two ways in which discrimination can be enforced. The discriminator may have some monopoly power vis-

à-vis the discriminatee, either in the form of direct government intervention, such as the South African laws that bar blacks from certain jobs, or because he controls some complementary factor of production. For example, if whites own all the physical capital or land, blacks must of necessity work for whites.

Alternatively, the discriminator may have a *taste* for discrimination. He may not want to associate with the discriminatee, and he may be willing to take whatever actions are necessary to avoid this association. If he refuses to shop in stores with black saleswomen, then the derived demand for black saleswomen falls, and their market wage falls.

DISCRIMINATION AGAINST BLACKS

To what extent does discrimination enter into income distribution in the United States? The most obvious instance has to do with blacks. The average black family has an income only 61.5 percent that of the average white family. In virtually every field, black earnings are less than white earnings in the same jobs. In itself, of course, such facts do not prove that wage discrimination exists. An apologist for the differentials in wages could claim that there is a real difference in the marginal productivity of whites and blacks. In that case the question is whether there has been discrimination at a more basic level; for instance, in the access to human capital.

Only a few years ago, it would have been simple to demonstrate that blacks were systematically prevented from acquiring equal skills or gaining access to jobs on equal terms. Their marginal productivity was lower because they were forced into the bottom jobs of society—unable to gain admission into many colleges, kept out of high-wage trades, and simply condemned by their own past poverty from accumulating the money needed to buy an education that would allow them to compete.

This picture is now changing in some important respects. As Table 17·3 shows, median incomes of black families are now much closer to those of white families, especially among younger workers. This change is the result of a substantial lowering of barriers against blacks entering many professions and occupations.

Income of black and other nonwhite families with college educations has risen dramatically: in 1961 only 28 percent of these educated families made more than $10,000; in 1970, 67 percent passed the $10,000 mark. Even more striking, in 1961 only 3.4 percent of college-educated nonwhites made over $15,000; in 1970, 48 percent made $15,000 or more. Discrimination against blacks is probably most serious today in low-paying, not high-paying jobs. The color of poverty in the United States is still disproportionately black.

DISCRIMINATION AGAINST WOMEN

A second major area of discrimination militates against women. Table 17·4 shows the same categories of occupations listed in Table 17·2 and compares women's pay (on a full-time, year-round basis) to that of men. As the table shows, women typically earn substantially less than men in all occupations. A portion of this differential may stem from women withdrawing from the labor force to have children and to nurture them in their early years, but there is no doubt that these "economic" reasons for pay differentials do not begin to account for the full differences we observe.

Statistics comparing men and women at age 35 show that the average *single* woman will be on her job another 31 years—which will be longer than her male counterpart will work—and the average *married* woman will work another 24 years. Second, the facts show that married women are *less* likely to leave one employer for another than men are (only 8.6 percent of employed women changed employers in a year compared with

TABLE 17 · 3
Median Income of Black Families as a
Percentage of Median Income of White Families

ALL BLACK FAMILIES	Under 35	35–44	Husband's age 45–54	55–64	65 and over
1959 (57%)	62%	60%	55%	51%	57%
1972 (76%)	85	76	71	59	72

Source: Bureau of the Census

THE CIRCULAR CAUSATION OF POVERTY

We must always be very careful before we impute poverty to any single source. One of the authors, sitting in a Ph.D. exam, was questioning a candidate about a dissertation on poverty. It seemed there were many causes for poverty, all impressively substantiated with econometric evidence.

"But if you had to single out one cause as the *most* important," asked the examiner, "which would it be?"

The candidate hemmed and hawed. There was skill. There was health. There was culture.

There was native ability. But if he *had* to choose, he would say that education—or rather, the lack of it—was the greatest contributory factor in poverty. Most poor people simply didn't have the knowledge to enable them to get high-paying jobs.

"And why didn't they have the education?" asked the examiner.

That was easy. Education was expensive. Poor people couldn't afford private schools. The need for income was so great that they dropped out of school early to earn money.

"I see," said the examiner. "People are poor because they are uneducated. They are uneducated because they haven't the money to buy education. So, *poverty causes poverty.*"

TABLE 17 · 4
Male and female earning differentials, 1971

Occupation	Women	Men
Physicians and surgeons	n.a.	over $25,000
Engineers, salaried	n.a.	14,509
Sales workers	$4,485	10,650
Primary and secondary school teachers	8,126	9,913
Self-employed managers	4,269	9,280
Clerical workers	5,696	9,124
Manufacturing operatives	4,901	7,966
Service workers	4,159	7,111
Laborers, nonfarm	4,548	6,866

11 percent of men). And finally, U.S. Public Health data reveal that on the average, women are absent from work only 5.3 days per year, a fractionally *better* record than men's.

Discrimination against women is beginning to melt away. The Women's Liberation movement has won important court battles to establish the right of equal pay for equal work, as well as equal rights to jobs, regardless of sex. Just by way of comment on our unconscious "sexist" attitudes, we might bear in mind that the United States has been unusually slow in admitting women to a full range of professional and occupational tasks. Only about 14 percent of our doctors are women, for example, whereas in West Germany 20 percent, and in the

U.S.S.R. 70 percent, are women. Perhaps even more surprising, in Sweden 70 percent of overhead crane operators are women, an occupation virtually unknown to women in the U.S.[1] These surprising percentages at both ends of the "social" scale leave little doubt that women *could* earn a great deal more than they do, if the barriers of discrimination were removed.

WAGE CONTOURS

Another distributional problem arises because individuals have strong feelings about how much they are "entitled" to. In our competitive model, each individual is a rational income maximizer, out for himself. At no time does any supplier of labor look at the wages of *other* laborers, except to determine whether or not he is being paid the competitive rate of return.

Yet, in real life individuals do not focus narrowly on their own wages but look around themselves at the entire wage structure. A policeman looks at firemen's wages, at truckers', perhaps even at schoolteachers' incomes. This interest in other incomes has different labels: psychologists might call it envy; sociologists might call it relative deprivation; economists have labeled it *wage contours*.

Now, in simple competitive economics, what other groups of workers are paid is irrelevant to what you should be paid; it is all a matter of relative marginal productivities. But in the real world, groups of workers look at other groups ("reference groups," in sociologists' terms) to see if they are being paid fairly. Studies in this area reveal that people strongly feel that their economic benefits should be proportional to their costs (i.e., their efforts, hardships, and the like), and that *equals should be treated equally*.

[1] National Industrial Conference Board, *Record*, Feb. 1971, p. 10.

Since there are various types of "costs" in different situations, and different rewards (income, esteem, status, power), the problem of defining "equals" immediately arises. To what group of people do you compare yourself, to determine whether you are being treated relatively "equally" and "proportionally"?

Reference groups seem to be both stable and restricted—people look at groups that are economically close to themselves. This helps explain why very large inequalities in the distribution of economic rewards seem to cause relatively little dissatisfaction, whereas small inequalities can raise a great commotion. A policeman does not "expect" to make as much as a movie actor—Hollywood is not his reference group. He does expect to make as much as, or more than, a sanitation worker, perhaps as much as a teacher. Changes in the pay of his reference groups therefore bring immediate changes in his estimation of the fairness of his own pay.

INTERNAL LABOR MARKETS

As a result of these comparisons, wage contours lead groups strongly to resist changes in relative wage rates. Even the person benefitting from a change in relative wages may feel that it is unfair that he receive more than someone else. He does not think that he is putting out more effort. Consequently, labor economists have discovered that wage rates among many groups move together, almost independently of movements in the supply and demand curves in each occupation.

Within individual firms these feelings are particularly strong. In labor negotiations, *relative wage differentials* become as important as the general increase in wages and have led to what has been called *internal labor markets*—the labor markets that are internal to the firm. Often they are rigidly segregated from the general labor market.

The market wage for ditchdiggers may be $2.00 per hour; but a ditchdigger working for General Motors may make $3 per hour, since he is the beneficiary of a wage-setting negotiation establishing group norms that elevate ditchdigging wages above the market wage for this skill.

Even without unions, such group feelings can be effective because production is not the atomistic process that we have been describing. It requires *teamwork*. In addition, the output of a worker—skilled or not—is not fixed but partly under his own control. Unlike a machine, he must be motivated to achieve his maximum output, and part of this motivation is having an acceptable team-wage structure. As a result, it is probably more realistic to think that workers are paid in proportion to their average team productivity, than to believe that each individual worker is paid in accordance with his or her marginal productivity.

PROPERTY

Last, we must inquire into the effects of property on the distribution of income. Here we enter a field where, as we have noted, marginal productivity theory has little to say, because the theory does not explain the distribution of wealth.

Nevertheless we should be familiar with the pattern of wealth distribution, because it obviously plays an important role in total income distribution. Table 17·5 gives us an overview of the ownership of U.S. wealth.

Note that whereas Table 17·1 showed us that the top 20 percent of all families had 42 percent of all *income*, Table 17·5 shows that the top 18.7 percent of all families have 76.2 percent of all *wealth*—stocks and bonds, houses, cars, bank accounts. Why is the distribution of wealth so much more unequal than that of income? The primary reason is the institution of inheritance. Family fortunes, once made, can be transmitted relatively intact, from one generation to another. Taxes on wealth (estate taxes) have had little effect in diminishing this transference, and in any event are less heavy than income taxes, owing to the presence of many estate tax loopholes.

This unequal distribution of wealth has a dual effect on the distribution of income. First, the possession of wealth makes it easier to attain a high marginal productivity, because it often requires large outlays of money to gain the specialized skills that result in high productivity. Second, the possession of large wealth brings income in the form of land, rents, interest, and profits. Property income is very unimportant as a proportion of total income for most Americans, but it be-

TABLE 17·5
U.S. Distribution of Family Wealth, 1962

	Percent of total families, by wealth	Percent of total family wealth	Approximate size of wealth holdings per family*
	Lowest 25.4	0.0	Less than $999
	Next 31.5	6.6	Less than $7,500
	Next 24.4	17.2	Less than $25,000
	Top 18.7	76.2	Over $50,000
	(Top 7.5)	(59.1)	
	(Top 2.4)	(44.4)	
	(Top 0.5)	(25.8)	Over $100,000

*Estimated.

THE PERPETUATION OF INCOME DIFFERENCES

Recall our anecdote on how poverty causes poverty (p. 251). The converse is that high incomes cause high incomes. The table below shows the advantage accruing to children of upper-income families in acquiring the college education that leads in turn to higher incomes for themselves.

Family income (1965)	% 1966 high-school graduates starting college in 1967
Under $3,000	19.8%
$3,000–$4,000	32.3
$4,000–$6,000	36.9
$6,000–$7,500	41.1
$7,500–$10,000	51.0
$10,000–$15,000	61.3
over $15,000	86.7

Another example of the tendency of different income and social levels to reproduce their respective economic levels can be seen in the table below, which compares the occupations of sons with those of their fathers, for 1962.

Father's occupation when son was 16	Son's occupation in March 1962		
	White collar	Blue-collar	Farm
White-collar	71.0%	27.6%	1.5%
Blue-collar	36.9	61.5	1.6
Farm	23.3	55.2	21.6

The table shows considerable movement out of one's parent's group into other groups. But it also shows that the great majority of white-collar families produce white-collar sons, that the preponderence of blue-collar families produce blue-collar sons, and that most young farmers come from farming families.

comes the main source of income at the topmost levels.

MARGINAL PRODUCTIVITY THEORY RECONSIDERED

All these difficulties explain why the distribution of total income cannot be completely explained by marginal productivity theory and why we must be very careful in assuming that the theory adequately accounts for variations in earnings—extremes of riches and poverty aside.

But we should stop now to consider another criticism that will pave the way to our next chapter, where we shall consider ways of altering income distribution. You will remember that marginal productivity theory seemed to establish that in a fully competitive market, the levels of remuneration of the factors would be "just," in that each

factor would be compensated fully for the contribution it made to output. We remember that all factors in a competitive market earned their marginal revenue products. What could be fairer than that?

The question is trickier than you might think. Of course it makes us ask "What is *fair*?"—a question we will come back to at the end of the coming chapter. But it also seems to imply that there is one, and only one, distribution of income that would qualify as "fair." The impression given by marginal productivity theory is that there is one pattern of wages and rents and interest that would be established once and for all if a perfect market prevailed. But that is not so.

Once again an illustration is the best way to understand the problem. Suppose that we decide through our political system to establish a much more equal distribution of income—perhaps imposing wage floors and taxing high incomes much more steeply than we do. Such a shift would change demand patterns, since poor families have a higher relative demand for food, and rich families have a higher relative demand for luxury goods. Our redistribution of income will therefore shift the demand curve for food to the right and move that for luxuries to the left.

But higher demand curves for food products will lead to higher food prices and to higher incomes for those factors of production specialized in the food industry. Conversely, those factors of production specialized in luxury industries will face a lower demand curve and lose income. Moreover, unspecialized factors of production will also be affected. Let's assume that the food industry is capital-intensive—it uses a lot of capital per unit of output—while the luxury industries are labor-intensive, using little capital per unit of output. Such a shift in demand will therefore mean an increase in the price of capital relative to the price of labor. Not only will people with farming skills become richer relative to those with luxury skills, but all capitalists will become richer relative to all laborers.

THE PROBLEM OF THE STARTING POINT

Now, both before and after these shifts, we can imagine factors being remunerated according to their marginal revenue products. Both worlds can be in equilibrium. But the distribution of the two worlds will be different—not only regarding those who happen to be in one industry or another, but also regarding the share of income going to capital and to labor!

Thus the *marginal productivity theory of distribution can be consistent with many kinds of income distribution, depending on the condi-* *tions from which we begin.* So, too, the distribution of income can change—still in accord with marginal productivity—depending on changes in "taste" or in technology. These new distributions of income can become more or less unequal, can favor one group or class over another.

Marginal productivity theory therefore essentially takes as a "given" the starting point of the circular flow and ignores exogenous changes that can alter that flow. We are therefore thrown back on the unanswered questions of how the initial starting point came into being or of how changing tastes or technology enters the system. History or the political process or the ill-understood course of technical development thus become the critical determinants of the income distribution process, even assuming that the market system operates perfectly. Marginal productivity theory could then describe the process of change, but not the starting point or the critical factors at work in altering the circular flow.

Moreover, each such different society could "justify" its different distribution of income by appealing to marginal productivity. This means that ultimately we must choose between alternative income distributions—none justifiable "once and for all"—because each reflects a given set of historical, cultural, and technical forces. There is no escape from making a value judgment at least once, before we agree that the market process should thereafter decide income distribution, whatever that determination may be.

WHAT IS LEFT OF MARGINAL PRODUCTIVITY?

Have our objections removed marginal productivity altogether as a useful concept? We must be careful in framing an answer. Certainly the theory cannot completely explain value in explaining the facts of income distribution across the nation. Moreover, we have seen that it cannot be used as the basis for justifying any particular income distribution, even if there were no market imperfections.

Yet, there remains a useful place for the idea of marginal productivity when we apply it to specific and limited situations. When we seek to explain differentials in earnings where there are relatively few barriers to mobility, marginal productivity gives us a useful first approximation to finding why the market works the way it does. So, too, when we try to predict the effects of actions such as unionization or minimum wage laws, we need a theory that relates factor costs and factor employments—and marginal productivity analysis again provides a useful point of entry. Thus economists continue to use (and to teach) how diminishing returns are reflected in factor outputs and how factor outputs affect marginal revenue products, *because this sheds light on how entrepreneurs behave*. But economists are very wary of suggesting that marginal productivity theory should be lifted from this specialized use and applied wholesale to explain—much less to justify—the way incomes are actually distributed among households.

KEY WORDS

Institutional basis of riches and poverty

Market imperfections

Ignorance, luck, time lags, monopolies

Discrimination

Wage contours

Property

Marginal productivity theory

Multiple solutions

Value judgments

CENTRAL CONCEPTS

1. Income distribution is highly unequal in the United States, and marginal productivity theory sheds little light on the extremes of this distribution. *Riches and poverty are to be explained by institutional facts that are beyond marginal productivity analysis.*

2. One of the basic problems of marginal productivity theory is the widespread existence of *market imperfections*.

3. Market imperfections arise from many sources. *Ignorance, luck, time lags, monopolies*—all result in factor rewards that are greater or less than their long-term equilibrium values.

4. A particularly important market imperfection is *discrimination*, especially against blacks and women. Discrimination may result from the exercise of monopoly power or because of a "taste" for discrimination.

5. *Wage contours* describe the very important process by which people judge the fairness of their income shares. The *reference group* they use for judging the fairness of their pay is relatively limited, often to their fellow workers.

6. *Property* is another source of income differentials. Property is much more unequally distributed than income. It affects income distribution through the advantages wealthier children have in gaining skills, and through the property income that is concentrated in relatively few families.

7. Marginal productivity theory implies a distribution of income in which all factors are remunerated in accordance with their marginal revenue products; that is, their contribution to output.
The theory has two weaknesses. First it ignores or assumes away all *market imperfections*. Second, it overlooks the original starting point of the economy, and the fact that changes in taste or technology will yield different income distributions, even though each one rewards its factors according to the MRPs. Thus *there is an inescapable value judgment to be exercised in choosing which distribution we prefer.*

QUESTIONS

1. In Southern Rhodesia about 65 percent of all income goes to the top 5 percent of families, compared with about 20 percent in most European countries or in the U.S. Can you explain this in terms of marginal productivity? How else would you explain it?

2. How do you account for the fact that the percentage distribution of income is stable in Western countries over long periods of time?

3. Going back to the Table 17·3, Median Annual Earnings, how much of the different occupational incomes do you think can be ascribed to marginal productivity differences? Most? None? Some? What other elements enter the picture?

4. How would you determine how much it would pay you to overcome ignorance in looking for the best-paying job; e.g., how much time should you spend reading all the want ads in all the papers in all cities?

5. Which of the following do you consider to be market imperfections: (1) certification by state boards before anyone can practice medicine, (2) limitations on the number of students admitted to medical school, (3) costs of a medical education.

6. Give an example of a "taste" for discrimination against women. How about women airplane pilots? Bartenders? Surgeons?

7. Suppose it could be demonstrated that the marginal productivity of blacks in a given occupation was less than that of whites. Would this *explain* their lower rates of pay? Would it *justify* it? What would you have to know before you could claim that no market imperfections were at work?

8. Explain why marginal productivity theory does not shed light on the income of a wealthy stockholder. Does it shed light on the income of a wealthy baseball player?

9. Suppose that consumers' tastes shift in favor of vodka rather than expensive wine. Trace the likely shifts in the distribution of income.

10. Why would employers be willing to tolerate wage contours among their employees? What, if anything, can they do about it?

Changing the Distribution of Income

18

CAN WE CHANGE THE DISTRIBUTION OF INcome? Of course we can. Should we? That is a more difficult question. When we speak of deliberately trying to change the distribution of income, our purpose is usually to make it fairer. By *fairer,* we generally mean more equal, although not always; sometimes we say it is not fair that certain groups, such as school teachers, do not get higher incomes, even though they are already receiving incomes that are above the median for the society. In the discussion that follows, we will largely be concerned with ways of making income distribution "fairer" by making it more equal. At the end of the chapter we will take up the question of the value judgments that lie behind this decision. But first, let us concern ourselves with ways and means.

A person who wants to change income distribution can choose among four basic ways of going about his task. He may try to change the marginal productivities of individuals and then let them fend for themselves on the market. Or he may try to limit the

workings of the market, so that certain people will receive larger or smaller incomes even if their marginal productivities remain the same. Or he may let marginal productivities and markets alone and intervene by the mechanism of taxes and transfers, rearranging the rewards of society according to some principle of equity. Finally, he can introduce some system of rewards wholly different from market-determined rewards. This is surely the boldest and most far-reaching method, but it is one about which economists have very little to say. Accordingly, we will confine ourselves to the first three methods.

1. Changing productivities

Assuming that low marginal productivity is a basic reason—if not the only reason—for low incomes, someone who wants to change income distribution would do well to begin by boosting the marginal productivity of the least skilled and trained. How is that to be done? By and large, by giving *education,*—a generalized skill, *training,* or a specialized skill to those who lack them.

A glance at the Table 18·1 should make it clear that there is a strong functional relationship between education and lifetime earnings.

TABLE 18 · 1
Education and Average Lifetime Earnings

	Lifetime earnings (1968)
Elementary school	
0–7 years	$196,000
8 years	258,000
High school	
1–3 years	294,000
4 years	350,000
College	
1–3 years	411,000
4 years	586,000

In Chapter 17, we saw that the proportion of children of upper income groups who go to college is much larger than that of children of lower income groups. The reasons are not difficult to understand. Poor people not only lack the money to go to college, but they cannot afford the opportunity cost. That is, their low income gives them a much higher time preference for present income over future income. They go to work because they can't "afford" not to—even though their choice will turn out to be ill-advised from a long-term point of view.

Programs to invest in human capital can compensate for these economic constraints. Very low (or zero) interest loans, for example, allow poor persons to invest in themselves up to the same point that a wealthier person does. In 1971 the federal government spent some $1.5 billion in educational aid and $1.9 billion in manpower programs designed to teach skills, part of a program to build up our stock of human capital at the bottom of the income scale. State governments also contributed to similar programs.

Alternatively, we can think of investments in the poor simply as efforts to overcome inequality because our values incline us to favor equality. In this case we may invest in low-productivity individuals beyond the point where the return on our investment, in terms of their future incomes, equals the rate of interest (the general measure for the opportunity cost of the money we put into any capital endeavor). We may deliberately undertake these programs to remedy handicaps imposed by other circumstances—ill health, disabilities, an unfortunate family background—because we believe that equality is a more valuable social goal than the "natural" inequalities of different human abilities.

Education is not the only way of lifting marginal productivity. Investments in the health of low-income persons can also increase their earning power. Investments in housing may improve marginal productivity by giving poor persons a "stake" and an incentive to succeed. Investments in programs to overcome barriers to mobility of all kinds enable persons to put productivities to better use: a highly-skilled worker who cannot afford to move to a city where his talents are in demand, or who does not know about jobs where his abilities could be best put to use, or who is barred from certain jobs because of discrimination, is prevented from using the market mechanism to his best advantage.

2. Intervening in the market

A quite different way of going about the task of changing income distribution is to intervene in the market. Here two methods warrant consideration, because they are widely used. One is to establish minimum wages. Another is to use the power of unions to set wage "floors."

GINI COEFFICIENTS

How does one measure inequality? How does one determine, between two unequal income distributions, which is *more* unequal? One widely used way of measuring inequality—or more accurately, of measuring the distribution of any variable, such as income, age, weight, I.Q.—is to use a geometrical device called a Lorenz curve.

We construction a Lorenz curve by drawing a box, as illustrated in the graph, with the vertical and horizontal axes depicting the variables in which we are interested, measured on a percentage scale. Thus in the diagram, the vertical axis shows percentages of total income from zero to 100, and the horizontal axis shows percentages of all families, from zero to 100.

Next, we rank families in terms of the variable we wish to measure (income in this case), from lowest to highest, and then indicate on the graph the relationship between a given percentage of families and the corresponding percentage of total income. Above, for example, the dots show that in the U.S., in 1972, the first (lowest income) 20 percent of all families received just over 5 percent of all income, the first 40 percent about 17

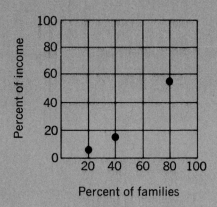

percent of all income, and the first 80 percent of families about 58 percent of income. If we fill in enough of these relationships, we get a "curve"—a so-called Lorenz curve—of income distribution.

Now comes the Gini coefficient. What would a perfectly equal distribution of income look like on a Lorenz diagram? It would be a "curve" in which each successive percentage group of families enjoyed the same percentage of income: the first 20 percent getting 20 percent of income; the first 40 percent, 40 percent of income; the first 60 percent, 60 percent of income. What would a

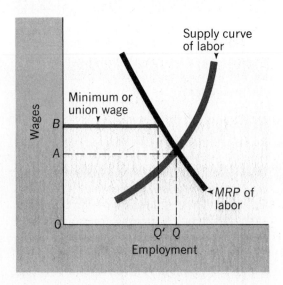

FIGURE 18 · 1
Minimum or union wage floors

curve of "perfect equality" look like? It would be a straight line, from one corner to the other.

And how about "perfect" inequality? It would show the last (richest) family getting *all* the income, and the first 99.9 . . . percent getting none. We show both these distributions in the chart below, together with a Lorenz curve that

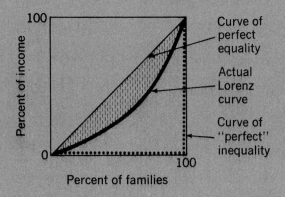

roughly depicts the actual distribution in America today.

Probably you have already seen what the Gini coefficient will measure. It is the ratio be-

tween the dashed area outlined by the Lorenz curve and the full area under the 45° line. In the case of perfect equality, the area enclosed by the Lorenz curve is zero, and the coefficient would also be zero. In the case of "perfect" inequality, the area enclosed by the curve, as shown by the dotted line, is virtually as great as the area under the 45° line, so that the Gini coefficient is 1.0. Thus, the closer to zero, the less the inequality of any given distribution.

In actual fact, the Gini coefficient for the United States in 1970 was slightly less than .40. This is about the same coefficient as we find in most West European countries, and it is probably considerably less than the coefficients of many under-developed countries, where the distribution of income is highly bi-modal (see p. 677). Whether or not the "Gini" is rising or falling within the United States (or whether it is rising or falling between groups such as blacks and whites) is a complex problem, about which opinions differ. Probably whatever movements exist are small and slow. The usefulness of the Lorenz curve with its Gini coefficient is that it enables us to speak about these changes in a precise manner, and therefore to compare our findings in a way that would not otherwise be possible.

Actually, both methods are very similar, so far as their economic consequences are concerned. In Fig. 18·1, we show a market equilibrium wage *OA*, and a higher wage *OB*. This higher wage could be imposed by law or by union contract. In both cases the effect on the employer is the same. Instead of employing *OQ* amount of labor, he will employ *OQ'*.

What will be the effects of such a wage floor? We must now differentiate carefully between the effects on those who will remain employed and the effects on labor as a whole. Clearly, for those employed at the new higher wage, the minimum wage or the union wage will be an unalloyed success.

But how about the income of labor as a whole? Formerly its income was *OA* wages multiplied by *OQ* employment. After the new wage, total labor income is *OB* × *OQ'*. Will labor as a whole gain or lose? That depends on two other elements. One of these is the ease with which the employer can "displace" labor with capital. This is largely a technical or an engineering consideration. If a minimum wage (or a union contract) pushes up the wages of labor, the employer may be able to use machinery that was formerly too expensive. In that case, he will substitute capital for labor and may reduce employment sharply. On the other hand, if he cannot use machines in place of labor, he will have to maintain his work force more or less unchanged and simply bear the higher price.

But that is only half the answer. Suppose that he does hold onto his labor force, paying higher wages. This will increase his costs and his selling prices. Now everything hinges on the elasticity of demand for his product. If demand is very elastic, his sales will diminish sharply at higher prices, and he will be forced to cut back on output and employment. If demand is inelastic, the public will pay the higher prices charged by the employer, and the wage increase or the minimum wage will have but little effect on employment.

uct is price-inelastic, we would expect that labor would benefit.*

3. Taxes and transfers

A third means of altering income distribution is to tax high incomes and to subsidize low ones. Taxes and subsidies (or *transfers*, as they are called) are used by all governments to redistribute incomes. Let us examine them carefully.

To begin, it is helpful to classify taxes into three types, so far as their effect on in-

TABLE 18 · 2
Percent of Household Income Paid as Tax, 1968

Incomes	All taxes	Federal tax		State and local taxes	
		Income	Soc. Sec.	Property	Sales
Under $2,000	50.0%	1.2%	7.6%	16.2%	6.6%
$2,000 – 3,999	34.6	3.5	6.5	7.6	4.9
$4,000 – $5,999	31.0	5.3	6.7	4.8	4.1
$6,000 – $7,999	30.1	6.5	6.8	3.8	3.6
$8,000 – $9,999	29.2	7.4	6.2	3.6	3.3
$10,000 – $14,999	29.8	8.7	5.8	3.6	2.9
$15,000 – $24,999	30.0	9.9	4.6	3.6	2.4
$25,000 – $49,999	32.8	12.9	2.5	2.7	1.8
$50,000 +	45.0	19.8	1.0	2.0	1.1

Thus it is difficult to say, without examining each industry's technical characteristics and market demand, whether raising wages by union action or by law will result in more income going to labor as a whole. If labor is easily replaced by capital and if demand is elastic, an attempt to raise the income of labor by direct action may have serious adverse consequences. If labor is not easily displaced and if demand for the prod-

*There is a further consideration that we should raise, although we cannot go into it fully here. A minimum wage law, or a nationwide union push for higher wages, will affect not only an employer's costs but also his income, because it will give rise to an increase in total demand. Therefore, to a certain extent, the rise in employers' cost curves will be offset by a rightward shift in their demand curves. The extent of this shift will vary from industry to industry. A rise in minimum wages may have an immediate effect on the sales of clothing. It is not likely to have any immediate effect on the sales of turbogenerators.

come distribution is concerned. We call a tax *progressive* if it takes an increasing proportion of income as income rises, so that a rich man pays a larger fraction of his income in taxes than a poor man does. We call a tax *proportional* if it takes an equal fraction of the income of a rich or a poor person; and we call it *regressive* if it takes a larger fraction of a low income than of a high one. Someone who wants to make income more equally distributed will want a progressive tax system. Someone who wants it more unequally distributed will want a regressive system.

What kind of system do we actually have? That is not an easy question to answer for two reasons. First, we have more than one "system." Second, it is not always possible to know who ultimately pays a tax.

TAX SYSTEMS

Table 18·2 shows the confusion of tax "systems," the proportion of incomes at different levels that are paid for various federal, state, and local taxes. Notice that the overall tax "system" is more or less proportional, with the exception of its regressive twist on the lowest incomes and its progressive bite at the top. As we can see, looking at the other columns, this weight at the top and bottom is due to the progression of some taxes (mainly the federal income tax) and the regression of others (property and sales taxes).

The table makes clear that a considerable change in income distribution could be brought about simply by removing some of the regressive taxes in our system. Social Security taxes, for example, could be made much more proportional by removing the ceiling on salaries that now imposes a cut-off above roughly $13,000. (In 1968, the ceiling was only $10,000, which is why the Social Security tax turned regressive at that level.) Sales taxes could be made less regressive by allowing the first $2,000 of consumption to

be free of tax or by imposing higher taxes on luxuries mainly consumed by high-income families, and lower taxes on goods which are consumed by the poor as well as the rich.

TAX INCIDENCE

But the effect of taxes on income distribution is complicated not only because of our variety of tax systems. The problem is made more obscure because it is not always clear who actually pays a given tax.

This seems paradoxical. Doesn't the person who writes out a check to the government or pays a bill with a tax on it obviously pay the tax? Not necessarily. Let's take the case of a sales tax. If a state levies a tax of 5 percent on some commodity, the cost of the commodity rises by this amount. This is the same thing as an upward shift in its supply curve, as in Fig. 18·2. As a result, the price rises from P_1 to P_2.

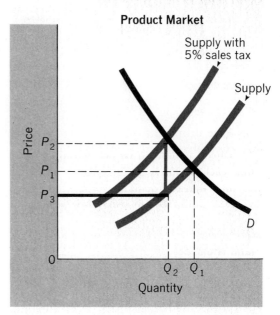

FIGURE 18·2
The incidence of sales taxes

Product Market

HUGGER MUGGER
IN THE TAX DEPARTMENT

There are two ways of not paying an income tax. One is *tax evasion:* just not filing a tax return or failing to report income on a return you do file. This is a route that has many advantages but one large disadvantage: if caught, you go to jail. The other way is *tax avoidance*. This has many disadvantages—it requires expensive legal and accounting advice, it may involve complicated paperwork, etc.—but it has one big advantage: it is perfectly legal.

Tax avoidance is the art of finding loopholes that make it possible to enjoy income without paying tax. For example, if you own tax exempt bonds, your income from these bonds is legally exempt from tax. Or if you own a large yacht on which you are forced to cruise while talking customers into buying your goods, you may be able to persuade the Internal Revenue Service that you must deduct part of the cost of the yacht from your income as a "necessary" business expense, thereby legally reducing your income considerably.

Philip M. Stern has written an eye-opening book about tax avoidance, *The Rape of the Taxpayer*. You might be interested in a table on p. 67 of that book, showing the number of individuals who paid *no* income taxes in 1969.

Income group	Number paying zero tax
Above $1,000,000	56
Above $500,000	117
Above $200,000	301
Above $100,000	761

According to Stern's data, the total amount of legal tax avoidance in the United States in 1973 amounted to $77 *billion*.

Now look carefully at the consequences. First, as we saw in Chapter 9, you will see that the price has risen *less* than the sales tax: the price is up from P_1 to P_2; the sales tax has raised costs from P_3 to P_2. Second, you will notice that the producer's income has gone down for two reasons. First, his sales have dropped from Q_1 to Q_2. Second, the net revenue that he retains after remitting the tax (P_3) has also gone down from P_1, the original sales price. As a consequence, his profits have fallen.

Who has now paid the tax? The consumer has paid some—the rise in price from P_1 to P_2, combined with the decrease in his consumption of the good. The producer has paid some—the drop in profits we have just analyzed. But this is not all. Because the producer's income has fallen, so has his demand for factors of production. The prices of the factors and the quantitites employed will fall. Hence they too bear part of the tax.

Thus to determine the distribution of income before and after taxes is a complicated operation. The problems of tax incidence strain the competence of not only a beginning student but also professional economists. The corporate income tax is a good example. Who ultimately pays this tax—the consumer, the capitalist, the worker? No one knows.*

*It seems only proper to add that of all taxes, the one whose incidence seems least open to doubt is the personal income tax. This is because the government taxes you on your *net* income, after you deduct legitimate expenses. To the extent that we are money maximizers, we will try to earn as large a *net* income as possible. The only way to pay less taxes would be to make less net income—and have more leisure.

TRANSFERS

Still another means of redistributing incomes consists of transfer payments or subsidies— the "transfer" of income from one person to another. Welfare, Social Security, crop-support payments, subsidies to the merchant marine are all transfer payments of one kind or another. When we study macro-economics we will find that transfers play an important role in government expenditure.

Here, however, we want to look into their effect on income. In recent years, much discussion has focused on the *negative income tax,* a plan to alleviate poverty. As the words suggest, the government *pays* these "taxes" to families under a given poverty line.

In 1972, the Department of Labor Statistics defined as "poor" an urban family of four having less than $4,300 annual income. Suppose that we decided to transfer income to all families having less than that amount. This would have cost about $12 billion, roughly one percent of GNP.

This seems like a simple method of eliminating poverty. Suppose, however, that we also decided *not* to help people who were no longer not in officially defined poverty, so that we reduced the amount of aid we gave by any income a poor family earned. If a family earned $2,300, it would then get $2,000 in assistance. If it earned $3,300, it would get $1,000. If it earned $4,300, it would get nothing.

This is the same as a 100 percent tax on all earnings under $4,300. Why should any family bother to earn any sum less than $4,300 if its welfare is immediately reduced by that amount? Would it not be sensible— rational—to stay on relief, rather than take a low-paying and probably unpleasant job?

WORK INCENTIVE PROGRAMS

Unfortunately, this is the way many welfare programs have been administered in the past.

Not surprisingly, they have resulted in strong disincentives to work. Accordingly, economists now propose that assistance programs should have work incentives built into them.

This means we must tax below-poverty earnings at less than 100 percent. To make the arithmetic simple, let's tax at 50 percent—actually a very high tax rate since it is the top rate currently paid on the highest salaries earned in the United States. If a family earns $1,000, it must then pay $500 in taxes. But this now leaves it with an income of $4,800 after taxes ($4,300 in assistance plus $1,000 in earnings, less $500 in taxes).

Note, however, that we are now making transfer payments to people who are *above* our poverty line of $4,300. We will continue to make such transfer payments until family earnings reach $8,600. At this point, a family's budget would look like this:

Transfer income	$4,300
Earned income	8,600
Tax on earnings	−4,300
Net income	**$8,600**

ECONOMIC AND POLITICAL PROBLEMS

This presents us with two problems. The first is economic. Under the plan we have just examined, a "poor" family earning $8,600 would have an advantage over a family that was not deemed "poor" and that earned the same amount, because the second family would owe taxes on all its $8,600 income and would therefore end up less well off than the income-supported family. To remove this inequity, we would have to pay family allowances to *all* families and remove all taxes on nonpoor families' earnings up to $8,600. This would involve massive transfer payments or a very great increase in the income tax rates above $8,600.

The second problem is political. Could we persuade families whose incomes were above $8,600—that is, families who were neither "in" nor "near" poverty—to bear the entire income tax burden because they recognized the unfairness of programs that locked unfortunate families into low brackets with a 100 percent tax on any earnings they made?

There is no economic answer to this question. What is at stake is essentially a political choice between two patterns of income distribution. That choice will be exercised through political programs that will favor one group or another—those at the bottom of the scale or those who are sufficiently affluent to be in the higher taxpaying brackets. This problem is complicated because the "pyramid of income distribution" is changing, as Fig. 18 · 3 shows.

This upward shift is welcome news, but it has its complex political implications. The middle-class affluence we seem to be headed for is a hard-earned, not an idle affluence. Today, three-quarters of all families in the $15,000 bracket have reached that level be-

cause two or even three members of the family work. At the same time, the proportion of the poor is shrinking: in terms of 1970 dollars, families with less than $5,000 will have diminished by 1980 to only a fifth of the population.

Hence the question: will an America that is increasingly "middle class" wish to help those who are left behind? Will the growth of political strength in the middle ranges make it more difficult to vote money to aid those at the bottom? Few questions have as much importance in determining the quality of our national existence in the coming years.

Value Judgments in Economics

The politics of income distribution finally brings us face to face with a problem that we raised at the outset of this chapter—indeed at the beginning of this book. This is the matter of value judgments, of deciding what is right, and who is to say what is right.

FIGURE 18 · 3

**The changing pyramid of income distribution
(total households each year = 100%; based on 1970 dollars)**

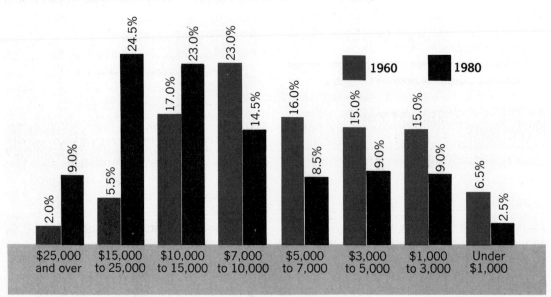

EQUITY VS. EFFICIENCY

Does economics have anything to say about value judgments? Traditionally, economists have been extremely cautious about making one kind of value judgment—that concerning *equity*. By equity, we mean justice: an equitable distribution of income is a just distribution, and an inequitable one an unjust distribution.

Of course this raises the question of what we mean by *just*. Philosophers have argued over the meaning of that word for millennia, but economists have politely bowed out of the discussion. They have refused to say what they thought was just, or else they have taken the position that as economists they had no special claim to knowledge about justice. They have therefore traditionally declared their own preferences to be no more than those of any citizen and have equally carefully tried to keep them out of their economic analyses.

On the other hand, economists have been quick to claim the validity of their analyses regarding *efficiency*. By and large they have distinguished between efficiency and equity, in economic terms, as follows. Problems of equity involve matters such as the distribution of income. Here the concept of justice enters, so that the economist has refused to say which of two or more income distributions is better. Efficiency, on the other hand, involves problems of minimizing input and maximizing output. Most people claim that a larger output is preferable to a smaller one, or that we want to get as much output as possible for a given input. Therefore economists have freely offered advice about efficiency, without feeling that they were getting involved in value judgments.

PARETO OPTIMALITY

The clearest example of an efficiency criterion is called Pareto Optimality, named after its

inventor, the brilliant Italian sociologist and economist Vilfredo Pareto (1848–1923).

Pareto gave us a definition of efficiency that has an immediate, intuitive appeal. Suppose that we have a society composed, let us say, of you and me, each of us enjoying a certain condition of well-being. Now let us say that a third person appraises our mutual condition; and seeing that you are very poor and that I am very rich, he decides that our mutual condition will be improved if some of my possessions are transferred to you.

How could he justify his decision? Essentially he could not, unless he made the value judgment that our more equal new condition was "better" than our former less equal condition. But he cannot *know* that. Perhaps he is wrong. For all he knows, my increased enjoyment of still more wealth might be greater than your decreased enjoyment of less wealth.

But, said Pareto, suppose that we can arrange matters so that one of us gains *while the other remains exactly as he was before.* Now we no longer have to make dubious calculations. No one has lost; someone has gained. Our second combined position is unambiguously better than the first. Moreover, Pareto added, there is a second way to accomplish the same end. Our social situation may be altered so that you gain and I lose, perhaps by giving you some of my income. But if you can *compensate me for my loss in some other way—say, by giving me some of your profits—so that I declare myself as satisfied as I was before,* once again there has been an unambiguous gain in welfare. We have improved our efficiency in a way that seems quite free of any interjection of values.

THE ECONOMICS OF EXCHANGE

This clever principle is obviously relevant to a market system. Suppose that you have a vast tract of land; I have huge piles of money. If you willingly accept some of my money for your land, and I willingly part with that money in exchange for your land, we should move in the direction of Paretian Optimality. We would not enter into the exchange unless the utility of the money was greater to you than the utility of the land you will sell. Similarly, I would not offer the money unless your land was worth more to me than the sum I must pay. By voluntary exchange, we will arrive at a redistribution of wealth that will make both of us better off by our own estimation of what we want or that will, at worst, leave one of us just as well off, while the other is better off.

Thus an economy of exchange, in which we can trade goods and services, should lead us in the direction of greater efficiency—more utility for at least one person, and less for none. It is not difficult to show that a perfectly competitive system should lead to Pareto Optimality. In such a system, all available output would have been attained, given the original starting point. Eventually no one could further improve his lot unless someone else was made less well off.

EFFICIENCY AS A VALUE: MORE IS BETTER

Is Pareto's principle correct? Does it establish a criterion for efficiency that is *quite free of all value judgments?* Economists used to think so. But recently we have come to see that underlying the idea of efficiency are two value judgments every bit as arbitrary and personal, every bit as imbued with concepts of justice and ethics, as the ideas of equity.

One of these is the notion that *more is better.* This is a notion that underlies the conception of economic man, to whom we have continually referred in our analysis of maximizing. Yet, we have been careful to describe maximizing as a general description

of how men *do* behave, not how they *ought to* behave in some moral sense.

Is more always better? Not necessarily. More may be dangerous, as we shall see when we study the long-term implications of industrial growth in Chapter 34. Or "more" may be unpleasant, vulgar, greedy. Or "more" may be meaningless in terms of utilities. Think of someone who does not want to be richer—let us say a person who has found that the greatest happiness in life comes from divesting himself of his worldly goods, as Jesus or Thoreau preached. Such a man has the happiness of experiencing no budget constraint. Economic goods have become free goods to him, with zero price. A Thoreau can be inefficient—can deliberately follow a course that lessens his possessions—without becoming less satisfied.

Thus the idea that "more is better" is not a *fact.* It is a *judgment.* It may be an accurate description of human affairs, but it cannot be an accurate prescription for human affairs—at least not without injecting a value judgment into the end we propose.

INDIVIDUAL PREFERENCES

The second value judgment on which Paretian efficiency rests is the notion of the *autonomy of our individual preferences.* Central to the Paretian idea is the belief that we make up our own minds, free of any influence: that we know what is best for ourselves.

Do we? The more we study the workings of society, the less convincing becomes the idea of the autonomous individual. We are all creatures of our family training, our group beliefs, our culture. Our "wants" and "needs," our desires and drives are shaped in a social mold from which none of us can wholly escape. We are bombarded by advertising, moved by political exhortations, influenced by philosophic beliefs.

Thus the value judgment hidden in

Pareto's assumption about individual choice is that "autonomous" choice would be *better* than social choice. But if, in fact, individuals are never wholly autonomous, then the line between individual and social choice becomes blurred. We can no longer claim that "free choice" is always better than socially imposed choice, because there is an element of social choice even in the "freest" individual decision.

EFFICIENCY AND POLICY

Thus we are skeptical nowadays about claiming that Pareto Optimality is a "value-free" concept. This makes us skeptical as well, of claims that a society of free exchange will, by its voluntary activities, unerringly move in the direction of increased social well-being. This is all the more true because most actual social changes do not fulfill the Paretian principle that no one should be made worse off (or should be compensated if he is made worse off). Almost all important economic policies deliberately seek to benefit one group *at the expense of another.* Laws that protect labor unions hurt employers. Laws that protect industries (such as tariffs) harm consumers. Laws that improve the condition of the poor do so at the expense of the rich. We are plunged into a sea of equity judgments every time we make economic policy decisions, even if those decisions aim at efficiency.

THE DILEMMA OF VALUES

If economics is inextricably entangled with equity, does that reduce all analysis to a question of sheer preference? You prefer one policy, I another—who is to say which is better?

To some extent, economics and all social science are forever entangled with just such problems. As an economist, you may take the position that the best distribution of income entrusts one man with virtually all the nation's wealth, while the rest of the people are his slaves. Perhaps you will say that this accords with your judgment of what is best. There is no way of "proving" your values wrong, just as there is no way of proving wrong the values of another economist who plumps for a totally equal distribution of income. At best we might be able to analyze the effect of both distributions on output; but even that, as we have seen, is a value judgment, insofar as we declare our preference for one size of income over another!

THE TEST OF SOCIAL VALUES

Yet, there is no reason to conclude from this that one economic prescription will be judged as good as another, for the policies of economists are subject in real life to a test much more severe than that of the private value judgments of the economist himself. They are tested by the *collective judgments of society.* Some policies will gain the approval and support of large numbers of people. Others will earn their dislike, even their hatred.

Social values are a fact of life as real as supply and demand curves—perhaps even more easily identifiable in real life. Like supply and demand curves, social values can change. But at any time they operate as strong forces within society, rejecting some economic policies, endorsing others. Thus we can hope to improve our feelings of social welfare by advocating policies that accord with social values. We can also, of course, try to change social values so that they will accord with our policies. An economist in favor of slavery will seek to convince people that slavery is best; an egalitarian economist will write tracts advocating his program of equality. In all likelihood, the values of society will change only slowly, paying scant heed to extreme proposals. Our policies will be deemed better or worse, insofar as they succeed in bringing about changes that accord with these slowly changing values.

Equity

But we can do more than recognize the fact of social value judgments. We can also try to think systematically about the problem of equity. This will not teach us what is "right" and "wrong," but it will lead us to understand better the nature and implications of the value judgments we make.

EQUALITY AND INEQUALITY

Behind economic equity judgments are two possible starting points. One is the belief, rooted in Greek philosophy, that men are by nature *unequal*. The other is the belief that men are *equal*—a point of view that made its forceful appearance into Western thought in the eighteenth century, with the work of Jean Jacques Rousseau.

Now this contradiction in starting points is not a disagreement about facts. Of course human beings are not exactly alike. The disagreement is about the nature of the arguments that one must bring to bear to justify social policies. Someone who believes that men are inherently unequal will demand a justification for any policy whose objective is to make men more equal. Since the policy goes against his fundamental assumption, he begins with the need to be "shown" that *some* equality might produce a better society than the existing condition of inequality.

For one who begins from the position of Rousseau, just the opposite is true. Since his premise is that men are equal, he will have to be shown that *some* inequality will produce a better society. Suppose we believe, in accordance with the prevailing value judgments, that a rising level of national output would be a good thing. We would then have

to demonstrate that output would be higher if there were *some* inequalities in rewards than if there were none. We might do so by claiming that higher rewards for higher output will tempt people to work harder. We would then have justified our admission of inequality because it was a means to the end of higher output. It also follows that to advocate such a policy we would have to rank our preference for output *above* our preference for equality.

RULES FOR EQUITY

Most Americans—indeed, most people in the modern world—subscribe to the idea that men are equal rather than unequal. This means that there is an underlying bias in favor of equality in their social values. We hear of policies in every nation that seek to diminish the differences between rich and poor. We hear of very few policies that openly advocate greater inequality. Even policies that support greater inequality—for example tax loopholes for millionaires—are "justified" in terms of their ultimate effect in raising the incomes of *all,* presumably lessening poverty.

But given this starting bias in favor of equality, we need some understanding of the kinds of exceptions we may make to the general rule. That is, we need to know, and to look carefully into, the arguments in favor of inequality. As we shall see, there are four of them:

1. We agree that inequality is justified if everyone has a "fair chance" to get ahead.

Most of us do not object to inequality—in fact, we generally favor it—if we are convinced that the race was run under fair conditions, with no one handicapped at the start.

What are fair conditions? That is where the argument becomes complex. Are large inheritances "fair"? Most Americans agree that *some* inheritance is fair but that taxes should prevent the full passage of wealth from generation to generation. What about inheritances of talent? No one is much concerned about this. Inheritances of culture? We are beginning to get exercised over the handicaps that are "inherited" by persons born in the slum or to nonwhite parents.

2. "We agree to inequality when it is the outcome of individual preferences."

If the outcome of the economic "game" results in unequal incomes, we justify these inequalities when they accord with different personal desires. One man works harder than another, so he is "entitled" to a larger income. One chooses to enter the law and makes a fortune; the other chooses to enter the ministry and make do with a small income. We acquiesce in these inequalities to the degree that they appear to mirror individual preferences.

3. We abide by inequality when it reflects merit.

Merit is not quite the same as fairness or personal preference. It has to do with our belief in the propriety of higher rewards when they are "justified" by a larger contribution to output. Here is the belief that underpins the idea that factors should be paid their marginal revenue product. We do not object to factors receiving different remunerations in the market, because we can show that each factor contributes a different amount to total output.

This is, of course, only a value judgment. Suppose there are two workers, side by side, on the assembly line. One is young, strong, unmarried and very productive. The other is older, married, has a large family and many expenses—and less productive. Should the first be compensated more highly than the second? We find ourselves in a conflict of values here. Our bias toward equality tells us no. Our exception for merit tells us yes. There is no correct solution for this or any other problem involving value judgments.

Once again, social values prevail, sometimes paying the younger man more than the older, sometimes both the same, sometimes the older man more.

4. Finally, we agree in violating the spirit of equality when we are convinced that inequality is for the "common good."

The common good is often translated into practical terms of gross national product. Thus we may agree to allow unequal rewards because we are convinced (or persuaded) that this inequality will ultimately benefit us all, by raising all incomes as well as the incomes of those who are favored.

Here we are again allowing ourselves to be swayed by the idea that more is better. But it is possible that we can define the common good in other terms that may also justify some inequality. We may establish environmental safety and quality as the highest goals of society and then decide that certain inequalities of rewards are the most expeditious ways of achieving this common good. For instance, we might reward persons who behaved in an environmentally favorable way, by lowering their taxes, or by subsidizing them. The point remains that it *is* the common good that justifies the departure from the value norm of equality.

The difficult question here is to define the common good. There are many conceptions of what such an objective should be. All incorporate value judgments. Even the common good of survival, which might justify giving a larger reward to those who must be entrusted with survival, is a value judgment. Do all societies deserve to survive? Was Nazi Germany justified in seeking survival at all costs?

Equity and Economics

These general principles do not describe the way we *should* think about inequality. They

TOWARD AN OPERATIONAL DEFINITION OF FAIRNESS

Can we specify with certainty what a "fair" income distribution would look like? Of course not. But suppose we ask the question differently. Can we specify a distribution of income that would accord with what most people think is "fair"? Here we put the question in such a way that we could test the results, for example by an opinion poll.

If we took such a poll, most persons in the United States would probably agree that existing income is not "fairly" distributed. They consider it unfair that some people are as poor as they are and others are as rich as they are.

But suppose we asked whether the public would approve of an income distribution that had the same "shape" as that for one group in which the more obvious advantages and disadvantages of the real world were minimized. *That group consists of the white adult males who work full-time and full-year.* In general, these workers suffer minimally or not at all from the handicaps of race, sex, age, personal deficiencies, or bad economic policies. By examining their earnings rather than their incomes we can eliminate the effects of inherited wealth. Might not such a

standard appeal to many people as constituting an "operational" definition of a just income distribution?

Since the poll has never been taken, we cannot answer the question. But we can examine what income distribution would look like under such a dispensation. The results are shown in the table. It is interesting to note that this standard of fairness, if applied, would reduce the dispersion of income by 40 percent.

Annual earnings (in thousands of dollars)	Distribution of income in accordance with "fairness" standard	Actual distribution of income, 1970
$ 0–1	1.7%	10.4%
1–2	1.3	8.3
2–3	1.5	6.9
3–4	3.0	6.8
4–5	4.4	6.2
5–6	6.8	6.7
6–7	8.6	7.0
7–8	10.5	7.8
8–10	19.7	13.2
10–15	27.9	17.7
15–20	11.2	6.8
25 & over	3.3	2.3

are an attempt to describe the way we do think about it—the arguments that we commonly hear or raise ourselves, to defend an unequal distribution of goods and services, or of wealth.

Each of these arguments, as we can see, poses its own tangled problems. And there is every reason that they should be tangled, for the distribution of incomes poses the most perplexing of all economic problems to any society. At one extreme, it criticizes all the privileges and inequalities that every society displays, forcing us to explain to ourselves why one man should enjoy an income larger than he can spend, while another suffers from

an income too small to permit him to live in decency. At the other extreme, it forces us to examine the complications and contradictions of a society of absolutely equal incomes, where each individual (or family?) received the same amount as every other, regardless of differences in his physical capacities, his life situation, his potential contribution to society.

Both extremes pose economic as well as moral problems. A society of "total inequality" would probably work very poorly—most slave societies have. A society of "complete equality" would also probably work poorly,—attempts in early Soviet history to reward all workers alike resulted in a near breakdown in production.

Thus we have to compromise, find reasons to support income distributions that are neither completely equal or completely unequal. Here is where we lean partly on our actual knowledge of the effects of income distribution on work and output, and partly on our value systems that define allowable exceptions to our basic rule that societies should seek equality as their goal, not inequality. As our values change—and we are now living in a period when values seem to be changing rapidly—we accord different weights to the various arguments by which we traditionally justify unequal incomes.

THE UNAVOIDABLE PROBLEM

Economic equity is complex, confusing, and often disconcerting. It makes us uneasy, not only because it confronts us with the often shaky presumptions we make about equity, but because we recognize that no solution will ever be found that can be demonstrated to be superior to all others. Economic efficiency, as we have been at some pains to show, is also a disguised belief in economic equity of a certain sort. Thus in the end we are thrown back on the moral promptings that are the very foundation of social life, but that are nonetheless the most vulnerable of any of society's institutions.

Yet, however difficult or dismaying, there is no escape from trying to specify what you mean by equity. If you fail to make such an effort, you will only end up with a hodgepodge of feelings about equity, many of them contradictory or unsatisfying. This does not mean that you can work out a scheme of economic equity that will be free of contradictions or uncertainties. But at least you will have understood why certain contradictions are inherent in the problem of equity, and you will be in a position to act with both intelligence and purpose when you are forced—as we are all forced—to decide which economic policies are better and which are worse.

KEY WORDS

Ways of changing distribution

Improving productivity

Intervening in the market

Taxes and transfers

Progressive

Proportional

Regressive

CENTRAL CONCEPTS

1. **There are four ways of changing income distribution: changing marginal productivities and allowing the market to work; intervening in the market; altering rewards by taxes or transfers; and restructuring the social system. Economics deals with the first three.**

2. **Productivities can be changed by** *improving the education and skills of low-productivity workers.* **We call this** *investing in human capital.* **All expenditures to lessen discrimination and improve mobility also enhance productivity.**

3. **There are two main ways of intervening in the labor market:** *minimum wages and unions.* **Both efforts attempt to establish wages that are above market equilibrium wages. These methods increase the incomes of those who remain employed, but may or may not increase labor income as a whole.**

4. *Taxes and transfers* **also rearrange factor rewards. Taxes can be classified as** *progressive, proportional, and regressive,* **depending on whether they take a growing, equal, or diminishing fraction as incomes rise. Taxes as a method of redistribution are made complex because the** *incidence of taxation* **is often difficult to calculate.**

Transfer payments

5. *Transfer payments* are a means of increasing the incomes of some individuals by "transferring" to them the taxes of others. *The negative income tax and work incentive program are two major plans for transfer payments.*

If we transfer incomes only to those below some defined "poverty line," we effectively tax away any earnings they make.

If we tax away only a part of earned income, the cost of the transfer program is much greater.

Negative income tax

Work disincentive

Equity

6. Economists have always distinguished between statements about *equity—justice*—and *efficiency*. They have sought to refrain from making equity statements but have not hesitated to make efficiency statements.

Efficiency

7. The most famous efficiency statement is called Pareto Optimality. It specifies an improvement in efficiency as one in which (a) *at least one person is better off and no one is less well off* or (b) *the person who is made less well off by a change can be compensated until he is at least as well off as before the change.*

Pareto Optimality

More is better

8. Actually, Pareto Optimality contains two hidden value judgments. One is that *more is better.* The other is that *individual (autonomous) choice is better than social choice.*

Autonomous choice

9. Public policy decisions impose an inescapable problem of values. These policies are judged to be "better" or "worse" according to whether they reflect prevailing social values. We are of course free to change these values to accord with our own.

Social values

Equality

10. Behind equity decisions lie two different starting points. They are our beliefs in *whether men are equal or unequal.* If, as most modern men, we believe in their equality, we are forced to *justify policies that favor inequality.*

"Fair" competition

11. American society makes general justifications for the existence of inequality when:

● *competition for place is fair*

Individual preferences

● individuals *choose economic actions that they know will lead to unequal results*

Merit

● inequality accords with *merit*

● inequality is for the *common good*

Common good

12. *Each of these commonly accepted rules for equity poses problems.* The purpose of thinking about equity is not to work out "solutions" to the problems of economic equity. It is to become aware of the complexity and even contradictoriness of these value problems. Its purpose is not to make us "right," but to make us thoughtful.

Problems of equity

QUESTIONS

1. Do you think the present educational system works to reduce or to maintain the structure of inequality in rewards? How would you suggest changing it, if you wanted less inequality? More?

2. What institutional barriers do you think are most responsible for inequalities in our income distribution? Let's assume that discrimination against blacks and women is on your list. How do you suggest lessening this?

3. Explain the effects of a minimum wage law on the total income of labor, assuming that (1) machines can be easily substituted for labor, and (2) public demand for the good is elastic. Under what conditions of substitutability and demand is a union most likely to increase the payrolls going to its members?

4. Why is the sales tax regressive? Why do you think a property tax is regressive? (Hint: landlords add the tax to rents. The demand for housing is inelastic. Rents rise. Show this in diagrams.)

5. Explain why a tax on net incomes cannot be "shifted." Will it affect the income earner's behavior?

6. Suppose that we instituted a family allowance program that insured an annual income of $5,000 per family, and we taxed all earned income at one third. How high a total income could a family earn before its tax payments to the government exceeded the government's payments to it?

7. Which of the following statements are value judgments:

Economic competition produces low prices.
Economic competition produces welfare.
Economic competition produces poor mental health.
Economic competition is inferior to economic cooperation.

8. Individual A has 100 acres, and B has $1000. If I give A $100 from B and B 10 acres from A, can I claim to have increased total well-being? Suppose they exchange voluntarily. Then what?

9. Why is "More is better" a value judgment? Why is it not a simple statement of fact? Can you think of instances in which more is not better?

10. Do you consider yourself an "autonomous" individual? Can you imagine a situation in which you might be better off if your choices were made for you by someone else than by yourself? How about the choices of a child? An uninformed consumer?

11. If you favored inequality, what sorts of arguments might appeal to you as reasons for endorsing policies designed to *decrease* inequality?

12. Do you think there should be any limit on the amount of money an individual should be allowed to make? If so, why? If not, why not?

13. What do you feel about merit as the basis for reward? Would you pay a young unmarried worker as much as an older worker with a family if their productivities were the same? Why or why not? In either event, what are the values underlying your choice?

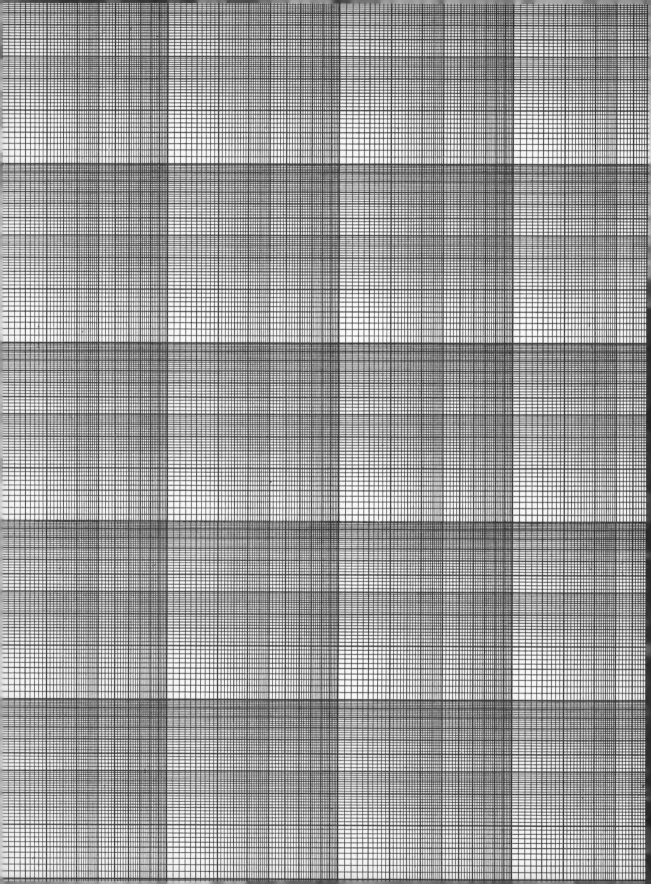

4

PROSPERITY AND RECESSION~ THE ECONOMICS OF THE MACRO SYSTEM

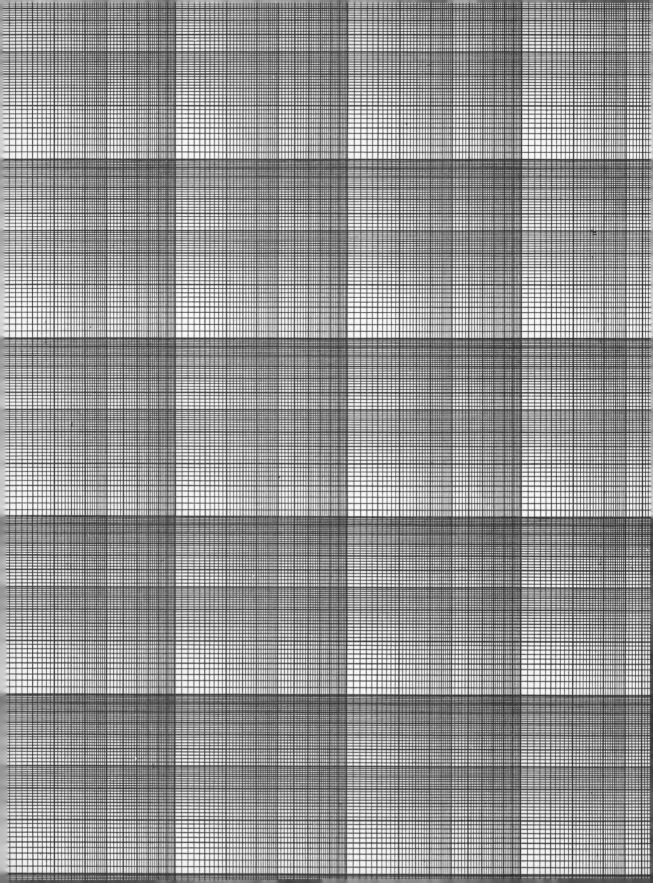

Wealth and Output

19

WHAT IS "MACROECONOMICS?" The word derives from the Greek *macro* meaning "big," and the implication is therefore that it is concerned with bigger problems than in microeconomics (*micro* = small). Yet microeconomics wrestles with problems that are quite as large as those of macroeconomics. The difference is really not one of scale. It is one of approach, of original angle of incidence. *Macroeconomics begins from a viewpoint that initially draws our attention to aggregate economic phenomena and processes,* such as the growth of total output—recall Adam Smith's model. Microeconomics begins from a vantage point that first directs our analysis to the workings of the marketplace, again as we saw in Adam Smith. Both views are needed to comprehend the economy as a whole, just as it takes two different lenses to make a stereophoto jump into the round. But we can learn only one view at a time, and now we turn to the spectacle of the entire national economy as it unfolds to the macroscopic gaze.

What does the economy look like from this perspective? The view is not unlike that from a plane. What we see first is the fundamental tableau of nature—the fields and forests, lakes and seas, with their inherent riches; then the diverse artifacts of man—the cities and towns, the road and rail networks, the factories and machines, the stocks of half-completed or unsold goods; and finally the human actors themselves with all their skills and talents, their energies, their social organization.

Thus our perspective shows us a vast panorama from which we single out for our special attention those elements and activities having to do with our overall economic performance. In fact, once more like Adam Smith, we concentrate on those processes that give rise to the *wealth* of our nation.

National Wealth

What is the wealth of a nation? In Table 19·1 we show the most recent inventory of our national wealth. Note that it consists of the value of those physical objects we noted from our aerial overview—land, buildings, equipment, and the like. Yet, a closer examination of the table reveals some odd facts.

To begin with, it does not include *all* our material goods. Such immense economic treasures as the contents of the Library of

TABLE 19 · 1

U.S. National Physical Wealth 1968 Value

	Billions of dollars, rounded
Structures	
Residential	$ 683
Business	394
Government	460
Equipment	
Producers (machines, factories, etc.)	377
Consumers durables (autos, appliances)	234
Inventories, business	216
Monetary gold and foreign exchange	14
Land	
Farm	152
Residential & business	419
Public	144
Net foreign assets	66
Total	**3,159**

Congress or the Patent Office cannot be accurately valued. Nor can works of art, nor military equipment—not any of them included in the total. Much of our public land is valued at only nominal amounts. Hence at best this is the roughest of estimates of the economic endowment at our disposal.

What is more important, the table omits the most important constituent of our wealth: the value of skills and knowledge in our population. If we estimate the value of those skills for 1968, they come to $5.15 trillion—more than the value of the material equipment with which they work! For reasons that we explain in the following box, "Human Wealth," we do not usually include human wealth along with physical wealth, although we shall return again and again to this all-important element of our economic system. But here we shall familiarize ourselves with the material side of our national balance sheet, leaving the human side for later.

CAPITAL

One portion of the endowment of a nation's physical wealth has a special significance. This is its national *capital*—the portion of its productive wealth that is *man-made* and therefore *reproducible.* If we look back at the table, we can see that our own national capital in 1968 consisted of the sum total of all our structures, our producers' equipment and our consumer durables, our inventories, our monetary gold and foreign assets—$2,744 billion in all.

We can think of this national capital as consisting of whatever has been preserved out of the sum total of everything that has ever been produced from the very beginning of the economic history of the United States up to a certain date—here December 31, 1968. Some of that capital—inventories for example—might be used up the very next day. On the other hand, inventories might also be increased. In fact, our national capital changes from date to date, as we do add to our inventories or to our stocks of equipment or structures, etc., or more rarely, as we consume them and do not replace them. But at any date, our capital still represents *all that the nation has produced*—yesterday or a century ago—*and that it has not used up or destroyed.*

The reason that we identify our national capital within the larger frame of our wealth is that it is constantly changing and usually growing. Not that a nation's inheritance of natural resources is unimportant; indeed, the ability of a people to build capital depends to no small degree on the bounties or obstacles offered by its geography and geology—think of the economic limitations imposed by desert and ice on the Bushman and the Eskimo. But the point in singling out our

HUMAN WEALTH

Why do we not include the value of human skills in our inventory of wealth? The reason is that all our inventory consists of *property* that can be sold; that is, marketable goods. When our economy included slaves, they were part of our wealth; but in today's market system, men are not property. They can sell their labor but not themselves. See box, p. 18.

Ideally, our inventory of wealth should therefore include the "asset value" of that labor, or the human capital that gives rise to the various tasks, skilled and unskilled, that men perform. How could we estimate that value? The method is much the same as that used to estimate the value of a machine. If a lathe produces a flow of output worth, say, $1,000 a year, we can "capitalize" the value of that flow of output to arrive at the current value of the machine itself. On page 237 we have explained how this process of capitalization works.

In the same way we can capitalize the flow of output of humans. The value of human output is measured by the *incomes* that the factor of production labor earns. In 1968 that stream of income was worth $515 billion. If we capitalize it at 10 percent—a rough and ready figure that is comparable to the rate at which we might capitalize many assets—the value of our human capital was therefore $5.15 trillion.

capital is that it represents the portion of our total national endowment over which we have the most immediate control. As we shall later see, much of a nation's current economic fortune is intimately related to the rate at which it is adding to its capital wealth.

WEALTH AND CLAIMS

There remains to be noted one more thing before we leave the subject of wealth. Our table of national wealth omits two items that would be the very first to be counted in an inventory of personal wealth: bank accounts and financial assets such as stocks or bonds or deeds or mortgages. Why are these all-important items of personal wealth excluded from our summary of national wealth?

The answer to this seeming paradox is that we have already counted the *things*—houses, factories, machines, etc.,—that constitute the real assets behind stocks, bonds, deeds, and the like. Indeed these certificates tell us only who *owns* the various items of our national capital. Stocks and bonds and mortgages and deeds are *claims* on assets, but they are not those assets in themselves. The reality of General Motors is its physical plant and its going organization, not the shares of stock that organization has issued. If by some curious mischance all its shares disintegrated, General Motors would still be there; but if the plants and the organization disintegrated instead, the shares would not magically constitute for us another enterprise.

So, too, with our bank accounts. The dollars we spend or hold in our accounts are part of our personal wealth only insofar as they command goods or services. The value of coin or currency as "objects" is much less than their official and legal value as money. But most of the goods over which our money exerts its claims (although not, it must be admitted, the services it also buys) are already on our balance sheet. To count our money as part of national wealth would thus be to count a claim as if it were an asset, much as in the case of stocks and bonds.

Why, then, do we have an item for monetary gold in our table of national wealth? The answer is that foreigners will accept gold in exchange for their own real assets (whereas they are not bound to accept our dollar bills) and that, therefore, monetary gold gives us a claim against *foreign* wealth.* In much the same way, the item of *net foreign assets* represents the value of all real assets, such as factories, located abroad and owned by U.S. citizens, less the value of any real wealth located in the United States and owned by foreigners.

REAL WEALTH VS. FINANCIAL WEALTH

Thus we reach a very important final conclusion. *National wealth is not quite the same thing as the sum of personal wealth.* When we add up our individual wealth, we include first of all our holdings of money or stocks or bonds—all items that are excluded from our national register of wealth. The difference is that as individuals we properly consider our own wealth to be the *claims* we have against one another, whereas as a society we consider our wealth to be the stock of material *assets* we possess, and the only claims we consider are those that we may have against other societies.

National wealth is thus a *real* phenomenon, the tangible consequence of past activity. Financial wealth, on the other hand —the form in which individuals hold their wealth—is only the way the claims of ownership are established vis-à-vis the underlying real assets of the community. The contrast between the underlying, slow-changing reality of national wealth and the overlying, sometimes fast-changing financial representation of that wealth is one of the differences between economic life viewed from the vantage point of the economist and that same life seen through the eyes of a participant in the process. We shall encounter many more such contrasts as our study proceeds.

*Gold has, of course, a value in itself—we can use it for jewelry and for dentistry. However, in the balance sheet of our national wealth, we value the gold at its formal international exchange price, rather than merely as a commodity.

The Flow of Production

WEALTH AND OUTPUT

But why is national wealth so important? Exactly what is the connection between the wealth of nations and the well-being of their citizens?

The question is not an idle one, for the connection between wealth and well-being is not a matter of direct physical cause and effect. For example, India has the largest inventory of livestock in the world, but its contribution to Indian living standards is far less than that of our livestock wealth. Or again, our national capital of goods (or skills) in 1933 was not significantly different from that in 1929, but one year was marked by widespread misery and the other by booming prosperity. Clearly then, the existence of great physical wealth by itself does not guarantee—it only holds out the possibility of—a high standard of living. It is only insofar as physical wealth interacts with the working population that it exerts its enormous economic leverage, and this interaction is not a mechanical phenomenon that we can take for granted, but a complex *social* process, whose motivations we must explore.

As the example of Indian livestock indicates, local customs and beliefs can effectively sterilize the potential physical benefits of wealth. Perhaps we should generalize that conclusion by observing that the political and social system will have a primary role in causing an effective or ineffective use of existing wealth. Compare the traditional hoarding of gold or gems in many backward societies with the possibility of their disposal to produce foreign exchange for the purchase of machinery.

In a modern industrial society, we take for granted some kind of effective social and political structure. Then why do we at times make vigorous use of our existing material assets and at other times seem to put them to little or no use? Why do we have "good times" and "bad times"? The question directs our attention back to the panorama of society to discover something further about its economic operation.

INPUTS AND OUTPUTS

This time, our attention fastens on a different aspect of the tableau. Rather than noticing our stock of wealth, we observe the result of our use of that wealth, a result we can see emerging in the form of a *flow of production.*

How does this flow of production arise? We can see that it comes into being as man combines his energies and his skills with his natural and man-made environment. We have already briefly described the long and painful history of his social and technical attempts to combine those energies and the environment successfully.

Hence we take for granted the fact that men organize their struggle with nature according to the rules of a market process whose operation was the main subject matter of Part Three. Now we want to know what happens to the flow of output emerging under our eyes from thousands of enterprises as their entrepreneurs hire the factors of production and combine their services as inputs to yield a saleable output.

It may help us picture the flow as a whole if we imagine that each and every good and service that is produced—each loaf of bread, each nut and bolt, each doctor's call, each theatrical performance, each car, ship, lathe, or bolt of cloth—can be identified in the way that a radioactive isotope allows us to follow the circulation of certain kinds of cells through the body. Then if we look down on the economic panorama, we can see the continuous combination of land, labor, and capital giving off a continuous flow of "lights" as goods and services emerge in their saleable form.

INTERMEDIATE GOODS

Where do these lights go? Many, as we can see, are soon extinguished. The goods or services they represent are *intermediate goods* that have been incorporated into other products to form more fully finished items of output. Thus from our aerial perspective we can follow a product such as cotton from the fields to the spinning mill, where its light is extinguished, for there the cotton disappears into a new product: yarn. In turn, the light of the yarn traces a path as it leaves the spinning mill by way of sale to the textile mill, there to be doused as the yarn disappears into a new good: cloth. And again, the cloth leaving the textile mill lights a way to the factory where it will become part of an article of clothing.

FINAL GOODS: CONSUMPTION

And what of the clothing? Here at last we have what the economist calls a *final* good. Why "final"? Because once in the possession of its ultimate owner, the clothing passes out of the active economic flow. As a good in the hands of a consumer, it is no longer an object on the marketplace. Its light is now extinguished permanently—or if we wish to complete our image, we can imagine it fading gradually as the clothing "disappears" into the utility of the consumer. In the case of consumer goods like food or of consumer services like recreation, the light goes out faster, for these items are "consumed" when they reach their final destination.*

We shall have a good deal to learn in later chapters about the macroeconomic behavior of consumers. What we should notice in this first view is the supreme importance of this flow of production into consumers'

---

*In fact, of course, they are not *really* consumed but remain behind as garbage, junk, wastes, and so on. Economics used to ignore these residuals, but it does so no longer. Recall our discussion in Chapter 15.

hands. By this vital process, the population replenishes or increases its energies and ministers to its wants and needs; it is a process that, if halted for very long, would cause a society to perish. That is why we speak of consumption as the ultimate end and aim of all economic activity.

A SECOND FINAL GOOD: INVESTMENT

Nevertheless, for all the importance of consumption, if we look down on the illuminated flow of output we see a surprising thing. Whereas the greater portion of the final goods and services of the economy is bought by the human agents of production for their consumption, we also find that a lesser but still considerable flow of final products is not. What happens to it?

If we follow an appropriate good, we may find out. Let us watch the destination of the steel that leaves a Pittsburgh mill. Some of it, like our cotton cloth, will become incorporated into consumers' goods, ending up as cans, automobiles, or household articles of various kinds. But some steel will not find its way to a consumer at all; instead, it will end up as part of a machine or an office building or a railroad track.

Now in a way, these goods are not "final," for they are used to produce still further goods or services—the machine producing output of some kind, the building producing office space, the rail track producing transportation. Yet there is a difference between such goods, used for production, and consumer goods, like clothing. The difference is that the machine, the office building, and the track are goods that are used by business enterprises as part of their productive equipment. In terms of our image, these goods slowly lose their light-giving powers as their services pass into flows of production, but usually they are replaced with new goods before their light is totally extinguished. That is why we call them *cap-*

ital goods or *investment goods* in distinction to consumers' goods. *As part of our capital, they will be preserved, maintained, and renewed, perhaps indefinitely. Hence the stock of capital, like consumers, constitutes a final destination for output.**

GROSS AND NET INVESTMENT

We call the great stream of output that goes to capital *gross investment.* The very word *gross* suggests that it conceals a finer breakdown; and looking more closely, we can see that the flow of output going to capital does indeed serve two distinct purposes. Part of it is used to replace the capital—the machines, the buildings, the track, or whatever—that has been used up in the process of production. Just as the human agents of production have to be replenished by a flow of consumption goods, so the material agents of production need to be maintained and renewed if their contribution to output is to remain undiminished. We call the part of gross investment, whose purpose is to keep society's stock of capital intact, *replacement investment,* or simply *replacement.*

Sometimes the total flow of output going to capital is not large enough to maintain the existing stock; for instance, if we allow inventories (a form of capital) to become depleted, or if we simply fail to replace worn-out equipment or plant. This running-down of capital, we call *disinvestment,* meaning the very opposite of investment: instead of maintaining or building up capital, we are literally consuming it.

Not all gross investment is used for replacement purposes, however. Some of the flow may *increase* the stock of capital by

*We might note that some products, like automobiles, possess characteristics of both consumption goods and capital goods. We call such goods *consumer durables;* and unlike ordinary goods (such as food) held by consumers, we include them in our inventory of national wealth (see Table 19·1).

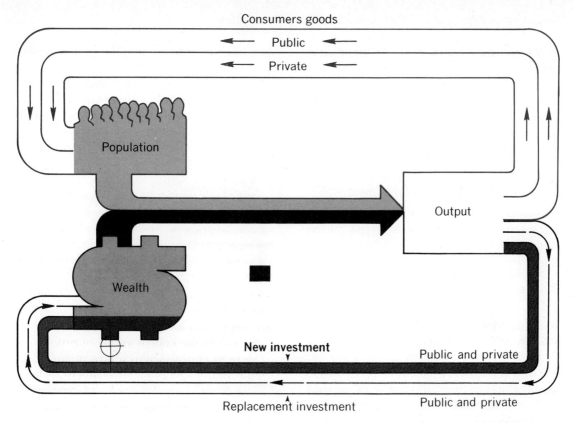

Consumers goods

Public

Private

Population

Output

Wealth

New investment

Public and private

Replacement investment

Public and private

FIGURE 19 · 1
The circular flow, view I

adding buildings, machines, track, inventory, and so on.* If the total output consigned to capital is sufficiently great not only to make up for wear and tear, but to increase the capital stock, we say there has been *new* or *net investment,* or *net capital formation.*

Sometimes it helps to have a homely picture in mind to keep things straight. The difference between replacement and net investment is made clear by using as an example the paving of streets. Each year some streets wear out, and we have to repave them to keep them passable. This is clearly *investment,* but it does not add to our ability to enjoy surface transportation. Hence it is solely *replacement* investment. Only when we

build *additional* streets do we undertake new or net investment. Also, sometimes we build a new street but allow an old one to deteriorate beyond usability. Here we have to offset the new investment with the *disinvestment* in our now unpassable road. Whether or not we have any net investment in street building as a whole depends on whether we have added or subtracted from street capacity when we consider gross investment and the disinvestment together.

CONSUMPTION AND INVESTMENT

A simple diagram may help us picture the flows of output we have been discussing. Figure 19·1 calls to our attention these important attributes of the economic system:

*Note carefully that *increased* inventory is a form of investment. Later this will take on a special importance.

1. It emphasizes the essential circularity, the self-renewing, self-feeding nature of the production flow.

This circularity is a feature of the macroeconomic process to which we will return again and again.

2. It illumines a basic choice that all economic societies must make: a choice between consumption and investment.

At any given level of output, consumption and investment are, so to speak, rivals for the current output of society.

3. It makes clear that society can invest (that is, add to its capital) only the output that it refrains from consuming.

The economic meaning of *saving,* as our diagram shows, is to release resources from consumption so that they can be used for the building of capital. Whether they *will* be so used is a matter that will occupy us through many subsequent chapters.

4. It shows that both the consumption flow and the investment flow can be split between public uses and private uses in any manner that the nation sees fit.

But the only way to increase public consumption and investment is to refrain from private consumption and investment.

5. Finally, it reveals that output is the nation's "budget constraint."

It indicates the total quantity of goods and services available for all public and private consumption and investment uses. More goods and services might be desired, but they do not exist.

Gross National Product

There remains but one preliminary matter before we proceed to a closer examination of the actual determinants of the flow of production. We have seen that the annual output of the nation is a revealing measure of its well-being, for it reflects the degree of interaction between the population and its wealth. Later we shall also find output to be a major determinant of employment. Hence it behooves us to examine the nature and general character of this flow and to become familiar with its nomenclature and composition.

We call the dollar value of the total annual output of final goods and services in the nation its gross national product. The gross national product (or GNP as it is usually abbreviated) is thus nothing but the dollar value of the total output of all consumption goods and of all investment goods produced in a year. We are already familiar with this general meaning, but now we must define GNP a little more precisely.

FINAL GOODS

We are interested, through the concept of GNP, in measuring the value of the *ultimate* production of the economic system—that is, the total value of all goods and services *enjoyed by its consumers or accumulated as new or replacement capital.* Hence we do not count the intermediate goods we have already noted in our economic panorama. To go back to an earlier example, we do not add up the value of the cotton *and* the yarn *and* the cloth *and* the final clothing when we compute the value of GNP. That kind of multiple counting might be very useful if we wanted certain information about our total economic activity, but it would not tell us accurately about the final value of output. For when we buy a shirt, the price we pay includes the cost of the cloth to the shirtmaker; and in turn, the amount the shirtmaker paid for his cloth included the cost of the yarn; and in turn, again, the seller of yarn included in his price the amount he paid for raw cotton. Embodied in the price of the shirt, therefore, is the value of all the intermediate products that went into it. Thus in figuring the value for GNP, we add only the values of all final goods, both for consumption and for investment purposes. Note as well that GNP includes only a given year's production of goods and services. Therefore sales of used car dealers, antique dealers, etc., are not included, because the value of these goods was picked up in GNP the year they were produced.

TYPES OF FINAL GOODS

In our first view of macroeconomic activity we divided the flow of output into two great streams: consumption and gross investment. Now, for purposes of a closer analysis, we impose a few refinements on this basic scheme.

First we must pay heed to a small flow of production that has previously escaped our notice. This is the net flow of goods or services that leaves this country; that is, the total flow going abroad minus the flow that enters. This international branch of our economy will play a relatively minor role in our analysis for quite a while; we will largely ignore it until Chapter 26, then again until Part Five. But we must give it its proper name: *net exports*. Because these net exports are a kind of investment (they are goods we produce but do not consume), we must now rename the great bulk of investment that remains in this country. We will henceforth call it *gross private domestic investment.*

By convention, gross private domestic investment refers only to investments in physical assets such as factories, inventories, homes. Personal expenditures on acquiring human skills, as well as expenditures for regular use, are considered *personal consumption expenditures*—the technical accounting term for *consumption*. As these accounting terms indicate, *public* consumption and in-

vestment are included in neither personal consumption expenditures nor gross private domestic investment. Here is our last flow of final output: all public buying of final goods and services is kept in a separate category called *government purchases of goods and services.*

C + I + G + X

We now have four streams of "final" output, each going to a final purchaser of economic output. Therefore we can speak of gross national product as being the sum of personal consumption expenditure (C), gross private domestic investment (I), government purchases (G), and net exports (X), or (to abbreviate a long sentence) we can write that

$$GNP \equiv C + I + G + X$$

This is a descriptive identity that should be remembered.

It helps, at this juncture, to look at GNP over the past decades. In Fig. 19·2 we show the long irregular upward flow of GNP from 1929 to the present, with the four component streams of expenditures visible. Later we will be talking at length about the behavior of each stream, but first we need to be introduced to the overall flow itself.

STOCKS AND FLOWS

One final point should be made about our basic identity. All through our discussion of GNP we have talked about *flows* of output. We do so to distinguish GNP, a "flow concept," from wealth or capital (or any asset) that is a *stock,* or a sum of wealth that exists at any given time.

A moment's reflection may make the distinction clear. When we speak of a stock of business capital or of land or structures, we mean a sum of wealth that we could actually

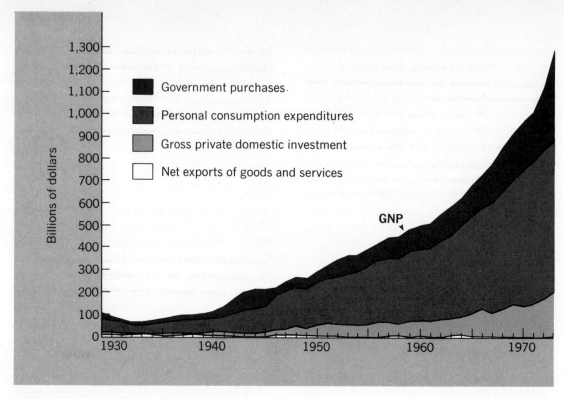

FIGURE 19·2
GNP and components, 1929–1973

inspect on a given date. GNP, however, does not "exist" in quite the same way. If our gross national product for a year is, say $1.5 trillion, this does not mean that on any day of that year we could actually discover this much value of goods and services. *Rather, GNP tells us the average annual rate, for that year, at which production was carried out; so that if the year's flow of output had been collected in a huge reservoir without being consumed, at the end of the year the volume in the reservoir would indeed have totaled $1.5 trillion.* GNP is, however, constantly being consumed as well as produced. Hence the $1.5 trillion figure refers to the value of the *flow of production over the year* and should not be pictured as constituting a given sum of output existing at any moment in time.

Cautions about GNP

GNP is an indispensable concept in dealing with the performance of our economy, but it is well to understand the weaknesses as well as the strengths of this most important single economic indicator.

1. GNP deals in dollar values, not in physical units.

That is, it does not tell us how many goods and services were produced, but only what their sales value was. Trouble then arises when we compare the GNP of one year with that of another, to determine whether or not the nation is better off. For if prices in the second year are higher, GNP will appear higher, even though the actual volume of output is unchanged or even lower!

We can correct for this price change very easily when all prices have moved in the same degree or proportion. Then it is easy to speak of "real" GNP—that is, the current money value of GNP adjusted for price changes—as reflecting the actual rise or fall of output. The price problem becomes more difficult, however, when prices change in different degrees or even in different directions, as they often do. Then a comparison of "real" GNP from one year to the next, and especially over a long span of years, is unavoidably arbitrary to some extent.

Figure 19·3 shows us the previous totals

for GNP corrected as best we can for price changes. In this chart, 1963 is used as the "base," and the GNP's of other years use 1963 prices, so that price changes are eliminated to the greatest possible extent. One can, of course, choose any year for a base. Choosing a different year would alter the basic dollar measuring rod, but it would only slightly change the profile of "real" year-to-year changes.

2. Changes in the quality of output may not be accurately reflected in GNP.

The second weakness of GNP also involves its inaccuracy as an indicator of "real" trends over time. The difficulty revolves

FIGURE 19 · 3
GNP in constant and current prices, 1929–1973

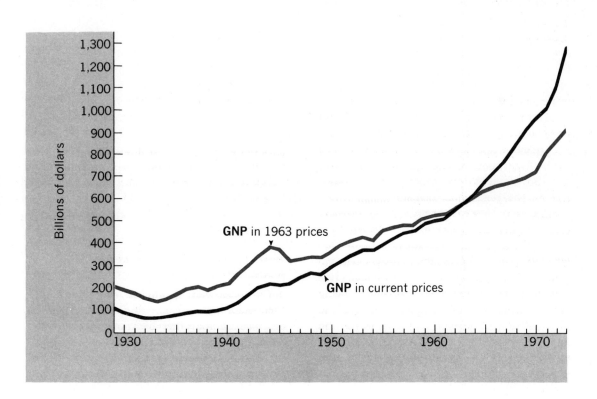

REAL AND CURRENT GNP

How do we arrive at a figure for "real" GNP? *The answer is that we "correct" the value of GNP (or any other magnitude measured in dollars) for the price changes that affect the value of our dollars but not the real quantitites of goods and services our dollars buy.*

We make this correction by applying a *price index*. Such an index is a series of numbers showing the variation in prices, year to year, from a starting or *base year* whose price level is set at 100. Thus if prices go up 5 percent a year, a price index starting in year one will read 105 for year two, 110+ for year three (105 × 1.05), 115.5 for year four, and so on.

In correcting GNP we use a very complex price index called a GNP *price deflator*. This index, constructed by the Department of Commerce, allows for the fact that different parts of GNP, such as consumers goods and investment goods may change in price at different rates. The present price deflator uses GNP price levels in 1958 as a "base." In 1973, the value of the deflator was 153.9.

Now let us work out an actual example. *To arrive at a corrected GNP, we divide the current GNP by the deflator and then multiply by 100.* For example, GNP in current figures was $977 billion for 1970; $1,055 billion for 1971; and $1,155 billion for 1972. The deflator for those years was 135, 142, and 146 respectively. Here are the results:

around the changes in the utility of goods and services. In a technologically advancing society, goods are usually improved from one decade to the next, or even more rapidly, and new goods are constantly being introduced. In an urbanizing, increasingly high-density society, the utility of other goods may be lessened over time. An airplane trip today, for example, is certainly highly preferable to one taken twenty or thirty years ago; a subway ride is not. Television sets did not even exist 40 years ago.

Government statisticians attempt to correct for changes in the quality of goods and services. Committees composed of government statisticians and industry representatives meet to decide on the extent to which price increases represent quality improvements. It is very difficult to determine whether these committees over- or underadjust for quality improvements. In the 1950s these committees counted the cost of putting "fins" on cars as a quality improvement rather than as a price increase. The fins did not affect the performance of the car, but they were thought to improve its beauty.

Completely new goods, such as the picture phones that have been demonstrated in some parts of the country, present an even more difficult problem. Clearly, a picture phone is not an ordinary telephone. Yet, how much of a quality improvement is it? Since there is no satisfactory answer to this question, picture phones will be valued at the price they are sold in any given year. If picture phones fall in price as they are introduced into the mass market, an evaluation of GNP in 1975 prices will give a much higher "weight" to picture phones than an evaluation of GNP in 1990 prices. This is a prime reason why base years and deflating formulas are periodically reconsidered.

3. GNP does not reflect the purpose of production.

A third difficulty with GNP lies in its blindness to the ultimate use of production. If in one year GNP rises by a billion dollars, owing to an increase in expenditure on education, and in another year it rises by the same amount because of a rise in cigarette production, the figures in each case show the same

$$\frac{\$977}{135} = \$7.23 \times 100 = \$723 \text{ billion}$$

$$\frac{\$1055}{142} = \$7.44 \times 100 = \$744 \text{ billion}$$

$$\frac{\$1155}{146} = \$7.93 \times 100 = \$793 \text{ billion}$$

Thus the "real value" of GNP in 1972 was $793 billion, *in terms of 1958 prices*, rather than the $1,115 billion of its current value. Two things should be noted in this process of correction. First, the "real value" of any series will differ, depending on the base year that is chosen. For instance, if we started a series in 1972, the "real value" of GNP for that year would be $1,115, the same as its money value.

Second, the process of constructing a GNP deflator is enormously difficult. In fact there is no single "accurate" way of constructing an index that will reflect all the variations of prices of the goods *within* GNP. To put it differently, we can construct different kinds of indexes, with different "weights" for different sectors, and these will give us differing results. The point then is to be cautious in using corrected figures. Be sure you know what the base year is. And remember that complex indexes, such as the GNP deflator, are only approximations of a change that defies wholly accurate measurement.*

*For a fuller discussion of price indexes and related problems, you might look into Part Seven, p. 677ff.

amount of "growth" of GNP. Even output that turns out to be wide of the mark or totally wasteful—such as the famous Edsel car that no one wanted or military weapons that are obsolete from the moment they appear—all are counted as part of GNP.

The problem of environmental deterioration adds another difficulty. Environmentalists concerned about the adverse impact of growth on our quality of life sometimes advocate Zero Economic Growth as a solution to our ecological ills. Unfortunately, the environmental problem is much more complicated than a "go" or "no go" decision on GNP. Some types of GNP growth directly contribute to pollution—cars, paper or steel production, for example. Other types of GNP growth are necessary to stop pollution—sewage disposal plants or the production of a clean internal combustion engine. Still other types of GNP have little impact on the environment; most personal services fall into this category.

The real problem therefore involves selecting the types of economic growth that are compatible with environmental safety. The sheer measure of GNP tells us nothing with respect to such a purpose. For example, our conventional measure of GNP makes no allowances for the harmful goods and services that are often generated by production. All forms of pollution and congestion are essentially *negative outputs*. They diminish individual pleasure or utility; they should be *subtracted* from GNP. Yet under our accounting procedures, they are included in GNP! So too, we fail to factor out of GNP those expenditures taken to repair the damage caused by other elements in the total. For instance, the cleaning bills we pay to undo damage caused by smoke from the neighborhood factory become part of GNP, although cleaning our clothes does not increase our well-being; it only brings it back to what it was in the first place.

These costs of cleaning up the harmful effects of economic growth are just one of a large number of *defensive expenditures* in GNP. Defensive expenditures are designed to prevent bad things from happening or to offset the impacts of adverse circumstances, rather than to cause good things to happen.

IMPUTED INCOMES

Imputed rents are calculated by determining the rent that an individual would have to pay for his own home if he did not own it. This hypothetical rent is added into the GNP to maintain consistency in the treatment of the nation's housing stock. The produce grown in personal vegetable gardens is handled in a similar fashion.

Given the precedent of imputed rents, why haven't government statisticians imputed the value of the housewifes services into the GNP? Statisticians can find market wages for cleaning women, babysitters, maids, cooks, and other servants. But is a wife or mother simply a combined cook, maid, and babysitter—a servant? While imputed values could be placed on these services, most wives and mothers would object strenuously (so would most husbands and children). As a result, government statisticians, and the politicians who hire them, do not want to get involved in placing a value on someone else's mother or wife. Only research economists are willing to be so presumptuous.

Imputing the value of housewives services also opens a Pandora's Box of problems with respect to other self-produced consumption goods. Men provide services in their homes as carpenters, bartenders, and do-it-yourselfers. Should imputed values also be placed on these services? Theoretically the answer is yes, but as with the housewife, the practical problems of assigning an actual value serves as a deterrent.

In addition to environmental expenditures, other major examples include military and police expenditures, flood control, repair bills, many medical outlays. These outlays are not desired in their own right; they are simply forced on us by man-made circumstances.

4. GNP does not include most goods and services that are not for sale.

Presumably GNP tells us how large our final output is. Yet it does not include one of the most important kinds of work and sources of consumer pleasure—the labor of wives in maintaining their households. Yet, curiously, if this labor were paid for—that is, if we engaged cooks and maids and baby sitters instead of depending on our wives for these services, GNP *would* include their services as final output, since they would be purchased on the market. But the labor of wives being unpaid, it is excluded from GNP.

The difficulty here is that we are constantly moving toward purchasing "outside" services in place of home services—laundries, bakeries, restaurants, etc., all perform work that used to be performed at home. Thus the process of *monetizing* activity gives an upward trend to GNP statistics that is not fully mirrored in actual output.

A related problem is that some parts of GNP are paid for by some members of the population and not by others. Rent, for example, measures the services of landlords for homeowners and is therefore included in GNP, but what of the man who owns his own home and pays no rent? Similarly, what of the family that grows some part of its food at home and therefore does not pay for it? In order to include such items of "free" consumption into GNP, the statisticians of the Commerce Department add an "imputed" value figure to include goods and services like these not tallied on a cash register.

5. GNP does not consider the value of leisure.

Leisure time is not only enjoyable in its own right, but it is also *necessary* in order to consume material goods and services. A boat without the time to use it is of little consumption value. Over the years, individual enjoyment and national well-being go up because each individual has more leisure time to spend as he pleases, *but leisure time does not show up in GNP.*

Leisure time has not been integrated into

measured GNP, since economists have not managed to find either a good technique for measuring its extent or for placing a value upon it. What should be subtracted from the 24-hour day to indicate hours of leisure? Presumably, leisure hours are those hours that give pleasure or utility. All hours that create pain or disutility should be subtracted.

Then what about work? Does it create utility or disutility? Does it create pleasure or pain? In economic theory, work is considered to be pain and should be subtracted. Yet this is clearly often incorrect. Many people enjoy their work. Even those who basically dislike their jobs often enjoy *parts* of work. Are those parts leisure or work?

And then, what about the maintenance activities—eating, sleeping, washing, etc.—that are necessary to keep the human body alive? Are they pleasure or pain? They are necessary to both consumption and work, yet they are neither pure leisure nor painful work. If you were asked to divide your own day into the hours that give pleasure (utility) and the hours that give you pain (disutility), how would you divide your day? Most of us would find many ambiguous hours that we could not really categorize one way or another. What you cannot do for yourself, economists cannot do for you.

As a result, people tend to focus on hours of paid labor on the assumption that leisure must be going up if hours of paid work are going down. While this assumption is probably correct, subtracting hours of paid work from the total quantity of available hours is obviously a very crude measure of the quality of leisure. For example, if people move to the suburbs and spend two hours a day commuting, their leisure has decreased, although their paid hours of labor have not.

6. The GNP does not indicate anything about the distribution of goods and services among the population.

Societies can and do differ in how they allocate their production of purchasable goods and services among their populations. A pure egalitarian society might allocate everyone the same quantity of goods and services. Many societies establish minimum consumption standards for individuals and families. Few deliberately decide to let someone starve if they have the economic resources to prevent such a possibility. Yet to know a nation's GNP, or even to know its average (per capita) GNP, is to know nothing about how broadly or how narrowly this output is shared. A wealthy country can be composed mainly of poor families. A poor country can have many wealthy families.*

GNP AND ECONOMIC WELFARE

These problems lead economists to treat GNP in a skeptical and gingerly manner, particularly insofar as its "welfare" considerations are concerned. Kenneth Boulding has suggested that we relabel the monster *Gross National Cost* to disabuse ourselves once and for all of the notion that a bigger GNP is necessarily a better one. Paul Samuelson suggests a new measure—Net Economic Welfare or NEW—to supplement GNP, the difference being mostly the maintenance or defensive or negative outputs we have mentioned. Economists Tobin and Nordhaus propose MEW—Measure of Economic Welfare—for much the same purposes, subtracting the outputs that contribute nothing to the sum of individuals' utilities and adding back other sums, mainly housewives' services, that are conventionally omitted. (We might note in passing that MEW, as calculated by Tobin and Nordhaus, grows much less rapidly per capita than does GNP. From 1929 to 1965, real per capita GNP mounted at 1.7 percent per year; MEW, at 1.1 percent.)

All these doubts and reservations should instill in us a permanent caution against using

---

*See the section on averages and bimodal distributions, Part Seven, for more on this important statistical problem.

GNP as if it were a clearcut measure of social contentment or happiness. Economist Edward Denison once remarked that perhaps nothing affects national economic welfare so much as the weather, which does not get into the GNP accounts. Hence, because the U.S. has a GNP per capita that is higher than that of say, the Netherlands, life is not consequently better here; it may be worse. In fact, by the indices of longevity, health, quality of environment, or ease of retirement, it probably *is* worse.

Yet, with all its shortcomings, *GNP is still the simplest way we possess of summarizing the overall level of market activity of the economy.* If we want to summarize its welfare, we had better turn to specific social indicators of how long we live, how healthy we are, how cheaply we provide good medical care, how varied and abundant is our diet, etc.—none of which we can tell from GNP figures alone. But we are not always interested in welfare, partly because it is too complex to be summed up in a single measure. For better or worse, therefore, GNP has become the yardstick used by most nations in the world; and although other yardsticks are sure to become more important, GNP will be a central term in the economic lexicon for a long time to come.

KEY WORDS

Macroeconomics

Wealth

Capital
Claims

Production
Consumption

Replacement
investment

Net and gross
investment

Gross national
product
  Consumption
  Gross domestic
  private investment
  Government
  purchases
  Net exports

CENTRAL CONCEPTS

1. **Macroeconomics is an approach to economic problems** through the study of certain aggregate processes.

2. **We begin the study of macroeconomic processes by observing how a flow of output comes from human resources interacting with physical resources.**

3. **We note that capital consists of real things,** and **not of the financial claims** against those things.

4. **The flow of output shows us the circular nature of the process of production.** From the interaction of population and wealth emerges a stream of goods going back to replenish consumers (consumption goods) and a stream to replenish and add to our capital wealth (gross investment).

5. **A study of the flow of output and its division into consumption and investment emphasizes the essential choice that must be made between these** two basic uses of output.

6. **The investment flow can be subdivided into two: one flow** replaces or renews capital that has been worn out or used up. This is replacement investment. The other flow **adds to the stock of capital wealth,** and is called new or net capital or investment. The two flows together are called gross investment.

7. **The name for the total flow of output is** gross national product. It is divided into four categories:

  ● *Consumption (C),* or the goods and services going to consumers.
  ● *Gross private domestic investment (I),* or that portion of output going to private businesses as replacements for, and additions to, their real domestic capital. It also includes housing.
  ● *Government purchases (G),* or those goods and services (both consumers' and capital) bought by all public agencies.
  ● *Net exports (X),* the net outflow of goods and services to other countries.
  ● **The formula GNP $\equiv$ C + I + G + X conveniently summarizes these four subdivisions.**

$GNP \equiv C + I$
$+ G + X$

8. Note that GNP counts only the dollar value of *final goods and services* in each of these categories. Intermediate goods or services are not counted, since their value is included in the value of final goods.

9. GNP is an indispensable concept in macroeconomics. It is a measure of the purchasable goods and services that the economy is producing. *It is not a measure of economic welfare.* To make judgments about economic welfare remember that:

* GNP deals in dollar values and not physical units. This leads to problems in adjusting for price changes, so that "real" GNP accurately reflects changes in output.

Pitfalls of GNP
  Final goods
  Prices
  Quality
  Use
  Defensive output
  Output not for sale
  Leisure
  Distribution

* GNP may not accurately reflect quality changes.
* GNP does not reveal the purpose or usefulness of production.
* GNP accounts include defensive expenditures that are made not to create utility but to prevent disutility, and "negative" outputs are not subtracted from it.
* GNP figures do not include output that is not for sale.
* GNP does not place a value upon leisure time.
* GNP does not reveal the distribution of output.

QUESTIONS

1. Why is capital so important a part of national wealth? Why is money not considered capital?

2. What is meant by the "circularity" of the economic process? Does it have something to do with the output of the system being returned to it as fresh inputs?

3. What is meant by net investment? by gross investment? What is the difference?

4. Write the basic identity for GNP and state carefully the exact names of each of the four constituents of GNP.

5. Suppose we had an island economy with an output of 100 tons of grain, each ton selling for $90. If grain is the only product sold, what is the value of GNP? Now suppose that production stays the same but that prices rise to $110. What is the value of GNP now? How could we "correct" for the price rise? If we didn't, would GNP be an accurate measure of output from one year to the next?

6. Now suppose that production rose to 110 tons but that prices fell to $81. The value of GNP, in terms of current prices, has fallen from $9,000 to $8,910. Yet, actual output, measured in tons of grain, has increased. Can you devise a price index that will show the change in real GNP?

7. Presumably, the quality of most products improves over time. If their price is unchanged, does that mean that GNP understates or overstates the real value of output?

8. When more and more consumers buy "do-it-yourself" kits, does the value of GNP (which includes the sale price of these kits) understate or overstate the true final output of the nation?

9. What is a public consumption good? A public investment expenditure? How would you classify the following: Blue Cross Insurance; airport flight controllers; museum fees? Are these public or private? Consumption or investment? It's not always easy to draw the line, is it?

10. What is an intermediate good, and why are such goods not included in the value of GNP? Is coal sold to a utility company an intermediate good? Coal sold to a consumer? Coal sold to the army? What determines whether a good will or will not be counted in the total of GNP?

11. A bachelor pays a cook $100 a week. Is this part of GNP? He then marries her and gives her an allowance of $100 a week. Allowances do not count in GNP. Hence the *measure* of GNP falls. Does welfare fall?

12. Do you think that we should develop measures other than GNP to indicate changes in our basic well-being? What sorts of measures?

The Growth of Output

MACROECONOMICS, WE SAID IN CHAPTER 2, when we were examining Adam Smith's *Wealth of Nations* is essentially about growth. At the center of its focus is the question of how an economy expands its output of goods and services (or if it fails to expand them, why growth does not take place). In our last chapter, we began to analyze this process by familiarizing ourselves with the way our stock of wealth interacts with our labor force to yield a flow of output that we call gross national product. In this chapter we are going to push forward by learning about the underlying trends and causes of growth in the American economy. That will set the stage for the work that still lies ahead, when we will narrow our focus down to the present and inquire into the reasons for the problems of our macrosystem—unemployment and inflation, booms and busts.

THE PRODUCTION POSSIBILITY CURVE

How much output can an economy produce, utilizing all its factors? If the economy produced only a single good, such as wheat, the answer would be some number of bushels

that we would discover by utilizing to the hilt every available acre, every tractor, every farmer's labor.

But obviously, economies produce many kinds of goods. Thus we cannot answer the question in terms of a single figure, but in terms of a range of possibilities, depending on which goods we produce. It would be difficult to represent this range of possibilities in a simple graph, so we abstract the range of possible outputs to two goods, say grain and milk, and we then show what *combinations* of outputs are possible, using all factors. As we learned in Chapter 3, we call such a schedule of alternative possibilities a *production possibility curve,* and we draw such a curve in Fig. 20·1.

THE EFFICIENCY FRONTIER

The production-possibility curve shows us a number of things. First it makes vivid the material meaning of the word *scarcity*. Any point outside the frontier of the curve is unattainable for our community, given its present resources. This is obviously true of point *X*. But look at point *Y*. This is an output that

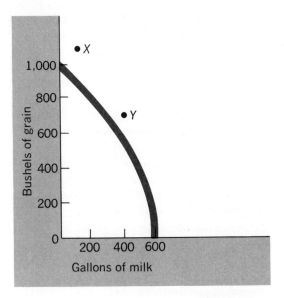

FIGURE 20 · 1
Production possibility curve

filled spinnaker sail. Any place on the sail represents some combination of consumption, investment, and government spending that is within the reach of the community. Any place "behind" the *efficiency frontier* represents a failure of the economy to employ all its resources. It is a graphic depiction of unemployment of men or materials.

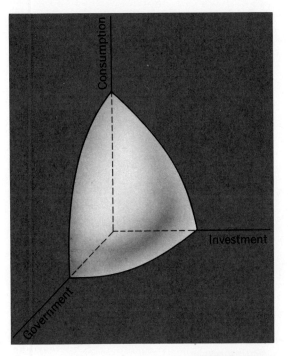

FIGURE 20 · 2
A production possibility surface

represents roughly 700 bushels of grain and 400 gallons of milk. Either one of these goals, taken separately, lies well within the production possibilities of the economy. What the curve shows us is that *we cannot have both at the same time.* If we want 700 bushels of grain, we must be content with less than 400 gallons of milk: and if we want 400 gallons of milk, we will have to settle for about 600 bushels of grain.

Such a two-commodity diagram may seem unreal. But remember that "milk" and "grain" can stand for consumption and investment (or any other choices available to an economy). In fact, with a little imagination we can construct a three-dimensional production-possibility *surface* that will show us the limits imposed by scarcity on a society that divides its output among three uses: such as consumption, investment, and government. Figure 20·2 shows what such a diagram looks like.

Note how the production-possibility surface swells out from the origin like a wind-

THE LAW OF INCREASING COST

One last point deserves clarification before we move on. The alert student may have noticed that all the production-possibility curves have bowed shapes. The reason for this lies in the *changing efficiency* of our resources as we shift them from one use to another. This is a phenomenon we first met in Chapter 3 and further explained in Chapter 11 (see box, p. 143).

We call this changing efficiency, represented by the bowed curve, the *law of increasing costs*. Note that it is a law imposed by nature, rather than behavior. For what would it mean if the curve connecting the two points of all-out grain or milk production were a straight line as in Fig. 20·3? It would mean that as we shifted resources from one use to the other, we would always get exactly the same results: the last man and the last acre put into milk would give us exactly as much milk at the loss of exactly as much grain, as the first man and the first acre.

Such a straight-line production possibility curve is said to exhibit *constant returns to specialization*. Except perhaps in a very simple economy, where a population might choose between hunting or fishing, constant returns to specialization is an unrealistic assumption, for it implies that there is no difference from one man, or acre, to another, or that it made no difference as to the *proportions* in which factors, even if they were homogenous, were combined. That is a very unrealistic assumption. Men and land (and any other resource) *are* different and different products *do* utilize them in different proportions. Hence, as we shift them from one use to another, assuming that we always choose the resources best suited for the job, society's efficiency changes. At first we enjoy a very low cost in terms of what we must give up for what we get; thereafter, we pay an increasing cost. Although the shapes of production-possibility curves may have considerably different contours, the unevenness of nature's gifts make most of them bowed, or concave from below.

FIGURE 20 · 3
Constant returns to specialization

FIGURE 20 · 4
Shifts in the production frontier

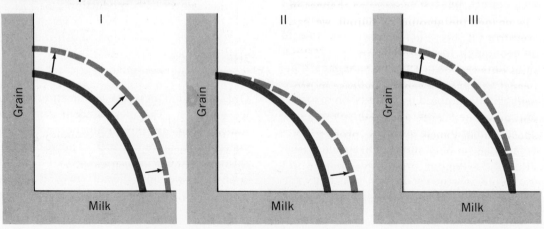

BEYOND THE FRONTIER

Our acquaintance with production possibility curves and their efficiency frontiers brings home the importance of economic growth. Consider point A in the diagram (right). It shows us an economy that is allowing attainable output to go to waste because it is not properly using the productive powers it possesses. Such an economy may expand output, but this is not really "growth" in the proper sense.

But now consider the shaded area outside the frontier. This represents combinations of goods that we cannot enjoy because we do not have the requisite productive capacity. We can enter this area only if we grow, thereby pushing our frontier outward.

This is by no means an exercise in mere imagination. For instance in 1960, a rather conservative Commission on National Goals established by President Eisenhower laid down a set of "goals," ranging from enhanced private consumption to a variety of improved public programs, to be achieved in some 15 areas of economic activity by 1975. A few years ago the National Planning Association "costed out" these goals to show how many of them were within reach. Its conclusion was that if output continued

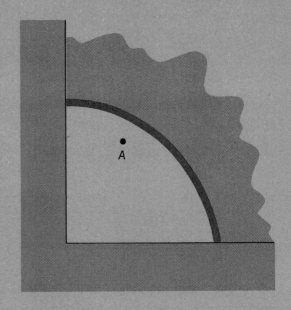

to grow at 4 percent a year (14 percent *faster* than the long-run 3.5 percent rate), our GNP in 1975 would fall short by $150 billion of achieving the aspirations of the Eisenhower Commission. Our production possibility curve was simply not moving out fast enough to place the Commission's targets within our efficiency frontier.

TECHNICAL PROGRESS

One last point about the shapes of production possibility curves. They are not static. As we know, technical progress or changes in skills change the amount of output we can derive from a given quantity of input. These changes move the production possibility curves outward, as in Fig. 20·4. Notice that in Panel 1 we enjoy a general increase in productivity that enables us to have more of both grain and milk (or any other two goods); but in Panels 2 and 3, productivity has increased in only one area of production, so that the maximum output available to us for one good stays the same, whereas the maximum for the other good increases.

Structural Requirements for Growth

An acquaintance with production possibility curves has given us a sense of the importance of growth. But it does not explain how an economy actually grows, how the production possibility frontiers move out. That is the question to which we will devote the remainder of this chapter.

We already know part of the answer from our study of GNP. An economy grows because it allocates a portion of its output to investment (public as well as private), rather than consuming the entire output it produces. But how does a society actually find the resources to devote to capital-building? In a rich, industrialized economy such as ours, the answer seems simple. We use our industrial equipment to make more industrial equipment. To put it differently, we use the people working in the steel and other capital-goods industries to make more steel and more capital goods.

Building Capital

But the matter-of-factness of our answer hides the real structural problem of growth. For when we look abroad or backward in time and ask how a poor society grows, the question is not so easy to answer. Such societies do not have steel mills waiting for orders, nor labor forces that are already deployed in capital-goods industries. How do they, then, create capital?

The process is by no means an obvious one. Suppose we have a very poor society (like an extremely underdeveloped nation) in which 80 percent of the population tills the soil, equipped with so little by way of capital —mere spades and hoes—that it produces only enough to maintain itself and the remaining 20 percent of society.

Who are the other 20 percent? In reality, of course, they might be government officials, landlords, and others, but we will simplify our model by assuming that the whole 20 percent is occupied in making the simple spades and hoes (the capital goods) used by the consumption-goods sector. Like the farmers, the toolmakers labor from dawn to dusk; and again like the farmers, they are so unproductive that they can produce only enough capital to replace the spades and hoes that wear out each year.

TWIN TASKS OF THE CAPITAL SECTOR

Now how could such a society grow? If we look again at the capital-goods sector, we

find a clue. For unlike the consumption-goods sector, we find here not one but two distinguishable kinds of economic activity going on. On the agricultural side, everyone is farming; but on the capital-goods side, not everyone is making the spades and hoes with which the agricultural laborers work. No matter how we simplify our model, we can see that *the capital-goods sector must carry out two different tasks*. It must turn out spades and hoes, to be sure. But part of the capital-goods labor force must also turn out a different kind of capital good—a very special kind that will produce not only spades and hoes, but also more of itself!

Is there such a kind of equipment? There is indeed, in a versatile group of implements known as *machine tools*. In our model economy, these may be only chisels and hammers that can be used to make spades, hoes, and more chisels and hammers. In a complex industrial system, machine tools consist of presses and borers and lathes that, when used ensemble, not only make all kinds of complicated machines but can also recreate themselves.

THE CAPITAL SECTOR

Thus we encounter the unexpected fact that there is a strategic branch of capital creation at the core of the whole sequence of economic growth.*

How does growth now ensue? Our model enables us to see that it is not simply a matter of bringing in peasants from the fields to make more spades and thereby to

---

*This raises the perplexing question of how the machine-tool industry *began*, since it needs it own output to grow. The answer is that it evolved as a special branch of industrial production during the industrial revolution when, for the first time, machinery itself began to be made by machinery instead of by hand. A key figure in the evolution of the machine-tool industry was Maudslay, whose invention of the screw-cutting lathe was "one of the decisive pieces of standardization that made the modern machine possible." (Lewis Mumford, *Technics and Civilization*. New York: Harcourt, 1963, p. 209.)

increase their productivity, for they will not be able to make spades until there is an increased output of spade-making tools. Before spades can be made, chisels and hammers must be made. Before textile or shoemaking or food-processing or transportation equipment can be made, machine tools must be made.

Thus at the core of the growth process—whether in a very backward nation or a highly industrialized one—we can see *two* great structural shifts that must take place:

1. Within the capital sector to increase its own productive capacity

2. From the consumption sector to the capital sector, to man the growing volume of equipment emerging from the enlarged capital sector. *

Can we actually trace this process in real life? The shifts *within* the capital sector are not always easy to see, because there is usually some excess capacity in the machine tool branch. By running overtime, for instance, it can produce *both* more machine tools *and* more hoes and spades. Yet if we examine a society in the process of rapid industrialization, such as the U.S.S.R. (or for that matter, the United States in its periods of rapid wartime industrial buildup), we can clearly see the importance of this critical branch in setting a *ceiling* on the overall pace of industrial expansion.

When we turn to the second shift, from the consumption sector into the capital sector, the movement in real life is very apparent. Table 20·1, for instance, shows us the proportion of the population engaged in agriculture for a number of industrialized nations at an early and a late stage of their transformations.

*How will these new factory workers be fed? Obviously, food must be diverted from the country to the city. In Chapter 39, when we discuss underdevelopment, we will examine how this can be done.

TABLE 20 · 1
Labor Force in Agriculture

PERCENT OF LABOR FORCE IN AGRICULTURE		
	Early 19th century	1971
France	63% (1827)	13
Great Britain	31 (1811)	3
Sweden	63 (1840)	8
United States	72 (1820)	4

Sources: Colin Clark, *The Conditions of Economic Progress* (London: Macmillan, 1960), pp. 512, 514, 518; B. R. Mitchell, *Abstract of British Historical Statistics* (Cambridge: Cambridge University Press, 1962), pp. 60–61, and *Basic Statistics of the Community.*

Here we see in reality the internal emigration that takes place in the industrializing process (note that by 1811 Britain was already well on the road). Over the course of the nineteenth and twentieth centuries, these countries have lost two-thirds to four-fifths of their erstwhile farmers—not all to capital-building alone, of course, but to the whole industrial and commercial structure that capital-building makes possible. In this way, the process of economic growth can be seen in part as a great flow of human and material resources from simple consumption goods output to a hierarchy of industrial tasks — a flow that is even more dramatic in real life than we might have divined from our imaginary model.

THE HISTORICAL RECORD

Thus, behind the phenomenon of growth—of production possibility frontiers moving outward—we encounter a hidden structural shift of the greatest importance. Once the shift has taken place, however—once a large capital-building sector has been established—the process of adding to the stock of capital is greatly simplified. Indeed, we now get that long process of gradually increasing output that provides us with our ordinary starting point in the study of growth. In Fig. 20·5 we see the American experience from the middle of the nineteenth century, in terms of real per capita GNP in 1929 prices.

How regular has been our average rate of growth? The answer: astonishingly constant, whether we take an average over the past thirty-odd years since the Great Depression, or whether we go back to the earliest reliable statistics and calculate our growth rate since the 1870s (or even 1830s). As the chart shows, the swings are almost all contained within a range of 10 percent above or below the trend. The trend itself comes to about 3.5 percent a year in real terms, or a little over 1.5 percent a year per capita.

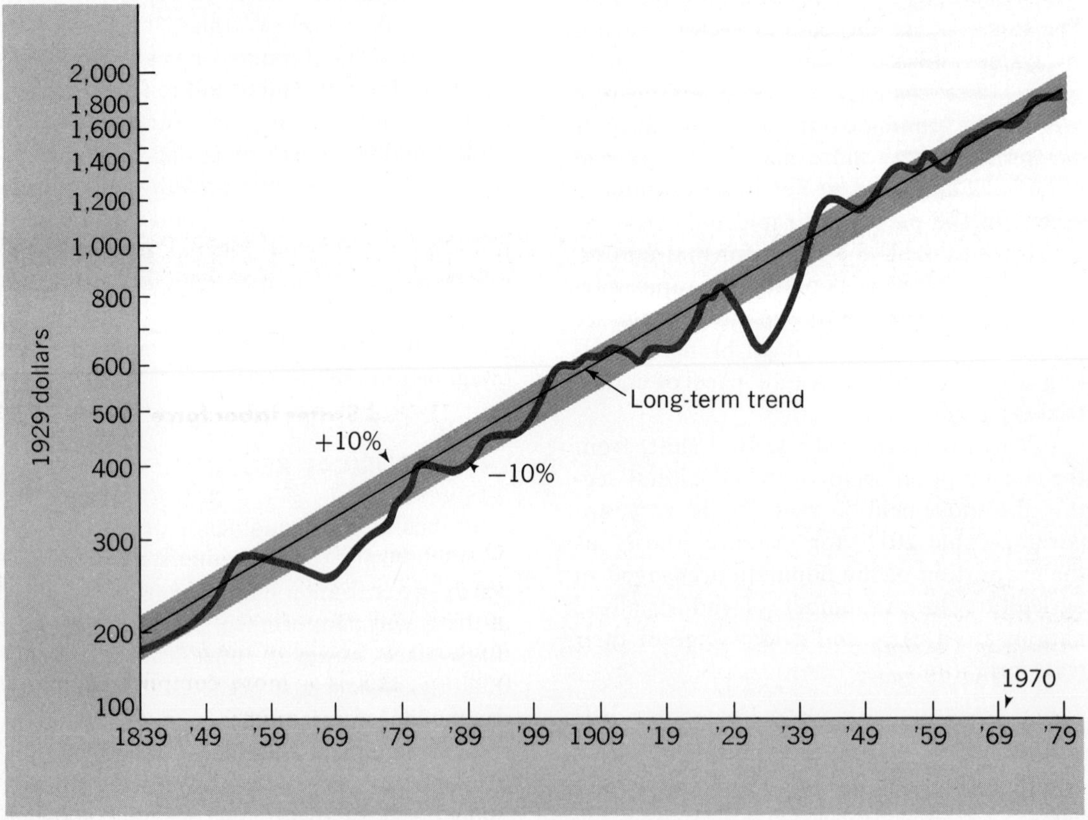

The Sources of Growth

How do we account for this long steady ascent? The answer takes us to a deeper consideration of the *sources of growth* in terms of the contribution of labor and capital (including land). Clearly, our growth reflects the fact that labor and capital have cooperated to bring about a rising stream of final goods and services. But this rising output could be the result of either or both of two quite separate trends:

1. The quantity of labor and capital that we use may be growing; that is, the expansion of our production frontier may simply reflect a growth in the sheer volume of "inputs."

FIGURE 20 · 5
Trend in real GNP per head, 1839–1973

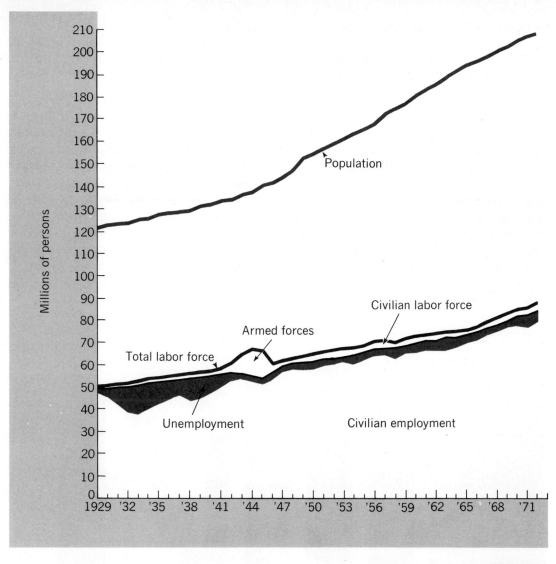

FIGURE 20·6
United States labor force, 1929–1973

2. *Labor and capital may be increasing in productivity.* **If each working individual or each "unit" of land or capital can produce more goods and services over time, output will rise, even if the total number of working individuals or the value of the stock of capital remains unchanged.**

Hence the next question on the agenda is to learn about the long-term trends in the quantity of labor and capital and their respective productivities.

EMPLOYMENT AND PRODUCTION

Output depends on work, and work depends on people working. Thus, the first source of growth that we study is the rise in the sheer numbers of people in the *labor force*. As we will see, this is a more complicated matter than might at first appear.

Figure 20·6 gives us a picture of the population and the labor force over the past almost half-century. As we would expect, the size of the force has been rising because our population has been rising. Yet this trend is more puzzling than might at first appear. For one might expect that as our society grew richer and more affluent, fewer people would seek employment. But that is not the case. Looking back to 1890 or 1900, we find that only 52 out of every 100 persons over 14 sought paid work. Today about 60 out of every 100 persons of working age seek employment. Looking forward is more uncertain; but if we can extrapolate (extend) the trend of the past several decades to the year 2000, we can expect perhaps as many as 65 persons out of 100 to be in the labor market by that date.

PARTICIPATION IN THE LABOR FORCE

How can we explain this curious upward drift of the labor force itself? The answer is to be found in the *different labor participation trends* of different ages and different sexes. Figure 20·7 shows this very clearly.

Note that the overall trend toward a larger participation rate for the entire population masks a number of significant trends.

FIGURE 20 · 7
Participation rates

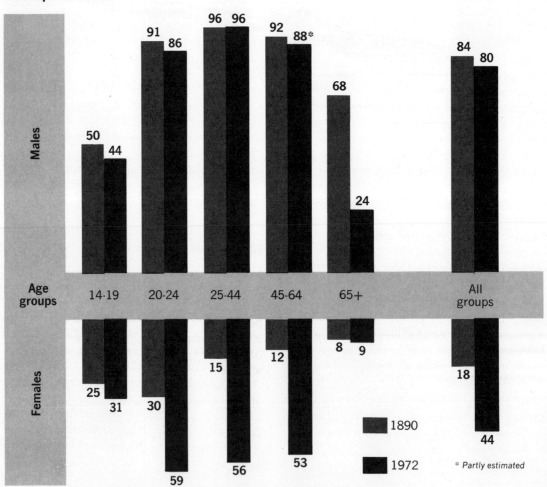

1. *Young males entering the labor force are older than those who entered in the past*.

A larger number of young men remain in high school now or go on to college. Only a third of elementary school pupils now go on to college, but the ratio is steadily growing.

2. *Older males show a dramatic withdrawal from the labor force*.

The reason is the advent of Social Security and private pension plans. It is probable that the proportion of older males in the labor force will continue to fall as the retirement age is slowly reduced.

3. *Counterbalancing this fall in male participation is a spectacular rise in total female participation. Indeed, the overall trend toward an increasing search for work within the population at large is entirely the result of the mass entrance of women into the labor force*.

This surge of women into the labor market reflects several changing factors in the American scene (many of these changes can be found abroad, as well). One factor is the growth of nonmanual, as contrasted with manual, jobs. Another is the widening cultural approval of working women and working wives—it is the amazing fact that the average American girl who marries today in her early twenties and goes on to raise a family will nevertheless spend *twenty-five years* of her life in paid employment after her children are grown. Yet another reason for the influx of women is that technology has released them from household work. And finally there is the pressure to raise living standards by having two incomes within the household.

4. *Actually, we must view the growing number of females in the labor force as part of a very old economic phenomenon whose roots we traced back to the Middle Ages—the monetization of work*.

The upward trend of female participation does not imply an increasing amount of labor performed within society. Rather, it measures a larger amount of *paid* labor. In the 1890s, many persons worked long and hard hours on a family farm or in a family enterprise, and above all within a household, *without getting paid* and, therefore, were not counted as members of the "labor force." To a very considerable extent, the rising numbers of female participants in the labor force mirror the transfer of these unpaid jobs onto the marketplace where the same labor is now performed in an economically visible way. There is every likelihood this process will continue.

These are not, of course, the only factors that bear on the fundamental question of how many persons will seek work out of a given population. The drift from country to city, the decline in the number of hours of labor per day expected of a jobholder, the general lengthening of life, the growth of general well-being—all these changes bear on the decision to work or not. *Overall, what the complex trends seem to show is that we are moving in the direction of a society where employment absorbs a larger fraction of the life (but not of the day) of an average woman, and a diminishing fraction of the life and of the day of an average man*.

HOURS OF WORK

In addition to deciding whether to participate in the labor force, individuals decide how much labor they wish to contribute as members of the labor force. That is, they must decide how many hours of work they wish to offer during a week or how many weeks they wish to work in a year.

Had we asked this question in the days of Adam Smith, it would have been relatively simple to answer. Wages were so close to subsistence that someone in the labor force

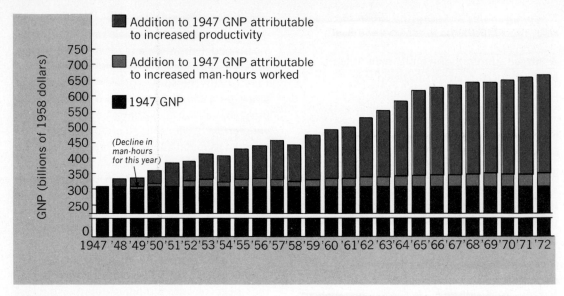

FIGURE 20 · 8
Source of GNP increases, 1947–1972

was obliged to work extremely long hours to keep body and soul together. Paid vacations were unknown to the employees of the cotton mills; unpaid vacations would have been tantamount to starvation.

But with the slow rise in productivity, working men and women gradually found their income rising above "subsistence," and a new possibility came into being: the possibility of deliberately working less than the physical maximum, *using part of their increased productivity to buy leisure for themselves instead of wages.* Thus, beginning in the early nineteenth century we find that labor organizations (still very small and weak) sought to shorten the workweek. In England, in 1847 a signal victory was won with the introduction of the Ten (!) Hour Day as the legal maximum for women and children. In America, in the prosperity of the 1920s, the 48-hour week finally became standard. More recently, the two-day weekend has become the general practice; and

now we hear talk of the coming of the three-day weekend.*

Thus the total supply of labor-time has not risen as fast as the labor force, because a decline in average hours has offset the rise in participation rates and population. On balance, the total supply of labor-hours has increased, but the supply of labor-hours *per employee,* male and female, has fallen.

LABOR PRODUCTIVITY

As we have seen, we can trace part of our long-term growth to increases in the total supply of man-hours of production. But this is by no means the most important source of growth. Far outpacing the growth in the sheer volume of labor-time has been the increase in the amounts of goods and services that each hour of labor-time gives rise to.

*Recall our discussion of leisure on p. 231. The trend to shorter hours may now be at an end: we do not yet know.

MEASURING PRODUCTIVITY

As the accompanying table shows, the average increase in productivity of 3.5 percent masks wide swings from year to year. Compare 1950, when productivity per man grew at 9.2 percent, with 1956, when it actually declined by 0.1!

But there is a caution here. These sharp ups and downs do not so much reflect real variations in output per manhour as they do the way in which we measure productivity. *Productivity is measured by dividing total output by total man-hours.* When recessions occur and output falls, businesses reduce their labor forces as much as possible, but they find that there are considerable numbers of overhead laborers who cannot profitably be let go simply because output is down. If General Motors' production falls by 25 percent, it does not reduce the working time of its president by 25 percent.

Hence, in recession years, a smaller output is divided by a number of man-hours that has been "kept high." The underlying normal growth in productivity of the labor force may still be occurring, but it is masked by the overhead labor that is not reduced as much as output. In booming years, just the opposite occurs. Output increases faster than employment, since the company does not need to add overhead as rapidly as output. Result: *year-to-year productivity figures must be interpreted with great care.*

	Productivity index (G.N.P./man-hour)	Per cent change in productivity per man-hour in the private economy
1947	100	
1948	103.4	3.4%
1949	105.9	2.4
1950	115.7	9.2
1951	121.0	4.6
1952	124.5	2.9
1953	130.2	4.6
1954	133.7	2.7
1955	139.5	4.3
1956	139.5	−0.1
1957	143.3	2.7
1958	146.6	2.3
1959	152.5	4.0
1960	154.2	1.1
1961	158.4	2.7
1962	166.8	5.3
1963	172.3	3.3
1964	178.7	3.7
1965	184.6	3.3
1966	193.1	4.6
1967	197.0	2.0
1968	204.9	4.0
1969	207.1	1.1
1970	206.8	1.0
1971	214.6	4.1
1972	222.2	3.8
1973	227.9	2.9

Technology Review, March, 1971.

Economists measure the productivity of the labor force by dividing the total output of goods by the total number of man-hours. Figure 20·8 shows us the wide margin by which changes in labor productivity outweigh changes in labor-time as a source of increased output.

Over the postwar period, the *average* increase in productivity per man-hour has been growing at about 3½ percent a year. At that rate, productivity per man-hour doubles in just under 20 years. Of course, this increase varies from one sector to another; over the last two decades it increased by 80 percent in manufacturing and *tripled* in agriculture.

Sources of Labor Productivity

What is the explanation for this tremendous and persistent increase in the ability of labor to turn out goods? Here are the most important answers.

1. Growth of human capital

By human capital, as we know, we mean the skills and knowledge possessed by the labor force. The measurement of "human capital" is fraught with difficulties, but the difficulty of measuring it does not permit us to ignore this vital contributory element in labor productivity. Ferenc Jánossy, a Hungarian economist, has suggested a vivid imaginary experiment to highlight the importance of skills and knowledge.

Suppose, he says, that the populations of two nations of the same size could be swapped overnight, so that 50 million Englishmen would awake to find themselves in, say, Nepal, and 50 million Nepalese would find themselves in England. The newly transferred Englishmen would have to contend with all the poverty and difficulties of the Nepalese economy; the newly transferred Nepalese would confront the riches of England. Yet the Englishmen would bring with them an immense reservoir of literacy, skills, discipline, and training, whereas the Nepalese would bring with them the very low levels of "human capital" that are characteristic of underdeveloped countries. Is there any doubt, asks Jánossy, that growth rates in Nepal with its new skilled population would in all likelihood rise dramatically, and that those of England would probably fall catastrophically?

One way of indicating in very general terms the rising "amount" of human capital is to trace the additions to the stock of education that the population embodies. Table 20·2 shows the change in the total number of years of schooling of the U.S. population over the past three quarters of a century, as well as the rise in formal education per capita. While these measures of human capital are far from exact or all-inclusive, they give some dimensions to the importance of skills and knowledge in increasing productivity.

2. Shifts in the occupations of the labor force

A second source of added productivity results from shifts in employment from low productivity areas to high productivity areas. If workers move from occupations in which their productivity is low relative to other occupations in which output per manhour is high, the production possibility curve of the economy will move out, even if there are no increases in productivity *within* the different sectors.

A glance at Table 20·3 shows that very profound and pervasive shifts in the location of labor have taken place. What have been the effects of this shift on our long-term ability to produce goods?

The answer is complex. In the early years of the twentieth century, the shift of

TABLE 20 · 2
Stock of Education, U.S.

	1900	1971
Total man-years of schooling embodied in population (million)	228	991
Percent of labor force with high-school education or more	6.4%	66.9%
Percent of high-school graduates entering college	17%	40%

	1900	1972
Agriculture, forests, and fisheries	38.1	4.2
Manufacturing, mining, transportation, construction, utilities	37.7	36.3
Trade, government, finance, professional and personal services*	24.2	59.5

TABLE 20 · 3
*Percent
Distribution of all
Employed
Workers*

Source: Calculated from *Historical Statistics*, p. 74; also from *Statistical Abstract*.

*It is customary to include transportation and utilities among the third, or service, area of activities. In this analysis, however, we group them with goods-producing or goods-handling activities, to highlight the drift into "purely" service occupations. Since domestic servants, proprietors, and the self-employed are omitted (owing to inadequate statistics), the table under-represents the labor force in the service and trade sector.

labor out of agriculture into manufacturing and services probably increased the overall productivity of the economy, since manufacturing was then the most technologically advanced sector. In more recent years, however, we would have to arrive at a different conclusion. Agriculture is now a highly productive but very small sector, in terms of employment. Moreover, the proportion of the labor force employed in manufacturing is roughly constant, up or down only a few percentage points, year to year, from its long-term level of 35 to 40 percent of all workers.

Today, growth in employment takes place mainly in the congeries of occupations we call the service sector: government, retail and wholesale trade, professions such as lawyers, accountants, and the like. The growth of output per capita is least evident in these occupations.* *Thus the drift of labor into the service sector means that average GNP per worker is growing more slowly today than if labor were moving into manufacturing or agriculture.*

Why is this growth-lowering shift taking

*It is only proper to note that we cannot measure productivity of output in the service sector nearly so unambiguously as in the goods sector, and there is no doubt that the *quality* of many services has increased substantially. Compare, for example, the "productivity" of a surgeon operating for appendicitis in 1900, 1930, and 1960. On the other hand, insofar as we are interested in increases of measurable output per capita, there seems little doubt of the considerable superiority of the goods-producing branches of the economy.

place? The reason has to do with the changing pattern of demand in an affluent society. There seems to be a natural sequence of wants as a society grows richer: first for food and basic clothing, then for the output of a wide range of industrial goods, then for recreation, professional advice, public administration, and enjoyments of other services. We will meet these trends again and study their implications when we examine the problems of automation and inflation.

3. Economies of large-scale production

A third source of increasing labor productivity is the magnifying effect of mass production on output. As we have seen, when the organization of production reaches a certain critical size, especially in manufacturing, economies of scale become possible. Many of these are based on the possibility of dividing complex operations into a series of simpler ones, each performed at high speed by a worker aided by specially designed equipment. It is difficult to estimate the degree of growth attributable to these economies of size. Certainly during the era of railroad-building and of the introduction of mass production, they contributed heavily to growth rate. In a careful study of the contemporary sources of U.S. growth, Edward F. Denison estimates that economies of large-scale production today are responsible for about one-tenth of our annual rate of productivity increase.

MASS PRODUCTION IN ACTION

Allan Nevins has described what mass production techniques looked like in the early Ford assembly lines.

Just how were the main assembly lines and lines of component production and supply kept in harmony? For the chassis alone, from 1,000 to 4,000 pieces of each component had to be furnished each day at just the right point and right minute; a single failure, and the whole mechanism would come to a jarring standstill.... Superintendents had to know every hour just how many components were being produced and how many were in stock. Whenever danger of shortage appeared, the shortage chaser—a familiar figure in all automobile factories—flung himself into the breach. Counters and checkers reported to him. Verifying in person any ominous news, he mobilized the foreman concerned to repair deficiencies. Three times a day he made typed reports in manifold to the factory clearing-house, at the same time chalking on blackboards in the clearing-house office a statement of results in each factory-production department and each assembling department.[1]

Such systematizing in itself resulted in astonishing increases in productivity. With each operation analyzed and subdivided into its simplest components, with a steady stream of work passing before stationary men, with a relentless but manageable pace of work, the total time required to assemble a car dropped astonishingly. Within a single year, the time required to assemble a motor fell from 600 minutes to 226 minutes; to build a chassis, from 12 hours and 28 minutes to 1 hour and 33 minutes. A stopwatch man was told to observe a 3-minute assembly in which men assembled rods and pistons, a simple operation. The job was divided into three jobs, and half the men turned out the same output as before.

[1] *Ford, the Times, the Man, the Company* (New York: Scribner's, 1954), I, 507.

4. Increases in the quantity and quality of capital

A fourth basic reason for the rising productivity of labor again harks back to Adam Smith's growth model. It is the fact that each additional member of the labor force has been equipped with at least as much capital as previous members; and that all members of the labor force have worked with a steadily more productive stock of capital.

We call the first kind of capital growth a *widening* of capital. It consists of matching additional workers with the same amounts and kinds of equipment that their predecessors had. The streams of additional part-time women workers coming into offices and stores, for example, would not be able to match the productivity of those who preceded them if they did not also get typewriters, cash registers, or similar equipment.

But we must also notice a *deepening* of capital as a source of increased labor productivity. This means that each worker receives *more* capital equipment over time. The ditch digger becomes the operator of a power shovel; the pencil-and-paper accountant uses a computer.

Over the long course of economic growth, increased productivity has required the slow accumulation of very large capital stocks per working individual. Thus investment that increases capital per worker is, and will probably continue to be, one of the most effective levers for steadily raising output per worker. But unlike the steady widening of capital, the deepening of capital is not a regular process. Between 1929 and 1947 there was no additional capital added per worker! This was, of course, a time of severe depression and thereafter of enforced wartime stringencies. Since 1947, the value of our stock of capital per worker has been growing at about 2.7 percent a year. As we shall see immediately following, however, the *size* of this additional stock of capital is of less crucial importance than the *productivity* of that capital—that is, its technological character.

5. *Technology*

We have just mentioned the fifth and last main source of increases in productivity—technology. During the past half-century, GNP has consistently grown faster than can be accounted for by increases in the work force or the size of the capital stock. (Even during the 1929–1947 era, for instance, when capital stock per worker remained fixed, the output of GNP per worker grew by 1.5 percent a year.) Part of this "unexplained" increase can be attributed to some of the sources of growth we have itemized above—mainly education and training and economies of scale. But contemporary economic investigation increasingly attributes the bulk of the "bonus" rate of growth to the impact of new technology.

The term is, admittedly, somewhat vague. By "new technology" we mean new inventions, innovations of a productivity-enhancing kind, the growth of knowledge in the form of research and development, changes in business organization or in techniques of management, and many other activities. What is increasingly apparent, however, is that the search for new products and processes is the main force behind much productivity-enhancing investment. Thus while investment has become less important for growth simply as a means of adding sheer quantities of capital to the labor force (although that is still a very important function,

particularly in construction), it remains the strategic variable as the carrier of technological change.

CHANGING SOURCES OF GROWTH

It is time to sum up what we have covered. We have seen that long-term growth proceeds from two sources: *more* input and more *productive* input, and we have been concerned with studying some of the main facets of both kinds of growth. Perhaps we can summarize our findings in Fig. 20·9, comparing the sources of growth in two eras of our past.

Note the declining importance played by increases in numbers of workers or sheer dollar value of capital, and the increasing importance of the "intangibles" of education and technology. To a number of observers, this shift implies that we have been slowly moving into a new phase of industrial organization in which productivity will more and more reflect the application of scientific knowledge, rather than the brute leverage of mechanical strength and power. Whether this "postindustrial" society will grow at a faster or slower rate than in the past is a question that we will not be able to answer for many years.

WHAT WE DO NOT KNOW ABOUT LONG-RUN GROWTH

Last, we must take cognizance of an important fact. We have learned a good deal about the sources of growth in the United States, but we have not really unlocked the secret of the historical trajectory of that growth. In fact, we can now see that this trajectory was the result of crosscurrents of many kinds. The potential stimulus to growth of a rising participation rate was dampened by a decline in the numbers of hours worked per year. Increases in productivity of labor in manufacturing and agriculture were offset by a

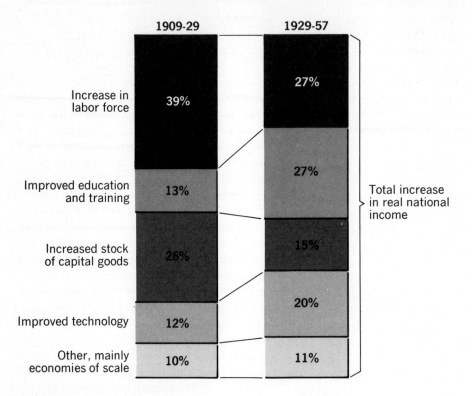

	1909-29	1929-57
Increase in labor force	39%	27%
Improved education and training	13%	27%
Increased stock of capital goods	26%	15%
Improved technology	12%	20%
Other, mainly economies of scale	10%	11%

Total increase in real national income

FIGURE 20·9
Sources of U.S. economic growth

shift of labor into the "low productivity" service sector.

The overall effect of these complex trends is the "steady" rate of 3.5 percent growth evidenced in the United States for many years. But we can now see that this steady rate was really the outcome of many contrary trends. Is there any underlying reason why the growth of GNP maintained such an even pace, or why that pace was 3.5 percent per year?

Not so far as we know. Other nations have different long-run growth rates; and those growth rates are not always as steady as those of the United States, by any means. Furthermore, within the United States, the steadiness of the average rate conceals a great deal of variation in shortrun rates as we shall see in our next chapter. The fact is, then, that we can describe but cannot really explain why our growth has followed the pattern shown in Fig. 20·9. This remains a profound problem for economists and economic historians.

DIFFERENT KINDS OF KNOWLEDGE

In thinking about technology and growth, it helps to differentiate among scientific knowledge, engineering knowledge, and economic knowledge. The relationship can best be understood if we look at the accompanying figure. Here we assume that knowledge can be arranged along a continuum from the least productive technologies to the most productive. On the extreme left are those techniques we have discarded; for example, water mills or treadmills for the production of energy. Next we come to the range of techniques

From an economist's point of view, the level of productivity in an economy depends not only on the location of all these techniques and frontiers, but on the distribution of plants *within* the bell curve. A high-productivity economy will have its curve of plants to the right of a low-productivity economy. Moreover, within that curve, its working equipment will be "bunched" toward the right-hand edge of *best-practice* plants; a low-productivity economy will have the opposite distribution. Incidentally, this is one reason why productivity is very high in industrial nations that have been severely damaged by war but have re-

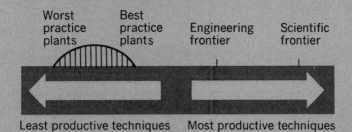

Worst practice plants   Best practice plants   Engineering frontier   Scientific frontier

Least productive techniques   Most productive techniques

in use. Here is a "bell curve" of plants, beginning with those that are still in use but almost obsolete—say, old-fashioned utilities—to the newest plant and equipment, perhaps nuclear power plants. Here we reach the *economic frontier*, the limit of knowledge that can be profitably used.

Still further to the right is another frontier—the limit of *engineering knowledge*. For instance, breeder reactors, still in "pilot plant stage" might be located near this point. Then to the far right is the boundary of *scientific knowledge*—for instance, fusion power—where our theoretical knowledge has not yet passed into the stage of engineering feasibility.

built their capital stock. Their factories will tend to incorporate the very newest and best in techniques, whereas an economy that was spared the damages of war will retain in use many older plants that still manage to show a small profit.

This consequence may help explain why the change in R&D expenditures has not significantly boosted American postwar productivity. *These expenditures have mainly moved out our engineering and scientific frontiers, without altering the distribution of actual techniques in use.* Most R&D in recent decades has been aimed at military and space technology, which has so far had only minor application to civilian use.

KEY WORDS	CENTRAL CONCEPTS

1. Total output possibilities for an economy cannot be represented simply by a given physical maximum. They involve choices that we represent by *production possibility curves* or surfaces.

2. The bowed shape of production-possibility curves reflects the *law of increasing costs*. This basic economic relationship describes the fact that our efficiency changes as we shift resources from one use to another, and that typically we experience a decreasing efficiency (with the consequence of increasing costs) as we move more and more resources into the production of any given item.

3. Production possibility curves also make vivid the idea of an *efficiency frontier*. An economy that has not reached its frontier has unemployed or underutilized factors. *An economy cannot reach any point beyond a frontier unless it grows, thereby moving its frontier outwards.*

4. Growth in output derives from *investment that adds to our stock of capital wealth*. In an industrialized society, this investment is achieved by utilizing an existing capital-goods sector. But in a poor society, the process is much more difficult.

5. In such a poor society, we can see that the *capital-goods sector consists of two subsectors*: one producing capital goods that, in turn, make the equipment used in the consumption-goods sector; and the other making machine tools, capital goods that can create more capital goods.

6. *Thus growth requires two shifts*: (1) the machine-tool capacity of the capital-goods sector must be enlarged and (2) resources must be shifted from the consumption sector to the capital goods sector.

7. United States' growth has been shown a *persistently steady rate* of 3.5 percent—1.5 percent per capita—for well over a century. From *year to year*, however, the *rate of growth is very uneven.*

8. Growth results from two sources: *more inputs* and *more productive inputs* of both labor and capital.

9. *Participation* in the working force has slowly increased over the long run. This is the net result of three forces: the *large-scale entry of women* into the labor force after childbearing, partly offset by the *later entry and earlier retirement of males.*

10. Higher participation rates have also been accompanied and offset by declines in the hours of work per year, per week, per day.

11. *Increases in labor productivity have been much more important than increases in labor hours* in accounting for growth. The main sources of increased productivity have been:
    - Growth of human capital.
    - Shifts in labor occupation (recently these shifts into services have slowed growth).
    - Economies of large-scale production.

    - Increases in the quantity and quality of capital. This involves both the *widening* and *deepening* of capital.
    - Technology.

    Technology and human capital have become increasingly important as sources of growth.

12. These factors present crosscurrents, some enhancing growth, some dampening it. We have no explanation of why the 3.5 average growth rate has been so steady.

QUESTIONS

1. Set up a production possibility curve for an economy producing food and steel. Show how there exists combinations of goods that cannot be produced, *although the quantity of either good alone is within the reach of economy.*

2. Describe carefully the difference between the law of diminishing returns and the law of increasing cost.

3. What kind of economy might display constant returns to specialization? Would a very simple, low-technology economy show such a straight-line efficiency frontier if its choice was, for example, to hunt or fish? Would this depend on the abundance of game or fish?

4. Explain why an economy would not want to operate to the left of its efficiency frontier.

5. What should an economy do if it does not *want* as many goods as it can have by being on its efficiency frontier? (Hint: remember that the efficiency frontier assumes that all factors of production are fully employed.)

6. Describe carefully the shifts in labor needed to begin the process of growth in a very simple economy. Are such shifts visible in an advanced economy?

7. What is the special property of machine tools? Is there a very simple article of output that is also capable of the dual uses of machine tools? How about the seed corn held back by a farmer?

8. How do you account for the fact that there are more people per hundred who want to work today than there were 70 years ago, when the nation was so much poorer? How much does the monetization of labor have to do with this? How much is it a change in life styles, especially for women? What do you expect for the very long run—say 100 years from now?

9. What is extensive investment? What is intensive investment? Which is more conducive to growth?

10. Why is productivity so important in achieving growth? What are its main sources? What would you recommend as a long-term program to raise American productivity? Asian productivity?

From Growth to Fluctuations

WITH SOME UNDERSTANDING of how long-term growth emerges—and with a chastening sense of the limits of our understanding—it is time to change the focus of our lens. When we do, a second problem of macroeconomics emerges, as important as those we have been studying. It is concerned not with the course of long-term growth, but of *short-term fluctuations*—of boom and bust, business cycles, recessions and unemployment, inflation.

Take the years 1895 to 1905, very smooth-looking in Fig. 20·5, in our preced-

TABLE 21·1
U.S. Rates of Growth 1895–1905

1895–1896	− 2.5%	1900–1901	+11.5%
1896–1897	+9.4	1901–1902	+ 1.0
1897–1898	+2.3	1902–1903	+ 4.9
1898–1899	+9.1	1903–1904	− 1.2
1899–1900	+2.7	1904–1905	+ 7.4

Source: *Long Term Economic Growth* (U.S. Dept. of Commerce, 1966), p. 107.

ing chapter. As Table 21·1 reveals, those years were, in fact, anything but steady.

Or examine a more recent period, not year by year, but in groups of years. As we can see in Fig. 21·1 the rate of growth has varied greatly over the last fifty years.

BUSINESS CYCLES

This extraordinary sequence of ups and downs, rushes of growth followed by doldrums, introduces us to a fascinating aspect of the subject of growth—*business cycles*. For if we inspect the profile of the long ascent carefully, we can see that its entire length is marked with irregular tremors or peaks and valleys. Indeed, the more closely we examine year-to-year figures, the more of these tremors and deviations we discover, until the problem becomes one of selection: which vibrations to consider significant and which to discard as uninteresting.

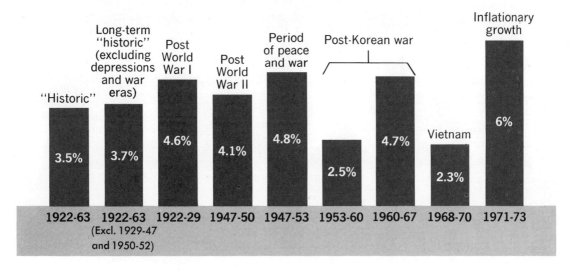

FIGURE 21 · 1
Short-term variations in the rate of growth

The problem of sorting out the important fluctuations in output (or in statistics of prices or employment) is a difficult one. Economists have actually detected dozens of cycles of different lengths and amplitudes, from very short rhythms of expansion and contraction that can be found, for example, in patterns of inventory accumulation and decumulation, to large background pulsations of seventeen or eighteen years in the housing industry, and possibly (the evidence is unclear) swings of forty to fifty years in the path of capitalist development as a whole.

Generally, however, when we speak of "the" business cycle we refer to a wavelike movement that lasts, on the average, about eight to ten years. In Fig. 21 · 2 this major oscillation of the American economy stands forth very clearly, for the chartist has eliminated the underlying tilt of growth, so that the profile of economic performance looks like a cross section at sea level rather than a cut through a long incline.

REFERENCE CYCLES

In a general way we are all familiar with the meaning of business cycles, for the alternation of "boom and bust" or prosperity and recession (a polite name for a mild depression) is part of everyday parlance. It will help us study cycles, however, if we learn to speak of them with a standard terminology. We can do this by taking the cycles from actual history, "superimposing" them, and drawing the general profile of the so-called *reference* cycle that emerges. It looks like Fig. 21 · 3. This model of a typical cycle enables us to speak of the "length" of a business cycle as the period from one peak to the next or from trough to trough. If we fail to measure from *similar* points on two or more cycles, we can easily get a distorted picture of short-term growth—for instance, one that begins at the upper turning point of one cycle and measures to the trough of the next. Much of the political charge and countercharge about

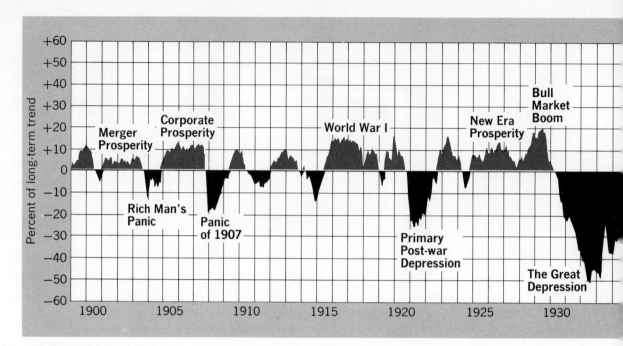

FIGURE 21 · 2
The business cycle

FIGURE 21 · 3
The reference cycle

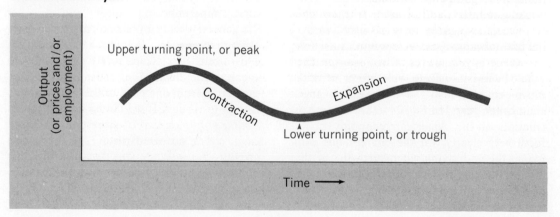

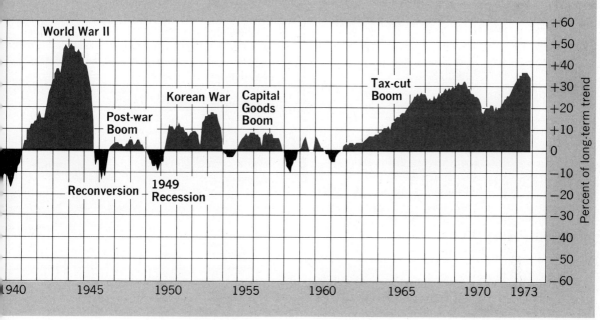

Percent of long-term trend

From Cleveland Trust Company

growth rates can be clarified if we examine the starting and terminating dates used by each side.

CAUSES OF CYCLES

What lies behind this more or less regular alternation of good and bad times?

Innumerable theories, none of them entirely satisfactory, have been advanced to explain the business cycle. A common business explanation is that waves of optimism in the world of affairs alternate with waves of pessimism—a statement that may be true enough, but that seems to describe the sequence of events rather than to explain it. Hence economists have tried to find the underlying cyclical mechanism in firmer stuff than an alternation of moods. One famous late-nineteenth-century economist, W. S. Jevons, for example, explained business cy-

cles as the consequence of sunspots—perhaps not as occult a theory as it might seem, since Jevons believed that the sunspots caused weather cycles that caused crop cycles that caused business cycles. The trouble was that subsequent investigation shows that the periodicity of sunspots was sufficiently different from that of rainfall cycles to make the connection impossible.

Other economists have turned to causes closer to home: to variations in the rate of gold mining (with its effects on the money supply); to fluctuations in the rate of invention; to the regular recurrence of war; and to yet many other factors. There is no doubt that many of these events can induce a business expansion or contraction. The persistent problem, however, is that none of the so-called underlying causes itself displays an inherent cyclicality—much less one with a periodicity of eight to ten years.

THE GREAT DEPRESSION

The Great Depression was probably the most traumatic economic episode in American life. GNP fell precipitously from $104 billion in 1929 to $56 billion in 1933. Unemployment rose from 1.5 million to 12.8 million: one person in every four in the labor force was out of work. Residential construction fell by 90 percent; nine million savings accounts were lost as banks closed their doors. Eighty-five thousand businesses failed. Wages fell to 5 cents an hour in sawmills, 6 cents in brick and tile manufacturing, 7½ cents in general contracting. As the stock market crashed, $30 billion in financial assets vanished. By 1932 nearly one in five of all Detroit schoolchildren was officially registered as seriously undernourished.

What caused the Great Depression? To this day we do not have a wholly convincing account. In part it was, of course, the consequence of a general decline in capital formation: investment expenditures fell by 88 percent from 1929 to 1933. But underlying this collapse were a number of contributory factors. Farm incomes had been steadily falling for years. The distribution of income was worsening, with profits booming at the same time that wage income was basically unchanged. Compounding and aggravating these weaknesses in the economy was a devastating collapse in credit. Whole structures of companies, pyramided one atop the other, fell like so many houses of cards when the stock market fell. And to worsen matters still further, the monetary authorities pursued policies of "prudence" and "caution" that unwittingly weakened the economy still further.

Can another Great Depression devastate the economy? Most economists would doubt it. Most bank accounts are today insured by the Federal Deposit Insurance Corporation, so that the wholesale wiping out of household assets would not happen again. The stock market, although still subject to wide swings, is unlikely to drag households or businesses into insolvency, because stocks can no longer be bought on the thin "margins" (partial payments) characteristic of the 1930s. Most important of all, the sheer size of government expenditure today makes a total collapse almost impossible. Moreover, if a severe depression were to begin, government would pursue policies of expansion either unknown or unthinkable in those days of laissez-faire economics. We will be studying all these matters shortly.

Then how do we explain cycles? Economists today no longer seek a single explanation of the phenomenon in an exogenous (external) cyclical force. Rather, they tend to see cycles as our own eye first saw them on the growth curve, *as variations in the rate of growth that tend to be induced by the dynamics of the growth process itself.*

FROM CAPACITY TO UTILIZATION

With this change in focus we bring into play a different set of concepts. In our survey of the long-term growth of GNP we were essentially dealing with the slow expansion of the *capacity* of the economy to produce goods and services. Now, however, as we concentrate on the immediate problems of boom and recession, it is not so much the capacity of the economy but *the degree of utilization* of the economy that becomes all-important. Hence, just as in microeconomics, when long-term price changes depended more on supply considerations than on demand considerations, and short-run price fluctuations depended more on demand shifts than supply shifts, so in dealing with the flow of total output, the potential supply of GNP was the main concern for long-run analysis, whereas in seeking to understand the short-run changes in GNP, changes in demand will be of primary interest.

Nevertheless, even in the determination of long-run prices, we will remember that demand was always present as one of the "blades of the scissors," as Alfred Marshall put it. So, too, in dealing with short-run changes in GNP, we see that supply is always present as one of the fundamental forces that determines how large output will actually be. It is, however, the *short-run forces of supply* that will occupy us in studying fluctuations of GNP. We will learn about them in this chapter, setting the stage for a much more extended examination of the forces of demand in the chapters to come.

THE SHORT-RUN SUPPLY OF GNP

What do we mean by the short-run supply forces for GNP? Let us first understand what we do *not* mean. A short-run supply function does not mean the relation between the quantity of GNP and its "price," as it might if we were talking about the short-run supply function for shoes. Nor does it mean the long-run changes in labor force or capital stock or productivity, which we learned about in our last chapter.

What does it mean? The idea is simple enough. The short-run supply function of GNP *tells us about the degree to which we utilize the existing capacity for production of our economy*. It tells us how much we can increase GNP in the short run by varying the degree of use of our existing labor and capital supplies; and it tells us how much GNP may fall if we fail to make use of these supplies.

Unemployment

That last point opens up the first obvious avenue for study. The supply of GNP that our stock of resources can deliver will certainly depend on the degree to which the productive capacity of labor and capital is used—or conversely, the degree to which labor and capital is unemployed and unutilized.

Usually when we think of unemployment we picture it as a kind of residual—the total labor force minus the number who are actually at work. If we look back at Fig. 20 · 6, we can see the colored band that represents this residual, a band that expanded to terrifying dimensions in the Great Depression (when one of every *four* members of the labor force was unable to find work) and that almost disappeared in the war years from 1942 to 1944.

AN ELASTIC LABOR FORCE

But when we look more closely into the actual phenomenon of unemployment we discover something quite surprising. It is that "unemployment" is not simply the difference between the number of people working and a fixed labor force, but the difference between the number working and an elastic, changeable labor force. Moreover we also discover that the size of the labor force is directly affected by the number of jobs, increasing as employment increases, shrinking as it falls.

The result is seemingly paradoxical. It is that employment and unemployment can both rise and fall at the same time, something that would be impossible if the labor force itself were fixed.

Table 21 · 2, on the next page, shows this curious phenomenon.

Notice that between 1969 and 1970, employment and unemployment *both* rose. The same phenomenon appeared between 1966 and 1967. How can this be?

The answer to the apparent paradox lies in the short-run responsiveness of the labor supply to the ease of finding work. In good times when jobs are plentiful, more youths

TABLE 21 · 2
Labor Force 1966–1972

| | SHORT-RUN CHANGES (IN MILLIONS) | | | | | | |
	1966	1967	1968	1969	1970	1971	1972
Number in civilian labor force	75.8	77.3	78.7	80.7	82.7	84.1	86.5
Civilian employment	72.9	74.4	75.9	77.9	78.6	79.1	81.7
Unemployment	2.9	3.0	2.8	2.8	4.1	5.0	4.8

and women will seek work. The whole labor force will then temporarily expand; and since not all of it may find work, both employment and unemployment may show increases. The reverse is true in a year of recession. What happens then is that many will be discouraged by bad times and "withdraw" from the labor force, remaining in school or in the household. As a result, the number of unemployed will then be smaller than if the larger labor force of a boom year had continued actively looking for work.

MEANING OF UNEMPLOYMENT

The concept of a variable participation rate for labor in the short run helps to elucidate the meaning of unemployment.

Clearly, unemployment is not a static condition, but one that varies with the *short-run participation rate* itself. Technically, the measure of unemployment is determined by a household-to-household survey conducted each month by the Bureau of the Census among a carefully selected sample.* An "unemployed" person is thereupon defined not merely as a person without a job—for perhaps such a person does not *want* a job—but as someone who is "actively" seeking work but is unable to find it. Since, however, the number of people who will be seeking work

*Sampling is an important statistical tool. If you would like to learn more about it, consult Section IV, Part Seven.

will rise in good times and fall in bad times, figures for any given period must be viewed with caution.

As employment opportunities drop, unemployment will not rise by an equivalent amount. Some of those looking for work when job opportunities are plentiful will withdraw from the labor force and become part of *hidden unemployment*. When job opportunities expand, these "hidden unemployed" will reenter the labor force, so that unemployment will not fall as fast as employment rises. Thus the ups and downs in the measured unemployment rate reflect the state of the economy, but the swings are not as large as they would be if the term "unemployment" measured the hidden unemployed.

WORKING HOURS

Like short-run participation rates, *hours of labor are also sensitive to the ease of finding work.* As jobs become more available, hours of work per employee tend to lengthen. This occurs partly because employees are able to work overtime, or even to "moonlight," taking on a second job to augment their incomes. It also occurs because employers find it cheaper to lengthen hours, even at overtime rates, than to hire and train new employees.

These cautions by no means invalidate the statistics of unemployment, but they warn us against taking those statistics at face value. In appraising the seriousness of a given rate of unemployment, the economist first looks at participation rates and hours of work, and then at unemployment rates, before he judges what fraction of the "long-run" potential labor supply is being utilized.

CAPACITY UTILIZATION

Is there a counterpart to unemployment affecting capital? There is—up to a point. We call it the rate of *capacity utilization.* This rate measures the relation between the maximum output that could be attained with a given capital stock and the actual output that is being produced by this stock. We usually use this rate for manufacturing industries, since adequate statistics for capacity output do not exist for many industries, such as construction or agriculture. But when we hear, for example, that steel plants are running at, say, 90 percent of capacity, the unused 10 percent is the capital counterpart of unemployment.

What about the elasticity of supply of capital? Does the stock of capital respond to economic conditions, in the way that the labor force does? To a certain extent it does. When demand is brisk, old plant destined for scrapping may be used for a time. Land that was taken out of cultivation can be brought back. Decayed buildings can be pressed into temporary service. Therefore there is a certain elasticity in the stock of capital, comparable to that of labor. Because the measurement of this flexibility of capital supply is so difficult, however, we do not make use of this elasticity when we speak of capital utilization.

CAPITAL OUTPUT RATIOS

There is a closely related concept that links short-term fluctuations in labor productivity with short-term fluctuations in capital productivity. (See box, "Measuring Productivity," Chapter 20.) Just as we can speak of a long-run productivity of labor, in terms of the relation between labor inputs and labor outputs, so there are, as we know, long-run trends in the productivity of capital. We can describe this productivity as the ratio between capital and output—*the amount of capital that it requires to create a unit of output.*

In the long run, this *capital-output* ratio changes only slowly as technology increases the productive power of capital. But in the short run, capital-output ratios are capable of sharp fluctuations much like those of labor. The amount of output that a "unit" of capital—say a factory or a machine—produces can swing sharply from month to month or year to year, especially if the factory or machine is underutilized and then brought up to "rated" capacity, or if it is pushed beyond the output rate it was designed for.

THE AGGREGATE PRODUCTION FUNCTION

Our discussion of the elasticity of supply of labor and capital has helped us understand how we can vary the amount of GNP we produce from our historically determined stock of capital and population. As the degree of utilization or employment changes, so will the amount of output that our factors of production turn out.

Now we must ask a larger question about this variable capacity to produce. Can we construct a supply curve for GNP as a whole, in the short run? We can indeed, and we call such a curve an *aggregate production function.* As the words suggest, it expresses the total supply capabilities of the economy as we alter the degree of utilization of all factors in it.

What does such an aggregate production

function look like? Figure 21·4 shows us three possibilities.*

First, a technical note. You will see that we have put output on the horizontal axis and input on the vertical axis. In microeconomics, we generally put output on the vertical axis and input on the horizontal (see Fig. 11·1, p. 141). This is simply a convention, but it explains why the curves turn in different directions than they would in microanalysis.

more output as we add inputs; in Panel III, less and less.

ECONOMIES OF SCALE

What accounts for these different shapes? The answer is not, as we might think, the law of variable proportions, because we are varying all factor inputs, not just one. Nor is the reason the law of increasing cost, because we are not shifting from one kind of output to

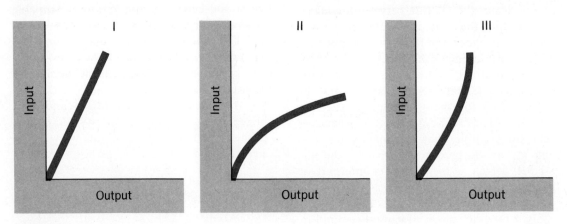

FIGURE 21·4
Possible production functions

Second, let us look at the three curves. In Panel I we depict a production function that gives us a constant proportional increase in output as we add "bundles" of labor and capital, let us say ½ percent more output for each 1 percent more in combined inputs. In Panel II, we get proportionally more and

*We depict the curves for simplicity's sake without calibrating the graph. Usually production functions are graphed on log-log paper to show the rates of change of inputs and outputs, rather than their amounts.

another. Rather, the constraint that determines the shape of the curve has to do with *economies of scale*. As we add more and more of all inputs, we might encounter economies of size, or beyond a certain scale of output, diseconomies of size.

Which of the three curves actually portrays the production function of the United States? The curve is probably a combination of all three. If we were to begin from a very low level of output, we would likely enjoy economies of scale at first. Then after a "nor-

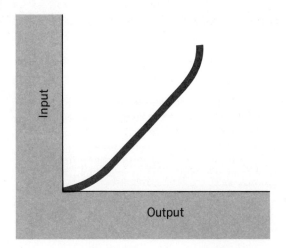

FIGURE 21 · 5
The aggregate production function

FIGURE 21 · 6
Shifts in the production function

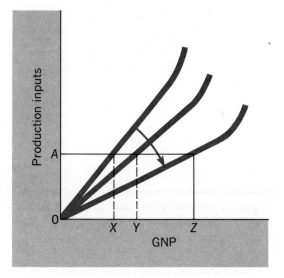

production function we might encounter diseconomies of scale (or we might simply run out of the highest-quality factors and be forced to resort to lower grade, relatively unproductive factors). Thus our aggregate production function looks like Fig. 21·5. We will be mainly interested in its upper portions.

We should note, however, that the slope of an aggregate production function need not remain fixed. As the productivity of all factors increases over time, we can attain the same total output with a smaller bundle of inputs, or what is the same thing, more output from the same bundle of inputs. This results in a clockwise shift in the production function as shown in Fig. 21·6.

Note that as our productivity grows, OA inputs give us larger and larger GNPs: first OX, then OY, then OZ.

AGGREGATE SUPPLY CURVE

The aggregate production function gives us what we set out to discover—a comprehension of the supply curve for GNP.* When we speak of the changing slope of such a production function, we are of course dealing in long-run time periods; but even in the short run we can think of the supply curve as *tending* downward as the productivity of labor and capital gradually increase.

What the aggregate production function therefore shows is how much GNP the economy can produce as it varies its inputs. It thereby gives us half of the information we need to arrive at a determination of the year-to-year, or month-to-month, fluctuations in GNP. We now know something about the shape of supply. The next problem is to discover what the demand for GNP will be.

mal" level of production were reached, we would enter a long stretch of constant returns to scale. At the very upper end of the

---

*Another means of studying total output is called input-output analysis. This is explained in a special supplement to this chapter.

KEY WORDS

CENTRAL CONCEPTS

Business cycles

Degree of
utilization

Short-run
participation
rates

Unemployment

Capital output
ratios

Aggregate
production
function

Constant returns
to scale

1. We turn now from growth to fluctuations in growth called *business cycles*. A typical business cycle has four phases: expansion, an upper turning point, contraction, and a lower turning point.

2. Business cycles introduce us to the concept of the *degree of utilization* of the economy. We are interested now in the relation between factor inputs and GNP in the short run, rather than in long-run growth trends.

3. Business cycles and degree of utilization introduce us to the idea of *unemployment* and *underutilization*. Here we note that the labor force is not a fixed number, but varies in size as the number of available jobs changes; i.e., there are changes in short-run *participation rates*.

4. As a result of flexible participation rates, the idea of *unemployment* becomes complex. *Unemployment reflects changes in the number of people looking for work, as well as changes in the number of available jobs.* Hours of labor are also responsive to demand.

5. Capital stock can also be *underutilized* and *capital output* ratios can change, as the stock of capital is left idle or pushed beyond its normal design capacity.

6. An *aggregate production function is the economy's supply curve.* It shows us the increase in the output of a given selection of goods if we increase inputs with "bundles" of capital and labor. Up to the point of "full employment and utilization," production functions tend to be straight lines, meaning that each proportionate increase in bundles of inputs yields an unchanging proportionate increase in output. We call this *constant returns to scale.*

7. The slope of the production function will move downward as the productivity of the inputs changes.

QUESTIONS

1. How can employment and unemployment increase at the same time?

2. If there were a severe shortage of labor, what would the probable effect be on the length of the work week? The work year?

3. Can you think of some mechanism that would produce regular business cycles? How about the tendency of most capital goods to wear out in about ten years? Assuming that capital goods were originally bought in a "bunched" fashion, could that produce a wavelike pattern? How about cycles in birth rates? (Or do you suppose these are *caused* by business cycles?)

4. What is meant by an aggregate production function? How is it different from a production possibility curve?

5. Under what conditions would an aggregate production function show an increasing slope? A decreasing slope?

6. What does it mean when we state that the U.S. economy mainly has a straight-line aggregate production function, at least up to "capacity utilization"?

7. Draw a series of aggregate production functions for an economy in which pollution problems were becoming an ever more burdensome problem.

Special Supplement on Input-Output

Input-output is another means of understanding the production function. It is an analytical procedure developed during the last two decades under the leadership of Wassily Leontieff of Harvard University, who won the Nobel prize for his efforts.

Input-output analysis is an effort to clarify the way the economy literally fits together in terms of the flows of goods from one producer to another or from the last producer to the final buyer. In our normal aggregative way of looking at GNP, we do not see the immensely complex interaction of production flows down the various "stages" of production. All these flows are ignored as we concentrate on *final* production. Input-output analysis concentrates on *all* production, final or intermediate. It thereby gives us a much more detailed understanding of the linkages of output than we can get from normal GNP analysis.

Input-output analysis begins by classifying production into basic inputs or industries. Today the Department of Commerce operates with an input-output table that lists 87 different industries, such as livestock and livestock products, ordnance and accessories, household appliances, amusements. These 87 industries are listed one below the other. Then the output of each industry is placed in a "cell" or "cells" corresponding to the industries to which it is sold. An actual input-output table or *matrix* is too large to be shown here. Instead, Table 21·3 gives us a look at a model of such a matrix for an extremely simple hypothetical economy.

What is such a matrix good for? First, let us read across the rows of the table, to trace where output goes. For example, of the total wheat crop of 600 (thousand bushels), 100 are kept back to sow next year's crop, none go to the machine or auto industry, and 500 are used for food (and sold to labor). Machines have a different pattern. Forty machines are produced. Ten are used in harvesting wheat, 5 are used in making more machines (machine tools), 25 go to the auto industry, none are sold to labor. Automobiles are sold to wheat farms (trucks), used by the machinery and auto industry, as trucks or vehicles for salesmen, and sold in large numbers to consumers. Labor is used by all producers, including labor itself (barbers, lawyers, teachers).

This shows us the flow of production "horizontally" through the economy. But we can also use the table to trace its "vertical" distribution. That is, we can see that the production of 600 "units" of wheat (last figure in the top row) required *inputs* (the column under wheat) of 100 units of wheat, 10 machines, 5 automobiles (trucks), and 20 units of labor. To make 40 machines, it takes no wheat, 5 machines, 10 autos, and 30 units of labor. The production of 68 automobiles needs 25 machines, 3 cars, and 60 labor units. To "produce" 120 "units" of labor—to feed and sustain that much labor—takes 500 units of wheat, 50 cars, 10 units of personal services.

Thus our input-output analysis enables us to penetrate deeply into the interstices of the economy. But more than that, it *enables us to calculate production requirements* in a way that far exceeds in accuracy any previously known

TABLE 21 · 3

	Wheat	Machines	Automobiles	Labor	Total
Wheat (000 bushels)	100	0	0	500	600
Machines (units)	10	5	25	0	40
Automobiles (units)	5	10	3	50	68
Labor (000 man-years)	20	30	60	10	120

method. Suppose, for example, that the economy wanted to double its output of autos. Forget for a moment about economies of scale. To begin with, we can see that it will need 25 more machines, 3 additional autos, and 60 more units of labor.

But that is only a list of its *direct demands*. There is also a long series of *indirect demands*. For when the auto industry buys five additional machines, the machine industry will have to increase its output by one-eighth. This means it will need one-eighth more inputs of machines, autos, and labor. But in turn this sets up still further requirements. To "produce" more labor will require more outputs of wheat and cars. To produce more wheat will require still further output of machines and autos. Thus a whole series of secondary, indirect demands spread out through the economy, each generating still further demands.

Input-output analysis uses a technique known as *matrix algebra* to sum up the total effects of any original change. This is not a subject that we will explore here. It is enough to understand how the matrix enables us to calculate production requirements, very much in the manner of an aggregate production function, but in finer detail.

We should note one difficulty with input-output analysis. When we took our example of doubling auto output, we assumed that there would be no changes in the proportions of inputs required to double output and that the input "mixes" for the other industries would be unaffected by increases in their outputs. This assumption of *fixed production coefficients* is not in accord with reality. Increases in output, such as a doubling of auto output, not only usually lead to economies of scale, but may also result from wholly new techniques. Input-output analysis has no way of handling or predicting these kinds of changes. At best it gives us a picture of the production requirements of an economy under the assumption that production methods and products are fixed, although we know they are not.

Nonetheless, no more powerful tool has yet been developed to examine the interactions of the economic system. Input-output analysis is used more and more, not only by government planners or economists, but by large corporations that want to calculate how changes in various sectors of the economy affect demand for their products. Input-output tables enable them to do this because they show the indirect as well as direct demands that economic changes generate.

The Demand
for Output

22

So far, we have talked about GNP from three points of view. First we familiarized ourselves with the actual process of production itself—the interaction of the factors of production and the accumulated wealth of the past as they cooperated to bring a flow of output into being. Next we examined the forces that swelled that volume of output over time, mainly the increase in skills and capital equipment and technology that are responsible for our long-term trend of growth. Then we looked into the immediate causes of short-run fluctuations of output, from the point of view of supply, acquainting ourselves with the idea of a supply curve of GNP that we call the aggregate production function.

Now let us turn to the vital question of demand—to the forces that induce the factors of production to generate output. In particular, we want to look into the question of fluctuations of demand that, together with fluctuations in supply, will determine the actual amount of GNP produced at any moment.

Output and Demand

Let us start with a basic question—at once very simple and surprisingly complex. *How do we know that there will be enough demand to buy the amount of output that the factors produce?* Once we understand that, we will be well on the way to unlocking the puzzle of macroeconomies.

The question leads us to understand a fundamental linkage between demand and output. For how does output actually come into existence? Any businessman will give you the answer. He will tell you that the crucial factor enabling him to run his business is *demand* or *purchasing power;* that is, the presence of buyers who are willing and able to buy some good or service at a price he is willing to accept.

But how does demand or purchasing power come into existence? If we ask any buyer, he will tell us that his dollars come to him because they are part of his *income* or his cash receipts. But where, in turn, do the dollar receipts or incomes of buyers come from? If we inquire again, most buyers will tell us that they have money in their pockets because in one fashion or another they have contributed to the process of production; that is, because they have helped to make the output that is now being sold.

Thus output is generated by demand—and demand is generated by output! Our quest for the motive force behind the flow of production therefore leads us in a great circle through the market system. Here is the circular flow we first met in Chapter 7, this time approached from a macro rather than micro perspective. We can see this in Fig. 22·1.

At the top of the circle we see payments flowing from households to firms or government units (cities, states, federal agencies, etc.), thereby creating the demand that brings forth production. At the bottom of the circle, we see more payments, this time flowing from

firms or governments back to households, as businesses hire the services of the various factors in order to carry out production. Thus we can see that there is a constant regeneration of demand as money is first spent by the public on the output of firms and governments, and then in turn spent by firms and governments for the services of the public.

AN ECONOMIC MODEL

Let us begin by examining this chain of payments and receipts as a model of the macro system.

Our model, to begin with, will be a very simple one. We must simplify it, at first, by ruling out some of the very events to which we will later turn as the climax of our study. For instance, we shall ignore changes in *people's tastes,* so that we can assume that

FIGURE 22 · 1

The circular flow, view II

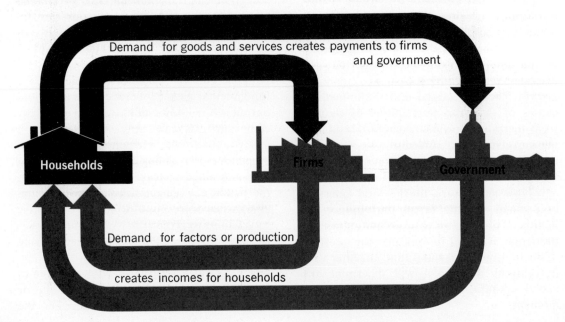

everyone will regularly buy the same kinds of goods. We shall ignore differences in the *structure of firms* or *markets,* so that we can forget about differences in competitive pressures. We shall rule out *population growth* and, even more important, *inventive progress,* so that we can deal with a very stable imaginary world. For the time being, we will exclude even *saving* and *net investment* (although of course we must permit replacement investment), so that we can ignore growth. Later, of course, we are to be deeply concerned with just such problems of dynamic change. But we shall not be able to come to grips with them until we have first understood an economic world as "pure" and changeless as possible.

COST AND OUTPUT

The very abstract model we have created may seem too far removed from the real world to tell us much about its operation. But if we now go back to the circle of economic activity in which payments to firms, governments, and factors become their incomes, and in turn reappear on the marketplace as demand, our model will enable us to explain a very important problem. *It is how an economy that has produced a given GNP is able to buy it back.*

This is by no means a self-evident matter. Indeed, one of the most common misconceptions about the flow of economic activity is that there will not be enough purchasing power to buy everything we have produced—that somehow we are unable to buy enough to keep up with the output of our factories. So it is well to understand once and for all how an economy can sustain a given level of production through its purchases on the market.

We start, then, with an imaginary economy in full operation. We can, if we wish,

imagine ourselves as having collected a year's output, which is now sitting on the economic front doorstep looking for a buyer. What we must now see is whether it will be possible to *sell* this gross national product to the people who have been engaged in producing it. In other words, *we must ask whether enough income or receipts have been generated in the process of production to buy back all the products themselves.*

Costs and Incomes

How does production create income? Businessmen do not think about "incomes" when they assemble the factors of production to meet the demand for their product. They worry about *cost.* All the money they pay out during the production process is paid under the heading of *cost,* whether it be wage or salary cost, cost of materials, depreciation cost, tax cost, or whatever. Thus it would seem that the concept of cost may offer us a useful point of entry into the economic chain. For *if we can show how all costs become incomes,* we will have taken a major step toward understanding whether our gross national product can in fact be sold to those who produced it.

It may help us if we begin by looking at the kinds of costs incurred by business firms in real life. Since governments also produce goods and services, this hypothetical firm should be taken to represent government agencies as well as business firms. Both incur the same kinds of costs; only the labels differ.

Table 22 · 1, a hypothetical expense summary of General Output Company, will serve as an example typical of all business firms, large or small, and all government agencies. (If you examine the year-end statements of any business, you will find that their costs all fall into one or more of the cost categories shown.)

TABLE 22 · 1

General Output Company Cost Summary

Wages, salaries, and employee benefits	$100,000,000
Rental, interest, and profits payments	5,000,000
Materials, supplies, etc.	60,000,000
Taxes other than income	25,000,000
Depreciation	20,000,000
Total	$210,000,000

FACTOR COSTS AND NATIONAL INCOME

Some of these costs we recognize immediately as payments to factors of production. The item for "wages and salaries" is obviously a payment to the factor *labor*. The item "interest" (perhaps not so obviously) is a payment to the factor *capital*—that is, to those who have lent the company money in order to help it carry on its productive operation. The item for rent is, of course, a payment for the rental of *land* or natural resources from their owners.

Note that we have included profits with rent and interest. In actual accounting practice, profits are not shown as an expense. For our purposes, however, it will be quite legitimate and very helpful to regard profits as a special kind of factor cost going to entrepreneurs for their risk-taking function. Later we shall go more thoroughly into the matter of profits.

Two things strike us about these factor costs. First, it is clear that they represent payments that have been made to secure production. In more technical language, they are payments for factor inputs that result in commodity outputs. All the production actually carried on within the company or government agency, all the value it has added to the

economy has been compensated by the payments the company or the agency has made to land, labor, and capital. To be sure, there are other costs, for materials and taxes and depreciation, and we shall soon turn to these. But whatever production or assembly or distribution the company or agency has carried out during the course of the year has required the use of land, labor, or capital. Thus *the total of its factor costs represents the value of the total new output that General Output by itself has given to the economy*.

From here it is but a simple step to add up *all* the factor costs paid out by *all* the companies and government agencies in the economy, in order to measure the total new *value added* by all productive efforts in the year. This measure is called *national income*. As we can see, it is less than gross national product, for it does not include other costs of output; namely, certain taxes and depreciation.

FACTOR COSTS AND HOUSEHOLD INCOMES

A second fact that strikes us is that *all factor costs are income payments*. The wages, salaries, interest, rents, etc., that were costs to the company or agency were income to its recipients. So are any profits, which will accrue as income to the owners of the business.

Thus, just as it sounds, national income means the total amount of earnings of the factors of production within the nation. If we think of these factors as constituting the households of the economy, we can see that *factor costs result directly in incomes to the household sector*. Thus, if factor costs were the only costs involved in production, the problem of buying back the gross national product would be a very simple one. We should simply be paying out to households, as the cost of production, the very sum needed to buy GNP when we turned around to sell it. But this is not the case, as a glance at the General Output expense summary shows. There are other costs besides factor costs. How shall we deal with them?

COSTS OF MATERIALS

The next item of the expense summary is puzzling. Called payments for "materials, supplies, etc.," it represents all the money General Output has paid, not to its own factors, but to other companies for other products it has needed. We may even recognize these costs as payments for those intermediate products that lose their identity in a later stage of production. How do such payments become part of the income available to buy GNP on the marketplace?

Perhaps the answer is already intuitively clear. When General Output sends its checks to, let us say, U.S. Steel or General Electric or to a local supplier of stationery, each of these recipient firms now uses the proceeds of General Output's checks to pay its own costs.

(Actually, of course, they have probably long since paid their own costs and now use General Output's payment only to reimburse themselves. But if we want to picture our model economy in the simplest way, we can imagine U.S. Steel and other firms sending their products to General Output and waiting until checks arrive to pay their own costs.)

And what are those costs? What must U.S. Steel or all the other suppliers now do with their checks? The answer is obvious. They must now reimburse their own factors and then pay any other costs that remain.

Figure 22·2 may make the matter plain. It shows us, looking back down the chain of intermediate payments, that what constitutes material costs to one firm is made up of factor and other costs to another. Indeed, as we unravel the chain from company to company, it is clear that all the contribution to new output must have come from the contribution of factors somewhere down the line, and that *all the costs of new output—all the value added—must ultimately be resolvable into payments to land, labor, and capital.*

Another way of picturing the same thing is to imagine that all firms or agencies in the

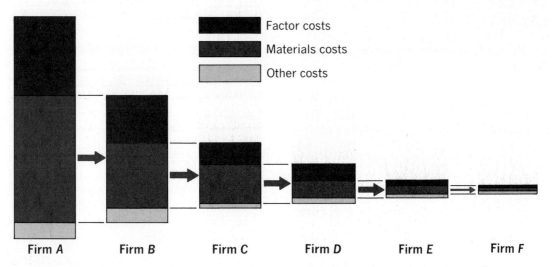

FIGURE 22 · 2
How materials costs become other costs

Factor costs

Materials costs

Other costs

Firm *A*   Firm *B*   Firm *C*   Firm *D*   Firm *E*   Firm *F*

country were bought up by a single gigantic corporation. The various production units of the new supercorporation would then ship components and semifinished items back and forth to one another, but there would not have to be any payment from one division to another. The only payments that would be necessary would be those required to buy the services of factors—that is, various kinds of labor or the use of property or capital—so that at the end of the year, the supercorporation would show on its expense summary only items for wages and salaries, rent, and interest (and as we shall see, taxes and depreciation), but it would have no item for materials cost.

We have come a bit further toward seeing how our gross national product can be sold. *To the extent that GNP represents new output made during the course of the year, the income to buy back this output has already been handed out as factor costs, either paid at the last stage of production or "carried along" in the guise of materials costs.*

But a glance at the General Output expense summary shows that entrepreneurs incur two kinds of costs that we have still not taken into account: taxes and depreciation. Here are costs employers have incurred that have not been accounted for on the income side. What can we say about them?

TAX COSTS

Let us begin by tracing the taxes that General Output pays, just as we have traced its materials payments.* In the first instance, its taxes will go to government units—federal, state, and local. But we need not stop there. Just as we saw that General Output's checks

*For simplicity, we also show government agencies as taxpayers. In fact, most government units do *not* pay taxes. Yet there will be hidden tax costs in the prices of many materials they buy. No harm is done by treating government agencies like taxpaying firms in this model.

to supplier firms paid for the suppliers' factor costs and for still further interfirm transactions, so we can see that its checks to government agencies pay for goods and services that these agencies have produced—goods such as roads, buildings, or defense equipment; or services such as teaching, police protection, and the administration of justice. General Output's tax checks are thus used to help pay for factors of production—land, labor, and capital—that are used in the *public sector.*

In many ways, General Output's payments to government units resemble its payments to other firms for raw materials. Indeed, if the government *sold* its services to General Output, charging for the use of the roads, police services, or defense protection it affords the company, there would be *no* difference whatsoever. The reason we differentiate between a company's payment to the public sector and its payments for intermediate products is important, however, and worth looking into.

The first reason is clearly that with few exceptions, the government does *not* sell its output. This is partly because the community has decided that certain things the government produces (education, justice, or the use of public parks, for instance) should not be for sale but should be supplied to all citizens without direct charge. In part, it is also because some things the government produces, such as defense or law and order, cannot be equitably charged to individual buyers, since it is impossible to say to what degree anyone benefits from—or even uses—these communal facilities. Hence General Output, like every other producer, is billed, justly or otherwise, for a share of the cost of government.

There is also a second reason why we consider the cost of taxes as a new kind of cost, distinct from factor payments. It is that when business firms have finished paying the factors, they have not yet paid all the sums that employers must lay out. *Some taxes, in other words, are an addition to the cost of production.*

INDIRECT VS. DIRECT TAXES

These taxes—so-called *indirect taxes*—are levied on the productive enterprise itself or on its actual physical output. Taxes on real estate, for instance, or taxes that are levied on each unit of output, regardless of whether or not it is sold (such as excise taxes on cigarettes), or taxes levied on goods sold at retail (sales taxes) are all payments that entrepreneurs must make as part of their costs of doing business.

Note that not all taxes collected by the government are costs of production. Many taxes will be paid, not by the entrepreneurs as an expense of doing business, but by the *factors* themselves. These so-called *direct* taxes (such as income taxes) are *not* part of the cost of production. When General Output adds up its total cost of production, it naturally includes the wages and salaries it has paid, but it does not include the taxes its workers or executives have paid out of their incomes. Such direct taxes transfer income from earners to government, but they are not a cost to the company itself.

In the same way, the income taxes on the profits of a company do *not* constitute a cost of production. General Output does not pay income taxes as a regular charge on its operations but waits until a year's production has taken place and then pays income taxes on the profits it makes *after* paying its costs. If it finds that it has lost money over the year, it will not pay any income taxes—although it will have paid other costs, including indirect taxes. *Thus direct taxes are not a cost that is paid out in the course of production and must be recouped, but a payment made by factors (including owners of the business) from the incomes they have earned through the process of production.*

TAXES AS COST

Thus we can see two reasons why taxes are handled as a separate item in GNP and are not telescoped into factor costs, the way materials costs are. One reason is that taxes are a payment to a *different sector* from that of business and thus indicate a separate stream of economic activity. But the second reason, and the one that interests us more at this moment, is that *certain taxes*—indirect taxes—*are an entirely new kind of cost of production, not previously picked up.* As an expense paid out by entrepreneurs, over and above factor costs (or materials costs), these tax costs must be part of the total selling price of GNP.

Will there be enough incomes handed out in the process of production to cover this item of cost? We have seen that there will be. The indirect tax costs paid out by firms will be received by government agencies who will use these tax receipts to pay income to factors working for the government. Any direct taxes (income taxes) paid by General Output or by its factors will also wind up in the hands of a government. Thus all tax payments result in the transfer of purchasing power from the private to the public sector, and when spent by the public sector, they will again become demand on the marketplace.

DEPRECIATION

But there is still one last item of cost. At the end of the year, when the company is totting up its expenses to see if it has made a profit for the period, its accountants do not stop with factor costs, material costs, and indirect taxes. If they did, the company would soon be in serious straits. In producing its goods, General Output has also used up a certain amount of its assets—its buildings and equipment—and a cost must now be charged for this wear and tear if the company is to be able to preserve the value of its physical plant intact. If it did not make this cost allowance, it would have failed to include all the resources that were used up in the process of production, and it would therefore be overstating its profits.

Yet, this cost has something about it clearly different from other costs that General Output has paid. Unlike factor costs or taxes or materials costs, depreciation is not paid for by check. When the company's accountants make an allowance for depreciation, all they do is make an entry on the company's book, stating that plant and equipment are now worth a certain amount less than in the beginning of the year.

At the same time, however, General Output *includes* the amount of depreciation in the price it intends to charge for its goods. As we have seen, part of the resources used up in production was its own capital equipment, and it is certainly entitled to consider the depreciation as a cost. Yet, it has not paid anyone a sum of money equal to this cost! How, then, will there be enough income in the marketplace to buy back its product?

REPLACEMENT EXPENDITURE

The answer is that in essence it has paid depreciation charges to itself. Depreciation is thus part of its gross income. Together with after-tax profits, these depreciation charges are called a business's *cash flow*.

A business does not *have to* spend its depreciation accruals, but normally it will, *to maintain and replace its capital stock*. To be sure, an individual firm may not replace its worn-out capital exactly on schedule. But when we consider the economy as a whole, with its vast assemblage of firms, that problem tends to disappear. Suppose we have 1,000 firms, each with machines worth $1,000 and each depreciating its machines at $100 per year. Provided that all the machines were

bought in different years, this means that in any given year, about 10 percent of the capital stock will wear out and have to be replaced. It's reasonable to assume that among them, the 1,000 firms will spend $100,000 to replace their old equipment over a ten-year span.*

This enables us to see that insofar as there is a steady stream of replacement expenditures going to firms that make capital goods, there will be payments just large enough to balance the addition to costs due to depreciation. As with all other payments to firms, these replacement expenditures will, of course, become incomes to factors, etc., and thus can reappear on the marketplace.

ANOTHER VIEW OF COSTS AND INCOMES

Because it is very important to understand the relationship between the "selling price" of GNP and the amount of income available to buy it back, it may help to look at the matter from a different point of view.

This time let us approach it by seeing how the economy arranges things so that consumers and government and business, the three great sectors of final demand, are provided with enough purchasing power to claim the whole of GNP. Suppose, to begin with, that the economy paid out income only to its factors and priced its goods and services accordingly. In that case, consumers could purchase the entire value of the year's output,

---

*But what if the machines *were* all bought in one year or over a small number of years? Then replacement expenditures will *not* be evenly distributed over time, and we may indeed have problems. This takes us into the dynamics of prosperity and recession, to which we will turn in due course. For the purpose of our explanatory model, we will stick with our (not too unrealistic) assumption that machines wear out on a steady schedule and that aggregate replacement expenditures therefore also display a steady, unfluctuating pattern.

but business would be unable to purchase any portion of the output to replace its worn-out equipment. (Also it raises the awkward question of how we would pay factors working for the government, since government agencies would have very little income.)

That would obviously lead to serious trouble. Hence we must arrange for business to have a claim on output and for government factors to be paid for their services. The latter is simple. By imposing direct (income) taxes on factors, we divert income from the private to the public sector. And by imposing indirect taxes on output, we price output above its factor cost, thus making it impossible for consumers to claim the entire output.

In exactly the same way, business also reserves a claim on output by pricing its products to include a charge for depreciation. By so doing, it again reduces the ability of consumers to buy back the entire output of the economy, while it gives business the purchasing power to claim the output it needs (just as taxes give purchasing power to government). Now, after paying direct and indirect taxes and depreciation, the consumer is finally free to spend all the remainder of

accrue to government and business, but also as the means by which the output of the economy is made available to two important claimants besides private households.

THE THREE STREAMS OF EXPENDITURE

Our analysis is now essentially complete. Item by item, we have traced each element of cost into an income payment, so that we now know there is enough income paid out to buy back our GNP at a price that represents its full cost. Perhaps this was a conclusion we anticipated all along. After all, ours would be an impossibly difficult economy to manage if somewhere along the line purchasing power dropped out of existence, so that we were always faced with a shortage of income to buy back the product we made. But our analysis has also shown us something more unexpected. We are accustomed to thinking that all the purchasing power in the economy is received and spent through the hands of "people"—usually meaning households. Now we can see that this is not true. There is not only one, but there are *three* streams of incomes and costs, all quite distinct from one another although linked by direct taxes).

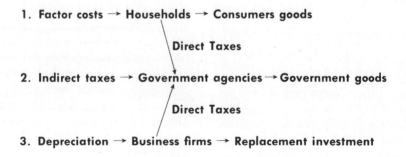

1. **Factor costs** → **Households** → **Consumers goods**
   Direct Taxes
2. **Indirect taxes** → **Government agencies** → **Government goods**
   Direct Taxes
3. **Depreciation** → **Business firms** → **Replacement investment**

his income without danger of encroaching on the output that must be reserved for public activity and for the replacement of capital.

In other words, we can look at taxes and depreciation not merely as "costs" that the consumer has to pay or as "incomes" that

The one major crossover in the three streams is the direct taxes of households and business firms that go to governments. This flow permits governments to buy more goods and services than could be purchased with indirect taxes alone.

THE THREE FLOWS

To help visualize these three flows, imagine for an instant that our money comes in colors (all of equal value): black, orange, and brown. Now suppose that firms always pay their factors in orange money, their taxes in brown money, and their replacement expenditures in black money. In point of fact, of course, the colors would soon be mixed. A factor that is paid in orange bills will be paying some of his orange income for taxes; or a government agency will be paying out brown money as factor incomes; or firms will be using black dollars to pay taxes or factors, and brown or orange dollars to pay for replacement capital.

But at least in our mind we could picture the streams being kept separate. A brown tax dollar paid by General Output to the Internal Revenue Service for taxes could go from the government to another firm, let us say in payment for office supplies, and we can think of the office supply firm keeping these brown dollars apart from its other receipts, to pay its taxes with. Such a brown dollar could circulate indefinitely, from government agencies to firms and back again, helping to bring about production but never entering a consumer's pocket! In the same way, a black replacement expenditure dollar going from General Output to, let us say, U.S. Steel could be set aside by U.S. Steel to pay for *its* replacement needs; and the firm that received this black dollar might, in turn, set it aside for its own use as replacement expenditure. We could, that is, imagine a circuit of expenditures in which black dollars went from firm to firm, to pay for replacement investment, and never ended up in a pay envelope or as a tax payment.

There is a simple way of explaining this seemingly complex triple flow. Each stream indicates the existence of a *final taker* of gross national product: consumers, government, and business itself.* Since output has final claimants other than consumers, we can obviously have a flow of purchasing power that does not enter consumers' or factors' hands.

The Completed Circuit of Demand

The realization that factor-owners do not get paid incomes equal to the total gross value of output brings us back to the central question of this chapter: can we be certain that we will be able to sell our GNP at its full cost? Has there surely been generated enough purchasing power to buy back our total output?

*We continue to forget about net exports until Chapter 26. We can think of them perfectly satisfactorily as a component of gross private investment.

We have thus far carefully analyzed and answered half the question. *We know that all costs will become incomes to factors or receipts of government agencies or of firms making replacement items.* To sum up again, factor costs become the incomes of workers, managements, owners of natural resources and of capital; and all these incomes together can be thought of as comprising the receipts of the household sector. Tax costs are paid to government agencies and become receipts of the government sector. Depreciation costs are initially accrued within business firms, and these accruals belong to the business sector. As long as worn-out capital is regularly replaced, these accruals will be matched by equivalent new receipts of firms that make capital goods.

THE CRUCIAL ROLE OF EXPENDITURES

What we have not yet established, however, is that these sector receipts will become sector expenditures. That is, we have not demonstrated that all households will now *spend* all their incomes on goods and services, or that government units will necessarily *spend* all their tax receipts on public goods and services, or that all firms will assuredly *spend* their depreciation accruals for new replacement equipment.

What happens if some receipts are not spent? The answer is of key importance in understanding the operation of the economy. A failure of the sectors to spend as much money as they have received means that some of the costs that have been laid out will *not* come back to the original entrepreneurs. As a result, they will suffer losses. If, for instance, our gross national product costs $1 trillion to produce but the various sectors spend only $900 billion in all, then some entrepreneurs will find themselves failing to sell all their output. Inventories of unsold goods will begin piling up, and businessmen will soon be worried about overproducing. The natural thing to do when you can't sell all your output is to stop making so much of it, so that businesses will begin cutting back on production. As they do so, they will also cut back on the number of people they employ. As a result, businessmen's costs will go down; but so will factor incomes, for we have seen that costs and incomes are but opposite sides of one coin. As incomes fall, the expenditures of the sectors might very well fall further, bringing about another twist in the spiral of recession.

This is not yet the place to go into the mechanics of such a downward spiral of business. But the point is clear. *A failure of the sectors to bring all their receipts back to the marketplace as demand can initiate profound economic problems. In the contrast between an unshakable equality of output and incomes on the one hand and the uncertain connection between incomes and expenditures on the other, we have come to grips with one of the most important problems in macroeconomics.*

THE CLOSED CIRCUIT

We shall have ample opportunity later to observe exactly what happens when incomes are not spent. Now let us be sure that we understand how the great circle of the economic flow is closed when the sectors *do* spend their receipts. Figure 22·3 shows how we can trace our three streams of dollars through the economy and how these flows suffice to buy back GNP for its total cost. For simplicity, we assume that there are no direct taxes.

We can trace the flow from left to right. We begin on the left with the bar representing the total cost of our freshly produced GNP. As we know, this cost consists of all the factor costs of all the firms and government units in the nation, all the indirect tax costs incurred during production, and all the depreciation charges made during production. The bar also shows us the amount of money demand our economy must generate in order to buy back its own output.

FROM GNP TO GNI

The next bars show us the transmutation of costs into sector receipts for householders, government units, and business firms (who retain their own depreciation accruals). This relationship between costs and sector receipts is one of *identity*—all costs *must* be receipts. Hence we use the sign ≡ to indicate that this is a relation of identities—of definitional differences only. If we use GNI to stand for gross national income (the gross incomes of all the sectors), then:

$$GNP \equiv GNI$$

That is an identity to be remembered—and *understood!*

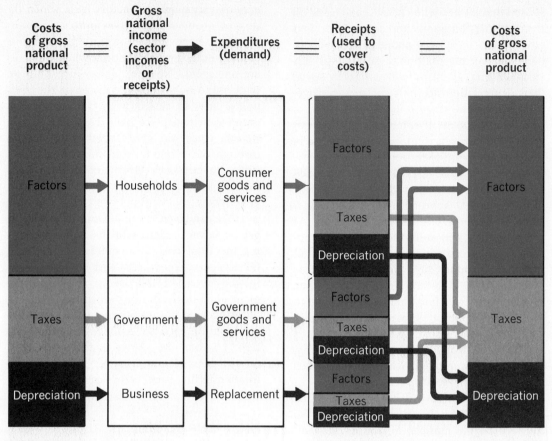

FIGURE 22 · 3
The circular flow, view III

INCOMES AND EXPENDITURES

Thereafter we notice the crucial link. We assume that each sector dutifully spends all its receipts, as it is supposed to. Our household sector buys the kinds of goods and services householders do in fact buy—consumption goods and services. Our government sector buys government goods and services, and our business sector buys replacement investment. This time we use an arrow (→) because this is emphatically *not* a relationship of identity. Our sectors may not spend all their in-

comes. Later we will see what happens if they don't.

Now note the next bar. Here we see what happens to these expenditures when they are received by the firms that make consumer goods or by the firms or individuals who make goods and services bought by governments or by the manufacturers of capital equipment. Each of these recipients will use the money he has received to cover factor payments, taxes, and depreciation for his own business. (What we show in our diagram are not these costs for each and every firm but the aggregate costs for all firms selling to each sector.)*

We are almost done. It remains only to aggregate the sector costs; that is, to add up all the factor costs, all the taxes, and all the depreciation accruals of *all* firms and government agencies—to reproduce a bar just like the one we started with. A circle of production has been completed. Firms and government units have received back, on the marketplace, a sum just large enough to cover their initial costs, including their profits for risk. The stage is set for another round of production, similar to the last.

GNP as a Sum of Costs and a Sum of Expenditures

Our bar graph also enables us to examine again the concept of gross national product, for now we can see that GNP can be looked at in one of two ways. We can think of measuring a year's gross national product as a *sum of all the costs incurred to make a year's output:* factor costs, indirect taxes, and depreciation. Or we can think of measuring the same GNP as the *sum of the expenditures that bought this output;* that is, consumption expenditure, government expenditure, and gross private investment expenditure. Since the final output is one and the same, we can see that the two methods of computing its value must also be the same.

TWO WAYS OF MEASURING GNP

An illustration may make it easier to grasp this identity of the two ways of measuring GNP. Suppose once again that we picture the

*Recall that for ease of exposition we are treating government agencies like firms and therefore show them as taxpayers.

economy as a gigantic factory from which the flow of production emerges onto a shipping platform, each item tagged with its selling price. There the items are examined by two clerks. One of them notes down in his book the selling price of each item and then analyzes that price into its cost (as income) components; factor cost (including profit), indirect taxes, and depreciation. The second clerk keeps a similar book in which each item's selling price is also entered, but his job is to note which sector—consumer, government, business investment, or export—is its buyer. Clearly, at the end of the year, the two clerks must show the same value of total output. But whereas the books of the first will show that total value separated into various costs, the books of the second will show it analyzed by its "customers"; that is, by the expenditures of the various sectors.

But wait! Suppose that an item comes onto the shipping platform without an order waiting for it! Would that not make the sum of costs larger than the sum of expenditures?

The answer will give us our final insight into the necessary equality of the two measures of GNP. For what happens to an item that is not bought by one of the sectors? It will be sent by the shipping clerk into inventory *where it will count as part of the business investment of the economy!* Do not forget that increases in inventory are treated as investment because they are a part of output that has not been consumed. In this case it is a very unwelcome kind of investment; and if it continues, it will shortly lead to changes in the production of the firm. Such dynamic changes will soon lie at the very center of our attention. In the meantime, however, *the fact that unbought goods are counted as investment*—as if they were "bought" by the firm that produced but cannot sell them—establishes the absolute identity of GNP measured as a sum of costs or as a sum of expenditures.

GNP AND GNI

To express the equality with the conciseness and clarity of mathematics, we can write, as we know:

$$GNP \equiv GNI$$

We already know that:

$$GNP \equiv C + I + G + X$$

and

$$GNI \equiv F + T + D$$

where *C, I, G,* and *X* are the familiar categories of expenditure, and *F, T,* and *D* stand for factor costs (income to land, capital, and labor), indirect taxes, and depreciation. Therefore, we know that:

$$C + I + G + X \equiv F + T + D$$

It is important to remember that these are all accounting identities, true by definition. The *National Income and Product Accounts,* the official government accounts for the economy, are kept in such a manner as to make them true.* As the name implies, these accounts are kept in two sets of "books," one on the products produced in the economy and one on the costs of production, which we know to be identical with the incomes generated in the economy. Since both sets of accounts are measuring the same output, the two totals must be equal.

NNP AND NATIONAL INCOME

It is now easy to understand the meaning of two other measures of output. One of these is called *net national product* (NNP). As the name indicates, it is exactly equal to the gross national product minus depreciation. GNP is

*There is a special supplement on these accounts at the end of this chapter.

used much more than NNP, since the actual measures of depreciation are very unreliable. The other measure, national income, we have already met. It is *GNP minus both depreciation and indirect taxes.* This makes it equal to the sum of factor costs only. Figure 22·4 should make this relationship clear. The aim of this last measure is to identify the net income that actually reaches the hands of factors of production. Consequently, the measure is sometimes called the *national income at factor cost.*

THE CIRCULAR FLOW

The "self-reproducing" model economy we have now sketched out is obviously still very far from reality. Nevertheless, the particular kind of unreality that we have deliberately constructed serves a highly useful purpose. An economy that regularly and dependably buys back everything it produces gives us a kind of bench mark from which to begin our subsequent investigations. We call such an economy, whose internal relationships we have outlined, an economy in *stationary equilibrium,* and we denote the changeless flow of costs into business receipts, and receipts back into costs, a *circular flow.*

We shall return many times to the model of a circular flow economy for insights into a more complex and dynamic system. Hence it is well that we summarize briefly two of the salient characteristics of such a system.

1. A circular flow economy will never experience a "recession."

Year in and year out, its total output will remain unchanged. Indeed, the very concept of a circular flow is useful in showing us that an economic system can maintain a given level of activity *indefinitely,* so long as all the sectors convert all their receipts into expenditures.

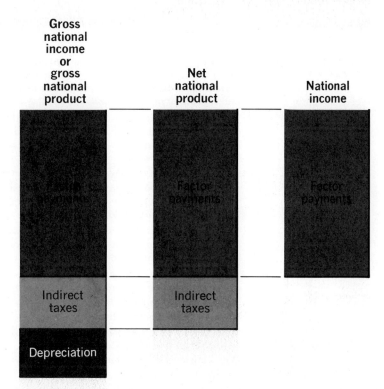

FIGURE 22 · 4
GNP, NNP, and NI

2. A circular flow economy also will never know a "boom."

That is, it will not grow, and its standard of living will remain unchanged. That standard of living may be high or low, for we could have a circular flow economy of poverty or of abundance. But in either state, changelessness will be its essence.

THE GREAT PUZZLE

What we have demonstrated in this chapter is an exceedingly important idea. There *can* always be enough purchasing power generated by the process of output to buy back that output.

Yet we all know, from our most casual acquaintance with economics, that in fact there is not always enough purchasing power

around, or that on occasions there is too much purchasing power. With too little, we have slumps and recessions; with too much, booms and inflation.

Hence the circular flow sets the stage for the next step in our study of macro-economics. If there *can be* the right amount of purchasing power generated, why isn't there? Or to put the question more perplexingly: if there *can be* enough purchasing power to buy *any* size output, small or large, what determines how large purchasing power will actually be, and therefore how large output will actually be?

These questions point the way for the next stage of our investigation. We must study the workings of demand much more realistically than heretofore, by removing some of the assumptions that were necessary to create a model of a circular flow system.

KEY WORDS	CENTRAL CONCEPTS

Output and demand

1. A great *circle of payments and receipts constantly renews the demand* that keeps the economy going. Income is pumped out to the participants in the production process, who then spend their incomes to buy the output they have helped to make, thereby returning money to the firms that again pump it out.

Model

2. We seek to elucidate this circle by constructing a *model*—an economy stripped of everything that is not essential to the process we seek to understand. Our model simplifies and therefore highlights the relationships that underlie the macroeconomic process.

Costs

3. We use our model to show *how an economy can in fact buy back all of its own output*—how enough demand can be created to keep the economy going. To do so, we must show that *every item of cost to business firms can become demand* for them on the marketplace.

Factor payments

4. The first item of costs is *factor payments* (wages, rent, profits, interest). These payments for factor services are *the source of household income.* In turn, households can spend their incomes for *consumption goods.*

Indirect taxes

5. *Indirect taxes* are costs paid by firms to government agencies. In turn, these agencies spend their tax receipts for *government purchases.*

(Note that direct taxes—income taxes—are not a cost of production, but a levy imposed on factors after they have been paid.)

Direct taxes

6. *Depreciation costs* are costs incurred by *businesses* for the wear and tear of capital. They are accrued by business firms, who, in turn, are able to use them to purchase *replacement investment.*

7. *Materials* costs are payments to other firms. Ultimately, they can all be broken down into factor costs, indirect tax costs, and depreciation costs.

Depreciation

8. *The total value of output (GNP) can be seen not only as a sum of expenditures* on different kinds of output $(C + I + G + X)$, *but as a sum of costs* $(F + T + D)$, or gross incomes incurred in making this output. This gives us the important identity:

Material costs

$$\text{GNP} \equiv \text{GNI}$$

GNP ≡ *GNI*

9. *All costs thus become sector receipts:* factor costs become household incomes; indirect tax costs become government receipts; depreciation costs become business receipts. In turn, *these sector receipts can be returned to the market as new expenditures* (demand).

10. This gives us *three streams of cost—income—expenditure (demand):*

Sector receipts

      Cost   Income   Expenditure
1. Factor costs→Households→Consumer goods
                 (Direct taxes)

Three streams of GNP

2. Indirect taxes→Government agencies→Government goods
                 (Direct taxes)

3. Depreciation→Business firms→Replacement investment

Expenditure

11. *The key linkage is that between receipts and expenditure. All costs must become receipts. But not all receipts need necessarily be spent.*

12. When all receipts are spent, then we have a perfect *circular flow.* In such a situation, the economy generates an unchanging flow of demand and therefore experiences neither boom, recession, nor growth.

Circular flow

QUESTIONS

1. How can a model elucidate reality when it is deliberately stripped of the very things that make reality interesting?

2. Why do we need a model to show that an economy can buy back its own production?

3. What are factor costs? What kinds of factor costs are there? To what sector do factor costs go?

4. What are direct taxes? What are indirect taxes? Which are considered part of production costs? Why?

5. To whom are materials costs paid? Why are they not part of the sum total of costs in GNP?

6. What is depreciation? Why is it a part of costs? Who receives the payments or accruals made for depreciation purposes?

7. Show in a carefully drawn diagram how costs become income or receipts of the different sectors.

8. Show in a second diagram how the incomes of the various sectors can become expenditures.

9. Why is the link between expenditure and receipt different from that between receipt and expenditure?

10. What is meant by a circular flow economy? Why does such an economy have neither growth nor fluctuation?

11. Explain the two different ways of looking at GNP and write the simple formula for each. Why is GNP the same thing as GNI?

12. Can we have demand without expenditure?

Supplement on the National Income and Product Accounts

The official accounts of the U.S. economy, the National Income and Product Accounts, illustrate all of the strengths and weaknesses of economic statistics. These statistics are published every quarter, but in the July issue of the *Survey of Current Business* (a publication of the U.S. Department of Commerce) a detailed set of accounts is given for the previous year. We will examine only the most basic tables. The official accounts present thousands of different numbers in dozens of formats as different ways of examining the economy. Any good economist should become familiar with these tables.

As we know, there are two prime ways of viewing the economy in the circular flow. One can look at *costs* (or *incomes*) or at *outputs*. Table 22·2 presents broad categories of each of these accounts. To avoid the confusion of calling the same number two different things—the gross national product and the gross national income or

costs—the National Income and Product Accounts do not use the term gross national income, but refer to the sum of both columns as the gross national product.

Many of the terms in the table are now familiar to us. But let us examine a few that need special attention.

First, note (on the output side) that *change in inventories*, not inventories, enters the GNP. Since GNP measures goods and services produced during a given year, it takes into account goods that have been made and put into inventory only during that year. Goods held over in inventory from last year do not count.

What happens, then, if we sell goods out of last year's inventory and do not replace them? From the point of view of overall output, this is exactly the same as if a business firm did not replace its worn-out capital. By convention, we count this diminution in the level of our capital

TABLE 22·2
The National Income and Product Accounts: 1973

Output		Income	
Personal consumption expenditures	804.0	1. Compensation of employees	785.2
Durable goods	130.8	2. Proprietors income	84.2
Nondurable goods	335.9	3. Rental income	25.1
Services	337.3	4. Corporate profits and inventory valuation	
Gross private domestic investment	202.1	adjustment	109.4
Nonresidential structures and		5. Net interest	50.4
equipment	136.2	6. National income	1054.3
Residential structures	58.0	7. Indirect business taxes and nontax liability	117.8
Change in inventories	8.0	8. Business transfer payments	4.9
Net exports of goods and services	5.8	9. Statistical discrepancy	2.9
Exports	102.0	10. Minus: subsidies less current surplus of	
Imports	96.2	government enterprises	0.4
Government purchases	277.1	11. Net national product	1179.5
Federal	106.6	12. Capital consumption allowances	109.6
defense	73.9	13. Gross national (income) product	1289.1
nondefense	32.7		
State and local	170.5		
Gross national product	1289.1		

stock as a fall in the total value of output. Hence, the item "change in inventories" is the only item on the product side of GNP that can have a negative value. We cannot produce less than zero consumers goods or government output, but we can produce less goods for inventories than we need to maintain their levels.

The first strange term on the income side is the *inventory valuation adjustment* (no. 4 on the list). Remember than the GNP accounts attempt to measure incomes that are produced this year. If a corporation made a good last year but did not sell it, it will add that good to its inventories. If the good is actually sold this year when prices and costs have risen, measured corporate profits will be higher than they would have been if the good had been produced in the year in which it was sold. The good is sold for this year's higher prices, but the cost of making it is last year's lower cost. As a result, measured profits are higher because of production undertaken in the past. If we are trying to measure the profits that are produced this year, we must subtract these extra profits; and that is what the "valuation adjustment" attempts to do. (If prices and costs were falling, we would make a similar adjustment to increase this year's true profits.)

Indirect business taxes (no. 7) are already known to us. They are the sales and excise taxes (gas taxes, liquor taxes, etc.) that are assessed on the value of some products. *Nontax liabilities* (no. 7) refer to the public fees (licences, etc.) that are collected from businesses. *Business transfer payments* (no. 8) are corporate gifts to nonprofit institutions, consumer bad debts, and a few other minor payments.

For the moment, let's skip the statistical discrepancy and move on to *subsidies less current surplus of government enterprises* (no. 10). This entry refers to the profits and losses of TVA, state liquor stores, and other such government businesses. When the government loses money on one of its enterprises, these losses must be subtracted from the other income flows, because a loss means that incomes have been paid out, but no corresponding product has been produced.* If the losses were not subtracted, the gross national

*If a private business loses money, this shows up in negative corporate profits or proprietor's income. Thus, this is really a government counterpart to these private income categories.

product would not equal the gross national income.

This loss is an example of a true but misleading number. The loss occurs because some activities that the ordinary individual would not consider a business are counted as a business. The agriculture support programs of the federal government are such a "business." When the government pays the farmer $2.00 for a bushel of corn and then sells it for $1.50, it incurs a loss. The purpose is to raise farm incomes through subsidies, but this shows up in the GNP accounts as though it were an accidental loss. Most government businesses that are designed to make money do in fact make money.

Subtracting subsidies less surpluses is just a convention. One could just as easily add surpluses less subsidies. In the former case you are subtracting a positive number; in the latter case you are adding a negative number.

Depreciation Charges in the National Income and Product Accounts

The charges for *capital consumption allowances* (no. 12) illustrate the care that must be used in interpreting any economic statistics. Real economic depreciation is a measure of how fast capital is wearing out, but depreciation charges are also very important in determining the profits on which taxes are levied. Depreciation charges are subtracted from gross corporate income to obtain taxable profits. Income taxes are levied only on profits, not upon depreciation charges.

Since World War II the federal government has repeatedly lowered business income taxes by raising the allowable depreciation charges, thereby lowering taxable profits. While the statutory corporate profit tax rate has remained constant at approximately 50 percent, the effective tax rate (tax collections as a percent of gross corporate income) has fallen from 39 percent in 1953 to 27 percent in 1973, because allowable depreciation has grown so rapidly.

Tax allowances for depreciation are now so large that they are far above real economic depreciation. Yet the GNP accounts use the tax laws rather than actual studies of real depreciation to estimate depreciation charges in the National Accounts. Why? There is a law stating that the U.S. Treasury must adjust depreciation charges (which enter into the calculation of taxable profits) to reflect real economic depreciation. To have the U.S. Department of Commerce use different estimates of real economic depreciation would be to accuse the U.S. Treasury of illegal actions. As a result, the two must use identical measures of depreciation—those allowed by the tax code, and not necessarily those called for by economic analysis. The net national product is consequently not a very accurate measure of what it is intended to measure—the output of goods and services that can be used after restoring our national capital to its previous level.

Now for the *statistical discrepancy* (no. 9). As we have seen, theoretically the gross national product must equal the gross national income. In practice, however, if you gave one group of statisticians the task of estimating the gross national product and another the task of estimating the gross national income, they would not come up with identical numbers. Why? Each side of the accounts is subject to measurement errors, and there is no reason why the errors should be identical. The statistical discrepancy is an estimate of these errors. Since the GNP and the GNI must be equal, we add a term (positive or negative) to the income side of the accounts to make them equal. The statistical discrepancy is simply the number that will make the final sum of both sides of the accounts the same.

Production of GNP

We have examined only the most basic table of GNP; and as we have said, there are dozens more. One of them is worth looking at. It shows who *produces* GNP, rather than who buys it or who earns it.

TABLE 22 · 3
Production of GNP, 1973

Private gross national product	1141.6
Business	1090.9
Farm	47.7
Nonfarm	1043.2
Households and institutions	41.1
Rest of the world	9.6
Gross government product	147.5

NOTE: Institutions are nonprofit private institutions such as universities and hospitals. They are added together with households, since neither attempts to earn profits. The *household* production of GNP includes the services of domestic servants and others.

Notice also that in this table, consumption and investment disappear, since all sectors—farms, households, governments, etc.—both consume and invest.

Many other tables are to be found in the Department of Commerce publications, including tables that show in much finer detail the large figures that appear on these tables. For anyone doing research on the activity of the economy, the annual July issue of the *Survey of Current Business* is indispensable. But before using these tables, one must understand the problems and pitfalls we have discussed in this introduction to the U.S. Income and Product Accounts.

Saving and
Investment

23

LET US RETURN FOR A MOMENT to our original perspective, an aerial view of the economic flow. We will remember that we could see the workings of the economy as an interaction between the factors of production and their environment, culminating in a stream of production—some private, some public—that was used in part for consumption and in part for the replacement or the further building up of capital. In our model of a circular flow economy we saw how such an economy can be self-sustaining and self-renewing, as each round of disbursements by employers found its way into a stream of purchasing power just large enough to justify the continuation of the given scale of output.

The Meaning of Saving

Yet we all know that such a circular flow is a highly unreal depiction of the world. In-deed, it omits the most important dynamic factor of real economic life—the steady accumulation of new capital (and the qualitative change in the nature of the capital due to technology) that characterizes a *growing* economy. What we must now investigate is the process by which society adds each year to its stock of real wealth—and the effect of this process on the circuit of production and purchasing.

We begin by making sure that we understand a key word in this dynamic analysis—*saving*. We have come across saving many times by now in this book, and so we should be ready for a final mastery of this centrally important economic term. In Chapter 19, "Wealth and Output," we spoke of saving in *real* terms as the act by which society relinquished resources that might have been used for consumption, thereby making them available for the capital-building stream of output. Now we must translate that underlying real meaning of saving into terms corresponding with the buying and selling, paying and receiving discussed in the preceding chapter.

What is saving in these terms? It is very simply *not spending all or part of income for consumption goods or services.** It should be very clear then why saving is such a key term. In our discussion of the circular flow, it became apparent that expenditure was the critical link in the steady operation of the economy. If saving is not-spending, then it would seem that saving could be the cause of just that kind of downward spiral of which we caught a glimpse in our preceding chapter.

And yet this clearly is not the whole story. We also know that the act of investing—of spending money to direct factors into the production of capital goods—requires an act of saving; that is, of not using that same money to direct those factors instead into the production of consumers goods. *Hence, saving is clearly necessary for the process of investment.* Now, how can one and the same act be necessary for economic expansion and a threat to its stability? This is a problem that will occupy us during much of the coming chapters.

GROSS VS. NET SAVING

It will help us understand the problem if we again have recourse to the now familiar diagram of the circular flow. But this time we must introduce into it the crucial new fact of net saving. Note *net* saving. Quite unnoticed,

we have already encountered saving in our circular flow. In our model economy, when business made expenditures for the replacement of capital, it used money that *could* have been paid in dividends to stockholders or in additional compensation to employees. Before a replacement expenditure was made, someone had to decide not to allocate that money for dividends or bonuses. Thus, there is a flow of saving—that is, of nonconsumption—even in the circular flow.

But this saving is not *net* saving. Like the regular flow of replacement investment itself, the flow of saving that finances this replacement serves only to maintain the existing level of capital wealth, not to increase it. Hence, just as with investment, we reserve the term *net saving* for saving that makes possible a rise in the total of our capital assets.

Gross and net saving are thus easy to define. *By gross saving we mean all saving, both for replacement and for expansion of our capital assets, exactly like gross investment. By net saving, we mean any saving that makes possible an increase in the stock of capital, again exactly as in the definition of net investment.*

We have already seen that an economy can maintain a circular flow when it saves only as much as is needed to maintain its capital. But now suppose that it saves more than that, as is shown in Fig. 23·1. Here householders save a portion of their incomes, over and above the amount saved by business to insure the maintenance of its assets.*

---

*Note "for consumption goods or services." Purchasing stocks or bonds or life insurance is also an act of saving, even though you must spend money to acquire these items. What you acquire, however, are assets, not consumption goods and services. Some acts of spending are difficult to classify. Is a college education, for instance, a consumption good or an investment? As we know, it is probably better thought of as an investment, even though in the statistics of GNP it is treated as consumption.

*Figure 23·1 represents all net saving as occurring in households, but it should be emphasized that a large fraction of this household savings actually takes place in corporations. When a corporation saves money, it is retaining earnings in the name of its owners—individuals who belong to households. Corporate saving can be thought of as household saving, since it ultimately belongs to households, but it is not directly under the control of households in the same sense that personal savings is under their control. We will discuss corporate saving in a later chapter.

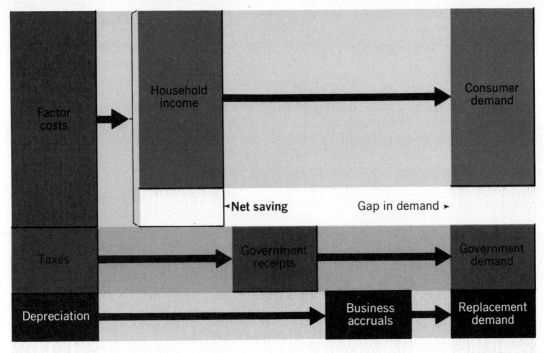

FIGURE 23 · 1
The demand gap

THE DEMAND GAP

What we see is precisely what we would expect. *There is a gap in demand introduced by the deficiency of consumer spending.* This means that the total receipts of employers who make consumer goods will be less than the total amounts they laid out. It begins to look as if we were approaching the cause of economic recession and unemployment.

Yet, whereas we have introduced net saving, we have forgotten about its counterpart, net investment. Cannot the investment activity of a growing economy in some way close the demand gap?

THE DILEMMA OF SAVING

This is indeed, as we shall soon see, the way out of the dilemma. But before we trace the way investment compensates for saving, let us draw some important conclusions from the analysis we have made up to this point.

1. Any act of saving, in and by itself, creates a gap in demand, a shortage of spending. Unless this gap is closed, there will be trouble in the economic system, for employers will not be getting back as receipts all the sums they laid out.

2. If the gap is caused by saving that is implicit in depreciation, it can be closed by replacement expenditures. But if it is caused by net saving, over and above the flow needed to maintain the stock of capital, it will require net investment to be closed.

3. The presence of a demand gap forces us to make a choice. If we want a dynamic, investing economy, we will have to be prepared to cope with the problems that net saving raises. If we want to avoid these problems, we can close the gap by urging consumers or corporations not to save. Then we would have a dependable circular flow, but we would no longer enjoy economic growth.

The Offset to Savings

How, then, shall we manage to make our way out of the dilemma of saving? The previous diagram makes clear what must be done. If a gap in demand is due to the savings of households, then *that gap must be closed by the expanded spending of some other sector*. There are only two other such sectors: government or business. Thus in some fashion or other, the savings of one sector must be "offset" by the increased activity of another.

But how is this offset to take place? How are the resources that are relinquished by consumers to be made available to entrepreneurs in the business sector or to government officials? In a market economy there is only one way that resources or factors not being used in one place can be used in another. Someone must be willing and able to hire them.

Whether or not government and business *are* willing to employ the factors that are not needed in the consumer goods sector is a very critical matter, soon to command much of our attention. But suppose that they are willing. How will they be able to do so? How can they get the necessary funds to expand their activity?

INCREASING EXPENDITURE

There are six principal methods of accomplishing this essential increase in expenditure.

1. The business sector can increase its expenditures by *borrowing* the savings of the public through the sale of new corporate bonds.

2. The government sector can increase its expenditures by *borrowing* savings from the other sectors through the sale of new government bonds.

3. The business sector can increase its expenditures by attracting household savings into partnerships, new stock, or other *ownership or equity*.

4. Both business and government sectors can increase expenditures by *borrowing* additional funds from commercial banks.*

5. The government sector can increase its expenditures by *taxing* the other sectors.

6. Both business and government sectors can increase their expenditures by drawing on *accumulated past savings*, such as unexpended profits or tax receipts from previous years.

The first four methods above have one attribute that calls them especially to our attention. They give rise to *claims* that reveal from whom the funds have been obtained and to whom they have been made available, as well as on what terms. Bonds, corporate or government, show that savings have been borrowed from individuals or banks or firms by business and government units. Shares of stock reveal that savings have been obtained on an equity (ownership) basis, as do new partnership agreements. Borrowing from banks gives rise to loans that also represent the claims of one part of the community against another.

We can note a few additional points about claims, now that we see how many of them arise in the economy. First, many household savings are first put into banks and insurance companies—so-called financial intermediaries—so that the transfer of funds from households to business or government may go through several stages; e.g., from household to insurance company and then from insurance company to corporation.

Second, not *all* claims involve the offsetting of savings of one sector by expenditures of another. Many claims, once they have arisen, are traded back and forth and bought and sold, as is the case with most stocks and bonds. These purchases and sales involve the *transfer of existing claims*, not the creation of new claims.

*Actually, they are borrowing from the public through the means of banks. We shall learn about this in Chapter 30.

Finally, not every new claim necessarily involves the creation of an asset. If A borrows $5 from B, bets it on the races, and gives B his note, there has been an increase in claims, but no new asset has been brought into being to match it.

PUBLIC AND PRIVATE CLAIMS

Now let us look at Fig. 23·2. This time we show what happens when savings are made available to the business sector by direct borrowing from households. Note the claim (or equity) that arises.

If the government were doing the borrowing, rather than the business sector, the diagram would look like Fig. 23·3, p. 354. Notice that the claim is now a government bond.

We have not looked at a diagram showing business or government borrowing its funds from the banking system. (This process will be better understood when we take up the problem of money and banking, in Chapter 30.) The basic concept, however, although more complex, is much the same as above.

COMPLETED ACT OF OFFSETTING SAVINGS

There remains only a last step, which must now be fully anticipated. We have seen how it is possible to offset the savings in one

FIGURE 23 · 2
"Transfer" of savings to business

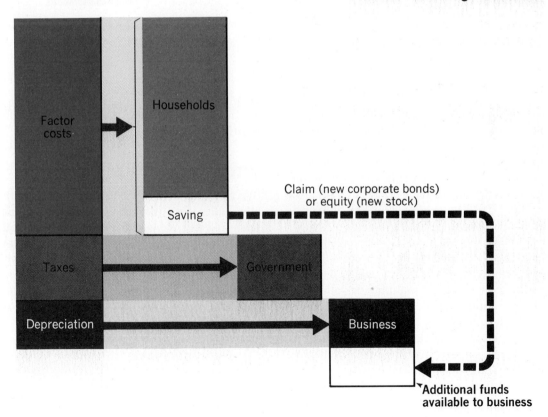

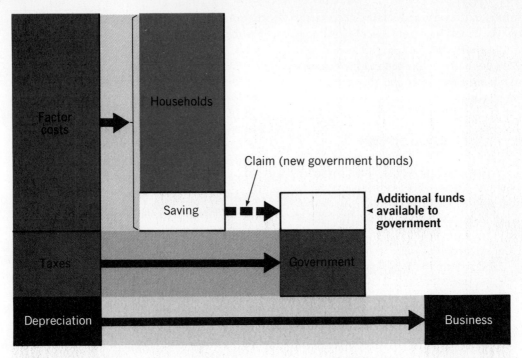

Claim (new government bonds)

Additional funds
available to
government

FIGURE 23 · 3
"Transfer" of savings to government

sector, where they were going to cause an ex-penditure gap, by increasing the funds available to another sector. It remains only to *spend* those additional funds in the form of additional investment or, in the case of the government, for additional public goods and services. The two completed expenditure cir-cuits now appear in Fig. 23·4.

While Fig. 23·4 is drawn so that the new investment demand or new government de-mand is exactly equal to net saving, it is im-portant to understand that there is nothing in the economic system guaranteeing that these demands will exactly equal net saving. The desire for new investment or new government goods and services may be either higher or lower than new saving. The need to regulate these new demands so that they will equal net savings is an important objective of *fiscal*

and monetary policies, a problem we will study later.

INTERSECTORAL OFFSETS

We shall not investigate further at this point the differences between increased public spending and increased business investment. What we must heed is the crucial point at issue: *if saving in any one sector is to be offset, some other sector (or sectors) must spend more than its income. A gap in demand due to in-sufficient expenditure in one sector can be compensated only by an increase in demand— that is, in expenditure—of another.*

Once this simple but fundamental point is clearly understood, much of the mystery of macroeconomics disappears, for we can then begin to see that an economy in move-

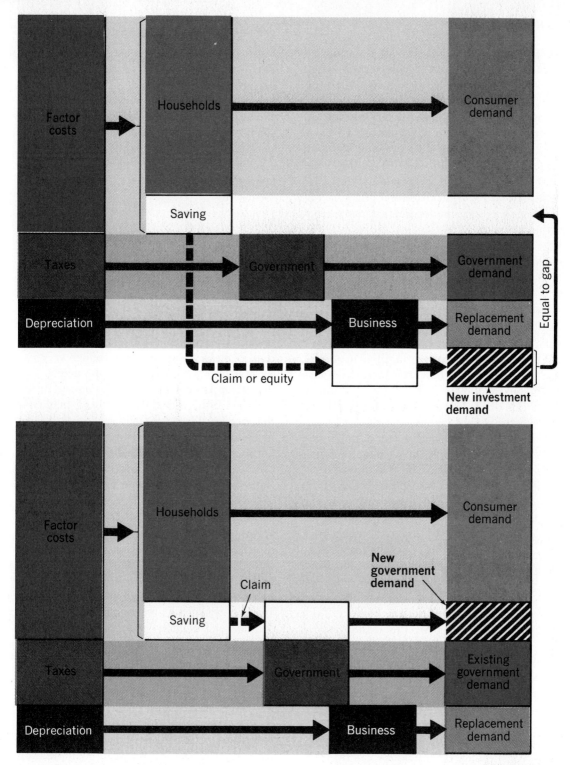

FIGURE 23 · 4
Two ways of closing the demand gap

ment, as contrasted with one in a stationary circular flow, is one in which sectors must *cooperate* to maintain the closed circuit of income and output. In a dynamic economy, we no longer enjoy the steady translation of incomes into expenditure which, as we have seen, is the key to an uninterrupted flow of output. Rather, we are faced with the presence of net saving and the possibility of a gap in final demand. Difficult though the ensuing problems are, let us not forget that net saving is the necessary condition for the accumulation of capital. *The price of economic growth, in other words, is the risk of economic decline.*

REAL AND MONEY SAVING

This central importance of saving in a growing economy will become a familiar problem. At this juncture, where we have first encountered the difficulties it can pose, we must be certain that we understand two different aspects that saving assumes.

One aspect, noticed in our initial overview of the economy, is the decision to relinquish *resources* that can be redeployed into capital-building. This is the real significance of saving. But this "real" aspect of saving is not the way we encounter the act of saving in our ordinary lives. We think of saving as a *monetary* phenomenon, not a "real" one. When we save, we are conscious of not using all our incomes for consumption, but we scarcely, if ever, think of releasing resources for alternative employments.

There is a reason for this dichotomy of real and money saving. In our society, with its extraordinary degree of specialization, the individuals or institutions that do the actual saving are not ordinarily those that do the actual capital-building. In a simple society, this dichotomy between saving and investing need not, and usually does not, occur. A farmer who decides to build new capital—for example, to build a barn—is very much

aware of giving up a consumption activity— the raising of food—in order to carry out his investment. So is an artisan who stops weaving clothing to repair his loom. Where the saver and the investor are one and the same person, there need be no "financial" saving, and the underlying real phenomenon of saving as the diversion of activity from consumption to investment is immediately apparent.

SAVERS AND INVESTORS

In the modern world, savers and investors are sometimes the same individual or group —as in the case of a business management that spends profits on new productive capacity rather than on higher executive salaries. More often, however, savers are not investors. Certainly householders, though very important savers, do not personally decide and direct the process of capital formation in the nation. Furthermore, the men and materials that households voluntarily relinquish by not using all their incomes to buy consumers goods have to be physically transferred to different industries, often to different occupations and locations, in order to carry out their investment tasks. This requires funds in the hands of the investors, so that they can tempt resources from one use to another.

Hence we need an elaborate system for directly or indirectly "transferring" money saving into the hands of those who will be in a position to employ factors for capital construction purposes. Nevertheless, underlying this complex mechanism for transferring purchasing power remains the same simple purpose that we initially witnessed. Resources that have been relinquished from the production of consumption goods or services are now employed in the production of capital goods. Thus, *saving and investing are essentially real phenomena*, even though it may take a great deal of financial manipulation to bring them about.

A final important point. *The fact that the decisions to save and the decisions to invest are lodged in different individuals or groups alerts us to a basic reason why the savings-investment process may not always work smoothly.* Savers may choose to consume less than their total incomes at times when investors have no interest in expanding their capital assets. Alternatively, business firms may wish to form new capital when savers are interested in spending money only on themselves. This separation of decision-making can give rise to situations in which savings are not offset by investment or in which investment plans race out ahead of savings capabilities. In our next chapters we will be investigating what happens in these cases.

Transfer Payments and Profits

We have talked about the transfer of purchasing power from savers to investors, but we have not yet mentioned another kind of transfer, also of great importance in the overall operation of the economy. This is the transfer of incomes from sector to sector (and sometimes within sectors).

TRANSFERS

As we already know, income transfers (called *transfer payments*) are a very useful and important means of reallocating purchasing power in society. Through transfer payments, members of the community who do not participate in production are given an opportunity to enjoy incomes that would otherwise not be available to them. Thus Social Security transfer payments make it possible for the old or the handicapped to be given an "income" of their own (not, to be sure, a currently *earned* income), or unemployment benefits give purchasing power to those who cannot get it through employment.

Not all transfers are in the nature of welfare payments, however. The distribution of money *within* a household is a transfer payment. So is the payment of interest on the national debt.* So is the grant of a subsidy to a private enterprise, such as an airline, or of a scholarship to a college student. Any income payment that is not earned by selling one's productive services on the market falls in the transfer category.

It may help to understand this process if we visualize it in our flow diagram. Figure 23·5, p. 358, shows two kinds of transfers. The upper one, from government to the household sector, shows a typical transfer of incomes, such as veterans' pensions or Social Security; the transfer below it reflects the flow of income that might be illustrated by a payment to agriculture for crop support. Transfers *within* sectors, such as household allowances, are not shown in the diagram.

One thing we may well note about transfers is that they can only *rearrange* the incomes created in the production process; they cannot increase those incomes. Income, as we learned in the last chapter, is inextricably tied to output—indeed, income is only the financial counterpart of output.

Transfer payments, on the other hand, are a way of arranging individual claims to production in some fashion that strikes the community as fairer or more efficient or more decorous than the way the market process allocates them through the production process. As such, transfer payments are an indispensable and often invaluable agency of social policy. But it is important to understand that no amount of transfers can, in itself, increase the total that is to be shared. That can happen only by raising output itself.

---

*As we know, the payment of interest on corporate debt is not considered a transfer payment, but a payment to a factor of production. Actually, much government interest should also be thought of as a factor payment (for the loan of capital for purposes of public output); but by convention, all government interest is classified as a transfer payment.

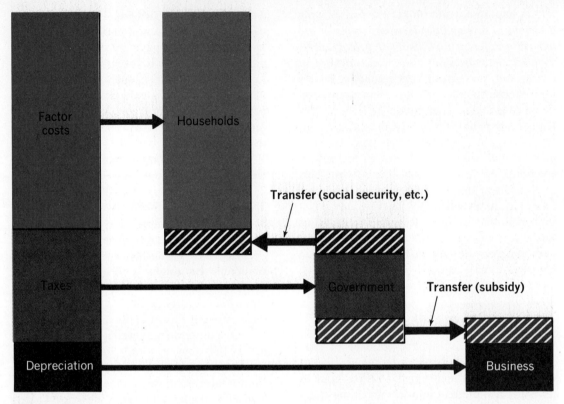

FIGURE 23 · 5
Transfer payments

TRANSFER PAYMENTS AND TAXES

We have mentioned, but only in passing, another means of transferring purchasing power from one sector to another: taxation. Heretofore, however, we have often spoken as though all government tax receipts were derived from indirect taxes that were added onto the cost of production.

In fact, this is not the only source of government revenue. Indirect taxes are an important part of state and local revenues, but they are only a minor part of federal tax receipts. Most federal taxes are levied on the incomes of the factors of production or on the profits of businesses after the other factors have been paid.

Once again it is worth remembering that the government taxes consumers (and businesses) because it is in the nature of much government output that it cannot be *sold*. Taxes are the way we are billed for our share —rightly or wrongly figured—of government production that has been collectively decided upon. As we can now see, taxes—both on business and on the household sector—also finance many transfer payments. That is, the government intervenes in the distribution process to make it conform to our politically expressed social purposes, taking away some incomes from certain individuals and groups, and providing incomes to others. Figure 23·6 shows what this looks like in the flow of GNP. (Note that the business sector is drawn with profits, as our next section will explain.)

As we can see, the exchanges of income between the household and the government sectors can be very complex. Income can flow from households to government units via taxation, and return to the household sector via transfer payments; and the same two-way flows can take place between government and business.

PROFITS AND DEMAND

The last diagram has already introduced a new element of reality in our discussion. Taxes on business *income* presuppose that businesses make *profits.* Let us see how these profits fit into the savings-investment process.

During our discussion of the circular flow, we spoke of profits as a special kind of factor cost—a payment to the factor *capital.* Now we can think of profits not merely as a factor cost (although there is always a certain element of risk-remuneration in profits), but as a return to especially efficient or forward-thinking firms who have used the investment process to introduce new products or processes ahead of the run of their industries. We also know that profits accrue to powerful firms who exact a semimonopolistic return from their customers.

FIGURE 23 · 6
Transfers and income taxes

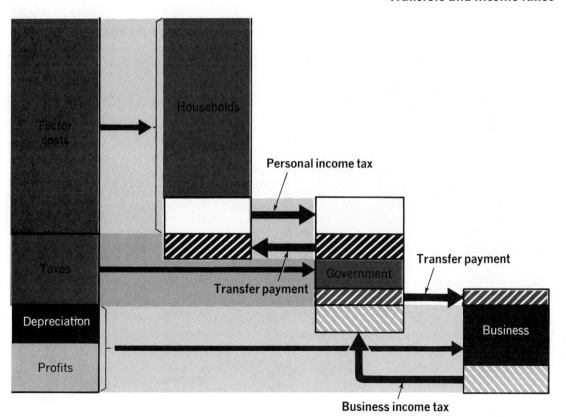

What matters in our analysis at this stage is not the precise explanation we give to the origin of profits, but a precise explanation of their role in maintaining a "closed-circuit" economy in which all costs are returned to the marketplace as demand. A commonly heard diagnosis for economic maladies is that profits are at the root of the matter, in that they cause a "withdrawal" of spending power or income from the community. If profits are "hoarded," or kept unspent, this can be true. In fact, however, profits are usually spent in three ways. They may be

1. Distributed as income to the household sector in the form of dividends or profit shares, to become part of household spending

2. Spent by business firms for new plant and equipment

3. Taxed by the government and spent in the public sector

All three methods of offsetting profits appear in Fig. 23·7.

Thus, we can see that profits need not constitute a withdrawal from the income stream. Indeed, unless profits are adequate, businesses will very likely not invest enough to offset the savings of the household sector. They may, in fact, even fail to make normal replacement expenditures, aggravating the demand gap still further in this way.

Thus the existence of profits, far from being deflationary—that is, far from causing a fall in income—is, in fact, essential for the

FIGURE 23·7
Profits in the circular flow

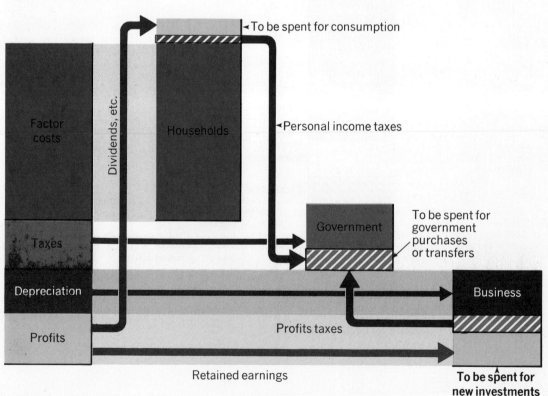

maintenance of a given level of income or for an advance to a higher level. Nonetheless, there is a germ of truth in the contentions of those who have maintained that profits can cause an insufficiency of purchasing power. *For unless profits are returned to the flow of purchasing power as dividends that are spent by their recipients or as new capital expenditures made by business or as taxes that lead to additional public spending, there will be a gap in the community's demand.* Thus we can think of profits just as we think of saving— an indispensable source of economic growth or a potential source of economic decline.

SAVING, INVESTMENT, AND GROWTH

We are almost ready to leave our analysis of the circle of production and income and to proceed to a much closer study of the individual dynamic elements that create and close demand gaps. Before we do, however, it is well that we take note of one last fact of the greatest importance. In offsetting the savings of any sector by investment, we have closed the production and income circuit, much as in the stationary circular flow, but there is one crucial difference from the circular flow. Now we have closed the flow by diverting savings into the creation of *additional* capital. Unlike the stationary circular flow where the handing around of incomes did no more than to maintain unchanged the original configuration of the system, in our new dynamic saving-and-investing model *each closing of the circuit results in a quantitative change— the addition of a new "layer" of capital.*

Hence, more and more physical wealth is being added to our system; and thinking back to our first impressions of the interaction of wealth and population, we would expect more and more productiveness from our human factors. With complications that we shall have to deal with in due course, *growth* has entered our economic model.

KEY WORDS	CENTRAL CONCEPTS

Saving

1. The critical element missing from the concept of the circular flow is *saving*. The key question is how an economy can buy back all its output when some of its receipts are saved rather than returned to the market through expenditure.

Demand gap

2. Saving poses a dilemma. On one hand, it breaks the circular flow and creates a *demand gap*. On the other hand, if there is *no saving*, there can be *no investment*.

3. The answer to the dilemma is that *saving*, although essential for growth, *must be offset by additional expenditure*. This means that another sector must spend more than its income.

Inter-sectoral cooperation

4. In six main ways, a sector can spend more than its income:
   - The *business sector can borrow* from the household sector.
   - The *government sector can borrow* from the household sector.
   - The *business sector* can attract household savings into *equities*.
   - Business and government can borrow from the *commercial banks*.
   - Governments can *tax* other sectors.
   - Business and government can spend *past savings*.

Real saving

5. Although saving involves money, it is essentially a "real" process (as is investment). That is, its real meaning is that resources are released from consumption. The acts of releasing resources (saving) and the acts of employing them (investment) are usually performed by different groups in modern society.

Transfer payments

6. *Transfer payments*, from one sector to another or within one sector, play an important part in *redistributing* income, but do not increase the total GNP.

7. Profits can be returned to the expenditure flow by being: (1) paid out as dividends, etc., to the household sector, where they can be used for consumption; (2) spent by business firms for new investment; or (3) taxed by the government and spent by it.

Profits and expenditure

8. Saving thus requires investment to assure that all payments will be returned to the market as demand. Note, however, that the process of *investment adds to capital* and thereby increases productivity. Saving and investment are therefore an integral part of the *process of growth*.

Growth

QUESTIONS

1. What do we mean by a demand gap? Show diagrammatically.

2. How is a demand gap filled by business investment? Show diagrammatically.

3. Why is saving indispensable for growth?

4. Can we have planned business investment without saving? Saving without planned business investment?

5. Draw carefully a diagram that shows how savings can be offset by government spending.

6. How is it possible for a sector to spend more than its income? How does it get the additional money?

7. What is a transfer payment? Draw diagrams of transfers from government to consumers, from government to business. Is charity a transfer? Is a lottery?

8. Diagram the three ways in which profits can be returned to the expenditure flow. What happens if they are not?

9. Why is a problem presented by the fact that those who make the decision to invest are not the same people?

10. In what way is a stationary circular flow economy different from an economy that saves and invests?

Consumption Demand

With a basic understanding of the crucial role of expenditure and of the complex relationship of saving and investment behind us, we are in a position to look more deeply into the question of the determination of gross national product. For what we have discovered heretofore is only the *mechanism* by which a market economy can sustain or fail to sustain a given level of output through a circuit of expenditure and receipt. Now we must try to discover the *forces* that dynamize the system, creating or closing gaps between income and outgo. What causes a demand for the goods and services measured in the GNP? Let us begin to answer that question by examining the flow of demand most familiar to us—consumption.

The Household Sector

Largest and in many respects most important of all the sectors in the economy is that of the nation's households—that is, its families and single-dwelling individuals (the two categories together called consumer units) considered as receivers of income and transfer payments or as savers and spenders of money for consumption.

How big is this sector? In 1973 it comprised some fifty-one million families and some fourteen million independent individuals who collectively gathered in $1,025 billion in income and spent $828 billion.* As Fig.

*The Department of Commerce has recently redefined some categories of the national income accounts, and the word *consumption* today applies, strictly speaking, only to personal expenditures for goods and services. Included in total consumer spending, however, are sizeable amounts for interest (mainly on installment loans) and for remittances abroad, neither of which sums are included in the amount for goods and services. The proper nomenclature for the total of consumer spending (goods and services plus interest and remittances) is now *personal outlays*. We shall, however, continue to use the simpler term, *consumption,* although our figures will be those for personal outlays.

Note, also, that the compilation of these figures is a time-consuming process in which earlier estimates are frequently subject to revision. Hence, figures for the components of consumption or, for that matter, for almost all magnitudes in the economic process are apt to vary slightly in successive printed statistics until, eventually, the "final" figures are arrived at. Note our caution about preliminary data in the tinted box "Cautions about Data," Chapter 6, and also in Chapter 42, p. 674.

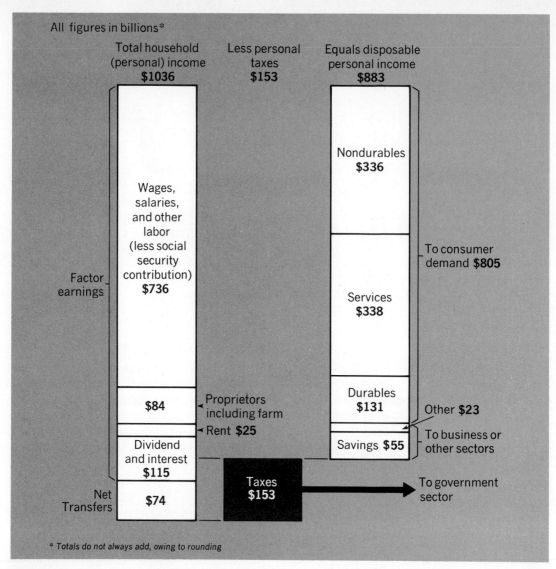

FIGURE 24 · 1
Consumption sector 1973

24·1 shows, the great bulk of receipts was from factor earnings, and transfer payments played only a relatively small role. As we can also see, we must subtract personal tax payments from household income (or *personal income* as it is officially designated) before we get *disposable personal income*— income actually available for spending. It is from disposable personal income that the crucial choice is made to spend or save.

Much of this chapter will focus on that choice.

SUBCOMPONENTS OF CONSUMPTION

Finally we note that consumer spending itself divides into three main streams. The largest of these is for *nondurable* goods, such as food and clothing or other items whose economic life is (or is assumed to be) short. Second largest is an assortment of expenditures we call consumer *services,* comprising things such as rent, doctors' or lawyers' or barbers' ministrations, theater or movie admissions, bus or taxi or plane transportation, and other purchases that are not a physical good but work performed by someone or some equipment. Last is a substream of expenditure for consumer *durable* goods, which, as the name suggests, include items such as cars or household appliances whose economic life is considerably greater than that of most nondurables. We can think of these goods as comprising consumers' physical capital.

There are complicated patterns and interrelations among these three major streams of consumer spending. As we would expect, consumer spending for durables is extremely volatile. In bad times, such as 1933, it has sunk to less than 8 percent of all consumer outlays; in the peak of good times in the early 1970s, it came to nearly double that. Meanwhile, outlays for services have been a steadily swelling area for consumer spending in the postwar economy. As a consequence of the growth of consumer buying of durables and of services, the relative share of the consumer dollar going to "soft goods" has been slowly declining. It is interesting to note, for example, that between 1960 and 1972 consumer spending for food and tobacco fell from 30 percent of all consumption to 22 percent, and that expenditures for apparel fell from 12 percent to 10 percent. Conversely, consumer spending on recreation and foreign travel and remittances climbed from 6 percent to more than 7 percent.

CONSUMPTION AND GNP

These internal dynamics of consumption are of the greatest interest to someone who seeks to project consumer spending patterns into the future—perhaps as an aid to merchandising. But here we are interested in the larger phenomenon of the relationship of consumption as a whole to the flow of gross national product.

Figure 24·2 shows us this historic relationship since 1929. Certain things stand out.

1. Consumption spending is by far the largest category of spending in GNP.

Total consumer expenditures—for durable goods such as automobiles or washing machines, for nondurables like food or clothing, and for services such as recreation or medical care—account for approximately two-thirds of all the final buying in the economy.

2. Consumption is not only the biggest, but the most stable of all the streams of expenditure.

Consumption, as we have mentioned, is *the* essential economic activity. Unless there is a total breakdown in the social system, households will consume some bare minimum. Further, it is a fact of common experience that even in adverse circumstances, households seek to maintain their accustomed living standards. Thus consumption activities constitute a kind of floor for the level of overall economic activity. Investment and government spending, as we shall see, are capable of sudden reversals; but the streams of consumer spending tend to display a measure of stability over time.

3. Consumption is nonetheless capable of considerable fluctuation as a proportion of GNP.

Remembering our previous diagrams, we can see that this proportionate fluctuation must reflect changes in the relative importance of investment and government spending. And indeed this is the case. As investment spending fell in the Depression, consumption bulked relatively larger in GNP; as government spending increased during the war,

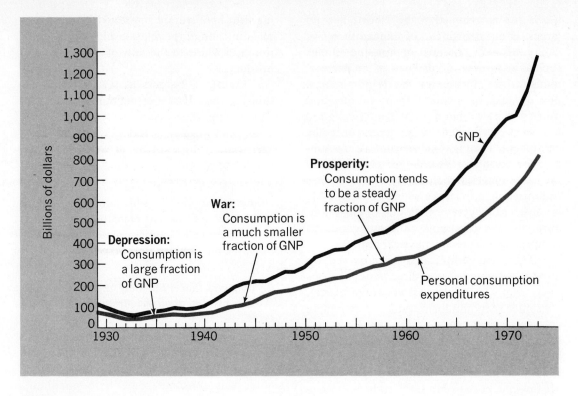

FIGURE 24 · 2
Consumption and GNP, current prices

consumption bulked relatively smaller. The changing *relative* size of consumption, in other words, reflects broad changes in *other* sectors rather than sharp changes in consuming habits.

4. Despite its importance, consumption alone will not "buy back" GNP.

It is well to recall that consumption, although the largest component of GNP, is still *only* two-thirds of GNP. Government buying and business buying of investment goods are essential if the income-expenditure circuit is to be closed. During our subsequent analysis it will help to remember that consumption expenditure by itself does not provide the only impetus of demand.

Saving in Historic Perspective

This first view of consumption activity sets the stage for our inquiry into the dynamic causes of fluctuations in GNP. We already know that the saving-investment relationship lies at the center of this problem, and that much saving arises from the household sector. Hence, let us see what we can learn about the saving process in historic perspective.

We begin with Fig. 24·3, showing the relationship of household saving to disposable income—that is, to household sector incomes after the payment of taxes.

What we see here are two interesting facts. First, during the bottom of the Great

Depression there were *no* savings in the household sector. In fact, under the duress of unemployment, millions of households were forced to *dissave*—to borrow or to draw on their old savings (hence the negative figure for the sector as a whole). By way of contrast, we notice the immense savings of the peak war years when consumers' goods were rationed and households were urged to save. Clearly, then, the *amount* of saving is capable of great fluctuation, falling to zero or to negative figures in periods of great economic distress and rising to as much as a quarter of income during periods of goods shortages.

In Fig. 24·4, p. 368, we are struck by another fact. However variable the amounts, the savings *ratio* shows a considerable stability in "normal" years. This steadiness is particularly noteworthy in the postwar period. From 1950 to the present, consumption has ranged between roughly 92 to 95 percent

FIGURE 24 · 3
Saving and disposable income

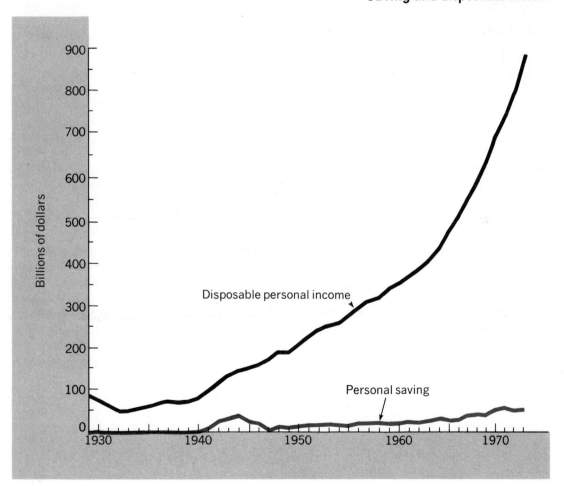

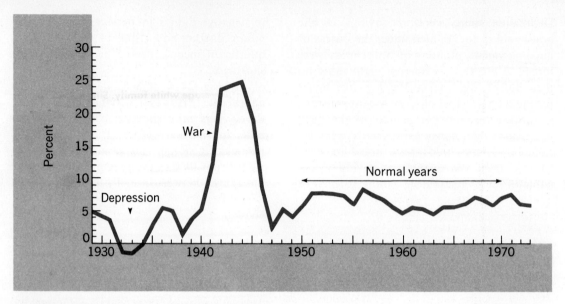

FIGURE 24 · 4
Saving as percent of disposable income

of disposable personal income—which is, of course, the same as saying that savings have ranged roughly between 8 percent and 5 percent. If we take the postwar period as a whole, *we can see that in an average year we have consumed a little more than 94 cents of each dollar of income and that this ratio has remained fairly constant even though our incomes have increased markedly.*

LONG-RUN SAVINGS BEHAVIOR

This stability of the long-run savings ratio is an interesting, important phenomenon and something of a puzzling one, for we might easily imagine that the savings ratio would rise over time. Statistical investigations of cross sections of the nation show that rich families tend to save not only larger amounts, but larger *percentages* of their income, than poor families do. Thus as the entire nation has grown richer and as families have moved from lower income brackets to higher ones, it

seems natural to suppose that they would also take on the higher savings characteristics that accompany upper incomes.

Were this so, the economy would face a very serious problem. In order to sustain its higher levels of aggregate income, it would have to invest an ever larger *proportion* of its income to offset its growing ratio of savings to income. As we shall see in our next chapter, investment is always a source of potential trouble because it is so much riskier than any other business function. If we had to keep on making proportionally larger investments each year to keep pace with our proportionally growing savings, we should live in an exceedingly vulnerable economic environment.

Fortunately, we are rescued from this dangerous situation, because our long-run savings ratio, as we have seen, displays a reassuring steadiness. In fact, there has been no significant upward trend in the savings ratio for the nation's households since the

SHORT-RUN VS. LONG-RUN
SAVINGS BEHAVIOR

How do we reconcile the stability of the long-run savings ratio with the fact that statistical studies always reveal that rich families save a larger percentage of their incomes than do poor families? As the nation has moved, en masse, into higher income brackets, why has it not also saved proportionately more of its income?

The explanation for the long-run stability of savings behavior revolves around the importance of *relative* incomes, or "keeping up with the Joneses", in consumption decisions. If a family earned $11,000 in 1940, it was a wealthy family with an income far above the average. It could save a large fraction of its income and still have more than other families in the community had to spend on consumption. By 1973 the family with an $11,000 annual income was simply an average family. To keep up with consumption standards of other families in the community, it needed to spend a large fraction of its income. As a result, the savings rates for families with $11,000 gradually fell over time as the families changed from wealthy to average.

The same relative income effect is seen in the savings rates of black families. For any given income level, the average black family saves more than the average white family. Since black family incomes are lower than white family incomes, any given income has a higher relative position among blacks than it does among whites. To keep up with their peer group, whites must consequently spend more than blacks.

As a result of these and still other motivations, savings behavior in the long run differs considerably from that in the short run. Over the years, American households have shown a remarkable stability in their rate of overall savings. Its importance has already been mentioned. In a shorter period of time, however—over a few months or perhaps a year—households tend to save higher fractions of increases in their incomes than they do in the long run. The very great importance of this fact we shall subsequently note.

mid-1800s, and there may have been a slight downward trend.*

The Consumption-Income Relationship

What we have heretofore seen are some of the historical and empirical relationships of consumption and personal saving to income. We have taken the trouble to investigate these relationships in some detail, since they are among the most important causes of the gaps that have to be closed by investment. But the statistical facts in themselves are only a halfway stage in our macroeconomic investigation. Now we want to go beyond the facts to a generalized understanding of the behavior that gives rise to them. Thus our next task is to extract from the facts certain behavioral *relationships* that are sufficiently regular and dependable for us to build into a new dynamic model of the economy.

If we think back over the data we have examined, one primary conclusion comes to mind. This is the indisputable fact that the *amount* of saving generated by the household sector depends in the first instance upon the income enjoyed by the household sector. Despite the stability of the savings ratio, we have seen that the dollar volume of saving in the economy is susceptible to great variation, from negative amounts in the Great Depression to very large amounts in boom times.

*Economists maintain a certain tentativeness in their assertions about long-run trends, since the statistical foundation on which they are based is inevitably subject to some error and uncertainty.

Now we must see if we can find a systematic connection between the changing size of income and the changing size of saving.

PROPENSITY TO CONSUME

There is indeed such a relationship, lying at the heart of macroeconomic analysis. We call it the *consumption function* or, more formally, the *propensity to consume,* the name invented by John Maynard Keynes, the famous English economist who first formulated it in 1936.* What is this "propensity" to consume? It means that the relationship between consumption behavior and income is sufficiently dependable so that we can actually *predict* how much consumption (or how much saving) will be associated with a given level of income.

We base such predictions on a *schedule* that enables us to see the income-consumption relationship over a considerable range of variation. Table 24·1 is such a schedule, a purely hypothetical one, for us to examine.

TABLE 24 · 1
A Propensity to Consume Schedule

BILLIONS OF DOLLARS		
Income	Consumption	Savings
$100	$80	$20
110	87	23
120	92	28
130	95	35
140	97	43

One could imagine, of course, innumerable different consumption schedules; in one society a given income might be accompanied by a much higher propensity to consume (or propensity to save) than in another. But the

*More about Keynes in the box on p. 418. Note that his name is pronounced "Kanes," not "Keenes."

basic hypothesis of Keynes—a hypothesis amply confirmed by research—was that the consumption schedule in all modern industrial societies had a particular basic configuration, despite these variations. The propensity to consume, said Keynes, reflected the fact that on the average, men tended to increase their consumption as their incomes rose, but not by as much as their income increased. In other words, as the incomes of individuals rose, so did both their consumption *and their savings.*

Note that Keynes did not say that the proportion of saving rose. We have seen how involved is the dynamic determination of savings ratios. Keynes merely suggested that in the short run, the *amount* of saving would rise as income rose—or to put it conversely again, that families would not use *all* their increases in income for consumption purposes alone. It is well to remember that these conclusions hold in going down the schedule as well as up. Keynes' basic "law" implies that when there is a decrease in income, there will be some decrease in the *amount of saving,* or that a family will not absorb a fall in its income entirely by contracting its consumption.

What does the consumption schedule look like in the United States? We will come to that shortly. First, however, let us fill in our understanding of the terms we will need for our generalized study.

AVERAGE PROPENSITY TO CONSUME

The consumption schedule gives us two ways of measuring the fundamental economic relationship of income and saving. One way is simply to take any given level of income and to compute the percentage relation of consumption to that income. This gives us the *average propensity to consume.* In Table 24·2, using the same hypothetical schedule as before, we make this computation.

TABLE 24 · 2
Calculation of the Average Propensity to Consume

BILLIONS OF DOLLARS		Consumption divided by income (Average propensity to consume)
Income	Consumption	
$100	$80	.80
110	87	.79
120	92	.77
130	95	.73
140	97	.69

The average propensity to consume, in other words, tells us how a society at any given moment divides its total income between consumption and saving. It is thus a kind of measure of long-run savings behavior, for households divide their incomes between saving and consuming in ratios that reflect established habits and, as we have seen, do not ordinarily change rapidly.

MARGINAL PROPENSITY TO CONSUME

But we can also use our schedule to measure another very important aspect of saving behavior: the way households divide *increases* (or decreases) in income between consumption and saving. This *marginal propensity to consume* is quite different from the average propensity to consume, as the figures in Table 24·3 (still from our original hypothetical schedule) demonstrate.

Note carefully that the last column in Table 24·3 is designed to show us something quite different from the last column of the previous table. Take a given income level— say $110 billion. In Table 24·2 the average propensity to consume for that income level is .79, meaning that we will actually spend on consumption 79 percent of our income of $110 billion. But the corresponding figure

TABLE 24 · 3
Calculation of the Marginal Propensity to Consume

BILLIONS OF DOLLARS			Change in consumption	Marginal propensity to consume = Change in consumption ÷ change in income
Income	Consumption	Change in income		
$100	$80	—	—	—
110	87	$10	$7	.70
120	92	10	5	.50
130	95	10	3	.30
140	97	10	2	.20

opposite $110 billion in the marginal propensity to consume table (24·3) is .70. This does *not* mean that out of our $110 billion income we somehow spend only 70 percent, instead of 79 percent, on consumption. It *does* mean that we spend on consumption only 70 percent *of the $10 billion increase* that lifted us from a previous income of $100 billion to the $110 billion level. The rest of that $10 billion increase we saved.

As we know, much of economics, in micro- as well as macroanalysis, is concerned with studying the effects of *changes* in economic life. It is precisely here that marginal concepts take on their importance. When we speak of the average propensity to consume, we relate all consumption and all income from the bottom up, so to speak, and thus we call attention to behavior covering a great variety of situations and conditions. But when we speak of the marginal propensity to consume, we are focusing only on our behavior toward *changes* in our incomes. Thus the marginal approach is invaluable, as we shall see, in dealing with the effects of short-run fluctuations in GNP.

A SCATTER DIAGRAM

The essentially simple idea of a systematic, behavioral relationship between income and consumption will play an extremely important part in the model of the economy we shall soon construct. But the relationships we have thus far defined are too vague to be of much use. We want to know if we can extract from the facts of experience not only a general dependence of consumption on income, but a *fairly precise method of determining exactly how much saving will be associated with a given amount of income.*

Here we reach a place where it will help us to use diagrams and simple equations rather than words alone. So let us begin by transferring our conception of a propensity to consume schedule to a new kind of diagram directly showing the interrelation of income and consumption.

The *scatter diagram* (Fig. 24·5) shows precisely that. Along the vertical axis on the left we have marked off intervals to measure total consumer expenditure in billions of dollars; along the horizontal axis on the bottom we measure disposable personal income, also in billions of dollars. The dots tell us, for the years enumerated, how large consumption and income were. For instance, if we take the dot for 1966 and look directly below it to the horizontal axis, we can see that disposable personal income for that year was roughly $510 billion. The same dot measured against the vertical consumption axis tells us that consumption for 1966 was a little more than $475 billion. If we now divide the figure for consumption by that for income, we get a value of 93.1 percent for our propensity to consume. If we subtract that from 100, our propensity to save must have been 6.9 percent.*

Returning to the diagram itself, we notice that the black line which "fits" the trend of the dots does not go evenly from corner to corner. If it did, it would mean that each amount of income was matched by an *equal* amount of consumption—in other words, that there was no saving. Instead, the line leans slightly downward, indicating that as income goes higher, consumption also increases, but not by quite as much.

Does the chart also show us our marginal propensity to consume? Not really. As we know, our short-run savings propensities are higher than our long-run propensities. This chart shows our "settled" position, from

---

*It is difficult to read figures accurately from a graph. The actual values are: disposable income, $512 billion; consumption, $479 billion; average propensity to consume, 93.4 percent.

For more information on "fitting" a line to such a graph, see Chapter 42, Part 7.

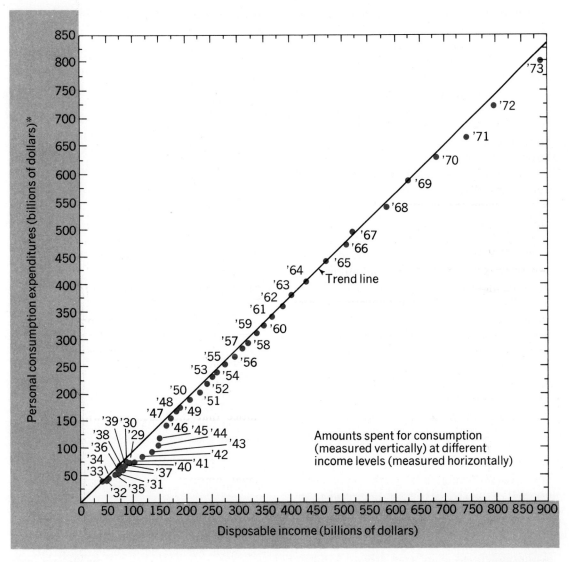

FIGURE 24 · 5
United States' propensity to consume, 1929–1973

year to year, after the long-run, upward drift of spending has washed out our marginal (short-run) savings behavior.

Nevertheless, if we look at the movement from one dot to the next, we get some notion of the short-run forces at work. Dur-

ing the war years, for instance, as the result of a shortage of many consumer goods and a general exhortation to save, the average propensity to consume was unusually low. That is why the dots during those years form a bulge below the trend line. After the war, we can also see that the marginal propensity to consume must have been very high. As a matter of fact, for a few years, consumption actually rose faster than income, as people used their wartime savings to buy things that were unavailable during the war. Between 1946 and 1947, for example, disposable income rose by some $9.8 billion, but personal outlays rose by almost $18 billion! By 1950, however, the consumption-income relationship was back to virtually the same ratio as during the 1930s.

IN SIMPLE MATHEMATICS

There is another way of reducing to short-hand clarity the propensity to consume. For obviously, what we are looking for is a functional relationship between income (Y), the independent variable, and consumption (C), the dependent variable. In the mathematical language now familiar to us, we write

$$C = f(Y)$$

and we want to discover what f looks like.

Highly sophisticated and complex formulas have been tried to "fit" values of C and Y. Their economics and their mathematics both are beyond the scope of this book. But we can at least get a clearer idea of what it means to devise a *consumption function* by trying to make a very simple one ourselves. If we look at Fig. 24·5 on p. 373, we can see that during the Depression years, at very low levels of income, around $50 billion, consumption was just as large as income itself. (In some years it was actually bigger; as we have seen, there was net dissaving in 1933.) Hence, we might hypothesize that a consumption function for the United States might have a fixed value representing this "bottom," plus some regular fraction designating the amount of income that would be saved for all income over that amount.

A GENERALIZED CONSUMPTION FUNCTION

This is a very important hypothesis. It enables us to describe the consumption function as an amount that represents rock-bottom consumption, to which we add additional consumption spending as income rises. If a is the "bottom," and subsequent spending out of additional income is $b(Y)$, where b represents this spending "propensity," we can now write the consumption function as a whole as:

$$C = a + b(Y)$$

We have seen that a is $50 billion, and we know that our actual spending propensity, b, is about 94 percent. Therefore, we can get a *very rough* approximation of consumption by adding $50 billion to 94 percent of our disposable income over $50 billion. In 1973, for example, disposable income was $883 billion. If we add $50 billion to .94 (883 − 50), we get $833. Actual consumption in 1973 was $828 billion.*

Let the reader be warned, however, that devising a reliable consumption function is much more difficult than this simple formula would indicate. The process of translating economics into *econometrics*—that is, of finding ways to represent abstract theoretical relationships in terms of specific empirical relations—is a very difficult one. Nonetheless, even our simple example gives one an idea of what the economist and the econometrician hope to find: a precise way of expressing functional interrelations (like those between consumption and income), so that the relations will be useful in making predictions.

*Would you like to know a little more about the mathematics of the consumption function? Look at Chapter 42, Section V.

INDIVIDUAL VERSUS AGGREGATE CONSUMPTION

Here an important warning is in order. The consumption function should not be taken as a representation of individual consumption patterns. Individual preferences vary enormously, and a wide variety of random factors causes individuals to purchase different commodities at different times. But the task of predicting the consumption expenditures of a large group of people is much easier than the task of predicting the consumption of any individual in the group. The random factors that make individual predictions difficult average out in a large group. Some individuals will spend more than we would predict on the basis of their income, but others will spend less.

AGE

In addition to random disturbances, systematic factors other than incomes influence consumption. Age is a critical variable. When individuals marry and establish households of their own they are confronted with the need to acquire many consumer durables. As a result, young families are apt to have low or even negative savings rates—they consume more than they earn by using installment payments. When individuals reach middle age, they have already incurred those expenditures necessary to raise a family and are starting to think about retirement. Their consumption propensities fall, and their savings propensities rise.

If the age distribution of the population were changing rapidly, the marginal propensity to consume would not be as stable and constant as it is. Age can be ignored only in the aggregate, since the age distribution of the population does not change rapidly.

PASSIVITY OF CONSUMPTION

Throughout this chapter we have talked of the dynamics of consuming and saving. Now it is important that we recall the main conclusion of our analysis, *the essential passivity of consumption as an economic process.* Consumption spending, we will recall, is a function of income. This means it is a *dependent* variable in the economic process, a factor that is acted *on*, but that does not itself generate spontaneous action.

To be sure, it is well to qualify this assertion. We have earlier paid special attention to the long-term stability of the national savings ratio and pointed out that one cause of this stability was a general upward tendency of consumption, as families "learned" to spend their rising incomes. This dynamic, although slow-acting, behavioral trend has exerted a strong background force on the trend of the economy. Then, too, there have been occasions, the most famous being the years just following World War II, when consumption seemed to generate its own momentum and—as we have seen—raced out ahead of income. But this was a period when wants were intense, following wartime shortages, and when huge amounts of wartime savings were available to translate those wants into action. During the normal course of things, no matter how intense "wants" may be, consumers ordinarily lack the spendable cash to translate their desires into effective demand.

This highlights an extremely important point. Wants and appetites *alone* do not drive the economy upward; if they did, we should experience a more impelling demand in depressions, when people are hungry, than in booms, when they are well off. Hence the futility of those who urge the cure of depressions by suggesting that consumers should buy more! There is nothing consumers

CONSUMER CREDIT

What about consumer credit, someone will ask. Aren't many families in debt up to their ears? Doesn't the ability to buy "on time" enable consumers as a group to spend *more* than their incomes?

Consumer credit indeed enables families to spend a larger amount than they earn as incomes or receive as transfers, for short periods of time. Nonetheless, consumers do not use credit to spend more than their total receipts: *some consumers do*, but consumers as a group do not. We know this is true because the *value of all consumption spending includes purchases that are made on credit*, such as cars or many other kinds of items bought on household loans or on installment. But this total spending is still less than the total receipts of the consumer sector. Thus there is net household saving, even though purchases are made on credit.

Would there be more saving if there were no credit? In that situation, many families would put income aside until they had accumulated enough to buy cars, refrigerators, houses, and other big items. During the period that they were saving up to buy these goods, their savings rates would certainly be higher than if they had consumer credit at their disposal. But after they had bought their "lumpy" goods, their savings rates would again fall, perhaps below the level of a consumer credit economy, which tempts us to buy lumpy items and to perform our saving through installment payments.

As a result, we would expect to find high savings rates in an economy where desires for lumpy items were increasing but where consumer credit was not available. Economists cite this as one explanation of the fact that Japanese families have savings rates that are more than three times as high as American families, even though Japanese incomes are lower. In Japan you cannot "buy now, pay later"; so you save now and buy later.

would rather do than buy more, if only they could. Let us not forget, furthermore, that consumers are at all times being cajoled and exhorted to increase their expenditures by the multibillion dollar pressures exerted by the advertising industry.

The trouble is, however, that consumers cannot buy more unless they have more incomes to buy with. It is true, of course, that for short periods they can borrow or they may temporarily sharply reduce their rate of savings; but each household's borrowing capacity or accumulated savings are limited, so

that once these bursts are over, the steady habitual ways of saving and spending are apt to reassert themselves.

Thus it is clear that in considering the consumer sector we study a part of the economy that, however ultimately important, is not in itself the source of major changes in activity. Consumption mirrors and, as we shall see, can magnify disturbances elsewhere in the economy, but it does not initiate the greater part of our economic fortunes or misfortunes.

KEY WORDS

Consumption

Disposable income

CENTRAL CONCEPTS

1. Consumption is the largest source of economic activity, and accordingly the largest absolute source of demand within the economy. Nonetheless, *consumption alone will not create enough demand to buy all of the nation's output.*

2. Consumption in absolute amounts is capable of wide fluctuations, but *the relation of consumption to disposable income is relatively stable.*

Consumption
function

3. Over the long run (since the mid-1800s), *the fraction of disposable income that has been saved seems to have been more or less unchanged.* This has prevented the economy from facing the problem of a growing proportion of saving. In the short run, the ratio of saving to increases in income is apt to be higher than over the long run.

Schedule

4. We call the relation between saving and disposable income the *consumption function.* We can represent this function (relationship) in a *schedule* showing the division of disposable income, at different levels of income, between consumption and saving.

Average
propensity to
to consume

5. The consumption schedule shows that the *amount of consumption rises as income rises, but not by as much as income.* Therefore the amount of saving also rises as income rises.

6. From the consumption schedule we can derive two ratios. One shows us the relation between the *total income* and the *total consumption* of any period. We call this the *average propensity to consume.* The other shows us the relationship between the *change in income and the change in consumption* between two periods. This is called the *marginal propensity to consume.*

Marginal
propensity
to consume

7. The average propensity to consume shows us how people behave with regard to consumption and saving over the *long run.* The marginal propensity to consume shows us how they behave over the *short run.*

8. Consumption is generally regarded as a *passive economic force,* rather than an initiating active one. It is acted on by changes in income. Thus we generalize the force of consumption by saying that it is a *function of income:* $C = a + b(Y)$, where a is the "bottom" and b is the marginal propensity to consume.

$C = a + b(Y)$
"Bottom"

QUESTIONS

1. What are the main components of consumption? Why are some of these components more dynamic than others?

2. "The reason we have depressions is that consumption isn't big enough to buy the output of all our factories." What is wrong with this statement?

3. What do you think accounts for the relative stability of the savings ratio over the long run? Would you expect the savings ratio in the short run to be relatively stable? Why or why not?

4. What is meant by the consumption function? Could we also speak of a savings function? What would be the relation between the two?

5. Suppose that a given family had an income of $8,000 and saved $400. What would be its average propensity to consume? Could you tell from this information what its marginal propensity to consume was?

6. Suppose the same family now increased its income

to $9,000 and its saving to $500. What is its new average propensity to consume? Can you figure out the family's marginal propensity to consume?

7. Draw a scatter diagram to show the following:

Family income	Savings
$4,000	$ 0
5,000	50
6,000	150
7,000	300
8,000	500

From the figures above, calculate the average propensity to consume at each level of income. Can you calculate the marginal propensity to consume for each jump in income?

8. How do you read $S = f(Y)$? From what you know of the propensity to consume, how would you describe the relation of S to Y?

9. Why can't we cure depressions by urging people to go out and spend?

25

Investment Demand

IN STUDYING THE BEHAVIOR of the consumption sector, we have begun to understand how the demand for GNP arises. Now we must turn to a second source of demand—investment demand. This requires a shift in our vantage point. As experienced consumers, we know about consumption, but the activity of investing is foreign to most of us. Worse, we are apt to begin by confusing the meaning of investment, as a source of demand for GNP, with "investing" in the sense familiar to most of us when we think about buying stocks or bonds.

INVESTMENT: REAL AND FINANCIAL

We had best begin, then, by making certain that our vocabulary is correct. *Investing, or investment, as the economist uses the term in describing the demand for GNP, is an activity that uses the resources of the community to maintain or add to its stock of physical capital.*

Now this may or may not coincide with the purchase of a security. When we buy an ordinary stock or bond, we usually buy it from someone who has previously owned it, and therefore our personal act of "investment" becomes, in the economic view of things, merely a *transfer* of claims without any direct bearing on the creation of new wealth. A pays B cash and takes his General Output stock; B takes A's cash and doubtless uses it to buy stock from C; but the transactions between A and B and C in no way alter the actual amount of real capital in the economy. Only when we buy *newly issued* shares or bonds, and then only when their proceeds are directly allocated to new equipment or plant, does our act of personal financial investment result in the addition of wealth to the community. In that case, A buys his stock directly (or through an investment banker) from General Output itself, and not from B. A's cash can now be spent by General Output for new capital goods, as presumably it will be.

Thus, much of investment, as economists

see it, is a little-known form of activity for the majority of us. This is true not only because real investment is not the same as personal financial investment, but because the real investors of the nation usually act on behalf of an institution other than the familiar one of the household. The unit of behavior in the world of investment is typically the business *firm*, just as in the world of consumption it is the household. Boards of directors, chief executives, or small-business proprietors are the persons who decide whether or not to devote business cash to the construction of new facilities or to the addition of inventory; and this decision, as we shall see, is very

different in character and motivation from the decisions familiar to us as members of the household sector.

The Investment Sector in Profile

Before we begin an investigation into the dynamics of investment decisions, however, let us gain a quick acquaintance with the sector as a whole, much as we did with the consumption sector.

Figure 25·1 gives a first general impression of the investment sector in a recent year. Note that the main source of gross

FIGURE 25 · 1
Business sector 1973

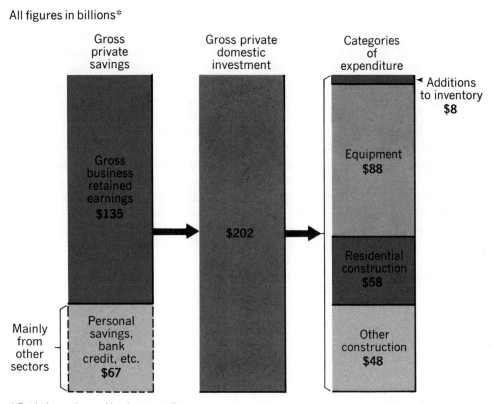

All figures in billions*

* Totals do not always add, owing to rounding

private domestic investment expenditure is the retained earnings of business; that is, the expenditures come from depreciation accruals or from profits that have been kept in the business. However, as the next bar shows, gross investment *expenditures* are considerably larger than retained earnings. The difference represents funds that business obtains in several ways.

1. It may draw on cash (or securities) accumulated out of retained earnings or depreciation accruals of previous years.

2. It may obtain savings from the household sector by direct borrowing or by sale of new issues of shares of stock or indirectly via insurance companies or savings banks or pension funds, and so on.

3. It may borrow from commercial banks.

4. The difference also represents investment in housing, which is not typically financed by corporate earnings but by consumers, borrowing from banks.

The last two sources of funds we will not fully understand until we reach Chapter 30 when we study the money mechanism. But our chart enables us to see that most gross investment is financed by business itself from its *internal* sources—retained earnings plus depreciation accruals—and that external sources play only a secondary role. In particular, this is true of new stock issues, which, during most of the 1960s and early 1970s, raised only some 3 to 8 percent of the funds spent by the business sector for new plant and equipment.

CATEGORIES OF INVESTMENT

From the total funds at its disposal, the business sector now renews its worn-out capital and adds new capital. Let us say a word con-

cerning some of the main categories of investment expenditure.

1. Inventories

At the top of the expenditure bar in Fig. 25·1, we note an item of $8.0 billion for *additions to inventory*. Note that this figure does not represent total inventories, but only *changes* in inventories, upwards or downwards. If there had been no change in inventory over the year, the item would have been zero, even if existing inventories were huge. Why? Because those huge inventories would have been included in the investment expenditure flow of *previous* years when they were built up.

Additions to inventories are capital, but they need not be additions to capital *goods*. Indeed, they are likely to include farm stocks, consumer goods, and other items of all kinds. Of course, these are goods held by business, and not by consumers. But that is the very point. We count inventory additions as net investment because they are output that has been produced but that has not been consumed. In another year, if these goods pass from the hands of business into consumers' hands, and inventories decline, we will have a negative figure for net inventory investment. This will mean, just as it appears, that we are consuming goods faster than we are producing them—that we are disinvesting.

Inventories are often visualized as completed TV sets sitting in some warehouse. While some inventories are completed goods sitting in storage, most are in the form of goods on display in stores, half-finished goods in the process of production, or raw materials to be used in production. When a steel company adds to its stock of iron ore, it is adding to its inventories.

Investments in inventory are particularly significant for one reason. Alone among the investment categories, inventories can be *rapidly* used up as well as increased. A positive figure for one year or even one calendar quarter can quickly turn into a negative figure the next. *This means that expenditures for inventory are usually the most volatile element of any in gross national product.* A glance at Fig. 25·2 shows a particularly dramatic instance of how rapidly inventory spending can change. In the second quarter of 1960, we were *disinvesting* in inventories at an annual rate of almost $5 billion. Two quarters later, we were building up inventories—*investing* in inventories—by roughly the same amount. Thus, within a span of six months, there was a swing of almost $10 billion in spending. Rapid inventory swings, although not quite of this magnitude, are by no means uncommon.

As we shall see more clearly later, this volatility of investment has much significance for business conditions. Note that while inventories are being built up, they serve as an offset to saving—that is, some of the resources released from consumption are used by business firms to build up stocks of inventory capital. But when inventories are being "worked off," we are actually making the demand gap bigger. As we would expect, this can give rise to serious economic troubles.

2. Equipment

The next item in the expenditure bar (Fig. 25·1) is more familiar: $88 billion for *equipment.* Here we find expenditures for goods of a varied sort—lathes, trucks, generators,

FIGURE 25 · 2
Inventory swings and GNP

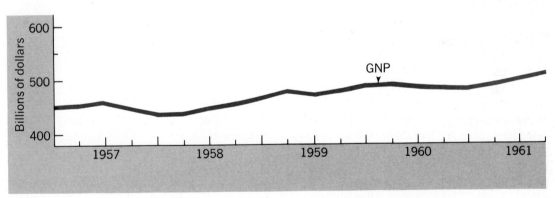

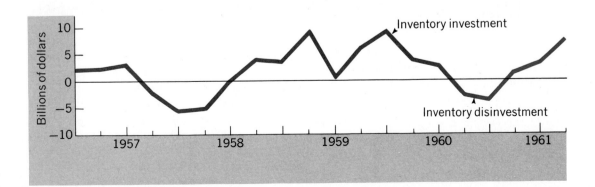

computers, office typewriters.* The total includes both *new equipment* and *replacement equipment,* and we need a word of caution here. Exactly what does it mean to "replace" a given item of equipment? Suppose we have a textile loom that cost $100,000 and that is now on its last legs. Is the loom "replaced" by spending another $100,000, regardless of what kind of machine the money will buy? What if loom prices have gone up and $100,000 no longer buys a loom of the same capacity? Or suppose that prices have remained steady but that owing to technological advance, $100,000 now buys a loom of double the old capacity?

From an economic perspective, *replacement is the dollar amount that would be necessary to buy the same productive capacity.* But this is an amount that is seldom known with great accuracy. It may not be possible to buy new equipment with exactly the same productive capacity as old equipment. Often businesses replace whole factories rather than individual pieces of equipment. These new factories are likely to have very different configurations of equipment as well as different productive capacities. Such problems make the definition of "replacement" an accountant's headache and an economist's nightmare. At the moment there isn't even a generally accepted estimate of replacement investment. We need not involve ourselves deeper in the question, but we should note the complexities introduced into a seemingly simple matter once we leave the changeless world of stationary flow and enter the world of invention and innovation.

3. Construction—residential

Our next section on the expenditure bar (Fig. 25·1) is total *residential construction.* Why do we include this $58 billion in the investment sector when most of it is represented by new houses that householders buy for their own use?

Part of the answer is that most houses are built by business firms, such as contractors and developers, who put up the houses *before* they are sold. Thus the original expenditures involved in building houses typically come from businessmen, not from households. Later, when the householder buys a house, he takes possession of an existing asset, and his expenditure does not pump new incomes into the economy but only repays the contractor who *did* contribute new incomes.

Actually, this is a somewhat arbitrary definition, since, after all, businessmen own *all* output before consumers buy it. However, another reason for considering residential construction as investment is that, unlike most "consumer goods," houses are typically maintained as if they were capital goods. Thus their durability also enters into their classification as investment goods.

Finally, we class housing as investment because residential purchases "behave" very much like other items of construction. Therefore it simplifies our understanding of the forces at work in the economy if we classify residential construction as an investment expenditure rather than as a consumer expenditure.

4. Other construction—plant

Last on the bar, $48 billion of *other construction* is largely made up of the "plant" in "plant and equipment"—factories and stores and private office buildings and warehouses. (It does not, however, include public construction such as roads, dams, harbors, or public buildings, all of which are picked up under government purchases.) It is interesting to note that the building of structures, as represented by the total of residential construction plus other private construction, accounts for over half of all investment expenditure, and this total would be further swelled

*But *not* typewriters bought by consumers. Thus the same good can be classified as a consumption item or an investment item, depending on the use to which it is put.

if public construction were included herein. This accords with the dominant role of structures in the panorama of national wealth we first encountered in Chapter 19. It tells us, too, that swings in construction expenditure can be a major lever for economic change.

Investment in Historic Perspective

With this introduction behind us, let us take a look at the flow of investment, not over a single year, but over many years.

In Fig. 25·3, several things spring to our notice. Clearly, investment demand is not nearly so smooth and unperturbed a flow of spending as consumption. Note that gross investment in the depths of the Depression virtually disappeared—that we almost failed to *maintain,* much less add to, our stock of wealth. (Net investment was, in fact, a negative figure for several years.) Note also investment was reduced during the war years as private capital formation was deliberately limited through government allocations.

Two important conclusions emerge from this examination of investment spending:

FIGURE 25 · 3
Gross private domestic investment, 1929–1973

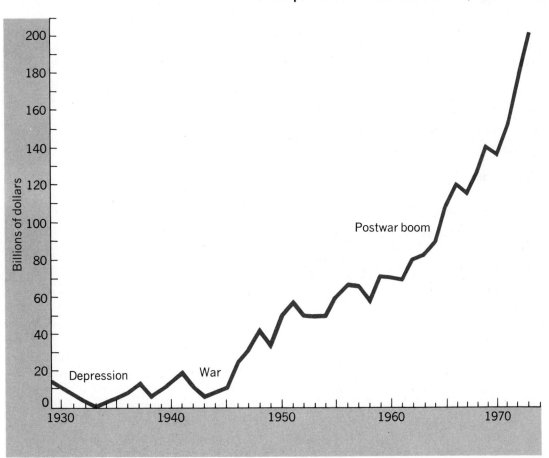

First, as we have already seen, investment spending contains a component—net additions to inventory—that is capable of drastic, sudden shifts. This accounts for much of the wavelike movement of the total flow of investment expenditure.

Second, investment spending as a whole is capable of more or less total collapses, of a severity and degree that are never to be found in consumption.

The prime example of such a collapse was, of course, the Great Depression. From 1929 to 1933, while consumption fell by 41 percent, investment fell by *91 percent,* as we can see in Fig. 25·3. Similarly, whereas consumption rose by a little more than half from 1933 to 1940, investment in the same period rose by *nine times.*

IMPORTANCE OF INVESTMENT

This potential for collapse or spectacular boom always makes investment a source of special concern in the economic picture. But even the tendency toward inventory fluctuations, or toward milder declines in other capital expenditures, is sufficient to identify investment as a prime source of economic instability. As we have said before, there is often a tendency among noneconomists to equate all buying in the economy with consumer buying. Let us never lose sight of the fact that the maintenance of, and addition to, capital is also a part of GNP spending and that a considerable part of the labor force depends for its livelihood on the making of investment goods. At the bottom of the Great Depression in 1933, it was estimated that one-third of total unemployment was directly associated with the shrinkage in the capital goods industry.

We shall want to look more closely into the reasons for the sensitivity of investment spending. But first a question must surely have occurred to the reader. For all its susceptibility to change, the investment sector is, after all, a fairly small sector. In 1973, total expenditures for gross private domestic investment came to only about one-seventh of GNP, and the normal year-to-year variation in investment spending in the 1960s and 1970s is only about 1 to 2 percent of GNP. To devote so much time to such small fluctuations seems a disproportionate emphasis. How could so small a tail as investment wag so large a dog as GNP?

The Multiplier

The answer lies in a relationship of economic activities known as the *multiplier.* The multiplier describes the fact that *additions to spending (or diminutions in spending) have an impact on income that is greater than the original increase or decrease in spending itself.* In other words, even small increments in spending can *multiply* their effects (whence the name).

It is not difficult to understand the general idea of the multiplier. Suppose that we have an island community whose economy is in a perfect circular flow, unchanging from year to year. Next, let us introduce the stimulus of a new investment expenditure in the form of a stranger who arrives from another island (with a supply of acceptable money) and who proceeds to build a house. This immediately increases the islanders' incomes. In our case, we will assume that the stranger spends $1,000 on wages for construction workers, and we will ignore all other expenditures he may make. (We also make the assumption that these workers were previously unemployed, so that the builder is not merely taking them from some other task.)

Now the construction workers, who have had their incomes increased by $1,000, are very unlikely to sit on this money. As we know from our study of the marginal pro-

INTENDED AND UNINTENDED
INVENTORY INVESTMENT

Changes in inventories reflect both the desires of firms and their planning errors. *Investment in inventories can therefore be intended or unintended.* If an automobile manufacturer expects to sell 1 million cars and is disappointed in this expectation, actual sales will lag behind expected sales. Since production was planned on the basis of expected sales, it will exceed actual sales until production can be readjusted to the new sales levels. Autos built but not sold will be added to inventories. In this case, inventories go up not because the firm wanted more inventories, but because it made a mistake.

If actual sales exceed expected sales, the reverse will happen. Inventories fall. But again, the firm did not lower inventories because lower inventories were more profitable, but by mistake. Low inventories can be extremely unprofitable. If certain models are not available or in limited supply, the auto maker will find that he is losing potential sales to a competitor that can give the consumer the kind of car he wants at the time he wants it.

These unintended changes in inventories enter into the statistics of GNP, just as do intended changes in inventories. But they obviously have very different significances regarding future economic behavior.

---

pensity to consume, they are apt to save some of the increase (and they may have to pay some to the government as income taxes), but the rest they will spend on additional consumption goods. Let us suppose that they save 10 percent and pay taxes of 20 percent on the $1,000 they get. They will then have $700 left over to spend for additional consumer goods and services.

But this is not an end to it. The sellers of these goods and services will now have received $700 over and above their former incomes, and they, too, will be certain to spend a considerable amount of their new income. If we assume that their family spending patterns (and their tax brackets) are the same as the construction workers, they will also spend 70 percent of their new incomes, or $490. And now the wheel takes another turn, as still *another* group receives new income and spends a fraction of it—in turn.

CONTINUING IMPACT OF RESPENDING

If the newcomer then departed as mysteriously as he came, we would have to describe the economic impact of his investment as constituting a single "bulge" of income that gradually disappeared. The bulge would consist of the original $1,000, the secondary $700, the tertiary $490, and so on. If everyone continued to spend 70 percent of his new income, after ten rounds all that would remain by way of new spending traceable to the original $1,000 would be about $38. Soon, the impact of the new investment on incomes would have virtually disappeared.

But now let us suppose that after our visitor builds his house and leaves, another visitor arrives to build another house. This time, in other words, we assume that the level of investment spending *continues* at the higher level to which it was raised by the first expenditure for a new house. We can see that the second house will set into motion precisely the same repercussive effects as did the first, and that the new series of respendings will be added to the dwindling echoes of the original injection of incomes.

In Fig. 25·4, we can trace this effect. The succession of colored bars at the bottom of the graph stands for the continuing injec-

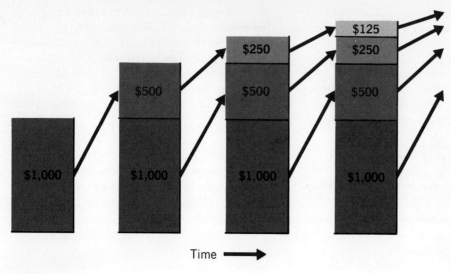

FIGURE 25 · 4
The multiplier

tions of $1,000 as new houses are steadily built. (Note that this means the level of new investment is only being maintained, not that it is rising.) Each of these colored bars now generates a series of secondary, tertiary, etc., bars that represent the respending of income after taxes and savings. In our example we have assumed that the respending fraction is 50 percent.

Let us now examine the effects of investment spending in a generalized fashion, without paying attention to specific dollar amounts. In Fig. 25·5, we see the effects of a single, *once-and-for-all* investment expendi-

FIGURE 25 · 5
Once-over and continuing effects of investment

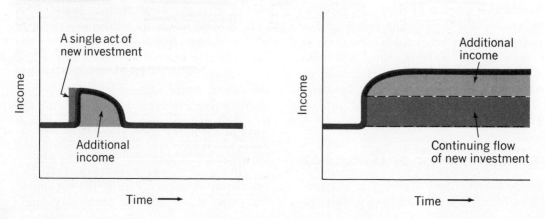

ture (the stranger who came and went), contrasted with the effects of a *continuing* stream of investment.

Our diagrams show us three important things:

1. A single burst of investment creates a bulge of incomes larger than the initial expenditure, but a bulge that disappears.

2. A continuing flow of investment creates a new higher permanent level of income, larger than the investment expenditures themselves.

3. A continuing flow of new investment creates a higher income level that gradually levels out.

the total addition to income due to the re-spending of that $1,000 is $3,000, we have a multiplier of 3; if the total addition is $2,000, the multiplier is 2.

What determines how large the multiplier will be? The answer depends entirely on our marginal consumption (or, if you will, our marginal saving) habits—that is, on how much we consume (or save) out of each dollar of additional income that comes to us. Let us follow two cases below. In the first, we will assume that each recipient spends only one-half of any new income that comes to him, saving the rest. In the second case, he spends three-quarters of it and saves one-quarter.

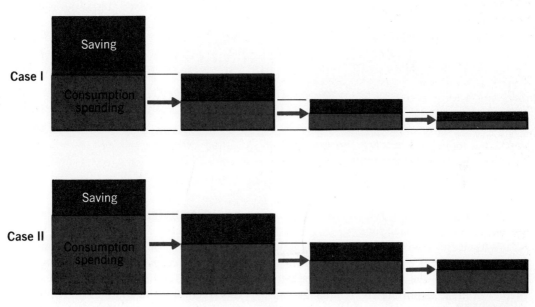

FIGURE 25 · 6
Comparison of two multipliers

MARGINAL PROPENSITY TO SAVE

We can understand now that *the multiplier is the numerical relation between the initial new investment and the total increase in income.* If the initial investment is $1,000 and

It is very clear that the amount of income that will be passed along from one receiver to the next will be much larger where the marginal propensity to consume is higher. In fact, we can see that the total amount of new incomes (total amount of boxes above)

MATHEMATICS FOR THE CURIOUS

How do we know that the multiplier will be 4, if the marginal propensity to save is ¼ ? Most of us "intuitively" see that the sum of respending hinges on the savings fraction, and we take on faith the simple formula that tells us how to calculate that sum by taking the reciprocal of the mps and multiplying it by the change in spending.

But some students may want to go beyond faith, to understanding. Here is a simple mathematical demonstration that the multiplier formula is "really" true.

What we are trying to get at, with the multiplier formula, is the *sum of a series*, in which an initial term is multiplied again and again by some number that is less than 1 (and greater than 0). Suppose the initial term is $10 and the number-less-than-one is .8. Then we want to know the sum of the following problem:

$$10 + .8(10) + .8[.8(10)]....$$

This is the same as if we wrote:

$$10 + .8(10) + .8^2(10) + .8^3(10) ... + ...8^n(10)$$

If we think of .8 as designating the marginal propensity to consume, we are looking for the sum of an initial new expenditure of $10, of which $8 will be spent in the first round (.8 × $10); $6.40 ($.8^2$ × $10) in the second round; $5.12 ($.8^3$ × $10) in the third, and so on. From the textbook, we "know" that this sum is found by taking the mps, which is .2, or $\frac{1}{5}$, and multiplying the original expenditure by its reciprocal. Thus, $10 × 5 = $50. Now let's prove it.

We can restate our multiplier series in simple algebra by calling the initial term a and the number-less-than-one (.8 above) b. Then the series looks like this:

$$a + b \cdot a + b^2 \cdot a... + ...b^n \cdot a$$

where b^n stands for the fraction spent on the last (nth) round.

Suppose we call the sum of this series S. Now we are going to perform a truly magical (but perfectly legitimate) mathematical trick. We will first write the formula we have just described, and below it we will write the *same* formula, after we have multiplied both sides of the equation by b.

must be mathematically related to the proportion that is spent each time.

What is this relationship? The arithmetic is easier to figure if we use not the consumption fraction, but the *saving fraction* (the two are, of course, as intimately related as the first slice of cake and the remaining cake). If we use the saving fraction, the *sum of new incomes is obtained by taking the reciprocal of* (i.e., inverting, or turning upside down) *the fraction we save*. Thus, if we save ½ our income, the total amount of new incomes generated by respending will be ½ inverted, or 2—twice the original increase in income. If we save ¼, it will be the reciprocal of ¼ or 4 times the original change.

BASIC MULTIPLIER FORMULA

We call the fraction of new income that is saved the *marginal propensity to save* (often abbreviated as mps). As we have just seen, this fraction is the complement of an already familiar one, the marginal propensity to consume. If our marginal propensity to consume is 80 percent, our marginal propensity to save must be 20 percent; if our mpc is three-quarters, our mps must be one-quarter. In brief, mps + mpc ≡ 1.

Understanding the relationship between the marginal propensity to save and the size of the resulting respending fractions allows us to state a very simple (but very important)

$$S = a + b \cdot a + b^2 \cdot a \ldots + \ldots b^n a$$
$$b \cdot S = b \cdot a + b^2 \cdot a \ldots + \ldots b^{n+1} \cdot a$$

We have strung out the second equation so that terms such as $b \cdot a$ lie underneath their counterparts in the first equation.

Now we subtract the second equation from the first. All the terms that are under one another just disappear. This leaves us:

$$S - b \cdot S = a - b^{n+1} \cdot a$$

Next we factor out S on the left side, giving us $S(1-b)$, and divide both sides by $(1-b)$. The result:

$$S = \frac{a - b^{n+1} \cdot a}{(1-b)}$$

We are almost at the end. Now we examine what happens as the exponent n approaches infinity. Remember that by definition b is a number less than 1, so that with each successive increase in the exponent, b becomes *smaller*. Thus we can assume that the final term approaches zero, as its exponent approaches infinity. That is to say, it "vanishes." This is very convenient because it leaves us with the much simpler formula:

$$S = \frac{a}{1-b}$$

Do you see the connection with the multiplier? The term b was the fraction (.8) by which we constantly multiplied the initial sum ($10). *Thus this fraction was exactly like the marginal propensity to consume!* Therefore, $1-b$ must be the difference between 1 and the mpc (or .2). We know this is the mps. Therefore we can write mps in place of $1-b$; and while we are about it, we can write $10, or ΔI, or any other number in place of a.

Hence our formula becomes translated into economic terms and looks like this:

$$S = \frac{\Delta I}{\text{mps}}$$

The term S stood for the sum of the series. An economist will call it ΔY since this is the sum of the additional incomes generated by each round of spending.

ΔY is therefore $10 \div .2$ (or $50). And that is why the formula below is *really* true.

formula for the multiplier:

change in income = multiplier ×
change in investment

Since we have just learned that the multiplier is determined by the reciprocal of the marginal propensity to save, we can write:

$$\text{multiplier} = \frac{1}{\text{mps}}$$

If we now use the symbols we are familiar with, plus a Greek letter Δ, delta, that means "change in," we can write the important economic relationship above as follows:

$$\Delta Y = \frac{1}{\text{mps}} \times \Delta I$$

Thus, if our mps is $\frac{1}{4}$ (meaning, let us not forget, that we save a quarter of increases in income and spend the rest), then an increase in investment of $1 billion will lead to a total increase in incomes of $4 billion

$$(\$4 \text{ billion} = \frac{1}{1/4} \times \$1 \text{ billion}$$

Note that the multiplier is a complex or *double* fraction:

$$\text{it is } \frac{1}{1/4} \text{ and } not \ \frac{1}{4}.$$

If the mps is $\frac{1}{10}$, \$1 billion gives rise to incomes of \$10 billion; if the mps is 50 percent, the billion will multiply to \$2 billion. And if mps is 1? This means that the entire increase in income is unspent, that our island construction workers tuck away (or find taxed away) their entire newly earned pay. In that case, the multiplier will be 1 also, and the impact of the new investment on the island economy will be no more than the \$1,000 earned by the construction workers in the first place.

LEAKAGES

The importance of the size of the marginal savings ratio in determining the effect that additional investment will have on income is thus apparent. Now, however, we must pass from the simple example of our island economy to the more complex behavioral patterns and institutional arrangements of real life. The average propensity to save (the ratio of saving to disposable income) runs around 6 to 7 percent. In recent years, the *marginal* propensity to save (the ratio of additional saving to increases in income) figured over the period of a year has not departed very much from this figure. If this is the case, then, following our analysis, the multiplier would be very high. If mps were even as much as 10 percent of income, a change in investment of \$1 billion would bring a \$10 billion change in income. If mps were nearer 6 percent—the approximate level of the average propensity to save—a change of \$1 billion would bring a swing of over \$16 billion. Were this the case, the economy would be subject to the most violent disturbances whenever the level of spending shifted.

In fact, however, the impact of the multiplier is greatly reduced because the successive rounds of spending are damped by factors other than personal saving. One of them we have already introduced in our imaginary island economy. This is the tendency of *government taxation* to "mop up" a fraction of income as it passes from hand to hand. This mopping-up effect of taxation is in actuality much larger than that of saving. For every dollar of change in income, federal taxes will take about 30 cents, and state and local taxes another 6 cents.

Another dampener is the tendency of respending to swell *business savings* as well as personal incomes. Of each dollar of new spending, perhaps 10 cents goes into business profits, and this sum is typically saved, at least for a time, rather than immediately respent.

Still another source of dampening is the tendency of consumers and businesses to increase purchases from abroad as their incomes rise. These rising *imports* divert 3 to 4 percent of new spending to foreign nations and accordingly reduce the successive impact of each round of expenditure.

All these withdrawals from the respending cycle are called *leakages,* and the total effect of all leakages together (personal savings, business savings, taxes, and imports) is to reduce the overall impact of the multiplier from an impossibly large figure to a very manageable one. The combined effect of all leakages brings the actual multiplier in the United States in the 1970s to a little more than 2 over a period of 2 years.*

To be sure—and this is very important—all these leakages *can* return to the income stream. Household saving can be turned into capital formation; business profits can be invested; tax receipts can be disbursed in gov-

---

*In dealing with the multiplier equation ($\Delta Y = \dfrac{1}{mps} \times \Delta I$), we can interpret mps to mean the total withdrawal from spending due to all leakages. This brings mps to around $\frac{1}{2}$, and gives us a multiplier of 2.

It is interesting to note that the leakages all tend to increase somewhat in boom times and to decline in recessions, which results in a slightly larger multiplier in bad times than in good.

ernment spending programs; and purchases from foreign sellers can be returned as purchases *by* foreigners. What is at stake here is the regularity and reliability with which these circuits will be closed. In the case of ordinary income going to a household, we can count with considerable assurance on a "return expenditure" of consumption. In the case of the other recipients of funds, the assurance is much less; hence we count their receipts as money that has leaked out of the expenditure flow, for the time being.

THE DOWNWARD MULTIPLIER

The multiplier, with its important magnifying action, rests at the very center of our understanding of economic fluctuations. Not only does it explain how relatively small stimuli can exert considerable upward pushes, but it also makes much clearer than before how the failure to offset a small savings gap can snowball into a serious fall in income and employment.

For just as additional income is respent to create still further new income, a loss in income will not stop with the affected households. On the contrary, as families lose income, they cut down on their spending, although the behavior pattern of the propensity to consume schedule suggests that they will not cut their consumption by as much as their loss in income. Yet each reduction in consumption, large or small, lessens to that extent the income or receipts of some other household or firm.

We have already noted that personal savings alone do not determine the full impact of the multiplier. This is even more fortunate on the way down than on the way up. If the size of the multiplier were solely dependent on the marginal propensity to save, an original fall in spending would result in a catastrophic contraction of consumption through the economy. But the leakages that cushion the upward pressure of the multiplier also cushion its downward effect. As spending falls, business savings (profits) fall, tax receipts dwindle, and the flow of imports declines. We shall discuss this cushioning effect when we look into the government sector.

All of these leakages now work in the direction of mitigating the repercussions of the original fall in spending. The fall in business profits means that less will be saved by business and thus less withdrawn from respending; the decline in taxes means that more money will be left to consumers; and the drop in imports similarly releases additional spending power for the domestic market. Thus, just as the various leakages pulled money away from consumption on the way up, on the way down they lessen their siphoning effect and in this way restore purchasing power to consumers' hands. As a result, in the downward direction as in the upward, the actual impact of the multiplier is about 2, so that a fall in investment of, say, $5 billion will lower GNP by $10 billion.

THE MULTIPLIER AND INVESTMENT

Even with a reduced figure, we can now understand how a relatively small change in investment can magnify its impact on GNP. If the typical year-to-year change in investment is around $10 billion to $20 billion, a multiplier of 2 will produce a change in GNP of $20 billion to $40 billion, by no means a negligible figure. In addition, as we shall shortly see, the multiplier may set up repercussions that feed back onto investment. But more of that momentarily. First let us make three final points in regard to the multiplier.

1. Other multipliers

We have talked of the multiplier in connection with changes in investment spending. *But we must also realize that any original change in any spending has a multiplier effect.*

We have used investment as the "trigger" for the multiplier because it is, in fact, a component of spending that is likely to evidence *large* and *sudden* changes. But an increase in foreigners' purchases of our exports has a multiplier effect, as does an increase in government spending or a decrease in taxes, or a spontaneous increase in consumption itself due to, say, a drop in the propensity to save. Any stimulus to the economy is thus not confined to its original impact, but gives a series of successive pushes to the system until it has finally been absorbed in leakages. We shall come back to this important fact in our next chapter.

2. Idle resources

Finally, there is a very important proviso to recognize, although we will not study its full significance until Chapter 31. This is the important difference between an economy with idle resources—unemployed labor or unused machines or land—and one without them.

For *it is only when we have idle resources that the responding impetus of the multiplier is useful.* Then each round of new expenditure can bring idle resources into use, creating not only new money incomes but *new production and employment.* The situation is considerably different when there are no, or few, idle men or machines. Then the expenditure rounds of the multiplier bring higher money incomes, but these are not matched by increased real output.

In both cases, the multiplier exerts its leverage, bringing about an increase in total expenditure larger than the original injection of new spending. In the case without idle resources, however, the results are solely *inflationary,* as the increased spending results in higher incomes and higher prices, but not in higher output. In the case where idle resources exist, we can avoid this mere "money" multiplication and enjoy a rise in output as a result of our increased spending. Indeed, we can even speak of the *employment*

multiplier in situations where there is considerable unemployment, meaning by this the total increase in employment brought about by a given increase in spending. We shall return in subsequent chapters to a fuller scrutiny of the difference between the case of idle and of fully employed resources, but we must bear the distinction in mind henceforth.

3. The importance of time lags

Last we must distinguish between the multiplier as a mathematical relationship and the multiplier in real life.

In equations, the multiplier is "instantaneous." If investment rises by $10 billions and the multiplier is 2, we "instantly" have a $20 billion rise in output. In actuality, the successive "rounds" of spending display very important *time lags.* Investment expenditures of $10 billion will first show up as increased sales of businesses. Businesses usually will draw down on inventories rather than immediately increasing production (and factor incomes), to hedge against the possibility that the increase is only temporary. This leads to a smaller increase in incomes, other than profits, than might be expected. And for the same hedging reason, businesses are unlikely at first to use their additional profits to pay higher incomes or to finance new investment.

Moreover, incomes that do go to consumers are also not instantaneously spent. One recent study has shown that families spent only 66 cents out of each dollar of new income in the first three months during which they received that income. Only gradually did their spending propensities build up to "normal." And even when they *did* spend their additional incomes, the businesses that enjoyed larger sales were again likely to display the cautious hedging attitudes we have described. That is another reason why the multiplier, 2 years after an investment increase, is in fact only about 2.

KEY WORDS

CENTRAL CONCEPTS

Investment

1. The investment sector is made up not of households and their activities, but of *business firms adding to their capital assets.* By and large, these additions to business capital are financed out of *internal funds* (retained earnings and depreciation accruals) rather than from external sources (borrowing or new equities). The main categories of investment expenditure are additions to inventory, new equipment, residential housing, and other construction, mainly plant.

Instability

2. The main characteristic of all investment expenditure is its *potential instability.* In times of serious recession, net investment can virtually cease. Even in ordinary times, inventory investment is capable of drastic changes.

Inventories

3. Changes in investment (or in any other kind of spending) are given larger economic impact because of the *multiplier effect.* This arises because incomes received from a new investment (or any other source) are *partly respent,* giving rise to additional new incomes which, in turn, are respent.

Multiplier

4. A single "burst" of investment creates a bulge in incomes that disappears over time; but a *continuing level of new investment creates a continuing higher level of new incomes.*

5. The size of the multiplier depends on the fraction of additional income spent for consumption by each new recipient. *The more the spending* (or the less the saving) *the greater will be the multiplier.*

Marginal propensity to save

6. *We calculate the multiplier by taking the reciprocal of the marginal propensity to save.* This gives us the important formula:

$\Delta Y = \dfrac{1}{mps} \times \Delta I$

$\Delta Y = \dfrac{1}{mps} \times \Delta I$ (change in income = multiplier × change in investment)

Leakages

7. The *size of mps is determined by leakages.* There are four main leakages:
   ● Saving
   ● Taxation
   ● Business profits
   ● Imports
   Total leakages in the U.S. amount to about one-half of increases in income. Therefore the U.S. multiplier is about 2.

8. *Each of these leakages takes money out of the "automatic" respending circuit of consumption.* Money going into leakages *can* return to the economy via additional investment, but it does not do so as reliably as money that stays in the consumption flow.

Downward multiplier

9. Magnifying the effects on income of a fall in investment, the *multiplier works downward,* as well as upward.

Full employment

10. The multiplier will have very different economic effects, depending on whether or not the economy is *fully employed.*

Time lags

11. The multiplier acts over *time,* not instantaneously.

QUESTIONS

1. If you buy a share of stock on the New York Stock Exchange, does that always create new capital? Why, or why not?

2. Why are additions to inventory so much more liable to rapid fluctuation than are other kinds of investment?

3. Why do we face the possibility of a total collapse of investment, but not of consumption?

4. Draw a diagram showing the multiplier effect of a $1,000 expenditure when the marginal propensity to save is one-tenth. Draw a second diagram, showing the effect when the marginal propensity to consume is nine-tenths. Are the diagrams the same?

5. Compare two multiplier diagrams: one where the marginal propensity to save is one-quarter; the other where it is one-third. The *larger* the saving ratio, the larger or smaller the multiplier?

6. Calculate the impact on income if investment rises by $10 billion and the multiplier is 2. If the multiplier is 3. If it is 1.

7. Income is $500 billion; investment is $50 billion. The multiplier is 2. If inventories decline by $10 billion, what happens to income?

8. Draw a diagram showing what happens to $1 billion of new investment given the following leakages: mps 10 percent; marginal taxation 20 percent; marginal propensity to import 5 percent; marginal addition to business saving 15 percent. What will be the size of the second round of spending? the third? the final total?

9. If the marginal propensity to consume is three-quarters, what is the size of the marginal propensity to save? If it is five-sixths? If it is 70 percent?

10. What is the formula for the multiplier?

The Motivation of Investment

26

THE INHERENT INSTABILITY OF INVESTMENT, and the multiplier repercussions that arise from changes in investment, begin to give us an understanding of the special importance of the business sector in determining the demand for GNP. In our next chapter we shall look into equally special characteristics of government demand before assembling the demand functions of all the sectors, to match them against the supply of GNP.

But before we proceed to that goal, we must learn something further about the nature of investment demand—in particular, about the motivations that give rise to it—for if we compare the underlying behavioral drives that impel consumption and investment, we can see a fundamental difference of the greatest significance.

UTILITY VS. PROFIT

Consumption demand, we remember, is essentially directed at the satisfaction of the individual—at providing him with the "utili-ties" of the goods and services he buys. An increasingly affluent society may not be able to say that consumer expenditure is any longer solely geared to necessity, but at least it obeys the fairly constant promptings of the cultural and social environment, with the result that consumer spending, in the aggregate, fluctuates relatively little, except as income fluctuates.

A quite different set of motivations drives the investment impulse. Whether the investment is for replacement of old capital or for the installation of new capital, the ruling consideration is not apt to be the personal use or satisfaction that the investment yields to the owners of the firm. Instead, the touchstone of investment decisions is *profit*.

Figure 26·1 (p. 396) shows corporate profits since 1929 and their division into retained earnings, dividends, and taxes. What is strikingly apparent, of course, is the extreme fluctuation of profits between prosperity and recession. Note that corporations as a whole lost money in the depths of the De-

HAVE CORPORATE PROFITS CHANGED?

Have corporate profits changed as a percent of GNP? The answer is not easily given, since the federal government has chosen to reduce corporation income taxes by increasing depreciation allowances rather than lowering nominal tax rates. (See supplement on National Income and Product Accounts at the end of Chap. 22.) In 1929, corporate gross cash flow (profits plus depreciation allowances) amounted to 12.4 percent of GNP; in 1947, they amounted to 11.4 percent, from 1965 to 1972 they averaged 10.9 percent.

Profits after tax were 8.3 percent of GNP in 1929, 8.7 percent in 1947, and fell to 5.4 percent from 1965 to 1972. Thus it appears that corporate profits have been falling as a share of the total economy, at least over the last decade or two. This does not mean that profit rates of every corporation have necessarily fallen: some have; some have not. And we should note that corporate profits have, of course, increased sharply as a percent of GNP from their level in the early 1900s, but the percentage increased because corporations themselves were not a dominant form of business organization until after World War I.

FIGURE 26 · 1
Profits, taxes, and dividends

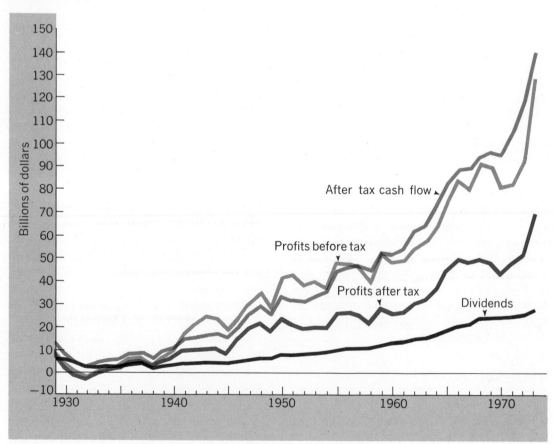

pression years, but that even in the lush postwar period, the swings from year to year have been considerable (compare 1958 and 1959).

EXPECTATIONS

The chart shows us how corporate profits looked to businessmen when the books were tallied at the end of each year. But the results of his last year's operation, although very important, is not the main thing that motivates a businessman to invest. Primarily, he is interested in the profits he expects from *next year's* operations. His view is never backward, but always forward.

Note the important stress on *expectations*. One firm may be enjoying large profits on its existing plant and equipment at the moment; but if it anticipates no profits from the sale of goods that an *additional* investment would make possible, the firm will make no additions to capital. Another firm may be suffering current losses; but if it anticipates a large profit from the production of a new good, it may launch a considerable capital expenditure.

There is a sound reason for this anticipatory quality of investment decisions. Typically, the capital goods bought by investment expenditures are expected to last for years and to pay for themselves only slowly. In addition, they are often highly specialized. If capital expenditures could be recouped in a few weeks or months, or even in a matter of a year or two, or if capital goods were easily transferred from one use to another, they would not be so risky and their dependence on expectations not so great. But it is characteristic of most capital goods that they *are* durable, with life expectancies of ten or more years, and that they tend to be limited in their alternative uses, or to have no alternative uses at all. You cannot spin cloth in a steel mill or make steel in a cotton mill.

The decision to invest is thus always forward-looking. Even when the stimulus to build is felt in the present, the calculations that determine whether or not an investment will be made necessarily concern the flow of income to the firm in the future. These expectations are inherently much more volatile than the current drives and desires that guide the consumer. Expectations, whether based on guesses or forecasts, are capable of sudden and sharp reversals of a sort rare in consumption spending. Thus in its orientation to the future we find a main cause for the volatility of investment expenditures.

Induced and Autonomous Investment

One kind of profit expectation, and the investment that stems from it, derives from *an observed rise in current consumption spending,* as a result of higher incomes.

Many business firms decide to invest because they must expand their capacity to maintain a given share of a growing market. Real estate developers who build to accommodate an already visible suburban exodus, or supermarkets that build to serve a booming metropolis, or gas stations that must be built to serve a new highway, or additions to manufacturing capacity that must be made because existing facilities cannot keep up with demand—these are all examples of what we call *induced investment.*

THE ACCELERATION PRINCIPLE

When rising incomes and consumption lead to induced investment, the relationship is called the *acceleration principle* or the *accelerator.* The name springs from the fact that the amount of induced investment depends upon the rate of growth of the economy. An economy that is not growing has no induced

TABLE 26 · 1
A Model of the Accelerator

Year	Sales	Existing capital	Needed capital (2 × sales)	Replacement investment	Induced new investment (2 × addition to sales)	Total investment
1	$100	$200	$200	$20	—	$20
2	120	200	240	20	$40	60
3	130	240	260	20	20	40
4	135	260	270	20	10	30
5	138	270	276	20	6	26
6	140	276	280	20	4	24
7	140	280	280	20	—	20
8	130	280	260	—	—	0
9	130	260	260	20	—	20

investment. Also, an economy that has unutilized capacity will not have induced investment.

Table 26·1 is a model that explains this phenomenon. It shows us an industry whose sales rise for six years, then level off, and finally decline. We assume it has no unused equipment and that its equipment wears out every ten years. Also, we will make the assumption that the capital-output ratio (see p. 323) in this industry is 2; that is, it requires a capital investment of $2 to produce a flow of output of $1.

Now let us see the accelerator at work.

In our first view of the industry, we find it in equilibrium with sales of, let us say, 100 units, capital equipment valued at 200 units, and regular replacement demand of twenty units, or 10 percent of its stock of equipment. Now we assume that its sales rise to 120 units. To produce 120 units of goods, the firm will need (according to our assumptions) 240 units of capital. This is 40 units more than it has, so it must order them. Note that its demand for capital goods now shoots from 20 units to 60 units: 20 units for replace-

ment as before, and 40 new ones. Thus investment expenditures *triple,* even though sales have risen but 20 percent!

Now assume that in the next year sales rise further, to 130 units. How large will our firm's investment demand be? Its replacement demand will not be larger, since its new capital will not wear out for ten years. And the amount of new capital needed to handle its new sales will be only 20 units, not 40 as before. Its total investment demand has *fallen* from 60 units to 40.

What is the surprising fact here? It is that *we can have an actual fall in induced investment, though sales are still rising!* In fact, as soon as the *rate of increase* of consumption begins to fall, *the absolute amount* of induced investment declines. Thus a slowdown in the rate of improvement in sales can cause an absolute decline in the orders sent to capital goods makers. This helps us to explain how weakness can appear in some branches of the economy while prosperity seems still to be reigning in the market at large.

Now look at what happens to our model in the eighth year, when we assume that sales slip back to 130. Our existing capital (280 units) will be greater by 20 units than our needed capital. That year the industry will have no new orders for capital goods and may not even make any replacements, because it can produce all it needs with its old machines. Its orders to capital goods makers will fall to zero, even though its level of sales is 30 percent higher than at the beginning. The next year, however, if sales remain steady, it will again have to replace one of its old machines. Its replacement demand again jumps to 20. No wonder capital goods industries traditionally experience feast or famine years!

There is, in addition, an extremely important point to bear in mind about the accelerator. *Its upward leverage usually takes effect only when an industry is operating at or near capacity.* When an industry is not near capacity, it is relatively simple for it to satisfy a larger demand for its goods by raising output on its underutilized equipment. Thus, unlike the multiplier, which yields its effects on output only when we have unemployed resources, the accelerator yields its effects only when we do *not* have unemployed capital.

AUTONOMOUS INVESTMENT

Not all investment is induced by prior rises in consumption. A very important category of investment is that undertaken in the expectation of a profit to be derived from a *new* good or a *new* way of making a good. This type of investment is usually called *autonomous* investment.

In autonomous investment decisions, prior trends in consumption have little or nothing to do with the decision to invest. This is particularly the case when new technologies provide the stimulus for investment. Then the question in the minds of the man-

agers of the firm is whether the new product will create *new* demand for itself.

Technological advance is not, however, the only cause for autonomous investment, and therefore we cannot statistically separate autonomous from induced investment. With some economic stimuli, such as the opening of a new territory or shifts in population or population growth, the motivations of both autonomous and induced investment are undoubtedly present. Yet there is a meaningful distinction between the two, insofar as induced investment is sensitive and responsive to sales, whereas autonomous investment is not. This means that induced investment, by its nature, is more foreseeable than autonomous investment.

At the same time, both spontaneous and induced investments are powerfully affected by the overall investment "climate"—not alone the economic climate of confidence, the level and direction of the stock market, etc., but the political scene, international developments, and so on. Hence it is not surprising that investment is often an unpredictable component of GNP, and thus a key "independent" variable in any model of GNP.

BUSINESS CYCLES AGAIN

Perhaps you noticed a wavelike movement in the investment induced by changing consumption. Our understanding of the accelerator, combined with the multiplier, now enables us to gain a much deeper insight into the mechanism of the *business cycle.*

Let us assume that some stimulus, such as an important industry-building invention, has begun to increase investment expenditures. We can easily see how such an initial impetus can generate a cumulative and self-feeding boom. As the multiplier and accelerator interact, the first burst of investment stimulates additional consumption, the additional consumption induces more investment,

THE STOCK MARKET AND INVESTMENT

How does the stock market affect business investment? There are two direct effects. One is that the market has traditionally served as a general barometer of the expectations of the business-minded community as a whole. We say "business-minded" rather than "business," because the demand for and supply of securities mainly comes from securities dealers, stockbrokers, and the investing public, rather than from nonfinancial business enterprises themselves. When the market is buoyant, it has been a signal to business that the "business climate" is favorable, and the effect on what Keynes called the "animal spirits" of executives has been to encourage them to go ahead with expansion plans. When the market is falling, on the other hand, spirits tend to be dampened, and executives may think twice before embarking on an expansion program in the face of general pessimism.

This traditional relationship is, however, greatly lessened by the growing power of government to influence the trend of economic events. Businessmen once looked to the market as the key signal for the future. Today they look to Washington. Hence, during the past decade when the stock market has shown wide swings, business investment in plant and equipment has remained basically steady. This reflects the feelings of corporate managers that government policy will keep the economy growing, whatever "the market" may think of events.

A second direct effect of the stock market on investment has to do with the ease of issuing new securities. One of the ways in which investment is financed is through the issuance of new stocks or bonds whose proceeds will purchase plant and equipment. When the market is rising, it is much easier to float a new issue than when prices are falling. This is particularly true for certain businesses—A.T. & T. is a prime example—that depend heavily on stock issues for new capital rather than on retained earnings.

and this in turn reinvigorates consumption. Meanwhile, this process of mutual stimulation serves to lift business expectations and to encourage still further expansionary spending. Inventories are built up in anticipation of larger sales. Prices "firm up," and the stock market rises. Optimism reigns. A boom is on.

What happens to end such a boom? There are many possible reasons why it may peter out or come to an abrupt halt. It may simply be that the new industry will get built, and thereafter an important stimulus to investment will be lacking. Or even before it is completed, wages and prices may have begun to rise as full employment is neared, and the climate of expectations may become wary. (Businessmen have an old adage that "what goes up must come down.") Meanwhile, per-

haps tight money will choke off spending plans or make new projects appear unprofitable.

Or investment may begin to decline because consumption, although still rising, is no longer rising at the earlier *rate* (the acceleration principle in action). We have already noticed that the action of the accelerator, all by itself, could give rise to wave-like movements in total expenditure. The accelerator, of course, never works all by itself, but it can exert its upward and downward pressures within the flux of economic forces and in this way give rise to an underlying cyclical impetus.

It is impossible to know in advance what particular cause will retard spending—a credit shortage, a very tight labor market, a saturation of demand for a key industry's products (such as automobiles). But it is all too easy to see how a hesitation in spending can turn into a general contraction. Perhaps warned by a falling stock market, perhaps by a slowdown in their sales or an end to rising profits, businessmen begin to cut back. Whatever their initial motivation, what follows thereafter is much like the preceding expansion, only in reverse. The multiplier mechanism now breeds smaller rather than larger incomes. Downward revisions of expectations reduce rather than enhance the attractiveness of investment projects. As consumption decreases, unemployment begins to rise. Inventories are worked off. Bankruptcies become more common. We experience all the economic and social problems of a recession.

But just as there is a "natural" ceiling to a boom, so there is a more or less "natural" floor to recessions.* The fall in inventories, for example, will eventually come to an end: for even in the severest recessions, merchants and manufacturers must have *some* goods on their shelves and so must eventually begin stocking up. The decline in expenditures will lead to easy money, and the slack in output will tend to a lower level of costs: and both of these factors will encourage new investment projects. Meanwhile, the countercyclical effects of government fiscal policy will slowly make their effects known. Sooner or later, in other words, expenditures will cease falling, and the economy will tend to "bottom out."

The Determinants of Investment

The profit expectations that guide investment decisions are largely unpredictable. But there exists one influence on investment decisions

*In retrospect, the tremendous and long-lasting collapse of 1929 seems to have been caused by special circumstances having to do mainly with speculation and monetary mismanagement.

that seems to offer a more determinable guide. This is the influence of the *rate of interest* on the investment decisions of business firms.

INTEREST COSTS

The rate of interest should offer two guides to the investing firm. If the businessman must borrow capital, a higher rate of interest makes it more expensive to undertake an investment. For huge firms that target a return of 15 to 20 percent on their investment projects, a change in the interest rate from 7 to 8 percent may be negligible. But for certain kinds of investment—notably utilities and home construction—interest rates constitute an important component of the cost of investment funds. To these firms, the lower the cost of borrowed capital, the more the stimulus for investment. The difference in *interest costs* for $1 million borrowed for twenty years at 7 percent (instead of 8 percent) is $200,000, by no means a negligible sum. Since construction is the largest single component of investment, the interest rate therefore becomes an important influence on the value of total capital formation.

A second guide is offered to those businessmen who are not directly seeking to borrow money for investment, but who are debating whether to invest the savings (retained earnings) of their firms. This problem of deciding on investments introduces us to an important idea: the discounting of future income.

DISCOUNTING THE FUTURE

Suppose that someone gave you an ironclad promise to pay you $100 a year hence. Would you pay him $100 *now* to get back the same sum 365 days in the future? Certainly not, for in parting with the money you are suffering an *opportunity cost* or a cost that can be measured in terms of the opportunities that your action (to pay $100 now) has foreclosed for you. Had the going rate of interest been 5 percent, for example, you could have loaned your $100 at 5 percent and had $105 at the end of the year. Hence, friendship aside, you are unlikely to lend your money unless you are paid something to compensate you for the opportunities you must give up while you are waiting for your money to return. Another way of saying exactly the same thing is that we arrive at the *present value* of a specified sum in the future by discounting it by some percentage. If the discount rate is 5 percent, the present value of $100 one year in the future is $100 ÷ 1.05, or approximately $95.24.

This brings us back to our businessman who is considering whether or not to make an investment. Suppose that he is considering investing $100,000 in a machine that he expects to earn $25,000 a year for 5 years, over and above all expenses, after which it will be worthless. Does this mean that the expected profit on the machine is therefore $25,000—the $125,000 of expected earnings less the $100,000 of original cost? No, it does not, for the expected earnings will have to be discounted by some appropriate percentage to find their present value. Thus the first $25,000 to be earned by the machine must be reduced by some discount rate; and the second $25,000 must be discounted *twice* (just as $100 to be repaid in *two* year's time will have to yield the equivalent of *two* years' worth of interest); the third $25,000, three times, etc.*

*The formula for calculating the present value of a flow of future income that does not change from year to year is:

$$\text{Present value} = \frac{R}{(1+i)} + \frac{R}{(1+i)^2} + \ldots + \frac{R}{(1+i)^n}$$

where R is the annual flow of income, i is the interest rate, and n is the number of years over which the flow will last.

Clearly, this process of discounting will cause the present value of the expected future returns of the machine to be less than the sum of the undiscounted returns. If, for example, its returns are discounted at a rate of 10 percent, the businessman will find that the present value of a five-year flow of $25,000 per annum comes not to $125,000 but to only $94,700. This is *less* than the actual expenditure for the machine ($100,000). Hence, at a discount rate of 10 percent, the businessman would not undertake the venture.

On the other hand, if he used a discount rate of 5 percent, the present value of the same future flow would be worth (in round numbers) $109,000. In that case, the machine *would* be a worthwhile investment.

INTEREST RATES AND INVESTMENT

What rate should our businessman use to discount his future earnings? Here is where the rate of interest enters the picture. Looking out at the economy, the businessman sees that there is a whole spectrum of interest rates, ranging from very low rates on bonds (usually government bonds) where the element of risk is very small, to high rates on securities of the same maturity (that is, coming due in the same number of years) where the risk is much greater, such as "low-grade" corporate bonds or mortgages. Among this spectrum of rates, there will be a rate at which *he* can borrow—high or low, depending on his credit worthiness in the eyes of the banking community. By applying that rate he can discover whether the estimated future earning from his venture, properly discounted, is actually profitable or not.

MARGINAL EFFICIENCY OF INVESTMENT

There is still another way in which the interest rate should help determine the volume of investment, although it is really only another

GOVERNMENT AND INDIVIDUAL INVESTMENT

Does the process of discounting earnings and comparing them with costs apply to public investment or to individuals investing in, say, education?

It may. A government contemplating an investment in roads, parks, or police stations does not expect to show a financial profit, but it does expect a flow of benefits—a kind of *social profit*. These benefits can often be roughly measured in terms of their financial worth, and the public institution can then compare the discounted value of these expected benefits against their costs.

In similar fashion, an individual contemplating a personal investment, such as acquiring a new skill, may make a similar calculation. He estimates the future increase in earnings that he expects from his training, discounts this sum, and compares it with the cost of undertaking the investment. Of course, individuals do not always act with the precision of "economic man." Nonetheless, the idea of discounting future returns helps give analytic clarity to the reason why a 20-year-old person will willingly accept the cost of becoming a doctor or an engineer, whereas a 55-year-old will not. For the younger person, the investment is expected to pay off (quite aside from the pleasures of the increased skills themselves). For the older person, it is not. An older person may go to school for pleasure, but not for profit.

version of the same idea. The businessman can calculate the rate of discount that would just suffice to make a particular investment pay. *This rate is called the marginal efficiency of investment.* In our illustration above, we saw that over a five-year period, a rate of 10 percent proved the machine to be unprofitable and that one of 5 percent made it show an expected profit. There must be some rate, between these two, that would just make the discounted value of the expected returns equal the cost of the $100,000 machine; if we figure it out, the rate turns out to be approximately 8 percent.*

*How did we arrive at the figure of 8 percent? By using the formula in the preceding footnote and solving for the value of i (the discount rate) that would make the sum of the terms on the right—the discounted future earnings—just equal to the *cost* of the machine, which we put on the left. The mathematics is a bit complicated, but the idea is not. Clearly, if we expect an investment to yield very high returns (R in the formula), it will have a very high marginal efficiency—that is, it will take a very high discount rate (i in the formula) to bring the sum of future earnings down to the level of cost. On the other hand, if the investment is expected to give only small returns, it will have a low marginal efficiency, for only a low discount rate will keep the discounted value of future earnings from falling below costs.

What has this to do with the rate of interest? The answer is that the businessman very likely has a choice among different investment projects from which he anticipates different returns. Suppose he ranks those projects, as we have in Fig. 26·2, starting with the most profitable (A) and proceeding to the least profitable (G). How far down the list should he go? The rate of interest gives the answer. Suppose the rate (for projects of comparable risk) is shown by OX. Then all his investment projects whose marginal efficiency is higher than OX (investments A through D) will be profitable, and all those whose marginal efficiency falls below OX (E through G) will be discarded or at least postponed.

Note that if the interest rate falls, more investments will be worthwhile; and that if it rises, fewer will be. As the figure on the right shows in generalized form, a fall in the rate of interest (e.g., from OX to OY) induces a rise in the quantity of investment (from OC to OG).

Increases in autonomous investment or induced investment can be represented as the

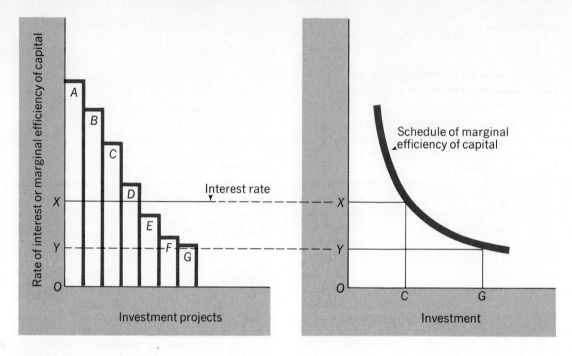

FIGURE 26 · 2
Marginal efficiency of capital

marginal efficiency of capital schedule shifting to the right. New opportunities, either arising from the development of new goods and services or because of increasing sales of old goods, mean that a given amount of capital can earn a higher rate of return or that more investment will occur at any given rate of interest. Draw in a new marginal efficiency curve in Fig. 26 · 2 and prove this to yourself.

Thus, whether we figure interest as a cost or as a guideline against which we measure the expected returns of a capital investment, we reach the important conclusion that *low interest rates should encourage investment spending*—or in more formal language, that *investment should be inversely related to the rate of interest*. To be sure, the fact that a given investment, such as project *B* above, has a marginal efficiency higher than the in-

terest rate is no guarantee that a business actually will undertake it. Other considerations —perhaps political, perhaps psychological— may deter management, despite its encouraging calculations. But assuredly a business will not carry out a project that yields less than the interest rate, because it can make more profit by lending the money, at the same degree of risk, than by investing it.

THE INTEREST RATE PUZZLE

Perhaps the reader will have remarked on a certain cautionary tone in this discussion of the effect of interest rates on investment. We have spoken of the effects that changing interest rates *should have* on business investment, as an opportunity cost or as a money payment or as a guideline against which to measure the marginal efficiency of investment.* We have not spoken of the effect that interest rates *do* have, because extensive empirical investigations have failed to disclose the effects we would expect.

Here we come to another puzzle in economics—an area in which our analysis, based on what rational profit-maximizing entrepreneurs should do, fails to square with what we find they actually do. Let us take a moment to explore this intriguing and unsolved problem.

THE INVESTMENT FUNCTION

We are now familiar with the consumption function, that relates income to consumption. If our analysis is to be believed, there should also be an *investment function* relating investment to the rate of interest. That is, we should be able to specify that for each percentage point fall in interest, investment rises by such-and-such a percent. We would expect the function to show a curve like the hypothetical one in Fig. 26·3.

In fact, when econometricians first began to inquire into the interest-investment relationship, they found exactly this kind of relation between interest rates and residential construction. As they expected, when it became cheaper to borrow or take out a mortgage, home-building increased. But to their consternation, when they investigated the relation between interest rates and plant and equipment investment, no such relationship appeared. Worse, the data seemed to show a "wrong" relationship: when interest rates went up, plant and equipment investment also went up! Figure 26·4 shows the kind of relation that research established between

*Economists also speak of the marginal efficiency of *capital*. Technically speaking, this is a somewhat different concept. But the two terms are often used interchangeably.

plant and equipment investment (shown as a proportion of GNP) and interest rates, *i*.

THE ELUSIVE INTEREST-RATE INVESTMENT FUNCTION

Did this mean that our theory is wrong in some fundamental sense? Econometricians

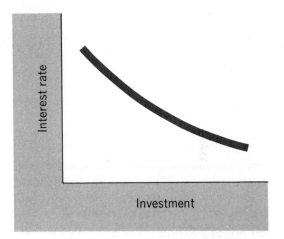

FIGURE 26 · 3
The hypothetical interest—investment function

FIGURE 26 · 4
The econometric interest—investment function

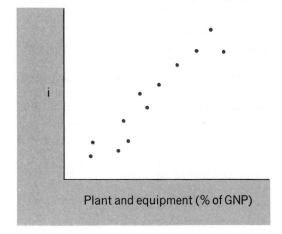

have tried a number of ways to make it come out right. One method was to correct the money rate of interest to the *real rate of interest*. The real rate of interest is the money rate reduced by the rate of inflation. If you get 5 percent on a savings bank deposit, but prices rise by 5 percent, your real interest return is zero. So too if businessmen could borrow at 8 percent, but prices rose by 5 percent, their real interest cost was only 3 percent. Unhappily, when money interest rates were corrected for inflation, the expected investment function still did not appear.

Numerous other attempts have also been made to "specify" an investment function that would reconcile the observed phenomenon of investment perversely rising with interest rates. Econometricians have struggled to incorporate after-tax profit rates and many other possible influences into their investment function term, but all to no avail. *No reliable interest rate—investment function has yet been devised.* Good investment functions have been found for explaining and predicting plant and equipment investment in terms of the rate of growth of the economy, the degree of capacity utilization, and the rate of return on capital investments; but the econometricians who build these functions have not been able to include interest rates in the expected manner.

OTHER EXPLANATIONS

Is there no way, then, of explaining a phenomenon that seems to fly in the face of common sense as well as theory? One plausible explanation has been advanced. Historically, interest rates rise during periods of rapid growth. This happens for two reasons. During these periods, the demand curve for *induced* investment shifts to the right. Therefore, even if higher interest rates tend to discourage autonomous investment, this effect may be overridden by the accelerator taking hold elsewhere.

Second, periods of rapid growth push economies toward full employment. Governments thereupon deliberately raise interest rates through the money mechanism to try to cool off the economy. In this complex of cross currents we can have the curious parallel of higher interest rates and higher investment in plant and equipment, but we can see that the influence of interest rates alone is difficult—even impossible—to isolate. Therefore we continue to assume that *if* we could isolate those effects, they would show the negatively sloped investment function we use in economic theory. We make this assumption because it accords with the premises of maximization, and because we believe we can explain away the seeming disconfirmation of our theory in real life. Nevertheless, the interest rate—investment relationship remains something of a puzzle and a source of discomfiture to economists.

The Export Sector

Before we go on to the problem of public demand, we must mention, if only in passing, a sector we have so far largely overlooked. This is the foreign sector, or more properly the sector of net exports.

If we lived in Europe, South America, or Asia, we could not be so casual in our treatment of foreign trade, for this sector constitutes the very lifeline of many, perhaps even most, countries. Our own highly self-sustained economy in which foreign trade plays only a small quantitative (although a much more important qualitative) role in generating total output is very much the exception rather than the rule.*

*In Chapter 36 we shall see, however, that international currency problems can play a very important role in our economic affairs.

In part, it is the relatively marginal role played by foreign trade in the American economy that allows us to treat it so cavalierly. But there is also another problem. The forces that enter into the flows of international trade are much more complex than any we have heretofore discussed. Not alone the reactions of American consumers and firms, but those of foreign consumers and firms must be taken into account. Thus comparisons between international price levels, the availability of foreign or domestic goods, credit and monetary controls, exchange rates —a whole host of other such considerations

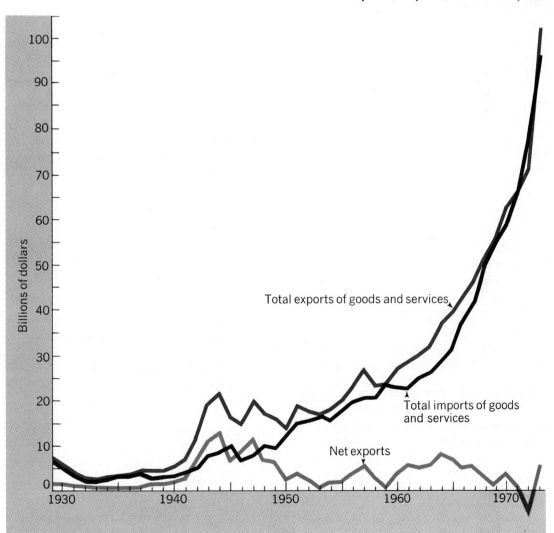

FIGURE 26 · 5
Exports, imports, and net exports

—lie at the very heart of foreign trade. To begin to unravel these interrelationships, one must study international trade as a subject in itself, and that we will defer until Part Five. Nevertheless, we should try to understand the main impact of foreign trade on the demand for GNP, even if we cannot yet investigate the forces and institutions of foreign trade as thoroughly as we might like.

IMPACT OF FOREIGN TRADE

We must begin by repeating that our initial overview of the economic system, with its twin streams of consumption and investment, was actually incomplete. It portrayed what we call a "closed" system, an economy with no flows of goods or services from within its borders to other nations, or from other nations to itself.

Yet such flows must, of course, be taken into account in computing our national output. Let us therefore look at a chart that shows us the main streams of goods and services that cross our borders, as well as a table of the magnitudes in our benchmark years (see Fig. 26.5, p. 407).

First a word of explanation. Exports show the total value of all goods and services we sold to foreigners. Imports show the total value of all goods and services we bought from foreigners. Our bottom line shows the net difference between exports and imports, or the difference between the value of the goods we sold abroad and the value we bought from abroad. This difference is called *net exports,* and it constitutes the net contribution of foreign trade to the demand for GNP.

If we think of it in terms of expenditures, it is not difficult to see what the net contribution is. When exports are sold to for-

eigners, their expenditures add to American incomes. Imports, on the contrary, are expenditures that we make to other countries (and hence that we do not make at home). If we add the foreign expenditures made here and subtract the domestic expenditures made abroad, we will have left a net figure that will show the contribution (if any) made by foreigners to GNP.

THE EXPORT MULTIPLIER

What is the impact of this net expenditure on GNP? It is much the same as net private domestic investment. If we have a rising net export balance, we will have a net increase in spending in the economy.

Conversely, if our net foreign trade balance falls, our demand for GNP will decline, exactly as if the demand for domestic investment fell. Thus, even though we must defer for a while a study of the actual forces at work in international trade, we can quickly include the effects of foreign trade on the level of GNP by considering the net trade balance as a part of our investment demand for output.

One point in particular should be noted. If there is a rise in the net demand generated by foreigners, this will have a *multiplier effect,* exactly as an increase in investment will have. Here is, in fact, the parable of an individual visiting an island (p. 384) come to life. Additional net foreign spending will generate new incomes which will generate new buying; and decreased net foreign spending will diminish incomes, with a similar train of secondary and tertiary effects. We will look into this problem again when we study the foreign trade difficulties of the United States in Chapter 37.

KEY WORDS

CENTRAL CONCEPTS

Expectations

1. The motivation for investment expenditure is not personal use or satisfaction, but *expected profit*. Note that investment is always geared to forward profit *expectations* rather than to past or present results.

Induced investment

2. We distinguish between two kinds of investment motivation. When investments are made to meet an expected demand arising from present or clearly indicated changes in consumption, we speak of *induced investment*. When investment is stimulated by developments (such as inventions) that have little relation to existing trends, we speak of *autonomous investment*.

Autonomous investment

3. Induced investment is subject to the *acceleration principle*, which describes how a given increase in C can give rise to a proportionally larger increase in *I*. The acceleration principle also shows us that the absolute level of induced investment can fall, even though the level of C is still rising. Thus it helps explain the onset of recessions. Note that the acceleration principle "takes hold" only when an industry is at or near full utilization.

Acceleration principle

4. The interaction of the multiplier and the accelerator helps elucidate the mechanism of *business cycles*.

Business cycles

5. The rate of interest should affect investment in two ways. (1) It directly affects *the cost of investment* and (2) it gives us a rate (for investments with risks similar to the one we are contemplating) that allows us to calculate whether the *discounted future returns of our investment are greater than its cost*.

Rate of Interest

6. We can also discover the rate of discount that just makes the expected returns on an investment equal to its cost. This rate is called the *marginal efficiency of investment*. Unless the marginal efficiency of investment is greater than the rate of interest, an investment will not be undertaken.

Discounting future income

7. Note that whether we consider interest as a cost or a guideline against which to compare the marginal efficiency of capital, *the lower the interest rate, the larger the* volume of investment should be.

Marginal efficiency of investment

8. The *investment function* is highly complex. It depends partly on changes within *the system, such as the accelerator, and partly on changes in* "external" *events, above all those affecting expected profits.*

9. Efforts to discover an investment function have not shown that plant and equipment spending declines when interest rises, although housing construction does. Corrections for the *real rate of interest*, and for other possible causes, have failed to give us an expected negatively sloped investment function.

Investment function

10. It is possible to explain this anomaly in terms of the *crosscurrents* that affect interest and investment during a boom. The induced investment during a boom causes a rise in the price of money. The actions of government also deliberately raise interest rates to lessen inflation. In these circumstances it is impossible to detect the actual effect of the rate on plant and equipment spending. We assume that our negatively sloped function is correct because it makes sense, but we cannot demonstrate it in fact.

Export multiplier

Real rate of interest

11. The *export sector* affects the demand for GNP by any excess of exports, or foreign purchases, over imports, or purchases from foreigners. It is therefore very similar to investment demand as an influence on the level of GNP. We can speak of an *export multiplier* exactly as we have spoken of an investment multiplier.

Export sector

QUESTIONS

1. Discuss the difference in the motivation of a consumer buying a car for pleasure and the same person buying a car for his business.

2. Which of the following are induced and which autonomous investment decisions: a developer builds homes in a growing community; a city enlarges its water supply after a period of water shortage; a firm builds a laboratory for basic research; an entrepreneur invests in a new gadget.

3. What is the basic idea of the acceleration principle? Describe carefully how the acceleration principle helps explain the instability of investment.

4. What is meant by "discounting" the value of an expected return? If the rate of interest were 10 percent, what would be the *present value* of $100 due a year hence? What would be its present value two years hence? (HINT: the first year's discounted value has to be discounted a *second time*.)

5. Assume that it costs 7 percent to borrow from a bank. What is the minimum profit that must be expected from an investment before it becomes worthwhile? Could we write that $I = f(r)$ where r stands for the rate of interest? What would be the relation between a change in r and I? Would $I = f(r)$ be a complete description of the motivation for investment? Why should future costs as well as profits be discounted?

6. If inflation is proceeding at 3 percent a year and the banks charge 7 percent for loans, what is the real rate of interest? What is meant by this "real" rate?

7. Why doesn't the accelerator work when there is idle equipment? What significance does this have for the flow of investment as the economy moves from a position of underutilization to one of high utilization?

8. Do you think it is valid to assume that the investment function is negatively related to interest rates, even though we can't show this statistically?

9. Explain how exports stimulate income. Does this mean that imports are bad? Are savings bad?

Government Demand

27

WE TURN NOW TO THE LAST OF THE MAIN sources of demand for GNP—the government. As before, we should begin by familiarizing ourselves with its long historical profile. Figure 27·1 (p. 412) at once shows the signal fact that will underlie the discussion in this chapter. It is that up to 1940 the government was almost insignificant as a source of economic demand. More important, the New Deal (1933–1940) and the postwar era marked a turning point in the *philosophy* of government, from a passive to an active force in macroeconomic affairs. In Europe, government has played a substantial economic role for a longer period; but in Europe as well as America, the deliberate *public management* of demand is a modern phenomenon on which this chapter will focus.

Government in the Expenditure Flow

Before we begin our analysis, let us take a closer look at a recent year, to help us fit the government sector into the flow of national expenditure. Figure 27·2 (p. 413) has the familiar bars of our flow diagram. Note that indirect taxes, totaling some $118 billion in 1973, amounted to almost 10 percent of the value of GNP. As can be seen, however, income taxes on households and businesses are much more important than indirect taxes in providing total government revenues. (What the diagram does not show is that about two-thirds of the indirect taxes are state and local in origin: property taxes, excise taxes, motor vehicle and gasoline taxes, and others. Income taxes and Social Security contributions constitute about nine-tenths of the income of the federal government.)

On the expenditure side, we see that state and local purchases of goods and services are more important than federal purchases in providing public demand; however, since two-thirds of all transfer payments are federal in origin, total federal *expenditures* (as contrasted with purchases of goods and services) run about one-fifth higher than all state and local expenditures.

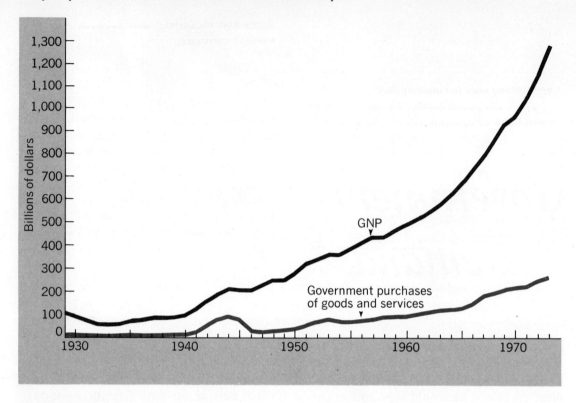

FIGURE 27 · 1
GNP and the government sector

Finally, it is worth reminding ourselves of the different significance and impact of public purchases and transfers. Public purchases of goods and services, whether they originate with local or federal government, require the use of land, labor, and capital. They thus contribute to GNP. Transfer payments, on the other hand, do not increase output. They are simply a reallocation of income, from factors to various groups of the community in the business sector or the household sector. Transfers, therefore, do not require new production and therefore do not add to GNP.

GOVERNMENT SECTOR IN HISTORICAL PERSPECTIVE

How large does the public sector bulk in the total flow of GNP? Let us again try to put a perspective into our answer by observing the trend of government purchases over the years.

We have already pointed out the striking change from prewar to postwar years. The government sector, taken as a whole, has changed from a very small sector to a very large one. In 1929, total government purchases of goods and services were only half of total private investment spending; in 1973, total government purchases were almost 40 percent *larger* than private investment. *In terms of its contributions to GNP, government is now second only to consumption.*

Thus, the public sector, whose operation we will have to examine closely, has become a major factor in the economy as a whole. Let us begin by learning to distinguish carefully among various aspects of what we call "government spending." As we shall see, it is very easy to get confused between "expenditures" and "purchases of goods and services"; between federal spending and total government spending (which includes the states and localities); and between war and nonwar spending.

1. Government expenditures vs. purchases of goods and services

When we speak of government spending, we must take care to specify whether we mean total *expenditures of the government,* which include transfer payments, or *purchases of goods and services by the government,* which represent only actual economic activity performed for, and bought by, the government. In the latter category we include all "production" that owes its existence to public demand, whether from federal, state, or local

FIGURE 27 · 2
Government sector, 1973

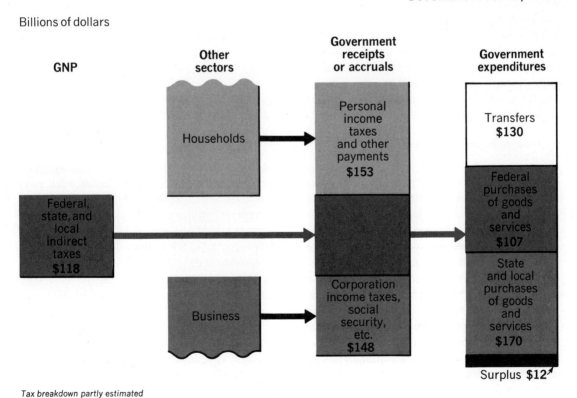

Billions of dollars

Tax breakdown partly estimated

agencies; in the former we include activities performed for the government *plus* transfer payments made by government, at all levels, as part of the redistribution of income. Thus, under "purchases" we include items such as arms and education and police and roads; under "expenditures" we count all these plus Social Security; interest on the debts of localities, states, and the federal government; welfare; and other such transfers.

The distinction is important in terms of the relative bulk of what we call government spending. The purchases of goods and services by all government agencies amounted in 1973 to about $277 billion (of which, as Fig. 27·2 shows, the federal government accounted for just over $100 billion). *The term "G" in our GNP equation stands for these total purchases.* The larger "expenditure" category came to $407 billion. Thus government purchases were the direct cause of the production of about 21 percent of GNP itself, whereas government expenditures amounted in all to not quite one-third of GNP. Remember that a rise in transfers does not increase GNP, so that you must be careful not to use "expenditures" and "purchases" indiscriminately.

2. Federal vs. state and local spending

In dealing with the public sector we must also be careful to distinguish between expenditures or purchases that originate with the federal government and those that stem from state and local agencies. As we noted in Fig. 27·2, state and local spending for goods and services is *larger* than federal purchasing. This is the consequence of the rise of an urbanized, motorized, education-minded society that has imposed vast new burdens on state and local authorities: the supervision of vehicular traffic alone requires the employment of roughly one out of every ten state and local employees, and the support of education now runs to $65 billion a year. These have been increasing during the last decade, and now, annual state and local spending for such goods and services runs about 60 percent ahead of federal purchases.

On the other hand, federal expenditures, *including transfers,* make *total* federal spending larger than total state-and-local spending. In 1973, for example, federal expenditures, including transfers such as Social Security, interest on the debt, various subsidies, grants to the states, etc., brought total federal outlays of all kinds to more than double the amount it spent for goods and services alone.

3. Welfare vs. warfare

Most of the rise in federal purchases of goods and services is the result of our swollen armaments economy, a problem so important that we will look into it separately in a supplement to this chapter. Defense spending in 1973 amounted to about 30 percent of our federal expenditures of all kinds including transfers, and to a much larger fraction—not quite 75 percent—of federal purchases of goods and services. In contrast, note that federal purchases of nonwar goods and services as a percent of GNP are actually smaller than in the prewar days and have shown only a slight rise during the last decade.

Meanwhile, social welfare expenditures of all kinds and of all government agencies (federal, state, and local), including such payments as Social Security, health and medical programs, public education, public housing, welfare assistance, etc., have risen from about 10 percent of GNP in the mid-1930s to about 19 percent today. This is not a large percentage by international standards. In 1971 (the last year for which we have comparative figures) at least 4 other nations spent a higher proportion of their GNP on education than we did. Other social welfare spending (excluding education) amounted to about 13 percent of our GNP, compared with an average of more than 15 percent among the in-

PUBLIC AND PRIVATE BUYING

It is important to realize that government buying can be divided into consumption and investment expenditures, just as private expenditures are. In 1972, for example, governments—federal, state, and local—spent $56 billion for structures and durable goods such as roads, schools, parks, sewage disposal plants and the like, as well as $69 billion on manpower training programs and education to upgrade human skills. These are all *public investment* programs.

Governments also spend large sums on *public consumption;* that is, on providing goods that are enjoyable or necessary for the public at large. Streets are swept, zoos operated, bombers flown, criminals caught.

Why do we separate government consumption and investment from private consumption and investment? The immediate answer is that the money is spent by some government agency rather than by a household or a firm. But there is a deeper reason behind this. It is that a *political decision* has been made to put certain types of expenditures into the hands of the public authorities.

This decision varies from nation to nation. Some countries, like the U.S., have private airlines. Others, such as most of the nations of Europe, have public airlines. In the old days, roads were private; today roads are public, although occasionally one finds a privately owned road. (Note, by the way, that we could not utilize our private consumption of automobile travel unless we simultaneously "consumed" the public road on which we travel.) All nations provide public defense, justice, administration; most provide some public health; a few provide public entertainment. Ideology draws the line, not only between socialist and capitalist governments, but within socialist and capitalist governments: there is a large private agricultural sector in Yugoslavia, a very small one in Russia; many municipally owned power stations in Europe; far fewer here.

What is important to realize is that government expenditure is not a form of economic activity different from consumption or investment. It is the same kind of economic activity, undertaken collectively, through a public agency, rather than privately.

TABLE 27 · 1
Federal Nondefense Purchases

Selected years	1929	1933	1940	1960–65	1966	1967	1968	1969	1970	1971	1972	1973
Percent of GNP	1.0*	3.0*	4.0	2.1	2.2	2.3	2.5	2.5	2.4	2.5	2.6	2.6

*Estimated.

dustrialized nations of Europe. It is noteworthy that in 1971 the average monthly Social Security check per married couple came to just over $132; in Scandinavian countries the payments, compared to average earnings, were roughly twice as generous as ours.

The Main Tasks of Government

The forms and functions of government spending are so complex that it may help us if we now step back and simplify the picture. Basically the federal government has three major economic functions. Measured in terms of expenditures, its largest responsibility lies in the conduct of *international affairs*. Here we find expenditures for defense, foreign aid, veterans' expenditures, military research including space exploration. In 1972 this was 41 percent of all federal spending.

Second, the federal government writes checks in the form of *transfer payments* to individuals and businesses. Here are the farm subsidies, subsidies for the merchant marine, and the very large outflow for Social Security and other welfare. In all, this adds up to another 42 percent of federal expenditure.

Third, the federal government writes checks, in the form of *grants-in-aid* to states and local governments. This accounts for 15 percent of federal outlays. The remainder of federal spending—2 percent—represents direct federal government operating costs and various miscellaneous functions.

It will help us review the main outlines of government spending if we look at Fig. 27·3. The first chart shows us the strikingly different *sources of funds* that flow to the federal and to state and local governments. Note the much heavier reliance of the federal government on income taxes, and the corresponding dependence of state and local governments on indirect taxes. The middle chart shows us the difference in the division of activity between federal and other governments by kinds of payments. But this table obscures a still more basic division, which we see in the third chart. Here we contrast the functions of federal and state and local governments. Now the importance of the three main functions of the federal government clearly emerges.

The Economics of the Public Sector

So far we have been mainly concerned with problems of a definitional kind—in finding out what the government does. Now we want to examine the public sector from a different angle; namely, its unique *economic* character. And here the appropriate place to begin seems to be in the difference in *motivations* that guide public, as contrasted with private, spending.

We recall that the motivations for the household sector and the business sector are lodged in the free decisions of their respective units. Householders decide to spend or save their incomes as they wish, and we are able to construct a propensity to consume schedule only because there seem to be spending and saving patterns that emerge spontaneously from the householders themselves. Similarly, business firms exercise their own judgments on their capital expenditures, and as a result we have seen the inherent variability of investment decisions.

But when we turn to the expenditures of the public sector, we enter an entirely new area of motivation. It is no longer fixed habit or profit that determines the rate of spending, but *political decision*—that is, the collective will of the people as it is formulated and expressed through their local, state, and federal legislatures and executives.

As we shall soon see, this does not mean that government is therefore an entirely unpredictable economic force. There are regularities and patterns in the government's economic behavior, as there are in other sectors. Yet the presence of an explicit political will that can direct the income or outgo of the sector *as a whole* (especially its federal component) gives to the public sector a special significance. *This is the only sector whose expenditures and receipts are open to deliberate control.* We can exert (through public action) very important influences on the behavior of households and firms. But we cannot directly alter their economic activity in the manner that is open to us with the public sector.

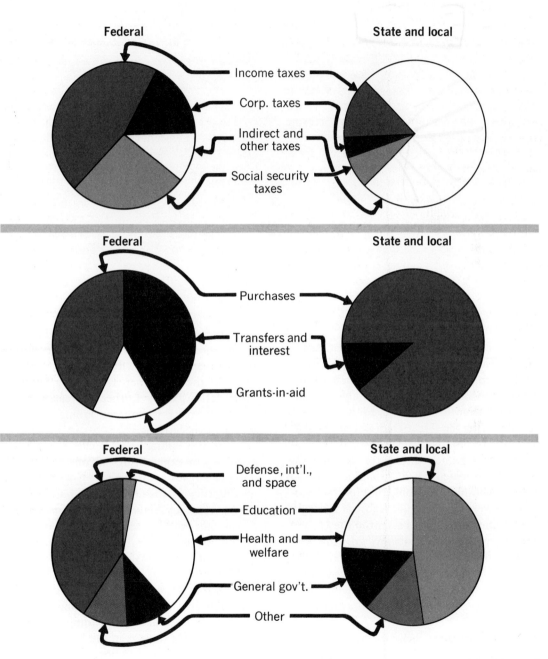

FIGURE 27 · 3
Federal, state, and local finances

JOHN MAYNARD KEYNES

Few economists have left so deep a mark on their own times as John Maynard Keynes, and few have roused such passions, pro and con. It is difficult now, when (as a famous conservative economist has said) "We are all Keynesians," to recall the impact of Keynes's seminal book, *The General Theory of Employment, Interest, and Money*, when it appeared in 1936. Yet there were debates in the halls of academe in which voices shook and faces became empurpled over questions such as whether or not savings and investment were equal, as Keynes *defined* them to be! (We shall come to that question shortly.)

What made Keynes so controversial? Partly it was the economic philosophy that lay half explicit, half implicit in his great book—a philosophy of active government intervention. In a period when the reigning philosophy in many circles was still laissez faire, this was reason enough for Keynes's disturbing impact.

But perhaps another reason was Keynes's personality. Inordinately gifted, he was successful at a dozen things: a brilliant mathematician, a major diplomat, a great collector of modern French art, a dazzlingly skillful investor and speculator—here was one theoretical economist who *did* make a lot of money—a fascinating speaker, a consummate stylist. Keynes was not one to wear these talents modestly, and his wit was savage. Sir Harry Goshen, chairman of a Scottish bank, once deplored a Keynesian proposal and urged that things should be allowed to take "their natural course." "Is it more appropriate to smile or rage at these artless sentiments?" Keynes asked. "Best perhaps to let Sir Harry take *his* natural course."

FISCAL POLICY

The deliberate use of the government sector as an active economic force is a relatively new conception in economics. Much of the apparatus of macroeconomic analysis stems essentially from the work of John Maynard Keynes during the Great Depression. At that time his proposals were regarded as extremely daring, but they have become increasingly accepted by both major political parties. Although the bold use of the economic powers of the public sector is far from commanding unanimous assent in the United States today, there is a steadily growing consensus in the use of fiscal policy—that is, the deliberate utilization of the government's taxing and spending powers—to help insure the stability and growth of the national economy.

The basic idea behind modern fiscal policy is simple enough. We have seen that economic recessions have their roots in a failure of the business sector to offset the savings of the economy through sufficient investment. If savings or leakages are larger than intended investment, there will be a gap in the circuit of incomes and expenditures that can cumulate downward, at first by the effect of the multiplier, thereafter, and even more seriously, by further decreases in investment brought about by falling sales and gloomy expectations.

But if a falling GNP is caused by an inadequacy of expenditures in one sector, our analysis suggests an answer. Could not the insufficiency of spending in the business sector be offset by higher spending in another sector, the public sector? Could not the public sector serve as a supplementary avenue for the "transfer" of savings into expenditure?

As Fig. 27·4 shows, a demand gap can indeed be closed by "transferring" savings to the public sector and spending them. The diagram shows savings in the household sector partly offset by business investment and partly by government spending. It makes clear that at least so far as the mechanics of the economic flow are concerned, the public sector can serve to offset savings or other leakages equally as well as the private sector.

How is the "transfer" accomplished? It can be done much as business does it, by offering bonds that individuals or institutions may buy with their savings. Unlike business, the government cannot offer stock, for it is not run as a profit-making enterprise. However, government has a source of funds quite different from business; namely, *taxes. In effect, government can "commandeer" purchasing power in a way that business cannot.*

TAXES, EXPENDITURES, AND GNP

We shall look more carefully into the question of how the government can serve as a kind of counterbalance for the private economy. But first we must discover something about the normal behavior of the public sector; for despite the importance of political decisions in determining the action of the public sector, and despite the multiplicity of government units and activities, *we can nonetheless discern "propensities" in government spending and receiving*—propensities that play their compensating role in the economy quite independently of any direct political intervention.

The reason for these propensities is that both government income and government outgo are closely tied to private activity. Government receipts are derived in the main from taxes, and taxes—direct or indirect—

FIGURE 27 · 4
Public expenditure and the demand gap

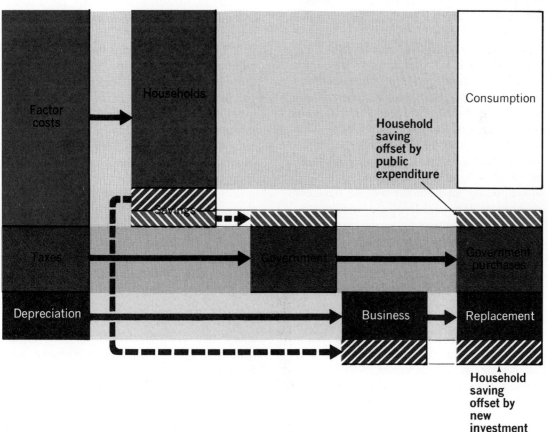

tend to reflect the trend of business and personal income. In fact, we can generalize about tax payments in much the same fashion as we can about consumption, describing them as a predictable function of GNP. To be sure, this assumes that tax *rates* do not change. But since rates change only infrequently, we can draw up a general schedule that relates tax receipts and the level of GNP. The schedule will show not only that taxes rise as GNP rises, but that they rise *faster* than GNP.

Why faster? Largely because of the progressive structure of the federal income tax. As household and business incomes rise to higher levels, the percentage "bite" of income taxes increases (see Table 18·2). Thus as incomes rise, tax liabilities rise even more. Conversely, the tax bite works downward in the opposite way. As incomes fall, taxes fall even faster, since households or businesses with lowered incomes find themselves in less steep tax brackets.

Government expenditures also show certain "propensities," which is to say, *some government spending is also functionally related to the level of GNP.* A number of government programs are directly correlated to the level of economic activity in such a way that spending *decreases* as GNP *increases,* and vice versa. For instance, unemployment benefits are naturally higher when GNP is low or falling. So are many welfare payments at the state and local level. So, too, are disbursements to farmers under various agricultural programs.

AUTOMATIC STABILIZERS

All these automatic effects taken together are called the *automatic stabilizers* or the *built-in stabilizers* of the economy. What they add up to is an automatic government counterbalance to the private sector. As GNP falls because private spending is insufficient, taxes

decline even faster and public expenditures grow, thereby automatically causing the government sector to offset the private sector to some extent. In similar fashion, as GNP rises, taxes tend to rise even faster and public expenditures decline, thereby causing the government sector to act as a brake.

The public sector therefore acts as an automatic compensator, even without direct action to alter tax or expenditure levels, pumping out more public demand when private demand is slowing and curbing public demand when private demand is brisk.

How effective are the built-in stabilizers? It is estimated that the increase in transfer payments plus the reduction in taxes offset about 35¢ of each dollar of original decline in spending. Here is how this works. Suppose that private investment were to fall by $10 billion. If there were no stabilizers, household spending might fall by another $10 billion (the multiplier effect), causing a total decline of $20 billion in incomes.

The action of the stabilizers, however, will prevent the full force of this fall. First, the reduction in incomes of both households and firms will lower their tax liabilities. Since taxes take about 35¢ from each dollar, the initial drop of $10 billion in incomes will reduce tax liabilities by about $3.5 billion. Most of this—let us say $3 billion—is likely to be spent. Meanwhile some public expenditures for unemployment insurance and farm payments will rise, pumping out perhaps $1 billion into the consumption sector, all of which we assume to be spent by its recipients.

Thus, the incomes of firms and households, having originally fallen by $10 billion, will be offset by roughly $4 billion— $1 billion in additional transfer incomes and $3 billion in income spent by households because their taxes are lower. As a result, the decline in expenditure will be reduced from $10 billion to about $6 billion (actually $6.5 billion, according to the calculations of the

Council of Economic Advisers). The multiplier will then indeed affect this net reduction in spending, so that total incomes in the economy will fall by twice as much, or $13 billion in all. But this is considerably less than the fall of $20 billion that would otherwise have taken place.

This is certainly an improvement over a situation with no stabilizers. Yet if the drop in investment is not to bring about some fall in GNP, it will have to be *fully* compensated by an equivalent increase in government spending or by a fall in taxes large enough to induce an equivalent amount of private spending. This will require public action more vigorous than that brought about automatically. Indeed, it requires that the government take on a task very different from any we have heretofore studied, the task of "demand management," or acting as the *deliberate* balancing mechanism of the economy.*

Demand Management

How does the government manage demand? It has three basic alternatives. It can

1. *increase or decrease expenditures*

2. *raise or lower taxes*

3. *alter its monetary policy*

We have already looked into the mechanics of the first option in Fig. 27·4, where we showed that government expenditure fills a demand gap exactly like private expenditure. It follows that a decrease in government

---

*Although we should note that different kinds of private and public spending programs may have different multipliers if they go to different spending groups. A government public works program that uses unskilled labor is apt to have a larger initial repercussion on GNP than a private investment project in computers. Additional transfer expenditures may also have initial multiplier effects different from direct purchases of goods and services. And finally, different tax structures will cause changes in GNP to affect private spending differently.

spending will also create a decrease in final demand, just as a drop in the spending of any other sector.

Our diagram did not show the direct effect of tax changes, simply because it is difficult to draw such a diagram clearly. But it is not difficult to understand the effect of a tax change. When the government lowers taxes it diminishes the transfer of income from households or firms into the public sector. Households and firms therefore have more income to spend. Contrariwise, in raising taxes, a government withdraws spending power from households and firms. As a result, we can expect that private spending will fall.

OFFSETS AND REPERCUSSIONS

The direct effects of changes in expenditures and taxes are thus easy to picture. (We will postpone for a few chapters the use of monetary policy, but we can already anticipate that the purpose of monetary policy is also to induce households or firms to spend more or less money, much as changes in taxes.

A few cautions are in order, however. When government expenditures or taxes are changed, the government wants to raise or lower total demand by a certain amount. Therefore the fiscal authorities must take into account the repercussions that their own policies may have. For example, an increase in government expenditures financed by borrowing may bring about a rise in private expenditures because the announcement of a policy of higher spending may spur businessmen into taking more aggressive investment action. If that is the case, government spending plus additional private spending may raise total demand by more than the government originally desired! At the other extreme, it is possible that the announcement of a government spending program may "scare off" business spending. This was probably the case in the early days of the New Deal, although it is much less likely to happen today, when business has come to welcome most government spending.

TIME LAGS

Second, there is a long delay between the adoption of a new tax or expenditure policy and the realization of its effects. Increased expenditures or new tax proposals have to move through Congress, often a time-consuming process. In addition, if expenditures require capital construction, it may take months, even years, before spending really gets rolling. Thus by the time the new expenditures begin to give their boosting effect, the economic situation may have changed in a way that makes those expenditures unwelcome. So, too, with expenditure cuts. It takes a long time to turn off most government programs. By the time the spending ceases, we may wish it were still with us!

Some economists have therefore suggested that we should have "stockpiles" of approved expenditure projects and "standby" authority to permit the President to raise or lower tax rates, within stated limits, in order to speed up the process of demand management. Other countries have successfully used such expenditure "stockpiles" as a means of accelerating the demand management process, and the last several U.S. administrations have sought—so far in vain—for executive power to adjust tax rates. Most economists would probably favor both proposals, but neither is yet an actuality. As a result, *very long time lags must be taken into account in the normal process of demand management.*

TAX CUTS VS. EXPENDITURES

Which of these two methods of managing demand—taxes or spending—is preferable?

The question basically asks us which we need more: public goods or private goods. But there are a number of technical economic criteria that we must also bear in mind.

First, tax cuts and expenditures tend to favor different groups. Tax cuts benefit those who pay taxes, and expenditures benefit those who receive them. This simple fact reveals a good deal about the political and economic pros and cons of each method. Tax cuts help well-to-do families and are of little direct benefit to poor families whose incomes are so low that they pay little or no income taxes. Expenditure programs *can* benefit these disadvantaged groups or areas—for example, by slum clearance in specific cities, training programs, or simply higher welfare payments. Expenditure programs can also help special groups, such as military or road contractors, or middle-income families who usually benefit from housing programs.

The difference, then, is that tax programs have a widespread impact, whereas expenditure programs tend to have a concentrated impact: *tax cuts or increases are diffused across the economy, exerting their influences on different income strata, whereas expenditure programs are often concentrated geographically or occupationally.* (Some expenditure programs, such as Social Security or medical aid, can have a broad "horizontal effect" as well.)

Second, *expenditure programs tend to be more reliable as a means of increasing demand, whereas tax programs tend to be effective in decreasing demand.* The reason is clear enough. If the government wishes to increase final demand and chooses to lower taxes, it makes possible a higher level of private spending, but there is no guarantee that firms or households will in fact spend all their tax savings. Indeed, the marginal propensity to consume leads us to be quite certain that firms and households will not spend all their tax reductions, at least for a time. Thus if the government wants to increase demand by say $7 billion, it may have to cut taxes by about $10 billion.

On the other hand, tax increases are a very reliable method of decreasing demand. Individuals or firms *can* "defy" tax increases and maintain their former level of spending by going out and borrowing money or by spending their savings, but it is unlikely they will do so. If the government tries to hold back total demand by cutting its own expenditure programs, however, there is the chance that firms and individuals will undo the government's effort to cut demand by borrowing and spending more themselves.

There is no magic formula that will enable us to declare once and for all what policy is best for demand management. It is often impossible to raise taxes for political reasons, in which case a decrease in expenditures is certainly the next best way to keep total demand from rising too fast. So too, it may be impossible to push through a program of public expenditure because public opinion or congressional tempers are opposed to spending. In that case, a tax cut is certainly the best available way to keep demand up if the nation is threatened with a recession.

GOVERNMENT-MADE CYCLES

Thus the management of demand is fraught with difficulties. One of them stems from the fact that by virtue of its size, government is itself the source of much economic instability. Government is not just a "balance wheel" in the economy, but a primary source of demand; and when that demand changes, we are likely to experience booms or recessions.

For example, the recessions of 1954 and 1957–1958 were both caused by cuts in military spending. No other government (or private) spending rose to fill the gap, and the

economy suffered a decline as a result. Again, a recession was "made in Washington" in 1960–1961, when the Eisenhower administration tried to balance the federal government's budget by cutting expenditures. Not surprisingly, total demand fell and the economy slumped. Yet another instance arose in 1969–1970, when the Nixon administration tried to curb inflation by trimming government spending. Once again the economy responded by slowing down its growth—alas, without curing its inflationary ills.

FISCAL DRAG

We shall return at length to the problem of inflation, but let us pursue a little further the difficulties of managing demand. One of these is the problem known as *fiscal drag,* a problem that arises from the same mechanism that gives rise to the automatic stabilizers. We have seen that most taxes depend on the level of income and that the federal government tends to increase its tax collections faster than income grows. As a result, if the government maintains a more or less "do nothing" policy, it will gradually collect more and more taxes, while its expenditures remain constant.

This would lead to a chronic, rising surplus in the federal budget. What this surplus means, in macroeconomic terms, is that the government is taking income away from the household and business sectors and failing to spend it. Hence such a surplus could seriously hold back the economy from attaining its maximum output. Thus the government may have to declare a "fiscal dividend" by cutting taxes or increasing expenditures, if it is to prevent a slowdown.

RUOs

Still another problem of demand management is exactly the converse of the one we have just mentioned. It has been called RUOs, or *relatively uncontrollable outlays.* This problem refers to the fact that certain kinds of expenditure gather a momentum of their own and defy sensible management. Military outlays were for many years sacrosanct, and they steadily rose, whatever their economic consequences. So were farm supports. More recently, Social Security expenditures have been rising more rapidly than the growth of GNP, because Social Security payments have become an important political issue.

Under the influence of RUOs, the federal budget can run badly into deficit, even though considerations of demand management would call for a cut in government spending. For example, in 1970 we incurred a federal deficit of $13 billion, and in 1971 a deficit of $22 billion—not because of deliberate planning to offset demand gaps elsewhere, but largely because we could not turn off certain spigots of government spending. According to a study made by the Brookings Institution, such relatively uncontrollable outlays now constitute over 70 percent of the federal budget, and their importance is going up, not down.

GRANTS-IN-AID

A different problem of demand management relates to the fiscal aid the federal government gives to the states, so-called grants-in-aid. Grants-in-aid have risen from $12.7 billion in 1966 to $42 billion in 1973, and they may rise further. *Because grants-in-aid encourage state and local spending, they are one of the most important macroeconomic instruments controlled by the federal government.*

Grants-in-aid are of three kinds. *Categorical grants* are made for specifically designated purposes; for example, grants for highway expenditure. *Block grants* are designated not for specific purposes but for broad cate-

gories of purpose; an example is a grant for "educational purposes." *Unrestricted grants* may be used for any purpose. The controversial plan to distribute a fixed percentage of federal revenue to states and localities under a "revenue-sharing" plan is, in part, such an unrestricted grant.

The distinction among these types of grants is clearer from a legal than from an economic point of view. Consider a locality that was going to build a new road. If it receives a categorical grant for that purpose, the community can use revenues, formerly intended for road building, for another purpose: a categorical grant has become in fact an unrestricted grant. So, too, a block grant for education may free up some funds formerly intended for education but that can now be spent in another area. The "catch" in these grants is that the federal government often attaches a matching provision to its grant, so that the locality has to spend some of its revenues on, say, roads or education, to qualify for federal aid. Thus a promise of $1 in federal aid for a given purpose may induce state and local authorities to undertake programs that lead to a much larger increase in total expenditure than the federal outlay.

REVENUE SHARING

The Nixon administration has strongly urged that many categorical grants be replaced by a system of automatic *revenue sharing.* Proponents of this plan stress two advantages. One is that the federal government is a much more effective taxgatherer than the states or cities. By sharing its revenues with states and localities it will give them spending abilities they could not otherwise enjoy. Second, the plan accords with the general philosophy of those who would like to see economic power deconcentrated and brought back toward local government.

Opponents of the plan fear that the more-or-less unrestricted granting of funds may lead to state and local tax cuts instead of to additional state and local spending. These cuts may occur because states tend to compete with one another for new industry, each state trying to keep its taxes lower than its neighbors. When all states do this, no one gains a competitive advantage, but the result is too little spending for domestic purposes. The danger is that the assurance of federal aid will take the pressure off states and localities, who will use their revenue shares to cut down on local taxation, thereby cutting down as well on local programs. Proponents of revenue sharing reply that the share of revenue going to each state can be determined by formulas that will reward states for making a strong local tax effort. Opponents then charge that this will benefit rich states that can afford higher taxes and penalize poor ones that cannot.

To these complex considerations, there must be added another. Grants-in-aid and revenue sharing force us to confront the difficult question of what level of government should make various expenditure and tax decisions. There is disagreement over which functions are best "reserved" for the federal government (e.g., highways?); over how much intervention the federal government should have in economic activities traditionally reserved for the states (education?); and over what sorts of strictly local programs should be aided by categorical grants (such as local police forces). These are matters for political determination, but their effects on the level and distribution of public demand can be very considerable.

THE RESPONSIBILITY OF PUBLIC DEMAND

All these considerations point out how difficult it is to conduct demand management as smoothly in practice as in textbooks. There was a time, not too long ago, when economists talked rather glibly of "fine-tuning" the economy. That was in the first flush of triumph of the *idea* of managed demand, before the hard realities of fiscal drag and RUOs and other problems had been fully faced. Economists are a good deal more modest in their claims these days.

Nevertheless, the basic idea of using the government as a balancing mechanism for the economy remains valid, however difficult it may be to realize the perfect balance in fact. It is valid because the federal sector is the only sector whose operations we can collectively control. There is no way for business to determine how much it should spend as a sector, no way for consumers to concert their activity. More important, even if there were such a way, business and consumer actions might not accord with the needs of the macroeconomy. Only the public sector can act consciously on behalf of the public interest; only the public sector can attempt to reconcile the needs of all groups. However exasperating or inefficient or clumsy public demand management may be, it remains a major accomplishment, both in theory and fact, of twentieth-century economics.

KEY WORDS

Government
purchases vs.
government
expenditures

CENTRAL CONCEPTS

1. **The** *public sector* **derives its income from three main sources: indirect taxes (mainly for state and local governments), personal income taxes, and corporate taxes. Expenditures for** *goods and services* **are roughly equally divided between federal government and state and local government, but the federal government is the larger source of** *transfer expenditures.*

Fiscal policy

Automatic stabilizers

Demand management

Expenditures

Taxes

Monetary policy

Time lags

Expenditures vs. tax cuts

Fiscal drag

RUOs

Grants-in-aid

2. Comparing 1970 with the 1920s, we find that the public sector has grown considerably as a proportion of GNP. Federal purchases have grown largely for defense purposes. Total expenditures have grown because of much larger defense spending and, in recent years, larger spending for welfare, including such items as education, social security, and welfare assistance.

3. The critical differentiating factor between the public and the private sectors is that the public sector can be deliberately employed as an instrument of *national economic policy*.

4. The use of government spending and taxing to achieve economic stability is called *fiscal policy*. *Automatic stabilizers* lessen the momentum of booms and cushion the impact of recessions. The stabilizers arise from the progressive incidence of income taxation, from expenditure programs geared to unemployment, and so on.

5. The effort to use the government as a balance wheel in the economy is called *demand management*. Demand management uses three main instruments. It seeks to
   - raise or lower aggregate demand by altering the *level of public expenditures*
   - influence aggregate demand by raising or lowering *taxes*
   - affect final demand through *monetary policy*

6. Demand management is complicated for many reasons. One of them is that changes in expenditures or taxes may induce augmenting or offsetting changes in the private sector. A second difficulty is that long *time lags* occur between policy and its implementation.

7. It is difficult to choose categorically between expenditures and taxes as a preferred means of control. Tax changes tend to benefit or hurt taxpayers and are of little immediate concern to the nonincome-taxpaying poor. Expenditures can be focused on special groups or areas. *Generally speaking, tax increases are more effective than expenditure decreases as a means of reducing demand; and expenditure increases are surer than tax cuts as a method of increasing demand.*

8. Other difficulties of demand management are the threat of *fiscal drag*, the problem of *relatively uncontrollable outlays* (RUOs), and the use of *grants-in-aid*. Grants-in-aid, in particular, are an important tool of demand management, but they present many problems with regard to the effect of various grant-in-aid programs and the desirable location of political decision over different kinds of economic activity.

QUESTIONS

1. What are the main differences between the public and the private sectors? Are these differences economic or political?

2. Show in a diagram how increased government expenditure can offset a demand gap. Show also how decreased government taxation can do the same.

3. What is meant by the automatic stabilizers? Give an example of how they might work if we had an increase in investment of $20 billion and the multiplier were 2; and if the increase in taxes and the decrease in public expenditure associated with the boom in investment were $3 billion and $1 billion, respectively.

4. What do you consider a better way of combating a mild recession—tax cuts or higher expenditures? Why? Suppose we had a deep recession, then what would you do?

5. In what sorts of economic conditions should the government run a surplus?

6. Suppose the government cuts taxes by $10 billion and also cuts its expenditures by the same amount. Will this stimulate the economy? Suppose it raises its expenditures and also raises taxes? Would this be a good antirecession policy?

Special Supplement on the Military Subeconomy

The Department of Defense (DOD) is the largest planned economy outside the Soviet Union. Its property—plant and equipment, land, inventories of war and other commodities—amounts to over $200 billion, equal to about 7 percent of the assets of the entire American economy. It owns 39 million acres of land, roughly an area the size of Hawaii. It rules over a population of more than 3 million—direct employees or soldiers —and spends an "official" budget of roughly $75 billion, a budget three-quarters as large as the entire gross national product of Great Britain.

This makes the DOD richer than any small nation in the world and, of course, incomparably more powerful. That part of its assets represented by nuclear explosives alone gives it the equivalent of six tons of TNT for every living inhabitant of the globe, to which must be added the awesome military power of its "conventional" weapons. The conventional explosives dropped in Indochina *before* the extension of the war to Laos amounted to well over 3 million tons, or 50 percent *more* than the total bomb tonnage dropped on all nations in both European and Pacific theaters during World War II.

The DOD system embraces both people and industry. In the early 1970s the people included, first, some 2.5 million soldiers deployed in more than 2,000 bases or locations abroad and at home, plus another half million civilians located within the United States and abroad. No less important are some 2 million civilian workers who are directly employed on war production, in addition to a much larger number employed in the secondary echelon of defense-related output. This does not include still further millions who owe their livelihood to the civilian services they render to the military. Symptomatic of the pervasive influence of military spending is the rise in unemployment in Connecticut to 8.1 percent of the labor force when a first small effort was made in 1971 to cut back on military output. What would be the effect of a major contraction in defense spending is a problem we will have to defer for later consideration, but no one denies that such a major contraction—*unless counteracted by*

vigorous expansion of civilian spending—could plunge this country into a severe depression.

The Web of Military Spending

The web of DOD expenditures extends to more areas of the economy than one might think. One expects Lockheed Aircraft, with 88 percent of its sales to the government, to be a ward of the DOD, which explains why the government agreed to rescue the company from bankruptcy in 1971 with an enormous advance of funds. One does not expect the DOD to show up (from 1960 to 1967) as the source of some 44 percent of the revenues of Pan Am or to be, through its Post Exchange (PX) system, the third largest marketing chain in the country, just after Sears and the A&P; or to be a major factor in the housing industry, spending more for military housing between 1956 and 1967 than the total spent by the federal government on all other public housing.

All in all, some 22,000 firms are prime contractors with the DOD, although the widespread practice of subcontracting means that a much larger number of enterprises—perhaps 100,000 in all—look to defense spending for a portion of their income. Within the main constituency, however, a very few firms are the bastion of the DOD economy. The hundred largest defense contractors supply about two-thirds of the $40-odd billion of manufactured deliveries; and within this group, an inner group of ten firms by themselves accounted for 30 percent of the total. Incidentally, many of these largest contractors owe not only their sales to the military economy, but a considerable portion of their capital as well. As of 1967, defense contractors used $2 billion worth of government-owned furniture and office machines, $4.7 billion worth of government-owned materials, and over $5 billion worth of government plant and equipment, on all of which they were allowed to make profits just as if they were using their own property.

The military establishment, as the largest single customer in the economy, not only supports a central core of industry but also penetrates that core with 2,072 retired military officers who were employed in 1969 by the 95 biggest contractors (the 10 largest firms averaging 106 officers each). Meanwhile, the establishment has a powerful political arm as well. In the early 1970s the DOD employed more than 300 lobbyists on Capitol Hill and (a conservative estimate) some 2,700 public relations men in the U.S. and abroad. This close political relationship undoubtedly has some bearing on the Pentagon's requests for funds being given, until recently, only the most cursory congressional inspection, leading among other things to cost "overruns" of $24 billion on 38 weapons systems.

This is not to say, of course, that the United States does not need a strong defense capability today. But there is no question that the Pentagon subeconomy—some would say "substate"—has become a major element in, and a major problem for, American capitalism. Indeed, the questions it raises are central: how important is this subeconomy to our economic vitality? How difficult would it be to reduce? Can our economy get along without a military subsector?

Military Dependency

Let us begin by reviewing a few important facts.

At the height of the Vietnam War in 1968, more than 10 percent of our labor force was employed in defense-related work. As the Vietnam War gradually decelerated, this percentage fell to about 7 percent. Defense expenditures, meanwhile, dropped from their peak of nearly $80 billion to just under $75 billion in 1973.

These global figures do not, however, give a clear picture of the strategic position of defense spending within the economy. For the problem is that war-related spending and employment are not distributed evenly across the system but are bunched in special areas and industries. In a survey made in 1967, the Defense Department found that 72 employment areas depended on war output for 12 percent or more of their employment and that four-fifths of these areas were communities with labor forces of less than 50,000. This concentration of defense activity is still a fact of economic life in the mid 1970s. The impact of a cutback on these middle-sized communities can be devastating.

In addition, defense-employment is concen-

trated among special skills as well as in a nucleus of defense-oriented companies. In the late 1960s, about one scientist or engineer out of every five in private industry was employed on a defense-related job. Thirty-eight percent of all physicists depended on war-work. Twenty-five percent of all sheet-metal workers, the same proportion of pattern-makers, and 54 percent of all airplane mechanics worked on defense projects. These proportions have declined, but "defense" is still a major employer of these skills. And as we have already remarked, there is a core of companies dependent on military spending for their very existence. These are not usually the largest companies in the economy (for whom, on the average, defense receipts amount to about 10 percent of total revenues) but the second echelon of corporations: of 30 companies with assets in the $250 million to $1 billion range, 6 depended on war spending for half their incomes, and 7 depended on it for a quarter of theirs.

Thus a cutback in defense spending will be felt very sharply in particular areas, where there may be no other jobs available, or among occupational groups who have no alternative employment at equivalent pay or in companies that are "captives" of the DOD. Such companies and areas naturally lobby hard for defense expenditure on which their livelihood depends. So do their representatives in Congress. It is this interweaving of economic and political interests that makes the problem of defense cutbacks so difficult.

Conversion Possibilities

Yet a cutback is not economically impossible. What is needed is a program of retraining, relocation, income support, and conversion aid that will cushion the inevitable shock of a decline in military spending. In 1971 the National Urban Coalition estimated that we would need to spend about $4 billion over two years to move and retrain the bulk of the personnel who would be displaced if we cut our military budget by one third (in real terms) by 1976.

The sum is not large compared with the saving in national resources that would thereby be affected. *What is crucial, economically, is that the decline in war-related spending be offset by increases in peace-related spending, to be sure that the overall level of demand would remain high enough to act as a magnet, attracting the displaced workers to other jobs.* The Coalition therefore proposes a broadscale attack on many social problems that would raise total (local, state, and federal) government expenditures on goods and services by over 50 percent to assure that the conversion would in fact be a smooth one.

Could such a vast conversion actually be carried out? That is a problem we will return to again in Chapter 41. But at this juncture it is important to point out that there is no insuperable *economic* problem posed by conversion, whatever the political or psychological problems. After World War II, defense spending fell from 37.5 percent of GNP (compared with only 7.4 percent today) to 6.6 percent in the short space of two years. During this period 10 million men and women were demobilized from the armed forces and added to the labor force. Yet unemployment from 1947 to 1949 rose only from 3.6 percent of the labor force to 5.5 percent. Clearly the problem today is much less formidable than it was then.

Opportunity Cost of Armaments

The difference between the two periods is that post-World War II conversion was largely the result of a hugh pent-up civilian demand. Today there is no evidence of unfilled consumer needs. The Vietnam War did not impose shortages on the American people. Thus, unless we were willing to undertake an ambitious program of public spending or tax cuts the conversion will almost certainly generate severe unemployment—and consequent political pressures to maintain military spending.

Meanwhile, it is clear that the armaments economy is not only costing us a vast sum of money, but is also weighing heavily on the Russians. The annual military expenditure of the U.S. and the U.S.S.R. together amounts to about two-thirds of the entire output of the billion peoples of Latin America, Southeast Asia, and the Near East. If we add in the arms expenditures of the rest of the world, we reach the total of more than $50 for every inhabitant of the globe. Not only

is this an opportunity cost of tragically large dimensions, if we consider the uses to which those resources might otherwise be applied, but even in terms of a strictly military calculus, much of it is *total waste, since neither side has been able to gain a decisive advantage despite its enormous expenditures.*

The Prisoners' Dilemma

How does such a senseless course commend itself to national governments? Needless to say, the answers lie deep in the web of political, economic, and social forces of our times. But analytical reasoning can nonetheless throw some light on the matter by unraveling a peculiar situation involving two prisoners, each of whom knows something about the other and who are being interrogated separately. If *both* prisoners remain silent, both will get very light terms, since the evidence against each is not conclusive. If one prisoner squeals on the other, he will get off scot-free as a reward for turning state's evidence, while the other prisoner gets a heavy term. But if *both* prisoners squeal, both will get severe terms, convicted by each other's testimony.

Now, if the two prisoners could confer—and more important, if they trusted each other absolutely—they would obviously agree that the strategy of shared silence was the best for each. But if the prisoners are separated or unsure of each other's trustworthiness, each will be powerfully tempted to rat on the other in order to reduce his own sentence. As a result, since the two are equally tempted, the outcome is likely to end in *both* sides ratting—and in both sides getting heavy punishment. Thus the pursuit of individual "self-interest" lures the two prisoners into a strategy that penalizes them both.

Something very like a Prisoners' Dilemma afflicts the nations of the world, especially the two superpowers, America and Russia. Although a policy of limited arms spending would clearly be to their mutual best advantage, their distrust of each other leads each to try to get ahead of the other. This state of mutual suspicion is then worsened when special interest groups in each nation deliberately play on the fears of the public. The end result is that both spend vast sums on armaments that yield them no advantage, while the world (including them) suffers an opportunity cost on an enormous scale. Only with a frank and candid exploration of shared gains and losses can both prisoners hope to get out of their dilemma. Prospects for a massive conversion of the U.S. arms economy may hinge on our understanding of this central military and political reality of our time.

Sources: Seymour Melman, *Pentagon Capitalism* (New York: McGraw Hill, 1970); Leonard Rodberg and Derek Sherer, *The Pentagon Watchers* (Garden City: Doubleday, 1970); Adam Yarmolinsky, *The Military Establishment* (New York: Harper, 1971); *Economic Report of the President* (1971).

Deficit Spending 28

UP TO THIS MOMENT, we have been analyzing the public sector in terms of its effect on the demand for GNP. Now we are going to take a brief but necessary respite from our systematic examination of the various sources of demand for output. The use of the public sector as a source of deliberate demand management poses a question that we must understand before we can comfortably resume our inquiry. This is the question of the government debt.

Any government that uses its budget as a stabilizing device must be prepared to spend more than it takes in in taxes. On occasion it must purposefully plan a budget in which outgo exceeds income, leaving a negative figure called a *deficit*.

That raises a problem that alarms and perplexes many people. Like a business or consumer, the government cannot spend money it does not have. Therefore it must *borrow* the needed funds from individuals, firms, or banks in order to cover its deficit. Deficit spending, in other words, means the spending of borrowed money, money derived from the sale of government bonds.

DEFICITS AND LOSSES

Can the government safely run up a deficit? Let us begin to unravel this important but perplexing question by asking another: can a private business afford to run up a deficit?

There is one kind of deficit that a private business *cannot* afford: a deficit that comes from spending more money on current production than it will realize from its sale. This kind of deficit is called a *business loss;* and if losses are severe enough, a business firm will be forced to discontinue its operations.

But there is another kind of deficit, although it is not called by that name, in the operations of a private firm. This is an excess of expenditures over receipts brought about by spending money on *capital assets.* When the American Telephone and Telegraph Company or the Exxon Corporation uses its own savings or those of the public to build a new plant and new equipment, it does not show a "loss" on its annual statement to stockholders, even though its total expenditures on current costs and on capital may have been greater than sales. Instead, expenditures are divided into two kinds, one relating current costs to current income, and the other relegating expenditures on capital goods to an entirely separate "capital account." Instead of calling the excess of expenditures a deficit, they call it investment.*

DEBTS AND ASSETS

Can A.T.&T. or Exxon afford to run deficits of the latter kind indefinitely? We can answer the question by imagining ourselves in an economic landscape with no disturbing changes in technology or in consumers' tastes, so that entrepreneurs can plan ahead with great safety. Now let us assume that in this comfortable economy, Exxon decides to build a new refinery, perhaps to take care of the growing population. To finance the plant, it issues new bonds, so that its new asset is matched by a new debt.

Now what about this debt? How long can Exxon afford to have its bonds outstanding?

The answer is—forever!

Remember that we have assumed an economy remaining changeless in tastes and techniques, so that each year the new refinery can turn out a quota of output, perfectly confident that it will be sold; and each year it can set aside a reserve for wear and tear, perfectly confident that the refinery is being properly depreciated. As a result, each year the debt must be as good as the year before— no better and no worse. The bondholder is sure of getting his interest, steadily earned, and he knows that the underlying asset is being fully maintained.

Admittedly, after a certain number of years the new factory will be worn out. But if our imaginary economy remains un-

---

*Investment does not *require* a "deficit," since it can be financed out of current profits. But many expanding companies do spend more money on current and capital account than they take in through sales and thereby incur a "deficit" for at least a part of their investment.

changed and if depreciation accruals have been properly set aside, when the old plant gives out, an identical new one will be built from these depreciation reserves. Meanwhile, the old debt, like the old plant, will also come to an end, for debts usually run for a fixed term of years. The Exxon Corporation must now pay back its debtholders in full. But how? The firm has accumulated a reserve to buy a new plant, but it has not accumulated a second reserve to repay its bondholders.

Nevertheless, the answer is simple enough. When the bonds come due in our imaginary situation, the Exxon Corporation issues *new* bonds equal in value to the old ones. It then sells the new bonds and uses the new money it raises to pay off the old bondholders. When the transaction is done, a whole cycle is complete: both a new refinery and a new issue of bonds exist in place of the old. Everything is exactly as it was in the first place. Furthermore, as long as this cycle can be repeated, such a debt could safely exist in perpetuity! And why not? Its underlying asset also exists, eternally renewed, in perpetuity.

REAL CORPORATE DEBTS

To be sure, not many businesses are run this way, for the obvious reason that tastes and techniques in the real world are anything but changeless. Indeed, there is every reason to believe that when a factory wears out it will *not* be replaced by another costing exactly as much and producing just the same commodity. Yet, highly stable businesses such as the Exxon Corporation or A.T.&T. do, in fact, continuously "refund" their bond issues, paying off old bonds with new ones, and never "paying back" their indebtedness as a whole. A.T.&T., for instance, actually increased its total indebtedness from $1.1 billion in 1929 to $24.1 billion in 1972. Exxon ran up its debt from $170.1 million in 1929 to

$2.7 billion in 1973. And the credit rating of both companies today is as good as, or better than, it was in 1929.

Thus some individual enterprises that face conditions of stability similar to our imaginary situations do actually issue bonds "in perpetuity," paying back each issue when it is due, only to replace it with another (and, as we have seen, *bigger*) issue.

TOTAL BUSINESS DEBTS

Most strong individual businesses can carry their debts indefinitely, and the business sector *as a whole* can easily do so. For although individual businesses may seek to retire their debts, as we look over the whole economy we can see that as one business extinguishes its debt, another is borrowing an even larger sum. Why larger? Because the *assets* of the total business sector are also steadily rising.

Table 28·1 shows this trend in the growth of corporate debt.*

Note that from 1929 through 1940, corporate debt *declined*. The shrinkage coincided with the years of depression and slow recovery, when additions to capital plant were small. But beginning with the onset of the postwar period, we see a very rapid increase in business indebtedness, an increase that continues down to our present day.

If we think of this creation of debt (and equity) as part of the savings-investment process, the relationship between debts and assets should be clear. Debts are claims, and we remember how claims can arise as the financial counterpart of the process of real capital formation. Thus, rising debts on capital account are a sign that assets are also increasing. It is important to emphasize the *capital account*. Debts incurred to buy capital assets are very different from those incurred to pay current expenses. The latter have very little close connection with rising wealth, whereas when we see that debts on corporate capital account are rising, we can take for granted that assets are probably rising as well. The same is true, incidentally, for the ever-rising total of consumer debts that mirror a corresponding increase in consumers' assets. As our stock of houses grows, so does our total mortgage debt; as our personal inventories of cars, washing machines, and other appliances grow, so does our outstanding consumer indebtedness.

GOVERNMENT DEFICITS

Can government, like business, borrow "indefinitely"? The question is important enough to warrant a careful answer. Hence,

TABLE 28 · 1
Corporate Net Long-Term Debt*

Year	1929	1933	1940	1950	1966	1967	1968	1969	1970	1971	1972
Billions of dollars	47	48	44	60	231	258	286	324	360	402	447

*Maturity over one year.

*We do not show the parallel rise in new equities (shares of stock), since changes in stock market prices play so large a role here. We might, however, add a mental note to the effect that business issues new stock each year, as well as new bonds. During the 1960s and early 1970s, net new stock issues have ranged from about $2 to $9 billion per annum.

let us begin by comparing government borrowing and business borrowing.

One difference that springs quickly to mind is that businesses borrow in order to acquire productive assets. That is, matching the new claims on the business sector is additional real wealth that will provide for larger output. From this additional wealth, business will also receive the income to pay interest on its debt or dividends on its stock. But what of the government? Where are its productive assets?

We have already noted that the government budget includes dams, roads, housing projects, and many other items that might be classified as assets. During the 1960s, federal expenditures for such civil construction projects averaged about $5 billion a year. Thus the total addition to the gross public debt during the 1960s (it rose from roughly $239 billion in 1960 to $470 billion in 1973) could be construed as merely the financial counterpart of the creation of public assets.

Why is it not so considered? Mainly because, as we have seen, the peculiar character of public expenditures leads us to lump together all public spending, regardless of kind. In many European countries, however, public capital expenditures are sharply differentiated from public current expenditures. If we had such a system, the government's deficit on capital account could then be viewed as the public equivalent of business's deficit on capital account. Such a change might considerably improve the rationality of much discussion concerning the government's deficit.

SALES VS. TAXES

But there is still a difference. Private capital enhances the earning capacity of a private business, whereas most public capital, save for such assets as toll roads, does not "make money" for the public sector. Does this constitute a meaningful distinction?

We can understand, of course, why an individual business insists that its investment must be profitable. The actual money that the business will pay out in the course of making an investment will almost surely not return to the business that spent it. A shirt manufacturer, for instance, who invests in a new factory cannot hope that the men who build that factory will spend all their wages on his shirts. He knows that the money he spends through investment will soon be dissipated throughout the economy, and that it can be recaptured only through strenuous selling efforts.

Not quite so with a national government, however. Its income does not come from sales but from taxes, and those taxes reflect the general level of income of the country. Thus any and all that government lays out, just because it enters the general stream of incomes, redounds to the taxing capacity or, we might say, the "earning capacity" of government.

How much will come back to the government in taxes? That depends on two main factors: the impact of government spending on income via the multipler, and the incidence and progressivity of the tax structure. Under today's normal conditions, the government will recover about half or a little more of its expenditure.* But in any event, note that the government does not "lose" its money in the way that a business does. Whatever goes into the income stream is always *available* to the government as a source of taxes; but whatever goes into the income stream is not necessarily available to any single business as a source of sales.

This reasoning helps us understand why federal finance is different from state and local government finance. An expenditure made by New York City or New York State is apt to be respent in many other areas of the country. Thus taxable incomes in New

---

*We can make a rough estimate of the multiplier effect of additional public expenditure as 2 and of the share of an additional dollar of GNP going to federal taxes as about $\frac{1}{3}$ (see p. 420). Thus $1 of public spending will create $2 of GNP, of which 65¢ will go back to the federal government.

York will not, in all probability, rise to match local spending. As a result, *state and local governments must look on their finances much as an individual business does.* The power of full fiscal recapture belongs solely to the federal government.

The National Debt

INTERNAL AND EXTERNAL DEBTS

This difference between the limited powers of recoupment of a single firm and the relatively limitless powers of a national government lies at the heart of the basic difference between business and government deficit spending. It helps us understand why the government has a capacity for financial operation that is inherently of a far higher order of magnitude than that of business. We can sum up this fundamental difference in the contrast between the *externality of business debts* and the *internality of national government debts.*

What do we mean by the externality of business debts? We simply mean that business firms owe their debts to someone distinct from themselves—someone over whom they have no control—whether this be bondholders or the bank from which they borrowed. Thus, to service or to pay back its debts, business must transfer funds from its own possession into the possession of outsiders. If this transfer cannot be made, if a business does not have the funds to pay its bondholders or its bank, it will go bankrupt.

The government is in a radically different position. Its bondholders, banks, and other people or institutions to whom it owes its debts belong to the same community as that whence it extracts its receipts. In other words, the government does not have to transfer its funds to an "outside" group to pay its bonds. It transfers them, instead, from some members of the national community over which

it has legal powers (taxpayers) to other members of the *same* community (bondholders). The contrast is much the same as that between a family that owes a debt to another family, and a family in which the husband has borrowed money from his wife; or again between a firm that owes money to another, and a firm in which one branch has borrowed money from another. *Internal debts do not drain the resources of one community into another, but merely redistribute the claims among members of the same community.*

PROBLEMS OF A NATIONAL DEBT

A government cannot always borrow without trouble, however. Important and difficult problems of money management are inseparable from a large debt. More important, the people or institutions from whom taxes are collected are not always exactly the same people and institutions to whom interest is paid, so that servicing a government debt often poses problems of *redistribution of income.* For instance, if all government bonds were owned by rich people and if all government taxation were regressive (i.e., proportionately heavier on low incomes), then servicing a government debt would mean transferring income from the poor to the rich. Considerations of equity aside, this would also probably involve distributing income from spenders to savers and would thereby intensify the problem of closing the savings gap.

In addition, a debt that a government owes to foreign citizens is *not* an internal debt. It is exactly like a debt that a corporation owes to an "outside" public, and it can involve payments that can cripple a nation. Do not forget that the internality of debts applies only to *national* debts held as bonds by members of the same community of people whose incomes contribute to government revenues.

PERPETUAL PUBLIC DEBTS

Can a national government therefore have a perpetual debt? We have seen that it can. To be sure, the debt must be constantly refunded, much as business refunds its debts, with new issues of bonds replacing the old. But like the business sector, we can expect the government debt in this way to be maintained indefinitely.

Will our public debt grow forever? That depends largely on what happens to our business debts and equities. If business debts and equities grow fast enough—that is, if we are creating enough assets through investment—there is no reason why government debts should grow. Government deficits, after all, are designed as *supplements* to deficits. The rationale behind public borrowing is that it will be used only when the private sector is not providing enough expenditure to give us a large enough GNP to provide reasonably full employment.

Nonetheless, the prospect of a rising national debt bothers many people. Some day, they say, it will have to be repaid. Is this true? It may aid us to think about the problem if we try to answer the following questions:

1. *Can we afford to pay interest on a rising debt?*

The capacity to expand debts, both public and private, depends largely on the willingness of people to lend money, and this willingness in turn reflects their confidence that they will be paid interest regularly and will have their principal returned to them when their bonds are due.

We have seen how refunding can take care of the repayment problem. But what about interest? With a private firm, this requires that interest costs be kept to a modest fraction of sales, so that they can easily be covered. With government, similar financial prudence requires that interest costs stay well within the taxable capacity of government. The figures in Table 28·2 give us some perspective on this problem today.

It can be seen that interest is a much higher percentage of federal revenues than of corporate revenues. But there is a reason for this. Corporations are supposed to maximize their revenues; the government is not supposed to maximize its tax income. Hence we must also judge the size of the federal interest cost in comparison with the size of GNP, the total tax base from which the government can draw. Finally, we should know that interest as a percentage of all federal receipts has remained very steady in recent years, and it is actually much lower than in the 1920s, when interest costs amounted to about 20 to 30 percent of the federal budget.

2. *Can we afford the burden of a rising debt?*

What is the "burden" of a debt? For a firm, the question is easy to answer. It is the *interest cost* that must be borne by those who owe the debt. Here, of course, we are back to the externality of debts. *The burden of a debt is*

TABLE 28·2
Debt and Interest Costs

	Net interest ($ billions)	Interest as proportionate cost
Nonfinancial corporations (1973)	$18.8	2.8 percent of gross corporate revenues
Federal government (1973)	14.4	{ 5.5 percent of receipts { 1.1 percent of GNP

the obligation it imposes to pay funds from one firm or community to another.

But we have seen that there is no such cost for an internal debt, such as that of a nation. The *cost* of the debt—that is, the taxes that must be levied to pay interest—becomes *income* to the very same community, as checks sent to bondholders for their interest income. Every penny that the debt costs our economy in taxes returns to our economy as income.

The same is also true of the principal of the debt. The debts we owe inside the nation we also *own* inside the nation—just as the I.O.U. a husband owes his wife is also an I.O.U. owned by the family; or, again, just as an amount borrowed by Branch A of a multibranch firm is owed to Branch B of the same firm.

There is a further point here. Internal debts are debts that are considered as financial *assets* within the "family." Nobody within A.T.&T. considers its debts to be part of the assets of the firm, but many thousands of people in the U.S. consider the country's debts to be their assets. Indeed, everyone who owns a government bond considers it an asset. Thus in contrast to external debts, paying back an internal debt does not "lift a burden" from a community, because no burden existed in the first place! When a corporation pays off a debt to a bank, it is rid of an obligation to an outside claimant on its property. But when a husband pays his wife, the *family* is no richer, any more than the *firm* is better off if one branch reimburses another. So, too, with a nation. If a national debt is repaid, the national economy is not rid of an obligation to an outside claimant. We would be rid only of obligations owed to one another.

REAL BURDENS

This is not to say—and the point is important—that government spending is costless. Consider for a moment the main cause of government spending over the past fifty years: the prosecution of three wars. There was surely a terrific cost to these wars in lives, health, and (in economic terms) in the use of factors of production to produce guns instead of butter. But note also that all of this cost is irrevocably and unbudgeably situated in the past. The cost of all wars is borne during the years when the wars are fought and must be measured in the destruction that was then caused and the opportunities for creating real wealth that were then missed. The debt inherited from these wars is no longer a "cost." Today it is only an instrument for the transfer of incomes within the American community.

So, too, with debts incurred to fight unemployment. The cost of unemployment is also borne once and for all at the time it occurs, and the benefits of the government spending to combat unemployment will be enjoyed (or if the spending is ill-advised, the wastes of spending will be suffered) when that spending takes place. Afterward, the debt persists as a continuing means of transferring incomes, but the debt no longer has any connection to the "cost" for which it was incurred.

Costs, in other words, are *missed opportunities,* potential well-being not achieved. Debts, on the other hand (when they are held within a country) only transfer purchasing power and do not involve the nation in giving up its output to anyone else.

INDIRECT EFFECTS

Does this mean that there are no disadvantages whatsoever in a large national debt?

We have talked of one possible disadvantage, that of transferring incomes from spenders to savers, or possibly of transferring purchasing power from productive groups to unproductive groups. But we must pay heed to one other problem. This is the problem a rising debt may cause indirectly, but none-

PERSONAL DEBTS
AND PUBLIC DEBTS

In view of the fact that our national debt today figures out to approximately $2,200 for every man, woman, and child, it is not surprising that we frequently hear appeals to "common sense," telling us how much better we would be without this debt, and how our grandchildren will groan under its weight.

Is this true? We have already discussed the fact that internal debts are different from external debts, but let us press the point home from a different vantage point. Suppose we decided that we would "pay off" the debt. This would mean that our government bonds would be redeemed for cash. To get the cash, we would have to tax ourselves (unless we wanted to roll the printing presses), so that what we would really be doing would be transferring money from taxpayers to bondholders.

Would that be a net gain for the nation? Consider the typical holder of a government bond —a family, a bank, or a corporation. It now holds the world's safest and most readily-sold paper asset from which a regular income is obtained. After our debt is redeemed, our families, banks, and corporations will have two choices: (1) they can hold cash and get *no* income, or (2) they can invest in other securities that are slightly *less* safe. Are these investors better off? As for our grandchildren, it is true that if we pay off the debt they will not have to "carry" its weight. But to offset that, neither will they be carried by the comfortable government bonds they would otherwise have inherited. They will also be relieved from paying taxes to meet the interest on the debt. Alas, they will be relieved as well of the pleasure of depositing the green Treasury checks for interest payments that used to arrive twice a year.

theless painfully, *if it discourages private investment*.

This could be a very serious, real cost of government debts, were such a reaction to be widespread and long-lasting. It may well be (we are not sure) that the long drawn-out and never entirely successful recovery from the Great Depression was caused, to a considerable extent, by the adverse psychological impact of government deficit spending on business investment intentions. Business did not understand deficit spending and interpreted it either as the entering wedge of socialism (instead of a crash program to save capitalism) or as a wastrel and a harebrained economic scheme. To make matters worse, the amount of the government deficit (at its peak $4 billion), while large enough to frighten the business community, was not big enough to begin to exert an effective leverage on total demand, particularly under conditions of widespread unemployment and financial catastrophe.

Today, however, it is much less likely that deficit spending would be attended by a drop in private spending. A great deal that was new and frightening in thought and practice in the 1930s is today well-understood and tested. World War II was, after all, an immense laboratory demonstration of what public spending could do for GNP. The experience of recent years gives good reason to believe that deficit spending in the future will not cause a significant slowdown in private investment expenditure.

THE PUBLIC SECTOR AGAIN IN PERSPECTIVE

We have spent enough time on the question of the debt. Now we must ask what is it that close examination of the problems of government finance reveals, making them look so different from what we expect. The answer is largely that we think of the government as if it were a firm or a household, when it is actually something else. *The government is a sector;* and if we want to think clearly about it, we must compare it, not to the maxims and activities of a household or a firm, but to those of the entire consumer sector or the entire business sector.

Then we can see that the government sector plays a role not too dissimilar from that of the business sector. We have seen how businesses, through their individual decisions to add to plant and equipment, act in concert to offset the savings of consumers. The government, we now see, acts in precisely the same way, except that its decisions, rather than reflecting the behavior of innumerable entrepreneurs in a search for profit, reflect the deliberate political will of the community itself.

Persons who do not understand the intersectoral relationships of the economy like to say that business must "live within its income" and that government acts irresponsibly in failing to do so. These critics fail to see that business does *not* live within its income, but borrows the savings of other sectors and thus typically and normally spends more than it takes in from its sales alone. By doing so, of course, it serves the invaluable function of providing an offset for saving that would otherwise create a demand gap and thereby precipitate a downward movement in economic activity.

Once this offsetting function is understood, it is not difficult to see that government, as well as business, can serve as a "spender" to offset savings, and that in the course of doing so, both government and business typically create new assets for the community.

PUBLIC AND PRIVATE ASSETS

Finally, we have seen something else that gives us a last insight into government spending. We have seen that the creation of earning assets is indispensable for business, because each asset constitutes the means by which an individual business seeks to recoup its own investment spending. But with the government, the definition of an "earning asset" can properly be much larger than with a business firm. The government does not need its assets to make money for itself directly, for the government's economic capability arises from its capacity to tax *all* incomes. So far as government is concerned, then, all that matters is that savings be turned into expenditures, and thereby into taxable incomes.

As a result, government can and should be motivated—even in a self-interested way—by a much wider view of the economic process than would be possible or proper for a single firm. Whereas a firm's assets are largely its capital goods, the assets of a nation are not only capital wealth but the whole productive capacity of its people. Thus government expenditures that redound to the health or well-being or education of its citizens are just as properly considered asset-building expenditures as are its expenditures on dams and roads.

KEY WORDS

Deficits

Government borrowing

Government debt

Assets and debts

Internal vs. external debts

Redistribution

Debt burden

CENTRAL CONCEPTS

1. The use of the government budget as a deliberate antirecession instrument leads to *deficits*. These deficits are financed by *government borrowing* and hence lead to government debt.

2. The government debt can be thought of in the same way as much private debt: as the *financial counterpart of assets.*

3. *All debts, as long as their underlying assets are economically productive, can be maintained indefinitely* by being refunded when they come due.

4. In a progressive and growing economy, debts increase as assets rise. Debts are only a way of financing the growth in assets.

5. National governments have the power of fiscal recapture of any money spent by them—a power not available to state governments or even to the largest businesses. Hence national governments are in a fundamentally different position regarding the safety of their domestically-held debts. This difference is expressed in the concept of *internal debts* versus *external ones.*

6. Domestically-held national debts do not lead to bankruptcy. They do present important and difficult problems of *monetary management*, and they can also result in the *redistribution of income*. These are *real burdens* of the debt.

7. *Repaying the debt would not lift a burden from the economy*. Taxes would decrease (because the debt need no longer be serviced), but income would also decrease (because interest would no longer be paid). Former government bondholders would have to find another acceptable financial asset.

8. The confusion with which the public debt is often viewed arises from a failure to understand that *the government is not a "household" but a sector*, fully comparable to the business sector in its intersectoral operations.

QUESTIONS

1. In what ways is a government deficit comparable to business spending for investment purposes? In what ways is it not?

2. If the government is going to go into debt, does it matter if it spends money for roads or for relief? For education or for weapons? Is there any connection between the use to which government spending is put and *the economic analysis of deficit spending?* Think hard about this: suppose you could show that some spending increased the productivity of the country and that other spending didn't. Would that influence your answer?

3. What is meant by the internality of debts? Is the debt of New York State internal? The debt of a country like Israel?

4. What relation do debts generally have to assets? Can business debts increase indefinitely? Can a family's? Can the debt of all consumers?

5. What are the real burdens of a national debt?

6. Trace out carefully all the consequences of paying back the national debt.

7. How would you explain to someone who is adamantly opposed to socialism that government deficit spending was (a) safe and (b) not necessarily "socialistic"? Or do you think it is not safe and that it is socialistic?

The Determination of GNP

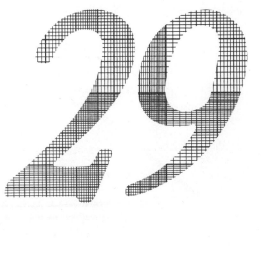

WE HAVE REACHED THE DESTINATION toward which we have been traveling for many chapters. We are finally in a position to understand how the forces of supply and demand determine the actual level of GNP that confronts us in daily life—"the state of the economy" that affects our employment prospects, our immediate well-being, our state of satisfaction or dissatisfaction with the way things are going. To repeat what we have earlier said, we are not interested now in the "historical" level of gross national product—the height to which the underlying processes of growth have carried us—but in the degree of utilization of that total capacity, the fluctuations of "good times and bad times" that we worry about from one year to the next.

THE SUPPLY OF GNP

As we know, that short-run level of GNP is determined by the outcome of two opposing tendencies of supply and demand, just as the level of prices and quantities in a marketplace is "set" by the counterplay of these forces. In Chapter 21 we summed up the short-run supply of GNP in terms of an aggregate production function, and Fig. 29·1 (next page) refreshes our memory of that function. *It shows us the rising curve of total potential output available to us from utilizing larger and larger quantities of labor and capital.* Note that the curve finally reaches a maximum—technically when every man and woman is working to the exhaustion point.

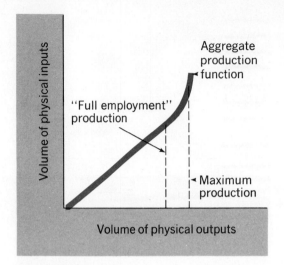

FIGURE 29 · 1
The production function

The normal level of "full" employment is well below that maximum.

THE DEMAND FOR GNP

Now what of the demand function? How much physical output will the factors of production buy as we vary output from levels where employment is very low to levels of full employment?

We will postpone for a moment a detailed discussion of the demand function, to emphasize one aspect. Unlike a curve showing quantities demanded and prices, this demand curve will *slope upward* since we are relating quantities demanded and *incomes*. This enables us to draw a generalized demand curve for all output, which we show in Fig. 29·2 (right).

EMPLOYMENT AND EQUILIBRIUM

Like the market for any single good or service, the market for all goods and services will find its equilibrium where the total quan-

tity of goods demanded equals that supplied. But now we must note something of paramount importance. While the economy will automatically move to this equilibrium point, the *point need not bring about the full employment of the factors of production, particularly labor*. In Fig. 29·2, the economy at equilibrium produces a GNP indicated by GNP_e, which corresponds with an employment of inputs given by I_e.

It could very well be, however, that if all factors of production were fully employed, inputs would be I_F. If they *were* all employed, a "full employment" GNP_F would be produced. *Thus, even though the economy is in equilibrium, it will not necessarily be at a point of full employment.* All we can say about the equilibrium point—exactly as in a market for goods and services—is that it is a level of output toward which the system will move, and from which it will not budge unless the supply curve or the demand curve shifts. It is not necessarily the *right* level—in this case, the level that results in "full" (or even high) employment. It may indeed be a very unsatis-

FIGURE 29 · 2
Supply and demand for physical output

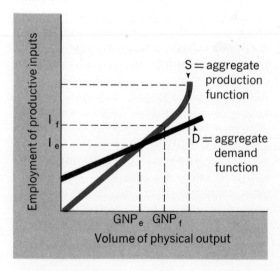

factory level, such as that in the Great Depression.

From Employment to Expenditure

Up to this point, our analysis has been in physical terms—units of inputs (employment) and units of output (actual production). Now we want to analyze the supply and demand for GNP in ordinary money terms, as the dollar value of inputs and the dollar value of outputs. This will enable us to speak of the determination of GNP in terms of *the supply of incomes* and *the demand for output*.

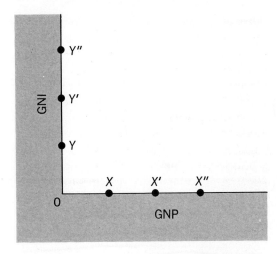

FIGURE 29 · 3
Identity of GNI and GNP

THE SUPPLY OF INCOMES

What do we mean by the supply of incomes? We mean the total payments that will be made to the factors of production as different quantities of them are employed. If we think in terms of the aggregate production function, we want to know how much income, rather than how much employment, will be associated with different levels of output.

Here is where the identity GNP ≡ GNI comes into play in a surprising way. For if we recall Chapter 22 (p. 339), we know that each and every output will give rise to a sum of incomes that is exactly equal to (identical with) the value of that output. If we now translate that identity into graphical form, we see that the supply curve of income must have the shape of a 45° line, as shown in Fig. 29·4.

Figure 29·3 shows us why this is so. The axes of the graph are GNI and GNP. If output is at point X along the GNP (output) axis, the corresponding point on the income axis must be at point Y where $OY = OX$. This is also true of points X' and Y', X'' and Y'', and so on.

From these points, it is easy to draw a

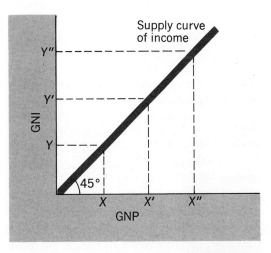

FIGURE 29 · 4
Supply curve of income

supply curve of income that relates GNP and GNI. As you can see, it will lie at a 45° slope from the origin, as Fig. 29·4 shows.

REAL AND MONEY GNP

This 45° line has a special characteristic that we should clearly understand. No matter what the shape of the aggregate production function in real terms—whether we are dealing with a highly efficient or very inefficient economy—the supply curve of *income* will always be a 45° slope. This is because the supply curve of income tells us only that there will always be a supply of incomes equal to the value of output (GNI ≡ GNP), whereas the supply curve of real goods and services looks "behind" the *value* of output and income to the measurement of *quantities* of output and the *quantities* of inputs. Thus the 45° line represents a definitional, rather than a functional, relationship.

THE DEMAND CURVE FOR OUTPUT

If the supply curve of income always has the same slope, why do we bother with it? The answer is that it enables us to highlight the crucial role played by the volume of expenditures in bringing about the equilibrium level of GNP. We can see this if we now add a demand curve for output to our 45° diagram. This is a curve that will be made up of the demands of the various sectors of the economy—households, business, and government. It is therefore a curve that sums up the demand function for consumption, investment, and public expenditure.

In the determination of GNP, the 45° supply curve plays a passive role very much like the totally inelastic (vertical) supply curve we discussed on page 104. In both cases, demand plays the active role in establishing the outcome—of GNP on the one hand, price on the other. But in both cases, supply is a necessary consideration: the other blade of scissors.

Now in real life, as we know, the demand functions of the business and government sectors are exceedingly complex. But in this schematic representation of how supply and demand bring a given money value of GNP into being, we can ignore those difficulties and simply assume that the demand

functions for *G* and *I* are "given" for any moment, although they will certainly shift up or down over time. The crucial demand function then becomes the expenditure of the household sector. And here we have an important clue. From our study of the propensity to consume, we know that this function starts at a "bottom" and slopes upward. Moreover, the upward slope is determined by the marginal propensity to consume, which is less than 1. Therefore the slope must be less than 45°.

We can now put together these components to form a demand function for total output. If we follow steps 1 through 4 in Fig. 29·5 (right) we can see how the demand side of GNP emerges. Figure 29·6 then combines the supply and the demand curves into a graph that shows us the forces that establish an equilibrium level for GNP.

This equilibrium, let us remember, tells us the money value of GNP brought about by the flow of demand against supply. It might, for example, indicate that this equilibrium value of GNP was $1.5 trillion. It does *not* tell us whether $1.5 trillion is a *good* size for GNP, any more than an equilibrium price of $20 for a commodity tells us whether that is a good or bad price from the viewpoint of buyers, producers, or the economy at large. We will spend a number of later chapters discussing problems such as inflation, employment, and growth, all of which critically examine the social usefulness of whatever GNP supply and demand have established.

THE CIRCULAR FLOW

Before we can discuss these problems, however, we must be sure we understand the nature of the GNP-establishing process, just as we must understand how a market works before we can get into issues of social policy connected with a market. Let us therefore take another step and connect our supply-

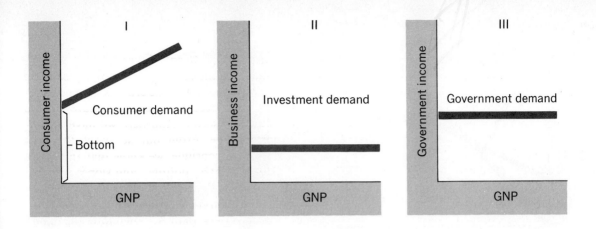

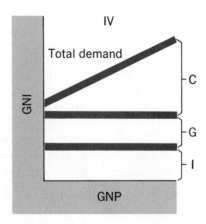

FIGURE 29 · 5
The components of the demand for GNP

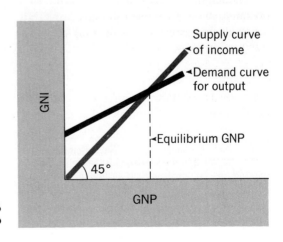

FIGURE 29 · 6
Supply and demand for GNP

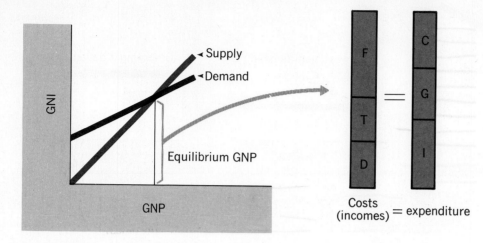

FIGURE 29 · 7
Equilibrium GNP

Costs (incomes) = expenditure

demand equilibrium for GNP with our previous discussion of the circular flow. In Fig. 29·7, let us take out a thin slice of output at equilibrium and examine it under a magnifying glass.

As always, GNP can be analyzed into its component factor, indirect tax, and depreciation costs ($F + T + D$) and into its component expenditures or demands ($C + G + I$). From our earlier discussion we recall that *all costs become incomes*. Therefore our supply curve of incomes must also represent the costs of that supply: we could relabel the GNI axis Gross National Cost. What we see in our diagram is that at equilibrium, total demand for GNP equals total cost for GNP, much like our circular flow (p. 340).*

THE DEMAND GAP

Now let us examine another slice of GNP that is above equilibrium, as in Fig. 29·8.

*But not exactly like our circular flow. We are now dealing with an economy that has saving and investing, profits and losses. Therefore we must think of the cost category "depreciation" (D) as including profits, and the expenditure category "investment" (I) as including net investment as well as replacement. We also continue to forget about exports, for simplicity.

Here the cost of GNP is represented by OA. It is composed of $F + T + D$, and as we know, it is *identical* with the incomes that are the other side of all costs. Thus OA is the cost of GNP measured in terms of gross national income. But now the demand for GNP, OB, is less than OA. The difference consists of the demand gap, with which we are familiar.

If we had taken a slice below equilibrium, what would have been the situation? Now the demand curve lies above the supply curve. That is, total expenditures would have been larger than total costs ($\equiv$ incomes). *We would have had a demand surplus, rather than a demand gap.* You might try drawing such a slice and seeing what the relation of $F + T + D$ would be to $C + G + I$.

THE MOVEMENT TO EQUILIBRIUM

Now let us trace the forces that would push GNP toward the position of equilibrium. At the level of GNP that lies above equilibrium, entrepreneurs and public agencies would have paid out larger sums as costs ($\equiv$ incomes) than they would receive back as sales ($\equiv$ demand). Sales would be below the level

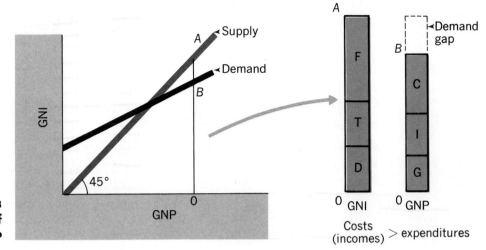

FIGURE 29 · 8
Analysis of equilibrium GNP

of expectations that led to the employment of the factors in the first place. The first result would be a piling up of unsold inventories. Quickly, however, production plans would be revised downward. Fewer factors would be employed. With the fall in employment, incomes would fall, and as incomes fell, so would demand.

The analysis is exactly reversed if GNP is below equilibrium. Now demand ($C + G + I$) is greater than costs or incomes ($F + T + D$). Entrepreneurs will meet this extra demand out of inventories, and they will begin to plan for higher output, hiring more factors, and embarking on investment programs, thereby raising costs and incomes.

Note that in both cases, demand does not change as rapidly as income. In the first case it does not fall as fast as income; in the second it rises more slowly than income. This is the result of the marginal propensity to consume, which, as we have seen, reflects the unwillingness of households to raise or lower their consumption spending as much as any change in their incomes.

Assuming for the moment that G and I remain unchanged, we can see that as the employment of factors increases or decreases, total demand must come closer to total costs or incomes. *If there is a demand gap, income will fall more rapidly than demand, and the gap will close. If there is a demand surplus, income will rise faster than demand and the surplus will gradually disappear.* In both cases, the economy will move toward equilibrium.

THE MOVEMENT OF EQUILIBRIUM

If we now introduce changes in G and I, we can see that *the equilibrium point itself may move.* As the economy enters a downward spiral, investment spending may fall, outbalancing the supportive action of the automatic stabilizers. If this is the case, then the equilibrium level of GNP will move leftward, and the recession may not halt until we reach a very low level of GNP. This is, in fact, exactly what happens when a severe recession causes investment to fall, and the economy does not "bottom out" until GNP has fallen substantially, bringing with it considerable unemployment. Figure 29·9 (p. 448) shows us this process schematically.

Let us begin at a level of GNP indicated by output *OA*. A demand gap exists, and the level of output begins to fall toward *OB*,

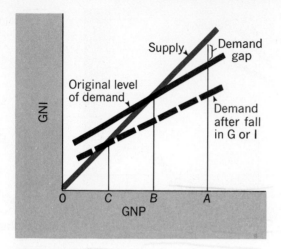

FIGURE 29 · 9
A change in equilibrium GNP

which is an equilibrium level at the *original level of demand*. But now the fall in GNP adversely affects I as well as C, so that the demand schedule for output shifts downward to the dotted line. Hence the economy will not settle at output OB but will continue downward until OC, where once again the demand for output equals the supply of output.

THE EXPANSION PROCESS

Just the opposite course of events helps us explain an upward movement. Suppose our economy "began" in equilibrium at output OC, following a severe recession. Now let us suppose that a rise in demand takes place. This could be the consequence of a burst of autonomous investment or simply the result of brighter expectations or the consequence of deficit spending or any combination. If you will extend the line at OC up to the new demand curve (the upper line, this time), you can see that demand for output ($C + I + G$) is now larger than the costs of output ($F + T + D$).

As a result, entrepreneurs will find their receipts rising. They will add factors, rehiring labor that has been let go during the recession and adding to their stock of inventories or equipment. The economy will begin to move toward the equilibrium depicted by output OB.

Once again, however, we must be careful not to imagine that the equilibrium point is fixed. As the economy moves so will autonomous investment and government spending and taxing. Hence the final equilibrium level may be less than, equal to, or greater than OC, depending on further shifts in the demand curve. But the *process* by which an equilibrium level of GNP is reached is always indicated by the relationship between the supply curve of GNI and the demand curve for GNP.

Another View of Equilibrium

LEAKAGES AND INJECTIONS

So far we have examined the determination of equilibrium by comparing the total supply of incomes and the total demand for goods and services at different levels of GNP. But there is another way of understanding the same idea, if we approach it from the familiar vantage point of savings and investment. We have seen that savings—or rather, all *leakages*—bring about potential demand gaps. We can now add a similar general category for *all increases in expenditure*—net investment, additional government spending, rising exports, or even a spontaneous jump in consumption—and call these *injections*. Equilibrium will then be determined by the interplay of these leakages and injections.

In Fig. 29·10, we show these injections as a solid band in Panel I. In Panel II, we show leakages as the triangle that represents the difference between incomes received (always the 45° line) and expenditure. (Note

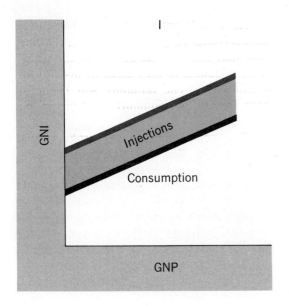

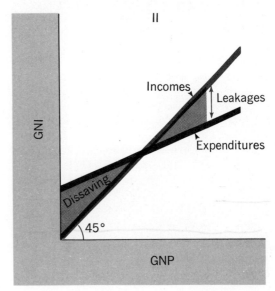

FIGURE 29 · 10
Injections and leakages

FIGURE 29 · 11
Equilibrium of GNP

that below a certain level, there is *dissaving*—a kind of negative leakage—as consumers strive to maintain their living standards by spending more than their incomes, drawing on accumulated savings, or as businesses maintain their depreciation flows by drawing on previous years earnings.)

If we superimpose the two diagrams (see Fig. 29·11), we can see that the equilibrium level of GNP can be looked at not only as the point at which total demand ($C + I + G$) equals total incomes ($F + T + D$), but also as the point where total leakages (household savings, business savings, additional taxes, additional imports) just equal total injections.

At any GNP higher than the equilibrium, the flow of leakages will be larger than that of injections, and a demand gap will push GNP back toward equilibrium. At any point below equilibrium, the band of injections is wider than the triangle of leakages, and the economy will move in the other direction, "up" toward equilibrium. Do not

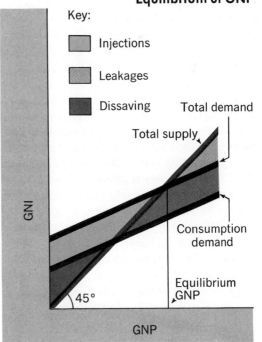

INTENDED AND UNINTENDED S AND I

The careful reader may have noted that we speak of *intended* savings and *intended* investment as the critical forces in establishing equilibrium. This is because there is a formal balance between *all* saving and investment (or all leakages and all injections) at every moment in the economy.

This sounds very strange. Are there not demand gaps, when saving is not offset by investment? Have we not just shown a schedule in which S was not equal to I at every level of income? How then can saving and investment be identities?

The answer is unexpectedly simple. Both saving and investment are made up of *intended and unintended* flows. I may intend to save a great deal, but if my income falls, my actual savings may be very small. As an entrepreneur, I may intend to invest nothing this year; but if sales are poor, I may end up with an unintended investment in unsold inventories. Thus, through fluctuations in incomes, profits, and inventories, people are constantly saving and investing more or less than they intended. These unintended changes make *total* savings equal to (identical with) *total* investment, whereas obviously the intended portions of saving or investment may be unequal.

Economists speak of the difference between intended and unintended activities as *ex ante* and *ex post*. *Ex ante* means "looking forward;" *ex post* means "looking backward." Ex ante savings and investment (or leakages and injections) are usually not equal. But at each and every moment,

SCHEDULES AND TIME

There is a problem inherent in diagrams such as these. One tends to interpret them as if they showed the movement of the economy over time—as if the horizontal axis read 1970, 1971, 1972, and so on. But this is not what the diagram is intended to show. As in the case of the determination of prices, we are trying to build a model to show the forces at work in *a given period of time*. Our diagrams refer to the crosscurrents during each and every moment of time for the economy. If we wanted to show time itself, we would have to introduce a whole ladder of demand and supply schedules showing the components of demand or supply at different dates. Such a diagram would he hopelessly complicated, so we must use our imaginations to portray time, and we must remember that in the graphs we abstract from time to reveal the underlying principles that are constantly at work determining the level of GNP.

forget that the demand curve can always shift, moving equilibrium to the right or left.

SAVING AND INVESTMENT

Equilibrium is always a complicated subject to master, so let us fix the matter in our minds by going over the problem one last time. Suppose that we are going to predict the level of GNP for an island community by means of a questionnaire. To simplify our task, we will ignore government and exports, so that we can concentrate solely on consumption, saving, and investment.

We begin by interrogating the island's business community about their intentions for next year's investment. Now we know that some investment will be induced and that, therefore, investment will partly be a result of the island's level of income; but again for simplification, we assume that businessmen have laid their plans for next year. They tell us they intend to spend $30 million for new housing, plant, equipment, and other capital goods.

Next, our team of pollsters approaches a carefully selected sample of the island's householders and asks them what their consumption and savings plans are for the coming year. Here the answer will be a bit

ex post savings and investment *will* be equal because someone will have been stuck with higher or lower inventories or greater or lesser saving than he intended ex ante.

The strict balance between the formal accounting meanings of saving and investment and the tug-of-war between the active forces of *intended* saving and investment are sources of much confusion to students who ask why the terms are defined in this difficult way. In part we owe the answer to Keynes, who first defined *S* and *I* as identities. Since then the usage has become solidified because it is useful for purposes of national accounting.

For our purposes, we must learn to distinguish between the formal, ex post identity between total saving and investment (or between all leakages and all injections) and the active, ex ante difference between *intended* savings and investment (or *intended* saving, *intended* imports, *intended* business saving, etc., and *intended* additional expenditures of all kinds).

What matters in the determination of GNP are the *actions* people are taking—actions that lead them to try to save or to invest or that make them struggle to get rid of unintended inventories or to build up desired inventories. These are the kinds of activities that will be moving the economy up and down in the never-ending "quest" for its equilibrium point. The fact that at each moment ex post savings and investment are identical from the viewpoint of the economy's balance sheet is important only insofar as we are economic accountants. As analysts of the course of future GNP, we concentrate on the inequality of ex ante, intended actions.

[handwritten: Ex ante = Intended / Ex post = Unintended]

---

disconcerting. Reflecting on their past experience, our householders will reply: "We can't say for sure. We'd *like* to spend such-and-such an amount and save the rest, but really it depends on what our incomes will be." Our poll, in other words, will have to make inquiries about different possibilities that reflect the island's propensity to consume.

Now we tabulate our results, and find that we have the schedule in Table 29·1.

TABLE 29 · 1

Income	Consumption	Saving	Investment
	(in millions)		
$100	$75	$25	$30
110	80	30	30
120	85	35	30

INTERPLAY OF SAVING AND INVESTMENT

If we look at the last two columns, those for saving and investment, we can see a powerful cross play that will characterize our model economy at different levels of income, for the forces of investment and saving will not be in balance at all levels. At some levels, the propensity to save will outrun the act of purposeful investment; at others, the motivations to save will be less than the investment expenditures made by business firms. In fact, our island model shows that at only one level of income—$110 million—will the saving and investment schedules coincide.

What does it mean when intended savings are greater than the flow of intended investment? It means that people are *trying* to save out of their given incomes a larger amount than businessmen are willing to invest. Now if we think back to the exposition of the economy in equilibrium, it will be clear what the result must be. The economy cannot maintain a closed circuit of income and expenditure if savings are larger than investment. This will simply give rise to a demand gap, the repercussions of which we have already explored.

But a similar lack of equilibrium results if intended savings are less than intended investment expenditure (or if investment spending is greater than the propensity to save).

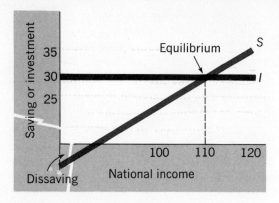

FIGURE 29 · 12
Saving and investment

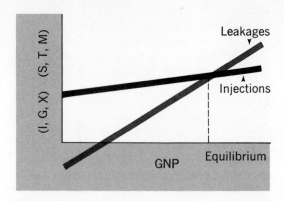

FIGURE 29 · 13
Leakages and injections

Now businessmen will be pumping out more than enough to offset the savings gap. The additional expenditures, over and above those that compensate for saving, will flow into the economy to create new incomes— and out of those new incomes, new savings.

Income and output will be stable, in other words, only when the flow of intended investment just compensates for the flow of intended saving. Investment and saving thus conduct a tug of war around this pivot point, driving the economy upward when intended investment exceeds the flow of intended saving; downward when it fails to offset saving.
In Fig. 29·12 we show this crosscurrent in schematic form. Note that as incomes fall very low, householders will *dissave*.

INJECTIONS VS. LEAKAGES

We can easily make our graph more realistic by adding taxes (T) and imports (M) to savings, and exports (X) and government spending to investment. The vertical axis in Fig. 29·13 now shows all leakages and injections. And just to introduce another feature of the real world, we will tilt the injection line upward, on the assumption that induced investment will be an important constituent of total investment. The leakages curve will

not be exactly the same shape as the savings curve, but it will reflect the general tendency of savings and imports and taxes to rise with income.

TWO DIAGRAMS, ONE SOLUTION

Our new diagram also enables us to see the connection between our first approach to equilibrium as the supply and demand for GNP, and our present analysis of equilibrium in terms of leakages and injections.

As Fig. 29·14 shows, the injections curve and the leakages curve in our previous diagram are exactly those we have already encountered in Fig. 29·11.

Notice that the shaded triangle showing leakages is transposed to the second diagram. There it looks "flatter" only because we now ignore consumption and therefore put the leakage triangle on a horizontal base instead of a sloping one. Our "fixed" injection schedule is also exactly the same in both diagrams, although once again, in the lower figure it is "flatter" because it no longer sits on a sloping C line. Now it should be obvious why the equilibrium intersection point must be at the same figure on the GNP axis of both diagrams.

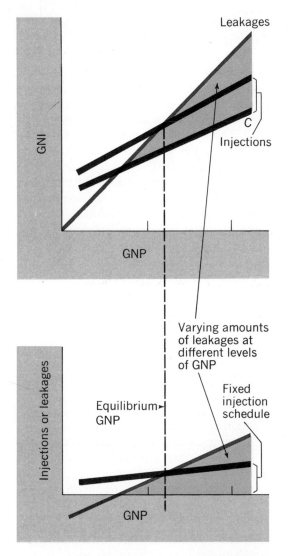

FIGURE 29 · 14
Two methods of showing equilibrium

The Multiplier

There remains only one part of the jigsaw puzzle to put into place. This is the integration of the *multiplier* into our analysis of the determination of GNP.

We remember that the essential point about the multiplier was that changes in in-

vestment, government spending, or exports resulted in larger changes in GNP because the additions to income were respent, creating still more new incomes. Further, we remember that the size of the multiplier effect depended on the marginal propensity to consume, the marginal propensity to tax, and the marginal propensity to buy imports as GNP rises. Now it remains only to show how this basic analytic concept enters into the determination of equilibrium GNP.

Let us begin with the diagram that shows injections and leakages, and let us now draw a new line showing an increase in injections (Fig. 29·15). Notice that the increase in GNP is larger than the increase in injections. *This is the multiplier itself in graphic form.*

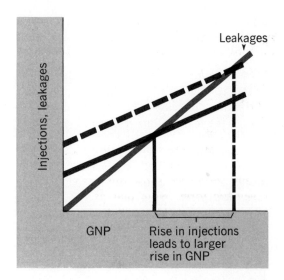

FIGURE 29 · 15
Multiplier in graphic form

We can see exactly the same result in our diagram of the supply and demand for GNP. Notice how a rise in the demand for GNP (a rise in injections) leads to a larger rise in the output of GNP (see Fig. 29·16).

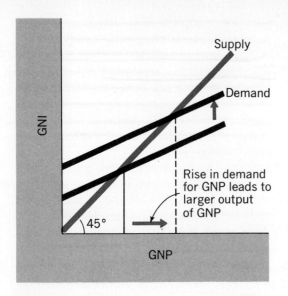

FIGURE 29 · 16
Multiplier and GNP

Rise in demand for GNP leads to larger output of GNP

45°

THE SLOPE OF THE LEAKAGE CURVE

Both diagrams also show that the relation between the original increase in injections and the resulting increase in GNP depends on the *slope* of the leakage line. Figure 29·17 shows

us two different injection-GNP relationships that arise from differing slopes.

Notice how the *same* increase in spending (from *OA* to *OB* on the injections axis) leads to a much smaller increase in panel I GNP (from *OX* to *OY*), where the leakage slope is high, than in panel II (from *OX'* to *OY'*), where the slope is more gradual.

Why is the increase greater when the slope is more gradual? The answer should be obvious. The slope represents the marginal propensity to save, to tax, to import—in short, all the marginal propensities that give rise to leakages. If these propensities are high—if there are high leakages—then the slope of the leakage curve will be high. If it is low, the leakage curve will be flat.

A LAST LOOK AT EQUILIBRIUM

Thus we finally understand how GNP reaches an equilibrium position after a change in demand. Here it is well to reiterate, however, that the word "equilibrium" does not imply a static, motionless state. Nor does it mean a desired state. We use the word only to denote the fact that *given* certain behavior

FIGURE 29 · 17
Two multipliers

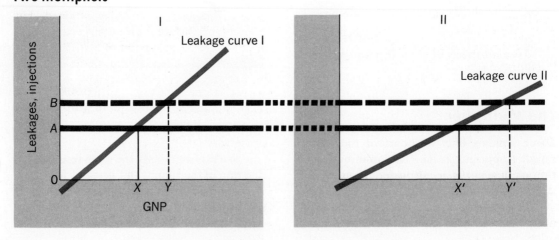

THE MULTIPLIER ONCE AGAIN

You can see the relation between our multiplier analysis and our graphical analysis by thinking about the following two examples.

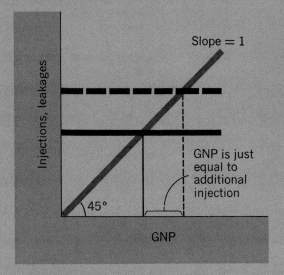

1. Suppose that the leakage fraction is 1; in other words, that we absorb *all* increases in income in additional savings, taxes, imports. What will the multiplier be? We know that the multiplier is $1/mps$. If $mps = 1$, then the multiplier fraction will be 1, and the increase in income will be 1 times the injection. In graphical terms, this looks like the accompanying figure.

The leakage curve shows that each dollar of additional GNP leads to another dollar of leakage. Hence the increase in GNP arising from an increase in injections is exactly equal to the original increase in injections. The multiplier is unity.

2. Now suppose that the leakage fraction is .5. The multiplier, once again, is $1/.5$ or 2. In the second figure, we show the same relationship in graphical terms.

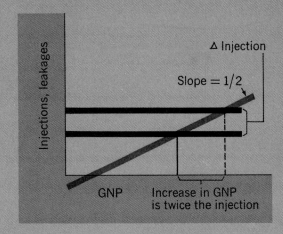

patterns, there will be a determinate point to which their interaction will push the level of income; and *so long as the underlying patterns of injections and leakages remain unchanged, the forces they exert will keep income at this level.*

In fact, of course, the flows of spending and saving are continually changing, so that the equilibrium level of the economy is constantly shifting, like a Ping-Pong ball suspended in a rising jet of water. Equilibrium can thus be regarded as a target toward which the economy is constantly propelled by the push-pull between leakages and injec-

tions. The target may be attained but momentarily before the economy is again impelled to seek a new point of rest. What our diagrams and the underlying analysis explain for us, then, is not a single determinate point at which our economy will in fact settle down, but the *direction* it will go in quest of a resting place, as the dynamic forces of the system exert their pressures.

THE PARADOX OF THRIFT

The fact that income must always move toward the level where the flows of intended saving and investment are equal leads to one of the most startling—and important—paradoxes of economics. This is the so-called paradox of thrift, a paradox that tells us that the *attempt to increase intended saving* may, under certain circumstances, lead to a *fall in actual saving.*

The paradox is not difficult for us to understand at this stage. An attempt to save, *when it is not matched with an equal willingness to invest or to increase government expenditure,* will cause a gap in demand. This means that businessmen will not be getting back enough money to cover their costs. Hence, production will be curtailed or costs will be slashed, with the result that incomes will fall. As incomes fall, savings will also fall, because the ability to save will be reduced. Thus, by a chain of activities working their influence on income and output, the effort to *increase* savings may end up with an actual *reduction* of savings.

This frustration of individual desires is perhaps the most striking instance of a common situation in economic life, the incompatibility between some kinds of individual

behavior and some collective results. An individual farmer, for instance, may produce a larger crop in order to enjoy a bigger income; but if all farmers produce bigger crops, farm prices are apt to fall so heavily that farmers end up with less income. So too, a single family may wish to save a very large fraction of its income for reasons of financial prudence; but if all families seek to save a great deal of their incomes, the result—unless investment also rises—will be a fall in expenditure and a common failure to realize savings objectives. The paradox of thrift, in other words, teaches us that the freedom of behavior available to a few individuals cannot always be generalized to all individuals.*

---

*The paradox of thrift is actually only a subtle instance of that type of faulty reasoning called the fallacy of composition. The fallacy consists of assuming that what is true of the individual case must also be true of all cases combined. The flaw in reasoning lies in our tendency to overlook "side effects" of individual actions (such as the decrease in spending associated with an individual's attempt to save more, or the increase in supply when a farmer markets his larger crop) which may be negligible in isolation but which are very important in the aggregate. See p. 65.

KEY WORDS

Supply and demand for GNP

Equilibrium GNP

GNP ≡ GNI
45° line

Upward-sloping demand curve

CENTRAL CONCEPTS

1. The level of GNP is determined by the interplay of two forces—the (short-run) supply of GNP and the demand for GNP. The first is the aggregate production function; the second is the sum of the consumption, investment, and government demands for output.

2. *The equilibrium level of GNP need not correspond to full employment. Thus the fact that GNP is in equilibrium does not mean that it is at a socially satisfactory level.*

3. GNP is determined not only by the supply and demand for output, but *also by the interplay of the supply of income and the demand for GNP stemming from an expenditure of income. Since GNP ≡ GNI, we can convert the aggregate supply function into a 45° line showing that all levels of GNP are matched by an identical GNI.*

4. *We can also depict the demand for goods arising from the expenditure of income. This demand curve is made up of the demands of the various sectors. Generally it slopes upward, since it relates quantities demanded and incomes, not prices.*

Moreover, it does not begin at the origin, but at some level representing "the bottom"—the expenditures that would have to be made even if no output were made and no incomes earned.

The meaning of equilibrium

5. At equilibrium levels of GNP $F+T+D$ (cash flow including profits) $= C+I+G+X$. At nonequilibrium levels there would be a demand gap, or a demand surplus.

Expenditures vs. costs

6. At levels of GNP other than equilibrium there will be forces that push GNP toward equilibrium. These forces arise because expenditures are greater than costs (or incomes), leading to a rise in GNP; or because expenditures are less than costs, leading to curtailment of production and a fall in GNP. *Equilibrium is reached because demand rises or falls less rapidly than income.*

Leakages Injections

7. *The simplest model of an economy in equilibrium is one in which intended S = intended I.* We can enlarge this model to include the equality of all intended *leakages* and *injections*.

Changing equilibrium

8. Equilibrium is not a static concept. Like the equilibrium price in a market, *equilibrium levels of GNP constantly change* as the forces of supply and demand change.

Slope of the curve of leakages

9. The multiplier shows us the relation between an increase in injections and the resulting increase in GNP. We can show it graphically as well as algebraically. In a graph, the *critical fraction* mps *is depicted by the slope of the leakage curve.*

Paradox of thrift

10. The desire to save can be frustrated by a failure of GNP to be high enough to permit households or businesses to save. Indeed, the attempt to save, by lowering expenditures, may actually drive income down and reduce the flow of saving. This is called the *paradox of thrift*.

QUESTIONS

1. Explain equilibrium in terms of the demand and supply for output, using an aggregate production function and a demand function for output. Why does the supply function begin at the origin? Why doesn't the demand function?

2. Draw an equilibrium diagram and indicate the volume of GNP it implies. Can you tell from the diagram if this is a full-employment GNP? What information would you need to draw in the various ranges of employment on the GNP axis?

3. Why is GNP an identity with GNI? Be sure you understand this.

4. Why does the "curve" of an identity always have a 45° slope? Demonstrate this by plotting a curve that relates the number of bachelors (horizontal axis) with the number of unmarried men (vertical axis).

5. Describe the "scenario" by which GNP is "pushed"

from a point above equilibrium back to equilibrium. Do the same for a GNP below equilibrium.

6. What are the components of the aggregate demand function? Can you write the function for the consumption portion of this total demand function? Can we write a plausible investment function? A government expenditure function? Is it reasonable to assume that these functions are "fixed" in the short-run period in which we are interested? Can we use an aggregate demand function, even though we do not know its precise shape, to highlight the process of adjustment of GNP?

7. Can you show why there is no demand gap at equilibrium? Remember: instead of "depreciation," use cash flow.

8. Now show how a shift in the demand function will bring about a new equilibrium point.

9. Draw a diagram showing the interplay of leakages

and injections, with injections as a layer of expenditures above consumption, and leakages as a triangle. Can you explain the triangle? How do you define the top line of the triangle? The bottom line? What is the difference?

10. Now show the interplay in the simplest form between savings and investment. Enlarge the saving/investment diagram to a leakage/injection diagram. Relate it to the diagram above.

11. Show how the multiplier affects the size of changes in GNP, according to the slope of the leakage curve. What does this slope represent? Relate the slope to the *mps*.

12. What is the paradox of thrift? Can you turn it upside down? Suppose no one wanted to save, and everyone tried to spend all his income. What would happen to total income? What would probably happen to saving?

13. Suppose that an economy turns out to have the following consumption and saving schedule (in billions):

Income	Saving	Consumption
$400	$50	$350
450	55	395
500	60	440
550	70	480
600	85	515

Now suppose that firms intend to make investments of $60 billion during the year. What will be the level of income for the economy? If investment rises to $85 billion, then what will be its income? What would be the multiplier in this case?

14. What is the difference between the 45° supply curve of income and the aggregate production function?

Money

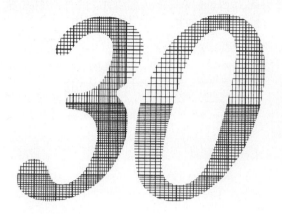

WE HAVE ALMOST COMPLETED our analysis of the major elements of macroeconomics, and soon we can bring our analysis to bear on some major problems of the economy. But first there is a matter that we must integrate into our discussion. This is the role that money plays in fixing or changing the level of GNP, along with the other forces that we have come to know.

Actually, we have been talking about money throughout our exposition. After all, one cannot discuss expenditure without assuming the existence of money. But now we must look behind this unexamined assumption and find out exactly what we mean when we speak of money. This will entail two tasks. In this chapter we shall investigate the perplexing question of what money *is*—for as we shall see, money is surely one of the most sophisticated and curious inventions of human society. Then in our next chapter, once we have come to understand what currency and gold and bank deposits are and how they come into being, we will look into the effect that money has on our economic operations.

The Supply of Money

Let us begin, then, by asking—"What is money?" Coin and currency are certainly money. But are checks money? Are the deposits from which we draw checks money? Are savings accounts money? Stamps? Government bonds?

The answer is a somewhat arbitrary one. From the spectrum of possible candidates, we reserve the term *money* for those items used to make *payments.* This means that we include cash in the public's possession and checking accounts, because we pay for most things by cash or check. Surprisingly, it means that we do not usually count savings accounts, since we have to draw "money" *out* of our savings accounts, in the form of cash, or have it transferred to our checking account, if we want to use our savings accounts to make expenditures. So, too, we have to sell government bonds to get money.*

---

*Some economists do count savings accounts as money. By convention, cash plus checking accounts are called the M_1 measure of money; and M_1 plus savings accounts are called the M_2 measure of money.

CREDIT CARDS

Money serves as a mechanism for storing potential purchasing power and for actually purchasing goods and services. Since cash and personal checks are the principal means for making these purchases, money has come to be defined as cash outside banks plus checking accounts. But what about credit cards. Shouldn't they be considered money?

Credit cards clearly can be used to make purchases, so that they appear on the surface to have a vital attribute of money. But a moment's reflection shows that in fact they *substitute* for cash or checks in which payment is finally made. The moment you pay your credit card bill, or the moment the credit card company pays the local merchant, the credit card is replaced by standard money. *Thus credit cards play the role of money only to the extent that credit bills are unpaid!*

CURRENCY

Money, then, is mainly currency and checking accounts. In 1973, for example, our total money supply was $260 billion, of which $60 billion was currency in the hands of the public, and $200 billion was the total of checking accounts (or demand deposits, as they are also called).

Of the two kinds of "money," currency is the form most familiar to us. Yet there is a considerable mystery even about currency. Who determines how much currency there is? How is the supply of coins or bills regulated?

We often assume that the supply of currency is "set" by the government that "issues" it. Yet when we think about it, we realize that the government does not just hand out money, and certainly not coins or bills. When the government pays people, it is nearly always by check.

Then who does fix the amount of currency in circulation? You can answer the question by asking how you yourself determine how much currency you will carry. If you think about it, the answer is that you "cash" a check when you need more currency than you have, and you put the currency back into your checking account when you have more than you need.

What you do, everyone does. The amount of cash that the public holds at any

time is no more and no less than the amount that it *wants* to hold. When it needs more—at Christmas, for instance—the public draws currency by cashing checks on its own checking accounts; and when Christmas is past, shopkeepers (who have received the public's currency) return it to their checking accounts.

Thus the amount of currency we have bears an obvious, important relation to the size of our bank accounts, for we can't write checks for cash if our accounts will not cover them.

Does this mean, then, that the banks have as much currency in their vaults as the total of our checking accounts? No, it does not. But to understand that, let us follow the course of some currency that we deposit in our banks for credit to our accounts.

BOOKKEEPING MONEY

When you put money into a commercial bank,* the bank does not hold that money for you as a pile of specially earmarked bills or as a bundle of checks made out to you from some payer. The bank takes notice of your deposit simply by crediting your "ac-

*A commercial bank is a bank that is empowered by law to offer checking services. It may also have savings accounts. A savings bank has only savings accounts and may not offer checking services.

In this role credit cards are not unique. Any unpaid bill or charge account is like money, in that you are able to purchase goods and services in exchange for your personal IOU. In a sense, each person is able to "print" money to the extent that he can persuade people to accept his IOUs. For most of us, that extent is very limited.

From an economist's point of view, the value of all outstanding trade credit (unpaid bills, unpaid charge accounts, or credit cards) *should* be considered money. It is not included in the official statistics for two reasons. First, it is difficult or impossible to figure how much trade credit is outstanding at any moment. Second, fluctuations in trade credit do not have a big impact on the economy. Ordinarily, the value of trade credit does not vary much, and therefore trade credit does not give rise to substantial changes in the effective money supply.

count," a bookkeeping page recording your present "balance." After the amount of the currency or check has been credited to you, the currency is put away with the bank's general store of vault cash and the checks are sent to the banks from which they came, where they will be charged against the accounts of the people who wrote them.

There is probably no misconception in economics harder to dispel than the idea that banks are warehouses stuffed with money. In point of fact, however, you might search as hard as you pleased in your bank, but you would find no money that was yours other than a bookkeeping account in your name. This seems like a very unreal form of money; and yet, the fact that you can present a check at the teller's window and convert your bookkeeping account into cash proves that your account must nonetheless be "real."

But suppose that you and all the other depositors tried to convert your accounts into cash on the same day. You would then find something shocking. There would not be nearly enough cash in the bank's till to cover the total withdrawals. In 1973, for instance, total demand deposits in the United States amounted to about $200 billion. But the total amount of coin and currency held by the banks was only $6.5 billion!

At first blush, this seems like a highly dangerous state of affairs. But second thoughts are more reassuring. After all, most of us put money into a bank because we do *not* need it immediately, or because making payments in cash is a nuisance compared with making them by check. Yet, there is always the chance—more than that, the certainty—that some depositors *will* want their money in currency. How much currency will the banks need then? What will be a proper reserve for them to hold?

FEDERAL RESERVE SYSTEM

For many years, the banks themselves decided what reserve ratio constituted a safe proportion of currency to hold against their demand deposits (the technical name for checking accounts). Today, however, most large banks are members of the Federal Reserve, a central banking system established in 1913 to strengthen the banking activities of the nation. Under the Federal Reserve System, the nation is divided into twelve districts, each with a Federal Reserve Bank owned (but not really controlled) by the member banks of its district. In turn, the twelve Reserve Banks are themselves coordinated by a seven-man Federal Reserve Board in Washington. Since the members of the board are appointed for fourteen-year terms, they constitute a body that has been purposely established as an independent, nonpolitical monetary authority.*

One of the most important functions of the Federal Reserve Board is to establish reserve ratios for different categories of banks, within limits set by Congress. Historically these reserve ratios have ranged between 13 and 26 percent of demand deposits for city banks, with a somewhat smaller reserve ratio for country banks. Today, reserve ratios are determined by size, and they vary between 18 percent for the largest banks and 8 percent for the smallest. The Federal Reserve Board also sets reserve requirements for "time" deposits (the technical term for savings deposits). These range from 3 to 5 percent. Do not forget, however, that time deposits do not count—or directly serve—as "money."

THE BANKS' BANK

Yet here is something odd! We noticed that in 1973 the total amount of deposits was $200 billion and that banks' holdings of coin and currency were only $6.5 billion. This is much less than the 18 percent—or even 8 percent —reserve against deposits established by the Federal Reserve Board. How can this be?

The answer is that cash is not the only reserve a bank holds against deposits. Claims on other banks are also held as its reserve.

What are these claims? Suppose, in your account in Bank A, you deposit a check from someone who has an account in Bank B. Bank A credits your account and then presents the check to Bank B for "payment." By "payment" Bank A does not mean coin and currency, however. Instead, Bank A and Bank B settle their transaction at still *another*

bank where both Bank A and Bank B have their own accounts. These accounts are with the twelve Federal Reserve Banks of the country, where all banks who are members of the Federal Reserve System (and this accounts for banks holding most of the deposits in our banking system) *must* open accounts. Thus at the Federal Reserve Bank, Bank A's account will be credited, and Bank B's account will be debited, in this way moving reserves from one bank to the other.†

In other words, *the Federal Reserve Banks serve their member banks in exactly the same way as the member banks serve the public.* Member banks automatically deposit in their Federal Reserve accounts all checks they get from other banks. As a result, banks are constantly "clearing" their checks with one another through the Federal Reserve System, because their depositors are constantly writing checks on their own banks payable to someone who banks elsewhere. Meanwhile, *the balance that each bank maintains at the Federal Reserve—that is, the claim it has on other banks—counts, as much as any currency, as part of its reserve against deposits.*

In 1973, therefore, when demand deposits were $200 billion and cash in the banks only $6.5 billion, we would expect the member banks to have had heavy accounts with the Federal Reserve banks. And so they did —$33 billion in all. Thus, total reserves of the banks were $39.5 billion ($6.5 billion in cash plus $33 billion in Federal Reserve accounts), enough to satisfy the legal requirements of the Fed.

FRACTIONAL RESERVES

Thus we see that our banks operate on what is called a *fractional reserve system.* That is, a certain specified fraction of all demand deposits must be kept "on hand" at all times

*This has resulted, on occasion, in sharp clashes of viewpoint with the Treasury Department or the Bureau of the Budget where fiscal and economic policy is formulated by each administration. There is some disagreement over whether the nation is better served by a Federal Reserve that can impede an economic policy it disagrees with or one that is bound to assist the economic aims of each incumbent administration. Generally speaking, however, the Federal Reserve gives in to any strongly held view of the administration, although it is legally independent of the executive.

†When money is put into a bank account, the account is credited; when money is taken out, the account is debited.

in cash or at the Fed. The size of the minimum fraction is determined by the Federal Reserve, for reasons of control that we shall shortly learn about. It is *not* determined, as we might be tempted to think, to provide a "safe" backing for our bank deposits. For under *any* fractional system, if *all* depositors decided to draw out their accounts in currency and coin from all banks at the same time, the banks would be unable to meet the demand for cash and would have to close. We call this a "run" on the banking system. Needless to say, runs can be terrifying and destructive economic phenomena.*

Why, then, do we court the risk of runs, however small this risk may be? What is the benefit of a fractional banking system? To answer that, let us look into our bank again.

LOANS AND INVESTMENTS

Suppose its customers have given our bank $1 million in deposits and that the Federal Reserve Board requirements are 20 percent, a simpler figure to work with than the actual one. Then we know that our bank must at all times keep $200,000, either in currency in its own till or in its demand deposit at the Federal Reserve Bank.

But having taken care of that requirement, what does the bank do with the remaining deposits? If it simply lets them sit, either as vault cash or as a deposit at the Federal Reserve, our bank will be very "liquid," but it will have no way of making an income. Unless it charges a very high fee for its checking services, it will have to go out of business.

And yet there is an obvious way for the bank to make an income, while performing a valuable service. The bank can use all the cash and check claims it does not need for its

reserve to make *loans* to businessmen or families or to make financial *investments* in corporate or government bonds. It will thereby not only earn an income, but it will assist the process of business investment and government borrowing. Thus the mechanics of the banking system lead us back to the concerns at the very center of our previous analysis.

Inside the Banking System

Fractional reserves allow banks to lend, or to invest in securities, part of the funds that have been deposited with them. But that is not the only usefulness of the fractional reserve system. It works as well to help enlarge or diminish the supply of investible or loanable funds, as the occasion demands. Let us follow how this process works. To make the mechanics of banking clear, we are going to look at the actual books of the bank—in simplified form, of course—so that we can see how the process of lending and investing appears to the banker himself.

ASSETS AND LIABILITIES

We begin by introducing two basic elements of business accounting: *assets* and *liabilities*. Every student at some time or another has seen the balance sheet of a firm, and many have wondered how total assets always equal total liabilities. The reason is very simple. Assets are all the things or claims a business owns. Liabilities are claims against those assets—some of them the claims of creditors, some the claims of owners (called the Net Worth of the business). Since assets show everything that a business owns, and since liabilities show how claims against these self-same things are divided between creditors and owners, it is obvious that the two sides of the balance sheet must always come to exactly the same total. The total of assets and the total of liabilities are an identity.

---

*A "run" on the banking system is no longer much of a threat, because the Federal Reserve could supply its members with vast amounts of cash. We shall learn how, later in this chapter.

T ACCOUNTS

Businesses show their financial condition on a *balance sheet* on which all items on the left side represent assets and all those on the right side represent liabilities. By using a simple two-column balance sheet (called a "T account" because it looks like a T), we can follow very clearly what happens to our bank as we deposit money in it or as it makes loans or investments.

EXCESS RESERVES

Now we recall from our previous discussion that our bank does not want to remain in this very liquid, but very unprofitable, position. According to the law, it must retain only a certain percentage of its deposits in cash or at the Federal Reserve—20 percent in our hypothetical example. All the rest it is free to lend or invest. As things now stand, however, it has $1 million in reserves—$800,000 more

ORIGINAL BANK

Assets	Liabilities
$1,000,000 (cash and checks)	$1,000,000 (money owed to depositors)
Total $1,000,000	**Total $1,000,000**

We start off with the example we have just used, in which we open a brand new bank with $1 million in cash and checks on other banks. Accordingly, our first entry in the T account shows the two sides of this transaction. Notice that our bank has gained an asset of $1 million, the cash and checks it now owns, and that it has simultaneously gained $1 million in liabilities, the deposits it *owes* to its depositors (who can, after all, take their money out any time).

As we know, however, our bank will not keep all its newly-gained cash and checks in the till. It may hang on to some of the cash, but it will send all the checks it has received, plus any currency that it feels it does not need, to the Fed for deposit in its account there. As a result, its T account will now look like this:

than it needs. Hence, let us suppose that it decides to put these *excess reserves* to work by lending that amount to a sound business risk. (Note that banks do not lend the excess reserves themselves. These reserves, cash and deposits at the Fed, remain right where they are. Their function is to tell the banks how much they may loan or invest.)

MAKING A LOAN

Assume now that the Smith Corporation, a well-known firm, comes in for a loan of $800,000. Our bank is happy to lend them that amount. But "making a loan" does not mean that the bank now pays the company in cash out of its vaults. Rather, *it makes a loan by opening a new checking account for the firm* and by crediting that account with $800,000.

ORIGINAL BANK

Assets		Liabilities	
Vault Cash	$ 100,000	Deposits	$1,000,000
Deposit at Fed	900,000		
Total	**$1,000,000***	**Total**	**$1,000,000**

*If you will examine some bank balance sheets, you will see these items listed as "Cash and due from banks." This means, of course, cash in their own vaults plus their balance at the Fed.

(Or if, as is likely, the Smith firm already has an account with the bank, it will simply credit the proceeds of the loan to that account.)

Now our T account shows some interesting changes.

serves are now sufficient to cover the Smith Corporation's account as well as the original deposit accounts. A glance reveals that all is well. We still have $1 million in reserves against $1.8 million in deposits. Our reserve

ORIGINAL BANK

Assets		Liabilities	
Cash and at Fed	$1,000,000	Original deposits	$1,000,000
Loan (Smith Corp.)	800,000	New deposit (Smith Corp.)	800,000
Total	**$1,800,000**	**Total**	**$1,800,000**

There are several things to note about this transaction. First, our bank's reserves (its cash and deposit at the Fed) have not yet changed. The $1 million in reserves are still there.

Second, notice that the Smith Corporation loan counts as a new asset for the bank because the bank now has a legal claim against the company for that amount. (The interest on the loan is not shown in the balance sheet; but when it is paid, it will show up as an addition to the bank's cash.)

Third, deposits have increased by $800,000. Note, however, that this $800,000 was not paid to the Smith firm out of anyone else's account in the bank. It is a new checking account, one that did not exist before. As a result, the supply of money is also up! More about this shortly.

THE LOAN IS SPENT

Was it safe to open this new account for the company? Well, we might see whether our re-

ratio is much higher than the 20 percent required by law.

It is so much higher, in fact, that we might be tempted to make another loan to the next customer who requests one, and in that way further increase our earning capacity. But an experienced banker shakes his head. "The Smith Corporation did not take out a loan and agree to pay interest on it just for the pleasure of letting that money sit with you," he explains. "Very shortly, the company will be writing checks on its balance to pay for goods or services; and when it does, you will need every penny of the reserve you now have."

That, indeed, is the case. Within a few days we find that our bank's account at the Federal Reserve Bank has been charged with a check for $800,000 written by the Smith Corporation in favor of the Jones Corporation, which carries its account at another bank. Now we find that our T account has changed dramatically to look like this:

ORIGINAL BANK

Assets		Liabilities	
Cash and at Fed	$ 200,000	Original deposits	$1,000,000
Loan (Smith Corp.)	800,000	Smith Corp. deposits	0
Total	**$1,000,000**	**Total**	**$1,000,000**

SECOND BANK

Assets		Liabilities	
Cash and at Fed	$800,000	Deposit (Jones Corp.)	$800,000
Total	**$800,000**	**Total**	**$800,000**

Let us see exactly what has happened. First, the Smith Corporation's check has been charged against our account at the Fed and has reduced it from $900,000 to $100,000. Together with the $100,000 cash in our vault, this gives us $200,000 in reserves.

Second, the Smith Corporation's deposit is entirely gone, although its loan agreement remains with us as an asset.

Now if we refigure our reserves we find that they are just right. We are required to have $200,000 in vault cash or in our Federal Reserve account against our $1 million in deposits. That is exactly the amount we have left. Our bank is now fully "loaned up."

EXPANDING THE MONEY SUPPLY

But the banking *system* is not yet fully loaned up. So far, we have traced what happened to only our bank when the Smith Corporation spent the money in its deposit account. Now we must trace the effect of this action on the deposits and reserves of other banks.

We begin with the bank in which the Jones Corporation deposits the check it has just received from the Smith Corporation. As the above T account shows, the Jones Corporation's bank now finds itself in exactly the same position as our bank was when we opened it with $1 million in new deposits, except that the addition to this "second generation" bank is smaller than the addition to the "first generation" bank.

As we can see, our second generation bank has gained $800,000 in cash and in deposits. Since it needs only 20 percent of this for required reserves, it finds itself with $640,000 excess reserves, which it is now free to use to make loans as investments. Suppose that it extends a loan to the Brown Company and that the Brown Company shortly thereafter spends the proceeds of that loan at the Black Company, which banks at yet a third bank. The following two T accounts show how the total deposits will now be affected.

As Fig. 30·1 makes clear, the process will not stop here but can continue from one bank to the next as long as any lending power

SECOND BANK
(after Brown Co. spends the proceeds of its loan)

Assets		Liabilities	
Cash and at Fed	$160,000	Deposits (Jones Corp.)	$800,000
Loan (to Brown Co.)	640,000	Deposits (Brown Co.)	0
Total	**$800,000**	**Total**	**$800,000**

THIRD BANK
(after Black Co. gets the check of Brown Co.)

Assets		Liabilities	
Cash and at Fed	$640,000	Deposit (Black Co.)	$640,000
Total	**$640,000**	**Total**	**$640,000**

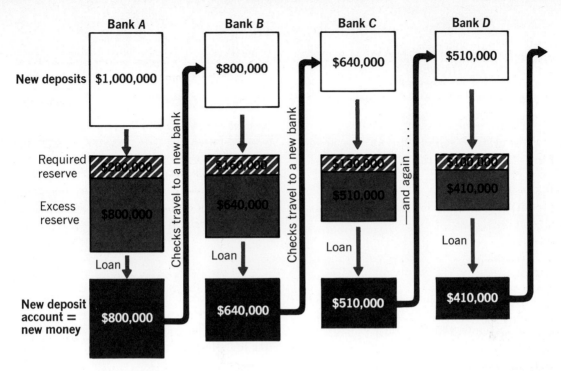

Bank A $1,000,000 | **Bank B** $800,000 | **Bank C** $640,000 | **Bank D** $510,000

New deposits

Required reserve: $200,000 / $160,000 / $130,000 / $100,000

Excess reserve: $800,000 / $640,000 / $510,000 / $410,000

Checks travel to a new bank — and again . . . . .

Loan

New deposit account = new money: $800,000 / $640,000 / $510,000 / $410,000

FIGURE 30 · 1
Expansion of the money supply

remains. Notice, however, that this lending power gets smaller and smaller and will eventually reach zero.

Expansion of the Money Supply

If we now look at the bottom of Fig. 30·1, we will see something very important. Every time any bank in this chain of transactions has opened an account for a new borrower, *the supply of money has increased.* Remember that the supply of money is the sum of currency outside the banking system (i.e., in our own pockets) plus the total of demand deposits. As our chain of banks kept opening new accounts, it was simultaneously expanding the total check-writing capacity of the economy. Thus, money has materialized, seemingly out of thin air.

Now how can this be? If we tell any banker in the chain that he has "created"

money, he will protest vehemently. The loans he made, he will insist, were backed at the time he made them by excess reserves as large as the loan itself. Just as we had $800,000 in excess reserves when we made our initial loan to the Smith Corporation, so every subsequent loan was always backed 100 percent by unused reserves when it was made.

Our bankers are perfectly correct when they tell us that they never, never lend a penny more than they have. Money is not created in the lending process because a banker lends money he doesn't have. *It is created because you and I generally pay each other by checks that give us claims against each other's bank.* If we constantly cashed the checks we exchanged, no new money would be created. But we do not. We deposit each other's checks in our own bank accounts; and in doing so, we give our banks more reserves than they need against the deposits we have just made. These new excess

467

MONEY AND DEBT

All this gives us a fresh insight into the question of what money is. We said before that it is whatever we use to make payments. But what do we use? The answer is a surprising one. We use *debts*—specifically, the debts of commercial banks. Deposits are, after all, nothing but the liabilities that banks owe their customers. Furthermore, we can see that one purpose of the banking system is to buy debts from other units in the economy, such as businesses or governments, in exchange for its own debts (which are money). For when a bank opens an account for a business to which it has granted a loan or when it buys a government bond, what else is it doing but accepting a debt that is *not* usable as money, in exchange for its deposit liabilities that *are* usable as money. And why is it that banks create money when they make loans, but you or I do not, when we lend money? Because we all accept bank liabilities (deposits) as money, but we do not accept personal or business IOUs to make payments with.

reserves make it possible for our banks to lend or invest, and thereby to open still more deposit accounts, which in turn lead to new reserves.

LIMITS ON THE EXPANSION

This all sounds a little frightening. Does it mean that the money supply can go on expanding indefinitely from a single new deposit? Wouldn't that be extremely dangerous?

It would of course be very dangerous, but there is no possibility that it can happen. For having understood how the supply of money can expand from an original increase in deposits, we may now understand equally well what keeps an expansion within bounds.

1. Not every loan generates an increase in bank deposits.

If our bank had opened a loan account for the Smith Corporation at the same time that another firm had paid off a similar loan, there would have been no original expansion in bank deposits. In that case, the addition of $800,000 to the Smith account would have been exactly balanced by a decline of $800,000 in someone else's account. Even if that decline would have taken place in a different bank, it would still mean that the nation's total of bank deposits would not have risen, and therefore no new money would

have been created. Thus, *only net additions to loans have an expansionary effect.* We will shortly see how such net additions arise in the first place.

2. There is a limit to the rise in money supply from a single increase in deposits.

As Fig. 30·1 shows, in the chain of deposit expansion each successive bank has a smaller increase in deposits, because each bank has to keep some of its newly gained cash or checks as reserve. Hence the amount of *excess* reserves, against which loans can be made, steadily falls.

Further, we can see that the amount of the total monetary expansion from an original net increase in deposits is governed by the size of the fraction that has to be kept aside each time as reserve. *In fact, we can see that just as with the multiplier, the cumulative effect of an increase in deposits will be determined by the reciprocal of the reserve fraction.* If each bank must keep one-fifth of its increased deposits as reserves, then the cumulative effect of an original increase in deposits, when it has expanded through the system, is five times the original increase. If reserves are one-fourth, the expansion is limited to four times the original increase, and so on.

If M is the money supply, D is net new deposits, and r is the reserve ratio, it follows that:

$$\Delta M = \frac{1}{r} \times \Delta D$$

Notice that this formula is exactly the same as that for the multiplier.*

3. The monetary expansion process can work in reverse.

Suppose that the banking system as a whole suffers a net loss of deposits. Instead of putting $1 million into a bank, the public takes it out in cash. The bank will now have too few reserves, and it will have to cut down its loans or sell its investments to gain the reserves it needs. In turn, as borrowers pay off their loans, or as bond buyers pay for their securities, cash will drain from other banks who will now find *their* reserves too small in relation to their deposits. In turn, they will therefore have to sell more investments or curtail still other loans, and this again will squeeze still other banks and reduce their reserves, with the same consequences.

Thus, just as an original expansion in deposits can lead to a multiple expansion, so an original contraction in deposits can lead to a multiple contraction. The size of this contraction is also limited by the reciprocal of the reserve fraction. If banks have to hold a 25 percent reserve, then an original fall of $100,000 in deposits will lead to a total fall of $400,000, assuming that the system was fully "loaned up" to begin with. If they had to hold a 20 percent reserve, a fall of $100,000 could pyramid to $500,000.

4. The expansion process may not be fully carried through.

We have assumed that each bank in the chain always lends out an amount equal to its excess reserves, but this may not be the case. The third or fifth bank along the way may have trouble finding a credit-worthy customer and may decide—for the moment,

*Why is ΔM determined by multiplying ΔD by $1/r$? Same reason as the multiplier. See box on pp. 388–89.

anyway—to sit on its excess reserves. Or borrowers along the chain may take out cash from some of their new deposits and thereby reduce the banks' reserves and their lending powers. Thus the potential expansion may be only partially realized.

5. The expansion process takes time.

Like the multiplier process, the expansion of the money supply encounters many "frictions" in real life. Banks do not instantly expand loans when their reserves rise; bank customers do not instantly spend the proceeds of bank loans. The time lags in banking are too variable to enable us to make an estimate of how long it takes for an initial increase in new deposits to work its way through the system, but the time period is surely a matter of months for two or three "rounds."

WHY BANKS MUST WORK TOGETHER

There is an interesting problem concealed behind this crisscrossing of deposits that leads to a slowly rising level of the money supply. Suppose that an imaginary island economy was served by a single bank (and let us forget about all complications of international trade, etc.), and suppose that this bank, which worked on a 20 percent reserve ratio, was suddenly presented with an extra one million dollars worth of reserves—let us say newly mined pure gold. Our bank could, of course, increase its loans to customers. By how much? By *five million dollars!*

In other words, our island bank, all by itself, could use an increase in its reserves to create a much larger increase in the money supply. It is not difficult to understand why. Any borrower of the new five million, no matter where he spent his money on the island, would only be giving his checks to someone who also banked at the single, solitary bank. The whole five million, in other words, would stay *within* the bank as its deposits, although the identity of those depositors would, of course, shift. Indeed, there is no reason why such a bank should limit its expansion of the money supply to five million. As long as the "soundness" of the currency was unquestioned, such a bank could create as much money as it wanted through new deposits, since all of those deposits would remain in its own keeping.

The imaginary bank makes it plain why ordinary commercial banks *cannot* expand deposits beyond their excess reserves. Unlike the monopoly bank, they must expect to *lose* their deposits to other banks when their borrowers write checks on their new accounts. As a result they will also lose their reserves, and this can lead to trouble.

OVERLENDING

This situation is important enough to warrant taking a moment to examine. Suppose that in our previous example we had decided to lend the Smith Corporation not $800,000 but $900,000, and suppose as before that the Smith Corporation used the proceeds of that loan to pay the Jones Corporation. Now look at the condition of our bank after the Smith payment has cleared.

Our reserves would now have dropped to 10 percent! Indeed, if we had loaned the

One way that a bank may repair the situation is by borrowing reserves for a short period (paying interest on them, of course) from another bank that may have a temporary surplus at the Fed; this is called borrowing federal funds. Or a bank may quickly sell some of its government bonds and add the proceeds to its reserve account at the Fed. Or again, it may add to its reserves the proceeds of any loans that have come due and deliberately fail to replace these expired loans with new loans. Finally, a bank may borrow reserves directly from its Federal Reserve Bank and pay interest for the loan. We shall shortly look into this method when we talk about the role of the Federal Reserve in regulating the quantity of money.

The main point is clear. A bank is safe in lending only an amount that it can afford to lose to another bank. But of course one bank's loss is another's gain. That is why, by the exchange of checks, the banking system can accomplish the same result as the island monopoly bank, whereas no individual bank can hope to do so.

INVESTMENTS AND INTEREST

If a bank uses its excess reserves to buy securities, does that lead to the same multiplication effect as a bank loan?

ORIGINAL BANK

Assets		Liabilities	
Cash and at Fed	$ 100,000	Original deposits	$1,000,000
Loan (Smith Corp.)	900,000	Smith Corp. deposit	0
Total	**$1,000,000**	**Total**	**$1,000,000**

company $1,000,000 we would be in danger of insolvency.

Banks are, in fact, very careful not to overlend. If they find that they have inadvertently exceeded their legal reserve requirements, they quickly take remedial action.

It can. When a bank buys government securities, it usually does so from a securities dealer, a professional trader in bonds.* Its

*The dealer may be only a middleman, who will in turn buy from, or sell to, corporations or individuals. This doesn't change our analysis, however.

check (for $800,000 in our example) drawn on its account at the Federal Reserve will be made out to a dealer, who will deposit it in his bank. As a result, the dealer's bank suddenly finds itself with an $800,000 new deposit. It must keep 20 percent of this as required reserve, but the remainder is excess reserve against which it can make loans or investments as it wishes.

Is there a new deposit, corresponding to that of the businessman borrower? There is: the new deposit of the securities dealer. Note that in his case, as in the case of the borrower, the new deposit on the books of the bank has not been put there by the transfer of money from some other commercial bank. The $800,000 deposit has come into being through the deposit of a check of the Federal Reserve Bank, which is not a commercial bank. Thus it represents a new addition to the deposits of the private banking system.

Let us see this in the T accounts. After our first bank has bought its $800,000 in bonds (paying for them with its Federal Reserve checking account), its T account looks like this.

Here there are excess reserves of $640,000 with which additional investments can be made. It is possible for such new deposits, albeit diminishing each time, to remain in the financial circuit for some time, moving from bank to bank as an active business is done in buying government bonds.

YIELDS

Meanwhile, however, the very activity in bidding for government bonds is likely to raise their price, and thereby lower their rate of interest.

This is important to understand. A bond has a *fixed* rate of return and a stated face value. If it is a 4 percent, $1,000 bond, this means it will pay $40 interest yearly. If the bond now sells on the marketplace for $1,100, the $40 yearly interest will be less than a 4 percent return ($40 is only 3.6 percent of $1,100). If the price should fall to $900, the $40 return will be more than 4 percent ($40 is 4.4 percent of $900). Thus the *yield* of a bond varies inversely—in the other direction—from its market price.

ORIGINAL BANK

Assets		Liabilities	
Cash at Fed	$ 200,000	Deposits	$1,000,000
Government bonds	800,000		
Total	**$1,000,000**	**Total**	**$1,000,000**

As we can see, there are no excess reserves here. But look at the bank in which the seller of the government bond has deposited the check he has just received from our bank:

When the price of government bonds changes, all bond prices tend to change in the same direction. This is because all bonds are competing for investors' funds. If the yield on

SECOND BANK

Assets		Liabilities	
Cash	$800,000	New deposit of bond seller	$800,000
Total	**$800,000**	**Total**	**$800,000**

"governments" falls, investors will switch from governments to other, higher yielding bonds. But as they bid for these other bonds, the prices of these bonds will rise—and their yields will fall, too!

In this way, a change in yields spreads from one group of bonds to another. A lower rate of interest or a lower yield on government securities is quickly reflected in lower rates or yields for other kinds of bonds. In turn, a lower rate of interest on bonds makes loans to business look more attractive. Thus, sooner or later, excess reserves are apt to be channeled to new loans as well as new investments. Thereafter the deposit-building process follows its familiar course.

Controlling the Money Supply

We have now seen how a banking system can create money through the successive creation of excess reserves. But the key to the process is the creation of the *original* excess reserves, for without them the cumulative process will not be set in motion. We remember, for example, that a loan will not result in an increase in the money supply if it is offset by a decline in lending somewhere else in the banking system; neither will the purchase of a bond by one commercial bank if it is only buying a security sold by another. *To get a net addition to loans or investments, however, a banking system—assuming that it is fully loaned up—needs an increase in its reserves.* Where do these extra reserves come from? That is the question we must turn to next.

ROLE OF THE FEDERAL RESERVE

In our example we have already met one source of changes in reserves. When the public needs less currency, and it deposits its extra holdings in the banks, reserves rise, as we have seen. Contrariwise, when the public wants more currency, it depletes the banks'

holdings of currency and thereby lowers their reserves. In the latter case, the banks may find that they have insufficient reserves behind their deposits. To get more currency or claims on other banks, they will have to sell securities or reduce their loans. This might put a very severe crimp in the economy. Hence, to allow bank reserves to be regulated by the public's fluctuating demand for cash would seem to be an impossible way to run our monetary system.

But we remember that bank reserves are not mainly currency; in fact, currency is a relatively minor item. Most reserves are the accounts that member banks hold at the Federal Reserve. Hence, if these accounts could somehow be increased or decreased, we could regulate the amount of reserves—and thus the permissible total of deposits—without regard to the public's changing need for cash.

This is precisely what the Federal Reserve System is designed to do. Essentially, the system is set up to regulate the supply of money by raising or lowering the reserves of its member banks. When these reserves are raised, member banks find themselves with excess reserves and are thus in a position to make loans and investments by which the supply of money will increase further. Conversely, when the Federal Reserve lowers the reserves of its member banks, they will no longer be able to make loans and investments, or they may even have to reduce loans or get rid of investments, thereby extinguishing deposit accounts and contracting the supply of money.

MONETARY CONTROL MECHANISMS

How does the Federal Reserve operate? There are three ways.

1. Changing reserve requirements

It was the Federal Reserve itself, we will remember, that originally determined how much in reserves its member banks should hold against their deposits. Hence by changing that reserve requirement for a given level of deposits, it can give its member banks excess reserves or can create a shortage of reserves.

In our imaginary bank we have assumed that reserves were set at 20 percent of deposits. Suppose now that the Federal Reserve determined to lower reserve requirements to 15 percent. It would thereby automatically create extra lending or investing power for our *existing* reserves. Our bank with $1 million in deposits and $200,000 in reserves could now lend or invest an additional $50,000 without any new funds coming in from depositors. On the other hand, if requirements were raised to, say, 30 percent, we would find that our original $200,000 of reserves was $100,000 short of requirements, and we would have to curtail lending or investing until we were again in line with requirements.

Do not forget that these new reserve requirements affect *all* banks. Therefore, changing reserve ratios is a very effective way of freeing or contracting bank credit on a large scale. But it is an instrument that sweeps across the entire banking system in an undiscriminating fashion. It is therefore used only rarely, when the Federal Reserve Board feels that the supply of money is seriously short or dangerously excessive and needs remedy on a countrywide basis. For instance, in early 1973, the board raised reserve requirements one-half percent for all banks, partly to mop up excess reserves and partly to sound a general warning against what it considered to be a potentially dangerous inflationary state of affairs.

2. Changing discount rates

A second means of control uses interest rates as the money-controlling device. Recall that member banks that are short on reserves have a special privilege, if they wish to exercise it. They can *borrow* reserve balances from the Federal Reserve Bank itself and add them to their regular reserve account at the bank.

The Federal Reserve Bank, of course, charges interest for lending reserves, and this interest is called the *discount rate.* By raising or lowering this rate, the Federal Reserve can make it attractive or unattractive for member banks to borrow to augment reserves. Thus in contrast with changing the reserve ratio itself, changing the discount rate is a mild device that allows each bank to decide for itself whether it wishes to increase its reserves. In addition, changes in the discount rate tend to influence the whole structure of interest rates, either tightening or loosening money.*

Although changes in the discount rate can be used as a major means of controlling the money supply and are used to control it in some countries, they are not used for this purpose in the U. S. The Federal Reserve Board does not allow banks to borrow whatever they would like at the current discount rate. The discount "window" is a place where a bank can borrow small amounts of money to smooth out short-run fluctuations in deposits and loans, but it is not a place where banks can borrow major amounts of money to expand their lending portfolios. As a result, the discount rate serves more as a signal of what the Federal Reserve would like to see happen than as an active force in determining the total borrowings of banks.

3. Open-market operations

Most frequently used, however, is a third technique called open-market operations. This technique permits the Federal Reserve Banks to change the supply of reserves by

---

*When interest rates are high, money is called tight. This means not only that borrowers have to pay higher rates, but that banks are stricter and more selective in judging the credit worthiness of business applications for loans. Conversely, when interest rates decline, money is called easy, meaning that it is not only cheaper but literally easier to borrow.

buying or selling U.S. government bonds on the open market.

How does this work? Let us suppose that the Federal Reserve authorities wish to increase the reserves of member banks. They will begin to buy government securities from dealers in the bond market; and they will pay these dealers with Federal Reserve checks.

Notice something about these checks: *they are not drawn on any commercial bank!* They are drawn on the Federal Reserve Bank itself. The security dealer who sells the bond will, of course, deposit the Fed's check, as if it were any other check, in his own commercial bank; and his bank will send the Fed's check through for credit to its own account, as if it were any other check. *As a result, the dealer's bank will have gained reserves, although no other commercial bank has lost reserves.* On balance, then, the system has more lending and investing capacity than it had before. In fact, it now has *excess* reserves, and these, as we have seen, will spread out through the system. *Thus by buying bonds, the Federal Reserve has, in fact, deposited money in the accounts of its members, thereby giving them the extra reserves that it set out to create.*

Conversely, if the authorities decide that member banks' reserves are too large, they will sell securities. Now the process works in reverse. Security dealers or other buyers of bonds will send their own checks on their own regular commercial banks to the Federal Reserve in payment for these bonds. This time the Fed will take the checks of its member banks and charge their accounts, thereby reducing their reserves. *Since these checks will not find their way into another commercial bank, the system as a whole will have suffered a diminution of its reserves.* By selling securities, in other words, the Federal Reserve authorities lower the Federal Reserve ac-

counts of member banks, thereby diminishing their reserves.*

ASYMMETRIC CONTROL

How effective are all these powers over the money supply? The Federal Reserve Board's capacity to control money is often compared to our ability to manipulate a string. If the Federal Reserve Board wishes to *reduce* the money supply, it can increase the discount rate or sell bonds. Sooner or later, this tends to be effective. If banks have free or excess reserves, they will not immediately have to reduce their lending portfolios; but eventually, by pulling on the string hard enough, the Fed can force a reduction in bank loans and the money supply.

The Federal Reserve Board's capacity to increase the money supply is not equally great. It can reduce reserve rates and buy bonds, but it cannot *force* banks to make loans if they do not wish to do so. Banks can, if they wish, simply increase their excess reserves. Normally, banks wish to make loans and earn profits; but if risks are high, they may not wish to do so. Such a situation occurred in the Great Depression. Banks

---

*Isn't this, some bright student will ask, really the same thing as raising or lowering the reserve ratio? If the Fed is really just putting money into member bank accounts when it buys bonds and taking money out when it sells them, why does it bother to go through the open market? Why not just tell the member banks that their reserves are larger or smaller?

Analytically, our student is entirely right. There is, however, cogent reason for working through the bond market. It is that the open-market technique allows banks to *compete* for their share of the excess reserves that are being made available or taken away. Banks that are good at attracting depositors will thereby get extra benefit from an increase in the money supply. Thus, rather than assigning excess reserves by executive fiat, the Fed uses the open market as an allocation device.

Finally, open-market operations allow the Fed to make very small changes in the money supply, whereas changes in reserve requirements would be difficult to adjust in very fine amounts.

piled up vast reserves rather than make loans, since the risks of defaults were too high to make most loans an attractive economic gamble. In terms of our analogy, the Federal Reserve Board can pull, but it cannot push on its string of controls.

STICKY PRICES

We are almost ready to look into the dynamics of money, in our next chapter, but we must examine a question that we have heretofore passed over in silence. We have taken for granted that we need a larger supply of money in order to expand output. But why should we? Why could we not grow just as well if the supply of money were fixed?

Theoretically we could. If we cut prices as we increased output, a given amount of money (or a given amount of expenditure) could cover an indefinitely large real output. Furthermore, as prices fell, workers would be content not to ask for higher wages (or would even accept lower wages), since in real terms they would be just as well or better off.

It is not difficult to spot the flaw in this argument. In the real world, prices of many goods cannot be cut easily. If the price of steel rose and fell as quickly and easily as prices on the stock exchange, or if wages went down without a murmur of resistance, or if rents and other contractual items could be quickly adjusted, then prices would be flexible and we would not require any enlargement of our money supply to cover a growing real output.

In fact, as we know, prices are extremely "sticky" in the downward direction. Union leaders do not look with approval on wage cuts, even when living costs fall. Contractual prices cannot be quickly adjusted. Many big firms, as we saw in Chapter 14, administer their prices and carefully avoid price competition: note, for example, that the prices of many consumer items are printed on the package months before the item will be sold.

Thus we can see that a fixed supply of money would put the economy into something of a straitjacket. As output tended to increase, businessmen would need more money to finance production, and consumers would need more money to make their larger expenditures. If businessmen could get more money from the banks, all would be well. But suppose they could not. Then the only way a businessman could get his hands on a larger supply of cash would be to persuade someone to lend it to him, and his persuasion would be in the form of a higher rate of interest. But this rising interest rate would discourage other businessmen from going ahead with their plans. Hence the would-be boom would be stopped dead in its tracks by a sheer shortage of spending power.

A flexible money supply obviates this economic suffocation. The fact that banks can create money (provided that they have excess reserves) enables them to take care of businesses that wish to make additional expenditures. The expenditures themselves put additional money into the hands of consumers. And the spending of consumers in turn sends the enlarged volume of purchasing power back to business firms to complete the great flow of expenditure and receipt.

Paper Money and Gold

Finally, let us clear up one last mystery of the monetary system—the mystery of where currency (coin and bills) actually comes from and where it goes. If we examine most of our paper currency, we will find that it has "Federal Reserve Note" on it; that is, it is paper money issued by the Federal Reserve System. We understand, by now, how the public gets these notes: it simply draws them from its checking accounts. When it does so, the commercial banks, finding their supplies of vault cash low, ask their Federal Reserve district banks to ship them as much new cash as they need.

And what does the Federal Reserve Bank do? It takes packets of bills ($1 and $5 and $10) out of its vaults, *where these stacks of printed paper have no monetary significance at all,* charges the requisite amount against its member banks' balances, and ships the cash out by armored truck. So long as these new stacks of bills remain in the member banks' possession, they are still not money! But soon they will pass out to the public, where they will be money. Do not forget, of course, that as a result, the public will have that much *less* money left in its checking accounts.

Could this currency-issuing process go on forever? Could the Federal Reserve ship out as much money as it wanted to? Suppose that the authorities at the Fed decided to order a trillion dollars worth of bills from the Treasury mints. What would happen when those bills arrived at the Federal Reserve Banks? The answer is that they would simply gather dust in their vaults. *There would be no way for the Fed to "issue" its money unless the public wanted cash.* And the amount of cash the public could want is always limited by the amount of money it has in its checking accounts.

THE GOLD COVER

Are there no limitations on this note-issuing or reserve-creating process? Until 1967 there *were* limitations imposed by Congress, requiring the Federal Reserve to hold gold certificates equal in value to at least 25 percent of all outstanding notes. (Gold certificates are a special kind of paper money issued by the U.S. Treasury and backed 100 percent by gold bullion in Fort Knox.) Prior to 1964 there was a further requirement that

the amount of gold certificates also be sufficient to give a 25 percent backing as well to the total amount of member bank deposits held by the Fed. Thus the legal obligation not to go beyond this 25 percent gold cover provided a strict ceiling on the amount of member bank reserves the Federal Reserve system could create or on the amount of notes it could ship at the request of its member banks.

All this presented no problem in, say, 1940, when the total of member bank reserves plus Federal Reserve notes came to only $20 billion, against which we held gold certificates worth almost $22 billion. Trouble began to develop, however, in the 1960s when a soaring GNP was accompanied by a steadily rising volume of both member bank reserves and Federal Reserve notes. By 1964, for example, member bank reserves had grown to $22 billion, and outstanding Reserve notes to nearly $35 billion. At the same time, for reasons that we shall learn more about in Part Five, our gold stock had declined to just over $15 billion. With $57 billion in liabilities ($22 billion in member bank reserves plus $35 billion in notes) and only $15 billion in gold certificates, the 25 percent cover requirement was clearly imperiled.

Congress thereupon removed the cover requirement from member bank reserves, leaving all our gold certificates available as "backing" for our Federal Reserve notes. But even that did not solve the problem. Currency in circulation continued to rise with a record GNP until it exceeded $40 billion in 1967. Our gold stock meanwhile continued to decline to $12 billion in that year and threatened to fall further. The handwriting on the wall indicated that the 25 percent cover could not long be maintained.

There were basically two ways out of the dilemma. One would have been to change the gold cover requirements from 25 percent to,

GOLDFINGER AT WORK

Some years ago a patriotic women's organization, alarmed lest the Communists had tunneled under the Atlantic, forced an inspection of the gold stock buried at Fort Knox. It proved to be all there. An interesting question arises as to the repercussions, had they found the great vault to be bare. Perhaps we might have followed the famous anthropological example of the island of Yap in the South Seas, where heavy stone cartwheels are the symbol of wealth for the leading families. One such family was particularly remarkable insofar as its cartwheel lay at the bottom of a lagoon, where it had fallen from a canoe. Although it was absolutely irretrievable and even invisible, the family's wealth was considered unimpaired, since everyone knew the stone was there. If the Kentucky depository had been empty, a patriotic declaration by the ladies that the gold really *was* in Fort Knox might have saved the day for the United States.

say, 10 percent. That would have made our gold stock more than adequate to "back" our paper money (and our member bank deposits, too).*

The second way was much simpler. *It was simply to eliminate the gold cover entirely.* With very little fuss, this is what Congress did in 1967.

GOLD AND MONEY

Does the presence or absence of a gold cover make any difference? From the economist's point of view it does not. Gold is a metal with a long and rich history of hypnotic influence, so there is undeniably a psychological usefulness in having gold "behind" a currency. But unless that currency is 100 percent convertible into gold, *any* money demands an act of faith on the part of its users. If that faith is destroyed, the money becomes valueless; so long as it is unquestioned, the money is "as good as gold."

Thus the presence or absence of a gold backing for currency is purely a psychological problem, so far as the value of a domestic currency is concerned. In Chapter 37 we will look into its international significance. But the point is worth pursuing a little further. Suppose our currency *were* 100 percent convertible into gold—suppose, in fact, that we used only gold coins as currency. Would that improve the operation of our economy?

A moment's reflection should reveal that it would not. We would still have to cope with a very difficult problem that our bank deposit money handles rather easily. This is the problem of how we could increase the supply of money or diminish it, as the needs of the economy changed. With gold coins as money, we would either have a frozen stock of money (with consequences that we shall trace in the next chapter), or our supply of money would be at the mercy of our luck in gold-mining or the currents of international trade that funneled gold into our hands or took it away. And incidentally, a gold currency would not obviate inflation, as many countries have discovered when the vagaries of international trade or a fortuitous discovery of gold mines increased their holdings of gold faster than their actual output.

MONEY AND BELIEF

As we cautioned at the outset, money is a highly sophisticated and curious invention.

*Actually as we shall see in the box on the gold standard on p. 581—the gold never really backed our currency, since no American was legally permitted to buy gold bullion.

477

At one time or another nearly everything imaginable has served as the magic symbol of money: whales' teeth, shells, feathers, bark, furs, blankets, butter, tobacco, leather, copper, silver, gold, and (in the most advanced nations) pieces of paper with pictures on them or simply numbers on a ledger page. In fact, anything is usable as money, provided that there is a natural or enforceable scarcity of it, so that men can usually come into its possession only through carefully designated ways. Behind all the symbols, however, rests the central requirement of faith. Money serves its indispensable purposes as long as we believe in it. It ceases to function the moment we do not. Money has well been called "the promises men live by."

But the creation of money and the control over its supply is still only half the question. We have yet to trace how our money supply influences the flow of output itself—or to put it differently, how the elaborate institutions through which men promise to honor one another's work and property affect the amount of work they do and the amount of new wealth they accumulate. This is the subject to which our next chapter will be devoted.

KEY WORDS

Money

Deposits

Currency

Federal Reserve Banks

Federal Reserve System

Fractional reserves

Excess reserves

Deposit creation

Loans

Reserve ratios

CENTRAL CONCEPTS

1. Money is defined as whatever we use to make *payments*. As such, in modern economies, the most important constituents of money are *currency outside the banking system and demand deposits (checking accounts)*.

2. Currency flows into, and is drawn out of, checking accounts. The total amount of checking accounts, however, far exceeds the actual currency held in banks.

3. Banks are forced, by law, to hold reserves against stated fractions of their demand deposits. For most banks these reserves can be either in *vault cash* or in accounts at a *Federal Reserve Bank*.

4. There are *twelve Federal Reserve Banks* that service their member banks exactly as the member banks service the public. The Reserve Banks are coordinated by a policy-making Board of Governors (Federal Reserve Board) in Washington. The Board is empowered to change reserve ratios for city or country banks, within legally established limits, and to take other actions to control the supply of money.

5. The function of the reserves established by the Federal Reserve Board is not to ensure the "safety" of the currency, but to provide a means of *controlling the supply of money*.

6. Any reserves of a commercial bank over and above those imposed by the Federal Reserve are called *excess reserves*. Commercial banks earn profits by lending or investing amounts equal to their excess reserves.

7. When a bank makes a loan, it opens an account in the name of the borrower. This account is a *net addition to total deposits and is therefore new money*. Thus bank lending can increase the supply of money. Investing in government bonds is also likely to lead to new demand deposits.

8. New deposits created by loans are typically drawn on by checks that go into other banks. Here they also give rise to excess reserves and to the possibility of *further deposit creation through more loans or investments*.

9. The total amount of new money that the banking system can create depends on the *reserve ratio*. The size of credit expansion is determined by the reciprocal of the reserve ratio.

Credit expansion

10. It is only the banking *system* that can expand the money supply up to the limit imposed by the reciprocal of the reserve ratio. A single bank can lend only up to the amount that it is prepared to "lose." *Hence each individual bank lends only an amount that is fully covered by its excess reserves.*

Discount rate

11. The Federal Reserve System controls the ability of the banking system to expand the supply of money by *controlling the amount of its reserves.* It can do so in three ways:

   ● By changing *reserve ratios*

Monetary controls

   ● By changing the *discount rate*, as a "signal" to the banking community

   ● By *open-market operations*

Open-market operations

12. The most commonly used method is open-market operations. This is a means of controlling the size of reserves by *purchases and sales of government bonds on the open market.* When the Federal Reserve System buys bonds, it issues in payment its own checks, which enter the commercial banks and are added to their reserves. This gives the commercial banks excess reserves and enables them to make additional loans or investments. Selling bonds brings checks from commercial bank accounts to the Federal Reserve Banks, and thereby lowers the reserve accounts of member banks. This reduces their ability to make loans or investments. Fed powers are more effective in contracting than in expanding the money supply.

Sticky prices

13. A flexible money supply is needed because of *sticky prices.*

Gold "cover"

14. There is no longer any gold backing required behind member bank reserves or behind Federal Reserve notes. The amount or percentage of gold cover is essentially arbitrary. *Gold plays only a symbolic role in a national monetary system.* The true value of money ultimately reposes in the faith men have in it.

QUESTIONS

1. Why do we not count cash in the tills of commercial banks in the money supply? Why don't we include savings accounts?

2. When you deposit currency in a commercial bank, what happens to it? Can you ask for your particular bills again? If you demanded to see "your" account, what would it be?

3. What determines how much vault cash a bank must hold against its deposits? Would you expect this proportion to change in some seasons, such as Christmas? Do you think it would be the same in worried times as in placid times? In new countries as in old ones?

4. Is currency the main reserve of a bank? Do reserves ensure the safety of a currency? What function do they have?

5. What are excess reserves? Suppose a bank has $500,000 in deposits and that there is a reserve ratio of 30 percent imposed by law. What is its required reserve? Suppose it happens to hold $200,000 in vault cash or at its account at the Fed. What, if any, is its excess reserve?

6. If the bank above wanted to make loans or investments, how much would it be entitled to lend or invest?

7. Suppose its deposits increased by another $50,000. Could it lend or invest this entire amount? Any of it? How much?

8. If a bank lends money, it opens an account in the name of the borrower. Now suppose the borrower draws down his new account. What happens to the reserves of the lending bank? Show this in a T account.

9. Suppose the borrower sends his check for $1,000 to someone who banks at another bank. Describe what happens to the deposits of the second bank. If the reserve ratio is 20 percent, how much new lending or investing can it do?

10. If the reserve ratio is 20 percent, and the original addition to reserves is $1,000, what will be the total potential amount of new money that can be created by the banking system? If the ratio is 25 percent?

11. What is the difference between a banking system and a single competitive bank? Can a single bank create new money? Can it create more new money than an amount equal to its excess reserves? Can a banking system create more money than its excess reserves?

12. Suppose that a bank has $1 million in deposits, $100,000 in reserves, and is fully loaned up. Now suppose the Federal Reserve System lowers reserve requirements from 10 percent to 8 percent. What happens to the lending capacity of the bank?

13. If the discount rate rises from 5 percent to 6 percent, does that affect the willingness of banks to lend? How?

14. The Federal Reserve Banks buy $100 million in U.S. Treasury notes. How do they pay for these notes? What happens to the checks? Do they affect the reserves of member banks? Will buying bonds increase or decrease the money supply?

15. Now explain what happens when the Fed sells Treasury notes. Who buys them? How do they pay for them? Where do the checks go? How does payment affect the accounts of the member banks at the Federal Reserve Banks?

16. Why do you think gold has held such a place of prestige in the minds of men?

Money and the Macro System

IN OUR PRECEDING CHAPTER, we found out something about what money is and how it comes into being. Now we must turn to the much more complicated question of how money works—the level of output. What happens when the banks create or destroy deposits? Can we directly raise or lower incomes by altering the quantity of money? Can we control inflation or recession by using the monetary management powers of the Federal Reserve System? These extremely important questions will be the focus of discussion in this chapter.

The Quantity Theory of Money

QUANTITY EQUATION

One relation between money and economic activity must already have occurred to us. It is that the quantity of money must have something to do with *prices*. Does it not stand to reason that if we increase the supply of money, prices will go up, and that if we

decrease the amount of money, prices will fall?

Something very much like this belief lies behind one of the most famous equations (really identities) in economics. The equation looks like this:

$$MV \equiv PT$$

where

M = *quantity of money* (currency outside banks plus demand deposits)

V = *velocity of circulation,* or the number of times per period or per year that an average dollar changes hands

P = *the general level of prices,* or a price index

T = *the number of transactions made in the economy* in a year, or a measure of *physical output*

If we think about this equation, its meaning is not hard to grasp. What the quantity equation says is that the amount of *ex-*

penditure (*M* times *V*, or the quantity of money times the frequency of its use) equals the amount of *receipts* (*P* times *T*, or the price of an average sale times the number of sales). Naturally, this is an identity. In fact, it is our old familiar circular flow. What all factors of production receive (*PT*) must equal what all factors of production spend (*MV*).

Just as our GNP identities are true at every moment, so are the quantity theory of money identities true at every instant. They merely look at the circular flow from a different vantage point. And just as our GNP identities yielded useful economic insights when we began to inquire into the functional relationships within those identities, so the quantity theory can also shed light on economic activity if we can find functional relationships concealed within its self-evident "truth."

THE ASSUMPTIONS OF THE QUANTITY THEORY

To move from tautologies to operationally useful relationships, we need to make assumptions that lend themselves to investigation and evidence. In the case of the GNP ≡ *C* + *G* + *I* + *X* identity, for instance, we made a critical assumption about the propensity to consume, which led to the multiplier and to predictive statements about the influence of injections on GNP. In the case of *MV* ≡ *PT*, we need another assumption. What will it be?

The crucial assumptions made by the economists who first formulated the quantity theory were two: (1) the velocity of money— the number of times an average dollar was used per year—*was constant;* and (2) transactions (sales) *were always at a full-employment level.* If these assumptions were true, it followed that the price level was a simple function of the supply of money:

$$P = \frac{V}{T} \cdot M$$

$$P = kM$$

where *k* was a constant defined by *V/T*.

If the money supply went up, prices went up; if the quantity of money went down, prices went down. Since the government controlled the money supply, it could easily regulate the price level.

TESTING THE QUANTITY THEORY

Is this causal relation true? Can we directly manipulate the price level by changing the size of our stock of money?

The original inventors of the quantity equation, over half a century ago, thought this was indeed the case. And of course it *would* be the case if everything else in the equation held steady while we moved the quantity of money up or down. In other words, if the velocity of circulation, *V*, and the number of transactions, *T*, were fixed, changes in *M* would have to operate directly on *P*.

Can we test the validity of this assumption? There is an easy way to do so. Figure 31·1 shows us changes in the supply of money compared with changes in the level of prices.

A glance at Fig. 31·1 answers our question. Between 1929 and 1973, the supply of money in the United States increased over eightfold while prices rose only a little more than twofold. Clearly, something *must* have happened to *V* or to *T* to prevent the eightfold increase in *M* from bringing about a similar increase in *P*. Let us see what those changes were.

CHANGES IN V

Figure 31·2 gives us a first clue as to what is wrong with a purely mechanical interpre-

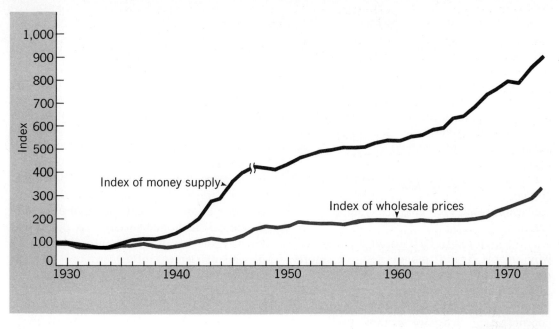

FIGURE 31 · 1
Money supply and prices

FIGURE 31 · 2
Money supply and velocity

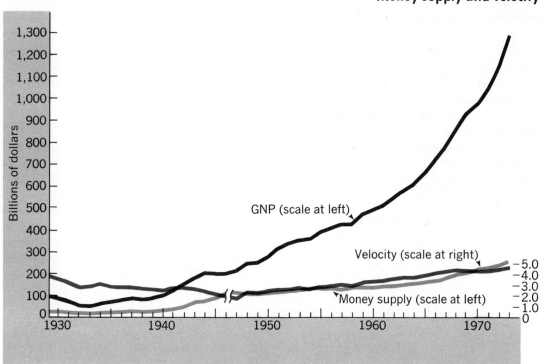

tation of the quantity theory. In it we show how many times an average dollar was used to help pay for each year's output.* We derive this number by dividing the total expenditure for each year's output (which is, of course, the familiar figure for GNP) by the actual supply of money—currency plus checking accounts—for each year. As the chart shows, the velocity of money fell by 50 percent between 1929 and 1946, only to rise again to the 1929 level over the postwar years.

We shall return later to an inquiry into why people spend money less or more quickly, but it is clear beyond question that they do. This has two important implications for our study of money. First, it gives a very cogent reason why we cannot apply the quantity theory in a mechanical way, asserting that an increase in the supply of money will *always* raise prices. For if people choose to spend the increased quantity of money more slowly, its impact on the quantity of goods may not change at all: whereas if they spend the same quantity of money more rapidly, prices can rise without any change in *M*.

Second and more clearly than we have seen, the variability of *V* reveals that money itself can be a destabilizing force—destabilizing because it enables us to do two things that would be impossible in a pure barter economy. We can:

1. delay between receiving and expending our rewards for economic effort

2. spend more or less than our receipts by drawing on, or adding to, our cash balances

*Note that final output is not quite the same as *T*, which embraces *all* transactions, including those for intermediate goods. But if we define *T* so that it includes only *transactions that enter into final output, PT* becomes a measure of gross national product. In the same way, we can count only those expenditures that enter into GNP when we calculate *MV*. It does no violence to the idea of the quantity theory to apply it only to final output, and it makes statistical computation far simpler.

The Classical economists used to speak of money as a "veil," implying that it did not itself play an active role in influencing the behavior of the economic players. But we can see that the ability of those players to vary the rate of their expenditure—to hang onto their money longer or to get rid of it more rapidly than usual—makes money much more than a veil. Money (or rather, people's wish to hold or to spend money) becomes an independent source of change in a complex economic society. To put it differently, the use of money introduces an independent element of uncertainty into the circular flow.*

CHANGES IN T

Now we must turn to a last and perhaps most important reason why we cannot relate the supply of money to the price level in a mechanical fashion. This reason lies in the role played by *T*; that is, by the volume of output.

Just as the early quantity theorists thought of *V* as essentially unvarying, so they thought of *T* as a relatively fixed term in the quantity equation. In the minds of nearly all economic theorists before the Depression, output was always assumed to be as large as the available resources and the willingness of the factors of production would permit. While everyone was aware that there might be minor variations from this state of full output, virtually no one thought they would be of sufficient importance to matter. *Hence the quantity theory implicitly assumed full employment or full output as the normal condition of the economy.* With such an assumption, it was easy to picture *T* as an unimpor-

*Technically, the standard economic definition of money is that it is both a means of exchange and a store of value. It is the latter characteristic that makes money a potentially disturbing influence.

WHY THE OLD QUANTITY THEORISTS ERRED

Modern economists can easily show that the velocity of money is not constant and that the volume of transactions (GNP) is not always at full employment. But it should not be thought that the originators of the quantity theory were stupid or too lazy to look up the basic data. Most of the numbers on which economists now rely were simply not in existence then. The national income, for example, was not calculated until the early 1930s, and GNP was not "invented" until the early 1940s. You cannot calculate the velocity of money unless you know the national income or the gross national product.

Neither did the original quantity theorists have accurate measures of unemployment or capacity utilization. They used the only method available to them: direct observation of the world, a method that is notoriously inaccurate when one's view is much smaller than "the world." In addition, believing in marginal productivity analysis, they could not see how any worker could be unemployed unless he wanted to be, or unless he (or his union) foolishly insisted on a wage higher than his marginal product. The idea of mass involuntary unemployment required the idea of an equilibrium output that would be less than a full-employment output, an idea completely foreign to pre-Keynesian thought.

tant term in the equation and to focus the full effect of changes in money on *P*.

The trauma of the Great Depression effectively removed the comfortable assumption that the economy "naturally" tended to full employment and output. At the bottom of the Depression, real output had fallen by 25 percent. Aside from what the Depression taught us in other ways, it made unmistakably clear that changes in the volume of output (and employment) were of crucial importance in the overall economic picture.

OUTPUT AND PRICES

How does our modern emphasis on the variability of output and employment fit into the overall question of money and prices? The answer is very simple, but very important. We have come to see that *the effect of more money on prices cannot be determined unless we also take into account the effect of spending on the volume of transactions or output.*

It is not difficult to grasp the point. Let us picture an increase in spending, perhaps initiated by businessmen launching a new

investment program or by the government inaugurating a new public works project. These new expenditures will be received by many other entrepreneurs, as the multiplier mechanism spreads the new spending through the economy. But now we come to the key question. What will entrepreneurs do as their receipts increase?

It is at this point that the question of output enters. For if businessmen are operating factories or stores *at less than full capacity,* and if there is an *employable supply of labor available,* the result of their new receipts is almost certain to be an increase in output. That is, employers will take advantage of the rise in demand, to produce and sell more goods and services. They may also try to raise prices and increase their profits further; but *if their industries are reasonably competitive,* it is doubtful that prices can be raised very much. Other businessmen with idle plants will simply undercut them and take their business away. An example is provided by the period 1934 through 1940, when output increased by 50 percent while prices rose by less than 5 percent. The reason, of course, lay in the great amount of unemployed resources, making it easy to expand output without price increases.

PRICES AND EMPLOYMENT

Thus we reach a general conclusion of the greatest importance. *An increase in spending of any kind tends to result in more output and employment whenever there are considerable amounts of unemployed resources.* But this is no longer true when we reach a level of high employment or very full plant utilization. Now an increase in spending *cannot* quickly lead to an increase in output, simply because the resources for more production are lacking. The result, instead, will be a rise in prices, for no firm can lose business to competitors when competitors are unable to fill additional orders. Thus the corollary of our general conclusion is that *additional spending from any source is inflationary when it is difficult to raise output.*

FULL EMPLOYMENT VS. UNDEREMPLOYMENT

It is impossible to overstress the importance of this finding for macroeconomic *policy.* Policies that make sense when we are fully employed may make no sense when we are badly underemployed, and vice versa.

To spend more in the public or in the private sector is clearly good for an economy that is suffering from underutilized resources, but equally clearly inflationary and bad for an economy that is bumping up against the ceiling of output. Similarly, to balance budgets or run budget surpluses makes little sense when men are looking for work and businessmen are looking for orders, but it is the course of wisdom when there are no idle resources to absorb the additional expenditure.

One of the main differences between con-temporary economic thought and that of the past is precisely this sharp division between policies that make sense in full employment and those that make sense in conditions of underemployment. It was not that the economists of the past did not recognize the tragedy of unemployment or did not wish to remedy it. It was rather that they did not see how an economy could be in *equilibrium* even though there was heavy unemployment.

The dragging years of the Great Depression taught us not only that output could fall far below the levels of full utilization, but—and perhaps this was its most intellectually unsettling feature—that an economy could be plagued with unemployed men and machines for almost a decade and yet not spontaneously generate the momentum to reabsorb them. Today we understand this condition of unemployment equilibrium, and we have devised various remedial measures to raise the equilibrium point to a satisfactory level, including, not least, additional public expenditure. But this new understanding must be balanced with a keen appreciation of its relevance to the underlying situation of employment. Remedies for an underemployed economy can be ills for a fully employed one.

INFLATION AND PUBLIC FINANCE

We can see that the conclusion we have reached puts a capstone on our previous analysis of deficit spending. It is now possible to add a major criterion to the question of whether or not to use the public sector as a supplement to the private sector. That criterion is whether or not substantially "full" employment has been reached.

If the economy is operating at or near the point of full employment, additional net public spending will only add more MV to a situation in which T is already at capacity and where, therefore, P will rise. But note

MAXIMUM VS. FULL EMPLOYMENT

What is "full" employment? Presumably government spending is guided by the objectives of the Employment Act of 1946, which declares the attainment of "maximum employment" to be a central economic objective of the government.

But what is "maximum" employment? Does it mean zero unemployment? This would mean that no one could quit his job even to look for a better one. Or consider the problem of inflation. Zero unemployment would probably mean extremely high rates of inflation, for reasons we will look into more carefully later. Hence no one claims that "full" employment is maximum employment in the sense of an absence of *any* unemployment whatsoever.

But this opens the question of how much *unemployment* is accepted as consistent with "maximum" employment. Under Presidents Kennedy and Johnson, the permissible unemployment rate was 4 percent. Under President Nixon the permissible unemployment rate rose to a range of 4.5 to 5 percent, largely because inflation had worsened. Hence the meaning of "full employment" is open to the discretion of the economic authorities, and their policies may vary from one period to another.

that this conclusion attaches to more than additional *public* spending. When full employment is reached, additional spending of any kind—public or private, consumption or investment—will increase MV and, given the ceiling on T, affect P.

A different conclusion is reached when there is large-scale unemployment. Now additional public (or private) spending will result not in higher prices, but in larger output and higher employment. Thus we cannot say that public spending in itself is "inflationary." Rather, we must see that *any kind of additional spending can be inflationary in a fully employed economy.*

Money and Expenditure

We have almost lost sight of our subject, which is not really inflation (we will come back to that in Chapter 32) but how money affects GNP. And here there is an important point. How does an increased supply of money get "into" GNP? People who have not studied economics often discuss changes in the money supply as if the government "put" money into circulation, mailing out dollar bills to taxpayers. The actual connection between an increase in M and an increase in MV is much more complex. Let us look into it.

INTEREST RATES AND THE TRANSACTIONS DEMAND FOR MONEY

From our previous chapter, we know the immediate results of an increased supply of money, whether brought about by open-market operations or a change in reserve ratios. *The effect in both cases is a rise in the lendable or investible reserves of banks.* Ceteris paribus, this will lead to a fall in interest rates as banks compete with one another in lending their additional unused reserves to firms or individuals.

As interest rates decline, some firms and individuals will be tempted to increase their borrowings. It becomes cheaper to take out a mortgage, to buy a car on an installment loan, to finance inventories. Thus, as we would expect, the demand curve for "spending money," like that for most commodities, slopes downward. As money gets cheaper, people want to "buy" (borrow) more of it. To put it differently, the lower the price of money, the larger the quantity demanded.

We speak of this demand curve for money to be used for expenditure as the *transactions demand for money.*

FINANCIAL DEMAND

But there is also another, quite separate source of the demand for money. This is the demand for money for *financial purposes,* to be held by individuals or corporations as part of their assets.

What happens to the demand for money for financial purposes as its price goes down? Financial demand also increases, although for different reasons. When interest rates are high, individuals and firms tend to keep their wealth as fully invested as possible, in order to earn the high return that is available. But when interest rates fall, the opportunity cost of keeping money idle is much less: if you are an investor with a portfolio of $10,000 and the rate of interest is 7 percent, you give up $700 a year if you are very "liquid" (i.e., all in cash); whereas if the interest rate is only 3 percent, your opportunity cost for liquidity falls to $300.

LIQUIDITY PREFERENCE

Economists call this increased willingness to be in cash as interest rates fall *liquidity preference.* The motives behind liquidity preferences are complex—partly speculative, partly precautionary. But they act in all cases to make us more and more willing or eager to be in cash when interest rates are low, and less and less willing when rates are higher. Thus the financial demand for cash, like the transactions demand, is a downward sloping demand curve.

We can now put together the two demand curves for money and add the supply curve of money—the actual stock of money available. The result looks like Fig. 31·3.

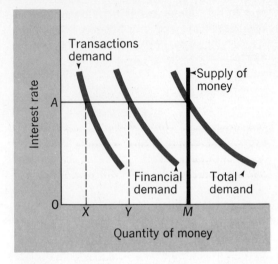

FIGURE 31 · 3
Transactions and financial demands for money

Our diagram shows us that at interest rate OA, there will be OX amount of money demanded for transactions purposes and OY amount demanded for liquidity purposes. The total demand for money will be OM (= $OX + OY$), which is just equal to the total supply.

CHANGING THE SUPPLY OF MONEY

Now let us suppose that the monetary authorities reduce the supply of money. We show this in Fig. 31·4. Now we have a curious situation. The supply of money has declined from OM to OM'. But notice that the demand curve for money shows that firms and individuals want to hold OM, at the given rate of interest OA. *Yet they cannot hold amount* OM, *because the monetary authorities have cut the supply to* OM'. What will happen?

The answer is very neat. As bank reserves fall, banks will "tighten" money— raise lending rates and screen loan applications more carefully. Therefore individuals and firms will be competing for a reduced supply of loans and will bid more for them. At the same time, individuals and firms will feel the pinch of reduced supplies of cash and will try to get more money to fulfill their liquidity desires. The easiest way to get more money is to sell securities, to get out of bonds and into cash. *Note, however, that selling securities does not create a single additional dollar of money. It simply transfers money from one holder to another. But it does change*

and firms would be holding more money than they wanted at the going rate of interest. They would try to get out of money into bonds, sending bond prices up and yields

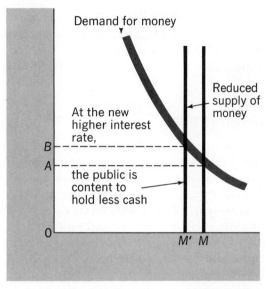

FIGURE 31 · 5
Determination of new equilibrium

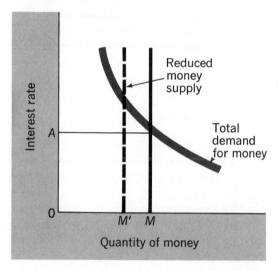

FIGURE 31 · 4
Reducing the supply of money

the rate of interest. As bonds are sold their price falls, and as the price of bonds falls, the interest yield on bonds rises (see p. 471).

Our next diagram (Fig. 31·5) shows what happens. As interest rates rise, the public is content to hold a smaller quantity of money. Hence a new interest rate, *OB,* will emerge, at which the public is *willing* to hold the money that there *is to hold.* The attempt to become more liquid ceases, and a new equilibrium interest rate prevails.

Suppose the authorities had increased the supply of money. In that case, individuals

FIGURE 31 · 6
Increasing the supply of money

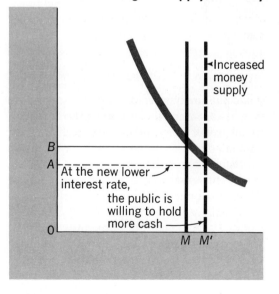

down. Simultaneously, banks would find themselves with extra reserves and would compete with one another for loans, also driving interest rates down. As interest rates fell, firms and individuals would be content to hold more money either for transactions or liquidity purposes, until a new equilibrium was again established. Fig. 31·6 shows the process at work.

THE DETERMINATION OF INTEREST RATES

This gives us the final link in our argument. We have seen that interest rates determine whether we wish to hold larger or smaller balances, either for transactions or financial (liquidity) purposes. But what determines the interest rate itself? We can now see that the *answer is the interplay of our demand for money and the supply of money.*

Our demand for money is made up of our transactions demand curve and our financial (liquidity) demand curve. The supply of money is given to us by the monetary authorities. The price of money—interest— is therefore determined by the demand for, and supply of, money, exactly as the price of any commodity is determined by the demand and supply for it.

MONEY AND EXPENDITURE

What our analysis enables us to see, however, is that once the interest rate is determined, it will affect the use to which we put a given supply of money. Now we begin to understand the full answer to the question of how changes in the supply of money affect GNP (and prices). Let us review the argument one last time.

1. Suppose that the monetary authorities want to increase the supply of money. They will lower reserve ratios or buy government bonds on the open market.

2. Banks will find that they have larger

reserves. They will compete with one another and lower lending rates.

3. Individuals and firms will also find that they have larger cash balances than they want at the going rate of interest. They will try to get rid of their extra cash by buying bonds, thereby sending bond yields down.

4. As interest rates fall, both as a result of bank competition and rising bond prices, the new, larger supply of money will find its way into use. *Part of it will be used for additional transactions purposes, as individuals and firms take advantage of cheaper money and increase their borrowings. Part of it will be used for larger financial balances, as the public's desire for liquidity grows with falling interest rates.*

THE PROCESS IN DIAGRAM

We can see the process very clearly in Fig. 31·7. We begin with *OM* money supply and

FIGURE 31 · 7
Using money for two purposes

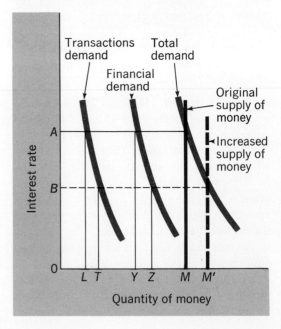

a rate of interest OA: as we can see, OL amount of money is held for liquidity purposes, and OY for transactions purposes. Now the stock of money is increased to OM'. The interest rate falls, for the reasons we now understand, until it reaches OB. At the new interest rate, liquidity balances have increased to OT, and transactions balances to OZ.

Exactly the same process would take place, in reverse, if the stock of money were decreased from OM' to OM. Can you see that the decreased supply of money will result partly in smaller transactions balances and partly in smaller liquidity balances? Do you understand that it is the higher rate of interest that causes the public to hold these smaller balances?

THE LIQUIDITY EFFECT

We have traced the circuitous manner in which a change in M "gets into" GNP. But there is yet another route that bypasses the rate of interest entirely. The "monetarist" school suggests that increases in M directly affect our spending habits, *even though interest rates remain unchanged.*

The monetarists suggest that changes in the supply of money directly affect our spending propensities because changes in the money supply alter our portfolios. A portfolio describes the way in which we hold our assets—in cash, savings accounts, checking accounts, various kinds of bonds, stocks, real estate, etc. When the money supply is altered, for example through open-market operations, the government is tempting the public to shift its portfolios into cash. But there is no reason to believe that the public wants this much cash; if it *had* wanted it, it would not have held the bonds (at their going rate of interest) in the first place. Therefore the public will seek to reduce its undesired cash holdings by buying real assets—cars, homes, inventories, and other things.

Most economists are willing to add this *liquidity effect* to the *interest rate effect,* so that monetary policy is believed to affect the economy both through its impact on the price of money and also directly through its impact on our portfolio preferences. What remains in doubt is the degree of influence that should be attributed to interest or to liquidity.

THE MODERN QUANTITY THEORY

We are now in a position to reformulate the quantity theory. Modern proponents of the theory recognize that economies do not always operate at full employment and that the velocity of money changes (we can see that liquidity preferences must be closely related to velocity). Hence they do not argue that an increase in the quantity of money is mechanically reflected in a proportionate rise in prices.

Instead, they contend that the demands for money for transaction purposes and liquidity purposes are *calculable functions,* just as consumption is a calculable function of income. The variables on which the demand for money depend are very complex —too complex to warrant explanation here. What is important is the idea that the relation between an increase in money supply and in transactions and financial demand can be estimated, much as the propensity to consume is estimated.

The Art of Money Management

We finally have all the pieces of the puzzle. We understand the curiously complex way in which changes in the supply of money affect changes in the expenditures of the public. It remains only to consider one aspect of the problem: the art of managing the supply of money so that the *right* increases in the supply of money will be forthcoming at the right time.

Why "art"? Is not the task of the monetary authority very clear? By increasing the supply of money, it pushes down interest rates and encourages expenditure. Hence all it has to do is to regulate the quantity of money to maintain a level of spending that will keep us at a high, but not too high, level of employment.

We have already seen some of the reasons why things are not that simple. The effect of interest rates on investment expenditure, as we previously learned, is obscure. So is the effect of liquidity on expenditure. We know that unwanted liquidity will encourage spending, but there is a time lag involved, and this lag may vary considerably at different phases of the business cycle. To add to the problem, the Federal Reserve Board can control the money supply with an eye on interest rates, or it can control it with an eye on liquidity effects, but it cannot do both at the same time. As a result, sometimes the board seems to focus entirely on the "price effect" of interest rates, and at other times on the liquidity effect of money supply. When two policies clash and there is no scientific means of judging between them, we trust to good sense, or to a "feel" of the economy. Hence the need for an "art" of money management.

SHIFTING LIQUIDITY PREFERENCES

Still another difficulty enforces the need for artful control. Suppose, for example, that the Federal Reserve creates excess reserves, in the expectation that interest rates will go down and that new loans will be pumped into investment. But suppose that at the same time, the public's "liquidity preferences" are rising because investors feel nervous and want to be more liquid. Then the shift in the quantity of money, as shown in Fig. 31·8, will be offset by a shift in liquidity preferences, and the rate of interest will not change

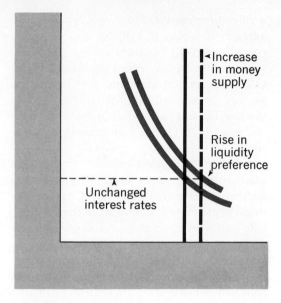

FIGURE 31 · 8
A shift in liquidity preference

at all! The new money will simply wind up in larger financial cash holdings, and none will be available for more transactions.

In other words, an attempt by the monetary authorities to drive down the rate of interest in order to encourage expenditure may be frustrated if the public uses all the additional funds for liquidity. At the bottom of the Great Depression, for example, banks had huge excess reserves because businessmen would not risk expenditure for new capital projects. People had an insatiable desire for liquidity, and no attempted reductions of the rate of interest could persuade them to spend the money they held for security.

In the same way, an attempt to raise interest rates and to halt price inflation by making credit tight may come to naught if the public reacts to higher interest rates by giving up its liquidity, thereby making funds available to others to finance increased transactions expenditure. Or take another instance: if the Fed tries to lower interest rates by increasing *M*, the effort may result in a general expectation of inflation and a movement out of bonds into stocks. In that case, interest rates, instead of falling, will go up! This actually happened in 1968.

CREDIT CRUNCHES

Still another difficulty of monetary management lies in *credit crunches*. These are sharp curtailments in the growth of credit. The reason for the designation "crunch" is that these curtailments do not impose their effects evenly across the economy. Tight money is not a serious deterrent to most large corporations with high credit ratings and large cash reserves of their own, but it may seriously impair the ability of states and municipalities to borrow, and it is apt to exact a real toll on residential building, where interest is a major item of cost. Even within the relatively unaffected corporate sector, a credit squeeze can hurt some companies very badly. The crunch in 1969–1970 almost caused the collapse of the Chrysler Corporation, one of the largest industrial enterprises in the nation.

Since that time, in an effort to prevent future crunches, special government lending intermediaries have been making loans to residential builders and to states and municipalities, if these borrowers are in trouble. We will have to wait and see how well they work, for the problem is that to some extent these intermediaries must take actions against the direction set by the Fed. When the Fed is trying to cut back credit, the intermediaries will be trying to extend credit.

MONETARY AND FISCAL POLICY

All these problems of monetary management help us understand why economists are generally reluctant to entrust the overall regulation of the economy to monetary policy alone. There is too much slippage between changes in the money supply and changes in expenditure; too little reliability as to the effects of changes in *M* on desired changes in *MV*.

Thus we look for our overall controls to both monetary policies and fiscal policies. Few economists today would rely solely on the money mechanism to move the general economy. Instead, they seek a combination of monetary and fiscal policies—easy money and more government spending (or tax cuts), or tight money and a public budgetary surplus.

KEY WORDS

Quantity equation

Velocity of circulation

Full employment vs. under-employment

CENTRAL CONCEPTS

1. **The quantity equation MV ≡ PT is an *identity saying that expenditures* (MV) *equal receipts* (PT). It was originally intended as a functional relation between M and P.**

2. **This statement would be true if V and T were fixed. In fact, *the velocity of circulation* (or GNP/M) changes during the cycle.**

3. **Even more important, short-run changes in output can be very marked. This is contrary to the *expectation of the early quantity theorists that the economy would always operate at full employment.***

4. **When a competitive economy is operating at substantially less than full employment, an increase in M, or in MV, leads to a rise in output rather than prices. *This distinction between the effects of additional expenditure, public or private, is one of the central conceptions of modern macroeconomics.***

5. How does an increase in M become a larger MV? One critical link is the *interest rate,* or the price of money. As the price of money falls, ceteris paribus, the demand for it increases.

6. There are two kinds of demand: demand for money for *transactions purposes* and for *financial* (speculative or precautionary) *purposes.* Demand for the former rises when we need money to finance larger purchases. Demand for the latter rises when we want to be more liquid.

7. *Changes in the supply of money leave the public with larger or smaller amounts of cash than it wants at going interest rates.* It will try to get out of, or into, cash by buying or selling bonds. As bonds are sold or bought, their yield changes; and as yields change, the public changes its willingness to hold cash. At the same time, changes in interest rates are brought about by banks trying to get rid of, or to conserve, their increased or decreased reserves. Thus, *changes in interest rates will lead to changes in transactions balances and financial balances.*

8. *The rate of interest is determined by the supply of, and the demand for, money.* Monetary authorities determine the supply; transactions and financial desires determine demand.

9. Changes in M will affect changes in expenditure (MV) in ways that depend on the shapes and positions of the transactions and liquidity preference schedules. *Note that an increase or decrease in M will usually be only partly reflected in a changed MV, since liquidity balances will also expand or contract.*

10. Monetarist theorists also believe that changes in M directly affect expenditure by altering our *portfolios.* This *liquidity effect* leads us to convert unwanted "excess liquidity" into real assets by spending our extra cash.

11. Modern monetary theory no longer ties increases in M directly to P. It asserts that there are *calculable functions for transactions and financial purposes* that enable us to relate changes in M to changes in MV.

12. Monetary management is an art, for many reasons. The effects of interest rate changes on expenditure are unclear. So are the effects of M on MV. Shifting liquidity preference schedules make the results of monetary action hard to predict.

13. As a result, most economists advocate a close coordination of fiscal and monetary policies, rather than reliance on either one alone.

QUESTIONS

1. Why is the quantity equation a truism? Why is the interpretation of the quantity equation that M affects P not a truism?

2. Suppose you are paid $140 a week and you put it in the bank. On each of the seven days of the week, you spend one-seventh of this sum. What is your average balance during the week? Now suppose that you spend the whole sum on the first day of the week. Will your average balance be the same? What is the relation between velocity of circulation and size of average balances?

3. What considerations might lead you, as a businessman, to carry higher cash balances? Could these considerations change rapidly?

4. The basic reason why the original quantity theorists thought that M affected P was their belief that V and T were fixed. Discuss the validity of this belief.

5. Why is the level of employment a critical determinant of fiscal policy?

6. If employment is "full," what will be the effects of an increase in private investment on prices and output, supposing that everything else stays the same?

7. In what way can an increase in excess reserves affect V or T? Is there any certainty that an increase in reserves will lead to an increase in V or T?

8. Suppose that you had $1,000 in the bank. Would you be more willing to invest it if you could earn 2 percent or 5 percent? What factors could make you change your mind about investing all or any part at, say, 5 percent? Could you imagine conditions that would make you unwilling to invest even at 10 percent? Other conditions that would lead you to invest your whole cash balance at, say, 3 percent?

9. Suppose that the going rate of interest is 7 percent and that the monetary authorities want to curb expenditures and act to lower the quantity of money. What will the effect be in terms of the public's feeling of liquidity? What will the public do if it feels short of cash? Will it buy or sell securities? What would this do to their price? What would thereupon happen to the rate of interest? To investment expenditures?

10. Suppose that the monetary and fiscal authorities want to encourage economic expansion. What are the general measures that each should take? What problems might changing liquidity preference interpose?

11. Do you unconsciously keep a "liquidity balance" among your assets? Suppose that your cash balance rose. Would you be tempted to spend more?

12. Show in a diagram how a decrease in the supply of money will be reflected in lower transactions balances and in lower financial balances. What is the mechanism that changes these balances?

13. Do you understand (a) how the rate of interest is determined; (b) how it affects our willingness to hold cash? Is this in any way different from the mechanism by which the price of shoes is determined or the way in which the price of shoes affects our willingness to buy them?

The Problem
of Inflation

32

HOW COULD WE SPEND TWO CHAPTERS ON money and not talk about inflation? The explanation is not that we have downgraded the problem. On the contrary, we consider it so important that we are devoting this entire chapter to it. Moreover, there was another reason to defer a discussion of inflation until now. By and large, our macroeconomic chapters up to this point have covered problems about which economic knowledge is fairly secure. Now and again we have pointed out unresolved or ill-understood issues, but in the main we were on firm ground.

Beginning with inflation, that is no longer the case. We are moving now into areas where economists are very tentative and unsure. As we shall see, we know something about each of the problems of the subsequent chapters, but not nearly enough.

INFLATION IN RETROSPECT

Inflation is both a very old problem and a very new one. If we look back over history, we discover many inflationary periods. Dio-

cletian tried (in vain) to curb a Roman inflation in the fourth century A.D., between 1150 and 1325 the cost of living in medieval Europe rose fourfold; between 1520 and 1650 prices again rose between 200 and 400 percent, largely as a result of gold pouring into Europe from the newly opened mines of the New World. In the years following the Civil War, the South experienced a ferocious inflation, and prices in the North doubled; during World War I, prices in the United States doubled again.

Let us focus closer on the U.S. experience up to 1950 (Fig. 32·1). Two things should be noted about this chart. *First, wars are regularly accompanied by inflation.* The reasons are obvious enough. War greatly increases the volume of public expenditure, but governments do not curb private spending by an equal amount through taxation. Invariably, wars are largely financed by borrowing, and the supply of money and the total amount of spending, public and private, rises rapidly.

Second, inflations have always been rela-

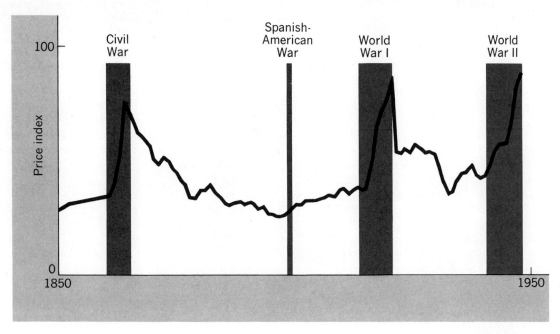

FIGURE 32 · 1
Inflation in perspective

tively short-lived in the past. Notice that prices fell during the periods 1866 to 1900 and 1925 to 1933, and that the long secular trend, although generally tilted upward, is marked with long valleys as well as sharp peaks.

RECENT INFLATIONARY EXPERIENCE

Now examine Fig. 32·2, which shows the record of U.S. price changes since 1950. Once again we notice that the outbreak of war has brought price rises, albeit relatively small ones. This is because the financing of the Korean and Vietnam wars, exactly as the preceding larger wars, did not sufficiently curtail private spending. But in a second vital regard, contemporary experience is different from that of the past. The peaks of inflationary rises have not been followed by long gradual declines. Instead, inflation seems to have become a chronic element in the economic situation, a lingering fever that has defied economic ice packs and economic antibiotics.

FIGURE 32 · 2
Wholesale prices since 1950

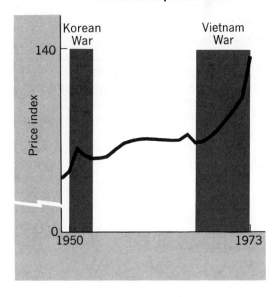

HYPERINFLATIONS

Hyperinflations are among the most destructive economic experiences that a modern economic society can undergo. In the German hyperinflation of the 1920s, for example, prices rose so rapidly that hotels and restaurants with foreign guests would not reveal the price of a meal until the diner had finished; then they would determine the "going" value of marks at that moment. Inflation mounted until a common postage stamp cost 9 billion marks, and a worker's weekly wage came to 120 trillion marks. Newspapers and magazines of the period showed people bringing home their weekly pay in wheelbarrows—billions and billions of marks literally worth less than the paper they were printed on.

Hyperinflations have also occurred in Hungary in 1923 and in China after World War II. They are really psychological—even pathological—phenomena, rather than strictly economic ones. That is, they signal a collapse of faith in the vitality and viability of the economy. Farmers typically hoard foodstuffs, rather than accept payment in currencies that they fear will be only so much wallpaper in a matters of weeks. Merchants and manufacturers are unable to make contracts, since suppliers ask for enormous prices in anticipation of price rises to follow. Shopkeepers are reluctant to sell to customers because this means giving up the true wealth of goods for the spurious wealth of paper money that no one trusts. Thus there is flight from all paper currency and a scramble to get into goods or into commodities such as gold, in which people retain faith. The scramble means that the demand curve for goods or gold shoots out to the right, and the supply curve moves to the left as those who own goods or

WORLDWIDE INFLATION

Before we attempt to explain this perplexing phenomenon, one further fact of great significance should be noted. *It is that inflation has been a worldwide experience.* It has ravaged underdeveloped countries, where prices have often risen by 20 to 50 percent per year. And it has appeared in every industrialized nation, *even though those nations did not participate in the Korean or Indochinese wars.* As Table 32·1 shows, the United States inflationary problem has been much *less* severe than the European or Japanese experience, except for a brief period from 1968–1969. In other words, contemporary inflation seems to be a new kind of economic problem that has appeared—and resisted attempts to remedy—in all industrialized nations.

TABLE 32 · 1
Worldwide Inflation

PRICE RISES IN INDUSTRIALIZED COUNTRIES

	Average annual percentage		
	1959–69	1968–69	1969–72
U.S.	2.2	5.1	4.5
Australia	2.4	2.8	5.3
Canada	2.4	4.3	3.7
France	3.7	5.7	5.8
Italy	3.6	2.5	5.1
Japan	5.0	4.9	6.1
Sweden	3.7	2.6	6.9
Switzerland	3.0	2.5	5.6
U.K.	3.4	5.1	7.6
W. Germany	2.4	2.6	4.9

The ABC of the Inflationary Process

SUPPLY AND DEMAND ONCE AGAIN

What do we know about inflations? A good place to start is by refreshing our memories of how an individual price rises. As we have seen, prices go up when demand curves shift

gold are reluctant to sell them for paper money. Meanwhile, governments find their expenses skyrocketing and are forced to turn to the printing presses as the only way to collect the revenues they require. Finally, people find that they must *barter* goods, as in primitive economic societies.

The only cure for a hyperinflation is the abandonment of the currency in which everyone has lost faith, and the institution of a new currency that people can be once again induced to believe will serve as a reasonably stable "store of value." For example, in 1958 General de Gaulle stopped an incipient runaway French inflation when he simply announced that there would be a new franc worth one hundred of the old, deteriorating francs. Because of de Gaulle's extraordinary prestige, Frenchmen willingly changed their old 100-franc notes for new one-franc coins and then stopped trying to get "out" of money and into

goods. The same magic feat was performed in the 1920s in Germany when the government announced that there would be a new mark "backed" by land. People believed that, and hyperinflation stopped.

There is a curious aspect of hyperinflation that we might stop to notice. Why do workers trundle their wages home in wheelbarrows? Why doesn't the government simply print trillion-mark notes, so that a man's wage would fit into his pocketbook? The answer is purely bureaucratic. The printing presses are busy turning out notes in denominations that would have been suitable for the price level of, say, six months earlier. No one dares give orders for denominations that might meet needs when the notes will actually be issued. Why? Because an order to print, say, trillion-mark notes instead of billion-mark notes would be construed as *inflationary!*

to the right or when supply curves shift to the left. This happens all the time in innumerable markets. Are these price rises "inflationary"?

The question begins to sharpen our understanding of what we mean by inflation. In a price rise that takes place in the way we have described, a new stable equilibrium price is established. In an inflationary situation there is no stable outcome, since prices continue to rise. This generally means that demand curves must be continuously moving outward and that supply curves must be upward sloping. Moreover, this unstable process is not taking place in a single market (such as a continuing rise in the price of a desired commodity like gold), but in *all* or nearly all markets at the same time.

CHANGES IN TOTAL DEMAND

What could bring about such a multimarket shift? Part of the problem we already understand. The continuing rightward shift in demand curves is the result of *a continuous rise in the volume of expenditure.* More and

more dollars are earned and spent. This in turn means that the supply of money must be increasing, or that the velocity of circulation must be constantly increasing, or both. MV must be rising if the national price level is rising. You cannot have inflation unless demand curves in most markets are shifting to the right; and this in turn cannot occur unless money incomes and expenditures are rising in the economy.

THE SUPPLY CONSTRAINT: BOTTLENECKS

But this is only half the picture. There is also the question of supply. A rightward shift of a demand curve will not cause prices to rise in a market if the supply curve shifts outward to meet it, or if the industry is producing under conditions of constant or decreasing cost. Thus, corresponding to the rightward shift in total demand must come a gradual upward tilt in supply curves, as Fig. 32·3 shows.

As demand moves from D_1 to D_2, prices hardly rise at all. But the shift from D_2 to D_3 brings a sharp increase because we run into

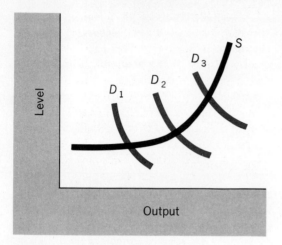

FIGURE 32 · 3
The bottleneck supply curve

bottlenecks, where output cannot be further increased, except at much higher cost. These bottlenecks, moreover, may begin to exert their constricting influence before the economy as a whole can be considered in a condition of overall full employment. Thus as demand curves for various goods and services move outward, we will experience price rises in some industries even though there is unused capacity in others, or even though considerable unemployment exists. If these industries bulk large in the pattern of production or in consumer budgets—for example, if we hit bottlenecks in steel or food output—the general price level will begin to rise.

DEMAND PULL VS. COST PUSH

Economists sometimes talk about the two processes that enter into inflation as *demand pull* or *cost push*. Demand pull focuses attention on forces that are causing thousands of demand curves to move to the right—for example, policies of easy money or expansionary fiscal policy. Cost push emphasizes the supply side, with cost curves moving to the left or sloping sharply upward.

Cost-push analyses often concentrate on the wage level as a prime causative agency for inflation. Of course, rising wages can be a source of higher costs. But it is important to distinguish between increases in *wage costs per hour* and increases in *wage costs per unit.* Wages may rise; but if *labor productivity keeps pace, cost per unit will not rise.*

Corresponding to cost push from rising wage costs per unit, there is cost push from higher profits, an argument frequently put forward by labor. Again we must distinguish between higher profits for the company and higher profits per unit. The latter may occur if the increase in demand outruns the increase in productive capacity, strengthening the market power of large companies.

THE PHILLIPS CURVE

To determine the inflationary pressure in thousands of markets, each with a different configuration of supply and demand curves, would be an impossible task. Hence, the English economist A.W. Phillips has sug-

FIGURE 32 · 4
The Phillips curve

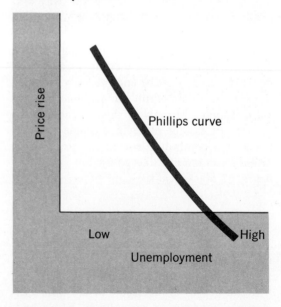

gested that we show the general relationship between higher wages and/or profits per unit and the degree of capacity utilization. As industries move toward higher rates of utilization, their costs per unit begin to mount more steeply. This is because they are operating in ever tighter markets, especially for labor. Hence Phillips has drawn a curve showing the over-all relationship between unemployment, a good indicator of capacity utilization, and the rise in prices. As Fig. 32·4 shows, the lower unemployment falls, the higher is the tendency toward inflation.

THE TRADE-OFF DILEMMA

The Phillips curve does more than point out a general relationship between employment and price rises. It also points up a fundamental dilemma. The dilemma is that we cannot choose one target for unemployment and another target for the price level *independent of each other*. We cannot, for example, "decide" to have 2 percent inflation and 2 percent unemployment, because we do not know how to reconcile low unemployment and low inflation rates. *We have to trade off unemployment against price stability.* The dilemma thus imposes a cruel choice. Governments must choose between alternatives, *both* of which are painful and costly. Before we investigate the means by which they seek to bring about whatever choice they finally make, let us inquire into the costs of these alternatives.

The Economic Costs of Unemployment

THE UNEMPLOYMENT OPTION

Suppose the government decides to lessen inflation by increasing the rate of unemployment. What does this cost?

We can begin with a fairly straightforward estimate of the losses that result from unemployment. For each percentage point of increase in the official unemployment figures, United States gross national product falls by about 3 percent, or roughly $40 billion. The reduction in GNP is much larger than the increase in officially measured unemployment because (as we have seen on pp. 321–22), the labor force and the hours worked per week both shrink as employment opportunities shrink.

Is this, then, the cost of choosing the unemployment "option"? Not quite. For the losses of output and income are by no means equally distributed. If increasing the unemployment rate from 4.5 to 5.5 percent meant that each person found himself unemployed for 14 instead of only 11 out of 250 working days, the income losses would be evenly shared among the labor force. On the other hand, if the additional one percent of unemployment meant that one percent of the working force was permanently unemployed all year, then the costs of unemployment would be entirely concentrated on this group.

If most people were asked which of the two ways of bearing the costs of unemployment were more equitable, they would probably choose the first. But in fact, the way unemployment is actually shared is closer to the second. Joblessness tends to concentrate in certain groups, especially in cyclical downswings. For instance, when the white unemployment rate rises by 1 percentage point, black unemployment rises by 2 percentage points. So, too, in 1971 when the average rate of unemployment for the labor force was more than 6 percent, the rate among married men was only 3.3 percent, whereas among young workers (16 to 19 years old) it was more than 15 percent. For white workers as a whole it was 4.5 percent; for Negro and Puerto Rican workers, 8.2 percent; among black teenagers it reached as high as 42 percent.

EASING THE COST OF UNEMPLOYMENT

To some extent, this concentrated cost of unemployment is offset by transfer payments made to some of the unemployed. Because these payments come from taxes that are mainly paid by employed persons, transfers help to some small degree to spread the burden. Unfortunately, unemployment compensation in the United States is far from generous. Although the systems differ widely among the states, in general they provide for payment of one-half of an unemployed person's income, up to some maximum weekly benefit that ranges from $105 in Washington, D.C., to $49 in Mississippi. There is also a limit on how many weeks' unemployment compensation can be claimed; usually 20 weeks. After that, there is welfare, which also varies from state to state.

OTHER COSTS

There are costs to unemployment other than those directly incurred by the unemployed. Discouraged workers who withdraw from the labor force do not count as "unemployed," as we know, but the incomes they "voluntarily" give up are also economic losses. These hidden unemployed are frequently housewives who have other means of support; but because they are not working full or part time, as they would if there were jobs available, their families will have fewer dollars to spend. Another source of loss is the decrease in overtime, often an important supplement to the earnings of blue-collar workers who are paid on an hourly basis.

Capitalists also bear some of the loss of unemployment. When the aggregate demand for goods and services falls, output declines too. As a result, profits fall; and since managers are usually reluctant to let overhead office staff go, profits may fall more rapidly

than output. As in the case of the labor force, the losses borne by firms are not evenly shared. Some industries, such as durable goods, tend to bear more than their share. Others, such as household staples, are relatively depression-proof.

The Economic Costs of Inflation

As we have seen, the costs of unemployment tend to be concentrated among certain groups or industries, rather than diffusely shared by all. Now, what of the costs of inflation? Are they also concentrated? Do they resemble the costs of unemployment?

INFLATION AND INCOME

Let us begin to answer this very important question by reviewing an important, familiar fact. It is our old identity between gross national product and gross national income. We recall that all costs of GNP must become the incomes of the factors of production. This gives us the knowledge that whenever the monetary value of GNP goes up, so must the monetary value of gross national income. This is true whether the increase in GNP results from increased output or higher prices, for in either case the factors of production must receive the monetary value of the output they have produced.

This provides a very important point of departure in comparing unemployment and inflation. *For the nation as a whole, inflation cannot decrease the total of incomes.* Here it differs sharply from unemployment. Whenever unemployment increases, there is a loss in potential GNP. But whatever the price at which GNP is sold, the real output of the nation remains the same—at least up to the point at which hyperinflation threatens to destroy the system itself.

A ZERO SUM GAME

Discussions of inflation tend to overlook this fact. They often speak of the losses that inflation brings to this or that group. Of course inflation *can* lower an individual's standard of living by raising the prices he or she must pay. *But since the total GNI is equal to GNP, one's person's loss must be transferred as a gain to another.* This in no way lessens the importance or even the dangers of inflations. But it makes clear that unlike unemployment, in which there are losers but no gainers, in inflation every loss is offset by an equal gain. We call this a zero sum game. What is important, then, is to try to weigh the benefits that inflation gives to some, against the losses it inflicts on others.

Who gains by inflation?

Speculators are one group. During an inflation everything does not rise by the same amount or at the same rate. Some items will shoot up; others lag behind, just as in a booming stock market not all stocks share alike in the rise. Speculators are those lucky or skillful individuals who have bought goods that enjoy the sharpest rises. This may be land, gold, or (in hyperinflations) food. Indeed, one of the reasons that hyperinflations result in a collapse of GNP is that more and more persons are *forced* to become speculators, seeking to sell or trade possessions for basic necessities. As individuals spend more and more time trading or scrounging, they begin to spend less and less time at their regular jobs. Thus output begins to fall, and we have the strange coexistence of an economic collapse and soaring prices.

MODERATE INFLATIONS

Hyperinflations are, fortunately, very rare. Nations have inflated as much as 200 or 300 percent a year without experiencing the breakdown and disorganization of a real runaway inflation.

In nonrunaway inflations, individuals will, of course, be gainers or losers, depending on whether their incomes go up faster or slower than prices. But we are interested in whether major *groups* can be classified as winners or losers. Usually one hears of three such groups: fixed-income receivers, labor, and capital. Let us look at each.

FIXED-INCOME RECEIVERS

By definition, anyone who lives on a fixed income, such as an annuity, must be a loser in inflation. Therefore retirees, as a group, should be badly penalized when prices continually rise. Curiously, they are not. Undoubtedly, there exist the much advertised widows and orphans living off small pensions, but they are probably a very small group. Most retirees live on (or depend largely on) Social Security, and this is quite another story. Social Security benefits have been periodically hiked up by Congress, so that a typical recipient in 1973 was well ahead of the game in terms of the purchasing power of benefits he received. Moreover, in 1973, Congress added to Social Security a cost-of-living escalator clause that automatically adjusted Social Security payments to compensate for cost-of-living increases. Welfare recipients have also had their benefits periodically increased, and they too, have not come out behind in the race since 1950.

LABOR AND CAPITAL

What about labor and capital? Table 32·2 gives us a breakdown of the income shares of strategic groups as a percentage of GNP for the last 20 years.

We must be cautious in reading this table, as we shall see, but it shows some im-

TABLE 32 · 2

Income Shares as Percentage of GNP

Year	Capital income Cash flow (corp. profits + depreciation) and interest income	Labor income (wages and other income)	Proprietors' income	Rental income	Transfer income
1950	19.5%	52.8%	13.1%	3.3%	5.3%
1960	19.3	56.1	9.1	3.1	5.6
1972	19.8	57.9	6.5	2.2	9.0

portant results. The first is that corporate cash flow income has barely changed during the last twenty-plus years. Many factors besides inflation influence corporate profits, but it is clear that inflation by itself has not been enough to increase profits.

Second, labor income has increased as a share of GNP. Here again a caution is needed: note the decline in proprietors' income in the third column. Much of that decline is the result of proprietors' giving up independent establishments to work for large enterprises. Their proprietors' income has become "wages." But if we add labor income and proprietors' income, we find that their combined share was just under 65.9 percent in 1950 and 64.4 percent in 1972—a trifling decline. Landlords' incomes have evidently fallen considerably as a share of GNP, as we would expect, since rents always lag behind rising prices. And transferees, as we have already seen, were strong gainers from the process.

WHY IS INFLATION SUCH A PROBLEM?

The figures raise an important question. The major single income-receiving group in the nation—wage and salary earners—has gained. The poorest stratum has gained. The corporate section—big business—has neither gained nor lost. *Why then the fuss about inflation?* Why is it not a popular economic process?

There are several plausible reasons. One is that *the losing groups, such as small proprietors and landlords, are politically influential and articulate and make their losses known to a much greater extent than the winners (labor or transferees) make their gains known.*

A second reason is that inflation is worrisome just because it affects everyone to some degree. Unlike the concentrated loss of unemployment, the diffuse gains-and-losses of inflation touch us all. *They give rise to the fear that the economy is "out of control"—which indeed, to a certain extent, it is.*

Third *is the lurking fear that moderate inflation may give way to a galloping inflation.* The vision of a hyperinflation is a specter that chills everyone, and not one to be lightly dismissed. There is always the chance, albeit a small one, that a panic wave may seize the country and that prices may skyrocket while the production of goods declines.

Fourth, *the effect of inflation on incomes is not always mirrored in its effects on assets.* Some assets, such as land, typically rise markedly during inflation because they are fixed in supply and because people, forgetting about the land booms and crashes of the past, want to get into something that has "solid value." But many middle-class families have their assets in savings banks or in government bonds, and they watch with dismay as the value of their savings declines. For as interest rates rise, bond prices fall, and

INFLATIONARY SCOREBOARD

James Kuhn of the Columbia School of Business has compiled an interesting scoreboard of winners during inflation. His main findings:

WAGE-EARNERS: average hourly earnings up 176 percent from 1950 to 1972; consumer prices up 74 percent. Between 1965 and 1972, wage earners' average hourly earnings were up 50 percent, although prices rose 33 percent during the period.

SALARIES (nonproduction workers, managers, technicians, etc): up 52 percent from 1965 to 1972, one-third more than prices.

SERVICE WORKERS: earnings up over 50 percent 1965–1972, again a third more than prices.

GOVERNMENT EMPLOYEES: earnings up 47 percent 1965–1971; prices up 28 percent in this period.

MORTGAGE HOLDERS (about one-half of all families): the median mortgage in 1971 was $18,000. Obviously, mortgage holders gain, because they enjoy rising incomes with which to make *fixed* monthly payments.

SOCIAL SECURITY RECIPIENTS: average monthly benefits up from $72 in 1950 to $132 in 1971 *in 1971 dollars*.

DIVIDEND PAYMENTS: roughly unchanged as a percentage of national income (2.8 percent).

---

stock prices do not typically stay abreast of the price level, although most people think they do. For instance, in late 1968 there was a stock market fall that brought stocks below their levels of 1966–1967; and again in early 1973, stock prices fell by a fifth, while consumer prices rose at record rates.

Last, *there is a kind of myopia that inflation produces. We blame inflation for reducing our real incomes by raising prices, but we do not give it credit for raising our incomes to pay those higher prices.* That is, for 364 days in the year we are aware of price rises as part of the inflationary process, but we do not connect the income increase we receive on the 365th day (if we get a yearly hike in pay) as being part of that selfsame process.

INFLATION VS. UNEMPLOYMENT AGAIN

Can we now compare the costs of unemployment and inflation? Three points stand out from our analysis.

1. *From the point of view of the economy as a whole, unemployment is more costly than moderate inflation.* That is because unemploy-

ment results in less production, whereas inflation does not.

2. *From the point of view of winners and losers, unemployment is also more costly than inflation.* For unlike inflation, unemployment is not a zero sum game. The losers in unemployment—the direct unemployed and the hidden unemployed—are not matched by winners who benefit from unemployment.

3. *The psychological impact of moderate inflation is probably much greater than that of moderate unemployment.* This is because we all feel affected by the inflationary process, whereas few of us worry too much about the unemployed—unless they happen to be ourselves.

Can we conclude from this that a nation should prefer inflation to unemployment—that a wise and humane policy would deliberately trade off a quite high degree of inflation in exchange for a low rate of unemployment? Some economists might recommend such a course, but it is doubtful that the nation as a whole would vote for it. Rightly or wrongly, most citizens feel that inflation is at least as great a danger as moderate unemployment, and they are not likely to support policies that reduce the joblessness of others in exchange for higher prices (or even higher prices and higher incomes!) for themselves.

The XYZ of Inflation

Until this point we have largely been concerned with conveying knowledge—what we know about inflation. Now we must change our tack. The rest of this chapter is mainly concerned with conveying ignorance. There is a great deal that we do not know about inflation, and a student should be as aware of our ignorance as of our knowledge.

THE ELUSIVE PHILLIPS CURVE

Let us begin with the Phillips curve. Consider Fig. 32·5.

The colored line shows that there is undoubtedly a *general* tendency for inflation to be inversely correlated with unemployment. When unemployment is *very* high, inflation is low, and vice versa. That much of the Phillips analysis remains. But now look at the scatter of dots within the tinted band representing the range from 3.5 to 6 percent unemployment. Compare 1964 and 1970, for instance. In 1964 an unemployment rate of just over 5 percent was associated with a rate

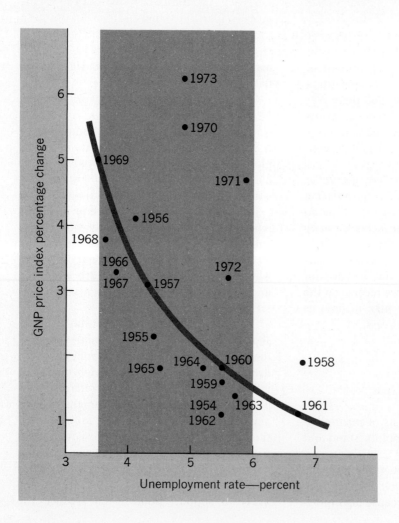

FIGURE 32 · 5
The unemployment–inflation relation

of inflation of less than 2 percent. In 1970 an unemployment rate of under 5 percent was associated with an inflation of over 5 percent.

What we have, in other words, is a relationship that is subject to such wide variations that economists have not been able to establish a function that enables them to *predict* what rate of unemployment will actually accompany a given inflation rate. This means that we cannot "target" a mix of inflation and unemployment, as our schematic diagram on p. 500 suggested, *because we do not know where the Phillips curve will be located.* Like a demand curve, it exists in our minds as a useful organizing concept, but unlike a demand curve, it is extremely difficult to locate this relationship in the real world.

A THEORY OF COINCIDENCES

Let us explore our ignorance from another angle. Consider the inflationary situation of 1973. We can easily describe many of its causes. There were poor crops and bad weather around the world. For some reason, anchovies did not appear off the coast of Peru. The Nixon administration sold a large fraction of the American grain crop to the Russians. There was a gasoline shortage because of a lack of refinery capacity and an Arab oil embargo. Lumber prices were up because foreigners were buying American lumber, made cheaper by the devaluation of the dollar. So the explanations go. Each and every price increase can be described in terms of shifts in demand and supply.

When we step back and look at the entire picture, however, this is no more than a theory of coincidences. Each price increase is caused by a particular event, not by the interrelated workings of the system. Weather, President Nixon's foreign policy, the vagaries of ocean currents, the gyrations of the foreign exchange market become the villains of the

piece. These descriptions may be correct, but they provide no means of avoiding or controlling inflation. Worse, they suggest that since inflation is caused by coincidences, we should run the economy at the highest possible employment rates, since the degree of inflation this will breed is largely in the lap of the gods.

Could this be the case? Economists are reluctant to admit it. To begin with, it would mean a declaration of *total* ignorance about inflation, and we do not feel totally ignorant. We are quite sure that running the economy at 2 percent unemployment would increase the rate of inflation, even though we do not know by how much. Second, we cannot quite believe that all the incidents actually add up to inflation. Are all nations all over the world having coincidences all the time? Why didn't they have similar coincidences in the past?

AN IMPASSE FOR THEORY

The net result is that economists do not have a satisfactory *theory* of inflation. We know a lot about various aspects of inflation and something about how to dampen it (as we will see in our next section). But we do not have a clear understanding of what causes inflation or how to cure it.

At the root of our problem is the question of expectations. We are all familiar with late afternoon commentators who tell us why the stock market went up or down as a result of a Presidential speech. Often these commentators have the text of the President's speech early that morning. Why don't they go on the air and tell us what the effect of the speech *will* be? The answer is that often neither they nor anybody else can predict how a statement will affect people's outlooks. Will an announcement of stiff price controls plunge businessmen into gloomy expectations of falling profits? Or will the same speech restore confidence and send the market up?

Our ignorance about the effect of pronouncements on the market is paralleled by our ignorance about the effects of actual developments on peoples' actions in the marketplace. Will an announcement that the cost-of-living index is up over last month send people on a buying spree, in anticipation of next month's expected increase? Or will it result in a tightening of budgets for the very same reason?

Thus, our inability to predict the course of inflation stems in large part from those very uncertainties about expectations we discussed in Chapter 4. Until we learn to anticipate expectations or to influence expectations, the movements of the marketplace—and of that gigantic marketplace called the economy—are likely to take us by surprise.

Controlling Inflation

Can we stop inflation? Of course—at a price. For example, wage and price controls with terrific penalties for noncompliance would surely dampen inflation. So would a decision not to increase the money supply at all. The trouble is that no one is willing to pay so great a price—the price of a police state in the first case, or of massive unemployment in the second. Therefore, we seek to control inflation with *politically acceptable* measures, and these have not been too successful.

Since World War II every major government has tried to keep the lid on inflation, but with scant results. In the United States, for example the Kennedy administration inaugurated the idea of "wage-price guidelines" that were intended to serve as an official index of productivity, to which it was hoped labor leaders would conform their demands. But the wage-price guidelines allowed for increases of only about 3.4 percent per year and were not very effective. Some unions whose productivity had gone up more rapidly than 3.4 percent claimed that they had the *right* to higher wages; other unions simply

ignored the guidelines; and most companies tried only halfheartedly to stick by them in wage bargaining sessions, because they counted on inflation itself to help them pass along any cost increases in higher prices.

Another tactic used by subsequent administrations was called "jawboning." This was simply the use of public pressure, usually through Presidential statements, to get companies (or less frequently unions) to scale down price increases. For instance in 1971, Bethlehem Steel suddenly announced that it was about to raise steel prices by 12 percent. Angry comments from the White House, coupled with stories carefully leaked to the newspapers about the possibility of relaxing steel import quotas, brought the welcome response of a much smaller (6 percent) increase from Bethlehem's main competitors. In the end, Bethlehem had to back down to the 6 percent mark.

WAGE AND PRICE CONTROLS

Still, 6 percent is a stiff price increase. Hence more and more pressure built up for the imposition of direct controls over wages and prices, which were finally instituted by the Nixon administration in 1971.

The problem with controls is that they are administratively clumsy, hard to enforce, and almost invariably evaded in one way or another. A wage and price freeze always catches someone at a disadvantage: a union that was just about to sign an advantageous but perhaps deserved contract, a store whose prices were at "sale" levels on the day that the freeze is announced, a business whose costs increase as a result of a rise in the prices of imported goods. Thus price controls lead to endless adjustment and adjudication. Moreover, unless there is a general sentiment of patriotic "pulling together" (as during World War II or the Korean war, when controls were fairly successful in repressing inflation), controls tend to lead to black or

ESCALATORS AND "INDEXING"

One way of mitigating the cost of inflation is to build escalator clauses into wage contracts, thereby guaranteeing workers that their wage rates will be automatically brought abreast of the cost-of-living index every six months or so. (This is also called "indexing" wages, since they are tied to a cost-of-living index.)

Such a policy clearly removes any loss of purchasing power for those who hold jobs, except for the brief period between adjustments. Then why not urge their widespread adoption?

Two main reasons are given by economists. One is that escalators would tend to be applied in big industries or government employments where contracts are negotiated and would not be applied in small businesses where employees are usually unorganized. Therefore it would result in a relative weakening of those whose wages were at the bottom of the scale.

The second objection is that built-in escalators would remove any incentive on the part of labor to bring the inflationary process to a halt. As long as labor's real wage was protected, what difference would it make whether inflation went faster or slower? In turn, this would remove any willingness to acquiesce in lower wage settlements or in wage guidelines or wage controls, to bring inflation to a halt.

Some economists would urge that "indexing" be applied much more broadly than in wage escalators. Professor Milton Friedman, for instance, would like to see us "index" savings accounts, bonds—indeed, all money contracts.

Such a policy would certainly mitigate the costs of inflation on fixed-money contractors, such as savings bank depositors. The worry, as with wage escalation, is that it might lower the last resistance to inflation, opening the door to accelerating price rises.

---

gray markets or to downgrading the quality of the "same" goods that are sold at fixed prices.

Not surprisingly, efforts to control wages tend to be more successful than efforts to control prices. This is because the government has an ally to help it enforce its wage regulations, namely, employers. Employers know what their employees make and have an interest in keeping wages within the legal limits. Consumers, on the other hand, usually do not know what products cost and cannot demand that stores refrain from marking up items. Renters are perhaps in the best position as consumers to enforce price controls on rents, but even they are in a weak position when they look for vacant apartments.

As a result, when controls are imposed, they tend to work fairly well for a short period of time on the wage side, while prices continue to creep upward. After a certain point, consumer resentment explodes in the form of demands from workers that their

wage restraints be loosened. In nation after nation, this process has been repeated.

INCOMES POLICY

There is, of course, an alternative. It is to curb inflation from the demand side rather than from the supply side, by imposing severe taxes that will bring the expansion of purchasing power to a halt. Once that is accomplished, the demand curve will stop its rightward movement, and a stable price level should be reached.

The problem here is obvious enough. Bringing the expansion of incomes to a halt means imposing severe taxes on the middle levels of American society where the bulk of purchasing power is concentrated. Unfortunately for anti-inflationary policy, increasing taxes on this group is political dynamite. Perhaps a truly effective tax increase could be imposed on the middle and upper middle class, if receivers of very large incomes were taxed at least as severely. This means closing the loopholes that have been stubbornly and successfully defended for many years.

It is a hard fact of political life that no government in any Western nation, including the most socially progressive such as Sweden, has succeeded in winning the political support necessary to impose a fully effective tax program. Perhaps people prefer inflation and relatively lower taxes to price stability and very high taxes. If this is the case, there is nothing the economist can do—except warn that inflation will continue.

STOP-GO

As a result of these partly economic, partly political difficulties, anti-inflationary policies here and abroad take on an aspect of "stop-go." When prices begin to rise too fast, remedial measures are put into effect. The money supply is tightened to curb spending. Governments trim their budgets. Taxes may be increased somewhat. Wage and price controls are put into effect.

For a time, these "stop" measures succeed. But then pressures mount in the opposite direction. High interest rates cut into home building. A slowdown in investment causes unemployment to rise. Tight government budgets mean that programs with important constituencies have to be cut back; Army bases are closed; social assistance programs abandoned. Businessmen chafe under controls. Hence pressures mount for a relaxation. The red light changes to green. The money supply goes up again; investment is encouraged; public spending resumes its former upward trend; controls are taken off. Before long the expected happens: prices begin to move ahead too rapidly once more, and the pendulum starts its swing in the opposite direction.

INFLATION AS A WAY OF LIFE

Does this mean that inflation has become a chronic fact of life, uncurable except at levels of unemployment that would be socially disastrous or by the imposition of severe and unpopular wage, price, and income controls?

Probably. What economists hope for is that a middle ground can be reached where the frictions generated by inflation and the real damage done by unemployment can be reduced to reasonable proportions. A price rise of 1 or 2 percent a year, for example, can be fairly easily tolerated, since a year-to-year increase in the quality of goods can be said to justify such an increase. Similarly, 3 or 4 percent unemployment (which includes voluntary unemployment and a hard core of "unemployables") is also socially acceptable, especially since both kinds of unemployment can be remedied by generous policies of unemployment compensation or by programs for retraining labor.

But it is one thing to announce such a goal and another to attain it. The simple fact remains that in no industrialized nation has anything like such an acceptable balance been achieved. In most nations, the claims of high employment quite properly take priority over those of inflation, and the rate of annual price increase has accordingly ranged from roughly 4 to 10 percent a year. As we saw at the beginning of this chapter, in this chronicle of inflationary failure, the record of the United States is by no means the worst. Unhappily, its record in combating unemployment, as we will see in our next chapter, is by no means the best.

A Problem of Affluence?

So we end our discussion of policy on much the same tentative and unsatisfying note as our discussion of theory. Yet perhaps it will help to put the unsolved problem of inflation into some perspective if we conclude with a few hypotheses of our own.

The critical fact about inflation, as we have said, is that it is a *process,* not a once-for-all jump in prices. This leads us to ask what changes might be sufficiently powerful and pervasive to bring such a process into being virtually around the world. Here are a few possibilities.

1. The shift to services

One striking fact that we notice in all industrialized nations is the movement of an ever larger fraction of their work forces into the service industries. In the U.S., as we have seen, over 60 percent of the labor force works in offices, shops, classrooms, municipal and state and federal buildings, producing the "services" that are ever more in demand in a highly urbanized, high-consumption society.

This movement has a powerful inflationary result. Productivity in the service sector tends to grow much more slowly than in the industrial or agricultural sectors. But wages in the service sectors tend to stay abreast of the wage levels set by the great industrial unions. Professor William Baumol has suggested that this major fact of life has inflationary repercussions that must be taken into account. For example, between 1962 and 1972 the consumer price index rose by 38 percent. This overall trend, however, concealed the fact that the price index of manufactured goods rose by only 27 percent, whereas the price index of services rose by 54 percent.

2. Increasing power in the marketplace

One of the most striking differences between modern inflations and those of the past is that in former days, inflationary peaks were followed by long, slow, deflationary declines, as prices and money wages both fell, particularly in times of recession. It was not uncustomary in the early years of this century for a large company to announce an across-the-board wage cut. Indeed, that happened frequently during the Great Depression. Certainly, prices frequently declined, partly as a result of technological advance, partly through the sporadic outbreak of cutthroat competition.

All that seems a part of the past beyond recall. Since World War II, most prices and wages in every nation have shown a "ratchet" tendency. They can go up, but they rarely or never come down. This characteristic is probably due to the increasing presence of oligopolistic market structures, to stronger trade unions, and to a business climate in which wage cuts and price wars are no longer regarded as legitimate economic policies. These changes may have salutary social consequences, but they undoubtedly add to our inflationary propensities.

3. The expansionist influence of governments

A third change is equally visible throughout the Western world and Japan. It is the much larger role played by the public sector in generating demand. This does not mean that government spending is inherently inflationary. We have seen that *any* kind of spending can send prices up once we reach an area of reasonably full employment. Rather, the presence of large government sectors and the knowledge that governments are dedicated to policies of economic growth affect expectations in an inflation-producing way.

In the old days, when governments were minor contributors to GNP, and when things such as deficit spending were unknown, businesses expected bad times as well as good and planned accordingly. Investment programs tended to be short range in character; and at the first sign of a storm, sails were furled. Thus, in the typical business cycle, when an upper turning point was reached, private spending in the business sector dwindled, pulling down incomes and lessening the pressure on prices.

That has also changed. Corporations, labor leaders, and the public now expect governments to prevent recessions. Accordingly, they no longer trim their sails at the first sight of trouble on the horizon. The willingness to maintain spending serves to set a floor under the economy, adding to the ratchet-like movement of incomes and prices.

4. The effects of affluence

A last suggestion is closely related to the previous ones. It is that the staying power of labor is vastly strengthened compared with prewar days. Only a generation ago, a strike was essentially limited by the meager savings of working families or the pittance of support that unions could offer. There was no unemployment compensation, no welfare, no large union treasuries. Today, strikes are backed by very substantial staying power, and both corporations and municipalities know it. Thus there is a tendency to settle for higher wages than would be granted if the employer felt that by waiting a few weeks he could enforce a better bargain.

Add to that a change in the expectations of the public that stem directly from a more affluent society. The strikes of teachers, garbage men, transportation workers, and other groups that have added their impetus to the inflationary surge of all industrial nations are very much a modern phenomenon, in that many of these groups of workers were formerly resigned to accepting low wages. A policeman in New York City, 1920, did not think that he had a "right" to earn as much as a worker in the Ford plant in Detroit; a policeman today sees no reason why he should not. In microeconomic language, wage contours are broadening (see p. 252). In an affluent society, where personal aspirations are encouraged and the constraints of poverty are lessened, the established "pecking order" of an old-fashioned society gives way to a free-for-all, in which each group tries to exploit its economic strength to the hilt. While this may be good for the group concerned and may lead to a more equitable distribution of income, it also lends its momentum to the forces that push our society toward a seemingly unstoppable inflation.

A LAST WORD

It seems then that we will have to live with inflation for a very long period. In fact, only two things seem capable of arresting it. One of these is a fearful economic collapse, comparable to the worldwide depression of the 1930s. This seems unlikely and is certainly undesirable. The other is the rise of a social consensus as to what an equitable distribution of income would be, whether imposed by direct income controls or by tax measures. This would surely be the ideal way of bringing inflation to a halt, but alas, it is an objective that has so far defied our efforts.

KEY WORDS

Inflation

Hyperinflation

Inflationary process

Demand pull/ cost push

Wage costs per hour and per unit

Phillips curve trade-off

Costs of unemployment

Inflation costs

Redistribution

Income groups

Inflationary fears

Myopia

What we do not know

CENTRAL CONCEPTS

1. Inflation is an ancient phenomenon, often associated with wars, but usually followed by periods of declining prices. The distinguishing feature of modern inflation is that *it has been experienced in all industrialized nations,* whether or not involved in war, and that it has not shown the traditional peaks and valleys.

2. *Hyperinflations* are episodes marked by a *flight from currency into goods.* They are periods in which the intangible trust in money as a store of value disappears. Only the issuance of a new monetary unit, in which faith can be placed, will restore price stability (see box, pp. 498–99).

3. Inflation differs from an ordinary increase in prices because it is not a movement from one point of equilibrium to another, but a *continuing process.*

4. This inflationary process may derive from a continuing increase in demand— "*demand pull*"; or it may derive from increases in cost—"*cost push.*"

5. Inflation is characterized by a tendency for *wage costs to outpace productivity* increases. We must be careful to distinguish wage costs per hour (which can be compensated by increased output per hour) from wage costs *per unit.*

6. The *Phillips* curve represents the general tendency of a fully employed economy to encounter increasing costs per unit. It shows us a general "*trade-off*" between *unemployment and inflation.*

7. This trade-off presents us with a difficult dilemma: comparing the costs of unemployment with those of inflation. *The costs of unemployment* are relatively straightforward. They consist in the *loss of income and output.* This loss of income is concentrated largely on the minority who are unable to find work. It can be eased by unemployment compensation, but even then, costs remain, such as the loss of income to the hidden unemployed who withdraw from the labor force, or the losses to certain industries from underproduction.

8. The economic costs of inflation are more complex. Because GNI ≡ GNP, inflation (which raises the money value of GNP) never lowers the income of the entire nation, as unemployment does. Instead, it *redistributes the increases in the value of GNP,* so that some people's real incomes rise and others fall. *Speculators* may gain substantially in inflations, if they succeed in buying goods that will rise in money value.

9. It is difficult to single out groups, rather than individuals, who win or lose in inflation. Fixed income recipients would of course be losers, but Social Security retirees have experienced increases in benefits that more than compensate for increases in prices. Labor as a whole has probably gained. Business profits have not grown.

10. Why is inflation so critical a political issue? The reasons seem to be these: (1) the losers include *politically articulate groups;* (2) *inflation is experienced by all,* not just by a minority; (3) we worry about the possibility of a *hyperinflation,* (4) the *value of assets,* particularly middle-class assets such as savings accounts or small investments, *may fall;* and (5) a *myopia* blinds us as to the effect of inflation on our incomes as well as on the prices we pay.

11. There is much we do not know about inflation. *The Phillips curve, for example, seems to shift around, so that it gives us no predictive power within the normal range of unemployment (3 to 7 percent).* Coincidences seem to affect many prices, but we have no unified theory of why these incidents all push prices up, not down. *In fact, we lack a coherent and cogent theory of the inflationary process.* This is intimately connected with our lack of understanding about the effect of expectations on behavior.

Expectations

Controls

Incomes policy

Stop-go

Chronic inflation

12. Efforts to control inflation have included "guidelines," admonitions ("jaw-boning"), and *wage and price controls*. None have been very successful. *Incomes policy*, through tax measures, have been difficult or impossible to impose. The result has been a *stop-go* pace, in which controls are applied and inflation slowed; and then as business weakens, controls are relaxed and output (and inflation) increase.

13. Inflation seems to be a chronic condition of contemporary industrialized societies. We suggest that the root causes may be found in the following attributes of these societies: (1) *a shift to service occupations*, with lower productivity combined with wage-setting by high-productivity industries; (2) *increasing power in the marketplace* for both corporations and unions; (3) *the expanded role of government*, which buoys expectations and encourages an expansionist attitude; and (4) a change in the staying power of labor and in its aspirations in *a more affluent society*.

QUESTIONS

1. Distinguish between a change in prices in an individual market and an inflationary change. What is meant by calling inflation a "process"?

2. What is meant by "demand pull"? By "cost push"? Can you have an inflationary *process* if costs are not rising?

3. What explanations can we give for the increase in demand? For the rise in costs?

4. What kind of event might give rise to a hyperinflation in the United States? Might a defeat in war trigger such an event? A victory? If we were to experience a runaway inflation, what measures would you counsel?

5. What is the importance of productivity in determining whether wage increases add to costs? Suppose that wages go up by 5 percent and that productivity goes up by 4 percent. Will there be an increase in costs per unit? Might this increase be absorbed by a fall in profits? Under what conditions is it likely to be passed on to consumers?

6. Is a war always inflationary? Does it depend on how it is financed? How should it be financed to minimize inflation? Does the same reasoning apply to an investment boom? (HINT: since corporations cannot tax, they must depend on savings for their expenditures.)

7. What is meant by the Phillips curve? Why is it a concept that is useful for a general understanding of the inflationary process? Why is it of little use for prediction?

8. Suppose that you could add up the costs of unemployment in terms of income lost. Suppose that you could add up the losses incurred just by those groups who are left behind in inflation (forget about the "winners"). Suppose further, that the losses imposed by inflation were greater than those imposed by unemployment. Does this mean that inflation is necessarily a worse economic disaster than unemployment? Must personal values enter into such a calculation? (If you simply compare amounts, is this also a value judgment?)

9. What measures can alleviate the costs of unemployment? The costs of inflation?

10. Why have measures to control inflation been so unsatisfactory? Why are price controls more difficult to monitor than wage controls?

11. Suppose that unemployment remains at about 5 percent and inflation rises to 10 percent a year. What measures would you propose?

12. What would you suggest as criteria for a "fair" income policy?

The Problem of 33 Unemployment

AGAIN AND AGAIN IN OUR STUDY OF MACRO-economics we have come up against the problem of unemployment. And no wonder. Unemployment bulks large in our studies, because it is a problem peculiarly associated with market systems. In traditional economies we do not encounter unemployment in the same form as we do in a market system. Peasants or nomadic tribesmen or serfs may be very unproductive, extremely poor, or reduced to idleness because their institutions exclude them from land or from the ownership of animals or other wealth, but unemployment in these societies is a static condition that results from existing institutions. We do not find periods of high and low unemployment in these societies (good and bad harvests perhaps excepted), but a more or less unchanged proportion of workers to the total population. So, too, in command societies, the labor force may be poorly allocated or underutilized, but here again, unemployment appears as a fault of the planning

mechanism, not as a result of the malfunction of the market machinery.

By way of contrast, the single most important social task that a market system entrusts to its "machinery" is that of finding acceptable work for the members of the society. It is also a task that the market does not perform altogether satisfactorily. From time to time, as we know, severe depressions have wracked market systems, causing unemployment to rise to heights that not only inflicted great damage on people who were unable to find work, but threatened the political and social stability of the system. In recent years, for example, one of the causes of racial unrest has been the failure of the market mechanism to provide jobs for the adolescent members of the labor force, especially among blacks, thereby contributing to the violence and unrest of the cities. Hanging over the politics of American and European societies are the memories of the trauma of the 1930s —a trauma in which mass unemployment

THE RECORD

The table (right) gives us important statistics. The terrible percentages of the Great Depression need no comment. Rather, let us pay heed to the level of unemployment in the 1960s. Here the record is mixed. During the early years of the decade we were troubled with persistent levels of unemployment much too high to be healthy. In the later years, unemployment declined sharply, but there is the discomfort of tracing much of this decline to an increase in war spending.

Part of our unemployment results from deliberate efforts made in 1970 by the Nixon administration to restrain inflation by "trading off" a limited rise in joblessness for a fall in the rate of price rise. As of the early 1970s, this trade-off was, unfortunately, more successful in producing unemployment than in slowing inflation.

Unemployment in the U.S.

Year	Unemployed (thousands)	Percent of civilian labor force
1929	1,550	3.2
1933	12,830	24.9
1940	8,120	14.6
1944	670	1.2
1960–65 av.	4,100	5.5
1966	2,875	3.8
1967	2,975	3.8
1968	2,817	3.6
1969	2,832	3.5
1970	4,085	4.9
1971	4,993	5.9
1972	4,840	5.6
1973	4,304	4.9

was a major factor in producing fascism in Europe and a crisis of faith in the United States.

Here we take one more look at the problem of unemployment—not from the viewpoint of aggregate demand, with which we are now familiar, but from another angle. We want to consider the question of *technological unemployment*—unemployment caused by the impact of machines. We want also to consider what remedies we have to combat unemployment, whatever the cause.

The Problem of Automation

For years, men have feared the effect of machinery on the demand for labor. The first-century Roman emperor Vespasian turned down a road-building machine, saying "I must have work for my poor." Shortly after Adam Smith's time, revolts of workers led by a mythical General Ludd smashed the hated and feared machines of the new textile manufacturers, which they believed to be

stealing the very bread from their mouths. To this very day, we call such antimachine attitudes "Luddite." David Ricardo, the great English economist of the early nineteenth century, shocked conventional opinion by adding a chapter "On Machinery" to the last edition of his famous *Principles of Political Economy,* in which he maintained that machines *could* displace labor, and therefore the capitalist's investment in equipment was not always in the interest of the working class.

In our own day, this lurking fear of machinery has focused on that extraordinary technology that "reads" and "hears" and "thinks" and puts human dexterity to shame. We call it automation. One of its characteristics is a feedback loop, by which the machine corrects itself, rather like a thermostat maintains a constant temperature. When such machines can make an engine block for a car almost without human supervision, it is understandable that thoughtful men should worry. There is a well-known story of Henry

Ford II showing his newly automated engine plant to the late Walter Reuther, the famed head of the United Auto Workers, and asking, "Well, Walter, how will you organize these machines?" Reuther replied, "How will you sell them cars?"

Does automation impose a wholly new and dangerous threat to employment? Let us begin to answer the question by using the tried and true method of supply and demand.

MACHINES AND SUPPLY

What has the introduction of machinery done to the supply of labor? We have already answered that question in part in Chapter 20. The effect of machinery has been vastly to increase the *productivity* of labor: a man with a tractor is incomparably more productive than a man with a shovel, not to speak of one with his bare hands.

But the effect of capital on productivity has not been evenly distributed among all parts of the labor force. On the contrary, one of the most striking characteristics of technology has been its *uneven entry* into production. In some sectors, such as agriculture, the effects of technology on output have been startling. Between 1880 and today, for instance, the time required to harvest an acre of wheat on the Great Plains has fallen from twenty hours to three. Between the late 1930s and the mid-1960s, the manhours needed to obtain a hundredweight of milk were slashed from 3.4 to 0.9; a hundredweight of chickens from 8.5 to 0.6.

Not quite so dramatic but also far-reaching in their effect have been technological improvements in other areas.

Table 33·1 shows the increase in productivity in five important branches of mining and manufacturing over the last two decades.

By way of contrast to the very great degree of technological advance in the primary and secondary sectors, we must note again

TABLE 33 · 1

Index of Output per Man-Hour for Production Workers

Industry	1950	1970
Coal mining	100	274
Railroad transportation	100	272
Steel	100	141
Paper and pulp	100	201
Petroleum refining	100	292

the laggard advance in productivity in the tertiary sector of activity. Output per man-hour in trade, for instance, or in education or in the service professions such as law or medicine or, again, in domestic or personal services such as barbering or repair work or in government has not increased nearly so much as in the primary and secondary sectors.

In the previous chapter, we noticed the inflationary consequences of this lag. Now we note the uneven entry of technology to be one main cause behind the overall migration of employment we observed in Chapter 20, (p. 308). Had we not enjoyed the enormous technical improvements in agriculture or mass production, but instead discovered vastly superior techniques of government services (in the sense of increasing the man-hour output of, say, policemen or firemen or teachers), the distribution of employment might look very different.

THE INFLUENCE OF DEMAND

These strikingly different rates of increase in productivity begin to suggest a way of analyzing the effects of automation, or for that matter, any kind of labor-saving machinery. Clearly, capital equipment increases the *potential output* of a given number of workers. But will increased output be ab-

AUTOMATION AT WORK

Some years ago one of the authors of this book, visiting the New York Federal Reserve Bank, saw automation in a striking application. One of the floors of the immense Renaissance palace that is the home of the New York Fed is devoted to sorting checks that its member banks send there for deposit, so that these checks can be returned to the banks on which they were drawn.

On one entire side of this block-long floor, hundreds of women sitting at desks used a foot mechanism to operate circular files, into which, with lightning rapidity, they sorted the checks in front of them. On the other side of the building was a single long machine, down which checks were whisked on a kind of belt mechanism. As the checks flew past, the magnetic ink numerals on each check were "read" by a machine, and the check was accordingly dropped into the appropriate slot as it traveled along the machine, much as an IBM card sorter works.

That one electronic check-reading machine performed as many "sorts" per hour as the small army of semimechanized check sorters on the other side of the aisle.

sorbed through expanded demand, allowing the workers to keep their jobs? Let us look at the question first in its broadest scope. We have seen that productivity has increased fastest in agriculture, next in manufacturing, least in services. What has happened to demand for the output of these sectors?

We can see this shift in demand in Table 33·2.

TABLE 33 · 2
Domestic Demand for Output

	PERCENTAGE DISTRIBUTION OF DEMAND*	
	1899–1908	1972
Primary sector (agriculture)	16.7%	3.2%
Secondary sector (mining, construction, manufacturing)	26.0	41.0
Tertiary sector (transportation, communication, government, other services)	57.2	55.7

*The figures actually show the *income originating in these sectors*. But we know from our macroeconomic studies that this income must have been produced by the expenditures—the demand—of society as it was distributed among the sectors.

What we see here is a shift working in a direction different from that of supply. As the productivity and potential output of the agricultural sector has risen, the demand for agricultural products has not followed suit but has lagged far behind. Demand has risen markedly for output coming from the secondary sector, but has remained roughly unchanged, in percentage terms, for the output of the tertiary service sector.

THE SQUEEZE ON EMPLOYMENT

If we now put together the forces of supply and demand, it is easy to understand what has happened to employment. The tremendous increase in productivity on the farm, faced with a shrinking proportionate demand for food, created a vast army of redundant labor in agriculture. Where did the labor go? It followed the route indicated by the growth of demand, migrating from the countryside into factory towns and cities where it found employment in manufacturing and service occupations. As Table 33·3 shows, the distribution of employment has steadily moved out of the primary and secondary, into the tertiary sector.

What we see here is the crucial role of technology in distributing employment among its various uses. As income rose,

TABLE 33 · 3
Distribution of Employment

	PERCENT DISTRIBUTION OF ALL EMPLOYED WORKERS	
	1900	*1972*
Primary sector	38.1%	4.6%
Secondary sector	37.7	32.0
Tertiary sector	24.2	63.4

purchasing power no longer used for food was diverted to manufactured goods and homes. We would therefore expect that the proportion of the labor force employed in these pursuits would have risen rapidly. Instead, as we can see, it has fallen slightly. This is because technological improvements entered the secondary sector along with manpower, greatly increasing the productivity of workers in this area. Therefore a smaller portion of the national labor force could satisfy the larger proportional demands of the public for output from this sector.

Most important of all is the service sector. As Table 33·2 shows us, the public has not much changed the share of income that it spends for the various outputs we call services. But technological advances have not exerted their leverage as dramatically here as elsewhere, so that it takes a much larger fraction of the work force to produce the services we demand.

IMPORTANCE OF THE TERTIARY SECTOR

The conclusion, then, is that the demand for labor reflects the interplay of technology (which exerts differing leverages on different industries and occupations at different times) and of the changing demand for goods and services. Typically, the entrance of technology into industry has a twofold effect. The first is to raise the *potential* output of the industry, with its present labor force. The second is to enable the costs of the industry to decline, or its quality to improve, so that actual demand for the product will increase. But normally, the rise in demand is not great enough to enable the existing labor force to be retained along with the new techniques. Instead, some labor must now find its employment elsewhere.

There are exceptions, of course. A great new industry, such as the automobile industry in the 1920s, will keep on expanding its labor force despite improved technology, for in such cases demand *is* sufficiently strong to absorb the output of the new technology, even with a growing labor force. Then too, there is the exception of capital-saving technology, making it possible for an industry to turn out the same product with a cheaper capital equipment, thereby making it attractive to expand production and to hire more labor.

But taking all industries and all technological changes together, the net result is unambiguous. As Table 33·4 reveals, technology has steadily increased our ability to create goods, both on the farm and on the factory floor, more rapidly than we have wished to consume them, with the result that employment in these areas has lagged behind output.

TABLE 33 · 4
Output and Employment Indices

	1950	*1971*
Agricultural output	100	131
Agricultural employment	100	44
Manufacturing output	100	205
Manufacturing employment	100	120
1950 = 100		

THE SERVICE SECTOR

Forty percent of all consumer spending, roughly a quarter of GNP, consists of purchases of "services." We have already seen (p. 48) that a service is a utility that we purchase without buying the actual physical object from which it flows (a movie admission, not a movie film and a projector; a subway ride, not the subway).

Here it is worthwhile stressing that services are not, as we often think, just the personal ministrations of individuals, such as lawyers, fortune-tellers, accountants, and others. Many services that we buy are heavily capital-intensive; that is, use large quantities of capital per worker. To take the two examples above, movie theaters require large capital investments; subways certainly do. Wholesale and retail trade are also services, but both require large investments in land and plant; electricity and steam are services that flow from vast capital agglomerations; telephone calls are services that utilize billions of dollars worth of capital. Even travel needs aircraft, ships, hotels.

The point is that we must differentiate, when we speak of services, between those services that require large capital investments and those that do not. Strictly speaking, the division between goods and services is of less interest, from the point of view of employment, than the division between labor-intensive and nonlabor-intensive occupations, but we do not have the information we need to analyze the latter.

Note how agricultural output has increased rapidly in this period, while agricultural employment has shrunk by over 50 percent; and notice that whereas manufacturing output has more than doubled, employment in manufacturing is up by only 20 percent.

During this same period, however, our total civilian labor force increased by over 20 million. Where did these millions find employment? As we would expect, largely in the service sector. Figures for employment in various parts of the service sector appear in Table 33·5. We might note that comparable shifts from agriculture "through" manufacturing into services are visible in all industrial nations.

THE IMPACT OF INVENTION ON COST

Our discussion of the forces of supply and demand enables us now to look more closely into the effect of introducing an individual labor-saving invention. To be sure, not all inventions by any manner of means are labor-saving. A profit-maximizing entrepreneur will not introduce a technical change into his process of production unless it lowers cost, but this cost can be capital-saving or land-saving as well as labor-saving. Fertilizer is land-saving, for example; miniaturization is often capital-saving; the oxygen process of making steel is labor-using but cost-cutting (i.e., it is relatively capital-saving).

TABLE 33 · 5
Service Employments 1950, 1972

	1950	1972	Percent increase 1950–1972
	(in millions)		
Trade	9.4	15.8	68
Services	5.4	12.3	128
Government	6.0	13.3	122
Finance and other	1.9	3.9	105
Total tertiary sector	22.7	45.3	100

LABOR-SAVING INVENTIONS

But the cost-cutting inventions that interest us here are *labor-saving* inventions or innovations, changes in technique or technology that enable an entrepreneur to turn out the same output as before, with less labor, or a larger output than before, with the same amount of labor.

Do such inventions "permanently" displace labor? Let us trace an imaginary instance and find out.

We assume in this case that an inventor has perfected a technique that makes it possible for a local shoe factory to reduce its production force from ten men to eight men, while still turning out the same number of shoes. Forgetting for the moment about the possible stimulatory effects of buying a new labor-saving machine,* let us see what happens to purchasing power and employment if the shoe manufacturer simply goes on selling the same number of shoes at the same prices as before, utilizing the new lower-cost process to increase his profits.

Suppose our manufacturer now spends his increased profits in increased consumption. Will that bring an equivalent increase in the total spending of the community? If we think twice we can see why not. For the increased spending of the manufacturer will be offset to a large extent by the decreased spending of the two displaced workers.

Exactly the same conclusion follows if the entrepreneur used his cost-cutting invention to lower the price of shoes, in the hope of snaring a larger market. Now it is *consumers* who are given an increase in purchasing power equivalent to the cut in prices. But again, their gain is exactly balanced by

---

*This is not an unfair assumption. The labor-saving technology might be no more than a more effective arrangement of labor within the existing plant, and thus require no new equipment; or the new equipment might be bought with regular capital replacement funds.

the lost purchasing power of the displaced workers.

INCOMES VS. EMPLOYMENT

Thus we can see that the introduction of labor-saving machinery does not necessarily imperil *incomes;* it merely shifts purchasing power from previously employed workers into the hands of consumers or into profits. But note also that *the unchanged volume of incomes is now associated with a smaller volume of employment. Thus the fact that there is no purchasing power "lost" when a labor-saving machine is introduced does not mean that there is no employment lost.*

Is this the end to our analysis of labor-displacing technology? It can be. It is possible that the introduction of labor-saving machinery will have no effect other than that of the example above: transferring consumer spending from previously employed labor to consumers or to entrepreneurs. But it is also possible that an employment-generating secondary effect may result. The entrepreneur may be so encouraged at the higher profits from his new process that he uses his profits to invest in additional plant and equipment and thereby sets in motion, via the multiplier, a rise in total expenditure sufficient to reemploy his displaced workers. Or in our second instance, consumers may evidence such a brisk demand for shoes at lower prices that, once again, our employer is encouraged to invest in additional plant and equipment, with the same salutary results as above. Do not fall into the trap of thinking that the new higher demand for shoes, will, *by itself,* suffice to eradicate unemployment. To be sure, shoe purchases may now increase to previous levels or even higher. But unless their incomes rise, consumer spending on other items will suffer to the exact degree that spending on shoes gains.

The moral is clear. *Labor-saving technology can offset the unemployment created by its immediate introduction only if it induces sufficient investment to increase the volume of total spending to a point where unemployment is fully absorbed.*

IMPACT OF AUTOMATION

It is in connection with our foregoing discussion that the much talked-of threat of *automation* becomes most meaningful. In the main, automation is clearly a cost-cutting and labor-saving kind of technology, although it has important applications for the creation of new products as well.

But one aspect of automation requires our special attention. It is the fact that automation represents the belated entry of technology into an area of economic activity that until now has been largely spared the impact of technical change. This is the area of service and administrative tasks that we have previously marked as an important source of growing employment. Thus the threat inherent in the new sensory, almost humanoid, equipment is not only that it may accelerate the employment-saving effects in the secondary (manufacturing) sector. *More sobering is that it may put an end to the traditional employment-absorptive effects of the tertiary service and administrative sector.*

What could be the implications of such a development? In simplest terms, it means that in the future, fewer people would be needed to produce the same quantity of services. The absorption of labor from agriculture and manufacturing into the ever-expanding service sector would now slow down or come to a halt, since the service sector could increase its output without hiring a proportionate increase in workers.

This *could,* of course, mean massive unemployment. But it need not. Just as our imaginary society limited its demand to agri-cultural goods and solved its labor problem by cutting the workweek, so a society that no longer needed to add labor as fast as its demands rose could easily solve the unemployment problem by more or less equitably sharing among its members the amount of labor it *did* require. To be sure, this raises many problems, not least among them the wage adjustments that must accompany such a reapportionment of hours. But it makes clear that, essentially, the challenge of automation is one of finding a new balance in our attitudes toward work, leisure, and goods, and an equitable means of sharing work (and income, the reward for work).

NEW DEMANDS

But now let us suppose that an inventor patents a new product—let us say a stove that automatically cooks things to perfection. Will such an invention create employment?

We will suppose that our inventor assembles his original models himself and peddles them in local stores, and we will ignore the small increase in spending (and perhaps in employment) due to his orders for raw materials. Instead, let us fasten our attention on the consumer who first decides to buy the new product in a store, because it has stimulated his demand.

Will the consumer's purchase result in a *net* increase in consumer spending in the economy? If this is so—and if the new product is generally liked—it is easy to see how the new product could result in sizeable additional employment.

But will it be true? Our consumer has, to be sure, bought a new item. *But unless his income has increased, there is no reason to believe that this is a net addition to his consumption expenditures.* The chances are, rather, that this unforeseen expenditure will be balanced by lessened spending for some other item. Almost surely he will not buy a regular stove. (When consumers first began buying television sets, they stopped buying as many radios and going to the movies as often.) But even where there is no direct competition, where the product is quite "new," everything that we know about the stability of the propensity to consume schedule leads us to believe that *total* consumer spending will not rise.

Thus we reach the important conclusion that new products do not automatically create *additional* spending, even though they may mobilize consumer demand for themselves. Indeed, many new products emerge onto the market every year and merely shoulder old products off. Must we then conclude that demand-creating inventions do not affect employment?

EMPLOYMENT AND INVESTMENT

We are by no means ready to jump to that conclusion. Rather, what we have seen enables us to understand that if a new product is to create employment, it must give rise to new *investment* (and to the consumption it induces in turn). If the automatic stove is successful, it may induce the inventor to borrow money from a bank and to build a plant to mass-produce the item. If consumer demand for it continues to rise, a very large factory may have to be built to accommodate demand. As a result of the investment expenditures on the new plant, GNP rises, consumers' incomes rise, and more employment will be created as they spend their incomes on various consumer items.

To be sure, investment will decline in those areas that are now selling less to consumers. At most, however, this decline can affect only their replacement expenditures, which probably averaged 5 to 10 percent of the value of their capital equipment. Meanwhile, in the new industry, an entire capital structure must be built from scratch. We can expect the total amount of investment spending to increase substantially, with its usual repercussive effects.

When we think of a new product not in terms of a household gadget but in terms of the automobile, airplane, or perhaps the transistor, we can understand how large the employment-creating potential of certain kinds of inventions can be. Originally the automobile merely resulted in consumer spending being diverted from buggies; the airplane merely cut into railroad income; the transistor, into vacuum tubes. But each of these inventions became in time the source of enormous investment expenditures. The automobile not only gave us the huge auto plants in Detroit, but indirectly brought into being multibillion-dollar investment in highways, gasoline refineries, service stations, tourism—all industries whose impact on employment has been gigantic. On a smaller, but still very large scale, the airplane gave rise not alone to huge aircraft building plants, but to airfields, radio and beacon equipment industries, international tourism, etc., whose employment totals are substantial. In turn, the transistor offered entirely new design possibilities for miniaturization and thus gave many businesses an impetus for expansion.

INDUSTRY-BUILDING INVENTIONS

What sorts of inventions have this industry-building capacity? We can perhaps generalize by describing them as inventions that are of sufficient importance to become "indispensable" to the consumer or the manufacturer, and of sufficient mechanical or physical variance from the existing technical environment to necessitate the creation of a large amount of supporting capital equipment to integrate them into economic life.

Demand-creating inventions, then, can indeed create employment. *They do so in-*

*directly, however—not by inducing new consumer spending, but by generating new investment spending.**

Unfortunately, there is no guarantee that these highly employment-generative inventions will come along precisely when they are needed. There have been long periods when the economy has not been adequately stimulated by this type of invention and when employment has lagged as a result.

AUTOMATION AND EMPLOYMENT

Will the technology of automation be industry-building or labor-saving? *We do not know.* It is possible that the computer, the transistor, the myriad new possibilities in feed-back engineering will play the same role as the automobile and the railroad, not only giving rise to an enormous flow of investment, but opening new fields of endeavor for other new industries that will also expand. If this is the case, the demand for labor will grow fast enough to match the increase in the productivity of labor, and there will be nothing unusual to worry about.

But it is also possible that the labor-saving impact of automation will make itself felt primarily in the tertiary sector—in government, retail and wholesale trade, in banking and finance, rather than in manufacturing or agriculture. If so, as we have seen, it will introduce a hitherto unfelt pressure on employment in that sector, "freeing" labor to seek employment elsewhere.

Where? The answer must be in some fast-growing sector of industry. But which one? The employment-generating effects of

---

*We should mention one effect of demand-creating inventions on consumption. It is probable that without the steady emergence of new products, the long-run propensity to consume would decline instead of remaining constant, as we have seen in Chapter 24. In this way, demand-creating technology is directly responsible for the creation of employment, by helping to keep consumer spending higher than it would be without a flow of new products.

the growth of private industry have been uneven: between 1950 and 1960, for example, new jobs in the private sector accounted for only one out of every ten new jobs in the nation; from 1960 to 1972 the private sector supplied 7 out of 10 jobs. In both cases, the rest were accounted for in the public or not-for-profit sector (private schools and universities, hospitals and the like). Thus, if private industry performs in the future as poorly as in the 1950s, automation *could* present us with a very serious problem. The question is: what could we do about it?

Combating Unemployment

This question leads us to consider the whole problem of combating unemployment. Because unemployment comes from more than one source, there is more than one answer to the problem. There is also less than one answer, because, as we shall see, it is easier to prescribe the remedies for unemployment than to apply them.

1. Increasing demand

We have learned that as a general rule, anything that increases the total demand of society is apt to increase employment. This is particularly true when unemployment tends to be widespread, both in geographic location and industrial distribution. *The expansion of GNP, whether by the stimulation of private investment or consumer spending or government expenditure or net exports is generally the single most reliable means of creating more employment.*

There is, however, a problem here. Doubtless, a vast amount of employment could be created if aggregate demand were enlarged; for instance, the systematic reconstruction of our cities, a task that is becoming an increasingly pressing necessity, could by itself provide millions of jobs for decades. So could the proper care of our rapidly-growing older population, or the provision of really first-class education for large segments of the population that lack it.

The problem is that these programs require the generation of large amounts of *public demand,* and this requires the prior political approval of the electorate. If this political approval is not forthcoming, the generation of additional demand will have to be entrusted to the private sector—that is, to individual entrepreneurs in search of a profit.

Can private enterprise, without a massive public investment program, generate sufficient demand to bring about full and lasting employment? One cannot be dogmatic about such questions, but, as we have seen, there is at least some historical reason for caution in assuming that the private sector, unaided by public programs of investment, will be able to offer as much employment as the growing labor force demands. *The likelihood is that a large enough aggregate demand will require the substantial use of public expenditure, whether for urban renewal and welfare services or for other ends.*

2. Wage policy

But suppose that a very high aggregate demand is maintained, through public or private spending. Will that in itself guarantee full employment?

The answer is that it will not if the spending creates only higher incomes for workers who are already employed, rather that new incomes for workers who are unemployed. This is because much of the increase in wages is apt to be saved, in the short run, rather than spent. Thus if we want to maximize the employment-creating effect of spending, whether private or public, *we need to hold back wage raises at least until the unemployment has fallen to a socially acceptable level.* (This applies particularly to unemployment among groups such as teen-agers.)

But if raising wages can impede the process of job creation, can cutting wages encourage it? The question is not a simple one, for lower wages set into motion contrary economic stimuli. On the one hand, lower wages cut costs and thereby tempt employers to add to their labor force. On the other hand, lower wages after a time will result in less consumption spending, and will thus adversely affect business sales. The net effect of a wage cut thus becomes highly unpredictable. If businessmen feel the positive gains of a cut in costs before they feel the adverse effects of a cut in sales, employment may rise —and thereby obviate the fall in consumption spending. On the other hand, employers may *expect* that the wage cut will lead to lower sales, and their pessimistic expectations may lead them to refrain from adding to their labor forces, despite lower costs. In that case, of course, employment will not rise. On balance, most economists today fear the adverse demand effects of wage-cutting more than they welcome the possible job-creating effects.

It seems, then, that maintaining wages in the face of an economic decline and restraining wage rises in the face of an economic advance is the best way of encouraging maximum employment. It is one thing, however, to spell out such a general guideline to action and another to achieve it. *To maintain wages against an undertow of falling sales requires a strong union movement. But once times improve, this same union movement is hardly likely to exercise the self-restraint needed to forego wage raises, so that additional spending can go into the pockets of the previously unemployed.* This poses another dilemma for a market society in search of a rational high employment policy, and there is at this moment no solution in sight.

3. Remedying structural unemployment

Not all unemployment is due to insufficient demand. Some can be traced to "structural" causes—to a lack of "fit" between the existing labor force and the existing job opportunities. For instance, men may be unem-

THE GHETTO SKILL MIX

A sad example of the lack of fit between the skills demanded by employers and those possessed by the labor force is to be found in the ghetto, where typically the labor force is badly undertrained. A recent study by the First National City Bank shows this situation in New York City.

In only one category—unskilled service—was the prospective demand for labor roughly in line with the skills available. This meant a reasonable employment prospect for maids, restaurant workers, bellhops, and the like—among the lowest-paid occupations in the nation. As for the common laborer, who comprised over half the "skill pool" of the New York ghetto, his outlook was bleak indeed—less than one percent of new jobs would open in that area. Conversely, for the widest job

	Occupational distribution of ghetto unemployed, 1968	Estimated job openings, 1965–1975
White collar	13.6%	65.7%
Craftsmen	2.8	7.4
Operatives	14.7	7.7
Unskilled personal service	16.6	18.6
Laborers	52.3	0.6

market in the city—the white-collar trades that offer two-thirds of the new jobs—the ghetto could offer only one-seventh of its residents as adequately trained. If these figures have any meaning, it is that ghetto poverty is here to stay, short of a herculean effort to rescue the trapped ghetto resident.

ployed because they do not know of job opportunities in another city, or because they do not have the requisite skills to get or hold jobs that are currently being offered. Indeed, it is perfectly possible to have structural unemployment side-by-side with a lack of manpower in certain fields.

One particularly important kind of structural unemployment comes from *labor-displacing* (as compared with labor-saving) inventions. Inventions can radically alter the specific skills demanded from the labor force. The labor-saving invention of our hypothetical shoe factory, for instance, might have been a mechanical cutter operated by one man who replaced three former hand cutters. Perhaps one of the hand cutters could be retrained to operate the mechanical cutter, but it is often easier and cheaper to hire a new, young employee and to let the former employees go. In this case, their skills have become obsolete. Even if shoe sales increase, the laid-off employees will not be put back to work at their former tasks. Such workers may find themselves permanently unem-

ployed or reduced to lower incomes because they are forced to accept unskilled work.

A sharp debate has raged in the United States concerning the importance of structural reasons (as contrasted with a general deficiency of demand) in accounting for the present level of unemployment. Many observers have pointed out that the unemployed are typically grouped into certain disprivileged categories: race, sex, age, lack of training, and unfortunate geographic location. The aged and the young, the black and the unskilled, and the Massachusetts textile worker are not quickly pulled into employment by a general expansion of demand. The broad stream of purchasing power passes most of them by and does not reintegrate them into the mainstream of the economy. Hence stress is increasingly placed on measures to assist labor mobility, so that the unemployed can move from distressed to expansive areas, and on the retraining of men for those jobs offered by a technologically fast-moving society.

Retraining is, unfortunately, much easier

when it is applied to relatively few persons than when it is proposed as a general public policy affecting large numbers of unemployed. Then the question arises: for what jobs shall the unemployed be trained? *Unless we very clearly know the shape of future demand, the risk is that a retraining program will prepare workers for jobs that may no longer exist when the workers are ready for them.* And unless the *level* of future demand is high, even a foresighted program will not effectively solve the unemployment problem.

Most economists would suggest a combination of measures to combat structural unemployment. One of them is *a much more effective job-finding system* than we now have —a computerized "job bank" has been proposed. Another is *a more generous program of unemployment benefits* to give people time to look for work they want, rather than force them to settle for work they can do. *Training schools* are very valuable, especially if they are connected with a program of *public demand,* so that trainees know that jobs exist. A recent survey made by the Urban Coalition in 34 cities has estimated that over a million potential new jobs exist in the public sector— in education, welfare administration, environment control, library services, police and fire protection, hospital and sanitation improvement, etc.

4. Reducing the supply of labor

Next, the possibility exists of attacking unemployment not from the demand side, but from the supply side, by cutting the workweek, lengthening vacations, and using similar measures. Essentially, the possibility held out by shortening the workweek is that a more or less fixed quantity of work will then be shared among a larger number of workers. This is entirely feasible and possible, provided that *the decrease in hours is not offset by an increase in hourly pay rates.* In other words, once again a rational wage policy holds the key between success and failure. Shorter hours, coupled with higher hourly wage rates, will merely raise unit costs (unless productivity rises quickly enough to compensate). This will certainly not contribute to increased employment. Shorter hours *without* increased hourly rates, on the other hand, may make it necessary for the employer to hire additional help in order to continue his established level of output.

Shortening hours of work can be a policy of despair. If people do not wish to change their working habits—the number of hours per week or the number of years in their lifetimes—then the cure for unemployment is surely to expand the demand for labor and not to diminish its supply. If private demand is inadequate to this task, then, as we have said, public demand may serve the purpose instead.

But an attack on unemployment that seeks to reduce the supply of labor, rather than to expand the demand for it, need not be a program of retreat. It can also become part of a deliberate and popularly endorsed effort to reshape the patterns and the duration of work as it now exists. Thus it may be possible to reduce the size of the labor force by measures such as subsidies that would induce younger people to remain longer in school or by raising Social Security to make it attractive for older people to retire earlier. Such policies can be useful not only in bringing down the participation rate and thus reducing "unemployment," but in affecting changes in the quality of life that would find general public approval.

5. The government as employer of last resort

Finally, we can effectively limit the problem of unemployment by formally adopting the policy of using the public sector as "employer of last resort."

This proposal, first advanced by the National Commission on Technology, Auto-

THE IMPORTANCE OF BEING THE RIGHT AGE

A special kind of structural unemployment arises because the age composition of the labor force changes, sometimes flooding the market with young untrained workers, sometimes with older workers. Take the group aged 14 to 24. This includes those who are finishing their educations, as well as those who have finished and are entering the work force. The "cohort" as a whole increased in numbers by roughly 8 to 10 percent from decade to decade in the period 1890 to 1960.

Then in the 1960s an explosion occurred. The so-called baby boom in the years immediately following World War II began to enter these age ranks. In the decade of the 1960s, the 14-to-24-year-old group increased by *52 percent*. In the 1970s it will increase by a "normal" 11 percent; in the 1980s it will *decline* by 8 percent. We can confidently predict these changes, because the members of this age group are already born.

Beginning in mid1980s, however, the rate of growth of the labor force will be very slow, except for women. Job prospects should then be very bright.

---

mation, and Economic Progress,[1] does not mean the use of the public sector as a kind of vast "work relief" program. On the contrary, the purpose is to use public employment primarily as a positive force in establishing a higher quality of life, an objective that will require a substantial expansion of much public service activity along the lines mentioned above in the proposals of the Urban Coalition. Only as a secondary objective is "the employer of last resort" to provide jobs for those who fail to find satisfactory work in the private sector or in the current career lines of public services.

THE U.S. AND EUROPEAN EXPERIENCE

As evidence of what can be done by a more vigorous attack on unemployment, using many of the above techniques, the performance of the United States in recent fairly "low unemployment" years can be compared with that of various European nations.

The lesson of Table 33·6 is clear. A tendency to generate unemployment may be an unavoidable consequence of a market sys-

[1] *Technology and the American Economy* (Washington, 1966), p. 37.

TABLE 33 · 6
Unemployment Rates, Selected Countries,* 1971

United States	5.9%
West Germany	0.7
France	2.1
Netherlands	1.4
United Kingdom	2.8
Norway	(1970) 0.8
Sweden	2.5
Denmark	1.1

*Definitions of unemployment vary slightly between the United States and European nations, but not enough to affect the basic conclusions revealed by the data.

tem, *but it is not an irremediable evil. It is possible to run a capitalist system at high levels of employment—admittedly, paying the price of a considerable degree of inflation.* The problem, then, is to bring about the political changes in the United States that may be the prerequisite for introducing the kinds of employment programs that have proved their worth in capitalism abroad. That may involve a very difficult problem of public education and persuasion, but it is a problem different from a defeatist admission that the economic system itself cannot be made to work adequately.

KEY WORDS

*Technological
unemployment*

Automation

*Uneven entry of
technology*

*Supply and
demand for labor*

Service sector

*Labor-saving
inventions*

*Industry-building
inventions*

*Aggregate
demand*

Wage policy

Wage cutting

*Structural
unemployment*

Retraining

Job information

*Reducing the
supply of labor*

*Employer of
last resort*

*European
experience*

CENTRAL CONCEPTS

1. Unemployment has been an endemic problem of market societies. We have already discussed this problem from the point of view of aggregate demand. Here we look into the important problem of *technological unemployment*.

2. *Automation* refers to a new category of machines with unusually flexible, self-correcting feedback mechanisms.

3. Machines affect both the supply of and the demand for labor. Changes on the supply side reflect the effect of machines in increasing productivity. The important point is the *uneven entry* of machines into the economy, greatly raising productivity in agriculture, raising it somewhat less (but also substantially) in manufacturing, having least effect in the services sector.

4. Employment depends on the demand for labor as well as the supply. We note that the demand for the products of the agricultural and manufacturing sectors has not risen as fast as the increase in productivity, resulting in a displacement of labor from those sectors.

5. Labor has moved, over many decades, into the services sector. The *special significance of automation is that it represents the entrance of technology into this previously labor-absorbing sector.*

6. *Labor-saving inventions* do not create a loss in income, but shift incomes previously earned by labor to other workers or to consumers (in terms of declines in price) or to profits. The inventions will create unemployment *unless these shifts give rise to investment that will create demand for additional labor.*

7. Employment-creating inventions give rise to new industries based on new demands. Automobiles are a prime example of such industry-building inventions. We do not know if automation will have such an effect on investment.

8. Unemployment can be combated in several ways. *Widespread unemployment is best attacked by increases in aggregate demand.*

9. *Wage policy* is a critical consideration in combating unemployment. If wages rise faster than productivity, employed labor will benefit; but the rise will deter the hiring of additional labor. Cutting wages is an uncertain and potentially dangerous mode of increasing employment, since the adverse effects of decreased consumption may more than offset lower costs.

10. *Structural unemployment* is an important source of joblessness. It results from a bad match between existing skills and the structure of work. *Labor-displacing inventions* refer to changes in technology that make existing skills obsolete.

11. The basic remedy for structural unemployment is retraining, but it is difficult to carry out unless one knows the skills that will be required in the future. Retraining is most successful when combined with programs of *public employment*, since their extent and skill requirements can be known in advance. Increasing *job information* is also very useful in reducing unemployment.

12. The *supply of labor can also be reduced*, thereby alleviating unemployment. This occurs when we lengthen the span of schooling or lower retirement ages.

13. The government can serve as an "*employer of last resort*," guaranteeing useful careers—not make-work jobs—if private demand is inadequate.

14. Recent European experience seems to indicate that unemployment can be held to lower levels than in the U.S.

QUESTIONS

1. What do you think accounts for the shift in demand from primary to secondary and tertiary products? In particular, what do you think is the reason for the steady growth of services as a kind of output that society seems to want?

2. Suppose that technology in the 1890s had taken the following turn: a very complex development of machines and techniques for improving public and private supervisory and administrative techniques, very clever devices that performed salesmen's and clerks' services, but almost no improvement in agricultural techniques. What would the distribution of the labor force probably look like?

3. Suppose that an inventor puts a wrist radio-telephone on the market. What would be the effects on consumer spending? What would ultimately determine whether the new invention was labor-attracting or labor-saving?

4. Suppose that another new invention halved the cost of making cars. Would this create new purchasing power? What losses in income would have to be balanced against what gains in incomes? What would be the most likely way that such an invention could increase employment? Would employment increase if the demand for cars were inelastic, like the demand for farm products—that is, if people bought very few more cars despite the fall in prices?

5. Unemployment among the black population in many cities in the late 1960s was worse than it was during the Great Depression. What steps would you propose to remedy this situation?

6. Would raising wages, and thereby creating more consumption demand, be a good way to increase employment?

7. Do you believe that there exists general support for large public employment-generating programs? Why or why not? What sorts of programs would you propose?

8. How would you encourage private enterprise to create as many *jobs* as possible?

9. Do you think that the computer, on net balance, has created unemployment? How would you go about trying to ascertain whether your hunch was accurate? Would you have to take into account the indirect effects of computers on investment?

10. What is the difference between a labor-saving and a labor-displacing invention? Which is more difficult to cope with?

11. How much inflation would *you* willingly accept, to lower unemployment to, say, 3 percent?

Problems of Economic Growth

34

FROM OUR FIRST INTRODUCTION to Adam Smith's *Wealth of Nations,* the issue of growth has been a central preoccupation of our macroeconomic investigations. Indeed, no sooner did we leave the imaginary circular flow of a static system, allowing investment to enter the picture, than growth became *the* problem of macroeconomics—too little growth bringing unemployment, too much growth pushing us toward inflation.

In this last chapter, growth once more comes into the foreground. For there are problems of growth that we have not yet met —as we shall see, problems of very great importance. In this chapter we shall deal with three of them: growth and stability; growth and the quality of life; and growth and its "limits." Perhaps these issues are more crucial and more complex than any others in the book.

The Stability of GNP

We met the issue of stability when we learned about business cycles in Chapter 21, but we have not yet explored an important aspect of the problem. Formerly, we looked into cycles as a problem involving fluctuations in demand. Now we want to investigate stability as a problem involving supply as well as demand. In particular, we want to learn more about the degree to which we utilize the *potential output* of the economy.

For even in the doldrums of recession (except for a very few severely depressed years) the economy manages to lay down a net increment of wealth in the form of investment, and this investment then adds its leverage to that of the entire stock of capital with which society works. So, too, every year, the labor force tends to grow as population increases, adding another component of potential input to the economic mechanism.

This does not mean, as we well know, that the economy will therefore automatically utilize its full resources. Unused men and unused machines are very well known to us in macroeconomic analysis. But the steady addition to the factors of production does mean that the *potential* of the economy will

THE DIFFERENCE THAT GROWTH RATES MAKE

The normal range in growth rates for capitalist economies does not seem to be very great. How much difference does it make, after all, if output grows at 3 percent or 5 percent or even 8 percent?

The answer is: an amazing difference. This is because growth is an *exponential* phenomenon —one involving a percentage rate of growth on a steadily rising base.

Recently, Professor Kenneth Boulding

pointed out that before World War II no country sustained more than 2.3 percent per capita growth of GNP whereas since World War II at least one country—Japan—has achieved and held a per capita growth rate of 8 percent. Boulding writes: "The difference between 2.3 and 8 percent may be dramatically illustrated by pointing out that [at 2.3 percent] children are twice as rich as their parents—i.e., per capita income approximately doubles every generation—while at 8 percent per annum, children are six times as rich as their parents."

be steadily rising. *The possibility of growth is thus introduced into the system by the accumulation of its basic instruments of production, both human and material.* If we multiply the slowly rising hours of total labor input by a "productivity coefficient" that reflects, among other things, technology and capital, we can derive a trend line of *potential GNP*. The question we then face is to see how much of this potential volume of output we will actually produce.

ACTUAL VS. POTENTIAL GNP

As Fig. 34·1 shows, all through much of the 1950s, 1960s and early 1970s, potential output ran well ahead of the output we actually achieved. Indeed, between 1958 and 1962 the amount of lost output represented by this gap came to the staggering sum of $170 billion. Even in 1972, a prosperous year, we could have added another $55 billion to GNP— $1,000 per family—if we had brought unemployment down from the actual level of 5.6 percent to 4 percent.

DEMAND VS. CAPACITY

The problem of a potential growth rate opens an aspect of the investment process that we

have not yet considered. Heretofore, we have always thought of investment primarily as an income-generating force, working through the multiplier to increase the level of expenditure. Now we begin to consider investment also as a *capacity-generating* force, working through the actual addition to our plant and equipment to increase the productive potential of the system.

No sooner do we introduce the idea of capacity, however, than a new problem arises for our consideration. If investment increases potential output as well as income, the obvious question is: will income rise fast enough to buy all this potential output? Thus at the end of our analysis of macroeconomics we revert to the question we posed at the beginning, but in a more dynamic context. At first, we asked whether an economy that saved could buy back its own output. Now we must ask whether an economy that grows can do the same.

MARGINAL CAPITAL-OUTPUT RATIO

The question brings us to consider a concept that we met earlier (p. 323) but have not yet put to much use. The *marginal capital-output ratio*, as the formidable name suggests, is not

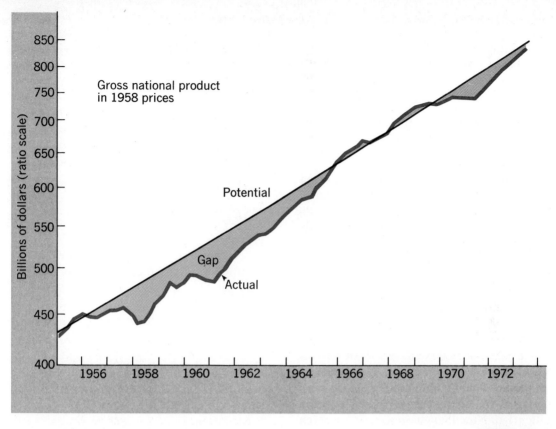

FIGURE 34 · 1
Actual and potential GNP

a relationship that describes behavior, as the multiplier does. It describes a strictly technical or engineering or organizational relationship between an *increase in the stock of capital and the increase in output that this new capital will yield.*

Note that we are not interested in the ratio between our entire existing stock of capital (most of which is old) and the flow of total output, but only in the ratio between the *new* capital added during the period and the *new* output associated with that new capital. Thus the marginal capital-output ratio directs our attention to the *net investment* of the period and to the *change in output* of the period. If net investment was $60 billion and the change in output yielded by that invest-

ment was $20 billion, then the marginal capital-output ratio was 3.

INCOME VS. OUTPUT

The marginal capital-output ratio gives us a powerful new concept to bring to bear on the problem of attaining and maintaining a high, steady rate of growth, for we can now see that the problem of steady growth requires the balancing out of two different economic processes. Investment raises productive capacity. *Increases* in investment raise income and demand. What we must now do is investigate the relationship between these two different, albeit related, economic variables.

Let us begin with a familiar formula

that shows how a change in investment affects a change in income. This is

$$\Delta Y = \frac{1}{mps} \times \Delta I,$$

which is nothing but the multiplier. For brevity, we will write it

$$\Delta Y = \frac{1}{s}\, \Delta I.$$

Now we need a new formula to relate I, the rate of new investment (not ΔI, the *change* in new investment), and ΔO, the change in dollar output. This will require a symbol for the marginal capital-output ratio, a symbol that expresses how many dollars' worth of capital it takes to get an additional dollar's worth of output. If we use the symbol σ (sigma), we can write this formula as $\sigma = I/\Delta O$.

For example, if we have $10 of new investment, and output rises by $5, then σ (the marginal capital-output ratio) must be 2 ($10 ÷ $5). If output rises by only $3, σ is 3.3 (10 ÷ $3). Note that the *lower* the marginal capital-output ratio, the *higher* the productivity of new investment.

BALANCED GROWTH IN THEORY

We now have two formulas. The first tells us by how much *income* will rise as investment grows. The second tells us the relationship between new investment and new *output*. Now, by a simple arithmetical operation, we can use our second formula to tell us the amount by which output will increase as investment increases. All we have to do is multiply both sides of the formula by ΔO and then divide both sides by σ to get $\Delta O = I/\sigma$.

This last formula tells us that if new investment is, say $10, and the marginal capital-output ratio is, say, 2.5, then the increase in output will be $4—that is, $10 ÷ 2.5.

We are ready to take the last and most important step. We can now discover *by how much investment must rise each year, to give us the additional income we will need to buy the addition to output that has been created by this selfsame investment.*

Our formulas enable us to answer that question very clearly. Increased income is ΔY. Increased output is ΔO. Since $\Delta Y = \frac{1}{s} \Delta I$, and $\Delta O = \frac{I}{\sigma}$ then ΔY will equal ΔO if $\frac{1}{s} \Delta I = \frac{I}{\sigma}$. This is the formula for balanced growth.

If we now multiply both sides of the equation by s, and then divide both sides by I, we get

$$\frac{\Delta I}{I} = \frac{s}{\sigma}$$

BALANCED GROWTH IN FACT

What does this equation mean? It tells us what *rate of growth of investment* ($\Delta I/I$) is needed to make $\Delta Y = \Delta O$. In words, it tells us by what percentage investment spending must rise to make income payments keep pace with dollar output. That rate of growth is equal to the marginal savings ratio divided by the marginal capital-output ratio. Suppose, for instance, that the marginal savings ratio (including leakages) is $\frac{1}{3}$ and that the marginal capital-output ratio (σ) is 3. Then s/σ is $\frac{1}{3} ÷ 3$ (or $\frac{1}{9}$), which means that investment would have to grow by $\frac{1}{9}$ to create just enough income to match the growing flow of output. If the rate of investment grew faster than that, income and demand would tend to grow ahead of output and we would be pushing beyond the path of balanced growth into inflation. If the rate of growth of investment were smaller than that, we would be experiencing chronic overproduction with falling prices and sagging employment.

What is the rate at which investment should rise for balanced growth in the United

States? To determine that, we would have to deal with tricky statistical problems of marginal capital-output ratios; we would have to calculate *net* investment—not easy to do; and marginal leakages would have to behave as tamely as they do in textbooks—not always the case. Moreover, to include public as well as private capital formation in the terms "investment" and "marginal capital output ratio" would also greatly complicate our computations.

Therefore we shall sidestep here the difficult empirical problems posed by the requirements for balanced growth and concentrate on the general issue that the formulation opens up. For the purpose of our discussion is to explain that there is a complex relationship between the growth in income and the growth in output. Our analysis shows that we can have a *growth gap* if our leakages are too high, so that increases in injections do not generate enough new purchasing power, or if our marginal capital-output ratio is too low, so that a given amount of investment increases our potential output (our capacity) too fast.*

POLICY FOR BALANCED GROWTH

Suppose that we are not generating income fast enough to absorb our potential output. Suppose, to go back to Fig. 34·1, that we have a persistent growth gap similar to that of the late 1950s. How can we bring the economy up to its potential?

*One question may have occurred to the reader. Aren't incomes and outputs *identities?* Isn't it true that GNI ≡ GNP? Then how can a growth "gap" occur? The answer lies in our familiar ex ante and ex post perspectives (see box pp. 450–51). Ex post, incomes are always the same as outputs. Ex ante they are not. The question, then, is not whether or not GNI will be equal to GNP, but *whether their identical values will be equal to potential output.* Our formula for balanced growth tells us which critical variables must be taken into account when we ask whether ex ante spending plans will bring us to a level of income and output that corresponds with *potential production.*

Our formula for balanced growth gives us the answer. The relationship between the growth of incomes and the growth of output depends above all on the rate at which investment *increases*. Thus, if expenditures fall short of the amount needed to absorb potential output, the answer is to raise the rate of investment or, perhaps more realistically, to raise the rate of growth of all expenditures, public and private. Conversely, of course, if we find ourselves pushing over the trend line of potential growth into inflation, as in the late 1960s, the indicated policy is to lower the rate of growth of investment (or of investment and government and consumption) to bring the flow of rising incomes back into balance with the rise of output.

Thus the critical element in balancing a growing economy is not the *amount* of investment needed to fill a given demand gap, but the *increase* in investment (or other injection-expenditure) needed to match a growing output capacity with a large enough demand.

PROBLEMS OF BALANCED GROWTH

But our analysis also points up an endemic problem for an expanding market system. *It is that the growth of incomes and the growth of output depend on unrelated variables.* There is nothing in the income side of the picture that ensures an appropriate magnitude for the marginal capital-output ratio. Nor is there anything on the capacity side that matches a given marginal capital-output ratio with an appropriate savings rate for the multiplier we want.

All this sets our previous discussion of fiscal and monetary policy into a more complex but more realistic framework. In trying to keep the economy on a path of balanced growth, we are trying to align two different sets of forces: one, largely technological, regulates the rate at which our capacity rises; one, largely behavioral, regulates the savings decision. This does not tell us what we want to know—exactly how fast to increase the supply of money or exactly how large our government expenditures should be—but it helps explain why a growing economy displays instability *just because it is growing.*

Growth and the Quality of Life

The second problem of growth is quite different from the first—philosophic rather than technical, imprecise and value-ridden rather than clear-cut and analytic. Yet it is no less pressing than the problem of stability. It is the question of *why growth fails to bring benefits that are proportionate to the increase in output.*

Perhaps an illustration will bring the problem home in full force. We are often fond of stating that we are twice as rich in material output as our parents, four times as rich as our grandparents, perhaps ten times richer than our Pilgrim forebears. Why then are we not twice, four, ten times as contented, satisfied, happy as they? Is it possible that growth is a process that somehow fails to offer deep and lasting satisfactions? If so, why?

THE DISBENEFITS OF GROWTH: EXTERNALITIES

We can only speculate , but speculations are better than not thinking. So let us begin by calling attention to one reason for the disappointing effects of growth. It is the possibility that *growth is the cause of many of those externalities* whose unpleasant attributes we have examined in Chapter 15.

In that chapter, we simply assumed the externalities were "there." Now we must recognize that many of them are the direct consequence of growth itself. The smoke, the virulent chemicals, the noise, the poisonous run-offs that make modern life disagreeable and even dangerous are all side effects of the industrial processes on which growth has depended. One reason why material growth fails to bring equal gains in well-being is that we add up the gains from growth but omit the losses it inflicts.

FAMILIARITY AND CONTEMPT

A second reason has less to do with the disutilities that flow from growth-generating processes, than with the utilities that the products of growth give us.

A new product is brought onto the market—let us say the first telephone. It is a source of excitement, wonder, joy. We are acutely conscious of the gain in well-being the flows from this miraculous instrument. We are able to talk with a sick friend; we close a business deal in a few seconds; our sense of well-being expands along with our gain in material possessions.

But now a few years pass. We are used to the telephone; we take it for granted. We call a sick friend and get a busy signal; we lose a business deal because our phone is out of order; the constant ringing of the phone is a nuisance, an interruption; we wish we could get rid of the thing, but we cannot.

This process of an initial surge, followed by a gradual decline in utilities, seems to be a common fact of life. It is probably true for total income as well as for individual goods: we are happy for a while if our income rises, then we get used to the new level and take it for granted. A cut in income that left us *above* our previous level would leave us much less happy than we formerly were.

What we encounter here is a more dynamic version of the notion of "wants" than economic theory deals with, but possibly a truer one. If our marginal utility curves are constantly moving leftward, as we become familiar with new goods or with higher levels of income, it follows that the gains in satisfaction from increasing our material wealth may be great—but will be short-lived.

MONEY AND HAPPINESS

There is still a third possible reason why growth has failed to bring a sense of increased well-being proportionate to the increase in wealth. "For most Americans," writes economist Richard Easterlin, "the pursuit of happiness and the pursuit of money come to the same thing. More money means more goods (inflation aside), and this means more of the material benefits of life." But, Easterlin asks, "What is the evidence on the relation between money and happiness? Are the wealthier members of society usually happier than the poorer? Does raising the incomes of all increase the happiness of all?"[1]

Easterlin has assembled data on the relation between wealth and happiness from some 30 surveys conducted in 19 developed and underdeveloped countries. The results are interesting—and paradoxical. In all societies, more money for the individual *is* reflected in a greater degree of happiness reported by the individual. However—and here is the paradox—raising the incomes of all does not increase the happiness of the entire society!

THE PARADOX OF WEALTH

How can we explain this paradox? Let us first examine the findings of the surveys. In every one of thirty separate surveys (eleven within the United States between 1946 and 1970; nineteen in other countries, including three communist nations), happiness was positively correlated with income. For ex-

[1]Richard Easterlin, "Does Money Buy Happiness?" *The Public Interest* (Winter, 1973).

ample, in a survey conducted among American families in December 1970, not much more than one-quarter of those in the lowest income group (under $3,000 per year) described themselves as being "very happy," whereas about half the members of the highest group (over $15,000) reported that they were "very happy." Moreover, in each intermediate income group, the proportion of "very happy" people rose steadily.

This is an impressive statistical association, although as Easterlin points out, we have to be careful that we are not dealing with "wrong-way causation" (Chap. 42, p. 692). It could be that "happier" people owe their good fortune to factors such as a better state of health or perhaps "better genes" and that their superior health or genetic endowments enable them to *make* more money. There is no doubt that some of the association between income and happiness comes from this causal relation. But another bit of evidence allows us to retain considerable confidence in the original hypothesis that money "causes" happiness. When individuals in all income groups are questioned about the reasons for their happiness or unhappiness, they typically mention three factors: family relations, health, and economic concerns. By removing one major source of worry—economic concerns—and by alleviating the problem of coping with ill-health, money can be a direct source of happiness in itself.

RELATIVE WEALTH

If this is the case, then should it not follow that the wealthier a nation becomes, the higher will be its sum total of happiness? Should we not expect to find that the overall proportion of persons calling themselves "very happy" would rise significantly as the real incomes of all groups in the nation rise?

As we have already learned, this is *not* the answer. In fact, jumping to that conclusion is another instance of the fallacy of composition—assuming that what is true for the individual must be true for society as a whole. According to the survey data, Easterlin writes, "Rich countries are not typically happier than poorer ones." Moreover, in the one nation—the United States—in which we can make comparisons over time, we find that the "profile" of happiness was not much different in 1970 from what it had been in the late 1940s, even though real income per capita, after allowance for taxes and inflation, was over 60 percent higher.

How can we explain this paradoxical outcome? The answer seems to lie in the way people estimate their degree of material well-being. They do not measure their income or possessions starting from zero. Instead, people evaluate their place on the scale of wealth by *comparing themselves with others.* Thus it is not our "absolute" level of material well-being but our relative level that determines whether or not we feel "rich" or "poor." That relative level of well-being, in turn, depends on the distribution of income; and income distribution, as we know, is slow to evidence major shifts.

WEALTH AND VALUES

This situation has two major consequences for our own society. One is that poverty, with its associated unhappiness, cannot be eradicated by simply raising the incomes of the bottom portion of the population along with those of everyone else. The families that make up the bottom groups will continue to feel "poor" and therefore unhappier than the well-to-do. *Only a change in the pattern of income distribution could be expected to eliminate the feeling of poverty and its associated unhappiness.* By the same consequence, however, a more equal distribution of income would reduce the feeling of being "rich" and its associated happiness! Whether a nation with a more equal distribution of income

would be "happier," en masse, therefore depends on our value judgments—specifically on the relative importance we would assign to an increase in the happiness of the bottom groups and a decrease in the happiness of top groups.

Second, Easterlin's conclusions suggest that we are locked in a "hedonic treadmill." We are all engaged in an effort to acquire wealth, in the expectation that it will bring happiness; but unlike the race in *Alice in Wonderland,* the race for wealth has all winners but no prizes. Does this have something to do with the chronic feeling of dissatisfaction—the complaints about the "rat race"—that are so much a part of our culture?

THE NEGLECT OF WORK

Another possibility is that growth has brought a change in the nature of the "felt work experience" that has added disutilities that largely offset the utilities of added income.

Objectively, it seems that most work experience has become easier during the last hundred years. The rigors and horrors of early industrial capitalism have largely vanished; the work day is shorter; the tasks we do are less dirty or dangerous. But with the rise in income has also come a change in aspirations and expectations. Conditions that would have been regarded as a great improvement a generation ago are as much taken for granted as the new commodities we first yearn for and then toss on the dust heap. Work becomes an experience that we no longer uncomplainingly accept, but that we regard critically. And we do not seem to like what we see.

Economics pays little heed to the felt experience of work. It concentrates on our welfare as consumers, much less on our welfare as producers. It describes as a "rising standard of living" the increasing flow of goods

GALBRAITH ON SOCIAL BALANCE

"The family which takes its mauve and cerise, air-conditioned, power-steered, and power-braked automobile out for a tour passes through cities that are badly paved, made hideous by litter, blighted buildings, billboards, and posts for wires that should long since have been underground. They pass on into a countryside that has been rendered largely invisible by commercial art.... They picnic on exquisitely packaged food from a portable icebox by a polluted stream and go on to spend the night at a park which is a menace to public health and morals. Just before dozing off on the air mattress, beneath a nylon tent, amid the stench of decaying refuse, they may reflect on the curious unevenness of their blessings. Is this, indeed, the American genius?"—*The Affluent Society* (Boston:Houghton Mifflin, 1958), p. 253.

and services we consume, but does not pay much heed to the "standard of work experience" in the factory and office, with its machine-like pace, its clock-regulated rhythms.

This is because economics assumes that the end of economic activity is consumption. But suppose the end is production? Suppose that the main "output" of an economy were considered not the quality of its goods but the quality of its working life? Then the growth in output might be almost irrelevant to the feelings of well-being of a society.

SOCIAL BALANCE

Consider one last possibility. *It is that our well-being lags behind our growth in output because we are producing the wrong output.* We are piling up mountains of toothpaste while we ignore the possibility of providing dental care to those who cannot afford it. We are "growing" in terms of military output but not in terms of the output of public housing. Thus, we are disappointed in the effects of growth because we do not even begin to live up to the ideal of an economy in which the marginal utility of expenditures for all purposes is equal.

This state of affairs comes about for two reasons. First, the market caters to wealth, not need, obeying the demands that emanate from the existing distribution of income.

Thus growth caters to the marginal utilities of those consumers whose total utilities are already much more "saturated" than those consumers who cannot fulfill their utilities through purchasing power.

Second, our society has a very imperfect means (if indeed any at all) of judging the marginal satisfactions of collective consumption versus private consumption. John Kenneth Galbraith has described our society as one of "private affluence and public squalor." To the extent that this is true, growth fails to yield proportionate satisfactions because it lacks "social balance"; that is, a condition in which marginal utilities of expenditures for all purposes have been brought to equality.

GROWTH AND WELL-BEING

These speculations and hypotheses are difficult to test empirically (perhaps with the exception of Easterlin's findings), but they offer plausible reasons for the widely shared feeling that material growth does not yield proportionate increments in satisfaction. If true, they also suggest that growth could bring more satisfaction if means were found to eliminate some of the causes of its failure to do so.

In particular it seems reasonable to assume that the gap between output and satisfaction could be narrowed *to some extent* if we took pains to minimize the externalities associated with growth, including not least, the experience of work; if we lessened the unequal distribution of income; and if we strove for a distribution of output that tried to equalize the marginal utilities of private and public output.

Such a task is a political rather than an economic one. The estimation of marginal utilities belongs to the sphere of value judgments, where the economist bows to the political decision-making process by which societies express their preferences. But the economic considerations we have raised may help us formulate those values more thoughtfully than if we simply assume that growth, in and of itself, will bring a sense of well-being along with an increase in material wealth. Perhaps the ultimate faith of all social science is that thoughtfulness and knowledge can ultimately bring their rewards in the quality of life we experience.

The Limits to Growth

We have been considering the question of why growth yields so much less net utilities than its gross utilities—a question that asks us to reconsider the assumption that "more is always better." But we have kept until last a much more fundamental question. It is whether growth, in the form in which we have experienced it, is a process that can be maintained much longer. Good or bad, growth has been the central tendency of our society for well over two centuries. Now we must ask whether the end to that tendency lies ahead.

MATERIAL CONSTRAINTS

Growth of the kind that has characterized the history of capitalism and of industrial socialism is a process that converts the raw materials of the planet into commodities that

men use for their consumption or for further production. Thus along with growth has come a steady rise in the volume of raw materials that man has extracted from the planet, and an even more rapid increase in the energy he has harnessed, both for the extraction and the processing of those raw materials.

A continuation of growth therefore depends on the continuing availability of the raw materials and energy essential for expanded industrial production. What is the outlook for these resources for the future?

Table 34·1 shows a recent estimate of the number of years that presently known and estimated future resources would supply us at present growth rates.

TABLE 34 · 1
Global Resource Availability

	Years of global resource availability at present growth rates	
Resource	If present resources stocks are used	If resource stocks are quintupled
Aluminum	31	55
Coal	111	150
Copper	21	48
Iron	93	173
Lead	21	64
Manganese	46	94
Natural gas	22	49
Petroleum	20	50
Silver	13	42
Tin	15	61
Tungsten	28	72

Source: Meadows, et al., *The Limits to Growth* (Washington, D.C.: Potomac Associates, 1972).

The figures are sobering, to say the least. Yet we must be careful before we take them

THE EXPONENTIAL FACTS OF LIFE

Exponential growth is a startling phenomenon. It is illustrated in the famous parable about the farmer who has a lily pond in which there is a single lily that doubles its size each day. Suppose that after a year the pond is completely covered. How long did it take for the pond to be *half* covered? The answer is—364 days. In the last day the doubling lily will completely fill the pond.

Exponential examples such as these must always be used with great care. This is because their mathematical logic does not take into effect the feedback mechanisms that inhibit the explosive behavior of exponential series. Long before the lily covered half the pond it would prob-ably have used up the nutrient matter in the pond and ceased growing. Long before the horrendous projections of exponential population growth, in which human beings will stand on one another's shoulders in a few centuries, feedbacks would have slowed down or halted or reversed population trends. Exponential trends show the *potential* growth of variables, but this potential is rarely realized.

One last point. There is a convenient way of figuring how long it takes for any quantity to double, if we know its exponential growth rate. *It is to divide the growth rate into the number 70.* Thus if population is growing at 2 percent a year, it will double in 35 years (70 ÷ 2). If the growth rate rises to 3 percent, the doubling time drops to 23+ years (70 ÷ 3).

at face value. First, the *presently known reserves include only those deposits of minerals that are available with today's technology.* They do not include minerals that exist in levels of concentration that are not "economic"; that is, too costly to utilize with existing techniques. But techniques change, and with them, the volume of "economic" resources. Consider the fact that taconite, the main source of iron today, was not even considered a resource in the days when the high-grade ores of the Mesabi Range provided most of our ore.

In some cases, such as oil, economic resources may be exhausted within a relatively short time. But in general, resources probably exist in quantities sufficient to support industrial growth for centuries, especially if we consider the gigantic amounts of minerals locked into the earth's crust or present in "trace amounts" in its seas. Given the technology and the energy, we could literally "mine the seas and melt the rocks" to provide ourselves with "unlimited" resources. Moreover, the necessary technology does not seem impossible to attain. Much of it already lies within the frontier of scientific knowledge.

THE EXPONENTIAL PROBLEM

This reassuring consideration must, however, be balanced against two less assuring facts. The first is that the rate of use of resources rises with frightening rapidity because growth is an *exponential* process. Today, global industrial output is rising at about 7 percent a year, thereby doubling every 10 years. If we project this rate of growth for another fifty years, it follows that the rate of use of resources would have doubled five times (assuming that today's technology of industrial production is essentially unchanged). Thus 50 years hence we would need 32 times as large a volume of material inputs as today. A century hence, when output would have doubled ten times, we would need over a thousand times the present volume of output —and this gargantuan volume of extraction would still be relentlessly doubling.

Does this mean that we will run out of resources? We are constantly making resources, learning to use less concentrated forms of minerals and finding wholly new materials, such as plastics. Thus we may be able to expand our supply of resources exponentially, along with our use of them.

THE TECHNOLOGICAL FACTOR

All this, however, requires the appropriate technology. It is one thing to mine the seas and the rocks on paper, and another to move and refine the millions of tons of water and earth in fact. Where will we get the energy to undertake these huge labors? It is one thing to run a nuclear fusion plant on paper; another to bring it on stream. (No controllable fusion process yet exists, except in the most advanced laboratories.)

Thus we are essentially engaged in a race between technology and the exponentially rising demands for raw materials. Technology enters this race in many ways. It may enable us to recycle existing wastes, so that we do not need to extract as much new materials. It may enable us to get more usable resources from a given quantity of raw material. It may open up new modes of production that enable us to shift production techniques away from materials that are becoming scarce (and therefore expensive), to those that remain abundant and therefore cheap. It gives us new sources of energy that enable us to use materials that are now too "low grade" for economic production.

One primary question in estimating the "limits" to growth is therefore the rate at which we will develop the appropriate technology. Unfortunately, the link between research and development and economically usable technology is not clearly understood. We do not really know whether we will find an appropriate technology to permit us, for example, to run a vast private automobile fleet in the year 2000, or whether we will be able to turn out the high quality steels in the volume that would support industrialization on a global scale fifty years hence. More important, we do not know if a technology that permitted us to "mine the seas and melt the rocks" will be perfected—or whether such a technology would be compatible with other ecological and environmental considerations.

THE HEAT PROBLEM

Here we reach another problem of great importance. To extract resources on the scale required to sustain industrial growth on its present path will require the application of tremendous amounts of energy. But the production of this energy, if it uses the combustion of conventional fuels or nuclear power (including fusion power) is associated with the generation of heat. This man-made heat is the most serious environmental barrier we face.

Today the amount of heat that all our man-made energy adds to the natural heat of the sun falling on our atmosphere is trivial. But the exponential increase in this man-made heat would sooner or later pose an insurmountable problem. As two well-known economists have written:

Present emission of energy is about 1/15,000th of the absorbed solar flux. But if the present rate of growth continued for 250 years, emissions would reach 100% of the absorbed solar flux. The resulting increase in the earth's temperature would be about 50°—a condition totally unsuitable for human habitation.[1]

DOOMSDAY?

Projections such as these easily give rise to Doomsday attitudes, beliefs that we are racing hell-bent on an unalterable disaster course. This is not what economists who are concerned about the growth problem have in mind. As we have already indicated, the technology required to keep us on the present growth path may not materialize in time, thereby averting or postponing for a long period a climatic catastrophe. Sources of energy such as solar power generators or wind machines may utilize the existing heat and en-

[1] Robert U. Ayres and Allen V. Kneese, *Economic and Ecological Effects of a Stationary State,* (Washington, D.C.: Resources for the Future). Reprint No. 99 (December 1972), p. 16.

ergy in the atmosphere to provide us with substantial amounts of power that do not pour man-made heat into the air. Most important of all, a shift in the direction of economic effort away from the encouragement of wasteful industrial growth, toward the growth of services or collective consumption goods (mass transit in place of automobiles), may greatly reduce our need to increase our energy inputs.

For all these and still other reasons, a Doomsday attitude is not warranted. But that is not the same thing as saying that an attitude of extreme caution about "limitless" industrial growth is also not warranted. On the contrary, *it seems likely that industrial growth, with its heavy dependency on resources and man-made energy, cannot continue indefinitely.* If the technology of industry does not change substantially, the threshold of a dangerous degree of climatic change would be reached in less than 150 years. At that point, growth would have to cease immediately. In all likelihood, the warnings of scientists or the difficulties of sustaining growth or ecological side effects will impose a rein on industrial expansion considerably before that time, although it is impossible to specify time horizons with any degree of assurance.

ZERO GROWTH?

Thus the long-term outlook for industrial societies implies that very low or zero industrial growth lies ahead. But here it is very important to recognize that even zero industrial growth in itself is not an answer to all problems. *A zero growth society may still pollute the atmosphere.* It may still be adding unacceptable amounts of heat or other harmful wastes, even though the flow of those pollutants into the air is not increasing over time. Thus zero industrial growth is not an end in itself, any more than positive growth was an acceptable end in itself. Zero growth

is useful only insofar as it leads to *zero pollution,* including heat pollution.

From this, another conclusion follows. A society may still enjoy a growth in output if that output does not contribute to pollution. A society that draws its powers from the wind and tides and sun can safely enlarge its volume of energy. A society that increases its technical efficiency, perhaps by recycling, can permit growth to occur if that growth will not be ecologically harmful. As we have many times emphasized, a society that seeks to grow through the enrichment of the *quality* of output need not impinge on the constraints of ecological safety.

Problems of Slow Growth

Can a zero pollution society be attained? We do not know. Probably some pollution is inescapable under any system of production, certainly under an industrial system. Thus even though zero industrial growth is not a sufficient condition to bring us to (or near) a zero pollution society, *the probabilities are great that a very low pollution society would be a society in which the rate of growth of industrial output was also very low.* It might in fact be a negative rate—a decline in industrial output being compensated by a rise in service outputs.

GLOBAL INEQUALITY

Two problems must be squarely faced if we contemplate the consequences of such a very slow-growing economy. The first has to do with the unequal distribution of income (and resources) among the nations of the world. As we will see in Chapter 39, the underdeveloped nations today "enjoy" standards of living that are far below even the poorest levels in the advanced countries. Thus the prospect of an enforced slowdown in the rate of industrial output raises the specter of *an international struggle for resources,* as the poor countries attempt to build up the framework of modern industrial structures, and the developed nations continue along their present course.

The question of an impending limit to industrial growth therefore poses a major economic problem for international relations. How are the remaining easily available resources of the world to be shared? As the advanced nations continue to build up their industrial systems, will they leave the underdeveloped regions on a permanently lower level of well-being? Our analysis poses this fundamental question, but cannot answer it.

STATIONARY CAPITALISM?

Second, we must ask whether a very slow rate of industrial growth—not to mention a zero or negative rate—is compatible with capitalism. As economists from Adam Smith through Karl Marx down to the present day have pointed out, such a "stationary state" would pose very great difficulties for a capitalist system. As expansion ceased, competition among enterprises would lead to a falling rate of profit. This would directly undercut the main source of income of the capitalist class. Worse, the cessation of net investment might set into motion a downward spiral of incomes and employment that would plunge capitalism into severe depression.

Could capitalism be rescued from this fate by a much larger volume of public spending that would not add to environmental danger? Could it give up industrial expansion in exchange for programs of education, the arts, greatly improved care for the old and the very young? Could it manage the political problem of income distribution in a situation in which the growth of material output was very small, or in which material output per capita actually fell because

of the demands of the underdeveloped countries?

Once again, we do not know. The impending environmental pressures surely indicate that major changes will be needed to make the economic system of the still distant future compatible with environmental constraints. Whether capitalism can make that adjustment—or whether industrial socialism can make it better—are questions that our analysis raises, but cannot answer.

A SPACESHIP ECONOMY

What is certain, however, is that *all* industrial systems, socialist as well as capitalist, will sooner or later have to change their attitudes toward growth.

For in the long run there is no alternative to viewing the earth itself as a spaceship (in Kenneth Boulding's phrase) to whose ultimately finite carrying capacity its passengers must adjust their ways. From this point of view, production itself suddenly appears as a "throughput," beginning with the raw material of the environment and ending with the converted material of the production process, which is returned to the environment by way of emissions, residuals, and so on. In managing this throughput, the task of producers is not to maximize "growth," but to do as little damage to the environment as possible during the inescapable process of transformation by which man lives. If "growth" enters man's calculations in this period of rationally controlled production, it can be only insofar as he can extract more and more "utility" from less and less material input; that is, as he learns to economize on the use of the environment by recycling his wastes and by avoiding the disturbance of delicate ecological systems.

Such a spaceship economy is probably still some distance off, although by no means so far away that our children or grandchildren may not encounter its problems. Much depends on the rate at which the Third World grows in population and productivity and on the technological means of lessening pollution in the advanced countries. Not least, a true spaceship earth would require a feeling of international amity sufficiently great so that the industrialized peoples of the world would willingly acquiesce in global production ceilings that penalized them much more severely than their poorer sister nations.

These longer perspectives begin to make us aware of the complexity of the problem of growth. Growth is desperately needed by a world that is, in most nations, still desperately poor. Yet, growth is already begin-

ning to threaten a world that is running out of "environment." If growth inevitably brings environmental danger, we shall be faced with a cruel choice indeed. Today we have only begun to recognize the problems of pollution-generating growth, and we are engaged in devising remedies for these problems on a national basis. Ahead lies the much more formidable problem of a world in which growth may encounter ecological barriers on a worldwide scale, bringing the need for new political and economic arrangements for which we have no precedent. The true Age of Spaceship Earth is still some distance in the future, but for the first time the passengers on the craft are aware of its limitations.

KEY WORDS

CENTRAL CONCEPTS

1. This chapter is concerned with three main problems: the *stability* of the growth process; the relation between growth and the *quality of life*; and the *limits* to growth.

Potential output

Growth gap

2. We must distinguish between actual output—GNP—and *potential output*, or potential GNP. The latter is the output available from the full utilization of the growing stock of labor and capital. The difference between potential and actual output is the *growth gap*.

Marginal capital-output ratio.

3. *The problem of a high steady rate of growth is to match our increase in capacity with an equal increase in purchasing power.* Increases in capacity depend on the marginal capital-output ratio, σ, which describes the relationship between an increase in the stock of capital and the increased flow of output from that new stock. Increases in demand result from the multiplier, which magnifies the income effect of a *rise in investment*.

Balanced growth
$\Delta I/s$
I/σ

4. *Balanced growth* occurs when the increase in purchasing power ($\Delta I \times 1/s$) is equal to the increase in capacity (I/σ). This formulation suggests a basic problem. There is no inherent mechanism to balance the critical terms $\Delta I/s$ and I/σ.

Quality of life
Externalities

5. A second major problem relates to growth and the *quality of life*. Why do we not seem to enjoy increases in well-being proportional to those in income and output? One hypothesis is that *externalities*, which yield pervasive disutilities, are integrally connected with the process of growth.

Diminishing utilities

6. A second hypothesis is that the utilities we derive from new products—or from additional income—rapidly diminish as we become accustomed to these new goods. *The gain in utilities from growth is real but short-lived.*

Happiness and relative incomes	7. A third hypothesis is the empirical evidence that the happiness that increased incomes brings is related to *our relative position on the income scale*. As all get richer, but the hierarchy of incomes is unchanged, we do not all feel "happier."
Social balance	8. Last is the possibility that our *social balance* is very far removed from the ideal of a society in which marginal utilities of all expenditures are equal. This is partly the consequence of existing income distribution, partly of a lack of any mechanism for estimating the marginal utilities of collective consumption vs. private consumption.
Criteria for well-being and growth	9. All this suggests the need for different criteria if growth is to bring more well-being and more material wealth. Specifically, society must take into account the disutilities it has formerly ignored, must seek a better social balance among the kinds of output in which it is realized, and should seek to minimize the hierarchies of income distribution that vitiate the gains from improving one's actual material position.
Exponential growth	10. The last and most serious problem concerns the *limits to growth*. Here the problem is to match our *exponential* rate of resource use with ample supplies of materials and energy.
Technology and resource availability	11. It is possible that *technology may make available much larger reserves of resources than now exist or may allow us to use resources much more efficiently. The principal long-run problem here is the heat pollution that combustion (including nuclear energy) adds to the atmosphere.*
Heat pollution **Doomsday predictions**	12. Doomsday predictions of a catastrophic collapse due to resource exhaustion or pollution must be treated with great care. However, the facts of resource exhaustion and heat pollution warn us that *present-day industrial growth cannot continue indefinitely.*
	13. Zero industrial growth, however, is not a sufficient condition for environmental safety. *Zero pollution is the necessary objective.*
Zero growth vs. zero pollution **Stationary state**	14. Zero pollution would probably imply slower rates of industrial growth. This will bring *severe problems for the international distribution of wealth and may pose major problems for capitalism in a stationary state.*

QUESTIONS

1. What is meant by the marginal capital-output ratio? Is it connected with behavior? Do you think it is susceptible to social control?

2. Write the formula for changes in capacity and for changes in income. Show algebraically the conditions under which increased income matches increased output.

3. What is the difference between our original "demand gap" and a growth gap? Explain how the remedy in one case is a given amount of investment; in the other, a *change in the rate of growth* of investment.

4. Do you think it would be possible to establish empirically the change in well-being, as contrasted with income? Could we use social indicators for this purpose? Do you think that statistics on health, longevity, crime rates, etc., shed light on this matter? Surveys on "happiness"?

5. What externalities can you think of that are a direct consequence of growth?

6. Do you think the conditions of work have deteriorated? Has factory labor become more onerous or less during the last 50 years? What about commuting? What role do expectations play in the experience of work?

7. What value judgments are implicit in the suggestion that growth might be associated with greater well-being if the criteria suggested in the text were followed?

8. How would you go about measuring the amount of iron ore "available"? At what level of concentration would you draw the line?

9. Why does the process of combustion add heat to the atmosphere but not the utilization of wind power?

10. Is a stationary state (i.e., a state without economic growth) without environmental effects? Describe ways in which we would have growth without environmental deterioration.

11. What policies do you think should be followed to avoid a severe environmental threat to industrial society?

12. Do you think that *global* growth control is a realistic goal today? If not, what do you foresee as the long-run scenario of world economics and politics?

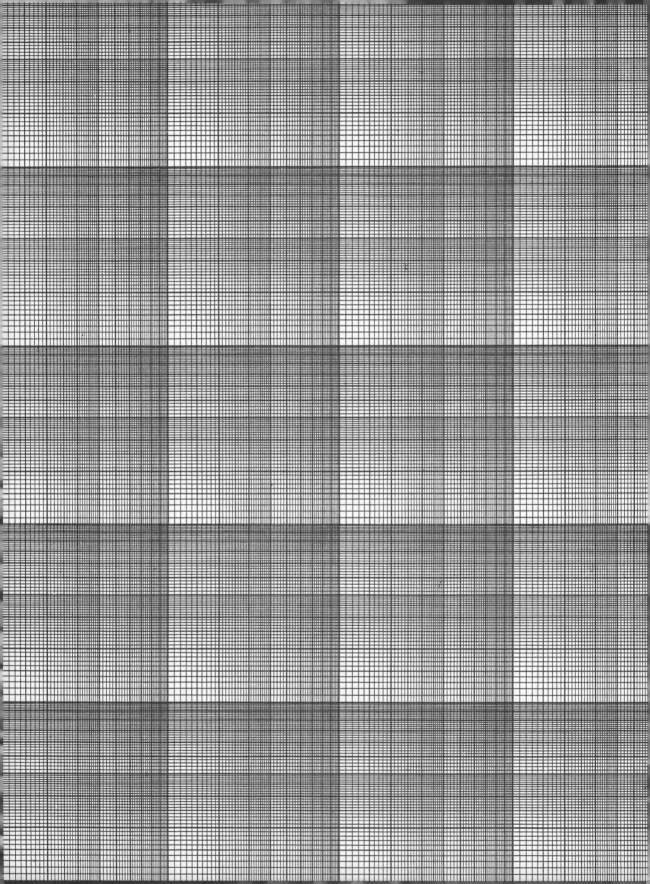

5

INTERNATIONAL
ECONOMICS

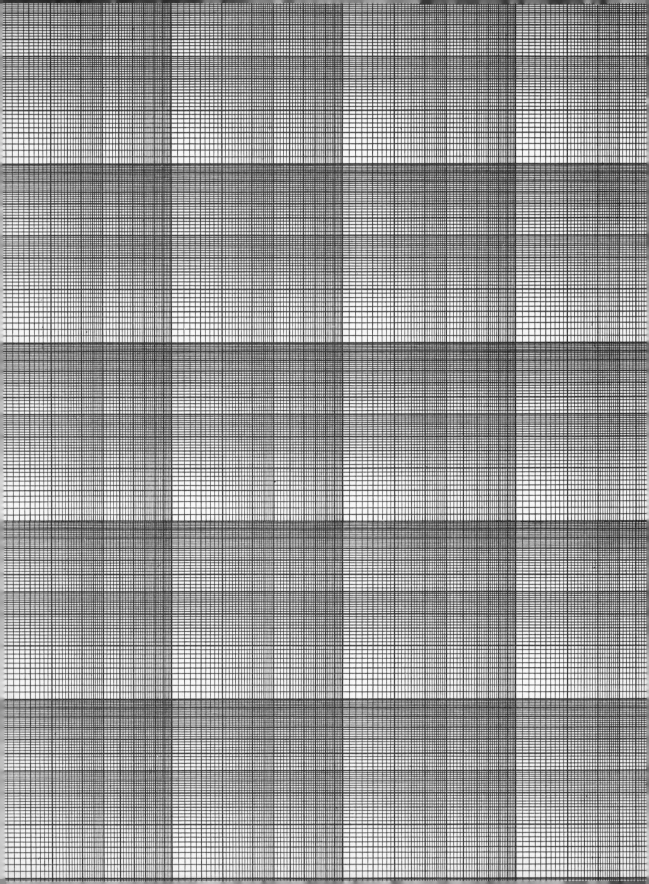

35

The Gains from Trade

AMERICANS HAVE HAD THE FORTUNE to be extraordinarily sheltered from the currents of international economics that wash against other shores. British students are brought up knowing about exports and imports because a quarter of their national income derives from foreign trade. A Canadian knows about international economics because one Canadian dollar in five is earned or spent beyond Canadian borders. Any educated person in the underdeveloped world will tell you that the future of his country is critically affected by the exports it sells to the developed world and the capital it brings in from that world. Only the United States, among the nations of the West, is generally unconcerned and uneducated about foreign trade, for the general opinion is that we are relatively self-sufficient and could, if we had to, let the rest of the world go hang.

Could we? It is true that only about 6 percent of our gross national product is sold overseas and only 4 or 5 percent of the value of GNP is bought overseas. Yet it is worth considering what would happen to our own economy if some mischance severed our ties with the rest of the world.

The first impact would be the loss of certain critical products needed for industrial production. In the earlier years of the country, we were inclined to treat our natural resources as inexhaustible, but the astounding rate of our consumption of industrial raw materials has disabused us of that notion. Today the major fractions of our iron ore, our copper, and our wood pulp come to us from abroad. Ninety percent of the bauxite from which we make aluminum is imported. Ninety-four percent of the manganese needed for high-tempered steels, all our chrome, virtually all our cobalt, the great bulk of our nickel, tin, platinum, asbestos, a rising fraction of our petroleum is foreign-bought. Many of these materials are so strategically important that we stockpile them against temporary disruption, but in a few years the stockpiles would be used up and we should be forced to make radical changes in some of our technology.

Then there would be other losses, less statistically impressive but no less irksome to consumers and industry: the loss of Japanese cameras, of British tweeds, of French perfume, of Italian movies, of Rolls Royce engines, Volkswagen cars, Danish silver, Indian jute and madras. Coffee and tea, the very mainstays of civilized existence, would no longer be available. Chocolate, the favorite flavor of a hundred million Americans, would be unobtainable. There would be no bananas in the morning, no pepper at supper, no Scotch whiskey at night. Clearly, shutting down the flow of the imports into America, however relatively self-sufficient we may be, would deal us a considerable blow. One can imagine what it would mean in the case of, say, Holland, where foreign products account for as much as 45 percent of all goods sold in that country.

But we have still not fully investigated the effects of international trade on the United States, for we have failed to consider the impact of a collapse of our exports. The farm country would feel such a collapse immediately of course, for a fifth of our cotton, almost a quarter of our grains, and more than a quarter of our tobacco go overseas. Mining country would feel it because a fifth of our coal and a third of our sulphur are sold abroad. Manufacturing enterprises in cities scattered all over the nation would feel the blow, as a quarter of our metalworking machinery and of our textile machinery, a third of our construction and mining machinery could no longer be sold overseas—not to speak of another thirty to forty industries in which at least a fifth of output is regularly sold to foreign buyers. In all, some three million to four million jobs, three-quarters of them in manufacturing or commerce, would

cease to exist if our foreign markets should suddenly disappear.

Many of those jobs would be replaced by new industries that would be encouraged if our overseas markets and sources of supply vanished. If we could not buy watches or watch parts in Switzerland, we would make more of them here. If we could not sell machine tools to the world, we would no doubt try to use our unemployed skills to make some product or service that could be marketed at home—perhaps one of the items we no longer imported. With considerable effort (especially in the case of strategic materials) we *could* readjust. Hence the question: Why don't we? What is the purpose of international trade? Why do we not seek to improve our relative self-sufficiency by making it complete?

THE BIAS OF NATIONALISM

No sooner do we ask the question of the aims of international trade than we encounter an obstacle that will present the single greatest difficulty in learning about international economies. This is the bias of nationalism—the curious fact that relationships and propositions that are perfectly self-evident in the context of "ordinary" economics suddenly appear suspect, not to say downright wrong, in the context of international economics.

For example, suppose that the governor of an eastern state—let us say New Jersey—wanted to raise the incomes of his constituents and decided that the best way to do so was to encourage some new industry to move there. Suppose furthermore that his wife was very fond of grapefruit and suggested to him one morning that grapefruit growing would be an excellent addition to New Jersey's products.

The governor might object that grapefruit needed a milder climate than New

Jersey had to offer. "That's no problem," his wife might answer. "We could protect our grapefruit by growing them in hothouses. That way, in addition to the income from the crop, we would benefit the state from the incomes earned by the glaziers and electricians who would be needed."

The governor might murmur something about hothouse grapefruit costing more than ordinary grapefruit, so that New Jersey could not sell its crop on the competitive market. "Nonsense," his wife would reply. "We can subsidize the grapefruit growers out of the proceeds of a general sales tax. Or we could pass a law requiring restaurants in this state to serve state grapefruit only. Or you could bar out-of-state grapefruit from New Jersey entirely."

"Now, my dear," the governor would return, "in the first place, that's unconstitutional. Second, even if it weren't, we would be making people in this state give up part of their incomes through the sales tax to benefit farmers, and that would never be politically acceptable. And third, the whole scheme is so inefficient it's just downright ridiculous."

But if we now shift our attention to a similar scene played between the prime minister of Nova Jersia and his wife, we find some interesting differences. Like her counterpart in New Jersey, the wife of the prime minister recommends the growing of hothouse grapefruit in Nova Jersia's chilly climate. Admittedly, that would make the crop considerably dearer than that for sale on the international markets. "But that's all right," she tells her husband. "We can put a tariff on foreign grapefruit, so none of the cheap fruit from abroad will be able to undersell ours."

"My dear," says the prime minister after carefully considering the matter, "I think you are right. It is true that grapefruit in Nova Jersia will be more expensive as a result of the tariff, but there is no doubt that a tariff

looks like a tax on them and not on us, and therefore no one will object to it. It is also true that our hothouse grapefruit may not taste as good as theirs, but we will have the immense satisfaction of eating our *own* grapefruit, which will make it taste better. Finally, there may be a few economists who will tell us that this is not the most efficient use of our resources, but I can tell them that the money we pay for hothouse grapefruit— even if it is a little more than it would be otherwise—stays in our own pockets and doesn't go to enrich foreigners. In addition to which, I would point out in my television appearances that the reason foreign grapefruit are so cheap is that foreign labor is so badly paid. We certainly don't want to drag down the price of our labor by making it compete with the cheap labor of other nations. All in all, hothouse grapefruit seems to me an eminently sensible proposal, and one that is certain to be politically popular."

SOURCE OF THE DIFFICULTY

Is it a sensible proposal? Of course not, although it will take some careful thinking to expose all of its fallacies. Will it be politically popular? It may very well be, for economic policies that would be laughed out of court at home get a serious hearing when they crop up in the international arena. Here are some of the things that most of us tend to believe.

Trade between two nations usually harms one side or the other.

Rich countries can't compete with poor countries.

There is always the danger that a country may sell but refuse to buy.

Are these fears true? One way of testing their validity is to see how they ring in our ears when we rid them of our unconscious national bias by recasting them as propositions in ordinary economics.

Is it true that trade between businesses or persons usually harms one side or the other?

Is it true that rich companies can't compete with poor ones?

Is it true that one company might only sell but never buy—not even materials or the services of factors of production?

What is the source of this curious prejudice against international trade? It is not, as we might think, an excess of patriotism that leads us to recommend courses of action that will help our own country, regardless of the effect on others. For, curiously, the policies of the economic superpatriot, if put into practice, would demonstrably injure the economic interests of his own land. The trouble, then, springs from a root deeper than mere national interest. It lies in the peculiarly deceptive problems posed by international trade. What is deceptive about them, however, is not that they involve principles that apply only to relations between nations. *All the economic arguments that elucidate international trade apply equally well to domestic trade.* The deception arises, rather, for two reasons:

1. International trade requires an understanding of how two countries, each dealing in its own currency, manage to buy and sell from each other in a world where there is no such thing as international money.

2. International trade requires a very thorough understanding of the advantages of and arguments for trade itself.

Gains from Trade

In a general way, of course, we are all aware of the importance of trade, although we have hardly mentioned it since the opening pages of our book. *It is trade that makes possible the division and specialization of labor on which our productivity is so largely based.* If we could not exchange the products of our specialized labor, each of us would have to be wholly self-supporting, and our standard of living would thereupon fall to that of subsistence farmers. Thus trade (international or domestic) is actually a means of *increasing productivity,* quite as much as investment or technological progress.

GAINS FROM SPECIALIZATION

The importance of trade in making possible specialization is so great that we should take a moment to make it crystal clear. Let us consider two towns. Each produces two goods, wool and cotton; but Wooltown has good grazing lands and poor growing lands, while Cottontown's grazing is poor, but growing is good. Suppose, moreover, that the two towns had equal populations and that each town employed half its people in cotton and half in wool. The results might look like Table 35·1.

TABLE 35 · 1
Unspecialized Production: Case I

	Wooltown	Cottontown
Wool production (lbs)	5,000	2,000
Cotton production (lbs)	10,000	20,000

As we can see, the same number of grazers in Wooltown turn out two-and-one-half times as much wool as they do in Cottontown, whereas the same number of cotton farmers in Cottontown produce double the amount of cotton that they do in Wooltown. One does not have to be an economist to see that both towns are losing by this arrangement. If Cottontown would shift its wool-workers into cotton, and Wooltown would shift its cotton farmers into wool, the output of the two towns would look like Table 35·2 (assuming constant returns to scale).

TABLE 35 · 2
Specialized Production

	Wooltown	Cottontown
Wool output	10,000	0
Cotton output	0	40,000

Now, if we compare total production of the two towns (see Table 35·3), we can see the gains from specialization.

TABLE 35 · 3
The Gain from Specialization

	Mixed	Specialized	Gain from specialization
Wool output	7,000	10,000	3,000
Cotton output	30,000	40,000	10,000

In other words, specialization followed by trade makes it possible for both towns to have more of both commodities than they had before. No matter how the gains from trade are distributed—and this will depend on many factors, such as the relative elasticities of demand for the two products—both towns will gain, even if one gains more than the other.

UNEQUAL ADVANTAGES

If all the world were divided into nations, like Wooltown and Cottontown, each producing for trade only a single item in which it has a clear advantage over all others, international trade would be a simple matter to understand. It would still present problems of international payment, and it might still inspire its prime ministers of Nova Jersias to forego the gains from trade for political reasons that we will examine at the end of this chapter. But the essential rationale of trade would be simple to understand.

Unfortunately for the economics student as well as for the world at large, this is not the way international resources are distributed. Instead of giving each nation at least one commodity in which it has a clear advantage, many nations do not have such an advantage in a single product. How can trade possibly take place under such inauspicious circumstances?

To unravel the mystery, let us turn again to Cottontown and Wooltown, but this time call them Supraville and Infraville, to designate an important change in their respective abilities. Although both towns still enjoy equal populations, which are again divided equally between cotton and wool production, in this example Supraville is a more efficient producer than Infraville in *both* cotton and wool, as Table 35·4 shows.

TABLE 35 · 4
Unspecialized Production: Case II

	Supraville	Infraville
Wool output	5,000	3,000
Cotton production	20,000	10,000

Is it possible for trade to benefit these two towns when one of them is so manifestly superior to the other in every product? It seems out of the question. But let us nonetheless test the case by supposing that each town began to specialize.

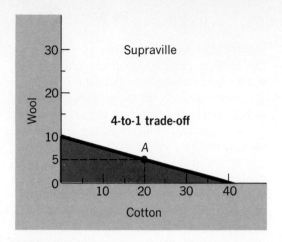
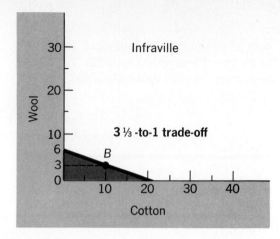

FIGURE 35·1
Production possibilities in the two towns before trade

TRADE-OFF RELATIONSHIPS

But how to decide which trade each town should follow? A look at Fig. 35·1 may give us a clue. The production-possibility diagrams are familiar to us from Chapter 20, where we used them to clarify the nature of scarcity and economic choice. Here we put them to use to let us see the results of trade.

What do the diagrams show? First, they establish maximums that each town could produce if it devoted all its efforts to one product. Since we have assumed that the labor force is divided, this means that each town could double the amount of cotton or wool it enjoys when it divides its workers fifty-fifty. Next, a line between these points shows the production frontier that both towns face.* We see that Supraville is located at point *A* where it has 5,000 lbs of wool and 20,000 lbs of cotton, and that Infraville is at

*Why are these lines drawn straight, not bowed as in Chapter 20? As we know, the bowing reflects the law of increasing cost, which makes the gains from a shift in resource allocation less and less favorable as we move from one extreme of allocation to another. Here we ignore this complication for simplicity of exposition. We have also ignored the problem of variable returns when we assumed that each town could double its output of cotton or wool by doubling its labor force.

B, where it has 3,000 lbs of wool and 10,000 lbs of cotton.

But the diagrams (and the figures in the preceding table, on which they are based) also show us something else. It is that each town has a different "trade-off" relationship between its two branches of production. When either town specializes in one branch, it must, of course, give up the output of the other. *But each town swaps one kind of output for the other in different proportions,* as the differing slopes of the two p-p curves show. Supraville, for example, can make only an extra pound of wool by giving up 4 pounds of cotton. That is, it gets its maximum potential output of 10,000 lbs of wool only by surrendering 40,000 lbs of cotton. Infraville can reach its production maximum of 6,000 lbs of wool at a loss of only 20,000 lbs of cotton. *Rather than having to give up 4 lbs of cotton to get one of wool, it gives up only 3.3 lbs.* Thus, in terms of how much cotton it must surrender, wool is actually cheaper in Infraville than in Supraville!

Not so the other way round, of course. As we would expect, cotton costs Supraville less in terms of wool than it costs Infraville. In Supraville, we get 40,000 lbs of cotton by relinquishing only 10,000 lbs of wool—a loss of a quarter of a pound of wool for a pound of the other. In Infraville, we can get the maximum output of 20,000 lbs of cotton only by a surrender of 6,000 lbs of wool—a loss of $\frac{1}{3}$ lb of wool rather than $\frac{1}{4}$ lb of wool for each unit of cotton.*

COMPARATIVE ADVANTAGE

Perhaps the light is beginning to dawn. Despite the fact that Supraville is more productive than Infraville in terms of output per man in both cotton and wool, it is *relatively* more productive in cotton than in wool. And despite the fact that Infraville is absolutely less productive than Supraville, man for man, in both cotton and wool, it is *relatively* more productive in wool. To repeat, it requires a smaller sacrifice of wool to get another pound of cotton in Infraville than in Supraville.

We call this kind of relative superiority *comparative advantage.* It is a concept that is often difficult to grasp at first but that is central to the reason for trade itself. When we speak of *comparative* advantage, we mean, as in the case of Supraville, that among *various* advantages of one producer or locale over another, there is one that is better than any other. *Comparatively* speaking, this is where its optimal returns lie. But just because it must abandon some lesser opportunity, its trading partner can now advantageously devote itself in the direction where *it* has a comparative advantage.

This is a relationship of logic, not economics. Take the example of the banker who is also the best carpenter in town. Will it pay him to build his own house? Clearly it will not, for he will make more money by devoting all his hours to banking, even though he

then has to employ and pay for a carpenter less skillful than himself. True, he could save that expense by building his own house. But he would then have to give up the much more lucrative hours he could be spending at the bank!

Now let us return to the matter of trade. We have seen that wool is *relatively* cheaper in Infraville, where each additional pound cost only 3.3 lbs of cotton, rather than 4 lbs as in Supraville; and that cotton is *relatively* cheaper in Supraville, where an additional pound costs but $\frac{1}{4}$ lb of wool, instead of $\frac{1}{3}$ lb across the way in Infraville. Now let us suppose that each side begins to specialize in the trade in which it has the comparative advantage. Suppose that Supraville took half its labor force now in wool and put it into cotton. Its output would change as in Table 35·5.

TABLE 35 · 5
Supraville

	Before the shift	After the shift
Wool production	5,000	2,500
Cotton production	20,000	30,000

Supraville has lost 2,500 lbs of wool but gained 10,000 lbs of cotton. Now let us see if it can trade its cotton for Infraville's wool. In Infraville, where productivity is so much less, the entire labor force has shifted to wool output, where its greatly inferior productivity can be put to best use. Hence its production pattern now looks like Table 35·6.

TABLE 35 · 6
Infraville

	Before the shift	After the shift
Wool	3,000	6,000
Cotton	10,000	—

*It takes long practice to master the arithmetic of gains from trade. Practice on questions 3 through 7 will help. It is more important, at this point, to "get the idea" than to master the calculations.

Infraville finds itself lacking 10,000 lbs of cotton, but it has 3,000 *additional* lbs of wool. Clearly, it can acquire the 10,000 lbs of cotton it needs from Supraville by giving Supraville *more* than the 2,500 lbs of wool it seeks. As a result, both Infraville and Supraville will have the same cotton consumption as before, but there will a surplus of 500 lbs of wool to be shared between them. As Fig. 35·2 shows, *both towns will have gained by the exchange, for both will have moved beyond their former production frontiers* (from *A* to *A'* and from *B* to *B'*).

This last point is the crucial one. If we remember the nature of production-possibility curves from our discussion of them in Chapter 20, any point lying outside the production frontier is simply unattainable by that society. In Fig. 35·2, points *A'* and *B'* do lie beyond the pre-trade *p-p* curves of the two towns, and yet trade has made it possible for both communities to enjoy what was formerly impossible.

OPPORTUNITY COST

Comparative advantage gives us an important insight into all exchange relationships,

for it reveals again a fundamental economic truth that we have mentioned more than once before. It is that *cost, in economics, means opportunities that must be foregone.* The real cost of wool in Supraville is the cotton that cannot be grown because men are engaged in wool production, just as the real cost of cotton is the wool that must be gone without. In fact, we can see that the basic reason for comparative advantage lies in the fact that opportunity costs vary, so that it "pays" (it costs less) for different parties to engage in different activities.

If opportunity costs for two producers are the same, then it follows that there cannot be any comparative advantage for either; and if there is no comparative advantage, there is nothing to be gained by specializing or trading. Suppose Supraville has a two-to-one edge over Infraville in *both* cotton and wool. Then, if either town specializes, neither will gain. Supraville may still gain 10,000 lbs of cotton for 2,500 lbs of wool, as before, but Infraville will gain only 2,500 lbs of wool (not 3,000) from its shift away from cotton. Thus, the key to trade lies in the existence of *different* opportunity costs.

Are opportunity costs usually different

FIGURE 35 · 2
Production possibilities in the two towns after trade

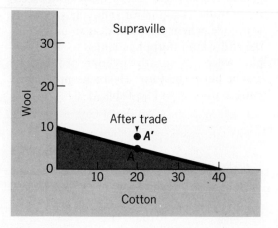

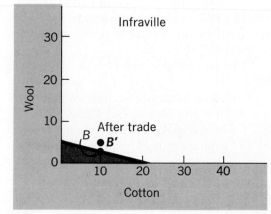

from country to country or from region to region? For most commodities they are. As we move from one part of the world to another—sometimes even short distances—climate, resources, skills, transportation costs, capital scarcity, or abundance all change; and as they change, so do opportunity costs. There is every possibility for rich countries to trade with poor ones, precisely because their opportunity costs are certain to differ.

EXCHANGE RATIOS

But we have not yet fully understood one last important aspect of trade—the *prices* at which goods will exchange. Suppose that Supraville and Infraville do specialize, each in the product in which it enjoys a comparative advantage. Does that mean they can swap their goods at any prices?

A quick series of calculations reveals otherwise. We remember that Supraville needed at least 2,500 lbs of wool for which it was going to offer some of its extra production of cotton in exchange. But how much? What price should it offer for its needed wool, in terms of cotton?

Suppose it offered 7,500 lbs of cotton. Would Infraville sell the wool? No, it would not. At home it can grow its own 7,500 lbs of cotton at a "cost" of only 2,273 lbs of wool, for we recall that Infraville traded off one pound of wool for 3.3 lbs of cotton (7,500 ÷ 3.3 = 2,273).

Suppose, then, that Infraville counter-offered to sell Supraville 2,500 lbs of wool for a price of 12,000 lbs of cotton. Would Supraville accept? Of course not. This would mean the equivalent of 4.8 lbs of cotton for a pound of wool. Supraville can do better than that by growing her own wool at her own trade-off ratio of only 4 lbs to one.

We begin to see, in other words, that the price of wool must lie between the trade-off ratios of Infraville and Supraville. Infraville

wants to import cotton. If it did not trade with Supraville, it could grow its own cotton at the cost of one pound of wool for every 3.3 lbs of cotton. Hence, for trade to be advantageous, Infraville seeks to get *more* cotton than that, per pound of wool.

Supraville is in the opposite situation. It seeks to export cotton and to import wool. It could make its own wool at the sacrifice of 4 lbs of cotton per pound of wool. Thus it seeks to gain wool for a *lower* price than that, in terms of cotton. Clearly, any ratio between 3.3 and 4.0 lbs of cotton per pound of wool will profit both sides.

THE ROLE OF PRICES

Let us put this into ordinary price terms. Suppose that cotton sells for 30¢ per pound. Then wool would have to sell between 99¢ and $1.20 (30¢ × 3.3 and × 4) to make trade worthwhile.* Let us say that supply and demand established a price of $1.10 for wool. Supraville can then sell its 10,000 lbs of extra cotton production at 30¢, which will net it $3,000. How much wool can it buy for this sum? At the going price of $1.10 per lb, 2,727 lbs. Therefore Supraville will end up with the same amount of cotton (20,000 lbs) as it had before specialization and trade, and with 227 *more* lbs of wool than before (2,500 lbs produced at home plus 2,727 lbs imported from Infraville—a total of 5,227 lbs). It has gained by trade an amount equal to the price of this extra wool, or $249.70.

How has Infraville fared? It has 3,273 lbs of wool left after exporting 2,727 lbs to Supraville from its production of 6,000 lbs, and it also has 10,000 lbs of cotton imported from Supraville in exchange for its wool ex-

---

*Needless to say, these prices are used for illustrative purposes only. And once again, let us reassure you: these calculations are easy to follow but not easy to do by yourself. Familiarity will come only with practice.

HIGH WAGES?

How do you tell whether a country is a high-wage country or a low-wage country? If German workers are paid 8 marks per hour, are their wages high or low compared to ours? Clearly, you cannot tell without knowing the exchange rate. If 8 marks can be traded for $1 then German workers are paid the equivalent of $1 per hour. We can then compare the German rates *relative* to the American. Given the average American wage of $4 per hour, we would conclude that German wages are low. If, however, 1 mark can be traded for 1 dollar, then German workers earn the equivalent of $8 per hour. In this case, German workers are highly paid *relative* to American workers. As a result we cannot really tell whether a country is high-wage or low-wage until we understand exchange rates and what determines them. More on this in the next chapter.

---

ports. Thus it, too, has a gain from trade—the 273 lbs of wool (worth $300.30) over the amount of 3,000 lbs that it would have produced without specialization and trade. In brief, *both* sides have profited from the exchange. To be sure, gains need not be distributed so evenly between the trading partners. If the price of wool had been $1.00, trade still would have been worthwhile, but Supraville would have gained almost all of it. Had the price of wool been $1.19, both sides again would have come out ahead, but now Infraville would have been the larger beneficiary by far. The actual price at which wool would sell would be determined by the supply and demand schedules for it in both communities.

The Case for Free Trade

Would the prime minister of Nova Jersia be convinced by these arguments? Would his wife? They might be weakened in their support for hothouse grapefruit, but some arguments would still linger in their minds. Let us consider them.

1. "Our workers cannot compete with low-wage workers overseas."

This is an argument one hears not only in Nova Jersia but also in the United States, where competition from sweatshop labor in Hong Kong is often cited at tariff hearings. And indeed, labor in Hong Kong is paid only a fifth of what it is paid here. Will not American labor be seriously injured if we import goods made under these conditions?

There is, of course, no doubt that an American textile worker who loses his job because of low-priced imports *will* be hurt. We shall come back to him later. But note that he would also be hurt if he lost his job because of regular domestic competition. Why do we feel so threatened when that competition comes from abroad?

Because, the answer goes, foreign competition isn't based on American efficiency. It is based on exploited labor. Hence it pulls down the standards of American labor to its own low level.

There is an easy reply to this argument. The reason Hong Kong textile labor is paid so much less than American textile labor is that *average* productivity in Hong Kong is so much lower than *average* productivity in America. To put it differently, the reason that American wages are high is that we use our workers in industries where their productivity is very high. If Hong Kong, with its very low productivity, can undersell us in textiles, then this is a clear signal that we must move our factors of production out of textiles into other areas where their contribution will be greater; for example, in the production of machinery. It is no coincidence that machinery—one of the highest wage industries in America—is one of our leading exports, or that more than 75 percent of our manufactured exports are produced by industries paying hourly wage rates above the national average for all manufacturing industries. In fact, all nations tend to export the goods that are produced at the highest, not lowest, local wages! Why? Because those industries employ their labor most effectively.

This fact opens our eyes to another. Why is it that the American garment industry is worried about competition from Hong Kong, but not the American auto industry or the electrical machinery industry or the farm equipment industry? After all, the manufacturers of those products could also avail themselves of low wages in Hong Kong.

The answer is that American manufacturers can make these products at much lower cost in America. Why? Because the technical skills necessary to produce them are available in the U.S., not in Hong Kong. Thus, if Hong Kong has a comparative advantage over us in the garment trade, we have a comparative advantage over her in many other areas.*

But suppose Hong Kong accumulated large amounts of capital and became a center for the manufacture of heavy equipment, so that it sold *both* garments and electrical generators more cheaply than we sold them. We are back to Supraville and Infraville. There would still be a *comparative* advantage in one or more of these products in which we would be wise to specialize, afterward trad-

ing with Hong Kong for our supplies of the other good.*

2. *"Tariffs are painless taxes because they are borne by foreigners."*

This is a convincing-sounding argument advanced by the prime minister of Nova Jersia (and by some other prime ministers in their time). But is it true? Let us take the case of hothouse grapefruit, which can be produced in Nova Jersia only at a cost of fifty cents each, whereas foreign grapefruit (no doubt produced by sweated labor) can be unloaded at its ports at twenty-five cents. To prevent his home industry from being destroyed, the prime minister imposes a tariff of twenty-five cents on foreign grapefruit—which, he tells the newspapers, will be entirely paid by foreigners.

This is not, however, the way his political opponent (who has had a course in economics) sees it. "Without the tariff," he tells his constituency, "you could buy grapefruit for twenty-five cents. Now you have to pay fifty cents for it. Who is paying the extra twenty-five cents—the foreign grower or you? Even if not a single grapefruit entered the country, you would still be paying twenty-five cents more than you have to. In fact, *you are being asked to subsidize an inefficient domestic industry.* Not only that, but the tariff wall means they won't ever become efficient because there is no pressure of competition on them."

Whether or not our economic candidate will win the electoral battle, he surely has the better of the argument. Or does he? For the prime minister, stung by these unkind remarks, replies:

---

*If every industry must have a comparative advantage in one country or another, how can there be steel industries (or any other) in more than one country? The answer, quite aside from considerations of nationalism, lies in *transportation costs,* which compensate for lower production costs in many products and thereby allow a relatively inefficient industry to supply a home market.

*Newspapers in Southeast Asia carry editorials seeking protection from American imports because, they say, we do not use labor in our production, and it is unfair to ask its citizens to compete with our machines that do not have to be paid wages.

ROBINSON CRUSOE

The beclouding effect of national bias on our thinking was never more charmingly or effectively presented than in this argument by Frédéric Bastiat, a delightful exponent of mid-nineteenth-century classical economic ideals, in a little book entitled *Social Fallacies.*[1]

In Bastiat's book, Robinson Crusoe inhabits an island with Friday. In the morning, Crusoe and Friday hunt for six hours and bring home four baskets of game. In the evening, they garden for six hours and get four baskets of vegetables. But now let Bastiat take over:

One day a canoe touched at the island. A good-looking foreigner landed and was admitted to the table of our two recluses. He tasted and commended very much the produce of the garden, and before taking leave of his entertainers, spoke as follows: "Generous islanders, I inhabit a country where game is much more plentiful than here, but where horticulture is quite unknown. It would be an easy matter to bring you every evening four baskets of game, if you will give me in exchange two baskets of vegetables."

At these words, Robinson and Friday retired to consult, and the debate that took place is too interesting not to be reported in extenso.

FRIDAY: What do you think of it?
ROBINSON: If we close with the proposal, we are ruined.
FRIDAY: Are you sure of that? Let us consider.
ROBINSON: The case is clear. Crushed by competition, our hunting as a branch of industry is annihilated.
FRIDAY: What matters it, if we have the game?
ROBINSON: Theory! It will no longer be the product of our labour.
FRIDAY: I beg your pardon sir; for in order to have game we must part with vegetables.
ROBINSON: Then, what shall we gain?
FRIDAY: The four baskets of game cost us six hours' work. The foreigner gives us them in ex-

[1] Translated by Frederick James Sterling (Santa Ana, Calif.: Register Publishing, 1944), pp. 203f.

3. "But at least the tariff keeps spending power at home. Our own grapefruit growers, not foreigners, have our money."

There are two answers to this argument. First, the purchasing power acquired by foreigners can be used to buy goods from efficient Nova Jersia producers and will thus return to Nova Jersia's economy. Second, if productive resources are used in inefficient, low-productivity industries, then the resources available for use in efficient, high-productivity industries are less than they otherwise would be, and the total output of the country falls. To keep out foreign grapefruit is to lower the country's real standard of living. The people of Nova Jersia waste time and resources doing something they do not do very well.

4. "But tariffs are necessary to keep the work force of Nova Jersia employed."

This is the time to remember our investigation of macroeconomic policies. As we learned in macroeconomics, the governments of Nova Jersia and every other country can use fiscal and monetary policies to keep their resources fully employed. If textile workers become unemployed, governments can expand aggregate demand and generate domestic job opportunities in other areas.

CLASSICAL ARGUMENT FOR FREE TRADE

Are there no arguments at all for tariffs? As we shall see, there are some rational arguments for restricting free trade. *But all of these arguments accept the fact that restrictions depress world incomes below what they would be otherwise. If world production is to be maximized, free trade is an essential in-*

change for two baskets of vegetables, which cost us only three hours' work. This places three hours at our disposal. . . .

ROBINSON: You lose yourself in generalities! What should we make of these three hours?

FRIDAY: We would do something else.

ROBINSON: Ah! I understand you. You cannot come to particulars. Something else, something else—that is easily said.

FRIDAY: We can fish, we can ornament our cottage, we can read the Bible.

ROBINSON: Utopia! Is there any certainty we should do either the one or the other? . . . Moreover there are political reasons for rejecting the interested offers of the perfidious foreigner.

FRIDAY: Political reasons!

ROBINSON: Yes, he only makes us these offers because they are advantageous to him.

FRIDAY: So much the better, since they are for our advantage likewise. . . .

ROBINSON: Suppose the foreigner learns to cultivate a garden and that his island should prove more fertile than ours. Do you see the consequences?

FRIDAY: Yes; our relations with the foreigner would cease. He would take from us no more vegetables, since he could have them at home with less labour. He would bring us no more game, since we should have nothing to give him in exchange, and we should then be in precisely the situation that you wish us in now. . . .

The debate was prolonged, and, as often happens, each remained wedded to his own opinion. But Robinson possessing a great influence over Friday, his opinion prevailed, and when the foreigner arrived to demand a reply, Robinson said to him: "Stranger, in order to induce us to accept your proposal, we must be assured of two things: the first is, that your island is no better stocked with game than ours, for we want to fight only with equal weapons. The second is, that you will lose by the bargain. For, as in every exchange there is necessarily a gaining and a losing party, we should be dupes, if you were not the loser. What have you got to say?"

"Nothing," replied the foreigner; and, bursting out laughing, he regained his canoe.

gredient. Free trade must therefore be considered a means of increasing GNP, a means not essentially different from technological improvement in its effect on output and growth. We may not want to maximize GNP, but we need to understand that to advocate restrictions on trade is to advocate lower real incomes.*

The Case for Tariffs

Are *all* arguments against tariffs? Not quite. But it is essential to recognize that these arguments take full cognizance of the inescapable

*These arguments apply cogently to developed countries. They are less persuasive when applied to underdeveloped countries, as we shall see when we discuss imperialism in Chapter 39.

costs of restricting trade. They do not contest the validity of the theory of free trade, but the difficulties of its application. Let us familiarize ourselves with them.

MOBILITY

The first difficulty concerns the problem of mobility. Explicit in Bastiat's case is the ease with which Crusoe and Friday move back and forth between hunting and gardening. Implicit in the case of Supraville and Infraville is the possibility of shifting men and resources from cotton to wool production. But in fact it is sometimes exceedingly difficult to move resources from one industry to another.

Thus when Hong Kong textiles press hard against the garment worker in New York, it is scant comfort to him to point to the higher wages in the auto factories in Detroit. He has a lifetime of skills and a long-established home in New York, and he does not want to move to another city where he will be a stranger and to a new trade in which he would be only an unskilled beginner. He certainly does not want to move to Hong Kong! Hence, the impact of foreign trade often brings serious dislocations that result in persistent local unemployment, rather than in a flow of resources from a relatively disadvantaged to a relatively advantaged one. If Crusoe had suggested that it was very difficult (perhaps because of the noonday sun) to work in the gardens in the morning when they usually went hunting, Friday would have been harder put for a reply.

TRANSITION COSTS

Second, we have seen that free trade is necessary to maximize world incomes and that it increases the incomes and real living standards of each country participating in trade. *But this does not mean that it increases the income and real living standards of each individual in each country.* Our New York textile worker may find himself with a substantial reduction in his income for the rest of his life. He is being economically rational when he resists "cheap" foreign imports and attempts to get his congressman to impose tariffs or quotas.

There is, it should be noted, an answer to this argument—an answer, at any rate, that applies to industrial nations. Since the gains from trade are generally spread across the nation, the real transition costs of moving from one industry, skill, or region to another should also be generally spread across the nation. This means that government (the taxpayers), rather than the worker or businessman, should bear the costs of relocation and retraining. In this way we spread the costs

in such a manner that a few need not suffer disproportionately to win the benefits of international trade that are shared by many.

We should also be aware of the possibility that transition costs may actually exceed the short-term benefits to be derived from international trade. Transition costs thus place a new element in the system, since the standard analysis of competitive systems —national or international—ignores them. A country may be wise to limit its international trade, if it calculates that the cost of reallocating its own factors is greater than the gains to be had in higher real income. Remember, however, that transition costs tend to be short-lived and that the gains from trade tend to last. Thus it is easy to exaggerate the costs of transition and to balk at making changes that would ultimately improve conditions.

FULL EMPLOYMENT

Third, *the argument for free trade rests on the very important assumption that there will be substantially full employment.*

In the days of the mid-nineteenth century when the free trade argument was first fully formulated, the idea of an underemployment equilibrium would have been considered absurd. When Crusoe asks what use they should make of their free time, Friday has no trouble replying that they should work or enjoy their leisure. But in a highly interdependent society, work may not be available, and leisure may be only a pseudonym for an inability to find work. In an economy of large enterprises and "sticky" wages and prices, we know that unemployment is a real and continuous object of concern for national policy.

Thus, it makes little sense to advocate policies to expand production via trade unless we are certain that the level of aggregate demand will be large enough to absorb that production. *Full employment policy therefore becomes an indispensable arm of trade policy.* Trade gives us the potential for maximizing production, but there is no point in laying the groundwork for the highest possible output, unless fiscal and monetary policy are also geared to bringing about a level of aggregate demand large enough to support that output.

NATIONAL SELF-SUFFICIENCY

Fourth, *there is the argument of nationalism pure and simple.* This argument does not impute spurious economic gains to tariffs. Rather, it says that free trade undoubtedly encourages production, but it does so at a certain cost. This is the cost of the vulnerability that comes from extensive and extreme specialization. This vulnerability is all very well within a nation where we assume that law and order will prevail, but it cannot be so easily justified among nations where the realistic assumption is just the other way. Tariffs, in other words, are defensible because they enable nations to attain a certain *self-sufficiency*—admittedly at some economic cost. Project Independence, the United States' effort to gain self-sufficiency in energy, is exactly such an undertaking.

When Crusoe argued that trade might cease, Friday properly scoffed. But the argument is much more valid for an economy of complex industrial processes and specialized know-how that cannot be quickly duplicated if trade is disrupted. In a world always threatened by war, self-sufficiency has a value that may properly override considerations of ideal economic efficiency. The problem is to hold the arguments for "national defense" down to proper proportions. When tariffs are periodically adjusted in international conferences, an astonishing variety of industries

(in all countries) find it possible to claim protection from foreign competition in the name of national "indispensability."

INFANT INDUSTRIES

Equally interesting is the nationalist argument for tariffs advanced by so-called infant industries, particularly in developing nations. These newly-formed or prospective enterprises claim that they cannot possibly compete with the giants in developed countries while they are small; but that if they are protected by a tariff, they will in time become large and efficient enough no longer to need a tariff. In addition, they claim, they will provide a more diversified spectrum of employments for their own people, as well as aiding in the national transition toward a more modern economy.

The argument is a valid one if it is applied to industries that have a fair chance of achieving a comparative advantage once grown up (otherwise one will be supporting them in infancy, maturity, and senility). Certainly it is an argument that was propounded by the youthful industries of the United States in the early nineteenth century and was sufficiently persuasive to bring them a moderate degree of protection (although it is inconclusive as to how much their growth was ultimately dependent on tariff help). And it is being listened to today by the underdeveloped nations who feel that their only chance of escaping from poverty is to develop a nucleus of industrial employment at almost any cost in the short run.

PRODUCERS' WELFARE

Finally there is an argument that comes down to desired life styles and the quality of life. Economists tend to think entirely of consumers' welfare and to ignore producers' welfare. By definition, work is a "disutility" that must create pain. If it didn't, the bribe of wages would not have to be paid to get people to work after subsistence was met. But in fact, the quality of an individual's productive life may be as important to him as, or more important than, the quality of his consumptive life. Individuals can and do choose to have lower standards of consumption in exchange for a job that they enjoy. Whole countries may make the same choice.

Assume for the moment that the U.S. has a comparative advantage in agricultural production vis-à-vis France, but Frenchmen enjoy being farmers. In a world of free trade, Frenchmen would be driven out of farming. They would work in the cities and have more goods and services than they would have on their farms. But they would no longer be able to enjoy their farms. Is it irrational for France to place high tariffs and quotas on American agricultural exports in this case? Clearly not. The only irrationality occurs when countries pretend that such actions do not impose costs and when they do not tell their populations that the whole country (farmers and nonfarmers) must reduce its material standard of living so that some can enjoy their work.

The problem of producers' welfare—the quality of work rather than consumption—is one with which neither economists nor society has adequately come to grips. It may become a key area in raising real standards of living.

THE BASIC ARGUMENT

Thus there are arguments for tariffs, or at least rational counterarguments against an extreme free trade position. Workers *are* hurt by international competition; and in the default of proper domestic plans for cushioning these blows, modest tariffs can buffer the pains of redeployment. Free trade *does* require a level of high employment; and when unemployment is already a national problem, tariffs may protect additional workers from losing their jobs. Strategic industries and development-stimulating industries *are* sometimes essential and may require protection from world competition. People may enjoy doing jobs they do not do as efficiently as other jobs. All these arguments are but qualifications to the basic proposition on which the economist rests his case for the freest *possible* trade, but they help to define "possible" in a realistic way.

Nonetheless it may help if we sum up the classical argument, for there is always a danger that the qualifications will take precedence over the main argument.

1. Free trade brings about the most efficient possible use of resources, and any interference with free trade lessens that efficiency.

Note that international trade is in no way different from interregional domestic trade in this regard. We recognize that we would suffer a loss in higher costs or smaller output by imposing restrictions on the exchange of goods between New York and Chicago. We suffer the same loss when we interfere with the exchange between New York and Hong Kong, whether by tariffs, quotas, or other means.

2. When international trade brings frictional problems, such as unemployment in an industry that cannot meet foreign competition, the answer is not to block the imports but to cure the unemployment by finding better uses for our inefficiently used resources.

Once again, international trade is no different in this regard from domestic trade. When low-price textiles from the South cause unemployment in New England, we do not prevent the sale of southern goods. We try to

ADAM SMITH ON FOREIGN TRADE

"Each nation has been made to look with an invidious eye upon the prosperity of all the nations with which it trades, and to consider their gain as its loss. Commerce, which ought naturally to be, among nations, as among individuals, a bond of union and friendship, has become the most fertile source of discord and animosity. The capricious ambition of kings and ministers has not, during the present and preceding century, been more fatal to the repose of Europe, than the impertinent jealousy of merchants and manufacturers. The violence and injustice of the rulers of mankind is an ancient evil, for which, I am afraid, the nature of human affairs can scarce admit of a remedy. But the mean rapacity, the monopolizing spirit of merchants and manufacturers, who neither are, nor ought to be, the rulers of mankind, though it cannot perhaps be corrected, may very easily be prevented from disturbing the tranquillity of anybody but themselves."—*Wealth of Nations* (Modern Library ed.), p. 460.

find new jobs for New Englanders, in occupations in which they have a comparative advantage over the South.

3. The purpose of all trade is to improve the well-being of the consumer by giving him the best and cheapest goods and services possible. Thus imports, not exports, represent the gains from trade.

The whole point of trade is to exchange things that we make efficiently for other things in which our efficiency is less. Anything that diminishes imports will reduce our standard of living, just as anything that blocks a return flow of goods from Chicago to New York will obviously reduce the benefit to New Yorkers of trade with Chicago.

In our last section, when we turn to the troubles of the underdeveloped world, we will see some of these matters illustrated not in textbook example, but in reality. However, it is encouraging that since 1948 the total value of world exports has risen from $54 billion to over $250 billion, and that the volume of world trade has been increasing at the rate of 6 percent a year since the 1960s.*

*A considerable part of the impetus to the growth of world trade must be credited to the spread of more rational—i.e., lower and fewer—tariff barriers. The General Agreement on Tariffs and Trade (GATT), an international body formed in 1947 to work for wider world trade, has succeeded in steadily reducing tariff levels and in dismantling import quotas. It is pleasant to record that the United States initially played a major role in this movement. During the 1930s we had the unenviable reputation of being one of the most restrictive trading nations in the world, but our tariff wall has been far reduced since those irresponsible days. Today our average level of duties on dutiable imports is roughly 10 percent, compared with 53 percent in 1930; and in addition, a third of all our imports are admitted duty-free. On the negative side, however, it must be noted that we continue to discriminate against imports that affect our manufacturing interests. For example, coffee comes in free, but not instant coffee, so that the underdeveloped nations who would like to process coffee within their own economies are gravely disadvantaged. Moreover, in recent years there has been a revival of U.S. protectionism, resulting in "voluntary" agreements on the part of foreign producers to restrict certain kinds of exports, such as textiles, to us, and (until very recently) in quotas on oil imports. The lessons of free trade continue to persuade economists more than businessmen.

KEY WORDS	CENTRAL CONCEPTS

KEY WORDS   CENTRAL CONCEPTS

Gains from trade

1. Imports and exports constitute small but strategic branches of economic activity for the United States. Precisely *the same arguments of economic rationality apply to them as to all purely domestic economic exchanges.* It is mainly the bias of nationalism that hinders us from applying the same reasoning in both cases.

Specialization

2. The gains from trade essentially arise *from the specialization and division of labor that trade makes possible.* Trade is a means of *increasing productivity.*

Comparative advantage

3. Trade is obviously advantageous when each of the two trading partners has a clear superiority in the production of one item. It can also be advantageous whenever a *comparative advantage exists*—that is, whenever one partner, although superior to the other in the production of all products, is relatively superior in some. By the logic of the case, the inferior partner must be relatively superior in the production of the others, and output will be increased if each devotes its energies to its advantaged products.

Trade-offs

Opportunity cost

4. We can tell which product is relatively favored for each trading partner by calculating the *trade-off relationship* represented by the productivity curve. This shows us the *opportunity cost* of producing each product in terms of the output of other products that must be foregone.

5. Products must exchange *at prices that lie between the ratios established by the trade-off possibilities.* Neither country will accept in exchange a product on terms (that is, at prices) that are less favorable than the terms it has available by devoting its own resources to the production of the item in question.

Low-wage arguments

6. The arguments against free trade often stress the dangers of low-wage competition. This argument overlooks the fact that *low wages are a symptom of low productivity.* Generally it is high-wage (i.e., high-productivity) items that are exported. It also ignores the fact that low-wage countries are generally deficient in capital and have trouble competing with capital-using products from nations in which capital is abundant.

Tariff

7. *Tariffs are a subsidy to a domestic industry that cannot meet competition from abroad.*

Factor mobility

8. One remediable cost of free trade is the necessity for *factor mobility.* Government can help bear relocation and retraining costs in cases where free trade imposes severe strains of readjustment.

Free trade

9. *Free trade assumes the existence of full employment, and a policy of encouraging free trade must be accompanied by one encouraging maximum output.*

Infant industries

10. *National self-sufficiency* and the encouragement of *infant industries* during their early years provide rational arguments for tariffs, although it is not easy to prevent these arguments from being indiscriminately applied.

11. The basic argument for free trade is that it brings about the *most efficient use of resources.* The purpose of all trade is to improve the well-being of the consumer. *Thus, imports, not exports, represent the gain from trade.*

Transition cost

12. Since the gains from international trade benefit the whole society, a case can be made for governments to assume the real *transition costs* that must be incurred to move factors of production from one place to another. If this does not happen, specific individuals are apt to suffer large reductions in income as the result of free trade.

Producers' welfare

13. *Perhaps the most compelling argument for restrictions on trade is that they are necessary to maximize producers' welfare.* People often like doing jobs in which they do not have a comparative advantage. As long as the costs of this option are known, it may be perfectly sensible to sacrifice consumption goods to get more enjoyment from work.

QUESTIONS

1. Is it true that a colossal nation like the United States can trade with a tiny one like Honduras to the benefit of both? Can it also trade with an industrial, small nation, like Holland? What products do we buy from and sell to each?

2. What do we mean when we say that trade is "indirect production"?

3. Suppose that two towns, Coaltown and Irontown, have equal populations but differing resources. If Coaltown applies its whole population to coal production, it will produce 10,000 tons of coal; if it applies them to iron production, it will produce 5,000 tons of iron. If Irontown concentrates on iron, it will turn out 18,000 tons of iron; if it shifts to coal, it will produce 12,000 tons of coal. Is trade possible between these towns? Would it be possible if Irontown could produce 24,000 tons of iron? Why is there a comparative advantage in one case and not in the other?

4. In which product does Coaltown have a comparative advantage? How many tons of iron does a ton of coal cost her? How many does it cost Irontown? What is the cost of iron in Coaltown and Irontown?

5. Draw a production-possibility diagram for each town. Show where the frontier lies before and after trade.

6. If iron sells for $10 a ton, what must be the price range of coal? Show that trade cannot be profitable if coal sells on either side of this range.

7. What is the opportunity cost of coal to Irontown? Of iron to Irontown?

8. Is it true that American watchmakers face unfair competition from Swiss watchmakers because wages are lower in Switzerland? If American watch workers are rendered unemployed by the low-paid Swiss, what might be done to help them—impose a tariff?

9. Is it true that mass-produced, low-cost American watches are a source of unfair competition for Switzerland? If Swiss watchmakers are unemployed as a result, what could be done to help them—impose a tariff? Is it possible that a mutually profitable trade in watches might take place between the two countries? What kinds of watches would each probably produce?

10. Are the duties on French wines borne by foreigners or by domestic consumers? Both? What, if any, is the rationale for these duties?

11. Do you believe that there should be a tariff on steel products because steel is essential for national defense? Should we refuse to buy low-cost Russian turbogenerators because the domestic industry needs support?

12. Why do imports represent the gains from trade and not exports?

13. How would you go about estimating the transition costs if we were to abolish the tariff on all wines and spirits? Who would be affected? What alternative employment would you suggest for the displaced labor? The displaced land?

14. What kinds of criteria would be useful in applying the argument of producers' welfare over that of consumers? Would it be easy to apply in the case of wines and spirits? In the case of a nation such as Mexico, in which agriculture is a way of life?

The Mechanism of International Transactions

36

We have learned something about one of the sources of confusion that surrounds international trade—the curiously concealed gains from trade itself. Yet our examples of trade have thus far not touched on another source of confusion—the fact that international trade is conducted in two (or sometimes more) currencies. After all, remember that Infraville and Supraville both trade in dollars. But suppose Infraville were Japan and Supraville America. Then how would things work out?

Foreign Exchange

The best way to find out would be to price the various items in Japan and America (assuming that Japan produces both wool and cotton, which she does not). Suppose the result looked like the table above, right.

TABLE 36 · 1

	United States	Japan
Price of wool (lb)	$1.10	¥ 300
Price of cotton (lb)	.30	¥ 100

What would this tell us about the cheapness or dearness of Japanese products compared with those of the U.S.? Nothing, unless we knew one further fact: *the rate at which we could exchange dollars and yen.*

Suppose you could buy 400 yen for a dollar. Then a pound of Japanese wool imported into America (forgetting about shipping costs) would cost 75¢ (¥300 ÷ 400), and a pound of Japanese cotton in America would cost $0.25 (¥100 ÷ 400). Assuming that these are the only products that either country makes for export, here we have a case in which Japan can seemingly undersell America in everything.

But now suppose the rate of exchange were not 400 to one but 250 to one. In that event a pound of Japanese wool landed in America would cost $1.20 (¥300 ÷ 250); and a pound of cotton, $0.40. At this rate of exchange everything in Japan is more expensive than the same products produced in the United States.

The point is clear. *We cannot decide whether foreign products are cheaper or dearer than our own until we know the rate of exchange,* the number of units of their currency we get for ours.

MECHANISM OF EXCHANGE: IMPORTS

How does international exchange work? The simplest way to understand it is to follow through a single act of international exchange from start to finish. Suppose, for example, that we decide to buy a Japanese camera directly from a Tokyo manufacturer. The price of the camera as advertised in the catalog is ¥20,000, and to buy the camera we must therefore arrange for the Japanese manufacturer to get that many *yen*. Obviously, we can't write him a check in that currency, since our own money is in dollars; and equally obviously we can't send him a check for dollars, since he can't use dollars in Tokyo any more than we could use a check from him in yen.

Therefore, we go to our bank and ask if it can arrange to sell us yen to be delivered to the Tokyo manufacturer. Yes, our bank would be delighted to oblige. How can it do so? The answer is that our bank (or if not ours, another bank with whom it does business) keeps a regular checking account in its own name in a so-called correspondent bank in Tokyo. As we might expect, the bank in Tokyo also keeps a checking account in dollars in *its* own name at our bank. If our banker has enough yen in his Tokyo account,

he can sell them to us himself. If not, he can buy yen (which he will then have available in Japan) from his correspondent bank in exchange for dollars which he will put into their account here.

Notice that two currencies change hands —not just one. Notice also that our American banker will not be able to buy yen unless the Japanese banker is willing to acquire dollars. And above all, note that banks are the intermediaries of the foreign exchange mechanism because they hold deposits in foreign banks.

When we go to our bank to buy ¥20,000, the bank officer looks up the current exchange rate on yen. Suppose it is 385. He then tells us that it will cost us $51.95 (20,000 ÷ 385) to purchase the yen, plus a bank commission for his services. We write the check, which is deducted from our bank balance and added to the balance of the Tokyo bank's account in this country. Meanwhile, the manufacturer has been notified that if he goes to the Tokyo bank in which our bank keeps its deposits of yen, he will receive a check for ¥20,000. In other words, the Tokyo bank, having received dollars in the United States, will now pay out yen in Japan.

EXPORTS

Exactly the opposite is true in the case of exports. Suppose that we were manufacturers of chemicals and that we sold a $1,000 order to Tokyo. In Japan, the importer of chemicals would go to his bank to find out how many yen that would cost. If the rate were 385, it would cost him ¥385,000, which he would then pay to the Japanese bank. The bank would charge his account and credit the yen to the Tokyo account of an American bank with which it did business, meantime advising the bank here that the transaction had taken place. When the appropriate papers arrived from Japan, our U.S. bank would then take note of its increased holdings of yen and pay the equivalent amount in dollars into our account.

FOREIGN EXCHANGE

Thus the mechanism of foreign exchange involves the more or less simultaneous (or anyway, closely linked) operations of two banks in different countries. One bank accepts money in one national denomination, the other pays out money in another denomination. Both are able to do so because each needs the other's currency, and each maintains accounts in the other country. *Note that when payments are made in international trade, money does not physically leave the country.* It travels back or forth between American-owned and foreign-owned bank accounts *in America.* The same is true in foreign nations, where their money will travel between an American-owned account there and the account of one of their nationals. *Taken collectively, these foreign-owned accounts (including our own overseas) are called "foreign exchange." They constitute the main pool of moneys available to finance foreign trade.*

Exchange Rates

Thus the mechanism of foreign exchange works through the cooperation of banks. But we must go beyond an understanding of the mechanism to see the actual forces of supply and demand at work. And this is confusing because we have to think in two money units at the same time.

BUYING AND SELLING MONEY

We are used to thinking of the price of shoes in terms of dollars. We don't turn around and ask what is the price of dollars in terms of shoes, because consumers don't use shoes to buy dollars.

When we buy pounds or francs or yen, however, we are buying a commodity that is indeed usable to buy the very money we are using. Dollars buy francs and marks and yen; and marks, francs and yen buy dollars. We will have to bear this in mind when we seek to understand the supply and demand curves for international exchange.

Now let us consider an exchange market, say the market for yen (Fig. 36·1). The

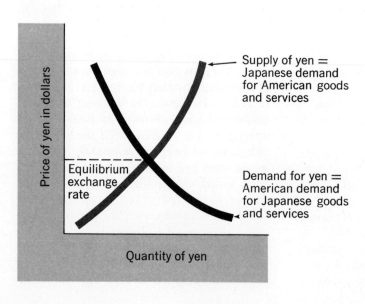

FIGURE 36·1
The market for exchange

demand curve for yen is easy to understand. It shows us that we will want to acquire larger amounts of yen as they get cheaper. Why? Because cheap yen means relatively cheaper Japanese goods and services. *Really our demand curve for foreign exchange is a picture of our changing demand for foreign goods and services as these goods get cheaper or dearer because the money we use to buy them gets cheaper or dearer.*

Now the supply curve. We can most easily picture it as the changing willingness and ability of Japanese banks to offer yen as we pay high or low prices for yen. (There is a better way of explaining the supply curve, but it takes some hard thought. Those who want to penetrate the mysteries of foreign exchange should look at the box, pp. 574-75.)

EQUILIBRIUM PRICES

What is important is that our diagram shows that there is an equilibrium price for yen that just clears the market. At that price, the amounts of yen that Americans want are exactly equal to the amounts of yen that Japanese want to supply. If we look through the "veil of money," we can see that at this price the value of all Japanese goods and services that we will buy must also be equal to the value of all American goods and services that they will buy!

APPRECIATION AND DEPRECIATION OF EXCHANGE RATES

From this, a very important result follows. Suppose that a United States importer wants to buy Japanese automobiles priced at ¥1 million per car. He goes to the bank to finance the deal. Here he has an unpleasant surprise. His banker tells him that exchange is very "tight" at the moment, meaning that the banker's own yen balances in Japan are very small. As a result, the American banker can no longer offer yen at the old price of,

say, 350 to the dollar. The Japanese banks with whom he does business are insisting on a higher price for yen—offering only 325 or perhaps even 300 yen for a dollar. Because of supply and demand, the yen has risen in price, or *appreciated;* and the dollar has fallen in price, or *depreciated.*

The importer now makes a quick calculation. At 350 yen to the dollar, a Japanese car that costs ¥1 million will cost him $2,857 (¥1,000,000 ÷ 350). At an exchange rate of 300, it would now cost him $3,333 (¥1,000,000 ÷ 300). The new higher price is too steep for the American market. He decides not to place the order. Exactly the opposite situation faces the Japanese importer. Suppose he wants to buy a $50,000 IBM computer. How much will it cost him *in yen* if he has to pay 350 yen for a dollar? 300 yen?

The principle is very clear. *Movements in exchange rates change relative prices among countries.* At different relative prices, imports will rise or fall, as will exports. If the price of the dollar falls, American exports will be increased and its imports diminished. If the price of the yen rises, Japanese exports will fall and its imports will rise.

Thus a moving exchange rate will automatically bring about an equilibrium between the demand for, and the supply of, foreign exchange, exactly as a moving price for shoes will bring about an equality between the value of the dollars offered for shoes and the value of the shoes offered for dollars! In one case as in the other, there may be time lags. *But the effect of a moving price in both cases is to eliminate "shortages" and "surpluses"—that is, to bring about a price at which quantity demanded (of a particular currency or any other commodity) equals quantity supplied.*

The Balance of Payments

How, then, can we account for the deficit we hear so much about in the U.S. balance of

ANOTHER LOOK AT
THE EXCHANGE PROBLEM

Let us trace the exchange process once more, very carefully. The chart of the New York market shows the demand for English pounds in dollars. When it costs $3 to buy £1, our demand is for one million pounds (we can think of them as com-

modities, like one million shoes). This is point **A**. When the price falls to $2 for £1, our demand rises to 2 million pounds, point **B**. The broken line **AB** is our demand curve for pounds.

Now we move to the London market on the right. We are going to show that the New York demand curve **AB** becomes a London supply curve **A'B'**. To do so, remember that when it costs

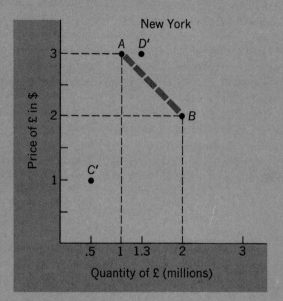

New York

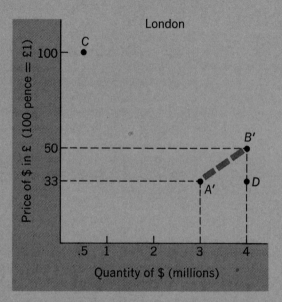

London

---

payments? To understand the answer, we first have to understand what we mean by the "balance" of payments. We don't speak of a "balance" of payments in, say, a market for shoes. Why, then, is there one in the market for foreign exchange?

DISAGGREGATING THE BALANCE OF PAYMENTS

The first part of the answer lies in an important attribute of this market. In a shoe market, all buyers want shoes, presumably to wear. In an exchange market, there are many

kinds of buyers (or sellers) who want to buy or sell exchange for different purposes. That is, the so-called balance of payments represents supplies and demand for foreign exchange by *different groups* in each economy. When all supplies and all demands are added together, the two totals must balance because we then have an identity: Purchases ≡ Sales (i.e., the purchases of any currency, such as dollars, must equal the sales of that currency). But they need not balance for any particular group in the economy. As a result, deficits and surpluses refer to groups that are demanding more foreign exchange than they supply, or supplying more than they demand.

$3 to buy £1, from the London point of view the price of $1 is 33 pence (one-third of a pound). What is the supply of dollars at this price? It is equal to the number of dollars spent for pounds in New York. At the $3 price, we bought one million pounds. Our supply of dollars is therefore $3 million. This gives us point A' in the London market.

It is now simple to get point B'. When £1 falls to $2 in New York, $1 rises to 50 pence (one-half pound) in London. How many dollars are supplied at this price? We can see in the New York diagram that we bought £2 million at $2 each, spending $4 million. Hence in the London market, we locate point B' at a price of 50 pence and a quantity of dollars equal to 4 million.

Now we have a demand curve in New York and a supply curve in London. We need a supply curve in New York and a demand curve in London. We'll start in London, with a high price for dollars. Point C shows us that when it costs £1 to buy $1, the demand for dollars is small—only $.5 million are demanded. But this point on the demand curve also gives us a supply of pounds: .5 million "units" of dollars at £1 each, or a total of £.5 million. Back in New York this shows up as point C'. (Remember: $1 = £1.)

Now back to London. The price of dollars falls to 33 pence or ⅓ of a pound. At that price, suppose Britishers demand $4 million, point D. To buy $4 million at 33 pence each, Britishers will have to spend 132 million pence, or £1.32 million. This gives us the supply of pounds in the New York market at the price that corresponds to $1 = 33 pence. This price is $3 = 100 pence (one pound). Point D' locates the supply curve at that price. We suggest you draw the two new curves: CD, the demand for dollars, and C'D', the supply of dollars.

Each panel now has an equilibrium price. In London it is a little over 33 pence, say 37 pence. *But the New York price must be the very same price, expressed in dollars instead of pounds.* If 37 pence = $1 in London, then in New York £1 must equal $2.70 ($1.00 ÷ .37). And if we look at the equilibrium price in New York, so it does.

This is not really surprising. The price of pounds in dollars is the same thing as the price of dollars in pounds "upside down." It is as if pounds were shoes and we were saying that a pair of shoes that cost $10 is the same thing as 10 dollars costing 1 pair of shoes. But it takes a while to get used to the idea of two markets in which supply and demand are linked, as in the case of international exchange. With a little practice, the mystery begins to evaporate.

ITEMS IN THE U.S. BALANCE OF PAYMENTS

Let us learn more about these groups by examining the actual balance of payments for the United States for 1972 (Table 36·2). We begin with some obvious and self-explanatory figures—the exports and imports of *merchandise*. As we can see, in 1972 exporters sold $48.8 billion, earning that many dollars (foreign buyers had to supply us with dollars to that amount). U.S. importers bought $55.7 billion worth of foreign goods, supplying that many dollars to the foreign exchange market. On net balance, the merchandise trade showed a balance of −$6.9 billion—a deficit arising from an excess of imports over exports. This is a net supply of dollars to the foreign exchange market.

The second group of items consists of supplies and demands for exchange to pay for *services* rather than goods. In our table we note a few of these major transactions. Note that *military transactions* gave rise to a large demand for foreign exchange to pay for expenses at U.S. bases abroad, the wind-up of the Indochina war, and the like. *Travel and transportation* mainly shows us that American tourists were demanding more foreign currencies to voyage abroad than foreigners were demanding dollars to travel

TABLE 36 · 2
The United States Balance of Payments, 1972 (billions of dollars)

1. *Merchandise*			
Exports	+$48.8		
Imports	−55.7		
Balance on merchandise	−6.9		
2. *Services*			
Military transactions	−3.6		
Travel & transportation	−2.9		
Investment income	+7.9		
Other	+0.9		
Balance on services	+2.3		
Balance on goods and services			−4.6
3. *Unilateral transfers*			
Remittances	−1.6		
Government transfers	−2.2		
Balance on transfers	−3.8		
Balance on current account			−8.4
4. *Long-term capital transactions*			
Private	−0.1		
Government	−1.3		
Balance on long-term capital	−1.4		
5. *Short-term capital (net balance)*	−0.5		
Official reserve transactions balance			−10.3
6. *Financed by:*			
Liquid liabilities held by foreign governments	+10.1		
Nonliquid liabilities held by foreign governments	+0.2		
Reserve assets	0.0		
	+10.3		

here. It also shows the net balance between U.S. payments for foreign carriers (for example, a flight on Lufthansa or the charter of a Greek freighter) and foreign payments for U.S. transportation (flights on PanAm or cargo on a U.S. owned ship).

More interesting is the item for *investment income*. This reflects the flow of profits from U.S. companies in foreign nations to their home offices in the United States, minus the flow from foreign companies in the United States to their home offices abroad. When IBM in Italy sends profits back to its U.S. headquarters, it buys dollars with its local bank balances of lire, creating a demand for dollars. When Nestlé sends profits back to *its* headquarters country, it uses its dollar balances to buy Swiss francs. From this large inflow of earnings we must subtract a small outflow of government interest payments going abroad. When we net out these flows, we can see that investment income was a strong source of dollars for the United States in 1972, amounting to $7.9 billion.

TWO PARTIAL BALANCES

If we now sum up all items on the merchandise account and all items on service account we get the so-called balance of goods and services. In 1972, this showed a deficit (or net supply) of $4.6 billion. Evidently we have not earned enough on exports or from investment income, etc., to offset our demands for imports, for travel, for military expenditures abroad, and the like. We will come back to the reasons for our foreign exchange problems later. Here we want only to explain what these partial balances mean.

Hence we move to two further items, under the category of *unilateral transfers.* Here we find remittances, or the sums that persons residing in America send to private individuals abroad, less any sums coming the other way from Americans residing abroad and sending their pay home. The pay that an American working abroad might send home would be a remittance that would earn us dollars; the sums sent home by a Britisher working in the United States would require the purchase of pounds. As we can see, remittances cause a further deficit in our accounts.

This is augmented by *government unilateral transfers*—sums "sent abroad" by the government for foreign aid, emergency relief, and so on. Of course, these sums are not actually sent abroad; rather, the U.S. government opens a dollar account for the recipient nation, which then uses these dollars. But in using them, the recipient country again sells dollars for other currencies.

Summing up again, we now reach a new partial balance—*the balance on current account,* which showed a deficit (net supply) of $8.4 billion in 1972.

ITEMS ON CAPITAL ACCOUNT

The next items reflect supplies and demands for foreign exchange associated with capital investments (not *income* from these investments, which we have already counted). This may include investment by U.S. companies in plant and equipment abroad, less investment by foreign companies in plant and equipment here; or purchases of foreign long-term securities by Americans less American stocks or bonds bought by foreigners. These long-term *private capital* flows cost us a net $0.1 billion in 1972. That small outflow was further augmented, however, by *government long-term capital transactions*—the purchases of foreign government securities by the U.S. government, less any purchases of U.S. bonds by foreign governments. The net balance on both private and public long-term capital account gave rise to a further deficit or net supply of $1.4 billion.

Last, we reach the item called *short-term capital.* This is actually an amalgam of several different flows that we do not show in detail. The most important of these consists of the transfer from one country to another of private balances, belonging to individuals or companies, that are moved about in response to interest rates or for speculative reasons. The treasurer of a multinational company may "park" his extra cash in Sweden one year and in the United States the next, depending on where he can earn more interest in short-term securities or special bank accounts. Some individuals and even some small governments move their bank balances from country to country in search of the best return or in anticipation of a move in exchange rates that will benefit them. This movement of short-term capital tends to be volatile and can on occasion give rise to speculative "flights" from one nation to another. In 1971, for example, when there was a general distrust of the American dollar, well over $7 billion was withdrawn from American accounts and "sent abroad," creating a deficit of that amount on short-term capital account.

SUMMING UP THE ACCOUNTS

As we have seen, very different motivations apply to these different actors on the foreign exchange markets. Exports and imports reflect the relative price levels and growth of output of trading countries. Tourism is also affected by prices abroad, as well as by the relative affluence of different countries. Flows of corporate earnings arise from investments made in the past. Long-term private capital items reflect estimates of the *future* earning power of investments home or abroad. Short-term capital is guided by interest rates and speculative moods. Government flows hinge largely on foreign policy decisions.

Whatever the different motives affecting these flows, each gives rise to supplies of, or demands for, dollars. Thus we can sum up the net outcome of all these varied groups to discover the overall demand and supply for dollars. This summing up is called the *official reserve transactions balance*. In 1972, as we can see, this balance was in deficit by $10.3 billion. That is, United States groups, in toto, demanded $10.3 billion more in foreign exchange than they supplied. Turning the coin over, we can say that foreigners demanded $10.3 billion less in U.S. goods and services and financial assets of all kinds than they sold to us.

FINANCING THE BALANCE

What we have traced thus far are the various groups whose economic (or political) interests caused them to supply dollars to or demand dollars on the exchange market. But we have arrived at a curious stopping point.

After we have summed up all the action of the groups, we find that 10.3 billion more dollars were supplied than were demanded. How can we square that with the identity of total purchases and sales? For every dollar that was sold there must have been a dollar bought. Something must be missing in our accounts.

It is. What is missing is a series of transactions that enable the actions we have described to take place. The fact that these various actions resulted in the sale of $10.3 billion more dollars than foreigners wanted for the various purposes we have enumerated means that someone else must have bought the otherwise unwanted dollars. And someone else did. These buyers were the central banks of the world.

CENTRAL BANKS

All nations have central banks, which serve two purposes. One of them is the exercise of various controls over the domestic money supply in their own countries: when we studied macroeconomics we saw how our central bank, The Federal Reserve, acts on this behalf. But all central banks also serve another function. *They are agencies of their governments, who buy or sell foreign exchange, making their own currencies available to foreigners when they buy foreign exchange and absorbing their own currency from foreigners when they sell foreign exchange.*

How do central banks acquire the capacity for these transactions in foreign exchange? The answer is that private banks in all countries have the option of transferring their own supplies of foreign exchange to their central bank, receiving payment in their own currency. For example, let us suppose that the Chase Manhattan Bank finds itself with large and unwanted supplies of francs. It can exchange these francs for dollars with the Federal Reserve. The Chase Manhattan

Bank will then get a dollar credit at the Federal Reserve, and the Federal Reserve will be the owner of the francs formerly belonging to Chase. In the same way, the Bank of Yokohama or Barclay's Bank or the Swiss Bank can exchange their holdings of dollars for yen or pounds or Swiss francs, in each case receiving a credit at their central bank in their own currencies and transferring their holdings of dollars to their government bank.

HOW CENTRAL BANKS WORK

Thus central banks are the holders of large amounts of foreign exchange, which they acquire indirectly from the activities of various groups in their own nations. The central banks are therefore the last "group" whose own actions must balance out the unbalanced flows we have heretofore examined. There are two ways in which this can be done.

1. Gold flows

For many years, any balances "left over" were settled by the shipment of gold from one central bank to another. For example, all through the early 1960s, the United States balanced its accounts partly by selling gold to cover any deficit in its Official Reserve Transactions Balance. The sale of gold was exactly like an export. Foreign central banks paid us in dollars from their holdings of dollar exchange, and this dollar inflow offset any deficit of dollars arising from other transactions. (Recently, a new kind of "paper gold" called Special Drawing Rights has also served as another *reserve asset* available to balance accounts. We will learn more about SDRs in our next chapter.) For reasons that we will also investigate there, gold shipments have been discontinued since the international monetary crisis of 1971. Notice in our overall balance of payments for 1972, that no reserve assets were sold to offset the deficit on official Reserve Transactions.

2. Holding reserve currencies

The second means by which the central banks can balance out the difference between demand and supply is to hold a foreign currency *as if it were a reserve asset*. This is exactly what the central banks of the world have (reluctantly) agreed to do in the case of the United States. The central banks of France, Germany, Japan and other nations have allowed their dollar holdings to mount as a "reserve currency" without converting those holdings into gold.

Thus the major balancing item consists of increases in holdings of dollars owned by foreign governments. In 1972, as Table 36·2 shows, these holdings (called liquid liabilities held by foreign governments) increased by $10.1 billion. (The nonliquid liabilities in the table refer to long-term bonds that foreign governments bought with their dollars.)

How does this increase in dollar holdings formally balance out the accounts? *The answer is that increased dollar holdings are counted in the overall balance of payments as a short-term credit for the United States.* They are, after all, foreign claims on U.S. wealth that have been "loaned" to us. In the official books they count as a "plus" item that offsets the "minus" items.

THE IMPORTANCE OF LIQUIDITY

Our analysis has shown us a very important fact. We can run an unbalanced foreign exchange account only if we can "finance" it by one or the other of the two means described above. Suppose, for example, that we had no gold and that foreign central banks refused to hold any more dollars. Then an American importer or tourist who went to buy francs or marks would soon discover that there weren't any, because no bank would accept any more dollars. Since there weren't any, he could not finance his import or his trip abroad.

The balance of payments would then be brought into balance at the cost of a lower level of international transactions. Americans would have to do with fewer Toyotas. Fewer Americans could visit Paris. No doubt more Fords would be sold instead and Yellowstone Park would be more crowded. But the level of consumer well-being would be lower than if trade could have occurred; and the total of world production, as we saw in our last chapter, would suffer because countries could not take full use of their comparative advantages.

Thus the willingness of central banks to hold one another's currencies and the quantity of reserve assets that they can use to "settle up" are of the greatest importance in determining the level of world trade. That is the meaning of the phrase that is often heard as to the importance of having enough "liquidity" in the world. Not having enough liquidity means that the ability to finance imports is crippled because a country has no gold or SDRs or because no central bank is willing to accept its currency in the way that dollars have been accepted. This absence of liquidity is particularly difficult for poor nations that desperately need imports and cannot pay for them.

The Price of Exchange

Somehow, in this description of the supply and demand for exchange, we have lost sight of the *price* of exchange. When Americans pile up a large deficit, does this not mean that at the going exchange rate, commerce and government find that their demand for foreign exchange exceeds the supply? Shouldn't this change the exchange rate itself, exactly as an "excess demand" for shoes at a given price will raise the price of shoes?

FIXED EXCHANGE RATES

This is exactly what would happen if the price of exchange were free to vary in the way that the price of shoes varies. *But what is different about exchange markets, until recently, has been that they have not been free to move.* Exchange rates have been "fixed" by international agreement. These fixed rates were made effective by the actions of central banks which stood ready to "supply exchange" to cover any deficits arising from trade or other transactions or to buy exchange to offset any surplus.

In the case of the United States, for reasons that we will look into in our next chapter, this fixed rate was too high; that is, appreciably higher than the equilibrium rate that would have established itself had the dollar been allowed to find a value equating supplies and demands for it. The result of the higher-than-equilibrium price was a "surplus" of dollars, exactly as a higher-than-equilibrium price creates a surplus of commodities in any market. *It was this surplus that was absorbed by foreign central banks who agreed to hold dollars.* An analogy would be a willingness of shoe buyers to use their "reserve assets," such as their savings, to buy and hold the "surplus" shoes that suppliers would put onto a market at a higher-than-equilibrium price.

WHY FIXED RATES?

Why have exchange rates been fixed? We will look into some of the practical advantages of a fixed rate in our next chapter, along with the disadvantages of a fixed rate. But the basic cause must be sought in history and psychology. For centuries, the only commodities that have universally commanded the magic of belief have been gold and silver. No nation in the past would accept the curious pieces of paper that another nation called

THE GOLD STANDARD

Until the Great Depression, the international monetary system was run on a gold standard, under which any citizen at any time could demand gold for paper money. This led to two problems. One was the risk of a "run" on gold in times of panic. It was this that provided the rationale for the gold "backing" of currency in the original Federal Reserve System.

The second problem was that anyone at any time could convert his money into gold and then ship the gold abroad to buy francs or marks or any other currency, if that was profitable. This international ebb and flow of gold kept all currencies tightly tied together. The difficulty was that the gold link among currencies made it im-

possible for any nation to launch an expansionary program *if the rest of the world was experiencing a recession*. As a result of its expansion program, its prices would rise. As prices rose, its citizens found it profitable to turn their money into gold, to send the gold abroad and to buy cheaper foreign goods or assets. This drained gold from the expanding economy, caused credit to contract, and promptly brought the boom to a halt.

After World War II, nations used a gold-exchange standard. Under this standard, gold was reserved for foreign exchange use. The U.S. Treasury sold gold only to foreign official holders of dollars, such as central banks. No gold was sold to foreign private citizens or to domestic U.S. citizens. As we shall see, even this attempt to safeguard the system did not work.

money. This led nations to "declare" the value of their paper monies in terms of their gold "content" and to agree that any foreign holder of its paper money (or of a checking account) could "redeem" that money in gold. In the United States, the value of a dollar from 1933 until very recent years was 1/35th of an ounce of gold.*

*Nations that owned very little gold declared the value of their currencies in terms of a major "reserve currency" such as dollars or pounds.

Numerous suggestions have been put forward by economists for other international standards of value, and many economists have urged that exchange rates should be cut loose entirely from any "fixed" value —that they should fluctuate like any other price. Until recently, however, these proposals have been stubbornly resisted by most governments. But that leads us into the problems discussed in our next chapter.

<table>
<tr><td>

KEY WORDS

Exchange rates

Foreign exchange

*Price of
foreign exchange*

*Freely moving
rates*

*Balances in the
balance of
payments*

*Official reserve
transactions
balance*

*Role of
central banks*

Liquidity

Reserve assets

*Reserve
currencies*

*Fixed exchange
rates*

</td><td>

CENTRAL CONCEPTS

1. We cannot compare international prices until we know the *exchange rates* of nations' currencies.

2. *Foreign exchange* arises from international transactions. In these transactions, importers or other demanders of foreign goods or services pay money to banks in their own nations. This money is then credited to the accounts of foreign banks, and the foreign banks pay the exporter or the supplier of the good or service in the money of his country. *The total of all foreign-owned bank accounts in all nations is the sum of foreign exchange.*

3. The price of exchange in a free market depends on demands for, and supplies of, exchange. Demands for exchange rise as the price of foreign money falls. Supplies of exchange rise as the price of foreign money rises. Two currencies are thus exchanged, and their prices must always be the reciprocal of each other.

4. *In a market with free exchange rates, there can be no surplus or deficit of exchange.* The price of exchange will move to the level at which quantities demanded and supplied are equal.

5. The balance of payments is actually made up of the activities of different groups engaged in international transactions. These groups generate different supplies and demands for exchange for their various purposes. The most important balances within the balance of payments are balances on *merchandise account,* on *current account,* on *capital account* (long and short run), and the *official reserve transactions balance.* The net balancing item is provided by transactions of central banks.

6. Central banks acquire private domestic banks' unwanted holdings of foreign exchange. They then "settle" any imbalance in the transactions between nations either by (1) *selling gold or some other reserve asset* (such as SDRs) and/or (2) *holding the currency of the country in deficit as a "reserve currency."*

7. The ability of a nation to import more than it exports essentially hinges on two things: (1) its supplies of reserve assets, such as gold; and (2) the willingness of other countries' central banks to hold its currency. The pool of reserve assets and reserve currencies is called *liquidity.* It is a prime determinant of the volume of international trade.

8. *Fixed exchange rates* were made possible by the long established habit of regarding gold (and to a lesser extent silver) as a commodity of international value. *Under fixed exchange rates, surpluses or deficits can occur in exchange markets,* just as in any market in which prices are kept above or below equilibrium levels. *It has been the function of central banks to absorb any surpluses from these markets or to supply any shortages in them.*

</td></tr>
</table>

QUESTIONS

1. If you wanted to buy a Swiss watch and discovered that it cost 200 francs, what would you need to know to discover if it were cheaper or more expensive than a comparable American watch? Suppose that the price of francs was 20¢ and the American watch cost $50? What if the price of francs rose to 30¢?

2. If you now bought the watch, to be sent to you, how would you pay for it? What would happen to your bank check? How would the Swiss watchmaker be paid?

3. Suppose that the Swiss, in turn, now decided to buy an American radio that cost $40. He finds the rate of exchange is 5 Swiss francs to the dollar. Explain how he makes payment.

4. Now suppose that the rate of exchange rises for the Swiss, so that he has to pay 6 francs for a dollar. What happens to the price of the radio in Swiss terms? Suppose the rate cheapens, so that he pays only 4 francs? Now what is the price of the radio to him?

5. How is an exchange rate determined in a free market? Can you explain why the demand for a foreign currency increases as its price decreases? Why the supply increases?

6. Show that the appreciation of the mark versus the franc is the same thing as the depreciation of the franc versus the mark.

7. Why can there be no surplus or deficit in a market with free exchange rates?

8. Show the relation between a "deficit" in the balance of payments and a surplus in a commodity market.

9. What is actually meant by the balance of payments? Is it possible for all flows of international expenditure and receipt to "unbalance"? Then why is the balance of payments a problem?

10. What is there to prevent a country from subsidizing its exports so that it could undersell the whole world in everything? Assume that the United States tried this policy. Explain carefully what would happen under a policy of (a) flexible exchange rates and (b) fixed exchange rates.

11. Explain why central banks are so important in international trade. Suppose there were no central banks. Could a fixed exchange system work? A flexible exchange system?

The International Monetary Problem

37

IN OUR LAST CHAPTER WE LEARNED SOMEthing about the technical meaning of a balance of payments deficit or surplus. But we have not begun to look into the question of why the United States ran such a large deficit in 1972 or the means by which such deficits might be remedied. Hence in this chapter we shall put our newly-gained knowledge to use by looking into the background and prospects for the great international monetary crisis of the early 1970s.

THE DETERIORATION OF TRADE

Figure 37·1 gives us a first clue to the problem. It shows the irregular but eventually precipitous fall in our earnings on current account. At the beginning of the period we were earning almost $1.5 billion in exchange. By the middle of 1972 we were running a deficit at almost the same rate. What was the cause of this fall?

We can find the answer by examining the

TABLE 37·1
Balance on Current Account

	Exports	Imports (billions)	Net balance*
1967	$30.7	$26.9	$+3.8
1968	33.6	33.0	+0.6
1969	36.4	35.8	+0.6
1970	42.0	40.0	+2.1
1971	42.8	45.4	−2.7
1972	48.8	55.7	−6.8

*Figures are rounded.

main components of our balance on current account. The largest of these are our transactions on merchandise account. Here we find an immediate reason for the deterioration of the balance on current account.*

*The balance of payments took an abrupt turn for the better in 1973. We will nonetheless focus on 1972 because it highlights the problems in which we are interested.

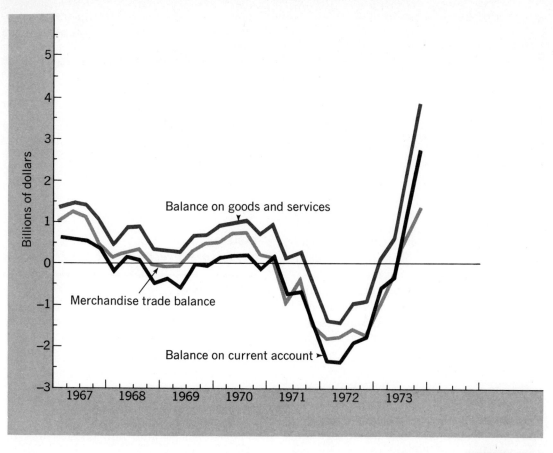

Billions of dollars

Balance on goods and services

Merchandise trade balance

Balance on current account

1967 1968 1969 1970 1971 1972 1973

FIGURE 37 · 1
Balance on current account

TRAVEL AND TRANSPORTATION

What was the cause of this tremendous decline in merchandise earnings? Let us postpone that question for the moment, until we have looked further into the history of the recent past; for sorry as is the record on merchandise, the overall balance on current account is worsened if we add in private services. At the beginning of the period, Americans were spending, net, $1.7 billion abroad for travel and transportation. By the end of the period, this figure had risen to $2.8 billion. Here, an enormous increase in tourism played a central role. A million and a half Americans traveled abroad in 1960 (creating a demand for foreign exchange). By 1971, this number had risen to 5.7 million.

MILITARY EXPENDITURES

These two large drains on American exchange were augmented by a third, this time originating from government rather than private persons. Every year, during the six-year period we are examining, the government ran up a large deficit for military purposes. In 1965 we were spending $3.0 billion in net foreign exchange to support U.S. military activities abroad. By 1972 this had risen to $4.7 billion, largely because of the Vietnam War. Notice that this is not the full cost of our military activities. Much of the war and nonwar military expenditure, such as the maintenance of U.S. troops in foreign bases, is paid in dollars and creates no demand for foreign exchange. The bombers we build or

EURODOLLARS

A student in international economics sooner or later hears about a mysterious currency called Eurodollars. Eurodollars are simply European bank accounts denominated in dollars rather than in the currency of the country. They represent a pool of funds that can be borrowed—*a pool of funds that are essentially unregulated by any government.* Since they are not held in the U.S., they are outside of the jurisdiction of the U.S. government. Since they are held in dollars and not in local currency, they are outside the jurisdiction of the local government.

Their main impact is to make it much more difficult for any country to control its own monetary policies. Suppose a firm wants to invest during a period when its own government is restrict-

fly and the pay for soldiers (in U.S. dollars) create no exchange problems. But inextricably connected with foreign military activity is a need to make large expenditures in other currencies. Non-U.S. personnel must be paid in their currencies, not in ours. Supplies such as food or local supplies must also be paid for in foreign currencies.

INVESTMENT INCOME

If we add up the deficits for 1972 (the worst year) on merchandise account, travel and transportation, and military, we get a total of $13 billion—far larger than the actual deficit on current account for that year of some $8 billion. What accounted for the difference?

Part of the difference is to be found in the other items we learned about in our last chapter, some of which helped and some of which hurt our balance on current account. During the six-year period, for example, our remittances abroad increased, worsening our balance, whereas our sales of other services (such as insurance on cargoes), improved, helping the situation. But the main source of dollars to offset the huge total of our merchandise, travel, and military expenditures was a large and growing income from United States private investments abroad, mainly the flow of profits of U.S. corporations. Table 37·2 shows the substantial rise in these sources of foreign exchange for the United States (remember that every time a foreign

TABLE 37 · 2
Net Private Investment Income (billions)

1967	1968	1969	1970	1971	1972
$+5.8	$6.2	$5.8	$6.4	$9.0	$9.8

branch of a U.S. company sends profit home, it must buy dollars).*

TRENDS ON CAPITAL ACCOUNT

We have still to look into the activities on capital account which, as we know, also enter into our official reserve transactions balance —the "balance" being achieved by sales of gold or other reserve assets or by holdings of currency by foreign central banks.

Table 37·3 shows the trends in both long-term and short-term capital.

What do these figures show? We see, first, that there has been a steady drain on the balance of payments from long-term government transactions. These largely reflect foreign aid loans, which are counted as claims against the U.S., although they are mainly spent *in* the U.S., creating additional exports.

*This figure is larger than the item for investment income of + $7.9 in Table 36·2, because the inflow of *private* earnings was reduced by an outflow of U.S. government interest payments. In Table 36·2 we show private-plus-public investment income; in Table 37·2, private income only.

ing the money supply and making it difficult to obtain loans. Unable to borrow in local currency, the firm borrows Eurodollars and exchanges these borrowed dollars for local currency. It now has the funds that it wished to have for investment purposes, and its government has been frustrated in its efforts to retard lending. In the United States credit crunch of 1969–1970, large U.S. firms were substantial borrowers of Eurodollars, to circumvent the Federal Reserve Board's policy of making it harder and harder to obtain local loans. Eurodollars are therefore one of the reasons for the uneven impact of monetary policies that we noted earlier.

Finally, Eurodollar accounts serve as a ready source of funds for speculating on international exchange rates. They are highly mobile and acceptable everywhere.

		1967	1968	1969	1970	1971	1972
TABLE 37 · 3 ***Long- and Short-Term Capital Flows (billions)***	**Long term**						
	Gov't.	$−2.4	$−2.2	$−1.9	$−2.0	$−2.4	$−1.3
	Private	−2.9	+1.2	−0.1	−1.4	−4.1	−0.1
	Short term	+1.2	+3.5	+8.8	−6.0	−7.8	−0.5

More important, we see in the next column an irregular deficit earned on long-term private account. This is partially the result of companies purchasing foreign exchange to build plants and equipment abroad —a flow that ran between $1 billion and $2 billion all during the early 1960s. It is also partly the result of companies and individuals investing in long-term foreign securities in the early 1970s. They invested because they thought growth prospects were good for these companies, or because they wanted to protect themselves against a change in the value of the dollar, or were speculating that such a change would take place.

For example, a person who bought a World Bank Bond in Swiss francs in 1970 would have paid the fixed rate of exchange, then about 25¢ per franc. If the dollar then fell in price, so that each franc now cost 35¢ instead of 25¢, an owner of a 10,000 franc bond that cost him $2,500 would be able to sell the bond for $3,500. (After the very large outflow of private capital in 1971, the United States placed a tax of 18 percent on purchases of most foreign securities, to deter just such speculative transactions.) This tax has since been removed.

SHORT-TERM TRENDS

Finally, we come to the column of short-term capital flows. As we know, these are sums that travel from nation to nation in search of profitable short-term investment. They are partly guided by interest rates, partly by speculative considerations of the kind we have just considered. Note how volatile this item is. Between 1969 and 1970, there was a difference of $14 billion—$8 billion entering this country in the first year, $6 billion leaving it the next. Here was another serious cause for the crisis that finally erupted in 1971.

Now let us sum up our partial balances, to look at the trend in the official reserve transactions balance. As we can see in Table 37·4, this balance worsened seriously in 1970, "collapsed" in 1971, and remained badly in deficit in 1972.

	1967	1968	1969	1970	1971	1972	TABLE 37 · 4
	$-3.4	$+1.6	$+2.7	$-9.8	$-29.8	$-10.3	**Official Reserve Transactions Balance (billions)**

REASONS FOR THE COLLAPSE

Can we account for this spectacular decline in our 1970–1972 Official Reserve Transactions Balances? As we can see, the reasons are complex rather than simple. Let us summarize them:

1. A deterioration in trade

Traditionally we have been large earners of foreign exchange from the international flow of merchandise. In the early 1960s, for example, our merchandise balance ran as high as $7 billion in our favor. Indeed, the net deficit on merchandise account that we suffered in 1971 was the first such deficit since 1895 and was itself a signal to the international community that something was drastically wrong.

Why had our trade balance eroded? There is more than one answer to the question. In part we must seek the reason in a spurt of inflation in the U.S. in the early 1970s, tending to price American goods out of many markets. But even more important than this was a steady rise in the productivity of European and Japanese competitors, enabling them to undersell Americans, not only abroad but in America itself. For example, during the late 1960s we became heavy importers of steel and autos from Europe and Japan, whereas before World War II we were always big steel and auto exporters.

Associated with this change in productivity was another fact, connected with American merchandising. Japanese and European producers anticipated important consumer trends and stole important marches on American producers in many fields. The business acumen of foreign producers and their rapidly increasing productivity were largely responsible for the rise of the Volkswagen, the Toyota, the Honda, the French and Italian sports cars, and the Japanese radio, electronic, and optical businesses.

Thus the reasons behind the deteriorating merchandise balance must be sought in real, not monetary factors. It was the international competition for productivity and for markets that was the background for United States international monetary troubles. In this competition the United States fared badly. As a result, its exports lagged, and its imports boomed.

2. The military

We have already noted the importance of military expenditures in creating the deficit. Unlike the deterioration of payments on business account, we cannot explain this adverse item in our accounts by economic or business reasons. The decision to maintain a large American presence abroad and to sustain our Indochina military involvement was a political decision. The subsequent failure of our war efforts to bring speedy or decisive victory was another reason for a gradual loss of faith in American foreign economic policy that led to the eventual monetary crisis.

3. Capital movements

A third reason for the collapse lay in the capital markets. As we have mentioned, long-term private capital flowed out of the nation in a steady stream during the early 1960s, but this was counterbalanced by a rising flow of earnings from that capital, which began to exert a powerful corrective influence on the balance of payments in the mid1960s.

Much more disruptive were the speculative flows of capital that created vast de-

GOLD "FLOWS"

The gold is *not* usually shipped abroad, as one might think. Instead, it is trucked to a vault many feet below the street surface in the Federal Reserve Bank of New York, where it is stacked in dull yellow bricks about the shape (but half the thickness) of a building brick. It is possible to visit this vault, which now holds some $13 billion of foreign gold, neatly separated into bins assigned to different countries. To see this modern equivalent of Montezuma's treasure is an astonishing sight. Gold may well be, as many have said, a kind of international psychosis, but its power over the imagination, no doubt the result of its traditional association with riches, is still remarkable. It is amusing to note that the Federal Reserve Bank, as custodian of this foreign gold, once suggested to its binholders that they might save a considerable sum if, instead of actually weighing the bricks and moving them from bin to bin whenever gold was bought and sold, both parties agreed to move the gold just on the books, the way bank balances are moved about. All governments demurred. They wanted the actual gold bricks in their bins. Hence, when gold moves from nation to nation to help settle up accounts, it is still actually pushed across the floor of the Fed's vault and carefully piled in the proper bin.

mands for foreign exchange in 1970 and 1971. These flows were in large part the consequence of a growing disbelief in the stability of the exchange rates that had been established after World War II and in a growing belief in an inevitable decline in the international value of the dollar. By "fleeing" abroad, speculative capital flows created enormous demands for foreign exchange that then made the position of the dollar increasingly untenable. The flight of capital was thus an example of a phenomenon economists call a *self-fulfilling prophecy*—a prophecy that results in actions that cause the prophecy to come true.

THE GOLD DRAIN

How did the United States meet its worsening foreign exchange situation? As we have seen, there are only two ways to finance a deficit: by selling gold (or other reserve assets) or by persuading foreign central banks to hold the currency of the deficitary nation, in this case dollars.

Figure 37·2 shows that this is exactly what happened in the case of the United States. Beginning in the 1960s, our gold stock

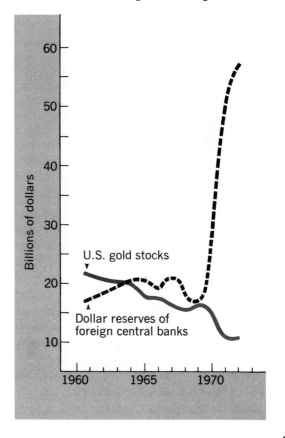

FIGURE 37 · 2
Changes in U.S. gold stocks

U.S. gold stocks

Dollar reserves of
foreign central banks

steadily declined, and the dollar reserves of foreign central banks rose. As we can see, as early as 1964 the holdings of dollars were so large that if foreign central banks had exercised their legal claims to gold, the entire gold stock of the nation would have been wiped out!

This would have been a major economic catastrophe—not because we would have no gold "behind" our currency, but because we would have cheated foreign nations out of hundreds of millions or billions worth of purchasing power. For if the gold supply ran out, foreigners could no longer repatriate their dollar holdings at a known exchange rate. Instead, their dollar holdings would now have to find their own price vis-à-vis their own currencies; and since dollars are held in very large quantities, this price might be very low.

Take, for example, a Netherlands bank that had allowed $1 million to remain in America, because it was confident that it could always get 28,571 ounces of gold for it (at $35 per ounce) *with which it could then purchase other currencies whose value in gold was also firmly fixed*. Suddenly it would find that its million-dollar deposit would be worth only as many ounces of gold as the market decreed—very possibly much less than the quantity it thought it owned. In turn, the amounts of other currencies available to the Netherlands bank would also fall, since it would have less gold to buy them with. *Thus countries that had cooperated with us by allowing their dollars to remain here would be penalized.* They would never again allow dollars to pile up in U.S. banks.

Not less important in staying the hand of the central banks who might have claimed our gold was the fear of a terrible slump in trade that could have followed. Since World War II, the dollar had been *the* unit for settling balances among nations. If the dollar were no longer—at least in theory—conver-

tible into gold, a main instrument for international "liquidity" would be wiped out. Unless another means of settling balances was put in its place, a straitjacket of gold could be placed on the volume of international trade, with an almost certain consequence of a drastic fall in the volume of trade.

THE GOLD RUSH OF 1967

Just because the consequences of a "run" on gold would be so disastrous, the central banks of foreign nations agreed not to demand their dollar balances in gold. This did not, however, prevent *private* speculators abroad from taking their holdings of dollars and converting them into gold. For many years there has been a perfectly legal market in gold in London and elsewhere, where private individuals could exchange their holdings of dollars or other currencies for gold bullion. The gold they bought was sold to them at the fixed price of $35 an ounce by an international gold pool comprised of the main goldholding nations. In the pool, the United States, as the largest holder of gold in the world, was committed to providing 59 percent of all bullion supplied to the market.

In March, 1968, the gold pool suddenly faced a crisis. Alarmed at the shrinkage of U.S. gold reserves, private speculators converged on the London market to convert their dollars into bullion. On the first day of that month, the pool, which had normally sold 3 to 4 tons of gold per day, suddenly found itself obliged to sell 40 tons. A week later the demand had risen to 75 tons. On March 13, it was 100 tons; the next day, 200 tons.

At this rate of drainage, the United States Treasury was being forced to put $1 million of gold into the pool every two to three minutes. Officials in Washington nervously figured that if the gold hemorrhage were not checked, the nation's entire gold reserves would be used up in a few weeks. To prevent such a crisis, the world's central bankers hurriedly convened in Washington and after a weekend of continuous conferences, announced that the gold pool was to be discontinued and that a new "two-price" system would immediately begin. All *official* holders of dollars (i.e., governments and central banks) would still be able to buy gold at $35 an ounce from the Treasury. But there would no longer be any effort to maintain the price of gold at an "official" level in the private market by supplying whatever gold was needed there at that price. Instead, there would be no sales from any national reserve of gold into the private market, and "private" gold would be allowed to find its own price.*

The system worked fairly well for a while. But the continuing imbalance of American accounts made it clear that such a means of avoiding an international currency crisis was at best a stopgap.

Curing the Balance of Payments Deficit

Before we look into the next episode of the international monetary crisis of the early 1970s, let us take a moment to reflect on the problem. As we have seen, the United States had run a persisting deficit on its official reserve transactions balance—a deficit that had caused a drastic fall in its holdings of gold and an even more marked increase in

the holdings of dollars by foreign central banks.

This raises the question of what means are available for curing such a deficit. To put it differently, it asks us to examine the various ways in which we could cut down the surplus of dollars that foreigners were acquiring, over and beyond the amount of dollars they were spending.

Basically there are six possibilities.

1. The classical medicine

One cure for a persisting balance of payments deficit is the bitter purge of the "classical remedy." This consists of a stiff application of higher interest rates and restrictive fiscal policies, whose purpose is to deflate the economy—forcing down our price level until our exports look more attractive to foreign buyers, and lowering our incomes so that our imports fall markedly.

The classical remedy can be very effective. But it is also very expensive, for *essentially it aims to cure our international ills at the expense of increased domestic ills.* The price of a smaller GNP is unemployment, and lower incomes. Thus the classical remedy puts a priority on our international economic relationships over our internal prosperity. It is hardly surprising that it is a remedy few countries are eager to apply.

2. Restraints on capital flows

A second remedy is much less deflationary. It aims to correct the balance of payments deficit by measures that seek to discourage American spending abroad on capital account. One of these is the imposition of a tax on the purchase of foreign securities. Another is a general plea to business to defer inessential plans for overseas expansion. A third is the deliberate raising of short-term interest rates at home to attract foreign short-term capital here.

All these measures (which we have

---

*By mid1974, the price of gold on these private markets reached the astronomical price of $175 per ounce. This was far beyond any reasonable calculation of the equilibrium price of the dollar. It was simply an index of the capacity of that extraordinary metal to command belief in itself and, of course, evidence of the international private speculators' continuing feelings that the existing exchange rate structure was still untrustworthy.

taken at various times) aim at restraining the net outflow of American capital. In the short run, there is no doubt that they can help alleviate our deficit on international account. The problem arises when we ask if this is a wise *long-run* policy, for there is a useful purpose served when American capital goes overseas—useful not only to the nations to which it brings wealth, but to us as well. Our capital exports soon earn income that returns to us as a "plus" item on current account. In a normal year this investment income from abroad is larger than our capital exports abroad. Thus if we stifle capital exports to save dollars today, we jeopardize a larger flow of dollars to ourselves tomorrow.

3. Correcting the government overseas deficit

The difficulties of taking more than short-term action in the capital market bring us to the next obvious candidate for paring down: the heavy overseas deficit incurred on government account. For the past decade, this government balance has cost us a net deficit of between $2 billion and $3 billion. Here, it seems is a likely cure for the balance of payments problem. What can be done about it?

One often suggested "remedy" is to reduce our foreign aid expenditures. But the overwhelming bulk of these expenditures does not constitute any dollar drain, *because foreign aid is largely used to buy American goods*. It thereby results in U.S. exports. If we cut back our foreign aid expenditures, we would therefore cut back on the volume of American production that is transferred to the underdeveloped world, but we would hardly improve our dollar drain at all. Moreover, the suggestion that the richest nation on earth curtail its aid to the poorest and most miserable for the sake of saving the few hundred millions of dollar drain that is the net "foreign exchange cost" of foreign aid seems like the last and not the first measure

to be taken in curing the balance of payments.

What of reducing our military expenditures abroad? That would be a much clearer net saving of dollars, for there is much less linkage between the dollars we spend to sustain troops or bases or actual fighting abroad and the *sale* of American goods to those areas. To the extent that we are able to reduce our overseas commitments, wherever we maintain a large and expensive armed presence, we will certainly contribute greatly to the solution of our payments problem. The problem, however, is not an economic one—although it has profound economic consequences. The flexibility of our military budget is based on political considerations about which the economist, whatever his private views, has no professional competence.

4. Flexible exchange rates

Finally, might it not be possible to avoid the problem of the deficit altogether by flexible exchange rates? We will remember that if exchange rates could move freely there would never be a "deficit" in the first place, since the price of exchange, like any price, would always clear the market.

The difficulty with a freely moving exchange rate is that is poses considerable difficulties for the smooth flow of trade and capital. The reasons are not difficult to understand. Most international transactions are not concluded immediately across a counter, but extend over weeks or even months between the time that a sale is agreed upon and the time when the goods arrive and payment is due. If exchange rates changed during this period, either the importer or the exporter could be severely penalized. If you agreed to buy a camera for ¥20,000, thinking it would cost $50, and when the camera arrived you found that yen had advanced in price, so that the camera cost $70, you would hardly be pleased. And although it is possible to insure oneself to some extent against exchange variations by buying "forward exchange," most traders would rather not deal in exchange rates that are likely to alter over the course of a transaction. More important, international investors who put money overseas for long periods have no way of protecting themselves against changes in rates and are even more concerned about the risks of flexible exchange.

In addition, many monetary authorities fear that fluctuating rates would lead to speculative purchases and sales of foreign currencies just for the purpose of making a profit on swings in their price—and that these speculative "raids" would have the effect of self-fulfilling prophecies in still further aggravating those swings.

5. Devaluation

Another course of action must be apparent. It is to "devalue" the dollar—that is, to change the relationship of the dollar to gold, so that the Treasury would ask not $35 for an ounce of gold, but $45 or $75 or any other large sum. Instantly this new exchange rate would make American prices cheaper abroad —it would be the classical medicine painlessly taken. But better yet: with devaluation, our remaining gold stock would immediately be worth a great many more dollars. Therefore, any deficit on the balance of payments could be easily settled by shipping a much smaller quantity of gold. The crisis would be over.

Or would it? We have already seen the problems that would ensue if the United States were unable to meet its obligations in gold at all. But devaluing would in fact be tantamount to very much such a state of affairs. Countries that had held their dollar balances secure in the conviction that they would always be worth so much in gold (and thereby in other currencies), would now find

themselves shortchanged by the amount of the devaluation.

The problem, in other words, is that devaluation is a very difficult course for a country to take when its own currency serves many other nations as part of their international reserves. For then, devaluation does not merely affect the value of *our* reserves, but of *theirs* too. Our gold stock is worth more; their dollars are worth less.

But this is not the only difficulty with devaluation. Changing the price of one's currency by fiat is always a two-edged sword. A nation devalues, among other reasons, to increase its exports and to discourage imports. But other nations may not take kindly to this prospect of increased international competition. Hence they may devalue also, leaving international price relationships just where they were. The era of the 1930s saw a number of such competitive devaluations that brought little but international friction.

Last, we must take into account an effect of devaluation that applies less cogently to us than to nations that are heavily dependent on foreign commerce. It is that *devaluation raises the costs of imports to countries that have devalued.* If the dollar now buys fewer yen, Japanese cameras will cost more. If imports are an important part of that country's standard of living, this means that devaluation will raise domestic prices.

Thus devaluation looks much more attractive superficially than it does on careful examination. This does not mean that it is therefore never useful. There are occasions when it is plain that a nation's exchange rate is hopelessly out of line with other countries and that none of the milder medicines will suffice to restore a balance of payments. In that case there is a choice to be made. Either trade must be allowed to fall to a much lower level, or the difficult step of devaluation must be made, once again putting the country on a competitive footing. Great Britain, for example, after World War II, devalued its prewar pound, which was obviously far out of line with postwar realities, from $4.00 to $2.80 and then, when it was unable to cure a persistent balance of payments gap that was draining away its reserves, had to devalue again in 1967, from $2.80 to $2.40. Underdeveloped nations on numerous occasions have devalued to bring their inflated price levels back to some sort of parity with world market levels.

6. Creating new reserve assets: SDRs

A sixth method is more imaginative. It consists of finding some reserve asset other than gold that will enable central banks to "cover" any deficits. As we have seen, for many years the dollar itself was such an asset. Many foreign nations, especially in the less affluent nations, counted their holdings of dollars as if they were "as good as gold." And as long as other nations were willing or even eager to hold these dollar balances (which, after all, earned interest, whereas gold holdings did not), the dollar itself was such a reserve asset.

But with the loss of faith in the ability of the United States to maintain the fixed exchange value of dollars, dollars ceased to become an acceptable reserve currency. Hence the effort was made to create a new reserve currency.

This effort was mounted by the International Monetary Fund (IMF), an institution which serves as a kind of central bank for many nations. Into the IMF, subscriber nations deposit both gold and their own currencies. From the bank, they can borrow gold or other currencies to meet temporary shortages of reserve assets. In turn, the fund serves as a kind of monitor of exchange adjustments, because it will not lend reserves unless it approves of the borrower's exchange rates and overall economic policies.

To meet the growing crisis of liquidity, the IMF in 1970 took the very important step of creating a new reserve asset called *Special Drawing Rights* or *SDRs*. These new assets were, in fact, created out of thin air (as all money ultimately is); but because SDRs had the backing of the fund, they were just as "good" as gold in settling international accounts.

SDRs are an important step in breaking the gold psychosis. The difficulty is that the different countries of the world cannot agree on how many Special Drawing Rights should be created each year or on how this international money should be distributed to the different countries of the world. Within each country, the government decides how much money to supply and how it should be inserted into the national economy. There is no comparable way to solve the same problem at the international level.

More important, however, is that reserves—no matter how large—cannot ultimately solve the problem of *persistent* deficits or surplus. They can lengthen the period before the problem becomes acute, but they do not solve the problem itself.

The Great Monetary Muddle

Now let us return to the situation of worsening tensions that we have traced up to 1971. Between 1968 and August 1971, none of the six measures above was effectively applied. The United States was unwilling to swallow the classical medicine. It tried to restrain capital outflows but with scant success, as our figures for capital movements show. It did not cut down overseas government spending—in fact, it stepped up its military expenditures. It did not allow the fixed exchange system to lapse and a new flexible exchange system to take over. It steadfastly opposed raising the number of dollars that an ounce of gold would buy (devaluation). It was unable to create a major addition to reserve assets, which in any case would not have solved the problem of persistent deficits.

The Crisis of August 1971

Inevitably, therefore, the U.S. balance of payments deficit worsened, and foreign holdings of dollars grew. As they grew, so did fears that the U.S. would eventually be *forced* to devalue as the only way out of its dilemma. This brought worries that foreign governments who held dollars would suffer large losses.

To forestall this possibility, in August 1971 the British government apparently asked for a guarantee that our government would compensate the British Treasury for any losses on its dollar reserves in case we devalued. This placed the American government in a quandary. If it granted the English request, it would be forced to make similar concessions to other governments. This would have transferred the losses from devaluation from foreign governments to ourselves, since we would have had to give foreign central banks additional dollars to compensate for the fall in purchasing power of their "old" dollars.

Faced with this cost, and with the rapidly worsening climate of confidence, the United States government chose a drastic step. *It announced that it would no longer sell any gold to foreign central banks at any price.* In effect, it severed the tie of dollars to gold. Since dollars could no longer be "valued" in gold, in effect the United States allowed dollars to find their own market price. The dollar was "floated." To show that the United States meant business in correcting its balance of payments deficits, it imposed a temporary 10 percent tax on all imports.

PAINFUL OPTIONS

This immediately led to acrimonious debates as to what should be done next. The trouble was that all possibilities were painful. If nothing were done, foreign countries would be "stuck" with unwanted dollar holdings. If the dollar were formally devalued, then foreign currencies would appreciate. In that case, American goods would become cheaper, and foreign goods more expensive for Americans. Foreign producers who competed with American exporters, (such as French farmers) feared they would be flooded with cheap American goods; and foreign exporters, such as automobile makers, feared they might be priced out of America's enormous market.

Moreover, if the dollar were devalued in terms of gold, by raising the number of dollars that an ounce of gold was worth, this action would reward all the less cooperative countries in the world who had refused to allow their reserves to be held in American dollars, while penalizing the very countries that had worked with us by agreeing not to exercise their option to exchange dollars for gold.

THE FIGHT TO HOLD FIXED RATES

Out of such conflicts of interest no happy solution could emerge, and none did. After long negotiations, European countries and Japan agreed to revalue (appreciate) their currencies vis-à-vis the dollar; the United States agreed to devalue its dollar in terms of gold, changing the price from $35 per ounce to $42, *and the nations of the world attempted to return to a new system of fixed exchange rates,* albeit rates that were different from those of the immediate past.

This patchwork lasted only a short time. Many individuals and corporations doubted that the exchange adjustments were large enough to correct the persisting deficit in America's balance of payments. Accordingly, they sought to buy currencies that were most likely to rise in value in the expected next round of devaluation. Once again a self-fulfilling prophecy fulfilled itself. During January and February of 1973, German reserves rose by $8 billion, or almost 50 percent, as speculators sought to buy German marks, the most obviously undervalued international currency (the Germans had been running persistent balance of payments *surpluses*). The German government then decided that it did not want its already huge dollar balances to rise further and announced that the mark would be allowed to "float." Once again the world was forced off a fixed exchange rate system and onto a floating one.

And once again the world tried to find a system of fixed rates that would hold up. The dollar was devalued again; calm was restored for a few weeks; but the rush into marks resumed, and the governments of the world were again forced to abandon their fixed rates and to allow their currencies to float.

Figure 37·3 shows the gyrations of currencies during the whole postwar period. Note the dramatic fall in most currencies vis-à-vis the dollar just after the war, and the swing upward during the 1970 crisis years.

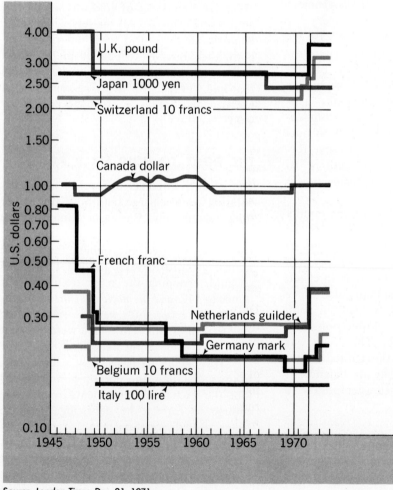

FIGURE 37 · 3
How parities have changed

Source: *London Times*, Dec. 21, 1971.

OUTLOOK FOR THE FUTURE

From the economists' point of view, the future lies between two uncertainties. Under flexible or floating exchange rates, business is subject to constant small uncertainties about what foreign exchange will cost. Under fixed rates, it is subject to infrequent but much larger and more disruptive uncertainties. Economists tend to favor the constant small risks; but somewhat surprisingly, many bankers and businessmen seem to prefer infrequent larger ones.

It seems likely therefore that attempts will be made to find a stable set of fixed rates, which will be held in place, as before, by the cooperation of the central banks.

There is talk of a compromise between the two systems—"semi-fixed" exchange rates with a "band" of 2 or 3 percent on either side, within which exchange rates will be free to fluctuate without central bank intervention. From the accumulated experience of the last decade, it seems unlikely that a system of fixed rates or even fixed rates with limited free fluctuations will last. The world is changing so rapidly, in terms of different national rates of economic growth and relative production costs, that no fixed exchange system seems viable. It is possible that new arrangements will be worked out, with fixed rates for close trading partners such as the members of the European Community (the Common Market), and floating rates among all of Europe and the U.S.; but this requires that Europe coordinate its monetary systems, still a step for the future.

THE LESSONS OF THE PAST

Can we sum up the lessons of these past difficult years? A few generalizations seem warranted.

1. The balance of payments is subject to very rapid improvement or deterioration.

In these years of dollar surpluses, we forget that the 1950s were years of dollar *shortages.* The worry then was that there were not enough dollars to finance world trade needs, not that there was a surplus of unwanted dollars.

Such changes in balances of payments can come about in two ways. First, the devaluations and revaluations of currencies take time to exert their effects. It has been a rule of thumb that a 1 percent change in the value of the dollar brings a change of about $1 billion in the balance of payments, after a period of a year or so. As of late 1973 the cheaper U.S. dollar had already restored a surplus in our basic balance of payments. This was a much quicker response than anyone had expected. If international capital movements once again bring short-term capital in large amounts to the United States, we could quickly find ourselves back in a dollar shortage position, with demands for dollars outrunning supplies. An untoward political event—for example, another crisis such as the Arab oil embargo—could trigger such a move. *Thus the international monetary situation has the capability of rapid change, and today's "solution" may be tomorrow's problem.*

2. The deterioration in our balance of payments suggests that we will have to move to a new economic relationship vis-à-vis the rest of the world.

During the troubled years of the early 1970s there was much talk that America was being priced out of markets all over the world. It *was* being priced out, but this was the result of fixed exchange rates that made the dollar too expensive. Now suppose that "equilibrium" fixed rates are found or that the world's currencies float. What is likely to be the role of America in world trade?

That brings us back to the question of comparative advantage. Until recent years, the trading strength of the United States lay in its high productivity in manufactures. It then seemed that we would always be the supplier of the world's least expensive steel, automobiles, and heavy machinery. Today that comparative advantage is waning and has perhaps disappeared. Curiously, our productivity advantage now appears to be greatest in the field of agriculture, the products that were the source of America's economic strength before it became a manufacturing nation. America's economic position in the world, according to some observers, will be less and less that of an exporter of industrial goods, and more and more an exporter of agricultural products and of special services and products such as aircraft, computer systems, and management expertise itself, in which we still are an unchallenged world leader.

A further consideration is that in a more "normal" world, where the American balance of payments was not constantly undermined by very large military expenditures, America would enjoy a steady large inflow of dollars from its foreign investments. This may mean that our long-term equilibrium position lies in accepting a permanent deficit on trade account offset by a permanent surplus on investment account.

A final point to remember is the lesson we have learned in Chapter 36. Under a system of free exchange rates (or under a fixed exchange rate that was really an "equilibrium" rate), every nation must have a comparative advantage in *some* product, barring only the unlikely case of two nations with identical endowments and therefore no wish to trade. It is impossible for the United States or any other country to be unable to sell anything because another nation, or "the rest of the world", can produce *everything* more cheaply. A moment's reflection on the case of Infraville and Supraville makes it clear that it is always possible to engage in mutually advantageous trade, assuming that tariffs or quotas or other impediments do not prevent the international specialization of labor from taking place.

What we cannot easily tell is *where* a nation's comparative advantage will lie, given free exchange rates. It may lie in areas that the nation does not wish to develop. For example, it may lie in the field of agriculture for the United States, whereas considerations of national economic policy may lead us to seek encouragement for other areas of activity. Here is the problem of Crusoe persuading Friday to give up hunting, which he enjoys, and to go in for gardening, which he does not. We shall consider this difficulty in our next point below.

3. Pressures for protectionism are likely to mount.

Perhaps the greatest danger of the present situation is that it exposes business and labor to a larger flow of imports. Efforts have already been made to lessen this inflow of goods by "voluntary" agreements from Japan and other nations to limit their exports to our markets. We have seen that the initial American devaluation was accompanied by a 10 percent tax on imports—since repealed.

But there is a strong possibility that we will respond to our changing international position by using tariffs or quotas, trying to protect American producers against "cheap foreign imports." Economists have traditionally been staunch supporters of free trade as the means of maximizing world production. That position was largely formulated in the days when it was taken for granted that maximizing world production also meant maximizing American production. If a new stable world trade arrangement calls for a shift out of American industrial products, this enthusiasm for free trade may wane.

Even if agricultural products become firmly established as America's most comparatively advantaged export, there will remain the difficult problem of unemployment that might develop in, say, the American automobile or steel industry. It hardly seems likely that factory workers would agree to return to farm work or that jobs would exist for them in the highly mechanized farm sector. There is a risk, therefore, that we will choose the easiest course of action; namely, restricting foreign competition to protect these workers and businesses, instead of making bold efforts to find new employments, for example in the provision of much needed public services, where international competition is virtually nonexistent.

In other words, *consumers' welfare is likely to be sacrificed to producers' welfare.* Whether this sacrifice is wise or not depends, in the end, on our value judgments with regard to the well-being of consumers and producers. Here no economic calculus can be applied. The decision must be made on political and moral grounds.

KEY WORDS

Trends in different balances

Self-fulfilling prophecies

Gold drain

Two-tier system

Classical medicine

Curbing capital flows

Curbing government outflow

Flexible exchange rates

Devaluation

SDRs

CENTRAL CONCEPTS

1. America's international trade position has been deteriorating for several years. Our merchandise balance of trade fell to a deficit in the early 1970s. Military expenditures drained dollars in large amounts. Capital flows and flights further weakened our position. Only a steady inflow of investment earning permitted the United States to hold the dollar fixed as long as it did.

2. Capital flights arise when individuals or firms expect a currency to deteriorate and seek investment in other currencies. These flows can give rise to *self-fulfilling prophecies.*

3. The result of our deteriorating accounts was a heavy gold drain and a large rise in dollar balances held by foreign central banks.

4. The continuing pile up of unwanted dollar balances and the loss of gold gave rise to currency "scares" as early as the late 1960s. At that time (1968) a makeshift arrangement tried to stem the gold flow by agreeing that central banks would only sell gold to one another, and not to outside speculators. This "two-tier" price system slowed but did not stop the trend of events.

5. There are six basic ways of "curing" a persistent deficit in a balance of payments:
   - The "classical" medicine requires that a country attempt to lower its price level by severe restrictive monetary and fiscal measures. *This cure requires the acceptance of substantial unemployment.*
   - Taxes or other measures can seek to *limit capital flows.* This imposes the danger of curtailing future inflows of investment income.
   - *Correcting government deficits* in international affairs. Here the problem rests on military and political policy considerations.
   - *Flexible exchange rates are another cure.* No deficit can occur when prices are free to move. But flexible exchange prices introduce a considerable *element of short-run risk* into international transactions.
   - *Devaluation is a means of cheapening a currency by reducing the "gold content" of that currency.* It imposes a cost on countries that have agreed to allow their reserve assets to remain in the currency that is devalued. It also raises import prices.
   - New reserve assets such as *SDRs (Special Drawing Rights)* are a means of giving nations additional liquidity. They do not change the forces at work that produce a persistent deficit.

*The crises of
the 1970s*

6. **None of these measures was successfully applied in the early 1970s, and the world saw a succession of devaluations forced upon the United States.** *Repeated attempts to reestablish fixed rates failed, and a floating system was imposed on the world by virtue of the failure to create an arrangement of stable fixed rates.*

*Fixed vs.
flexible rates*

7. **Economists by and large tend to favor the continuing minor risks of flexible rates rather than the sporadic major risks of currency crises. But business and bankers prefer the fixed system to the floating one. As of now, the outlook is for efforts to reconstruct a fixed system, against very considerable difficulties.**

8. **The lessons of the past are important:**

*Rapid changes in
balance of
payments*

- *Changes in the balance of payments can arise rapidly.* **Changes in the relative price levels of nations and in their growth rates, plus changes in expectations of international investors, make any "solution" open to failure.**

- *The American role in international trade is still uncertain.* **Possibly we will have to depend on agricultural exports and highly specialized manufactures for our comparative advantage in the coming years. Our erstwhile advantage in heavy manufactures seems likely to diminish. We may also run a "normal" merchandise deficit, offset by investment income.**

*New areas of
comparative
advantage*

- *Protectionist sentiment is likely to mount.* **If the American international economic position is endangered in international competition, we may well see a recrudescence of tariff and quota measures designed to protect American producers. Needless to say, this will be at the cost of American consumers.**

9. **There is no economic rationale for preferring the interests of one over the other. The decision must be made on political grounds.**

Protectionism

QUESTIONS

1. Give a general account for the reasons behind the growing imbalance in the American balance of payments.

2. Why is short-term capital a source of such potential difficulty? Explain what is meant by a self-fulfilling prophecy. Can you think of an example in domestic affairs? Suggestion: a rumor to the effect that a bank might not be able to meet its claims. Can you think of others?

3. What are the difficulties of devaluation as a policy for curing a persistent deficit in international payments? Would devaluation be more or less welcome to a nation such as Japan that imports a large proportion of its raw materials, than in the U.S., where imports are small in relation to production?

4. What are the difficulties of flexible exchange rates? Would they affect merchandise trade more than long-run investments?

5. Why do you think that gold has retained its "value" in the eyes of international speculators?

6. Suppose that the only cure for a persistent deficit were the classical remedy. Suppose that a country refused to swallow the cure. What would be the results?

7. Suppose that Japan becomes a nation in which the most modern technology is wedded to "cheap" labor. Is it possible that Japan could undersell everyone in everything? Will there be no comparative advantage left to anyone? Think this problem through very carefully. It is crucial.

8. Suppose that Japanese and European productivity give their auto industry a decisive edge over ours. What policy would you favor and why: (1) allowing foreign cars to take over the American market and allowing the big American car makers to fold up shop, (2) imposing tariffs or quotas on car imports, (3) making an effort to relocate the capital and labor in the American auto industry? Assuming that you chose the last possibility, where would you try to locate these factors of production? Suppose the labor pool in Detroit did not want to move to those areas in which you had found work for them? How would you then resolve the dilemma?

The Multinational Corporation

WE HAVE LOOKED FAIRLY CAREFULLY into the advantages and disadvantages of free trade as a means of organizing the economic intercourse of nations, and we shall have something more to say about it when we study the problems of the underdeveloped world. But in this chapter we must add a new dimension to our picture of international economics. Everything we have heretofore learned has had to do with the international *exchange* of goods and services. What we must now consider is a revolutionary development in international economics, based not on exchange, but on the *international production* of goods and services. To understand what this means we must learn about a new institution that has sprung up as the most important agency for international production: the multinational corporation.

THE MULTINATIONAL CORPORATION

What is a multinational corporation? Essentially, we mean a corporation that has producing branches or subsidiaries located in more than one nation. Take PepsiCo, for example. PepsiCo does not ship its famous product around the world from bottling plants in the United States. It *produces* Pepsi Cola in 512 plants located in 114 countries. When you buy a Pepsi in Mexico or the Philippines or Israel or Denmark, you are buying an American product that was manufactured in that country.

PepsiCo is a far-flung, but not a particularly large multinational corporation: in 1969 it was only the 119th largest company. More impressive by far is the Ford Motor Company, a multinational that consists of a network of 60 subsidiary corporations, 40 of them foreign-based. Of the corporation's total assets of $8 billion, 36 percent are invested in 27 foreign nations; and of its 388,000 employees (as of 1966) 150,000 were employed outside the U.S. And if we studied the corporate structures of GM or IBM or the great oil companies, we would find that they, too, are multinational companies with substantial portions of their total wealth invested in productive facilities outside the United States.

Multinational corporations differ from international corporations in that they produce goods and services in more than one country, while international corporations produce most of their goods and services in one country, even though they sell them in many. Volkswagen is a large international corporation, since it sells Volkswagens all over the world; but it is not much of a multinational corporation, since most of its actual production is concentrated in Germany.

If we broaden our view to include the top 100 American firms, we find that 62 have such production facilities in at least 6 nations. Moreover, the value of output that is produced overseas by the largest corporations by far exceeds the value of the goods they still export from the United States. In 1970, sales of foreign affiliates of U.S. multinational firms (which means their wholly or partially owned overseas branches) came to over $115 billion. In the same year, our total exports of manufactures amounted to $47 billion, only 41 percent as much as American firms produced abroad.

INTERNATIONAL DIRECT INVESTMENT

Another way of establishing the spectacular rise of international production is to trace the increase in the value of U.S. foreign direct investment; that is, the value of foreign-located, U.S.-owned plant and equipment (*not* U.S.-owned foreign bonds and stocks). In 1950 the value of U.S. foreign direct investment was $11 billion. In 1972 it was over $94 billion. Moreover, this figure, too, needs an upward adjustment, because it includes only the value of American dollars invested abroad and not the additional value of foreign capital that may be controlled by those dollars. For example, if a U.S. company has invested $10 million in a foreign enterprise

whose total net worth is $20 million, the U.S. official figures for our foreign investment take note only of the $10 million of American equity (ownership) and not of the $20 million wealth that our equity actually controls. If we include the capital controlled by our foreign direct investment as a whole, the value of American overseas productive assets may be as large as $250 billion. In general, something between a quarter and a half of the real assets of our biggest corporations are located abroad.

THE "AMERICAN CHALLENGE"

The stunning expansion of American corporate production overseas, especially in Europe, has given rise to what has been called *The American Challenge*. In a book with that title, written in 1967, Jean-Jacques Servan-Schreiber described the "take over" of French markets by dynamic American firms that had seized 50 percent of Europe's semiconductor market, 80 percent of its computer market, and 95 percent of its integrated circuit production. Servan-Schreiber feared that the American Challenge portended the Americanization of Europe's fastest growing industries. "Fifteen years from now," he wrote, "it is quite possible that the world's third greatest industrial power, just after the United States and Russia, will not be Europe, but American industry in Europe."[1]

Servan-Schreiber argued that Europe must fend off this American threat by the formation of equally efficient, large pan-European corporations. But he failed to see that the movement toward the internationalization of production was not a strictly American but a truly multinational phenomenon. If the American multinationals are today the most imposing (of the world's biggest 500 corporations, 306 are American),

[1] J.-J. Servan-Schreiber, *The American Challenge* (New York: Atheneum, 1968), pp. 3, 13.

THE TOP 15 MULTINATIONAL COMPANIES, 1971

Company	Total sales (billions of dollars)	Foreign sales as percentage of total	Number of countries in which subsidiaries are located
General Motors	$28.3	19%	21
Exxon	18.7	50	25
Ford	16.4	26	30
Royal Dutch/Shell*	12.7	79	43
General Electric	9.4	16	32
IBM	8.3	39	80
Mobil Oil	8.2	45	62
Chrysler	8.0	24	26
Texaco	7.5	40	30
Unilever*	7.5	80	31
ITT	7.3	42	40
Gulf Oil	5.9	45	61
British Petroleum*	5.2	88	52
Philips Gloeilampenfabrieken*	5.2	NA	29
Standard Oil of California	5.1	45	26

*Not a U.S. firm.

they are closely challenged by non-American multinationals. Philips Lamp Works, for example, is a huge Dutch multinational company with operations in 68 countries. Of its 225,000 employees, 167,000 work in nations other than the Netherlands "home" nation. Royal Dutch/Shell is another vast multinational, whose "home" is somewhere between the Netherlands and the United Kingdom (it is jointly owned by nationals of both countries): Shell *in the United States* ranks among "our" top 20 biggest companies. Another is Nestlé Chocolate, a Swiss firm, 97 percent of whose $2 billion revenues originate outside Switzerland.

THE INTERNATIONAL CHALLENGE

Indeed, if we take the ten leading capital-exporting nations together (including the United States), we find that in 1967 their combined exports came to over $130 billion, but their combined overseas production amounted to well over $240 billion. For 1970,

Judd Polk, economist for the International Chamber of Commerce, estimates that total international production—U.S. production abroad, foreign production here, and foreign production in other foreign countries—may account for as much as *one-sixth* of the total value of all world output, and a much higher fraction of the world output of industrial commodities.

Because we do not have complete statistics on many aspects of international production, the full extent of this new form of economic relationship is still uncertain. But a number of economists have made projections based on the continuance of the rapid growth of international production. Professor Harold Perlmutter, for example, predicts that by 1988 most noncommunist trade will be dominated by 300 large corporations, of which 200 will be American, and that these corporations will account for roughly half of total world industrial output. Perhaps the reader will recall Professor Berle's remark that some corporations "can be thought of only in somewhat the way we have heretofore thought of nations." A glance at Table 38·1, comparing the GNPs of various countries with the sales of selected multinational corporations, shows that Berle's statement was not just rhetoric.

Cautionary note: Corporate sales are not the equivalent of GNPs. The table vastly overstates the relative importance of corporations with respect to manpower: Portugal, for instance, has a population of over 9 million, whereas GM employs far less than 1 million. On the other hand, the table understates the economic strength of corporations: GM can borrow a great deal more easily than Portugal; and it controls *all* of its receipts, whereas Portugal gets only the taxes from its GNP. Nonetheless, the table makes it clear that Berle's comparison was not a wholly fanciful one.

EMERGENCE OF MULTINATIONAL PRODUCTION

Why has the multinational phenomenon arisen? After all, it is cheaper to export goods or to license production abroad than to establish a branch in a faraway nation and encounter troubles and risks, which we will look into. Why, then, has production itself leaped overseas?

The initial reason arises from a charac-

teristic of the firm, to which we have heretofore paid only passing attention. This is the drive for expansion that we find in nearly all capitalist enterprises.* This "logic" of expansion has driven firms, from early times, to expand their market overseas. Samuel Colt, the inventor of the first "assembly line" revolver, opened a foreign branch in London in the mid-1850s (and promptly failed). But in those same years, American entrepreneurs were already successfully pushing a railway line to completion across the Panama isthmus, and by the 1870s the Singer Sewing Machine was gaining half its revenues from overseas production and exports.

MOTIVES FOR OVERSEAS PRODUCTION

But what drives a firm to *produce* overseas rather than just sell overseas? One possible answer is straightforward. A firm is success-

---

*Alfred Chandler has shown in a brilliant book, *Strategy and Structure* (Cambridge, Mass.: M.I.T. Press, 1962), that the typical domestic firm went through a series of "logical" changes in organization, growing from the single-product, single-plant firm (in which every operation was supervised personally by the founder-owner) to the multidivisional, multiproduct enterprise in which a tiered organizational structure became necessary to superintend the strategic requirements of national geographic scope and increasing technical complexity.

TABLE 38 · 1

GNP of Various Countries Compared with Sales of Selected Multinational Corporations (billions of current dollars, 1971)

Netherlands	$39.4	**Royal Dutch Shell**	**$12.7**
Belgium	31.7	Greece	10.8
General Motors	**28.3**	**General Electric**	**9.4**
Switzerland	26.3	**IBM**	**8.3**
Exxon	**18.7**	**Mobil**	**8.2**
Denmark	18.2	**Chrysler**	**8.0**
Austria	17.8	**Unilever**	**11.5**
Ford	**16.4**	**Texaco**	**7.5**
Norway	15.6	Portugal	7.3

ful at home. Its technology and organizational skills give it an edge on foreign competition. It begins to export its product. The foreign market grows. At some point, the firm begins to calculate whether it would be more profitable to organize an overseas production operation. By doing so, it would save transportation costs. It may be able to evade a tariff by producing goods "behind" a tariff wall. It may be able to take advantage of lower wage rates. Finally, it ceases shipping goods abroad and instead exports capital, technology, and management—and becomes a multinational.

Or calculations may be more complex. By degrees, a successful company may change its point of view. First it thinks of itself as a domestic company, perhaps with a small export market. Then it builds up its exports and thinks of itself as an international company with a substantial interest in exports. Finally its perspective changes to that of a multinational, considering the world (or substantial portions of it) to be its market. In that case, it may locate plants abroad *before* the market is fully developed, in order to be firmly established abroad ahead of its competition.*

In the multinational boom of the 1950s and 1960s, still other considerations may have played a role. For American corporations, one factor was probably our overvalued dollar, which made it possible to buy or build foreign plant and equipment cheaply. In fact, many experts believe that the dollar devaluation of the early 1970s will slow down the multinational thrust of American companies, although by the same reasoning it would stimulate the thrust of foreign companies, now able to buy American dollars more cheaply. The added element of risk

*The internal dynamics that send some firms overseas, but not others, are by no means wholly understood. The internationalization of production is much more widely spread in some industries, such as glass, than in others, such as steel. Drugs are widely produced on an international basis; machine tools are not.

that results from floating exchange rates, however, may deter both American and foreign firms from investing large sums abroad, since domestic value may fall if the exchange rate fluctuates against them.

THE ECONOMICS OF MULTINATIONAL PRODUCTION

Whether or not the multinational boom continues at its past rate, the startling rise of multinationals has already changed the face of international economic relationships. One major effect has been a dramatic shift in the *geographic location* and the *technological character* of international economic activity.

The shift away from exports to international production has introduced two changes into the international economic scene. One change is a movement of foreign investment away from its original concentration in the underdeveloped areas of the world toward the richer markets of the developed areas. Fifty years ago, in the era of high imperialism, most of the capital leaving one country for another flowed from rich to poor lands. Thus foreign investment in the late nineteenth and early twentieth centuries was largely associated with the creation of vast plantations, the building of railways through jungles, and the development of mineral resources.

But the growth of the multinational enterprise has coincided with a decisive shift away from investment in the underdeveloped world to investment in the industrial world. In 1897, 59 percent of American foreign direct investment was in agriculture, mining, or railways, mainly in the underdeveloped world. By the end of the 1960s, our investment in agriculture, mining, and railways, as a proportion of our total overseas assets, had fallen to about 20 percent; and its geographical location in the backward world came to only 36 percent of all our overseas direct investments. More striking, almost three-quarters of our huge rise in direct investment during the decade of the 1960s had been in the developed world; and the vast bulk of it has been in manufacturing (and oil) rather than in plantations, railroads, or ores. *Thus the multinational companies are investing in each others' territories rather than invading the territories of the underdeveloped world.* This is not to say that they do not wield great power in the background regions, as we shall see, but their thrust of expansion has been in other industrial lands, not in the unindustrialized ones.

The second economic change is really implicit in the first. It is a shift away from "heavy technology" to "high technology" industries—away from enterprises in which often vast sums of capital were associated with large, unskilled labor forces, as in the building of railways or plantations—toward industries in which capital is perhaps less strategic than research and development, skilled technical manpower, and sophisticated management techniques typical of computer, petrochemical, and other new industries. Table 38·2 sums up the overall shift.

Note the dramatic shift away from Latin America and away from transport, mining, and agriculture into Europe and manufacturing, a shift that would be even more accentuated if we were not still dependent on oil as a major source of the world's energy. If nuclear power or the fuel cell displace oil within the next two decades, we can expect a still more rapid decline in investment in the backward areas (especially in the Near East), and a proportionately still larger concentration of foreign direct investment in manufacturing.

PROBLEMS FOR POLICY MAKERS

Multinationals have not only changed the face of international economic activity, but also have added considerably to the problem of controlling domestic economies. Assume that a country wants to slow down its economy through monetary policies designed to

TABLE 38 · 2
Size and Distribution of U.S. Foreign Direct Investment

	1929	1950	1972
Total (millions)	**$7,528**	**$11,788**	**$94,031**
Percent distribution, by market			
Canada	27	30	27
Europe	18	14	33
Latin America	47	41	18
Asia, Africa, other	8	15	22
Percentage distribution, by industrial sector			
Manufacturing	24	31	42
Petroleum	15	29	28
Transport and utilities	21	12	N.A.
Mining	15	9	8
Trade	5	7	N.A.
Agriculture	12	5	N.A.
Other	8	6	N.A.

reduce plant and equipment spending. A restrictive monetary policy at home may be vitiated by the ability of a multinational to borrow *abroad* in order to finance investment at home. Conversely, a monetary policy designed to stimulate the home economy may end up in loans that increase production in someone else's economy. *Thus the effectiveness of national economic policy making is weakened.* Moreover, it is not easy to suggest that monetary policies should be coordinated among countries, since the economic needs of different countries may not be the same: what is right for one country at a given time may be wrong for another.

A second problem has already come to our attention when we considered the international movement of short-term capital. *Multinationals are almost inevitably thrust into the role of international currency speculators.* They can quickly move billions of dollars from one country to another—and as profit maximizers, they are motivated to carry out such movements. A multinational that allowed its balances to remain in a currency that seemed overpriced would be deliberately courting a loss. Thus the multinationals aggravate short-term capital flows, sometimes forcing the hand of governments. As we have seen, this contributed to the currency problems of the early 1970s.

TAX PROBLEMS

Difficult problems also emerge in the area of taxation. Multinationals often produce components in one country and finished goods in another. When they ship components across national boundaries, they are in a position to charge high or low prices for these components. If they charge high prices, they will create profits for the component manufacturing country; if they charge low prices, profits will be higher in the component receiving country. (This is exactly like the international oil companies, whose pricing strategies we looked into in Chapter 13.)

Since there are different tax laws in different countries, these pricing policies will be largely influenced by taxes rather than by other criteria. As a result, countries may not receive their fair share of total corporation tax payments. Indeed, countries may be forced to compete with one another in their tax rates, much as the various states in the United States who seek to attract industry by special tax arrangements. In one case as in the other, this is a zero-sum game, where the winner is the corporation who ends up paying less taxes than it should, and where the loser is the individual taxpayer who must pick up the burden.

COMPLICATIONS IN THE BALANCE OF PAYMENTS

Last, there is the exceedingly difficult problem of determining whether a multinational helps or hurts its own country's balance of payments. Consider what happens when a corporation begins to produce abroad. To begin with, it will hurt the balance of payments by buying foreign exchange in order to build or buy foreign plant and equipment. Then it will further increase the balance of payments deficit by ceasing to export its goods, as production starts up abroad. To add insult to injury, it may begin to export goods *from* its new foreign facility back into the United States, increasing our imports. Doubtless it will increase the deficit on travel as its executives travel around the world.

But that is only half the picture. The plants abroad will often require special machinery or parts or services from home, resulting in the export of goods or services that would not otherwise have been exported. Most important of all, the multinational will be sending home a steady and probably rising total of dividends. As we have seen, this homeward flow of profits is a major "plus" item in the balance on current account.

Can we then add up all the minuses and all the pluses and calculate whether a multinational is a net help or hindrance? It seems impossible to do so. The fact of overseas production changes the problem so markedly that we can no longer compare the situation under multinationals with that which preceded it. Foreign production is likely to be much larger than exports would have been. Therefore foreign-earned profits of the corporation may also be larger. It is really impossible to say whether, for example, GM earns more foreign exchange as a multinational than it would as a large exporter.

What does seem probable is that *domestic* employment will be less when a company goes multinational. GM will no longer employ American workers to make cars for export; instead, it employs workers in Germany. On the other hand, many American workers may work for a foreign company—Lever Brothers or BP Petroleum or Olivetti. If these companies were not also multinationals, they would be employing labor in their home countries rather than here.

POLITICAL ECONOMICS OF MULTINATIONAL ENTERPRISE

All these considerations make it difficult or impossible to draw up a balance sheet of the pluses and minuses of multinationals strictly in economic terms. Indeed, the problems posed by the multinationals make it necessary to view international production from a perspective that is different from the one we take toward international trade.

As we have seen, the "classical" conception of international trade was based on the familiar model of a competitive world in which factors of production were free to move about to find their points of highest return *within* their own national territories. The result, as our discussion of Infraville and Supraville revealed, was a final equilibrium in which each nation discovered the best allocation of resources both with regard to its own resources and to the advantageous exchanges it could make with other nations. In this final equilibrium, political considerations might enter in the form of tariffs or other trade barriers, but these were always considered by economists as the product of national nearsightedness, so that in theory—and to some extent in practice—it was possible to use the notion of *an equilibrium determined entirely by economic considerations.*

Not so, under the peculiar conditions of international production. For now the unit of economic activity is no longer the tiny "factor of production" subject to market forces, but a giant corporation capable of the maneuvers, tactics, and strategies characteristic of oligopolistic, rather than competitive, economic units. *Within* a given country, as we have seen, oligopolies tend to "settle down" to a more or less steady division-of-the-market, usually by tacit agreement not to engage in price warfare. But this division-of-the-market becomes rudely disturbed when a newcomer from the "outside" establishes his production unit within the home market of another nation. Moreover, the new division-of-the-market between the invader and the established giants will not be determined solely by the economic growth potentials of the various contestants. *There is an inescapable political element that will also play an important, perhaps determinative, role.*

CREATING PAN-NATIONAL ENTERPRISES

This political element has two aspects. On the one hand, it depends on the ability of smaller nations, such as those we find in Europe, to relinquish their feelings of national pride sufficiently to allow the creation of big enough pan-European enterprises to challenge the (American) invader successfully. (Here is Servan-Schreiber's problem in a nutshell.) The cost of IBM's research and development, for example, is more than the entire sales of its largest British competitor. To stand up against IBM, therefore, it may be necessary for British companies to merge with French or German ones, to form a competitor with the economies of scale, the ability to generate finance, the command over technical talent that will put it in a league with IBM. But that will require a *political* decision on the part of England or France or Germany to give up exclusive national control over "their" biggest computer companies.

A few such genuinely international companies have come into existence. Royal Dutch/Shell and Unilever are truly binational enterprises run by boards of directors that represent both Dutch and English directors and an effort is being made to create more such genuinely "European" enterprises. Whether or not this effort will be successful, it is too early to tell. But surely the critical element in the final division of the market among the multinationals will depend fully as much on the politics of international merger as on the economics of market tactics.

NATIONAL PREROGATIVES

There is, moreover, a second political element that enters into the new multinational thrust. This is the question of the extent to which foreign governments will *permit* "foreign" corporations to operate within their borders.

This problem has its counterpart in the political decision on how much to tax (by tariffs) the goods that foreign companies sell to a given nation. But the problem of permitting the entry of a foreign-owned production unit is much more important—and much more difficult to resolve—because this may be the only way that a nation can "import" the *technology* and the *productivity* that the multinational invader will bring with it.

The French government, for example, has been extremely uneasy over the virtual preemption of its high technology computer industries by American firms. Thus when General Electric sought to buy a 20 percent interest in Machines Bull, a leading French manufacturer of computer and desk machinery, the French government balked at the "Americanization" of the firm and forbade the transaction. That was in 1962. Within a very few years it became apparent that without an infusion of American technology, Machines Bull could not stand up to the competition of IBM, and the French government unhappily acquiesced in the American "takeover." But then it *was* a takeover, for G.E. demanded (and got) 50 percent control of Machines Bull.*

Thus, the pan-national thrust of economic activity unavoidably hinges on political decisions concerning the national independence of the economies into which the multinationals seek to move. Canadians, for example, have recently awakened to the fact that they own only *15* percent of their "own" industry. All the rest is foreign owned. Americans alone control 46 percent of Canadian manufacturing, 58 percent of its oil and gas, and almost 100 percent of its auto industry. The next biggest slice is owned by various European, mainly British, interests. Canada is thus a minority stockholder in its own economy, a situation that has led to strong sentiments to block further foreign ownership. But there would, of course, be a real economic price—a diminished rate of growth—for such a blockage.

---

*An investment that did not work out very successfully, by the way, for G.E.

HOST AND HOSTAGE

This conflict between the jealous claims of nation-states who seek to retain national control over productive activity within their own borders and the powerful thrust of pan-national corporations for new markets in foreign territories introduces profound tensions into the political economics of multinational production. On the one hand, the multinational is in a position to win hard bargains from the "host" country into which it seeks to enter (as in the case of France and G.E.), because the corporation *is* the main bearer of new technologies and management techniques that every nation seeks. Therefore, if one country—say France—refuses to give a would-be entrant the right to come in (and possibly to cause financial losses to its established firms), the multinational may well place its plants, with their precious economic cargo of productivity, in another country, leaving the recalcitrant nation the loser in the race for international growth.

On the other hand, the power is by no means entirely one-sided; for once a multinational *has* entered a foreign nation, it becomes a *hostage* of the host country. It is now bound by the laws of that country and may find itself forced to undertake activities that are "foreign." In Japan, for example, it is an unwritten law that workers engaged by giant corporations are *never* fired, but become permanent employees. Japan has been extremely reluctant to allow foreign capital to establish manufacturing operations on Japanese soil, to the great annoyance of foreign companies. But if, as now seems likely, Japan is opened to American and European capital, we can be sure that American or European corporations will be expected to behave in the Japanese way with their employees. This will not be an easy course to follow, since these corporations are not likely to receive the special support that the Japanese government gives to its own big firms.

Or take the problem of a multinational that is forced by a fall in demand to cut back the volume of its output. A decision made along strictly economic lines would lead it to close its least profitable plant. But this may bring very serious economic repercussions in the particular nation in which that plant is located—so serious that the government will threaten to take "action" if the plant is closed. What dictates shall the multinational then follow: those of standard business accounting or those of political accounting?

Or consider the multinational seeking to expand or to alter its operations in an underdeveloped country. This, too, may lead to friction, for as former Undersecretary George Ball has candidly asked: "How can a national government make an economic plan with any confidence if a board of directors meeting 5,000 miles away can, by altering its pattern of purchasing and production, affect in a major way the country's economic life?"

As we have seen in Peru, Bolivia, and Libya, this incompatibility of aims may become so great that the underdeveloped country eventually seizes and nationalizes the local assets of the multinational.

MULTINATIONALS AND WORLD ORDER

Is there a resolution to this conflict between the business rationality of the multinational corporation and the political priorities of the nation-state? At this stage in the development of both institutions, none is in sight. The very idea of pan-national production is itself so new that we lack even a conceptual model of how to deal with its problems, much less a set of practical rules and regulations to follow. Take something so simple—but so important—as the location of the "head office" of the corporation to which a nation would make representations if the action of an international corporation were contrary to national interest. From one nation to another, the legal definition of the head office differs: in the U.S. and the U.K., it is the place where the company is formally incorporated; in Morocco it is the location of the "registered" home office; in France, Germany, and Belgium, it is the main center of management; in Italy and Egypt it is the place of principal business activity. Which of these will respond when a country in search of legal redress calls out: "Will the real head office please stand up?"

Or take the question of the patriotic accountability of the multinational enterprise. Suppose a company wishes to move its profits from country A to country B, but that country A has balance-of-payments difficulties. Is the company bound to obey the wishes of A? Suppose the government in B also has balance-of-payments difficulties and desires the company to import capital? What should the corporation do? In 1966, for example, the United States government asked corporations not to export capital, lest our precarious balance of payments be worsened. But the Ford Motor Company decided that its long-term interests required the purchase of British Ford; so despite objections of the U.S. government, Ford exported $600 million of capital to make the purchase. Was Ford "unpatriotic," if its actions were in the long-term interests of the company? Was it similarly "unpatriotic" (from a British point of view) for a group of English investors to export British capital to finance the building of the Pan Am skyscraper in New York?

Or suppose a multinational company, in one of its foreign plants, undertakes work that is integral to the defense of that nation and that the foreign policy of the nation in question brings its military strength to bear against another country where the multinational is also located? Is the company supposed to "take sides"?

UNRESOLVED QUESTIONS

There are no answers to such questions, only speculations. Here are some of them.

Speculation 1. Will the conglomerate serve as a means of mediation between the demands of business production efficiency and political "national" control? For example, why didn't Ford build the Pan Am building, and why didn't the British investors put their pounds into autos? Both sides would have told you (in 1966) that they weren't "in the business" of real estate or cars, as the case may be. But the rise of the conglomerates opens new possibilities in this direction. If, in the future, the controlling center of Ford becomes (as ITT has already become) a capital-allocation office rather than the headquarters of an auto firm, it is possible that Ford *will* build buildings and that British realtors *will* build automobiles. In that case, the big corporation may surrender some of its multinationality in exchange for multi-industrial coverage.

Speculation 2. Will the multinationals, if they continue their growth, constitute the skeleton of a new form of world order? Some economists and businessmen see the rise of a pan-national system of oligopolies controlled by boards of directors that represent many nations (rather than mainly one, as is now the case), whose operations will pave the way for a much more pragmatic, down-to-earth, effective system of world production and distribution than the present competition of hostile, suspicious, and dangerous nation-states. Such an international rationalization of production, they believe, could bring a "businessman's peace," in which the big companies serve to accelerate world growth, to introduce efficiency into the backward areas, and to assert the logic of economic performance over the outmoded rivalries of jealous nation-states.

Speculation 3. Will the multinationals, on the contrary, serve to heighten the tension of a world which, for better or worse, must continue to use the powerful appeal of nationhood? How can giant corporations, necessarily dedicated to profit-making, adjust their operations to the often unprofitable needs of national development? How can corporations, jockeying for market position, provide the basis for a stable world economic system? The opponents of the multinationals see in them not a force for progress but only a means of imposing a calculus of profits on a world whose needs at the moment often demand an entirely different set of fundamental values.

AN UNWRITTEN ENDING

It is much too early to determine which of these arguments will eventually be proven correct. Perhaps the safest guess is that all will be, at least in our lifetimes. The big corporations are likely to continue to go in both multinational and conglomerate directions. To some extent they will be the international carriers of efficiency and development, especially in the high technology areas for which they seem to be the most effective form of organization. But if the power of the nation-state will be challenged by these international production units, it is not likely to be humbled by them. There are many things a nation can do that a corporation cannot, including, above all, the creation of the spirit of sacrifice necessary both for good purposes such as development and for evil ones such as war.

Perhaps all we can say at this stage of human development is that both nation-states and huge corporations are necessary, in that they seem to be the only ways in which we can organize mankind to perform the arduous and sustained labor without which humanity itself would rapidly perish. Perhaps after the long age of capital accumulation has finally come to an end and sufficient capital is available to all peoples, we may be able to think seriously about dismantling the giant enterprise and the nation-state, both of which overpower the individual with their massive organized strength. However desirable that ultimate goal may be, in our time both state and corporation promise to be with us, and the tension between them will be part of the evolutionary drama of our period of history.

KEY WORDS

Multinational corporations

Foreign direct investment

Internationaliza-tion of production

CENTRAL CONCEPTS

1. The basic conception of international trade has entered a new stage marked by *international production carried on by the multinational corporation.*

2. The natural impetus of the drive for growth has pushed most large U.S. corporations into multinational operations. As a result, our exports of manufactures are today only 40 percent of our overseas production of manufactured commodities. In fact, *overseas production is now considerably larger than the exports of all U.S. goods.*

3. As a result of the thrust of the multinationals, U.S. foreign direct investment has risen to approximately $94 billion of foreign direct American investment and may be much larger than that if the value of foreign capital controlled by American equity is included.

4. The rise of the American multinational in Europe has been very rapid and has led to a European fear of an "American challenge." However justified in Europe, this fear masks the fact that *European (and Japanese) foreign investment is rising as rapidly as American—and much of it in America.*

5. The prospect is for a genuine internationalization of production within the next few decades. A few hundred large firms, of which the majority will probably be American, are apt to control a very large fraction of the production of industrial commodities on a worldwide basis.

Shifts in foreign investment

6. *The rise of multinational production has resulted in striking shifts in the geographic location and technological character of international investment.* The trend has been away from investment in underdeveloped nations to investment in the markets of high-consumption countries; and away from low-skilled labor-intensive or capital-intensive industries to high skill-intensive industries.

High technology

7. *The incursion of the multinationals across national boundaries has raised sharp new problems of political economics.* An equilibrium in international production will depend on political decisions as well as economic forces. These are the
   - Ability of smaller nations to abandon national prerogatives in order to form effective international corporations to challenge American or other companies
   - Willingness of countries to allow foreign companies to dominate industries that they may consider vital to their national interests
   - Ability of most nations to import new high technology if they deny entry to the multinationals

Economic interdependence

8. Multinational firms make it difficult for different countries to pursue independent economic policies. They are forced into a state of economic interdependency, since multinational firms can avoid many of the controls that limit national firms.

Conflicts of the multinational corporation

9. Equally perplexing problems are posed for the multinationals themselves. *If they are often in a position to drive hard bargains before entrance, once they have committed themselves, they are "hostages."* To which of the many countries where they may be located do they owe allegiance if the interests of those countries clash? To what extent should the dictates of profit determine their international allocations of manpower and capital? To what extent can a multinational "defy" and to what extent must it "bow before" the demands of a host country?

10. There are no answers to the questions above. *We still have no established rules or procedures or laws to cover this new mode of organizing the world's production.* Nor do we know if the growth of multinational enterprise will serve as a force for order and rationality or for friction and tension.

QUESTIONS

1. How many products produced in America can you identify as "foreign"? (You might start with the detergents and soaps produced by Lever Brothers, the office machinery produced by Olivetti-Underwood, the gas and oil refined by Shell.)

2. Can you suggest a hypothesis that might be testable as to why certain industries seem to go overseas more rapidly than others? Why autos and tractors but not washing machines? Why plate glass but not sheet steel? (Little is known about this, so you might become famous if you come up with an idea that tests out.)

3. Do you think a nation is right to exclude a foreign company from producing certain products? If you could get high-speed computers only by allowing a Japanese company to produce them in the U.S., would you still keep out the Japanese company?

4. What do you think is the duty of a company that is active in nations A, B, and C, when all have balance-of-payments problems and request domestic corporations not to make unnecessary international payments? Suppose the headquarters of the company is in A and the stockholders want the profits from B and C repatriated?

5. Can you draw up a plan for a company that would be privately owned and managed but not officially "headquartered" in any one nation? From whom would it receive its charter? Under what laws will it operate so far as the top management is concerned?

6. Do you think that the rise of multinational production opens the way for a more rational world or a more divided one?

7. Explain how companies can make their profits show up in different countries, depending upon where it is most advantageous to have profits.

8. Multinational companies pose various difficulties for economic policy makers. Which difficulties do you think are most important and why?

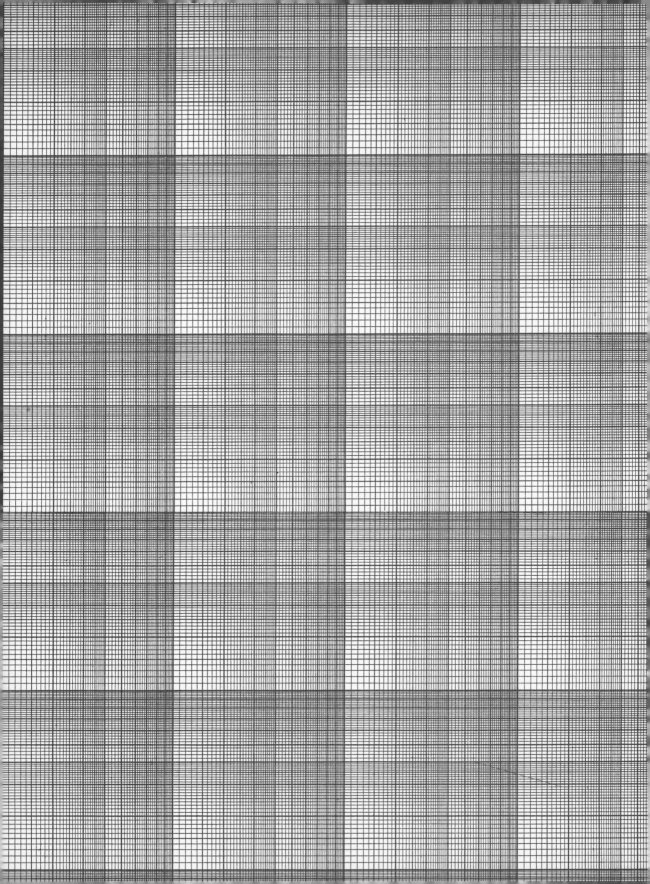

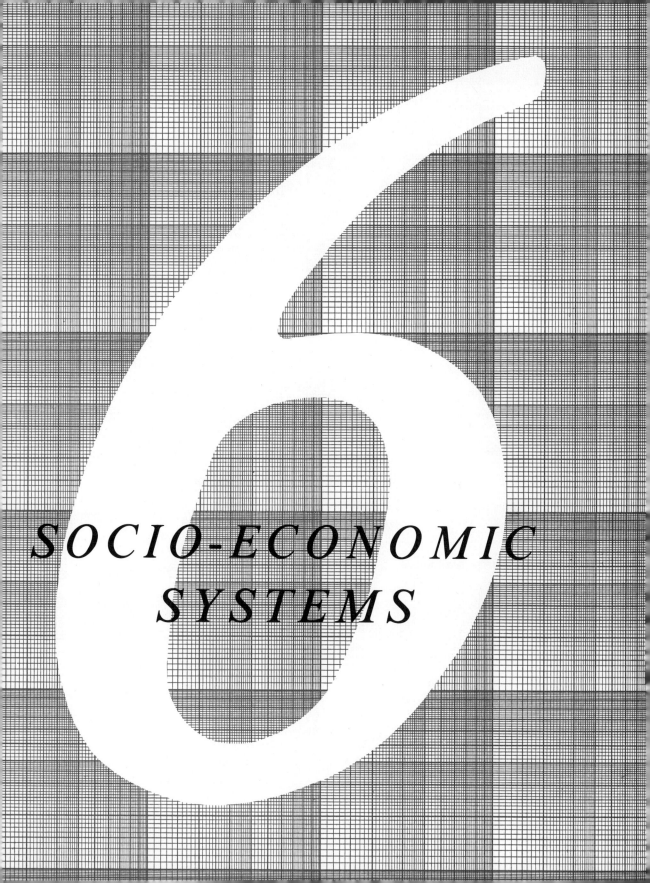

6

SOCIO-ECONOMIC SYSTEMS

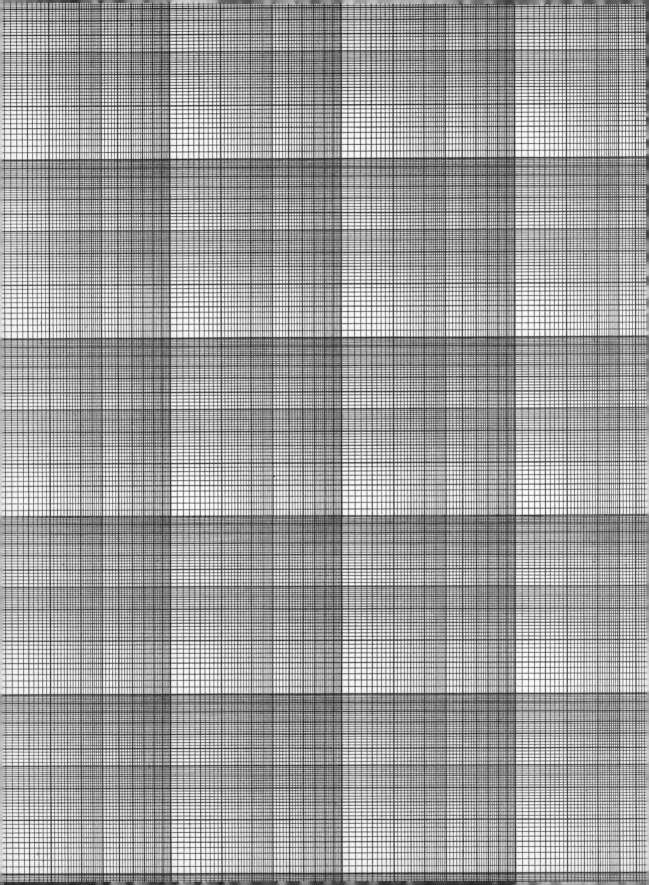

The Underdeveloped World

WE MUST BEGIN THIS VERY IMPORTANT chapter with a sobering realization. It is that our account of the long sweep of Western economic advance has ignored the economic existence of four out of five human beings. Mere parochialism was not, however, the reason for this concentration on Western progress. Rather, it was the shocking fact that, taken in the large, *there was no economic progress in the rest of the world.*

This is not to say that tides of fortune and misfortune did not mark these areas, that great cultural heights were not achieved, and that the political or social histories of these regions do not warrant interest and study. Yet the fact remains that the mounting tide of *economic* advance that has engaged our attention was a phenomenon limited to the West. It is no doubt something of an over-simplification, but it is basically true to claim that in Asia, Africa, South America, or the Near East, economic existence was not ma-terially improved for the average inhabitant from the twelfth—and, in some cases, the second—to the beginning of the twentieth century. Indeed, for many of them it was

worsened. A long graph of non-Western ma-terial well-being would depict irregular rises and falls but an almost total absence of cu-mulative betterment.

The near end of such a graph would show the standard of living of three-quarters of the human race who inhabit the so-called underdeveloped areas today. Most of this mass of humanity exists in conditions of poverty that are difficult for a Westerner to comprehend. When we sum up the plight of the underdeveloped nations by saying that a billion human beings have a standard of liv-ing of "less than $100 a year," and that another, more fortunate, billion people enjoy in a year one-quarter to one-half the income a typical American family spends in a single *month,* we give only a pale statistical mean-ing to a reality that we can scarcely grasp.

BACKGROUND TO UNDERDEVELOPMENT

Why are the underdeveloped nations so piti-ably poor? Only a half-century ago it was common to attribute their backwardness to

SNAPSHOTS OF UNDERDEVELOPMENT

DAMASCUS: in famine years the children of the poor examine the droppings of horses to extract morsels of undigested oats.

CALCUTTA: 250,000 people have no home whatsoever; they live, eat, defecate, mate, and die in the streets.

HONG KONG: large numbers of families live in floating villages that tourists like to photograph. A family of six, eight, or ten occupies a home approximately the size of a rowboat.

CALI, COLOMBIA: When the rains come, the river rises and the sewers run through the homes of the poor.

HYDERABAD: child labor employed in sealing the ends of cheap bracelets is paid eight cents per gross of bracelets.

KATMANDU, Nepal: Life expectancy is between 35 and 40 years. Tuberculosis is chronic. One hears people coughing themselves to death at night.

NEW DELHI: "Oh, sir! Someone has dropped ice cream on your shoes! I will clean them." The tourist finds himself propped against a building, with two boys each shining one shoe. "Oh, sir! Your laces are frayed. See, they break! I will sell you a new pair." The tourist buys the new pair. As he leaves, an Old India Hand says to him: "Have to watch those little beggars. Fling mud on your shoes." These are the tactics which poverty generates.

geographic or climatic causes. The underdeveloped nations were poor, it was thought, either because the climate was too debilitating or because natural resources were lacking. Sometimes it was just said that the natives were too childlike or racially too inferior to improve their lot.

Bad climates may have had adverse effects. Yet, many hot areas have shown a capacity for sustained economic growth (for example, the Queensland areas of Australia), and we have come to recognize that a number of underdeveloped areas, such as Argentina and Korea, have completely temperate climates. So, too, we now regard the lack of resources in many areas more as a *symptom* of underdevelopment than a cause—which is to say that in many underdeveloped areas, resources have not yet been *looked for*. Libya, for instance, which used to be written off as a totally barren nation, has been discovered to be a huge reservoir of oil. Finally, little is heard today about native childishness or inherent inferiority. (Perhaps we remember how the wealthy classes in Europe similarly characterized the poor not too many cen-

turies ago.) Climate and geography and cultural unpreparedness unquestionably constitute obstacles to rapid economic growth—and in some areas of the globe, very serious obstacles—but there are few economists who would look to these disadvantages as the main causes of economic backwardness.

Why then are these societies so poor?

The answer takes us back to the first chapter of our book. These are poor societies because they are *traditional* societies—that is, societies which have developed neither the mechanisms of command nor of the market by which they might launch into a sustained process of economic growth. Indeed, as we examine them further we will have the feeling that we are encountering in the present the anachronistic counterparts of the static societies of antiquity.

Why did they remain traditional societies? Why, for instance, did Byzantium, which was economically so advanced in contrast with the Crusaders' Europe, fall into decline? Why did China, with so many natural advantages, not develop into a dynamic economic society? There are no simple, or even fully satisfactory, answers. Perhaps the absence of economic progress elsewhere on the globe forces us to look upon our Western experience not as the paradigm and standard for historic development, but as a very special case in which various activating factors met in an environment peculiarly favorable for the emergence of a new economic style in history. The problem is one into which we cannot go more deeply in this book. At any rate, it is today an academic question. The dominant reality of our times is that the backward areas are now striving desperately to enter the mainstream of economic progress of the West. Let us examine further their chances for doing so.

CONDITIONS OF BACKWARDNESS

Every people, to exist, must first feed itself; there is a rough sequence to the order of demands in human society. But to go beyond existence, it must achieve a certain level of efficiency in agriculture, so that its efforts can be turned in other directions. What is tragically characteristic of the underdeveloped areas is that this first corner of economic progress has not yet been turned.

Consider the situation in that all-important crop of the East, rice. Table 39·1 shows the difference between the productivity of rice fields in the main Asiatic countries and those of the United States and Australia and Japan.

What is true of rice can be duplicated in most other crops.* It is a disconcerting fact

TABLE 39 · 1
Rice Production

(100 kilograms per hectare)	1971
U.S.	52.0
Australia	73.8
Japan	52.5
India	17.1
Indonesia	22.0
Thailand	19.7
Philippines	17.2

that the backward peasant nations that depend desperately on their capacity to grow food cannot even compete in these main products with the advanced countries: Louisiana rice undersells Philippine rice, California oranges are not only better but cheaper than Indonesian oranges.

Why is agriculture so unproductive? One apparent reason is that the typical unit of agricultural production in the underdeveloped lands is far too small to permit efficient farming. "Postage stamp cultivation" marks the pattern of farming throughout most of Asia and a good deal of Africa and South America. John Gunther, reporting the situation in India over a generation ago, described it vividly. It has not changed materially since that time.

There is no primogeniture in India as a rule, and when the peasant dies his land is subdivided among all his sons with the result that most holdings are infinitesimally small. In one district in the Punjab, following fragmentation through generations, 584 owners cultivate no less than 16,000 fields; in another, 12,800 acres are split into actually 63,000 holdings. Three-quarters of the holdings in India as a whole are under ten acres. In many parts of India the average holding is less than an acre.[1]

---

*Table 39·1 shows only the productive differentials of equal *areas* of land. When we consider that a single American farmer tends up to a hundred times as large an acreage as a peasant in an underdeveloped area, the difference of output *per man* would be much more striking. The "Green Revolution" (discussed near the end of the chapter) has improved the situation, but we do not yet have the data we need to determine outputs for the new rice strains. Despite the improvement, a vast gulf still separates U.S. agricultural productivity from that of the underdeveloped nations.

[1]*Inside Asia* (New York: Harper, 1939), p. 385.

In part, this terrible situation is the result of divisive inheritance practices which Gunther mentions. In part, it is due to landlord systems in which peasants cannot legally own or accumulate their own land; in part, to the pressure of too many people on too little soil. There are many causes, with one result: agriculture suffers from a devastatingly low productivity brought about by grotesque man/land ratios.

These are, however, only the first links in a chain of causes for low agricultural productivity. Another consequence of these tiny plots is an inability to apply sufficient capital to the land. Mechanical binders and reapers, tractors and trucks are not only impossible to use efficiently in such tiny spaces, but they are costly beyond the reach of the subsistence farmer. Even fertilizer is too expensive: in much of Asia, animal dung is used to provide free fuel rather than returned to the soil to enrich it.

This paralyzing lack of capital is by no means confined to agriculture. It pervades the entire range of an underdeveloped economy. The whole industrial landscape of a Western economy is missing: no factories, no power lines, no machines, no paved roads meet the eye for mile upon mile as one travels through an underdeveloped continent. Indeed, to a pitiable extent, an underdeveloped land is one in which human and animal muscle power provide the energy with which production is carried on. In India in 1953, for instance, 65 percent of the total amount of productive energy in the nation was the product of straining man and beast.[2] The amount of usable electrical power generated in *all of India* would not have sufficed to light up New York City.

SOCIAL INERTIA

A lack of agricultural and industrial capital is not the only reason for low productivity. As we would expect in traditional societies, an endemic cause of low per capita output lies in prevailing social attitudes. Typically, the people of an underdeveloped economy have not *learned* the "economic" attitudes that foster rapid industrialization. Instead of technology-conscious farmers, they are tradition-bound peasants. Instead of disciplined workers, they are reluctant and untrained laborers. Instead of production-minded businessmen, they are trading-oriented merchants.

For example, Alvin Hansen reports from his observations in India:

Agricultural practices are controlled by custom and tradition. A villager is fearful of science. For many villagers, insecticide is taboo because all life is sacred. A new and improved seed is suspect. To try it is a gamble. Fertilizers, for example, are indeed a risk.... To adopt these untried methods might be to risk failure. And failure could mean starvation.[3]

In similar vein, a UNESCO report tells us:

In the least developed areas, the worker's attitude toward labour may entirely lack time perspective, let alone the concept of productive investment. For example, the day labourer in a rural area on his way to work, who finds a fish in the net he placed in the river the night before, is observed to return home, his needs being met....[4]

An equally crippling attitude is evinced by the upper classes, who look with scorn or disdain upon business or production-oriented careers. More than a decade ago, UNESCO

[2]Daniel Wit and Alfred B. Clubok. "Atomic Power Development in India," *Social Research,* Autumn 1958, p. 290.

[3]*Economic Issues of the 1960s* (New York: McGraw-Hill, 1960), pp. 157–58.

[4]*Report on the World Social Situation,* UNESCO, March 9, 1961, p. 79.

reported that of the many students from the underdeveloped lands studying in the United States—the majority of whom come from the more privileged classes—only 4 percent were studying a problem fundamental to all their nations: agriculture.[5] This has not changed over time.

All these attitudes give rise to a *social inertia* that poses a tremendous hurdle to economic development. A suspicious peasantry, fearful of change that might jeopardize the slim margin yielding them life, a work force unused to the rhythms of industrial production, a privileged class not interested in production—all these are part of the obdurate handicaps to be overcome by an underdeveloped nation.

FURTHER PROBLEMS: POPULATION GROWTH

Many of these problems resemble the pre-market economies of antiquity. But in addition, the underdeveloped lands face an obstacle with which the economies of antiquity did *not* have to cope: a crushing rate of population increase that threatens to nullify their efforts to emerge from backward conditions.

Only a few figures are needed to make the point. Let us begin with our southern neighbor, Mexico. Today, Mexico has a population equal to that of New York State, Pennsylvania, New Jersey, and Connecticut. Thirty years from now, if Mexico's present rate of population increase continues, it will have as many people as the present population of these four states *plus* the rest of New England, *plus* the entire South Atlantic seaboard, *plus* the entire West Coast, *plus* Ohio, Indiana, Illinois, Michigan, and Wisconsin. Or take the Caribbean and Central American

area. In some thirty years, at present growth rates, that small part of the globe will outnumber the entire population of the United States. South America, now 5 percent less populous than we, will be 200 percent larger than our present population. India could then number a billion souls.

We have already seen one result of the relentless proliferation of people in the fragmentation of landholdings. But the problem goes beyond mere fragmentation. Eugene Black, formerly president of the International Bank for Reconstruction and Development (the World Bank) has written that in India a population equivalent to that of all Great Britain has been squeezed out of any landholding whatsoever—even though it still dwells in rural areas.[6] Consequently, population pressure generates massive and widespread rural poverty, pushing inhabitants from the countryside into the already overcrowded cities. Five hundred families a day move into Jakarta from the surrounding Javanese countryside, where population has reached the fantastic figure of 1,100 per square mile.

Even these tragic repercussions of population growth are but side effects. The main problem is that population growth adds more mouths almost as fast as the underdeveloped nations manage to add more food. They cancel out much economic progress by literally eating up the small surpluses that might serve as a springboard for faster future growth.

Ironically, this population "explosion" in the underdeveloped countries is a fairly recent phenomenon, attributable largely to the incursion of Western medicine and public health into the low income areas. Prior to World War II, the poorer countries held

[5]*Ibid.*, p. 81.

[6]Eugene R. Black, *The Diplomacy of Economic Development* (Cambridge: Harvard University Press, 1960), p. 9.

A PRIMER ON DEMOGRAPHY

Anyone interested in development soon comes up against the formidable problems of population growth, and this quickly leads to the field of demography, the study of the behavior of populations. Demography is a fascinating and immensely important subject, which we cannot go into in this book, but we must take a moment to introduce the student to some of its elements.

Let us begin by looking at two imaginary countries, both having zero population growth. We will call them Westernland and Easternland because the facts they represent, although somewhat exaggerated, are approximated by conditions in the developed Western and the underdeveloped Eastern (and Southern) nations.

First consider Westernland. The demographic facts are simple. Of 2,000 population (50 percent women), all marry. Each couple has two children, 1 girl, 1 boy. None die. At age 26 the original population has reproduced itself, and a new cycle can begin with no increase in numbers.

Now Easternland. The original population is again 2,000. But only 900 marriages take place, owing to various restrictions of caste, taboos, ill-

	Westernland	Easternland
Population, age 26	2,000	2,000
Number of marriages	1,000	900
Total births	2,000	3,780
Infant deaths	0	880
Number surviving to age 1	2,000	2,900
Child and young adult deaths	0	900
Population, age 26	**2,000**	**2,000**

nesses, and so forth. The 900 marriages now produce 4.2 children each. This is lower than the number they would produce if some of the husbands and wives did not die before the years of child-bearing were finished. It is roughly the family size we find in many underdeveloped areas. However, of the 3,780 births, 880 die during the first year and 900 in the next 25 years—again, "age-specific" mortality rates closely corresponding with the facts of the poorer regions. Hence we reach a new generation, at age 26, that is only 2,000 strong, just as in the case of Westernland. *There is no population growth.*

There is, however, an enormous potential for population growth in the frustrated fertility of their population growth in check because death rates were nearly as high as birth rates. With DDT and penicillin, death rates have plunged dramatically. In Ceylon, for example, death rates dropped 40 percent in one year following the adoption of malaria control and other health measures. As death rates dropped in the underdeveloped areas, birth rates, for many reasons, continued high, despite efforts to introduce birth control. In the backward lands, children are not only a source of prestige and of household labor for the peasant family, but also the only possible source of "social security" for old age. The childless older couple could very well starve. As parents or grandparents, they are at least assured of a roof over their heads.

Is there a solution to this problem? The mood of demographers has swung between black despair and cautious hope over the past decades. New birth control methods have, from time to time, offered the chance for dramatic breakthroughs. Poor birth control programs have repeatedly dashed these hopes.

Today most demographers expect the population explosion to continue for at least another generation. Part of the problem lies in the disproportionately large numbers of the population under age 15 in the poorer regions—young people who are themselves the result of the population flood. Even if each girl who comes of child-bearing age were to adopt birth control, the ever larger

Easternland. One hundred additional marriages could take place. The total number of children born to the 2,000 potential marriages would be 5,250 if neither adult died and if prevailing family size were achieved by all. If all infant, childhood, and young adult deaths were prevented, 1,780 additional marriages could result. In that case, the population at the end of 26 years would not be 2,000 but 5,250—a rate of increase reminding us of Central America.

Now two important demographic terms. *The net reproduction rate* tells us the number of daughters who will be born to 1,000 girl babies by the end of their child-bearing years, *assuming that existing fertility and mortality rates prevail.* In both Westernland and Easternland this is 1,000, or a net reproduction rate of 1.00. Each girl baby in each land will eventually be succeeded by one potential mother at age 26. *Gross reproduction rates* tell us the number of female offspring, *ignoring mortality and other wastage,* such as failure to marry. In Westernland it is also 1.00. In Easternland it is 2,625 (half of 5,250), or a rate of 2.625. *The difference between the two rates is therefore the measure of frustrated fertility in a population.*

The table shows us that a population in equilibrium, with zero population growth, can be big with potential increase. It also reveals the surprising number of factors that enter into population growth rates. Age at marriage is one: if that age were 20 instead of 26, the average family in Easternland would have over 6 people. It calls attention to celibacy as an element in the population equation, a factor of some importance in certain areas. It points out the role of premature death, both in limiting the size of families that will be achieved, and in winnowing out the number of survivors who will eventually become parents. And of course it highlights the all-important factor of desired or traditional family size in determining how large the rate of population growth will be. Clearly, the limitation of population growth is a good deal more complicated than just birth control, important as the latter is.

Anyone wishing to learn more about this crucially important subject might look into E. A. Wrigley, *Population and History* (New York: McGraw-Hill, 1969) from which this example has been taken. This is an engrossing and easily accessible book on a subject every economist should know something about.

number of such child-bearers will result in rapid population growth for another generation, despite a declining birth rate. And in all probability, the newly wed girls of the next generation in most parts of the world will *not* practice birth control. We will have to await the arrival of effective governments before population control programs can be expected to succeed. More about that later.

NINETEENTH-CENTURY IMPERIALISM

This gives us a brief introduction to underdevelopment as it exists today. Before we turn to the problem of how this condition can be remedied, we must inquire into one more question. Why did not the market society, with all its economic dynamism, spread into the backward areas?

The answer is that the active economies of the European and American worlds *did* make contact with the underdeveloped regions, beginning with the great exploratory and commercial voyages of the fifteenth and sixteenth centuries. Until the nineteenth century, unfortunately, that contact was little more than mere adventure and plunder. And then, starting in the first half of that century and gaining momentum until World War I, came that scramble for territory we call the Age of Imperialism.

What was this imperialism? It was, in retrospect, a compound of many things: militarism, jingoism, a search for markets and for sources of cheap raw materials to feed growing industrial enterprises. Insofar as the colonial areas were concerned, however, the first impact of imperialism was not solely that of exploitation. On the contrary, the incursion of Western empires into the backward areas brought some advantages. It injected the heavy doses of industrial capital: rail lines, mines, plantation equipment. It brought law and order, often into areas in which the most despotic personal rule had previously been the order of the day. It introduced the ideas of the West, including, most importantly, the idea of freedom, which was eventually to rouse the backward nations against the invading West itself.

Yet if imperialism brought these positive and stimulating influences, it also exerted a peculiarly deforming impulse to the underdeveloped—indeed, then, totally undeveloped —economies of the East and South. In the eyes of the imperialist nations, the colonies were viewed not as areas to be brought along in balanced development, but essentially as immense supply dumps to be attached to the mother countries' industrial economies. Malaya became a vast tin mine; Indonesia, a huge tea and rubber plantation; Arabia, an oil field. In other words, the direction of economic development was steadily pushed in the direction that most benefited the imperial owner, not the colonial peoples themselves.

The result today is that the typical underdeveloped nation has a badly lopsided economy, unable to supply itself with a wide variety of goods. It is thereby thrust into the international market with its one basic commodity. For instance, in South America we find that Venezuela is dependent on oil for some 90 percent of its exports; Colombia, on coffee for three-quarters of its exports; Chile, on copper for two-thirds of its foreign earnings; Honduras, on bananas for half of its foreign earnings. On the surface, this looks like a healthy specialization of trade. We shall shortly see why it is not.

Economic lopsidedness was one unhappy consequence of imperialism. No less important for the future course of development in the colonial areas was a second decisive influence of the West: its failure to achieve political and psychological relationships of mutual respect with its colonial peoples. In part, this was no doubt traceable to an often frankly exploitative economic attitude, in which the colonials were relegated to second-class jobs with third-class pay, while a handful of Western whites formed an insulated and highly paid managerial clique. But it ran deeper than that. A terrible color line, a callous indifference to colonial aspirations, a patronizing and sometimes contemptuous view of "the natives" runs all through the history of imperialism. It has left as a bitter heritage not only an identification of capitalism with its worst practices, but a political and social wariness toward the West which deeply affects the general orientation of the developing areas.

IMPERIALISM TODAY

What about imperialism today? Certainly it has changed. The naked power grabs are in the past, when imperialism often meant only the acquisition of territory that would look good on a map. In the past, also, are the seizures of raw materials on the unfair terms characteristic of mineral empires built in the late nineteenth century. Less prominent, as well, are attitudes of racial "superiority," so infuriating to peoples whose culture was often of far greater delicacy and discrimination than that of the West.

Thus, the nature of this imperialism is now changing, partly under the pressures exerted by a restive Third World, partly as a result of developments within the advanced nations themselves. The rise of the so-called multinational corporation, for example, puts the problem of the economic relationship of advanced and backward countries in a new light. The backward nations are extremely wary of the power inherent in a giant U.S. or European corporation. Yet they also want some of the things the multinationals offer. Big multinationals pay higher wages, keep more honest books, provide better working conditions and fancier career opportunities, and bring in more technological expertise than do the domestic enterprises of the "host" nation.

The result is that the problem of imperialism in our day, at least so far as the United States is concerned, has taken an unexpected turn. On the economic side of the question, the *danger now is as much that the big companies will bypass the backward nations as that they will dominate them.*

Meanwhile, the political element of imperialism seems to be diminishing. The erstwhile capitalist empires of Germany, Belgium, Netherlands, England (with a few exceptions), have disappeared. Portugal, for all its ruthless domination of Angola, is an archaic remnant of the past, not an entering wedge of the future.* What is left is a strong effort on the part of the United States to preserve its ideological and political influence, particularly in Latin America and Southeast Asia, but the debacle of the American Vietnam policy indicates that the prospects for a "successful" imperialism of this kind are limited, at best.

The Engineering of Development

Up to this point we have concentrated our attention mainly on the background of un-

*The 1974 overthrow of the Portugal government emphasizes this point.

derdevelopment. Now we must ask a more forward-looking, more technically "economic" question: How can an underdeveloped nation emerge from its poverty?

From what we have learned, we know the basic answer to this question. The prerequisite for economic progress for the underdeveloped countries today is not essentially different from what is was in Great Britain at the time of the industrial revolution or what it was in Russia in 1917. To grow, an underdeveloped economy must build capital.

How is a starving country able to build capital? When 80 percent of a country is scrabbling on the land for a bare subsistence, how can it divert its energies to building dams and roads, ditches and houses, railroad embankments and factories that, however indispensable for progress tomorrow, cannot be eaten today? If our postage-stamp farmers were to halt work on their tiny unproductive plots and go to work on a great project like, say, the Aswan Dam, who would feed them? Whence would come the necessary food to sustain these capital workers?

Here is our grim "model" of a peasant economy from Chapter 20 come to life, and we will remember from that model that when consumption could not be cut, growth could not ensue. Still, when we look again at the underdeveloped lands, the prospect is not quite so bleak as that. In the first place, these economies *do* have unemployed factors. In the second place, we find that a large number of the peasants who till the fields are not feeding themselves. They are, also, in a sense, taking food from one another's mouths.

As we have seen, the crowding of peasants on the land in these areas has resulted in a diminution of agricultural productivity far below that of the advanced countries. Hence the abundance of peasants working in the fields obscures the fact that *a smaller number of peasants, with little more equipment*

ZERO MARGINAL PRODUCTIVITY?

The suggestion that output could be increased if there were fewer people working on the land implies that the "last" people working must have *negative* marginal productivities—that they must, in fact, duplicate the seemingly far-fetched example on pp. 139f., where the addition of still more workers actually causes output to drop.

Do they really have negative marginal productivities? There has been considerable discussion about this in the professional literature and quite extensive investigations in the field. When we look at a peasant family at work, it seems impossible to find a case where an extra pair of hands does not produce some increment to the crop. The difficulty arises from the fact that *average* productivities are so low—in part a consequence of low productivity.

To unscramble that sentence, as a result of the low average productivity of peasant workers, diets are meager, and men and women cannot work nearly as successfully as if they were well-fed. Thus, taking men off the farm may permit the remaining members of the household to eat a little more (they cannot eat a lot more, or they would eat up all their productivity gains), and the result of eating more may, in turn, enhance their productive ability. Food thus becomes a capital good. Eating is investment, in conditions of extreme poverty.

—perhaps even with no more equipment— could raise a total output just as large. One observer has written: "An experiment carried out near Cairo by the American College seems to suggest that the present output, or something closely approaching it, could be produced by about half the present rural population of Egypt."[7] Here is an extreme case, but it can be found to apply, to some degree, to nearly every underdeveloped land.

Now we begin to see an answer to the predicament of the underdeveloped societies. In nearly all of these societies, there exists a disguised and hidden surplus of labor which, if it were taken off the land, could be used to build capital. Most emphatically, this does not mean that the rural population should be literally moved, en masse, to the cities where there is already a hideous lump of indigestible unemployment. It means, rather, that the inefficient scale of agriculture conceals a reservoir of both labor and the food to feed that labor if it were elsewhere employed. By raising the productivity of the tillers of the soil, a work force can be made available for the building of roads and dams, while this "transfer" to capital building need not result in a diminution of agricultural output.

This rationalization of agriculture is not the only requirement for growth. When agricultural productivity is enhanced by the creation of larger farms (or by improved techniques on existing farms), *part of the ensuing larger output per man must be saved.* In other words, the peasant who remains on the soil cannot enjoy his enhanced productivity by raising his standard of living and eating up all his larger crop. Instead, the gain in output per cultivator must be siphoned off the farm. It must be "saved" by the peasant cultivator and shared with his formerly unproductive cousins, nephews, sons, and daughters who are now at work on capital-building projects. We do not expect a hungry peasant to do this voluntarily. Rather, by taxation or exaction, the government of an underdeveloped land must arrange for this indispensable transfer. Thus in the early stages of a *successful* development program there is apt to be no visible rise in the individ-

[7] Ragnar Nurkse, *Problems of Capital Formation in the Underdeveloped Countries* (New York: Oxford University Press, 1958), p. 35, fn. 2.

ual peasant's food *consumption,* although there must be a rise in his food *production.* What is apt to be visible is a more or less efficient—and sometimes harsh—mechanism for assuring that some portion of this newly added productivity is not consumed on the farm but is made available to support the capital-building worker. This is a problem that caused the Russian planners such trouble in the early days of Soviet industrialization.

What we have just outlined is not, let us repeat, a formula for immediate action. In many underdeveloped lands, as we have seen, the countryside already crawls with unemployment, and to create, overnight, a large and efficient farming operation would create an intolerable social situation. We should think of the process we have just outlined as a long-term blueprint which covers the course of development over many years. It shows us—as did our earlier model—that the process of development takes the form of a huge internal migration from agricultural pursuits, where labor is wasted, to industrial and other pursuits, where it can yield a net contribution to the nation's progress.

PROBLEM OF EQUIPMENT

Our model also showed us that capital-building is not just a matter of freeing hands and food. Peasant labor may construct roads, but it cannot, with its bare hands, build the trucks to run over them. It may throw up dams, but it cannot fashion the generators and power lines through which a dam can produce energy. In other words, what is needed to engineer the great ascent is not just a pool of labor. It is also a vast array of industrial equipment.

How is this equipment obtained? In our model, by expanding the machine-tool—that is, the capital-equipment-building—subsector. But an underdeveloped economy does not have a capital-equipment-building sector and cannot take the time *painfully* to create one. Consequently, *in the first stages of industrialization, before the nucleus of a self-contained industrial sector has been laid down, a backward nation must obtain its equipment from abroad.*

This it can do in one of three ways. (1) It can buy the equipment from an industrialized nation by the normal process of *foreign trade.* Libya, for example, can sell its oil and use the foreign currency it receives to purchase abroad the tractors, lathes, and industrial equipment it needs. (2) It can receive the equipment by *foreign investment* when a corporation in an advanced nation chooses to build in a backward area. This is the route by which the United States got much of its capital from Britain during the nineteenth century, and it is the means by which the underdeveloped nations themselves received capital during their colonial days. (3) It may receive the foreign exchange needed to buy industrial equipment as a result of a grant or a loan from another nation or from a United Nations agency such as the World Bank. That is, it can buy industrial equipment with *foreign aid.*

TRADE PROBLEMS

Of these three avenues of industrialization, the most important is foreign trade. In all, the underdeveloped nations earn over $70 billion a year from foreign trade. Not all of this, by any manner of means, however, is available for *new* industrial capital. A lion's share of export earnings, unfortunately, must go to pay for indispensable imports—replacements of old equipment, or even food—or to pay interest on loans contracted with the industrialized world.

In addition, another problem plagues the underdeveloped nations in foreign trade. We have seen how international trade is the

HUMAN CAPITAL AGAIN

An allied problem of no less importance arises from the lack of technical training on which industrialization critically depends. At the lowest level, this is evidenced by appalling rates of illiteracy (up to 80 or 90 percent) which make it impossible, for instance, to print instructions on a machine or a product and expect them to be followed. And at a more advanced level, the lack of expert training becomes an even more pinching bottleneck. Before its destructive civil war, United Nations economists figured that Nigeria alone would need some 20,000 top-level administrators, executives, technicians, etc., over the next ten years and twice as many subordinates. On a world-wide scale, this implies a need for at least 700,000 top-level personnel and 1,400,000 second-level assistants. Not 1 percent of these skilled personnel exists today in the poor countries, and to "produce" them will be a task of staggering difficulty. Yet, without them it is often impossible to translate development plans into actuality.

means by which a great international division of labor can be achieved—that is, by which productivity can be enhanced in all trading countries, by enabling each to concentrate on those products in which it is most efficient.

With the underdeveloped nations, however, this international division of labor has worked badly. First, as we have seen, their structural backwardness has prevented them from developing their productivities even in their main occupational tasks. Second, most of them suffer from another problem. As sellers of raw commodities—usually only one raw commodity—typically, they face a highly inelastic demand for their goods. Like the American farmer, when they produce a bumper crop, prices tend to fall precipitously, and demand does not rise proportionately. At the same time, the industrial materials they buy in exchange tend to be firm or to rise in price over the years.

TERMS OF TRADE

Thus the "terms of trade"—the actual *quid pro quo* of goods received against goods offered—have usually moved against the poorer nations, who have given more and more coffee for the same amount of ma-

chinery. In 1957 and 1958, when commodity prices took a particularly bad tumble, the poor nations actually lost more in purchasing power than the total amount of all foreign aid they received. In effect, they subsidized the advanced nations! As another example, it has been estimated that falling prices cost the African nations more, in the first two decades since World War II, than all foreign funds given, loaned, or invested there.

It is possible—we do not yet know— that tightening markets in resources may reverse this trend. The last few years have seen enormous sums flowing into the coffers of mid-Eastern governments, many of whom may become lenders, not borrowers, on the international capital markets. If the world resource picture worsens, along the lines we sketched out in Chapter 34, the underdeveloped countries may find themselves the beneficiaries of inelastic demand curves, and the developed nations may be the ones complaining about the terms of world trade.

That is, however, still in the future. Today foreign exchange reserves are still very scarce for all but the oil-producing countries. Hence the underdeveloped raw material producers are looking for commodity stabilization agreements, not altogether dissimilar from the programs that have long supported

American farm prices. Recently, the Western nations have recognized the need for some such device if the underdeveloped countries are to be able to plan ahead with any assurance of stability.

Another possibility lies in the prospect of encouraging diversified exports from the underdeveloped nations—handicrafts, light manufactures, and others. The difficulty here is that these exports may compete with the domestic industry of the advanced nations: witness the problems of the American textile industry in the face of textile shipments from Hong Kong. No doubt a large source of potential earnings lies along this path, but it is unlikely to rise rapidly as long as the advanced nations refuse to allow the backward countries equal access to their own markets.

LIMITATIONS ON PRIVATE FOREIGN INVESTMENT

A second main avenue of capital accumulation for the backward nations is foreign investment. Indeed, before World War II, this was *the* source of their industrial wealth. Today, however, it is a much diminished avenue of assistance, for reasons we have learned in our exploration of the multinational corporation. The former capital-exporting nations are no longer eager to invest private funds in areas over which they have lost control and in which they fear to lose any new investments they might make. For reasons that we have discussed, many of the poorer nations view Western capitalism with ambivalence. They need capital, technology, and expertise; but the arrival of a branch of a powerful corporation run by faraway "headquarters" looks to them like another form of the domination they have just escaped. As a result, foreign investment is often hampered by restrictive legislation in the underdeveloped nations, even though it is badly needed. Con-

sequently, not much more than $3 billion to $4 billion a year from all the advanced nations goes overseas as foreign investment into the underdeveloped world.

Another difficulty is that Western corporations partially offset the growth-producing effects of their investments by draining profits out of the country. In the period 1950–1965, for example, the flow of income remitted from Latin America to the United States was $11.3 billion, three times larger than the flow of new capital into Latin America. In 1971, income of $4.1 billion was transmitted to the United States, and only $0.9 billion was sent back to Latin America. This pattern of economic flows should not be misinterpreted as implying that foreign investment is a "negative" influence: the plant and equipment that the West has sent abroad remains in the underdeveloped world, where it continues to enhance the productivity of labor, or perhaps to generate exports. But the *earnings* on this capital are not typically plowed back into still more capital goods, so that their potential growth-producing effect is far from realized.

THE CRUCIAL AVENUE OF AID

These considerations enable us to understand the special importance that attaches to the third channel of capital accumulation: foreign aid. Surprisingly, perhaps, in the light of the attention it attracts, foreign aid is not a very large figure. International assistance, from *all* individual nations and from the UN and its agencies, has run at a rate of about $6 billion to $7 billion per year through the 1960s and is now *diminishing*. This is an insignificant contribution to the gross output of the noncommunist underdeveloped world.

It is, however, a substantial fraction—perhaps as much as 15 percent—of the *gross investment* of that world. Thus, foreign aid makes possible the accumulation of industrial capital much faster than could be accomplished solely as a result of the backward lands' export efforts or their ability to attract foreign private capital.* To be sure, an increase in foreign earnings or in private capital imports would have equally powerful effects on growth. But we have seen the difficulties in the way of rapidly increasing the receipts from these sources. For the near future, foreign aid represents the most effective channel for *quickly* raising the amount of industrial capital which the underdeveloped nations must obtain.

Foreign aid, particularly from UN sources, is also an extremely important source of *technical assistance* which enables the backward regions to overcome the handicaps imposed by their lack of skilled and trained personnel. For the near term, this may be even more important than the acquisition of the industrial capital itself in promoting growth.

ECONOMIC POSSIBILITIES FOR GROWTH

Against these handicaps, can the underdeveloped nations grow? Can the terrible conditions of poverty be relegated to the past? Economic analysis allows us to ask these questions systematically, for growth depends on the interplay of three variables.

1. Rate of investment that an underdeveloped nation can generate

As we know, this depends on the proportion of current effort that it can devote to capital-

creating activity. In turn, the rate of saving, the success in attracting foreign capital, the volume of foreign aid—all add to this critical fraction of effort on which growth hinges.

2. Productivity of the new capital

The saving that goes into new capital eventually results in higher output. But not all capital boosts output by an equal amount. A million-dollar steel mill, for example, will have an impact on GNP very different from that of a million-dollar investment in schools. In the short run, the mill may yield a higher return of output per unit of capital investment; in the long run, the school may have the edge. But in any event, the effect on output will depend not merely on the amount of investment, but on the marginal capital-output ratio of the particular form of investment chosen.

3. Population growth

Here, as we know, is the negative factor. If growth is to be achieved, output must rise faster than population. Otherwise, per capita output will be falling or static, despite seemingly large rates of overall growth.

THE CRITICAL BALANCE

With these basic variables, is growth a possibility for the backward lands? We can see that if investment were 10 percent of GNP and if each dollar of new investment gave rise to a third of a dollar of additional output,* a 10 percent rate of capital formation would yield a 3.3 percent rate of growth (10 percent × one-third). This is about equal to population growth rates in the nations with the highest rates of population income.

---

*Note "makes possible." There is some disturbing evidence that foreign aid may displace domestic saving, so that an underdeveloped country receiving aid may relax its own efforts to generate capital. Much depends on the political will of the recipient country. (See K. B. Griffin and J. L. Enos, "Foreign Assistance: Objectives and Consequences," *Economic Development and Cultural Change,* April, 1970.)

---

*This seems to be *roughly* what the marginal capital-output ratio of new investment in the underdeveloped areas may be.

The trouble is that most of the backward nations have investment rates that are closer to 5 than to 10 percent of GNP. In that case, even with a marginal capital-output ratio of one-half, growth rates would not be enough to begin a sustained climb against a population growth of 2.5 percent (5 percent × ½ = 2.5 percent). And this gloomy calculation is made gloomier still when we confront the fact that the labor force is rising faster than the population as a whole, as vast numbers of children become vast numbers of workers. In the 1960s in Latin America it was estimated that at least *25 percent* of the working-age population was unemployed. In the decade since then, unemployment as a percent of the labor force seems to have increased in virtually every underdeveloped country.[8]

ECONOMIC PROSPECTS: THE "GREEN REVOLUTION"

Nobody can confront these economic realities and make optimistic forecasts for the developing countries. Yet the situation is not hopeless. For many nations, an increase in their capital formation rates of 50 or 100 percent—a difficult but by no means impossible task, especially if, like the oil producers of the Middle East, they become beneficiaries of a resource squeeze—should bring them to the point of cumulative growth. What is then needed is a program of encouragement to small industry, to absorb the flood of unemployed while the essential industrial core is being built, and an all-out effort to bring about manageable rates of population growth. These are also difficult but, again, not impossible undertakings.

Meanwhile a vitally important break-through has been scored on the agricultural front with the development of new crops that can literally double and triple yields. This so-called Green Revolution, in effect in parts of India, has somewhat relieved the specter of mass famine, which haunted the food officials of the world during the 1960s when they looked ahead to the 1970s. But the Green Revolution will require vast amounts of fertilizer, huge irrigation projects, and much administration. It is far from a simple "solution" to the development problem, but it may offer vital time for a critical effort to stop the juggernaut of population growth and to institute an all-out developmental effort.

SOCIAL AND POLITICAL PROBLEMS

Thus it is possible, not only in theory but in actuality, to foresee a long, slow developmental climb. But this is not the end of our analysis, for it is impossible to think of development only in terms of economics. As we have seen in the case of Western growth, *economic development is nothing less than the modernization of an entire society.* When we talk of building capital or redirecting agriculture, we must not imagine that this entails only the addition of machines and farm equipment to a peasant society. It entails the conversion of a peasant society into another kind of society. It means a change in the whole tenor of life, in the expectations and motivations, the environment of daily existence itself.

We have already noted some of the changes that economic development imposes on a society. Illiterate peasants must be made into literate farmers. Dispirited urban slum-dwellers must be made into disciplined factory workers. Old and powerful social classes, which have for generations derived their wealth from feudal land tenure, must be deprived of their vested rights. New managerial attitudes must be implanted in new

[8]See William Thiesenhausen, "Latin America's Employment Problem," *Science,* March 1971; also H. Oshima in *Economic Development and Cultural Change,* January 1971.

elites. Above all, the profligate generation of life, conceived in dark huts as the only solace available to a crushed humanity, must give way to a responsible and deliberate creation of children as the chosen heirs to a better future.

These changes will *in time* be facilitated by the realization of development itself. A growing industrial environment breeds industrial ways. The gradual realization of economic improvement brings about attitudes that will themselves accelerate economic growth. A slowly rising standard of living is likely to quicken the spread of birth control, as it did in the West.

All these changes, as we have said, may take place in time. But it is time itself that is so critically lacking. The changes must begin to take place now—today—so that the process of development can gain an initial momentum. Given the momentum of population growth, the transition from a backward, tradition-bound way of life to a modern and dynamic one cannot be allowed to mature at its own slow pace. Only an enormous effort can inaugurate, much less shorten, the transition from the past into the future.

COLLECTIVISM AND UNDERDEVELOPMENT

These sobering considerations converge in one main direction. They alert us to the fact that *in the great transformation of the underdeveloped areas, the market mechanism is apt to play a much smaller role than in the comparable transformation of the West during the industrial revolution.*

We will recall how lengthy and arduous was the period of apprenticeship through which the West had to pass in order for the ideas and attitudes, the social institutions and legal prerequisites of the market system to be hammered out. When the industrial revolution came into being, it exploded within a historic situation in which market institutions, actions, customs had already become the dominant form of economic organization.

None of this is true in the underdeveloped nations today. Rather than having their transition to a market society behind them, many of those nations must leap overnight from essentially tradition-bound and archaic relationships to commercialized and industrialized ones. Many of them are not even fully monetized economies. None of them have the network of institutions—and behind that, the network of "economic" motivations—on which a market society is built.

Hence it is not difficult to foresee that the guiding force of development is apt to be tilted in the direction of central planning. Regardless of the importance of private enterprise in carrying out the individual projects of development, the driving and organizing force of economic growth will have to be principally lodged with the government.

POLITICAL IMPLICATIONS

But the outlook indicates more than a growth of economic command. Implicit also in the harsh demands of industrialization is the need for strong political leadership, not only to initiate and guide the course of development, but to *make it stick*. For it is not only wrong, but dangerously wrong, to picture economic development as a long, invigorating climb from achievement to achievement. On the contrary, it is better imagined as a gigantic social and political earthquake. Eugene Black, ex-president of the World Bank, soberly pointed out that we delude ourselves with buoyant phrases such as "the revolution of rising expectations" when we describe the process—rather than the prospect—of development.[9] To many of the

[9]*The Diplomacy of Economic Development,* p. 9.

people involved in the bewildering transformations of development, the revolution is apt to be marked by a loss of traditional expectations, by a new awareness of deprivation, a new experience of frustration. For decades, perhaps generations, a developing nation must plow back its surplus into the ugly and unenjoyable shapes of lathes and drills, conveyor belts and factory smokestacks. Some change toward betterment is not ruled out, particularly in health, basic diet, and education; but beyond this first great step, material improvement in everyday living will not—cannot—materialize quickly.

As a consequence, many of the policies and programs required for development, rather than being eagerly accepted by all levels of society, are apt to be resisted. Tax reform, land reform, the curtailment of luxury consumption are virtually certain to be opposed by the old order. In addition, as the long march begins, latent resentments of the poorer classes are likely to become mobilized; the underdog wakens to his lowly position. Even if his lot improves, he may well feel a new fury if his *relative* well-being is impaired.

SOCIAL STRESSES

These considerations enable us to understand how social tensions and economic standards can rise at the same time. And this prospect, in turn, enables us to appreciate the fearful demands on political leadership, which must provide impetus, inspiration, and, if necessary, discipline to keep the great ascent in motion. The strains of the early industrial revolution in England, with its widening chasm between the proletariat and capitalist, are not to be forgotten when we project the likely course of affairs in the developing nations.

In the politically immature and labile areas of the underdeveloped world, this exercise of leadership typically assumes the form of "strong-man" government. In large part, this is only the perpetuation of age-old tendencies in these areas; but in the special environment of development, a new source of encouragement for dictatorial government arises from the exigencies of the economic process itself. Powerful, even ruthless, government may be needed, not only to begin the development process, but to cope with the strains of a *successful* development program.

It is not surprising, then, that the political map reveals the presence of authoritarian governments in many developing nations today. The communist areas aside, we find more or less authoritarian rule in Egypt, Pakistan, Burma, South Korea, Indonesia, and the succession of South American junta governments. From country to country, the severity and ideological coloring of these governments varies. Yet in all of them we find that the problems of economic development provide a large rationale for the tightening of political control. At least in the arduous early stages of growth, some form of political command seems as integral to economic development as the accumulation of capital itself.

THE ECOLOGICAL PROBLEM

To this endless list of problems one last one must now be added. As we have seen in Chapter 34, there may well be limits to the amount of industrialization the planet can sustain. These limits are likely to impose a ceiling on material output in the underdeveloped areas that would leave them far below Western standards.

This problem reinforces the political argument above. For it is clear that the free play of market forces, *insofar as these are permitted to develop along the lines of traditional capitalism,* would lead rapidly to an ecological impasse. Instead, the development of the backward nations, once an initial

CHINESE ECONOMICS

Until very recent years, information about the Chinese economy percolated into the United States only through roundabout sources, and our knowledge of how the Chinese economy operates was scanty indeed.* Recently, however, a number of American economists were invited to visit China. One of them, David Gordon, has written this brief report of his impressions.

The Chinese have begun the long road toward industrialization. As many have reported in the American press during the recent burst of interest about China, the economy has made amazing progress in raising the basic standard of living of the entire population. Chinese planners now feel that agricultural production has grown rapidly enough to permit the diversion of substantial resources into industrial development. They are increasing the levels of investment in heavy industry, and they are also stepping up production in their key light industries. (Many Chinese are now able to buy the three most desired consumer durables: bicycles, sewing machines, and transistor radios.)

Although many other underdeveloped economies have begun to industrialize at analogous points in their economic histories, the Chinese industrial development strategy has some unique and unprecedented features. The recent Cultural Revolution (1966–1968) has had a profound effect on that strategy. It involved some fundamental struggles over power and ideology, eventually resolved through the consolidation of Chairman Mao Tse-tung's power and the clarification of the country's ideological direction. The forces which triumphed during the Cultural Revolution were waging ideological battle against Liu Shao-chi, Mao's former heir apparent and leading party official. They accused Liu Shao-chi of "revisionism" and of being a "capitalist-roader." In particular, they argued that Liu

Shao-chi's economic development policies would lead China down the same economic paths already traveled by both the Western capitalist countries and the developed socialist nations like Russia. (The Chinese accuse the Russians of having abandoned socialism; they say the Soviet Union is the leading model of "state capitalism," or "bureaucratic capitalism.")

The problem with both these versions of economic development, according to the anti-Liu faction, is that the economy develops unevenly. The most advanced sectors become more advanced. Those workers with the greatest skills become even more skilled. Organizations become ever more hierarchical and bureaucratic. In the pursuit of more and more economic growth, justified by arguments of efficiency, private capitalism and "state capitalism" both reinforce inequalities in the already uneven patterns of development.

Particularly since the rout of the Liu forces during the Cultural Revolution, the Chinese have been intensifying their efforts to promote a path of industrial development aimed at overcoming the inequalities inherited from history. They are trying to ensure that all regions and all people share both in the process and the fruits of increasing industrialization. For example, some regions in the country inherited a considerable industrial capacity (they had automobile and tractor factories in Shanghai). By conventional Western economic criteria, it would probably be more efficient to build on that existing capacity. Instead, the Chinese are encouraging the dispersion of industrial investment throughout the country. Four-fifths of the people live in the country; therefore, much of the new industrial building is occurring in rural areas where the bulk of the people live. Since many of these people have neither industrial skills nor large stocks of capital for investment, the scale of industry will initially be quite small and the technology quite primitive.

We saw a living, breathing, fuming illustration of these economic policies in rural Tsunhua County in the province of T'ang-shan (to the east of

---

*One of the best sources of information is a book by Barry Richman, a Canadian economist, *Industrial Society in China* (New York: Random House, 1969).

Peking). Following the new economic guidelines, the county government had begun to develop its own, admittedly crude, heavy industries. In 1970 the county began to build an iron and steel factory. Relying on what everyone in China calls "indigenous methods," the inexperienced local workers designed and built the iron and steel plant themselves. The plant was financed from the county's own "accumulation fund," earned from profits in other county-sponsored production. Almost 400 workers are now employed in the factory.

The plant looked primitive indeed. It comprised a series of open-air sheds, all small. The furnaces were tiny. The ores were conveyed almost entirely by horse-drawn carts. Molten steel was drawn and processed through simple machines that looked like clothes wringers; the long red-hot metal bars were pulled through with simple iron tongs. As the bars grew longer and longer, they would whip and lash as they came through the machine. The workers, wearing only rubber gloves for protection, would dodge the glowing metal like smaller Muhammad Alis, dancing nimbly like butterflies to escape its sting.

One of the members of our group was an industrial economist, schooled in American corporate technological perspectives. After we finished touring the plant, he whispered that he thought the plant was a joke. "What a waste of money," he said, "to build such a primitive plant when you have better technologies available."

Many of us thought otherwise. There are many reasons why that little iron and steel plant made a great deal of sense. Building local industry on a small scale means that many peasants who would otherwise never leave their rural pursuits will participate in the process of industrialization. The fact that local people designed and built the plant gives them pride in its products and provides a strong incentive for them to work as hard as they can in production. Having involved themselves this far, they have started expanding their economic horizons. In order to be completely self-reliant in iron and steel production, for in-

stance, they have begun exploring for iron ore in the local hills, hoping to find additional deposits that they could mine themselves. They have found 23 different ores, including iron, which they are now extracting. Building on a small scale has also taken advantage of crude scrap materials that could be incorporated into indigenous plant designs but could not possibly be built into more technologically advanced factory structures. And the local base of production has guaranteed, to a certain extent, that the products of their heavy industry are oriented to their own needs.

Gordon's comments point up the important difference between the criteria of "Western" economics and those of the Chinese. The Chinese have established as basic goals the development of self-reliant communities, with as diverse a range of occupations as possible, rather than industrial complexes that achieve high standards of economic efficiency at the cost of the human problems associated with extreme regional and occupational specialization. In our terms, they are deliberately trading off a certain amount of efficiency for the attainment of social goals.

It should be added, of course, that this emphasis on rural diversification also adds an element of safety against military attack and may have an even more down-to-earth justification in saving the transportation costs that would be involved in moving men and materials to massive industrial centers. Nonetheless, with all these considerations, there is no doubt that the Chinese are attempting to modernize in a new way—a way that seeks to avoid the bureaucratic characteristics of big business and oversized ministries.

What remains to be seen is whether small-scale rural-based industry can achieve enough efficiency to allow China to develop rapidly enough to satisfy its leaders and its people. We will have to wait several years to find that out. If the Chinese succeed, they may be a pattern of development for much of the remainder of the underdeveloped world, now in a schizophrenic state of mind—eager to begin the "Western" industrialization process and yet fearful of the social costs that progress has imposed on the West.

momentum has been attained, will have to be planned to economize on materials and production to a degree unknown in the West. The absurd waste of resources involved in providing transportation through the proliferation of private automobiles instead of public conveyances; the encouragement of individual ownership of washing machines and television sets and domestic conveniences—the very center of the Western ideal of a "high standard of living"—will probably be impossible to duplicate on a global scale.

What will be required in their place is a new pattern of *public consumption*, perhaps even a wholly new conception of what is meant by an advanced society. All this will be required not as a matter of ideological preference, but as one of long-run necessity. The ecological barriers of resource availability and the absorption capacity of the earth pose truly staggering problems for the underdeveloped nations, once they manage to escape from the stagnation that still characterizes most of them. No one knows how these constraints of nature will be translated into the realities of social life, but it is certain that the relatively laissez-faire attitudes of Western capitalism will not be adequate to the task.

KEY WORDS

Underdevelopment

Population explosion

Industrialization

Imperialism

Foreign aid

CENTRAL CONCEPTS

1. *Underdevelopment* constitutes the economic environment for the vast majority of mankind. It is ascribable in part, perhaps, to bad climates or inadequate resources, but in the main it springs from the inability of traditional societies to mount sustained programs of investment and change.

2. The main attributes of underdevelopment are *very low levels of productivity*, especially in agriculture where man/land ratios often impose highly inefficient scales of production. No less important, but more difficult to correct, are deep-seated *attitudes of inertia* on the part of the population. And constituting a main obstacle to a development effort are very high rates of *population growth*.

3. The development effort requires a *shift from consumption to investment activity*. This necessitates a prior increase in agricultural productivity, accompanied by measures that will transfer the food surplus from the peasant cultivator to workers on capital projects. This is an exceedingly difficult task to carry out.

4. In addition to shifting resources from agriculture to capital-building tasks, a nucleus of *industrial equipment* must be brought in. This can be done by *international trade*, by *foreign investment*, or by *foreign aid*.

5. The channel of international trade is a difficult one for many backward nations, as a consequence of *imperialism*. Many underdeveloped nations have "lopsided" economies that sell one raw commodity in markets where demand is typically inelastic. Foreign investment has not produced much capital for the backward nations, partly because of their own suspicions of Western nations, partly because the West has not reinvested its earnings there. *Foreign aid* thus becomes a small but crucial avenue for the transfer of funds and skills. It should be noted that skill is often essential to acquire before funds can be absorbed.

Rate of investment

6. **Development hinges on three economic variables: the** *rate of investment*, **the** *capital-output ratio* **(the size of additional output yielded by net investment) and the** *rate of population growth*. **With a 10 percent net investment rate and a one-third capital output ratio, it should be possible to begin growth even against a 3 percent population increase.**

Marginal capital-output ratio

7. **The problem is that the economic effort is not detachable from the whole array of** *social and political problems*. **To mount a development program requires that many traditional institutions and ways be discarded for new and untried ones. This process of change occurred over several centuries in the West, but the exigencies of the population crisis make it necessary to compress it within a few generations today.**

Ecological constraints

8. **Constraints imposed by resource limitations impose ceilings on the material attainments open to the underdeveloped nations. At least in the long run, they will be forced to develop resource allocation patterns very different from ours.** *Planned economies* **are not a matter of choice, but necessity, for their futures.**

QUESTIONS

1. In what ways do you think underdeveloped countries are different from the American Colonies in the mid-1600s? Think of literacy, attitudes toward work and thrift, and other such factors. What about the relationship to more advanced nations in each case?

2. Why do you think it is so difficult to change social attitudes at the lowest levels of society? At the upper levels? Are there different reasons for social inertia at different stations in society?

3. Does the United States have a population problem? Will population growth here affect economic or social aspects of life more? Do you think we should adopt an American population control policy? What sort of policy?

4. Review your acquaintance with the model of a peasant economy on pp. 300–301. Is this model applicable to China today? Was it applicable to the American Colonies? What differences do population pressure and social inertia impart to the solutions implied by the model?

5. Many economists have suggested that all advanced nations should give about 1 percent of their GNP for foreign aid. In the U.S., that would mean a foreign aid appropriation of $10 billion. Actually we appropriate about $2 billion. Do you think it would be practicable to suggest a 1 percent levy? How would the country feel about such a program?

6. What are the main variables in determining whether or not growth will be self-sustaining? If net investment were 8 percent of GNP and the capital output ratio were $\frac{1}{4}$, could a nation grow if its rate of population increase were $2\frac{1}{4}$ percent? What changes could initiate growth?

7. What do you think is the likelihood of the appearance of strong-arm governments and collectivist economies in the underdeveloped world? For the appearance of effective democratic governments? For capitalist economies? Socialist ones? Is it possible to make predictions or judgments in these matters that do not accord with your personal preferences?

8. If you had to plan the 50-year development of a nation like India, and if you knew that it would be essential to economize on the production and consumption of minerals, how would you suggest that patterns of consumption be changed to effect major resource savings? To what extent would such changes impose further changes in the pattern of social life?

From Market to Planning

WE CAN SEE THE BEGINNING of worldwide economic development as a genuine watershed in human history. An active and dynamic form of economic life, until recently the distinctive characteristic of the industrial West, is about to be generalized over the face of the globe. The process of diffusion will take generations, but it marks a profound, irreversible, and truly historic alteration in the economic condition of man.

Yet, if the process of economic growth is henceforth to be carried out on a global scale, it is also clear that there will be significant change in the auspices under which this process is likely to unfold. As we have seen, it is command rather than the market system which is in the ascendant as the driving force in the underdeveloped regions. And when we combine the geographic extent of these regions with those where communism has become firmly entrenched, it seems that command now bids fair to become *the* dominant means of organizing economic activity on

this planet, as tradition was, not very long ago.

But again there is a difference. During the centuries when tradition held sway over most of the world, the economies run by the market system were the locus of progress and motion. Today one cannot with assurance say the same; nor will we be able to, in the future, for a preeminent motive of the rising economies of command is to *displace* the market societies as the source of the world's economic vitality.

Does this mean that economic history now writes finis to the market system? Does it mean that the market, as a means of solving the economic problem, is about to be relegated to the museum of economic antiquities or at best limited to the confines of North America, Western Europe, and Japan? The question brings to a focus our continuing concern with the market system. Let us attempt in these last pages to appraise its prospects.

Stages of Economic Development

We might well begin an appraisal by taking a last survey of the array of economic systems that mark our times. It is, at first glance, an extraordinary assortment: we find, in this third quarter of the twentieth century, a spectrum of economic organization that represents virtually every stage in economic history, from the earliest and most primitive. But at second look, a significant pattern can be seen within this seemingly disordered assemblage. The few remaining wholly traditional economies, such as those of the South Seas or tribal Africa, have not yet begun to move into the mainstream of economic development. A much larger group of underdeveloped nations, in which institutions of economic command are now rising amid a still traditional environment, have just commenced their development efforts and are now coping with the initial problems preparatory to eventual all-out industrialization. Going yet further along we find the economies of iron command, such as China and to a lesser extent Russia. Here we find national communities that are (or recently were) wrestling with the gigantic task of rapid massive modernization. Finally, we pass to the market economies of the West, to encounter societies with their developmental days behind them, now concerned with the operation of high-consumption of economic systems.*

The categorization suggests a very important general conclusion. *The economic structures of nations today bear an integral relation with their stage of economic develop-*

*A glance at the back end papers reveals a map keyed to these economic differences. Needless to say, the map gives no more than a subjective interpretation of the extent of Tradition, Command, and Market in the various countries of the world.

ment. Acts of foreign intervention aside, the choice of command or market systems is not just the outcome of political considerations or ideologies and preferences. It is also, and perhaps primarily, the result of functional requirements that are very different at different levels of economic achievement.

INCEPTION OF GROWTH

We have already noted this connection in our discussion of the underdeveloped areas. Now, however, we can place what we have learned into a wider frame of reference. For if we compare the trend of events in the underdeveloped economies with the "equivalent" stage of development in Western history, we see a significant point of resemblance between the two. The emergence of command in the development-minded countries today has a parallel in the mercantile era, when the Western nations also received a powerful impetus toward industrialization under the organizing influence of the "industry-minded" governments of the seventeenth and early eighteenth centuries.

Thereafter, to be sure, the resemblance ceases. In the West, following the first push of mercantilism, it was the market mechanism that provided the main directive force for growth; in the underdeveloped lands, as we have seen, this influence is likely to be preempted to a much larger extent by political and economic command.

PRESENT VS. PAST

Three main reasons lie behind this divergence of paths. *First, the underdeveloped areas today start from a lower level of preparedness* than did the West in the seventeenth and eighteenth centuries. Not only have the actual institutions of the market not yet appeared in many backward lands, but the whole process of acculturation has failed to duplicate that of the West. In many ways—not all of them economic—the West was "ready" for economic development. A similar readiness is not in evidence in the majority of the backward lands today, with the result that development, far from evincing itself as a spontaneous process, comes about as the result of enforced and imposed change.

Second, the *West was able to mount its development effort in leisurely tempo.* This is not to say that its rate of growth was slow or that strong pressures did not weigh upon many Western countries, arousing within them feelings of dissatisfaction with their progress. Yet the situation was unlike that of the backward areas today. Here immense pressures, both of population growth and of political impatience, create an overwhelming need and desire for speed. As a result, the process of growth is not allowed to mature quietly in the background of history, as it did for much of the West, but has been placed at the very center of political and social attention.

Finally, the underdeveloped countries, who suffer from so many handicaps in comparison with the developmental days of the West, enjoy one not inconsiderable advantage. Because they are in the rear guard rather than the vanguard of history, they know where they are going. *In a manner denied to the West, they can see ahead of them the goal they seek to reach.* They do not wish to reach this goal, however, by retreading the painful and laborious path marked out by the West. Rather, they intend to shortcut it, to move directly to their chosen destination, by utilizing the mechanisms of command to bring about the great alterations that must be made.

Can economic command significantly compress and accelerate the growth process? The remarkable performance of the Soviet Union suggests that it can. In 1920 Russia was but a minor figure in the economic councils of the world. Today it is a country whose economic achievements bear comparison with those of the United States. If Soviet production continues to gain on American production at the rate of the last ten years, in little more than another generation its total industrial output (although not its per capita output) will be larger than our own.

The case of China is less clearcut. Until the famine disaster of 1959–1960, Chinese economic growth was double or triple that of India; since then, perhaps because of the convulsions of the Cultural Revolution, its record is less easy to appraise. It may possibly have grown less rapidly than India in *quantitative terms;* but by the reports of all observers, its *qualitative* improvements in health, education, and welfare are strikingly better than those of India.

It is no doubt wise not to exaggerate the advantages of a command system. If it holds the potential for an all-out attack on backwardness, it also contains the possibilities of substantial failure, as in the disappointments of the planned Cuban economy. The mere existence of a will to plan is no guarantee that the plans will be well drawn or well carried out or reasonably well obeyed. Nonetheless, these caveats must be set against the dismal record of economies that continue to wallow in the doldrums of tradition or that undertake the arduous transition into modernity under the inadequate stimulus of half-hearted regimes and half-formed market systems. In this comparison of alternatives, the advantage seems unmistakably on the side of those backward societies that are capable of mustering a strong central economic authority.

Planning and Its Problems

What are the advantages, what are the problems of planning? The subject is large enough to fill many books, and we will not attempt to discuss the full economics of planning in this short chapter. But a few general remarks may serve as an introduction to the subject.

How is planning carried out? The question goes to the heart of the matter, for all planned economies have found their central difficulty in going from the vision of a general objective to the actual attainment of that objective in fact. It is one thing to plan for 6 percent growth, another to issue the directives to bring forth just the right amounts of (quite literally) hundreds of thousands of items, so that 6 percent growth will result.

SOVIET PLANNING

In the Soviet Union this complicated planning mechanism is carried out in successive stages. The overall objectives are originally formulated by the Gosplan, the State Planning Agency. The long-term overall plan is then broken down into shorter one-year plans. These one-year plans, specifying the output of major sectors of industry, are then transmitted to various government ministries concerned with, for example, steel production, transportation, lumbering, and so forth. In turn, the ministries refer the one-year plans further down the line to the heads of large industrial plants, to experts and advisers, and so on. At each stage, the overall plan is thus unraveled into its subsidiary components, until finally the threads have been traced as far back as feasible along the productive process—typically, to the officials in charge of actual factory operations. The factory manager of, for instance, a coking operation is given a planned objective for the next year, specifying the output needed from his plant. He confers with his production engineers, considers the condition of his machinery, the availability of his labor force, and then transmits his requirements for meeting the objective back upward along the

hierarchy. In this way, just as "demand" is transmitted downward along the chain of command, the exigencies of "supply" flow back upward, culminating ultimately in the top command of the planning authority (the Gosplan) itself.

SUCCESS INDICATORS

The coordination and integration of these plans is a tremendously complicated task. Recently the Soviets have adopted techniques of input-output analysis (see pp. 327–28), which have considerably simplified the problem. Even with input-output, however, the process is bureaucratic, cumbersome, slow, and mistake-prone. A Russian factory manager has very little leeway in what he produces or the combination of factors that he uses for production. Both inputs and outputs are carefully specified for him in his plan. What the manager *is* supposed to do is to beat the plan, by "overproducing" the items that have been assigned to his plant. Indeed, from 30 to 50 percent of a manager's pay will depend on bonuses tied directly to his "overfulfillment" of the plan, so that he has a very great personal incentive to exceed the output "success indicators" set for him.

All this seems sensible enough. Trouble comes, however, because the manager's drive to exceed his factory's quota tends to distort the productive effort from the receivers' point of view. For example, if the target for a textile factory is set in terms of yards of cloth, there is every temptation to weave the cloth as loosely as possible, to get the maximum yardage out of a given amount of thread. Or if the plan merely calls for tonnages of output, there is every incentive to skimp on design or finish or quality, in order to concentrate on sheer weight. A cartoon in the Russian satirical magazine *Krokodil* shows a nail factory proudly displaying its record output: one gigantic nail

PLANNING UNDER LENIN AND STALIN

Official literature of the Communist movement gives little guidance for running a socialist society. Marx's *Das Kapital*, the seminal work of communism, was entirely devoted to a study of capitalism; and in those few essays in which Marx looked to the future, his gaze rarely traveled beyond the watershed of the revolutionary act itself. With the achievement of the revolution, Marx thought, a temporary regime known as "the dictatorship of the proletariat" would take over the transition from capitalism to socialism, and thereafter a "planned socialist economy" would emerge as the first step toward a still less specified "communism."

What is the difference between *socialism* and *communism*? In the West, *socialism* implies an adherence to democratic political mechanisms, whereas *communism* does not. But within the socialist bloc there is another interesting difference of definition. Socialism there represents a stage of development in which it is still necessary to use "bourgeois" incentives in order to make the economy function; that is, people must be paid in proportion to the "value" of their work. Under communism, a new form of human society will presumably have been achieved in which these selfish incentives will no longer be needed. Then will come the time when society will be able to put into effect Karl Marx's famous description of communism: "From each according to his ability; to each according to his need."

In a true communist economy—the final terminus of economic evolution according to Marx —there were hints that the necessary but humdrum tasks of production and distribution would take place by the voluntary cooperation of all citizens, and society would turn its serious attention to matters of cultural and humanistic importance. Indeed, in a famous passsge in *State and Revolution*, Lenin described the activities of administering a socialist state as having been "*simplified* by capitalism to the utmost, till they have become the extraordinarily simple operations of watching, recording, and issuing receipts, within the reach of anybody who can read and who knows the first four rules of arithmetic."

Many of the problems of early Soviet history sprang from the total absence, on the part of its rulers, of any comprehension of the staggering difficulties of planning in fact rather than in thought. The initial Soviet attempt to run the economy was a disastrous failure. Under inept management (and often cavalier disregard of "bourgeois" concerns with factory management), industrial output declined precipitously; by 1920 it had fallen to *14 percent* of prewar levels. As goods available to the peasants became scarcer, the peasants, themselves, were less and less willing to acquiesce in giving up food to the cities. The result was a wild inflation followed by a degeneration into an economy of semibarter. For a while, toward the end of 1920, the system threatened to break down completely.

To forestall the impending collapse, in 1921 Lenin instituted a New Economic Policy, the so-called NEP. This was a return toward a market

suspended from an immense gantry crane. (On the other hand, if a nail factory has its output specified in terms of the *numbers* of nails it produces, its incentive to overfulfill this "success indicator" is apt to result in the production of very small or thin nails.)

PROFIT AS A SUCCESS INDICATOR

What is the way out of this kind of dilemma? A few years ago, a widely held opinion among the Russian planners was that more detailed and better integrated planning performed on a battery of computers would solve the problem. Few still cling to this belief. The demands of planning have grown far faster than the ability to meet them: indeed, one Soviet mathematician has predicted that at the current rate of growth of the planning bureaucracy, planning alone would require the services of the entire Russian population by 1980. Even with the most

system and a partial reconstitution of actual capitalism. Retail trade, for instance, was opened again to private ownership and operation. Small-scale industry also reverted to private direction. Most important, the farms were no longer requisitioned but operated as profit-making units. Only the "commanding heights" of industry and finance were retained in government hands.

There ensued for several years a bitter debate about the course of action to follow next. While the basic aim of the Soviet government was still to industrialize and to socialize (i.e., to replace the private ownership of the means of production by state ownership), the question was how fast to move ahead—and, indeed, *how* to move ahead. The pace of industrialization hinged critically on one highly uncertain factor: the willingness of the large, private peasant sector to deliver food for sustaining city workers. To what extent, therefore, should the need for additional capital goods be sacrificed in order to turn out the consumption goods that could be used as an inducement for peasant cooperation?

The argument was never truly resolved. In 1927 Stalin moved into command and the difficult question of how much to appease the unwilling peasant disappeared. Stalin simply made the ruthless decision to appease him not at all, but to *coerce* him by collectivizing his holdings.

The collectivization process solved in one swoop the problem of securing the essential transfer of food from the farm to the city, but it did so at a frightful social (and economic) cost. Many peasants slaughtered their livestock rather than hand it over to the new collective farms; others waged outright war or practiced sabotage. In reprisal, the authorities acted with brutal force. An estimated five million "kulaks" (rich peasants) were executed or put in labor camps, while in the cities an equally relentless policy showed itself vis-à-vis labor. Workers were summarily ordered to the tasks required by the central authorities. The right to strike was forbidden, and the trade unions were reduced to impotence. Speedups were widely applied, and living conditions were allowed to deteriorate to very low levels.

The history of this period of forced industrialization has left abiding scars on Russian society. It is well for us, nonetheless, to attempt to view it with some objectivity. If the extremes to which the Stalinist authorities went were extraordinary, often unpardonable, and perhaps self-defeating, we must bear in mind that industrialization on the grand scale has always been wrenching, always accompanied by economic sacrifice, and always carried out by the more or less authoritarian use of power.

We might note in passing that universal male suffrage was not gained in England until the late 1860s and 1870s. Aneurin Bevan has written: "It is highly doubtful whether the achievements of the Industrial Revolution would have been permitted if the franchise had been universal. It is very doubtful because a great deal of the capital aggregations that we are at present enjoying are the results of the wages that our fathers went without." (From Gunnar Myrdal, *Rich Lands and Poor*. New York: Harper, 1957, p. 46.)

complete computerization, it seems a hopeless task to attempt to beat the problem of efficiency by increasing the "fineness" of the planning mechanism.

Rather, the wind for reform in the Soviet Union is now blowing from quite another quarter. Led by economist E. G. Liberman, there is a growing demand that the misleading plan directives of weight, length, etc., be subordinated to a new "success indicator" independently capable of guiding the manager to results that will make sense from the overall point of view. And what is that overriding indicator? It is the *profit* that a factory manager can make for his enterprise!

We should note several things about this profit. To begin with, it is not supposed to arise from price manipulations. Factory managers must continue to operate with the prices established by planners; but they will now have to *sell* their output and *buy* their inputs, rather than merely deliver or accept them. This means that each factory will have to be responsive to the particular needs of its customers if it wishes to dispose of its output. In the same way, of course, its own suppliers will now have to be responsive to the factory's needs if the suppliers are to get the factory's business.

Second, the profit will belong not to the factory or its managers, but to the State. A portion of the profit will indeed be allocated for bonuses and other rewards, so that there is a direct incentive to run the plant efficiently, but the bulk of the earnings will be transferred to the State.

THE MARKET AS A PLANNING TOOL

Thus, profits are to be used as an efficiency-maximizing indicator, just as we saw them used in our study of microeconomics.

Indeed, to view the change even more broadly, we can see that the reintroduction of the use of profits implies a deliberate return to the use of the *market mechanism* as a means of achieving economic efficiency. Not only profits but also interest charges—a capitalist term that would have been heresy to mention in the days of Stalin—are being introduced into the planning mechanism to allow factory managers to determine for themselves what is the most efficient thing to do, both for their enterprises and for the economy as a whole.

The drift toward the market mechanism is still new in the Soviet Union, and we do not know how far it will ultimately progress. The objectives of the 1971–1975 Plan call for a much greater emphasis on consumer goods, but speak of "an extensive use of economic-mathematical methods," which implies something of a return to the computer rather than a rapid movement in the direction of freer trade. Nonetheless, there seems to be no doubt from which quarter the winds blow

most steadily. As Soviet economist A. Birman has put it: "Only three years ago, no one would have thought that there would be anything but the direct physical allocation of goods. Now economists talk of *torgovat* (trading) instead of *snabzhat* (allocating)."[1] Furthermore, the government has warmly endorsed the idea of "production associations,"—groups of geographically separate plants that coordinate their marketing, purchasing, research, and management, just like corporations. The market idea of economies of large-scale production has triumphed over the political idea of production organized by locality.

MARKET SOCIALISM

Meanwhile, the trend toward the market has proceeded much further in a large part of Eastern Europe, and above all in Yugoslavia. There, the market rules very nearly as supreme as it does in Western capitalist countries. Yet the Yugoslavs certainly consider themselves a socialist economy. As in the U.S.S.R., enterprise profits do not go to the "owners" of the business but are distributed as incentive bonuses or used for investment or other purposes under the overall guidance of the State. And again as in the U.S.S.R., the market is used as a deliberate instrument of social control, rather than as an institution that is above question. Thus, the main determination of investment, the direction of development of consumers' goods, the basic distribution of income—all continue to be matters established at the center as part of a planned economy. More and more, however, this central plan is allowed to realize itself through the profit-seeking operations of highly autonomous firms, rather than

---

[1] From Marshall I. Goldman, *The Soviet Economy: Myth and Reality* (Englewood Cliffs, N.J.: Prentice-Hall, 1968), p. 129.

through being imposed in full detail upon the economy.*

MARKET VS. PLAN

The drift of planning toward markets raises a question of fundamental importance. Why plan at all? Why not let the market take over the task of coordination that has proved such a formidable hurdle for industrial planners, for is not the market itself a "planning mechanism"?

After all, in the market, the signal of profitability serves as the guide for allocation of resources and labor. Entrepreneurs, anticipating or following demand, risk private funds in the construction of the facilities that they hope the future will require. Meanwhile, as these industrial salients grow, smaller satellite industries grow along with them to cater to their needs.

The flow of materials is thus regulated in every sector by the forces of private demand, making themselves known by the signal of rising or falling prices. At every moment there emanates from the growing industries a magnetic pull of demand on secondary industries, while, in turn, the growth salients themselves are guided, spurred, or slowed down by the pressure of demand from the ultimate buying public. And all the while, counterposed to these pulls of demand, are the obduracies of supply—the cost schedules of the producers themselves. In the cross fire of demand and supply exists a marvelously

*One of the most interesting aspects of Yugoslav socialism is its effort to introduce worker-run enterprises. At least in theory, the top authority in most enterprises is a Worker's Council, which has the right to *fire* the boss of the plant, and which also determines wage and bonus payments, plant investments, and so on. We do not yet know whether this economic democracy exists in fact as well as on paper. If it does, it may prove to be one of the important social developments of our time. See Paul Blumberg, *Industrial Democracy* (New York: Schocken Books, 1969).

sensitive social instrument for the integration of the overall economic effort of expansion.

ECONOMIES IN MID-DEVELOPMENT

This extraordinary integrative capacity of market systems returns us to the consideration of the suitability of various economic control mechanisms to different stages of development. We have seen that central planning is likely to be necessary to move stagnant, traditional economies off dead center. Once the development process is well under way, however, the relative functional merits of the market and the command mechanisms begin to change. After planning has done its massive tasks—enforcing economic and social change, creating an industrial sector, rationalizing agriculture—another problem begins to assume ever more importance. This is the problem of efficiency, of dovetailing the innumerable productive efforts of society into a single coherent and smoothly functioning whole.

In the flush period of mid-development, the market mechanism easily outperforms the command apparatus as a means of carrying out this complex coordinating task. Every profit-seeking entrepreneur, every industrial salesman, every cost-conscious purchasing agent becomes in effect part of a gigantic and continuously alert planning system within the market economy. Command systems do not easily duplicate their efforts. Bottlenecks, unusable output, shortages, waste, and a cumbersome hierarchy of bureaucratic forms and officials typically interfere with the maximum efficiency of the planned economy in midgrowth.

What we see here is not just a passing problem, easily ironed out. One of the critical lessons of the twentieth century is that the word *planning* is exceedingly easy to pronounce and exceedingly difficult to spell out. When targets are still relatively simple, and the priorities of action beyond dispute—as in the case of a nation wrenching itself from the stagnation of an ineffective regime—planning can produce miracles. But when the economy reaches a certain degree of complexity, in which the coordination of ten activities gives way to the coordination of ten thousand, innumerable problems arise, *because planned economies enjoy no "natural" congruence between private action and public necessity.*

Here is where the market comes into its own. As we know from our study of microtheory, each firm must combine its factors of production with one eye on their relative costs and the other on their respective productivities, finally bringing about a mix in which each factor is used as effectively as possible, given its cost. Thus in seeking only to maximize their own profits, the units in a market system inadvertently tend also to maximize the efficiency of the system as a whole.

PRIVATE AIMS, PUBLIC GOALS

Even more remarkable: one operating rule alone suffices to bring about this extraordinary conjunction of private aims and public goals. *That single rule is to maximize profits.* By concentrating on that one criterion of success and not by trying to maximize output in physical terms or by trying to live by a complicated book of regulations, entrepreneurs in a competitive environment do in fact bring the system toward efficiency. In other words, *profits are not only a source of privileged income, but also an enormously versatile and useful "success indicator" for a system that is trying to squeeze as much output as possible from its given inputs.*

Furthermore, the market mechanism solves the economic problem *with a minimum of social and political controls.* Impelled by the drives inherent in a market society, the individual marketer fulfills his public economic function without constant attention from the authorities. In contradistinction to his counterpart in a centralized command society, who is often prodded, cajoled, or even threatened to act in ways that do not appeal to his self-interest, the classical marketer obeys the peremptory demands of the market as a voluntary exercise of his own economic "freedom."

Thus it is not surprising that we find many of the motivating principles of the market being introduced into command societies. For as these societies settle into more or less established routines, they too can utilize the pressure of want and the pull of pecuniary desire to facilitate the fulfillment of their basic plans.

Economic freedom, as we know it in the West, is not yet a reality, or even an official objective, in any of these countries. The right to strike, for example, is not recognized, and nothing like the fluid consumer-responsive market system is allowed to exert its unimpeded influence on the general direction of economic development. But the introduction of more and more discretion at the factory level argues strongly that the principles of the market society are apt to find their place in planned societies at an appropriate stage of economic development.

HIGH CONSUMPTION ECONOMIES

Thus our survey of successive stages of development brings us to a consideration of Western economic society—that is, to the advanced economies that have progressed beyond the need for forced industrialization and now enter the stage of high consumption.

From our foregoing discussion, it is clear that the market mechanism finds its most natural application in this fortunate period of economic evolution. Insofar as the advanced Western societies have reached a stage in which the consumer is not only permitted but encouraged to impose his wants on the direction of economic activity, there is little doubt that the market mechanism fulfills the prevailing social purpose more effectively than any other.

Nonetheless, as we noted in Chapters 14 and 15, the market is not without its own grave problems, even in this regard. For one thing, *it is an inefficient instrument for provisioning societies*—even rich societies—*with those goods and services for which no "price tag" exists,* such as education or local government services or public health facilities.

A market society "buys" such public goods by allocating a certain amount of taxes for these purposes. Its citizens, however, tend to feel these taxes as an exaction in contrast with the items they voluntarily buy. Typically, therefore, a market society underallocates resources to education, city government, public health or recreation, since it has no means of "bidding" funds into these areas, in competition with the powerful means of bidding them into autos or clothes or personal insurance.

A second and perhaps even deeper-seated failing of the market system is its application of a strictly economic calculus to the satisfaction of human wants and needs. As we said before, the market is an assiduous servant of the wealthy, but an indifferent servant of the poor. Thus it presents us with the anomaly of a surplus of luxury housing existing side-by-side with a shortage of inexpensive housing, although the social need for the latter is incontestably greater than the former. Or it pours energy and resources into the multiplication of luxuries for which the wealthier classes offer a market, while allowing more basic needs of the poor to go unheeded and unmet.

Finally, these shortcomings are aggravated by *the inability of the market system to cope with the externalities that, as we have seen, are inextricable from the workings of the market system.* But there is no need to spell out further the deficiencies of the market. In one form or another, all are indicative of a central weakness of the market mechanism: *its inability to formulate effective stimuli or restraints other than those that arise from the marketplace itself.*

So long as the public need roughly coincides with the sum of the private interests to which the market automatically attends, this failing of the market system is a minor one. But in an advanced economic society, it tends to become ever more important. As primary wants become satisfied, the public aim turns toward stability and security, objectives not attainable without a degree of public control. As technological organization becomes more complex and massive, again a public need arises to contain the new agglomerations of economic power. So, too, as wealth increases, pressure for education, urban improvement, welfare, and the like comes to the fore, not only as an indication of the public conscience, but as a functioning requirement of a mature society. And finally, the public stimulus and management of continued growth take on increased political urgency as the ecological problems of industrial societies multiply.

We have already paid much attention to the rise of planning in the advanced market societies as a corrective force to deal with just such problems. Now we can go so far as to generalize the economic meaning of this trend. *Planning arises in the advanced market societies to offset their inherent goal-setting weaknesses, just as the market mechanism arises in advanced command societies to offset their inherent motivational weaknesses.* In other words, planning and market mechanisms, in those societies which have begun

CONSERVATIVE PLANNING

All through Europe we see a reliance on planning that is both greater and more outspoken than anything we have encountered in the United States. In our own country, we have arrived at a consensus as to fiscal and monetary policy as the proper implements for achieving a stable and satisfactory rate of growth. But in most European nations, there is visible a further commitment to planning as a means of achieving publicly determined patterns of resource allocation as well as adequate rates of growth.

In France, for example, a central planning agency, working in consultation with Parliament and with representatives of industry, agriculture, labor, and other groups, sets a general plan for French growth—a plan that not only establishes a desired rate of expansion but determines whether or not, for example, the provincial cities should expand faster or slower than the nation as a whole, or where the bulk of new housing is to be located, or to what degree social services are to be increased. Once decided, the plan is then divided into the various production targets needed for its fulfillment, and their practicality is discussed with management and labor groups in each industry concerned.

From these discussions arise two results. First, the plan is often amended to conform with the wishes or advice of those who must carry it out. Second, the general targets of the plan become part of the business expectations of the industries that have helped to formulate them. To be sure, the government has substantial investment powers that can nudge the economy along whatever

to enter the stage of high consumption, are not mutually incompatible. On the contrary, they powerfully supplement and support one another.

CONVERGENCE OF SYSTEMS

What seems to impend at the moment, then, is a *convergence of economic mechanisms* for the more advanced societies. In the planned economies, the market is being introduced to facilitate the smoother achievement of established objectives; while in the market economies, a degree of planning is increasingly relied upon to give order, stability, and social direction to the outcome of private activity.

This does not imply that the two major systems today are about to become indistinguishable. The convergence of economic mechanisms may blur but not obliterate the basic distinctions between them. Nor does the convergence of mechanisms in itself portend profound changes in the larger social structures of socialism and capitalism. A gradual rapprochement of the economic mechanisms should not lead us to hasty con-

clusions about the rebirth of "capitalism" in the Soviet Union or the advent of "socialism" in the United States. This is a matter into which we will look more deeply in our last chapter.

COMMON PROBLEMS

There is another way in which the phenomenon of convergence reveals itself, in addition to that of a coming together of economic mechanisms. *It is the appearance of similar problems in advanced industrial societies.* When we examine capitalism and socialism, we will pay special attention to the problems that separate and distinguish these two kinds of societies. Here it is important to realize that they are also bound together by certain common difficulties.

What is the nature of these overarching problems? As we would expect, they stem from the very technical capability and social organization that bring similar economic mechanisms into being. Three problems in particular seem of major importance.

path has been finally determined. But in the main, French "indicative" planning works as a *self-fulfilling prophecy*—the very act of establishing its objectives sets into motion the behavior needed to realize them.

In England, Germany, the Netherlands, Italy, and Scandinavia, we see other forms of government planning, none so elaborately worked out as the French system, but all also injecting a powerful element of public guidance into the growth and disposition of their resources.

The plans have not been wholly successful. Inflation has been the curse of Europe to an even greater degree than it has here; and nothing like a successful "incomes policy" has been worked out in any nation. But considerable success has been attained in the allocation of resources for public purposes through planning, and in the shaping of the general contours of national development.

What is beyond dispute is that a basic commitment to planning seems to have become an integral part of modern European capitalism. *Note, however, that all these planning systems utilize the mechanism of the market as a means for achieving their ends.* The act of planning itself is not, of course, a market activity; but the realization of the various desired production tasks for industry is entrusted largely to the pull of demand acting on independent enterprises. Thus the market has been utilized as an instrument of social policy.*

*Anyone who wishes to learn more about the important subject of European planning should read Andrew Shonfield, *Modern Capitalism: the Changing Balance of Public and Private Power* (New York: Oxford University Press, 1969).

1. Control over technology

One of the most important attributes of modern history is lodged in a striking difference between two kinds of knowledge: the knowledge we acquire in physics, chemistry, engineering, and other sciences, and that which we gain in the sphere of social or political or moral activity. The difference is that knowledge in some sciences is cumulative and builds on itself, whereas knowledge in the social sphere does not. The merest beginner in biology soon knows more than the greatest biologists of a century ago. By way of contrast, the veteran student (or practitioner) of government, of social relations, of moral philosophy is aware of his modest stature in comparison with the great social and moral philosophers of the past.

The result is that all modern societies tend to find that their technological capabilities are constantly increasing, while the social and political and moral institutions by which those capabilities are controlled cannot match the challenges with which they are faced. Television, for example, is an immense force for cultural homogenization; medical technology changes the composition of society by altering its age groups and life expectancy; rapid transportation vastly increases mobility and social horizons; and the obliterative power of nuclear arms casts a pervasive anxiety over all of life. All these technologically-rooted developments fundamentally alter the conditions and problems of life, but we do not know what social, political, and moral responses are appropriate to them. *As a result, all modern societies—socialist and capitalist—experience the feeling of being at the mercy of a technological and scientific impetus that shapes the lives of their citizens in ways that cannot be accurately foreseen nor adequately controlled.*

2. The problem of participation

The second problem derives from the first. Because advanced societies are characterized by high levels of technology, they are necessarily marked by a high degree of organization. The technology of our era depends on the cooperation of vast masses of men, some at the levels of production, some at the levels of administration. The common undergirding of all advanced industrial or "post-industrial" societies lies not alone in their gigantic instrumentalities of production, but in their equally essential and vast instrumentalities of administration, whether these be called corporations, production ministries, or government agencies.

The problem is then how the citizen is to find a place for his individuality in the midst of so much organization; how he is to express his voice in the direction of affairs, when so much bureaucratic management is inescapable; how he is to "participate" in a world whose technological structure calls for ever more order and coordination. This is a matter which, like the sweeping imperative of technology, affects both capitalism and socialism. In both kinds of societies, individuals feel overwhelmed by the impersonality of the work process, impotent before the power of huge enterprises—above all, the state itself—and frustrated at an inability to participate in decisions that see more and more beyond any possibility of personal influence.

No doubt much can be done to increase the feeling of individual participation in the making of the future, especially in those nations that still deny elementary political freedoms. *But there remains a recalcitrant problem of how the quest for increased individual decision making and participation can be reconciled with the organizational demands imposed by the technology on which all advanced societies depend.* This is a problem that is likely to trouble societies—capitalist or socialist—as long as technology itself rests on integrated processes of production and requires centralized organs of administration and control.

3. The problem of the environment

As we saw in Chapter 34, all industrial nations face an era in which exponential growth is beginning to absorb resources at rates faster than we may be able to provide them with new technologies; and all industrialized societies—indeed, the whole world—may soon be entering an era in which environmental limitations on energy or the ability to absorb heat will impose a slowdown on rates of growth.

Moreover, we stand at a period in history when underdeveloped nations are belatedly making their own bid for a share in the rising output per capita that has until now mainly been confined to advanced nation-states.

In this period of long-run economic stringency, industrial socialist and capitalist nations again seem likely to share common problems—not alone in bringing about a controlled slowdown in output, but in achieving social harmony under conditions that no longer allow their citizens to look forward to ever-higher standards of material consumption. Here, too, similar social and political problems may override differences in economic institutions and ideologies.

CONVERGENCE AND HISTORY

In a larger sense, then, "convergence" brings us beyond economics to the common human adventure in which economic systems are only alternate routes conducting humanity toward much the same general direction and destination. Perhaps it is well that we end our survey of economic history with the recognition that the long trajectory of the market system does not bring us to a terminus of social history, but only to a state in which some kinds of problems—the pitifully simple problems of producing and distributing goods—begin to be solved, only to reveal vastly larger problems in the very technology and organization that prepared the means for solving them.

But those are problems for the future.

Meanwhile it is the present that absorbs us, for it is in the present that we have our personal encounter with history. And so we cannot end our study of economics until we have considered a question to which we have often glancingly referred but are only now ready to confront. This is the question of America today, and more specifically, of American capitalism today. What are its prospects, its portents?

KEY WORDS

Stages of development

Mercantilism

Central planning

Collectivization

Success indicators

Indicative planning

Mid-development

High consumption economies

Market and goals Convergence

CENTRAL CONCEPTS

1. The spectrum of economic systems in the world corresponds generally to their *stages of economic development.*

2. The inception of growth seems to require a political stimulus. In the West, this was provided by *mercantilism*, after which the market took over the task of growth. The West was favored over the present underdeveloped areas by its ability to mount a *leisurely development*, and by its *vanguard position* vis-à-vis the rest of the world.

3. Today, *command economies* are attempting to push their societies off dead center and to initiate the process of growth. For these tasks, command seems a more appropriate system than the market.

4. *Central planning*, such as that used by the U.S.S.R., requires the formulation of *highly detailed and accurately interlocked subplans*, any one of which, if wrongly designed or unmet, can wreck the larger one. *Inefficiency* has been the plague of the central planning system, although the system did succeed in achieving its larger objectives. The problem has been that "success indicators" have led to misallocation of resources.

5. As a result, we see the *introduction of the market mechanism*, using profit as the success indicator, into the European communist economics. *The market is being introduced into socialism just as planning is being introduced into capitalism.*

6. Among the many European planning mechanisms, the most elaborate is the French system of "*indicative planning.*" A centrally formulated plan is discussed among representatives of industry, labor, and other groups, and in the process of discussion and amendment becomes part of the general expectations of these groups. This leads to the requisite investment needed to bring the plan about. Thus the system works as a *self-fulfilling prophecy.* (See box, pp. 648–49).

7. Economies in *mid-development*, such as the Soviet Union, are characterized by mixtures of command and market systems. Command is best suited for massive economic and social reallocations, but not to the problems of running a smoothly integrated high-output economy. Bureaucratic inefficiency, and the absence of a congruence between private interest and public requirement, introduce many difficulties into the planning mechanism. Here the market begins to achieve a new relevance.

8. *High consumption economies* are naturally suited to market guidance. However, these economies now suffer from the inability of the market to formulate public needs. *Thus, planning arises in market systems to offset their inherent goal-setting weaknesses, just as the market arises in command systems to offset their inherent motivational weaknesses.*

9. Thus we see a *convergence of systems* in the more advanced nations. However, a convergence of economic systems does not mean that "capitalism" and "socialism" are becoming indistinguishable.

*Cumulative vs.
noncumulative*

Participation

Environment

10. **Convergence seems to mean, also, that all advanced industrial societies share certain overarching problems: (1) a noncumulative social or political or moral understanding trying to control the effects of a cumulative technology; (2) finding a place for the individual who wants to participate in a system that demands large-scale, coordinated processes of production and administration; and (3) coping with the problems of the environment. These problems manifest themselves differently in capitalism and industrial socialism, but in neither does a solution seem to be in sight.**

QUESTIONS

1. How do you account for the simultaneous existence in the world of such radically different economic systems?

2. In what ways was the early growth experience of the West unlike the present growth prospects of the backward world? Do you think these differences will have a substantial effect on the choice of economic systems for the backward countries?

3. Discuss the difference in social goals and priorities between a nation that is just beginning its development and one in mid-development. Between one in mid-development and one at a stage of high mass consumption. What is the relevance of planning techniques for each of these stages of development? Would you expect the techniques of planning to be similar in all stages? What differences would you look for?

4. What is meant by the congruence of self-interest and public requirement in a market system? Is this what we mean by the "invisible hand"? How can this congruence be reconciled with the fact that the market has no means of establishing public priorities.

5. What are the advantages of the market system for economic freedom? Do you think the market system is also productive of political freedom? Draw a scatter diagram showing on one axis the degree of planning and on the other axis your estimate of relative political freedoms for the following nations: Sweden, England, U.S., France, South Africa. Is there much, if any, relationship between the two variables?

6. Do you think there can be a convergence of economic systems without a convergence of social and political systems?

7. What specific technological processes seem to defy social control? How about the effect of television? Urbanization? The arms race? Can you think of others?

8. Why do you think that knowledge accumulates more easily in science than it does in the areas of morality or politics or social activities?

9. Do you think that bureaucracy is an avoidable aspect of industrial society? If so, how? If not, what problems does it pose?

10. Do you think that the freedom to choose one's profession, to go to graduate school "to see how you like it," to become a hippie, etc., makes life easier or more difficult? Do you think your parents would agree?

Is Capitalism
the Problem?

41

AT THIS FINAL STAGE IN OUR INQUIRY, as in its first stage, let us speak as two individuals, because what we have to say should not be given the authority that rightly or wrongly seems to adhere to statements made in the impersonal third person of textbook prose. What follows is no more than the fruit of personal reflection, offered not to settle matters, but only to make explicit our convictions and to give student and instructor an opportunity to discuss the most important social issue with which economics is ultimately concerned. As the title of this chapter indicates, that ultimate problem is capitalism itself, and the extent to which it must be held responsible for the troubles of our country.

DEFINITIONS: WHAT IS CAPITALISM? SOCIALISM?

We had better begin with definitions. If we are now to ask whether America's troubles are due to capitalism, we should know what we mean by that crucial word.

It is surprisingly difficult to find a succinct definition of capitalism.[1] But all shades of opinion, from right to left, would agree that its essential characteristics are these:

1. The legal right to private ownership of the means of production.

Under capitalism, the capital equipment of society is owned by a minority of individuals (capitalists) who have the right to use this property for private gain.

2. The market determination of distribution.

Capitalism relies primarily on the market system, not only to allocate its resources among various uses, but also to establish the levels of

---

[1] One of the few clear-cut definitions is in Max Weber's *General Economic History* (New York: The Free Press, 1950), Chapters 22 and 30. The Marxian definition is more diffuse, referring at different times to an *economic system,* to the *bourgeois society* built on that system, and occasionally to a *stage of Western civilization.* See George Lichtheim, *Marxism* (New York: Praeger, 1961), p. 164, note 4.

income (such as wages, rents, profits) of different social classes.

What is the corresponding definition of socialism? As we might expect, it is something of a mirror image of capitalism. "In its primary usage," writes Paul M. Sweezy, a leading Socialist theoretican, "the term 'socialism' means a social system which is differentiated from other social systems by the character of its property relations.... Capitalism recognizes a relatively unrestricted right of private ownership in the means of production, while socialism denies this right and reserves such ownership to public bodies."[2]

Thus Sweezy, like most socialists, makes the crucial distinction between capitalism and socialism the question of *property ownership*—to which most socialists would also add that socialism, unlike capitalism, depends primarily on *planning,* rather than on the market, both for its overall allocation of resources and for its distribution of income.

IDEAL TYPES VS. REAL CASES

These definitions are what the sociologist Max Weber called "ideal types." They are meant to summarize and abstract out of the enormous variety of actual institutions and historical experiences those essential elements that make up a pure model of the institution or activity in which we are interested. We dealt with such an ideal type when we studied the operation of a "market" under pure competition, knowing full well that no market was in fact ever free of all imperfections. In the same fashion, the emphasis on public vs. private property, and on market vs. planned distribution, serves to sharpen our conception of the "irreducible" elements of capitalism and socialism that are to be dis-

covered behind their many variations in actual history.

But no sooner do we create these ideal types than we find ourselves in something of a quandary. For the question then arises as to what practical function these models of capitalism and socialism serve. For example, if one asks a dedicated humanitarian socialist if "socialism" is better than "capitalism," he will unhesitatingly tell you that it is, basing his reply on the superiority of public over private ownership and on the preference to be accorded to planning over the market. But the same humanitarian socialist recoils in horror at the repressiveness of Russia and looks with approval on the humaneness of (capitalist) Denmark. How does he reconcile this contradiction? By telling you that Russia is not "really" socialist, but only a grim travesty of socialism; and that Denmark is not "really" capitalist but a modified socialist version of capitalism. Yet, unquestionably, the Soviet Union has public ownership of property and a thorough-going system of planning, and Denmark has private ownership of property and a general market determination of incomes and outputs.

The point of this disconcerting confrontation is clear. It is that the elements that all agree are decisive in defining capitalism and socialism as "ideal types" do not tell us very much about the societies that display those characteristic elements. As a matter of fact, thinking about the differences among capitalist nations—compare Sweden and the Union of South Africa—or among socialist countries—contrast Russia and China—we begin to wonder if the words *capitalism* and *socialism* mean anything at all.

CAPITALISM AND SOCIALISM AS ECONOMIC SYSTEMS

The answer is that the terms *do* mean something, although perhaps less than we often assume. As we shall see, there are crucial

---

[2] Paul M. Sweezy, *Socialism* (New York: McGraw-Hill, 1948), p. 3.

areas of life to which they add little if any understanding. But there are other areas where they add a good deal, and it is to these that we now direct our attention.

The first such area should be obvious. It is that of economics proper. *Capitalism and socialism as ideal types identify for us a series of economic problems that we find among all members of each type.*

What are these problems? For capitalism, we have but to refer to the micro and macro sections of this text. Disequilibrium, instability, misallocation of resources, and inequality of incomes are results of the economic process in *every* society that has private ownership of property and a market determination of prices. Whether we look to Japan or Sweden, the Union of South Africa or the United States, we see similar tendencies toward too much or too little growth, inflation or unemployment, a struggle between the private and the public sector, and a highly uneven division of incomes between the property-owning and the working classes. These are problems as specific to capitalism as the problems of guild life were specific to feudalism.

Can we apply the same general finding to socialism? To a certain extent our comparison is muddied by the fact that so many socialist systems are still in (or only very recently out of) a period of forced economic growth. Hence we do not really have "mature" socialisms to compare with mature capitalisms.

Nonetheless, there seems to be a set of common economic problems built into socialism in much the same fashion as the problems that are intrinsically part of the capitalist mechanism. As we would expect, these are problems of public ownership and planning —in particular, the problem of controlling unwieldy state bureaucracies and avoiding inefficient production and distribution directives. Indeed, one of the most brilliant socialist economic theoreticians, the late Oskar

Lange, wrote presciently in 1938: *"The real danger of socialism is that of the bureaucratization of economic life. . . ."*[3]

CAPITALISM AND SOCIALISM AS POLITICAL SYSTEMS

We shall return to a consideration of what these deep-rooted problems imply for capitalism and socialism.* But first let us ask if we find a parallel to the economic attributes of the two systems in their political characteristics. Some eminent political economists, Milton Friedman prominent among them, believe that we do. Friedman believes that the *existence of a market mechanism separate from the state provides the necessary basis for a free political life.* He does not insist that every capitalist society is therefore a free one —indeed, he specifically singles out Fascist Italy, Nazi Germany, Spain, and Tsarist Russia as societies whose capitalist economic structures did *not* provide political freedom. But he is firm in his conviction that the existence of capitalism is a *necessary,* even if not a sufficient, condition for political freedom.

Friedman spells out his argument forcefully.

One feature of a free society is surely the freedom of individuals to advocate and propagandize openly for a radical change in the structure of society. . . . It is a mark of the political freedom of a capitalist society that men can openly advocate and work for socialism. . . . How could the freedom to advocate capitalism be preserved and protected in a socialist society?

In order for men to advocate anything, they must in the first place be able to earn a living. This already raises a problem in a socialist society, since all jobs are under the direct control of political

[3] Oskar Lange and Fred M. Taylor, *On the Economic Theory of Socialism* (New York: McGraw-Hill, 1956), p. 109.

*It is equally obvious that the phenomenon of convergence in the economic mechanisms of the two systems tends to bring the problems of one system into the other.

authorities. It would take an act of self-denial whose difficulty is underlined by experience in the United States after World War II with the problem of "security" among Federal employees, for a socialist government to permit its employees to advocate policies directly contrary to official doctrine.

But let us suppose this act of self-denial to be achieved. For advocacy of capitalism to mean anything, the proponents must be able to finance their cause—to hold public meetings, publish pamphlets, buy radio time, issue newspapers and magazines and so on. How could they raise the funds?[4]

Does Friedman's logic establish a second fundamental dividing line between capitalism and socialism? It may. There *is* a certain refuge for the dissident individual in the free-for-all of the market that may be missing in socialism for the very reasons that Friedman suggests. Indeed, as M. A. Adelman has written, one main objection to socialist society is that it too much resembles the company town under capitalism, where one employer has power over everyone within his jurisdiction.[5]

Yet the issue does not have the clarity of the economic differences between capitalist and socialist societies. Certainly it weakens Friedman's position that among capitalist societies—that is, in the spectrum of systems relying on private property and the market system—*there seems to be no correlation whatsoever between the degree of political freedom and the use of "socialist" limitations of property rights or the market mechanism.* Indeed, the most "socialistic" capitalisms, such as the Scandinavian bloc, England, or New Zealand, are probably the world's freest political societies.

[4] *Capitalism and Freedom*, pp. 10, 16–18.
[5] Cited in *The Business Establishment*, ed. Earl Cheit (New York: Wiley, 1964), p. 217, note 17.

Second, it should be pointed out that in the provision of a *guaranteed income* to all citizens, there is available a protection against the very kind of political vulnerability Friedman warns us about. As we have seen, the United States seems to be moving slowly toward such a system, and there is no reason why a socialist system could not adopt one very easily. A guaranteed income under socialism would free the dissident citizen from political pressure, just as private employment offers him his freedom under capitalism.

To be sure, such a system also requires that a socialist state value political liberty so highly that it would not cut off the income of an agitator for capitalism. Certainly no socialist state today displays any such easy tolerance—just as very few capitalist states allow complete liberty for agitators against capitalism. In all societies there is a natural "defensive" reaction against hostile ideas, and there is no reason to suppose that socialist authorities will be more tolerant of dissent than are capitalist ones. Indeed, in the present stage of historical development it is doubtful that socialist governments in general can be expected to be as politically tolerant as capitalist ones, partly because many socialist economies are in the throes of forced growth, and partly because most socialist governments feel insecure in a world still dominated by capitalist power.

Whether socialism will continue to be less politically free than the most liberal capitalisms is a forecast that cannot be made at present. As we have seen, it is possible to guard against the company-town aspect of socialism by means of a guaranteed income. But that is not to say that even a wealthy and self-assured socialist society will *wish* to put political liberty, with all its risks, high on its agenda. In the final analysis, we suspect that the degree of political liberty will vary from nation to nation under socialism, as it does under capitalism; and that the *economic* framework will have very little to do with the quality of political life in either system.

VALUES AND LIFE STYLES UNDER CAPITALISM AND SOCIALISM

This complex relationship between economic structures and political liberty brings us deeper into the question of how "capitalism" as an ideal type differs from "socialism." Specifically it leads to the question of whether we can discover a basis for *social* differences between the two systems—differences in values or life styles that might find their origins in the economic systems characteristic of each.

At first glimpse, there seems no relationship whatsoever between the two. Indeed, if we were able to compare "socialist" Yugoslavia with "capitalist" Italy we would probably find many more *similarities* of social tone, cultural life, and general atmosphere, than if we compared "socialist" Yugoslavia with "socialist" China. And in the same way, "capitalist" West Germany probably more closely resembles "socialist" East Germany in general life style than it does "capitalist" Australia. *National* differences in culture and tradition seem so overwhelmingly important that any common social traits traceable to capitalism or socialism recede far into the background, if indeed they are discoverable at all.*

*Merely as one indication of this diversity, let us look into the much discussed matter of suicide rates. Does "welfare capitalism" bring a high suicide rate, as President Eisenhower once suggested? The evidence does not show it. Suicides per 100,000 population are higher (18.5 per 100,000) in Sweden than in the U.S. (10.4 per 100,000), but they are higher here than in Norway (7.9) or the Netherlands (6.6 per 100,000). Denmark, on the other hand, has an even higher rate—20.3 per 100,000—but it suffered from this rate *before* it became a welfare state. See Alan Gruchy, *Comparative Economic Systems* (Boston: Houghton, 1966), p. 436.

This emphasis on national differences is, we believe, of the greatest importance, and we will return to it in full measure when we inquire further into the problems of American capitalism. Yet at this stage of our inquiry, while we are still seeking to divide all capitalisms from all socialisms, perhaps one distinctive element should be noted. It is the presence within all socialist countries of an ideal not to be found in any capitalist country. This is the ideal of Socialist Man—of man transformed from the competitive, acquisitive being that he is (and that he is *encouraged* to be) under all property-dominated, market-oriented systems, into the cooperative human being who finds fulfillment in unselfishness and who presumably can develop only in the benign environment of a propertyless, nonmarket social system.

It need hardly be said that this ideal of Socialist Man remains an unrealized goal in any socialist society, and may perhaps even be an unrealizable goal, comparable to the ideal of a truly religious community vainly pursued by centuries of Christianity. Yet, however distant, however vulgarized in practice, however abused as a mere slogan for social manipulation, the conception of Socialist Man provides a spiritual basis for an ideology that is powerfully persuasive. Capitalist nations may be efficient, humane, democratic, permissive, creative, but they do not have a "vision" built on the elements of property ownership and the market—a vision comparable to that which is founded on the ideal of common ownership and sharing.

This absence of a capitalist ideal may strike some as a fact little to be regretted, in view of the human suffering that so many official ideals have brought in the past. But if we suspend our judgments and merely ask whether there is an ultimate difference between the worlds of socialist and capitalist values and goals, we think the answer must be yes, that "socialism" has a sense of high human mission—perhaps even redemption—that "capitalism," even at its best, cannot quite match.

PROBLEMS AND SOLUTIONS

Our discussion suggests that the ideal types of "capitalism" and "socialism" *are* useful because they indicate different kinds of problems that the two systems tend to generate. But we have not yet inquired into an extremely important question that our findings pose. Granted that capitalism and socialism have common problems, *does this mean that they all find similar solutions to these problems?*

To ask the question is to answer it. Obviously, different capitalisms respond to their economic and social and political problems in very different ways, as do different socialisms. Take capitalism as an example. Two well-known Marxist critics of capitalism have written that genuine planning or resolute action to provide housing would be impossible in America because "such planning and such action ... will never be undertaken by a government run by and for the rich, as every capitalist government is and must be."[6] They have obviously concentrated on the lack of an effective social sector in the U.S., and overlooked the planning and housing undertaken by Norway, Sweden, Denmark, New Zealand, Netherlands, and other governments presumably run by and for the rich, since they are certainly countries where private ownership of the means of production prevails.

In the same way, because all capitalist systems do suffer from macroinstabilities, it does not follow that all therefore suffer from the same degree of unemployment. For example, during the 1960s when unemployment here and in Canada reached levels over 5 percent, in West Germany unemployment never rose over 1 percent of the labor force, and in New Zealand it was considerably less than that.

This same variety of responses can be found in socialist economies. Oskar Lange's diagnosis of bureaucracy has proved all too true within all socialisms, but some have responded with a reliance on market socialism (Yugoslavia or pre-invasion Czechoslovakia); others, in efforts to solve the problem with better computer-planning (U.S.S.R.); and still others, with a search for "moral incentives" (Cuba or China).

American Capitalism

All this has an obvious relevance to the central issue with which we began this chapter; for we can see now that whereas many of the problems that beset America undoubtedly have their roots in our capitalist institutions, the fact that we have coped with them so inadequately is not a matter that can be blamed on capitalism as such.

Take, for example, the question of social neglect that is so dismaying an aspect of American life. If we compare the United States with, say, Norway in terms of various indicators of social well-being, there is no doubt that we show up poorly. Infant mortality in the United States is a full 50 percent higher than it is in Norway. Norway spends a higher proportion of its GNP on education than we do, even though Norway has a much smaller per capita GNP. Norway has more hospital beds per thousand population than we do. It allocates a larger proportion of its GNP to social security expenditures than does the United States. Its cities are essentially free of all slums. Although Norwegian citizens have a much lower GNP than American citizens, "poverty" as a relative condi-

---

[6] Paul Sweezy and Paul Baran, *Monopoly Capital* (New York: *Monthly Review Press,* 1966), p. 300.

tion—or poverty as a human condition of neglect by society—has been virtually eliminated there.

If we are willing to grant on the basis of these and other indicators that Norway has a record of social welfare superior to that of the United States, we would expect large differences in the economic structures of the two countries; for instance, in their income distributions. Yet, as Table 41·1 shows, there are not!

TABLE 41 · 1

Pre-Tax Income Distribution:
United States and Norway

| | Percent of Income Going to Percentile Groups | |
	Norway 1963	U.S. 1972
Lowest 20%	4.5	5.4
Second 20%	12.1	11.9
Third 20%	18.5	17.5
Fourth 20%	24.4	23.9
Top 20%	41.5	41.4

Note that the years are not identical, but in view of the extreme slowness of movements in income distribution, this is not a serious consideration.

What the table shows is that the pre-tax distribution of income—that is the relative shares of rich and poor as determined by the market mechanism—is much the same in both nations. Indeed, in the United States the pre-tax share of the rich and the relative share of the second and third groups from the bottom (which includes the bulk of the working class) is vertually the same as in Norway.

But this surprising result would be totally altered if we now took into account the distribution of income *after* taxes and subsidies. In contrast to the generally nonprogressive structure of U.S. taxes, the Norwegian tax structure is one of the most pro-

gressive in the world and this progressive tax incidence is further emphasized by a system of subsidies to low income groups, which greatly lightens their net tax burden. The point, of course, is clear. It is that *the original capitalist dispositions of the marketplace do not leave an irretrievable mark on their societies, but can be radically modified by taxes and transfers.*[7]

CAPITALISM OR AMERICAN CAPITALISM

All this has a sobering, as well as an encouraging, implication. It is that much of what troubles America seems to be related to factors that, however much exacerbated by our economic system, cannot be uniquely attributed to capitalism as such. The low level of social services in America, the enormous role played by the military, the "rat-race" tempo of American life, the extent of our slums, the callous treatment of criminals, the obsession with "communism," and many other unlovely aspects of our social system are not predominant in many other capitalist systems.

The problem, in other words, resides more in those elements of our society that are *American* than in those that are capitalist. To put it differently, the significant question for us is to understand why capitalism here has not achieved the possibilities realized by capitalism elsewhere, rather than to compare the deficiencies of life in America with the presumed advantages that "socialism" might bring. For unless we understand and correct the failures of *American* capitalism, the likelihood is that a change of economic systems in this country might produce only an *American* socialism that would manifest many of the very failings of American capitalism.

[7] See Gruchy, *Comparative Economic Systems,* p. 338. See also his discussion of Swedish tax incidence, p. 402.

Can we identify the elements in America that have brought about the failure of capitalism here to match the social achievements of similar systems abroad? Here we move from the reasonably firm grounds of empirical evidence to the quicksand regions of social conjecture. There are, perhaps, no "right" or "wrong" answers to this crucial question—only answers that are more or less useful in helping us think about an immensely complex and elusive problem. But with all the difficulties, let us try to suggest some possible explanations for the failures of our system.

TWO HYPOTHESES THAT FAIL: SIZE AND HOMOGENEITY

We might start by asking what is the most striking difference between the United States and the more socially responsive capitalisms we have referred to. One such difference that immediately suggests itself is sheer *size*. Is the greater social neglect of American capitalism simply the result of the scale of our continental country? Are we too big to create a genuine sense of community?

It is certainly plausible that a large country, with its variations in regions and interests, is less apt to feel a sense of shared responsibility than is a small one. And yet at best, the explanation of size can be only a partial one. Canada, although geographically larger than the United States, has a considerably more highly developed social welfare program and "point of view" than we have. Australia is another example of a large, regionally variegated nation with a high level of social services. Perhaps more to the point, our density of population (which is the way that "size" becomes translated into human experience) is 25 percent *greater* than Sweden's and double that of New Zealand, so that our continental expanse does not physically separate man from man to a degree

that might explain the lack of our "communal feelings."

A second hypothesis seems more convincing. It would explain the failures of American capitalism by the extreme diversity of our population—the many ethnic and cultural groups that have obstructed the formation of a socially-minded single community comparable to the Danes or the Swedes.

Yet this hypothesis also loses much of its persuasiveness under scrutiny. Switzerland, for example, has three different language and nationality groups and yet maintains a very high level of communal concern. Canada, despite the political friction of its French- and English-speaking regions, has a well-developed social welfare system. Perhaps even more telling is the egregious social neglect in certain areas of the United States where there *are* strong ethnic, regional, and racial bonds—the rich whites of Appalachia have paid little attention to the decline of their poor white kinsmen.

THE RACE PROBLEM

Yet, if our social heterogeneity does not provide a fully adequate reason for our inadequate social performance, it does point in one direction that sharply differentiates American capitalism from capitalism in other nations.* This is the fact that America is burdened with a *race problem* that is closely entwined with many areas of its laggard social performance.

There is no need at this point to rehearse the statistics of black poverty and neglect. Instead, it is necessary to link our deep-seated prejudice against blacks with out national disregard of the (largely black) slums, our national neglect of the prisons (disproportionately filled with blacks), the low level

---

*With the exception of the Union of South Africa, which significantly also displays many of the same symptoms of social neglect that we do—and for the same reasons.

of welfare provision (roughly half of which goes to blacks). To put it bluntly, one very likely reason for the continued failure of American society to clean up its ghettos, improve its prisons, and liberalize antipoverty programs is the unwillingness of many white Americans to allow their hard-earned dollars to be taxed and spent in ways that will tend to benefit black Americans more than white ones.

THE TRADITION OF DEMOCRATIC INDIVIDUALISM

The race issue, so central a part of our history, reaches deep into the core of American social neglect, but it does not account for every aspect of our problems. Racism does not explain our reluctance to deal generously with white poverty, nor our militarism, nor our failure to cope effectively with the deterioration of the environment. To what special factors in American experience might we attribute these weaknesses of American capitalism?

The first answer we will suggest may be a surprising one. It is America's tradition of *democratic individualism*—a tradition born and nurtured on our frontiers and perpetuated in the image of the town meeting as the ideal, democratic mode of government.

This is surely an odd suggestion, in that it is precisely this tradition that is one of the proudest claims of American society—indeed, the genuinely democratic quality of American life continues to impress visitors even from the most socially advanced European nations.

Yet, there is a price to be paid for this heritage. It is the lack of a tradition of strong central government and *noblesse oblige,* the duty of the benevolent ruler to provide for his less fortunate subjects. Both these traditions are woven into the cultural traditions of most European nations and have helped them create *welfare* capitalisms.

Alexis de Tocqueville, the great nineteenth-century French sociologist, wrote perceptively about the difference in traditions when he visited America in the 1830s.

Aristocracy [he wrote] links everybody, from peasant to king, in one long chain. Democracy breaks the chain and frees each link. As social equality spreads, there are more and more people who, though neither rich nor powerful enough to have much hold over others, have gained or kept enough wealth and enough understanding to look after their own needs. Such folk owe no man anything and hardly expect anything from anybody. They form the habit of thinking of themselves in isolation and imagining that their whole destiny is in their own hands. Thus, not only does democracy make men forget their ancestors, it also clouds their view of their descendants, and isolates them from their contemporaries. Each man is forever thrown back on himself alone, and there is a danger that he may be shut up in the solitude of his heart.[8]

This is not to claim that European aristocracies were in fact more solicitous of their peoples in the nineteenth century than was democratic America. On the contrary, as the direction of immigration showed, quite the opposite was the case. Nonetheless, with eloquence and perception, Tocqueville points to a deep difference between democratically based and aristocratically based social systems. It is the presence, within the older governments of Europe, of a *legitimacy of authority* that, in the changed conditions of the twentieth century, made possible a much more vigorous and direct governmental attack on social problems than was possible in the American democratic-individualist environment.

But it is not only our tradition of small-scale local government that has handicapped us in dealing with large-scale, national problems. What Tocqueville alerts us to is an in-

[8] Alexis de Tocqueville, *Democracy in America* (New York: Harper, 1966), p. 478.

hibition in the *feeling of social responsibility* of a democratic society, compared with one of more aristocratic lineage. This suggests that one of the reasons that Americans have not developed an effective attack on social problems is that there is no popular support for the idea that government should help the needy. The result has not only been an anaesthetizing of the American social conscience, but a paralysis of the mechanism by which that conscience might have been best expressed.*

LACK OF SOCIAL DEMOCRATIC HERITAGE

To this possibly heretical suggestion, let us now add a second, which applies to America at the other end of the socio-political scale. It is that American capitalism, unlike its European counterparts, never developed much of a tradition of *democratic socialism as a reformist force.*

Many observers have commented on the failure of the socialist ideal to implant itself in America, a failure due in part to our economic success and in part to our political (if not economic) stress on equality, at least for whites. "On reefs of roast beef and apple pie," as the economic historian Werner Sombart put it, "socialistic utopias of every sort [were] sent to their doom."[9]

---

*Sociologist Seymour Martin Lipset gives a telling illustration of the difference in national attitude between our highly individualist tradition and that of a more aristocratic society. Both the United States and Canada have created national folk-heros connected with the frontier that was so important for both societies. How suggestive that America should have chosen the free-wheeling cowboy and that Canada should have picked the scarlet-clad Canadian mounted police, representative of central authority and law and order! See Lipset, *The First New Nation* (New York: Basic Books, 1963), p. 251.

[9]Quoted in Daniel Bell, *Marxian Socialism in the United States* (Princeton, N.J.: Princeton University Press, 1967), p. 4.

It has been customary for Americans to congratulate themselves on the absence of a native socialist movement, for socialism has certainly brought sharp conflicts and severe social tensions within many nations. Yet, it has also brought an impetus for social change and a widened agenda for discussion that has been missing from America. In Sweden, Norway, Denmark, England, Netherlands, New Zealand—in short, virtually everywhere that capitalism has achieved a high level of social welfare—the driving force for change has been a democratic socialist party, pushing against the limits of conservative capitalism. In America, with the important exception of the early New Deal, no such "socialistic" enhancement of the conception of what a society might achieve has ever significantly enlarged the conception of what "capitalism" meant. Hence measures such as the nationalization of industry where that might be useful, or the massive redistribution of income—both measures used by a number of capitalist nations—are avoided in America because they smack of "socialism."*

THE PROBLEM OF MILITARY POWER

There remains, however, one vast problem to which neither of the preceding suggestions directly applies. It is the question of the relationship of capitalism to the rise of an aggressive American military state.

Is militarism itself a specifically capital-

---

*The political scientist Robert A. Dahl has written that Americans are "half colorblind" when they think about problems such as how to control giant corporations. "An important reason," he writes, "is that our history has left us without a socialist tradition.... The consequence, I think, is a serious limit to our capacity for clearheaded consideration of how economic enterprises should be governed. Because we have no socialist tradition, our debates about economic institutions nearly always leave some major alternatives—chiefly 'socialist' alternatives—unexplored." (*After the Revolution?* New Haven: Yale University Press, 1970, p. 119).

ist phenomenon? The evidence of history would hardly indicate as much. There have been pacific capitalisms (Sweden, Switzerland) and aggressive ones (pre-World War II Germany, postwar United States); and the actions of the Soviet Union in its invasion of Hungary and Czechoslovakia or its border clashes with China are enough to indicate that capitalism has no monopoly on international violence.

Yet there is no doubt that in recent years American capitalism has acted as a military state that felt itself impelled to police the world. Since World War II, the United States has intervened in Lebanon, the Dominican Republic, Cuba, South Vietnam, Cambodia, and Laos; has clandestinely overthrown at least one government (Guatemala), and probably another (Iran); has actively supported repressive right-wing regimes in Greece and Brazil; and has "shown the flag" on some 50 occasions.

Whence derives this military commitment? The answer is not easy to give. In part, it lay in a genuine fear of postwar Soviet intentions, hardly reassured by Russia's general truculence, by its walling-off East Berlin, and by its anticapitalist rhetoric. Yet, compared with the reaction of many European nations, the American response to these threats was exaggerated—indeed, so exaggerated that it undoubtedly helped to provoke the very Soviet (and later Chinese) belligerency by which Americans could then justify their uncompromising stand.

The sources of this American response are many: clergymen who equated the rise of communism with the anti-Christ and who stirred the fears of fundamentalist religious groups; labor unions whose anticommunist feelings reached such irrational levels that longshoremen refused to unload Polish merchandise at the very time that Poland was trying to break *away* from the Soviet bloc; veterans organizations to whom patriotism

meant an uncritical support of American "might"; congressmen who subscribed to the belief that America must defend "freedom" everywhere, except perhaps for selected members of their own constituencies; military leaders who gradually enlarged the conception of American "defense" to the ability to wage two major wars and one minor war *simultaneously;* industrial circles for whom the arms economy provided a seemingly inexhaustible source of revenue; high government officials who saw America as the world's bulwark against "international communism," even after international communism had dissolved into mutually hostile nation-states; intellectuals and professors who applied the cool logic of "maximizing" national security but never questioned the premises on which security might be based.

In this multiplicity of causes, each of which has played its part in the American anticommunist stance, the role of "capitalism" is both central and elusive. Certainly capitalism all over the world has felt itself menaced by the rise of powerful socialist governments in Russia and China and the prospect of revolutionary socialism in the underdeveloped areas, much as the Catholic Church once felt menaced by the rise of Protestantism; or the British aristocracy, by the rise of democracy in France. Yet, as we know, Catholicism made its peace with Protestantism—differently, to be sure, in different countries—and the aristocracies made their peace with democracy—also differently in different countries. That private property and the market mechanism can also make a peace with public property and planning seems equally evident—but that, too, will take place differently in different countries.

In many capitalist nations that process of peacemaking seems now fairly far advanced. Various capitalist nations have long carried on trade freely with Russia, have begun to invest in communist nations, have recognized and traded with China, and while maintaining a healthy skepticism toward the national policies of these nations, have felt no need to mount an ideological crusade against them. Now, after the disastrous Indochinese War that has been the price America has paid for its blind anticommunism, we shall have to see if here, too, we cannot find a pragmatic mode of coexistence and become once again a citizen, not the policeman, of the world.

WHAT SHOULD BE DONE?

If our discussion points to any conclusion, it is that we must seek to change many aspects of American capitalism in the direction that capitalism in other nations has indicated. This does not mean, of course, that America could or should set out to "copy" the achievements of other lands. Institutions, like good wines, do not always travel easily. We shall have to find our own ways of coping with the problems of health and the cities and national planning and the arts and pollution and education—ways that may take inspiration from foreign accomplishments, but that must in the end reflect our own ways of doing things.

Yet it may help if we have at least a general idea of what such a major effort would look like. There is such a "preview" in the proposed budget of government expenditure for the late 1970s, drawn up by the National Urban Coalition. Envisaging a new program of National Health insurance, public service employment, family assistance for the poor, aid to students from preschool to college, grants for improving police departments, taxes to curb pollution, plus a wide range of other social services, the Coalition looks ahead to a total federal government expendi-

ture in 1976 of $354 billion,[10] compared with $206 billion in 1971.

Is such an enormous increase in expenditure imaginable in a capitalist society? Comparisons are helpful here. In 1967 the United States used 18.8 percent of its GNP for nonwar public purposes. The proposals above would raise the nonwar proportion of GNP used by government to only 20 percent, assuming that GNP will grow very rapidly (about 5 percent per year) under the many-sided stimulus of this new program.

But if GNP grew at only 3 percent, the "burden" of the proposed nonwar expenditures on GNP would be only 26 percent of GNP (and the addition of the defense portion would raise it to only 30 percent). In Table 41·2, compare the portion of GNP used by other nations for nonwar purposes only. Clearly, the *Counterbudget* proposals lie well within the range of actual experience in other capitalist societies.

TABLE 41 · 2
Nonwar Public Uses of GNP, Selected Nations (1967)

Canada	26.7%
France	31.6
West Germany	30.9
Italy	27.6
Sweden	36.8
United Kingdom	27.0

Source: *Counterbudget*, p. 313.

[10]*Counterbudget*, p. 300. (The 1976 figure includes $94 billion of purchases of goods and services, $50 billion of which are allocated for military expenditure, and $260 billion of transfers to individuals and states and localities.) These figures are in current dollars, based on the assumption that prices in 1976 will be only 21 percent higher than in 1971. The inflation rates of the 1972–1974 period would raise the Coalition's figures considerably.

The problem, then, is not one of economics, but of politics—of will. The program will require increased tax revenues, and the Coalition proposals call for graduated increases in taxes on incomes over about $12,000 plus the closing of many tax loopholes. Thus a family earning $13,000 a year would pay about $100 more in taxes; one earning $20,000 would pay roughly $1,000 more; one earning $35,000 might pay as much as $2,500 more.[11]

Will those influential Americans who constitute the 40 percent of all families above the $12,000 line assent to such an increase in their taxes to bring about a new America? Only time and effort will tell. Many Americans, particularly those who find themselves in the influential upper echelons in the nation, are not deeply disturbed by the problems of the remaining 60 percent, much less those of peoples many thousand miles away. Yet, others—including sizeable numbers who are aware of their good fortune—feel the urgency of the problems and the inadequacy of the American response. Indeed, if we judge by the past, the leadership for the reform of America's backwardness will come from those very groups who will have to make some sacrifice of their private income to achieve a greater public good.

THE MORAL CHALLENGE

Textbooks are not proper places for expression of political sentiment. They are not improper places, however, for giving vent to moral sentiments, provided these are properly identified. Hence let us end our long discussion of the origins and character and problems and promise of our society by saying that we believe that American capitalism stands at the threshold of a crucial decade. For almost a generation our society has been identified in the eyes of much of the world as a nation that is rich but indifferent; peace-loving in rhetoric but aggressive in behavior; boastful of its economic strength, blind to the misuse of that strength. Now the years are at hand when that image—and the realities that are uncomfortably close to the image—can be changed. We will continue to be a rich nation; we must cease to be in so many ways a poor one. We will continue to be a powerful nation; we must learn the limits of our power. We will continue to be an economic colossus, but our economic system must now become the undergirding for a good society.

In view of our heritage, our traditions, our deeply-rooted failings, this will not be an easy transformation for America to make. It may even prove to be an impossible one, although we strongly believe that this is not the case. But in any event, the all-important thing is to make the effort. The challenge is to change America today into a different and much better America tomorrow. In our view, this will be the most searching test of its character that American society has ever faced.

Central Concepts and Questions

All the previous chapters offered summaries at the end—key words to be learned and concepts that an instructor is likely to hold you responsible for. Not this chapter. This is not so much a chapter to be "studied" as it is to be thought about. This time you must pose your own questions.

[11]*Ibid.,* p. 10. (The figures have been roughly adjusted to 1975 values.)

7

QUANTITATIVE METHODS

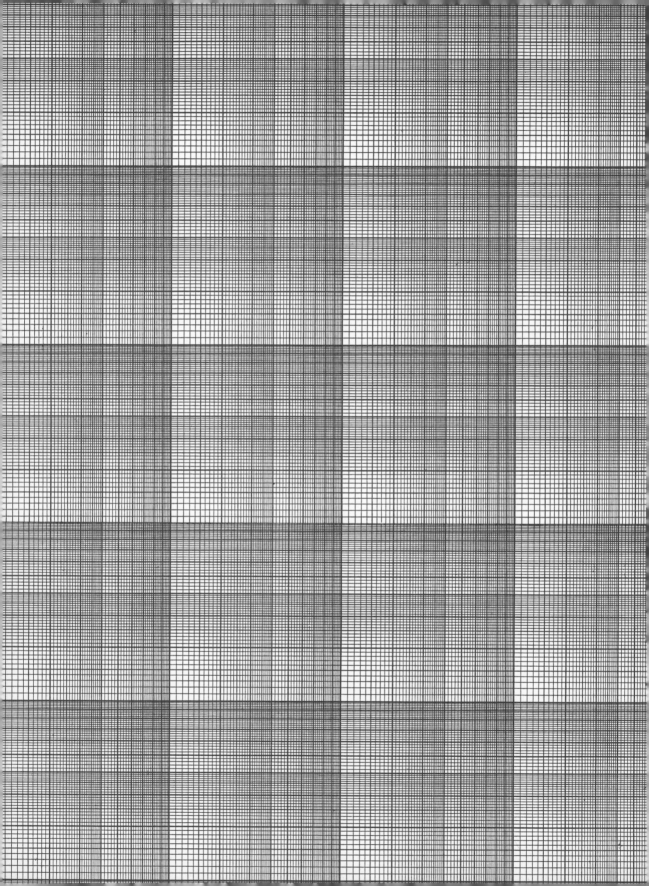

An Introduction to Statistics and Econometrics

42

STATISTICS ARE CENTRAL to the study of economics. Why? Because economics, more than any other social science, involves things that can be counted, concepts that can be quantified, activities that can be measured. Prices, wages, GNP itself, as we have learned, all express our ability to quantify certain aspects of social activity. This ability to quantify is often essential to our ability to analyze and control. The GNP accounts, invented in the 1930s, made it possible to develop macroeconomics. Without these accounts, only fuzzy impressionistic statements could be made about macroeconomic activity, just as only vague statements could be made about the velocity of money before we could actually measure output (see p. 485). If we could quantify "utility," we could measure welfare far better than we now do.

Hence an economist is constantly using statistics and *econometrics,* a branch of economics that combines statistical techniques with economic theory. To understand the techniques of statistics or econometrics you must learn much more than we are going to teach in this chapter. This is most emphatically not a mini-course that will substitute for a thorough study of both, but we hope that it will pave the way for further study by taking away some of the mystery surrounding the use and compilation of the numbers on which economics depends.

WHAT THIS CHAPTER IS ABOUT

You will find this longish chapter divided into eight sections, really eight sub-chapters. The first of them has to do with some overall *cautions and warnings.* It is easy to read and does not involve any "work," unless you call learning something useful "work." Everyone should take its message to heart. Statistics are indispensable for economics, but they are also the source of much trouble for the unwary amateur.

Section II is about *Distributions and Averages.* These are concepts we use every day, as economists or in everyday life. If ever there was a word loaded with problems, it is that deceptive word *average.* This is a section that will richly repay study, and it is basically easy to master.

Section III is about *Price Indexes.* This is a more specialized subject, into which we have already looked very briefly (p. 290). It takes a little calculation to work through the construction of a price index, but the results are worth the effort, especially if you are thinking of taking further courses in economics. Then you will really have to understand "weighting" and how to convert a price index in current dollars into one in constant dollars.

Section IV is about *Sampling,* a tremendously important subject for anyone interested in empirical research. We can give you only a general idea of the often complex considerations involved in sampling, but we hope you will emerge from these few pages with a much better understanding of the general problems of sampling than you probably had before you began.

Section V brings us to econometrics and specifically to *Functions.* We have already had an introduction to functional relationships in Chapter 5, but the subject is so useful that a careful review of the field is worthwhile, particularly since functions lie at the heart of econometrics.

This brings us to *Correlations and Regression,* section VI. Here is the core of econometrics. Part of the section, about regressions, introduces you to the general notion of the techniques of one very important branch of functional analysis. It does

not try to teach you how to make an actual regression analysis, but it should make clear what you are trying to do when you sit down with a desk calculator or use a computer to work out the complex formulae of econometric analysis.

Section VII is perhaps even more important. It is a series of warnings about the meaning of *causation and correlation.* In particular, it alerts you against easy conclusions that such-and-such is the "cause" of so-and-so, just because the arithmetic "looks right." Everyone ought to look at this section, even if you have only the haziest notion of how you go about finding a "least squares" coefficient.

From here we go to section VIII, *Forecasting.* As we said in Chapters 5 and 6, prediction is one of the most important—and troublesome—aims of economic analysis. Our brief review will serve to remind you of these problems before you find yourself forecasting on the basis of flimsy data or uncertain premises.

We don't suggest that you sit down to read this chapter as a single lesson. Use it as a reference source. Perhaps you have already looked into it, when the text suggested you look ahead into certain pages. Or read it in short takes, section by section, to get a feeling for the field of statistics and econometrics. If the chapter dispels some of the fog surrounding that field and tempts you to take a course in statistics, it will have amply served its purpose.

I. Some Initial Cautions

QUALITY VS. QUANTITY

We should immediately clear up two misconceptions about statistics. The first might be called "statistics worship." It finds expression in the view that if you can't count something, it doesn't count. The second is just the opposite. It scoffs at statistics because all the important things in life are qualitative and therefore can't be measured.

Both of these *are* misconceptions, and it is important to understand why. The statistics worshipper must realize that before anyone can count anything, he must *define* it, and that definition is ultimately an act of judgment, not measurement. And the statistics scoffer must realize that before he can bring his reason to bear on social problems involving good and evil, he must have at least some notion of how *large* a good or evil he is concerned with. Some of these matters we have touched on in Chapter 6, but it will do no harm to look into them again a little more carefully.

Take as an illustration the problem of poverty. When Professor Galbraith wrote *The Affluent Society* in 1958, he said that one family in thirteen in America was poor. By current official statistics, one in seven is poor. Does that mean that a larger fraction of Americans are poor today? Of course not. The answer is that today the United States statistical authorities' *definition* of poverty is different from the definition used by Galbraith, so that the "growing" percentage of the poor reflects nothing more than a changing conception of what poverty means.

Thus, the extent of poverty in America is a quantitative measure founded on a qualitative definition, and the person who places a blind faith in the "facts" must remember that many of these facts change as our definition of social problems change. But isn't that just the point, asks the statistical skeptic. Since definition is everything, the problem of poverty boils down in the end to what we *think* poverty is. What is important is not what the numbers tell us, but our moral judgments from which the numbers emerge.

Of course it is true that definitions *do* come first, but they establish the unit or yard-stick by which we measure the problem. Poverty may rise or fall in America if we use an unchanged yardstick. It can also rise or fall because we change its definition. But once we have made that choice of the yardstick, we use the definition to learn the magnitude of the problem *as we have defined it.* For until we know its magnitude as measured by the *same* yardstick, we cannot know how serious a problem it is, much less determine what action will be needed to remedy it.

PATTERNS OF MOVEMENT

Often, as with poverty, *the pattern of movement is as important as, or even more important than, the absolute value of the variable.* If poverty is rapidly diminishing, we may need a set of public programs very different from those needed if it seems to persist. So, too, with other measured magnitudes, such as GNP. Is GNP rising or falling, and how fast? It is no accident that most discussions of GNP in different countries tend to focus on comparative rates of growth. Changes in GNP (per capita) are probably a better measure of changes in economic welfare within a country than are efforts to measure the absolute level of GNP in one nation and to compare it with another.

Unemployment presents a similar problem. Different countries define unemployment in different ways, and the seriousness of any given level of unemployment can often be determined only by comparing the current level with its historical path. For this reason, any new statistical series should be used with caution. Its usefulness cannot be determined until we see how it moves over time and under different circumstances.

Sometimes definitions themselves are important in determining the pattern of movement of economic variables. Sometimes they are not. One of the techniques for investigating the importance of different definitions is to see if they imply that the economy is changing in different ways. If Professor Galbraith's definition of poverty yielded a pattern of movement different from the official definition (suppose poverty was growing on one definition and falling on the other) then the movement of poverty is *sensitive* to its definition. If this were true, one would want to be extremely careful when discussing poverty.

THE PROBLEM OF DEFINITIONS

The second lesson follows from the first. Words are slippery, even when no arbitrary judgment is involved. Any maker or user of statistics must be exceedingly sure of the exact meaning of the terms being used. In Chapter 6, p. 68f., we showed how tricky the words *family* or *household* could be, for, a household can be a single person! Another example of a treacherous definition is the one for—of all things—*motor vehicles.* If you look up the data in the *Statistical Abstract,* and read the footnotes, you will see that a "motor vehicle" includes a mobile trailer home. The result is that the motor vehicle statistics overstate the production of vehicles if you think of these as cars and trucks, and that correspondingly, the statistics for real housing investment (in which mobile trailer homes are not included) understate the value of residential construction.

UPDATING SERIES

Numbers, perhaps even more than words, have a magical authority inherent in them, but that does not mean that either are necessarily correct, even from the most impeccable sources. People make honest mistakes in gathering statistics, or mistakes creep in because of the difficulties in compiling statistics, some of which we will learn about later.

In Chapter 6, pp. 68–69, we also saw

how official figures, such as those for GNP can change from one edition of the *Statistical Abstract* to the next.

The reason is that many statistical series are refined and corrected as more complete data are collected, so that virtually all figures change until the "final" returns are in. Most economic statistics are necessarily based on fragmentary data and are gradually improved as fuller information is obtained. The latest statistics are *always* subject to revision and are often marked *prelim.,* or *est.,* to warn the reader that they will be subject to change. The changes are usually not large, but occasionally large enough to alter precise calculations based on earlier data.

At regular intervals, the various government offices concerned with the collection of statistical information bring out revised—sometimes drastically revised—series of data. These revisions not only establish new "bases" and "weights" for many indices (we will come to the meaning of these terms), but involve recalculation of many past series based on new concepts, on more sophisticated data-handling techniques, and so on. (One recent controversy, for example, concerns the problem of whether or not we have accurately measured the value of the private capital stock in the United States. Economist Robert Gordon has claimed that we have "mislaid" $45 billion of this stock by failing to make allowance for various government transactions, such as the very low prices at which some companies acquired plants built by the government during World War II and subsequently sold at nominal figures to private enterprise.*)

Of course no ordinary user of official statistics can be aware of all these pitfalls or can anticipate the revisions that may alter the numbers on which he is relying. An excess of

*See *American Economic Review,* June 1969, and rejoinder, *ibid.,* September 1970.

wariness would only paralyze the research that we must carry on with the only data we have at hand. But a healthy pinch of caution, allowing for moderate changes in the magnitudes at hand, has saved many a researcher from trying to prove a point by relying on very *small differences in magnitudes that may later disappear.*

MOUNTAINS OUT OF MOLEHILLS

The best recent example of misusing small differences is in the handling of the monthly reports on the rate of inflation. Monthly changes are multiplied by 12 to get the annual rate of inflation published in newspaper headlines. Let's assume that in one month the rate of inflation is 0.34 percent, and in the next month it is 0.36 percent. Since the Bureau of Labor and Statistics rounds its numbers to one place to the right of the decimal, these would be reported as a 0.3 percent increase and a 0.4 percent increase. When they are multiplied by 12, they yield a 3.6 percent rate of inflation versus a 4.8 percent rate of inflation.

In fact, the initial difference is probably well within sampling error (see below). This means that the two numbers 0.34 and 0.36 may be the same; they only appear different because of the errors inherent in any process of measurement and sampling. As a result, one should not place too much weight on month-to-month changes, but look at the pattern of movement over several months.

VISUAL DECEPTIONS

Often visual devices are used to display statistical data. These are particularly subject to misleading interpretations. Let's say we are trying to chart a crime wave over time. We have two graphs, displaying exactly the same numbers, but on different scales. In the first graph, it looks as if there were little or no change in crime; the second looks as if there were a horrendous acceleration. Can you see the reason for the difference?

FIGURE 42 · 1
Visual deceptions

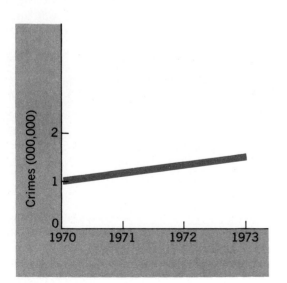

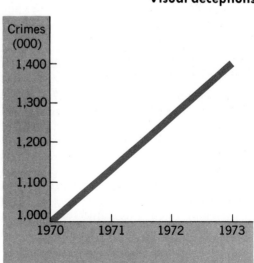

PERSPECTIVES ON DATA

Normally, the statistical "truth" about any phenomenon depends upon examining it from a number of different perspectives. Consider the much discussed question of the size of government in the U.S. economy. This is a matter we have already looked into in pp. 412f., but it is worth reconsidering here as a general problem in statistics. The size of government will depend largely on how we define *government*. Do we mean federal government or federal plus state and local? Do we mean purchases of goods and services or all expenditures, including transfer payments? Just as a review of the difference that these definitions make, consider the following "measures" of government size:

	Percent of GNP, 1973
Federal expenditures	20
Federal state and local expenditures	31
All federal purchases	8
All government purchases	21

All these statistics accurately indicate something about the relationship of "government" and the size of the economy. The important thing, therefore, is to use the figures that are appropriate to the problem you are investigating. Are you interested in federal or total government activity? In production, purchases, or in expenditure? In the *uses* of production or expenditure? (In the last case, you will need more data, since the figures above do not show, for example, welfare vs. warfare; subsidies or public investment, etc.)

USES OF STATISTICS

These are very general warnings, but not to be taken lightly. More than one researcher has been in serious trouble because he didn't read the caption at the top of a table of figures or because he failed to check on the most recent compilations of data or because he was fooled by numbers that did not adequately reflect the changing nature of the problem he was trying to measure.

Now, at the end, a word to redress the balance. We have stressed skepticism toward statistics because the general attitude of the beginning student is generally one of blind acceptance. But too much skepticism is perhaps worse than none at all. *Carefully defined and collected data, clearly labeled and competently used, are the only way we have of measuring very important facets of our social activity.* This book would be impossible to write without statistics, and economics would be severely crippled if we could not rely on numerical magnitudes. Statistics really are an integral part of economics. The thing is to be wary of their weaknesses while appreciating their great virtues and absolute necessity.

II. Distributions and Averages

Distributions and averages are essentially statistical devices for viewing aggregate phenomena from different perspectives. For example, GNP measures total output. Sweden had an aggregate output of $39 billion in 1972; Spain had a GNP of $42 billion; the U.S. one of $1,118 billion. These numbers indicate something about the potential economic power of each country—although not very much, since Spain's ability to wage war, for example, is probably less than that of a nation such as Sweden. But even less do the numbers tell us about the amounts of goods and services that are available to typical people in each nation. To learn about this, GNP needs further examination.

MEANS

The simplest procedure is to divide total GNP by the total population to obtain the "average" GNP per person. We call this arithmetical average the *mean:* in the U.S. in 1972 it was $1,118 billion divided by 209 million individuals, or $5,353 per capita. In Sweden the per capita (mean) GNP was $4,749; in Spain, $1,221. In common usage this is what we usually have in mind when we use the word *average,* but to the statistician it is only one of several ways to give meaning to that very important word.*

MEDIANS

Means can be very misleading "averages." Suppose that one person or a very few people had virtually all the income, and the rest had very little (not so far from reality in a case like Pakistan). The mean income would then tell us very little about the income of an individual chosen at random. Hence, statisticians often use another definition of "average" called the *median.* As the word suggests, the median income is the income of the middle individual. If we lined up the population in order of income and selected the person who stood midpoint in the line, that person's income would be the median. Half the country would have smaller incomes; half larger.

ANOTHER CAUTION: CHOICE OF UNITS

Here is a good place to interject cautions about data and definitions, once more. If we were to calculate the median income of *individuals* in the United States, the answer would be zero! This is because more than half of all individuals have zero incomes. They are children, nonworking females, older retirees, and others. (You can see that it makes a difference if we choose "income" or "income plus transfers.") But it obviously

*Here is a problem in perspectives on data. If we want "average" GNP to tell us something about welfare, we probably don't want to use GNP as the measuring rod, because too much GNP becomes corporate or government end-product. Better to use an aggregate such as personal income.

makes no sense to say that the median individual in the U.S. had no income. Therefore, we focus on *family* income, on the assumption that all members of a family share equitably in the income received by the family as a whole. Actually that assumption is not true: children have much less purchasing power than adults; but it serves our general purposes to make this assumption, as long as we know that we are using the data in a special way.

MEANS AND MEDIANS

Is there a systematic difference between means and medians? There is: *medians are almost always smaller than means.* A moment's thought tells us that if the median income is $7,100, then the range of income in the poorer half of the population must be smaller than the range in the half that is richer; 50 percent of the population must receive between $1 and $7,100, and 50 percent between $7,100 and the income of the richest person in the country. *This asymmetry in the distribution of income is at the root of the difference between the two averages.*

A SIMPLE ILLUSTRATION

Perhaps the principle involved is clear to you by now. If not, an arithmetical example may help.

Assume that there are two towns, A and B, and that each consists of five families. The following table represents the distribution of income.

A $4,000 $6,000 $7,000 $8,000 $10,000
B $4,000 $6,000 $7,000 $8,000 $15,000

In each town, the median is $7,000, the income of the family in the middle. But notice that in town A, the distribution of income is symmetrical; that is, the median family is $1,000 away from the second and fourth families, and $3,000 away from the first and fifth families. Note, however, that this is not true of town B. The income of the top (fifth) family is much further away from the median than the income of the first family.

In town A, the median should be equal to the arithmetic mean. And, of course, it is: total income of $35,000 of town A divided by 5 equals $7,000. But in town B, because of the imbalance of income toward the wealthier side, the arithmetic mean is higher than the median: total income of $40,000 of town B divided by 5 equals $8,000. So the mean is $1,000 greater than the median in B.

In its way, town B is a highly simplified version of the country as a whole. If we lined up all the approximately 54 million families in the U.S., from poorest to richest, and picked the middle family—that is, the family with 27 million families on either side—our designated *median* household would have an income of $11,116. If instead, we added up the *incomes* of all families, in any order, and divided by 54 million, the resulting *mean* income would be $12,625.

SKEWNESS

This asymmetry in distribution is called *skewness,* a term we shall make use of subsequently. A quick way of grasping skewness is to plot data on a graph. For example, in Fig. 42·2, *A* and *B* show two hypothetical distributions, one skewed and one not. In both graphs we measure income along the horizontal axis and the number of families along the vertical axis.

The nonskewed distribution is shown in *A*, where we can see that the distribution of income is just the same on one side of the median as on the other. For example, if the median income is $8,000, we note that 5 million families have an income of $6,000 and that 5 million families have an income of $10,000, each income being the same $2,000 distance from the median figure.

This kind of unskewed frequency distribution is called a *bell-shaped curve* or a *normal* curve. Graph *B*, on the other hand, is skewed to the right. Suppose that $8,000 is the median income for this population. We

FIGURE 42 · 2
Normal and skewed distributions

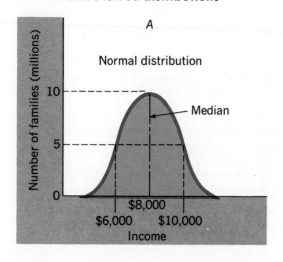

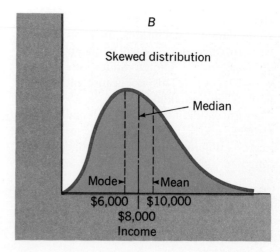

can see from the shape of the curve that there is no longer an even pairing of families on either side of this median. The incomes of the families on the right "tail" of the curve are further from the median than the incomes of those on the left "tail," and so the mean (the "old-fashioned" average) must be above the median (middle) income.

Here the mean, as we know from our definition above, is higher than the median (we suppose it to be $10,000 on the graph).

MODES

Furthermore, we can now speak of *another* kind of average, called a *mode,* which shows the particular income level on which there is the *largest number* of families. In our graph, this is represented by the figure of $6,000. Notice that in a bell-shaped distribution, the mode is the same as the median and the arithmetic mean, but this is true for only a normal, bell-shaped distribution.

WHICH AVERAGE TO USE?

If we want to express the average income in the United States, which average should we use? To know the complete truth, we would need to know the shape of the entire distribution of incomes, but we often need a summary measure. Ordinarily, we choose the median as the summary measure, because what we have in mind with the word *average* is indeed that income which is "middle-most." If we use the arithmetic mean, we distort the picture of the middle family, for we now show an amount of "average" income swollen by the presence of very large incomes at the upper end of the income register.

Sometimes, however, the mean is a better average than the median. This is the case when the distribution of income is approximately normal. Under these circumstances, the arithmetic mean has the advantage that it takes into account the *actual value* of the money incomes received by rich and poor families, whereas the median counts only the number of families on either side of the middle figure.

But bear in mind that in some kinds of distributions, the use of either average can be misleading. Say we had a country where 95 percent of the population received about $500, and the other 5 percent received more than $500, ranging up to $1 million. This is not unlike the situation in India. With such a violent skewness, we probably should use the *mode,* $500. It tells us more than either the mean or the median reveals about the distribution of income.

But suppose the upper 5 percent were all clustered around $10,000. Then we could separate the population into two main groups, one rich and one poor. Such a curve is likely to be *bimodal,* showing that a great many families cluster around one income level, and another group clusters around a wholly different level. We show such a bimodal distribution in Fig. 42·3, p. 678.

Statisticians are well aware of the difficulties of accurately representing such bimodal frequencies with "averages." Thus when you see well-conceived data that present averages—whether for IQs, academic grades, incomes, or other phenomena—you can be pretty confident that the data have only one mode and represent the kind of frequency distribution shown in Fig. 42·2.

III. Price Indexes

The problem of finding a useful way to represent "average" or typical incomes brings us to another problem when we want to reduce aggregate figures to their "real meaning." This has to do with price changes, a subject we are familiar with from our discussion in macroeconomics. In fact, in our first chapter of macroeconomics, pp. 290–91, we looked into the problem of how we change

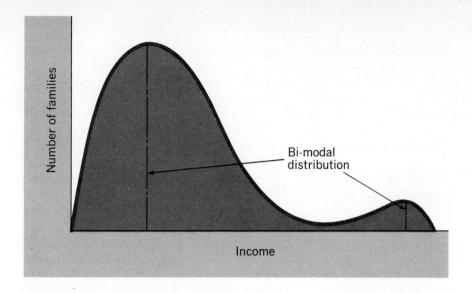

FIGURE 42 · 3
Bi-modal distribution

GNP in current dollars into GNP in constant dollars.

Here we want to investigate that problem a little more carefully. As we know, the device we use to discover changes in "real" purchasing power is a *price index*. There are many different kinds of price indexes, depending on what prices we want to measure: price indexes for wholesale commodities, for housing, for capital goods, for GNP.

Usually, when we try to discover changes in consumer well-being, we use the *Consumer Price Index,* constructed by the Department of Labor. It reflects the changing prices of a typical "market basket" of goods and services bought by an urban consumer over the period we want to investigate. By taking the prices in an initial or *base year* and the prices for the same commodities in other years, we arrive at a consumer price series, or an index of consumer purchasing power over that period.

BUILDING A PRICE INDEX

The idea of a price index is simple (although the actual statistical problems, such as collecting the data, are formidable). Let us quickly review the main steps.

Take an imaginary economy with only one consumer product, which we will call bread. Say that in the first year, which we will use as our base, bread sells for $2 per baker's dozen, and that in the second year its price is $4. What would the price index be for the second year? Elementary algebra gives us the answer:

$$\$2 \text{ is to } \$4 \text{ as } 100 \text{ is to } X$$

$$\frac{2}{4} = \frac{100}{x}$$

Solving by cross-multiplying, we get:

$$2x = 400$$
$$x = 200$$

Note that a price index is a percentage: in the above example, an index of 200 means that prices in that year are double the index (100) in the base year.

With 200 as our price index, we are now ready to compute real income in the second year in terms of base-year prices, just as we did earlier. Suppose a worker in a one-product economy was paid $5,000 the first year and $11,000 the second. What has happened to his real income? To find out, we divide this money income by the price index for that year, and then multiply the quotient by 100.

$$\frac{\$5,000}{100} \times 100 = \$5,000$$

(real income in Year 1)

$$\frac{\$11,000}{200} \times 100 = \$5,500$$

(real income in Year 2)

Thus, his real income has risen, but not by nearly so much as the sheer dollar increase indicates.

WEIGHTS

Now let's drop the assumption of a one-product economy and see what happens when there is more than one product. Assume an economy that consists of two products, bread and shoes. How would we compute a price index in this case?

Here we face a new problem. Before we can calculate *one* index for *two* products, we have to impute a proportional importance to each. For example, if more than half of consumer expenditures went for bread, then a doubling in its price would count more heavily than if the price of shoes doubled. The way we customarily deal with such a problem is *to take the amount of each product that the "typical" consumer purchased in the base year, and then compute the rise in living costs, under the assumption that the consumer will buy each product in the same proportion in succeeding years.*

Let us say that the consumer spent $100 in the base year and allotted the money in this way, between bread at $2 per unit and shoes at $10 per pair.

	No. units				
Product	*bought*	×	*Price*	=	*Total*
Bread	35 dozen	×	$2	=	$70
Shoes	3 pairs	×	$10	=	$30
					$100

In the second year, the price of bread doubles to $4 and the price of shoes is cut in half to $5. If the consumer plans to buy the same number of units of each, then his new budget must look like this:

	No. units				
Product	*bought*	×	*Price*	=	*Total*
Bread	35 dozen	×	$4	=	$140
Shoes	3 pairs	×	$5	=	$15
					$155

This gives us an aggregate expenditure for each year, from which we can compute the price index. Since it cost $100 to buy a typical basket of goods in the base year and $155 to buy *the same basket of goods* in the succeeding year, then the relevant index numbers are, of course, 100 and 155.

A more common way to go about this computation is *to derive "weights" from the proportional number of dollars spent on each product in the base year.* Since 70 percent of the typical consumer's money was spent on bread and 30 percent on shoes in 1967, the weights are .7 and .3, respectively.

In order to use the weights, however, we must first compute a separate price index for each product. For bread, which doubled in price, the index is 200 for the second year; for shoes, which halved in price, the index is 50. Now we multiply the index for each product by its corresponding weight and add up the results. For the base year:

Product	*Index*	×	*Weight*	=	*Total*
Bread	100	×	.7	=	70
Shoes	100	×	.3	=	30
					100

For the succeeding year:

Product	*Index*	×	*Weight*	=	*Total*
Bread	200	×	.7	=	140
Shoes	50	×	.3	=	15
					155

Again, our index numbers are 100 and 155.

Generally speaking, we assign weights by the value of purchases of the base year.* But what is more important is that the *same* weights must be used in the computation of each index number (or weights that are at least approximately the same). The consequences of doing otherwise are that an index number for one year would have little or no relation to an index number for another. If we are going to talk about fluctuations in a price level, then it must be a price level for the *same* basket of goods, as in the two products of our example, or for a basket of goods that is *relatively unchanging* in composition.

NEW GOODS

What do you do about the price of goods that do not exist in the base year? They have no base-year price. What price should they be multiplied by? There is no easy answer. Perhaps the good is a new product, such as a microwave oven, with an old function, cooking meals. In that case, government statisticians, in conjunction with an outside panel of experts, try to determine how much of the price of the new good reflects quality improvements and how much reflects price increases.

For example, suppose a conventional oven sold for $100 at the time the microwave oven was introduced at $150. Suppose further that the expert committee thought the microwave oven was 50 percent better than a conventional oven. Then the higher cost microwave oven would not show up as an increase in prices but as an increase in the quality of output. Each microwave oven produced would have the same effect on the real GNP as one and one-half conventional ovens; it would count as a $150 contribution to real GNP rather than a $100 contribution.

*For special purposes, there are also "chain indices," where weights are changed every year, as usage changes.

But suppose the committee thought that the microwave oven was *not* a quality improvement. Then each oven would show up as a $100 increase in the real GNP, just as conventional ovens. To the extent that people bought the higher price, but not higher quality microwave ovens, the prices of ovens in the GNP price indexes would go up, even though the price of conventional ovens did not change. In quality terms, you are being sold a conventional oven at a new higher price—$150.

NEW FUNCTIONS

But what can you do about products that serve completely new functions? Television sets were such an example. Since there was nothing to compare them with when they were introduced after World War II, these goods and services were evaluated at the prices for which they were first sold.

This creates a problem in measuring real growth. Most new goods are sold for very high prices when they are first introduced. Thereafter, their prices fall as they become mass produced and consumed. Let's assume that TV sets cost $500 when they were first sold in 1947. Each TV set produced would thus add $500 to the real GNP. If one million sets were produced in a year, this would add $500 million to the GNP. But let's now assume that the price of black-and-white sets fell to $100 and the government moved its base year forward to a year in which TVs cost only $100. Now the production of one million sets per year would add only $100 million to the real GNP, even though the same number had added $500 million when the government was using an earlier year as the base year.

This could be avoided by not moving the base year. But in that case the base year will rapidly become obsolete as new goods are introduced into the economy.

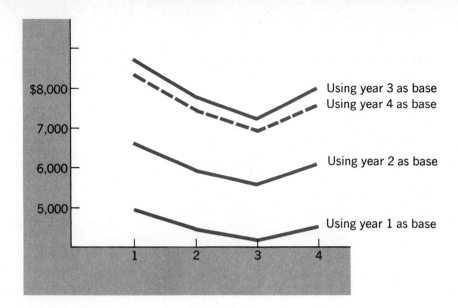

FIGURE 42 · 4
Shifting the base

As a result, corrections for price changes can be only an approximation of the truth. They should never be relied on "down to the last dollar" to make comparisons in real standards of living, especially if the periods being compared are widely separated in time. They are absolutely necessary to rid ourselves of the "money illusion," but they offer no more than thoughtful estimates—not precise measures—of what is happening to real standards of living over time.

APPLYING THE INDEXES

Constructing indexes is difficult; applying them is easy, and we have already given the basic rule on p. 290. But let us review it here, for completeness.

Suppose that the results of research give us the following data:

Year	Income in current $	Price index
1	$5,000	60
2	6,000	80
3	7,000	100
4	8,000	105

It now becomes very simple to compute the changes in real income. We divide the income in current dollars in each year by the price index and multiply by 100. The answers are:

Year	Real income
1	$5,000 ÷ 60 × 100 = $8,333
2	6,000 ÷ 80 × 100 = 7,500
3	7,000 ÷ 100 × 100 = 7,000
4	8,000 ÷ 105 × 100 = 7,619

CHANGING THE BASE

Note that in this case the base year is year 3. Would it make a difference if the base year were 1 or 2? Obviously, the numbers would change. If year 1 were the base, then the "real income" in year 1 becomes $5,000 instead of $8,333. *But the relative changes over time would not be different.*

Suppose we graph the profile of real income, using each year in turn as a base. Figure 42·4 shows us that we have in fact only shifted the location of the series on the vertical scale. The *shape* of the real income series is not thereby affected.

SPLICING

Changing the base suggests that we can "splice" one time series, with a certain "base," onto another. Indeed, we can and do. As the statistical agencies "update" their bases to prevent them from becoming obsolete, they usually splice the old series onto the new, by recomputing it on the new base. The only problem here is one of interpretation. By splicing one series to another we can get charts that extend far into the past. They tend to make us forget the difficulties of making "real" corrections with which we have been concerned. Remember when you see a long time series that it has probably been spliced many times and that its apparent unbroken continuity very likely masks many difficult problems of statistical definition. Use such series with care!

IV. Sampling

Most economic data, such as family income, is collected with sampling techniques rather than complete enumeration. Few things are more disconcerting to someone who begins to work with such data than to be told that most of the "facts" which he correlates or uses are based on *samples*— that is, on figures derived from observing or counting not the entire set of objects in which he is interested, but only a fraction of that population, and often a very small fraction at that. How can the statistician presume to be telling us the exact story, the student asks, when he really knows only a part of it?

The question is a very important one. We should understand the answer, for it is true that most economic data *are* derived from samples. And yet it is also true that sampling can provide extraordinarily accurate information about "populations" that are not counted—or even *countable!* For example, during World War II, American and British statisticians estimated the totals

for German war production by applying sampling methods to the serial numbers on captured equipment. After the war, many of their figures turned out to be just as accurate as the figures that the Germans compiled from the actual records of production itself. Moreover, the Allied figures were available sooner, since their method of counting took much less time.

THE LOGIC OF SAMPLING

What is the logic of sampling? It is based on a kind of reasoning that we use all the time. For example, we may complain that the streets in a city are dirty. Yet it is very unlikely that our complaint is based on an inspection of *all* the streets in the city or even a majority of them. Does this render our conclusion untrustworthy? It all depends. If we have seen a "fair" number of streets in all parts of the city, and all those we saw were dirty, we are quite justified in concluding that those we haven't seen would look much the same. But what is a "fair" number? What are the other criteria for selecting a sample that will enable us to generalize with a high degree of reliability about data that we have not observed?

SAMPLE SIZE

The first and most obvious requirement for a "fair sample" is its size. Yet, surprisingly, in making accurate statements based on samples, size turns out to be much less of a stumbling block than we might think. The U.S. Census Bureau, for example, on the basis of a sample of only about 5 percent of all families, gives a detailed and reliable description of the United States population. And if we want to study collections of data less complicated than those for the U.S. population, a sample much smaller than 5 percent will yield surprisingly good results. The Gallup and other polls forecast election results—on the whole quite accurately—on the basis of a sample of less than 100th of 1 percent of all voters.

How big a sample do we need to arrive at accurate estimates? That depends on two things: (1) how much we know about the larger "universe" of data that we are sampling, and (2) how accurate we want to be.

Suppose, for example, that we had two barrels of marbles, one filled with marbles of only two colors, white and black, and the other filled with an unknown number of colors. Obviously, the proportion of black to white marbles can be found from a much smaller sample than we would need for the second barrel, where a whole spectrum of colors and their various proportions are involved. Hence, the simpler the problem and the more clear-cut the data, the smaller the sample can be: it takes many fewer cases to establish the proportion of male to female births than to establish the pattern of average weight-to-height relation of infants at birth.

But suppose we have a "universe" that is well understood, such as our first barrel. How many marbles will we have to draw until we can make some kind of reliable statement about the *actual* proportion of black to white in the whole barrel? Here the answer depends on how accurate we want to be. Statisticians speak of "confidence intervals" to describe the fact that we can establish our own limits of reliability by increasing the size of a sample.
The "correct" sample size then turns out to depend on our need for accuracy. If we want to be 100 percent accurate, after all, we will have to check each and every marble, thereby raising the sample to the size of the population itself, assuming we make no enumeration errors—no easy task.

BIAS

Surprisingly, then, the problem of sample size is not so much of a difficulty as we might have thought, assuming that we know something about the characteristics of the population we are investigating. But suppose we do not. Suppose that we have traveled exclusively in the downtown streets of a city, not knowing there was a great slum just a few blocks away, and we based our judgments of the city as a whole on the cleanliness of the business district alone. Here we have encountered a much more serious problem. *Bias* has entered our calculations. By *bias* we mean that we have not planned our sampling technique in such a way that *each and every item in the population has an equal chance of being observed.*

Bias can enter sampling in the most unexpected ways. Suppose, for example, that the white marbles in our barrel were (unknown to us) slightly heavier than the black ones, so that when we shook up the barrel to be certain the marbles were fairly mixed, we were actually causing the white marbles to move toward the bottom. If our sample were taken from the top of the barrel, it would be biased in favor of black marbles. Or suppose that we chose our city streets absolutely by chance, but that we failed to take into account that the Sanitation Department visited different parts of the city on different days. Our sample could easily be biased in one direction or another.

Perhaps the most famous example of bias was the *Literary Digest* poll of the Roosevelt-Landon election in 1936. The magazine (long since defunct) sampled 2.3 million people, most of whom said they were going to vote for Alfred Landon. On this basis, the *Digest* predicted a landslide for the Republican candidate. In fact, as we all know, the election *was* a landslide—but for the other side.

What went wrong? Obviously, the problem was that the sample, although very large, was terribly biased. It was taken from subscription lists of magazines, telephone directories, and automobile registration lists.

In 1936, people who subscribed to magazines, had phones, and owned autos were
highly concentrated in the upper brackets of
income distribution. Furthermore, in that
year, income had a great deal to do with
party choice. The result has been used in statistics texts ever since as the perfect example
of bias.

CORRECTING FOR BIASES

How do we avoid biases? The answer is to
strive to choose our sample as randomly as
possible—that is, in such a way that every
unit in the population has an equal chance of
being in our sample. A purely random sample has the best chance to duplicate, in miniature, all the characteristics of the larger
"universe" it represents. Clearly, the more
complex that universe, the larger the random
sample will have to be to give us figures that
fall within respectable confidence intervals.
That is why the U.S. Bureau of the Census
uses a sample *as large as* 5 percent to collect
information about age, sex, race, income,
place of residence and many more attributes
of the population as a whole.

Getting a sample that is free of bias is by
no means easy. Statisticians spend much time
in devising ways to avoid the errors of bias
and in detecting unsuspected errors in the
work of others. Doesn't this mean, then, that
sampling is a technique we should view with
considerable suspicion? In fact, doesn't it
more or less cast serious doubt on a good
deal of what we think are "the facts"?

As we have said before, a healthy skepticism with regard to data is often very useful.
But skepticism is not at all the same thing as
a rejection of sampling as a statistical *technique*. Sampling is ultimately based on the
laws of probability about which we know
quite a bit. Hence, far from avoiding sampling, we should use our knowledge to perfect it.

Here are three reasons why sampling is
both essential and reliable:

1. Samples may be the only way of obtaining information.

Suppose that you are in charge of planning
the development program for a nation like
India and that you need to know many facts
about the birth rate, the average size of landholdings, or average incomes. *There may be
no possible way of obtaining this information
except by sampling.* Or suppose you are a historian interested in reconstructing the average length of life or the average family size of
some period far in the past. There may be no
possible method of gaining this information
except by sampling (for example, using the
data on gravestones), thereafter making the
best adjustment you can for the inevitable
biases that have crept into your sample (the
poorest people didn't get gravestones). Or
suppose your doctor wants to test your
blood. Would you want him to test *all* of it?

2. Samples may be cheaper than counting.

Even when you can count all the items in a
collection, it may not be worth your while
to do so. It would take an enormous expenditure of time and effort to interview each and
every householder in the U.S. about every
set of facts in the decennial census. It would
cost a fortune to test each and every light
bulb that General Electric makes. Sampling
gives us accurate enough data at a vast saving
in cost and time.

3. Sampling may be more accurate than counting.

Believe it or not, sampling is often more accurate than taking a full count. For example,
a census survey of every household would require the services of an army of census takers,
most of whom could not be trained in all the
pitfalls of interviewing. It is quite likely that
the inaccuracies of their mass survey would

be greater than those in the much smaller number of reports prepared by expert census takers who are careful to follow instructions to the letter. In full counts, immense amounts of data must be handled and transcribed. Clerical errors are apt to be fewer in smaller but more easily manipulated samples.

Sampling, in short, has its inescapable problems. But it is an absolutely indispensable and astonishingly reliable technique with which to deal with the overwhelming mass of data in the real world.

V. Econometrics. The Use of Functions

Now that we have an idea how sampling and other statistical techniques generate different types of economic data, we want to learn something about how data are used to test economic theories. This brings us to a relatively new use for statistics called *econometrics, a use that applies statistics to test various hypotheses about how the economy works.*

FUNCTIONAL RELATIONSHIPS

One of the central uses of econometrics is to discover functional relationships among the variables of the economic system. From Chapter 5, we are familiar with the general meaning of functional relationships, and we have again and again looked for, or applied, such relationships, as in the propensity to consume, a functional relationship between income and consumption.

Let us increase our basic knowledge by studying the consumption function further— not as a problem of empirical research, but as an exercise that will introduce us further to the vocabulary and techniques of econometrics.

For this purpose let us assume that the consumption function is this:

$$C = 2 + .8\,Y$$

This formula enables us to set up a table or schedule, relating C and Y, as follows:

If Y is equal to	*then C is equal to*			*shown on graph as point*
0	2 + .8(0)	= 2 + 0 =	2	A
10	2 + .8(10)	= 2 + 8 =	10	B
20	2 + .8(20)	= 2 + 16 =	18	C
40	2 + .8(40)	= 2 + 32 =	34	D
75	2 + .8(75)	= 2 + 60 =	62	E
100	2 + .8(100)	= 2 + 80 =	82	F

Let's now graph this relationship. Remember that each point in the two-dimensional graph represents *two* numbers; namely, C and Y.

FIGURE 42 · 5
Propensity to consume

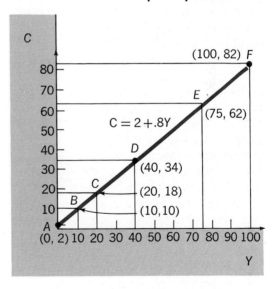

All this, of course, is very familiar. But now consider: *how do we know that the consumption function is a straight line?*

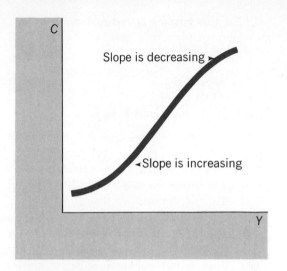

FIGURE 42 · 6
Nonlinear functions

The answer is that the word *straight* means that the line has an *unchanging slope*. If it were not a straight line, the slope would be increasing or decreasing from point to point, as Fig. 42·6 shows. By way of contrast, all lines of unchanging slope are always straight, although they need not all slope at the same angle. Figure 42·7 shows a variety of lines of *unchanging* slope, with very *different angles* of slope.

What is it about our formula that tells us that the (hypothetical) consumption function of $C = 2 + .8Y$ is a straight line? It is not the number 2, which simply tells us the value of C when $Y = 0$. This so-called *intercept* tells us where our line "begins" on the vertical axis, but it says nothing about its slope. That is determined by the second term. In this term, the number .8 is called the *coefficient* of Y, and shows the slope of the line. The "linearity" (straight line) aspect of the equation depends on the fact that Y is of the "first order"—that is, it is not squared or of a higher power. $C = 2 + .8Y^2$ would *not* be a linear function, but a curvilinear one. A

linear function is always of the form $Y = a \pm bx$.

COEFFICIENTS

The coefficient of the independent variable tells us how much the dependent variable will change for each unit of change of the independent variable. That is, .8 tells us that for each variation of $1 in income, consumption will change by 80¢. Given our unchanging coefficient, this relationship will remain the same, whether we are dealing with $1 or with $1 million. In each case, the change in C will be 80 percent the change in Y. If, in fact, the coefficient did change—if, e.g., we consumed a larger fraction of a small income than of a large one—the line would not be straight, and the equation that represented that line would not be an equation of the very simple form we have used.*

FIGURE 42 · 7
Linear functions

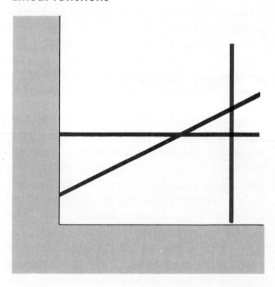

*The two numbers 2 and .8 in our formula are called *parameters*. A change in either or both parameters indicates a *shift* in the consumption line.

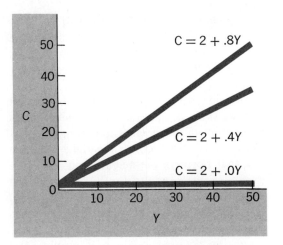

FIGURE 42 · 8
Changing coefficients

Coefficients, even of the simplest kinds, can represent all kinds of slopes. In Fig. 42·8 we show three coefficients that give rise to three slopes. The lowest line has a coefficient of zero, giving rise to a straight horizontal line parallel to the income axis. (Such a line depicts a situation in which we would not consume any more than a fixed amount, no matter how high our incomes—a highly implausible state of affairs.) The middle line shows a consumption function of approximately .4—a state of affairs in which we regularly consume 40¢ out of each additional $1 of income over the $2 intercept. The topmost line shows us our familiar coefficient of .8.

Note three final points.

1. The coefficient describes the slope of a line.

2. All lines described by unchanging coefficients are straight.

3. Econometrics commonly uses equations rather than graphs because equations can handle many more variables than we can conveniently show on a graph.

VI. Correlations and Regression

Up to this point we have spoken of functional relationships as if we knew the relationship among the variables in our equations. But one of the most difficult problems in actual economic research is to determine this relationship. In everyday language, we spend much time in econometric research trying to determine what affects what, and by how much. Once we have determined these relationships, it is relatively simple to put the results into formulas and to solve for the answers. Hence we are now going to turn to a fundamental technique of econometrics —correlation and regression—that will lead into the important problem of establishing reliable functional relationships.

THE ANALYSIS OF RELATIONSHIPS

We have come into contact with this problem in our study of the propensity to consume. There we were interested in examining the association between two variables, income and consumption. Much of econometrics is concerned with the kinds of *problems* exemplified by the income-consumption relationship, although the kinds of *activities* that econometrics investigates runs a vast gamut that includes price-quantity relations, interest rate-investment relationships, and wage-employment relationships.

What is it that we want to discover when we look into the relationship of two variables, such as income and consumption, wages and employment, or whatever? Generally, we seek to establish three things: (1) whether a change in one of these variables— say, income—or wages—is usually associated with a change in the second variable; (2) if there *is* such an association, whether a given change in one variable usually results in large or small change in the associated variable; and (3) once two variables have been shown to be linked in some way, just how precisely the relationship between the two can be described.

CORRELATION AND SCATTER DIAGRAMS

All of these problems use statistical techniques that we will not attempt to teach fully in this text. But for each of them there is also an intuitive representation that will make you familiar with the nature of these econometric questions.

The first such task, we have said, is to discover whether or not a relationship exists at all between two variables. The simplest way to test for the probable existence of such a relationship is to perform the operation we did in Chapter 24 ("A Diagram of the Propensity to Consume"), where we made a *scatter diagram* of income and consumption. In Fig. 42·9, we show two such scatter diagrams, each a visual representation of the association between two variables. On the left, we show a diagram that depicts the relationship of the heights and weights of a group of male adults; on the right we show one that portrays the heights and IQs of a group of male adults (both diagrams are fictitious).

The difference between the two diagrams is obvious. There is a clear-cut relationship between height and weight—the taller the man, the heavier he tends to be (with exceptions, of course). But there is no clear-cut relation between height and IQ: tall men are neither brighter nor more stupid than short men. In the first case, there is prima facie evidence that a correlation exists; in the second case there is none.

FIGURE 42 · 9
Scatter diagrams

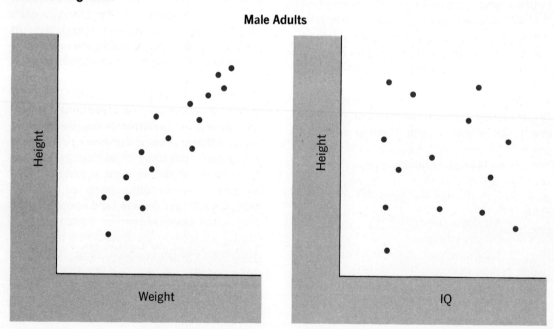

Male Adults

REGRESSION

Once we have some evidence that a correlation exists, we would like to find a way of describing it. That is, we would like to discover the value of one variable, if we know the value of the other, or the rate of change in one variable, given a change in the other. For example, we would like to know the *average change* in weight associated with an additional inch of height, or the *average change* in consumption associated with an additional dollar of income. (The last is, of course, our familiar marginal propensity to consume.)

When econometricians seek to reduce this relationship to a numerical magnitude, they speak of "running a regression" of one variable (such as height or consumption) on the other (weight or income). How do you run such a regression? The first problem, to which we shall turn in Section VII, is to establish the direction of causality—that is, whether we are regressing X on Y, or Y on X. The second problem is the procedure by which we measure the association in which we are interested. The details of computation are better reserved for a statistics course. We will just ask you to take on faith that there is a fairly simple arithmetical technique for deriving a regression line that will tell us what variable A will tend to be, given Variable B.

In Fig. 42·10, we show a very simple height-weight scatter diagram (it could be, of course an income-consumption scatter diagram) with *two* regression lines. Each line was drawn visually in an attempt to show the form of the relationship which the dots exhibit. Forget the colored lines for a moment. How do we know which is the better of the two, line A or B?

LEAST SQUARES

The answer is that we try to find the line that describes the relationship best; that is, the

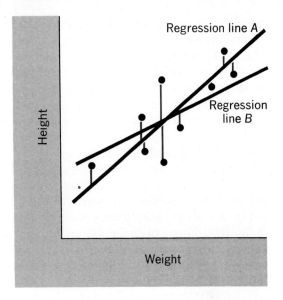

FIGURE 42·10
Least squares

line that lies closest to the dots. But how do we define "closest"? Here is where the colored lines enter. (1) From each dot we draw a vertical line to a given regression line (line A in our example). (2) We measure each "colored-line" distance and *square* it.* (3) We add together the squared distances.

In our diagram we have shown the distances of the dots from line A. Now we would have to do the same thing for line B, or for any other line we could draw. In the end, *we choose the line that has the smallest sum of squared distances as being the best fit.* We call this technique of finding such a line "fitting a regression by least squares." There are other ways of fitting lines, as well. We must leave these problems for a statistics

---

*Why do we square the distances? One answer is that this gives additional importance (weight) to any dots that lie considerably away from the regression line, thereby making our measure more sensitive to "exceptions" to what seems to be the rule. The mathematics of statistical theory is another, and more fundamental, reason.

course. But the basic idea of a "least squares" fit should now be plain.

What does the resulting line show? We know the answer from the propensity to consume (note the least squares line fitted on the diagram, Fig. 24·5). The regression coefficient, or the *slope* of the regression line, shows us the change in the dependent variable (C) associated with a change in the independent variable (Y). We know that this relationship can also be expressed (very roughly) in the formula on page 374, where the coefficient is $.94Y$. Thus once again we have shown that a diagram and an equation are only two ways of depicting the same thing—in this case, the relationship between consumption and income.*

CORRELATION COEFFICIENT

There remains one last problem. We know now what we mean by correlation and by a regression analysis. We even understand, in general, the criterion by which we fit a line to a group of "dots." But we do not know how we distinguish between two correlations to determine which is better.

Once again, an example will help make the answer clear. In Fig. 42·11 we show two regression lines, *each one of which is the best fit we can get by the least squares criterion.* Yet, clearly, the line on the right shows a better correlation of Y on X than the line on the left, simply because the dots lie directly on, or just off, the regression line, whereas in

---

*Note that we have been talking exclusively about *positive* coefficients; that is, about cases where an added inch of height is associated with an *added* amount of weight, or where an extra dollar of income is associated with *extra* consumption spending. We can also have negative regression coefficients, where the groups of "dots" would be downward sloping, and where the coefficient would have a minus sign in front of it. For example, if you graph weight on the vertical axis and life expectancy on the horizontal, you would find a downward sloping cluster of dots. Can you think of other such cases?

the second case they lie at some distance from it.

There is a mathematical way of describing the difference in "closeness of fit" of the two lines; it is called a *correlation coefficient.* A correlation coefficient is a number that tells us how closely our regression line fits the actual data. This coefficient always has a value between $+1$ and -1. If the correspondence is perfect—that is, if the dots lie exactly *on* the line, the correlation coefficient is either $+1$ or -1 (depending on whether the two variables move in the same direction or in directions opposite to each other). If the correlation coefficient is 0, then no correlation exists, as in the case of our example of height and IQ. Once again, we leave the calculation of the coefficient itself to a statistics course. But you can now see that the closer a correlation coefficient (for which we use the symbol r) is to $+1$ or -1, the stronger is the relationship between the two variables, and the better is the correlation.

One further piece of information. Statisticians usually use the *square* of the correlation coefficient (r^2) to describe the degree of association between two variables. In other words, a high r^2 is evidence of a "significant" relationship between two variables, and a low r^2 is evidence of the *lack* of such a relationship. This raises two questions: (1) by a "significant" correlation, do we imply that changes in one variable *cause* changes in the other? and (2) how high an r^2 is "significant"? The first of these questions we will examine in our next sections. The second is a technical matter that must be left for explanation in a course on statistics.

MULTIPLE CORRELATION

A last problem remains, to which we can only allude in passing in this introduction to correlation analysis. It is the question of how

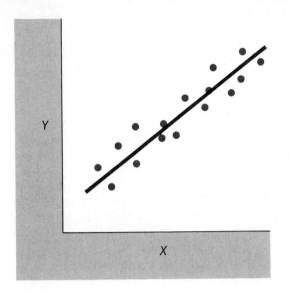
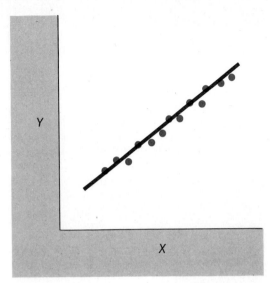

FIGURE 42 · 11
Closeness of fit

we deal with the problem of correlation when more than one variable is associated with changes in another. For instance, how do we figure out the relationship not merely between consumption and income, but between consumption and income *and* wealth *and* changes in prices *and* consumer expectations *and* still other variables that may influence consumption expenditures?

The answer is that we run what is called a *multiple regression analysis.* We do this by quantifying the relationship between the variable in which we are interested (the *dependent variable*) and the other variables (the *independent variables*), by techniques that enable us to discover the separate effect of each independent variable on the dependent variable. Once again, the techniques for doing this must be kept for a statistics course, but we should now be able to understand the meaning of the complicated formulas that emerge from multiple regression analysis.

Such a formula (purely imaginary in this case) might look like this:

$$C = .8Y + .3W - 2T$$

The purpose of such a multiple regression formula is to show us the different effects on our dependent variable, consumption (C), of various independent variables such as income (Y), wealth (W), taxes (T), and so on. What our particular formula tells us is that the net effect on consumption of $1 of added income is 80 cents, assuming that there is no change in wealth, taxes, etc.; that the net effect of an increase in wealth is 30 cents (again assuming that we have held the other variables constant); that the net effect of a rise in taxes is a *decline* in consumption (note the minus sign before the coefficient of .2), again with other variables held constant.

VII. Correlation and Causation

Regression techniques are widely used in econometric research, and we have been able only to indicate in very general terms the actual procedures that econometricians use. But the idea of correlation opens up for us a problem into which we must look very carefully.

691

This is the connection between *correlation and causation*. When we say that there is a high correlation between two variables—let us say cigarette smoking and heart disease—do we mean that cigarette smoking *causes* heart disease?

The question is obviously of very great importance. All sorts of disputes rage over the degree of "causation" that can be attributed to correlations. Econometrics cannot solve these disputes, because the word "cause"—as anyone knows who has ever looked into a book on philosophy—is a perplexing and elusive one. But econometrics can shed a strong light on some of the pitfalls associated with it. In particular, it makes us very cautious about declaring that such-and-such an event is the cause of another, simply because there is a high degree of correlation between them.

Here are a few examples to think about.

1. Wrong-way causation

It is a statistical fact that there is a positive correlation between the number of babies born in various cities of northwestern Europe and the number of storks' nests in those cities. Is this evidence that storks really do bring babies? The answer is that we are using a correlation to establish a causal connection in the wrong way. The true line of causation lies in the opposite direction: cities that have more children tend to have more houses, which offer storks more chimneys to build their nests in!

2. Spurious causation

It is a statistical fact that there has been a positive correlation all during the 1960s between the cost of living in Paris and the numbers of American tourists visiting there. Does that imply that American visitors are the cause of price increases in that city?

Here at least there is little danger of getting the causal links back to front; few people would argue that more Americans visit Paris *because* its prices are going up. But it would be equally difficult to argue that American tourists are the cause of rising Parisian prices, simply because the total amount of American spending is small in relation to the total amount of expenditure in Paris (with the exception of a few tourist traps).

The answer, then, is that the correlation is "spurious" in terms of causality, although it is "real" in terms of sheer statistics. The true explanation for the correlation is that the rising numbers of American visitors and the rising costs of living in Paris are both aspects of a worldwide expansion in incomes and prices. Neither is the "cause" of the other; both are the results of more fundamental, broader-ranging phenomena.

3. The problem of ceteris paribus

Finally, we must consider again the now familiar problem of *ceteris paribus*—the necessity of other things being equal. Suppose we correlate prices and sales, in order to test the hypothesis that lower prices "cause" us to increase the quantities we buy. Now suppose that the correlation turns out to be very poor. Does that disprove the hypothesis? Not necessarily. First we have to find out what happened to income during this period. We also have to find out what, if anything, happened to our tastes. We might also have to consider changes in the prices of other, competitive goods.

This problem affects all scientific tests, not just those of economics. Scientists cannot test the law of gravitation unless "other things" are equal, such as an absence of air that would cause a feather to fall much more slowly than Galileo predicted. But the trouble with the social sciences is that the "other things" are often more difficult to spot—or just to think of—than they are in the laboratory. For example, what are we to make of the fact that there is a positive correlation between shoe sizes and mathematical ability among school children? The answer, once we think about it, is that we have not held "other things" equal in one very important respect—the age of the school children. Of course older children, with bigger feet, will be able to do more arithmetic than little children. Hence the first thing to do is to see if there is a correlation between shoe sizes of children of *equal ages* and mathematical ability. Children of equal ages will also have different shoe sizes, but the correlation with problem-solving ability is hardly likely to be there.

WHAT CAN CORRELATION TELL US?

These (and still other) pitfalls make econometricians extremely cautious about using correlations to "prove" causal hypotheses. *Even the closest correlation may not show in which direction the causal influences are working;* for instance, the high correlation between the number of corporate mergers and the level of stock prices in the 1960s does not show whether mergers caused stock prices to move up, or whether booming prices encouraged companies to merge and float their new securities on a favorable market.

So, too, *the interconnectedness of the economic process often causes many series of data to move together.* In inflationary periods, for example, most prices tend to rise, or in depression many indexes tend to fall, without establishing that any of these series was directly responsible for a movement in another particular series.

And finally, econometricians are constantly on the lookout for factors that have not been held constant during a correlation, so that *ceteris paribus conditions were not in fact maintained.*

Is there an answer to these (and still other) very puzzling problems of correlation and causation? There is a partial answer. We cannot claim that a correlation is proof that a causal relationship exists. But a causal relationship is likely to exist when we can demonstrate a strong correlation, backed up by theoretical reasoning. That is, every valid economic hypothesis (or for that matter every valid hypothesis of any kind) *must* show a high and "significant" correlation coefficient between the "cause" and the "effect," provided that we are absolutely certain that our statistical test has rigorously excluded spurious correlations of various kinds and unsuspected "other things." Needless to say this is often very difficult and sometimes impossible to do with real data. A physicist can hold "other things equal" in his laboratory, but the world will not stand still just so an economist can test his theories. The net result is that correlations are a more powerful device for *disproving* hypotheses than for proving them.

All this has a very important moral. Because economists deal with quantifiable data, correlation analysis is one of the most important tools in the economist's kit. All the "laws" of economics, from supply and demand to the various relationships of macroeconomics, are constantly being subjected to highly sophisticated econometric tests, *because these are the only objective methods we have for establishing whether variable X is associated with variable Y.* Thus we constantly use correlations to buttress—but not to prove—our belief in economic relationships. Conversely, when careful correlation analysis fails to show high coefficients, there is good reason to look very skeptically at our economic hypotheses.

Hence, correlation is an indispensable part of economic science. But that is not the same thing as saying that it *is* economic science. Behind the tools of econometrics lies the process of *economic reasoning*—that is, the attempt to explain why people behave in certain ways on the marketplace, or how the hard realities of nature or technology shape and constrain the ongoing economic process. If the economic reasoning is faulty, econometrics will not set it straight: recall the puzzling inability to explain the interest rate's effect on investment (pp. 404–6). Here, in the end, we allowed the logic of theory to override the "evidence" of econometrics.

Yet, with all its problems, econometrics is the most penetrative means of empirical analysis that economics has yet developed, and our understanding of the workings of the economy would be terribly handicapped if we did not command its powerful techniques. But ultimately it is to economic theory that we look for that basic understanding itself.

Thus, the predictive tool of the econometrician consists of *a series of equations which depict the simultaneous interaction of those activities that the model-builder believes to be crucial in affecting the economy.* As can be imagined, this leads to extremely complex equations and systems of equations. The famous model of the Wharton School of Business of the University of Pennsylvania, for example, has some 150 equations, most of them with strings of multiple regression elements. To solve such a system of equations by pencil and paper would be impossible— or would take so long that the future would long ago have become history. Thus the computer with its lightning speeds is an integral part of the econometrician's equipment.

VIII. Forecasting

In addition to their role in testing economic hypotheses, correlations and regressions are used to build forecasting models. Indeed, this is one of the most important functions of econometrics.

How does an econometrician attempt to forecast? We know part of the answer from our introduction to econometrics. The first step is to assemble all available statistical information concerning important economic relationships. Suppose, for example, it is known from multiple regression that the demand for automobiles can be represented by an equation in which the demand for cars is related to changes in disposable income, to changes in car prices, to credit conditions, etc. This long and complicated equation will then become one element in a model of the economy that will also contain the demand for many other final products, as well as the relationships (also econometrically derived) between or among elements such as the quantity of money and the level of prices, the level of employment and consumer disposable income, and so on.

EXOGENOUS AND ENDOGENOUS VARIABLES

When we examine forecasting models, we find that the many variables they contain can be classified into two kinds. Many of the variables *depend* on other variables for their values; for example, automobile purchases depend on disposable income, car prices, and other variables. These are *dependent* variables in the model; that is, elements whose values will be determined by the action of the independent variables in the system.

These *exogenous* ("outside") variables must be introduced into the system to determine the values of the dependent variables. Where does the econometrician get these exogenous variables? Some of them may be known facts from which various kinds of future economic activity may be inferred by econometric analysis. For instance, the model-builder may introduce the known level of population growth as an exogenous variable on which the future level of home-building or car-buying may depend. Or exogenous variables may be based on surveys or on expected policy decisions, such as surveys of intended corporate investment or policy statements of the Federal Reserve. These and other exogenous variables are arranged along with the endogenous variables in a set of simultaneous equations that make up the econometric model and reflect the complex mechanism of the economy. The econometrician then solves the model, thus deriving values for his endogenous variables, such as consumer disposable income, automobile purchases, or GNP itself.

Here, of course, lies a central problem for econometric forecasting. We can never be certain of the accuracy of our all-important exogenous variables. We cannot predict with certainty, for instance, what the money supply will be, because we cannot know whether Federal Reserve policy will change. We cannot predict with certainty what business investment will be, because business expectations may change. To the extent that our exogenous variables are wrongly estimated, the values we derive from them for dependent elements in our model will also be wrong. In 1966, for example, the Council of Economic Advisers predicted that the rate of inflation was about to slow down. Actually, it accelerated. The mistake did not lie with the reasoning of the council or with its econometric model. It lay in the fact that the council assumed, on the basis of the data at hand, that the budget for 1967 was going to be balanced. What happened in fact was that a budgetary deficit of $9 billion was incurred to finance an expansion of the Vietnam War —a deficit that had not been included in the original budget estimates.

One way to get around such problems is to assume several sets of values for key "policy variables" and to run the model for each one. For example, a forecast put out by the Federal Reserve Bank of St. Louis has made three different predictions based on three different postulated rates of growth in the money supply. Having such a spectrum of predictions seems at first to be weaseling on the job. But actually it serves a very useful purpose. Such alternative models show us the probable *range* of our results and can therefore tell us, for instance, if the consumption of autos will be much or little affected if the money supply grows fast or slowly.

Second, because many of the key policy variables *cannot* be accurately predicted (who can foretell with certainty what next year's federal budget will be?) a spectrum of forecasts enables individuals or corporations to make their own guesstimates about these critical variables and then to examine what the probable configuration of *other* economic variables are apt to look like, given these key assumptions.

Every person or firm bases his economic decisions on some kind of implicit or explicit estimate of future economic conditions. To try to improve these decisions, econometric forecasting is increasingly used by large industrial corporations, by financial institutions, and by governments. The results of the big models at the Brookings Institution or the Wharton School are even featured in news articles. Hence it is important to ask how well the econometric predictions have fared.

PERFORMANCE

Most econometricians would be the first to admit that their performance has been uneven. The models have worked very well— have predicted changes in the economy quite accurately—as long as things were moving along fairly steadily in one direction. But when things changed direction, *which is just when we want our models to work best,* they have not been too reliable. The influential Brookings Model, for example, did not forecast the recessions of 1957–58, or of 1960–1961. In 1968, most econometricians predicted a recession for 1969, but it did not actually take place until 1970.

What accounts for this disappointing performance? Part of the trouble lies in the problems of "specifying" the model correctly—that is, of overcoming the numerous problems of econometrics that we have touched on in past supplements. The fact that different econometric models use different multiple regression coefficients for the same variables attests in itself to the fact that there is more than one way to depict certain relationships.

Second is the problem that econometricians, as we have also made clear, ultimately rely on economics; and economists are by no means of one mind about the relative importance of certain key exogenous variables, such as the influence of fiscal or monetary policy on the price level or on GNP. Hence our models misperform because we are still in disagreement about economic theory itself.

And last, our models misperform because we cannot predict the course of certain critical exogenous variables that lie beyond the reach of economic reasoning itself. For example, no economic theory could have predicted the degree of escalation of the Vietnam War. Yet that event dominated macroeconomic behavior in the late Sixties.

THE LIMITS OF PREDICTION

Does this mean that econometric forecasting is doomed to failure? Not necessarily. The need is rather to understand better what our models can do for us and what they cannot. One thing they can certainly do is to clarify the nature of the interactions of many economic activities, so that important relationships within the economic system can be specified with much greater accuracy. A second very important purpose of models is to make clear the economic *requirements* to reach given targets, a task that is not quite the same as that of prediction, but that is no less important regarding the future.

Equally certainly, however, we cannot use our models to give us predictions about such things as the date of the next stock market crash or even the level of GNP ten years hence. When matters of volatile *behavior* are concerned, such as the psychology of the stock market, an econometric model has nothing to go by. By definition, such unforecastable psychological changes *alter* the very coefficients on which predictive equations are built.

In the same way, long-term predictions are made hazardous by events such as international crises, technological breakthroughs, or political upheavals, all of which also upset the parameters on which all prediction is ultimately based.

Here, in the realm of the exogenous, the *science* of econometric forecasting ends, and the art of forecasting begins. For it is here that the shrewd guess of the economist or the hunch of the econometrician will lead him—in the face of the established patterns of the past—to postulate that things have shifted or are about to shift and therefore to regard the results of his model with more than the usual degree of caution, or to recast the parameters on which his model is based. At this point the economist or the econometrician certainly does not claim access to the occult, but humbly confesses that scientific methods, powerful as they are, still cannot describe the shape of many things to come.

CENTRAL CONCEPTS

I. SOME INITIAL CAUTIONS

Statistics are always best approached with caution. *Definitions* are often key to the interpretations we place on data. Data can be out-dated, misleading if not carefully identified, deceptive if "overinterpreted" (be wary of making mountains out of molehills).

Visual tricks using data to make a point or the failure to get a rounded perspective on facts are frequent sources of statistical error—or worse—statistical *deception*.

II. DISTRIBUTIONS AND AVERAGES

Learn to distinguish between *means*—the arithmetic average—and *medians*—the middle value of a series. Mean values are usually different from medians, because data are often *skewed*.

Modes are "averages" in terms of that value which has the largest number of units. *In a normal, bell-shaped distribution, mode, mean, and median will all be the same. This will not be the case in skewed distributions.*

We must be particularly careful to guard against *bi-modal* distributions—distributions in which there are two (or more) modal clusters.

III. PRICE INDEXES

Price indexes are used to "correct" current dollar values, in order to establish "real" changes in economic magnitudes.

Price indexes can be constructed from many kinds of data. The Department of Labor's *Consumer Price Index* is an index number that measures the changing cost of a "market basket" of typical consumer goods.

It is simple to construct an index if only one good is involved. When more than one good is in the market basket, we must assign *weights* to the various items. Each is given the degree of importance proportional to its value in the "market basket."

New goods present many difficult problems for the construction of an index. To the extent that the good fulfills a *new function*, the problem is magnified because, typically, new goods fall in price.

Price indexes, once constructed, are simple to apply: current dollar values are divided by the index and then multiplied by 100. The resulting "real" series will differ in its dollar values according to which year is used as a *base*, but the relative changes from year to year will be the same. Thus indexes *can be spliced*. Be wary, however, of ignoring the statistical difficulties of long-term time series, just because they are spliced into one continuous series.

IV. SAMPLING

The logic of sampling is based on *probability*. There are well-established laws of probability that enable us to determine certain characteristics of large "universes" of data by sampling techniques.

The most important objective of good sampling is to avoid *bias*. This is accomplished by aiming for a random sample. The size of the random sample will then depend on the size of the total "universe" and on the number of its characteristics that we wish to test for.

Econometrics

Sampling, surprisingly enough, may be *more* accurate than full enumeration, as well as cheaper and easier.

Functional relationships

V. ECONOMETRICS AND FUNCTIONAL RELATIONSHIPS

One main use of econometrics is the use of statistics to test economic hypotheses by statistical methods.

Another principal use is to establish the existence or nonexistence of functional relationships.

Linear graphs

Functional relationships are usually described by equations, rather than graphs, especially if the variables are numerous. *The slope of the line on all linear graphs will be determined by the coefficients* in the equation that describes the functional relationship.

Coefficients

Associations among variables

VI. CORRELATIONS AND REGRESSIONS

Correlations attempt to establish *associations among variables*, and regressions are the means by which we quantify these associations.

Scatter diagrams

Scatter diagrams give us a visual clue to whether correlations exist. But we test for such correlations by a method known as *least squares*.. This method gives us a *regression line* that is mathematically defined as the closest possible fit to a scatter of paired variables. The regression equation describes the nature and direction of the association between the two variables.

Least squares

Regressions do not tell us how closely the data cluster around a given line. To establish the "tightness" of fit, we need another measure called the *correlation coefficient*.

Regression line

A method known as *multiple correlation* enables us to quantify the relationship among more than two variables by holding all variables "fixed" while we test the correlation of the particular pair we wish to investigate.

Correlation coefficient

VII. CORRELATION AND CAUSATION

We must be very careful about attributing causal relationships to correlated variables. There are many pitfalls: *wrong-way correlation*, *spurious correlation*, improper attention to *ceteris paribus*. All can deceive us as to the existence or nonexistence of a causal connection between variables.

Multiple correlation

Causation

Correlations, when carefully interpreted, can often be useful in disproving hypotheses. They are less dependable as "proofs" of hypotheses.

VIII. FORECASTING

Wrong-way correlation

Forecasting is essentially an effort to *predict the movement of dependent variables on the basis of assumptions regarding the independent variables.*

The independent variables are *exogenous*, introduced into the system from outside. They are treated as stated values that will determine or limit the values of the dependent variables.

Spurious correlation

Because exogenous variables are intrinsically unpredictable, the forecasting powers of econometrics are inherently limited.

Exogenous variables

Econometrics models are still invaluable as a means of examining the interrelationships among various parts of the economy and testing theories that seek to explain those relationships.

QUESTIONS

1. If someone told you that the unemployment rate was 4 percent in West Germany and 5 percent in the U.S., would you conclude that unemployment was therefore higher in the U.S.? What cautions would you advise before coming to that conclusion?

2. Draw a graph that is *calculated* to deceive someone. Example: an alarming depiction of stock price changes in the last month.

3. What is the best average to describe the "average" IQ of a population? What do you mean by "best"? What characteristics would be shown by a mean that would not be shown by a median? Would a mode be useful? Suppose instead of IQ, you wanted to show the "average" stockholdings of the same population. What answers do you now give to the questions?

4. Describe a series that is likely to be bi-modal. Try to think of one that isn't in the text. The trick is to find something that typically comes in two or more "sizes" instead of being more or less evenly spaced over the range of all sizes.

5. Write down four median incomes and four price index numbers; figure out real income for each year. Shift bases and compare your results.

6. How would you handle the problem of fitting new cars into the Consumer Price Index? How much of their price is improved quality, how much is just inflation? How would you go about making such a decision?

7. Take a market basket with two goods: bread and wine. Bread costs 50¢; wine $5.00. In year 1, you spend $10 for bread, $5 for wine. Now bread falls to 25¢, wine rises to $6.00. Can you make a weighted index from this data? Do you need to know how much you spend for bread and wine in year 2?

8. Describe some data that could be more accurately surveyed by sampling techniques than by complete enumeration. Numbers of grains of sand on a beach? Name some economic possibilities.

9. What is meant by regression? By correlation? How would you decide if a high correlation was a prima facie case for causation? Describe a highly correlated pair of variables in which causation is uncertain. Are cigarette smoking and ill-health one? How would you find out? *Could* you find out "beyond all possibility of error"?

10. Why is the forecaster's life a hard one?

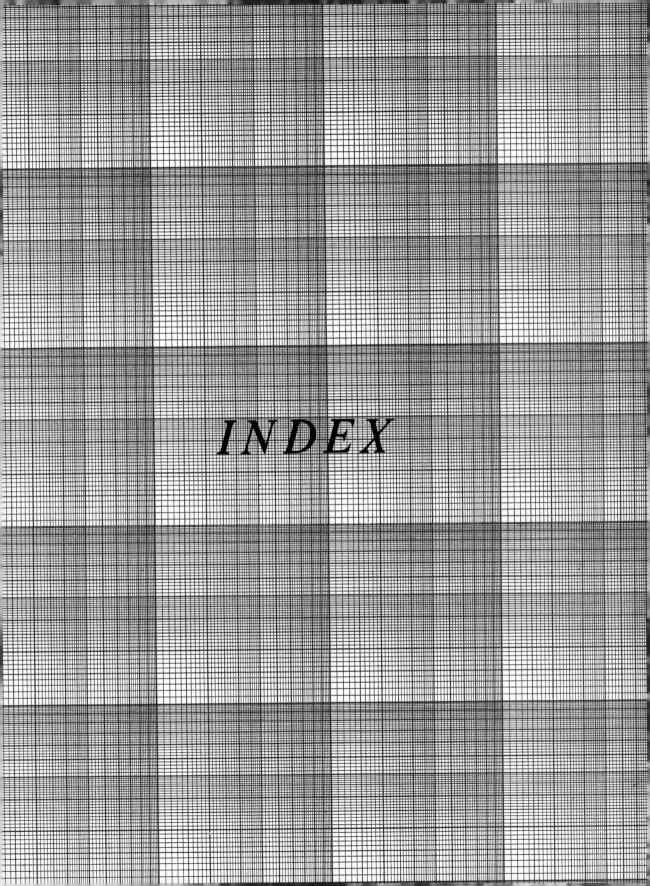

INDEX

Index

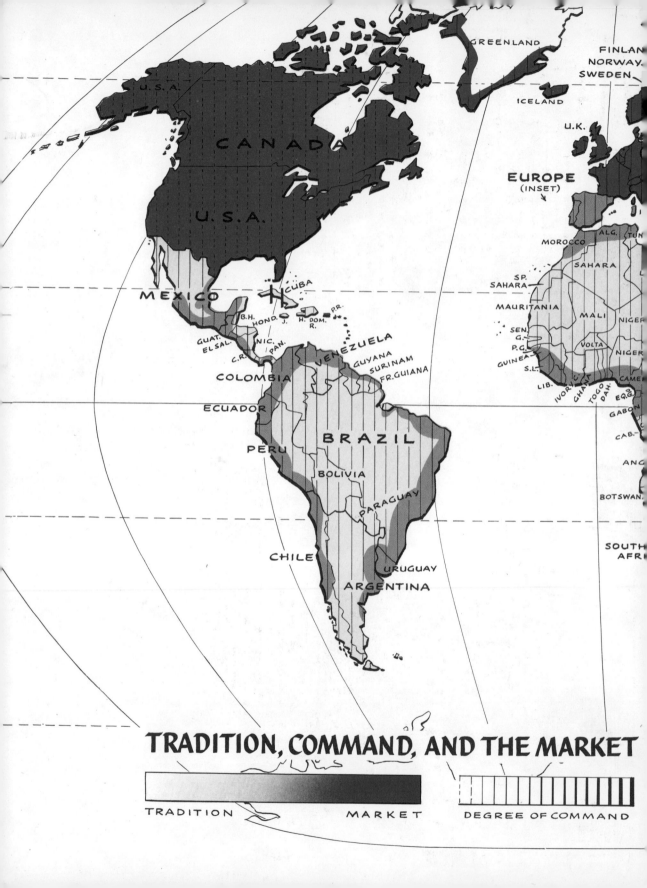

TRADITION, COMMAND, AND THE MARKET

TRADITION   MARKET   DEGREE OF COMMAND